Ecosystem-Based Adaptation
Approaches to Sustainable Management of Aquatic Resources

Ecosystem-Based Adaptation
Approaches to Sustainable Management of Aquatic Resources

Arvind Kumar
India Water Foundation, New Delhi, India

Elsevier
Radarweg 29, PO Box 211, 1000 AE Amsterdam, Netherlands
The Boulevard, Langford Lane, Kidlington, Oxford OX5 1GB, United Kingdom
50 Hampshire Street, 5th Floor, Cambridge, MA 02139, United States

Copyright © 2022 Elsevier Inc. All rights reserved.

No part of this publication may be reproduced or transmitted in any form or by any means, electronic or mechanical, including photocopying, recording, or any information storage and retrieval system, without permission in writing from the publisher. Details on how to seek permission, further information about the Publisher's permissions policies and our arrangements with organizations such as the Copyright Clearance Center and the Copyright Licensing Agency, can be found at our website: www.elsevier.com/permissions.

This book and the individual contributions contained in it are protected under copyright by the Publisher (other than as may be noted herein).

Notices
Knowledge and best practice in this field are constantly changing. As new research and experience broaden our understanding, changes in research methods, professional practices, or medical treatment may become necessary.

Practitioners and researchers must always rely on their own experience and knowledge in evaluating and using any information, methods, compounds, or experiments described herein. In using such information or methods they should be mindful of their own safety and the safety of others, including parties for whom they have a professional responsibility.

To the fullest extent of the law, neither the Publisher nor the authors, contributors, or editors, assume any liability for any injury and/or damage to persons or property as a matter of products liability, negligence or otherwise, or from any use or operation of any methods, products, instructions, or ideas contained in the material herein.

British Library Cataloguing-in-Publication Data
A catalogue record for this book is available from the British Library

Library of Congress Cataloging-in-Publication Data
A catalog record for this book is available from the Library of Congress

ISBN: 978-0-12-815025-2

For Information on all Elsevier publications
visit our website at https://www.elsevier.com/books-and-journals

Publisher: Candice Janco
Acquisitions Editor: Louisa Munro
Editorial Project Manager: Emily Thomson
Production Project Manager: Bharatwaj Varatharajan
Cover Designer: Greg Harris

Typeset by MPS Limited, Chennai, India

Contents

About the author xi
Preface xiii
Acknowledgment xvii

CHAPTER 1 Introduction 1
    1.1 Defining disaster 1
        1.1.1 Hazard 1
        1.1.2 Risk 2
    1.2 Definition of disaster 2
        1.2.1 Toward consensual definition 5
    1.3 Hazard-disaster linkages 7
    1.4 Disaster, vulnerability, and resilience 10
    1.5 Toward disaster risk reduction 12
    1.6 Hyogo framework for action 14
    1.7 Sendai framework for disaster risk reduction 15
    1.8 Ecosystem-based adaptation for disaster risk reduction 19
    1.9 Conclusion 25
        References 26
        Further reading 37

CHAPTER 2 Challenge of climate change 39
    2.1 Introduction 39
    2.2 Defining climate 42
        2.2.1 Climate and weather 43
    2.3 Climate system 44
        2.3.1 Atmosphere 44
        2.3.2 Hydrosphere 46
        2.3.3 Cryosphere 47
        2.3.4 The lithosphere 48
        2.3.5 The biosphere 49
    2.4 Defining climate change 51
        2.4.1 Definition of climate change 53
    2.5 Abrupt climate change 55
    2.6 Global warming 57
        2.6.1 Global warming versus climate change 58
        2.6.2 Global warming and Anthropocene 59
    2.7 Anthropogenic drivers of climate change 60
        2.7.1 Economic growth 61

            2.7.2 Urbanization 64
            2.7.3 Consumption 67
    2.8 Extreme weather events 68
            2.8.1 Extreme cold events 69
            2.8.2 Extreme heat events 70
            2.8.3 Droughts 72
            2.8.4 Extreme snow and ice storms 74
            2.8.5 Tropical cyclones 76
            2.8.6 Wildfires 81
    2.9 Impacts of climate change 82
    2.10 Conclusion 83
            References 84
            Further reading 103

CHAPTER 3 Ecosystem-based adaptation approach: concept and its ingredients 105
    3.1 Introduction 105
    3.2 Concept of ecosystem-based adaptation 107
            3.2.1 Defining ecosystem-based adaptation 107
    3.3 Ecosystem-based adaptation's ingredients 110
            3.3.1 Adaptation 110
    3.4 Ecosystem services 117
            3.4.1 Classification of ecosystem services 119
            3.4.2 Ecosystem services and biodiversity 123
    3.5 Ecosystem-based adaptation: costs and benefits 125
            3.5.1 Costs and benefits of adaptation 125
            3.5.2 Costs and benefits of ecosystem-based adaptation 126
    3.6 Conclusion 132
            References 133
            Further reading 141

CHAPTER 4 Coping with climate chang 143
    4.1 Introduction 143
    4.2 Mitigation 144
            4.2.1 Climate risk 145
            4.2.2 Internal determinants 147
            4.2.3 Excludable benefits 149
            4.2.4 Influence from above 149
            4.2.5 Harnessing mitigation 150
            4.2.6 International framework for climate change mitigation 153
    4.3 Adaptation 169

		4.3.1 Delineating adaptation 172
		4.3.2 Adaptation under Paris Agreement.. 173
		4.3.3 Limits and barriers to adaptation... 175
		4.3.4 Evaluating climate change adaptation 176
	4.4	Linkages between mitigation and adaptation 178
	4.5	Geoengineering and climate change.. 182
		4.5.1 Solar radiation management 185
		4.5.2 Carbon dioxide reduction... 188
	4.6	Loss and damage associated with climate change 197
		4.6.1 Defining loss and damage.. 198
		4.6.2 Warsaw International Mechanism for Loss and Damage 199
	4.7	Ecosystems approach and climate change... 203
	4.8	Conclusion 206
		References... 206
CHAPTER 5	**Toward water security** **235**
	5.1	Introduction 235
	5.2	Water availability 236
	5.3	Demand for water... 237
	5.4	Water quality 239
	5.5	Water scarcity... 241
		5.5.1 Physical water scarcity... 242
		5.5.2 Economic water scarcity 243
		5.5.3 Social water scarcity 244
		5.5.4 Technological water scarcity 245
	5.6	Water and climate change.. 247
		5.6.1 Water and extreme weather events.. 248
	5.7	Securing water 251
		5.7.1 Defining water security.. 251
		5.7.2 Gender and water security 255
		5.7.3 Integrated water resource management 257
		5.7.4 Water−energy−food security nexus 259
		5.7.5 Water security and nature-based solutions.. 267
	5.8	Conclusion 270
		References... 270
		Further reading 287
CHAPTER 6	**Toward sustainable food security** **289**
	6.1	Introduction 289
	6.2	Hunger 290
		6.2.1 Defining hunger 292

6.3	Toward food security	293
	6.3.1 Definition of food security and its evolution	293
6.4	Food security and climate change	300
	6.4.1 Climate change, climate drivers, and food security	305
	6.4.2 Climate change and food availability	307
	6.4.3 Climate change and access to food	310
	6.4.4 Climate change and food utilization	311
	6.4.5 Climate change and food stability	313
6.5	Adaptation options	314
	References	317
	Further reading	324

CHAPTER 7 Sustainable smart cities ... 325

7.1	Introduction	325
7.2	Urbanization and sustainable development	326
	7.2.1 Local sustainable development	330
7.3	Sustainable urban development	332
7.4	Livable cities	334
7.5	Eco-cities	335
7.6	Toward sustainable smart cities	345
	7.6.1 Sustainable concept	346
	7.6.2 "Smart" concept	347
	7.6.3 "Cities" concept	348
	7.6.4 Defining a smart city	349
	7.6.5 Dimensions of a smart city	353
7.7	Ecosystem-based adaptation for smart cities	384
7.8	Conclusion	386
	References	387
	Further reading	416

CHAPTER 8 Sustaining life below water ... 417

8.1	Introduction	417
8.2	Environmental stressors	418
	8.2.1 Ocean warming	422
	8.2.2 Global sea level rise	424
	8.2.3 Ocean acidification	429
	8.2.4 Ocean deoxygenation	432
8.3	Anthropogenic stressors	437
	8.3.1 Marine microplastic pollution	439
	8.3.2 Oil-spills pollution	444
	8.3.3 Overfishing	448

	8.3.4 Greenhouse gases	451
	8.3.5 Land-based sources of marine pollution	457
8.4	Ecosystem-based management for oceans	460
	8.4.1 Ecosystem-based approach versus ecosystem-based management	463
	References	466
	Further reading	498

CHAPTER 9 Preserving life on Earth 503

9.1	Introduction	503
9.2	Freshwater ecosystems	504
	9.2.1 Climate change and freshwater ecosystems	506
9.3	Conserving forests	507
	9.3.1 Deforestation	509
	9.3.2 Impact of droughts	511
	9.3.3 Forest degradation	513
	9.3.4 Wild and forest fires	515
	9.3.5 Forest fragmentation	517
9.4	Sustainable land use	519
	9.4.1 Land degradation	521
	9.4.2 Desertification	524
9.5	Conserving mountain ecosystems	527
9.6	Protecting biodiversity	531
9.7	Genetic resources	536
	9.7.1 Types of genetic resources	537
9.8	Conservation of wildlife	546
	9.8.1 Threats to wildlife species	547
9.9	Preventing invasive aliens species on land and water	552
	9.9.1 Defining invasive alien species	555
	9.9.2 Invasive alien species and climate change	556
	9.9.3 Regulatory framework for invasive alien species	557
	References	561
	Further reading	601

CHAPTER 10 Mainstreaming ecosystem-based adaptation 603

10.1	Introduction	603
10.2	Concept of mainstreaming	604
	10.2.1 Categories of mainstreaming climate change adaptation	605
10.3	Mainstreaming ecosystem-based adaptation into policy process	607
10.4	Entry points for mainstreaming ecosystem-based adaptation	609
	10.4.1 Mainstreaming ecosystem-based adaptation in national adaptation plan	610

- 10.4.2 Mainstreaming ecosystem-based adaptation into sectoral adaptation plan ... 612
- 10.4.3 Mainstreaming ecosystem-based adaptation into local and community planning processes ... 613
- 10.5 Mainstreaming ecosystem-based adaptation in G-20 countries ... 614
- 10.6 Mainstreaming ecosystem-based adaptation in Sweden ... 616
- 10.7 Mainstreaming ecosystem-based adaptation in Bangladesh ... 617
- 10.8 Mainstreaming ecosystem-based adaptation in India ... 618
  - 10.8.1 India's move toward climate change policy ... 620
  - 10.8.2 From water—energy—food nexus approach to ecosystem-based adaptation ... 621
  - 10.8.3 Mainstreaming/integrating ecosystem-based adaptation into Meghalaya ... 623
  - 10.8.4 Integrating ecosystem-based adaptation into Arunachal Pradesh ... 632
  - 10.8.5 Integrating ecosystem-based adaptation into the Indian Himalayan Region ... 632
- 10.9 Emerging lessons of mainstreaming ecosystem-based adaptation ... 633
- 10.10 Barriers to mainstreaming ecosystem-based adaptation ... 636
- 10.11 Conclusion ... 638
- References ... 639
- Further reading ... 645

CHAPTER 11 Conclusion ... 647
- 11.1 Natural disasters versus human-induced disasters ... 647
- 11.2 Tackling global warming ... 650
  - 11.2.1 Greenhouse gases ... 651
- 11.3 Global outlook for biodiversity ... 653
- 11.4 Global outlook for life on Earth ... 656
- 11.5 Global outlook for life below water ... 660
- 11.6 COVID-19 pandemic and climate change ... 665
- 11.7 Outlook for ecosystem-based adaptation ... 668
- References ... 670

Index ... 675

About the author

Dr Arvind Kumar is Governor World Water Council and founder president of India Water Foundation a non-profit organization and think tank established for generating a heightened public awareness at national level in India and sub-regional level in Asia, regarding water and its impact on human health, economic growth and environmental sustainability. He is a strategist and key-influencer in development sector, author, columnist, Water activist, and specializes in ecosystem-based adaptation, water-energy-food nexus, with specific emphasis on inter-linkages between water, environment and SDGs. He is Member Technical Advisory Committee for India's Third National Communication and Biennial Update Reports to UNFCCC and member of the 'National Wetlands Committee', ministry of environment forest and climate change, Government of India. He is Member Meghalaya State Water Resources Council and the State Council for Climate Change and Sustainable Development, government of Meghalaya. He is PhD Defense Studies has published over 400-plus research articles in reputed journals. He is also the Editor of online e-Magazine Focus Global Reporter and authored many books. Some of them worth mentioning are- published by SAC Dhaka in December 2015 entitled 'SAARC Outlook on Water-Energy-Food Nexus in SAARC Region', 'United Nations at 75 and Beyond' and upcoming book 'India at 75 and beyond' celebrating 75 years of India's independence.

Preface

Agenda 2030 is a plan of action for the prosperity of the people of planet Earth that also seeks to endeavor to heal and secure this planet by taking bold and transformative steps that are urgently needed to shift the world onto a sustainable and resilient path. This agenda also articulates the determination to protect the planet from degradation, sustainably managing its natural resources, and taking urgent action on climate change, so that it can support the needs of the present and future generations. This book is greatly inspired by major sustainable development goals (SDGs), such as tackling the problem of hunger (SDG-2) by ensuring food security, clean drinking water (SDG-6) by ensuring water security, sustainable cities (SDG-11) by moving towards sustainable smart cities, climate action (SDG-13) by understanding the magnitude of the challenge of climate change and suggesting means to cope with this problem, safeguarding life below water (SDG-14) by suggesting means and measures of sustain life below water, and protect life on Earth (SDG-15) by adhering to means and measures that help conserve life. The salient feature of this book is its emphasis on nature-based solutions, with specific emphasis on ecosystem-based adaptation (EbA), and it recommends mainstreaming EbA into national, provincial, and local level adaptation plans as a means to realize the goals of Agenda 2030.

Chapter 1 is in introductory form that focuses on the concept of disaster and its interlinkages with notions of risk and hazard. In an endeavor to define the concept of disaster by examining various definitions, it also takes into account efforts towards a consensual definition of disaster, and this chapter places reliance on the definition provided by UNISDR in 2017. While taking into consideration briefly linkages between hazard and disaster, the chapter proceeds to appraise linkages between disaster, vulnerability, and resilience. While tracing the evolution of the mechanism of disaster risk reduction, the chapter focuses on Hyogo Framework for Action (2005–15) and thereafter the adoption of Sendai Framework at the expiry of Hyogo Framework for Action in 2015. While assessing the successes and failures of the Sendai Framework for Disaster Risk Reduction (SFDRR) (2015–30) for years of its operationalization (2016–20), it also examines the prospects of applying EbA for disaster risk reduction, also termed as Eco-DRR approach.

Magnitude of the challenge of climate change as the greatest challenge confronting humankind is dealt with in Chapter 2 that takes an overview of the concept of climate change along with an attempt to delineate the concept of climate change, coupled with a brief description of distinction between weather and climate change. While providing a brief description of the main components of climate change—atmosphere, biosphere, cryosphere, hydrosphere, and lithosphere—an endeavor is made to examine various prevalent definitions of the concept of climate change and finally it focuses on the definitions provided by IPCC and UNFCCC that are widely used in the contemporary literature. Providing a brief appraisal of the level adaptation notion of abrupt climate change, this chapter proceeds to analyze the notion of global warming and its relationship with climate change and Anthropocene. While analyzing anthropogenic drivers of climate change, such as economic change, urbanization, and consumption, the focus of this chapter shifts to extreme weather events like extreme cold events, extreme heat events, droughts, extreme ice and snow storms, tropical storms, extra-tropical storms, and wildfires. While asserting that climate change is a risk management problem, the chapter concludes that there is a need for implementing the most cost-effective ways to limit risks to an accepted/agreed level, informed by the best scientific evidence along with quick action on adaptation and mitigation to ward off these risks.

Chapter 3 focuses on EbA approach as a concept along with a brief examination of major ingredients of the approach. Acknowledging the ever-increasing traction being garnered by EbA as one of the most effective nature-based adaptation measures to deal with the adverse impacts of climate change that come to be reckoned with as cost-effective measure yielding multiple benefits in the form of goods and services provided by ecosystems, the chapter proceeds to examine the concept of EbA along with its evolving definition along with a brief analysis of main ingredients of EbA, such as adaptation and need for adaptation, adaptive capacity, and limits to adaptation; ecosystem services, and the types of ecosystem services like provisioning services, regulating services, and sociocultural services. While delineating interconnectedness between ecosystem services and biodiversity, the chapter finally dwells on the theme of cost−benefit analysis of adaptation in general and EbA in particular.

Essentiality of tackling adverse impacts of the ongoing process of climate change comprises the main theme of Chapter 4 that takes into account pros and cons of mitigation and adaptation measures to deal with climate change. Conceding that international community has been engaged in in-depth exploration of the causes and outcomes of climate change impacts and concurrently devising strategies, plans, and programs to tackle the adverse impacts of climate change, starting with mitigation and adaptation approaches, the chapter focuses on the efforts of international community on having applied geoengineering approach comprising solar radiation management and carbon dioxide reduction approach. There is also growing support for loss and damage associated with climate change approach. Nevertheless, all these approaches are under implementation and have yielded some satisfactory outcomes to some extent. In its final analysis, the chapter suggests that a combination of these approaches is deemed appropriate in coping with the adverse impacts of climate change, because no single approach can provide a complete solution to cope with climate change impacts and each approach suffers from some type of lacuna either of governance or technology or finance or capacity building.

The substantive theme of water security constitutes the main focal point of analysis in Chapter 5 that describes water as a finite source despite the fact that two-thirds of the Earth is surrounded by water; nevertheless, only 2.5% is freshwater that is used for human consumption, agriculture, domestic, and industrial use. Burgeoning population and the resultant enhanced requirements of increased supply of freshwater for increased food production, industrial, and economic activities along with other socioeconomic requirements have exerted increased pressures on water resources that being finite are getting depleted in the wake of rapid pace of depletion of groundwater resources, pollution of surface water resources and speedier melting of glaciers in the wake of vagaries of climate change, thereby triggering water scarcity or water insecurity. While focusing on the current availability of water and growing demand for water resulting in the gap between supply and demand, the chapter also takes into considerations the issues of water quality and water scarcity and thereafter proceeds to analyze the impact of climate change on water resources. While analyzing the concept of water security, issue of gender and water security is also focused on in this chapter. While examining the integrated water resource management (IWRM) as a model of water resources management, emphasis is also placed on water-energy-food nexus approach to deal with water insecurity. Noting that water security is often regarded as a broader concept of IWRM, that is designed to balance resource protection and resource use, the analysis in this chapter demonstrates that IWRM is barely delineated as a perspective-way of discerning water, based on technical-scientific knowledge, while water security embraces a more inclusive and diverse way of perceiving water, with a greater emphasis on human values, ethics, and power. In the wake of growing importance being garnered by nature-based solutions (NBS) as a more viable and effective pathway to address water security, this chapter focuses on EbA approach, among the various NBS approaches, as a viable and cost-effective approach to ensure water security.

Focus on ending hunger by ensuring food security forms the main subject matter of Chapter 6 that takes into account burgeoning population and vagaries of climate change that have increasingly necessitated the urgency of ensuring food security. The notion of food security is explained by the concept of hunger and thereafter the chapter proceeds to define the notion of food security by tracing the evolution of the idea of food security. While analyzing the impacts of climate change on ensuring food security, the focus is on food availability vis-à-vis climate change, access to food vis-à-vis climate change, food quality vis-à-vis climate change, food utilization vis-à-vis climate change and food stability vis-à-vis climate change. While taking into account the emanation of greenhouse gases from agricultural production, focus is also on various adaptation options available to ward-off the adverse impacts of climate change, with specific emphasis on increasing preference for climate change adaptation (CCA) in conjunction with disaster risk reduction (DRR) or CCA +DRR approach.

Chapter 7 deals with smart cities, especially at a time when the cities are at the center as becoming home to growing majority of global population. While elaborating increasing penchant for smart cities, this chapter takes into account linkages between urbanization and sustainable development to evaluate the concept of sustainable urban development along with brief appraisal of prevalent notions of cities like livable cities, eco-cities and their related components that make city life worth living. Thereafter, the study proceeds to examine prospects of sustainable smart cities, with specific focus on its constituents like smart mobility, smart economy, smart living, smart people, smart governance, and smart environment. While assessing options available for cities to tackle the vagaries of climate change, the chapter seeks to present a case for EbA as a cost-effective, viable, and durable option to deal with adverse impacts of climate change. Lastly, it suggests the implementation of New Urban Agenda of the UN-Habitat in tandem with sustainable development goal-11 as a way out.

The theme of life below water is dealt with in Chapter 8. Biotic life below water entails life in oceans, seas, rivers, estuaries, lakes, and other bodies that store water and this biotic life below water comprises organisms, both big and small, ranging from large fish, mammals, reptiles, and other tiny species. The marine life is threatened by both environmental and anthropogenic stressors. Ocean warming, acidification, deoxygenation, sea-level rise are the major components of environmental stressors; and anthropogenic stressors include plastic pollution, oil-spills, overfishing, greenhouse gases, land-based sources of marine pollution usually comprising all inputs that are released directly into the marine environment via human activities in coastal settlements, nearby cities, and native industries and the consequential discharge and disposal of urban and industrial waste, litter, untreated sewage, and coastal agricultural installations like fish farms, etc. Besides, there are land-based pollution sources. While elaborating the concept of ecosystem-based management (EBM) that is based or predicated on making use of the natural ecosystem boundaries as a framework instead of being confined by political or administrative boundaries, the chapter also deal with the application of EBM in the context of oceans because EBM focuses on the maintenance or augmentation of ecological structure · and function, and the benefits that accrue to the society from healthy oceans This chapter also provides a brief overview of linkages between EBM and EbA and in its final analysis, it suggests that in order to sustain life below water, EBM can be applied to overcome environmental and anthropogenic stressors.

Substantive theme of life on Earth is dealt with in Chapter 9, with specific focus on human life and other biotic organisms inhabiting Earth that are endangered due to the vagaries of climate change, overexploitation and unsustainable use of natural resources like freshwater ecosystems, forests, genetic resources, wildlife and land-use, etc., and concurrently emphasis in also laid on specific remedial measures in respect of each category with regard to safeguard against the adverse impacts of climate change and other related threats that endanger those categories. While emphasizing on the vitality of natural ecosystems and the goods and services accruing from them for human and other biotic organisms, focus is also stressed on the conservation and sustainable use of natural resources to ward off adverse impacts of climate change and sustain the continuity of life-cycle on Earth.

Emphasis on mainstreaming EbA in programs and policies in the action plan at national, provincial levels is stressed in Chapter 10 that, while focusing on the devastating havoc being wreaked by the vagaries of climate change, also notes the need that has increasingly augmented the urgency of mainstreaming EbA into policies and programs at all levels. While elaborating the concept of mainstreaming, the analysis also takes cognizance of categories of mainstreaming climate change adaptation, such as programmatic mainstreaming, managerial mainstreaming, regulatory mainstreaming, organizational mainstreaming, and directed mainstreaming. While focusing on mainstreaming EbA into policy process, the study emphasizes on the entry points for EbA mainstreaming, with specific emphasis on national action plan, sectoral action plan, and mainstreaming EbA at local and community levels. Chapter proceeds to assess pragmatic aspects of mainstreaming EbA in some countries like G-20 nations, Sweden, Bangladesh, and India with specific focus on Meghalaya, along with emerging lessons from mainstreaming EbA based on the field experience and the impediments that hinder the mainstreaming of EbA into national, provincial and local policy levels. In the final analysis, emphasis is focused on frequent and regular exchange of fresh knowledge and expertise garnered through ongoing EbA mainstreaming projects in order to surmount prevalent barriers.

Chapter 11 is in the form of conclusion wherein the impact of the outbreak of COVID-19 pandemic is briefly assessed during 2020 on different sectors that have been covered in previous chapters because the coverage of the data in these chapters is limited to the closing part of 2019, prior to the outbreak of the global pandemic. In other words, this chapter sheds light on the future outlook for sectors focused on in this book. Emphasis is on statistical information and status of factors pertaining to disasters, global warming, biodiversity loss, life on Earth, and life below water. It also serves as an update for the previous chapters with specific focus on the future outlook for EbA.

Acknowledgment

Having an idea and turning it into a book is as hard as it sounds but far more rewarding than I could have ever imagined. The experience is both internally challenging and exhilarating. I know it is a lonely pursuit, and sometimes it is hard to see the forest for the trees. COVID-19 has shown us how instrumental our families are to our existence. I am deeply indebted to my family for giving me the unwavering support and bearing with me through this journey.

I would also like to extend my deepest gratitude to my editors Ms. Emily Thomson and Ms. Louisa Munro for their monumental patience throughout the tedious process of completion of this book thus far whose constant cooperation has been a source of inspiration for me and to Elsevier for giving me this opportunity to write a book on Ecosystem-based Adaptation.

Without the experiences and support from my peers and team at the India Water Foundation, this book would have not existed, so I thank them all. You have given me the opportunity to lead a great group of individuals. I acknowledge all who were part of my journey and a few I lost during the trail.

I cannot begin to express my thanks to my wife Shweta for having illuminating discussions and for standing beside me through it all. I hope one day to possess just a sliver of your superhuman productivity and incredible knowledge of the literary world.

Finally, I recognize and acknowledge you—the reader. If you do not read my books, I would not write them. So once again, Thank you!!

Introduction

CHAPTER 1

No geographic region of the globe, including the humans inhabiting that region, is immune from the risks of hazards or disaster-related losses in terms of human life, destruction of property and infrastructure (NAP, 2001, p. 1). The 21st century is characterized by unprecedented risks of disasters, both natural and man-made, the rapid recurrence of which has amplified the scale and complexity of those risks to human existence, from local to global levels. An unparalleled pace of change taking place in human lives has proved instrumental in envisaging transformational shifts in the realms of economic, environmental, geopolitical, societal, and technological systems that offer unique opportunities as well as enhanced implicit systemic risks. It is in this context that world economic forum (WEF), in its global risks report 2014, emphasizes that stakeholders across business, government, and civil society face an "evolving imperative" in discerning and managing "emerging global risks" which defy geographical boundaries (WEF, 2014).

1.1 Defining disaster

There has been a mushroom growth of definitions of the term "disaster" in the current literature on hazards and disasters. Until recently, the terms "hazard," "risk," and "disaster" continued to be used interchangeably, despite the fact that each term connotes different meanings (Cutter, 1993, 1994; Kunreuther & Slovic, 1996; Quarantelli, 1998a; Mileti, 1999). Prior to attempting a widely accepted definition of disaster, it seems appropriate to have a brief overview of interconnectedness between these terms.

1.1.1 Hazard

In the broadest sense of the term, a hazard is construed as threat to human lives and property. Apart from entailing the inherent capacity of occurrence, hazards also imply the probability of wielding real impact of an event on humans or places. Imbalanced interaction between social, technological and natural systems spurs the emergence of hazards. The earlier mechanism of classifying hazards in accordance with their origin, for instance, natural hazards like earthquakes and technological hazards like industrial/chemical hazards, is no longer popular with the experts of hazard/disaster literature in the wake of new developments like deforestation, adverse impacts of climate change, and its attendant implications, thereby making hazards to have more complicated origins. According to S.L. Cutter (2001, p. 3), it is almost inconceivable to have cognizance of hazards without taking into account the context (social, historic, political, environmental, etc.) within which hazards take place owing to interlinkages between hazards and societal interactions.

1.1.2 Risk

The United States Presidential/Congressional Commission on Risk Assessment and Risk Management (1997) has identified risk as the probability of an event taking place or the likelihood of a hazard occurring. According to Cutter (2001, p. 4), risk emphasizes the "estimation and quantification" of probability to decide adequate levels of "safety and acceptability" of a technology or subsequent way to deal with it. In other words, risk is a constituent of hazard.

Like hazards, occurrence of disasters is attributable to many sources like natural systems, social systems, technological failure, etc., and disaster as a singular event causes tremendous losses to human lives, property, infrastructure, and environmental ecosystems. Quarantelli (1998a) opines that there are multiple perspectives on what comprises a disaster. Experts during the decades of the 1980s and the 1990s (Hamilton & Viscusi, 1999; Hewitt, 1997; Whyte & Burton, 1980) emphasized on the importance of differentiation between hazard, risk, and disaster because it demonstrated the heterogeneity of perspectives on to recognize and assess environmental threats (risks), what to be done about them (hazards), and how to respond to them after their occurrence (disaster). Cutter (2001, p. 4) has viewed growing emphasis on hazard, risk and disaster as also "reflective" of diverse "disciplinary orientations" of researchers and practitioners.

Until recently, concern about risks formed within the purview of the health sciences, psychology, economics and engineering in terms of risks' quantification, mathematical attributes, and use in decision-making. Interests in hazards emanated from the fields of geography and geology, whereas disasters attracted the intellectual domain of sociology. With the nature of hazards, risks, and disasters becoming more complicated and "intertwined" and the field of hazards' research and management "more integrated," Cutter opines that, these differentiations "blurred as did the differentiation between origins as "natural," "technological," or "environmental"" (Cutter, 2001, p. 4).

1.2 Definition of disaster

Disasters, being highly complex events entailing "low probability—high impact," are usually defined by the event itself and the venue of its occurrence which envisages variations in the definition of disaster that generally "reflects the nature and focus of the organization or individuals defining it" (Singh & Swarnim, 2010, p. 1). The first systematic disaster study is said to be conducted by Prince (1920), and Carr (1932) is credited with introducing definition and context to disaster study. The trajectory of growth witnessed by disaster study during the decade of the 1950s gathered momentum with the founding of the Disaster Research Center (DRC) in the United States in 1963. The occurrence of highly perceptible and apocalyptic disasters along with the availability of funding for disaster research during the decades of the 1970s and 1980s proved instrumental in increasing inventory of disaster research findings (Tierney, Lindell, & Perry, 2001). Submission of summaries of 1000 empirical studies by Drabek (1986) coincided with the calls by Quarantelli (1987) to focus attention on issues of defining disasters. Quarantell's insistence on keeping the issue of disaster visible culminated in the production of special issues of journals and volumes dedicated to the disaster study (Perry & Quarantelli, 2005; Quarantelli, 1998a) and the resultant outcome was the emergence of increased number of formal definitions from different perspectives in the research literature.

Period commencing with the end of the Second World War and closing with the publication of Fritz's definition of disaster, categorized by Perry (2018, p. 5) as classical period for disaster research, was characterized by the emergence of some disaster studies often left the meaning of disaster implicit. Definitions emanating from these studies clearly referred to an agent as catalyst denoted by the use of the term "event" and mostly dealt with social disruption, with less emphasis on specific agents underlying disaster. However, some allusions were made to different agents connected with differing elements of experience like speed, duration, magnitude or scope of onset of disaster (Perry, 1985, p. 18). However, in his seminal work on disaster research, Fritz (1961b) argued that the "therapeutic community" emerged out of "the social disruption" itself. Thus during the classical period, disaster research, while focusing on event-centric assessment of disasters also dealt with some implications for social disruption. Framework of disaster research developed during the classical period was carried forward by social psychologists and some disaster researchers that yielded the vision of social interactions assisted by norms that could be turned futile by disasters. Disaster research developed during the classical period contained both researches based on field studies as well as hypothesis-based research, which proved instrumental in spurring subsequent theory and definitional endeavors in disaster literature.

Suggestion of Killian (1954, p. 67) in describing disasters as "disruption of the social order" resulting in physical destruction and death entailed that people had to deal with such an eventuality by departing "from the pattern of norm of expectations." In an identical manner, Wallace (1956, p. 1) described disasters as "extreme situations" entailing not only impact, but the risk of disruption of ordinarily "effective procedures for reducing certain tensions, together with a dramatic increase in tensions." On the basis of his studies of tornadoes in Texas, Moore (1958, p. 310), opined that disasters made people follow "new behavior patterns" as a "defining feature," nevertheless, he also affirmed that "the loss of life is an essential element." These three definitions of disasters formulated by Killian (1954), Wallace (1956), and Moore (1958) during the classical era, apart from sharing remarkable consistency with each other, each described disaster in terms of the effect or risk of an agent and each focused on social disruption.

Fritz (1961a, 1961b, p. 655) has defined disaster, as "an event, concentrated in time and space, in which a society, or a relatively self-sufficient subdivision of a society, undergoes severe danger and incurs such losses to its members and physical appurtenances that the social structure is disrupted and the fulfillment of all or some of the essential functions of the society is prevented." While encapsulating the sociological concept of disaster, Fritz perceived disaster as impacting the whole society or some subdivision of it and incorporated threat and actual effect and while doing so concurrently he underscored that essential functions of the society get prevented. While not distinct from past definitions in many respects, Fritz's definition did try to be specific about the role of social interactions in an eventuality of a disaster. In response to criticism from some critics (Quarantelli, 1984), Fritz (1961a) did add caveat of "time and space." Formulation of the notion of disaster by Fritz occurred in the backdrop of the Cold War being at its zenith and hence it was inevitable for disaster to be perceived in the social context of the Cold War. Added emphasis on threats from external attacks during the Cold War era could be said to have influenced the notion that disasters could be driven by agents and external to society or social group. Fritz's definition proved instrumental in spurring bulk of studies undertaken by large number of multidisciplinary and international body of disaster researchers (Quarantelli, 1987a).

For almost six decades now, definition of disaster formulated by Fritz in 1961 continues to inspire scholars of disaster research, directly or indirectly. Many of these scholars have either adopted Fritz definition verbatim or cited in their own research studies or adopted the basic traits of the definition to suit the contemporary understandings. Such examples include studies of bush fire disasters by Wettenhall (1975), research by Peacock, Bates, Dynes, De Marchi, and Pelanda (1987, p. 292) on social change and disaster, study of a nuclear power plant accident by Perry (1985), the review of flood studies by Perry and Lindell (1997), and cross-national studies of natural disasters by Lowendahl (2013, p. 11).

While adopting Fritz' definition, Drabek (1986, p. 7) incorporated the provision that "disasters are accidental or uncontrollable events, actual or threatened." Fritz definition is broadened by Buckle (2005, p. 179) by adding emphasis on the scale of social disruption, stating that there is a sense of substantial, irretrievable loss and damage, calling for "the need of long-term recovery." Likewise, Smith (2005, p. 301) suggested that disasters are events that bring about death and damage and cause "considerable social, political and economic disruptions." Period subsequent to Fritz's formulation of definition of disaster was characterized by the inclusion of changes and additions by disaster researchers, while retaining the major substance of the original definition, to add theoretical clarity and update for changes in the contemporary disaster research literature. Focus on the social order constitutes a major defining tenet of many disaster researchers in the existing disaster literature.

Emphasis of G.A. Kreps (1998, p. 34) in his description of disaster as "nonroutine event" that causes social disruption and physical damage and his focus on four key defining features of disaster that include forewarning, magnitude of impact, scope of impact, and duration of impact, undeniably bring his notion of disaster closer to Fritz. Similar proximity to Fritz can be discerned in Porfiriev (1998, p. 1) when he describes disaster as the "destabilization" of the "social" system pointed out by failure of routine functioning that calls for an intervention" to restore stability.

Emerging trends in recent literature on disaster research make it discernible that many scholars are shifting away their focus from agent-based conceptions or that are optimally dependent on perceptions of physical damage. Physical destruction still continues to be part of the quantum of disruption caused by the disaster, but not as the key characteristic of disaster. While drawing upon the fundamentals of Fritz's definition, Drabek (2013) in his work specifically points out that disasters are more social-centric rather than agent-centric. The notion of disaster being social-centric in origin and not agent-centric is gaining wider support and recognition in current disaster research. While contending that "disasters are not a function of agents, but are social in origin"; Gilbert (1998, p. 13) subsequently also perceived disasters originating from human agency. Emphasis on disaster as a social phenomenon pertaining to "social change" that is acknowledged across "social time" as a "radical change" in the prescriptive environment is underlined by Rosenthal (1998, p. 226). Inherent social nature of disaster has also been emphasized by Wisner, Kelman, and Gailliard (2014, p. 16).

Russell Dynes (1998, p. 109, 126) describes disaster as an occasion when the failure of norms impels a community to engage in extraordinary efforts designed to safeguard and benefit some social resource. Disaster is viewed by Rodriguez and Barnshaw (2006, p. 222) as "human-induced, socially constructed events" that form the part of the "social processes that characterize societies." McEntire (2015, p. 3) defines disasters as "deadly and destructive" social events that "injure and kill people, damage infrastructure and personal property" and complicate daily routine activities.

Disaster is perceived by Pescaroli and Alexander (2015, p. 5) as a situation that engenders a sequence of events in "human subsystems that result in physical, social and economic disruption."

1.2.1 Toward consensual definition

Devising a consensual definition of disaster has almost been an imponderable task in the wake of vast array of definitions worked out at different times by scholars from many different disciplines. Irrespective of the divergence of views on traits underlying the definitions of the phenomenon of disaster neglect of views pertaining to causes or consequences of this phenomenon, there has been increasing compatibility of observations among scholars in recent years about salient features of disasters. Undoubtedly, explicit differences are discernible between disciplines regarding focus on the phenomenon of disaster; nonetheless, these differences stemming from the diverse domains of disciplines have seemingly been instrumental in enriching the fecundity of research entailing models and theory-work about disasters across disciplines owing to cross- and interdisciplinary involvement. Despite prodigious variation with regard to the theory context in which the notion of disaster is defined and disagreement among scholars with respect to assigning number of defining features to the notion of disaster. Nevertheless, there is increasing agreement in contemporary literature on assigning minimum defining features to disasters.

At the cusp of the new millennium Quarantelli (2000, p. 682) tried to elaborate a consensus definition of disaster by asserting that disasters are "relatively sudden occasions when... the routines of collective social units are seriously disrupted and when unplanned courses of action" ought to be undertaken to cope. The modern-day scholars may not find much difficulty in dealing with this composite definition. Subsequent assertion by Quarantelli (2005, p. 339) that disasters must be construed as an inherently social phenomenon finds endorsement from many present-day scholars that the disaster is the key disruption in the social system, irrespective of its size, that entails the potential of rendering futile any pattern of prevalent social interaction. There is increasing acknowledgment by many scholars about close proximity between different patterns of social interactions and the disasters (Donner & Rodriguez, 2008).

A closer scrutiny of diverse definitions of the term disaster makes it discernible that various issues like the role of a hazard agent and the mechanism of gauging physical damage in disaster definition have seemingly been left out to help reach a wider consensual definition of disaster. Until the recent past it has been said that disasters are defined by social disruption rather than by hazard agent and as has been argued by Shaluf (2007, p. 687), agent-specific focus culminates in developing pseudo-typologies that endeavor to describe or group disasters into various groups like natural, man-made, public health, creeping, hybrid, etc.

Quarantelli (2005) argues that there is ample disaster research and theory to demonstrate that social scientific attention ought to focus classification along more fundamental theory-based lines.

According to Perrow (2007), despite the fact these classifications are not theory-based and are seldom used analytically, they still find place in various literatures. Significance of a proximal agent as an articulation of hazard processes while defining disasters continues to be emphasized by some scholars. Some these scholars, based in physical science outlook, especially the geologists, focus their work on hazard process and define disasters accordingly, as pointed out by

Abbott (2014) and Keller and DeVecchio (2014). Some definitions of disasters propounded by few social scientists recognize that the nature of physical agents affects features of disaster occasions and features, interalia, include level of fear, magnitude of impact, etc. It is also acknowledged that the behaviors are also affected by these physical agents during and following the disruption. According to Perry (2018, p. 15), "There has been, however, movement away from the contention that any agent "is" the disaster, but disagreement persists regarding the extent to which agents are central or peripheral features of disaster definitions."

A contention reflected in the classical era definitions of disasters that disasters occur or originate "outside" the nidus of social system finds rare allusion in contemporary literature perhaps in the wake of increasing recognition that all disasters are human-induced. A related issue is the contention that disasters originate "outside" the focal social system, which arises in some classical era definitions. This claim appears rarely in contemporary literature, probably owing to the growing acknowledgment that all disasters are human-caused. Ecological perspectives lay emphasis on the origin of disasters within the social system itself where causes are inherent in the social structure, social interactions, and the environment as a whole. Different interpretations are offered with regard to the role of physical damage in defining disasters and the fact that damage is not essentially a defining feature of disaster has been acknowledged by scholars since the classical era. Despite the controversy over whether damage should constitute a part of disaster definition or not, there prevails a semblance of some agreement that the extent of damage should be gauged not in terms of loss of lives or property but by the magnitude of the disruption and failure of the normative or cultural system (Perry, 2018, p. 15).

Assertion by Perry (1989, p. 354) that typologies offer a means of classifying occasions and findings to have more conceptually suitable comparisons finds support in Lukic et al. (2013) when they argue that disasters meaningful definitions of disaster can only be had within the categories of a classification scheme or typology. Many categories in a typology of "collective stress" were created by Barton (1969, 2005) and he subsequently (Barton, 1989) classified disaster types based upon a medium of four dimensions—scope of impact, speed of onset, duration of impact, and social preparedness—and described each cell in social and interactive terms. Kreps (1989) developed a complex system by exploring domains, tasks, resources and activities (DTRA). Elaboration of "crises" as a more general dimension incorporating disaster and to conceptualize other kinds of crises as well, has been facilitated by Boin (2005).

The corpus of available literature makes it discernible that only scant use of typological classification has been made by researchers till date to put their research in conceptual space. Nevertheless, Perry (2018) has argued that in the aftermath of the work of Boin (2005), the practice may see an upsurge. The theory-based proposal of a 10-point scale by Fischer (2003, p. 100) to measure the severity of disasters has also been rarely used by social scientists. Thinking and research of the geographers and anthropologists has mostly been dictated by the context in which hazards and disasters occur. Asserting that mounting contemporary stress on ecological perspectives may envisage new strategy for research and design thereby spurring adjustments of disaster definition, Perry (2018, p. 16) opines that ecological perspectives "embrace a macroscopic view that minimally should direct attention to the threat or risk environment."

Some research is said to have been undertaken in studying disasters in the context of the ambit of threats impacting the focal system or environment. Study of hazard perception in the context of three natural hazards—volcanoes, fires, and earthquakes—was undertaken by Perry and Lindell (2007). In their research study, Lindell and Hwang (2008) focused on natural hazards (flood and hurricane) with a toxic chemical release threat. The risk environment of communities threatened by multiple volcanoes is examined by Diefenbach, Wood, and Ewert (2015). Study of the release of hazardous materials in connection with natural disasters is also gaining prominence in literature (Sengul, Santella, Steinberg, & Cruz, 2012; Young, Balluz, & Mililay, 2004).

Perry (2018, p. 16) draws attention to the increasing usage of the term "cascading disasters" in the literature to characterize the broader vulnerability of a place. However, Kumasaki, King, Arai, and Yang (2016) point out that sometimes the usage of this term is narrow in scope, alluding to disasters that occur in time sequence and appear to be connected. Nevertheless, Pescaroli and Alexander (2015) have argued that conceptualization of the cascading disasters can be facilitated in broader terms, not "falling dominoes," that more efficiently captures the hazard and disaster context. However, Perry (2018, p. 17) opines that cascading disasters are "not a variant on the disruption (disaster), but a focus on the broader hazard and disaster environment and how that environment may be manifest in multiple disaster episodes that are in some way sequential or linked."

For the purpose of this book reliance can be placed on the definition provided by UNISDR (2017): "Disaster is a serious disruption of the functioning of a community or a society at any scale due to hazardous events interacting with conditions of exposure, vulnerability and capacity, leading to one or more of the following: human, material, economic and environmental losses and impacts."

1.3 Hazard-disaster linkages

Relatedness of hazards and disaster to many disciplines and professional communities has culminated in the emergence of multiple definitions of these two terms. In many cases, the terms hazards and disaster refer to natural hazards and disasters and are often used interchangeably; however, each term has a "precise and distinct meaning" (Paul, 2011, p. 1). Of the many definitions of hazards, only definition by Alexander (2000, p. 7) envisages a differentiation between hazards and disasters when he writes that "ä hazard is an extreme geophysical event that is capable of causing a disaster." Apparently, Alexander's definition does not indicate any distinction between hazard and disaster, but it does suggest that hazards may "transform into disasters" thereby to become "sequential" events (Paul, 2011, p. 2). For Thywissen (2006), every disaster starts with a hazard. Paul (2011, p. 2) opines that the fact of human actions often playing prime role in causing and/or exacerbating the impacts of extreme events gets ignored in Alexander's definition. Suggesting that hazards epitomize the "potential occurrence" of extreme "natural events," or "likelihood" to cause the severe adverse effects Tobin and Montz (1997) opine that disasters accrue from "actual" hazard events. Thus "a hazard is a threat not the actual event" (Paul, 2011, p. 2).

Hazards as source of threat to human lives, property, and environment is explained by Cutter (1993). Reiterating similar stance, Oliver (2001, p. 2) has defined a hazard "as a threat posed to people by the natural environment." A hazard is defined by the United Nations as an agent or threat that is "a potentially damaging physical event, phenomenon or human activity that may cause the loss of life or injury, property damage, social and economic disruption or environmental degradation" (UNISDR, 2004). While directly acknowledging the role of humans in causing/exacerbating hazards, this definition also distinguishes between a hazard and a disaster by including the word "potentially," thereby accounting for all possible hazard manifestations.

Undoubtedly, discernment of natural hazards entails many disciplines; nonetheless, geography and other geographical disciplines constitute the prime source of studying natural hazards. One emphasis is on examining the hazard processes that leads to the occurrence of earthquakes, tornadoes, floods, volcanic eruptions, and other identical events. Another stress is on natural disaster albeit within the framework of the processes allied to the hazard and this is termed as an all-inclusive approach that is occasionally construed in the backdrop of another attempt like resource management (Burton & Kates, 1964). It is opined by scholars like Burton, Kates, and White (1968) and Kates (1971) that natural hazards perceptions entail early and durable links to human ecology. Nevertheless, Burton, Kates, and White (1978) allude to the classic statement of the hazards approach. A disaster, within this framework, is often perceived as an extreme event that takes place when a hazard agent crisscrosses a human use mechanism. The resultant occurrence of disasters as part of normal environmental processes entails the potential of making such processes as focus of study. Perry (2018, p. 8) cites the example of the occurrence of an earthquake that is treated as a disaster if it affects the people, and an earthquake takes place as a consequence of seismic activity irrespective of the fact that whether the people are affected by it or not. In this context, Perry (2018, p. 8) opines: "At least in early formulations, the cause of a disaster is the extreme event and understanding disaster rests upon understanding the larger process (engaging both social science and natural science perspectives)."

The narrow focus on disaster events articulated in bulk of the definitions evolved during the classical period manifests a differentiation from the mega view of hazards researchers. Disasters are prone to become "an epiphenomenon" instead of becoming a focal point of definition and explanation, as pointed out by Quarantelli (2005, p. 342), when hazard cycles and agents are the nidus. Concomitantly, it is identically right, nevertheless, as asserted by Birkmann et al. (2014, p. 4), that "a disaster is but a moment or materialization of [important] underlying conditions." Each of the disciplines like sociology, geography, psychology, anthropology, etc., where disasters are studied is prone to manifest disciplinary interests in evolving definitions (Gaillard, 2016). Undeniably, research from each standpoint has made significant contribution to the corpus of literature on disasters.

In his attempt to define disaster as a component of the environmental process, Oliver (1980, p. 3) also delineates it as a process that takes place when human systems traverse with the hazard producing major "human hardship with significant damage." This explanation includes the crucial issue of a recurring environmental process along with the notion of somber social interruption and physical impairment. An endeavor by Susman, O'keefe, and Wisner (1983, p. 264) to define disaster as "the interface between an extreme physical event and a vulnerable human population," brings them closer to the view of traditional geographers in this regard. Exposition of disaster as events where "physical agents define the problem" is facilitated by Hewitt (1998, p. 77). Interestingly, Hewitt had argued in 1983 that disasters could be construed in terms of unannounced and unparalleled impacts that "derive from natural processes of events" (Hewitt, 1983, p. 10). It is opined by Peek, Mileti, Bechtel, and Churchman (2002, p. 512) that occurrence of disasters takes place when extreme events in the natural environment interact with "the natural, social and constructed environments." The view that interactions between human use systems and natural processes that generate negative impacts for people and the given environment spur the occurrence of disasters is pioneered by Paton and McClure (2013, p. 4).

Nevertheless, the damage caused to the systems supporting human life, particularly agriculture, infrastructure, etc., is counted among the consequences of disaster by these scholars who base their surmise on the assumption that such damage entails the potential of impacting human systems even if they are either far-fetched or otherwise protected. Spotlight on the traditional concern of hazards researchers with the system of hazard agents and the resultant impact of their interaction with human systems gets focused in each of these definitions. Conceding that while the major thrust of hazards perspectives dealt with hazards from natural resources, Perry (2018, p. 9) does not rule the possibility of using a hazards' view when "the nature of the underlying threat is human-generated by specifying the underlying force or process."

Some hazard researchers have, keeping in consonance with a macroscopic significance, adhered to an unambiguous focus on the nature of consequential impacts on social vulnerability. It has been opined by Alexander (1993, p. 4) that natural disasters can be construed as rapid "onset events" with tremendous effect of the "natural environment upon the socio-economic system." Elaborating this phenomenon in his subsequent writing, Alexander (2005, p. 29) averred that disasters could not be defined by fixed events but by "social constructs and these are liable to change." Here the emphasis by Alexander is on the fact that the disaster is not merely an event accruing from carrefour of human and natural systems but the social consequences of the event and these consequences are ever changing and divergent across groups. While conceding that disasters do arise from the intersection of the physical, built and social environments, Mileti (1999, p. 3) asserts that disasters are "social in nature." While emphasizing on the fact that human encroachment upon physical environment can be instrumental in creating disasters, thereby tracing the origin of disasters to a hazard, Mileti here also underscores the social emphasis while studying the event.

In their attempt to define a disaster as a situation entailing "a natural hazard which has consequences in terms of damage, livelihoods, economic disruption and/or casualties" that can surpass local capacity to deal with, Wisner, Gaillard, and Kelman (2012, p. 30) also sounded a warning that they did not intend to do away with events in small isolated places that are bereft of the choice of seeking resources from outside. While elaborating his notion of disasters in a vulnerability context, Cutter (2005, p. 39) maintained that the issue at hand is not disasters as events but rather human "vulnerability (and resiliency) to environmental threats and extreme events." Viewed in a broad sense, it can be surmised that each of these definitions points toward a focus on social contexts to changeable degrees. Undoubtedly, approaches dealing with hazards entail a long-lasting interest "consequences and vulnerability" as pointed out by Quarantelli (1998b), nevertheless, definitions worked out from this perspective have frequently incorporated social disruption as at any rate one prominent feature of disaster. The hazard researchers' proneness to move in this direction attests to the fact that they seemingly in agreement with sociological researchers "to place people and social relationships at the core of disaster study" (Perry, 2018, p. 9).

Interconnectedness between disasters and social systems has been focused by many researchers. Disaster is viewed by Dynes (1998, p. 13) as events accruing from failure of norms, spurring a community to adhere to unusual endeavors "to protect and benefit some social resources." AN identical perspective is enunciated by Rodriguez and Barnshaw (2006, p. 222) when they describe disaster as "human-induced, socially constructed events that are part of the social processes that characterize societies." Ability of disasters to strike "with such severity that the affected community has to respond by taking exceptional measures," has been emphasized by Carter (2008, p. 9). While viewing disasters in relationship to rudimentary hazards, McEntire (2015, p. 3) also emphasizes that disasters are important disruptive social events that call for alterations in routine behaviors. While looking at disaster as situation that "generate a sequence of events in human subsystems that result in physical, social and economic disruption," Pescaroli and Alexander (2015, p. 5) argue that the magnitude of the disruption is determined by the levels of vulnerability. Presence of event notion as intrinsic to the definition of a disaster is acknowledged by these authors and they also characterize disasters as social phenomena that are ingrained in the social structure itself. This aspect is amply demonstrated by Ait-Chellouche (2015, p. 423) when he describes disaster as "serious disruption of the functioning of the community following widespread human, material, economic or environmental losses."

While drawing upon human interactions and the perception that drives them, Jigyasu (2005) formulates his definition of disasters entirely in social systems. On the contrary, Horlick-Jones (1995, p. 311) describes disasters as "disruptions in cultural expectations" that gives rise to the impression about the inability of institutions in keeping threats under check. Articulating an almost identical perspective, Dombrowsky (2005) regards the disaster as social, engendered in social structure and can only be examined *via* that route. For Boin (2005, p. 159), disasters are ingrained in social structure and occur in the eventuality of disruption in the life sustaining functions of the system. For Hewitt (2016, p. 8), the key traits of disaster emanate from the "disruption of a significant part of society's productive activity and administrative functions." So, there is no exaggeration in surmising that many scholars have perceived the phenomenon of disaster in the social context and accordingly they have tried to mull definitions of disaster in the context of social change.

1.4 Disaster, vulnerability, and resilience

Undoubtedly, vulnerability and resilience do not constitute defining features of disaster definition; nonetheless, they are related to causes, conditions, or consequences of disasters (Quarantelli, Lagadec, & Boin, 2006). The available corpus of literature on disaster includes many studies and endeavors at theorizing to establish vulnerability and resilience as causes or effects of disasters; hence each notion merits brief appraisal here. According to Boin, Comfort, and Demchak (2010) notions of vulnerability and resilience have emerged as trendy catchwords not just in technical literatures, but also in popular narrative about politics, sports, and everyday pastimes. It has been opined by Gaillard (2010, p. 219) that each term glaring started becoming discernible in the disaster literature in the 1970s—vulnerability first (O'Keefe, Westgate, & Wisner, 1976), and resilience later (Torry, 1979). Subsequent period was characterized by frequent appearance of each concept in research and theory, especially by the scholars making use of hazard type perspectives (Singh-Peterson, Salmon, Baldwin, & Goode, 2015, p. 756).

Perry (2018, p. 13) opines that "the idea if not the term" vulnerability has been articulated in many historical discussions of disaster. In some cases, this term has been explicitly used by researchers in their disaster definitions. According to Wamsler (2014, p. 4), disasters emerge as a consequence of an interaction between "hazards and vulnerable conditions." It is opined by Bradshaw (2014, p. 34) that presence of disaster becomes discernible when "an individual or group is vulnerable to the impact of a natural or human-made hazard." Scholars like Konukcu, Mentese, and Kilic (2015, p. 14) hold the notion of vulnerability as a cause, condition or consequence of

1.4 Disaster, vulnerability, and resilience

disasters, or interrelated with extent of disruption, but not as a trait or component of the definition itself. An elaboration of relationship between human vulnerability and disasters is provided by Blaikie, Cannon, Davis, and Wisner (1994).

Noting the conceptual and operational ambiguity in the formal or conventional definition of vulnerability, Lindell (2013, p. 11−12), also emphasizes on the need for identifying identify which variables are indicators of vulnerability, which are proximal and distal causes, and which are simply correlates of vulnerability. In a similar vein, Alexander (2016, pp. 2−3) while describing vulnerability as a crucial concept for future research and practice, also emphasizes that attention should be accorded to both elucidation of conceptual relationship of vulnerability to disaster and comprehending the critical dimensions of vulnerability itself as a distinct concept. Construed in a broad perspective, bulk of the challenges presented by vulnerability as well as resilience emanate from the need to especially adapt it to disaster research and theory. While initiating the process of elucidating the relationships among the notions of disaster, vulnerability, and resilience, Aguirre (2007, p. 41) suggested that a great deal of scrutiny was called for by scholars to effectively integrate either concept into the discourse about disasters.

Arguing that disaster resilience is a reasonable extension of and complement to the notion of vulnerability, Zakour and Gillespie (2013, p. 73) postulate that resilience beguiles the capacity to minimize the effects of disasters *via* multiple potential mechanisms. In the opinion of Aguirre (2006), resilience is a useful concept for disaster researchers and they ought to continue to look for pellucidity and unanimity on issues of meaning and conditions. In the wake of numerous ostensibly different definitions and specifically illumination of the notion of "bouncing back from disasters," both Hayward (2013) and Aldunce, Beilin, John, and Howden (2014, p. 252) emphasize on seeking basic meaning consensus. An endeavor in quest of assimilating an eclectic assortment of perspectives—individual, community, institutional, and environmental—has been made by Paton (2006, p. 305). Undoubtedly, attempts have been made by Berkes and Ross (2013) to seek "common ground" between approaches grounded on social-ecological systems and those positioned in the psychology of individuals; nevertheless, Barrios (2014) opines that there remain unresolved issues for applications at the community level of analysis, for means to appraise resilience (Cutter, Ash, & Emrich, 2014) and for the relationship of resilience to public policy, especially disaster risk reduction (DRR) (Amundsen, 2012). While pinpointing the interrelatedness between resilience and vulnerability, Cutter (2016) emphasizes on the need for laying down not just the conceptual particulars of each concept but also the subtle relationship between them.

Conceding that resilience may come up in disaster definition explanations to the extent that it is conceptually perceived as a modifier of vulnerability or in the applied arena to the degree that resilience can be defined, learned, and implemented across disasters and communities (Leykin, Lahad, Cohen, Goldberg, & Aharonson-Daniel, 2016). Nonuse of resilience as an element of disaster definitions themselves to-date is explained by Perry (2018, p. 13): "probably because the thrust of the concept appears to be more as either a reaction to disasters (after the disruption) or as features of the unit of analysis (ecologically the community) that modify the magnitude of subsequent disasters (thereby before the focal disruption)."

Instinctively, the disruption that constitutes a defining feature of disasters entails the likelihood of being affected by the presence of resilience. It is in this backdrop that Klein, Nichols, and Thomalla (2003, p. 41) suggest that systematic use of the concept must anticipate further conceptualization that explicates the definition and elements of resilience and its relationship to other concepts such as vulnerability, as well as empirical verification of the projected relationships.

1.5 Toward disaster risk reduction

During the past four decades (1980–2019), there has been phenomenal increase in major disaster events. The period between 2000 and 2019, recorded 7348 disaster events claiming 1.23 million lives, affecting 4.2 billion people, incurring nearly $2.97 trillion in global economic losses (CRED-UNDRR, 2020). The previous 20 years, between 1980 and 1999, there occurred 4212 disasters that were linked to natural hazards globally claiming nearly 1.19 million lives and affecting 3.25 billion people and the accrual of economic losses were estimated at $1.63 trillion. In other words, there was a sharp increase in 2000–19 period over 1980–99 period in the occurrence of disasters, people affected, loss of human lives as well as economic losses. Nevertheless, much of the difference is discernible by a rise in climate-related disasters including extreme weather events that increased from 3656 climate-related events during the 1980–99 period to 6681 climate-related disasters in the period 2000–19 (CRED-UNDRR, 2020). For several years now, the world has been witnessing an increase in loss of human lives, augmentation in economic losses as well as increase in the number of people affected worldwide due to both natural and climate-related disasters (Pandey & Okazaki, 2003, p. 1; Tang, Wu, Ye, & Liu, 2019). A suite of factors like climate events, high population densities and substandard housing conditions were found to be rendering cities vulnerable to natural hazards, especially in Africa (Parnell, Simon, & Vogel, 2007, p. 357; Salami, von Medling, & Giggins, 2017). The physical, social, and economic losses incurred on account of natural hazards are more severe in developing countries (Botzen, Deschenes, & Sanders, 2019).

The effects of natural hazards also wield influence over socio-economic conditions, tradition, culture and environment of the affected communities (Pandey & Okazaki, 2005, p. 1; Mata-Lima, Alvino-Borba, Pinheiro, Mata-Lima, & Almeida, 2013). There has been a progressive shift in the approach to deal with the impacts caused by natural hazards (Innocenti & Albrito, 2011, p. 730), and one of the ways in which natural hazards and their impacts were broached has been on the context of disaster management (DM), especially during the 1980s when the DM approaches still only centered on the underlying risks that make people vulnerable (Van Riet & Van Niekerk, 2012, p. 38; UNISDR, 2015b). Disaster management is defined by the UNDP (1992, p. 21) as "the body of policy and administrative decisions and operational activities which pertains to the various stages of hazards at all stages." It comprises cycles that draw in prevention, mitigation, early warning and recovery (Gratwa & Bollin, 2002, p. 19). During this juncture of time when disaster management was operational, institutional hazard management practice had been mostly making use of a top–down strategy based on the supposition that exposure to natural hazards constitutes risk (Delica-Willison & Gaillard, 2012, p. 713). The ratiocinative and practices of the approach of disaster management were more directed toward reacting and responding to emergencies, managing emergencies and aftershock recovery (Innocenti & Albrito, 2011, p. 730). Given that all activities and resources in the DM approach were directed toward catastrophic events, this approach was confronted with some challenges (Ramanuja, 2015). Irrespective of the fact that the emphasis on the primary causes of hazards like risks and vulnerability was not taken into account in most of the cases (Van Riet & Van Niekerk, 2012, p. 38), DM was still regarded as the effective way of responding to hazards. Apart from exclusion of any involvement of communities at risk, this approach also paid no attention to underlying risk (Gratwa & Bollin, 2002, p. 19).

Decision of the United Nations vide Resolution 44/236 of 22 December 1989, to launch an International Decade for Natural Disaster Reduction (1990–99) commencing from 1 January 1990 proved instrumental in shift of emphasis from disaster management to DRR along with the recognition that DRR was a social and economic imperative (IDNDR, 1996). The international action initiated under the IDNDR, especially in the developing countries was designed to reduce loss of lives, poverty damage, social and economic disruption caused by natural hazards (UN-ECOSOC, 1999). Issues of major concern in disaster risk management (DRM) pertaining to education, capacity development, social impact and vulnerability, civil society and public-private partnership, economic and health aspects, land use planning and environmental protection were addressed under the aegis of the IDNDR (IDNDR, 1999). Convening of the world conference on Disaster Reduction at Yokohama, Japan in 1994 under the aegis of the IDNDR proved instrumental in the adoption of the Yokohama Strategy and Plan for a safer world (IDNDR, 1994). Yokohama Strategy is said to have been adopted in the backdrop of the realization that impacts of natural hazards in terms of human and economic losses continue to rise and societies are becoming more vulnerable to such hazards. The Yokohama Strategy proved instrumental in playing a pivotal role in emerging as a precursor to the establishment of the International Strategy for Disaster Reduction (ISDR) in 2000 under the auspices of the UN that aimed at the development of a global culture of DRR into the broader context of sustainable development and environmental considerations, and concomitantly, ISDR succeeded the IDNDR in 2000 (UNISDR, 2009a, 2009b, p. 20). An assortment of efforts was prioritized through the UNISDR to address the limitations of the disaster management (DM) approach by the humanitarian groups and those NGOs that were directly involved in helping the communities affected by hazards (Gaillard, 2007), it was only then that the significance of focusing on as to what makes the people susceptible was realized (Gratwa & Bollin, 2002, p. 20).

This cognition was also spurred by observing that apart from the happening of hazards and people's propinquity and exposure to hazards, people also suffered owing to prevailing socio-economic and political conditions that render them susceptible to these natural hazards (Delica-Willison & Gaillard, 2012, p. 713; Shaw, 2012, p. 4). This realization also proved instrumental in the shift of emphasis from an earlier approach that had been used to tackle natural hazards to proactive approaches as new mechanisms to deal with the impacts wrought by natural hazards. Proactive approaches aim at understanding as to why hazards occur by applying integrated holistic approaches to decrease the impacts of hazards as well as focusing on underlying risks (Bankoff, 2015). Deployment of more proactive approaches to tackle the effects of hazards resulted in the emergence of the notion of DRR that has come to replace the previously used top–down approach that mostly failed to reach community members who were worst affected by hazards (Srivastava et al., 2016; USAID, 2011). Under DRR, specific attention is paid to the people who are most affected by hazards by putting them at the center of all initiatives (Eiser et al., 2012, p. 5; Ferdinand, O'Brien, O'Keefe, & Jayawickrama, 2012, p. 85). People at the grassroots level are acknowledged under DRR *via* community-based organizations re allowed to respond to emergencies in a rapid, efficient and fair manner (Gratwa & Bollin, 2002, p. 19). Furthermore, DRR provides a platform wherein people can be encouraged to participate from the initial development of the program until its implementation (Patterson, Weil, & Patel, 2010, p. 127). Emphasis on members of the affected communities under the DRR mechanism gradually evolved in the official developed of an official approach that has come to be known as Community-Based Disaster Risk Reduction (CBDRR) (Shaw, 2012, p. 6).

1.6 Hyogo framework for action

There had started developing thinking in 2001 that there was lack of knowledge among many decision-makers that led them more focused on disaster relief rather than disaster reduction, and at the same time, they also lacked appropriate information with regard to the short-term socio-economic effects and long-term losses of disasters on vulnerable people and communities who could be saved if the focus was on reduction rather than relief (de Govet, Marti, & Osorio, 2006). In this backdrop, UNISDR (2001) emphasized that the focus of decision-makers ought to be on reducing natural disasters by dealing with vulnerability factors like dependence on agriculture, deforestation, forced displacements, and climate change. An example is cited in this regard in respect of Humboldt Center, a local community center in Nicaragua that followed a successful method to minimize the effects of disasters by examining soil factors in that region to decide the places where people could build houses and live safely (UNISDR, 2001). Thus emphasis on building the resilience of vulnerable people to face natural disasters became the focal point and main topic of discussion at the World Conference on Disaster Reduction held in Hyogo, Japan in 2005 (UN, 2005). This conference adopted a framework for action and that has become known as Hyogo Framework for Action (HFA) 2005–15. Characterized by a holistic strategic approach for disaster risk reduction, HFA has emphasized not only on natural hazards but also their interaction and relation with connected technological and environmental hazards (USAID, 2011). While building on the success that had been attained since Yokohama Strategy (1994) in terms of lessons learned, HFA identifies a number of gaps that need to be addressed (UNISDR, 2005). Some of these gaps, interalia, include: (1) Governance: organizational, legal, and policy framework; (2) risk identification, assessment, monitoring and early warning; (3) knowledge management and education; (4) reducing underlying risk factors; and (5) preparedness for effective response and recovery (UNISDR, 2005, p. 2).

Hailed as an international blueprint for disaster risk reduction, the HFA has a comprehensive goal of building the resilience of nations and communities to disasters toward 2015 (UNISDR, 2005). It accords primacy to five priorities for action toward achieving the following: making DRR a priority; knowing the risks and taking actions, that is, identifying, assessing and monitoring disaster risks, building the culture of resilience making use of all possible knowledge, innovation and as well increase understanding and awareness, minimizing risk factors via adequate risk management techniques; always be prepared and respond appropriately when necessary. These priority areas were further simplified by Benson and Twigg (2007) into five thematic areas: governance, knowledge and education; risk management and vulnerability reduction; risk assessment; disaster preparedness; and response. While delineating the subthemes of the themes that are relatively the characteristics of resilient communities, Zhou, Perera, Kayawickrama, and Adeniy (2014) have opined that the HFA identified some issues that are required to be taken into consideration while conducting the key activities anticipated under the priority actions, and these issues are: the importance of multihazard approach to all actions, gender perspective or influence as well as cultural diversity; community and volunteers' participation; and capacity building and technology transfer. HFA identified state governments, regional organizations and institutions and international organizations, including United Nations System and International Financial Institutions, as prime actors in this regard. There were prone to undergo modification, especially in the post-2015 edition in the wake of identification of the potential significant role of the community along with local and international private sector actors (Ki-Moon, 2013; UNISDR, 2013a, 2013b).

As the implementation deadline of Hyogo framework was drawing near, some pertinent context toward a successful post-2015 framework was identified by UNISDR (2013a, 2013b) and this emphasized on the need for better addressing underlying risk factors along with emphasis on engaging more actors like private sector and local government leaders. It was also demonstrated, as noted by Zhou et al. (2014), that bulk of the issues being raised now had been focused upon in international declarations made previously; therefore "it should be made clear if the issues are still being mentioned for the sake of reaffirmation or a step toward pointing out the gaps in how previous instruments handled for the issues" (Zhou et al., 2014). Apart from emphasis on cross-cutting issues like gender contributes to the effective implementation of the prevalent framework, it was also pointed out that an improved implementation and a follow-up mechanism was required (UNISDR, 2013a, 2013b). Irrespective of the continued relevance and utility of the of the existing Hyogo framework, a call for a deeper look, specifically into the priority action four had been made (Zhou et al., 2014). Undeniably, various tool had been deployed under the HFA with the focus on tracking progress in DRR implementation rom the regional to national scales; nevertheless, an identical reporting tool was not established to assess HFA's implementation at the local level, "suggesting limited consideration on the part of the international community to track the community-level impacts of these DRR strategies" (Tozier de la Poterie & Baudoin, 2015).

1.7 Sendai framework for disaster risk reduction

The HFA was a decade-long plan, effective from 2005 to 2015 and this period was geopolitically and economically presented worlds apart. The near-global economic crisis of 2007−08 represents a major watershed between the two UN global platforms on disaster reduction (UNCTAD, 2010; Economist, 2013). With austerity measures in force and acute shortage of employment becoming common in many countries, this financial crisis rendered people more vulnerable to an array of additional shocks and crisis like pensions, disability assistance and other factors in social safety nets being reduced in many countries along with knock-on effects rippling through the economies of poor and highly-indebted countries (Wisner, 2020). Global politics during this period was characterized by recurring incidents of political instability, outbreak of civil wars, resulting in exodus of millions of immigrants and refugees (Lindsay, 2015; Song, Lu, Liang, Wang, & Lin, 2017). The period from 2005 to 2015 was also marked by occurrence of disasters around the globe that contributed to human, economic, infrastructure and ecological losses, particularly in the most vulnerable and poorest countries (World Bank, 2012; Nicholson, 2014). In the wake of commitments to support · DRR made under HFA coming to an end in 2015, the third World Conference on Disaster Risk Reduction (WCDRR) was held in Sendai in Japan on March 14−18, 2015. While reviewing the HFA, the third WCDRR noticed some gaps in the HFA that interalia included: lack of early warning systems in economic and social vulnerabilities, dearth of systematic assessment of multihazards, ineffective implementation of HFA at the local level; and inadequate integration of DRR into the national and international policies of sustainable development (Zhou et al., 2014; UN-ESCAP, 2011). Accordingly, a new improved framework was adopted at the third WCDRR and it came to known as Sendai Framework for Disaster Risk Reduction (SFDRR) commencing from 2015 till 2030. The areas to which priority was accorded in SFDRR included: understanding disaster risk; strengthening disaster risk governance to manage disaster risk; investing in DRR for resilience; and enhancing disaster preparedness for effective response and to "Build Back Better" in recovery, rehabilitation and reconstruction UNISDR, 2015a, p. 20).

The SFDRR 2015—30 has specifically been launched to enhance DRR policy worldwide to augment human understanding of the complexity of disaster risk in current context (Faivre, Sgobbi, Happaerts, Raynal, & Schmidt, 2018). Considering the implementation of the SFDRR framework at the local level; nonetheless, the core spirit of the SFDRR is seemingly not adequately embraced across countries (Goniewicz & Burkle, 2019; Dube, 2020). Apart from these, the global change along with an array of changes have been instrumental in triggering multiple evolving risks in recent times, and pondering over these should spur mulling new commitments to humans' sustainable living, resilience and well-being (Dube, 2020). Causal links between climate change and DRR have already been focused on in research studies on climate change adaptation (CCA) and DRR, and that has been examined and determined by vulnerability, exposure and the strength of people to prepare, respond and recover from its ill-effects (Busayo, Kalumba, & Orimoloye, 2019). Nevertheless, it has seemingly become critical for envisaging a paradigm shift from hazard response to identification of risks, assessing and ranking them accordingly (Van der Keur et al., 2016; Mojtahedi & Oo, 2017). Emphasis of this shift relies on considering social factors impacting the local population and interpretation of the risk in tandem with their thresholds for action, and this underlines that "bringing a synergy between societal indicators of risk vis-à-vis strong policy implementation, for example, SFDRR and collective agency assistance supported by science and technology (S&T) skills, better socio-economic development can be attained" (Busayo, Kalumba, Afuve, Ekundayo, & Orimoloye, 2020). Instead of relying on disaster vulnerability approach, SFDRR emphasizes on a human-based approach and probable roles by stakeholders. Busayo and colleagues (2020) have called this approach a "bottom to up" that encourages germane institutions on the "requisite" to pursue the development of sustainable policies that would, in turn, forge ahead disaster risk evaluation, design of disaster monitoring plans, and its core implementation. The SFDRR is said to have been analyzed by many researchers in different fields, including identification of methods to implement it at various levels (Estrella, Renaud, Sudmeier-Rieux, & Nehren, 2016; de la Poterie & Baudoin, 2015).

The cognition of disaster risk has come to play a pivotal role in DRR. Assessment of hazard, vulnerability, exposure and capabilities of the populace to prognosticate, respond to, and to recover from its effects have been the focal point of investigation in the recent research studies (Shaw, Izumi, & Shi, 2016a, 2016b). In this backdrop, the emphasis of priority-1 under the SFDRR has shifted the paradigm from disaster management to disaster risk management, and this shift from mere hazard response to the identification, assessment and categorizing of vulnerabilities and risks assumed significance (Rahman & Shaw, 2015). This shift of emphasis in focus takes into consideration social factors, influencing local population's interpretation about risks and their thresholds for action (Shaw et al., 2016a, 2016b). It implies that there is need for identifying influencing societal determinants of risk in order to achieve better economic and social development pathways (Kubal, Haase, Meyer, & Scheuer, 2009). Priority-1 of the SFDRR has seemingly integrated S&T in DRR for sustainable future by having an understanding of disaster risk (Rumbach & Follingstad, 2018). An understanding of urban disasters can be helpful in unveiling the accumulating risk in the built environment and economy, especially in the aftermath of occurrence of a major disaster, and such risk is subject to variation from town to town and time to time depending on the development and economic activities in urban areas (Malalgoda, Amaratunga, & Haigh, 2014; Rahman & Fang, 2019).

Assessment, dissemination and communicating risk information are seemingly the essential components of the first priority under SFDRR. Risk information, involving a broader scale of data inputs relative to the prevalent hazard-based approach, also includes information on vulnerability, capacity, exposure of people and assets, hazard features and the environment (Cardona et al., 2012). While alluding to the inventory of elements in an area in which hazard events may occur (UNISDR, 2004, 2009b), exposure also refers to the property, infrastructure and people that entail the potential of being affected by a disaster; whereas vulnerability reflects the susceptibility of an individual, a community; assets or systems to the impacts of hazards (Bogardi & Birkmann, 2004; UNISDR, 2004, 2009b). Nevertheless, bulk of the information available and used to inform policy is barely focused on hazard likelihood (Aggarwal, 2016; Mahmood & Rahman, 2019). Apart from lacking in wider dissemination of information, risk information is also confronted with the challenge of funding with regard to risk assessment and mapping (OECD, 2012). Undeniably, there is still a gap in the standardized assessment methodologies and digital maps of natural resources in some of the flood plains are progressing with the avowed objective of improvement in standardization, including a focus on Sade risk assessment; nevertheless, particular details are yet to be evolved (Díez-Herrero & Garrote, 2020). Lamentably, adoption of this guidance is voluntary except it is funded by some organization and that denotes that provincial and local governments could follow risk assessments and mapping that are incomparable across jurisdictions (UNISDR, 2014).

Following gaps in the current research activities regarding First Priority of SFDRR have been summarized by Rahman and Fang (2019) along with suggestions, such as: less importance to international protocols; ignoring indigenous knowledge and approaches; lack of policy implementation and coordination; need for developing databases for public assess; free and open access to satellite-based fine resolution spatial data; need to improve national DRR knowledge management system; need for including DRR experts in decision-making process and policy formulation; and increased emphasis on augmenting investment in research and development along with capacity building. In other words, policies and practices pertaining to DRM are required to be based on an understanding of disaster risk in all its dimensions of vulnerability, capacity, exposure of persons and assets, hazard characteristics and the environment (Mahmood & Rahman, 2019). It has been suggested that such knowledge entail the potential of being leveraged for the purpose of predisaster risk assessment, prevention, and mitigation and for the implementation and development of apposite preparedness and effective response to disasters (Rahman & Fang, 2019). Countries along with regional and international organizations and other relevant stakeholders are called upon to take into account the key activities listed under Priority-1 of SFDRR in their approach to DRR and ought to implement them, as appropriate, while keeping in view respective capacities and capabilities in consonance with national laws and regulations (Street et al., 2018). Well-concerted international cooperation, an enabling international environment and means of implementation, especially in the wake of increasing global interdependence, are direly required to stimulate and contribute to promoting the knowledge, capacities and motivation for DRR (Foudi, Osés-Eraso, & Tamayo, 2015).

The relationship between civil society and the State is often described as one of "fierce friends and friendly enemies" (Wisner & Haghebaert, 2006), and in recent years, especially after the promulgation of the SFDRR in 2015, such tensions between the civil society and the State have seemingly increased as social unrest continues to sweep through many countries (Hribernik & Haynes, 2020). The SFDRR is committed and is also in quest of mobilizing the knowledge and skill of local communities and the people; nevertheless, in the eventuality of some countries locking out NGOs and other civil society organizations from the decision-making loop, it becomes difficult for the SFDRR to deal with the governments. With S&T having gained salience at the time of conclusion of the HFA in 2005, this S&T got catapulted to prominent place by the time of coming into being of SFDRR in 2015 and the targets of the SFDRR also started attracting appropriate focus accordingly. The targets envisaged in the SFDRR are as follow: (1) substantial reduction in global disaster mortality by 2030; (2) substantial reduction in the number of affected people globally by 2030; (3) Reduction in direct disaster-related economic losses in relation to global GDP by 2030; (4) substantial reduction in disaster damage to critical infrastructure by 2030; (5) increase in the number of countries with national and local DRR strategies by 2020; (6) enhancing international cooperation to provide adequate and sustainable support for developing countries to enable them to implement the disaster prevention framework by 2020; and (7) substantially increase people's opportunities to access multihazard early warning systems and disaster risk information and assessments by 2030 (SFDRR, 2015, p. 11). Lamenting that it is difficult to get the knowledge used in the "right place and the right time," that is sometimes regarded as the "last mile problem," Wisner (2020) questions as to how does a moderate system of communication, such as satellites, ocean buoys, radio transmitters, and computers get a warning conveyed to someone in a remote area like a village on the coast with no means of communication, such as no mobile phone, no radio, and maybe a bicycle. Making appropriate use of local knowledge in designing warning systems that are local, pertinent, and efficacious can also be regarded as the "first mile problem" (Gaillard & Kelman, 2018). Undeniably, progress has been achieved to some extent in the case of Bangladesh where largescale network of volunteers the Red Crescent in communicating cyclone warning to the people in coastal areas; nevertheless, this challenge still remains at a vaster scale for the implementation of the SFDRR (Roy & Kovordanyi, 2015). The question in the implementation of Sendai Framework remains as to whether or not the fixation on technology will be tempered by common-sense "mud on the boots" appreciation of the daily lives and situations, cultures, vernaculars and ways of thinking or communicating that constitute "the first mile" (Krueger, Bankoff, Cannon, Orlowski, & Schipper, 2015; Wisner, 2020).

Civil society is called upon to play an important role in the implementation of the SFDRR (Vaughan & Hillier, 2019). Global Network of Civil Society Organizations for Disaster Reduction (GNDR) conducted global surveys and compiled them under the title Views from the Front Lines (GNDR, 2019). At one time, these surveys comprised principal tools for holding UNISDR and the signatory governments to account for as well as make available a counterpoint to the HFA's accounting tool—HFA Monitor, and currently these surveys are being used by the Sendai Monitor (Wisner, 2020). Still, there is increasing need for enhanced global networking in implementing Sendai Framework (van Liekerk, Coetzee, & Nemakonde, 2020). Increased interaction between

UNDRR and civil society and other organizations is the need of the hour to speed up implementation of Sendai Framework (UN-GA, 2020). During the past five years of the operationalization of the Sendai Framework (2015–19), mechanism is said to have been put in place for filtration of data by gender and age for gauging damage assessments and for preparedness and risk reduction planning (UNDRR, 2019, 2020a), and to monitor impacts of climate change policy for their possible negative effects on women and girls (UNDRR, 2020b; Habtezion, 2016). Women's role is also acknowledged as skillful and capable leaders in DRR (UN Women, 2019). While lamenting that there is more to successful implementation than the mere fulfillment of countable targets, Wisner (2020) argues that neither the Hyogo Framework nor the Sendai Framework "encourages governments to think of disaster reduction in terms of human rights or challenges governments to stand up to private investors concerning megaprojects that threaten to create new risks, more is required."

1.8 Ecosystem-based adaptation for disaster risk reduction

Recent past years have progressively been characterized by focusing more attention on the convergence between DRR and CCA agendas conceptually and pragmatically at subnational, national and international levels (Mitchell et al., 2010). This has corresponded to the advent of "adaptation" as a vital component of the worldwide response to climate change and institutionalization of DRR as epitomized by the agreement of the 2005 HFA. Irrespective of substantial literature, both academic and policy-focused (Schipper & Pelling, 2006; Yamin, Rahman, & Huq, 2005), the 2009 Global Assessment Report on DRR indicated that an array of national processes for dealing with DRR and CCA are in vogue analogously and entail separate policy and institutional frameworks (UNISDR, 2009a, 2009b). DRR and CCA are two approaches, with some identical goals and activities but operating in different policy landscapes. Globally, DRR is the domain of United Nations Office for Disaster Risk Reduction (UNDRR) and CCA is the province of United Nations Framework Convention on Climate Change (UNFCCC). Admittedly, these two spheres—DRR and CCA—are distinct vastly on account of the policies and institutions involved in each; nevertheless, global agreements pertaining to both CCA and DRR allude a reference to each other. Irrespective of the fact that mandates of CCA and DRR are distinct, still there is some crossover as well as some differences that has spurred calls for and signs of convergence between these two domains (UNDRR, 2020a).

Natural hazards or weather or climate extremes, also called extreme, severe, rare or high-impact weather, or flood events (Stephenson, 2008), climate extremes, weather and climate variability (IPCC, 2012), constitute risks to numerous assets both natural and man-made, such as, ecosystems, the lives and livelihoods of people, communities, infrastructure, cultural heritage, the economy and societies in general (IPCC, 2012). The scale of the hazard along with its underlying factors often entails the potential of escalating to a disaster; "Severe alterations in the normal functioning of a community or a society due to hazardous physical events interacting with vulnerable social conditions, leading to widespread adverse human, material, economic, or environmental effects that require immediate emergency response to satisfy critical human needs and that may require external support for recovery" (IPCC, 2014: 1763). There is ample evidence to demonstrate that the altering likelihood of individual weather events is attributable to climate change (Stott et al., 2015;

Otto et al., 2018); however, the evidence is more indistinct as to whether climate change has added to the increasing socio-economic impacts and outcomes of climate-related hazards (IPCC, 2014; Pielke Jr. et al. 2008). Like any externality (Pigou), the factors contributing to climate change, and specifically the adverse repercussions of climate change, are relatively not the same factors that bear the consequences (Stern et al., 2006). As pointed out by Tol (2009, p. 29), "climate change is the mother of all externalities," it is a subtle and complex, substantially unpredictable and probably a societal challenge wherein endeavors to deal with are convoluted by multiple political and technological challenges (Knutti et al., 2016).

Major political initiatives worldwide are said to have been underway in recent years to facilitate reduction of the risks emanating from disasters and climate change, first though the 2005 HFA (UNISDR, 2005) and thenceforth, the adoption in 2015 of SFDRR (UNISDR, 2015a) that lay foundation for international efforts in multihazard DRR and DRM. The conclusion of the Paris Agreement on Climate Change in 2015 under the aegis of the UNFCCC epitomizes the global political will to tackle the menace of climate change (UNFCCC, 2015). Undoubtedly, the reduction of greenhouse gas (GHG) emissions from energy, land use and other sources, as core theme of climate change (IPCC, 2014), is presently hogging bulk of the academic, especially political attention; nonetheless, in practice, multiple challenges like the dearth of political ambition (Rogelj et al., 2018), want of cooperation and coordination in worldwide climate policy (Keohane and Victor, 2016; Nordhaus, 2015), absence of appropriate technologies to reach a required level of GHG emissions (Fuss et al., 2016), individual preferences and behavior (Cozzi & Petropoulos, 2019), etc., have been instrumental in maintaining an increasing rate of GHG emissions. Consequently, irrespective of the stated political will to kibosh the augmentation in GHG emissions since the first Conference of Parties (COP-1) to the UNFCCC in 1995, there has been steady increase in the GHG emissions (NASA, 2019).

In the wake of these developments, emphasis has been focused on the immediacy and challenges pertaining to policies and measures that are directed at lessening the socio-economic impacts of climate-induced events primarily through CCA (Adger et al., 2005; Hallegatte, 2009). Climate change mitigation and adaptation are widely acknowledged as efforts designed to lessen the impacts of climate change, with mitigation lessening human requirement for CCA and vice versa (Tol, 2004). Scientific and academic research has been instrumental in evolving models that comprise the complementarity of both mitigation and adaptation with regard to climate change for efficacious management of climate change risk, yielding in an economically optimal blending of mitigation and adaptation (Kane and Shogren, 2000).

Conversely, in order to facilitate prevention, reduction and embark on preparation for disaster risk and to respond to and recover from disasters linked to natural hazards and climate change, there are two, usually discretely addressed domains of research, policy and practice, and these are—DRM and CCA (IPCC, 2012, 2014). Reduction of impacts of weather and climate extremes the prime objectives of DRM and CCA in human systems *via* the exposure and vulnerability of people and assets at risk and to facilitate improvement in disaster response and recovery. This is accomplished by the development of improved and integrated policies and strategies, implementing measures via investing in technological development and innovations, and advancing the adoption of behavioral change at an individual level (Tol, 2005; IPCC, 2012). The advantage of CCA in terms of DRM is that it fetches a long-term perspective to the traditional DRM approach (Kelman et al., 2015; Hanemman, 2000; Rivera & Wamsler, 2014). Irrespective of the potential of enhancing the efficaciousness and effectiveness of DRM and CCA by means of their integration into research, governance and practice (Venton & La Trobe, 2008), integration is still in its early stage (Kelman et al., 2015). Undoubtedly, challenges underlying the lack of integration have been identified; nonetheless, no exact definition denoting the relationship between DRM and CCA and their joint integration or mainstreaming, in other fields is available within the integration literature. Besides, empirical evidence of DRM and CCA integration is also meager.

Reduction of vulnerability lies at the core of DRM and CCA; nevertheless, its accomplishment is usually confronted with low quality governance (UNISDR, 2015b) which is interlinked with other socio-cultural factors (Alexander & Davis, 2012). With the bulk of literature on governance challenges having focused on economically less developed countries, with little analysis on effective governance in developed countries, the literature also falls short on analysis on the complexity of DRM and CCA governance due to numerous, competing decision-making criteria entailing that effective implementation of DRM and CCA governance and implementation seldom implies cost-efficiency, leading to potential over-adaptation to disasters and climate change (Hanemann, 2000). Moreover, governance and improved DRM and CCA measures seldom negate the significance of the role of decision-making down to the level of individual in facilitating the reduction of impacts of natural hazards, disasters and climate change (Adger et al., 2005).

Common goal of reducing the effects of extreme events and enhancing resilience to disasters, specifically among vulnerable populations interconnects DRR to CCA (McVittie, Cole, Wreford, Sgobbi, & Beatriz Yordi, 2017), and these close linkages between DRR and CCA have resulted in close calls for the integration of CCA and DRR policy (Doswald & Estrella, 2015). This is amply demonstrated in recent developments in EU policies and strategies pertaining to DRR and CCA whereby in recent years the policy framework for CCA and DRR at the EU level has evolved significantly, with augmentation in synergies between the two policy spheres (McVittie et al., 2017), along with more systematic attention focused on the potential role of nature-based solutions (NBSs). The 2013 EU strategy on adaptation to climate change (European Commission, 2013) warrants the implementation of adaptation policies in tandem with and full coordination with disaster risk reduction. It also outlines numerous benefits accruing from ecosystem-based approaches, and their contribution to achieving its strategy objective of a climate-resilient Europe, including DRM and DRR. The EU Action Plan on the SFDRR (European Commission, 2016) envisages a coherent agenda for augmenting risk prevention, building resilience of societies and developing risk-proofed investments via different EU policies, flood risk management, water and biodiversity protection, Furthermore, an array of sectorial policies including the Floods Directive, the Biodiversity Strategy and the EU Green Infrastructure Strategy acknowledges the necessity for an integrated approach to risk management and therefore contributes directly or indirectly to strengthening the CCA and DRR agendas (Faivre et al., 2018).

Poorly-planned or unplanned socio-economic development in areas exposed to an array of hazards in many parts across the globe contributes to increase disaster risk (UNDRR, 2020). Climate change entails the potential of exacerbating the frequency and magnitude of hydrometeorological hazards (IPCC, 2012, 2014). Human life and assets are significantly impacted by a combination of ill-informed infrastructure projects like dams and dykes that contribute to floods, droughts and other hazard risks along with rapid pace of industrialization and economic growth, unplanned urbanization in exposed coastal and river areas along with the loss of ecosystem services (UNISDR, 2015b). Issues pertaining to global warming such as extreme weather and biodiversity loss have been ranked as the top five risks in terms of likelihood in the ensuing decade by WEF in its Global Risks Report for 2020 (WEF, 2020). Human security and ecosystem well-being are confronted by multiple complex challenges and systemic risks on account of these interacting processes. This calls for an urgent need for countries to have a better understanding of the impacts and associated risks of ecosystem decline and "to integrate ecosystem conservation and rehabilitation, and the sustainable use and management of natural resources in national DRR policies and plans (UNDRR, 2020c).

Environmental degradation is an invitation to disaster risk. This can be exemplified from the fact that lack of good vegetation cover on slopes can cause landslides under the spell of heavy rainfall, and concurrently, the absence of ecosystem services can result in the exacerbation of disaster impacts and affect recovery endeavors and livelihood recovery in the aftermath of a disaster. Conversely, disasters also affect ecosystems, triggering environmental degradation and losses that in turn enhance probability of risk. This highlights interconnectedness and interdependence between human well-being, ecosystems and altering risk patterns; therefore emphasizing the need for going in for NBSs like conservation, restoration and the sustainable use and management of natural resources (UNDRR, 2020c). Ecosystem-based approach (EbA) is gaining prominence among SBSs because of its ability to deliver important services that can address an assortment of risk factors.

With EbA having garnered ample recognition as a significant strategy for risk disaster risk reduction, the term ecosystem-based DRR (Eco-DRR) has gained wide traction. Eco-DRR is defined as "sustainable management, conservation and restoration of ecosystems to reduce disaster risk, with the aim to achieve sustainable and resilient development" (Estrella & Saalismaa, 2013). Restoration of coastal vegetation areas like mangroves to protect shorelines from surges, management of invasive alien species linked to land degradation endangering food security and water supplies and managing ecosystems to complement, protect and durability to investments in hard infrastructure are some examples of Eco-DRR.

The mechanism of Eco-DRR has been able to garner support in the international policy arena in multiple ways. Eco-DRR was referred to in the UNFCCC Cancun Adaptation Framework that adopted as part of the Cancun Agreements in 2010 (Ranger, Surminski, & Silver, 2011), and the Nairobi work program on impacts, vulnerability and adaptation to climate change (UNFCCC, 2007). Even IPCC's Special Report on Extreme Events investing in ecosystems, sustainable land management and ecosystem restoration and management (IPCC, 2012). Furthermore, the SFDRR 2015—30, while building on the HFA 2005—15, outlines seven global targets to be achieved over the next 15 years, prioritizing "ecosystem-based approaches…to build resilience and reduce disaster risk" (SFDRR, 2015). Massive support and endorsements have been garnered by Eco-DRR in the outcomes of regional DRR platforms of Asia, Africa, Latin America and Arab states, and of the European Ministerial meeting on DRR.

Eco-DRR measures are almost same as EbA activities. In many cases because both are designed to reduce disaster risk, and therefore, EbA can sometimes be construed as an example of Eco-DRR and vice versa. A review of commonalities and differences between EbA and Eco-DRR undertaken by Doswald and Estrella (2015) makes it discernible that in practice, it is problematic to make distinction between EbA and Eco-DRR because there more commonalities than differences between the two owing to the primary shared basic value of employing the ecosystem approach and enhancing the resilience of people and communities. Emphasis on EbA and participation of nongovernmental organizations (NGOs) in implementation, encouraging involvement of local communities along with assessment of vulnerabilities and risks are often the commonalities shared by both EbA and Eco-DRR (Doswald & Estrella, 2015). Involvement of indigenous people and local communities is usually projected as a salient principle of EbA and DRR implementation. Community-managed DRR that corresponds to community-based adaptation in disaster risk reduction, is an approach that entails the potential of helping the communities in identifying the hazards they are exposed to and work out effective measures to promote resilience (Fitzgibbon & Crosskey, 2013). Relatively new approaches bobbling up from broader adaptation and DRR practice, both EbA and Eco-DRR are in the initial stages of developing assessment, monitoring and evaluation methodologies. Nevertheless, policy frameworks that endorse EbA have been developed over the past many years.

Differences between EbA and Eco-DRR reflect those of general CCA and DRR activities. In the first place, operating under different policy forums, both EbA and Eco-DRR are often undertaken by different institutions to deal with different types of hazards, and employ different terminologies to convey similar terms and concepts (Doswald & Estrella, 2015). Undeniably, EbA relatively deal with climate-related hazards; nevertheless, there are instances of EbA interventions like implementing protection forests that accord stability to the soil to prevent landslides, both climate and nonclimate related. EbA interventions are often called upon to address slow-onset climate change impacts like altering perception patterns, rising mean temperatures, sea-level rise, etc., and that has not been a tradition focus in DRR. On the other hand, Eco-DRR accosts both nonclimate hazards like earthquakes, tsunamis, etc., and climate-related natural hazards like hurricanes and heat waves and other kinds of hazards. Eco-DRR is inclined to focus on rapid as well as slow-onset events wherefrom a system is expected to recover, instead of chronic and irreparable stressors to which systems must adapt, such as gradually warming temperatures, seal-level rise and melting of glaciers. Besides, Eco-DRR comprises constituents like early warning systems, preparedness and contingency planning, response, recovery and reconstruction and these have not often been the focus of EbA (Doswald & Estrella, 2015). Moreover, until recently, EbA and other forms of adaptation interventions have responded to both observed and projected climate trends, whereas DRR has only dealt with historical return rates of weather-related disasters to inform planning rather than projections for altering return rates in the future (Zissener, 2011).

The scale of damage and losses accruing from natural hazards can be vastly reduced despite increase in frequency and intensity of these events with the help of implementation of DRR strategies. Strategies adhered to in DRR frequently place reliance on man-made structures in order to protect communities, whereas nature-based approaches, especially EbA, are equal significant constituents of DRR (Sobey & Monty, 2016). Sound and adequately-managed ecosystems are prone to play a significant role in cost-effective DRR strategies. It has been pointed out by Sudmeier-Rieux, Ash, and Murti (2013) that there are four reasons as to why ecosystems matter to disaster risk reduction, and these are: (1) dependence of human well-being on ecosystems; (2) ecosystems providing cost-effective natural buffers against hazards; (3) resilience of ecosystems to extreme weather events; and (4) ecosystem degradation reduces the ability of natural systems to sequester carbon. Effective adoption of ecosystem-based solutions to DRR requires the need of understanding the differences in scope and approach between EbA and Eco-DRR. EbA deals with climate-related natural disasters and Eco-DRR addresses other types of disasters such as earthquakes (Doswald & Estrella, 2015). Another insidious difference that entails the potential of impacting the type of approach and measure adopted, pertains to the timeframe of the underlying analysis (Doswald & Estrella, 2015).

Preparation of planning for efficacious EbA calls for emphasis on long-term changes in frequency and intensity along with future potential impacts emanating from climate change, including slow-onset events, whereas, Eco-DRR addresses current risks and disasters (Mitchell & van Aalst, 2008). EbA and Eco-DRR at their intersection entail the potential of facilitating implementation in an array of sectors to address several hazards along with engendering additional ecosystem service benefits (CBD, 2019a). This scale of potential ecosystem service co-benefits denotes that EbA activities can add to both climate adaptation, and also be beneficial to multiple stakeholders. This can be exemplified from the fact the protection or restoration of wetlands and floodplains can safeguard against floods, including flashfloods (Doswald & Osti, 2011). Likewise, restoration of coastal wetlands, mangroves, and coral reefs can be helpful in reducing the susceptibility to and impact of storm surges (McIvor, Spencer, Möller, & Spalding, 2012). In an identical manner, green infrastructure and spaces, roof gardens, tree-lining in urban environments also entail the potential of facilitating reduction in the impacts of heat waves, while also augmenting water retention (Strosser, Delacámara, Hanus, Williams, & Jaritt, 2015). It can be possible to make a distinction between EbA and Eco-DRR measures at the level of individual measures; nonetheless, this distinction gets blurred when the approaches are considered cumulatively or in their entirety. According to Naumann et al. (2011), the protection and restoration of resilient ecosystems are among the most cost-effective means of restricting the range and adverse consequences of disasters and climate change, especially in view of all the co-benefits of the measures beyond augmenting resilience against hazards.

Eco-DRR comprises a combination of natural resources management approaches or the sustainable management of ecosystems, and DRR methods like early warning systems and emergency planning, so as to have more effective disaster prevention, reducing the impact of disasters on people and communities, and support disaster recovery (Sudmeier-Rieux, Nehren, Sandholz, & Doswald, 2019). Delivery of societal benefits in a fair and equitable manner, in a way that encourages transparency and participation at a larger scale is also facilitated by Eco-DRR. With Eco-DRR and EbA being acknowledged as integral constituents of risk reduction and CCA strategies, both approaches are deemed to emphasize the significance of biodiversity and ecosystems in minimizing risk and build on other of restricting practices like conservation and ecosystem restoration that are in quest of enhancing the resilience of ecosystems for the benefit of people (CBD, 2019b). Both are also said to function well with grey infrastructure either as a complement, a substitute, or a safeguard (IADB, 2020). While pointing out NBSs are key to meeting worldwide goals for climate change and sustainable development, Seddon et al. (2020) have called upon the ecosystem science community to work in tandem with policy makers to identify meaningful adaptation targets that are beneficial both for people and the ecosystems on which they depend.

Deployment of EbA for DRR has been increasing garnering endorsement and recognition. The 2015 Global Assessment Report for Disaster Risk Recognition (UNISDR, 2015b) included EbA approaches and laid emphasis on new approaches blending gray and green infrastructure to optimize ecosystem services. While taking into consideration the pluralistic nature of risk in various dimensions, at multiple scales and with numerous impacts, UNDRR's 0219 Global Assessment Report outlined environmental degradation as a primary aspect in creating risk, acknowledged the necessity of understanding systemic risks to people and the ecosystems, took notice of ecosystems as a core concept for encouraging the integration of DRR with SDGs, incorporated case-studies of DRR/CCA integration and recognized the significance of balanced ecosystems for livable communities in urban governance. EbA has been incorporated in IPCC's Fifth Assessment Report and the IPCC Special Report on the Ocean and Cryosphere in a Changing Climate (IPCC, 2019), wherein recognition is accorded to ecosystem-based measures and hybrid approaches combining ecosystems and built infrastructure as actions to reduce hazards. Actions aiming at reducing hazards, vulnerability, and exposure are required to be weighed against systems—human, ecological, economic, etc.—and action scales such as global, regional, national and subnational, etc. While demonstrating that more of current disaster events are associated with environmental degradation and climate change, the 2019 Asia-Pacific Disaster Report (UN-ESCAP, 2019) argues that risks get enhanced due to environmental degradation and that one of the strongest defenses against disasters is a healthy ecosystem. The Report highlights that recent advances in ecosystem-based approaches to DRR provide new and innovative solutions to reduce risk and vulnerability.

Apart from direct benefits accruing from ecosystems for DRR and CCA, ecosystems also bequeath an array of other socio-economic and environmental benefits for multiple stakeholders and those benefits can be instrumental in further reducing risks (Sudmeier-Rieux et al., 2019). The most vulnerable people in many countries depend on ecosystems for their livelihoods and resilience. Deployment of EbA in DRR and Eco-DRR is helpful in reducing social vulnerability and enhancement of people's resilience by sustaining livelihoods and delivering essential natural resources like food, water and materials (Renaud, Sudmeier-Rieux, Estrella, & Nehren, 2016). Other socio-economic benefits accruing from ecosystem-based approaches interalia include: carbon storage and sequestration, biodiversity conservation, and poverty alleviation. Accrual of multiple system-wide benefits warrants investing in NBSs crucial to facilitate reduction in disaster risk, adaptation to climate change, conserving natural resources, alleviating poverty, and attainment of sustainable development (UNEP, 2019). Available literature amply demonstrates that Eco-DRR/EbA provides effective cost-efficient and "no-regret" or low regret' solutions for reducing disaster risk and promoting resilience (IPCC, 2012; Sudmeier-Rieux et al., 2019). Eco-DRR is relatively a new concept for practitioners and policy makers, and as such dearth of knowledge on similar nature-based approaches, their efficaciousness and implementation process, erects a serious impediment to its adoption and scaling up. Such a scenario warrants increasing need for countries, specifically developing countries, to have cognizance of what is meant by Eco-DRR and how to make it operation in order to pave way for making it a key investment option for sustainable development (Sobey & Monty, 2016).

1.9 Conclusion

The notion of disaster has witnessed multiple definitions and visions in social scientific studies undertaken by scholars and researches over the decades, especially from the classical period onward. Over the years, the concept of disaster has been amply refined and defined. The initial phase was characterized by intensive research and theorizing designed to isolate what constituted the "disaster" from attendant causes, condition, and consequences. Mounting corpus of research literature helped the notion of disaster to move away from an agent-centric, damage-driven, and irrepressible event vision to be acknowledged as a social phenomenon with social disruption as the key defining feature. Conceptual refinements have enabled the phenomenon of disaster to be understood at individual, organizational, and social system levels of disruption and how these may differ or interact within the ambit of the notion of disaster. A social scientific vision of disasters calls for focusing on the fundamental dimensions of the concept independent of externalities that may comprise causes, conditions for, or consequences of disasters. Evolving a theory-basis for disaster research calls for appropriate knowledge of causes, conditions, and consequences and concurrently, it is equally significant to build a body of knowledge on shared understanding of the concept in its entirety. NBSs, especially the ecosystem-based approach, are garnering traction to address disaster risks, and available literature on DRR is increasingly acknowledging the utility of employing EbA for DRR and this approach is termed as Eco-DRR approach. Nevertheless, being a new concept, it is faced with the problem of lack of knowledge about Eco-DRR and related other nature-based approaches, their effectiveness, implementation process, etc.

References

Abbott, P. (2014). *Natural disasters* (9th ed.). New York: McGraw-Hill.
Adger, W. N., Arnell, N., & Thompkins, E. M. (2005). Successful adaptation to climate change across scales. *Global Environmental Change, 15*, 77–86.
Aggarwal, A. (2016). Exposure, hazard and risk mapping during a flood event using open source geospatial technology. *Geomatics, Natural Hazards and Risk, 7*(4), 1426–1441.
Aguirre, B. E. (2006). *On the concept of resilience*. Preliminary Paper #356. Newark, DE: Disaster Research Center.
Aguirre, B. E. (2007). Dialectics of vulnerability and resilience. *Georgetown Journal on Poverty Law and Policy, 14*(1), 39–59.
Ait-Chellouche, Y. (2015). *Mainstreaming and implementing disaster risk reduction in West Africa*. Addis Ababa, Ethiopia: United Nations Economic Commission for Africa.
Aldunce, P., Beilin, R., John, H. J., & Howden, M. (2014). Framing disaster resilience. *Disaster Prevention and Management, 23*(3), 252–270.
Alexander, D. A. (1993). *Natural disasters*. New York, USA: Chapman and Hall.
Alexander, D. A. (2000). *Confronting catastrophe: New perspectives on natural disasters*. New York: Oxford University Press.
Alexander, D. A. (2005). An interpretation of disaster in terms of changes in culture, society and international relations. In R. W. Perry, & E. L. Quarantelli (Eds.), *What is a disaster: New answers to old questions*. Philadelphia: Xlibris Publishers: 25–38.
Alexander, D., & Davis, I. (2012). Disater Risk Reduction: An Altenative Viewpoint. *International Journal of Disaster Risk Reduction, 2*, 1–5.
Alexander, D. A. (2016). The game changes. *Disaster Prevention and Management, 25*(1), 2–10.
Amundsen, H. (2012). Illusions of resilience? *Ecology and Society, 17*(4), 1–19.
Bankoff, G. (2015). 'Lahat para sa lahat' (everything to everybody). *Disaster Prevention and Management, 24*(4), 430–447.
Barrios, R. (2014). Here, I'm not at ease: Anthropological perspectives on community resilience. *Disasters, 38*(2), 329–350.
Barton, A. H. (1969). *Communities in disaster*. New York, NY, USA: Doubleday.
Barton, A. H. (1989). Taxonomies of disaster and macrosocial theory. In G. A. Kreps (Ed.), *Social structure and disaster* (pp. 346–350). Newark, DE, USA: University of Delaware Press.
Barton, A. H. (2005). Disaster and collective stress. In R. W. Perry, & E. L. Quarantelli (Eds.), *What is a disaster: New answers to old questions* (pp. 125–152). Philadelphia: Xlibris Publishers.

References

Berkes, F., & Ross, H. (2013). Community resilience. *Society and Natural Resources*, 26(1), 5−20.

Birkmann, J., Cardona, O., Carreno, M., Barbat, A., Pelling, M., Schneiderbauer, S., ... Welle, T. (2014). Theoretical and conceptual framework for the assessment of vulnerability to natural hazards and climate change in Europe. In J. Birkmann, S. Kienberger, & D. Alexander (Eds.), *Assessment of vulnerability to natural hazards* (pp. 1−20). London: Elsevier.

Blaikie, P., Cannon, T., Davis, I., & Wisner, B. (1994). *At risk: Natural hazards. People's vulnerability and disasters*. London: Routledge.

Bogardi, J., & Birkmann, J. (2004). Vulnerability assessment: The first step towards sustainable risk reduction. In D. Malzahn, & T. Plapp (Eds.), *Disasters and society—From hazard assessment to risk reduction* (pp. 75−82). Berlin, Germany: Logos Verlag:.

Boin, A. (2005). From crisis to disaster: Towards an integrative perspective. In R. W. Perry, & E. L. Quarantelli (Eds.), *What is a disaster: New answers to old questions* (pp. 153−172). Philadelphia: Xlibris Publishers.

Boin, A., Comfort, L., & Demchak, C. (2010). The rise of resilience. In L. Comfort, A. Boin, & C. Demchak (Eds.), *Designing resilience* (pp. 1−13). Pittsburgh: University of Pittsburgh Press.

Botzen, W. J. W., Deschenes, O., & Sanders, M. (2019). The economic impacts of natural disasters: A review of models and empirical studies. *Review of Environmental Economics and Policy*, 13(2), 167−188.

Bradshaw, S. (2014). Engendering development and disasters. *Disasters*, 30, 34−55.

Buckle, P. (2005). Mandated definitions, local knowledge and complexity. In R. W. Perry, & E. L. Quarantelli (Eds.), *What is a disaster: New answers to old questions* (pp. 173−200). Philadelphia: Xlibris Publishers.

Burton, I., & Kates, R. (1964). The perception of natural hazards in resource management. *Natural Resources Journal*, 3, 412−441.

Burton, I., Kates, R., & White, G. (1968). *The human ecology of extreme geophysical events*. Working Paper No. 1. Boulder, CO: University of Colorado, Boulder, Natural Hazards Center.

Burton, I., Kates, R., & White, G. (1978). *The environment as hazard*. New York, USA: Oxford University Press.

Busayo, E. T., Kalumba, A. K., Afuve, G. A., Ekundayo, O. Y., & Orimoloye, E. I. (2020). Assessment of the Sendai framework for disaster risk reduction studies since 2015. *International Journal of Disaster Risk Reduction*, 50, 101906.

Busayo, E. T., Kalumba, A. M., & Orimoloye, I. R. (2019). Spatial planning and climate change adaptation assessment: Perspectives from Mdantsane Township dwellers in South Africa. *Habitat International*, 90, 101978.

Cardona, O. D., van Aalst, M. K., Birkmann, J., Fordham, M., McGregor, G., Perez, R., ... Sinh, B. T. (2012). *Determinants of risk: exposure and vulnerability*. Managing the risks of extreme events and disasters to advance climate change adaptation. *A Special Report of Working Groups I and II of the Intergovernmental Panel on Climate Change (IPCC)* (pp. 65−108). Cambridge, UK: Cambridge University Press.

Carr, L. T. (1932). Disaster and the sequence-pattern concept of social change. *American Journal of Sociology*, 38, 207−218.

Carter, N. (2008). *Disaster management*. Manila, Philippines: Asian Development Bank.

Convention on Biological Diversity (CBD). (2019a). *Outreach into sectors: Integrating ecosystem-based approaches to climate change adaptation and disaster risk reduction*. Montreal, Canada: CBD Secretariat.

Convention on Biological Diversity (CBD). (2019b). *Voluntary guidelines for the design and effective implementation of ecosystem-based approaches to climate change adaptation and disaster risk reduction and supplementary information* (No. CBD Technical Series No. 93). Montreal, Canada: CBD Secretariat.

Cozzi, L. & Petropoulos, A. (2019). *Growing preference for SUVs challenges emissions reductions in passenger car market*. Web portal of iea.org, 15 October 2019. Available online.

Center for Research on the Epidemiology of Disasters − United Nations Office for Disaster Risk Reduction (CRED-UNDRRR). (2020). *Human cost of disasters: An overview of the last 20 years, 2000−2019*. Geneva, Switzerland: UNDRRR.

Challenges in risk assessment and risk management. In H. Kunreuther, & P. Slovic (Eds.), *Special Issue, Annals of the American Academy of Political and Social Science* (545, pp. 8−183).

Cutter, S. (2005). Are we asking the right question? In R. W. Perry, & E. L. Quarantelli (Eds.), *What is a disaster: New answers to old questions* (pp. 39−48). Philadelphia: Xlibris Publishers.

Cutter, S. (2016). Resilience to what? Resilience for whom? *The Geographical Journal, 182*(2), 110−113.

Cutter, S., Ash, K., & Emrich, C. (2014). The geographies of community disaster resilience. *Global Environmental Change, 29,* 65−77.

Cutter, S. L. (1993). *Living with risk: The geography of technological hazards.* London: Edward Arnold.

Cutter, S. L. (Ed.), (1994). *Environmental risks and hazards.* Englewood Cliffs, NJ: Prentice-Hall.

Cutter, S. L. (2001). The changing nature of risks and hazards. In S. L. Cutter (Ed.), *American hazardscapes: The regionalization of hazards and disasters.* Washington. D.C.: Joseph Henry Press.

de Govet, C. D. V., Marti, R. Z., & Osorio, C. (2006). Natural disaster mitigation. In D. T. Jamison, J. G. Breman, A. R. Measham, G. Alleyne, M. Claeson, D. B. Evans, P. Jha, A. Mills, & P. Musgrove (Eds.), *Disease control priorities in developing countries* (2nd edition, pp. 1147−1162). Washington, D.C.: World Bank.

de la Poterie, A. T., & Baudoin, M. A. (2015). From Yokohama to Sendai: Approaches to participation in international disaster risk reduction frameworks. *International Journal of Disaster Risk Science, 6*(2), 128−139.

Delica-Willison, Z., & Gaillard, J. C. (2012). Community action and disasters. In B. Wisner, J. C. Gaillard, & I. Kelman (Eds.), *The Routledge handbook of hazards and disaster risk reduction* (pp. 711−722). London: Routledge.

Diefenbach, A., Wood, N., & Ewert, J. (2015). Variations in community exposure to lahar hazards from multiple volcanoes in Washington State. *Journal of Applied Volcanology, 4*(4), 15−24.

Dombrowsky, W. R. (2005). Not every move is a step forward. In R. W. Perry, & E. L. Quarantelli (Eds.), *What is a disaster: New answers to old questions* (pp. 79−98). Philadelphia: Xlibris Publishers.

Donner, W., & Rodriguez, H. (2008). Population composition, migration and inequality. *Social Forces, 87*(2), 1089−1114.

Doswald, N., & Estrella, M. (2015). *Promoting ecosystems for disaster risk reduction and climate change adaptation: Opportunities for integration. Discussion Paper.* Nairobi: UNEP.

Doswald, N., & Osti, M. (2011). Ecosystem-based approaches to adaptation and mitigation—Good practice examples and lessons learned in Europe. *Report for the German Federal Agency for Nature Conservation (BfN), BfN-Skripten 306, by UNEP-WCMC.* Bonn: UNEP-WCMC.

Drabek, T. E. (1986). *Human system responses to disaster.* New York, USA: Springer.

Drabek, T. E. (2013). *The human side of disaster* (2nd ed.). Boca Raton, FL, USA: CRC Press.

Dube, E. (2020). The build-back-better concept as a disaster risk reduction strategy for positive reconstruction and sustainable development in Zimbabwe: A literature study. *International Journal of Disaster Risk Reduction, 43,* 101401.

Dynes, R. R. (1998). Coming to terms with community disaster. In E. L. Quarantelli (Ed.), *What is a disaster: Perspectives on the question* (pp. 109−126). London: Routledge.

Díez-Herrero, A., & Garrote, J. (2020). Flood risk analysis and assessment, applications and uncertainties: A bibliometric review. *Water, 12,* 2050.

Economist. (2013). The origins of the financial crisis: Crash course. *The Economist,* 7 September. Available online at: https://www.economist.com/schoolsbrief/2013/09/07/crash-course.

Eiser, R. J., Bostrom, A., Burton, I., Johnston, D. M., McClure, J., Paton, D., ... White, M. P. (2012). Risk interpretation and action: A conceptual framework for responses to natural hazards. *International Journal of Disaster Risk Reduction, 1*(0), 5−16.

Estrella, M., & Saalismaa, N. (2013). Ecosystem-based disaster risk reduction (Eco-DRR): An overview. In F. Renaud, K. Sudmeier-Rieux, & M. Estrella (Eds.), *The role of ecosystem management in disaster risk reduction* (pp. 26−54). Tokyo: UNU Press.

Estrella, M., Renaud, F. G., Sudmeier-Rieux, K., & Nehren, U. (2016). Defining new pathways for ecosystem-based disaster risk reduction and adaptation in the post-2015 sustainable development agenda. In F. G. Renaud, K. Sudmeier-Rieux, M. Estrella, & U. Nehren (Eds.), *Ecosystem-based disaster risk reduction and adaptation in practice* (pp. 553–591). Cham, Switzerland: Springer.

European Commission (2013). *An EU strategy on adaptation to climate change*, April 2013. Available at: https://ec.europa.eu/clima/sites/clima/files/docs/eu_strategy_en.pdf.

European Commission (2016). *Action plan on the Sendai Framework for disaster risk reduction 2015–2030: A disaster risk-informed approach for all EU policies*. Brussels, SWD (2016) 205 final/2. Available at: https://ec.europa.eu/echo/sites/echo-site/files/sendai_swd_2016_205_0.pdf.

Faivre, N., Sgobbi, A., Happaerts, S., Raynal, J., & Schmidt, L. (2018). Translating the Sendai framework into action: The EU approach to ecosystem-based disaster risk reduction. *International Journal of Disaster Risk Reduction, 32,* 4–10.

Ferdinand, I., O'Brien, G., O'Keefe, P., & Jayawickrama, J. (2012). The double bind of poverty and community disaster risk reduction: A case study from the Caribbean. *International Journal of Disaster Risk Reduction, 2,* 84–94.

Fitzgibbon, C., & Crosskey, A. (2013). Disaster risk reduction management in the drylands in the Horn of Africa. *Technical brief prepared by the Technical Consortium for Building Resilience to Drought in the Horn of Africa hosted by the CGIAR Consortium in partnership with the FAO Investment Centre*. Nairobi: International Livestock Research Institute.

Fischer, H. (2003). The critics Corner: The sociology of disaster. *International Journal of Mass Emergencies and Disaster, 21,* 91–108.

Foudi, S., Osés-Eraso, N., & Tamayo, I. (2015). Integrated spatial flood risk assessment: The case of Zaragoza. *Land use Policy, 42,* 278–292.

Fritz, C. E. (1961a). Disaster. In R. Merton, & R. Nesbit (Eds.), *Contemporary social problems* (pp. 651–694). New York, USA: Harcourt Publishers.

Fritz, C. E. (1961b). *Disaster and community therapy*. Washington, DC, USA: National Academy of Sciences.

Gaillard, J. C. (2007). Resilience of traditional societies in facing natural hazard. *Disaster Prevention and Management, 16,* 522–544.

Gaillard, J. C. (2010). Vulnerability, capacity and resilience. *Journal of International Development, 22,* 218–232.

Gaillard, J. C. (2016). Natural hazards and disasters. In D. Richardson, N. Castree, M. Goodchild, A. Kobayashi, W. Liu, & R. Marston (Eds.), *The international encyclopedia of geography* (pp. 863–871). Chichester: Wiley-Blackwell.

Gaillard, J. C., & Kelman. I. (2018). The first mile of warning systems: Who's sharing what with whom? *Humanitarian Practice Network*, 8 November 2018. Available at: https://odihpn.org/blog/first-mile-warning-systems-whos-sharing/.

Gilbert, C. (1998). Studying disaster. In E. L. Quarantelli (Ed.), *What is a disaster: Perspectives on the question* (pp. 11–18). London: Routledge.

Global Network of Civil Society Organizations for Disaster Reduction (GNDR). (2019). *Views from the frontline*. London: GNDR. Available from https://www.gndr.org/programmes/views-from-the-frontline.html.

Goniewicz, K., & Burkle, F. M. (2019). Challenges in implementing Sendai framework for disaster risk reduction in Poland. *International Journal of Environmental Research and Public Health, 16*(14), 2574.

Gratwa, W. & Bollin, C. (2002). *Disaster risk management*. Working Concept. Eschborn; GTZ.

Habtezion, S. (2016). *Gender and disaster*. New York: UNDP and Global Gender and Climate Alliance.

Hamilton, J. T., & Viscusi, W. K. (1999). *Calculating risks: The spatial and political dimensions of hazardous waste policy*. Cambridge, MA: MIT Press.

Hanemman, M. (2000). Adaptation and its measurement. *Climatic Change, 45*(3), 571–581.

Hayward, B. (2013). Rethinking resilience. *Ecology and Society*, *18*(4), 1−7.
Hewitt, K. (1983). The idea of calamity in a technocratic age. In K. Hewitt (Ed.), *Interpretations of calamity* (pp. 3−32). Boston, USA: Allen and Unwin.
Hewitt, K. (1997). *Regions of risk: A geographical introduction to disasters.* Essex, UK: Longman.
Hewitt, K. (1998). Excluded perspectives in the social construction of disaster. In E. L. Quarantelli (Ed.), *What is a disaster? Perspectives on the question* (pp. 75−91). London: Routledge.
Hewitt, K. (2016). *Regions of risk* (2nd ed.). London: Routledge.
Horlick-Jones, T. (1995). Modern disasters as outrage and betrayal. *International Journal of Mass Emergencies and Disasters*, *13*, 305−316.
Hribernik, M. & Haynes, S. (2020). *47 Countries witness surge in civil unrest—Trend to continue in 2020.* Versil Maplecroft, 20 January 2020. Available at: https://www.maplecroft.com/insights/analysis/47-countries-witness-surge-in-civil-unrest/.
Inter-American Development Bank (IADB). (2020). *Increasing infrastructure resilience with nature-based solutions (NBSs).* Washington, D.C.: IADB and ACCLIMATISE.
International Decade for Natural Disaster Reduction (IDNDR). (1994). *Yokohama strategy and plan of action for a safer world: Guidelines for natural disaster prevention, preparedness and mitigation.* Geneva, Switzerland: IDNDR Secretariat.
International Decade for Natural Disaster Reduction (IDNDR). (1996). *Towards practical and pragmatic natural disaster reduction by 2000: A policy document based upon observations and lessons leaned during 1990−1996.* Geneva, Switzerland: IDNDR.
International Decade for Natural Disaster Reduction (IDNDR). (1999). *Proceedings: IDNDR programme forum, 1999. 5−9 July 1999, Geneva.* Geneva, Switzerland: IDNDR Secretariat.
Innocenti, D., & Albrito, P. (2011). Reducing the risks posed by natural hazards and climate change: The need for a participatory dialogue between the scientific community and policy makers. *Environmental Science & Policy*, *14*(7), 730−733.
Intergovernmental Panel on Climate Change (IPCC). (2012). *Managing the risks of extreme events and disasters to advance climate change adaption: special report of the intergovernmental panel on climate change.* Cambridge, UK: Cambridge University Press.
Intergovernmental Panel on Climate Change (IPCC). (2014). *Climate change 2014: Synthesis report. Contribution of working groups I, II and III to the fifth assessment report of the intergovernmental panel on climate change. Intergovernmental panel on climate change.* Cambridge, UK: Cambridge University Press.
Intergovernmental Panel on Climate Change (IPCC). (2019). *Special report on the ocean and cryosphere in a changing climate. Intergovernmental panel on climate change.* Available online at IPCC portal.
Jigyasu, R. (2005). Disaster: A reality or construct? In R. W. Perry, & E. L. Quarantelli (Eds.), *What is a disaster: New answers to old questions* (pp. 49−59). Philadelphia: Xlibris Publishers.
Kates, R. W. (1971). Natural hazard in human ecological perspective. *Economic geography*, *47*(3), 438−451.
Keller, E., & DeVecchio, D. (2014). *Natural hazards* (4th ed.). Upper Saddle River, NJ, USA: Prentice Hall.
Kelman, I., Gaillard, J. C., Mercer, J., Lewis, J., & Carrigan, J. (2015). Island vulnerability and resilience: Combining knowledge for disaster risk reduction including climate change adaptation. In E. DeLoughrey, J. Didur, & J. Carrigan (Eds.), *Global Ecologies and the Environmental Humanities: Post-Colonial Approaches* (pp. 162−185). London: Routledge.
Killian, L. M. (1954). Some accomplishments and some needs in disaster study. *Journal of Social Issues*, *10*, 66−72.
Ki-Moon, B. (2013). *Disaster-related economic losses out of control.* Available at: http://www.un.org/News/Press/2013/sgsm15027.doc.htm.
Klein, R., Nichols, R., & Thomalla, F. (2003). Resilience to natural hazards: How useful is this concept? *Environmental Hazards*, *5*(1), 35−45.

Konukcu, B., Mentese, E., & Kilic, O. (2015). Assessment of social vulnerability against disasters. *WIT Transactions on the Built Environment, 150*, 13−23.

Kreps, G. A. (1989). Disaster and the social order. In G. A. Kreps (Ed.), *Social structure and disaster* (pp. 31−51). Newark, DE, USA: University of Delaware Press.

Kreps, G. A. (1998). Disaster as systemic event and social catalyst. In E. L. Quarantelli (Ed.), *What is a disaster: Perspectives on the question* (pp. 31−55). London: Routledge.

Krueger, F., Bankoff, G., Cannon, T., Orlowski, B., & Schipper, E. L. (2015). *Cultures and disaster*. London: Routledge.

Kubal, C., Haase, D., Meyer, V., & Scheuer, S. (2009). Integrated urban flood risk assessment−adapting a multicriteria approach to a city. *Natural Hazards and Earth System Sciences, 9*(6), 1881−1895.

Kumasaki, M., King, M., Arai, M., & Yang, L. (2016). Anatomy of cascading natural disasters in Japan. *Natural Hazards, 80*, 1425−1441.

Leykin, D., Lahad, M., Cohen, R., Goldberg, A., & Aharonson-Daniel, L. (2016). The dynamics of community resilience between routine and emergency situations. *International Journal of Disaster Risk Reduction, 15*(1), 125−131.

Lindell, M. K. (2013). Disaster studies. *Current Sociology Review, 61*(5−6), 797−825.

Lindell, M. K., & Hwang, S. (2008). Households' perceived personal risk and responses in a multi-hazard environment. *Risk Analysis, 28*(2), 539−556.

Lindsay, J. (2015). *The world events that mattered most in 2015*. The Atlantic, 22 December 2015. Available at: https://www.theatlantic.com/international/archive/2015/12/world-events-news-2015/421545/.

Lowendahl, B. (2013). *Strategic management of disasters: Italy, Japan and the United States*. Lisbon, Portugal: Universidade Catolica Portuguesa, Escola de Administração de Empresas.

Lukic, T., Gavrillo, M. B., Markovic, S. B., Komac, B., Zorn, M., Mladan, D., ... Prentovic, R. (2013). Classification of natural disasters between the legislation and the application: Experience of the Republic of Serbia. *Acta Geographica Slovenica, 53*(1), 150−164.

Mahmood, S., & Rahman, A. (2019). Flash flood susceptibility modelling using geo-morphometric and hydrological approaches in Panjkora Basin, Eastern Hindu Kush, Pakistan. *Environmental Earth Sciences, 78*(1), 43−58.

Malalgoda, C., Amaratunga, D., & Haigh, R. P. (2014). Challenges in creating a disaster resilient built environment. *Procedia Economics and Finance, 18*, 736−744.

Mata-Lima, H., Alvino-Borba, A., Pinheiro, A., Mata-Lima, A., & Almeida, J. A. (2013). Impacts of natural disasters on environmental and socio-economic systems: What makes the difference. *Ambiente & Sociedade, 16*(3), 45−64.

McEntire, D. (2015). *Disaster response and recovery*. Hoboken, NJ, USA: John Wiley.

McIvor, A. L., Spencer, T., Möller, I., & Spalding, M. (2012). Storm surge reduction by mangroves. *Natural Coastal Protection Series: Report 2. Cambridge Coastal Research Unit Working Paper 41*. London: The Nature Conservancy and Wetlands International.

McVittie, A., Cole, L., Wreford, A., Sgobbi, A., & Beatriz Yordi, B. (2017). Ecosystem-based solutions for disaster risk reduction: Lessons from European applications of ecosystem-based adaptation measures. *International Journal of Disaster Risk Reduction, 32*, 42−54.

Mitchell, T., van Aalst, M., & Villanueva, P. S. (2010). *Assessing Progress on Integration Disaster Risk Reduction Climate Change Adaptation in Development Processes*. Sussex, UK: Institute of Development Studies, University of Sussex.

Mileti, D. (1999). *Disasters by design: A reassessment of natural hazards in the United States*. Washington DC: Joseph Henry Press.

Mitchell, T., & van Aalst, M. K. (2008). *Convergence of disaster risk reduction and climate change adaptation: A review for department for international development (DFID)*. London, UK: DFID.

Mojtahedi, M., & Oo, B. L. (2017). Critical attributes for proactive engagement of stakeholders in disaster risk management. *International Journal of Disaster Risk Reduction, 21*, 35–43.

Moore, H. E. (1958). *Tornadoes over Texas*. Austin, TX, USA: University of Texas Press.

NAP (National Academic Press). (2001). *Disaster resilience: A national imperative*. Washington, D.C.: National Academic Press.

National Aeronautics and Space Administration (NASA) (2019). *Carbon dioxide-current measurement*, 20 October 2020. NASA Web Portal. Available online.

Naumann, S., Anzaldua, G., Berry, P., Burch, S., Davis, M., Frelih-Larsen, A., ... & Sanders, M. (2011). *Assessment of the potential of ecosystem-based approaches to climate change adaptation and mitigation in Europe*. Final report to the European Commission, DG Environment, Contract no. 070307/2010/580412/SER/B2, Ecologic Institute and Environmental Change Institute, Oxford University Centre for the Environment. Brussels: European Commission.

Nicholson, S. E. (2014). A detailed look at the recent drought situation in the Greater Horn of Africa. *Journal of Arid Environments, 103*, 71–79.

Organization for Economic Cooperation and Development (OECD). (2012). *Disaster risk assessment and risk financing: A G20/OECD methodological framework*. Paris: OECD. Available at: https://www.oecd.org/gov/risk/G20disasterriskmanagement.pdf.

O'Keefe, P., Westgate, K., & Wisner, B. (1976). Taking the naturalness out of natural disasters. *Nature, 260*, 566–567.

Oliver, J. (1980). The disaster potential. In J. Oliver (Ed.), *Response to disaster. North Queensland: Center for disaster studies* (pp. 3–28). James Cook University.

Oliver, S. (2001). Natural hazards: Some new thinking. *Geography review, 14*(3), 2–4.

Otto, F. F. E., van der Wiel, K., van Olderborgh, G. J., Philip, S., Kew, S. F., Uhe, P., & Cullen, H. (2018). Climate change increses the probability of hevy rains in Norther England and Souther Scotland like those of storm Desmond - a real time event attribution revisited. *Environment Research Letters, 13*, 024006.

Pandey, B., & Okazaki, K. (2003). *Community Based Disaster Management: Empowering Communities to Cope with Disasters*. Japan: United Nations Center for Regional Development.

Parnell, S., Simon, D., & Vogel, C. (2007). Global environmental change: Conceptualizing the growing challenge for cities in poor countries. *Area, 39*(2), 357–369.

Paton, D. (2006). Disaster resilience. In D. Johnson, & D. Paton (Eds.), *Disaster resilience* (pp. 3–10). Springfield, IL, USA: Charles C. Thomas.

Paton, D., & McClure, J. (2013). *Preparing for disaster*. Springfield, Illinois: Charles C. Thomas.

Patterson, O., Weil, F., & Patel, K. (2010). The role of community in disaster response: Conceptual models. *Population research and policy review, 29*(2), 127–141.

Paul, B. K. (2011). *Environmental hazards and disasters: Contexts*. Perspectives and management. West Sussex, UK: Willey-Blackwell.

Peacock, W., & Bates, F. L. (1987). Disasters and social change. In R. Dynes, B. De Marchi, & C. Pelanda (Eds.), *Sociology of disasters* (pp. 291–330). Milan, Italy: Franco Angeli.

Peek, L. A., & Mileti, D. S. (2002). The history and future of disaster research. In R. Bechtel, & A. Churchman (Eds.), *Handbook of environmental psychology* (pp. 511–524). Hoboken, NJ, USA: Wiley.

Perrow, C. (2007). *Next Catastrophe: Reducing Our Vulnerabilities to Natural, Industrial and Terrorist Disasters*. Princeto, NJ: Princeton Uiversity Press.

Perry, R. W. (1985). *Comprehensive emergency management*. Greenwich, CT, USA: JAI Press.

Perry, R. W. (1989). Taxonomy, classification and theories of disaster. In G. A. Kreps (Ed.), *Social structure and disaster* (pp. 351–359). Newark, DE, USA: University of Delaware Press.

Perry, R. W. (2018). Defining disaster: An evolving concept. In H. Rodriguez, J. E. Trainor, & W. Donner (Eds.), Handbook of disaster research (pp. 3–20). Gewerbestrasse, Switzerland: Springer International Publishing.

Perry, R. W., & Lindell, M. K. (1997). Aged citizens in the warning phase of disasters. *International Journal of Aging and Human Development*, *44*(4), 257−267.
Perry, R. W., & Lindell, M. K. (2007). *Emergency planning*. Hoboken, NJ, USA: John Wiley.
Perry, R. W., & Quarantelli, E. L. (2005). *What is a disaster? New answers to old questions*. Philadelphia: Xlibris Publishers.
Pescaroli, G., & Alexander, D. (2015). A definition of cascading disasters and cascading effects. *Planet at Risk*, *3*(1), 1−9.
Porfiriev, B. N. (1998). Issues in the definition and delineation of disasters and disaster areas. In E. L. Quarantelli (Ed.), *What is a disaster: Perspectives on the question* (pp. 56−72). London: Routledge.
Prince, S. (1920). *Catastrophe and social change*. New York, NY, USA: Columbia University, Faculty of Political Science.
Quarantelli, E. L. (1984). Perceptions and reactions to emergency warnings of sudden hazards. *Ekistics: Reviews on the problems and science of human settlements*, *309*(6), 511−515.
Quarantelli, E. L. (1987). Presidential address: What should we study? *International Journal of Mass Emergencies and Disasters*, *5*, 7−32.
Quarantelli, E. L. (1987a). Disaster studies. *International Journal of Mass Emergencies and Disasters*, *5*(3), 285−310.
Quarantelli, E. L. (Ed.), (1998a). *What is a disaster? Perspectives on the question*. London: Routledge.
Quarantelli, E. L. (1998b). Epilogue: Where we have been and where we might go. In E. L. Quarantelli (Ed.), *What is a disaster: Perspectives on the question* (pp. 234−273). London: Routledge.
Quarantelli, E. L. (2000). Disaster research. In E. Borgatta, & R. Montgomery (Eds.), *Encyclopedia of sociology* (pp. 682−688). New York, NY, USA: Macmillan.
Quarantelli, E. L. (2005). A social science research agenda for the disasters of the 21st century. In R. W. Perry, & E. L. Quarantelli (Eds.), *What is a disaster: New answers to old questions* (pp. 325−396). Philadelphia: Xlibris Publishers.
Quarantelli, E. L., Lagadec, P., & Boin, A. (2006). A heuristic approach to future disasters and crises. In H. Rodriguez, E. L. Quarantelli, & R. Dynes (Eds.), *Handbook of disaster research* (pp. 16−41). New York, NY, USA: Springer.
Rahman, A., & Fang, C. (2019). Appraisal of gaps and challenges in Sendai Framework for disaster risk reduction priority 1 through the lens of science, technology and innovation. *Progress in Disaster Science*, *1*, 100006.
Rahman, A., & Shaw, R. (2015). Disaster and climate change education in Pakistan. In A. Rahman, A. N. Khan, & R. Shaw (Eds.), *Disaster risk reduction approaches in Pakistan* (pp. 315−336). Tokyo: Springer.
Ramanuja, N. (2015). Challenges in disaster management. *International Journal of Business*, *9*(1), 5−16.
Ranger, N., Surminski, S., & Silver, N. (2011). *Open questions about how to address "loss and damage" from climate change in the most vulnerable countries: A response to the Cancún adaptation framework*. London: Grantham Research Institute on Climate Change and the Environment.
Renaud, F. G., Sudmeier-Rieux, K., Estrella, M., & Nehren, U. (Eds.), (2016). *Ecosystem-based disaster risk reduction and adaptation in practice, advances in natural and technological hazards research*. Cham, Switzerland: Springer.
Rivera, C., & Wamsler, C. (2014). Integrating climate change adaptation, disaster risk reduction and urban planning: A review of Nicaraguan policies and regulations. *International Journal of Disaster Risk Reduction*, *7*, 78−90.
Rodriguez, H., & Barnshaw, J. (2006). The social construction of disasters: From heat waves to worst-case scenarios. Contemporary. *Sociology*, *35*(3), 218−223.
Rosenthal, U. (1998). Future disasters, future definitions. In E. L. Quarantelli (Ed.), *What is a disaster: Perspectives on the question* (pp. 146−159). London: Routledge.

Roy, C., & Kovordanyi, R. (2015). The current cyclone early warning system in Bangladesh: Providers' and receivers' views. *International Journal of Disaster Risk Reduction, 12*, 285−299.

Rumbach, A., & Follingstad, G. (2018). Urban disasters beyond the city: Environmental risk in India's fast-growing towns and villages. *International Journal of Disaster Risk Reduction, 34*, 94−107.

Salami, R. O., von Medling, J. K., & Giggins, H. (2017). Urban settlements' vulnerability to flood risks in African cities: A conceptual framework. *Jamba: Journal of Disaster Risk Studies, 9*(1), 370.

Schipper, L., & Pelling, M. (2006). Disaster risk, climate change and international development: Scope for, and challenges to, integration. *Disasters, 30.1*, 19−38.

Seddon, N., Daniels, E., Davis, R., Harris, R., HouJones, X., Huq, S., ... Wicander, S. (2020). Global recognition of the importance of nature-based solutions to climate change impacts. *Global Sustainability, 3*(e15), 1−12.

Sengul, H., Santella, N., Steinberg, L., & Cruz, M. (2012). Analysis of hazardous material releases due to natural hazards in the United States. *Disasters, 36*(4), 723−743.

Sendai Framework for Disaster Risk Reduction (SFDRR). (2015). Sendai framework for Disaster Risk Reduction, 2015−2030. Geneva, Switzerland: UNISDR. Available at: https://www.preventionweb.net/files/43291_sendaiframeworkfordrren.pdf.

Shaluf, M. (2007). An overview on the technological disaster. *Disaster Prevention and Management, 16*(3), 380−390.

Shaw, R. (2012). *Community Based Disaster Risk Reduction*. London, UK: Emerald Publishers.

Shaw, R., Izumi, T., & Shi, P. (2016a). Perspectives of science and technology in disaster risk reduction of Asia. *International Journal of Disaster Risk Science, 7*, 329−342.

Shaw, R., Izumi, T., Shi, P., Lu, L., Yang, S., & Ye, Q. (2016b). *Asia science technology status for disaster risk reduction*. Beijing, China: IRDR, Future Earth, and ASTAAG.

Singh, R. K., & Swarnim, K. (2010). *Encyclopedia of biological disaster management* (Vol. 1). New Delhi: Rajat Publications.

Singh-Peterson, L., Salmon, P., Baldwin, C., & Goode, N. (2015). Deconstructing the concept of shared responsibility for disaster resilience. *Natural Hazards, 79*, 755−774.

Smith, D. (2005). Through a glass darkly. In R. W. Perry, & E. L. Quarantelli (Eds.), *What is a disaster: New answers to old questions* (pp. 292−307). Philadelphia: Xlibris Publishers.

Sobey, M. N., & Monty, F. (2016). *Regional assessment on ecosystem-based disaster risk reduction and biodiversity in Oceania*. Gland, Switzerland: IUCN.

Song, T., Lu, D., Liang, Y., Wang, Q., & Lin, J. (2017). Progress in international geopolitical research from 1996 to 2015. *Journal of Geographical Sciences, 27*, 497−512.

Srivastava, N., Bharti, N., Tyagi, B., Saluja, S., Bhattacharya, P. K., Aslam, S., & Aggarwal, V. (2016). *Effects of various disaster management approaches: An evidence summary*. London: EPPI-Centre.

Stott, P. A., Chritidis, N., Otto, F. E. L., Sun, Y., Vanderlinden, J.-P., van Olderborgh, G. J., ... Zwiers, F. W. (2015). Attribution of extreme weather events and climate-related events. *WIREs, 7*(1), 23−41.

Street, R. B., Buontempo, C. C., Mysiak, J., Karali, E., Pulquério, M., Murray, V., & Swart, R. (2018). How could climate services support disaster risk reduction in the 21st century. *International Journal of Disaster Risk Reduction, 34*, 28−33.

Strosser, P., Delacámara, G., Hanus, A., Williams, H., & Jaritt, N. (2015). A guide to support the selection, design and implementation of natural water retention measures in Europe—Capturing the multiple benefits of nature-based solutions. *Final version, April 2015. NWRM Pilot Project Financed DG Environment*. Brussels: European Commission.

Sudmeier-Rieux, K., Ash, N., & Murti, R. (2013). *Environmental guidance note for disaster risk reduction: Healthy ecosystems for human security and climate change adaptation*. Gland, Switzerland: IUCN.

Sudmeier-Rieux, K., Nehren, U., Sandholz, S., & Doswald, N. (2019). *Disasters and ecosystems: Resilience in a changing climate—Source book*. Geneva, Switzerland: UNEP and TH Köln - University of Applied Sciences.

References

Susman, P., O'keefe, P., & Wisner, B. (1983). Global disasters, a radical interpretation. In K. Hewitt (Ed.), *Interpretations of calamity* (pp. 263–283). Boston: Allen and Unwin.

Tang, R., Wu, J., Ye, M., & Liu, W. (2019). Impact of economic development levels and disaster types on the short-term macroeconomic consequences of natural hazard-induced disasters in China. *International Journal of Disaster Risk Science, 10*, 371–385.

Thywissen, K. (2006). *Components of Risk- A Comparaive Glossary*. Bonn, Germany: United Nations University.

Tierney, K., Lindell, M., & Perry, R. W. (2001). *Facing the unexpected*. Washington, DC, USA: John Henry Press.

Tobin, G. A., & Montz, B. E. (1997). *Natural hazards: Explanation and integration*. New York: The Guilford Press.

Torry, W. I. (1979). Hazards, hazes and holes: A critique of the environment as hazard and general reflections on disaster research. *Canadian Geographer, 23*(4), 368–383.

Tozier de la Poterie, A., & Baudoin, M.-A. (2015). From Yokohama to Sendai: Approaches to participation in international disaster risk reduction frameworks. *International Journal of Disaster Risk Science, 6*, 128–139.

Twigg, J. (2007). *Characteristics of a disaster-resilient community*. UK: Department for International Development.

United Nations (UN). (2005). R*eport of the world conference on disaster reduction*. Kobe, Hyogo, Japan, 18–22 January 2005. A/CONF/206/6. 16 March 2005.

UN Women. (2019). *Promoting women's leadership in disaster risk reduction and resilience*. Available at: https://www.unwomen.org/en/news/stories/2019/5/news-promoting-womens-leadership-in-disasterrisk-reduction-and-resilience.

United Nations Conference on Trade and Development (UNCTAD). (2010). *The financial and economic crisis of 2008–2009 and developing countries*. Geneva, Switzerland: UNCTAD Secretariat.

United National Development Programme (UNDP). (1992). *An overview of disaster management*. Geneva: UNDP-DMTP.

United Nations Office for Disaster Risk Reduction (UNDRR). (2019). *Global assessment report on disaster risk reduction*. Geneva, Switzerland: UNDRR.

United Nations Office for Disaster Risk Reduction (UNDRR). (2020a). *Disaster risk reduction and climate change adaptation, pathways for policy coherence in sub-Saharan Africa*.

United Nations Office for Disaster Risk Reduction (UNDRR) (2020b). *Integrating disaster risk reduction and climate change adaptation in UN sustainable development cooperation framework*.

United Nations Office for Disaster Risk Reduction (UNDRR). (2020c). *Ecosystem-based disaster risk reduction: Implementing nature-based solutions for resilience*. Bangkok, Thailand: UNDRR Regional Office for Asia and the Pacific.

United Nations-Economic and Social Council (UN-ECOSOC) (1999). Final *report of the scientific and technical committee of the international decade for natural disaster reduction*.

United Nations Environment Programme (UNEP) (2019). *Green infrastructure nature's best defence against disasters*. UNEP Press Release, 17 May. Available at: UNEP Portal.

United Nations-Economic and Social Commission for Asia and the Pacific (UN-ESCAP) (2011). *Achievements and challenges in implementing the Hyogo Framework for Action in Asia and the Pacific*. E/ESCAP/CDR (2)/INF/8. 21 April 2011. Available at: https://www.unescap.org/idd/events/cdrr-2011/CDR2-INF8.pdf.

United Nations-Economic and Social Commission for Asia and the Pacific (UN-ESCAP). (2019). *Asia Pacific disaster report 2019: The disaster riskscape across Asia-Pacific - pathways for resilience, inclusion and empowerment*. Bangkok, Thailand: ESCAP.

United Nations Framework Convention for Climate Change (UNFCCC) (2015). *Paris agreement*. Available at: https://unfccc.int/sites/default/files/english_paris_agreement.pdf.

United Nations Framework Convention on Climate Change (UNFCCC) (2007). *Nairobi work program on impacts, vulnerability and adaptation to climate change.*

United Nations International Strategy for Disaster Reduction (UNISDR) (2015a). *Proceedings: Third UN conference on disaster risk reduction.* 14–18 March, Sendai, Japan. Available at: https://www.unisdr.org/files/45069_proceedingsthirdunitednationsworldc.pdf.

United Nations International Strategy for Disaster Reduction (UNISDR). (2001). *Countering disasters, targeting vulnerability.* Geneva: UNISDR.

United Nations International Strategy for Disaster Reduction (UNISDR). (2004). *Living with risk: A global review of disaster reduction initiatives.* Geneva: UNISDR.

United Nations International Strategy for Disaster Reduction (UNISDR. (2005). *Hyogo framework for action 2005–2015: Building the resilience of nations and communities to disasters.* Geneva: UNISDR.

United Nations International Strategy for Disaster Reduction (UNISDR). (2009a). *Terminology on Disaster risk reduction.* Geneva Switzerland: UNISDR.

United Nations International Strategy for Disaster Reduction (UNISDR). (2009b). *Global assessment of disaster risk reduction 2009.* Geneva: UNISDR.

United Nations International Strategy for Disaster Reduction (UNISDR). (2013a). *Global assessment of disaster risk reduction 2013.* Geneva: UNISDR.

United Nations International Strategy for Disaster Reduction (UNISDR). (2014). *Understanding risk: The evolution of disaster risk assessment since 2005. Background paper prepared for 2015 global assessment report on disaster risk reduction.* Geneva, Switzerland: UNISDR.

United Nations International Strategy for Disaster Reduction (UNISDR). (2015b). *Global assessment report on disaster risk reduction.* Geneva, Switzerland: UNISDR.

United Nations Office for Disaster Risk Reduction (UNISDR) (2017). Terminology. Available at: https://www.unisdr.org/we/inform/terminology.

United Nations International Strategy for Disaster Reduction (UNISDR). (2013b). *Towards the post-2015 framework for disaster risk reduction tackling future risks, economic losses and exposure.* Geneva: UNISDR.

United States Agency for International Development (USAID). (2011). *Introduction to disaster risk reduction.* Washington, D.C.: USAID.

Van der Keur, P., van Bers, C., Henriksen, H. J., Nibanupudi, H. K., Yadav, S., Wijaya, R., ... van Scheltinga, C. T. (2016). Identification and analysis of uncertainty in disaster risk reduction and climate change adaptation in South and Southeast Asia. *International Journal of Disaster Risk Reduction, 16*, 208–214.

van Liekerk, D., Coetzee, C., & Nemakonde, L. (2020). Implementing the Sendai Framework in Africa: Progress against the targets (2015–2018). *International Journal of Diaster Risk Reduction, 11*, 179–189.

Van Riet, G., & Van Niekerk, D. (2012). Capacity development for participatory disaster risk assessment. *Environmental Hazards, 11*(3), 213–225.

Vaughan, A., & Hillier, D. (2019). *Ensuring impact: The role of civil society organizations in strengthening world bank disaster risk financing.* London: Center for Disaster Protection.

Venton, P., & La Trobe, S. (2008). *Linking climate change adaptation and disaster risk reduction.* London: Institute of Development Stuies.

Wallace, A. F. C. (1956). *Human behavior in extreme situations.* Washington, DC, USA: National Research Council—National Academy of Sciences.

Wamsler, C. (2014). *Cities, disaster risk and adaptation.* London: Routledge.

(World Economic Forum (WEF). (2014). *The Global Risks Report 2014.* Geneva: World Economic Forum.

World Economic Forum (WEF). (2020). *Global Risks Report.* Geneva, Switzerland: WEF.

Wettenhall, R. L. (1975). *Bushfire disaster: An Australian community in crisis.* Sydney: Angus & Robertson.

Whyte, A. V. T., & Burton, I. (1980). *Environmental risk assessment. SCOPE 15.* New York: John Wiley & Sons.

Wisner, B. (2020). Five years beyond Sendai—Can we get beyond frameworks? *International Journal of Risk Science, 11*, 239–249.

Wisner, B., & Haghebaert, B. (2006). Fierce friends/friendly enemies: State/civil society relations. Paper presented at ProVention Consortium Forum 2006, 2–3 February 2006, Bangkok, Thailand. Available at: http://proventionconsortium.net/?pageid = 42.

Wisner, B., Gaillard, J. C., & Kelman, I. (2012). *Handbook of hazards and disaster risk reduction and management*. London: Routledge.

Wisner, B., Kelman, I., & Gailliard, J. C. (2014). Hazard, vulnerability, capacity, risk and participation. In A. Lopez-Carresi, M. Fordham, B. Wisner, I. Kelman, & J. C. Gaillard (Eds.), *Disaster management* (pp. 13–22). London: Routledge.

World Bank. (2012). *Disaster risk financing and insurance in Sub-Saharan Africa: Review and options for consideration*. Washington DC: The World Bank.

Yamin, F., Rahman, A., & Huq, S. (2005). Vulnerability, adaptation and climate disasters: A conceptual overview. *IDS Bulletin, 36*(4), 1–14.

Young, S., Balluz, L., & Mililay, J. (2004). Natural and technologic hazardous material releases during and after natural disasters. *Science of the Total Environment, 322*(1), 3–20.

Zakour, M., & Gillespie, D. (2013). *Community disaster vulnerability*. New York, NY, USA: Springer.

Zhou, L., Perera, S., Kayawickrama, J., & Adeniy, O. (2014). the implication of hyogo framework for action for disaster resilience education. *Procedia, Economics and Finance, 18*, 576–583.

Zissener, M. (2011). Solutions for those at risk in climate disasters. United Nations University Portal. 19 March 2011. Available online at: https://unu.edu/publications/articles/solutions-for-those-at-risk-in-climate-disasters.html#info.

Further reading

Bradshaw, S., & Fordham, M. (2013). *Women, girls and disaster: A review for DFID*. London: Department for International Development.

Janssen, M. A., Schoon, M. L., Ke, W., & Börner, K. (2006). Scholarly networks on resilience, vulnerability and adaptation within the human dimensions of global environmental change. *Global Environmental Change, 16*(3), 240–252.

Seager, J. (2014). Disasters are gendered: What's new? In A. Singh, & Z. Zommers (Eds.), *Reducing disaster: Early warning systems for climate change* (pp. 265–281). Cham, Switzerland: Springer.

Surianto, S., Alim, S., Nindrea, R. D., & Trisnantoro, L. (2019). Regional policy for disaster risk management in developing countries within the sendai framework: A systematic review. *Open Access Macedonian Journal of Medical Sciences, 7*(13), 2213–2219.

United Nations International Strategy for Disaster Reduction (UNISDR). (2011). *Global assessment of disaster risk reduction 2011*. Geneva: UNISDR.

CHAPTER 2

Challenge of climate change

2.1 Introduction

Phenomenon of climate change has come into reckoning of the international community, from being one of global risks to the humankind some years ago to be the greatest risk to humanity at the cusp of beginning of the third decade of the 21st century. About two decades back, climate change had been termed as one of the greatest "world risks" by Beck (1998) that humankind has brought forth.

Climate change that was viewed as a "source of risk" by the World Economic Forum (WEF) in its inaugural issue of Global Risk Report in 2006, came to be ranked among the "top five" perceived global risks, in terms of "impact" in WEF's global risks reports for 2014 and 2015 and environment-related risks have continued to dominate the WEF's global risks reports from 2017 to 2019 in a row, accounting for three of the top five risks by likelihood and four by impact (WEF, 2019). Risks stemming from climate change were regarded as "still emerging" by WEF's global risk report for 2006, on which the consensus about the nature of this risk and its resultant consequences for the international society and global economy were yet to be arrived at, despite a growing corpus of scientific evidence available and pointing to the gravity of the long-term challenge (WEF, 2006, p. 2). Alluding to the growing consensus over the phenomenon of climate change being real, WEF's 2006 global risk report cautioned that climate change entailed the potential of becoming "irreversible over the next 10–20 years" (WEF, 2006, p. 3).

In the wake of growing evidence in the scientific literature on climate change demonstrating that the risks emanating from climate change are primarily intertwined with other major risks originating from storms and ecosystem degradation to regulating and long-term energy uses (WEF, 2006, p. 5), the risk potential of climate change assumed added dimensions culminating in upscaling of risk profile of climate change. "Threat multiplier" effect of climate change caught the imagination of national security experts in the United States during the second half of the first decade of the 21st century (CNA, 2007) and it was also being described as an "accelerant of instability" (DoD, 2010). Description of climate change as "threat multiplier" and "accelerant of instability" denoted that climate change entails the potential of exacerbating other drivers of insecurity like water, food and energy. Concomitantly, it also demonstrates that risk does not emanate from climate change as such but from how climate change interacts with other environmental, socioeconomic and political factors.

Department of Defense of the United States in its 2014 Quadrennial Defense Review concisely described climate change as a "threat multiplier," meaning that it may exacerbate other threats to security (DoD, 2014). Notwithstanding acknowledgment of climate change as "threat multiplier" by the US military, risk assessments that dealt with international, regional and subnational risks frequently botched to capture the "multiplier" effect of climate change; rather on the contrary, the phenomenon of climate change was often regarded as an "environmental" factor with little to no connection to other risk factors in the socioeconomic, political and security realms (Werrell & Femia, 2015).

Nevertheless, WEF's global risk reports for 2014 and 2015 captured the multiplier effect of climate change by identifying climate change as a "perceived risk." In both reports, climate change ranked among the "top five" perceived global risks, in terms of "impact" in both these reports (WEF, 2014, p. 13; WEF, 2015, p. 9). This also demonstrates explicit relationship between three of the other top 10 risks—food crises, water crises and extreme weather events—with climate change. While asserting that global risks reports for 2014 and 2015 represent an important step forward in understanding and addressing systemic global risks such as climate change, Werrell & Femia (2015, p. 3) lament that continued viewing of climate change as an environmental risk, rather than as a broader societal, economic and geopolitical risk and separating it from other stresses like water, food security and extreme weather events could spur the societies and policy makers in underestimating the scope and scale of the risks. Therefore they regard climate change not merely about environment but also about the economy, security, geopolitics, and society as a whole.

Publication of the Fourth Assessment by the Intergovernmental Panel on Climate Change (IPCC) in 2007 that envisaged a comprehensive and thorough review of the science of climate change proved instrumental in generating worldwide awareness of the urgency of an international response to climate change. This also spurred the international scientific community globally to focus on the issues articulated in the Fourth Assessment Report of the IPCC. Subsequent successful convening of the climate change conference [Conference of Parties (CoP-15)] in Copenhagen in December 2009 under the aegis of the United Nations Framework Convention on Climate Change (UNFCCC) proved instrumental in bringing the climate change agenda from the laboratories and offices of the scientists and environmentalists onto the agenda of heads of governments. At the CoP-15, world leaders acknowledged the urgency of containing the human-induced global warming (GW) to a maximum of 2°C above pre−industrial levels (UNDP, 2010). The CoP-15 helped in reaching a significant pinnacle in humankind's "journey toward addressing one of the greatest global challenges of our time" (Rasmussen, Richardson, Steffan, & Liverman, 2011).

Corpus of literature on climate change either emanating from the international scientific community or international organizations like IPCC, appearing on the scene in post−Copenhagen period, emphasizes on the urgency of limiting GW and also cautions against the catastrophic impacts on individual or collective human lives, health, food system, ecosystems, biodiversity, economy, infrastructure, and society as a whole, accruing from extreme events like flooding, melting of glaciers, rise in sea levels, droughts, etc. Fifth Assessment Report of the IPCC published in 2014 envisaged that human influence on the climate change is evidently clear and noted that anthropogenic emissions of greenhouse gases (GHGs) have been highest in the human history with the result that climate changes occurring entail widespread impacts on human and natural systems (IPCC, 2014b, 2014c).

IPCC's Special Report on GW 1.5°C, released in Early October 2018 meeting a target of 1.5°C is possible to achieve; nevertheless, attaining this target calls for "deep emissions reductions" along with envisaging "rapid, far reaching and unprecedented alterations in all aspects of society." Pointing out that facilitation of limiting GW to 1.5°C compared with 2°C would result in reduction of challenging impacts on ecosystems, human health, and human well-being, the IPCC (2018) report further cautioned that a 2°C temperature increase would further aggravate extreme weather, rising sea levels and melting of Arctic ice, coral bleaching, and loss of ecosystems, among other impacts (IPCC, 2018).

Adverse impacts of climate change on human health, individually and collectively through vector diseases, famine, air pollution and water pollution culminating in water-borne diseases etc., are examined by Luber and Lemery (2015), Lemery and Auerbach (2017) and McMichael (2019). Climate change impacts all dimensions of food security and nutrition—food availability, food access, food utilization and food stability. Climatic conditions affect production of some staple crops, and variations in rainfall and temperatures impact crop yield as well as crop quality and quantity thereby denying access to quality food by the poor and entailing the potential of augmenting food insecurity (Behnassi, Pollman, & Gupta, 2018).

Climate change affects ecosystems in multiple ways: forcing the change in habitats where species live, undermining the capacity of the ecosystems to mitigate extreme events and disturbances, impacting the mountain and arctic ecosystems and species that are specifically sensitive to climate change and projected warming entail the potential of greatly enhancing the rate of extinction of species, especially in sensitive areas (Watanabe, Kapur, Aydın, Kanber, & Akça, 2019).

Climate change is affecting the habitats of several species, which must either adapt or migrate to areas with more favorable conditions. Climate change at global scale and its multifaceted impacts on ecosystem, society, economy has been witnessing the focus of the contemporary researches carried out by the scholars throughout the globe. The biodiversity in ecosystem is also facing a severe modification as well as a relentless decaying of varying magnitude due to this climatic variability at spatial and temporal references; and it has been making the biodiversity management process more complicated to address the effort of synergy between the issues of biodiversity conservation under the threat of global climate change and streamlining the utilization of biosphere resources for the benefit of mankind (Filho, Barbir, & Preziosi, 2019).

Climate change in its entirety and GW in particular, are said to wield a negative impact on biodiversity in three main ways: Firstly, climate change leads to enhancement in temperatures owing to climate change is known to be harmful to a number of organisms, especially those in sensitive habitats such as coral reefs and rainforests; secondly, the pressures emerging from climate entail the likelihood of leading to sets of responses in varied areas like phenology, range and physiology of living organisms; and thirdly, the impacts of climate change on biodiversity are reckoned to be realized in the short term in respect of some species and ecosystems, but also in the medium and long term in many biomes. Undeniably, some of these impacts may be irretrievable, if not addressed adequately (Filho et al., 2019). Infrastructure, comprising an interconnected web of systems, is alarmingly becoming vulnerable to the vagaries of climate change; and the critical infrastructure like transport, housing, electricity and water entailing the potential of being at risk in the wake of gradually rising sea levels and extreme temperatures. Coastal and low-lying areas in many developing countries entailing buildings, residential dwellings and other types of critical infrastructure are always at risk in the wake of rising sea levels that on an average record an increase rate of 2.6–2.9 mm per year (Ayyub, 2018).

Climate change poses a serious threat to the economies of both developed and developing countries in myriad ways, including increased flooding and storm damage, altered crop yields, lost labor productivity, higher crime, reshaped public-health patterns, and strained energy systems, among many other effects. Climate change affects local, regional and even national and global economies; concomitantly, economic choices also explicitly impact the greenhouse gas (GHG) emissions that drive climate change; thereby, there exists interrelationship between climate change and economics (Fistory & Pepyvrakis, 2016).

Myriad ways in which climate change wields multiple dangerous impacts on living organisms on Planet Earth not only makes climate change as a challenge of immense magnitude but also portending existential threat to human civilization. A report released on May 30, 2019 had cautioned that unless humankind takes "drastic and immediate" action to stop further deterioration in climate change, a blend of food production, water scarcity and extreme weather could culminate in a societal collapse worldwide and human civilization itself could be past the point of no return by 2050 (Spratt & Dunlop, 2019).

2.2 Defining climate

Defining notions of climate and climate change is nontrivial and contentious (Todorov, 1986, p. 259). There exists lack of unanimity of opinion and agreement among climate experts on the definition of climate (Stott & Kettleborough, 2002). Undoubtedly, in public and scientific discourses notions of climate and climate change are often applied loosely; nonetheless, it remains unclear yet as to what is exactly understood by them. Undeniably, various definitions of climate and climate change are discussed in the climate science literature; nevertheless, what is absent is a clear, distinct and comprehensive conceptual analysis of the different definitions and their "benefits and problems" (Lorenz, 1995). Climate is a potent enduring idea that often defies easy definition (Hulme, 2015). Certainly, geographers and Geography have long been familiar with the idea of climate (Barry, 2013); however, presently the notion of climate has come to be associated with the narrative of climate change and its scientific, political, economic, religious, social, and ethical dimensions (Hulme, 2009) thereby enabling the scientific community and common people to have a better understanding of the idea of climate itself.

According to Carey (2012), climate is an old but also a versatile idea entailing enormous power and utility; and certainly, it continues to acquire new powers (Hulme, 2011). Fleming and Jankovic (2011) credibly argue that climate has worked both as index and as agency. Use of climate as index facilitates description of the accumulated patterns of weather in places; as agent, climate is used as explanation for a wide range of physical and human outcomes. According to Hulme (2008), this dual function of climate has recurred throughout human cultural history and it works too in contemporary discourses about climate change. While ideas about climate are always situated in a time and in a place, ideas of climate also change (Agnew, 2014).

Generally, the scientific definition of climate commences with somewhat like the official wording used by the World Meteorological Organization (WMO): climate is "...a statistical description in terms of the mean and variability of relevant meteorological quantities over a period of time ranging from months to thousands or millions of years" (cited in Hulme, 2015). This description of climate orthodoxly places reliance on 30 years of weather data. However, Arguez and Vose (2011) contest this convention. Climate could also be construed a more wide-ranging scientific sense as a portrayal of the state and dynamics of the physical planetary system comprising five major components: "the atmosphere, the hydrosphere, the cryosphere, the lithosphere, and the biosphere, and the interactions between them. The climate system evolves in time under the influence of its own internal dynamics and because of external forcing such as volcanic eruptions, solar variation sand anthropogenic forcing" (IPCC, 2013a, 2013c, p. 1451).

2.2.1 Climate and weather

In common parlance, terms "climate" and "weather" are often used interchangeably denoting the same meaning. However, climate and weather are related but do not convey the same meaning. What differentiates climate from weather is time. Weather explains *pas seul* or variation of the atmosphere in a specific place over a short period of time. Weather denotes inclusion of conditions like changes in air pressure, temperature, humidity, clouds, wind and precipitation—rain, hail and snow. Besides, variation in weather can occur over the course of minutes, hours, days or weeks (Ormerod & McAfee, 2017).

On the other hand, climate is the normal or expected weather for a specific place or region that is characteristically assessed over a 30-year time period described as a normal. The time scale for climate is months, seasons, years, decades, centuries and even millennia. Climate is often regarded merely the long-term average temperature and precipitation that also incorporates variability and the probability of extreme events.

According to Cambridge dictionary, the term "weather" means "the conditions in the air at a particular time, such as wind, rain, or temperature" (Cambridge Dictionary, 2008, p. 1080). In other words, weather can be described the state of the atmosphere with respect to wind, temperature, cloudiness, moisture, pressure, etc. Broadly viewed, instant meteorological conditions that are measured accurately and statistically averaged over a period of time, usually engender the conventional definition. Nevertheless, climate is not weather. Weather has an imminence and evanescence that climate lacks. Apart from being continuously in flux, weather is always both evanesce and in renewal; and it can be seen and felt. However, the notion of climate spurs a feeling of constancy and regularity in the wake of disturbing and unpredictable weather. According to Hare (1966, p. 99–100), "Climate is the ordinary man's[sic] expectation of weather…there is a limit to the indignities that the weather can put upon him, and he can predict what clothes he will need for each month of the year." According to Solomon et al. (2007), "Climate in a narrow sense is usually defined as the average weather, or more rigorously, as the statistical description in terms of the mean and variability of relevant quantities over a time period ranging from months to thousands or millions of years. The classical period for averaging these variables is 30 years, as defined by the World Meteorological Organization. The relevant quantities are most often surface variables such as temperature, precipitation and wind. Climate in a wider sense is the state, including a statistical description, of the climate system" (Solomon et al., 2007, p. 942).

Weather and climate are primarily different and vary in their respective definitions (Werndl, 2016); however, "weather" is frequently defined as the state of the atmosphere at a point in time, while "climate" is the statistical distribution of weather aggregated over a period of time, nearly about three decades (Arguez & Vose, 2011). Both weather and climate are influenced by atmospheric and ocean processes; nevertheless, diverse facets have more or less significance depending on the time scale of interest. Chaotic behavior in the atmosphere entails potential of limiting weather predictions, and as Palmer (1993) points out, a weather forecast loses all skill beyond a lead time of approximately two weeks. Undoubtedly, precise observations of the atmosphere are essential to make adroit short-term weather and climate forecasts (Collins, 2002); nonetheless, these predictions are also prone to lose their importance for long-term future climate projections (Hawkins & Sutton, 2009). The future projections of weather and climate, along with human ability to comprehend past weather and climate, impact the kind of information that can be provided.

Undeniably, there is a dearth of research on layman's conceptualization of a climate system (Schreiner, Henriksen, & Kirkeby Hansen, 2005); nevertheless, research has also demonstrated that laymen often fail to comprehend that climate is the long-term changes in weather patterns (Pruneau, Moncton, Liboiron, & Vrain, 2001). Such a scenario often spurs laymen to think that climate change could be evident and directly experienced in short-term weather patterns (Gowda, Fox, & Magelky, 1997). Accordingly, many people get inclined to associate observed and experienced weather events as evidence supporting or refuting climate change.

2.3 Climate system

The climate system, as described in the Third Assessment Report of IPCC, is an interactive system comprising five major components—the atmosphere, the hydrosphere, the cryosphere, the land surface and the biosphere—that gets forced or influenced by various external forcing mechanisms, the most important of which is the Sun. Besides, the direct impact of human activities on the climate system is also considered an external forcing. The quantum of energy received from the Sun at any point on Earth's surface undergoes change on daily, yearly and longer-term cycles linked with Earth's movement via the solar system. This cyclical fluctuation in the input of solar energy, also called solar forcing, envisages changes in the surface environment—rise in temperatures during day and fall in the night as well as causing rise in temperatures during summer and fall in winter (Grotizinger & Jordan, 2014, p. 408). An understanding of each component of the climate system is deemed essential to comprehend the entire gamut of the climate system.

2.3.1 Atmosphere

Earth's atmosphere is the most unstable, the most mobile and rapidly changing part of the climate system. Composition of the atmosphere has undergone transformations along with the cycle of the evolution of the Earth. The atmosphere is layered and about three-fourths of its mass is concentrated in the layer closest to Earth's surface, known as the troposphere. Above the troposphere is the stratosphere, a dryer layer that extends to an altitude of about 50 km. The outer atmosphere, above the stratosphere, has no hasty or abrupt limit; it gradually gets thinner and disappears into outer space. The Irregular heating of Earth's surface by the Sun contributes to strong convection of the troposphere and the resulting "convection patterns in the troposphere, combined with Earth's rotation, set up a series of prevailing wind belts" (Grotizinger & Jordan, 2014, p. 408). Remarkably, heat energy from the warmer equatorial regions is carried by the spiral-like global circulation of air in the wind belts to the cooler polar regions. According to Grotizinger and Jordan, (2014, p. 408), the atmosphere is a mixture of gases, with nitrogen constituting 78% by volume, oxygen 21% by volume; and the remaining 1% comprises argon (0.93%), carbon dioxide (0.035%), and other minor gases (0.035%), including methane and ozone. Water vapor, concentrated in the troposphere in highly variable quantities, along with carbon dioxide (CO_2) are the major GHGs in the atmosphere (Ahlonsou et al., 2001, p. 88). Presence of trace gases like CO_2, methane (CH_4), nitrous oxide (N_2O), and ozone (O_3), also entailing the capacity to absorb and emit infrared radiation, play an essential role in the Earth's energy budget.

O_3, a highly reactive GHG produced principally by the ionization of molecular oxygen by ultraviolet radiation from the Sun, plays a vital role the Earth's energy budget. Most atmospheric ozone is found in the stratosphere, where it filters out incoming ultraviolet radiation, protecting the biosphere at Earth's surface from its potentially damaging effects. Apart from these gases, the atmosphere also contains solid and liquid particles, known as aerosols and clouds that interact with the incoming and outgoing radiations in a complex and spatially very variable manner. Aerosol is a suspension of airborne solid or liquid particles, with a typical size, that stays in the atmosphere for at least many hours. Aerosols, be of either natural or anthropogenic origin, entail the potential of impacting climate in several ways: "directly through scattering and absorbing radiation, and indirectly by acting as cloud condensation nuclei or ice nuclei, modifying the optical properties and lifetime of clouds" (Allwood, Bosetti, Dubash, Gómez-Echeverri, & von Stechow, 2014). Atmospheric aerosols, either natural or anthropogenic, emanate from two separate pathways— "emissions of primary particulate matter, and formation of secondary gaseous precursors" (Allwood et al., 2014) and most of the aerosols are of natural origins. Aerosol emanating from anthropogenic sources such as fossil and biofuel burning, mostly remain limited to populated regions, whereas aerosol originating from natural sources like desert dust, sea salt, volcanoes and the biosphere, are important in both hemispheres and likely dependent on climate and land-use change (Carslaw et al., 2010). Mahowald et al. (2010) opine that long-term trends in aerosols from natural sources are more difficult to identify due to international variability.

Water vapor in the atmosphere is the largest feedback component owing to its copiousness, responsible for nearly two-thirds of the greenhouse effect or natural warming. Rising temperatures change the water vapor feedback and engender more evaporation/transpiration, transferring water from the Earth's surface to the troposphere (IPCC, 2007a). Any enhancement in water vapor can further cause rise in the temperature differentially through the atmosphere (Solomon et al., 2010). After coming in contact with cooler temperatures, the water vapor gets condensed around particles in the troposphere forming clouds, and the clouds thus formed reflect solar energy, cooling the Earth, as well as simultaneously clouds assimilate infrared energy emitted by the Earth's surface, causing warming (Shepardsona, Niyogib, Roychoudhury, & Hirschd, 2012). Formation of clouds generally occurs in rising air, resulting in expanding and cooling thereby allowing the activation of aerosol particles into cloud droplets and ice crystals in supersaturated air. Cloud particles are generally larger than aerosol particles and composed mostly of liquid water or ice. Generally, it is cogitated that clouds will have a cooling effect on the Earth's climate (IPCC, 2007a). CO_2 and other GHGs remain present in the troposphere for longer periods spanning over hundreds of years, accordingly, even if there is any diminution in the emissions of GHGs taking place currently, their impact on GW will continue to last for years to come (IPCC, 2007a). Undoubtedly, no single weather event can be attributed to GW; nonetheless, long-term (spanning over 30 years) patterns or changes in weather may be caused by the warming of the troposphere (Rhein et al., 2013). Cloud and aerosol entail properties that are extremely variable in space and time. The short lifetime of cloud particles in "sub-saturated" air creates relatively sharp cloud edges and fine-scale variations in cloud properties, which is less typical of aerosol layers (Boucher et al., 2013). While the distinction between aerosols and clouds is generally apt and advantageous, it is not always explicit, thereby entailing likelihood of causing interpretational difficulties (Koren, Remer, Kaufman, Rudich, & Martins, 2007).

2.3.2 Hydrosphere

Hydrosphere is that component of the climate system that comprises all the liquid water on, over, and under Earth's surface and subterranean water, such as oceans, seas, rivers, freshwater lakes, underground water, etc. (IPCC, 2013a, 2013c). Nearly all of that liquid water—in the oceans, lakes, streams, and groundwater—constitute a mere 1% of the hydrosphere. However small, these continental components of the hydrosphere, though small, are reservoirs for moisture on land and serve as facilitator for transport system for returning precipitation and transporting salt and other minerals to the ocean, and as such are called upon to play a vital role in the climate system (Grotizinger & Jordan, 2014, p. 410). This vital role of the oceans in the climate system assumes significance because of the oceans' ability of controlling the quantum of GHGs, including CO_2, water vapor, and N_2O, along with heat in the atmosphere. According to Rhein et al. (2013), up to now bulk of the net energy enhancement in the climate system from anthropogenic radiative forcing (RF) has been in the form of ocean heat. This supplementary heat that accounts for nearly 60% is stored principally in the upper 700 m of the ocean (Johnson et al., 2016). According to Fahey, Doherty, Hibbard, Romanou, and Taylor (2017), ocean warming and climate-driven changes in ocean stratification and circulation change oceanic biological productivity and consequently CO_2 uptake; combined, these feedbacks impact the rate of warming from RF.

Marine ecosystems absorb CO_2 from the atmosphere in the similar manner that plants do on land. According to some scholars, nearly half of the world's net primary production (NPP) is by marine plants (Carr et al., 2006; Chavez, Messié, & Pennington, 2011; Falkowski et al., 2004). It is also claimed that Phytoplankton NPP supports the biological pump that facilitates transportation of organic carbon ranging from 2 to 12 GtC/year to the deep sea (Doney, 2010; Passow & Carlson, 2012) where it is kept away from the atmospheric pool of carbon for longer durations ranging from 200 to 1500 years. By virtue of the ocean being a vital carbon sink, climate-driven changes in NPP constitute a significant feedback owing to their potentiality to alter atmospheric CO_2 abundance and forcing. While referring to multiple links between RF-driven changes in climate, physical changes to the ocean, and feedbacks to ocean carbon and heat uptake, Fahey et al. (2017) have opined that fluctuations in ocean temperature, circulation, and stratification driven by climate change modify phytoplankton NPP, along with the increase in acidity of the ocean owing to absorption of CO_2 and that in turn entail the potential of impacting NPP and thus the carbon sink.

Occurrence of the bulk of surface evaporation and rainfall over the ocean enables the latter to assume a prominent role in the hydrological cycle, apart from being a significant carbon sink (Schanze, Schmitt, & Yu, 2010). Increase in the surface ocean salinity has been reported in areas of high salinity, such as the subtropical gyres, and diminution in surface salinity in areas of low salinity, such as the Warm Pool region, over decadal time scales (Good, Gregory, Lowe, & Andrews, 2013). This upsurge in stratification in select regions and mixing in other regions are feedback processes for the reason that they set in motion changed patterns of ocean circulation that entails the potential of influencing uptake of anthropogenic heat and CO_2. Augmented stratification constrains surface mixing, high-latitude convection, and deepwater formation, and in so doing theoretically enfeebling ocean circulations, especially the Atlantic Meridional Overturning Circulation (AMOC) (Kostov, Armour, & Marshall, 2014). Lesser deepwater formation and gradual overturning are linked with diminished heat and carbon sequestration at greater depths. Rahmstorf et al. (2015) have observed that future projections demonstrate that the strength of AMOC could prominently decline as the ocean warms and freshens and as upsurge in the Southern Ocean gets enfeebled owing to the storm track moving poleward (Rahmstorf et al., 2015).

While Fahey et al. (2017) have opined that such a deceleration of the ocean currents entail the likelihood of impacting the rate at which the ocean engrosses CO_2 and heat from the atmosphere; according to Rignot and Thomas (2002), enhanced ocean temperatures also quicken ice-sheet melt, especially for the Antarctic Ice Sheet where basal sea-ice melting is significant compared to surface melting due to colder surface temperatures. Undersea melting at tidewater margins, particularly in the case of the Greenland Ice Sheet, is also contributing to volume loss (van den Broeke et al., 2009). Sequentially, alterations in the cold and freshwater inputs get affected by changes in ice-sheet melt rates thereby impacting ocean stratification. This, in turn, entails the potential of impacting ocean circulation and the ability of the ocean to absorb more GHGs and heat (Enderlin & Hamilton, 2014). Increased sea-ice export to lower latitudes enhances local salinity anomalies such as the Great Salinity Anomaly (Gelderloos, Straneo, & Katsman, 2012) and hence to changes in ocean circulation and air—sea exchanges of momentum, heat, and freshwater that successively could impact the atmospheric distribution of heat and GHGs. Despite prevailing variations across climate model projections, there is still good agreement that in the future there will be increasing stratification, decreasing NPP, and a decreasing sink of CO_2 to the ocean via biological activity (Fu, Randerson, & Moore, 2016).

2.3.3 Cryosphere

Cryosphere comprises all regions on and beneath the surface of the Earth and ocean where water is in solid form, including sea ice, lake ice, river ice, snow cover, glaciers and ice sheets, and frozen ground, including permafrost (Planton, 2013). The role of the cryosphere in the climate system varies from that of the liquid hydrosphere owing to the fact that ice is comparatively immobile and because it reflects almost all of the solar energy that falls on it. Approximately one-third of the land surface is covered by seasonal snows, almost entirely (all but 2%) in the Northern Hemisphere. Melting snow constitutes the source of bulk of the freshwater in the hydrosphere. While the seasonal exchange of water between the cryosphere and the hydrosphere comprises an important process of the climate system; even bulk amounts of water are exchanged between the cryosphere and the hydrosphere during glacial cycles (Grotizinger & Jordan, 2014, p. 411). The cryosphere draws its importance to the climate system from its high reflectivity, also known as albedo, for solar radiation, its low thermal conductivity, its large thermal inertia and, especially, its critical role in driving deep ocean water circulation; and for the reason that the ice sheets store a large amount of water, variations in their volume are a potential source of sea level variations (Ahlonsou et al., 2001, p. 88).

In the wake of high reflectivity or albedo of snow and ice to solar radiation as compared to land surfaces and the ocean, loss of snow cover, glaciers, ice sheets, or sea ice accruing from climate warming lowers Earth's surface albedo. Accrual of such losses engender the snow—albedo feedback for the reason that subsequent increases in absorbed solar radiation lead to further warming as well as changes in turbulent heat fluxes at the surface (Sejas et al., 2014). For seasonal snow, glaciers, and sea ice, a positive albedo feedback takes place where light-absorbing aerosols are deposited to the surface, darkening the snow and ice and accelerating the loss of snow and ice mass (Yang, Xu, Cao, Zender, & Wang, 2015). The positive radiative feedback, for instance on the ice sheets of Antarctica and Greenland, gets further amplified by dynamical feedbacks on ice-sheet mass loss.

Explicitly, as continental ice shelves restrict the discharge rates of ice sheets into the ocean; any melting of the ice shelves fast-tracks the discharge rate, generating a positive feedback on the ice-stream flow rate and total mass loss (Joughin, Alley, & Holland, 2012). Warming oceans also spur augmentation of melting of basal ice—ice at the base of a glacier or ice sheet—and subsequent ice-sheet loss (Alley, Scambos, Siegfried, & Fricker, 2016). Feedbacks pertaining to ice-sheet dynamics happen on longer time scales than other feedbacks—many centuries or longer. Substantial ice-sheet melt can also cause changes in freshwater input to the oceans, entailing the potential of impacting ocean temperatures and circulation, ocean–atmosphere heat exchange and moisture fluxes, and atmospheric circulation (Masson-Delmotte et al., 2013).

In polar regions, the surface-albedo feedback wields substantial effect within the cryosphere (Taylor, Ellingson, & Cai, 2011); there is also an indication that the polar surface-albedo feedbacks could impact the tropical climate as well (Hall, 2004). Undeniably, changes in sea ice can also actuate arctic cloudiness; nevertheless, recent research makes it discernible that arctic clouds have responded to sea ice loss in fall but not summer (Taylor, Kato, Xu, & Cai, 2015). Asserting that such a trend implying important implications for future climate change, Fahey et al. (2017) opine that as an upsurge in summer clouds could counterbalance a component of the intensifying surface-albedo feedback, decelerating the rate of arctic warming.

2.3.4 The lithosphere

The lithosphere comprises the upper layer of the solid Earth, both continental and oceanic, consisting of all crustal rocks and the cold, mainly elastic part of the uppermost mantle. Volcanic activity, although part of the lithosphere, is not regarded as part of the climate system, but acts as an external forcing factor (Planton, 2013). The land surface is the most important part to the climate system because the composition of the land-surface wields immense impact on the manner of absorption or release of solar energy to the atmosphere. An upsurge in the land temperature proves instrumental in radiating back more heat energy into the atmosphere culminating in the enhanced water evaporation from the land surface and its subsequent entry into the atmosphere. Use of substantial energy in the evaporation process lets the land surface to cool; and accordingly, as the rates of evaporation are impacted by soil moisture and other related factors like vegetation cover and the surface flow of water, hence these play a significant role in controlling atmospheric temperatures (Grotizinger & Jordan, 2014, p. 411).

The manner of receipt of solar energy and its release into the atmosphere is controlled by vegetation and soil at the land surface. Some portion of the energy received from the Sun is returned as long-wave, also called infrared, radiation resulting in heating the atmosphere as the land surface gets warmed; and some portion of that energy serves to evaporate water, either in the soil or in the leaves of plants, fetching water back into the atmosphere. According to Ahlonsou et al. (2001, p. 89), the texture of the land surface, especially its roughness that is determined by both topography and vegetation, impacts the atmosphere dynamically as winds blow over the land's surface. Blowing wind also carries with it dust from the land surface into the atmosphere where it interacts with the atmospheric radiation.

Alterations in the land surface and vegetative cover imbibe the potential of impacting reflectivity or albedo and the carbon cycle (IPCC, 2007a). The process of deforestation, for instance, not only causes diminution in carbon sink; nonetheless, also envisages decline in the albedo of the land surface, enhancing the amount of energy that is absorbed from the Sun (Pielke et al., 2002). Furthermore, as soils warm in the Arctic region along with thawing of tundra, CO_2 and methane get released into the troposphere (IPCC, 2007a). A warming climate is prone to enhance the likelihood of increasing incidents of forest fires and insect infestations, also triggering release of more carbon dioxide into the troposphere as trees burn or die and decay (IPCC, 2007b). According to Ainsworth and Long (2005), enhanced levels of CO_2 could cause some plants to grow and help in the removal of CO_2 in the troposphere. Such "extra" plant growth could be restricted by the availability of water, nitrogen, and temperature, thereby increasing the likelihood of stabilizing or level out of the removal of CO_2 from the troposphere (IPCC, 2007b).

Despite the complex nature of this land−vegetation−carbon cycle, still it is believed that land/vegetation sinks would either deteriorate or become less efficient (Oren et al., 2001). Alterations in albedo and carbon cycling can cause augmentation in global temperatures, broadening the growing season in some regions (IPCC, 2007a, 2007b), resulting in two concomitant effects—increase in production in mid and high latitudes and decrease in production at lower latitudes (Parry, Rosenzweig, Iglesias, Fischer, & Livermore, 1999). According to Hibbard, Hoffman, Huntzinger, and West, 2017, historically some studies accounting for the contribution of the land cover to RF have focused on albedo forcing only while ignoring those from alterations in land-surface geographical properties like plant transpiration, evaporation from soils, plant community structure and function or in aerosols. Myhre et al. (2013) draw attention to recent assessments that have made a beginning in gauging the comparative contribution of land-use and land-cover change to RF in addition to albedo and/or aerosols. According to Ward, Mahowald, and Kloster (2014), recent reviews and estimates of the combined albedo and greenhouse gas RF for land-cover change make it discernible that these are estimated to account for 40% ± 16% of the human-caused global RF from 1850 to 2010. Interestingly, Ward and Mahowald (2015) have reported that the net RF owing particularly to fire—after accounting for short-lived forcing agents (O_3 and aerosols), long-lived GHGs, and land albedo change both now and, in the future—is projected to be near zero due to regrowth of forests which offsets the release of CO_2 from fire.

2.3.5 The biosphere

Biosphere—comprising terrestrial biosphere and marine biosphere—is an important component of climate system. The terrestrial biosphere is the part of the Earth system that consists of all ecosystems and living organisms, in the atmosphere, on land; and marine biosphere contains all ecosystems and living organisms in the oceans, including derived dead organic matter, such as litter, soil organic matter, and oceanic detritus (Planton, 2013). Undeniably, life is found everywhere on Earth; nonetheless, the amount of life at any location is dependent on local climate conditions. Both the marine and terrestrial biospheres are destined to play vital role by wielding a major impact on the atmosphere's composition. The uptake and release of GHGs is influenced by the biota. The photosynthetic process equips both marine and terrestrial plants (especially forests) to accumulate substantial amounts of carbon from carbon dioxide. It is in this way that the biosphere plays a pivotal role in the carbon cycle, as well as in the budgets of many other gases, such as methane and nitrous oxide (Ahlonsou et al., 2001, p. 89). Fossils, tree rings, pollen, etc., serve as preservers of records of the influence of climate on the biosphere and much of what is known in present times about past climates comes from such biotic indicators.

According to Grotizinger & Jordan, (2014, p. 412), the total energy contained and transported by biota is rather small on a global scale, "less than 0.1% of incoming solar energy is used by plants in photosynthesis and thus enters the biosphere." Other metabolic processes strongly link the biosphere to the other components of the climate system. Atmospheric temperature can be affected by terrestrial vegetation owing to the fact that plants absorb solar radiation for photosynthesis and release it as heat during respiration, and atmospheric moisture is capable of take up groundwater and release it as water vapor. Ability of organisms to absorb or release GHGs like CO_2 and methane (NH_4) equips them to regulate the atmospheric composition. Plants and algae transfer CO_2 from the atmosphere to the biosphere via the process of photosynthesis (Grotizinger & Jordan, 2014, p. 412).

Concentration of the global mean atmospheric CO_2 is regulated by emissions emanating from burning of fossil fuels, wildfires, and permafrost thaw balanced against CO_2 absorbed by the oceans and terrestrial biosphere (Ciais et al., 2013). According to Le-Quéré et al. (2016), hardly less than a third of anthropogenic CO_2 has been taken up during the past decade by the terrestrial environment, and another quarter by the oceans via the process of photosynthesis and through direct absorption by ocean surface waters. While explaining that the capacity of the land to continue absorb carbon dioxide is undefined and dependent on land-use management and on responses of the biosphere to climate change, Fahey et al. (2017, p. 92) further add that altered absorption rates impact atmospheric CO_2 abundance, forcing, and rates of climate change. Even ocean carbon-cycle changes in future climate scenarios are also highly unpredictable.

According to Wenzel, Cox, Eyring, and Friedlingstein (2016), elementary principles of carbon-cycle dynamics in terrestrial ecosystems suggest that augmented atmospheric CO_2 concentrations can directly enhance plant growth rates and, therefore, enhance carbon uptake (the "CO_2 fertilization" effect), nominally sequestering much of the added carbon from fossil-fuel combustion. Nevertheless, this effect is capricious; sometimes plants acclimate so that higher CO_2 concentrations no longer augment growth (Franks et al., 2013). Besides, "CO_2 fertilization" is often equipoised by other factors restricting plant growth, such as water and or nutrient availability and temperature and incoming solar radiation that can be modified by changes in vegetation structure. Large-scale plant mortality through fire, soil moisture drought, and/or temperature changes, as Fahey et al. (2017) opine, also affect successional processes that contribute to reestablishment and revegetation (or not) of disturbed ecosystems, changing the amount and distribution of plants available to uptake CO_2. According to Seppälä (2009), with sufficient disturbance, it has been argued that forests could, on net, turn into a source rather than a sink of CO_2.

Exact decision about future CO_2 stabilization scenarios is contingent on explicating the significant role that the land biosphere plays in the global carbon cycle and feedbacks between climate change and the terrestrial carbon cycle (Hibbard et al., 2017). According to Fahey et al. (2017), Earth System Models (ESMs) are enhancing the representation of terrestrial carbon-cycle processes, including plant photosynthesis, plant and soil respiration and decomposition, and CO_2 fertilization, with the latter based on the assumption that an increased atmospheric CO_2 concentration provides more substrate for photosynthesis and productivity. Recent advances in ESMs have begun to explain other significant factors such as nutrient limitations (Wieder, Cleveland, Smith, & Todd-Brown, 2015). ESMs that do include carbon-cycle feedbacks appear typically to overestimate terrestrial CO_2 uptake under the present-day climate (Anav et al., 2013; Smith et al., 2016) and underestimate nutrient limitations to CO_2 fertilization (Wieder et al., 2015).

2.4 Defining climate change

The term "climate change" has emerged as the defining narrative of the currently ongoing political, economic, socio-cultural and security-related discourses around the globe for multiple reasons. It is perhaps one of the most serious environmental issues with which today's humankind is faced with (Grover, 2010) though the issue is not new (Vlassopoulos, 2012). from the early 19th century till late 20th century, the discussion on the issue of climate change was exclusively confined within the scientific society (Vlassopoulos, 2012). It is reported to have first emerged on the public agenda in the mid- to late 1980s (Moser, 2010). Integration of discussions on climate change into public agenda has seemingly proved instrumental in the emergence of two camps—one camp is led by believers (referred here as protagonists) and the other camp is led by deniers (referred here as skeptics). The protagonists have often maintained that upshot of anthropogenic or human activities on world climate has reached to an alarming state and posing critical threats to physical, socioeconomic structures and even human existence (Spratt & Dunlop, 2019). On the other hand, the skeptics have also left no stone unturned in presented justly sufficient evidence to rebut the anthropogenic aspect of climate change.

Nevertheless, the hectic flurry of activities by the advocates of the climate change in debating over the urgency and dire need of addressing the issues pertaining to the climate change to look for appropriate methods of adaptation and mitigation over the past two decades through civil society, nongovernmental organizations (NGOs), forums of the United Nations and other international and regional forums, etc., by eliciting increasing public involvement in the climate change discourse seems to have spurred awareness-engendering regarding the potential risks and uncertainties attached to the issues related to the climate change. The resultant outcome is that the issues pertaining to climate change have been debated and problematized from diverse standpoints with increasing public participation, both nationally and globally.

"Climate change" refers to long-term change in the statistical distribution of weather patterns like temperature, precipitation, etc., over decades to millions of years of time. Climate on earth has been undergoing alterations on all time scales even since long before human activity could have played a role in its transformation (IPCC, 2007a). Nevertheless, the United Nations framework for Convention on Climate Change (UNFCCC) (UNFCCC, 1994) defined climate change as "a change of climate which is attributed directly or indirectly to human activity that alters the composition of the global atmosphere and which is in addition to natural climate variability observed over comparable time periods." However, the IPCC definition of climate change includes change due to natural variability alongside human activity (UNFCCC, 2011). Australian Government's Department of Climate Change and Energy Efficiency (DCCEE, 2012) in its website described climate change: "our climate is changing, largely due to the observed increases in human produced GHGs. GHGs absorb heat from the sun in the atmosphere and reduce the amount of heat escaping into space. This extra heat has been found to be the primary cause of observed changes in the climate system over the 20th century." Accordingly, in the ongoing environmental discourse different stakeholders have categorized climate change as primarily the change in modern climate increased by anthropogenic or human activities. The adverse human activities like burning fossil fuel and deforestation, etc., are attributed to bring likely change in some climatic aspects.

The term climate change, as articulated through definitions or policy propagation, is seemingly projected as a negative anthropogenic climate change in its present meaning, whereas at the onset it did not actually have focused on its damaging effect (Agrawala, 1998). The narrative is said to be set in motion by French mathematician and physician Jean Fourier in his article published in 1824 wherein he reported greenhouse effect, that's in fact at the core of the current ongoing climate debate (Agrawala, 1998). Nearly five decades later, Arrhenius (1896) published first calculation of GW from human emissions of CO_2 nevertheless Keeling was the first to measure precisely CO_2 in the Earth's atmosphere in 1960 (Weart, 2009). According to Vlassopoulos (2012), climatic variations were perceived accurately as a scientific issue until 1970, hence the debate mostly remained limited to the scientific community of climatologists and relevant research was fragmented into different academic and university endeavors only.

It is maintained that the growing inclination toward environmentalism in the early 1970s along with the climate change discourse, apart from engendering public doubts about the benefits of human activity for Earth, also proved instrumental in turning the "curiosity" about climate into "anxious concern" (Weart, 2010). Subsequent period since then has witnessed organization and convening of international cooperation programs and meetings on the concern about anthropogenic global degradation spreads in which apart from the scientific community, other concerned stakeholders including representatives from interested community have participated enthusiastically (Vlassopoulos 2012). The decade of the 1970s was seemingly characterized by convening of programs and meetings to explore and acknowledge the extent of anthropogenic climate change and convening of Global Atmospheric Research Program (GARP) organized by WMO and the International Council of Scientific Unions (ICSU) in 1974 (Flohn, 1977) and the first World Climate Conference (WCC) organized by the WMO in 1979, were the examples in this regard. Some significant events were held in the decades of the 1980s and the 1990s. Montreal Protocol of the Vienna Convention in 1987 imposes international restrictions on emission of ozone-destroying gases.

The decade of the 1990s was characterized by convening of two momentous events: the 1992 conference in Rio de Janeiro that produced UNFCCC and another 1997 International conference producing Kyoto Protocol that came into effect in 2005 and set targets for industrialized nations to reduce GHG emissions (Weart, 2009). Since the establishment of UNFCCC coming into force in 1994, various conventions, known as Conference of the Parties (COP), held under the aegis of the UNFCCC till the COP-17 held in November–December 2011 in Durban, South Africa, the participating countries continued to strive to negotiate the post–Kyoto agenda (Olmstead & Stavins, 2012) because Kyoto Protocol was to expire in 2012. The Kyoto Protocol's first commitment period started in 2008 and ended in 2012. A second commitment period was agreed in 2012, known as the Doha Amendment to the Kyoto Protocol (UNFCCC, n.d.a) was adopted at the COP 18 held from November 26 to December 8, 2012 in Doha, Qatar. During the second commitment period

2.4 Defining climate change

commencing from 2013 to 2020, the parties committed to reduce GHG emissions by at least 18% below 1990 levels. As of July 2019, 130 countries had accepted the Doha Amendment, while entry into force requires the acceptance of 144 countries. And of the 37 countries with binding commitments, 7 had ratified the Doha Amendment to Kyoto Protocol (UNFCCC, n.d.b).

The changes in the conceptualization of climate change, including its evolutionary epistemology over the years, as can be discerned from the discursive history of climate change, reveal humankind's relationship with the Earth (Bergthaller, Emmett, Johns-Puta, & Limmer-Kneitz, 2014). Rooted in research of the natural sciences, the study of the climate change phenomenon was utilized, having been rooted in the research of the natural sciences, in the 1970s largely extended support to the notion of environmental justice (Dunne, Kurki, & Smith, 2010). Early ideas of environmental sustainability were construed as discordant with development and modernity, also known as the limits-to-growth narrative (Dunne et al., 2010). Predictably, such skeptical views did not augur well in politics and there was a contempt or disregard for environmentalism in the public consciousness. Additionally, while environmental activists implored for earnestness and resources, whereas with the passage of time climate discourse has started assuming added dimensions. Undeniably, the earlier notion that the impacts of climate change are not often immediate and thus, the "tipping point" of the issue (Nixon, 2011), has now undergone alteration; nevertheless, the long-term impacts of climate adaptation and mitigation programs will often not be recognized immediately (Nixon, 2011).

During the closing decades of the twentieth century, the climate change research then led by scientists and NGOs started percolating itself beyond its natural scientific domain to the fields of political science, economics, and humanities (Bergthaller et al., 2014). With the establishment of the Intergovernmental Panel for Climate Change (IPCC) in the late 1980s, as the scientific body for UN climate research, the issue of climate change became institutionally endorsed by the UN and thus, legitimized on an international scale (McDonald, 2013). Such a research-led shift proved instrumental in orienting the notion of climate change from being "a single causal issue," to a "threat multiplier" endangering the fabric of human civilization (McDonald, 2013). Founding of the UNFCCC in 1992 as the official UN arm for climate governance (Agrawala, 1998) along with already established IPCC spurred independent research groups and NGOs to campaign for environmental responsibility with endorsements by politicians forthcoming (Dunne et al., 2010). While illustrating the advancements registered by the climate debate in recent decades, such developments also demonstrated that once neglected or sidelined discussion of climate change was gradually inching toward occupying center stage, placing the issue of environmental security at the center of the environmental challenge (Dyer, 2001).

2.4.1 Definition of climate change

Two globally accepted definitions of climate change—one by UNFCCC and the other by IPCC—serve as references. According to Pielke (2005, p. 549), there are severe discrepancies between what the scientific community under the IPCC regard as "climate change" and what constitutes "climate change" in the language of the climate change convention or UNFCCC. Under Article 1.2 of the Convention establishing the UNFCCC, the term "climate change" is defined as "a change of climate which is attributed directly or indirectly to human activity that alters the composition of the global atmosphere and which is in addition to natural climate variability observed over comparable time periods" (UNFCCC, 1992). The IPCC (2014a) has mooted a more inclusive definition, in contrast to the restrictive one in the UNFCCC. Climate change for the IPCC refers to a change in the state of the climate that can be identified by changes in the mean and/or the variability of its properties and which persist for an extended period, typically decades or longer and changes in the state of climate are identified through the use of statistical tests (IPCC, 2014a, p. 1255). It is important to note here that the definition of climate change proposed by the IPCC in 2014 is the outcome of some adjustments since the first IPCC Assessment Report in 1992, which were made as more scientific evidences were brought forward. The definition of climate change by the UNFCCC has, on the other hand, seemingly enjoyed a certain steadiness, undoubtedly thanks to its statutory nature, albeit it has been subjected to frequent criticism and calls for adjustments ever since. It has been observed by Hardy (2003, p. 4) that scientists are often inclined to use the UNFCCC's definition of climate change while referring to the postindustrial era, whereas they use the IPCC one for the pre–industrial times.

According to Gupta (2010, pp. 636–637), the construal of "the climate change definition" has evolved through initially taking into account environmental issues, and later on started broadening ever more to embrace developmental considerations. Undeniably, the UNFCCC attributed climate change solely to human activities, not taking into consideration any other probable causative factors; nonetheless, for the IPCC, climate change was the result of many factors, not only those of anthropogenic origin. To that end, the IPCC mentions for instance, the natural internal processes and the external forces such as the modulations of solar cycles and the volcanic eruptions (IPCC, 2014a, p. 1255) The UNFCCC having adopted such a restrictive approach in defining climate change had seemingly made it explicit from the beginning that it was of the view that all the efforts to be deployed in the sense of healing the climate were to be done with focus on anthropogenic activities alone. Pielke (2005) has opined that by way of appalling discourses, countries were put under pressure and persuaded to initiate quick climate change actions, despite the fact that there was not yet sufficient certainty on the human origin of climate change. Pielke (2005) further informs that many countries later discarded that initial "monolithic anthropogenic conception of the human's origins of climate change" and embraced "a more questionable conception of the matter, which henceforth accompanied the climate change discussions under the aegis of the UNFCCC" (p. 548).

Kiss and Sheldon (2000) explain that the UNFCCC did so, because its purpose was to prevent any harm to the climate system by way of regulating state actions that are influential to the global climate, whilst a legal instrument was not the proper tool to regulate or have any effect on any of the natural causes of climate change. Therefore the UNFCCC being a treaty had to focus on the anthropogenic causes, because they are the only ones that are subject to human re-adjustment and manipulation (Kiss and Sheldon, 2000, p. 512).

Generally, researchers in climate change often refer to both definitions because both are considered to be credible.

Many hostile phenomena such as the warming of the atmosphere and the ocean, violent storms and cyclones, an increase of sea levels, floods of low coastal areas, loss of biodiversity, degeneration of natural ecosystems, heats, etc. are listed among the consequences of climate change (IPCC, 2014d, p. 40). In its 2014 IPCC Synthesis Report, the IPCC identified five reasons of concern about additional temperature increases due to climate change (IPCC, 2014d, p. 70). These include the concern regarding the following:

- Unique and threatened ecosystems and species.
- The increase in the frequency and damage from extreme weather events.

- The greater climate change vulnerability of homes of poorer communities.
- The growing economic costs caused by the impacts acquired over time by increased atmospheric concentrations GHG.
- The growing possibility of the occurrence of large-scale singular events.

The avowed objective of the definitions of climate change is to elicit global cooperation in undertaking concrete actions to limit the adverse impacts of climate change and commit the countries to restrict the GW to the internationally accepted minimum level in accordance with the Paris Climate Change Agreement coming operational in 2020.

2.5 Abrupt climate change

Dynamics of alterations characterizing the ongoing process of climate change are receiving top priority of the national governments and international community across the globe for the redressal of catastrophic impacts of climate change; nevertheless, there prevails substantial uncertainty with regard to prospective changes whether being gradual, permitting natural ecosystems and societal infrastructure to adapt within a stipulated timeframe or "will some of the changes be more abrupt, crossing some threshold or 'tipping point' to change so fast that the time between when a problem is recognized and when action is required shrinks to the point where orderly adaptation is not possible?" (NRC, 2013, p. vii). Inevitability of "tipping points" taking place in the wake of occurring rapid changes having crossed the thresholds and culminating abrupt changes in the climate system is attested to by the past history of climate change. Admittedly, recent scientific research has enabled humankind to help lessen climatic uncertainty in two significant cases—probable abrupt changes in ocean deepwater formation; and release of carbon from frozen soils and ices in polar regions—nonetheless, the potential for abrupt changes in ecosystems, weather and climate extremes and critical groundwater supplies now seem more likely, severe and imminent (NRC, 2013, p. vii).

It can be discerned from past studies of ice cores massive alterations in climate could occur in a matter of few decades or even years. In this regard, Dansgaard, White, and Johnsen (1989) and Alley et al. (1993) have cited the examples of temperature changes of dozen degrees or more at local or regional levels or doubling and halving of precipitation rates or changes in dust concentrations by orders of magnitude as sufficient basis for abrupt climate change. Undoubtedly, scientific research in recent decades has enhanced human understanding of abrupt climate change in a significant way resulting in allaying some original fears; nevertheless, some new fears have cropped up.

During the closing part of the 1970s and in the early part of the 1980s, there had prevailed archetypal view that major climate change could be attributed, among other factors, to gradual shifts, triggered by the alterations in solar energy that go with likely variations in Earth's orbit around the Sun over thousands to tens of thousands of years (Hays, Imbrie, & Shackleton, 1976). It could be discerned from some early studies of rates of climate change, especially during the last glacial period and the transition from glacial to interglacial climates, that massive changes had occurred obviously in short periods of time (Coope, Morgan, & Osborne, 1971). Nevertheless, bulk of the paleoclimate records dating to back tens of thousands of years fell short of the sequential resolution to sort out yearly to decadal changes. However, the closing years of the 1980s witnessed this scenario beginning to change in the wake of scientific community's focus on examining events like the climate transition that took place at the end of the Younger Dryas about 12,000 years ago (Dansgaard et al., 1989), and the large swings in climate during the glacial period that came to be termed "Dansgaard–Oescher events" or "D–O events"; named after two of the ice core scientists who first studied these phenomena using ice cores (NRC, 2013, p. 23). At the initial stage of "D–O events"; many persons viewed these variations being too large and fast to be climate change; nonetheless, it was only after examining more ice cores that confirmed these variations (Anklin et al., 1993; Grootes, Stuiver, White, Johnsen, & Jouzel, 1993), and in many properties (Alley et al., 1993), including GHGs (Severinghaus & Brook, 1999) that they were widely acknowledged as real.

According to NRC (2013), first definitions of abrupt climate change were tied directly to the D–O events, which in turn were delineated by changes in temperature, precipitation rates, dust fallout, and concentrations of certain GHGs. The resulting outcome was that past reviews of abrupt change largely tended to concentrate on the physical climate system, and the potential for abrupt changes and threshold behavior were articulated principally in climatic terms. While providing the first systematic review of abrupt climate change, the US-based National Research Council (NRC) in its publication in 2002 defined abrupt climate change thus: "Technically, an abrupt climate change occurs when the climate system is forced to cross some threshold, triggering a transition to a new state at a rate determined by the climate system itself and faster than the cause. Chaotic processes in the climate system may allow the cause of such an abrupt climate change to be undetectably small" (NRC, 2002, p. 14).

Emphasizing that its definition of abrupt climate change explained in 2002, the NRC (2002) claimed that its definition assumed critical importance on two counts—first, its focus on climate system itself, a focus that is still used; and second, it engendered the probability of thresholds or tipping points being forced or hard-pressed by an imperceptibly small change in the cause of the shift. Emphasis was also focused on expanding this definition by placing abrupt climate change into a social context. Acknowledging that place this definition in a policy setting or public discourse called for some additional context, NRC (2002) pointed out that while many scientists gauged time on geological scales whereas most people cared more about changes and their potential impacts on society and ecological time scales. In this backdrop NRC (2002) observed: "From this point of view, an abrupt change is one that takes place so rapidly and unexpectedly that human or natural systems have difficulty adapting to it. Abrupt changes in climate are most likely to be significant, from a human perspective, if they persist over years or longer, are larger than typical climate variability, and affect subcontinental or larger regions. Change in any measure of climate or its variability can be abrupt, including change in the intensity, duration, or frequency of extreme events" (NRC, 2002, p. 14).

In the aftermath of elaboration of abrupt climate change by NRC in 2002, many papers on abrupt climate change were published subsequently; and some of these published papers contained definitions more focused on time (Clark, Pisias, Stocker, & Weaver, 2002), and others on the relative speed of the causes and reactions. Overpeck and Cole (2006) defined abrupt climate change as "a transition in the climate system whose duration is fast relative to the duration of the preceding or subsequent state." Lenton et al. (2008) formally introduced the concept of tipping point, defining abrupt climate change as "We offer a formal definition, introducing the term "tipping element" to describe subsystems of the Earth system that are at least subcontinental in scale and can be switched—under certain circumstances—into a qualitatively different state by small perturbations.

The tipping point is the corresponding critical point—in forcing and a feature of the system—at which the future state of the system is qualitatively altered."

A report of the US Climate Change Science Program (USCCSP) in late 2008 defined abrupt climate change as "A large-scale change in the climate system that takes place over a few decades or less, persists (or is anticipated to persist) for at least a few decades, and causes substantial disruptions in human and natural systems" (USCCSP, 2008). This simple definition directly focused attention on the impacts of change on natural and human systems and is considered significant because it directly links the physical climate system with human impacts. An increasingly interdisciplinary approach taken to studying abrupt climate change witnessed concurrent evolution in thinking, expanding from abrupt changes in the physical climate system to embrace abrupt impacts from climate change (NRC, 2012).

In 2013 the NRC has adopted a broader definition of abrupt climate change as thus: the term "abrupt climate change" as being abrupt changes in the physical climate system, and the related term, "abrupt climate impacts," as being abrupt impacts resulting from climate change, even if the climate change itself is gradual (but reaches a threshold value that triggers an abrupt impact in a related system) (NRC, 2013, p. 26). In this new definition, the NRC while embracing the wider concept of abrupt climate change elaborated in NRC's 2002 report, as well as the definition from its 2012 Climate and Social Stress Report, the NRC claims that its new definition apart from being critical, represents a broadening of the focus from just the physical climate system itself to also incorporate "abrupt changes in the natural and human-built world that may be triggered by gradual changes in the physical climate system" (NRC, 2013, p. 26).

2.6 Global warming

GW has come to be acknowledged as the most frequently used idiom in the current discourse on climate change and there is seemingly a semblance of consensus in the relevant scientific research literature (Cook et al., 2013, 2016). While stating that "warming of the climate system is unequivocal," the Fifth Assessment Report of the IPCC further observes that "It is extremely (95%–100%) likely that human influence has been the dominant cause of the observed warming since the mid-20th century" (IPCC, 2013a, 2013c). Such findings based on scientific research can apprise policy responses in tandem with other factors such as risk aversion, discounting of the future and assessments of the severity of future climate impacts (Richardson, Cowtan, & Millar, 2018).

Hulme, Nerlich, and Pearce (2013) citing Oxford English Dictionary (OED) note that the term "global warming" was first used in 1952 and the OED cited the local newspaper of Texas (USA) that published a report that scientists who were studying GW trends pointed out that not a single iceberg was cited last year south of Parallel-46. According to Russell and Landsberg (1971, p. 1312), first use of the term "global warming" in the journal Science came in June 1971. Since then it has emerged as the dominant term in climate change debates, especially in the United States. The third edition of the OED published in 2009 defines GW as "a long-term gradual increase in the temperature of the earth's atmosphere and oceans, spec. one generally thought to be occurring at the present time, and to be associated esp. with side effects of recent human activity such as the increased production of greenhouse gases." Nevertheless, Broecker (1975) has opined that climate scientists seem to have used the phrase in a more restrictive sense denoting the meaning as rise in global mean surface air temperature or "mean planetary temperature."

In this regard, Hulme et al. (2013) have opined that there could be a difference between general and popular understanding of GW. In the broad sense it may be referring to overall Earth system warming in both atmosphere and oceans, termed by Hulme et al. (2013) as GW-1; and the more precise scientific usage of GW in the narrow sense of surface air temperature warming, termed by Hulme et al. (2013) as GW-2; and concurrently Hulme et al. (2013) also hold the view that since a dichotomy entails the potential of leading to confusion. In other words, GW-1 entailing broad meaning refers to warming of the entire Earth system, atmosphere, cryosphere, and oceans; whereas GW entailing narrow meaning alludes to warming of global mean surface air temperature, that is, the lowest part of the atmospheric boundary layer. Subsequent period has witnessed scientific community increasingly adhering to narrow scientific meaning of GW as well as it has come to dominate policy discourses "around limiting GW to 2 degrees, determining the carbon budget and so on" (Hulme et al., 2013). Seemingly, Wigley, Jones, and Kelly (1981) drew attention to the risk of concentrating on surface air temperature rather than full atmospheric heat content, without mentioning about ocean heat. Interestingly, this narrow meaning of GW has become order of the day that rules the roost of scientific research and climate and environmental discourse.

While arguing that the term "climate change" appeared in the scientific literature prior to the term "global warming," Leiserowitz et al. (2014) have opined that the term "climate change" has been used more frequently in peer-reviewed articles for more than four decades. Referring to the interchangeable usage of both these terms in popular media, especially in the United States, Wilson (2000) notes that GW and climate change have different technical definitions. GW alludes to the increase in the Earth's average surface temperature since the Industrial Revolution, largely owing to the emission of GHGs (GHGs) emanating from the burning of fossil fuels and land-use change; nevertheless, climate change talks about the long-term change of the Earth's climate, including changes in temperature, precipitation, and wind patterns over a period of several decades or longer (IPCC, 2007). In the opinion of Li, Johnson, and Zaval (2011), undoubtedly scientists have made use of both terms in the peer-reviewed literature for decades; nonetheless, the scientific community commonly prefers the term climate change because this term encompasses a wider range of phenomena than just the increase in global surface temperatures. Citing the widely reported testimony of James Hansen at a Senate in 1988 in the United States wherein Hansen had stated that GW had reached "a level such that we can ascribe with a high degree of confidence a cause and effect relationship between the greenhouse effect and observed warming," Shabecoff (1988) claimed that the mass media in the United States widely adopted the term "global warming" after Hansen's testimony. Asserting that the news media in the United States often makes use of both terms interchangeably, with some sources saying that explicitly, Leiserowitz et al. (2014) note that for some media outlets this interchangeability may echo the assumption that the two terms convey not only the identical meaning for the general public, nonetheless entails the same emotional and political impact.

2.6.1 Global warming versus climate change

While differentiating between GW and climate change, Benjamin, Por, and Budescu (2017) express the opinion that both these terms arouse dissimilar interpretations and call attention to different facets of the changing global climate. A GW notion reminds tie-up with temperature increases, severe weather, greater concern, human causes, and negative affect, whereas an idea of climate change focuses on alterations in general weather patterns and the likelihood of natural fluctuations, and provides a fillip to reminiscences of non-heat-related consequences such as increased precipitation (Leiserowitz et al., 2014). While informing about ascertainment of differences in people's beliefs, perceptions and preferences of the phenomenon as an outcome of the term used, some researchers point out that climate change results in higher reported belies that climate change is occurring and will entail severe consequences (Leiserowitz et al., 2014).

Benjamin et al. (2017) have pointed out that people could construe the phenomenon differently under different frames. The first reason is that two terms deal with different facets of the same phenomenon. Plass (1956) is said to have devised the term "climatic change" as he theorized a strong historical relationship between carbon dioxide and the Earth's temperature and climate. The usage of the term "global warming" is attributed to Broecker (1975) who devised a simple model to forecast as to how global temperatures would alter with increased CO_2 in the atmosphere. Charney et al. (1979), in their research report named as Charney Report, envisaged an explicit distinction in usage of words for GW denoting the change in global temperatures and climate change as entailing other impacts like precipitation and moisture, as well. Scientific community has come to regard climate change to be a general term denoting constant variations in conditions overtime; while GW points to "one aspect of climate change" (United States Environment Protection Agency (US-EPA), 2013; cited in Benjamin et al., 2017) also entailing precise increases in air, surface or ocean temperature (IPCC, 2013a, 2013c).

Between 1880 and 2012, global average surface temperature warmed by 0.85°C (IPCC, 2013b), and it has now become an established fact that human influence on climate has been the dominant cause of observed warming since the mid-20th century. Many regions of the globe have already greater regional-scale warming, with 20%−40% of the world population having experienced over 1.5°C of warming in at least one season. Human and natural systems have already undergone profound changes in the wake of rising temperature, and these changes also include increases in droughts, floods, and some other types of extreme weather; sea level rise; and biodiversity loss; and the resultant outcome is that these changes are wreaking unprecedented risks to vulnerable persons, places and infrastructure (IPCC, 2012, 2014a; Mysiak, Surminski, Thieken, Mechler, & Aerts, 2016). The people living in low-and-middle-income countries are most vulnerable and affected people and many of them have experienced a diminution in food security and that can be partly ascribed to augmentation in poverty and poverty (IPCC, 2012).

2.6.2 Global warming and Anthropocene

The plethoric empirical evidence of the unparalleled rate and global scale of effect of human influence on the Earth System (Steffen et al., 2016) has impelled many scientists to call for a recognition of the fact that the Earth has been ushered into a new geological epoch: the Anthropocene (Gradstein, Ogg, Schmitz, & Ogg, 2012). Undeniably, rates of change in the Anthropocene are essentially evaluated over much shorter spans than those used to assess long-term baseline rates of change; nevertheless, contemporary challenges for direct comparison, they are still striking. According to some scholars (Bereiter et al., 2015), the increase in global CO_2 concentration since 2000 is about 20 ppm per decade, accounting for up to 10 times faster than any sustained rise in CO_2 during the past 800,000 years. The global average temperature has been witnessing an increase at a rate of 1.7°C per century since 1970, relative to a long-term decline over the past 7000 years at a baseline rate of 0.01°C per century (NOAA, 2016). These global-level rates of human-driven change far outnumber the rates of change driven by geophysical or biosphere forces that have changed the Earth System path in the past (Summerhayes, 2015; Foster, Royer, & Lunt, 2017); even abrupt geophysical events do not approach current rates of human-driven change.

Undoubtedly, Zalasiewicz et al. (2017) maintain that the process of formalizing the Anthropocene is ongoing; nonetheless, bulk of the majority members of the Anthropocene Working Group (AWG), set up by the Sub-commission on Quaternary Stratigraphy of the International Commission on Stratigraphy, have expressed unanimity on the following: (1) the Anthropocene has a geological merit; (2) it should follow the Holocene as a formal epoch in the Geological Time Scale; and (3) its onset should be defined as the mid-20th century (AWG (Anthropocene Working Group), 2019).

Potential indicators or pointers in the stratigraphic record comprise an assortment of novel manufactured materials of human origin that, as pointed out by Wolter, Hoerling, Eischeid, and Cheng (2016) serve as combined signals to "render the Anthropocene stratigraphically distinct from the Holocene and earlier epochs." Geological science community, according to Wolter et al. (2016), had in 1885 formally acknowledged that the Holocene epoch had begun 11,700 years ago with a more stable warm climate paving way for emergence of human civilization and growing human-nature interactions that have culminated in giving rise to the Anthropocene. In the opinion of Christian Schwägerl, considered an authority on Anthropocene, gave an interview to Rowan Hoope of New Scientist in February et al. 2015, the working group on the Anthropocene, a part of the International Union of Geological Sciences, favors a date around 1950 for the beginning of the Anthropocene epoch, because nuclear explosions and the commencement of modern consumerism began to have long-term effects on the biosphere (Hooper, 2015).

Brondizio et al. (2016) are in favor of employing the Anthropocene as a "boundary concept" to formulate climacteric insights into discerning the drivers, dynamics and specific challenges in finding pathways to realize the ambition of keeping global temperature well below 2°C while undertaking efforts toward and adapting to a 1.5°C warmer world. The Paris Agreement on Climate Change (PACC), concluded under the aegis of UNFCCC, acknowledges capability of humans to impact geophysical planetary processes (UNFCCC, 2015). While offering a structured discernment of the coalesce of past and present human–environmental relations, the Anthropocene also affords an opportunity to better envision the future to curtail drawbacks (Delanty & Mota, 2017; Pattberg & Zelli, 2016), along with the acknowledgment of the secernated responsibility and opportunity to limit GW and invest in prospects for climate-resilient sustainable development (Harrington, 2016). New vistas of opportunities are unfolded by the Anthropocene to raise questions with regard to the regional differences, social inequities, and uneven capacities and drivers of global social–environmental changes, thereby enabling to explore for solutions (Biermann et al., 2016). The Anthropocene also helps in linking incompatible influences of human actions on planetary functions to an irregular dispersal of impacts over and above the responsibility and response capacity to limit GW to no more than a 1.5°C rise above pre–industrial levels (Allen et al., 2018).

2.7 Anthropogenic drivers of climate change

"Anthropogenic" drivers of climate change allude to the human activities that cause climate change and the societal actions that impact and specify those actions (Rosa & Dietz, 2012). Apparently, human driving forces, in this regard, constitute the spectrum of attributes of societies that wield considerable impact on the global climate (Dietz, Rosa, & York, 2010; Rosa, Rudel, York, Jorgenson, & Dietz, 2015). There has been phenomenal augmentation in emissions and atmospheric concentrations of long-lived GHGs, specifically carbon emissions, since the pre−industrial era, owing largely to human activities connected with fossil-fuel use and agriculture; and other land-use changes entailing deforestation, etc., do make significant but relatively lesser contributions (IPCC, 2007, 2014a; USGCRP, 2017). The increasing level of emissions GHGs has become a significant component in proposed move of designating the "Anthropocene" epoch as a formal unit of the geologic time scale and this proposal has received massive attention in the scientific and public media (Finney, 2016, p. 4).

While conceding that long-term and short-term drivers of climate change are frequently explained differently within and between social sciences, Jorgenson et al. (2019, p. 1) are of the view that "long-term" may be applied to several decades or even longer periods, whereas "short-term" may allude to a period shorter than a year. Dietz (2017) has opined that both "long-term" and "short-term" drivers of climate change are engaged in persistent interaction. Economic growth, demographic growth and shifts, consumption, trade, social stratification and inequality, technology and land-use change and land transformation are identified as among the major drivers of climate change by social sciences.

2.7.1 Economic growth

The notion of economic development, distinguishable from human welfare that exclusively focuses on economic growth or the growth in gross domestic product (GDP) (Acemoglu, 2009), is a more comprehensive term incorporating social equity, poverty eradication, the meeting of basic human needs like access to health, education, and water and sanitation services, etc., along with the provision of physical infrastructure including housing, energy, transport and communications, etc., and the guarantee of essential political, economic, and social freedoms as espoused by Sen (2000). In an identical manner, the notion of economic development is broadly defined as the expansion of capacities that contribute to the advancement of society through individuals', firms', and communities' potential with specific focus on structural transformation by introducing changes in industrial structure—from an agriculture-based structure toward industry and services—social organization—moving from small-scale productive activities toward large-scale organizational structures, and the diversification of skills (Feldman, Hadjimichael, Lanahan, & Kemeny, 2016). Economic growth is one of the major drivers of climate change. Economic growth comprises both long-term and near-term factors that affect the timing and range of the driver's impact.

Interestingly, in order to evaluate the possibly changing impact of economic development on national-level carbon emissions, sociological research has applied longitudinal modeling techniques and statistical interactions (Thombs, 2018a). This corpus of research offers a sociological approach to analyses of a probable delinking of gross GDP and emissions (OECD, 2002), and normally concentrates on examining hypotheses deduced from social theories, especially ecological modernization theory and treadmill of production theory. Mol (2003) and Mol, Spaargaren, and Sonnenfeld (2014) have opined that ecological modernization theory indicates that mechanisms of modernization yield "added reflexivity" throughout the socioeconomic system. Jorgenson et al. (2019) maintain that technological development and environmental awareness, entailing close proximity with economic development, are often construed as primary components of modernization through "the greening of industrial" mechanisms including diminution in reduced fossil-fuel consumption, also culminating in more sustainable consumption.

The treadmill of production theory demonstrates as to how the persistent search for economic growth results in advanced economies being trapped on a "treadmill," where their prosperity is not improved by economic growth, nevertheless the outcomes of this pursuit of growth causes "massive, unsustainable environmental damages" (Curran, 2017). In other words, it is argued in favor of treadmill production theory that because market economies are based on enhancing profits via expansion, energy consumption and forms of pollution frequently expand culminating in the deterioration of overall environmental conditions (Gould, Pellow, & Schnaiberg, 2008). Consequently, economic development entails augmentation in resource use and the generation of waste accruing from the diverse stages of production processes, including increased fossil-fuel consumption leads to increased carbon emissions.

Jorgenson et al. (2019), while citing the research findings of Jorgenson and Clark (2010) and Thombs (2018a), have opined that decoupling of interactions between GDP per capita and time in longitudinal models of anthropogenic carbon emissions make discernible mixed support for ecological modernization theory treadmill of production theory; thereby, suggesting that both frameworks could gain from additional contemplation as to how international organization of production and "the structure of international trade influence the relationship between carbon emissions and economic development"(Jorgenson et al., 2019, p. 2). According to Jorgenson (2014), almost identical modeling techniques incorporating interactions between time and measures of GDP per capita have been applied in studies of the impact of economic growth "carbon intensity of human well-being" (CIWB)—a ration between per capita carbon emissions and a measure of human well-being—for samples of countries in the Americas, Europe, Oceania and Africa. Similar modeling techniques that include interactions between time and measures of GDP per capita have been used in studies of the effect of economic growth on the "carbon intensity of human well-being" (CIWB)—a ratio between per capita carbon emissions and a measure of human well-being—for samples of nations in the Americas, Europe, Oceania, Asia, and Africa (Jorgenson, 2014). Research findings by Jorgenson (2014) suggesting that the impact of GDP per capita on CIWB is fairly large, positive, and stable in magnitude through time for countries of North America, Europe, and Oceania, also find support from some recent studies (Schroder & Storm, 2018).

With the economy having taken a quantum jump from an "empty world" to a "full world" (Daly, 1973), it has become evident that conventional notion of growth is unable to sustain far into the future (UNEP, 2011) and with the stimulus for the social and political commitment to a penchant for a vision of sustainable growth gathers further momentum. Many reasons are attributable for this strong urge for economic growth that plays a "number of vital roles in modern society, including poverty eradication, the pursuit of social justice, the building of social solidarity, the defense of civic peace and the establishment of good governance" (UNEP, 2019, p. 36). Poverty eradication is regarded as the most significant role of economic development. As per World Bank data, nearly 783 million people, accounting for 10.7% of the global population, still live on less than US$1.90 per day, and 48.7% of the population lives on less than US$5.50 per day (World Bank, 2013). Poverty adversely impacts the lives of children as well. According to United Nations International Children Emergency Fund (UNICEF), approximately 22% of children are stunted and 7.5% are underweight worldwide (UNICEF, 2018a) whereas 264 million children along with adolescents, are incapable of entering or completing school education (UNICEF, 2018b), the majority of them being girls. Concurrently, as per UNICEF/WHO data, about 2.1 billion people lack adequate access to "safely-managed" water and 2.3 billion people lack basic sanitation facilities (UNICEF & WHO, 2017).

The poverty is not merely attributable to want of economic resources. Rather economic growth has come to be recognized as the only reliable method of alleviating poverty through redistributive policies and social security provisions for the poor so as to enable them reap benefits of fast and steady growth. Besides, the need for economic growth, especially in developing countries, is also felt to restrict the massive income gap that delineates developed countries as distinct from developing countries. The income gap between the developed and countries that had continued to widen even during the closing decades of the past 20th century, has interestingly started getting narrowed in the early two decades of the 21st century in the wake of surge in growth rates of developing countries. Inevitability of the growth for developing countries is supported even by the critics of the growth agenda (Jackson, 2009, p. 4). With their major critique focused on developed countries where the critics of the growth agenda usually argue that growth is neither essential nor desirable in such countries (Jackson 2009). Among the supporters of the growth agenda in developing countries, Friedman et al. (2005) recognizes the political role of growth in developing countries, while others also argue that along with growth, supporting fairness, social mobility and social solidarity also helps in garnering popular support for civic and international peace (Gartzke, 2007, p. 180).

The poverty agenda that has thus far remained unfinished has been accorded highest priority in the recently adopted 2030 Agenda under sustainable development goals, under SDG-1 dealing with the aim of ending poverty in all its forms everywhere (Kuenkel, 2019). In the wake of adverse impacts accruing from the vagaries of climate change culminating in narrowing down of the planetary boundaries, the prospects of development are increasing under threat. As the then Director-General of the International Labor Organization (ILO) in one of his pronouncements, while citing the founding principles of the ILO said: "Poverty anywhere is a threat to prosperity" (ILO, 2011), demonstrates the truth that global peace, stability and prosperity cannot be achieved when a significant part of humankind remains condemned to permanent deprivation and subservience under the bane of poverty.

The resultant slowdown in global growth in the aftermath of financial crisis (2007–08) had reportedly to do with moribund international trade, revival of the menace of trade wars, heightened policy uncertainty, and a diminishing of the main engine of global growth viz., emerging economies (World Bank, 2017, p. 3). In the wake of shrinking growth in global trade and widening income inequalities between countries, the emerging global scenario is best represented by famous Milanovic's global elephant curve (Weldon 2016). demonstrating that between 1988 and 2011, the incomes of the top 1% along with those in the 40–70 percentiles, supposedly to be in the developing economies, recorded a surge; while rising incomes in the bottom 10% and in the 80–90 percentiles, supposedly to be in middle class of the developed countries, recorded slow growth. Alvaredo, Chancel, Piketty, Saez, and Zucman (2017) have projected continued rise in income inequalities even in 2050 under both business-as-usual scenario as well as the US-style inequality scenario. Under the business-as-usual scenario, the income shares of the bottom 50% of the global population somewhat declines from nearly 10% currently (2017) to less than 9% in 2050 and the top 1% share rises from less than 21% currently (2017) to more than to more than 24% of world income in 2050 thereby reflecting a steep rise global inequality in this scenario, notwithstanding strong growth in emerging economies. Under the US-style scenario, the top 1% worldwide would earn close to 28% of global income by 2050 while the bottom 50% would earn close to 6% less than in 1980; and in this scenario this increase in the top 1% income share would largely be at the expense of bottom 50% (Alvaredo et al., 2017, p. 5).

There seemingly exists no upfront relationship between income inequality and the use of natural resources. is not straightforward. It is often argued in the classic economy that the poor have a higher tendency to consume than the rich (Caroll et al., 2017), and transferring income from the rich to the poo should therefore reduce the impact on the natural environment. Asserting that the question of whether higher levels of inequality are linked to environmental damage is not a new one, a recent UNESCAP publication notes the efforts of researchers for more than two decades to ascertain this relationship and this publication infers that the conclusions have been mixed one (UNESCAP, 2018, p. 49). It is discerned from some cross-country comparative studies that there is relationship between inequality and deforestation/biodiversity loss where more equal countries entail the tendency of having lower deforestation and impacts on biodiversity (Holland, Peterson, & Gonzalez, 2009; Islam, 2015; Koop & Tole, 2001). It also becomes evident from some studies that countries with high levels of inequality are prone to consume relatively high quantities of energy and carbon-intensive products, use more water and engender more waste (Dorling, 2017). High level of income inequality generates upward pressure on resources, both through the impact of visible consumption and out of the coercing of the middle class. Inequality entails the potential of escalating the conflict that in turn can adversely affect the environment; and Islam (2015) has pointed out, impacts of inequality on the environment get channelized through the consumption, investment and community outlets.

2.7.2 Urbanization

Urbanization is a discrete conduit through which demographic trends affect environmental resources. Endowed with higher incomes and consumption, greater access to political power, higher rates of economic growth, and, per capita, urban areas become well-equipped to exert a higher pressure on natural resources. Concomitantly, cities manifest better efficacy in the use of resources per unit of income generated and better potential for energy efficiency (Cottineau, Finance, Hatna, Arcaute, & Batty, 2018). No country throughout the globe has ever made the transition from poverty to middle/higher income status without experiencing a period of rapid urbanization because cities serve as the engines of growth; and, at the same time, urbanization is commonly associated with a lowering of fertility rates (Martine, Alves, & Cavenaghi, 2013). Currently, somewhat more than half of the world's population is living in urban areas, and it is a share that is likely to rise to 60% by 2030 and 66.4% by 2050 (Melchiorri et al., 2018).

Projections by UN-Habitat (United Nations Human Settlements Programme) predict that around 90% of the growth of cities will occur in low-income countries (UN-Habitat, 2014). While Africa is the world's most rapidly urbanizing region, whereas least population growth was recorded in European cities in the 1995–2015 period (UN-Habitat, 2016). The crucial factor attributable for these trends is neither fertility nor age structure (UN-Habitat, 2016). Noting that it took almost two centuries for the urban share of global population to grow from 3% to 50% to reach 3.5 billion people in 2010, projections by the United Nations caution that the coming decades are going to be more (UN-DESA, 2014). Undoubtedly, urban population is set to more than double over this century, as per projections of the United Nations; nonetheless, in all the ensuing centuries, world may add, at most, another billion or so; thereby, making the current global urbanization era not just immense, but also brief (Fuller & Romer 2014). Bhattacharya, Meltzer, Oppenheim, Qureshi, and Stern (2016) have opined that nevertheless, the world's infrastructure would be more than double in the next two decades; however, there would be very narrow window of opportunity to help plan and design new cities.

The potential for urban growth and its probable impact on natural resources can be ascertained with the help of the pattern of urbanization. Megacities, defined by the UN-Habitat as cities with more than 10 million population (UN-Habitat 2016, p. 7), have come to epitomize high end of urbanization; and interestingly, bulk of these megacities are located in the global South. There were 10 megacities housing 153 million people in 1990, accounting for 7% of the total urban population at that time, and by 2014, the number of these megacities rose to 28, with 453 million people, constituting 12% of the then world population (UN-DESA 2014). The number of megacities increased to 31 in 2016, of which 24 were located in the less-developed regions, the global South; of these, 6 were in China and 5 in India (UN-DESA, 2016); and in 2018, the number of megacities rose to 33 (UN-DESA, 2018).

Interestingly, approximately half of the world's population is living in small and medium cities that are recording fastest rates of growth of urban population (UN-DESA, 2014; UNESCAP & UN-Habitat, 2015). This segment of population is expected to "deliver nearly 40% of global growth by 2025, more than the entire developed world and emerging market megacities combined" (UN-Habitat, 2015a, p. 2). While appreciating the role of small and medium cities in in the global growth, Birkmann, Welle, Solecki, Lwasa, and Garschagen (2016) have also warned that these types of cities are more vulnerable to natural hazards than big cities and megacities. Small and medium cities, entailing complex systems, are becoming hubs of agglomeration economies, and these cities exhibit quantitative and qualitative changes in composition as they grow in size (Cottineau et al., 2018). Entailing the benefit of people clustering to minimize transport costs for goods, people and ideas, agglomeration economies serve as catalyst for boosting higher productivity that facilitates inflows of people, who play vital role in further augmenting productivity as well as envisaging a "positive feedback loop" that can be instrumental in multiplying the impact of external productivity factors and thereby culmination in the enhancement of urban production and wages (Zenghelis 2017).

Hamilton and Hartwick (2017) describe cities as a source of wealth creation, where wealth is appraised as the totality of natural, human, and physical assets. Natural capital, according to UNEP (2019), entails natural assets in their role of making available natural resource inputs and environmental services for economic production. Natural capital includes land, parks, green spaces, mineral and fossil fuels, solar energy, water and biodiversity, and the services provided by the interactions of all these elements in ecological systems. Human capital includes the population's education, knowledge and skills. Housing, infrastructure, industry and offices, etc., fall within the ambit of physical capital. Besides, there is also intangible capital that constitutes the combination of all intangibles of an organization such as human, structural and relational capital (Cruz-Cunha, Moreira, & Varajao, 2014); along with ideas and inspiration captured in forms incorporating research and development, patents, intellectual property rights, customer lists, brand equity, social capital and institutional governance. Intangible capital is perhaps the most important but "feeds off and interacts with the other forms of capital. It also provides the source for innovation and investment necessary to decouple growth from resource use and CO_2 emissions, in absolute levels as well as in rate of growth terms" (UNEP, 2019, p. 34).

Pollution, congestion, urban heat effect, ill-health, crime, informal settlements (slums), lack of affordability and waste, etc., are the attendant penalties of urbanization. Medium- and long-term costs of disorganized, unregulated and unplanned urban sprawl outweigh short-term benefits and the society at large along with the economy and the environment become more vulnerable to severe consequences. Unregulated urban habitats are prone to be more polluted, congested and ineffective in the utilization of resources and this attested to by the available evidence of nearly a third of the global urban population already living in slum-like pathetic plight without basic amenities and social protection [United Nations Population Fund 2010/2011 cited in UN Habitat III (United Nations Human Settlements Programme), 2015a, p. 3]. Pitiable plight is also reported about poor women residing in slums and they are especially vulnerable and confront various handicaps in having access to some of the benefits of urban living [United Nations Population Fund 2014 cited in UN Habitat III (United Nations Human Settlements Programme), 2015a, p. 2]. According to Lopez-Moreno, Pomeroy, Revuelto, and Vicente-Serrano (2012), about two-thirds of urban dwellers reside in cities where income disparities increased between 1980 and 2010 [Lopez Moreno et al., 2012, cited in UN Habitat III (United Nations Human Settlements Programme), 2015a, p. 1].

While reiterating close interconnectedness between carbon emissions and processes of economic growth, Jorgenson, Auerbach, and Clark (2014) have alluded to some recent sociological analyses of national carbon emissions that outline prominent sequential differences in the impacts of urbanization across regions. It also becomes discernible from related research that in developed countries with larger slum conglomerations, the overall impact of urbanization on carbon emissions gets stifled to some extent, assuming that households in urban slums suffer from structural disadvantages and usually fossil-fuel energy and other carbon-intensive goods are consumed by them in lesser quantity (McGee, Ergas, Greiner, & Clement, 2017). Some scholars are of the opinion that technical progress, when seen in historical context, has only moderately made up for supplementary emissions from economic growth in both domestic and external contexts (Jackson et al., 2016).

Urban expansion, absence of adequate public transport system and a lack access to basic amenities like water, waste collection and energy not only "offset the economic benefits of urban concentrations" and augment costs but become penalties depriving of "opportunity to prosper and also exacerbate urban poverty" (UNEP, 2019, p. 33). Alluding to disadvantages accruing from disorganized and unplanned urban growth, Floater and Rode (2014) mention that these disadvantages include: increased GHG emissions, alienation and social exclusion along with other attendant socioeconomic and environmental such as congestion, ill-health and crime. It is almost an acknowledged fact that tradeoffs of urban living outweighing the disadvantages thereof exert massive burden on urban governance structures (UNESCAP & UN-Habitat, 2015). Deficiency in technical capabilities to "lead a major urban development process" (UN-Habitat, 2012, p. XIV) on the part of many small and medium-sized cities render them vulnerable to suffer devolved responsibilities sans matching resources, obstructing their planning capacity (Frank & Martinez-Vazquez, 2014). The resultant outcome is the severe constraining of the urban governing structures to safeguard both natural resources and rights their citizens.

The recent wave of rapid urbanization followed by influx of people in developing regions, including Southern Asia and sub-Saharan Africa, entails the likelihood of placing great strain on urban institutional resources and infrastructure. Concomitantly, apart from presenting a challenge, these rapidly urbanizing areas also afford "the largest opportunities for future urban GHG emissions reduction [... because their ...] urban form and infrastructure is not locked-in" (Seto et al., 2014, p. 928). The transition to sustainable cities requires taking into consideration factors that wield impact on urban environment. These factors, apart from increasing population density, also include urban form, especially the pattern of urban physical infrastructure that is difficult to modify easily, and regulates land use, along with long-term transportation and energy demands (Güneralp et al., 2017; Seto et al., 2016). It is acknowledged that urban form patterns entail implications for energy consumption, GHG emissions, biodiversity (Salat, Chen, & Liu, 2014), water infrastructure (Farmani & Butler 2014), and land use and conversion of croplands (Bren d'Amour et al., 2016). Ramaswami, Russell, Culligan, Sharma, and Kumar, (2016, p. 940) have opined that urban form, "infrastructure design and socio-spatial disparities within cities are emerging as critical determinants of human health and well-being."

Keeping in view cities' vulnerability to environmental and climate impacts such as heat, water stress, and floods; while coastal cities face sea level rise, saltwater incursion and storm surges, UNEP's Sixth Global Environment Outlook for 2019 cautions that if future cities are "built over the next two or three decades on a resource-hungry, carbon-intensive model, based on sprawling urbanization, all hope of meeting ambitious resource and climate-risk targets will be lost. This could leave cities and countries struggling to meet their resource needs and unable to compete in global markets, with the stranding of physical and human assets" (UNEP, 2019, p. 34).

2.7.3 Consumption

Being the largest component of output, consumption is often construed as a major player in the expansion of in GHG emissions. Accordingly, Baudrillard (2017) has opined that amidst the emergence of consumer society actively practicing consumer culture, where human beings are increasingly leading a consumer-oriented way of life, the notion of consumption entails sufficient potential of being understood as a driver of emissions. Expenditure and investment patterns are somewhat affected by factors like Income, infrastructure, social organization, and culture; and in turn these factors wield direct impacts on climate change. Higher income and wealth, usually culminating in consumption of larger quantities of energy consumption, results in increased rates of carbon emissions. And accordingly, Leichenko and Solecki (2005) have argued that the urbanization of populations, especially in low- and moderate-income settings, is also connected with the development of high resource-consuming lifestyles.

Augmentation in income-level across countries leads to change in carbon-disparity level. Hubacek et al. (2017a), in their research conducted in the backdrop of geographical principles, have tried to demonstrate that ranking of countries from lowest to highest income order makes discernible extensive dissimilarity among lower-income countries; and with increase income there follows a decrease in disparity of carbon footprints. Undoubtedly, the decline in disparity of carbon footprints sets in as the countries grow richer; nevertheless, the average carbon footprint increases along with income. Socio-cultural contexts are also helpful in understanding consumption as a driver of climate change (Dietz, Stern, & Weber, 2013), and a close scrutiny of status consumption as well as that of status competition helps in appreciating socio-cultural context (Ehrhardt-Martinez & Schor, 2015). Status consumption is defined by Eastman and Eastman (2015) as the interest "a consumer has to improve one's social and/or self-standing through consumption of consumer products that may be conspicuous and that confer and symbolize status for the individual and surrounding significant others" (Eastman & Eastman, 2015, p. 3). Quest for status culminates in exacerbation of emissions as it spurs people to possess carbon-intensive goods and services like big homes, large vehicles and visit tourist spots to spend vacations along with possession of other luxury goods that are often looked upon as symbol of status owing to their social visibility (Schor, 1998). Consumption patterns also entail the potential of help decrease emissions; nevertheless, when green products such as hybrid vehicles or solar power installations are reckoned as indicators of high status (Griskevicus, Tybur, & Van Den Bergh, 2010).

Consumer habits assume significance because enhanced energy-intensive activities like frequent use of heating or cooling devices or frequent usage of water is likely to augment emissions. On the contrary, introducing modifications in these habits or adopting alternative options, such as according preference to public transportation over personal vehicle, can help in diminution of emissions (Ehrhardt-Martinez & Schor, 2015). Going in for green energy options, especially as rooftop solar photovoltaic systems, entail a sturdy spatial pattern of adoption yielding the discernment that "peer effects" could be a robust force in consumer choices. According to Graziano and Gillingham (2015), adoption usually takes place among neighboring habitats notwithstanding economic class and political party.

The notion of "lifestyle," as a component of consumption, is equally significant in appraising carbon emissions. The manners in which people lead their lives and their modes of consuming products get mirrored in the consumption patterns of societal groups dissimilar socioeconomic backgrounds and features like identity, education, employment or family status (Baiocchi, Minx, & Hubacek, 2010). Housing has come to be reckoned as one such important aspect of lifestyle-related choices (Huddart Kennedy, Krahn, & Krogman, 2014). People residing in suburban areas, also termed as suburbanites, particularly in developed countries, usually are in the habit of owning spacious and large homes equipped with massive heating and cooling facilities. Access to personal vehicles or public transportation for commuting distance, visiting recreation areas and city centers, public services and shops or malls, etc., are other significant neighborhood-specific that constitute lifestyle-linked determinants of carbon emissions (Baiocchi et al., 2010). According to UNEP (2016), A "sustainable lifestyle" is a cluster of habits and patterns of behavior embedded in a society and facilitated by institutions, norms and infrastructures that frame individual choice, in order to minimize the use of natural resources and generation of wastes, while supporting fairness and prosperity for all.

2.8 Extreme weather events

Human society has been affected by extreme weather events since the beginning of recorded history or perhaps even long prior to that. Humans, along with every other living organism on the Earth, have adapted to a certain degree of variability in the weather. Undeniably, extreme weather can wreak havoc in terms of loss of life and significant damage to property; nevertheless, people and virtually every other living organism have, at best to some degree, adapted to the sporadic extremes they experience within their normal climatic zone. Use of the fossil fuels by humans since the advent of Industrial Revolution has proved instrumental in commencing modifications in the

Earth's climate in ways that could hardly be imagined a century ago. Impacts of this alteration in climate are seemingly visible almost everywhere in terms of rise in average temperatures, changing precipitation patterns, melting of ice sheets and rising sea levels. The resultant outcomes of these climatic changes can be discerned from the availability and quality of water supplies, increasing food insecurity, and the impacts on the ecosystems on land and in the oceans (Titley, 2016, p. ix).

Notwithstanding such drastic changes, some people may still regard magnitude of climate change and its attendant risks as distant and remote possibility in both time and space owing to daily or seasonal variability of weather entailing the potential of masking the alterations in the overall climate. Hitherto, it is the occurrence of extreme events entailing greater frequency or increasing intensity to be experienced by the people that may propel them discern linages or interconnectedness between climate change and extreme events. While explaining the utility of understanding the scientific basis of the attribution of extreme events to changes in the climate system for satisfying the public curiosity, Titley (2016, p. x) further adds that such understanding can also make available useful information about the future risks of such events for the policy makers and official emergency managers to devise policy framework and precautionary measures accordingly. It is in this context that improved attribution and eventually prediction of extreme events assume significance in validating "an even more nuanced and sophisticated understanding of the climate system and will enhance scientists' ability to accurately predict and project future weather and climatic states" (Titley, 2016, p. x).

Major instances of extreme weather or climate events, inter alia, include, but are not limited to, heat waves, cold waves, floods, extreme precipitation, drought, tornadoes and tropical cyclones, etc. Albeit, extreme events occur occasionally, these are prone to have profound impacts in terms causing severe damage to infrastructure, affecting economy and human health as well as culminating in loss of lives (Melillo, Richmond, & Yohe, 2014; WHO, 2012; IPCC, 2013a, 2013c). In the wake of regional variability characteristics of climate, the definition of an extreme weather or climate event and its threshold is bound to diverge from location to location. In other words, an extreme value of a specific climate element in one location could be within the normal range in different location (WMO, 2018). There are also other pragmatic reasons, apart from natural and geographic reasons, for the varying definitions. WMO (2018) cites the case of defining heat waves in a heat-health warming system for which a heat wave is defined precisely in accordance with the potential impacts on human health.

It deems worthwhile to briefly appraise major extreme weather/climate events.

2.8.1 Extreme cold events

Undeniably, general description of extreme cold events is delineated in terms of temperature; nonetheless, impacts of such an extreme cold event can be compounded by wind, snow, and ice. Regarded as a meteorological event, a cold wave is usually characterized by steep fall in air temperature near the surface, leading to enormously scummy values, sharp rise of pressure and strengthening of wind speed, or associated with hazardous weather like frost and icing. Human health and agriculture are frequently subjected to terrible impacts of extreme cold event, also involving high heating costs, and can even culminate in mortality for human beings and livestock. WMO (2018) defines cold wave as a marked and unusual cold weather marked by "a sharp and significant drop of air temperatures near the surface (maximum, minimum and daily average) over a large area and persisting below certain thresholds for at least two consecutive days during the cold season."

The genuine temperatures that delineate a cold event diverge regionally and seasonally, nevertheless the temperatures during the course of such an event will be in the cold tail of the probability distribution of temperatures for a location or region and time of year. The event definitions most often are based on Daily temperatures most often constitute the basis of the event definitions, albeit although multiday or longer averages also have been used. Sillmann, Kharin, Zwiers, Zhang, and Bronaugh (2013), Sillmann, Kharin, Zhang, Zwiers, and Bronaugh (2013) have opined that the criteria can be either an absolute temperature threshold (e.g., 0°C, 0°F, −20°C), often randomly chosen, or a percentile value such as the 1-percentile or the 10-percentile criterion used in the Expert Team on Climate Change Detection and Indices (ETCCDI) ClimDEX database.

A combination of thermodynamics (cold air mass formation) and dynamics (the large-scale circulation, advection) tug or drive extreme cold events. Self-organizing maps derived from atmospheric re-analyses have been used by Horton et al. (2015) to demonstrate that both factors—thermodynamics and dynamics—have played roles in recent changes in extreme cold events. The surges in cold extreme daily minimum temperatures (i.e., warming) are usually higher than are the increases in extreme daily maximum temperatures, and there is no hint of enhanced variability of daily or monthly winter temperatures over the United States (Kunkel, Vose, Stevens, & Knight, 2015; Screen, Deser, & Sun, 2015). An analysis of the ETCCDI indices for 1948−2005 in four different atmospheric re-analyses and 31 Coupled Model Inter-comparison Project Phase 5 (CMIP5) models undertaken by Sillmann, Kharin, Zwiers, et al. (2013), Sillmann, Kharin, Zhang, et al. (2013) reveals a similar warming of the coldest temperatures over other land areas of the globe. The propensity for cold extremes to warm by more than hot extremes can also be evidenced from the findings of Collins et al. (2013) and Melillo et al. (2014).

Recent research (van Oldenborgh et al., 2015; Wolter et al., 2015) make it discernible that a few recent winters in 2012 and 2014 have been characterized by extreme cold events in eastern North America; nevertheless, such events have been less frequent along with their actual temperatures having been less extreme in the past few decades as compared to earlier decades of the 20th century. The decreased frequency of cold extreme emanates largely from the underlying increase of the mean temperature, not from the decreased variability (Wolter et al., 2015). Retreating Arctic sea-ice cover affords a good context of examining extreme cold events. In this regard, Screen et al. (2015) have, by suggesting diminishing Arctic ice cover but historically observed ocean temperatures outside of the Arctic in two different global climate models, argued that ice loss is linked to decreased probability of extreme cold events, as well as diminishing variability of temperature, approximately over the entire Northern Hemisphere land areas. On the one hand, some studies draw attention to impacts of sea-ice change on large-scale dynamics (Francis & Vavrus, 2015), whereas the indications remain rooted in the noise of natural variability (Barnes, Dunn-Sigouin, Masato, & Woollings, 2014), and, from the standpoint of extreme cold events, are overwhelmed by the underlying warming.

2.8.2 Extreme heat events

According to WMO (2018), an extreme heat event should be defined as "a period of marked unusual hot weather (maximum, minimum and daily average temperature) over a region persisting at least three consecutive days during the warm period of the year based on local (station-based) climatological conditions, with thermal conditions recorded above given thresholds." Attempts at delineating heat events over an assortment of time scales in the literature, from as little as 1 day to at least 1 year, have reportedly yielded some temperature anomalies that are distinguishable, especially between temperature anomalies of short duration comprising days, heat events, and those of longer duration consisting of weeks and longer warm anomalies (NASEM, 2016, p. 90). Owing to the fact that the temperature is a constant variable, the spatial range of a given heat event or warm anomaly is, to a certain extent defined one-sidedly and with the unfolding of the event can also undergo change via time. Characteristically, a latitude-longitude box is used, nevertheless occasionally single stations (King et al., 2015) or political boundaries are used. Undeniably, considerable attention has been focused in some studies on heat events over land, albeit some (Funk, Shukla, Hoell, & Livneh, 2015; Kam, Knutson, Zeng, & Wittenberg, 2015) have viewed warm sea surface temperatures (SSTs) anomalies over periods of seasons to years.

The effects of heat events and warm anomalies, for instance on human health, can be aggravated by high dew points, and also by high nighttime temperatures that sequentially are more probable in the event of high dew points (Gershunov & Guirguis, 2012). On the contrary, the amplitudes of the warm anomalies themselves can be enhanced by land-atmosphere feedbacks in case of low and this probably manifests a connection between drought and warm anomalies. Apart from their direct impacts, warm anomalies over both land and ocean also entail the potential of triggering other types of extreme events like droughts or wildfires.

Hartmann et al. (2013) in IPCC's Fifth Assessment Report have noted, "a large amount of evidence continues to support the conclusion that most global land areas analyzed have experienced significant warming of both maximum and minimum temperature extremes since about 1950." After having opined this, they seldom rule out the probability of human influence having contributed "to observed global scale changes in the frequency and intensity of daily temperature extremes since the mid-20th century, and likely that human influence has more than doubled the probability of occurrence of heat waves in some locations." Also noting that there has been more increase in minimum temperatures comparable to maximum temperatures, they have further emphasized that maps of changes make it discernible statistically significant increases in two indices of extreme temperatures in almost every land area since 1950. With regard to longer-term trends in heat waves and warm events in the United States, the US National Climate Assessment notes, as quoted in Walsh et al. (2014) that heat waves have usually with "western regions (including Alaska) setting records for numbers of these events in the 2000s. Most other regions in the country had their highest number of short-duration heat waves in the 1930s." Regarding future projections, Collins et al. (2013) are of the opinion: "It is also very likely that heat waves, defined as spells of days with temperature above a threshold determined from historical climatology, will occur with a higher frequency and duration."

Heat events have come to be reckoned unquestionably as the extreme weather events and having the longest history. In the aftermath of the 2003 European heat wave that claimed tens of thousands of excess deaths, Stott, Stone, and Allen (2004) came out with a seminal paper wherein the methods used served as the groundwork for subsequent studies in this field. According to NASEM (2016), of the events reported in the annual Explaining Extreme Events special issue of BAMS for 2014, heat events or warm anomalies occupy the largest share, being eight out of 32 for 2014. This may mirror the greater probability of successful attribution of heat waves, compared to other types of events, to human-induced climate change using existing models and data (NASEM, 2016, p. 92).

While suggesting that simulations of heat events and warm anomalies could derive gain from improvements in land-surface schemes in global and regional models, NASEM (2016, p. 92) laments that few studies embrace an assessment of the models' ability to simulate the important statistical properties of the event of interest. In their examples of a highly conditioned approach that explicitly could be applied to heat events, Trenberth et al. (2014) have not included heat events among their examples, starting perhaps with one of the most impactful events, like the Russian heat wave of 2010. According to NASEM (2016), heat events and warm anomalies could perhaps be the best candidates for evaluating the dependability and validity of attribution methods for the reason that the direct thermodynamic impacts on this sort of extreme event are commonly more forthright than, for example, heavy rainfall (NASEM, 2016, p. 93).

2.8.3 Droughts

Dai (2011) regards drought as a recurring extreme climate event over land characterized by below-normal precipitation over a period of months to years. Drought is a natural hazard and is one of the least understood and manageable phenomena impacting the world today (Wilhite, 2012). Comparing drought to aridity in arid areas, drought is a temporary dry spell phase, in contrast to the permanent aridity in arid regions. Drought happens in many parts of the globe, even in wet and humid areas and the reason for this is that drought is delineated as a dry period comparable to its local normal conditions and arid areas are vulnerable to drought owing to the fact that their rainfall amount is dependent on a few rainfall events (Sun, Solomon, Dai, & Portmann, 2006). Drought is defined by WMO (2018) as "a period of abnormally dry weather characterized by a prolonged deficiency of precipitation below a certain threshold over a large area and a period longer than a month." Regarded as a complex phenomenon, droughts embrace different amalgamations of atmospheric inputs; primarily precipitation and also temperature

Variations in global climate have contributed to enhancement of vulnerability to drought (Coles, Eslamian, Eslamian, & Eslamian, 2017). Burgeoning human population, coming of water resources under increasing pressure, industrial uses, agricultural production systems and environmental requirements are such significant factors that have defied universal definition of drought (BoM, 2014a, 2014b; Wilhite, 2012). Generally, a drought is described as a deficit of precipitation over a protracted period of time, say for agriculture production generally a season or more; causing water deficiency for some other activity; industry group, community, or environmental sector (NDCM, 2014a, 2014b). Drought is not strictly a physical phenomenon to be delineated rigorously in terms of climate variability. Accordingly, a drought is not merely designated as low rainfall; if it was, most of the arid and semiarid regions of the world, for instance, would be in almost everlasting drought (BoM, 2014a). In other words, drought can thus be ascertained by the equilibrium between water supply and demand. The impact of a drought is regulated by the interaction between a natural event such as less precipitation or water inputs than expected; and the demand directed at the water supply, with human activities ordinarily aggravating the impact of a drought (NDCM, 2014a, 2014b).

Apart from the conceptual definition of drought, Coles et al. (2017) have opined that the scale and inclemency of drought can be monitored in numerous ways and is usually dependent on the impact a drought wields on a definite activity or phenomenon and this, according to them, is often denoted as the operational definition of a drought and is monitored for example in terms of (Coles et al. 2017, p. 2):

- Rainfall deficiencies;
- The impact on primary industries, such as agricultural production;
- Groundwater recharge and streamflow;
- Social expectations, economics, and perceptions of water availability.

Consequently, drought has come to be defined in both conceptual and operational terms such that six general classes of drought have been previously recognized (USGS, 2014; Wilhite, 2012) as follows:

- Meteorological drought—Defined only in terms of precipitation deficiencies, in infrangible amounts, for a specified period.
- Climatological drought—Delineated in terms of precipitation insufficiencies, in percentages of normal values.
- Atmospheric drought—Outlined not only in terms of precipitation inadequacies but likely in terms of temperature, humidity, or wind speed.
- Agricultural drought—Defined primarily in terms of soil moisture and plant behavior.
- Hydrological drought—Delineated in terms of decrease in streamflow, diminution in lake or reservoir storage, and the lowering of groundwater levels (Subrahmanyam, 1967).
- Water management drought—Defined in terms of water supply shortages triggered by the failure of water management practices or facilities, such as an integrated water supply system and surface or subsurface storage, to overcome normal or abnormal dry periods and level the water supply throughout the year (Matthai, 1979).

Thus full description of a drought can only be delineated by depicting its several climatic, hydrologic, and operational elements (Williams-Sether, Macek-Rowland, & Emerson, 1994). Following an extensive research and review (Wilhite and Glantz, 1985), the six classifications suggested earlier were further revised into four basic approaches to measuring drought:

- Meteorological drought—A period of unusually dry weather amply protracted for the dearth of water to trigger serious hydrologic asymmetry in the affected area (Huschke, 1959).
- Agricultural drought—A climatic sashay involving a deficiency of precipitation adequate to adversely affect crop production or range production (Rosenberg, 1979)
- Hydrological drought—A period of below average water content in streams, reservoirs, groundwater aquifers, lakes, and soils (Vujica, Hall, & Salas, 1977).
- Socioeconomic and environmental drought—A period when the deteriorating water supply comparable to demand impacts human activities and ecosystem function to the threshold of failure and may be allied with elements of meteorological, hydrological, and agricultural drought (Wilhite, 1985).

The first three definitions earlier focus on ways to appraise drought and the fourth deals with drought in terms of the impact on supply and demand (Fuchs, Svoboda, Wilhite, & Hayes, 2014) with the fourth term being a variation on a water management drought (Matthai, 1979) and the engineers' drought (SBR, 1887). In such cases, as pointed out by Coles et al. (2017), socioeconomic definitions of drought have sprung up to pinpoint a link between the supply and demand of an economic good with elements of meteorological, hydrological, and agricultural drought. This proffer is distinguishable from the other types of drought in that it is dependent on the supply chain, while this supply chain is also climate or weather dependent. Consequently, many of the economic goods, such as water, forage, food grains, fish, and hydroelectric power, get impacted. Thus the relative supply and demand of goods will be dictated by the availability of water due to the natural variability of climate (ONRG, 2006; Wilhite, 2012).

While expressing "medium confidence" with regard to prognosis of future droughts over the 21st century owing to human impacts, the IPCC's 2012 Special report on Extremes had stated that droughts would intensify in the 21st century in some seasons and areas, as a result of decrease in precipitation and/or increased evapotranspiration. Applicability of these projections was deemed appropriate to regions including southern Europe and the Mediterranean region, central Europe, central North America, Central America and Mexico, northeast Brazil, and southern Africa (Seneviratne et al., 2012). Incongruity between different projections, accruing both from dissimilar models and from different indices of drought earned "low confidence" of IPCC's Special Report. Some experts are of the view that supplementary uncertainties emanate from soil moisture limitations on evapotranspiration, the effect of CO_2 concentrations on plant transpiration, observational uncertainties germane to interpretation of historical trends, and process representation in current land models (Greve et al., 2014). Some scholars in their drought-related attribution studies have used identical approach to those for heat (Funk et al., 2015).

In the wake of the ambiguity of results relating to the trends in segment of global area afflicted by drought (Hartmann et al., 2013), attribution studies seldom discern strong effect of anthropogenic climate change. Recent studies of the Colorado River by Vano, Das, and Lettenmaier (2012) and Vano et al. (2014) antedate striking effects on river flows associated with changes in temperature. A recently conducted study of eastern Brazil's recent drought by Otto et al. (2015) did not specifically identify anthropogenic contribution; relatively, it was attributed to a natural but uncommon digression of the South Atlantic Convergence Zone. While ambiguous or uncertain changes in probability and strength were discerned in some other studies (Barlow and Hoell, 2015), whereas Shiogama, Stone, Nagashima, Nozawa, and Emori (2013a) informed that their results were sensitive to bias correction.

While drought is acknowledged to be a complex phenomenon due to the many physical processes involved and the broad range of societal factors that influence its occurrence and intensity, some aspects of drought are influenced by temperature in ways that are better understood, and thus more amenable to attribution, than others. In particular, temperature exacerbates hydrological drought in some regions by increasing surface evaporation, so that increasing temperature causes an increasing risk of hydrological drought even if precipitation does not change (Diffenbaugh, Swain, & Touma, 2015). While acknowledging that droughts are attributable to multiple factors at different scales and contexts, NASEM (2016, p. 98) emphasizes that an area that needs further focus is understanding the leading factors that have historically been at the roots of drought in specific regions and watersheds.

2.8.4 Extreme snow and ice storms

Snow and ice storms, frequently followed by wind often characterize severe winter weather. In the absence of a universal criteria for delineating extreme snow or ice storms, NASEM (2016, p. 103), while citing the case of the US National Weather Service issuing frequent warnings about heavy snow and ice storms, also laments that the dearth of universal metrics for evaluating heavy snow and ice events obscures the analysis of trends and attribution studies. Effects of a snow or an ice storm get compounded by wind along with by the population of the area affected by the storm. Northern Hemisphere's overall snow cover has shrunk, partly owing to higher temperatures that shorten the time snow is on the ground (Derksen and Brown, 2012). Trends in heavy snow and ice events have attracted few studies; nevertheless, specifically over regional and larger spatial scales. A mixed evidence with regard to trends in the frequency and intensity of cold-season storms can be discerned for the entire Northern Hemisphere, irrespective of the fact that whether they generate snow and/or icy rain.

A northward swing in the primary tracks during winter in the overall storm frequencies is reflected in some studies (Seiler & Zwiers, 2015a, 2015b). O'Gorman (2014) moots the suggestion that the occurrence of extreme snowfalls in the coldest climates should increase with warming owing to increasing atmospheric water vapor, whereas it should decrease in warmer climates owing to reduced frequency of subfreezing temperatures, though by less than mean snowfall decreases. There have been few attribution studies of global or regional trends dealing with observations of extreme snow and ice events in the wake of the data limitations and the ambiguities in event definition. Nevertheless, there are some attribution studies dealing with specific events, inured on initial conditions in the atmosphere. Having simulated the western South Dakota blizzard of October 2013, Edward et al. (2014) have found no difference in accumulated snowfall, snow water equivalent, between pre–industrial counterfactual runs and present-day simulations. By making use of an ensemble of model simulations of recent winters, Añel et al. (2014) have come to the conclusion that heavier-than-normal snowfall seasons in the Spanish Pyrenees are not directly ascribable to anthropogenic forcing. Wang et al. (2015) have demonstrated that Himalayan blizzards such as the October 2014 event entail an enhanced probability of occurrence when tropical cyclones from the Bay of Bengal intermingle with stronger extratropical systems, and they have also extrapolated an "increased possibility" of such circumstances in the future.

Taking possible exception of a tropical cyclone connection in the study by Wang at el. (2015), it has been observed by NASEM (2016) that none of the event attribution studies point to anthropogenic climate change as a major factor in the heavy snow events. In the wake of the fact that the sample of case studies of extreme snow events examined thus far being too small, the probability of possible anthropogenic warming effects cannot be ruled out. Undoubtedly, trends in freezing rain events in the northern middle latitudes constitute major contenders for effects of anthropogenic warming (Klima and Morgan, 2015); nevertheless, systematic analyses of observed trends in freezing rain events have yet to be undertaken. While drawing attention to the fact that attribution of extreme snow and ice events suffers from a similar challenge as do some other extreme event types, NASEM (2016) laments that owing to this "the events are strongly governed by the atmospheric circulation, for which externally forced changes are uncertain" (p. 106). Owing to this reason, the possibility of attribution of extreme snow and ice storm events benefiting from an emphasis on the thermodynamic state during particular events cannot be rules out, as argued by Trenberth et al. (2014).

There are also severe convective storms (SCSs). Such storms are those that engender strong winds, hail, tornadoes, extensive lightning, or heavy precipitation, and SCRs usually occur over land. In meteorological parlance, the term "convection" refers to strong vertical motion—updrafts and downdrafts—spurred by resistance in the atmosphere. Identically, the term "severe" is characteristically used when some variables exceed specified thresholds—for example, wind speeds greater than 25 m/s or hailstones larger than 2 cm (Doswell, 2001). According to Tippett, Allen, Jensini, and Brooks (2015), the term "hazardous convective weather" also has been used to describe SCSs events. In both spatial extent and temporal duration, SCSs are small compared to many other extreme weather events. According to NASEM (2016), the most extreme hazards like tornadoes and large hail, are specifically localized and not well resolved by conventional meteorological observations, resulting in availability of aberrational datasets based on reports by amateur observers on the ground, especially in the United States; and it laments at the non-existence of good long-term data in this regard in much of the world (p. 118).

Detection of trends is difficult due to data heterogeneities, even scarcely available, make the task of detecting trends in SCSs cumbersome. Undeniably, observations of both tornadoes and hail in the United States exhibit significant increases over the latter half of the 20th century; nonetheless, these are extensively construed to be artifacts of increased frequency of reporting rather than actual meteorological trends (Brooks and Dotzek, 2007). According to Tippett et al. (2015), environmental variables predictive of tornado formation, for example, do not exhibit the trends that tornadoes themselves do. Admittedly, there are some constant indications in some literature about increased year-to-year variability, along with concentration of activity in fewer outbreaks of larger magnitude (Tippett, 2014), albeit there is no explicit link between this and climate change. Absence of detailed assessments of future projections of SCS activity in recent literature is probably attributable to the limited number of such studies. The 2012 IPCC Special Report on Extremes did take into account hail as distinct from other precipitation extremes, noting that "confidence" was still low for hail projections specially owing to absence of hail-specific modeling studies, and "a lack of agreement among the few available studies" (Seneviratne et al., 2012).

NASEM (2016) does not rule out the feasibility of highly conditioned approaches for SCS in present times, with the help of using either environmental indices or small-domain, high-resolution models forced by environmental conditions derived from larger-scale ones (NASEM, 2016, p. 120), as has been undertaken earlier for tropical cyclones (Knutson and Tuleya, 2004); and interestingly a small number of studies have already been undertaken by applying this methodology for future scenarios (Trapp and Hoogewind, 2016). However, NASEM (2016) admits that it is "not aware of any attribution studies of any kind for individual SCS events, whether single storms or outbreaks consisting of multiple storms."

2.8.5 Tropical cyclones

Tropical cyclone is regarded as a broadly defined term that is used to delineate those cyclonic disturbances—tropical depressions, tropical storms, hurricanes, typhoons, and cyclones—that spring up over the tropical and subtropical regions of the world's oceans. A cyclone can be defined as "a weather system in which winds move in a circular direction around a warm center of low barometric pressure. Unlike the larger extratropical cyclone, the tropical cyclone has no frontal systems, and its strongest winds are located near Earth's surface" (Longshore, 2008, p. 397). US National Oceanic and Atmospheric Administration(NOAA) has described tropical cyclone as "a rotating, organized system of clouds and thunderstorms that originates over tropical or subtropical waters and has a closed low-level circulation" (NOAA, 2011). Noting that tropical cyclones rotate counterclockwise in the Northern Hemisphere, NOAA (2011) has classified tropical cyclones as follow:

- Tropical Depression—A tropical cyclone with maximum sustained winds of 38 mph (33 knots) or less.
- Tropical Storm— A tropical cyclone with maximum sustained winds of 39−73 mph (34−63 knots).
- Hurricane—A tropical cyclone with maximum sustained winds of 74 mph (64 knots) or higher. Hurricanes are called typhoons in the western North Pacific; and similar storms in the Indian Ocean and South Pacific Ocean are called cyclones.
- Major Hurricane—A tropical cyclone with maximum sustained winds of 111 mph (96 knots) or higher, corresponding to a Category 3, 4 or 5 on the Saffir-Simpson Hurricane Wind Scale.

WMO maintains a Regional Specialized Meteorological Center in each region of the globe that is vulnerable to tropical cyclones, and determines when a given system is a tropical cyclone and also fixes its intensity from available observations. Conventionally the intensity of a tropical cyclone is construed to reflect its maximum sustained wind; however, this is only a "loose guide" to the prospective grimness of a particular storm's impacts. Hazards accompanying cyclones often embrace both coastal and freshwater flooding as well as winds. Undeniably, a particular tropical cyclone event could be delineated for attribution purposes by strong surge, precipitation, storm size, economic damage, or other variables; nonetheless, there are inadequate observations for some of these quantities (NASEM, 2016, p. 107). Satellite images are largely helpful in determining optimum sustained wind speed, with the usage of in-situ observations where available. However, uncertainties are important (Landsea and Franklin, 2013) and could be larger for other variables like storm surge in regions where automated tide measures are unavailable. Different variables may afford diverse level of severity of an event despite good observations. A storm even with feeble winds still entails the potential of triggering a major disaster owing to precipitation storm surge, or high vulnerability. According to NASEM (2016), merely observation-based methods have not been put in use to conduct event attribution studies on tropical cyclones. van Oldenborgh et al. (2015) are of the opinion that methods relying on extreme value theory are impractical for tropical cyclones. Being rare events, tropical cyclones seldom provide time series variables.

Synthesis studies, having made use of delineated thresholds of statistical importance against a null hypothesis of zero trend, characteristically make it discernible that no explicit detection of long-term trends can be had in tropical cyclone numbers, intensities, or integrated measures of activity (IPCC, 2014b, 2014c; Walsh et al., 2015). However, the frequency of the most intense storms could be an exception. Marginal significant increases are reported by some studies in the occurrence of category 4 and 5 storms (Kossin, Olander, & Knapp, 2013), whereas others make it discernible hitherto greater significance by spotting a sequential pattern of increase that more thoroughly resembles estimates of GHG-driven change rather than a mere linear trend (Holland and Bruyere, 2014). Explicit trends in recent decades are becoming discernible in some regions; for example, the Atlantic, where data are of highest quality, stands out (Emanuel, 2006). The attribution of these trends to specific causes still remains a debatable question; nevertheless, some have attributed these trends to natural variability and others to diminution in anthropogenic aerosol forcing (Mann & Emanuel, 2006). Kossin, Emanuel, and Vecchi (2014) have discerned a strong increase—both in the global and hemispheric means and in most individual basins—in the average latitude at which storms reach their utmost intensities.

Usage of new models has seemingly proved instrumental in facilitating emergence of a semblance of broad consensus with regard to the expected future trends of tropical cyclones and their levels of certainty (IPCC, 2013a, 2013c; Knutson et al., 2010; Walsh et al., 2015). With the warming of climate, tropical cyclones are prone to become more intense. NASEM (2016) finds "considerable confidence" in this inference in the wake of supporting evidence found in a wide range of numerical models as well as justification offered by theoretical understanding, predominantly since there is a well-established body of theory for the maximum potential intensity of tropical cyclones (Bryan & Rotunno, 2009). Admittedly, the rate of intensification per degree of global mean surface warming remains quantitatively inexact; however, because maximum potential intensities are predicted to rise (e.g., Camargo, 2013), future observations of tropical cyclones with intensities significantly higher than those observed in the past would be uniform with expectations in a warming climate, and attribution studies for such storms would have a firm basis in physical understanding (NASEM, 2016, p. 110).

Undeniably, the progression of a wide-ranging climate theory of tropical cyclones remains intangible (Walsh et al., 2015); nonetheless, recent advances in high-performance computing have enabled multidecadal simulations of climate models at tropical-cyclone-permitting resolutions. In tandem with conceptual models, such numerical models are the instrument of choice for examining prospective future changes in tropical cyclones (Wehner, Reed, & Zarzycki, 2017). Preceding research has assessed the effect of climate change on tropical storms via unrealistic representations of future climate through consistent augmentations in GHGs and SST (Walsh et al., 2015; Wehner et al., 2015) or more accurate but more extreme cases of warming by making use of the Representative Concentration Pathway (RCP4.5 or RCP8.5) scenarios (Bacmeister et al., 2018; Knutson et al., 2015). However, as found in their comparison of RCP4.5 to RCP8.5, Bacmeister et al. (2018) have discerned major uncertainties in the pattern of SST changes that also portend a significant challenge in precisely projecting future tropical storm frequency. Alterations in other significant features of tropical cyclone behavior are elusive. Both warmer climate conditions taken into account in present context reflect substantial changes in the poleward density of tropical storm tracks in comparison to the historical simulations, nonetheless the differences between them are not going probably to be highly significant. Likewise, changes in accumulated cyclonic energy (ACE), storm duration, track length and translational speed are complex with the differences clearly evident for only the most intense storms. Finally, some properties of tropical cyclones are not significantly altered in warmer climates, most notably the robust relationship between maximum wind speeds and central pressure minima.

2.8.5.1 Extratropical cyclones

Daily weather conditions in many parts of the globe are significantly influenced by extratropical cyclones through their associated wind and precipitation patterns (Dacre, Hawcroft, Stringer, & Hodges, 2012). Extratropical cyclones, also known as midlatitude storms, apart from a critical component of the global circulation (Chang, Lee, & Swanson, 2002), also facilitate transportation of vast quantities of moisture and energy. The route of these systems is accountable for much of the day-to-day changeability of weather in the midlatitudes, with cyclones and fronts fetching up to 90% of the precipitation (Hawcroft, Shaffrey, Hodges, & Dacre, 2012) including extremes, for instance, as exemplified by events above the 99th percentile (Catto & Pfahl, 2013) and triggering damage connected with strong winds (Leckebusch et al., 2006). Apart from having become a significant part of the IPCC process, the evaluation of extratropical storm tracks has also attracted sufficient recent interest in investigating not just the mean preferred storm locations but also the structure of extratropical cyclones in climate models and their associated cloud and precipitation features (Govekar, Jakob, & Catto, 2014). Nevertheless, Catto (2016) laments that these studies have mainly focused on the average properties of a large number of cyclones, without considering their salient features

Extratropical cyclones are Being a critical component of the atmospheric circulation system in the extratropical region, extratropical cyclones play a significant role in the maintenance of the energy equilibrium of the atmosphere by transferring energy and moisture poleward to lessen the meridional temperature gradient (Zhang, Xu, Ma, & Deng, 2019). Extratropical cyclones are mostly positioned in the Northern Hemisphere over Kuroshio in the Pacific, subtropical Northwest American coast, East Pacific, Southeast Greenland, and Barents sea in winter (Gulev, Zolina, & Grigoriev, 2001). Prevalence of an apparent correlation between Atlantic storm frequency and the North Atlantic Oscillation (NAO) is often discerned on the international time scale (Gulev et al., 2001). According to Gulev et al. (2001), frequency of the eastern Pacific cyclone is associated with the Pacific-North American (PNA) pattern. Undoubtedly, it was discerned by Zhang, Ding, and Li (2012) that the Arctic Oscillation (AO) index is certainly interrelated with the frequency of extratropical cyclones at the high latitudes in the Northern Hemisphere; nonetheless, a negative interrelatedness can be discerned at the low latitudes. Vicissitudes in long-term trends of extratropical cyclones over the Northern Hemisphere have garnered many attentions owing to GW (Wang, Feng, Chan, & Isaac, 2016).

It has been ascertained by Wang et al. (2016) that the enumerations of deep cyclones exhibit a general upsurge in both hemispheres over the past half century. The upsurge is more apparent in the Northern Hemisphere than in the Southern Hemisphere, and it is more palpable in winter than in summer. It has also been pointed by Chang (2017) that, by the closing of the 21st century, there will be a substantial intensification in the frequency of extreme cyclones in all four seasons irrespective of the definition, whereas the total number of cyclones demonstrates the opposite trend. It has also been discerned from many studies that there would be reduction in overall number of the boreal winter extratropical cyclones in the Northern Hemisphere during the winter by the closing part of 21st century (Eichler, Gaggini, & Pan, 2013). Suggesting that augmentation in near-surface heating at high altitudes could be a crucial cause in the diminution of extratropical cyclones, some studies also make it noticeable that, on the one hand, it leads to the decline of meridional temperature gradient and culminate in diminution of the existing potential energy for extratropical cyclones; whereas on the other hand, it works to undermine the atmospheric baroclinicity and inhibit the formation of cyclones (Eichler et al., 2013). Aside from this, the overall tendency of extratropical cyclones is intimately associated with the activities of strong cyclones that entail the disposition of moving poleward in the Atlantic Ocean (Wang, Swail, & Zwiers, 2006).

Noticeable multidecadal variability gets reflected through the statistics of observed events, often associated with extensive circulation patterns such as the NAO. Nevertheless, trends are occasionally reported in the literature, these trends are exceedingly sensitive to the period selected and to how the storms are defined. Contending that valuations of historical centennial timescale changes have to be based mostly on re-analyses, Krueger, Schenk, Feser, and Weisse (2013), do not rule out the possibility of these valuations containing long-term heterogeneities. A recent wide-ranging review for the North Atlantic and northwest Europe conducted by Feser et al. (2015), and for the US East Coast by Colle, Booth, and Chang (2015) make it explicitly evident that there is lack of no consensus on attributed trends in observations, at least in the Northern Hemisphere. Owing to competing factors, even the anticipated impact of human-induced climate change on extratropical cyclones is uncertain. The diminution reduction in pole-to-equator temperature gradient anticipated from polar amplification entails the likelihood of countermining cyclones; nevertheless, the surge in moisture is prone to strengthen them, as would the increase in upper tropospheric temperature gradient (O'Gorman, 2010). The surmise arrived at by the IPCC Fourth Assessment Report that that cyclones would be expected to strengthen was reportedly based on a study by Lambert and Fyfe (2006) that used minimum surface pressure as the index. The general anticipated deterioration in surface pressure at higher latitudes hence brought forth a trend that was not essentially associated with cyclone intensity. Consequently, the IPCC in its Fifth Assessment Report explicitly noted that future prognoses of extratropical cyclones were found to be unsure or uncertain (Christensen et al., 2013).

Additionally, the possibility of storm track positions changing locations has not been ruled out. A general escalation of the wintertime storm track over northern Europe in the CMIP5 models and an enfeebling of the Mediterranean storm track has been reported by Zappa, Shaffrey, and Hodges (2013); nonetheless, the confidence in this projection has remained unclear owing to the fact of lack of full comprehension of the entire gamut of related pertinent physical processes. Findings by Seiler and Zwiers (2015a, 2015b) report that explosive cyclones "rapidly intensifying low-pressure systems with severe wind speeds and heavy precipitation" get inclined to move poleward in the Northern Hemisphere, diminution in frequency owing to flagging baroclinicity, and somewhat surge in intensity. Hoskins and Woollings (2015) inform about multiple physical mechanisms that have been intended for triggering anthropogenic circulation vicissitudes at midlatitudes and their connection to weather extremes, and it is surmised by them that there is considerable incertitude with regard to what can be expected in the future. The Southern Hemisphere seems to have stronger human influence where it is said to be wielded via stratospheric ozone depletion. Some recent model-based attribution studies have discerned an ozone depletion influence on Southern Hemispheric extratropical cyclones and related extreme precipitation, discernible more explicitly in a poleward swing in the storm track (Grise, Son, Correa, & Polvani, 2014; Kang et al., 2013).

After having made use of a seasonal prediction system to evaluate the drivers of the extreme storminess over the central United States and Canada in winter 2013/14; Yang et al. (2015) did not find any proof of a human influence, but they did find a FAR in the range of 33%–75% due to the multiyear anomalous tropical Pacific winds. Augmentation in precipitations, cyclone intensity and low-level jet strength accruing from the latent heating over the US East Coast has been reported by Marciano, Lackmann, & Robinson (2015) and there was no assessment of human influence. A study Hurricane Sandy in 2012 by Lopeman, Deodatis, and Franco (2015) and discussion of long-term changes in New York City by Colle et al. (2015) reveal that the anthropogenic contribution to past storm increases was estimated to be small but projected to emerge as a potent factor, in terms of decreases in return periods during the course of 21st century.

According to NASEM (2016), it is viable in present times to run global models with a reasonable representation of extratropical cyclones and in that case the "main issue for event attribution, then, is to assess whether simulated anthropogenic changes in the large-scale circulation that affect the storm tracks are credible" (NASEM, 2016, p. 114). Likewise, Hoskins and Woollings (2015) have opined that this would be a challenge without a robust physical understanding of the processes controlling such changes, or a clear signature in observations.

2.8.6 Wildfires

Undeniably, wildfires are not acknowledged as meteorological events, still their proneness or likelihood and extent entail probability of being influenced by climatic factors. Wildfires are habitually enceinte and swiftly dispersible fires impacting forests, shrub areas, and/or grasslands. Occurrence of the incidents of wildfires takes place in many areas of the globe, particularly those with widespread forests and grasslands (Romero-Lankao et al., 2014). Despite the fact that bulk of wildfires are started by lightning, NASEM (2016) notes that "a substantial number are started by humans, especially near populated areas. The most common metric of wildfires is the area burned, either by a single wildfire or by all wildfires during a fire season in a particular region" (p. 115).

Attribution of wildfire trends and extreme events is convoluted by many factors. In the first place is the role of humans in kindling or ignition, fire suppression, and management of forests and other biomes (Gauthier, Bernier, Kuuluvainen, Shvidenko, & Schepaschenko, 2015). Second factor pertains to the role of lightning, wherein small-scale thunderstorms often prove instrumental in igniting large fire outbreaks (Struzik, 2017). Third factor accounts for the role of larger-scale weather that serves as a catalyst in the spread of wildfires and helps it grow into major events, particularly the role of winds and humidity in spreading fire and rain for dousing a fire breakout (Abatzoglou & Kolden, 2011). Fourth factor is related to the health of the forest and that in turn is closely linked to white pine bark beetle infestation. The increased frequency of extreme fire events and large-scale bark beetle mortality has been reported by studies in the recent past (Bentz et al., 2010), coupled with increasing likelihood of such events under climate change. In this regard, NASEM (2016) emphasizes on the need for wildfires attribution studies to focus on three time/space scales: "(1) individual large fires, which are controlled primarily by short-term weather patterns; (2) regional scale within-season extreme fire periods, which are driven by seasonal weather patterns; and (3) large fire seasons, which are regional-scale events resulting from climate teleconnections associated with persistent blocking ridges that cause extended fire seasons (with delayed season-ending rains)" (NASEM, 2016, p. 115).

Paucity of the availability of authentic data records is the major stumbling block in having an analysis of wildfire trends and extremes. Nevertheless, Jolly et al. (2015) inform that fire weather season lengths have exhibited profound increases during 1979–2013 across more than one-fourth of the Earth's vegetated surface, demonstrating a 19% augmentation in the global means fire weather season length. Association of heat and drought with wildfires as reflected in one of the earliest attribution studies undertaken by Gillett, Weaver, Zwiers, and Flannigan (2004) demonstrated that the augmentation of wildfire burn-areas in Canada during 1959–99 in tandem with anthropogenic summer warming. According to NASEM (2016), apart from the controls by climate and weather, the availability of fuels and hitherto, the state of vegetation have a bearing on individual fires, as well as overall fire season severity. Admittedly, it is pointed by Jolly et al. (2015) that longer fire seasons are caused by climate warming as can be evidenced from the recent observed augmentation in severe fire years in the western United States and Alaska, along with Brazil and Africa and some parts of Eurasia; nonetheless, NASEM (2016) points out that what is less evident is as to how "climate warming is driving changes in the atmospheric circulation and its teleconnections, resulting in persistent areas of high pressure that lead to large fire years on regional scales" (NASEM, 2016, p. 117).

2.9 Impacts of climate change

Climate change has come to be acknowledged as potential amplifier and multiplier of existing risks and engender new risks for natural and human systems (IPCC, 2014a). The distribution of these risks is not only unequal but also entails substantial hazards for deprived and disadvantaged people and communities. Vagaries of climate change affect universally almost at all levels of development. The risk of climate-related effects is an outcome of complicated exchanges or interactions between climate-related hazards and the vulnerability, exposure and adaptive capacity of human and natural systems. Any soaring in the rates and scales of warming and other alterations in the climate system, along with ocean acidification, enhance the risk of dangerous, persistent and in some cases irreparable damaging effects. There has already been rise in the annual global mean surface temperature at an average rate of 0.07°C per decade since 1880 and at an average rate of 0.17°C per decade since 1970 (NOAA, 2015). The Fifth Assessment Report of the IPCC has also reported about the prevalence of almost identical trends in SST, marine air temperature, sea level, tropospheric temperature, ocean heat content and specific humidity (IPCC, 2014a).

Apart from increases in temperature, there have also been augmentations in the frequency and chroma of wildfires that sequentially release GHGs. It becomes discernible from observations and climate model simulations about polar warming amplification accruing from several feedbacks in the climate system—the positive ice-albedo feedback being the strongest (Taylor et al., 2013). It becomes explicitly evident from the decreased level of ice cover about the presence of a darker surface that tends to indicate toward a reduced albedo, sequentially culminating in a robust intake of solar radiation and a further speeding up of warming. Vaughan et al. (2013) have reported that sea-ice extent of the Arctic is recording robust decrease, especially in summer, in response to the enhanced warming in the region. Nevertheless, trends from recent literature (Pithan and Mauritsen, 2014) lead to surmise that temperature feedbacks play a leading role, rendering surface-albedo feedback the second main contributor to Arctic amplification.

Climate change has been affecting the global water cycle, wielding impact on global-scale precipitation patterns over land, and on surface and subsurface ocean salinity, contributing to global-scale changes in frequency and intensity of daily temperature extremes since the mid-20th century. Alterations in the hydrological cycle are an anticipated consequence of anthropogenic climate change. Hegerl and colleagues (2015) have opined that the Clausius–Clapeyron relationship mulls a strong quasiexponential increase in water vapor concentrations with warming at about 6%–7% K-1 near the surface. This is in accordance with observations of change over the ocean (Chung, Soden, Sohn, & Shi, 2014) and land (Willett, Jones, Thorne, & Gillett, 2010) and with simulations of future changes (Allen and Ingram 2002) and presumes that on vast scales the relative humidity changes little, as normally expected (Allen and Ingram 2002; Sherwood, Roca, Weckwerth, & Andronova, 2010) and nearly seen in models (Collins et al., 2013). Nevertheless, virtual humidity changes in proximity areas may occur where large-scale circulation patterns alter or when moisture sources are limited over land (Vicente-Serrano et al., 2014).

Wide-ranging impacts on various ecosystems have become noticeable on account of alterations in the climate system. Change, climate change is exacerbating current pressures on land, water, biodiversity and ecosystems. Smith, Harrison, and Jordan (2011) have expressed the opinion that in case if there is an increase in the atmospheric CO_2 concentration from the current levels of 406 ppm to 450—600 ppm, resulting in greater than 2°C warming over the coming century, it could lead to several irretrievable impacts, including sea level rise. An elaboration of individual risks as well as overarching key risks on account of climate change, as explained by O'Neill et al. (2017), inter alia include risks to biodiversity, health, agriculture and so on, as well as risks of extreme events such as extreme precipitation and heat waves and risks to specific ecosystems such as mountain and Arctic, to name but a few.

While projecting that future climate will henceforth be dependent on the compounding of committed warming caused by past anthropogenic emissions, the impact of future anthropogenic emissions, natural climate variability and climate sensitivity, we are informed (UNEP, 2019, p. 46) that there are regions, especially at northern, mid- and high latitudes, that are already experiencing greater warming than the global average, with mean temperature rise exceeding 1.5°C in these regions and that these impacts entail implications for the quality and quantity of ecosystem services, as well as for patterns of resource use, their distribution and access across regions and within countries.

While signaling a warning that time is conking out to prevent the irreversible and dangerous impacts of climate change, recent studies have cautioned that lest a radical reduction in GHG emissions takes place, the world remains on a course to exceed the agreed temperature threshold of 2°C above pre—industrial levels, which would increase the risk of pervasive effects of climate change, beyond what is already seen and these effects could include: extreme events (including flooding, hurricanes and cyclones) leading to loss of lives and livelihoods, persistent droughts leading to loss of agricultural productivity and food insecurity, severe heat waves, changes in disease vectors resulting in increases in morbidity and mortality, slowdowns in economic growth, and increased potentials for violent conflict (IPCC, 2018; Mechler, Bouwer, Schinko, Surminski, & Linnerooth-Bayer, 2019; UNEP, 2019). The range, dispersal and severe nature of the impacts of climate change vary from country to country. Several islands, especially in the Pacific region, have faced multiple impacts in one season or annually in multiple years; and these impacts entail the potential of undermining food security mechanisms and systems, as well as social and economic progress in health and other areas (Handmer et al. 2019).

Hoeppe (2016) notes that twofold increase in frequency of climate-related loss events since 1980 is one sign of the probable impacts of climate change. UNEP (2019) points out that climate-related loss events are already estimated to have resulted in the loss of 400,000 lives and the imposition of a cost of US$1.2 trillion annually on the global economy, wiping 1.6% from global GDP.

The IPCC Report on GW informs that global economic damage accruing from GW is estimated to be $54 trillion in 2100 under a warming scenario of 1.5°C and $69 trillion under a warming scenario of 2°C (IPCC, 2018). The severity of these climate-related risks is greatest, at the current juncture as well as in the near future, especially for those segments of the population like coastal communities, people in agriculture and forest communities, and those people who are dependent on natural resources. The extent of potential damage accruing from the vagaries of climate change portends a major systemic risk to the future human well-being and the ecosystems on which we depend, especially for societies in less-developed, coastal island nations and less-resilient countries (OECD, 2017).

2.10 Conclusion

A scintilla of optimism is held out by the PACC that limiting warming by the end of the 21st century could help prevent more problems. It explicitly states the need for achieving a balance of emissions and removals in the second half of the century. The 2°C target assumes significance as a target to be achieved, so that it could help in minimizing the probability of more intense storms, protracted droughts, rising sea levels and other natural disasters that are being progressively reported (Munich, 2017). In order to brighten the prospects of staying below 2°C, and at realizable costs, it is imperative that emissions ought to register a drop or diminution by 40%–70% globally between 2010 and 2050, falling to zero by 2100 (IPCC, 2014a; Kroeze & Pulles 2015). The contemporary pathway of global annual and cumulative emissions of GHGs is not in consonance with the widely known and discussed goals of limiting GW to 1.5°C–2.0°C above pre–industrial levels. The temperature goals set in the PACC are likely to become almost unattainable, in case emissions are allowed to rise beyond 2020 or even remain at constant level. Holding up remedial action or enfeebled near-term policies could lead to enhance the mitigation challenges in the long-term. While referring to risks associated with exceeding 1.5°C GW by the end of the 21st century in terms of increases in the severity of projected impacts and in the adaptation needs, UNEP (2019) cautions that such a scenario is likely to make the achievement of many SDGs much more difficult. According to a recent World Bank Report (2016), the overall costs and risks of climate change comprise a forecast that some regions could witness decline in growth by as much as 6% of GDP by 2050. If the worst of the climate change-related risks are to be avoided, An OECD Publication (2017) suggests that the momentum and extent of the required economic transformation is to be maintained at unprecedented levels in order to avoid the worst of the climate-related risks. Thus climate change is a risk management problem necessitating as to how to find and implement the most cost-effective ways to limit risks to an accepted/agreed level, informed by the best scientific evidence along with quick action on adaptation and mitigation to ward off these risks (OECD, 2018).

References

Añel, J. A., López-Moreno, J. I., Otto, F. E. L., Vicente-Serrano, S., Schaller, N., Massey, N., ... Allen, M. (2014). The extreme snow accumulation in the western Spanish Pyrenees during Winter and Spring 2013. *Bulletin of American Meteorological Society*, S73–S76.

Abatzoglou, J. T., & Kolden, C. A. (2011). Relative importance of weather and climate on wildfire growth in interior Alaska. *International Journal of Wildland Fire*, 20(4), 479–486.

Acemoglu, D. (2009). *Introduction to modern economic growth*. Princeton, NJ: Princeton University Press.

Agnew, J. (2014). By words alone shall we know: Is the history of ideas enough to understand the world to which our concepts refer? *Dialogues in Human Geography*, 4, 311–319.

Agrawala, S. (1998). Structural and process history of the Intergovernmental Panel On Climate Change. *Climactic Change*, 39(4), 621–642.

Ainsworth, E., & Long, S. (2005). What have we learned from 15 years of free-air CO_2 enrichment (FACE)? A *meta*-analytic review of the responses of photosynthesis, canopy properties and plant production to rising CO_2. *New Phytologist*, 165, 351–CO72.

Allen, M. R., & Ingram, W. J. (2002). Constraints on future changes in climate and the hydrologic cycle. *Nature*, 419, 224–232.

Allen, M. R., Dube, O. P., Solecki, W., Aragón-Durand, F., Cramer, W., Humphreys, S., ... Zickfeld, K. (2018). *Framing and context*. Global Warming of 1.5°C. *An IPCC special report on the impacts of global warming of 1.5°C above pre-industrial levels and related global greenhouse gas emission pathways, in the context of strengthening the global response to the threat of climate change, sustainable development, and efforts to eradicate poverty*. Geneva: IPCC.

Alley, K. E., Scambos, T. A., Siegfried, M. R., & Fricker, H. A. (2016). Impacts of warm water on Antarctic ice shelf stability through basal channel formation. *Nature Geoscience, 9*, 290−293.

Alley, R. B., Meese, D. A., Shuman, C. A., Gow, A. J., Taylor, K. C., Grootes, P. M., . . . Zielinski, G. A. (1993). Abrupt increase in Greenland snow accumulation at the end of the Younger Dryas Event. *Nature, 362*(6420), 527−529.

Allwood, J. M., Bosetti, V., Dubash, N. K., Gómez-Echeverri, L., & von Stechow, C. (2014). *Glossary*. Climate change 2014: Mitigation of climate change. *Contribution of Working Group III to the Fifth Assessment Report of the Intergovernmental Panel on Climate Change*. Cambridge: Cambridge University Press.

Alvaredo, F., Chancel, L., Piketty, T., Saez, E., & Zucman, G. (2017). The elephant curve of global inequality and growth. In: *Working papers series n° 2017/20: 1.13*.

Anav, A., Friedlingstein, P., Kidston, M., Bopp, L., Ciais, P., Cox, P., . . . Zhu, Z. (2013). Evaluating the land and ocean components of the global carbon cycle in the CMIP5 earth system models. *Journal of Climate, 26*, 6801−6843.

Anklin, M., Barnola, J. M., Beer, J., Blunier, T., Chappellaz, J., Clausen, H. B., . . . Wolff, E. W. (1993). Climate instability during the last interglacial period recorded in the grip ice core. *Nature, 364*(6434), 203−207.

Arguez, A., & Vose, R. S. (2011). The definition of the standard WMO climate normal: The key to deriving alternative climate normals. *Bulletin of the American Meteorological Society, 92*(6), 699−704.

Arrhenius, S. (1896). On the influence of carbonic acid in the air upon the temperature of the ground. *Philosophical Magazine Series 5, 41*(251), 237−276. Available from https://doi.org/10.1080/14786449608620846. Available at.

Ayyub, B. M. (2018). *Climate-resilient infrastructure: Adaptive design and risk management*. Reston, VA: American Society of Civil Engineers (ASCE).

AWG (Anthropocene Working Group). (2019). What is the Anthropocene: Current definition and Status. Sub-commission on Quaternary Stratigraphy. Available online at: http://quatenary.straightgraphy.org/working-groups/anthropocene/

Bacmeister, J. T., Reed, K. A., Hannay, C., Lawrence, P. J., Bates, S. C., Truesdale, J. E., . . . Levy, M. N. (2018). Projected changes in tropical cyclone activity under future warming scenarios using a high-resolution climate model. *Climate Change, 146*(3−4), 547−560.

Baiocchi, G., Minx, J., & Hubacek, K. (2010). The impact of social factors and consumer behavior on CO_2 emissions in the UK: A panel regression based on input-output and geo-demographic consumer segmentation data. *Industrial Ecology, 14*(1), 50−72.

Barlow, M., & Hoell, A. (2015). Drought in the Middle East and Central-Southwest Asia during winter 2013/14 [in "Explaining extremes of 2014 from a climate perspective"]. *Bulletin of the American Meteorological Society, 96*(12), S71−S76.

Barnes, E. A., Dunn-Sigouin, E., Masato, G., & Woollings, T. (2014). Exploring recent trends in Northern Hemisphere blocking. *Geophysical Research Letters, 41*(2), 638−644.

Barry, R. G. (2013). A brief history of the terms 'climate and climatology'. *International Journal of Climatology, 33*, 1317−1320.

Baudrillard, J. (2017). *The consumer society: Myths and structures*. Los Angeles, CA: Sage.

Beck, U. (1998). *World risk society*. Cambridge: Polity Press.

Behnassi, M., Pollman, O., & Gupta, H. (2018). *Climate change, food security and natural resource management: Regional case studies from three continents*. Geneva: Springer.

Benjamin, D., Por, H.-H., & Budescu, D. (2017). Climate change vs global warming: Who is susceptible to the framing of climate change? *Environment and Behavior, 49*(7), 745−777.

Bentz, B. J., Régnière, J., Fettig, C. J., Hansen, E. M., Hayes, J. L., Hicke, J. A., ... Seybold, S. J. (2010). Climate change and bark beetles of the western United States and Canada: Direct and indirect effects. *Bioscience*, *60*, 602−613.

Bereiter, B., Eggleston, S., Schmitt, J., Nehrbass-Ahles, C., Stcoker, T. F., Fischer, H., ... Chappellaz, J. (2015). Revision of the EPICA Dome C CO_2 record from 800 to 600-kyr before present. *Geophysical Research Letters*, *42*(2), 542−549.

Bergthaller, H., Emmett, R., Johns-Puta, A., & Limmer-Kneitz, A. (2014). Mapping common ground: Ecocriticism, environmental history, and the environmental humanities. *Environmental Humanities.*, *5*(1), 261−276.

Bhattacharya, A., Meltzer, J., Oppenheim, J., Qureshi, Z., & Stern, N. (2016). *Delivering on sustainable infrastructure for better development and better climate*. Washington, DC: The Brookings Institution.

Biermann, F., Bai, X., Bondre, N., Broadgate, W., Arthur, C.-T., Dube, P., ... O'Brien, K. (2016). Down to Earth: Contextualizing the Anthropocene. *Global Environmental Change*, *39*, 341−350.

Birkmann, J., Welle, T., Solecki, W., Lwasa, S., & Garschagen, M. (2016). Boost resilience of small and midsized cities. *Nature*, *537*(7622), 605−608.

Brooks, H. E., & Dotzek, N. (2007). The spatial distribution of severe convective storms and an analysis of their secular changes. In H. Diaz, & R. Murane (Eds.), *Climate extremes and society* (pp. 35−53). Cambridge, UK: Cambridge University Press.

BoM (Bureau of Meteorology, Australia BoM). *Southern oscillation index, Climate glossary*. (2014b). <http://www.bom.gov.au/climate/glossary/soi.shtml>.

BoM (Bureau of Meteorology, Australia). *Drought statement-rainfall deficiencies*. (2014a). <http://www.bom.gov.au/climate/drought/drought.shtml> Accessed 01.11.14.

Boucher, O., Randall, D., Artaxo, P., Bretherton, C., Feingold, G., Forster, P., ... Zhang, X. Y. (2013). Clouds and aerosols. *Climate change 2013: The physical science basis. Contribution of Working Group I to the Fifth Assessment Report of the Intergovernmental Panel on Climate Change*. Cambridge: Cambridge University Press.

Bren d'Amour, C., Reitsma, F., Baiocchi, G., Barthel, S., Güneralp, B., Erb, K.-H., et al. (2016). Future urban land expansion and implications for global croplands. *Proceedings of the National Academy of Sciences*, *114*(34), 8939−8944.

Broecker, W. S. (1975). Climatic change: Are we on the brink of a pronounced global warming? *Science (New York, N.Y.)*, *189*(4201), 460−463.

Brondizio, E. S., O'Brien, K., Bai, X., Biermann, F., Steffen, W., Berkhout, F., ... Arthur Chen, C.-T. (2016). Reconceptualizing the Anthropocene: A call for collaboration. *Global Environmental Change*, *39*, 318−327.

Bryan, G. H., & Rotunno, R. (2009). The maximum intensity of tropical cyclones in axisymmetric numerical model simulations. *Monthly Weather Review*, *137*(6), 1770−1789.

Camargo, S. J. (2013). Global and regional aspects of tropical cyclone activity in the CMIP 5 models. *Journal of Climate*, *26*, 9880−9902.

Cambridge Dictionary. (2008). *Cambridge academic content dictionary*. Cambridge: Cambridge University Press.

Carey, M. (2012). Climate and history: A critical review of historical climatology and climate change historiography. *WIREs Climate Change*, *3*, 233−249.

Caroll, C., Roberts, D. R., Michalak, D. L., Lawler, J. L., Nielsen, S. E., Stalberg, D., ... Weng, T. (2017). Scale-dependent complementarity of climatic velocity and environmental diversity for identifying priority areas for conservation under climate change. *Global Change Biology*, *23*, 4508−4520.

Carr, M.-E., Friedrichs, M. A. M., Schmeltz, M., Noguchi Aita, M., Antoine, D., Arrigo, K. R., ... Yamanaka., Y. (2006). A comparison of global estimates of marine primary production from ocean color. *Deep Sea Research Part II: Topical Studies in Oceanography*, *53*, 741−770.

Carslaw, K. S., Boucher, O., Spracklen, D., Mann, G., Rae, J. G., Woodward, S., & Kumala, M. (2010). A review of natural aerosol interactions and feedbacks within the Earth system. *Atmospheric Chemistry and Physics*, *10*(4), 1701−1737.

Catto, J. L. (2016). Extratropical cyclone classification and its use in climate studies. *Review of Geophysics*, *54*, 486−520.

Catto, J. L., & Pfahl, S. (2013). The importance of fronts for extreme precipitation. *Journal of Geophysical Research Atmospheres*, *118*, 10,791−10,801.

Chang, E. K. (2017). Projected significant increase in the number of extreme extratropical cyclones in the Southern Hemisphere. *Journal of Climate*, *30*, 4915−4935.

Chang, E. K. M., Lee, S., & Swanson, K. L. (2002). Storm track dynamics. *Journal of Climate*, *15*, 2163−2183.

Charney, J. G., Arakawa, A., Baker, D. J., Bolin, B., Dickinson, R. E., Goody, R. M., & Wunsch, C. I. (1979). *Carbon dioxide and climate: A scientific assessment*. Washington, DC: National Academy of Sciences.

Chavez, F. P., Messié, M., & Pennington, J. T. (2011). Marine primary production in relation to climate variability and change. *Annual Review of Marine Science*, *3*, 227−260.

Christensen, J. H., Kumar, K. K., Aldrian, E., An, S.-I., Cavalcanti, I. F. A., de Castro, M., ... Zhou, T. (2013). *Climate phenomena and their relevance for future regional climate change. Climate change 2013: The physical science basis. Contribution of Working Group I to the Fifth Assessment Report of the Intergovernmental Panel on Climate Change*. Cambridge: Cambridge University Press.

Chung, E.-S., Soden, B., Sohn, B. J., & Shi, L. (2014). Upper-tropospheric moistening in response to anthropogenic warming. *Proceedings of National Academy of Sciences United States of America*, *111*, 11636−11641.

Ciais, P., Sabine, C., Bala, G., Bopp, L., Brovkin, V., Canadell, J., ... Thornton, P. (2013). *Carbon and other biogeochemical cycles*. In Climate change 2013: The physical science basis. *Contribution of Working Group I to the Fifth Assessment Report of the Intergovernmental Panel on Climate Change* (pp. 465−570). Cambridge: Cambridge University Press.

Clark, P. U., Pisias, N. G., Stocker, T. F., & Weaver, A. J. (2002). The role of the thermohaline circulation in abrupt climate change. *Nature*, *415*(6874), 863−869.

CNA (Center for Naval Analysis). *National security and the threat of climate change*. (2007). <https://www.cna.org/cna_files/pdf/national%20security%20and%20the%20threat%20of%20climate%20change.pdf>.

Coles, N. A., & Eslamian, S. (2017). Definition of drought. In S. Eslamian, & F. Eslamian (Eds.), *Handbook of drought and water scarcity: Principles of drought and water scarcity* (pp. 1−12). London: CRC Press.

Colle, B. A., Booth, J. F., & Chang, E. K. M. (2015). A review of historical and future changes of extratropical cyclones and associated impacts along the United States East Coast. *Current Climate Change Reports (Topical Collection on Extreme Events)*, *1*(3), 125−143.

Collins, M. (2002). Climate predictability on interannual to decadal time scales: The initial value problem. *Climate Dynamics*, *19*(8), 671−692.

Collins, M. R. K., Arblaster, J., Dufresne, J.-L., Fichefet, T., Friedlingstein, P., Gao, X., ... Wehner, M. (2013). *Long-term climate change: Projections, commitments and irreversibility*. Climate change 2013: The physical science basis. *Contribution of Working Group I to the Fifth Assessment Report of the Intergovernmental Panel on Climate Change* (pp. 1029−1136). Cambridge: Cambridge University Press.

Cook, J., Nuccitelli, D., Green, S. A., Richardson, M., Winkler, B., Painting, R., ... Skuce, A. (2013). Quantifying the consensus on anthropogenic global warming in the scientific literature. *Environmental Research Letters*, *L8*, 24024.

Cook, J., Oreskes, N., Doran., Peter, T., Anderegg., William, R. L., ... Rice, K. (2016). *Consensus on consensus: a synthesis of consensus estimates on human-caused global warming*, . *Environmental Research Letters* (L11, p. 48002).

Coope, G. R., Morgan, A., & Osborne, P. J. (1971). Fossil Coleoptera as Indicators of climatic fluctuations during last glaciation in Britain. *Palaeogeography, Palaeoclimatology, Palaeoecology.*, *10*(2−3), 87−101.

Cottineau, C., Finance, C., Hatna, E., Arcaute, E., & Batty, M. (2018). Defining urban agglomerations to detect agglomeration economies. Environment and Planning B: Urban Analytics and City. *Science (New York, N. Y.)*, *45*(2), 1−16.

Cruz-Cunha, M. M., Moreira, F., & Varajao, J. (2014). *Handbook of research on enterprise*. Hershey, PA: IGI Global.

Curran, D. (2017). The treadmill of production and the positional economy of production. *Canadian Review of Sociology*, *54*(1), 28−47.

Dacre, H. F., Hawcroft, M. K., Stringer, M. A., & Hodges, K. I. (2012). *An extratropical cyclone atlas—A tool for illustrating cyclone structure and evolution characteristics* (pp. 1497−1502). American Meteorological Society.

Dai, A. (2011). Drought under global warming: A review. *WIREs Climate Change*, *2*, 45−65.

Daly, H. E. (1973). *Toward a Steady State Economy*. San Francisco: W.H. Freeman.

Dansgaard, W., White, J. W. C., & Johnsen, S. J. (1989). The abrupt termination of the Younger Dryas climate event. *Nature*, *339*(6225), 532−534.

DCCEE (Department of Climate Change and Energy Efficiency). (2012). *Climate change in a nutshell*. Canberra: Australian Government.

Derksen, C., & Brown, R. (2012). Spring snow cover extent reductions in the 2008−2012 period exceeding climate model projections. *Geographical Research Letters*, *39*, L19504.

Delanty, G., & Mota, A. (2017). Governing the anthropocene. *European Journal of Social Theory*, *20*(1), 9−38.

Dietz, T. (2017). Human drivers of environmental change. *Annual Review of Environment and Resources*, *42*, 189−213.

Dietz, T., Rosa, E., & York, R. (2010). Human driving forces of global change: Dominant perspectives. In E. Rosa, A. Diekmann, T. Dietz, & C. Jaeger (Eds.), *Human footprints on the global environment: Threats to sustainability* (pp. 83−134). Cambridge: MIT Press.

Dietz, T., Stern, P., & Weber, E. (2013). Reducing carbon-based energy consumption through changes in household behavior. *Daedalus*, *142*(1), 78−89.

Diffenbaugh, N. S., Swain, D. L., & Touma, D. (2015). Anthropogenic warming has increased drought risk in California. *Proceedings of the National Academy of Sciences of the United States of America*, *112*(13), 3931−3936.

DoD (United States Department of Defense). (2010). *Quadrennial defense review report 2010*. Washington, DC: Department of Defense.

DoD (United States Department of Defense). (2014). *Quadrennial defense review 2014*. Washington, DC: Department of Defense.

Doney, S. C. (2010). The growing human footprint on coastal and open-ocean biogeochemistry. *Science (New York, N.Y.)*, *328*, 1512−1516.

Dorling, D. (2017). *The equality effect: Improving life for everyone*. Oxford: New Internationalist Publications Ltd.

Doswell, C. A. (2001). Severe Convective Storms—An Overview. *Meteorological Monographs*, *28*, 50.

Dunne, T., Kurki, M., & Smith, S. (2010). *International relations theories*. Oxford: Oxford University Press.

Dyer, H. (2001). Environmental security and international relations: The case for enclosure. *Review of International Studies*, *27*(3), 441−450.

Eastman, J. K., & Eastman, K. ,L. (2015). Conceptualizing a model of status consumption theory: An exploration of the antecedents and consequences of the motivation to consume for status. *The Marketing Management Journal*, *25*(1), 1−15.

Edwards, M., Bresnan, E., Cook, K. B., Heath, M., Helaouet, P., Nynam, C., ... Claire, W. (2014). Impacts of Climate Change on Plankton. *MCCIP Science Review*, 98−112.

Ehrhardt-Martinez, K., & Schor, J. B. (2015). Consumption and climate change. In R. Dunlap, & R. Brulle (Eds.), *Climate change and society: Sociological perspectives*. Oxford, England: Oxford University Press.

Eichler, T. P., Gaggini, N., & Pan, Z. (2013). Impacts of global warming on Northern Hemisphere winter storm tracks in the CMIP5 model suite. *Journal of Geophysical Research, 118*, 3919−3932.

Emanuel, K. (2006). Climate and tropical cyclone activity: A new model downscaling approach. *Journal of Climate., 19*(19), 4797−4802.

Enderlin, E. M., & Hamilton, G. S. (2014). Estimates of iceberg submarine melting from high-resolution digital elevation models: Application to Sermilik Fjord, East Greenland. *Journal of Glaciology, 60*, 1084−1092.

Fahey, D. W., Doherty, S. J., Hibbard, K. A., Romanou, A., & Taylor, P. C. (2017). *Physical drivers of climate change*. Climate Science Special Report: Fourth National Climate Assessment, Volume I (pp. 73−113). Washington, DC: United States Global Change Research Program.

Falkowski, P. G., Katz, M. E., Knoll, A. H., Quigg, A., Raven, J. A., Schofield, O., & Taylor, F. J. R. (2004). The evolution of modern eukaryotic phytoplankton. *Science (New York, N.Y.), 305*, 354−360.

Farmani, R., & Butler, D. (2014). Implications of urban form on water distribution systems performance. *Water Resources Management, 28*(1), 83−97.

Feldman, M., Hadjimichael, T., Lanahan, L., & Kemeny, T. (2016). The logic of economic development: A definition and model for development. *Environment and Planning C: Government and Policy, 34*, 5−21.

Feser, F., Barcikowska, M., Krueger, O., Schenk, F., Weisse, R., & Xia, L. (2015). Storminess over the North Atlantic and northwestern Europe—A review. *Quarterly Journal of the Royal Meteorological Society, 141*(687), 350−382.

Filho, W. L., Barbir, J., & Preziosi, R. (Eds.), (2019). *Handbook of climate change and biodiversity*. Cham: Springer.

Finney, S. C. (2016). The "Anthropocene" epoch: Scientific decision or political statement? *GSA Today: A Publication of the Geological Society of America., 26*(3−4), 4−10.

Fistory, F. R., & Pepyvrakis, E. (2016). *An introduction to climate change economics and policy*. London: Routledge.

Fleming, J. R., & Jankovic, V. (2011). Revisiting Klima. *Osiris, 26*(1), 1−11.

Floater, G., & Rode, P. (2014). Cities and the new climate economy: The transformative role of global urban growth. *The New Climate Economy*. Available online.

Flohn, H. (1977). Climate and energy: a scenario to a 21st century problem. *Climate Change, 1*(1), 5−20. Available online.

Foster, G., Royer, D., & Lunt, D. (2017). Future climate forcing potentially without precedent in the last 420 million years. *Nature Communication, 8*, 14845.

Francis, J. A., & Vavrus, S. J. (2015). Evidence for a waiver jet stream in response to rapid Arctic warming. *Environmental Research Letters., 10*(1), 14005.

Frank, J., & Martinez-Vazquez, J. (2014). *Decentralization and infrastructure in the global economy: From gaps to solutions*. Atlanta, GA: Georgia State University.

Franks, P. J., Adams, M. A., Amthor, J. S., Barbour, M. M., Berry, J. A., Ellsworth, D. S., . . . von Caemmerer, S. (2013). Sensitivity of plants to changing atmospheric CO_2 concentration: From the geological past to the next century. *New Phytologist, 197*, 1077−1094.

Friedman, C. S., Estes, R. M., Stokes, N. A., Burge, C. A., Hargove, G. S., Barber, B. J., . . . Reece, K. S. (2005). Herpes virus in juvenile Pacific oysters Crassostrea gigas from Tomales Bay, California, coincides with summer mortality episodes. *Diseases of Aquatic Organisms, 63*(1), 33−41.

Fu, W., Randerson, J. T., & Moore, J. K. (2016). Climate change impacts on net primary production (NPP) and export production (EP) regulated by increasing stratification and phytoplankton community structure in the CMIP5 models. *Biogeosciences., 13*, 5151−5170.

Fuchs, B. A., Svoboda, M. D., Wilhite, D. A., & Hayes, M. J. (2014). Drought indices for drought risk assessment in a changing climate. In S. Eslamian (Ed.), *Handbook of engineering hydrology* (Vol. 2). Boca Raton, FL: Taylor & Francis/CRC Press, *Modeling climate changes and variability*.

Fuller, B., & Romer, P. (2014). Urbanization as opportunity. *Policy research working paper*. Washington, D. C: World Bank.

Funk, C., Shukla, S., Hoell, A., & Livneh, B. (2015). Assessing the contributions of East African and West Pacific warming to the 2014 boreal spring East African drought [in "Explaining extremes of 2014 from a climate perspective"]. *Bulletin of the American Meteorological Society, 96*(12), S77−S82.

Gartzke, E. (2007). The capitalist peace. *American Journal of Political Science, 51*(1), 166−191.

Gauthier, S., Bernier, P., Kuuluvainen, T., Shvidenko, A. Z., & Schepaschenko, D. G. (2015). Boreal forest health and global change. *Science (New York, N.Y.), 349*(6250), 819−822.

Gelderloos, R., Straneo, F., & Katsman, C. A. (2012). Mechanisms behind the temporary shutdown of deep convection in the Labrador Sea: Lessons from the great salinity anomaly years 1968−71. *Journal of Climate, 25*, 6743−6755.

Gershunov, A., & Guirguis, K. (2012). California heat waves in the present and future. *Geophysical Research Letters, 39*, L18710.

Gillett, N. P., Weaver, A. J., Zwiers, F. W., & Flannigan, M. D. (2004). Detecting the effect of climate change on Canadian forest fires. *Geophysical Research Letters, 31*(18).

Good, P., Gregory, J. M., Lowe, J. A., & Andrews, T. (2013). Abrupt CO_2 experiments as tools for predicting and understanding CMIP5 representative concentration pathway projections. *Climate Dynamics, 40*, 1041−1053.

Gould, K., Pellow, D., & Schnaiberg, A. (2008). *The treadmill of production: Injustice and unsustainability in the global economy*. Boulder, CO: Paradigm.

Govekar, P., Jakob, C., & Catto, J. L. (2014). The relationship between clouds and dynamics in Southern Hemisphere extratropical cyclones in the real world and a climate model. *Journal of Geophysical Research Atmospheres, 119*, 6609−6628.

Gowda, M. V. R., Fox, J. C., & Magelky, R. D. (1997). Students' understanding of climate change: Insights for scientists and educators. *Bulletin of the American Meteorological Society, 78*(1), 2232−2240.

Gradstein, F. M., Ogg, J. G., Schmitz, M. D., & Ogg, G. M. (Eds.), (2012). *The geologic time scale*. Boston, MA: Elsevier BV.

Graziano, M., & Gillingham, K. (2015). Spatial patterns of solar photovoltaic system adoption: The influence of neighbors and the built environment. *Journal of Economic Geography, 15*(4), 815−839.

Greve, P., Orlowsky, B., Mueller, B., Sheffield, J., Reichstein, M., & Seneviratne, S. I. (2014). Global assessment of trends in wetting and drying over land. *Nature Geoscience, 7*(10), 716−721.

Grise, K. M., Son, S. W., Correa, G. J. P., & Polvani, L. M. (2014). The response of extratropical cyclones in the Southern Hemisphere to stratospheric ozone depletion in the 20th century. *Atmospheric Science Letters, 15*(1), 29−36.

Griskevicus, V., Tybur, J. M., & Van Den Bergh, B. (2010). Going green to be seen: Status, reputation and conspicuous conservation. *Journal of Personality and Social Psychology, 98*(3), 392−404.

Grootes, P. M., Stuiver, M., White, J. W. C., Johnsen, S., & Jouzel, J. (1993). Comparison of oxygen-isotope records from the Gisp2 and Grip Greenland ice cores. *Nature, 366*(6455), 552−554.

Grotizinger, John P., & Jordan, Thomas H. (2014). *Understanding Earth (7th ed)*. New York: W.H. Freeman & Company.

Grover, H. (2010). *Local response to global climate change: The role of local development plans in Climate Change management* (ProQuest dissertations and theses). Texas A&M University.

Gulev, S. K., Zolina, O., & Grigoriev, S. (2001). Extratropical cyclone variability in the Northern Hemisphere winter from the NCEP/NCAR reanalysis data. *Climate Dynamics, 17*, 795−809.

Güneralp, B., Zhou, Y., Ürge-Vorsatz, D., Gupta, M., Yu, S., Patel, P. L., et al. (2017). Global scenarios of urban density and its impacts on building energy use through 2050. *Proceedings of the National Academy of Sciences, 114*(34), 8945−8950.

Gupta, J. (2010). History of international climate change policy. *Wiley Interdisciplinary Reviews: Climate Change, 1*(5), 636−653.

Hall, A. (2004). The role of surface albedo feedback in climate. *Journal of Climate, 17,* 1550−1568.

Handmer, J., & Nalau, J. (2019). Understanding Loss and Damage in Pacific Small Island Developing States. In M. Reinhard, L. M. Bouwer, T. Schinko, S. Surminski, & J. A. Linnerooth-Bayer (Eds.), *Loss and damage from climate change: Concepts, methods and policy options* (pp. 365−380). Cham: Springer.

Hamilton, K., & Hartwick, J. (2017). Wealth and sustainability. In K. Hamilton, & C. Hepburn (Eds.), *National wealth: What is missing, why it matters?* Oxford, UK: Oxford University Press. Available online at https://oxford.universitypressscholarship.com/view/10.1093/oso/9780198803720.001.0001/oso-9780198803720-chapter-15.

Hardy, J. T. (2003). *Climate change: Causes, effects, and solutions.* London: John Wiley & Sons.

Hare, F. K. (1966). The concept of climate. *Geography (Sheffield, England), 51,* 99−110.

Harrington, C. (2016). The ends of the world: International relations and the anthropocene. *Millennium: Journal of International Studies, 44*(3), 478−498.

Hartmann, D. L., Klein Tank, A. M. G., Rusticucci, M., Alexander, L. V., Brönnimann, S., Charabi, Y., ... Zhai, P. M. (2013). Observations: Atmosphere and surface. Climate change 2013: The physical science basis. *Contribution of Working Group I to the Fifth Assessment Report of the Intergovernmental Panel on Climate Change* (pp. 159−254). Cambridge: Cambridge University Press.

Hawcroft, M. K., Shaffrey, L. C., Hodges, K. I., & Dacre, H. F. (2012). How much Northern Hemisphere precipitation is associated with extratropical cyclones? *Geophysical Research Letters, 39,* L24809.

Hawkins, E., & Sutton, R. (2009). The potential to narrow uncertainty in regional climate predictions. *Bulletin of the American Meteorological Society, 90*(8), 1095−1107.

Hays, J. D., Imbrie, J., & Shackleton, N. J. (1976). Variations in Earth's orbit: Pacemaker of the ice ages. *Science (New York, N.Y.), 194,* 1121−1132.

Hegerl, G., Black, E., Allan, R., Ingram, W., Polson, D., Trenberth, K., ... Zhang, X. (2015). Challenges in quantifying changes in the global water cycle. *Bulletin of the American Meteorological Society, 96,* 1097−1115.

Hibbard, K. A., Hoffman, F. M., Huntzinger, D., & West, T. O. (2017). *Changes in land cover and terrestrial biogeochemistry, . Climate science special report: Fourth national climate assessment* (Volume I). Washington, DC: United States Global Change Research Program: 277−302.

Hoeppe, P. (2016). Trends in weather related disasters—Consequences for insurers and society. *Weather and Climate Extremes, 11,* 70−79.

Holland, G., & Bruyere, C. L. (2014). Recent intense hurricane response to global climate change. *Climate Dynamics, 42*(3−4), 617−627.

Holland, T. G., Peterson, G. D., & Gonzalez, A. (2009). A cross-national analysis of how economic inequality predicts biodiversity loss. *Conservation Biology, 23*(5), 1304−1313.

Horton, R., Bader, D. A., Kushner, Y., Little, C., Blake, R., & Rosenzweig, C. (2015). New York City Panel on Climate Change 2015 Report: Climate observations and projections. *Annals of the New York Academy of Sciences, 1336,* 18−35.

Hooper, R. (2015). All hail the Anthropocene, the end of Holocene thinking. *New Scientist.* Available online.

Hoskins, B., & Woollings, T. (2015). Persistent extratropical regimes and climate extremes. *Current Climate Change Reports (Topical Collection on Extreme Events), 1*(3), 115−124.

Hubacek, K., Baiocchi, G., Feng, K., Munoz Castillo, R., Sun, L., & Xue, J. (2017a). Global carbon inequality. *Energy, Ecology and Environment, 2*(6), 361−369.

Huddart Kennedy, E., Krahn, H., & Krogman, N. (2014). Egregious emitters: Disproportionality in household carbon footprints. *Environment and Behavior*, 46(5), 535−555.

Hulme, M. (2008). The conquering of climate: Discourses of fear and their dissolution. *The Geographical Journal*, 174; 5−16.

Hulme, M. (2009). *Why we disagree about climate change: Understanding controversy, inaction and opportunity*. Cambridge: Cambridge University Press.

Hulme, M. (2011). Reducing the future to climate: A story of climate determinism and reductionism. *Osiris*, 26, 245−266.

Hulme, M. (2015). Climate and its changes: a cultural appraisal. *GEO: Geography and Environment*, 2(1), 1−11.

Hulme, M., Nerlich, B., & Pearce, W. (2013). Global warming is dead, long live global heating? *Making Science Public*. Available online.

Huschke, R. E. (Ed.), (1959). *Glossary of meteorology*. Boston, MA: American Meteorological Society.

ILO (International Labour Organization). *ILO Director-General address to the European Parliament*. (2011). Available online.

IPCC (Intergovernmental Panel on Climate Change). (2007). Climate change 2007: Synthesis report. *In Contribution of working groups, I, II and III to the Fourth Assessment Report of the Intergovernmental Panel on Climate Change*. Geneva: Intergovernmental Panel on Climate Change.

IPCC (Intergovernmental Panel on Climate Change). (2007a). Climate change 2007: The physical science basis. *Contribution of Working Group I to the Fourth Assessment Report of the IPCC*. Cambridge: Cambridge University Press.

IPCC (Intergovernmental Panel on Climate Change). (2007b). Climate change 2007: Impacts, adaptation and vulnerability. *Contribution of Working Group II to the Fourth Assessment Report of the IPCC*. Cambridge: Cambridge University Press.

IPCC (Intergovernmental Panel on Climate Change). (2012). Meeting report of the Intergovernmental Panel on Climate Change expert meeting on geoengineering. Potsdam: IPCC Working Group III Technical Support Unit, Potsdam Institute for Climate Impact Research.

IPCC (Intergovernmental Panel on Climate Change). (2013a). *Annex III: Glossary*. Climate change 2013: The physical science basis. *Contribution of Working Group I to the Fifth Assessment Report of the Intergovernmental Panel on Climate Change*. Cambridge: Cambridge University Press.

IPCC (Intergovernmental Panel on Climate Change). (2013b). *Summary for Policymakers*. Climate change 2013: The physical science basis. *Contribution of Working Group I to the Fifth Assessment Report of the Intergovernmental Panel on Climate Change* (pp. 3−29). Cambridge: Cambridge University Press.

IPCC (Intergovernmental Panel on Climate Change). (2013c). *Climate change 2013: The physical science basis. Contribution of Working Group I to the Fifth Assessment Report of the Intergovernmental Panel on Climate Change*. Cambridge: Cambridge University Press.

IPCC (Intergovernmental Panel on Climate Change). (2014a). *Climate change 2014: Impacts, adaptation, and vulnerability. Part A: Global and sectoral aspects. Contribution of Working Group II to the Fifth Assessment Report of the Intergovernmental Panel on Climate Change*. Cambridge: Cambridge University Press.

IPCC (Intergovernmental Panel on Climate Change) (2014b). *Climate change 2014: Impacts, adaptation and vulnerability: Part B: Regional aspects*. Contributions of Working Group II to the Fifth Assessment Report of the Intergovernmental Panel on Climate Change. Cambridge, UK: Cambridge University Press.

IPCC (Intergovernmental Panel on Climate Change). (2014c). Climate change 2014: Mitigation of climate change. *Contribution of Working Group III to the Fifth Assessment Report of the Intergovernmental Panel on Climate Change*. Cambridge: Cambridge University Press.

IPCC (Intergovernmental Panel on Climate Change). (2014d). Climate change 2014: Synthesis report. *Contribution of Working Groups I, II and III to the Fifth Assessment Report of the Intergovernmental Panel on Climate Change*. Geneva: IPCC.

IPCC (Intergovernmental Panel on Climate Change). (2018). *Summary for policymakers*. Global warming of 1.5°C. *An IPCC special report on the impacts of global warming of 1.5°C above pre-industrial levels and related global greenhouse gas emission pathways, in the context of strengthening the global response to the threat of climate change, sustainable development, and efforts to eradicate poverty*. Geneva: World Meteorological Organization Available at. Available from https://report.ipcc.ch/sr15/pdf/sr15_spm_final.pdf.

Islam, S. N. (2015). *Inequality and environmental sustainability*. UNDESA working paper no. 145. Available at http://www.un.org/esa/desa/papers/2015/wp145_2015.pdf.

Jackson, R. B., Canadell, J. G., Le Quéré, C., Andrew, R. M., Korsbakken, J. I., Peters, G. P., & Nakicenovic, N. (2016). Reaching peak emissions. *Nature Climate Change*, *6*, 7–10.

Jackson, T. (2009). *Prosperity without growth? The transition to a sustainable economy*. London: Sustainable Development Commission.

Johnson, G. C., Lyman, J. M., Boyer, T., Domingues, C. M., Ishii, M., Killick, R., ... Wijffels, S. E. (2016). [Global Oceans] Ocean heat content [in "State of the Climate in 2015"]. *Bulletin of the American Meteorological Society*, *97*, S66–S70.

Jolly, W. M., Cochrane, M. A., Freeborn, P. H., Holden, Z. A., Brown, T. J., Williamson, G. J., & Bowman, D. M. J. S. (2015). Climate-induced variations in global wildfire danger from 1979 to 2013. *Nature Communications*, *6*, 7537.

Jorgenson, A. (2014). Economic development and the carbon intensity of human well-being. *Nature Climate Change*, *4*, 186–189.

Jorgenson, A., & Clark, B. (2010). Assessing the temporal stability of the population/environment relationship: A cross-national panel study of carbon dioxide emissions, 1960–2005. *Population and Environment*, *32*, 27–41.

Jorgenson, A., Auerbach, D., & Clark, B. (2014). The (de-)carbonization of urbanization, 1960–2010. *Climatic Change*, *127*, 561–575.

Jorgenson, A. K., Fiske, S., Hubacek, K., Li, J., McGovern, T., Rick, T., ... Zycherman, A. (2019). Social science perspectives on drivers of and responses to global climate change. *WIREs Climate Change*, *10*(1), 1–17.

Joughin, I., Alley, R. B., & Holland, D. M. (2012). Ice-sheet response to oceanic forcing. *Science (New York, N.Y.)*, *338*, 1172–1176.

Kam, J., Knutson, T. R., Zeng, F., & Wittenberg, A. T. (2015). Record annual-mean warmth over Europe, the northeast Pacific, and the northwest Atlantic during 2014: Assessment of anthropogenic influence [in "Explaining extreme events of 2014 from a climate perspective"]. *Bulletin of the American Meteorological Society*, *96*(12), S61–S65.

Kang, S. M., Polvani, L. M., Fyfe, J. C., Son, S. W., Sigmond, M., & Correa, G. J. P. (2013). Modeling evidence that ozone depletion has impacted extreme precipitation in the austral summer. *Geophysical Research Letters*, *40*(15), 4054–4059.

Klima, K., & Morgan, M. G. (2015). Ice storm frequencies in a warmer climate. *Climatic Change*, *133*(2), 209–222.

Kiss, A. C., & Shelton, D. (2000). *International environmental law* (2nd ed.). New York: Transnational Publishers.

Knutson, T. R., & Tuleya, R. E. (2004). Impact of CO_2-induced warming on simulated hurricane intensity and precipitation: Sensitivity to the choice of climate model and convective parameterization. *Journal of Climate*, *17*(18), 3477–3495.

Knutson, T. R., McBride, J. L., Chan, J., Emanuel, K., Holland, G., Landsea, C., ... Sugi, M. (2010). Tropical cyclones and climate change. *Nature Geoscience*, *3*, 157–163.

Knutson, T. R., Sirutis, J. J., Zhao, M., Tuleya, R. E., Bender, M., Vecchi, G. A., ... Chavas, D. (2015). Global projections of intense tropical cyclone activity for the late twenty-first century from dynamical downscaling of CMIP5/RCP4.5 scenarios. *Journal of Climate*, *28*, 7203–7224.

Koop, G., & Tole, L. (2001). Deforestation, distribution and development. *Global Environmental Change, 11* (3), 193−202.

Koren, I., Remer, L. A., Kaufman, Y. J., Rudich, Y., & Martins, J. V. (2007). On the twilight zone between clouds and aerosols. Geophysics. *Research Letter, 34,* L08805.

Kossin, J. P., Emanuel, K. A., & Vecchi, G. A. (2014). The poleward migration of the location of tropical cyclone maximum intensity. *Nature, 509*(7500), 349−352.

Kossin, J. P., Olander, T. L., & Knapp, K. R. (2013). Trend analysis with a new global record of tropical cyclone intensity. *Journal of Climate, 26*(24), 9960−9976.

Kostov, Y., Armour, K. C., & Marshall, J. (2014). Impact of the Atlantic meridional overturning circulation on ocean heat storage and transient climate change. *Geophysical Research Letters, 41,* 2108−2116.

Kroeze, C., & Pulles, T. (2015). The importance of non-CO_2 greenhouse gases. *Journal of Integrative Environmental Sciences, 12*(1), 1−4.

Krueger, O., Schenk, F., Feser, F., & Weisse, R. (2013). Inconsistencies between long-term trends in storminess derived from the 20CR reanalysis and observations. *Journal of Climate, 26*(3), 868−874.

Kuenkel, P. (2019). *Stewarding sustainability transformations: An emerging theory and practice of SDG implementation.* Cham: Springer.

Kunkel, K. E., Vose, R. S., Stevens, L. E., & Knight, R. W. (2015). Is the monthly temperature climate of the United States becoming more extreme? *Geophysical Research Letters, 42*(2), 629−636.

Landsea, C. W., & Franklin, J. L. (2013). Atlantic Hurricane Database Uncertainty and Presentation of a New Database Format. *Monthly Weather Review, 141,* 3576−3592.

Lambert, S. J., & Fyfe, J. C. (2006). Changes in winter cyclone frequencies and strengths simulated in enhanced greenhouse warming experiments: Results from the models participating in the IPCC diagnostic exercise. *Climate Dynamics, 26*(7−8), 713−728.

Leckebusch, G. C., Koffi, B., Ulbrich, U., Pinto, J. G., Spangehl, T., & Zacharias, S. (2006). Analysis of frequency and intensity of European winter storm events from a multi-model perspective at synoptic and regional scales. *Climate Research, 31,* 59−74.

Leichenko, R. M., & Solecki, W. D. (2005). Exporting the American dream: Globalization and the creation of consumption landscapes in less developed country cities. *Regional Studies, 39,* 241−253.

Leiserowitz, A., Feinberg, G., Rosenthal, S., Smith, N., Anderson, A., Roser-Renouf, C., & Maibach, E. (2014). *What's in a name? Global warming vs. climate change.* New Haven, CT: Yale University Project on Climate Change Communication.

Lemery, J., & Auerbach, P. (2017). *Enviromedic: The impact of climate change on human health.* Lanham, MD: Rowman & Littlefield.

Lenton, T. M., Held, H., Kriegler, E., Hall, J. W., Lucht, W., Rahmstorf, S., & Schellnhuber, H. J. (2008). Tipping elements in the Earth's climate system. *Proceedings of the National Academy of Sciences of the United States of America, 105*(6), 1786−1793.

Le-Quéré, C., Andrew, R. M., Canadell, J. G., Sitch, S., Korsbakken, J. I., Peters, G. P., ... Zaehle, S. (2016). Global carbon budget 2016. *Earth System Science Data, 8,* 605−649.

Li, Y., Johnson, E., & Zaval, L. (2011). Local warming: Daily temperature change influences belief in global warming. *Psychological Science, 22*(4), 454−459.

Longshore, D. (2008). *Encyclopedia of hurricanes, typhoons and cyclones.* New York: The Facts on Files, Inc.

Lopez-Moreno, J. I., Pomeroy, J. W., Revuelto, J., & Vicente-Serrano, S. M. (2012). Response of snow processes to climate change: spatial variability in a small basin in the Spanish Pyrenees. *Hydrological Processes, 28*(17), 2637−2650.

Lopeman, M., Deodatis, G., & Franco, G. (2015). Extreme storm surge hazard estimation in lower Manhattan. *Natural Hazards, 78*(1), 355−391.

Lorenz, E. (1995). *Climate is what you expect* (pp. 1−33). Prepared for publication by NCAR. Unpublished, cited in Wendl, C. (2016). On defining climate and climate change. *The British Journal for the Philosophy of Science*, *67*(2), 337−364.

Luber, G., & Lemery, J. (Eds.), (2015). *Global climate change and human health: From science to practice*. San Francisco, CA: Jossey-Bass.

Mahowald, N. M., Kloster, S., Engelstaedter, S., Moore, J. K., Mukhopadhyay, S., McConnell, J. R., ... Zender, C. S. (2010). Observed 20th century desert dust variability: Impact on climate and biogeochemistry. *Atmospheric Chemistry and Physics.*, *10*(22), 10875−10893.

Mann, M. E., & Emanuel, K. A. (2006). Atlantic hurricane trends linked to climate change. *Eos Transactions AGU*, *87*(24), 233241.

Marciano, C. G., Lackmann, G. M., & Robinson, W. A. (2015). Changes in U.S. East Coast cyclone dynamics with climate change. *Journal of Climate*, *28*(2), 468−484.

Martine, G., Alves, J. E., & Cavenaghi, S. (2013). *Urbanization and fertility decline: Cashing in on structural change*. London: International Institute for Environment and Development Available at. Available from http://pubs.iied.org/pdfs/10653IIED.pdf.

Masson-Delmotte, V., Schulz, M., Abe-Ouchi, A., Beer, J., Ganopolski, A., González Rouco, J. F., ... Timmermann, A. (2013). *Information from paleoclimate archives. In* Climate change 2013: The physical science basis. *Contribution of Working Group I to the Fifth Assessment Report of the Intergovernmental Panel on Climate Change* (pp. 383−464). Cambridge: Cambridge University Press.

Matthai, H. F. (1979). Hydrologic and human aspects of the 1976−1977 drought. In: *United States Geological Survey professional paper 1130*. Washington, DC.

McDonald, M. (2013). Discourses of climate security. *Political Geography*, *33*, 42−51.

McGee, J., Ergas, C., Greiner, P., & Clement, M. (2017). How do slums change the relationship between urbanization and the carbon intensity of well-being? *PLoS One.*, *12*(12), e0189024.

McMichael, A. J. (2019). *Climate change and health of nations: Famines, fevers and the fate of populations*. New York: Oxford University Press.

Mechler, R., Bouwer, L. M., Schinko, T., Surminski, S., & Linnerooth-Bayer, J. A. (Eds.), (2019). *Loss and damage from climate change: Concepts, methods and policy options*. Cham: Springer.

Melchiorri, M., Florczyk, A., Freire, S., Schiavina, M., Pesaresi, M., & Kemper, T. (2018). Unveiling 25 years of planetary urbanization with remote sensing: Perspectives from the global human settlement layer. *Remote Sensing.*, *10*(5), 768.

Melillo, J. M., Richmond, T. C., & Yohe, G. W. (Eds.), (2014). *Climate change impacts in the United States: The third national climate assessment*. Washington, DC: United States Global Change Research Program.

Mol, A. (2003). *Globalization and environmental reform: The ecological modernization of the global economy*. Cambridge: MIT Press.

Mol, A., Spaargaren, G., & Sonnenfeld, D. (2014). Ecological modernization theory: Taking stock, moving forward. In S. Lockie, D. Sonnenfeld, & D. Fisher (Eds.), *The Routledge international handbook of social and environmental change* (pp. 15−30). New York: Routledge.

Moser, S. C.Communicating Climate Change. (2010). History, challenges, process and future directions. *Wiley Interdisciplinary Reviews: Climate Change*, *1*(1), 31−53.

Munich, R. E. *Natural disasters: The year in figures*. (2017). <https://natcatservice.munichre.com/events/1.filter = eyJ5ZWFyRnJvbSI6MTk4MCwieWVhclRvIjoyMDE3fQ%3D%3D&type = >.

Myhre, G., Shindell, D., Bréon, F.-M., Collins, W., Fuglestvedt, J., Huang, J., ... Zhang, H. (2013). *Anthropogenic and natural radiative forcing*. Climate change 2013: The physical science basis. *Contribution of Working Group I to the Fifth Assessment Report of the Intergovernmental Panel on Climate Change* (pp. 659−740). Cambridge: Cambridge University Press.

Mysiak, J., Surminski, S., Thieken, A., Mechler, R., & Aerts, J. (2016). Brief communication: Sendai framework for disaster risk reduction – Success or warning sign for Paris? *Natural Hazards and Earth System Sciences*, *16*(10), 2189–2193.

NASEM (National Academies of Sciences, Engineering, and Medicine). (2016). *Attribution of extreme weather events in the context of climate change*. Washington, DC: The National Academies Press.

NDCM (National Drought Mitigation Center). (2014a). *Drought basics—What is drought?*. Lincoln, NE: NDCM. Available at http://drought.unl.edu/DroughtBasics/WhatisDrought.aspx.

NDCM (National Drought Mitigation Center). (2014b). *Drought basics—Types of drought*. Lincoln, NE: NDCM. Available at http://drought.unl.edu/DroughtBasics/TypesofDrought.aspx.

Nixon, R. (2011). *Slow violence and environmentalism of the poor*. Cambridge, MA: Harvard University Press.

NOAA (National Oceanic and Atmospheric Administration). (2016). *Global temperature*. US NOAA. Available at https://www.ncdc.noaa.gov/sotc/global/201613.

NOAA (National Oceanic and Atmospheric Administration). *State of the climate: Global climate report for annual 2015*. (2015). <https://www.ncdc.noaa.gov/sotc/global/201513>.

NOAA (United States National Oceanic and Atmospheric Administration). *Tropical cyclone—A preparedness guide*. (2011). <https://www.weather.gov/media/zhu/ZHU_Training_Page/tropical_stuff/tropical_cyclone_brochure/TropicalCyclones.pdf>

NRC (National Research Council). (2002). *Abrupt climate change: Inevitable surprises*. Washington, DC: National Academic Press.

NRC (National Research Council). (2012). *Climate and social stresses: Implications for security analysis*. Washington, DC: National Academies Press.

NRC (National Research Council). (2013). *Abrupt impacts of climate change: Anticipating surprises*. Washington, DC: The National Academies Press.

O'Gorman, Paul A. (2014). Contrasting responses of mean and extreme snowfall to climate change. *Nature*, *512*(7515), 416–418.

O'Gorman, P. A. (2010). Understanding the varied response of the extratropical storm tracks to climate change. *Proceedings of the National Academy of Sciences of the United States of America*, *107*(45), 19176–19180.

O'Neill, B. C., Oppenheimer, M., Warren, R., Hallegatte, S., Kopp, R. E., Pörtner, H. O., et al. (2017). IPCC reasons for concern regarding climate change risks. *Nature Climate Change*, *7*(1), 28–37.

OECD (Organization for Economic Co-operation and Development). (2002). *Indicators to measure decoupling of environmental pressure from economic growth*. Paris, France: OECD.

OECD (Organization for Economic Cooperation and Development). (2017). *Investing in climate, investing in growth*. Paris: OECD.

OECD (organization of Economic Cooperation and Development). (2018). *Implementing the paris agreement: Remaining challenge and the role of OECD*. Paris: OECD.

Olmstead, S., & Stavins, R. N. (2012). Three key elements of a post-2012 international climate policy architecture. *Review of Environmental Economics and Policy*, *6*(1), 65–85.

ONRG (Ojos Negros Research Group). *Drought facts*. (2006). <http://ponce.sdsu.edu/three_issues_droughtfacts01.html>.

Ormerod, K. J., & McAfee, S. (2017). *Nevada's weather and climate. Fact sheet-17-04*. Las Vegas: University of Nevada.

Oren, R., Ellsworth, D. S., Johnsen, K. H., Phillips, N., Ewers, B. E., Maier, C., ... Katul, G. G. (2001). Soil fertility limits carbon sequestration by forest ecosystems in a CO_2-enriched atmosphere. *Nature*, *411*, 469–471.

Otto, F. E. L., Boyd, E., Jones, R. G., Cornforth, R. J., James, R., Parker, H. R., & Allen, M. R. (2015). Attribution of extreme weather events in Africa: a preliminary exploration of the science and policy implications. *Nature Climate Change, 132*, 531–543.

Overpeck, J. T., & Cole, J. E. (2006). Abrupt change in Earth's climate system. *Annual Review of Environment and Resources., 31*, 1–31.

Palmer, T. N. (1993). Extended-range atmospheric prediction and the Lorenz model. *Bulletin of the American Meteorological Society, 74*(1), 49–65.

Parry, M., Rosenzweig, C., Iglesias, A., Fischer, G., & Livermore, M. (1999). Climate change and world food security: A new assessment. *Global Environmental Change, 9*, S51–S67.

Passow, U., & Carlson, C. A. (2012). The biological pump in a high CO_2 world. *Marine Ecology Progress Series, 470*, 249–271.

Pattberg, P., & Zelli, F. (Eds.), (2016). *Environmental politics and governance in the Anthropocene: Institutions and legitimacy in a complex world.* London: Routledge.

Pielke, A. R. (2005). Misdefining "climate change": consequences for science and action. *Environmental Science & Policy, 8*(6), 548–561.

Pielke, R. A., Sr., Maryland, G., Betts, R. A., Chase, T. N., Eastman, J. L., Niles, J. O., ... Running, S. (2002). The influence of land-use change and landscape dynamics on the climate system: Relevance to climate change policy beyond the radiative effect of greenhouse gases. *Philosophical Transactions of the Royal Society, 360*, 1705–1719.

Pithan, F., & Mauritsen, T. (2014). Arctic amplification dominated by temperature feedbacks in contemporary climate models. *Nature Geoscience, 7*, 181–184.

Glossary. Annex III. In S. Planton (Ed.), Climate change 2013: The physical science basis. *Contribution of Working Group I to the Fifth Assessment Report of the Intergovernmental Panel on Climate Change.* Cambridge: Cambridge University Press.

Plass, G. N. (1956). The carbon dioxide theory of climatic change. *Tellus, 8*(2), 140–154.

Pruneau, D., Moncton, U., Liboiron, L., & Vrain, E. (2001). People's idea about climate change: A source of inspiration for the creation of educational programs. *Canadian Journal of Environmental Education, 6*(1), 58–76.

Rahmstorf, S., Box, J. E., Feulner, G., Mann, M. E., Robinson, A., Rutherford, S., & Schaffernicht, E. J. (2015). Exceptional twentieth-century slowdown in Atlantic Ocean overturning circulation. *Nature Climate Change, 5*, 475–480.

Ramaswami, A., Russell, A. G., Culligan, P. J., Sharma, K. R., & Kumar, E. (2016). *Meta*-principles for developing smart, sustainable, and healthy cities. *Science (New York, N.Y.), 352*(6288), 940–943.

Rasmussen, L. L. (2011). Foreword. In K. Richardson, W. Steffan, D. Liverman, et al. (Eds.), *Climate change: Global risks, challenges and decisions.* Cambridge: Cambridge University Press.

Richardson, M., Cowtan, K., & Millar, R. J. (2018). Global temperature definition affects achievement of long-term climate goals. *Environmental Research Letters, L13*, 054004.

Rhein, M., Rintoul, S. R., Aoki, S., Campos, E., Chambers, D., Feely, R. A., ... Wang, F. (2013). Observations: Ocean. Climate Change 2013: The Physical Science Basis. *Contribution of Working Group I to the Fifth Assessment Report of the Intergovernmental Panel on Climate Change* (pp. 255–316). Cambridge, UK: Cambridge University Press.

Rignot, E., & Thomas, R. H. (2002). Mass balance of polar ice sheets. *Science (New York, N.Y.), 297*, 1502–1506.

Romero-Lankao, P., Smith, J. B., Davidson, D. J., Diffenbaugh, N. S., Kinney, P. L., Kirshen, P., ... Ruiz, L. V. (2014). *North America*. Climate change 2014: Impacts, adaptation, and vulnerability. *Part B: Regional aspects. Contribution of Working Group II to the Fifth Assessment Report of the Intergovernmental Panel on Climate Change.* Cambridge: Cambridge University Press.

Rosa, E., & Dietz, T. (2012). Human drivers of national greenhouse gas emissions. *Nature Climate Change*, 2, 581–586.

Rosa, E., Rudel, T., York, R., Jorgenson, A., & Dietz, T. (2015). The human (anthropogenic) driving forces of global climate change. In R. Dunlap, & R. Brulle (Eds.), *Climate change and society: Sociological perspectives* (pp. 32–60). Oxford, England: Oxford University Press.

Rosenberg, N.J. (Ed.). (1979). Drought in the great plains—Research on impacts and strategies, In: *Proceedings of the workshop on research in great plains drought management strategies*, March 26–28, 1979, University of Nebraska, Lincoln, NE. Littleton, CO: Water Resources Publications.

Russell, C. S., & Landsberg, H. H. (1971). International environmental problems—a taxonomy. *Science (New York, N.Y.)*, 172(3990), 1307–1314.

Salat, S., Chen, M., & Liu, F. (2014). *Planning energy efficient and livable cities: Energy efficient cities*. Energy Sector Management Assistance Program. Available online.

SBR (Symons' British Rainfall). (1887). British Rainfall—A Monthly and Annual Acount of British Rainfall. London: G. SHield Printer, 1888. A copy of this is available in University of Dundde Archives.

Schanze, J. J., Schmitt, R. W., & Yu, L. L. (2010). The global oceanic freshwater cycle: A state-of-the-art quantification. *Journal of Marine Research*, 68, 569–595.

Schor, J. B. (1998). *The overspent American: Upscaling, downshifting and the new consumer*. New York: Basic Books.

Schreiner, C., Henriksen, E., & Kirkeby Hansen, P. J. (2005). Climate education: Empowering today's youth to meet tomorrow's challenges. *Studies in Science Education*, 41, 3–50.

Schroder, E., & Storm, S. (2018). *Economic growth and carbon emissions: The road to 'hothouse earth' is paved with good intentions. Working paper 84*. New York: Institute for New Economic Thinking.

Screen, J. A., Deser, C., & Sun, L. (2015). Reduced risk of North American cold extremes due to continued Arctic sea ice loss. *Bulletin of the American Meteorological Society*, 96(9).

Seiler, C., & Zwiers, F. W. (2015a). How well do CMIP5 climate models reproduce explosive cyclones in the extratropics of the Northern Hemisphere? *Climate Dynamics*, 46(3–4), 1241–1256.

Seiler, C., & Zwiers, F. W. (2015b). How will climate change affect explosive cyclones in the extratropics of the Northern Hemisphere? *Climate Dynamics*, 46(11–12), 3633–3644.

Sejas, S. A., Cai, M., Hu, A., Meehl, G. A., Washington, W., & Taylor, P. C. (2014). Individual feedback contributions to the seasonality of surface warming. *Journal of Climate*, 27, 5653–5669.

Sen, A. (2000). *Development as freedom*. New York: Anchor Books.

Seneviratne, S. I., Nicholls, N., Easterling, D., Goodess, C. M., Kanae, S., Kossin, J., ... Zhang, X. (2012). *Changes in climate extremes and their impacts on the natural physical environment*. Managing the risks of extreme events and disasters to advance climate change adaptation. *A special report of Working Groups I and II of the Intergovernmental Panel on Climate Change*. Cambridge: Cambridge University Press.

Seppälä, R. (2009). A global assessment on adaptation of forests to climate change. *Scandinavian Journal of Forest Research*, 24, 469–472.

Seto, K. C., Davis, S. J., Mitchell, R. B., Stokes, E. C., Unruh, G., & Ürge-Vorsatz, D. (2016). Carbon lockin: Types, causes, and policy implications. *Annual Review of Environment and Resources*, 41, 425–452.

Seto, K. C., Dhakal, S., Bigio, A., Blanco, H., Delgado, G. C., Dewar, D., et al. (2014). *Human settlements, infrastructure and spatial planning*. Climate change 2014: Mitigation of climate change. *Contribution of Working Group III to the Fifth Assessment Report of the Intergovernmental Panel on Climate Change*. Cambridge: Cambridge University Press.

Severinghaus, J., & Brook, E. (1999). Abrupt climate change at the end of the last glacial period inferred from trapped air in polar ice. *Science (New York, N.Y.)*, 286, 930–934.

Shabecoff, B. (1988). Global warming has begun, expert tells senate. *The New York Times*. Available at http://nyti.ms/1g7Z9bZ (24 January 1988).

Shepardsona, D. P., Niyogib, D., Roychoudhury, A., & Hirschd, A. (2012). Conceptualizing climate change in the context of a climate system: Implications for climate and environmental education. *Environmental Education Research, 18*(3), 323–352.

Sherwood, S. C., Roca, R., Weckwerth, T. M., & Andronova, N. G. (2010). Tropospheric water vapour, convection and climate. *Review of Geophysics, 48*, RG2001.

Shiogama, H., Stone, D. A., Nagashima, T., Nozawa, T., & Emori, S. (2013a). On the linear additivity of climate forcing-response relationships at global and continental scales. *International Journal of Climatology, 33*(11), 2542–2550.

Sillmann, J., Kharin, V. V., Zwiers, F. W., Zhang, X., & Bronaugh, D. (2013). Climate extremes indices in the CMIP5 multimodel ensemble: Part 2. Future climate projections. *Journal of Geophysical Research-Atmospheres, 118*(6), 2473–2493.

Sillmann, J., Kharin, V. V., Zhang, X., Zwiers, F. W., & Bronaugh, D. (2013). Climate extremes indices in the CMIP5 multimodel ensemble: Part 1. Model evaluation in the present climate. *Journal of Geophysical Research-Atmospheres, 118*(4), 1716–1733.

Smith, D. E., Harrison, S., & Jordan, J. T. (2011). The early Holocene sea level rise. *Quaternary Science Reviews, 30*(15–16), 1846–1860.

Smith, W. K., Reed, S. C., Cleveland, C. C., Ballantyne, A. P., Anderegg, W. R. L., Wieder, W. R., ... Running, S. W. (2016). Large divergence of satellite and Earth system model estimates of global terrestrial CO_2 fertilization. *Nature Climate Change, 6*, 306–310.

Solomon, S., Rosenlof, K. H., Portmann, R. W., Daniel, J. S., Davis, S. M., Sanford, T. J., & Plattner, G.-K. (2010). Contributions of stratospheric water vapor to decadal changes in the rate of global warming. *Science (New York, N.Y.), 327*(5970), 1219–1223.

Solomon, S., Qin, D., Manning, M., Chen, Z., Marquis, M., Averyt, K. B., ... Miller, H. L. (2007). Climate change 2007: The physical science basis. *Contribution of Working Group I to the Fourth Assessment Report of the Intergovernmental Panel on Climate Change*. Cambridge: Cambridge University Press.

Spratt, D., & Dunlop, I. (2019). *Existential climate-related security risk: A scenario approach*. Melbourne: National Centre for Climate Restoration.

Steffen, W., Leinfelder, R., Zalasiewicz, J., Waters, C. N., Williams, M., Summerhayes, C., ... Schellnhuber, H. J. (2016). Stratigraphic and Earth System approaches to defining the Anthropocene. *Earth's Future, 4*(8), 324–345.

Stott, P. A., & Kettleborough, J. A. (2002). Origins and estimates of uncertainty in predictions of twenty-first century temperature rise. *Nature, 416*, 723–726.

Stott, P. A., Stone, D. A., & Allen, M. R. (2004). Human contribution to the European heatwave of 2003. *Nature, 432*(7017), 610–614.

Struzik, E. (2017). *Firestorm: How wildfires will shape our futures*. Washington, DC: Island Press.

Subrahmanyam, V. P. (1967). Incidence and Spread of Continental Drought. *Reports on WMO/IHD Projects, 2*, World Meteorological Organization, International Hydrological Decade. Geneva, Switzerland: WMO.

Summerhayes, Coli P. (2015). *Earth's Climate Evolution*. West Sussex, UK: Wiley Blackwell.

Sun, Y., Solomon, S., Dai, A., & Portmann, R. (2006). How often does it rain? *Journal of Climate, 19*(6), 916–934.

Taylor, P. C., Cai, M., Hu, A., Meehl, J., Washington, W., & Zhang, G. J. (2013). A decomposition of feedback contributions to polar warming amplification. *Journal of Climate, 26*(18), 7023–7043.

Taylor, P. C., Ellingson, R. G., & Cai, M. (2011). Geographical distribution of climate feedbacks in the NCAR CCSM3.0. *Journal of Climate, 24*, 2737–2753.

Taylor, P. C., Kato, S., Xu, K.-M., & Cai, M. (2015). Covariance between Arctic sea ice and clouds within atmospheric state regimes at the satellite footprint level. *Journal of Geophysical Research Atmospheres, 120*, 12656–12678.

Thombs, R. (2018a). The transnational tilt of the treadmill and the role of trade openness on carbon emissions: A comparative international study, 1965–2010. *Sociological Forum, 33*(2), 422–442.

Titley, D. M. (2016). Preface. National Academies of Sciences, Engineering, and Medicine. 2016. Attribution of extreme weather events in the context of climate change. Washington, DC: The National Academies Press.

Tippett, M. K., Allen, J. T., Jensini, V. A., & Brooks, H. A. (2015). Climate and hazardous convective weather. Current Climate Change Reports, 1, 60–73.

Tippett, M. K. (2014). Changing volatility of U.S. annual tornado reports. Geographical Research Letters, 41 (19), 6956–6961.

Trapp, R. J., & Hoogewind, K. A. (2016). The Realization of Extreme Tornadic Storm Events under Future Anthropogenic Climate Change. Journal of Climate, 29(14), 5251–5265.

Todorov, A. V. (1986). Reply. Journal of Applied Climate and Meteorology, 25, 258–259.

Trenberth, K. E., Dai, A. G., van der Schrier, G., Jones, P. D., Barichivich, J., Briffa, K. R., & Sheffield, J. (2014). Global warming and changes in drought. Nature Climate Change., 4(1), 17–22.

UN-Habitat. (2012). State of the World's Cities, 2012–2013, Prosperity of Cities. Nairobi, Kenya: UN-Habitat.

UN-DEAS (United Nations-Department of Economic and Social Affairs). (2016). The UN Sustainable Development Goals Report 2016. New York: UN.

United States Environment Protection Agency (US-EPA). (2013). Report on the 2013 US Environmental Protection Agency (EPA) International Contamination Research and Development Conference. Washington, D.C.: US-EPA.

UN Habitat III (United Nations Human Settlements Programme). (2015a). Habitat III issue papers 1— Inclusive cities. Nairobi. Available online.

UN-DESA (United Nations-Department of Economic and Social Affairs). World's cities 2018, data booklet. (2018). Available online.

UN-DESA (United Nations-Department of Social and Economic Affairs). (2014). World urbanization prospects: 2014 Revision, highlights. New York.

UNDP (United Nations Development Programme). (2010). The outcomes of the Copenhagen: The negotiations and the accord. New York: UNDP.

UNEP (United Nations Environment Programme). (2011). Towards a green economy: pathways to sustainable development and poverty eradication. Nairobi: UNEP.

UNEP (United Nations Environment Programme). (2016). A framework for shaping sustainable lifestyles – Determinants and strategies. Nairobi: UNEP.

UNEP (United Nations Environment Programme). (2019). Global environment outlook—GEO-6: Healthy planet, healthy people. Nairobi: UNEP.

UNESCAP (United Nations Economic and Social Commission for Asia and the Pacific) & UN-Habitat (United Nations Human Settlement Programme). The State of Asian and Pacific cities 2015: Urban transformations. Shifting from quantity to quality. (2015). Available Online.

UNESCAP (United Nations Economic and Social Commission for Asia and the Pacific). (2018). Inequality in Asia and the Pacific in the era of the 2030 agenda for sustainable development. Bangkok: UNESCAP.

UNFCCC (n.d.a). The Doha amendment to the Kyoto Protocol. UNFCCC. Available at https://unfccc.int/files/kyoto_protocol/application/pdf/kp_doha_amendment_english.pdf.

UNFCCC (n.d.b). What is Kyoto Protocol?. UNFCCC. Available at https://unfccc.int/kyoto_protocol.

UNFCCC (United Nations Framework Convention on Climate Change). (1992). Text of the UNFCCC Treaty . United Nations. Available at https://unfccc.int/resource/docs/convkp/conveng.pdf.

UNFCCC (United Nations Framework Convention on Climate Change). Convention UNFCCC, Article 1 – Paragraph 2. (1994). <http://unfccc.int/essential_background/convention/background/items/2536.php>.

UNFCCC (United Nations Framework Convention on Climate Change). Fact sheet: Climate Change science— The status of Climate Change science today. (2011). <http://unfccc.int/files/press/backgrounders/application/pdf/press_factsh_science.pdf>.

UNFCCC (United Nations Framework Convention on Climate Change). *Paris Agreement.* (2015). <https://unfccc.int/sites/default/files/english_paris_agreement.pdf>.
UN-Habitat (United Nations Human Settlement Programme). (2014). Urbanization and sustainable development: Towards a new United Nations urban agenda. CEB/2014/HLCP-28/CRP.5. In: *New york agenda item 6: New UN urban agenda.* https://habnet.unhabitat.org/sites/default/files/oo/urbanization-andsustainable-development.pdf.
UN-Habitat (United Nations Human Settlement Programme). (2015a). *World population prospects: Key findings and advanced tables. The 2015 revision.* New York.
UN-Habitat (United Nations Human Settlement Programme). (2016). *World cities report 2016: Urbanization and development—Emerging futures.* Nairobi.
UNICEF (United Nations Children's Fund) & WHO (World Health Organization). (2017). *Progress on drinking water, sanitation and hygiene: 2017 Update and sustainable development goal baselines.* Geneva.
UNICEF (United Nations Children's Fund). (2018a). *Malnutrition.* <https://data.unicef.org/topic/nutrition/malnutrition/>.
UNICEF (United Nations Children's Fund). *Education.* (2018b). <https://www.unicef.org/education>.
USCCSP (United States Climate Change Science Program). (2008). *Synthesis and assessment product 3.4: Abrupt climate change.* Washington, DC: United States Department of the Interior.
USGCRP (United States Global Climate Research Program). (2017). *Climate science special report: Fourth national climate assessment* (Vol. I). Washington, DC: United States Global Change Research Program.
USGS (United States Geological Survey). (2014). *What is drought?.* North Dakota Water Science Center, USGS. Available at http://water.usgs.gov/drought/faqs/faq1.html.
van den Broeke, M., Bamber, J., Ettema, J., Rignot, E., Schrama, E., van de Berg, W. J., ... Wouters, B. (2009). Partitioning recent Greenland mass loss. *Science (New York, N.Y.), 326,* 984–986.
van Oldenborgh, G. J., Stephenson, D. B., Sterl, A., Vautard, R., Yiou, P., Drijfhout, S. S., ... van den Dool, H. (2015). Correspondence: Drivers of the 2013/14 winter floods in the UK. *Nature Climate Change, 5*(6), 490–491.
Vano, J. A., Udall, B., Cayan, D. R., Overpeck, J. T., Brekke, L. D., Das, T., ... Lettenmaier, D. P. (2014). Understanding uncertainties in future Colorado River streamflow. *Bulletin of the American Meteorological Society, 95*(1), 59–78.
Vano, J. A., Das, T., & Lettenmaier, D. P. (2012). Hydrologic sensitivities of Colorado River runoff to changes in precipitation and temperature. *Journal of Hydrometeorology, 13*(3), 932–949.
Vaughan, D. G., Comiso, J. C., Allison, I., Carrasco, J., Kaser, G., Kwok, R., et al. (2013). *Observations: Cryosphere.* Climate change 2013: The physical science basis. *Contribution of Working Group I to the Fifth Assessment Report of the Intergovernmental Panel on Climate Change* (pp. 317–382). Cambridge: Cambridge University Press.
Vicente-Serrano, S. M., Azorin-Molina, C., Sanchez-Lorenzo, A., Morán-Tejeda, E., Lorenzo-Lacruz, J., Revuelto, J., ... Espejo, F. (2014). Temporal evolution of surface humidity in Spain: Recent trends and possible physical mechanisms. *Climate Dynamics, 42,* 2655–2674.
Vlassopoulos, C. A. (2012). Competing definition of climate change and the post-Kyoto negotiations. *International Journal of Climate Change Strategies and Management, 4*(1), 104–118.
Vujica, Y., Hall, W. A., & Salas, J. D., (Eds.) (1977). Drought research needs. In: *Proceedings of the conference on drought research needs,* December 12–15, 1977, Colorado State University, Fort Collins, CO.
Walsh, J., Wuebbles, D., Hayhoe, K., Kossin, J., Kunkel, K., Stephens, G., ... Somerville, R. (2014). *Our changing climate. Climate change impacts in the United States: The third national climate assessment.* Washington, DC: United States Global Change Research Program.
Walsh, K. J. E., Camargo, S., Vecchi, G., Daloz, A. S., Elsner, J., Emanuel, K., ... Henderson, N. (2015). Hurricanes and climate: The United States CLIVAR working group on hurricanes, B. *American Meteorological Society, 96,* 997–1017.

Wang, X. L., Feng, Y., Chan, R., & Isaac, V. (2016). Inter-comparison of extra-tropical cyclone activity in nine reanalysis datasets. *Atmospheric Research, 181*, 133−153.

Wang, X. L., Swail, V. R., & Zwiers, F. W. (2006). Climatology and changes of extratropical cyclone activity: Comparison of ERA-40 with NCEP−NCAR reanalysis for 1958−2001. *Journal of Climate, 19*, 3145−3166.

Wang, X., Thompson, D. K., Marshall, G. A., Tymstra, C., Carr, R., & Flannigan, M. D. (2015). Increasing frequency of extreme fire weather in Canada with climate change. *Climate Change, 130*, 573−586.

Ward, D. S., & Mahowald, N. M. (2015). Local sources of global climate forcing from different categories of land use activities. *Earth System Dynamics, 6*, 175−194.

Ward, D. S., Mahowald, N. M., & Kloster, S. (2014). Potential climate forcing of land use and land cover change. *Atmospheric Chemistry and Physics, 14*, 12701−12724.

Watanabe, T., Kapur, S., Aydın, M., Kanber, R., & Akça, E. (Eds.), (2019). *Climate change impacts on basin agro-ecosystems*. Cham: Springer.

Weart, S. (2009). *The discovery of global warming*. American Institute of Physics. Available online.

Weart, S. (2010). *Introduction: A hyperlinked history of climate change science*. American Institute of Physics. Available online.

WEF (World Economic Forum). (2006). *The global risks report 2006*. Geneva: World Economic Forum.

WEF (World Economic Forum). (2014). *The global risks report 2014*. Geneva: World Economic Forum.

WEF (World Economic Forum). (2015). *The global risks report 2015*. Geneva: World Economic Forum.

WEF (World Economic Forum). (2019). *The global risks report 2019*. Geneva: World Economic Forum.

Wehner, M. F., Prabhat., Reed, K., Stone, D., Collins, W. D., & Bacmeister, J. (2015). Resolution dependence of future tropical cyclone projections of CAM5.1 in the United States CLIVAR Hurricane Working Group idealized configurations. *Journal of Climate, 28*, 3905−3925.

Wehner, M. F., Reed, K. A., & Zarzycki, C. M. (2017). High-resolution multi-decadal simulation of tropical cyclones, Chapter 8. In J. Colins, & K. Walsh (Eds.), *Hurricanes and climate change* (pp. 187−207). Springer.

Weldon, D. (2016). Globalisation is fraying: Look under the Elephant Trunk. *Bull Market*, 13, June 2016.

Wenzel, S., Cox, P. M., Eyring, V., & Friedlingstein, P. (2016). Projected land photosynthesis constrained by changes in the seasonal cycle of atmospheric CO_2. *Nature, 538*, 499−501.

Werndl, C. (2016). On defining climate and climate change. *The British Journal for the Philosophy of Science, 67*, 337−364.

Werrell, C. E., & Femia, F. (2015). Climate change as threat multiplier: Understanding the broader nature of the risk. In: *Briefer number 25*. Available online.

Wieder, W. R., Cleveland, C. C., Smith, W. K., & Todd-Brown, K. (2015). Future productivity and carbon storage limited by terrestrial nutrient availability. *Nature Geoscience, 8*, 441−444.

Wigley, T. M. L., Jones, P. D., & Kelly, P. M. (1981). Global warming? *Nature, 291*(5813), 285.

Wilhite, D. A. (2012). *Drought assessment management and planning: theory and case studies, business and economics*. Dordrecht: Springer Science and Business Media.

Willett, K. M., Jones, P. D., Thorne, P. W., & Gillett, N. P. (2010). A comparison of large-scale changes in surface humidity over land in observations and CMIP3 GCMs. *Environmental Research Letters, 5*, 025210.

Williams-Sether, T., Macek-Rowland, K. M., & Emerson, D. G. (1994). Climatic and hydrologic aspects of the 1988−92 drought and the effect on people and resources of North Dakota. In: *North Dakota State Water Commission, Water Resources Investigation 29*. Bismarck, ND.

Wilhite, D. A., & Glantz, M. H. (1985). Understanding the drought phenomenon: The role of definitions. University of Nebraska − Lincoln.

Wilson, K. M. (2000). Communicating climate change through the media: Predictions, politics, and perceptions of risk. In S. Allan, B. Adam, & C. Carter (Eds.), *Environmental risks and the media* (pp. 201−217). London: Routledge.

WMO (World Meteorological Organization). (2018). *Guidelines on the definition and monitoring of extreme weather and climate events*. Geneva: WMO. Electronic copy available online.

Wolter, K., Hoerling, M., Eischeid, J. K., van Oldenborgh, G. J., Quan, X.-W., Walsh, J. E., ... Dole, R. M. (2015). How unusual was the cold winter of 2013/14 in the upper Midwest? [in "Explaining extremes of 2014 from a climate perspective"]. *Bulletin of the American Meteorological Society*, 96(12), S10–S14.

Wolter, K., Hoerling, M., Eischeid, J. K., & Cheng, L. (2016). What history tells about 2015 US daily rainfall extremes. *Bulletin of the American Meteorological Society*, 97(12), S9–S13.

World Bank. (2016). *High and dry: Climate change, water, and the economy*. Washington, DC.

World Bank. *Poverty and equity*. (2013). <http://databank.worldbank.org/data/reports.aspx?source = poverty-and-equity-database>.

World Bank. *World development indicators*. (2017). <http://datatopics.worldbank.org/worlddevelopment-indicators/>.

World Health Organization (WHO). (2012). *Flooding and communicable diseases fact sheet*. WHO. Available at http://www.who.int/hac/techguidance/ems/flood_cds/en/.

Yang, S., Xu, B., Cao, J., Zender, C. S., & Wang, M. (2015). Climate effect of black carbon aerosol in a Tibetan Plateau glacier. *Atmospheric Environment*, 111, 71–78.

Zalasiewicz, J., Water, C. N., Wolfe, A. P., Barnosky, A. D., Cearreta, A., Edgeworth, M., ... Williams, M. (2017). Making the case for a formal anthropocene epoch: An analysis of ongoing critiques. *Newsletters on Stratigraphy*, 50(2), 205–226.

Zappa, G., Shaffrey, L. C., & Hodges, K. I. (2013). The ability of CMIP5 models to simulate North Atlantic extratropical cyclones. *Journal of Climate*, 26(15), 5379–5396.

Zenghelis, D. (2017). Cities, wealth and the era of urbanisation. In K. Hamilton, & C. Hepburn (Eds.), *National wealth: What is missing, why it matters*. Oxford University Press, Chapter 14.

Zhang, J., Xu, H., Ma, J., & Deng, J. (2019). Interannual variability of spring extratropical cyclones over the yellow, Bohai, and East China seas and possible causes. *Atmosphere*, 10(1), 40.

Zhang, Y. X., Ding, Y. H., & Li, Q. P. (2012). Interdecadal variations of extratropical cyclone activities and storm tracks in the Northern Hemisphere. *Chinese Journal of Atmospheric Sciences*, 36, 912–928.

Further reading

Albert, S., Robin, B., Nixon, T., Javier, L., Douglas, Y., Jillian, A., ... Alistair, G. (2017). Heading for the hills: climate-driven community relocations in the Solomon Islands and Alaska provide insight for a 1.5°C future. *Regional Environmental Change*, 18(8), 2261–2272.

Cattiaux, J., Vautard, R., Cassou, C., Yiou, P., Masson-Delmotte, V., & Codron, F. (2010). Winter 2010 in Europe: A cold extreme in a warming climate. *Geophysical Research Letters*, 37, L20704.

Fernandes, R., Zhao, H., Wang, X., Key, J., Qu, X., & Hall, A. (2009). Controls on Northern Hemisphere snow albedo feedback quantified using satellite Earth observations. *Geophysical Research Letters*, 36, L21702.

Francis, J. A., & Vavrus, S. J. (2012). Evidence linking Arctic amplification to extreme weather in mid-latitudes. *Geophysical Research Letters*, 39, L06801.

Hall, A., & Qu, X. (2006). Using the current seasonal cycle to constrain snow albedo feedback in future climate change. *Geophysical Research Letters*, 33, L03502.

Houghton, J. T. (2009). *Global warming: The complete briefing* (4th ed.). Cambridge: Cambridge University Press.

Inoue, J., Hori, M. E., & Takaya, K. (2012). The role of Barents sea ice in the wintertime cyclone track and emergence of a warm-Arctic cold-Siberian anomaly. *Journal of Climate*, 25(7), 2561–2568.

Morice, C. P., Kennedy, J. J., Rayner, N. A., & Jones, P. D. (2012). Quantifying uncertainties in global and regional temperature change using an ensemble of observational estimates: The HadCRUT4 data set. *Journal of Geophysical Research Atmospheres, 117*, D08101.

Sheffield, J., Wood, E. F., & Roderick, M. L. (2012). Little change in global drought over the past 60 years. *Nature, 491*(7424), 435–438.

Thornton, P. E., Lamarque, J.-F., Rosenbloom, N. A., & Mahowald, N. M. (2007). Influence of carbon-nitrogen cycle coupling on land model response to CO_2 fertilization and climate variability. *Global Biogeochemical Cycles, 21*, GB4018.

Wehner, M. F., Reed, K. A., Loring, B., Stone, D., & Krishnan, H. (2018). Changes in tropical cyclones under stabilized 1.5 and 2.0°C global warming scenarios as simulated by the Community Atmospheric Model under the HAPPI protocols. *Earth System Dynamics, 9*, 187–195.

Zhang, Y. C., & Rossow, W. B. (1997). Estimating meridional energy transports by the atmospheric and oceanic general circulations using boundary fluxes. *Journal of Climate, 10*, 2358–2373.

CHAPTER 3

Ecosystem-based adaptation approach: concept and its ingredients

3.1 Introduction

Ongoing process of climate change is increasingly casting its adverse impact on human beings and ecosystems alike, thereby emerging as the key stumbling block to the realization of sustainable development (Turetta, Walter Leal, & Leonardo Esteves de, 2018). Adverse consequences of climate change, especially through its major component of greenhouse gas (GHG) emissions, resulting in the unsustainable management of ecosystems entail potential of enhancing the vulnerabilities of the people and natural ecosystems (Baig, Rizvi, Josella, & Palanca-Tan, 2016). In the wake of inadequate levels of action to control the GHG emission level, World Bank (2013) has warned that a 4°C rise is imminent. In its annual Global Risk Report for 2013, the World Economic Forum, while highlighting GHG emissions as one of the five major risks confronting the global economy, also pointed out that global economic and environmental systems are concurrently under strain in the wake of adverse impact of climate change becoming more and more visible (WEF, 2013).

According to Girot, Ehrhart, and Oglethorpe (2012), the poor people and marginalized communities throughout the world have become the most vulnerable to the vagaries of climate change and have been pushed to the forefront of a changing climate change with more to lose and very little to buffer the blow. Increasing adverse impact cast by climate change on traditional coping mechanisms and means of food production have rendered the world's poor and marginalized communities more vulnerable to famines, droughts, floods, diseases, etc. (Girot et al., 2012).

The continuing process of climate change portends grave challenges to sustainable livelihoods and socioeconomic development, especially of the developing countries globally (Muthee, Mbow, Macharia, & Filho, 2017). Adverse impacts of climate change are visible in different sectors like environment, health, education, food security, energy (Andrade, Herrera, & Cazzolla, 2010), and inter alia these climatic effects are emerging as major risks to poor people and marginalized communities which feel deficiency of adequate financial, institutional, and technical capabilities to adapt to the vagaries of climate change (Munang, Andrews, Alverson, & Mebratu, 2014). Remarkably, the United Nations Framework Convention on Climate Change (UNFCCC) has pointedly warned that a temperature rise beyond 2°C entails the possibility of having damaging effects on crop production, water access, health, and economic development (UNFCCC, 2011a). Such a scenario requires "different players" such as governments, communities, institutions, and individuals to "recognize the urgency" of accosting social, environmental, and economic effects of climate change (Muthee et al., 2017, p. 205).

While acknowledging the vulnerability of the least developed countries (LDCs) to the vagaries of climate change, the UNFCCC has accordingly guided these countries to set up the National Adaptation Programs of Action (NAPA; Pramova, Locatelli, Brockhaus, & Fohlmeister, 2012).

NAPA is deemed as a political mechanism to help the LDCs in identifying and prioritizing their most urgent adaptation needs (Pramova et al., 2012), and the UNFCCC cautions that any delay in establishing NAPA can entail enhanced vulnerability and high costs later (UNFCCC, 2010a, 2010b). According to Muthee et al. (2017, p. 205), NAPA provides an "ideal starting point" for country-specific adaptation measures through adaptation projects. Adoption of the intended nationally determined contributionsintended nationally determined contributions (INDCs) by the countries party to the UNFCCC is expected to look after these countries' priority actions for voluntary contribution to mitigation endeavors as well as rendering support to adaptation needs in developing countries (UNFCCC, 2014).

Adverse impact of climate change on natural resources, ecosystems, and species entails the possibility of limiting alternatives for development at local and national levels and concomitantly the probability of augmentation in pressure on the residual terrestrial, freshwater, and marine habitats cannot be ruled out (Turetta et al., 2018, p. 193). Factors like erosion and salinization entail the potential of causing losses in land fertility and landscape-level productivity, thereby affecting the livelihoods of rural and coastal communities along with further minimizing prospects for sustainable development and heightening poverty in terms of decreased income opportunities (Girot et al., 2012).

Effective mechanism for combating the vagaries of climate change calls for more holistic approaches to adaptation that comply with principles of human well-being, efficient environmental management, and also suggest ways and means to engulf the unreal divide between adaptation approaches and those that encourage the role of local communities. It could be discerned from the Hyogo Framework for Action (2005–15), which referred to international acknowledgment, that efforts to reduce disaster risks "must be" integrated systematically into policies, plans, and programs for sustainable development and poverty reduction and supported through bilateral, regional, and international cooperation, including partnership. While alluding to interconnectedness between and mutually supportive objectives of sustainable development, poverty reduction, good governance, and disaster risk reduction, the Hyogo framework emphasized that in order to meet the challenges ahead, "accelerated efforts" ought to be made to build the necessary capacities at the community and national levels to "manage and reduce risk" (UISDR, 2005). In other words, the challenge of climate change warrants enhanced resilience of both social and ecological systems.

Emphasis on continuing mitigation efforts to control the rate and extent of climate change has been focused in the Fourth Assessment Report of the Inter-Governmental Panel on Climate Change (IPCC) released in 2007 (IPCC, 2007a, 2007b), which has also been reiterated in IPCC's Fifth Assessment Report (AR5) released in 2014 (IPCC, 2014a). Admittedly, substantive endeavors have been undertaken to mitigate the adverse impacts of climate change; nonetheless, even with existing mitigation efforts the climate "does and will continue" to vary, thereby, making adherence to adaptation as an inevitable strategy for adjusting human communities and ecosystems so that they become more resilient and capable of coping with the emerging vagaries of climate change (Baig et al., 2016). Adaptation has come to be regarded as a vital approach amongst arrays of strategies to deal with changing climate and its impacts. A report by the World Bank in 2009 had projected that countries were expected to invest in traditional options like infrastructure for coastal defenses and flood control, and new irrigation facilities and reservoirs for water shortages (World Bank, 2009). Concurrently, opinions with regard to these traditional options were gaining credence with regard to the likelihood of these options being costly and that such traditional options were prone to neglect conservation aspects of ecosystems and biodiversity (Baig et al., 2016).

Undoubtedly, there has recently emerged surge in interest veering round different adaptation approaches that are prone to play a significant role in enhancing the NAPA's capabilities in promoting adaptation and sustainable development; nonetheless, most of these adaptation approaches are either dispersed or narrow in scope. Nonetheless, ecosystem-based adaptation (EbA) approach is gaining salience at a faster pace as an effective, efficient and almost cost-effective adaptation approach.

3.2 Concept of ecosystem-based adaptation

Amid the increased attention being focused on adaptation approaches in recent years, EbA approach to climate change has earned more prominence and gained wider recognition and it is widely understood as the use of biodiversity and ecosystems to assist the people in their adaptation to the ill-effects of climate change (Mellmann, 2015). Credit for laying down the theoretical basis of the concept of EbA can be attributed to the fifth meeting of the parties of the Convention on Biodiversity (CBD) held in 2000, which defined ecosystem approach as "[…] a strategy for the integrated management of land, water and living resources that promotes conservation and sustainable use in an equitable way. Thus, the application of the ecosystem approach will help to reach a balance of the three objectives of the Convention: conservation; sustainable use; and the fair and equitable sharing of the benefits arising out of the utilization of genetic resources" (CBD, 2000, p. 103). With the publication of five-volume study by the Millennium Ecosystem Assessment (MEA) in 2005, the CBD's notion of ecosystem approach gained the nomenclature of "ecosystem-based management" which lost no time in gaining wide acceptance (Göhler et al., 2013, p. 1) from the scholarly community. Concurrently, the MEA had focused on the declining ecosystem health worldwide (Mercer, Kelman, Afghan, & Kurvits, 2012, p. 1909).

Viewing the concept of ecosystem-based management as a notion that, "[…] recognizes that plant, animal and human communities are interdependent and interact with their physical environment to form distinct ecological units called ecosystems" (UNEP GPA, 2006, p. 4) by the United Nations Environment Program (UNEP) has seemingly spurred the conceptualization of the EbA concept within "NGOs and intergovernmental organization circles," owing to its emphasis on ecosystem services (ESs) (Millemann, 2015, p. 4). According to Göhler et al. (2013, p. 1), the notion of EbA was looked upon as an instrument of harnessing natural solutions to deal with climate change. Preceding years have witnessed evolution of the concept of EbA as a significant link between the three Rio Conventions—the UNFCCC, the CBD, and the United Nations Convention to Combat Desertification (UNCCD). Nevertheless, of the three Rio Conventions, the CBD has contributed tremendously to the elaboration of the concept of EbA. Göhler et al. (2013, p. 1) have opined that the first introduction of the concept of EbA into the UNFCCC occurred at the COP 14 in 2008 in Poland, where this concept was pushed by NGOs like IUCN and TNC.

3.2.1 Defining ecosystem-based adaptation

Admittedly, there exists an array of different definitions of the concept of EbA; nevertheless, there is a lack of a common and widely accepted definition of the concept (UNFCCC-SBSTA, 2013, p. 6).

Elaboration of the notion of EbA in 2009 by the CBD is often said to be the most cited definition in existing scientific literature, which inter alia states,

"[...] Ecosystem-based adaptation is the use of biodiversity and ecosystem services as part of an overall adaptation strategy to help people to adapt to the adverse effects of climate change. Ecosystem-based adaptation uses the range of opportunities for the sustainable management, conservation, and restoration of ecosystems to provide services that enable people to adapt to the impacts of climate change. It aims to maintain and increase the resilience and reduce the vulnerability of ecosystems and people in the face of the adverse effects of climate change. [...]" (CBD, 2009, p. 41). Besides, "[...] Ecosystem-based adaptation, [...] can be cost-effective and generate social, economic and cultural co-benefits and contribute to the conservation of biodiversity" (CBD, 2009, p.10). Subsequently, at its 10th COP held in Nagoya, Japan the CBD slightly further elaborated the EbA definition video its decision X/33 wherein it is envisaged that the Conference of the Parties recognizes, that the ecosystem-based approach, "[...] may include sustainable management, conservation and restoration of ecosystems, as part of an overall adaptation strategy that takes into account the multiple social, economic and cultural co-benefits for local communities" (CBD, 2010a, 2010b, p. 3).

Construed in a broad perspective, the elaboration of the concept of EbA provided by the CBD continues to form the basis of other definitions articulated by various scholars from time to time, though with slight alterations. For Jones, Hole, and Zavaleta (2012), measures for EbA approach are designed to harness the capacity of nature "to buffer human communities against the adverse impacts of climate change through the sustainable delivery of ESs" and as such EbA is usually deployed in the form of "targeted management, conservation and restoration activities," and is generally focused on particular ESs. Scholars like Ahmmad, Nandy, and Husnain (2013) have tried to define EbA with a nonanthropocentric focus when they explain EbA referring to "use of natural resources through conservation and enhancing resilience of ecosystem to buffer the worst impacts of climate changes on species and well-being of community." While asserting that a definition of EbA with a nonanthropocentric focus is rather rare in the literature, Mellmann (2015, p. 5) has opined that that this fact demonstrates that in majority of the cases, the notion of EbA is perceived as anthropocentric.

Focus on the need for a shift from a purely anthropogenic perspective to one that encompasses both ecosystems and people when defining ecosystem-based approaches for adaptation, was emphasized by the participants of the UNFCCC's technical workshop of the Subsidiary Body for Scientific and Technological Advice in Bonn in 2013. The participants also stressed on the need for a conceptual detachment between ecosystem-based approaches for adaptation and adaptation for ecosystems, while acknowledging the larger role that ecosystems could play in actions to deal with climate change (UNFCCC-SBSTA, 2013, p. 6). Mellmann (2015, p. 5) considers this emphasis on the need to reassess the conceptual design of the EbA as the "beginning of a conceptual discussion over the concept" of ecosystem-based adaptation. While explaining that the concept of EbA is currently in its "early stages," Mellmann (2015) draws attention to the observation of Doswald et al. (2014, p. 185) that states that some organizations "still" conceptualize EbA as the adaptation of ecosystems to climate change, rather than the use of ecosystems for human adaptation to climate change. In the light of these observations, Mellmann (2015) emphasizes on the need for working further on the conceptual outline of the concept of EbA in order to bring its all aspects under one head.

EbA is also delineated as the approach that facilitates sustainable management, conservation, and restoration of ecosystems to provide the services that "allows people to adapt to climate change effects" (IUCN, 2015). Some scholars opine that EbA measures harness ESs and biodiversity as a component of the community adaptation strategies to combat adverse effects of climate change (Munang, Thiaw, & Rivington, 2013a). The EbA approach takes into consideration those adaptation projects that have both "ecosystem face" and "human face" on them (Muthee et al., 2017). The key role played by the ESs in the minimization of people's vulnerability to the effects of climate change is recognized by EbA (UNFCCC, 2011a).

EbA strategies are directed by many principles which also includes upgrading of multisectoral approach in managing ecosystems of various landscapes (Speranza, Kiteme, Ambenje, Wiesmann, & Makali, 2010). Promotion of collaboration and coordination between and among different sectors communities and players utilizing ESs is also facilitated by the EbA approach (Delica-Willison & Gaillard, 2011). According to Cadag and Gaillard (2012), the EbA acts at manifold spatial scales and landscapes such as local, subnational, national and region. Some scholars have emphasized on considering the complications of ecosystems like factors leading to vulnerability, geographical and political factors involved in the ecosystem management and transboundary nature of an ecosystem (Orlove, Roncoli, Kabugo, & Majugu, 2010). EbA's role in promoting involution, cultural relevance, accountability, and inclusiveness in project design and implementation is highlighted by Munang et al. (2013b).

Viewed in a global context, climate change entails wide-ranging ramifications and impacts which extend beyond ecosystems (Leal, 2001). While alluding to a specific salient feature of EbA, Locatelli and Pramova (2015) have opined that it entails the potential of providing a tangible contribution toward enhancing the resilience of the ecosystems to climate change, apart from harnessing ESs for adaptation of the people. EbA entails the potential of being applied to wide-ranging areas such as agriculture, forest management intervention, biodiversity conservation and management and water resource management, etc. (Muthee et al., 2017, p. 207). In the wake of multiple challenges being posed by the vagaries of climate change at different levels, there is a need for new and enhanced adaptation approaches for effectively combating the effects of climate change (Muthee et al., 2017, p. 207). Planning and allocation of resources for adaptation action requires participation of all stakeholders, including decision-makers, ecologists, and environmentalists (Vignola, Locatelli, Martinez, & Imbach, 2009).

According to (McVittie et al., 2017), EbA can be defined as measures or practices that facilitate utilization of natural or managed biophysical systems (ecosystems) and processes to realize adaptation objectives. Accordingly, multiple uses of EbAs can be had throughout an assortment of land use and land cover types, or what can be termed as adaptation sectors. On account of this inexplicit use of ecosystem or ecosystem-like processes, the effectuation of EbAs can be associated with the delivery of an array of other ESs. While arguing that these relationships are often positive entailing cobenefits or antagonistic (with inherent trade-offs), (McVittie et al., 2017) opine that "interest in the use of EbAs reflects a recognition that past or current land use or land use change has often focused on a narrow set of ecosystem services" such as food and timber production. The notion of EbA reportedly corresponds to an array of other terms or notions that delineate identical types of approaches and actions, even though often emanating from dissimilar policy perspectives (EU, 2016).

According to Naumann et al. (2011), ecosystem conservation and restoration, upkeep of natural areas and augmentation of biodiversity are identified as the most plebeian EbA activities currently in use. Nevertheless, EbAs can also be said to contain alterations in the management of what are otherwise rigorously looked after biophysical systems like agriculture and forestry. Such management activities entail the potential of lessening the adverse impacts of land use either by way of directly affecting an adaptation-specific ecosystem process such as controlling surface water flow or enhancing the broader ecological resilience of the system. Benefits across a wide array of ESs can be expected in both the cases. The increasing interest in EbAs across land uses and land covers mirrors a discernment that EbAs "offer cost-effective alternatives or complements to gray infrastructure" (McVittie et al., 2017, p. 5).

3.3 Ecosystem-based adaptation's ingredients

Adaptation, ESs, biodiversity and linkages between biodiversity and ESs altogether constitute the basic ingredients of EbA approach.

3.3.1 Adaptation

Usage of the term "adaptation" in common parlance as well as in the environment literature derives its origin from evolutionary biology where it denotes genetic or behavioral changes that permit organisms or species to subsist within changing and contending environments (Smit and Wandel, 2006). Meaning of the term "adaptation" depends on its usage in a particular system or discipline. Usage of "adaptation" in social systems connotes different meaning from its use in ecological systems. Redman and Kinzig (2003) are of the view that change largely takes place in ecological systems on evolutionary timescales on account of mutation and information made available by natural selection. Social systems are said to have inbuilt capability of sharing information quickly, and new behaviors and system features can grow within a single generation (Redman & Kinzig, 2003). Adaptation in the case of human understanding and foresight can be defensive or preemptive (Adger, Lorenzoni, & O'Brien, 2009a).

It has been amply demonstrated from research in archeology and other related fields that adaptation to socio-ecological systems that include climate variability, is not a phenomenon of recent origin for human societies but has been an integral part since ancient times (Thomsen, Smith, & Keys, 2012, p. 1). Human societies and communities entail a prolonged history of adjusting to the vagaries of climate change and weather impacts through an array of various practices that inter alia include: crop diversification, irrigation, water management, disaster risk management and insurance, etc. (IPCC, 2007a, 2007b, p. 65). Nevertheless, the risks emanating from vagaries of climate change in terms of droughts, heat waves, accelerated glacier retreat, storm surges, flooding, hurricane and typhoon intensity, as well as land erosion at the seaside, occurring more frequently in modern times, often transcend the geographic boundaries (IPCC, 2007a, 2007b).

Moser and Ekstrom (2010, p. 22026) have opined that the term "adaptation" entails a long and multidisciplinary history of investigation and that has envisaged divergence in its meanings subject to its use by field and practice. It has been pointed out by Thomsen et al. (2012) that the concept of adaptation lacks a consensus of definition in relation to climate change. Indeed, adaptation is an emerging concept which is being explored from different perspectives, with specific emphasis on its social and ethical dimensions (Mellmann, 2015). Jones and Boyd (2011) find fault with the method of examining the concept from this perspective. In this backdrop, Mellmann (2015) opines that the concept of adaption is still evolving.

Climate change adaptation is described by the IPCC in its Fourth Assessment Report as "adjustment in natural or human systems in response to actual or expected climatic stimuli or their effects, which moderates harm or exploits beneficial opportunities" (IPCC, 2007a, 2007b), a definition also accepted by the UNISDR (2009, p. 4). Under this definition, adaptation can plausibly include an array of actions or adjustments, from additions to or changes in physical or built structures, to changing industry practices (Bradshaw, Dolan, & Smit, 2004), to individual behavioral change (Grothmann & Patt, 2005), to changes in the processes of significant institutions dealing with climate policy. With regard to planning, as Niven (2014) suggests, adaptation accosts vital climate change-related risks within a designated geographical area in a cyclical process which involves: (1) assessing current risk and related vulnerability; (2) reviewing current practices, policies, and infrastructure; (3) assessing potential adaptation measures; (4) prioritizing and implementing measures; and (5) evaluating and managing implemented measures (Moser & Ekstrom, 2010). Pelling (2011) suggests that adaptation is a sociopolitical act that is closely connected to "contemporary," and with the possibility for "re-shaping future," power relations in society and concurrently, it also recognizes that different actors observe contrasting roles for adaptation.

Adaptation is identified either being proactive or reactive and the differentiation between proactive and reactive adaptation is often referred to in classifications or catalogs of social system adaptation (IPCC, 2014a). According to Berrang-Ford, Ford, and Paterson (2011), undoubtedly human systems are well equipped for proactive adaptation; nonetheless, the bulk of that adaptive activity remains reactive. While attributing this to difficulties in anticipating changes in local conditions, Mendelsohn and Dinar (2009) suggest that as long as actions are transient, most actors are well placed to react to climate change rather than expecting it. They do, nevertheless, emphasize on the significance of proactivity in enduring investments such as land use planning. Many scholars suggest occurrence of adaptation as a rare phenomenon merely in response to climate drivers alone (Rickards & Howden, 2012). While construing the process of adaptation in an agricultural context, it becomes discernible that an array of nonclimate factors impacts adaptation actions, which inter alia include: economic and regulatory conditions, social and technological imperatives, fluctuating labor and commodity prices, and the ongoing process of globalization (IPCC, 2014a).

The possibility of these nonclimate factors interacting with and overshadowing the impact of climate change in many cases cannot be ruled out (IPCC, 2014a). According to Dessai and Hulme (2007), this entails the potential of obscuring the definition of adaptation to climate change. Several definitions for adaptation have been "proposed" in the climate change literature, with "varying degrees" of reference to climate change and according to Gawith (2017), broadly adaptation can be defined as incorporating both ecological and social systems. Similar stance is reiterated by Adger et al. (2009b, p. 337) when they state, "in essence, adaptation describes adjustments made to changed environmental circumstances that take place naturally within biological systems and with some deliberation or intent in social systems." Inclusion of possible opportunities is a salient trait of a number of other definitions of adaptation and McCarthy, Canziani, Leary, Dokken, and White (2001) cite the example the IPCC definition of adaptation that is usually adopted in the climate change field, and it defines adaptation as "adjustment in natural or human systems in response to actual or expected climate stimuli or their effects, which moderates harm or exploits beneficial opportunities" (McCarthy et al., 2001, p. 928).

A detailed definition of adaptation proposed by Moser and Ekstrom (2010) is said to be an improvement over a number of weaknesses in the IPCC's definition. The fact of justification of adaptation by conditions or opportunities other than climate change is acknowledged by Moser and Ekstrom (2010) who also point out that despite the IPCC's definition assuming effectiveness, meaningful adaptations may also fail. The definition proposed by Moser and Ekstrom (2010) concedes that adaptation can vary from "coping strategies to system transformation" and it connects it to the concept of resilience (Gawith, 2017). Moser and Ekstrom (2010) propose the following definition: "Adaptation involves changes in social-ecological systems in response to actual and expected impacts of climate change in the context of interacting nonclimatic changes. Adaptation strategies and actions can range from short-term coping strategies to longer term, deeper transformations, aim to meet more than climate change goals alone, and may or may not be successful in moderating harm or exploiting beneficial opportunities" (Moser & Ekstrom, 2010, p. 22026).

Current studies on climate change adaptation are widely using definition of adaptation as provided by the IPCC in their 2014 AR5 which has defined adaptation as the "[...] process of adjustment to actual or expected climate and its effects. In human systems, adaptation seeks to moderate or avoid harm or exploit beneficial opportunities. In some natural systems, human intervention may facilitate adjustment to expected climate and its effects." (IPCC, 2014a, p. 5). Almost an identical definition is given by the UNFCCC that states: "Adaptation refers to adjustments in ecological, social, or economic systems in response to actual or expected climatic stimuli and their effects or impacts. It refers to changes in processes, practices, and structures to moderate potential damages or to benefit from opportunities associated with climate change" (UNFCCC, 2014; cited in Mellmann, 2015).

Adger et al. (2005a, 2005b) explore "good" or "desirable" aspects of adaptation. However, to know as to what constitutes "good" or "desirable" adaptation depends on the understanding of what people and societies value (Adger et al., 2005a, 2005b). Undoubtedly, the goals of adaptation are delineated by values; nonetheless, Ford and Berrang-Ford (2011, p. 19) observe: "there is surprisingly little research about what these values are, indeed, whose values they are and what they imply for adaptation outcomes." Dessai and Hulme (2007) regard these values as highly context-specific. Besides, the goals of adaptation pertaining to shared contexts are usually repugned (Adger et al., 2009b). Entailing the probability of differences between scales of successful adaptation, Adger et al. (2005a, 2005b) note that "successful private adaptation" may not align with collective adaptation and it is in this context that Adger et al. (2009b, p. 341) have observed: "The goal of adaptation will likely depend on who or what is adapting." In spite of having this knowledge about adaptation, assessment of adaptation goals is still faced with conceptual, methodological, and practical challenges. Gawith (2017) opines that discernment of the assortment of values held by actors and adaptation goals shaped by these values, constitutes a critical component of examining adaptation. However, Nicholas Stern (2007a, 2007b) cautions that quantification of these values and goals is a cumbersome task.

According to Jones et al. (2012), adaptation actions are usually delineated into so-called "hard" and "soft" approaches, with the former concentrating on engineered structures, and the latter on information and institutional capacity building. Measures adhered to under green or ecosystem-based approaches can be regarded either as a "third way" (Jones et al., 2012) or as constituent of an extensively redacted "soft" approach (Kithiia & Lyth, 2011). Ecological structures like wetlands, vegetation, forests, grasslands, waterbodies and components of "blue and green infrastructure" in cities such as parks, gardens, street trees and ponds, etc., render adaptation services (Niemelä et al., 2010), which inter alia include regulating services such as water, soil, local climate, and natural hazard regulation. Niven (2014) points out that further adaptation-relevant services, depending on the context, can include (but are not limited to) food and fiber provisioning, pest control and disease regulation, and preservation of genetic diversity.

3.3.1.1 Need for adaptation

While drawing attention to cautionary notes sounded by the IPCC's AR5 (2014) through its many projected changes in climate change if the world fails to quickly act on reducing GHG (IPCC, 2014a) along with signals of great ambitions articulated by many countries through the 2015 Paris Agreement on Climate Change (PACC) and the resulting slower progress on many of the issues from the 2015 PACC demonstrated at the 2016 COP-22 at Marrakesh, Nalau, Filho, Nalau, and Filho (2018) aver that the "adaptation policy space" is now globally recognized. Urgency for adaptation policy is further reinforced by occurrence of recent events like the year 2016 being declared the warmest year on record globally since the earth surface temperature records began in 1880 (NASA, 2017) and the extreme coral bleaching in 2016 in the Great Barrier Reef in Australia, where about 93% of the reefs were impacted (ARC, 2016). In the wake of frequent visitations by erratic rainfall patterns, floods, droughts, hurricanes, and other natural calamities becoming a recurring phenomenon, the narrative on how to adapt to global climate change is beginning to become a reality.

There has been growing international concern in recent years over the massive problem in relation to GHG emissions. In 2013 the IPCC in its report on Physical Science Basis disclosed that concentrations of the GHGs carbon dioxide (CO_2), methane (CH_4), and nitrous oxide (N_2O) exceeded the highest concentrations recorded in ice cores during the past 800,000 years (IPCC, 2013, p. 11). The report further reveals that rise in average concentration in the atmosphere is 40% for CO_2, 150% for CH_4 and 20% for N_2O in comparison to preindustrial times in 1750 (IPCC, 2013, p. 12). With the commencement of the burning of substantial amounts of fossil sunlight energy in the aftermath of the industrial revolution, humankind has been persistently tapping into this energy source as a result of which there has been an alarming increase in GHG concentrations in the atmosphere (Mellmann, 2015). Over the decades, deforestation and land degradation have also emerged as major sources of emissions.

The concentrations of the GHG over long periods of time in different parts of the globe have proved instrumental in severely disturbing radiative forcing thereby demonstrating unsuccessful attempts of humankind in controlling the output of emissions. It can be discerned from UNEP's annual Emission Gap Reports of preceding years, which provide a comparison of the present track of emissions confronting the humankind and the requirement of keeping global warming to an optimum of a 2°C average increase in temperature compared to preindustrial times with a likelihood of >66%, that the international community is lagging behind in attaining that target (UNEP, 2013, p. xii). There has been an ongoing debate on 2°C maximum increase in temperature, as agreed upon in the 2009 COP-15, also known as the Copenhagen Agreement; nevertheless, "we are not even on track for a maximum rise of 2°C" (Mellmann, 2015).

The UNEP's Emissions Gap Report of 2013 envisaged that expected emissions for the year 2020 were estimated to be 59 $GtCO_2$ per year (range: 56–60 $GtCO_2$ per year), under a business-as-usual scenario that already considers all the mitigation actions currently under way assuming that all of the pledges will be met. However, in order to stay below the 2°C, annual quantum of emissions should be about 44 GtCO$_2$ per year (range: 38–47 GtCO$_2$ per year), which leaves a gap of about 8–12 GtCO$_2$ per year (UNEP, 2013, p. xii). Undoubtedly, 2015 PACC emphasized on a long-term goal of keeping the increase in global average temperature to well below 2°C above preindustrial levels; nonetheless, UNEP's 2016 Gas Emissions Report revealed a steady increase, reaching approximately 52.7 GtCO$_2$ in 2014. It is further revealed from this report that the rate of global GHG emissions increase during the 2000–10 period was faster (2.2% per year) than during 1970–2000 period (about 1.3% per year) and during 2010–11 period it stood at 3.5% per year then showing a decline in 2012–13 being at 1.8% per year (UNEP, 2016, p. xii).

As per UNEP's gas Emissions Report 2017, the time series data for global total CO$_2$ and GHG emissions used for the Emissions Gap Reports have been updated since the 2016 report and according to the updated version, the total GHG emissions in 2014 stood at 51.7 GtCO$_2$e, and the estimate for global total CO$_2$ emissions in 2015 updated to 35.6 GtCO$_2$. According to this report, total global GHG emissions, including emissions from land use, land-use change and forestry (LULUCF) in 2016, are estimated at about 51.9 GtCO2e/year. However, the report laments that emissions still show a slowdown in growth in the past 2 years, with calculated increases of 0.9%, 0.2%, and 0.5% in 2014, 2015, and 2016, respectively (UNEP, 2017, p. xv). It further warns: "Looking beyond 2030, it is clear that if the emissions gap is not closed by 2030, it is extremely unlikely that the goal of holding global warming to well below 2°C can still be reached" (UNEP, 2017, p. xiii).

Failure in humankind's endeavors to mitigate GHG emissions to an acceptable level is "one reason" why adaptation assumes added emphasis (Mellmann, 2015). Efforts at developing adaptation plans and policies by integrating them into wider development plans at various levels of government in most of the countries suggests that governments are making efforts to integrate climate change considerations (IPCC, 2014b, p. 8), and according to Naumann et al. (2011, p. 45), recent years have witnessed increased acceptance and interest in climate change adaptation. The IPCC in its 2007 Fourth Assessment Report emphasized on adaptation to enhance the capacity of countries, regions, communities and social groups in ways sync with sustainable development (Adger et al., 2007, p. 737). The increased realization of the need for adaptation also becomes discernible from the mushroom growth in scientific publications on the subject, for which current authentic number is not available but according to a literature review undertaken in 2011 by Glick et al., it was reported that there had been a fivefold increase in climate change adaptation literature from 2007 to 2010 (Glick, Chmura, & Stein, 2011, p. 4).

3.3.1.2 Adaptive capacity

According to Adger et al. (2011), adaptation is the act of change whereas adaptive capacity ascertains whether adaptation can occur. Adaptive capacity constitutes one of the three components of the vulnerability framework developed by the IPCC and the other two are exposure and sensitivity (Cinner et al., 2013, 2015). It has been suggested by Cinner et al. (2013, p. 0.9) within this framework that adaptive capacity "is perhaps the component of vulnerability most amenable to influence, and may be a useful focus for adaptation planning." Scholars like Kellstedt, Zahran, and Vedlitz (2008), and Roser-Renouf and Nisbet (2008) have opined that people's inclination toward adaptation begins when they realize about the benefit of adaptation and risk of nonadaptive/maladaptive responses (Steel, Lovrich, Lach, & Fomenko, 2005). In other words, awareness among the people about the benefit of adaptive response can bolster their adaptive capacity and conversely, people employ their full force to adapt due to fear of loss from no adaptive response (Saroar & Routray, 2015).

A section of scholars, while asserting that cognition is essential but not sufficient, have argued that mere understanding of the need for adaptation cannot ensure any adaptive response sans people's ability to adapt (Adger et al., 2005a, 2005b). Blaikie, Cannon, Davis, and Wisner (1994) opine that availability of material resources at one's disposal determines his/her ability to adapt. Another group of scholars, while relying on personal motivation theory based on the literature of Maddux and Rogers (1983), have argued that people's adaptive capacity is also determined by psychological and behavioral factors. Those who accord priority to disposition of various physical resources opine that an individual bereft of these can at best initiate a maladaptive response (Wisner, Blaikie, Cannon, & Davis, 2004; Pelling & High, 2005; Allison et al., 2009). On the contrary, those according priority to psychological matters allude to a person's noesis about his/her ability to adapt and understanding about the efficacy of such adaptive responses (Blennow & Persson, 2009). Significance or vitality of past experience of adaptive response to demonstrate adaptive capacity against a fresh onslaught of disastrous event is also argued by the protagonists of adaptive behavioral factors (Grothmann & Reusswig, 2006; Saroar & Routray, 2012). The underlying emphasis here is on the fact that accumulated experience is helpful in initiating a new adaptive response. In this regard, Saroar and Routray (2015) opine that all arguments have their own merits.

Emphasis on the necessity for climate information communication as an important determinant of people's adaptive response has been focalized by some scholars (Saroar & Routray, 2010) with the argument that timely access to climate/weather information products can enable the people to prepare themselves to adapt. Timely access to information regarding sudden and rapid onset disastrous events can be helpful in enhancing adaptive capacity of the people. Equal emphasis needs to be stressed on physical environmental or spatial/location factors (Tol, Klein, & Nicholls, 2008). This assumes significance because some locations, owing to their geographical/morphological character are more prone to natural disasters and the people inhabiting such marginalized areas are often the disadvantaged groups entailing very little adaptive capacity (Molnar, 2010). Apart from these, several demographic and socioeconomic factors also often contribute to as cross cutting elements of adaptive capacity (Moser & Satterthwaite, 2009). In fact, all these factors contribute to adaptive capacity in one way or the other. According to Saroar and Routray (2015), a right mix of many of these factors can work effectively in a specific situation, even though the same combination may not prove effective while adapting to a very different situation.

According to Cinner et al. (2015), adaptive capacity differs between people and over time. Adaptive capacity is perceived in the climate change literature as a dormant trait of individuals and groups that directs their abilities to expect and respond positively to change (Whitney et al., 2017). Nelson (2011) and Cinner et al. (2015) have defined it as the sets of preconditions that enable occurrence of adaptation. While these prerequisites are often comprehended as different forms of capital, people's capacities to adapt frequently rely on features beyond those usually understood to be "capitals" (Mortreux & Barnett, 2017; Nelson, 2011). After reviewing the development of the concept of adaptive capacity in the literature, Mortreux and Barnett (2017) noted a distinct broadening of the concept into the behavioral and psychosocial sciences, including an evolving cognizance of the importance of aspects such as risk attitudes, trust, and place attachment.

After reviewing an array of approaches to the assessment of adaptive capacities and having ascertained huge variations in the approaches used, Whitney et al. (2017) arrived at the conclusion, "The variety of ways in which adaptive capacity is defined, applied, assessed, and measured reflect a diversity of interests, areas of expertise and rationales" (Whitney et al., 2017, p. 2). While conceding that certain attributes can be observed to augment adaptive capacity in a broad sense, these scholars affirm that many facets of adaptive capacity especially pertain to the change in question. While conducting research on adaptive capacities among primary resource users, Marshall, Park, Howden, Dowd, and Jakku (2013) and Cinner et al. (2013) have uncovered four vital and assessable dimensions of adaptive capacity: the management of risk; the ability to plan, learn, and reorganize; financial and psychological flexibility; and interest in undertaking change. According to Simões et al. (2017), different identifiable aspects and attributes of adaptive capacity can be seen as the enablers of adaptation.

3.3.1.3 Limits to adaptation

Increasing acceptance of the concept of adaptation in scientific and policy literature on climate change has also been instrumental in adaptation options, definitions, frameworks, and case studies to demonstrate adaptation processes and its implementation (Nalau et al., 2018). Ever-increasing frequency of international level negotiations on climate change through UNFCCC, specific focus is centered on mitigation, adaptation, and associated loss and damage (Warner, 2015), and this has catapulted the concept of "limits to adaptation" to garner increased attention from the policy and research communities in trying to ascertain as to why limits to adaptation emerge, how these can be tackled, and the kinds of ethical underpinnings that are required to be taken into consideration (Klein et al., 2014). Some scholars (Dow, Berkhout, & Preston, 2013; Klein et al., 2014) opine that sociocultural, economic, historical, and environmental contexts wield great influence on adaptation limits which come in many shapes and sizes. Notion of "adaptation limit" is still problematic because majority of the discussions taking place continue to focus on "potential future limits," and as Nalau et al. (2018, p. 2) have opined: "The evidence is still in the making and many adaptation limits remain unknown and poorly understood partly due to the complexity of interacting processes and factors, which together constitute a potential limit and partly due to the changing nature of soft limits as more options become available and hence open up new pathways that enable shifts in what constitutes a limit to adaptation."

Allusion to limits to adaptation finds mention in the IPCC's AR5 in terms of defining conceptual linkages between limits and transformational adaptation and with reference to specific ethical consideration and cross-linkages between concepts (Klein et al., 2014).

Three types of limits to adaptation—biophysical, sociocultural, and economic—have been identified by the AR5 (Klein et al., 2014, p. 923). Citing the evidence available in current literature, Klein et al. (2014) point out that currently more interest is being evinced in investigating adaptation limits within a specific sector, ecosystem or at particular species level. An adaptation limit is defined as "the point at which an actor's objectives cannot be secured from intolerable risks through adaptive actions" (Dow et al., 2013).

"A limit is said to be reached in the wake of inability of adaptation efforts to deliver an acceptable level of security from risks to the existing objectives and values and thwart the loss of vital characteristics, components, or services of ecosystems" (Klein et al., 2014, p. 919). Two kinds of limits to adaptation are identified, such as hard limits and soft limits. Hard limit entails no more availability of options; whereas soft limit is the one where options might not be accessible currently but could be obtainable in the future (Barnett et al., 2015; Dow et al., 2013; Klein et al., 2014). While differentiating between an adaptation constraint, some scholars point out that constraints can be overcome but a limit epitomizes a point where a radical change is called for in the event of non-availability of other options such as maintaining a particular livelihood or residing in a particular place (Barnett et al., 2015; Dow et al., 2013; Moser & Ekstrom, 2010). Scholars like Islam, Sallu, Hubacek, and Paavola (2014) consider limits as physical attributes such as more intense cyclones or hurricanes and regard constraints as socially constructed impediments like paucity of credit. According to Barnett et al. (2013), limits to adaptation "involve irreversible losses of things individuals care about, either due to climate change impacts or as outcomes of climate change policies." Nevertheless, it is argued by Warner (2015) that limits at the household scale can be construed in terms of trade-offs characterized by irrevocability and shrouded in ambiguity.

Socially constructed configuration of adaptation limits has long been acknowledged by many scholars (Klein et al., 2014). As per the version of Barnett et al. (2013), the limits to adaptation are "to some extent, in the eyes of the beholder," as perceptions; nevertheless, the notion as to what is a limit depends on values that obviously is distinguishable among groups and individuals (Klein et al., 2014). Notions of limits to adaptation, and "intolerable risk," as pointed out by Dow et al. (2013) and Klein et al. (2014), are steered by what is perceived to be at risk. According to Nalau and Handmer (2015), projection of climate change as a complex limitless problem is prone to significantly impact the types of management interventions and options that are essential to combat it.

Divergence of opinions on the notion of adaptation, because it is an evolving concept, pervades the discourse on adaptation limits as well. Multiscalar nature of limits spanning across diverse jurisdictions and times, as articulated by Klein et al. (2014), is prone to be impacted to a great degree by the interaction between both public and private actors and institutions. According to Dow et al. (2013), likelihood of a change and shift in adaptation limit is contingent upon concerted planning and sound decision-making and these authors also regard risk-based decision-making and adaptation limits as two analogous realms of investigation that are intimately colligated. Lack of unanimity over research and policy focusing on the notion of adaptation limits demonstrates "taking into consideration other processes than just climate change, which are already impacting ecosystems and/or increasing community vulnerability" (Nalau et al., 2018, p. 4).

3.4 Ecosystem services

Recent years have witnessed the concept of ES garnering salience in a wide spectrum of disciplines spanning both the natural and social sciences (Bennett et al., 2015; Pocock et al., 2016). Built on the benefit of hindsight of five-volume study by MEA in 2005 that crystallized multidisciplinary approaches into the concept of ES and supplemented by major advances in ecological research in these years, it has not only emerged as a vast and varied field of research but is to be reckoned with a field in its own right. Focus of research in ESs which was confined earlier to twin areas of monetary valuation of ESs and linking ESs to socioeconomic systems is now shifting to discern negative/positive consequences of human interventions and measuring impact of environmental factors, some of which are closely linked to biodiversity or ecosystem functioning, that could potentially affect ESs (Pocock et al., 2016).

Usage of the term "Ecosystem" was popularized by A.G. Tansley, a British biologist and plant ecologist and he envisaged this term as the key principle of the ecology (Tansley, 1935). Since then, it has come to gain increasing currency in scientific literature. An ecosystem entails "the structure of interrelationships of living beings to one another and to their inorganic environment" (Grunewald, Bastian, Karsten, & Olaf, 2015). An ecosystem, as defined by MEA (2005), is a "dynamic complex of plant, animal, and microorganism communities and the nonliving environment interacting as a functional unit. Humans are an integral part of ecosystems." The notion of ESs, described by Daily (1997, p. 3) as the conditions and processes "through which natural ecosystems and the species that make them up, sustain and fulfill human life" has witnessed rapid expansion in the scientific literature in recent years (Hubacek & Kronenberg, 2013). Availability of wide-ranging literature on ESs, along with their classification and valuation is attested to by Costanza et al. (1997), De Groot, Wilson, and Boumans (2002), and MEA reports (MEA, 2005).

The wave of globalization set in motion at the outset of the 1990s was followed by mounting demands of humankind on the limited resources of the planet resulting in upsetting the natural equilibrium, to be followed by huge loss of biodiversity and in the emergence of "the problem complex of energy and the climate" (Grunewald et al., 2015). With environmental degradation becoming part of the international environmental narrative, the notion of ESs also became the part of this narrative (Costanza et al., 1997). With the publication of MEA reports in 2005, attention came to be focused on ESs, which was further supplemented by The Economics of Ecosystems and Biodiversity (TEEB (The economics of Ecosystems and Biodiversity), 2008) studies (TEEB, 2009), and Rationalizing Biodiversity Conservation in Dynamic Ecosystems project or RUBICODE project funded by the European Commission to provide frameworks for aiding decision-making for biodiversity conservation (Harrison, 2010). The concept of ESs gathered further impetus in the aftermath of adoption of the Strategic Plan for 2011–20 at the 10th Conference of the Parties of the CBD in Nagoya on October 18–29, 2010, wherein the usage of the term ESs found frequent repetitions (CBD, 2010a, 2010b). Since then, it has come to gain sufficient salience.

The concept of ESs assumes importance owing to its ability to focus added emphasis on taking into consideration the role played by the ecological services—the services made available to humankind free of cost—in decision-making processes and ensuring sustainable use of land so as to avert overconsumption and dilapidation of natural resources. For Müller and Burkhard (2007), integrative, interdisciplinary and transdisciplinary lineament of the concept of ES along with its ability to connect environmental and socioeconomic elements constitute the core of appeal of ES concept.

Necessity for moving away from the notion of evaluating ES primarily in monetary terms toward utilizing a wide spectrum of indicators instead (UNEP-WCMC, 2011) has also been felt by the business community in the wake of fast depleting natural resources, reduced biodiversity and the degradation of ESs that not only portend mounting risk for business entities, investors and financial institutions but also the prospects of better financial opportunities that may unfold after tackling these problems. Inclusion of biodiversity and ES in the business models and core business strategies as a key stipulation for ensuring sustainable growth and success is gaining wider recognition (BESWS, 2010).

In the business community, too, there is a growing realization that scarcity of natural resources, reduced biodiversity, and the degradation of ES not only bear a growing level of risk for companies, investors, banks, and insurance companies, but also that solving these problems may open up opportunities of great financial significance. Leading companies are increasingly realizing that the maintenance and protection of nature is neither merely a marginal issue, nor is it something that can be dealt with by the commitment of volunteers. Rather, biodiversity and ES must be firmly rooted in their business models and core strategies, as a key precondition for ensuring sustainable growth and success (BESWS, 2010). Perception of nature as a productive force like capital and labor has enabled the ES approach to become not only the focal point of attention for the decision-makers in government and corporate sector and civil society but also of economic sciences which are now focusing on developing mechanism of economically evaluating ecosystems and the changes in them, and in resource economics, "the concepts of "external effects" and "economic total value" have been created for this purpose" (Grunewald et al., 2015, p. 2).

3.4.1 Classification of ecosystem services

The task of developing a distinct, widely accepted and all-embracing classification of ES is rendered difficult by the diverse and complex nature of ecosystems and the services supplied by them. There exist several proposals, classification systems and partly differing opinions with regard to the classification of ESs. According to Bastian, Grunewald, and Syrbe (2015, p. 45), "Depending on the goals of the assessment, spatial scales and specific decision-making context, they all show both strengths and weaknesses." Among the early scholars dwelling on the theme of ES classification, de Groot et al. (2002) delineated regulation, production, habitat, and information functions (or services). The habitat services were also identified by the TEEB study as a different category to emphasize the importance of ecosystems to provide habitat for migratory species and gene-pool "protectors" (TEEB, 2010). Building upon the definition of Costanza et al. (1997), the MEA (2005) provided a simple classification of ecosystems services that have been widely used in the international research and policy literature and these include Provisioning services, Regulating services, Cultural services, and Supporting services.

Prevalence of incongruity about the assignment of phenomena, which constitute the basis for the services of the three other classes, is pointed out by Bastian et al. (2015) and according to Müller and Burkhard (2007), this is applicable to supporting services or basic services. Bastian et al. (2015) consider supporting services an intermediate (analytical) stage and constitute a prerequisite for defining the other three groups of services. Scholars like Burkhard, Kroll, Müller, and Windhorst (2009) and Haines-Young and Potschin (2010) have also reiterated emphasis on treating supporting services differently from other ES wherein benefits directly accrue to the people.

According to Bastian, Haase, and Grunewald (2012), the split into productive (economic), regulating (ecological), and societal functions or services entails the benefit that it can be connected to both fundamental concepts of sustainability and risk using the established ecological, economic, and social development categories. Broadly speaking, the classification of ESs depends on the respective researcher; nevertheless, in general, three or four groups entailing a total of 15–30 functions or services are differentiated by Bastian et al. (2015). Undoubtedly, availability of information on appropriate indicators is a prerequisite; nonetheless, according to TEEB (2009), there are still gaps in the literature in this respect.

Building on the knowledge garnered by, de Groot et al. (2002) and Vandewalle et al. (2008) along with the addition of their own experiences and reflections, Grunewald et al. (2015) have classified 30 ESs according to three main categories: provisioning, regulation and sociocultural services—each with subdivisions, while providing name of the ES, its description with examples and indicators is also provided. These categories deem appropriate in the present context and are

Table 3.1 Provisioning services.

Name of the ecosystem services	Description	Examples
A. Food (provision of plant and animal materials)		
A.1 Food and forage plants A.2 Livestock A.3 Wild fruits and game A.4 Wild fish A.5 Aquaculture	Cultivated plants as food/forage for humans and animals Slaughter and productive Livestock Edible plants and animals from the wilderness Fishes and seafood caught in waters Fishes, shells or algae growing in ponds or farming installations	Cereals, vegetables, fruits, edible oil, hay Cattle, pigs, horses, Poultry Berries, mushrooms, Game Eels, herrings, shrimps, Shells Carps, shrimps, oysters
B. Renewable raw materials		
B.1 Wood and tree products B.2 Vegetable fibers B.3 Regrowing energy sources B.4 Other natural materials	Raw materials from trees in forests, plantations or agro-forest systems Fibers from herbaceous plants (from nature or cultivated) Biomass from energy crops and wastes Materials for industry, crafts, decoration, arts, souvenirs	Timber, cellulose, resin, natural rubber Cotton, hemp, flax, sisal Fire wood, charcoal, maize, rape, dung, liquid manure Leather, flavorings, pearls, feathers, ornamental fishes
C. Other renewable natural resources		
C.1 Genetic resources C.2. Biochemicals, natural medicine C.3. Freshwater	Genes und genetical information for breeding and biotechnology Raw materials for medicine, cosmetics and others to enhance health and well-being Clean water in ground- and surface waters, precipitation and in the underground for private, industrial and agricultural use	Seeds, resistance genes Etheric oils, tees, Echinacea, garlic, food supplements, leeches, natural crop protection products Rain, spring and fountain discharge, bank filtrate

From Bastian, O., Grunewald, K., & Syrbe, R.-U. (2015). Classification of ES. In K. Grunewald & O. Bastian (Eds.), Ecosystem services—Concept, methods and case studies (p. 47). Berlin: Springer.

adapted accordingly (Tables 3.1–3.3). Each category is described in a separate table along with the name of ES, its description and examples; however, indicators, as contained in the original tables, are excluded in each table.

3.4.1.1 Provisioning services

Multiple goods and services, ranging from oxygen and water to food and energy to medicinal and genetic resources, and materials for clothing and shelter, etc., are provided by ecosystems. These goods and services can be living organisms or biotic as well as nonliving organisms or abiotic. The goods and services falling in the category of living organism may include the products of living plants and animals and these are renewable biotic resources (Table 3.1). Abiotic resources

Table 3.2 Regulation services.

Name of ecosystem services	Description	Example
A. Climatologic and air hygienic services		
A.1 Air quality regulation A.2 Climate regulation A.3 Carbon sequestration A.4 Noise protection	Air cleaning gas exchange Impacts on the maintenance of natural climatic processes and on reducing the risks of extreme weather events Removing carbon dioxide from the atmosphere and relocation into sinks Reducing noise emissions by vegetation and surface forms	Filter effects (fine dust, aerosols), oxygen production Cold air production, humification, reducing temperature by the vegetation, weakening of extreme temperatures and storms Photosynthesis, fixation in the vegetation cover and in soils Noise protection effects of vegetation
B. Hydrological services		
B.1 Water regulation B.2 Water purification	Balancing impacts on the water level of watercourses and the height, duration, delay and avoiding floods, droughts and (forest) fires, protection against tidal flooding (e.g., by coral reefs, mangroves), water as transport medium, water power Filter effects, storage of nutrients, decomposition of wastes	Natural irrigation, soil storage, leaching/groundwater recharge Nitrogen retention, Denitrification, Self-purification of rivers and lakes
C. Pedological services		
C.1 Erosion protection C.2 Maintenance of soil fertility	Effects of vegetation on soil erosion, sedimentation, capping and silting Regeneration of soil quality by the edaphon (soil organisms), soil generation (pedogenesis) and nutrient cycles	Protection against landslides and avalanches, breaking winds Nitrogen fixation, waste decomposition, humus formation and accumulation
D. Biological services (habitat functions)		
D.1 Regulation of pests and diseases D.2 Pollination D.3 Maintenance of biodiversity	Mitigating influences on pests and the spread of epidemics Spread of pollens and seeds of wild and domestic plants Conservation of wild species and breeds of cultivated plants and livestock	Songbirds, lacewings, ladybirds, parasitic wasps, tics (Encephalitis) Honey and wild bees, bumblebees, butterflies, syrphid flies Refuge and reproduction habitats of wild plants and animals, partial habitats of migrating species, nursery spaces (e.g., spawning grounds for fishes), cattle breeds

From Bastian, O., Grunewald, K., & Syrbe, R.-U. (2015). Classification of ES. In K. Grunewald & O. Bastian (Eds.), Ecosystem services—Concept, methods and case studies (pp. 48–49). Berlin: Springer.

Table 3.3 Sociocultural services.

Name of Ecosystem services	Description	Example
A. Psychological–social goods and services		
A.1 Ethical, spiritual, religious values A.2 Esthetic values A.3 Identification A.4 Opportunities for recreation and (eco) tourism	Possibility to live in harmony with nature, Integrity of Creation, freedom of choice, fairness, generational equity Diversity, beauty, singularity, naturalness of nature and landscape Possibility for personal bonds and sense of home in a landscape Conditions for sports, recreation and leisure activities in nature and landscape	Bioproducts, sacred places Flowering mountain meadows, harmonious landscape Natural and cultural heritage, places of memory, traditional knowledge Accessibility, security, stimuli
B. Information services		
B.1 Education and training values, scientific insights B.2 Mental, spiritual and artistic inspiration B.3 Environmental indication	Opportunities to gain knowledge about natural interrelations, processes and genesis, scientific research and technological innovations Stimulating fantasy and inventiveness, inspiration in architecture, painting, photography, music, dance, fashion, folklore Gaining knowledge of environmental conditions, changes and threats by visually perceptible structures, processes and species	Natural soil profiles, functioning ecosystems, rare species, traditional land knowledge Impressive landscapes, mounts, rivers, cliffs, old trees Indication with lichens (air quality), indicator plants (site conditions)

Adapted with slight modification from Bastian, O., Grunewald, K., & Syrbe, R.-U. (2015). Classification of ES. In K. Grunewald & O. Bastian (Eds.), Ecosystem services—Concept, methods and case studies (p. 50). Berlin: Springer.

comprising raw materials beneath the earth's surface along with wind and energy cannot be allocated to any specific ecosystems and hence, according to Bastian et al. (2015), these are not to be considered ecosystem goods and services.

Human intervention leading to modification in ecosystems, as in the case of farmland, renders it difficult to distinguish between natural and human inputs in labor, material and energy to a service or good (Bastian et al., 2015). Biodiversity is the upshot of innumerable ecological and evolutionary events that have taken place over many scales in time and space due to genetic modifications facilitated by human intervention along with changes prodded by economic interests over many millennia. Natural resources and natural capital stocks are finite and, in some ways, analogous to financial capital in bank accounts (Haldane & May, 2011) and extraction of these resources is subject to a limit whereas overextraction either leads to depletion or crash of the system (Raffaelli, 2016). Some countries of Europe have well-regulated forest management system entailing a good stock-and-flow example of an ecosystem-yield approach that aspires for a sustainable timber provision.

3.4.1.2 Regulating services

Regulating services forked out via different coproduction processes are mostly beneficial for human well-being locally, while local, regional and global changes like pollution, landscape, fragmentation and climate warming, etc., impact upon many regulating services across multiple scales (Gill et al.,

2016). Raudsepp-Hearne, Peterson, and Bennett (2010) have opined that, on the whole, service diversity linked to biodiversity seems to be a good and reliable predictor for the delivery of regulating ESs. Human life is substantially conditioned by the biosphere and its ecosystems. Survival of human life on earth is exceedingly dependent on procedures such as energy transformation chiefly from solar radiation into biomass, storage and transfer of mineral material and energy in food chains, biogeochemical cycles, mineralization of organic matter in soils, climate regulation, etc. (Bastian et al., 2015). In contrast, interaction of abiotic factors with living organisms wields immense impact on these procedures or processes to galvanize them. Securing survival and smooth functioning of ecosystems, especially natural and quasinatural ecosystems, is essential to enable the people to continue to enjoy the benefits of these processes in the future. Indirect benefits accruing from Regulating services (Table 3.2) often renders them either to be ignored or neglected until they get damaged or lost, whereas according to De Groot et al. (2002) they constitute the basis for human life on earth. Regulating services inter alia include: Climatologic and air hygienic services, hydrological services, pedological services and biological services (Table 3.2).

3.4.1.3 Sociocultural services

Multiple opportunities are afforded by natural and seminatural ecosystems in terms of pleasure, inspiration, intellectual enrichment, esthetic delight and recreation. Such services, often nomenclature as "psychological—social" services or "sociocultural services" are as significant to the people as regulating and provisioning services. Nevertheless, such services are either get ignored or don't get fully appreciated. This can partially be attributed to the difficulty of evaluating these services economically, specifically in terms of money (Bastian et al., 2015). ESs also render information services in science and technology to ascertain knowledge about natural interrelations, environmental conditions and stimulus to fantasy and inventiveness (Table 3.3). The values of some of these ESs are seemingly prone to demonstrate the level or national of social organization. Take the example of zoological and botanical gardens that entail the potential of positively influencing the attitudes of the visitors (Williams, Jones, Gibbons, & Clubbe, 2015). Besides, ornamental plants, "charismatic species," or totemic animals like North American bald eagle, etc., are such organisms whose presence causes emotional changes in humans.

Bastian et al. (2015) have advised that any attempt to develop or evolve a commonly applicable classification system of ESs needs to be perceived carefully because it is not targeted to some extent and up to a large extent, ESs are the outcome of compound interactions of the biotic and abiotic environment, claims on utilization and anticipation of the users. It is worth mentioning here that adoption of any incompatible classification system is bound to yield unreliable results. It is in this backdrop that Bastian et al. (2015) insist on adopting a classification system that facilitates distinction between intermediate and final services along with benefits. Allusion is made to the common International Classification of Ecosystem Goods and Services (CICES) promoted by the European Environment Agency (EEA) and according to Haines-Young and Potschin (2010), the goal of CICES is, beginning with MEA (2005) to develop a new classification system that is compatible with already existing national accounting systems.

3.4.2 Ecosystem services and biodiversity

According to the CBD, biodiversity, or biological diversity means "the variability among living organisms from all sources including, inter alia, terrestrial, marine and other aquatic ecosystems and the ecological complexes of which they are part; this includes diversity within species, between species and of ecosystems" (CBD, 2010a, 2010b). This definition of biodiversity that is binding under international law, comprises diversity of species, the diversity of ecosystems, and genetic diversity. Usually, biodiversity and ESs are referred to in identical terms in view of the fact that diversity of ecosystem, biotic associations and landscapes are the components of biodiversity (TEEB, 2009). In view of the fact that biodiversity especially supports the functioning of ecosystems, Grunewald et al. (2015, p. 4) have opined that biodiversity can also be defined as an ES in its own right as the ES of providing biodiversity. Undoubtedly, concept of ESs and notion of biodiversity intersection or overlap each other; nonetheless, they are not similar or identical. Undeniably, any loss or degradation of biodiversity is prone to impact ES; nevertheless, no direct or linear relationship between the two can be presumed (IEEP, 2009).

By virtue of directly producing goods and enabling ecosystem functions or services (Mace, Norris, & Fitter, 2012), biodiversity can be said to be a key driver of ESs (Harrison et al., 2014). Identification of fields of high significance for biodiversity conservation and ES delivery can sometimes be facilitated (Egoh, Reyers, Rouget, Bode, & Richardson, 2009) that can entail potential for unfolding opportunities for comanagement (Sachs et al., 2009). Undoubtedly, CBD's 2014 Global biodiversity outlook describes this development (CBD, 2014) being instrumental in leading to enhanced policy-level emphasis on whole-ecosystem approaches to biodiversity conservation; nevertheless, some scholars like Schröter et al. (2014) argue that important debate still remains over the pertinence or relevance of ES approaches to biodiversity conservation, especially, as pointed by Cardinale et al. (2012), when our understanding of biodiversity and ES linkages still remains inadequate.

A direct or linear positive linkage between biodiversity and delivery of individual ES is not always apparent. Relationship between biodiversity and ES either takes divergent forms and shapes like nonlinear relationship (Gamfeldt et al., 2013) or exhibits assorted relationship (Cardinale et al., 2012) or be altogether nonexistent (Cardinale et al., 2012; Harrison et al., 2013). Presence of assorted relationships between biodiversity and ESs for individual ESs plays up the unevenness in the impact of biodiversity on a given ES in a specific context (Cardinale et al., 2012; Harrison et al., 2013). Besides, any type of dissimilarity in individual biodiversity−ES relationship can eventually result in trade-offs as well as synergies between multiple ESs (Cardinale et al., 2012).

While referring to the prevalence of such unevenness in the relationships between biodiversity and ESs, Bennett, Peterson, and Gordon (2009) and Nicholson et al. (2009) have pointed out that research in the realms of biodiversity and ESs has traditionally adopted a rather "top-down correlative" approach, with the implicit ecological mechanisms being neglected or ignored. The resultant impact has been that in spite of the ever-growing literature on biodiversity−ES, as pointed out by Duncan, Thompson, and Pettorelli (2015), we are still far away from comprehending the mechanisms influencing relationships between biodiversity and ESs. Cardinale et al. (2012) have opined that fields of biodiversity and ESs are widely recognized to be disconnected or detached from each other, seldom working together and usually being carried out in diverse contexts and at entirely different scales. Lamenting at the inability of extending current lessons leant from the growing corpus of emerging mechanistic research in the fields of biodiversity and ESs to studying the relationships between biodiversity and ESs, "We may thus be looking at too simplistic a picture of B−ES relationships and overlooking opportunities to link biodiversity and ES delivery via full functional pathways" (Duncan et al., 2015, p. 1).

3.5 Ecosystem-based adaptation: costs and benefits
3.5.1 Costs and benefits of adaptation

The mechanism of costs and benefits analysis of EbA is closely linked to the mechanism of costs and benefits of adaptation, an analysis of which plays a significant role in justifying the case for action, and for prioritizing available resources to deliver ample socioeconomic and environmental benefits. The information thus garnered bears relevance at the global level, as an input to the negotiations and deliberations on international financing needs. In equal terms, this information bears relevance for national adaptation plans (NPAs) to facilitate efficient, effective and equitable strategies; as well as for local and project level adaptation, as a vital input to facilitate appraisal. At a theoretical level, as Stern (2007a, 2007b) suggests, a common framework can be used or the analysis of costs and benefits at all three geographic levels, and this has garnered wider approval and acceptance. In the first instance, this framework makes an assessment of the impacts and economic costs of climate change, including slow onset trends and changes in extreme events; thereafter, it proceeds to assess the potential costs and benefits of adaptation to minimize these impacts. Usage of this information can also be made to assess the economic efficacy of adaptation to ascertain whether the economic benefits of adaptation outweigh the costs. Besides, this information also entails the potential of being used to compare alternative adaptation options. According to Chiabai et al. (2015), there is an additional step to be undertaken in this analysis, and this of assessing the residual impacts of climate change after adaptation, denoting that it will be rarely be completely effective—or even technically feasible or possible—to remove impacts completely. Striking a semblance of equilibrium between the costs of adaptation measures, the benefits of adaptation measures and Chaabi the residual impacts is perhaps the most effective or even economically optimal level of adaptation (Chiabai et al., 2015).

In the aftermath of the publication of the AR5 of the IPCC in 2014, the discussion around climate change has shifted toward a focus on risks (IPCC, 2014a), thereby leading to some changes in the terminology compared to the framework discussed above, with focus on future climate risks rather than impacts, and residual risks remaining after adaptation. In a risk framework, the costs of adaptation are equivalent to investments in risk reduction, and it has led to a "change in the framing around adaptation, moving away from the previous impact-assessment framework toward relative climate risk management" (Chiabai et al., 2015).

Over the years, a number of methods are said to have been developed to get estimates of the costs and benefits of adaptation (Watkiss & Hunt, 2010), though most of these have primarily used the impact-assessment method, while the approach is simple and forward, there are numerous challenges when putting this framework in practice (UNEP, 2014a, 2014b). In the first place, it is difficult to estimate the future impacts and economic costs of climate change, owing to the wide range of potential risks, the scientific and economic information available, data gaps and modeling impediments. And these issues are prone to get amplified when taking into account adaptation costs and benefits, particularly in the wake of the large number of potential adaptations available. Secondly, the uncertainty associated with future climate change in the wake of already existing challenges makes it more cumbersome. At a time when it is unclear as to what future emission pathway the world is on, and even if this were known, significant climate model uncertainty would remain (Chiabai et al., 2015). This uncertainty entails two outcomes; in the first instance, it makes it cumbersome to estimate the scale of the impacts of climate change and the benefits of adaptation; and secondly, it enhances the adaptation costs relative to a situation where it can be assumed to be able to predict the future.

Under the impact framework, the costs and benefits of adaptation are determined by the framework that is used and the objectives that are set forth—for instance, whether the optimal level is based on economic efficacy versus a defined level of acceptable risk. These vary with context, country and across stakeholder groups and this means that it is very problematic to make available a definitive cost of adaptation. Besides, the baseline assumptions and the future timescales under investigation in the impact framework are prone to lead to big variations in estimates, and as such, the option of discount rate (DR) is of specific relevance in this context because it impacts the weight put on benefits taking place in the future. There is another problematic issue with the existing adaptation deficit—the gap between the current state of a system and a state that reduces the adverse impacts from the existing climate variability—as adaptation to future climate change will be less effective if this deficit is not addresses first (Burton, 2012). Admittedly, this is a specific problem confronting the developing countries; nonetheless, even OECD countries also have adaptation deficits or are near to the limits of dealing with current climate variability (ASC, 2011). Mechanisms like trade and financial flows between and across countries entail the potential of spilling over the impacts of climate change in one area into other areas, and as such, these can only be modeled at a worldwide scale, concurrently entailing the potential of affecting costs reported at the regional, national or local scare.

Usage of the assumptions to deal with these challenges entails the potential of greatly impacting results and a consequence of this is that the results of any, and the estimates of the costs and benefits of adaptation they deliver, entail the potential to be ambiguous or misleading if viewed in isolation. Therefore, it is significant for any study to be transparent with regard to assumptions used and implications of these on potential decisions. One of the primary objectives of examining the costs and benefits of adaptation is to facilitate allocation of resources, to inform national adaptation planning by governments through to local decisions. Thus, the impact-assessment framework reflects a "stylized model of reality," that calculates technical costs that are used to estimate the reduction in future damages, and while "such studies are useful for raising awareness, and generating headline estimates of practical (early) adaptation as they are highly theoretical" (Chiabai et al., 2015). While arguing that the framing of adaptation, in responses to the multiple challenges of adaptation costing and other emerging issues, is moving away from a focus on science first and impact-assessment, Chiabai et al. (2015) note that now the emphasis is on the use of iterative climate risk management to consider uncertainty has emerged that focuses on climate and nonclimate risks as a dynamic set of risks, and identifies phased adaptation. Accordingly, usage of different approaches has made too difficult to compile and compare estimates. While emphasizing that studies now make use of different methods, objectives, metrics and assumptions, and often focus on different time periods, and are carried out at different scale and geographical resolution, and that no method is absolutely right or wrong and they all have strengths and weaknesses in accordance with the objectives of the exercise and the specific application, Chiabai et al. (2015) note a major difference between earlier and later studies, and according to him, the focus is on the state-of-the-art and key lessons, rather than providing absolute estimates of the costs of adaptation.

3.5.2 Costs and benefits of ecosystem-based adaptation

Economic perspective is the way that makes the case for EbA relative to other adaptation activities because EbA activities entail the potential of delivering multiple benefits for communities, and perhaps there might be cases where other adaptation activities, including engineered solutions, may provide more benefits for less cost. Nevertheless, there are cases where hard-engineering solutions for adaptation are essential, there are still many instances where nature-based approaches can provide cost effective and/or economically beneficial, as well as longer term solutions with a variety of cobenefits in terms of the goods and services provided by ecosystems (Baig et al., 2016). Emphasis on economic assessments of EbA approaches entail the potential of bringing to limelight the efficacy of EbA projects comparable to hard-engineering projects, especially when development projects, in tandem with conservation and risk reduction aspects are taken into consideration. There is a need to investigate the benefits of EbA activities relative to their costs in a number of ecological, institutional and social settings in order to understand as to when EbA is an economically preferable approach. Moreover, it can be beneficial to have an assessment of the cost-effectiveness of EbA approaches by comparing their costs to other available options. Admittedly, EbA is garnering increasing traction; nonetheless, it is a developing and evolving field, and as such, there is paucity of data with regard to the economic benefits accruing from EbA projects and whether these benefits exceed the costs of implementation, and this lack of information has "hindered investment into EbA options, especially in developing countries, where communities are likely to impacted the most by a changing climate" (Baig, et al., 2016).

The significance of assessing economic costs and benefits of adaptation was highlighted by UNFCCC in its Nairobi Work Program (NWP) on impacts, vulnerability, and adaptation to climate change that was launched in 2005. The key objective of the NWP is to assist all member countries party to the UNFCCC, in particular developing countries, including the LDCs and small island developing states, to help improve their understanding and assessment of impacts, vulnerability and adaption in order to enable these countries to make informed decisions on practical adaptation actions and measures to respond to climate change on a sound scientific, technical and socioeconomic basis, taking into account current and future climate change and variability (UNFCCC, 2005). The NWP was expected to play an important role in the UNFCCC process via engaging stakeholders, catalyzing targeted action and facilitating knowledge sharing and learning on adaptation. In 2010, UNFCCC in its report informed on the new action pledges by partners of the NWP in response to the call for action focusing on adaptation planning and practices in relation to climate related risks and extreme events. The report described progress made under different areas of work and how organizations, institutions, experts, communities and the private sector had been engaged in the activities of the second phase of the NWP up to December 2010 (UNFCCC, 2010a, 2010b).

In view of limited evidence pertaining to actual adaptation costs and benefits, (Kumar et al., 2010) proposed, according key research. priority to the economics of EbA. Until that period, research on adaptation costs was related to climate change impact studies, where the objective was to understand the expected value of avoided climate associated damages. Kumar et al., 2010 focused on studies by Agrawala and Fankhauser (2008), which revealed that information on costs and benefits of adaptation is limited with the exception of coastal protection. Moreover, there have been other studies that examine costs of adaptation and the benefits accruing thereof, in terms of diminished vulnerability and enhanced welfare. Kumar et al., 2010 have pointed out toward a small number of studies undertaken prior to 2010 and showing the benefits of EbA in economic terms against the cost of implementing those projects. Nevertheless, some studies have endeavored to present estimation of short-term adaptation costs based on extending protected areas, broader conservation measures and off-reserve measures (Baig et al., 2016), and such responses deal with current vulnerability and according to Watkiss, Downing, and Dyszynski (2010), the case can be made that they would result in increased resilience to climate change.

While emphasizing that it is equally important to understand as to how the benefits and costs are distributed among various groups and across genders that helps to explain why people must invest in EbA activities and how additional investment can be garnered, Baig et al. (2016) have opined that governments and investors invest in EbA when the benefits outweigh the costs. It can sometimes happen that in certain cases, the benefits and costs could go to the same person or group of people and if the benefits outweigh the costs, the same person or group of people would like to invest in EbA in terms of finances, time and effort. Interestingly, Baig et al. (2016) hypothesizes a situation where in some cases the benefits and costs might be dispersed among such a broad group of people that no single person would have an incentive to invest in EbA because the that person would bear the costs, while the benefits would be enjoyed by the larger group. There is also possibility of nonincorporation of gender consideration, such that while the work load of women could increase as and when project is implemented, they could be deprived of the benefits.

Thus, a prior information of the manner of distribution of the benefits and costs of EbA is helpful in identifying economic policies, such as subsidies and taxes that can align individual incentives to attain socially beneficial outcomes when they otherwise would not take place (Rizvi, Baig, & Verdone, 2015). Accordingly, it is not only significant to assess the environmental and social costs and benefits of adaptation but also its economic costs and benefits as well in order to engage in an informed planning process (UNFCCC, 2011b). It is equally important concurrently to have a comparative assessment of costs and benefits of EbA related projects to ascertain their effects on local communities, specifically the marginalized groups such as women, children and the elderly, and ecosystems, as well as on national economies, and the information or data thus accumulated will be helpful in promoting EbA related projects within countries and at the global forums (Rizvi et al., 2015).

In view of multiple uncertainties associated with future climate change along with the manner in which ecosystem goods and services get delivered, measuring costs and benefits of EbA options become difficult. Besides, future socioeconomic development also obstructs the identification and implementation of maximal adaptation options (UNFCCC, 2011b). Undoubtedly, there are some well-delineated baseline scenarios portraying the impacts wielded by climate change on natural capital stocks and ESs; accordingly, while costs of implementing EbA projects are available in some cases, there is generally no information with regard to economic benefits made available by the EbA projects as well as how these benefits are handed out or passed on. In the backdrop of such a scenario, Rizvi and colleagues (2015) opine that it becomes difficult to compare and contrast the costs and benefits of ecosystem-based adaptation. Lack of baselines entails the potential of rendering cost−benefit data and measurements embracing uncertainties associated with them (UNFCCC, 2011b). Undoubtedly, there has emerged increasing trend of undertaking economic valuation of ecosystem goods and services; nonetheless, there are still data gaps, often culminating in incomplete cost and benefit assessments. Short-term nature of projects often makes the measurement of economic benefits of ESs tending to be for short-term. There are identical difficulties in assessing the costs and benefits in the case of adaptation, specifically when looking at EbA and its related benefits that come as a host of products and services (Devisscher, 2010). According to IPCC's Fourth Assessment Report, adaptation costs are "the costs of planning, preparing for, facilitating, and implementing adaptation measures, including transition costs, and benefits are the avoided damage costs of the accrued benefits following the adaptation and implementation of adaptation measures" (IPCC, 2007a).

It is essential to assess the projected climate change impacts and the costs of various adaptation options to facilitate reaching estimates of the benefits of adaptation, including EbA, against a baseline scenario. It is equally significant to know that EbA or any other adaptation measure will not generally lead to wiping out of the adverse impacts of climate change; consequently, the cost of residual damage—the damage left over after implementation of option—must also be incorporated in the overall costs. According to the UNFCCC (2011b), the options with the highest net benefits are the ones that are required to be selected for implementation. Moreover, inclusion of cobenefits of nature-based solutions in the estimation of overall net benefits assumes significance to be able to get a holistic picture. There is an array of approaches that are used and implemented to examine economic benefits of goods and services and these same approaches "can" and "are" used to examine costs and benefits of adaptation options, including EbA. As per UNFCCC (2011b), there are three most prominent and commonly used approaches: (1) cost−benefit analysis (CBA); (2) cost-effective analysis; and (3) multicriteria analysis. Besides, there is also total economic valuation that is often used to examine the economic benefits made available by various ecosystems goods and services.

CBA entails calculating and comparing all of the costs and benefits and expressing them in monetary terms, and CBA is used when efficiency is the sole decision-making criterion for selecting adaptation measure, including EbA. CBA requires the distribution of costs and benefits to be taken into consideration, along with their aggregate values, with the result that an assortment of impacts can be compared by adhering to use a single metric (UNFCCC, 2011b). Nevertheless, in the case of EbA, it is relatively difficult to come to monetary conclusions using the CBA because many ecosystems' goods and services seldom have a market value, nonetheless, they may deliver innumerable unmeasured economic benefits. Moreover, CBA does not address the issues of gender equity and distribution of costs and benefits (UNFCCC, 2011b), and accordingly, this aspect needs to be taken into consideration when selecting the projects on the basis of CBA. Furthermore, many nature-based solutions are inclined to deliver benefits—economic and other—after a longer duration of time, and that generally comes after the project is completed. Baig et al. (2016) suggest that two approaches can be applied while undertaking CBA. In the first instance, CBA can be undertaken prior to initiating the project to help stakeholders understand the costs and benefits of different EbA activities. Secondly, there is a need to undertake detailed economic analyses of ongoing and completed projects with a view to comprehend and garner evidence that explains why EbA provides more economic benefits than other solutions and accordingly this can be extrapolated for national level EbA approaches, policies and strategies.

CBA of the adaptation measures takes into account both the private and external benefits accruing from these EbA interventions so as to augment the ESs to enable people adapt to the emergent impacts of climate change, and as such, CBA is an economic tool to facilitate better decision-making (GoN/UMDP, 2015). Given the fact that different ecosystems of a landscape make available provisioning, regulating, cultural and supporting services, accordingly, these services quantified and valued to have an estimation of the accrual of the benefits from the landscape so that various investments can be made to adapt to the changing climate scenario and to enhance the accrual of benefits from these ESs. CBA enables to have an assessment of the cost of these investments and thereafter endeavors to compare it with the outcome benefits as perceived in the enhanced ESs, and if the benefits outweigh the cost, it would make an economic sense to invest in further areas in the future, and as such, it helps in making suitable and effective policies, including the ones in adaptation. While offering alternative solutions to decision-makers, CBA also helps them in making robust decision when it comes to implementing projects to deal with the vagaries of climate change. It can be discerned from case studies of Amriso and Timur cultivation in Panchase in Nepal, undertaken by the Government of Nepal with assistance of the UNDP, that EbA interventions can provide better economic returns, and the research also demonstrates that the provision of quality seedlings and seeds by the government would provide more incentives to the farmers to plant these species (GoN/UNDP, 2015).

Rodgers (2014) finds CBA useful for four decision-making purposes: (1) evaluating a stand-alone EbA intervention ("no project" baseline), (2) evaluating EbA intervention(s) against alternative approach ("no project" baseline), (3) evaluating EbA as one component (climate-proofing measure) of a proposed investment project (baseline: project without EbA intervention), and (4) evaluating EbA as one of alternative approaches to climate-proofing an investment project (baseline: project without EbA intervention). Concurrently, UNFCCC (2011b) suggests important steps in undertaking a CBA, and that inter alia include: (1) Agree on the adaptation objective and identify potential adaptation options; (2) Establish a baseline; (3) Quantify and aggregate the costs over specific time periods; (4) Quantify and aggregate the benefits over specific time periods; (5) Compare the aggregated costs and benefits. In this regard, Venton (2010) has suggested three indicators that can be used to arrive at results. According to him, benefit-to-cost ratio signifies the level of Vinton benefit that will be accrued for every $1 of cost. A ratio greater than 1; therefore, is indicative of the fact that the project is worth investment from a financial perspective, whereas anything less than one symbolizes the fact of a negative return. Net present value takes the net benefit (benefit minus costs) annually and discounts these to their current value. Furthermore, if the result is greater than zero, that denotes that the benefits outweigh the costs. The higher the value, the greater the financial incentive for initiating the project. Usage of DR is facilitated to discount costs and benefits taking place in the future, as people often accord higher value on assets provided in the present and accord lower value on benefits accrual of which may take place further in the future.

An example of usage of CBA is provided by Gray and Srinidhi (2013) in their study of CBA of Watershed Development (WSD) for the Kumbharwadi rain-fed watershed in Maharashtra (India). The emerging trends indicate that the market benefits of this specific WSD inter alia include improved livestock and crop sales; avoided travel costs for migratory work and drinking water; avoided cost of government supplied water tankers; and improved fuelwood and fodder supplies. The nonmarket benefits, inter alia, included: Carbon sequestration; a cobenefit of the afforestation and reforestation intervention; improved biodiversity; pollination and water filtration; improved nutrition and health; increased enrollment in schools owing to improved livelihoods and resultant enhanced incomes; female empowerment; community development; and improved resilience to drought. The cumulative impact of these benefits is prone to help to minimize the vulnerabilities of the local communities to the vagaries of climate change. Baig et al. (2016) in their study on "Costs and Benefits of Ecosystem-based Adaptation: A Case Study of the Philippines," aver that the prime objective of the study is to make "case that undertaking analyses of economic costs and benefits of EbA options is important and must be undertaken to assist in decision and policy making, as well as demonstrating the real benefits (cobenefits) to human livelihoods and to natural resource management." This; however, should be done while considering their strengths and weaknesses and ensuring that aspects that often get neglected are included (Baig et al., 2016, p. 20).

Wide-ranging institutional, sociocultural, ecological, and economic benefits accruing from EbA approaches help in promoting restoration and protection of ecosystems thereby spurring emergence of healthy ecosystems (McGray, Hammill, & Bradley, 2007) that in turn entails the potential of serving as a natural barrier to adverse impacts of climate change like droughts, landslides, flooding, and extreme temperatures, among others (Andrade et al., 2010). Resilience of a healthy ecosystem serves as a bulwark against the vagaries of climate change thereby enabling the communities to continue to utilize the benefits accruing from ESs rendered by a healthy ecosystem (Falkenburg, Burnell, Connell, & Russell, 2010). According to Locatelli, Kanninen, Brockhaus, Murdiyarso, and Santoso (2008), EbA approach facilitates protection, restoration and management of ecosystems that in turn, apart from promoting conservation of biodiversity, also helps in building competency and capacity of the people to adapt to climate change variability (Mercer, Dominey-Howes, Kelman, & Lloyd, 2007), eventually contributing to sustainable development. The Nature Conservancy (TNC, 2009) has provided broad benefits accruing from adoption of EbA approach (Table 3.4).

According to Devisscher (2010), a salient characteristic of the EbA approaches is their applicability to virtually all types of ecosystems as well as at multiple scales—from the local to the national, regional, and international. According to TEEB (2009), along with achieving long-term and short-term priorities, EbA can also engender manifold environmental and societal benefits. Vignola et al. (2009) ascribe to the view that multisectoral and multiscale traits of EbA empowers with the capacity to assimilate an array of variety of disciplines, stakeholders, and institutions, so that they can function at various governance levels and can influence a number of decision-making networks thereby delivering benefits to diverse groups and communities.

McVittie et al., 2017 have opined that the evidence for cost-effectiveness of EbA is currently sparse, and difficult to quantify. What comprises an EbA can incorporate an array of actions, and many a times its efficacy at attaining the adaptation goal as well as the cobenefits may be indistinct. Current evidence with regard to efficacy of EbA banks on observations of contemporary conditions instead of future climate change impacts and such an eventuality is prone to lead to a significant incertitude regarding the future effectiveness of measures. Additionally, perhaps more than adaptation measures in other sectors, EbAs are "extremely locally specific," meaning that "what may be successful in one area may not be in another" (McVittie et al., 2017).

Table 3.4 Benefits accruing from ecosystem-based adaptation approach.

Restoring fragmented or degraded natural areas Protecting groundwater recharge Zones or restoration of floodplains Connecting expanses of protected forests, grasslands, reefs, or other habitats Protecting or restoring natural infrastructure such as barrier beaches, mangroves and coral reefs	Enhances critical ecosystem services such as water flow or fisheries provision Secures water resources so that entire communities can cope with drought Enables people and biodiversity to move to better or more viable habitats as the climate changes Buffers human communities from erosion and flooding

From TNC (The Nature Conservancy). Adapting to climate change—Ecosystem-based approaches for people and nature *(p. 4). (2009). <https://www.sprep.org/att/irc/ecopies/global/306.pdf>*.

The study by Muthee, Duguma, Nzyoka, and Minang (2021) evaluates the contributions of EbA practices to the water−energy−food (WEF) nexus balance, design practical pathways, and analyze barriers toward attainment of EbA−WEF balance. Analyzing data collected from 50 community forests spread across three regions in The Gambia (in West Africa), this case study established fourteen priority EbA practices and categorized then into four major groups in accordance with their application similarities. The anticipated ESs, inter alia, included: enhanced water resource conservation, food and feed production, enhanced energy supply, and improved community livelihoods to augment their resilience. Key enablers identified in this case study included a conducive policy framework institutional support, diverse incentives, information, knowledge and technology transfer. Nevertheless, climate and nonclimate barriers were cited as obstacles. The study concludes by outlining recommendations to surmount the established barriers. Several practical implications for decision-makers and policy planners were mooted to demonstrate as to how EbA practices can promote the WEF nexus and contribute to the development of livelihood assets and these include: increasing intersectoral collaboration across different institutions, policy frameworks, and departments to reduce policy and institutional conflicts in the water, energy and food/agricultural sectors, increased incentives via credit facilities, subsidies in farm inputs, enterprise and market development, building capital and promoting the different EbA practices. It is further emphasized that designing interventions like EbA practices ought to be based on current and future climate and adaptation scenarios to boost their sustainability. It accords equal emphasis on the need for deliberate endeavors from government agencies, development partners, and other stakeholders to engender awareness and build social and human capital among the parties involved in the use and management of natural resources.

3.6 Conclusion

Undoubtedly, within a short span of time EbA has come to be reckoned as a popular concept yielding numerous benefits, having garnered support from the scientific community and conservation agencies; nevertheless, it has also become imperative for this concept to prove its ability to deliver all the benefits attributed to it, or else, as Mellmann (2015) has cautioned, "the momentum could be lost rather quickly." Availability of few systematic studies on EbA's efficacy suggests that "the evidence base is thus lacking information" (Doswald et al., 2014, p. 186). This brings into focus the limitations and challenges of the concept of EbA. Lack of a commonly accepted definition of EbA, as pointed out by Doswald et al. (2014, p. 186), can be the reason as to why some decision-makers show reluctance in adhering to EbA along with considering other options.

The process of quantification is seemingly another knowledge gap affecting EbA concept. Limited base of evidence to prove that EbA can be a cost-effective option to conventional techniques dependent on gray or hard infrastructure make adherence to EbA doubtful for some decision-makers. According to Mellmann (2015), another key aspect surrounding the EbA concept is time and timescale. While pointing out that adaptation is a long-term process and success of an intervention could make itself visible after the implementation of the project, Mellmann (2015, p. 8) suggests: "It is important to have evidence regarding timescales." It is in this regard that Doswald et al. (2014) has mooted the suggestion for effective integration of knowledge, research, and monitoring on timescales.

References

Adger, W. N., Arnell, N. W., & Tompkins, E. L. (2005a). Successful adaptation to climate change across scales. *Global Environmental Change, 15*(2), 77–86.

Adger, W. N., Hughes, T. P., Folk, C., Carpenter, S. R., & Rockstrom, J. (2005b). Social-ecological resilience to coastal disasters. *Science (New York, N.Y.), 309*(5737), 1036–1103.

Adger, W. N., Brown, K., Nelson, D. R., Berkes, F., Eakin, H., Folke, C., ... Tompkins, E. L. (2011). Resilience implications of policy responses to climate change. *Wiley Interdisciplinary Reviews: Climate Change, 2*(5), 757–766.

Adger, W. N., Lorenzoni, I., & O'Brien, K. O. (2009a). *Adapting to climate change: Thresholds, values, governance*. Cambridge: Cambridge University Press.

Adger, W. N., Dessai, S., Goulden, M., Hulme, M., Lorenzoni, I., Nelson, D. R., Naess, L. O., Wolf, J., & Wreford, A. (2009b). Are there social limits to adaptation to climate change? *Climate Change, 93*, 335–354.

Adger, W. N., Agrawala, S., Mirza, M. M. Q., Conde, C., O'Brien, K., Pulhin, J., ... Takahashi, K. (2007). Assessment of adaptation practices, options, constraints and capacity. Climate Change 2007: Impacts, adaptation and vulnerability. In M. L. Parry, O. F. Canziani, J. P. Palutikof, P. J. van der Linden, & C. E. Hanson (Eds.), *Contribution of working group II to the fourth assessment report of the intergovernmental panel on climate change* (pp. 717–743). Cambridge: Cambridge University Press.

Agrawala, S., & Fankhauser, S. (Eds.), (2008). *Economic aspects of adaptation to climate change. Costs, benefits and policy instruments*. Paris: OECD.

Ahmmad, R., Nandy, P., & Husnain, P. (2013). *Unlocking ecosystem-based adaptation opportunities in coastal Bangladesh, Journal of Coastal Conservation* (17, pp. 833–840).

Allison, E. H., Perry, A. L., Badjeck, M. C., Adger, W. N., Brown, K., Conway, D., ... Dulvy, N. K. (2009). Vulnerability of national economies to the impacts of climate change on fisheries. *Fish and Fisheries, 10*, 173–196.

Andrade, A., Herrera, B., & Cazzolla, R. (2010). *Building resilience to climate change: Ecosystem based adaptation and lessons from the field*. Gland: IUCN.

ARC (Australian Research Council). (2016). *Coral reef studies. Only 7% of the Great Barrier Reef has avoided coral bleaching*. <https://www.coralcoe.org.au/media-releases/only-7-of-the-great-barrier-reef-has-avoided-coralbleaching>.

ASC (Adaptation Sub Committee). (2011). Research to identify potential low-regrets adaptation options to climate change in the residential buildings sector. *Commissioned by the Adaptation Sub Committee, July 2011*. London: Climate Change Committee.

Baig, S. P., Rizvi, A., Josella, M., & Palanca-Tan, R. (2016). *Cost and benefits of ecosystem based adaptation: The case of the Philippines*. Gland: IUCN.

Barnett, J., Evans, L. S., Gross, C., Kiem, A. S., Kingsford, R. T., Palutikof, J. P., ... Smithers, S. G. (2015). From barriers to limits to climate change adaptation: Path dependency and the speed of change. *Ecology and Society, 20*(3), 5. Available from https://www.ecologyandsociety.org/vol20/iss3/art5/.

Barnett, J., Mortreux, C., & Adger, W. N. (2013). Barriers and limits to adaptation: Cautionary notes. In S. Boulter, J. Palutikof, D. John Karoly, & D. Guitart (Eds.), *Natural disasters and adaptation to climate change* (pp. 223–235). Cambridge University Press.

Bastian, O., Grunewald, K., & Syrbe, R.-U. (2015). Classification of ES. In K. Grunewald, & O. Bastian (Eds.), *Ecosystem services—Concept, methods and case studies*. Berlin: Springer.

Bastian, O., Haase, D., & Grunewald, K. (2012). Ecosystem properties, potentials and services—the EPPS conceptual framework and an urban application example. *Ecological Indicators, 21*, 7–16. Available from https://esanalysis.colmex.mx/Sorted%20Papers/2012/2012%20DEU%20−3F%20Phys%202.pdf.

Bennett, E. M., Cramer, W., Begossi, A., Cundill, G., Díaz, S., Egoh, E., ... Martín-Lopez, B. (2015). Linking ecosystem services to human well-being: Three challenges for designing research for sustainability. *Current Opinion in Environmental Sustainability, 14*, 76−85.

Bennett, E. M., Peterson, G. D., & Gordon, L. J. (2009). Understanding relationships among multiple ecosystem services. *Ecological Letters, 12*, 1−11.

Berrang-Ford, L., Ford, J. D., & Paterson, J. (2011). Are we adapting to climate change? *Global Environmental Change, 21*(1), 25−33.

BESWS (Biodiversity and Ecosystem Service Work Stream). *Demystifying materiality: Hardwiring biodiversity and ecosystem services into finance*. (2010). <http://www.unepfi.org/fileadmin/documents/CEO_DemystifyingMateriality.pdf>

Blaikie, P., Cannon, T., Davis, I., & Wisner, B. (1994). *At risk: Natural hazards, people's vulnerability and disasters*. London: Routledge.

Blennow, K., & Persson, J. (2009). Climate change: Motivation for taking measure to adapt. *Global Environmental Change, 19*, 100−104.

Bradshaw, B., Dolan, H., & Smit, B. (2004). Farm-level adaptation to climatic variability and change: Crop diversification in the Canadian prairies. *Climatic Change, 67*, 119−141.

Burkhard, B., Kroll, F., Müller, F., & Windhorst, W. (2009). Landscapes' capacities to provide ecosystem services—A concept for land-cover based assessments. *Landscape Online, 15*, 1−22.

Burton, I. (2012). Climate change and the adaptation deficit. In: *Paper 3 of the proceedings of the international conference on climate change: Building the adaptive capacity*. Available at: http://projects.ca/climate/files/2012/10/Book-5.Paper3.pdf.

Cadag, J., & Gaillard, J. (2012). Integrating knowledge and actions in disaster risk reduction: The contribution of participatory mapping. *Area, 44*, 100−109.

Cardinale, B. J., Emmett Duffy, J., Gonzalez, A., Hooper, D. U., Perrings, C., Venail, P., ... Naeem, S. (2012). Biodiversity loss and its impact on humanity. *Nature, 486*, 59−67.

CBD (Convention on Biological Diversity). (2000). *Decisions adopted by the conference of the parties to the convention on biological diversity at its fifth meeting*, Nairobi, 15−26 May 2000.

CBD (Convention on Biological Diversity). (2009). Connecting biodiversity and climate change mitigation and adaptation: Report of the Second Ad Hoc Technical Expert Group on Biodiversity and Climate Change. In: *CBD. Technical series no. 41*.

CBD (Convention on Biological Diversity). (2010a). Decision adopted by the conference of the parties to the convention on biological diversity at its 10th meeting. In: *X/33 Biodiversity and climate change, UNEP/CBD/COP/DEC/X/33*.

CBD (Convention on Biological Biodiversity). (2010b). Global biodiversity outlook 3. Montreal: CBD Secretariat.

CBD (Convention on Biological Diversity). (2014). Global biodiversity outlook 4. Montreal: CBD Secretariat.

Chiabai, A., Hunt, A., Galarraga, I., Lago, M., Rouillard, J., Sainz de Murieta, E., ... Watkiss, P. (2015). Using cost and benefits to assess adaptation options. *ECONADAPT Project*. Bath: University of Bath.

Cinner, J. E., Huchery, C., Darling, E. S., Humphries, A. T., Graham, N. A. J., Hicks, C. C., ... McClanahan, T. R. (2013). Evaluating social and ecological vulnerability of coral reef fisheries to climate change. *PLoS One, 8*(9), e74321.

Cinner, J. E., Huchery, C., Hicks, C. C., Daw, T. M., Marshall, N., Wamukota, A., & Allison, E. H. (2015). Changes in adaptive capacity of Kenyan fishing communities. *Nature Climate Change, 5*, 872.

Costanza, R., Grasso, M., D'Arge, R., De Groot, R., Farber, S., Hannon, B., & Van Den Belt, M. (1997). The value of the world's ecosystem services and natural capital. *Nature, 387*.

de Groot, R. S., Wilson, M. A., & Boumans, R. M. (2002). A typology for the classification, description and valuation of ecosystem functions, goods and services. *Ecological Economics, 41*(3), 393−408.

Delica-Willison, Z., & Gaillard, J. (2011). Community action and disaster. In B. Wisner, J. C. Gaillard, & I. Kelman (Eds.), *The Routledge handbook of hazards and disaster risk reduction*. London: Routledge.

Dessai, S., & Hulme, M. (2007). Assessing the robustness of adaptation decisions to climate change uncertainties: A case study on water resources management in the East of England. *Global Environmental Change*, *17*(1), 59−72.

Devisscher, T. (2010). Ecosystem-based adaptation in Africa: Rationale, pathways and cost estimates. In: *Sectoral report for the adapt cost study*. Stockholm: Stockholm Environment Institute (SEI).

Doswald, N., Munroe, R., Roe, D., Giuliani, A., Castelli, I., Stephens, J., ... Reid, H. (2014). Effectiveness of ecosystem-based approaches to adaptation: Review of the evidence-base. *Climate and Development*, *6*(2), 185−201.

Dow, K., Berkhout, F., & Preston, B. L. (2013). Limits to adaptation to climate change: A risk approach. *Current Opinion in Environmental Sustainability*, *5*, 384−391.

Duncan, C., Thompson, J. R., & Pettorelli, N. (2015). The quest for a mechanistic understanding of biodiversity-ecosystem services. *Proceedings of the Royal Society B. Biological Sciences*.

Egoh, B., Reyers, B., Rouget, M., Bode, M., & Richardson, D. M. (2009). Spatial congruence between biodiversity and ecosystem services in South Africa. *Biological Conservation*, *142*(3), 553−562.

EU (European Union). (2016). Taking stock on ecosystem-based initiatives in the European Commission, draft report EU Commission.

Falkenburg, J., Burnell, W., Connell, D., & Russell, B. (2010). Sustainability in near-shore marine systems: Promoting natural resilience. *Sustainability*, *2*(8), 2593−2600.

Gamfeldt, L., Snäll, T., Bagchi, R., Jonsson, M., Gustafsson, L., Kjellander, P., ... Bengtsson, J. (2013). Higher levels of multiple ecosystem services are found in forests with more tree species. *Nature Communication*, *4*, 1340.

Gawith, D. (2017). *Estimating the Adaptation Deficit—An empirical analysis of the constraints on climate change adaptation in agriculture*. Cambridge: University of Cambridge, Unpublished dissertation.

Gill, R. J., Bal dock, K. C. R., Brown, M. J. F., Cresswell, J. E., Dicks, L. V., Fountain, M. T., ... Stone, G. N. (2016). Protecting an ecosystem service: Approaches to understanding and mitigating threats to wild insect pollinators. *Advances in Ecological Research*, *54*, 135−206.

Girot, P., Ehrhart, C., & Oglethorpe, J. (2012). *Integrating community and ecosystem-based approaches in climate change adaptation responses*. ELAN—Ecosystem & Livelihoods Networks. Available at http://cmsdata.iucn.org/downloads/a_eba_integratedapproach_15_04_12_0.pdf.

Glick, P., Chmura, H., & Stein, B. A. (2011). Moving the conservation goalposts: A review of climate change adaptation literature. *National Wildlife Federation*.

GoN/UNDP (Government of Nepal—United Nations Development Program). (2015). Cost-benefit of EbA interventions: Case studies from Panchase project area. In: *Cost benefit analysis report*. Kathmandu, Nepal.

Gray, E., & Srinidhi, A. (2013). *Watershed Development in India: Economic Valuation and Adaptation Considerations. Working Paper*. Washington, D.C.: World Resource Institute.

Grothmann, T., & Patt, A. (2005). Adaptive capacity and human cognition: The process of individual adaptation to climate change. *Global Environmental Change*, *15*, 199−213.

Grothmann, T., & Reusswig, F. (2006). People at risk of flooding: Why some residents take precautionary action while others do not. *Natural Hazards*, *38*, 101−120.

Grunewald, K., & Bastian, O. (2015). Ecosystem services (ES): More than just a vogue term? In G. Karsten, & B. Olaf (Eds.), *Ecosystem services—Concept, methods and case studies*. Springer-Verlag.

Göhler, D., Müller, F., Mytanz, C., Oliver, J., Renner, I., Riha, K., & Tscherning, K. (2013). *Ecosystem-based adaptation (EbA)*. Bonn: GIZ.

Haldane, A. G., & May, R. M. (2011). Systemic risk in banking ecosystems. *Nature*, *469*, 351−355.

Harrison, P. A. (2010). Ecosystem services and biodiversity conservation: An introduction to the RUBICODE project. *Biodiversity Conservation*, *19*, 2767−2772.

Harrison, P. A., Berry, P. M., Simpson, G., Haslet, J. R., Blicharska, M., Bucur, M., ... Turkelboom, F. (2014). Linkages between biodiversity attributes and ecosystem services: A systematic review. *Ecosystem Services*, 9, 191–203.

Harrison, P. A., Holman, I. P., Cojocaru, G., Kok, K., Kontogianni, A., Metzger, M. J., & Gramberger, M. (2013). Combining qualitative and quantitative understanding for exploring cross-sectoral climate change impacts, adaptation and vulnerability in Europe. *Regional Environmental Change*, 13(4), 761–780.

Hubacek, K., & Kronenberg, J. (2013). Synthesizing different perspectives on the value of urban ecosystem services. *Landscape and Urban Planning*, 109(1), 1–6.

IEEP–Institute for European Environmental Policy, Alterra, Ecologic, PBL–Netherland Environmental Assessment Agency und UNEP-WCMC (2009). Scenarios and models for exploring future trends of biodiversity and ecosystem services change. In: *Final Report to the European Commission, DG Environment on Contract ENV.G.1/ ETU/2008/0090r*.

IPCC (Intergovernmental Panel on Climate Change). (2007a). *Climate Change 2007: Synthesis Report. Contribution of Working Groups I, II and III to the Fourth Assessment Report of the Intergovernmental Panel on Climate Change*. Geneva, Switzerland: IPCC.

IPCC (Intergovernmental Panel on Climate Change). (2007b). Climate change 2007: Impacts, adaptation and vulnerability. *Contribution of Working Group II to the Fourth Assessment Report of the Intergovernmental Panel on Climate Change*. Cambridge: Cambridge University Press.

IPCC (Intergovernmental Panel on Climate Change). (2013). *Summary for policymakers*. Climate change 2013: The physical science basis. *Contribution of Working Group I to the Fifth Assessment Report of the Intergovernmental Panel on Climate Change*. Cambridge: Cambridge University Press.

IPCC (Intergovernmental Panel on Climate Change). (2014a). *Summary for policymakers*. Climate change 2014: Impacts, adaptation, and vulnerability. Part A: Global and sectoral aspects. *Contribution of Working Group II to the Fifth Assessment Report of the Intergovernmental Panel on Climate Change* (pp. 1–32). Cambridge: Cambridge University Press.

IPCC (Intergovernmental Panel on Climate Change). (2014b). *Climate Change 2014: Mitigation of Climate Change, Contribution of Working Group III to the Fifth Assessment Report of the Intergovernmental Panel on Climate Change*. Cambridge, UK: Cambridge University Press.

Islam, M. M., Sallu, S., Hubacek, K., & Paavola, J. (2014). Limits and barriers to adaptation to climate variability and change in Bangladeshi coastal fishing communities. *Marine Policy*, 43, 208–216.

IUCN (International Union for Conservation of Nature). (2015). *Nature based solutions for human resilience: A mapping analysis of IUCN's ecosystem-based adaptation projects*. Gland, Switzerland: IUCN.

Jones, H. P., Hole, D. G., & Zavaleta, E. S. (2012). Harnessing nature to help people adapt to climate change. *Nature Climate Change*, 2(7), 504–509.

Jones, L., & Boyd, E. (2011). Exploring social barriers to adaptation: Insights from Western Nepal. *Global Climate Change*, 21, 1262–1274.

Kellstedt, P. M., Zahran, S., & Vedlitz, A. (2008). Personal efficacy, the information environment, and attitudes toward global warming and climate change in the United States. *Risk Analysis: An Official Publication of the Society for Risk Analysis*, 28(1), 113–126.

Kithiia, J., & Lyth, A. (2011). Urban wild-scapes and green spaces in Mombasa and their potential contribution to climate change adaptation and mitigation. *Environment and Urbanization*, 23(1), 251–265.

Klein, R. J. T., Midgley, G. F., Preston, B. L., Alam, M., Berkhout, F. G. H., Dow, K., et al. (2014). Adaptation opportunities, constraints and limits. In C. B. Field, V. R. Barros, D. J. Dokken, et al. (Eds.), Climate change 2014: Impacts, adaptation, and vulnerability. Part A: Global and sectoral aspects. *Contribution of the working group II to the fifth assessment report of the intergovernmental panel on climate change* (pp. 899–943). Cambridge: Cambridge University Press.

Kumar, K. K. S., Shyamsundar, P., & Nambi, A. A. (2010). *The economics of climate change adaptation in India—Research and policy challenges ahead.* Policy Note Nr. 42–10, April 2010. The South Asian Network for Development and Environmental Economics [SANDEE]. Available at: https://www.epw.in/journal/2010/18/commentary/economics-climate-change-adaptation-india.ntml.

Leal, F. W. (Ed.), (2001). *The economic, social and political aspects of climate change.* Berlin: Springer.

Locatelli, B., Kanninen, M., Brockhaus, C., Murdiyarso, D., & Santoso, H. (2008). Facing an uncertain future: How forests and people can adapt to climate change. CIFOR (Center for International Forestry Research). *Forest Perspectives*, 5. Available from http://hal.cirad.fr/cirad-00699333/document.

Locatelli, B., & Pramova, E. (2015). *Ecosystem-based adaptation.* Indonesia: École thématique.

Mace, G. M., Norris, K., & Fitter, A. H. (2012). Biodiversity and ecosystem services: A multilayered relationship. *Trends in Ecology and Evolution*, 27, 19–26.

Maddux, J. E., & Rogers, R. W. (1983). Protection motivation and self-efficacy: A revised theory of fear appeals and attitude change. *Journal of Experimental Social Psychology*, 19(5), 469–479.

Marshall, N., Park, S., Howden, S. M., Dowd, A. M., & Jakku, E. (2013). Climate change awareness is associated with enhanced adaptive capacity. *Agricultural Systems*, 117, 30–34.

McCarthy, J. J., Canziani, O., Leary, N. A., Dokken, D. J., & White, K. S. (2001). Climate change 2001: Impacts, adaptation and vulnerability. *IPCC Working Group II.* Cambridge: Cambridge University Press, Intergovernmental Panel on Climate Change.

McGray, H., Hammill, A., & Bradley, R. (2007). *Weathering the storm: Options for framing adaptation and development.* Washington, DC: World Resources Institute (WRI).

McVittie, A., Cole, L., & Wreford, A. (2017). *Assessing adaptation knowledge in Europe: Ecosystem-based adaptation.* Brussels: European Union. Available at: http://ec.europa.eu/clima/sites/clima/fimes/adaptation/what/docs/ecosystem_based_adaptation_en.pdf.

Mellmann, N. (2015). Ecosystem-based adaptation. In: *Theory and practice—A case study of projects supported by the international climate initiative* (Unpublished Master's thesis). Uppsala: Department of Earth Sciences, Uppsala University. Available at http://www.diva-portal.org/smash/get/diva2:848265/FULLTEXT03.

Mendelsohn, R., & Dinar, A. (2009). *Climate change and agriculture.* Cheltenham: Edward Elgar Publishing.

Mercer, J., Dominey-Howes, D., Kelman, I., & Lloyd, K. (2007). The potential for combining indigenous and western knowledge in reducing vulnerability to environmental hazards in small island developing states. *Environmental Hazards*, 7(4), 245–256.

Mercer, J., Kelman, I., Afghan, B., & Kurvits, T. (2012). Ecosystem-based adaptation to climate change in Caribbean small island developing stats: Integrating local and external knowledge. *Sustainability*, 4, 1908–1932.

Millennium Ecosystem Assessment (MEA). (2005). *Ecosystems and human well-being: Current state and trends.* Washington: Island Press.

Müller, F., & Burkhard, B. (2007). An ecosystem based framework to link landscape structures, functions and services. In Ü. Mander, H. Wiggering, & K. Helming (Eds.), *Multifunctional Land Use* (pp. 37–63). Berlin, Germany: Springer.

Molnar, J. J. (2010). Climate change and societal response: Livelihoods, communities, and the environment. *Rural Sociology*, 75(1), 1–16.

Mortreux, C., & Barnett, J. (2017). Adaptive capacity: Exploring the research frontier. *Wiley Interdisciplinary Reviews: Climate Change*, 8(4), e467.

Moser, C., & Satterthwaite, D. (2009). *Towards pro-poor adaptation to climate change in the urban centers of low- and middle-income countries.* Discussion paper no. 3, IIED. Available at http://pubs.iied.org/pdfs/10564IIED.pdf.

Moser, S., & Ekstrom, J. A. (2010). A framework to diagnose barriers to climate change adaptation. *Proceedings of the National Academy of Sciences of the United States of America*, 107(51), 22026–22031.

Munang, R., Andrews, J., Alverson, K., & Mebratu, D. (2014). Harnessing ecosystem-based adaptation to address the social dimensions of climate change. *Environment*, *1*(56), 18−24.

Munang, R., Thiaw, I., & Rivington, M. (2013a). Ecosystem management: Tomorrow's approach to enhancing food security under a changing climate. *Sustainability*, *3*, 937−954.

Munang, R., Thiaw, I., Alverson, K., Mumba, M., Liu, J., & Rivington, M. (2013b). Climate change and ecosystem-based adaptation: A new pragmatic approach to buffering climate change impacts. *Current Opinion in Environmental Sustainability*, *5*, 1−5.

Muthee, K., Duguma, L., Nzyoka, J., & Minang, P. (2021). Ecosystem-based adaptation practices as a nature-based solution to promote water-energy-food nexus balance. *Sustainability*, *13*, 1142.

Muthee, K., Mbow, C., Macharia, G., & Filho, W. L. (2017). Ecosystem-based adaptation (EbA) as an adaptation strategy in Burkina Faso and Mali. In W. L. Filho, B. Simane, J. Kalangu, M. Wuta, P. Munishi, & K. Musiyiwa (Eds.), Climate change adaptation in Africa: Fostering resilience and capacity to adapt (pp. 205−215). Gland: Springer International Publishing.

Nalau, J., & Filho, W. L. (2018). Limits to adaptation. In J. Nalau, & W. L. Filho (Eds.), *Limits to Climate Change Adaptation*. Springer.

Nalau, J., & Handmer, J. (2015). When is transformation a viable policy alternative? *Environmental Science and Policy*, *54*, 349−356.

NASA (National Aeronautics and Space Administration). *NASA, NOAA data show 2016 warmest year on record globally.* (2017). <https://www.nasa.gov/press-release/nasa-noaa-data-show-2016-warmest-year-on-record-globally>.

Naumann, S., Anzaldua, G. Berry, P., Burch, S., Davis, M. K., Frelih-Larsen, A., . . . Sanders, M. (2011). Assessment of the potential of ecosystem-based approaches to climate change adaptation and mitigation in Europe. In: *Final report to the European Commission, DG Environment, Contract no. 070307/2010/580412/SER/B2*. Ecologic Institute and Environmental Change Institute, Oxford University Centre for the Environment.

Nelson, D. R. (2011). Adaptation and resilience: Responding to a changing climate. *Wiley Interdisciplinary Reviews: Climate Change*, *2*(1), 113−120.

Nicholson, E., Mace, G. M., Armsworth, P. R., Atkinson, G., Buckle, S., Clements, T., . . . Milner-Gulland, E. J. (2009). Priority research areas for ecosystem services in a changing world. *Journal of Applied Ecology*, *46*(6), 1139−1144.

Niemelä, J., Saarela, S.-R., Söderman, T., Kopperoinen, L., Yli-Pelkonen, V., Väre, S., & Kotze, D. J. (2010). Using the ecosystem services approach for better planning and conservation of urban green spaces: A Finland case study. *Biodiversity and Conservation*, *19*(11), 3225−3243.

Niven, L. (2014). *Harnessing the green and blue: An investigation of ecosystem-based adaptation measures in four southern Swedish coastal municipalities* (Unpublished Master's thesis). Sweden: Centre for Sustainability Studies, Lund University.

Orlove, B., Roncoli, C., Kabugo, M., & Majugu, A. (2010). Indigenous climate knowledge in Southern Uganda: The multiple components of a dynamic regional system. *Climatic Change*, *100*, 243−265.

Pelling, M. (2011). *Adaptation to climate change: From resilience to transformation*. London: Routledge.

Pelling, M., & High, C. (2005). Understanding adaptation: What can social capital offer assessments of adaptive capacity? *Global Environmental Change*, *15*(4), 308−319.

Pocock, M. J. O., Evans, D. M., Fontaine, C., Harvey, M., Julliard, R., McLaughlin, O., . . . Bohan, D. A. (2016). The visualisation of ecological networks, and their use as a tool for engagement, advocacy and management. *Advances in Ecological Research*, *54*, 41−85.

Pramova, E., Locatelli, B., Brockhaus, M., & Fohlmeister, S. (2012). Ecosystem services in the national adaptation programmes of action. *Climate Policy*, *12*(4), 393−409.

Raffaelli, D. (2016). Ecosystem structures and processes: Characterising natural capital stocks and flows. In M. Potschin, R. Haines-Young, R. Fish, & R. K. Turner (Eds.), *Routledge handbook of ecosystem services*. New York: Routledge.

References

Raudsepp-Hearne, C., Peterson, G. D., & Bennett, E. M. (2010). Ecosystem service bundles for analyzing trade-offs in diverse landscapes. *Proceedings of National Academy of Sciences, United States A, 107*, 5242−5247.

Redman, C. L., & Kinzig, A. P. (2003). Resilience of past landscapes: Resilience theory, society, and the Longue Durée. *Ecology and society, 7*(1).

Rickards, L., & Howden, S. M. (2012). Transformational adaptation: Agriculture and climate change. *Crop and Pasture Science, 63*(3), 240−250.

Rizvi, A. R., Baig, S., & Verdone, M. (2015). *Ecosystem-based adaptation: Knowledge gaps in making an economic case for investing in nature-based solutions for climate change*. Gland, Switzerland: IUCN.

Rodgers, C. (2014). Presentation on ecosystem based adaptation: Economic analysis. In: *At the inter-regional workshop on mainstreaming ecosystem-based approach to adaptation and accessing adaptation finance*, Kuala Lumpur, Malaysia. Stockholm: Stockholm Environment Institute (SEI).

Roser-Renouf, C., & Nisbet, M. (2008). The measurement of key behavioral science constructs in climate change research. *International Journal of Sustainability Communication, 3*(1), 37−95.

Sachs, J. D., Baillie, J. E. M., Sutherland, W. J., Armsworth, P. R., Ash, N., Beddington, J., . . . Jones, K. E. (2009). Biodiversity conservation and the millennium development goals. *Science (New York, N.Y.), 325*, 1502−1503.

Saroar, M., & Routray, J. K. (2015). Local determinants of adaptive capacity against the climatic impacts in coastal Bangladesh. In W. L. Filho (Ed.), *Handbook of climate change adaptation*. Berlin, Heidelberg: Springer-Verlag.

Saroar, M. M., & Routray, J. K. (2010). In situ adaptation against sea level rise (SLR) in Bangladesh: Does awareness matter? *International Journal of Climate Change Strategies and Management, 2*(3), 321−345.

Saroar, M. M., & Routray, J. K. (2012). Impacts of climatic disasters in coastal Bangladesh: Why does private adaptive capacity differ? *Regional Environmental Change, 12*(1), 169−190.

Schröter, M., van der Zanden, E. H., van Oudenhoven, A. P. E., Remme, R. P., Serna-Chavez, H. M., de Groot, R. S., & Opdam, P. (2014). Ecosystem services as a contested concept: A synthesis of critique and counter-arguments. *Conservation Letters, 7*, 514−523.

Simões, E., de Sousa Junior, W. C., de Freitas, D. M., Mills, M., Iwama, A. Y., Gonçalves, I., . . . Fidelman, P. (2017). Barriers and opportunities for adapting to climate change on the North Coast of São Paulo, Brazil. *Regional Environmental Change, 17*(6), 1739−1750.

Speranza, C., Kiteme, B., Ambenje, P., Wiesmann, U., & Makali, S. (2010). Indigenous knowledge related to climate variability and change: Insights from droughts in semi-arid areas of former Makueni District, Kenya. *Climatic Change, 100*, 295−315.

Steel, B., Lovrich, N., Lach, D., & Fomenko, V. (2005). Correlates and consequences of public knowledge concerning ocean fisheries management. *Coast Management, 33*(1), 37−51.

Stern, N. (2007a). *Economics of climate change: The stern review*. Cambridge: Grantham Research Institute.

Stern, N. (2007b). *The economics of climate change: The stern review*. Cambridge: University of Cambridge Press.

Tansley, A. G. (1935). The use and abuse of vegetational concepts and terms. *Ecology, 16*, 284−307.

TEEB (The economics of Ecosystems and Biodiversity). (2008). *An interim report* (p. 7). Brussels: European Commission. Available at: www.teebweb.org/media/2008/5/TEEB-Interim-Report_English.pdf.

TEEB. (2009). *The economics of ecosystems and biodiversity for national and international policy makers—Summary: Responding to the value of nature*. Available at: www.teebweb.org/wp-content/uploads/Study%20and%20Reports/Reports/National%20AND%international%20%Policy%20Making/TEEB%20for%20National%20%Policy%20Makrs%20report/TEEB%20National.pdf.

TEEB. (2010). In P. Kumar (Ed.), *The economics of ecosystems and biodiversity: Ecological and economic foundations*. London: Earthscan.

Thomsen, D. C., Smith, T. F., & Keys, N. (2012). Adaptation or manipulation? Unpacking climate change response strategies. *Ecology and Society, 17*(3), 20.

TNC (The Nature Conservancy). *Adapting to climate change—Ecosystem-based approaches for people and nature.* (2009). <https://www.sprep.org/att/irc/ecopies/global/306.pdf>.
Tol, R. S. J., Klein, R. J. T., & Nicholls, R. J. (2008). Toward successful adaptation to sea-level rise along Europe's coast. *Journal of Coastal Research, 24*(2), 432–442.
Turetta, A. P. D. (2018). An ecosystem approach to indicate agriculture adaptive strategies to climate change impacts: Managing vulnerability, fostering resilience. In F. Walter Leal, & F. Leonardo Esteves de (Eds.), Climate change in Latin America—Managing vulnerability, fostering resilience. (pp. 193–206). Gewerbestrasse: Springer International Publishing.
UNEP (United Nations Environment Program). (2006). Global Programme of Action (GPA) for the protection of the marine environment from land-based activities. In: *Ecosystem-based management—Markers for assessing progress.* The Hague: UNEP GPA.
UNEP (United Nations Environment Program). (2013). *The Emissions gap report 2013—A UNEP Synthesis Report.* Nairobi.
UNEP (United Nations Environment Program). (2014a). *The adaptation gap report—Executive summary.* Nairobi.
UNEP (United Nations Environment Program). (2014b). *The adaptation gap report 2014—A UNEP synthesis report.* Nairobi.
UNEP (United Nations Environment Program). (2016). *The emissions gap report 2016—A UNEP synthesis report.* Nairobi.
UNEP (United Nations Environment Program). (2017). *The emissions gap report 2017—A UNEP synthesis report.* Nairobi.
UNEP-WCMC (World Conservation Monitoring Centre of the United Nations Environment Programme). (2011). Developing ecosystem service indicators: Experiences and lessons learned from bus-global assessments and other initiatives. In: *Secretariat of the Convention on Biological Diversity, CBD technical series 58.*
UNFCCC (United Nations Convention on Framework for Climate Change). (2005). *Nairobi work program on impacts, vulnerability and adaptation to climate change.* Bonn, Germany: UNFCCC Secretariat.
UNFCCC (United Nations Convention on Framework for Climate Change). (2010a). *Progress made in implementing under the activities Nairobi Work Program on impacts, vulnerability and adaptation to Climate Change.* Bonn: UNFCCC Secretariat.
UNFCCC (United Nations Framework Convention on Climate Change). (2010b). *National adaptation programmes of action (NAPA).* Bonn: UNFCCC Secretariat.
UNFCCC (United Nations Framework Convention on Climate Change). (2011a). *Ecosystem-based approaches to adaptation: Compilation of information.* Bonn: UNFCCC.
UNFCCC (United Nations Framework Convention on Climate Change). (2011b). Assessing the costs and benefits of adaptation options: An overview of approaches. *The Nairobi Work Program on Impacts, Vulnerability and Adaptation to Climate Change. United Nations Framework Convention on Climate Change.* Bonn: UNFCCC Secretariat.
UNFCCC (United Nations Framework Convention on Climate Change). (2013). Report on the technical workshop on ecosystem-based approaches for adaptation to climate change. In: *Subsidiary Body for Scientific and Technological Advice (SBSTA), thirty-eighth session.* Bonn.
UNFCCC (United Nations Framework Convention on Climate Change). *Intended nationally determined contributions (INDCs).* (2014). <http://unfcc.int/focus/indc_portal/items/9766.php>.
UNISDR (United Nations Secretariat of the International Strategy for Disaster Reduction). (2009). *UNISDR terminology on disaster risk reduction.* Geneva: UNISDR.
Vandewalle, M., Sykes, M. T., Harrison, P. A., Luck, G. W., Berry, P., Bugter, R., ... Zobel, M. (2008). *Review paper on concepts of dynamic ecosystems and their services.* RUBICODE deliverable D2.1. Available at http://www.rubicode.net/rubicode/RUBICODE_e-conference_report.pdf.

Venton, C. C. (2010). *Cost benefit analysis for community based climate and disaster risk management: Synthesis report*. Tearfund and OXFAM America.

Vignola, R., Locatelli, B., Martinez, C., & Imbach, P. (2009). Ecosystems-based adaptation to climate change: What role for policy-makers, society and scientists? *Mitigation and Adaptation Strategies for Global Change, 14*(8), 691–696.

Warner, B. P. (2015). Understanding actor-centered adaptation limits in smallholder agriculture in the Central American dry tropics. *Agriculture and Human Values, 33*, 1–13.

Watkiss, P., Downing, T., & Dyszynski, J. (2010). *AdaptCost project: Analysis of the economic costs of climate change adaptation in Africa*. Nairobi: UNEP.

Watkiss, P., & Hunt, A. (2010). Review of adaptation costs and benefits estimates in Europe for the European environment state and outlook report 2010. *Technical Report prepared for the European Environmental Agency for the European Environment State and Outlook Report 2010*. Copenhagen: European Environment Agency.

WEF (World Economic Forum). (2013). *Global risks 2013* (8th ed.). An Initiative of the Risk Response Network, World Economic Forum 2013.

Whitney, C. K., Bennett, N. J., Ban, N. C., Allison, E. H., Armitage, D., Blythe, J. L., ... Yumagulova, L. (2017). Adaptive capacity: From assessment to action in coastal social-ecological systems. *Ecology and Society, 22*(2).

Williams, S. J., Jones, J. P. G., Gibbons, J. M., & Clubbe, C. (2015). Botanic gardens can positively influence visitors' environmental attitudes. *Biodiversity and Conservation, 24*, 1609–1620.

Wisner, B., Blaikie, P., Cannon, T., & Davis, I. (2004). *At risk: Natural hazards, people's vulnerability and disasters* (2nd ed.). London: Routledge.

World Bank. (2009). *Convenient solutions to an inconvenient truth: Ecosystems based approaches to climate change*. Washington, DC: Environment Department, World Bank.

World Bank. (2013). Turn down the heat: Climate extremes, regional impacts, and the case for resilience. *A report for the World Bank by the Potsdam Institute for Climate Impact Research and Climate Analytics*. Washington, DC: World Bank.

Further reading

Diamond, J. M. (2005). *Collapse: How societies choose to fail or succeed*. New York: Viking.

Nelson, G. C., Rosegrant, M. W., Koo, J., Robertson, R., Sulser, T., Zhu, T., ... D., L. (2010). *The costs of agricultural adaptation to climate change*. Washington, DC: The World Bank.

Nelson, K., Brummel, R., Jordan, N., & Manson, S. (2014a). Social networks in complex human and natural systems: The case of rotational grazing, weak ties, and eastern United States dairy landscapes. *Agriculture and Human Values, 31*(2), 245–259.

Nelson, G. C., Valin, H., Sands, R. D., Havlík, P., Ahammad, H., Deryng, D., ... Willenbockel, D. (2014b). Climate change effects on agriculture: Economic responses to biophysical shocks. *Proceedings of the National Academy of Sciences, 111*(9), 3274–3279.

Newitz, A. (2013). *Scatter, adapt, remember: How humans will survive a mass extinction*. Collingwood: Black Inc.

Pielke, R. A., Jr. (1998). Rethinking the role of adaptation in climate policy. *Global Environmental Change, 8*(2), 159–170.

UNISDR (United Nations Secretariat of the International Strategy for Disaster Reduction). (2005). *Hyogo Framework for Action 2005–2015: Building the resilience of nations and communities to disasters*. Geneva: UNISDR.

CHAPTER 4

Coping with climate chang

4.1 Introduction

Climate change has come to be reckoned with as a problem entailing global scope with localized impacts. These impacts are usually in the form of severe floods, decreasing water reserves, droughts, rapid pace of degradation in biodiversity and agricultural production (Hatfield et al., 2014), worsening living conditions (Luber et al., 2014), detrimental effect on economic development (Burke, Hsiang, & Miguel, 2015), rising sea levels and landscape changes (Adger, Barnett, Brown, Marshall, & O'Brien, 2013) in different parts of the globe such as Africa (Filho, Kalangu, Munishi, & Musiviwa, 2017), the Asia-Pacific region (Filho, 2015), Eastern Europe (Filho, Trbić, & Filipovic, 2019), Latin America (Filho & de Freitas, 2018), North America (Filho & Keenan, 2017), etc. Climate change entails multiple dimensions—scientific, political, economic, societal, and moral along with ethical questions—that make it a "global problem felt on local scales" (NASA, Website Portal).

Devastating impacts of climate change that wreak havoc on human and other biotic organisms have spurred many agencies to conjure up clarion calls for adhering to immediate actions to cope up with the vagaries of climate change to save humankind from climatic catastrophe. CRI (2019) explicitly warns that signs of escalating climate change can no longer be ignored on any continent or any region. IMF (2019) has lamented that the window of opportunity for containing climate change to manageable levels is closing rapidly. According to Inger Andersen (2019), Executive Director of the UNEP (United Nations Environment Program), world at this juncture is faced with a "stark choice" of either setting in motion the radical transformation needed now or face the consequences of a planet radically altered by climate change (Andersen, 2019, p. xiii). UNEP (2019a) cautions that even if all current unconditional commitments under the Paris Agreement were implemented, temperature was expected to rise by 3.2°C, bringing even wide-ranging and more destructive climate impacts. While bemoaning the widening gap in greenhouse gas (GHG) emissions during this decade between the COP-15 (Copenhagen Summit of 2009) and COP-25 (Spain/Chile in 2019), Christensen and Olhoff (2019) note that despite a decade of enhanced political and societal focus on climate change and the "milestone" Paris Agreement, GHG emissions "have not been curbed, and the emissions gap is larger than ever."

Similar warning signals of catastrophic impacts of climate change have come from IPCC's Special Report on Climate Change and Land (IPCC, 2019a) and IPCC's Special Report on The Ocean and Cryosphere in a Changing Climate Change (IPCC, 2019b). In an identical manner, IPBES (2019) has noted that with increase in GHG emissions, the climate change is already impacting nature from the level of ecosystems to that of genetics and these impacts are expected to increase over the coming decades, in some cases surpassing the impact of land and sea use change and other drivers. Striking a cautionary note, Patricia Espinosa, Executive Secretary of United

Nations Framework Convention on Climate Change (UNFCCC) says, "...(A)ctive participation by all of us—governments, businesses, investors, regions and more—is needed if we are to face and overcome the climate emergency we currently face" (Espinosa, 2019, p. iv). A report released in late September 2019 by World Meteorological Organization (WMO) exhibits the tell-tale signs of and impacts of climate change—such as sea-level rise, ice loss, and extreme weather—recording increase during 2015–19, which is set to be the warmest 5-year period on record (WMO, 2019a). Another WMO report informs about global atmospheric concentrations of GHGs having reached record levels in 2018 with carbon dioxide (CO_2) reaching 407.8 ± 0.1 parts per million, 147% of pre−industrial levels; along with global mean temperature for January to October 2019 rising $1.1°C \pm 0.1°C$ above pre−industrial levels; thereby, making 2019 perhaps to be the second or third warmest year on record (WMO, 2019a).

While unfolding latest critical data and scientific findings on the climate crisis, a recent UN publication entitled United for Science shows as to how climate change is already changing and it also highlights the far-reaching and dangerous impacts that will unfold for generations to come (WMO, 2019b, 2019c; WMO et al., 2019). Taking note of the fact of the continuous increase in global temperature from 2015 onward has culminated in impacts of climate change hitting harder and sooner than projected by climate assessments even a decade ago, this report calls for immediate and all-inclusive action encompassing deep decarbonization complemented by ambitious policy measures, protection and enhancement of carbon sinks and biodiversity, and effort to remove CO_2 from the atmosphere thereby meeting the requirements of the Paris Agreement on Climate Change (PACC).

Admonishing that time is running out, Pihl et al. (2019) denote: "Climate reality shows it, science states it and protesters around the globe express it with strikes... (to warn that) acting now is the only possible solution to solve the climate crisis."

In the wake of mounting scientific of adverse impacts accruing from the vagaries of climate change, focus is increasingly turning in the public, private and nonprofit domains to the viable options available for coping with climate change. Available literature comprising scientific evidence, policy analysis, and empirical research makes it discernible about the availability of four options—Mitigation, Adaptation, Geoengineering, and Loss and Damage—to deal with climate change.

4.2 Mitigation

Realization of the devastating aspects of climate change spurred prime attention to be focused on devising strategies that could mitigate climate change (Somorin et al., 2011). Mitigation includes man-made attempts at developing systems to alleviate GHG emissions or minimize the level of these gases after they have been emitted. Such a strategy encompasses options like harnessing clean technologies and improving energy-use efficiency by switching to renewable energy resources such as wind energy, solar, hydropower and biofuel generation, reducing emissions from reforestation and forest degradation (REDD +) and development of electric vehicles (UNFCCC, 2009b). One of the outcomes of the Kyoto Protocol (KP) is monetizing the mitigation of GHG emissions through the introduction of market-based mechanisms. The strategy of combating with dangerous outcomes of climate change initially focused on mitigation measures and with the founding of the UNFCCC in 1992 emphasis on mitigation gathered further momentum because of its according priority to stabilize GHG concentration at a level that could be instrumental in preventing dangerous anthropogenic interference with the climate system. A decade later the notion of adaptation started gaining currency in the climate change literature as a potent mechanism of reducing many of the adverse impacts of climate change and augment "beneficial impacts" (IPCC TAR, 2001). This amply demonstrated enhanced cognizance that the adverse impacts of climate change could not be entirely kiboshed merely with the aggressive adherence to mitigation effort in the strong torpors or inertias in the climate system will expose modern societies to some degree of warming irrespective of whatever efforts are put in to curb emissions.

In the meanwhile, the notion of mitigation has come to seen in the light of "homo economicus model" that lies at the core of most rational choice theories of human behavior (Levitt & List, 2008). According to Ostrom (2007), under the homo economicus model actors are well-equipped with information and appropriate preferences, and act accordingly to optimize the net worth of their expected returns. Act of mitigation at the municipal level requires cognizance of two important factors. First, no municipality, irrespective of its size or strength, is in a position to stop or delay the phenomenon of global climate change, owing to the fact that its emissions constitute on a fraction of the global total. Second, owing to the global nature of the system of climate change, mitigating local emissions cannot protect the local area from the impacts of emissions. Therefore, from the perspective of a classical political economy, local governments are not in a position to mitigate their emissions, as Stewart (2008) points out, unless (1) damages accruing from climate change are substantial; (2) their share of emissions is tremendous; and (3) the costs of substantial mitigation are properly low.

Decision-making in response to climate change can affect both individuals and groups at different places and times; and circumstances of these decisions entail uncertainty and disagreement that is "sometimes both severe and wide-ranging concerning not only the state of the climate and broader social consequences of any action or inaction on our part but also the range of actors available to us and what significance we should attach to their possible consequences" (Bradley & Steele, 2015, p. 799). This complex process of decision-making in response to climate change in tandem with the consequences arising thereof have led to focusing attention on four factors—climate risk; internal determinants; excludable benefits; and influence from above—and these call for brief elaboration.

4.2.1 Climate risk

Climate-related risks are engendered by a range of hazards and some of these are slow in their onset (like changes in temperature and precipitation leading to droughts, or agricultural losses), while others occur more suddenly, such as tropical storms and floods (UNFCCC, n.d.). human perceptions of risks are influenced by an array of factors that also embrace scientific information delineated by the professionals and authority, personal experiences, values, and worldviews (Dunlap, Liere, Mertig, & Jones, 2000). Decision-making levels like voting behavior, support of policy initiatives, and lifestyle decisions on an individual basis entail the probability of being influenced by the level of risk perception for individuals (Brody, Zahran, Vedlitz, & Grover, 2008). Attitudes and socioeconomic can constitute the basis of determining risk perceptions (Brody et al., 2008).

According to Slovic (1999), risk perceptions are also determined by the interactions of factors that an individual possesses, such as psychological, social, cultural, and political attributes. The knowledge of causes of climate change, the consequences brought upon by climate change, and the extent in which individuals feel that the effects of climate change will be harmful to their lifestyles are the determinants of the level of risk perception for climate change (Brody et al., 2008).

Psychometric paradigm and culture theory are considered as two main schools of thought in the research ambit of risk perception. The psychometric paradigm is utilized within risk analysis and risk perception studies as a means to have cognizance of as to why different people perceive types of risk in various ways (Siegrist, Keller, & Kiers, 2005). Focus of this approach is on the factors that affect risk perception of laypeople, or the general public, as opposed to experts within a particular hazard's field. The idea underlying risk perception concerning the psychometric paradigm is that a person's risk perception toward hazards is based on the qualitative characteristics of the hazards themselves (Pidgeon, 1998). Unambiguously, hazards can be ranked against each other on the basis of dimensions that pertain to their perceived risk. Nevertheless, most risk perception studies that make use of the psychometric paradigm emphasize exclusively on the differences in risk between hazards themselves (Bronfman, Cifuentes, & Gutierrez, 2008), some studies combine the dimensions within the psychometric paradigm with that of how demographics compare (Mumpower, Liu, & Vedlitz, 2016). The risk factor can be explained as apprehending fear toward a hazard because it is disastrous in nature or the hazard has unavoidable harm, while the personal exposure factor delineates the fact of risk perception attributable to personal experience with such hazard or fear toward that hazard on a firmly individual basis (Bubeck, Botzen, & Aerts, 2012). While exploring demographics vis-à-vis dread and personal exposure factors of the psychometric paradigm, a higher level of risk resembles with lower education, lower income, women, the young, and African Americans (Hakes & Viscusi, 2004).

Hiatus generated in the wake of inability of the psychometric paradigm in providing information on social and cultural influences on risk perception is engulfed by the Cultural Theory (Rippl, 2002) that focuses on the notion that risk perception is produced by cultural biases and global views (Bickerstaff, 2004). Risk perception observed by an individual gets powerfully impacted by the social and cultural groups that an individual is associated with (Kahan et al., 2012). One way the examination of Cultural Theory is facilitated within the risk perception literature is through the categorization of people within one of four groups-based concepts of which they are fearful: egalitarian, individualistic, hierarchic and fatalistic (Maibach, Leiserowitz, Roser-Renouf, & Mertz, 2011). Having a cognizance as to why some people find hazards as riskier than other persons do is not as easy as a study exclusively within the psychometric paradigm is inclined to be (De Groot, Steg, & Poortinga, 2013). New ecological paradigm factors like environmental beliefs and political party preferences have come to characterize within the ambit of the cultural theory (Amburgey & Thoman, 2011).

Undoubtedly, homo economicus model affords complete information and regimented preferences; nevertheless, laypeople don't—neither do institutions that lack the selective pressure and information-generating capabilities of the competitive market (Ostrom, 2007). Costly venture of seeking information and limited human capabilities of information-processing often render individuals to make choices based on insufficient knowledge of all probable options and their likely outcomes. Therefore, equipped with insufficient information and deficient information-processing capabilities, individuals are prone to commit mistakes in devising strategies aimed at realizing a set of goals (Ostrom, 2007, p. 31). Within the ambit of such a more realistic, limited-rational model of human behavior, mitigation outcomes from deficient human rationality; decision-makers and the public at large can misconstrue climate change and their ability to influence it (Krause, 2011b). This misapprehension entails the likelihood of resulting in mitigation when it is juxtaposed with a motive like vulnerability to the impacts of climate change. It has often been found that urban municipal areas are susceptible to the vagaries of climate change and rise in temperatures projected for urban areas are greater than global scale projections (Grimmond, 2007). Land cover modifications characterizing urban areas are responsible for this, like the heat-island effect (Yang, Qiana, Songa, & Zhenga, 2016). Comparable paucity of wind and deteriorating air quality in urban areas also reduce the ability of local population to cope with heat (Mills, 2007). Admittedly, other climate impacts like floods, droughts, hurricanes and fires, etc., are not specifically urban in nature; nonetheless, the density and concentration of population, infrastructure, and economic activity in urban areas is prone to render these areas at greater risks (Krause, 2011a). Industries like manufacturing, tourism, or agriculture that constitute the spine of urban economies, can also be vulnerable to climate change.

It is interesting to note that Zahran, Grover, Brody, and Vedlitz (2008a) and Zahran, Samuel, Vedlitz, Grover, and Miller (2008b) tested the hypothesis—climate risk predicts municipal mitigative action—using participation in the Cities for Climate Protection (CCP) climate network as the dependent variable. However, Zahran et al. didn't use cities as their unit of analysis, however; instead, they analyzed the metropolitan statistical areas (2008a) and counties (2008b) around the cities in question; and it could be discerned from their regression models that being located in a coastal area, anticipated changes in temperature, and prior deaths from natural hazards were all statistically important factors that enhanced the of CCP accession. Concurrently, other risk and vulnerable factors like precipitation levels and the percentage of land covered by forests and wetlands, were not found significant. While making use of a different variable—a count of the number of mitigation policies adopted—Pitt (2010a) ascertained that coastal proximity is not a statistically important predictor. Using yet another dependent variable—climate plan quality—Tang, Samuel, Quinn, Chang, and Wei (2010) came to the different conclusion that property damages occurring between 1995 and 2000 was a statistically important negative predictor. Usage of a slightly different measure for climate change—the number of times since 2000 that the adjacent country has been declared a weather-related federal disaster area—helped Krause (2011b) ascertain that it didn't significantly predict.

4.2.2 Internal determinants

The internal determinants are an assortment of demographic and political institutional characteristics that "may impact the extent of climate mitigation planning within a given municipality" (Pitt, 2010a, 2010b). The demographic features—population, income per capita, education, voting history, and college town status—are pepped up the findings of Zahran et al. (2008a, 2008b) that "civic capacity" variables such as income, education, and voting trends are major drivers of CCP membership, along with other past work that has linked these variables to the adoption of energy conservation or other environmental policies (Kahn, 2006). According to the internal determinants' theory, municipal mitigation is an outcome of the municipality's own political, economic, and social characteristics. For instance, local politicians and citizens can derive succor from municipal mitigation in the form of preference satisfaction, irrespective of whether or not that mitigation protects the climate; or, as Engel and Orbach (2008) denote, these people may value the "warm glow" feeling they derive from behaving altruistically. Their support may also mirror hyped benefits, undervalued costs, and/or costs that have been discounted exaggeratedly (Engel & Orbach, 2008). There may be an exaggerated account of benefits because climate change policies often define emissions reduction goals, they usually fail to point out the effect of those reduction on the global climate. This lop-sided demonstration of information can give rise to the impression that an elongate relationship exists between emissions and climate change, whereas the relationship is nonlinear and the effect of local mitigation is trivial (Woods & Potoski, 2010). Woods and Potoski, (2010) Undervaluation of costs may be mulled under the presumption that the efficacy of capital and technological spinoff would decrease costs once the mandatory reductions are in place (Ji & Darnall, 2017).

Political support is also a significant factor in predicting municipal mitigative action and it is often projected by supply-side theories of the provision of public goods that local governments having more financial and institutional capacity are better placed to respond to the demand for public goods (Ziblatt, 2008). Fiscal resources and institutional capacity of the local government are important factors in predicting municipal mitigative action. Support of general public as a key predictor of municipal mitigation is attested to by both qualitative and quantitative studies (Gerber, 2013; Krause, 2013). A strong causal role for environmental preferences is established by Millard-Ball (2012). Essentiality of measures of political orientation and environmental awareness as statistically significant predictors of CCP membership is noted by Zahran et al. (2008a, 2008b). Findings by Krause (2011a) make it discernible that political orientation bears a close link with the formulation of mitigation policies as obviously climate-protecting, and the institutionalization of climate protection as an explicit municipal aim. Nevertheless, Pitt (2010a) finds that voting history is not a statistically significant predictor of the number of mitigative policies municipalities adopt. Some studies also attest to the fact that political support is a necessary (if insufficient) condition for municipal mitigation action (Gerber, 2013; Tang, Wei, Quinn, & Zhao, 2012). Political opposition can play a significant role in influencing local government's mitigation policies (Schroeder & Bulkeley, 2009).

Findings by Zahran et al. (2008a, 2008b) show that the amount of automobile commuting and the percentage of workers employed in carbon-intensive industries significantly decreased the likelihood of accession to the CCP climate network. On the other hand, Krause (2011c) informs that a higher dependence on manufacturing in the local economy significantly decreases the probability of network membership, even though the magnitude of the impact is small. Harnessing an array of mitigation policies as the dependent variable renders; nonetheless, the influence of manufacturing on the local economy almost insignificant (Krause, 2011b). There is explicit evidence of political support as a significant factor in municipal mitigative policies (Dierwechter & Wessells, 2013). Quantitative support has been found using such measures as per capita income (Krause, 2011c; Zahran et al., 2008a, 2008b) and gross government expenditures (Vasi, 2006)—although in these cases, the dependent variable was climate network membership. Findings by Krause (2011b) show that income is a statistically significant and negative predictor when the number of mitigation policies is used as the dependent variable. Deployment of its staff by the local government as a measure of institutional capacity has found support in some qualitative studies (Pitt & Randolph, 2009; Pitt, 2010b). The role of nongovernmental organizations (NGOs), educational and other institutions is also considered important sometimes in assisting local governments with additional staff capacity and technological support (Knuth, Nagle, Steuer, & Yarnal, 2007). Quantitative studies using measures such as population (Gerber, 2013), staff (Krause, 2012), and educational institutions (Gerber, 2013; Krause, 2013) have also found support for the role of institutions in municipal mitigative policies. Nevertheless, findings by Pitt (2010a) present a notable exception in this regard when it points out that that neither income nor education are statistically significant predictors of the number of mitigation policies municipalities adopt.

4.2.3 Excludable benefits

Undoubtedly, local governments bear the brunt of the costs of their mitigation policies; nonetheless, they don't avail of the benefits to the global climate. Consequently, Olson (1971) predicted that only a separate and selective incentive—that is, an excludable benefit that is availed of locally—could inspire mitigative action and this surmise is in consonance with the presumption that "excludable benefits" predict municipal mitigative action. Meliorations in local air quality are a significant excludable benefit, because decrease in GHG emissions often has the effect of reducing ozone, nitrous oxide, sulfur dioxide, and other local air pollutants, as well (Romero-Lankao, 2007). Making use of measures of carbon monoxide and particulate matter (PM) pollution, Vasi (2006) informed that local air quality is not a statistically significant predictor—even though in Vasi's qualitative discussions, the perception that mitigation policies might improve local air quality was influential with local decision-makers. Admittedly, qualitative discussion by Pitt (2010b) pointed in the same direction; nonetheless, his quantitative assessment, like Vasi's (2011), found that poor air quality was an insignificant predictor. Pitt (2010a) defines poor air quality as the failure to meet one of the EPA's National Ambient Air Quality Standards. Interestingly, the findings of Zahran et al. (2008a, 2008b) are in line with Pitt and Vasi, but they make use of an unusual measure of local air pollution—hazardous air pollutants (HAP) emissions per capita. HAP emissions are an odd choice because the local air pollutants most commonly associated with GHG emissions are not listed as HAP (Houyoux, 2019).

4.2.4 Influence from above

Local mitigation policies or efforts can be influenced by central or federal government. Ostrom (2007) opines that institutional studies ought to include multiple levels owing to the fact that decisions about rules at any one level are made within the structure of rules at higher levels. States in the US and other federal countries can mandate municipal mitigation an example is cited of the California state as the only state having done so in the US. The fear or threat of mandates can spur or incentivize municipal mitigation and owing to this municipalities get an incentive to prepare or fall in line with the mandate and in the process relieve their uncertainty. However, within the US, California is the only state having legally binding emission caps (Krause, 2011c). While the office of the Attorney General in Californian has taken not only legal action against cities under the California Environmental Quality Act for violating the provisions of Act; nonetheless, it has also encouraged "cities to develop climate plans as a way to mitigate their GHG emissions" (Millard-Ball, 2012, p. 290). Mitigative actions of municipalities can also be influenced by State governments indirectly through enactment of legislation and adopting their own climate policies. Shipan and Volden (2006) conjecture a "snowball effect," such that state-level policies enable or spur the adoption of local-level policies, whereby state climate action may also incentivize local action by minimizing "the political and information-gathering costs associated with policy passage" (Krause, 2012, p. 2406); and a contrasting "pressure valve effect," whereby state-level policies lessen the probability of local-level initiatives by driving out pressure for governmental action or creating the impression that action has already been taken.

The presumption that state mitigation policies predict municipal mitigative action finds support in numerous qualitative studies that reveal that mitigation by municipalities is inspired and spurred by policies at higher levels of government (Kern & Alber, 2008). However, there is a paucity of quantitative studies in support of this presumption. On the one hand, findings of Tang et al. (2010) demonstrated that whether or not a state had passed climate legislation was a statistically significant, strongly positive predictor of the climate plan quality of that state's localities; on the other hand, Krause (2011c) makes use of multilevel modeling to find that state-level characteristics—including the presence of state GHG reduction targets and climate plans—do not have a statistically significant effect on the probability that municipalities will join the United States Mayors Climate Protection Agreement (USMCPA) climate network.

It indicated an increasing awareness that climate change could not be completely halted even with the aggressive mitigation effort on the strong inertias in the climate system will expose modern societies to some degree of warming no matter what they do to curb emissions. Furthermore, the constant difficulties encountered by international climate negotiations also make the implementation of aggressive mitigation even less optimistic, at least in the short term.

4.2.5 Harnessing mitigation

Climate change mitigation deals with actions that are designed to reduce or prevent emissions of GHG causing human-induced climate change. Mitigation of climate change can be attained by harnessing new technologies, promoting renewable energies, improving efficiency of older energy systems, or changing management practices or consumer behavior. According to the Intergovernmental Panel on Climate Change (IPCC, 2014a, 2014b), mitigation is the effort to control the human sources of climate change and their cumulative impacts, notably the emission of GHGs and other pollutants, such as black carbon particles, that also impact the planet's energy balance. Mitigation also includes efforts to enhance the processes that remove GHGs from the atmosphere, known as sinks. Mitigation of climate change is delineated as "a human intervention to reduce the sources or enhance the sinks of GHGs. This report also assesses human interventions to reduce the sources of other substances which may contribute directly or indirectly to limiting climate change, including, for example, the reduction of PM emissions that can directly alter the radiation balance (e.g., black carbon) or measures that control emissions of carbon monoxide, nitrogen oxides (NO_x), volatile organic compounds (VOCs) and other pollutants that can alter the concentration of tropospheric ozone (O_3) which has an indirect effect on the climate" (Allwood, Bosetti, Dubash, Gómez-Echeverri, & von Stechow, 2014). Owing to mitigation's potential in lowering the projected impacts of climate change along with the risks of extreme weather impacts, mitigation constitutes part of a broader policy strategy that encompasses adaptation to already occurring climate change impacts; therefore adaptation and mitigation are often considered holistically as two faces of the same effort to combat the negative impacts of climate change.

4.2 Mitigation

GHG emissions have grown at a rate of 1.5% annually in the last decade, stabilizing only briefly between 2014 and 2016; and total GHG emissions, including from land-use change, peaked at a record high of 55.3 GtCO2e in 2018 (UNEP, 2019a). Carbon Dioxide (CO_2) account for more than three-fourths of total GHG emissions and the rest GHG emissions comprise Methane (16%), Nitrous Oxide (6%) etc. (Fig. 4.1). Fossil CO_2 emissions, from energy use and industry, dominating total GHG emissions, rose by 2.0% in 2018, reaching a record 37.5 $GtCO_2$/year (UNEP, 2019a). CO_2 is the most diffused among GHG that gets released into the atmosphere via burning fossil fuels (e.g., coal, natural gas, and oil), solid waste, trees and wooden products, as well as an outcome of some chemical reactions like manufacturing of cement or glass (Alloisio & Borghesi, 2019; Lallanilla, 2019). Carbon sequestration or removal from the atmosphere is said to take place when plants as part of biological carbon cycle absorb it (Jain et al., 2012). Atmosphere can retain carbon dioxide for a protracted period, even spanning up to centuries and each additional quantity of carbon dioxide at present is bound to affect wellbeing of the people for decades/centuries to come

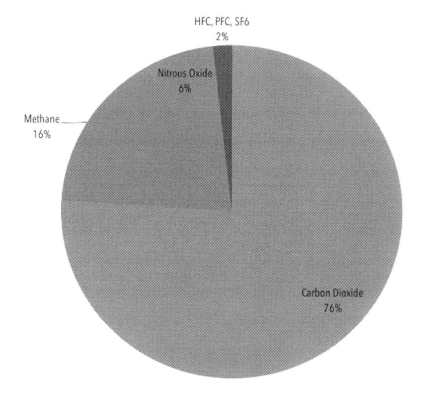

FIGURE 4.1

Global man-made greenhouse gas emissions by gas, 2015 EPA (United States Environmental Protection Agency).

Adapted from EPA (United States Environmental Protection Agency). Inventory of United States Greenhouse Gas and Sinks 1990–2015. *(2017).* <https://www.epa.gov/ghgemissions/inventory-us-greenhouse-gas-emissions-and-sinks-1990-2015>.

(Lindsey, 2019). Apart from CO₂, other GHGs such as methane (CH₄), dinitrogen monoxide (N₂O), hydrofluorocarbon, perfluorinated compounds, sulfur hexafluoride gases (SF₆), nitrogen trifluoride (NF₃) are also getting increasingly concentrated in the atmosphere entailing the probability of exacerbating the problem (WMO, 2019a).

Resultant trapping of excess heat in the atmosphere has culminated in the increase in the global average surface temperature and this phenomenon is known as 'global warming. According to Bosetti, Carraro, Massetti, and Tavoni (2014), average surface temperature increased by about 0.6°C during 1951–80 and 0.8°C with respect to the pre–industrial average. The global average surface temperature rose 0.6°C–0.9°C between 1906 and 2005 (NASA, 2010). Having recoded an increase at an average rate of 0.07°C per decade since 1880 and over twice that rate (+0.18°C) since 1981, the average temperature across global land and ocean surfaces was 0.95°C above the twentieth-century average of 13.9°C in 2019, making it the second-warmest year on record (Lindsey & Dahlman, 2020). Temperatures are projected to continue to rise for decades/centuries "because the climate system has a delayed response to the stock of the GHG and equilibrium temperature grows linearly with cumulative emissions of CO2" (Bosetti et al., 2014; NOAA, 2020). The CO₂ concentration in the atmosphere was around 350 parts per million (ppm) in 1972 and continued to register increase by around 1 ppm/year (Sachs, 2015). Presently, the CO₂ concentration in the atmosphere is recording an increase rather steadily at about 2 ppm/year. Admittedly, the global average atmospheric CO₂ in 2018 was 407.4 ppm, with a range of uncertainty of plus or minus 0.1 ppm; nonetheless, "Carbon dioxide levels today are higher than at any point in at least the past 800,000 years" (Lindsey, 2019).

According to the IPCC (2014a, 2014b), "mitigation scenarios in which it is likely that the temperature change caused by anthropogenic GHG emissions can be kept to less than 2°C relative to pre–industrial levels are characterized by atmospheric concentrations in 2100 of about 450 ppm CO₂eq (equivalent)." In contrast to traditional pollutants, stabilization of CO₂ concentrations can only be facilitated if global emissions peak and in the long-term decline toward zero. The lower the concentration at which CO₂ is to be stabilized, the sooner and lower the peak should be. Fundamental changes in the global energy system relative to a baseline scenario are required to stabilize concentration of GHG. As per the IPCC (2014a, 2014b), in mitigation scenarios reaching 450 ppm CO₂eq concentrations in 2100, CO₂ emissions from the energy supply sector decline over the next decades, reach 90% below 2010 levels between 2040 and 2070, and in many scenarios fall below zero thereafter. This concentration level "is possible thanks to consistent energy-efficiency improvements and almost quadrupling of the share of low and zero carbon energy technologies [from renewables, nuclear energy, and fossil energy with carbon dioxide capture and sequestration (CCS)] and of technologies aimed at negative emissions such as bioenergy with CCS (BECCS) by 2050" (Alloisio & Borghesi, 2019).

Ensuring a 50% chance of limiting global warming to 2°C, the IPCC (2014a, 2014b) suggests that the world can support a maximum carbon dioxide emission level, also known as carbon budget, of 3000Gt, of which an estimated 1970Gt had already been emitted prior to 2014. IEA (2015) mulls that in order to account for CO₂ emissions from industrial processes and land use, land-use change and forestry (LULUCF) over the rest of the 21st century leave the energy sector needs to be left with a carbon budget of just 980Gt. The energy sector is the largest contributor to global GHG, representing roughly two-thirds of all anthropogenic GHG, and CO₂ emissions from the sector have risen over the past century to ever higher levels; and energy-related CO₂ emissions rose 1.7% to a historic high of 33.1 GtCO$_2$ (IEA, 2019). Options for climate change mitigation in the energy sector need to be prudently worked out; and relatively climate change mitigation options also exist persist in the energy demand sector in terms of demand-side management and energy efficiency at household or business level or in transport. Energy extraction, conversion, storage, transmission and distribution processes constitute major components of energy supply sector. Among the different options available for climate change mitigation, some are designed to replace fossil fuel usage with technologies without direct GHG emissions (Goldman, Ungar, Capanna, & Simchak, 2012), Renewable and nuclear energy sources are such options (IAEA, 2018). Nevertheless, others options that aim at mitigating GHG emissions from the extraction, transport, and conversion of fossil fuels, inter alia, comprise increased efficiency, fuel switching (e.g., from coal to gas), and GHG capture, known as carbon capture and sequestration or CCS (Folger, 2013).

Energy efficiency has come into reckoning as a prime option for climate change mitigation. It can be described as the ratio of the desired energy output for a specific task or service to the energy input for the given energy conversion process (Nakićenović, Gilli, & Kurz, 1996). Other approaches often define energy efficiency in relative terms, such as the ratio of minimal energy needed by the current best practice technology to actual energy use, everything else being constant (Stern, 2012). Economic studies often use Energy intensity primarily construed as the ratio of energy use per dollar of GDP in the economic literature, is often used as an indicator of how effectively energy is applied to produce goods and services. Nevertheless, energy intensity is dependent on multiple factors other than technical efficiencies and is not an appropriate substitution of actual energy efficiency (Stern, 2012). IPCC (2014a, 2014b) acknowledges that outcomes of climate change mitigation measures are dependent on the extent to which appropriate efforts are undertaken for the implementation of climate change policies and measures. Success of such efforts is subject to the mitigation capacity of different countries that are differentiable in accordance with their economic development level. This is why the issue of burden sharing among countries assumes significance with regard to international cooperation on climate change (IPCC, 2014a, 2014b).

4.2.6 International framework for climate change mitigation

Deep preference heterogeneity and highly multidimensional nature of the complex phenomenon of climate change has vested the countries to opt for different preferences over the pace of GHG emissions mitigation, the extent of cumulative mitigation, and eventually the allocation of remaining GHG emissions consistent with any temperature target across countries (Cameron, 2011). Incessant concentration of globally mixing GHGs in the atmosphere has rendered anthropogenic climate change as global commons problem (Knopf, 2015). Therefore attaining significant progress in mitigating climate change calls for international cooperation. Admittedly, literature on climate change has extensively dealt with incentives to "free ride" on climate protection (Stavins, 2011), concomitantly it also mulls that in certain cases, efficient common property management of open-access resources can confine or even do away with excessive use or overuse (Wiener, 2009). Implementation of efficacious management at global level, by allocating rights to emit and providing disincentives for excessive use by imposing sanctions or pricing emissions is essential for effectual management of common property management of the atmosphere (Schrijver, 2016). Internalization of external costs such as those costs that are not included into market prices or through legal remediation can be helpful in augmenting production of public goods. Economic measures can combine external costs and benefits into prices, making available incentives for private actors to more optimally decrease external costs and enhance external benefits (Buchholz, Cornes, & Rübbelke, 2012). Legal remediation may comprise looking for injunctive relief or compensatory payments (IPCC, 2007a, 2007b).

Global nature of the complex phenomenon of climate change necessitates international cooperation in significantly mitigating climate change (UNFCCC, 2016). Cooperation at the global level entails the potential of addressing multiple challenges, several actors that entertain divergent perceptions of the costs and benefits of cumulative action, sources of emissions that are disproportionately disbursed; assorted climate change effects that are ambiguous and far-etched in time and space and mitigation costs that are at variance (IPCC, 2001, pp. 607−608). Absence of universal cumulative action may spur smaller groups of individual actors to embark on supplying public goods, especially if such actors are well familiarized with each other, anticipate recurring interactions, can keep nonmembers out, and can keep a tab on and sanction defiance in the form of either overconsumption or underproduction (Eckersley, 2012). An interesting aspect of climate change discourse is that it embraces both proponents and protagonists. Among the proponents some scholars are optimistic (Chen & Murthy, 2019; Roberts, 2019a, 2019b) regarding such "mini-lateralism" (Falkner, 2015) and others are more skeptical (Rensburg, 2015).

In the absence of a universally acceptable international governing body or world government, each country can be expected to voluntarily consent to be bound by any international agreement. Such international agreements, in order to be effective, ought to be enchanting enough to garner broad participation (Brousseau, Dedeurwaerdere, Jouvet, Willinger, & Willinger, 2012). In the absence of international cooperation, consideration of relationship between mitigation costs and climate trade-offs is seemingly a deficient stimulus for actors at any level to facilitate significant reduction of emissions. Nevertheless, behavioral research demonstrates that individuals are on certain occasions spurred to cooperate or even punish or admonish those who defy, to a scale bigger than strict national models project (Andreoni & Samuelson, 2006). This could provide a semblance of the backdrop for appraising some of the observed policies being deployed to minimize GHG emissions at the national, subnational, corporate and individual level. Some reduction in emissions can take place without cooperation under the presumption of rational action, owing to positive externalities of otherwise co-trade-offs or self-advantage actions, for example actions designed to minimize energy expenditures, augmenting the security of energy supply, decreasing levels of local air pollution, improving land-use conserving biodiversity (Seto, Guneralp, & Hutyra, 2012). Considerable attention is being focused in the literature on the cobenefits of climate change mitigation (UNEP, 2013b; UNECE, 2016).

With the emergence of climate change as one of the most defining issues in present times warranting immediate and prioritized action, it is being addressed in a growing number of global forums and institutions and across a wider spectrum of range of scales and these forums and institutions have been operating under the aegis of the UN, international institutions/organizations; international civil society organizations (ICSOs) and NGOs at global, regional, national, subnational and local levels (Bhandari, 2018; Mubaya & Mafongoya, 2017). Global climate change "Policy Architecture" alludes to "the basic nature and structure of an international agreement or other multilateral (or bilateral) climate regime" (Aldy & Stavins, 2010a). This term embraces the sense of durability and resilience, in respect of both policy structure and the institutions to implement, enforce and support that structure (Schmalensee, 2010), which is congenial and appropriate to the long-term nature of the problem of climate change. Differentiation in architectures of mitigation of international emissions can be had on the basis of the possible approaches—bottom-up approach and top-down approach—along with different mechanisms like market-based instruments versus command-and-control regulations being adopted (Aldy & Stavins 2007). The top-down approach typically comprises international climate agreements such as KP (UNFCCC, 1998) and PACC (UNFCCC, 2015a, 2015b). Bottom-up approach, according to Jaffe and Stavins (2009), serves as a link between independent national and regional tradable permit systems. Market-based instruments inter alia include subsidies, taxes, and/or emission trading systems (ETSs) like, cap-and-trade systems (Drummond, Caranci, & Tulk, 2007), while command-and-control mechanisms set particular limits for emissions and/or mandates on pollution control technologies to be used (Goers, Wagner, & Wegmayr, 2010).

The prevalent top-down climate policy architecture is exemplified by the coming into being of the UNFCCC in 1992 and the UNFCCC entered into force in 1994. The UNFCCC is styled as a framework convention wherein multilateral treaties are routinely worked out in a procedural way that starts with a "framework" agreement in which the parties recognize the existence of a problem or threat, and commit to cooperative action, sans undertaking substantive obligations (Zakkour, Scowcroft, & Heidug, 2014). Presently, 197 parties (196 countries and one regional economic integration organization) have ratified the UNFCCC (n.d.). Recognizing the specific needs and difficulties of developing countries, the Convention and introduced the principle of "common but differentiated responsibilities and respective capabilities." Taking into account the fact that the bulk share of historical emissions of GHGs originated from developed countries, the Convention introduced more comprehensive requirements for developed country Parties, which are listed in Annex I to the Convention; and various topics which are negotiated under the Convention like on the provision of support or on reporting distinguish between "Annex I Parties" and "non-Annex I Parties" (Moosmann, Urrutia, Siemons, Cames, & Schneider, 2019). The UNFCCC acknowledges the long-term impacts of GHG emissions by setting long-term environmental goal and a near-term goal for industrialized countries (the so-called Annex I countries as opposed to non-Annex I countries); and Annex I countries agreed to a nonbinding quantitative emission target aimed at stabilizing their GHG emissions at 1990 levels starting in 2000 (Aldy & Stavins 2007).

4.2.6.1 The Kyoto Protocol

The First Conference of the Parties (COP-1) of the UNFCCC in 1995 made the Parties recognize the inadequacy of the Convention's voluntary targets of stabilizing the GHG emissions of the developed countries at 1990 levels by the year 2000 (UNFCCC, 1995); and accordingly, the Parties initiated the process of negotiating legally binding targets of emissions reduction or limitation for the so-called Annex I countries (that is, developed countries and those with economies in transition). This complex process culminated at COP-3, giving birth to the KP to the UNFCCC in 1997 in Kyoto, Japan, a historic landmark in international environmental law (Oberthur & Ott 1999). With 192 parties agreeing on the terms of the KP, it entered into force in 2005. Nevertheless, the Unites States, one of the countries with highest emissions, did not ratify the KP. Under the Protocol, 37 industrialized countries and the European Community (EC) had committed to reducing their emissions by an average of 5% against 1990 levels over the 5-year period (2008–12). For this group of countries, reductions of 11% were projected for the first Kyoto commitment period from 2008 to 2012, provided policies and measures planned by these countries were put in place and these countries were also had to make use of the Protocol's flexible mechanisms in order to reach their collective emission reduction goal (UNFCCC, 2011a,b,c).

Within the first commitment phase (2008–12), the industrialized countries and the European EC were required to reduce their collective GHG emissions to an average of 5% below 1990 levels. Nevertheless, the 2007 report by the IPCC asserts, "the numerous mitigation measures that have been undertaken by many Parties to the UNFCCC and the entry into force of the KP in February 2005 are inadequate for reversing overall GHG emission trends" (IPCC, 2007a, 2007b). The Ad Hoc Working Group on Further Commitments for Annex I Parties under the Kyoto Protocol (AWG-KP) was established at the first meeting of the Parties to the KP held in 2005 (UNFCCC, 2005). Conference of the Parties to the UNFCCC held in Bali (Indonesia) in 2007 adopted Bali Plan of Action (UNFCCC, 2008a, 2008b) that envisaged a comprehensive process to enable the full, effective and sustained implementation of the Convention through long-term cooperative action, now, up to and beyond 2012. Accordingly, the Ad hoc Working Group on Long-term Cooperative Action (AWG LCA) was established, as part of the Bali Plan of Action, with participation of all UNFCCC Parties, including the United States (UNFCCC, 2008a, 2008b).

The Bali Plan of Action envisaged a process to "reach an agreement on long-term cooperative action up to and beyond 2012" (UNFCCC, 2008a, 2008b). In the wake of the Copenhagen Climate Conference convened in 2009 under the aegis of the UNFCCC, the intervening period between 2007 and 2009 had proved instrumental in heightening people's expectations from the Copenhagen Climate Change Conference, also known as COP-15 (UNFCCC, 2010). Nevertheless, "Copenhagen" did not succeed in working on the Bali Plan of Action in the absence of consensus on Copenhagen accords that contained an assortment of aspects that constituted the basis for following more successful negotiation rounds, included the statement that developing countries would implement mitigation actions and it was construed as a first "break" with the principle of 'common but differentiated responsibilities (JIN, 2010). The COP-16 held at Cancun (Mexico) in December 2010 yielded Cancun Agreements that mirrored a shift in the climate negotiations since "Copenhagen" from a top-down architecture wherein an all-embracing goal is translated into individual country targets, such as in the KP, to one wherein national commitments ought to combine a joint international effort (Kok et al., 2010); and member countries were called upon by the Cancun Agreements to frame national targets in the form of pledges to agree on international review procedures for these (UNFCCC, 2011a).

The climate change negotiations until the Copenhagen summit had tried to formulate an overall GHG emission reduction target based on the scientific evidence as reports of the Intergovernmental Panel on Climate Change (IPCC) in tandem with the precautionary tenets of the UNFCCC with individual country targets. Viewed in a broader perspective, Copenhagen and Cancun summits on climate change could be said to have demonstrated that climate policy making might be "more acceptable for countries if climate actions are embedded in domestic sustainable development objectives, especially when aiming at actively involving developing countries in global climate policy making" (van der Gaast & Monica, 2015). Renewed attempt at working out a plan for a legal agreement "applicable to all Parties" was essayed at the COP-17 held at Durban, South Africa in December 2011 and in this regard a new platform under the title the Ad Hoc Working Group on the Durban Platform for Enhanced Action (ADP) was launched and it comprised two workstreams (UNFCCC, 2011b). Accordingly, Workstream I focused on the proposed 2015 agreement and envisaged interim steps like a draft negotiating text not later than December 2014 at COP-20 in

Lima (Peru), with the objective of making available a negotiated text by May 2015, and that was, inter alia, to include work on mitigation, adaptation, finance, technology development and transfer, capacity building and transparency of action and support.

While concentrating on emission reductions prior to 2020, Workstream II focused on exploring possible options to minimize the huge gap between Parties' mitigation pledges and the pathways in accordance with limiting the increase of global average temperature below 2°C or 1.5°C above pre--industrial levels. The UNEP (2013) had cautioned that additional reductions were required in the range of 8–10 billion tons of CO_2 equivalent by 2020. Thus the Workstream II aimed at exploring "opportunities for actions with high mitigation potential, including those with adaptation and sustainable development co-benefits, ... scalable and replicable, with a view to promoting voluntary cooperation ... in accordance with nationally defined development priorities" (UNFCCC, 2012a, 2012b). In view of the first phase of commitment (2007–12) of the KP nearing its completion and conclusion of an all-embracing agreement still far away, negotiating the second phase of commitment under the KP was being contemplated; and concurrently it was opined that the adoption of the KP was hardly lauded as a panacea to the global climate problem, either at its inception or later (Boyd, 2010).

The Conference of the Parties at its 18th meeting (COP-18) held at Doha (Qatar) from November 26 to December 8, 2012 embarked on a new commitment phase under the KP, also called the second commitment phase, commencing from January 1, 2013 to December 31, 2020; as an interim measure until the new agreement was expected to enter into force in 2020 (UNFCCC, 2013c). Apart from delineating the second phase of the Protocol, the Doha Amendment also added nitrogen trifluoride to the list of GHGs covered, and facilitated the unilateral strengthening of commitments by individual parties; and for the second phase period, Annex 1 Parties committed to reduce GHG emissions by at least 18% below 1990 levels (Kolmuss, 2013). Not all countries included in the first phase of the KP decided to commit in the second phase and with Japan, New Zealand and Russia opting out of the second phase; the remaining countries that also included EU, Australia, Switzerland and Norway, jointly contributed no more than 15% of global emissions (Alessi, 2015). Other industrialized countries that did not form part of the second phase of KP included the US that did not ratify the first phase of the KP and Canada withdrew from the Protocol in 2012.

At the Doha round of negotiations on KP, the key issue of climate fiancé remained a core issue of contention between developing and industrialized countries and on account of the tough economic conditions, no collective mid-term commitment on scaled-up funding was made and only a few European countries could individually pledge increased finance (Morgan, 2011). Irrespective of its limitations, second phase of the KP was deemed to be a "success as it kept the only legally binding instrument under the UNFCCC alive" (Alessi, 2015), and "Countries with emission commitments from the KP emit on an average 7% less CO2 than similar countries that did not ratify the Protocol" (Grunewald & Martinez-Zarzoso, 2015); thereby, demonstrating flexible mechanism of the KP.

In the wake of publication of the Fifth Assessment Report (AR5) by the IPCC in 2013, prior to the convening of COP-19 in Warsaw (Poland), ample evidence was made available with regard to emission trends and estimates of the impacts of existing and proposed policies still entailed the potential of leading the average global temperature rise of 4°C above pre—industrial levels by 2100; and even if Parties to the UNFCCC facilitated full implementation of their pledges, the increase in temperature was projected to reach 3.3°C (Climate Action Tracker, 2011). The pattern of continuous consumption of fossil fuel energy in the previous decade seemingly provides the basis of higher estimate of temperature increase. A gap of 8 Gt carbon dioxide equivalent ($GtCO_{2e}$) was expected to remain even under the eventuality of the most stringent implementation of the pledges, as can be evidenced from Fig. 4.2 that also illustrates different scenarios.

Under the scenario of business-as-usual (BAU), the gap would be 14 $GtCO_2$eq/year; whereas under different cases of country pledges, the projected gap would be between 8 and 13 $GtCO_2$eq/year (cases 1–4); as shown in Fig. 4.2, and under the most ambitious scenario, the gap would be 8 $GtCO_2$eq/year (UNEP, 2013). Thus this ostensible "ambition gap" had emerged as one of the core issues of the contemporary climate change negotiations that then aimed at closing this gap by 2020. Undoubtedly, technical options were available at that stage to narrow this gap (Black, 2011); nonetheless, success depended on the political will to implement actions beyond the existing pledges (Alessi, 2015). According to UNEP (2013), the technical potential for emission reductions by 2020 were estimated to be about 17 +/− 3 $GtCO_2$eq at marginal costs below US$50–100/$tCO_2$eq, comfortably above the reductions needed to reach the 2°C range.

Irrespective of the issues raised by Fifth Assessment Report (IPCC, 2013), the COP-19 held at Warsaw (Poland) in November 2013 witnessed advances in negotiations that were meager in contents and focused more on issues pertaining to Parties' determination to reach a legally binding agreement. In the months prior to the convening of the COP-19 at Warsaw, Parties were perhaps had a clear vision of what was required to be agreed to in COP-19 and three issues were "key" for a positive outcome —finance; loss and damage; and advancing work toward the 2015 agreement (Violetti & Rodriguez, 2013). The COP-19 achieved some limited success in some areas like monitoring, reporting and verification for domestic action as well as adaptation. Another positive outcome of the COP-19 was the establishment of Warsaw International Mechanism (WIM) for Loss and Damage associated with Climate Change Impacts to address, to address cases such as extreme and slow onset events in vulnerable developing countries (UNFCCC, 2014). While agreeing on the rulebook for deforestation and forest degradation, the Parties at COP-19 also agreed on a timeline for the development of the 2015 agreement. One significant additional change envisaged by the Warsaw COP-19 that substantially transformed the process of the negotiations was the consensus reached that countries would submit their emission reduction pledges as "intended nationally determined commitments" (INDCs), and that international rules would ensure transparency and appropriate monitoring mechanisms on implementation (UNFCCC, 2014). The term "contribution" was introduced as a compromise of the term "commitment," to explain the proposed efforts of both developed and developing countries (Höhne, Ellermann, & Li, 2014). Accordingly, it was also decided at COP-19 that INDCs would be submitted "without prejudice to the legal status of the contributions" (UNFCCC, 2013c), permitting the distinction under the "Common but Differentiated Capabilities" principles (CC&ES, 2013). This negotiated settlement proved instrumental in paving way for ore significant advances in the forthcoming COP-20 and COP-21 scheduled to be held in Lima (Peru) and Paris (France).

Racing toward the goal of reaching a global legally binding agreement in 2015 necessitated the COP-20 held at Lima in 2014 to be entrusted with the herculean task of delivering solid groundwork. Among the important decisions taken at COP-20, one of the most significant was the inclusion of a first partial draft agreement text for COP-21 in the final declaration (UNFCCC, 2015a, 2015b) and it allowed the Parties to embark on negotiating in preparation of the Paris conference.

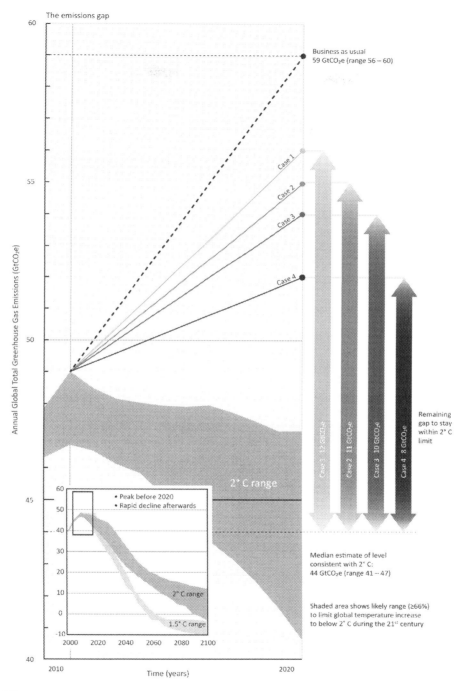

FIGURE 4.2

Emission scenarios for emission reductions, 2011–20 UNEP (United Nations Environment Programme), 2013.

From UNEP (United Nations Environment Programme) (2013). The Emission Gap report 2013: A UNEP synthesis report. Nairobi: UNEP. Available at: http://web.unep.org/sites/default/files/EGR2013/EmissionsGapReport_2013_high-res.pdf.

Lima negotiations also led to additional decisions that inter alia included: recognition of adaptation to have the same significance as mitigation, in line with the demands of developing countries; a formal recognition on the National Adaptation Plan as a way to deliver resilience with visibility tools and linking them to the Green Climate Fund (GCF); and countries reaching a new level of understanding on monitoring and verification measures on the delivery of their INDCs to climate change mitigation (UNFCCC, 2015a). Nevertheless, Lima negotiations left many areas still unresolved; and in the first place, it did not address the contributions of different countries in terms of emission reductions (Erbach, 2015). Nonetheless, the initial text from Lima made it discernible that the ADP Workstream I had been successful in producing at the climate negotiations at Geneva a full long draft negotiating text for the Paris negotiations, "meaning a much more comprehensive document than the Lima draft" (Alessi, 2015). The protracted journey of negotiations on climate change culminating in the signing of PACC is shown in Table 4.1.

4.2.6.2 Paris Agreement on Climate Change

Protracted negotiations toward the goal of arriving at a global legally binding agreement ripened into fruition at the COP-21 to the UNFCCC at Paris when the final draft was approved by the Parties and it came to be known as Paris Agreement or PACC in December 2015. Having entered into force on November 4, 2016, the PACC calls on the Parties to contrast climate change and to speed up and intensify action and investments required for a sustainable low-carbon future and to adapt to the increasing adverse impacts of climate change. The PACC epitomizes a significantly more ambitious shift in the recognition that the long-term temperature goal should be to "hold [...] the increase in the global average temperature to well below 2°C above pre-industrial levels and to pursue efforts to limit the temperature increase to 1.5C above pre-industrial levels" (UNFCCC, 2015b). The PACC envisages a different approach that is based on the notion that self-imposed voluntary commitments are more probably to be met rather than agreed through collective negotiations by the world community, and that manifested domestic progress, full transparency, nonparty engagement, a collective commitment to progression and ambition and regular review of the collective effort "are key to moving parties beyond no-regrets actions" (Doelle et al., 2017). Viewed in a broad spectrum, the relative merits of such approaches have been frequently dealt with in the realms of international relations and legal literature for some time (Stewart, Kingsbury, & Rudyk, 2009), and this had forecasted the ensuing shift to bottom-up approach.

The top-down approach contemplated in the KP has been based on the assumption that countries would always act in self-interest, necessitating a global agreement that aligns their self- interest with the global interest through binding commitments and strong compliance. The PACC is contemplated to be founded on the presumption that countries can be spurred into action in the global interest through managerial approaches that can help attain new norms of state behavior. Transparency, a clear enunciation of the collective goal, focus on the state of the science, collective progress being revisited periodically, and opportunities for interaction and information sharing, along with the flexibility to adjust to changing circumstances and science form the main constituents of this managerial approach (Bodansky, 2016). There is much common between the approaches of the KP and PACC because both are based on the idea that collective ambitious action on climate change could yield national and global economic benefits (Hisbaron, 2017), and both are based on strong transparency, "even if the transparency under Kyoto is focused on compliance, whereas the transparency in Paris is directed at progression of ambition" (Doelle et al., 2017).

4.2 Mitigation

Table 4.1 Timeline of global climate change negotiations (Hameso, 2012).

Period	Event	Objectives
1979	World Meteorological Organization (WMO) organized the First World Climate Conference (Geneva, Switzerland)	Assessing existing knowledge of how higher atmospheric GHG concentration levels could lead to average temperature increases.
1987	The Brundtland Commission Report Our Common Future (United Nations)	Reconciling traditional development objectives with environmental ones: promoting harmony among human beings and between humanity and nature.
1988	Formation of the Intergovernmental Panel on Climate Change (IPCC)	Provide governments with a clear scientific view of what is happening to the world's climate.
1990	IPCC & Second World Climate Conference call for Global Treaty	IPCC released First Assessment Report saying emissions resulting from human activities are substantially increasing concentration of GHGs leading to calls by IPCC and the second World Climate Conference for Global Treaty.
1992	UN Conference on Environment and Development (UNCED) negotiated UNFCCC	Agreement reached on the need for serious action to reduce man-made GHG emissions. Over 180 countries signed UNFCCC treaty.
1995	First Conference of Parties (COP-1) held in Berlin (Germany)	The Parties agreed to establish a process to negotiate strengthened commitments for developing countries thus laying groundwork for KP.
1997	Kyoto Protocol adopted	The third Conference of the Parties (COP-3) achieved an historical milestone with adoption of the KP, the world's first GHG emissions reduction treaty.
2000	Millennium Development Goals	Goal 7. Ensure environmental sustainability. Aims to reverse current trends in environmental degradation in order to sustain the health and productivity of the world's ecosystems.
2001	Marrakesh Accords reached	Marrakesh Accords reached at COP-7 set the stage for ratification of KP.
2004	UNDP's Climate Change Adaptation Policy Framework	Help developing country governments to prepare their National Adaptation Programs of Action (NAPAs).
2005	EU Emissions trading launched. KP entered into force	The European Union Emission trading Scheme, the first and largest emissions trading scheme in the world, launched as a major pillar of EU climate policy.
2006	The Subsidiary Body for Scientific and Technological Advice (SBSTA)	COP-12 held in Kenya, mandated the Subsidiary Body for Scientific and Technological Advice (SBSTA) to undertake a program to address impacts, vulnerability and adaptation to climate change.
2007	Developing countries produced their NAPAs	Specific adaptation options or measures identified.

(Continued)

Table 4.1 Timeline of global climate change negotiations (Hameso, 2012). *Continued*

Period	Event	Objectives
2009	Copenhagen UN climate summit	Deadline for binding global agreement on climate change missed. A new climate change treaty was not reached.
2010	Cancun Agreements dealing with climate change	The GCF, Technology Mechanism and the Cancun Adaptation Framework adopted at COP-16 held at Cancun in Mexico.
2011	Durban Summit	Formation of Durban Platform for Enhanced Action to negotiate legally binding emissions reduction by 2015. Any reductions take effect after 2020.
2012	Rio + 20 Earth Summit, June 2012 COP-18 held at Doha (Qatar) 26 November to 7 December 2012.	The United Nations Conference on Sustainable Development—or Rio + 20—took place in Rio de Janeiro, Brazil, resulted in a focused political outcome document that envisaged clear and practical measures for implementing sustainable development. Launched second commitment period of KP from 2013 to 2020.
2013	Warsaw Climate Change Conference COP-19	Governments took decisions to stay on track toward securing a universal climate change agreement in 2015. Also established the new WIM for Loss and Damage associated with Climate Change Impacts, to address cases such as extreme and slow onset events in vulnerable developing countries.
2014	The Global Commission on the Economy and Climate report UN Secretary-General's Climate Summit in November 2014 COP-20 at Lima (Peru) in December 2014.	The New Climate Economy is the Commission's project set up to provide independent and authoritative evidence on the relationship between actions which can strengthen economic performance and those which reduce the risk of dangerous climate change. Summit hosted to mobilize action and ambition on climate change in advance to COP-21 at Paris in 2015. Decisions taken to prepare the ground for a legally binding agreement. One of the most important was the incorporation of a first partial draft agreement text for COP-21 in the final that allows the Parties to start negotiating in preparation of the Paris conference.
2015	COP-21 at Paris (France)	Concluded Agreement on Climate Change, also known as the Paris Agreement, it not only sets a target of limiting a global temperature increase to "well below 2°C," but acknowledges the need to aim for a limit of 1.5°C, taking therefore into account the needs of the most vulnerable island nations.

Table 4.1 Timeline of global climate change negotiations (Hameso, 2012). *Continued*

Period	Event	Objectives
2016	COP-22 held at Marrakech (Morocco)	Adopted Marrakech Action Proclamation for Climate & Sustainable Development; Marrakech Partnership for Global Climate Action; and Progress Tracker.
2017	COP-23 was held in Bonn (Germany)	Significant progress was made on implementation guidelines for Paris Agreement, also called the Paris Rulebook. A rift began to emerge between developed and developing countries over pre-2020 action. In the end, Parties agreed to hold additional stocktaking sessions in 2018 and 2019 to review progress on reducing emissions, as well as produce two assessments on climate finance in 2018 and 2020.
2018	IPCC released *special report* on the impacts of *global warming of 1.5°C on 8 October 2018*. COP-24 held in Katowice (Poland)	The Special Report deals with the impacts of *global warming of 1.5°C* above pre–industrial levels and related *global* GHG emission pathways, in the context of strengthening the *global* response to the threat of *climate change*, sustainable development, and efforts to eradicate poverty. Agreement on the rules to implement Paris Agreement; some countries disagreed to "welcome" IPCC Special Report on 1.5°C. It was also agreed among the members that future pledges should cover a "common timeframe" from 2031.
2019	COP-25 held in Madrid (Spain)	Countries agreed to table their new and improved carbon curbing plans at COP26 in 2020. Developed nations agreed to honor their commitments on climate change before the end of 2020. However, countries failed to reach consensus on several most-contentious (technical and nontechnical) issues including, among others, negotiations related to Art. 6 of the Paris Agreement, financing of the global climate action, transparency reporting requirements and (common) time frames for climate commitments.

Compiled from Hameso, S. (2012, March 1). Development challenges in the age of climate change: The case of Sidama. In: Paper presented at the workshop on the economy of Southern Ethiopia. Hawassa: Ethiopian Economics Association; Alessi, M. (2015). History of the UN climate negotiations—Part 2—From 2011 to the present. Climate Policy Info Hub. UNFCCC website.

Debates on similar lines have seemingly occurred in many other international and domestic contexts with regard to efficacy of several regulatory approaches, with some opting for top-down command-and-control approaches, while others advocating managerial market-based or voluntary approaches; nevertheless, recent domestic debates have resulted in spurring a growing number of academics and practitioners calling for a mix of approaches that is customized to the circumstances (OECD, 2011; Tollefson, Zito, & Gale, 2012). The PACC has come to epitomize an experiment in the deployment of a more managerial approach that seeks to spur countries "beyond no-regrets actions" on climate change; and while retaining significant elements of the Kyoto approach, it nonetheless "represents a significant departure in that it seeks to raise ambition through careful management and norm-building" (Doelle et al., 2017).

Some of the important issues highlighted in the preamble of the PACC are recognized for the first time within a legally binding agreement under the UN climate regime; and these significant issues inter alia include: the imperative of a just transition of the workforce; the recognition of concepts such as "Mother Earth" and "climate justice"; the need to respect, promote, and consider human rights; emphasis on considering the right to health, the rights of indigenous peoples, local communities, migrants, children, persons with disabilities, and people in vulnerable situations; and the requirement of taking into consider the right to development, gender equality, the empowerment of women, and to intergenerational equity when taking action on climate change (Doelle et al., 2017; WRI, 2016).

Most of the paragraphs envisaged in the Preamble of the PACC contain parallel provisions in the operative part. This is seemingly pertinent because the juxtaposition of preambular paragraphs with operative provisions provides the former more binding character, according to well-established doctrine and law (Mbengue (n.d.)). Distinct feature of the PACC is that it goes beyond the realms of international climate and environmental law in multiple ways; specifically, it is the first multilateral environmental agreement (MEA) to comprise references to human rights, the rights of particular groups, gender equality, and intergenerational equity (Bach, 2016). Article 2 of the PACC incorporates the objective of keeping global average temperature increase "well below" 2°C, and aims at limiting this increase to 1.5°C can be said to be the major achievement of the Paris negotiations. Admittedly, the target of achieving 1.5°C as now perhaps emerged as the final standard against which the success of collective mitigation endeavors will be gauged. Besides, this long-term objective affords a significant basis for each country's future nationally determined contributions (NDCs), their justification on the basis of equity and the 5-year cycles of NDC communication and the global stocktake (Doelle et al., 2017; Thorgeirsson et al., 2017).

The provisions relating to mitigation are central to the PACC and Article 4 along with its subarticles emphasize on the collective mitigation endeavor represented by the NDCs assessed against the long-term temperature goal in the wake of successive NDCs ought to be consolidated every 5 years commencing rom 2025, thereafter to be followed by a global stocktaking exercise to be conducted every 2 years before each updated NDC is due. Articles 4.2–4.7 deal with individual and differentiated obligations with progression, support, flexibility and cobenefits, respectively (Winkler et al., 2017). Provisions regarding communicating information, time frames, recoding and accounting with regard to NDCs are provided from Articles 4.8– 4.14. Provision of Article 4.19 enjoins on all Parties to strive to formulate and communicate long-term low GHG emission development strategies. There is a disconnectedness between the aim of the PACC to keep increase in temperature well below 2°C and embark on efforts to limit it to 1.5°C (UNEP, 2015), and the aggregate effect of mitigation INDCs. According to Winkler et al. (2017), "In a regime where bottom-up pledges play a more central role, the strength of review and other top-down elements will be critical in determining whether the Paris Agreement is adequate or not. It is too early to tell."

Provisions of Article 4.5 of the PACC affirm significance of conserving and enhancing of sinks and reservoirs of GHGs, including forests. Most important sinks of carbon are land-based ecosystems like vegetation or living biomass, dead organic matter in litter and soils along with old soil carbon in wetland and permafrost soils. Nevertheless, land-based ecosystems, while subject to variations, are among the most important sinks of GHGs (Ciais et al., 2013, p. 470), after the use of fossil fuels and cement production. About one-fourth of anthropogenic GHG emissions globally come from the agriculture, forestry and other land-use (AFOLU) sector (Smith et al., 2014). Interestingly, activities in AFOLU sector can serve both as sources and sinks of emissions. Some case studies in the recent past have found that forests and tree- based ecosystems minimize the likelihood of social vulnerability to climate change, especially in developing countries (Pramova, Locatelli, Djoudi, & Somorin, 2012a, 2012b); and perhaps the probability of the greater vulnerability of developing countries could be part of the motivation for the PACC's emphasis in Article 4.5 on "the forest sector, where other sectors have not been referred so explicitly" (La Viña et al., 2017).

Voluntary cooperation of the Parties in implementation of their NDCs is recognized under the provisions of Article 6 of the PACC, with a view to allow for higher ambition and to promote sustainable development and environmental integrity. Principles for voluntary use of Internationally Transferred Mitigation Outcomes (ITMOs) toward NDCs are set out under Articles 6.2 and 6.3; a new centrally governed market mechanism is outlined under Articles 6.4, 6.5, and 6.6; and a framework for non-market approaches is envisaged under the provisions of Articles 6.8 and 6.9 (Hood, 2019; Howard et al., 2017). The issue of "rulebook" that had remained unresolved at the Paris negotiations in 2015 was resolved with the adoption of "rulebook" at the COP-24 at Katowice (Poland) and it contained a set of detailed guidance for implementing Article 6 of the PACC. The rulebook incorporates guidance under Article 6.2 for Parties making use of ITMOs, and rules, modalities, and procedures governing the Article 6.4 mechanism (UNFCCC, 2018a). ITMOs use a carbon dioxide equivalent [CO_2eq] metric for a new set of market provisions or other GHG mitigation outcomes that are delineated under article 6 of the PACC. These rules are set to come into force in 2020 and these have been designed to replace other prevalent forms of international carbon credits such as those announced under the Kyoto-era Clean Development Mechanism and Joint Implementation. Viewed in broad context, ITMOs still lack specific definition and "could take many forms including through linking ETSs across jurisdictions, investment in emission reduction projects, technology transfers and even credits from REDD + schemes" (Roth, Echeverria, & Gass, 2019). It is being argued by Roth et al. (2019), "Article 6 could be a useful way to channel technology, finance and capacity-building from developed to developing countries" that are defined under Article 6 of the PACC. Undoubtedly, Parties has initially agreed that rules for the implementation of the mechanisms would be worked out at COP-24; nevertheless, negotiations in this respect had proved cumbersome the and continued to be so at COP-25 at Madrid (Spain) with "Parties having different understandings on how the mechanisms are to be implemented" (TWN, 2019).

Article 7 of the PACC deals with adaptation and in the process, it acknowledges a global goal of increasing adaptive capacity, strengthening resilience and minimizing vulnerability to climate change (Pérez et al., 2017). The notion of loss and damage is delineated by Article 8 of the PACC as a "standalone" article and its placement in the Agreement separates it from adaptation, signaling that loss and damage carries with it a unique set of cross-cutting issues (Siegele et al., 2017). Both the notions of adaptation and loss and damage associated with climate change are discussed in more details in the succeeding pages of this chapter; therefore only brief reference to these notions is made here under the PACC.

Article 9 of the PACC envisages a broader understanding of climate finance that thus far had historically been confined to public financial flows from developed to developing country parties. Construed in this context, Article 9 is seemingly in line with Article 2.1(c) that envisages "making finance flows consistent with a pathway toward low GHG emissions and climate-resilient development" as the third objective of the PACC and it is an outcome of the recognition of the economic and financial reality that, in order to surmount the challenges posed by climate change, every single kind of available financial flows is required (Doelle et al., 2017). Keeping in view the past negotiations in respect of adequacy and status of previous commitments to mobilize climate finance of US$100 billion a year by 2020, the PACC calls upon developed countries to come up with a substantive roadmap for achieving the 2020 goal of US$100 billion. And Parties to the PACC were to conduct a facilitative dialog in 2016 to assess progress and as of November 2018, 183 Parties to the PACC have committed to "making finance flows consistent with a pathway toward low GHG emissions and climate-resilient development" as per Article 2.1c of the PACC (Whitley, Thwaites, Wright, & Caroline, 2018).

The PACC under Article 10 offers a new technology framework entailing significant potential for overlap with the prevalent Technology Mechanism (TM) and limited progress on the key major issues. TM consists of the Climate Technology Center Network (CTCN) and the Technology Executive Committee (TEC) (UNFCCC, 2015c). Construed in a broad context, Article 10 of the PACC along with other related decisions in this regard comprise a mix of actions that are "fairly concrete"; and the binding nature of the Article 10 makes it discernible that collective actions would be adhered to; nonetheless, these actions seem to be essentially limited in scope and manifest only a fraction of the overall "puzzle of how to support climate action in developing countries appropriately, although this can be seen as consistent with the overall emerging approach in the climate arena where it is increasingly looking as if much of the action will be outside the formal UNFCCC processes and institutions" (de Cornick & Sagar, 2017, p. 259). There is a dire need of accelerating and strengthening technological innovation to attain the goals of the PACC so that technology can be harnessed in delivering "environmentally and socially sound, cost-effective and better-performing climate technologies on a larger and widespread scale" (UNFCCC, 2017) by adhering to different innovation approaches.

Incorporation of capacity building under Article 11 of the PACC, for which efforts had been afoot since COP-13 at Bali, has been complemented with the establishment of the Paris Committee on Capacity Building (PCCB). The aim of capacity building, as envisaged in Article 11, is to augment the ability of developing countries to take effective action on climate change in the fulfillment of the objectives of the UNFCCC as well as of the PACC (UNFCCC, 2018b). While underlining the means through which capacity building is to be augmented under the PACC, Article 11 also outlines the roles and obligations of all Parties, as well as those "specific to developed and developing country parties respectively" (d'Auvergne et al., 2017). Undoubtedly, incorporation of capacity building in the PACC along with establishment of PCCB are laudable steps in enabling developing countries to enhance their respective capacity-building programs in accordance with Article 11 of the PACC; nevertheless, more time is required to attain the capacity building objectives and "Parties need to consider how best to equip the PCCB to fulfill its extended mandate, as effectively and efficiently as possible" (WRI, 2019).

Other more significant provisions incorporated in the PACC pertain to transparency under Article 13, global stocktake under Article 14 and compliance under Article 15. The notion of transparency imbibes building trust and letting the people know whether their objectives are being achieved. While raising the overall ambition of the climate change regime as whole and of the actions being reported, transparency helps in improving the "reliability of assessment that try to look at the big picture of how well we are doing collectively in tracking climate change" (Barakat et al., 2017). The provisions of transparency are envisaged under Article 13 of the PACC that apply to all Parties, with slight modest differentiation, primarily through a commitment to flexibility and support for developing countries (Doelle et al., 2017). Article 13 envisages general principles for the transparency framework and a path of moving on for enhancing the manner in which countries exchange information on progress attained in the realms of mitigation, adaptation, making available or receiving finance, advance technology development and transfer along with strengthening capacity building. It also focuses on the procedure for the subsequent review of such information, while the detailed modalities, procedures and guidelines for the enhanced transparency framework "remains to be developed" (Dagnet et al., 2017). Given the fact that climate finance is an essential enabler of enhanced climate action, transparency with regard to the scale and type of climate finance made available and mobilized by developed and other countries, and received by developing countries, assumes significance for national and international purposes (OECD, 2016). Interlinkages between capacity-building and transparency framework entail utility for developing countries because developing country Parties have different starting points and in the absence of capacity building some such Parties may have to strive hard to implement enhanced transparency framework as envisaged in the PACC (Dagnet, Nathan, Bird, Bouye, & Rocha, 2019a, 2019b).

The avowed objective of establishing a stocktaking process under Article 14 of the PACC is said to be another central element of the overall effort to ensure the goal of the Paris Agreement is met through the collective efforts of parties. The notion of global stocktaking encompasses mitigation, adaptation, means of implementation, and support. The first global stocktake would take place in 2023, in time for the revision of parties' NDCs by 2025 (UNFCCC, 2015b, Article 14.2). Aiming at enhancing national action and international cooperation, an initial stocktaking process among parties, called "facilitative dialog," was scheduled for 2018 (Doelle et al., 2017). The global stocktake aims at facilitating augmentation of ambition through three main functions that are designed to help parties in: (1) understanding how far they have collectively come in achieving their goals; (2) realizing what is still required collectively to reach them; and (3) be informed about possible options on how to augment their actions both nationally and internationally and thereby "hopefully be motivated to do more ... in this manner, the global stocktake can be compared to the engine of the Agreement, which is expected to drive the political momentum toward higher ambition and thus ensure dynamism" (Friedrich et al., 2017). This dynamism in turn is a precondition for a durable Agreement, since only through an upward dynamic can the Agreement deliver in the long term. Suggesting that the design of global stocktake ought not be locked in by December 2018, Northrop, Dagnet, Hohne, Thwaites, and Mogelgaard (2018) have informed that Parties should consider about upcoming decisions pertaining to interlinkages to processes and work programs that will provide input to the global stocktake and on the issue of agreeing on common time frames for NDCs. While advising the Parties to build on lessons learned from the Talanoa Dialogue (UNFCCC, 2018c), and each global stocktake to make next stocktake more impactful, Northrop et al. (2018) opine that a certain degree of "flexibility should therefore be built in to ensure that the mechanism can evolve over time."

Facilitating the establishment in Article 15 of a mechanism, consisting of an expert-based committee, to enable implementation and promote compliance under Article 15 of the PACC constitutes an essential component of the Paris Agreement's transparency and accountability regime. Consisting of twelve members with requisite technical expertise, with membership determined in a way similar to the facilitative branch of the compliance committee under the KP, this compliance committee is to be sensitive to national capabilities and circumstances of Parties in conducting its work (UNFCCC, 2015b, Article 15.2). The incorporation of such a mechanism in a MEA just like the Paris Agreement assumes significance in accomplishing vital functions, especially when it can accelerate implementation and extend support in garnering greater cooperation by—contributing to and maintaining trust among parties; addressing capacity problems and clarifying ambiguities; and preventing "free- riding" by promoting compliance with agreed obligations or norms (Dagnet et al., 2017). By agreeing to the inclusion of this mechanism in the Paris Agreement, the Parties have signaled the momentousness with which they intend to implement their obligations under the PACC.

The Modalities and Procedures for the effective operation of the Article 15 Committee foresee that the Compliance Committee builds up its rules of procedure for adoption by CMA3 in 2020. The rules of procedures will encompass more specific details on, for example, timelines, conflict of interest, role of the cochairs and reasoning in the decisions of the committee (Voigt, 2019; Zihua, Voigt, & Werksman, 2019). Viewed in a broad context, of all the elements and components that will require to be reflected through the modalities and procedures under the compliance mechanism, three issues—how individual Parties can be referred to the Committee; the measures available to the Committee; and appropriate overreaching operational guidance to the Committee to circumscribe its discretion and thereby provide reassurance to Parties—come to the fore as being key for striking the requisite equilibrium between each consensus among the Parties and ensuring a robust and effective committee that can help fulfill its purpose mandated by Article 15 (Oberthur & Northrop, 2018). Since the PACC is too operational in 2020, henceforth the success of the transparency, review, stocktaking, and compliance approach in the Paris Agreement in heightening ambition appropriately to meet the long- term goal will finally be dependent on "many factors outside the purview of the new climate regime, most notably the economic, political, and social circumstances in key member states" (Doelle et al., 2017).

The conclusion of the PACC in 2015 and its ratification and subsequent coming into force in November 2016 was deemed as a step toward transforming the paradigms of climate diplomacy because of its adoption of bottom-up structure for emission targets in the form of NDCs along with balancing of top-down provisions for strong global emission goals and key accountability provisions like reporting and review. While emphasizing that all Parties are required to be sensible in respect of the level of rigor and detail called for in the rules and provisions of the Paris Agreement, Stem (2018) suggests that the Parties perceptibly need to be strong enough to "support ambitious action but going overboard will make developing countries feel that the central bottom-up-top-down balance is getting lost ... The important thing is to get countries all in one effective system and get moving." With much of the Paris rulebook having been finalized at the Katowice COP-24, there were still many issues on the table at COP-25 at Madrid and one of that pertained to many developing countries wanting that in addition to reducing the future impacts of climate change, there needs to be accountability for past and present damages, with specific emphasis on wealthier countries allocating more funds for dealing with impacts of extreme weather events in poorer countries rather than simply funding projects that curb emissions and deployment of clean energy

(Irfan, 2019). Under the provisions of the Paris Agreement that call for 5-year cycles of global stocktaking in the hopes of "ratcheting up" ambition, Parties are being required to submit their next round of pledges by a UN meeting scheduled for in December 2020. At the outset of the 2020, most of the Parties have targets for 2030, and are being required to submit new, more ambitious goals for that year; nevertheless, they are also allowed to simply stick with their old targets and as such this makes 2020 "as a make-or-break year for climate policy" (Beeler, 2019). The process of US withdrawal from the Paris Agreement, path for which got paved in the aftermath of installation of the presidency of Donald Trump and was going to take final effect on November 4, 2020, has already caused some damage and in fact, "there are reasons to believe that the Paris Agreement is in bad shape, that in the 2020 it break down or even fall apart entirely" (Roberts, 2019a, 2019b). Despite downward spiral of dissent, dysfunction and disengagement along with articulation of pessimism over Paris Agreement, Noah (2019) still thinks it is better to have Paris Agreement than not to have it because "it facilitates international cooperation, has regular review conference, requires submissions of NDCs on a defined timetable, and requires submissions of inventories...what it cannot do is to command any particular result."

4.3 Adaptation

The notion of adaptation as a distinct response to climate change started gaining salience in the scientific literature in the mid-1990s onward, and there also ensued coevolution of adaptation policy, practice, and research, resulting in shaping and informing one another. Early years in the aftermath of the founding of the UNFCCC in 1992 witnessed scant attention being focused on adaptation for a variety of reasons, while mitigation received considerable attention. Article 2 of the UNFCCC dealing with the objective of this Convention delineated mitigation by alluding to actions adhered to for reducing emissions of GHGs in order to "prevent dangerous anthropogenic interference with the climate system" (UNFCCC, 1922, Article 2), adaption was not defined in the text of the Convention or UNFCCC. Consequently, the notion of adaptation connoting its meaning started finding articulation in the academic and policy discourses from time to time (Schipper & Burton 2008). In some quarters, apprehension was expressed that added focus on adaptation could divert attention from the international focus on mitigation (Burton, 1996; Pielke 1998), or that focus on adaptation from developing countries could be construed as admission of onus in causing climate change (Verheyen, 2002), entailing potential of "liability floodgates" (Klien et al., 2016).

Undoubtedly, with the publication of the Fifth Assessment report of the IPCC in 2014 that demonstrated the centrality of the occurrence of the anthropogenic climate change (IPCC, 2014a, 2014b, p. 48), there also came the realization of the need for climate adaptation ought to be underscored by scientific evidence that climate change is both occurring and that impacts are experienced; nevertheless, the similar level of centrality was missing in the early reports of the IPCC. According to Schipper (2006), this lack of scientific consensus on anthropogenic climate change muffled interest in adaptation and its resultant impact could be discerned in making adaptation almost a political nonstarter at the higher levels, and early references to adaptation in in official UNFCCC texts were often construed as intentionally vague (Burton, Huq, Lim, Pilifosova, & Schipper, 2002). The second half of the 1990s and early years of the first decade of the 21st century started witnessing adaptation research gradually focusing on better comprehension of the impacts of climate change on specific communities (Smith & Lazo, 2001), as a basis for identification of adaptation options. Admittedly, the formative phase of research work on adaptation witnessed fruitful outcome; nonetheless, it was realized that adaptation research "posed a unique set of questions and dilemmas beyond the scope of impact assessment" (Klein, Adams, Dzebo, Davis, & Siebert, 2016), and as also pointed out by Burton et al. (2002), the adaptation research produced during that phase had difficulty of encompassing significant policy conditions that impact adaptation outcomes, and was often incapable of taking into account the scale of adaptation options and tended to ignore the roles of different stakeholders in the pursuit of adaptation. Apart from being descriptive in nature, the adaptation research during this phase was also faced with political resistance and scientific uncertainty; and the priority of adaptation researchers at that stage was to "to better understand the future of climate impacts, to determine which impacts may be no longer avoidable, and begin to explore the possibility of adapting to those impacts" (Klein et al., 2016).

With the passage of time, especially during the 2000s, a shift of emphasis regarding climate change could be discerned as taking place. With the publication of Third Assessment Report by the IPCC in 2001 (IPCC, 2001) along with the convening of the COP-7 at Marrakech (UNFCCC, 2002), the phenomenon of climate change came to be acknowledged as a problem of development, in contrast to a global environmental issue that could impact all countries in equal proportion. It was explicitly evident from the contemporary descriptive research that developing countries were to be affected most by changing climate, and the Least Developed Countries were specifically subject to the vagaries of climate change (Adger, Huq, Brown, Conway, & Hulme, 2003). In the wake of establishment of three multilateral funds to manage adaptation funding—Adaptation Fund; Least Developed Country Fund; and the Special Climate Change Fund (Parker, Keenlyside, & Conway, 2014), along with the initiation of National Adaptation Programs of Action (NAPAs) established under the provisions of Article 4.9 of UNFCCC (1992) in 2001, enhanced efforts were made to bring adaptation to the climate agenda; nevertheless, the bulk of emphasis of the IPCC reports and negotiations conducted at the COPs remained confined to mitigation (Kates, 2000). Concurrently, in the wake of these developments, contemporary adaptation researchers also continued to explore multiple aspects of adaptation including what adaptation was, who was adapting, and how it was being adapted (Adger, Arnell, & Tompkins, 2005). Moreover, simultaneous endeavors were also underway in developing core concept relating to adaptation like sensitivity, vulnerability, resilience and adaptive capacity (Smit & Wandel, 2006). Growing corpus of research on adaptation also witnessed emergence of the notion of adaptive capacity evolving into a prime area for innovative research. Often construed as the capability of an individual or community to embark on an adaptive behavior, the notion of adaptive capacity is intrinsically social in nature, and epitomizes a crucial nexus between social, capital, politics, economics, and geophysical conditions and interestingly many studies have emerged on this aspect (Smith, Klein, & Huq, 2003). Questions pertaining to adaptation outcomes, probing technical and social aspects of identifying feasible and desirable outcomes of adaptation (Adger et al., 2003) along with focus on the likelihood of a desirable equilibrium between adaptation and mitigation efforts as well as adaptation and development also formed the focal points of adaptation research (Klein et al., 2005).

Publication of the Fourth Assessment Report in 2007 by the IPCC coincided with the lead-up to the convening of COP-13 in Bali (Indonesia). While stating that the warming of the climate is unequivocal and that climate change impacts are already occurring, IPCC (2007a, 2007b) report identified adaptation requirements in different sectors and regions. Besides, adoption of the Bali Action Plan (UNFCCC, 2008a, 2008b) as an outcome of the COP-13 proved instrumental in raising the political status of adaptation by incorporating adaptation as a key "pillar" of climate action, alongside mitigation, technology transfer and finance. Thus the Bali Action plan carved out a niche for adaptation, acknowledging its crucial significance for socioeconomic issues ranging from poverty to social marginalization and other alike factors. This also proved instrumental in spurring scientific research on climate to focus on varied aspects of adaptation from different perspectives. Apart from documenting observable changes in climate change and its attendant impacts, the IPPC (2007) report unveiled new awareness to the need for climate change adaptation (Ayers & Dodman, 2010). And since then, there has been no looking back, albeit past two decades have witnessed rapid proliferation of adaptation responses to climate change around the globe (Arnott, Moser, & Goodrich, 2016). A way has been paved for unfolding new vistas of more funding opportunities, improved scientific understanding, and enhanced public awareness has seemingly helped launch a mosaic of adaptation initiatives executive by government, international development, nonprofit and community organizations and seemingly some of the adaptation initiatives also comprise those entailing the potential of influencing human behavior and even decision-making (Vaughan & Dessai, 2014). Apart from this, adaptation initiatives have also been helpful in innovating new technologies and infrastructure (Schenk, Vogel, Maas, & Tavasszy, 2016); implementing management, governance, and institutional policies (Henstra, 2016); improving people's access to resources (Ribot, Mearns, & Norton, 2010); and minimizing exposure to environmental risk (Jordan 2015). Interestingly, these adaptation initiatives are focused on climate-vulnerable sectors such as agriculture, coasts, ecosystems, health, water, and urban systems.

There is little consensus on what constitutes as effective adaptation in practice and this is partially attributed to the fact that adaptation initiatives that are usually mulled and planned get rarely implemented (Mimura et al., 2014); and accordingly, academic literature has also been inclined to focus on these proposed activities. Barriers that routinely impede Adaptation efforts that are often are confronted with barriers inter alia include insufficient resources, prohibitive policies, competing or conflicting priorities for action, and uncertainty about future changes (Biesbroek, Klostermann, Termeer, & Kabat, 2013). Singular, exceptional and localized combinations of underlying context like politics, funding, motivation, power dynamics, and cultural values generally mold or influence adaptation efforts. Sometimes it becomes problematic to differentiate climate change from other activities, such as reducing risk to environmental issues or alleviating poverty and under such circumstances, complications are prone to arise for successful completion of adaptation efforts. The Special Report on Climate Change released by IPCC in 2018 has argued that including climate adaptation into sustainable development strategies would yield win-win solutions (Roy et al., 2018); whereas attempts at mainstreaming climate adaptation may appear efficient; nevertheless, efforts that address present development issues may run foul with actions addressing future climate risks or vice versa (Roy et al., 2018). Thus it is on account of these factors that it becomes difficult to disentangle a simple understanding of adaptation in practice.

While rummaging through the research literature on adaptation, it becomes discernible that a large part of this focuses on critical appraisal that points out toward failures and negative outcomes of adaptation (Eriksen, Aldunce, Bahinipati, & Ulsrud, 2011; Wise et al., 2014), disproportionate burdens of adaptation that befall more vulnerable populations (Shi et al., 2016), and disparity regarding as to who benefits from adaptation and who is bereft of it (Bäckstrand & Lövbrand

2006); thereby, making it explicit that climate change adaptation activities seldom work as intended. There is even insufficient insight into as to what comprises effective practices with regard to adaptation. One way suggested to analyze success of adaptation is to incorporate routine evaluation into the adaptation procedure (de Coninck et al., 2017). Interestingly, some scholars have mulled over frameworks to guide evaluation designs for adaptation (GEF, 2016; Vogel, Letson, & Herrick, 2017).

de Coninck et al. (2017) have discerned three major factors that challenge the implementation of adaptation as standard practice—uncertainty; lack of agreement; and attribution (p. 386). The first factor of uncertainty entailing actions and objectives that comprise adaptation practices has been analyzed in above paragraphs. The second factor of lack of agreement deals with best methods for gathering robust and longitudinal data and focuses on right indicators for monitoring. While common metrics to facilitate comparison of adaptation outcomes across projects and scales have been developed by agencies like UNFCCC (Craft & Fisher, 2016), whereas others have evolved pliable set of metrics and indicators to demonstrate accountability and effectiveness within projects (GIZ, 2017; Christiansen, Martinez, & Naswa, 2018). The third factor that questions effective implementation of adaptation takes into account the difficulty of attributing successful consequences to specific adaptation activities. Undoubtedly, evaluation practices are designed to show linear connection between activities and outcomes; nevertheless, adaptation is a cyclical process with no discrete end point (Smit & Pilifosova, 2001). Keeping in view the fact that there is little likelihood of the entire society becoming fully adapted, evaluation practices and the notion of success need to include the state of adaptation as a recurring and orbital process.

4.3.1 Delineating adaptation

Construed in common parlance, the term "adaptation" connotes adjusting behaviors, actions, and decisions within biological, social and built systems in response to climate change (Smit & Pilifosova 2001). According to Smit, Burton, Klein, and Wandel (2000), adjustments can be reactive to events that have taken place or preemptive in response to future events or conditions. Practically speaking, there is lack of clarity in the definition of adaptation (IPCC, 2014a, 2014b). Adaptation is always dependent upon the events or conditions to which it is reacting or anticipating. Adaptation is often coalesced with adaptation to environmental disasters or climate variability (FAO, 2015). Admittedly, the domain of climate change adaptation is the emerging field; nonetheless, it emerges from decades of research with regard to responses to environmental change (Liverman, 2015), and was framed as a key concept in hazard and risk studies and cultural ecology (Head, 2010). Applications of adaptation depend on the multiple and interconnecting ways wherein people know, experience and tackle climate change related issues. While arguing that at the core of this complexity lie the pluralities of how people understand the world (different epistemologies) and the pluralities of knowing what the world is (different ontologies), Goldman, Turner, and Daly (2018) write, "Such an approach challenges an assumption that most of us hold dear: that there is one reality out there, about which we can explore different perspectives" (Goldman et al., 2018, p. 3).

Adaptation community ought to know as to how to deal with coexistent realities of, experiences with and modes of gaining knowledge about climate change. Goldman et al. (2018) describe as to how climate change for pastoralists of Maasai in Tanzania means the extension of the cold season from one month—July only—to four months—June to September—and this conception of Maasai pastoralists about climate change differs from that of the Inuit harvesters of Canada who experience changes in ice conditions on account of Arctic warming trends (Ford et al., 2013). In other words, there is an array of ways to adapt behaviors in response to myriad ways of knowing, experiencing and understanding the vagaries of climate change. The challenge of defining basic goals of adaptation reflects the repugned nature of adaptation in practice. Enhancing adaptive capacity or the capacity to respond effectively to changing stresses and shocks to manage or minimizing risk is construed as one of the goals of adaptation (Engle, 2011). Enhancing resilience or the capability of a social or ecological to continue functioning when faced with shock and stress is another goal of adaptation (Nelson, Kokic, Crimp, Meinke, & Howden, 2010). Besides, reducing vulnerability or susceptibility to cause harm in the eventuality of exposure to an external hazard is also considered as the goal of adaptation (Yamin, Rahman, & Huq, 2005). Multiple disciplines like ecology, human geography, sustainability science, and risk management along with development give rise to these adaptation goals and these have seemingly come together under the canopy of adaptation research (Goldman et al., 2018). Interconnectedness between these goals of adaptation is pointed out by Smit & Wandel (2006). Strengthening of adaptive capacity by the populace of the Pacific Islands has enabled them to reduce vulnerability by evolving systems to share resources and labor that has helped the communities readying for and recover from drought and cyclones (de Coninck et al., 2018, p. 360).

Undoubtedly, these adaptation goals can be defined in a variety of ways; however, they are prone to become visibly generic when their application is facilitated sans context. Vulnerability is often construed as biophysical and/or socioeconomic exposure to risk (Fussel, 2007); nevertheless, conditions of exposure to risk get actively generated by past and present politico-economic, environmental, and social processes (Dooling & Simon, 2012). Citing the example of corn farmers in Mexico who are often confronted with extreme weather events and increased pest population, both impacts linked to climate change, Bellante (2017) narrates the plight of those farmers who lacked resources to adapt to changing climate change conditions resulting shift in their preferences from local to imported goods; thereby demonstrating that their exposure to climate risk is politically, economically and culturally generated through the convergence of international trade decisions, economies of scale and individual food choices and tis what O'Brien and Leichenko (2010) have termed as "double exposure," as farmers are made to face the effects of climate change and globalization. Mere or exclusive focus on addressing exposure to risk that contains symptoms of vulnerability without also addressing underlying or causal systemic issues that are regarded as the root cause of vulnerability, there is probability that adaptation initiatives may not yield effective outcomes in terms of reducing vulnerability (Dooling & Simon 2012). Therefore articulation of both the tangible and epistemological goals of adaptation initiatives is significant because these entail the potential of affording an assortment of choices and outcomes. The conceptualization of adaptation initiatives holds the key to shaping the rest of the adaptation process along with the selection and implementation of activities and how success is determined.

4.3.2 Adaptation under Paris Agreement

Due prominence has been accorded to adaptation in the PACC (referred as Paris Agreement). While acknowledging linkages between mitigation ambition and adaptation requirements and the need for a balanced allocation of financial resources to support mitigation and adaptation measures in developing countries, Paris Agreement also envisages international follow-up to set out priorities, plans, needs, and action of developing countries through the enhanced transparency framework and the global stocktake to further promote enhanced action (Suarez & Huang, 2015). Transparency rules are applied tenderly, albeit comprehensively for support and mitigation those rules are designed to address issues like clarity and tracking of progress toward attaining Parties' adaptation action to inform the global stocktake (Morgan, Dagnet, & Tirpak, 2015). Several key elements have been introduced in the Paris Agreement that comprises part of a "new regime" designed to attain the worldwide goal on adaptation through enhanced action. Prominent among these elements are "to-down" elements like the global goal on adaptation that is strengthened by its relation to the goal of temperature; the adaptation dimension of global stocktake and a "cycle of action for strengthening adaptation efforts regularly" by linking nationally determined adaptation priorities and needs to international mechanism framework, and processed that provide follow-up that seek to understand and enhance adaptation action (Morgan et al., 2015).

Adaptation is set out in Paris Agreement vide Article 7 and its subclauses. Most of the provisions pertaining to adaptation as envisaged in Article 7 do not offer any particular obligations and are phrased in terms of acknowledgment; and as such they are "unusual" in a treaty like this and entailing little legal weight per se; however, significance of these provisions lies less in their legal character rather in their ability to spell out a political dimension that "raises adaptation as a cornerstone of action under the Paris Agreement and a context for adaptation efforts" (Pérez et al., 2017). The long-term goal of adaptation, viz., "of enhancing adaptive capacity, strengthening resilience and reducing vulnerability to climate change, with a view to contributing to sustainable development, and ensuring an adequate adaptation response in the context of the temperature goal" is set out under Article 7.1 (UNFCCC, 2015b, p. 9). This goal builds on the purpose envisaged in Article 2.1 (b) and is focused on three dimensions in terms of adaptation to climate change: enhancing adaptive capacity, strengthening resilience and reducing vulnerability with a view to contributing to sustainable development, and ensuring an adequate adaptation response in the context of a temperature goal (UNFCCC, 2015b, p. 3). Inclusion of this provision is usually seen as a feature of new regime for adaptation efforts under the Paris Agreement, along with the input Parties would provide via the adaptation communication under Article 7.14; and the enhanced transparency framework under Article 13 of the Paris Agreement (Pérez et al., 2017).

Admittedly, the operationalization of the global goal on adaptation is not articulated in details under the Paris Agreement and is still to be resolved; nonetheless, the expression of the link between mitigation and adaptation, as part of this goal, makes this link a reference point for both mitigation and adaptation action. Besides, Decision 1/CP.21 (UNFCCC, 2016b) provides mandates to constituted bodies of the Convention to deliver on key aspects of the goal for the Conference of the Parties serving as the meeting of the Parties to the Paris Agreement (CMA) to consider or decide upon and these mandates include development of:

- modalities to recognize the adaptation efforts of developing countries (paragraph 41);
- review the work of adaptation-related institutional arrangements under the Convention (paragraph 42(a));
- methodologies for assessing adaptation needs with a view to assist developing countries (paragraph 42(b));

- methodologies and recommendation to facilitate the mobilization of support for adaptation in developing countries in the context of the limit to global average temperature increase (paragraph 45(a))22; and
- methodologies and recommendation to review the adequacy and effectiveness of adaptation and support referred to in Article 7.14(c) (paragraph 45(b)).

4.3.3 Limits and barriers to adaptation

Adaptation may not be able to avoid harm to social and ecological systems under even modest rates of global warming (Stafford-Smith et al., 2011), because much depends on the harshness of the climatic disruption and the sensitivity and resilience of the systems at risk (Dow et al., 2013a). Concepts of limits and barriers often formulate research on how adaptation is encumbered by several social and biophysical factors, or may fail to avoid catastrophic climate impacts. Inclusion of the concepts of barriers and limits to adaptation in the Fourth Assessment Report of the IPCC in 2007 proved instrumental in bringing these concepts in wider usage in the scientific literature (Adger et al., 2007) and subsequently inclusion of a chapter entitled *Adaptation Opportunities, Constraints, and Limits* in the IPCC Fifth Assessment Report (Klein et al., 2014) provided it further impetus of frequent usage in research. Cognizance of as to where, when, and how limits and barriers to adaptation emerge and arise has become a significant domain in climate change research (Palutikof, Barnett, Guitart, & Christoff, 2014a). It has been argued that advances in knowledge call for comparative studies that endeavor to seek to identify the underlying drivers of the barriers and limits to adaptation (Eisenack et al., 2014).

Adaptation as an ongoing process of adjusting to changes, with no end point, adaptation denotes that delineating successful adaptation is more about the sustainability of processes and the principles of fairness and equity than it is about measuring outcomes at any given point in time (Hurlimann et al., 2014). Thus successful adaptation is a matter of "socially and environmentally sustainable development pathways, including both social justice and environmental integrity" (Eriksen et al., 2011, p. 8). In the wake of myriad of definitions, synonyms, and typologies of barriers that impede adaptation (Biesbroek et al., 2013), barriers to adaptation can be defined as obstacles that can be overcome with concerted effort (Moser & Ekstrom 2010), and limits to adaptation can be construed as the points at which adaptation actions fail to protect things that stakeholders' value (Adger et al., 2009a, 2009b). Most of the explanations pertaining to the notion of "barriers to adaptation" aim at identifying them as factors that obstruct adaptation, albeit they are manageable if efforts are made. A systematic typology of barriers, also known as constraints, is provided by IPCC and that also includes knowledge, awareness and technology; the physical environment; biological tolerance; economic factors, financial factors; human resources; socio-cultural factors; and governance and institutional processes (Klein et al., 2014). The barriers to adaptation are often context-specific and generally they are denied in terms of obstructions to specific actions essential to meet the goals of specific actors that underline the problem of trade-offs among competing goals in adaptation decisions (Eisenack et al., 2014). A good deal of research on barriers to adaptation to sea-level rise in primary industries and local governments has been carried out within Australia where commonly identified barriers include a lack of or incapable leadership; inadequate knowledge of risks and responses; paucity of fiscal resources; difficulties in negotiating among competing values and goals; a lack of institutional support; and inadequate coordination across levels of government (Hurlimann et al., 2014; Marshall & Stokes 2014a). Besides, competing policy priorities among most actors is also seen as a major barrier to adaptation (Waters, Barnett, & Puleston, 2014).

Path dependency is seemingly one of the deep drivers of barriers to adaptation in some countries (Haasnoot et al., 2019). Path dependency denotes resistance to changing the way things have always been done, even if BAU has seemingly been increasingly maladaptive. The manifestation that opposition to things that have never been done or to focus on improving inefficient practices seems to reflect on much of the problem of social vulnerability to climate change that crop up owing to denial of important social freedoms and opportunities like imparting of primary education to girls or social protection in times of crises due to ignorance or prejudice (Barnett, 2009). Undoubtedly, path change is always feasible (Garrelts & Lange, 2011); nonetheless, its failure to commence for a prolonged period or if it records progress more gradually than climate change, in that eventuality path dependency may be delineated as a deeper cause of limits to climate change. In fact, ambiguity underlines the differentiation between barriers and limits to adaptation. Bulk of the hypothesized limits to adaptation is social constructs in two ways. In the first place, they come up owing to social processes that render groups' exposure to climate risks, restrain their adaptive capacities or hinder their adaptation responses (Ostrom, 2005). Second, the things that are valued by stakeholders either at risk of loss or damage are themselves the product of shared meanings (Dow, Berkhout, & Preston, 2013b). In this backdrop, Eisenack et al. (2014) argue that it is essential to ascertain the perspective of actors to understand as to how and far whom adaptation is constrained and limited. Essentiality of societal transformations for managing climate change is warranted by socially constructed nature of many barriers and causes of limits to adaptation. Large-scale changes can be instrumental in altering the social processes that hinder or limit adaptation or the shared meanings of what is at risk (Klein et al., 2014).

In the wake of the probability of some barriers becoming limits, some limits to adaptation can possibly be surmounted and transformed into barriers in tandem with new cultures, values, technologies o governance systems. Therefore socially determined limits to adaptation are sometimes also called as "soft" limits (Dow et al., 2013b; Klein et al., 2014). These "soft" limits are distinguishable from "hard" limits that occur when species, communities or ecosystems cannot adjust to new climate regimes in time to avert deterioration or degradation or collapse, as is the case with coral reef s being affected by ocean warming and acidification (Veron, 2011). There is important, albeit seldom acknowledged interplay between barriers and limits to adaptation. Time and path dependence can be helpful in altering some perceived barriers into limits when it becomes evident that they are insurmountable.

4.3.4 Evaluating climate change adaptation

Measuring success of adaptation is a complicated process in view of complex nature of conceptualizing adaptation and its objectives. The domain of adaptation envisages pragmatic methods to record progress and change. Two functions of evaluation are helpful for adaptation—formative (dealing with process) and summative (dealing with outcomes) (van Drooge & Spaapen, 2017). Formative evaluation is concerned with learning about process itself and envisages guidance for improving adaptation efforts. While focusing on adaptation outcomes, summative evaluation also analyzes progress and accountability. Launch of adaptation initiatives to be coincided with the establishment of formative and summative evaluation processes entails the potential of yielding plans, strategies, and activities that are more effective, efficient, unfailing, equitable and reduce the incidence of erroneous and maladaptive practices. Undoubtedly, frequent dependence of evaluation on quantitative and qualitative metrics or indicators demonstrates relationships between observable impacts of adaptation initiatives; nonetheless, as pointed by IPCC, determining the right metrics or indicators to measure progress across adaptation initiatives is not unambiguous, partly owing to the fact that different people and agencies have significantly diverse uses for evaluation (IPCC, 2014a, 2014b). Economic incentives spurred initial calls for evaluating climate adaptation projects coupled with financers' desire for standard metrics to have a comparative assessment of outcomes across projects globally to decide which project to be financed (Möhner, 2018). Undeniably, this motivation still comprises a potent driving factor, albeit evaluation can be used to valuate much more than economic efficacy. Evaluation can also be helpful to the people and organizations in ascertaining as to how activities are helping them adapt as well as who is the beneficiary. For instance, take the case fisheries where evaluation can be helpful to fisheries managers in determining as to how adaptation policies improve aquatic health and sustain operations concurrently also extending help to fisheries workers and coastal communities to discern as to how these policies affect their livelihoods and local food security (GEF, 2017).

Substantive indicators and metrics that exhibit impacts are prone to vary based on different values and goals; within a project, the goals and values of members of the community may not tally with those of funding organizations (FAO, 2018; GIZ, 2015). Evaluation frameworks are required to be balanced, or at least recognize, such competing needs for information. Evaluation research is confronted with another practical challenge of attribution or establishing causality between actions and outcomes. Effects can be inherent or acute; direct or indirect; and "cascading, cumulative, synergistic, connected, and distributional" (Moser & Boykoff, 2013, p. 17). Relative factors like poverty levels, development interventions or management policies are continually in flux, rendering it cumbersome to link improved conditions directly to a particular adaptation activity. The effect of a peculiar or singular activity remains uncertain without having multiple controlled experiments for comparison. Besides, quantification of direct effect is seldom feasible or meaningful (Christiansen et al., 2018). The adaptation community, in order to address the issue of attribution, can take a cue from a wide-ranging other evaluation practices such as global management and agricultural extension programs. Apart from conducting frequent outcome and project evaluations to appraise contributions toward attaining specific objectives, the UNDP also conducts impact evaluation, albeit occasionally, that takes into account all of a project's long-term impacts, both intended and nonintended; and in most cases, it is improbable to segregate the effect of one development initiative from other influencing factors (UNDP, 2019).

In view of the fact that climate change functions on longer timescales than most other kinds of projects that envisages another challenge for evaluating adaptation, the outcomes may not become discernible for longer period, spanning over years or even decades, if at all. According to Moser and Boykoff (2013, p. 18): "Publicly perceived success is achieved when an anticipated problem or impact does not occur—deaths prevented, damages avoided—yet proving this is the result of a policy or management intervention is often difficult." It is worth remembering that correlation does not imply causation in the case of adaptation success. Nevertheless, the challenges of evaluating adaptation are manageable and not often insuperable. It is often found more pragmatic by evaluators to focus on pursuing an activity's contributions toward a bigger goal, instead of attrition. It has been suggested by Mayne (1999) that reassessing what measurement can usefully mean, "Measurement in the public sector is less about precision and more about increasing understanding and knowledge. It is about increasing what we know about what works...and thereby reducing uncertainty" (Mayne, 1999, p. 50). Contribution analysis (Mayne 2008) envisages methods to ascertain whether or not expected outcomes occurred; whether or not a specific activity or process yielded those results; identify additional factors that impacted results; or disclose other explanations for why results occurred (Belcher, Suryadarma, & Halimanjaya, 2017). Anyhow, the process of systematic evaluation enhances clarity of understanding the links between activities and outcomes. The growing corpus of research literature has analyzed adaptation initiatives along with planning, implementation and outcomes (Vogel et al., 2017), and the results accruing from these studies envisage various components for good adaptation practices, strengthening institutional partnerships, engaging local individuals and communities in adaptation design, research and implementation; promoting sound leadership along with facilitating social learning. These results also reflect on primary barriers to adaptation that inter alia include lack of policies, human and financial resources, and institutional capacities to support the adaptive process (Bierbaum et al., 2013). In sum, the time is ripe for the adaptation community to inject new insight into prevalent evaluation practices along with emphasis on critical assessments to explore as to how success is defined, for whom and by whom.

4.4 Linkages between mitigation and adaptation

Indubitably, it is significant that mitigation and adaptation strategies are categorized separately; nevertheless, the IPCC has recognized that "there is a high confidence that neither adaptation nor mitigation alone can avoid all climate change impacts...adaptation and mitigation can complement each other and together can significantly reduce the risk of climate change" (IPCC, 2007, c:65). Whereas some researchers have endeavored to ascertain the optimal mix between adaptation and mitigation strategies (Kane & Shogren, 2000), others have ruminated that ascertaining the optimal mix such as the most effective and least expensive mixture of adaptation and mitigation initiatives might not be the best way out to deal with climate change (Klein et al., 2005; Swart & Raes, 2007). They emphasized that an understanding of the linkages between adaptation and mitigation ought to be prioritized. While developing an integrated framework in the agricultural sector, a study has acknowledged with a view to increase adaptive capacity in agriculture and other related sectors, important efforts are required to be undertaken in ascertaining knowledge about trade-offs and synergies between mitigation and adaptation (Jarvis et al., 2011). Similar suggestions were suggested by other studies in agriculture, water, land and urban planning sectors (Rizvi, Baig, Barrow, & Kumar, 2015a, 2015b; Thornton & Comberti, 2017).

Knowledge and understanding of the linkages between mitigation and adaptation is very significant to reduce the impact of mitigation strategy undermining an adaptation strategy or vice versa. Nevertheless, studies focusing on linking mitigation and adaptation strategies have been limited and not adequately documented (IPCC, 2007a, 2007b). Sectors that have been studied from these linkages' perspective include: agriculture (Smith, Gregory, & van Vuuren, 2010), building industry and urban development (Walsh et al., 2011), water resources (Pittock, 2008) and forestry (Damato et al., 2011). For example, switching over to low-carbon technologies for generating

4.4 Linkages between mitigation and adaptation

energy such as biofuel production enhances demand for water, ensuing in competition for water for food production and conflict with conservation of freshwater environment (Pittock, 2008). Additionally, research has also demonstrated that the synergistic impacts of integrating adaptation and mitigation are moderately low (Swart & Raes, 2007) and that there are more conflicts and trade-offs between adaptation and mitigation policies/strategies (Hamin & Gurran, 2009). Hamin and Gurran, (2009) cite the example of the United States and Australia where 50% of land-use mitigation and adaptation policies were found to be conflicting rather than complimentary. These outcomes are partially attributable to differences existing between the adaptation and mitigation strategies in various operational scales as demonstrated in Table 4.2.

Table 4.2 Differences between mitigation and adaptation.

Issue	Mitigation	Adaptation	References
Cause/effect	Cause	Consequences	Swart and Raes (2007)
Spatial	Global	Local	McEvoy et al. (2006); Swart and Raes (2007); Locatelli (2011)
Temporal	Long-term	Short-term	McEvoy et al. (2006); Swart and Raes (2007)
Beneficiaries	Global; other, that is, intergenerations	Local; only those who Implement it	McEvoy et al. (2006); Swart and Raes (2007); Locatelli (2011)
Sectors	Few; energy, agriculture, mining, forestry, transport, building industries, etc.	Wide, urban planning, water, agriculture, health, coastal zone settlement, etc.	Dang et al. (2003); McEvoy et al. (2006); Swart and Raes (2007); Locatelli (2011)
Decision-making/cooperation required	National initiatives in the context of international obligations	Local in the realm of local/regional economies and land managers	Dang et al. (2003); McEvoy et al. (2006); Swart and Raes (2007)
Incentives	Usually required	Often not required	Dang et al. (2003); McEvoy et al. (2006)
Urgency	Lower political urgency	Higher political urgency, especially in poor countries	Dang et al. (2003); McEvoy et al. (2006)
Equity	The free-rider effect; motivated in countries Less vulnerable to climate change	Unfair; the victims are not always responsible for causing climate change	Dang et al. (2003);
Uncertainty	Needs to be changed regularly to take into account new projections	Can wait till concrete evidence of climate change impacts is available	Dang et al. (2003)

Compiled from IPCC (Intergovernmental Panel on Climate Change). (2001). Climate change 2001: Mitigation: Contribution of working group III to the third assessment report of the Intergovernmental Panel on Climate Change. Cambridge: Cambridge University Press; Klein, R. J. T., Huq, S., Denton, F., Downing, T. E., Richels, R. G., Robinson, J. B., & Toth, F. L (2007). Inter-relationships between adaptation and mitigation. In: Climate change 2007: Impacts, adaptation and vulnerability. Contribution of Working Group II to the Fourth Assessment Report of the Intergovernmental Panel on Climate Change (pp. 745–777). Cambridge: Cambridge University Press; Locatelli, B. (2011). Synergies between adaptation and mitigation in a nutshell. Bogor: CIFOR (Center for International Forestry Research).

Prime emphasis needs to be focused on ascertaining the linkages between mitigation and adaptation strategies by fist determining the prevalent differences between them. These differences have been summarized in Table 4.2. The issue of cause and effect is seemingly one of the major differences whereupon mitigation is spurred by the desire to affect the cause of climate change; and thereby, focuses on reducing the GHG emissions or augmenting the GHG sink, whereas adaptation is triggered by the outcomes of climate change and, therefore, focuses on enhancing adaptive capacity and resilience to minimize prospects of vulnerability (Swart & Raes, 2007). In the realms of spatial and temporal differences and beneficiaries, mitigation wields wider impacts where can be experienced worldwide and in the long-term by future generations, whereas adaptation may benefit those who are engaged in its implementation and its impact may of short duration as is the case with diversification of crops in agriculture during drought season (Swart & Raes, 2007). Another area that differentiates mitigation from adaptation is the scope. Only few industries like coal, auto, energy, agriculture, mining, forestry and building industries, are usually involved in practicing mitigation strategies, while adaptation comprises a wider scope in terms of wider range of areas including urban planning, agriculture, water health and coastal zone settlements (Raven et al., 2018). As far as cooperation is concerned, mitigation entails wider scope because mitigation may require national initiatives in the context of international commitments, whereas adaptation may warrant local actions in the context of local or regional economies and land managers (Landauer, Juhola, & Soderholm, 2015).

Another factor that differentiates between mitigation and adaptation is that while the risk of climate change may differ from one region to another, it is the developing and poor countries that are most probably to be impacted the most and thus are spurred by greater political urgency to implement adaptation even without receiving any incentives, whereas developed countries are more interested in implementing mitigating strategies, generally in exchange for economic incentives (Thuy et al., 2014). From the equity perspective, the developed and newly industrialized countries including China and India are among the largest emitters of GHG; nevertheless, the levels of damaging effects may not be similar with some developed countries garnering net benefits (Kongsager, Locatelli, & Chazarin, 2016). Most of the developed countries have often emphasized on mitigation strategies to reduce dependence on fossil fuels and to switch over to clean, green sustainable technologies (Sreeramana & Shubhrajyotsna, 2016). On the other hand, poorer countries of the developing world have much lower rate of national GHG emissions per capita, with the exception of China and India, it is projected that these countries will be mostly impacted by the adverse effects of the vagaries of climate change and adaptation could be the only available option for these countries (Mertz, Halsnæs, Olesen, & Rasmussen, 2009).

Consequently, adaptation has come to garner greater attention, specifically its significance to developing countries that possess limited resources to adapt socially, financially, and technologically (UNFCCC, 2007). These adaptation approaches may comprise evolving practical solutions to help communities cope with floods, droughts and extreme weather events (UNFCCC, 2007). Finally, in the wake of uncertainty, mitigation is required to be changed frequently keeping in view new projection of atmospheric GHG, whereas the change in adaptation can wait until tangible evidence of climate effects is available (Dang et al., 2003). Apart from the prevalent differences between mitigation and adaptation, it is noteworthy that strategies to address climate change and the capability of reaching emission targets are subject to substantial variations from region to region (Ngum et al., 2019). Besides, pivotal role is played by factors like geographical location of the region, stage of economic development, population growth, availability of resources and finances, accessibility to green technology and industrial base for renewable energy.

Prior to the conclusion of Paris Agreement in 2015 (UNFCCC, 2015b), climate-related international policies were primarily focused on mitigation, as the need for adaptation was looked upon as a failure of mitigation or a way to undermine mitigation efforts (Locatelli, 2011). Concurrently, it was also contended that the putting into practice the mitigation-adaptation synergy measures would confront institutional and organizational complexity or obstacles at the international and national levels, and enforcing them together might be counterproductive (Locatelli, 2011). Nevertheless, in the wake of inevitability of the adverse impacts of climate change portending a risk for people and the planet, both mitigation and adaptation are essential and significant; mitigation endeavors to reduce future impacts and adaptation affects shot-term impacts (Wilbanks et al., 2007). Undeniably, mitigation and adaptation initiatives were hitherto managed separately, there is growing support for synergistic approaches to adaptation and mitigation that could yield massive benefits at multiple scales (Duguma, Minang, & van Noordwijk, 2014a).

Recent years have witnessed growing support in scientific literature for synergistic approach and Klein et al. (2005), representing some of the few pioneer literatures on synergy, has stated that synergies are produced when measures controlling atmospheric GHG contributions also reduce adverse impacts of climate change or vice versa. According to IPCC (2007a, 2007b), synergy alludes to the intersection of adaptation and mitigation so that "their combined effect is greater than the sum effect if implemented separately." Synergy between adaptation and mitigation has been described by Dugum and colleagues as an approach wherein both mitigation and adaptation initiatives are addressed "without any prioritization, mainly undertaken within a system-thinking context to address climate change issues" (Duguma et al., 2014a). It was posited by Klein et al. (2007) that synergies between adaptation and mitigation can enhance the efficiency of climate change measures, making it more lucrative funding agencies. In an identical manner it was argued by Dang et al. (2003) that if the equilibrium between adaptation and mitigation could be maintained, climate change policies may be socio-economically effective and may even contribute to sustainable development. Thus, even in the past decade, scientific literature made it discernible that synergy approach was gradually was gathering momentum for future climate policy.

Mitigation can be positively or negatively influenced by adaptation activities, and vice versa (Denton et al., 2014). Incentivizing activities that contribute to both climate objectives can enhance the efficacy of fund allocation and reduce trade-offs. The deficiency of consideration of mitigation in adaptation projects could result in enhanced GHG emissions, and that is one type of "maladaptation," as pointed out by Barnett and O'Neill (2010). Similarly, measures for GHG emission, without consideration of adaptation, entail the likelihood of underperforming owing to direct climate hazards like increasing flooding, as well as enhance the vulnerability and reduce the capacity of communities to adapt to a changing climate. The overall efficacy of global climate funding entails the prospects of being greatly reduced in the wake of frequent negative interactions between adaptation and mitigation. Concurrently, there are also opportunities that better integration could help narrow the adaptation funding gap with mitigation finance (Klein et al., 2005). Current level of flow of funds is projected to be far below any investment requirements for restricting climate change and its impact, and mainly for adaptation, as most climate funding supports mitigation (Buchner et al., 2013). Land-related activities such as agriculture and forestry explicitly manifest synergies and trade-offs between adaptation and mitigation (Locatelli, Fedele, Fayolle, & Baglee,

2015a, 2015b). Synergies and trade-offs are not only limited to mitigation and adaptation activities, albeit are also discernible among diverse types of mitigation and adaptation measures (Felgenhauer & Webster, 2013). Exploring and exploiting synergetic relationships can lessen costs and to balance the dual goals of mitigation and adaptation under limited resources (Zhao et al., 2018), and that is a prioritized practical issue for regions with high emissions and increased vulnerability (Fu, Zheng, & Wang, 2014). Given the fact that developing countries are assuming a more active role in climate change mitigation and concurrently realize their enhanced need for adaptation, "pursuing synergies between the two measures is gaining impetus as a viable option for their response to climate change" (Duguma et al., 2014a, 2014b).

4.5 Geoengineering and climate change

Geoengineering, often touted as the "third" response to climate change alongside mitigation and adaptation (The Royal Society, 2009), is still quite controversial (CSTCAAAS, 2010; Resnik & Vallero, 2011). It relies on two main approaches. The first approach is based on solar reflectance such as increasing the whiteness of clouds thus reflecting more radiation back into space and reducing the amount reaching the earth's surface (Vaughan & Lenton, 2011). The second approach is based on carbon capture, the most widely known being the carbon and capture storage technologies that have been proposed with injection of liquefied CO_2 deep underground (Vaughan & Lenton, 2011). Principles of geoengineering, while subject to controversy, are not indeed new and were first articulated during the Second World War with the objective of changing weather patterns with a view to elicit more favorable conditions on a regional scale (The Royal Society, 2009). Cloud-seeding has come to be reckoned with as one of the well-known techniques of geoengineering and it has been used to induce rain in drought-affected areas as well as a method to reduce the harshness of tropical storms (Rosenfeld, 2007). While rummaging through scientific debate on geoengineering it is discerned that there exist politico-ethical barriers that first need to be addressed (Virgoe, 2009). Nevertheless, there have been some somber proposals to conduct some studies on the pilot scale, while some has already been initiated, as has been the case with the Southern Ocean Iron Release Experiment (SOIREE), wherein about 3800 kg of iron sulfate was dumped into the Southern Ocean and it has resulted in an increase in growth of specific type of algae that in turn caused changes in cycling of carbon leading to a 10% drawdown of surface CO_2 (Boyd & Law, 2001).

The term "geoengineering" was coined in the 1970s by Cesare Marchetti and later formally published in the inaugural issue of the journal Climatic Change to describe a method for the "disposal" of atmospheric CO_2 through its injection into sinking thermohaline oceanic currents (Marchetti, 1977). Nevertheless, defining geoengineering is somewhat more complex than its dictionary meaning connotes, especially owing to the complex etymology of the term, entailing ambiguities as to what constitutes geoengineering, and how the term is used and competed with. Selected definitions of geoengineering are listed in Table 4.3. The definitions of geoengineering cited in Table 4.3 make it discernible that there is one element common to all these definitions is that for any action to constitute geoengineering that action ought to be in large scale. Undeniably, this action to be taken is regarded by most of the definitions, in relation to dealing with climate

4.5 Geoengineering and climate change

Table 4.3 Selected definitions of geoengineering.

Source	Definition of geoengineering
NAS (1992, p. 433)	Geoengineering proposals involve large-scale engineering of our environment in order to combat or counteract the effects of changes in atmospheric chemistry.
Keith (2000, p. 245, 247)	Geoengineering is the intentional large-scale manipulation of the environment…For an action to be geoengineering, the environmental change must be the primary goal rather than a side effect and the intent and effect of the manipulation must be large in scale, for example, continental to global, etc. Three core attributes will serve as markers of geoengineering: scale, intent and the degree to which the action is a countervailing measure
Barrett (2008, p. 45)	Geoengineering is to counteract climate change by reducing the amount of solar radiation that strikes the Earth…[not] by changing the atmospheric concentration of GHGs.
AMS (2009)	Geoengineering—deliberately manipulating the physical, chemical, or biological aspects of the Earth system [to reduce the risks of climate change].
The Royal Society (2009)	The deliberate large-scale intervention in the Earth's climate system, in order to moderate global warming.
OED (2008)	The deliberate large-scale manipulation of an environmental process that affects the Earth's climate, in an attempt to counteract the effects of global warming'
NAS (2015a, 2015b)	Climate geoengineering is "large scale, intentional interventions in earth and/or climate systems with the aim of reversing or reducing the rate or impacts of anthropogenic climate change."
Pasztor (2017, p. 20)	Geoengineering (also called climate engineering) is the deliberate, large-scale human intervention in the Earth's climate system to reduce the risks of climate change.

Compiled from the sources cited in the table.

change; nonetheless, it has also been used more sparingly to denote the manipulation of the environment at large. An attempt to offer a semblance of more specific definition of geoengineering has been made by Keith (2000) by envisaging three definitional markers: scale, intent and the extent to which an action is a counterbalancing measure. However, it has been pointed out by Fleming that latter two markers—intent and extent—ought not to be used to limit actions already delineated by their scales, and that could bring about unwanted as well as desirable counterbalancing ends. In fact, anthropogenic climate change in itself has been construed to be unintentional or inadvertent geoengineering (NAS, 1992).

The groundbreaking report by the Royal Society (2009) on geoengineering is hailed as furnishing the most commonly accepted definition of geoengineering having been reiterated by the Government of the United Kingdom (HoCSTC, 2010a, 2010b) and the IPCC (IPCC, 2010) amongst others.

Interesting, the concept of geoengineering has come to confront competition from relative terms like "climate engineering" (Bodansky, 1996); "climate modification" (McCormick & Ludwig, 1967); "Earth systems engineering" (Schneider, 2001); "planetary engineering" (Hoffert et al., 2002); and "climate remediation" (BPC, 2011). Moreover, this latter term, "climate remediation" presents a fascinating case as it signifies an effort to "rebrand" geoengineering. According to Sarewitz (2011), this term was selected by some to be in sync more conveniently alongside mitigation and adaptation, but it did not go unchallenged in its adoption. Proposals pertaining to geoengineering are usually divided in two subset categories that are in themselves contending with alternative terms. For instance, "carbon geoengineering" are those that aim at removing and sequestering CO_2 from the atmosphere, and have also been denoted as "carbon dioxide removal" (CDR) methods (Royal Society, 2009); "Negative Emissions Technologies" (NETs) (Kraxner, Nilsson, & Obersteiner, 2003); and most recently, "Greenhouse Gas Removal" (GGR) methods (Boucher et al., 2013); the latter two of which afford space for the incorporation of technologies that aim at looking for the removal of GHGs other than CO_2. Second, "solar geoengineering" proposals are those that try to enhance the reflection of sunlight back into space, and have also been referred as "solar radiation management" (SRM) methods (Royal Society, 2009) or, under another attempt to "rebrand" geoengineering, the less challenging 'Sunlight Reflection Methods (also, SRM) (SRMGI, 2011). Others have merely used "geoengineering" itself to refer exclusively to solar geoengineering proposals, and precisely stratospheric aerosols injection, overlooking carbon proposals in the definition altogether (Barrett, 2008). Geoengineering proposals have also been delineated along different lines of differentiation, including between those that look for enhancing Earth systems and those that embody "black-box" engineering (Rayner, 2011).

With geoengineering gaining salience as a policy response to climate change, some have seemingly been spurred to rethink the categorization of the proposals with regard to mitigation choices adaptations (Lin, 2013; Nassiry, Pickard, & Scott, 2017). While pondering over the opacities or ambiguities of not only geoengineering but that of mitigation and adaptation as well, Boucher et al. (2013) have brought to notice a number of overlapping concepts that entail the potential of confounding the policy discourses. Bucher et al. (2013: 29–30) present five new categories: anthropogenic emissions reductions; territorial or domestic removal of CO_2 and other long-lived GHGs (D-GGR); trans-territorial or transboundary removal of CO_2 and other long-lived GHGs (T-GGR); regional to planetary targeted climate or environmental modification (TCM); and adaptation and local targeted climate or environmental modification (CCAM). In a diverse categorization of responses to climate change, Heyward (2013) while retaining differentiations between mitigation, CDR, SRM, and adaptation, presents "rectification" (compensation) as type of responses for when the others have failed; and surmises that "research and debate [should] cease to be about 'geoengineering' and instead focus on the specific features of the proposed technologies, and the appropriate mix of [options]" (Heyward, 2013, p. 26), and urges for a rejection of the term. On the other hand, Cairns (2013, p. 3) contends that the ambiguities of the term offer "interpretative flexibility for articulating diverse interests within and across contested framings."

Presently, the best evidence to support or reject geoengineering mostly places reliance upon forecast models that frequently reveal complex and energetic changeability—from the probable or potential disruption of the monsoon cycle (Burns, 2011) to unsure impacts on multiple life systems like oceanic food webs (Lin, 2013). In other words, there is essentially no certain way of grasping full and complete cognizance and understanding of the impacts and effects of large-scale climate engineering initiatives. In that eventuality, some have espoused the potential for new models of governance, specifically those underlining experimentation, emanating from dire human need to mitigate climate change, although any kind or type of "techno-fix," particularly one that could endeavor to negate politics, has been extensively denied (Stilgoe, 2015, 2016; Wapner & Elver, 2016). In the wake of growing popularity of geoengineering amongst the scientific community, researchers and climate experts, the notion of postnormal science has often been used as a signifier and conceptual prism that facilitates contextualizing research in terms of significance of uncertainty and extended facts, civic engagement in the form of the concept of an extended peer community, and probable implementation of climate engineering measures emphasizing on how values might shape such initiatives (Chris 2016; Rayner, Matthias, & McGoey, 2015). Expectedly or unexpectedly, recent years have witnessed a proliferation of articles, research initiatives, meetings and workshops veering round the theme of geoengineering with the avowed aim of fostering a dialog on geoengineering within and amongst civic participants (Nassiry et al., 2017 Scheer & Renn 2014).

4.5.1 Solar radiation management

SRM is deemed as a projected method for "cooling the planet by reflecting a small percentage of incoming sunlight back into space before it can warm the Earth" (Chhetri et al., 2018). SRM is delineated as a set of geoengineering techniques that are designed to aim to "counter human-made climate change by artificially increasing the reflection of heat from sunlight (solar radiation) back into space" (ETC Group, 2017). SRM, also sometimes referred as solar climate engineering, has been accorded increasing attention as some are of the view that large-scale engineering initiatives might be the only viable means to deal with global warming, while such scenarios entail an array of uncertainties (Ban-Weiss & Caldeira, 2010; Budyko, 1977b; Keith, 2013). While Mount Pinatubo eruption in 1991 exponentially enhanced interest on SRM (Self et al., 1996), the idea of geoengineering using stratospheric aerosols to reflect solar radiation actually was articulated in 1977 by Mikhail Budyko, who argued "it may be feasible to modify the aerosol layer of the stratosphere in the near future" (Budyko, 1977a, p. 242). In honor of the groundbreaking work of Mikhail Budyko, some have begun to calling SRM proposals "Budyko's Blanket," and supporters for this approach often uphold the notion that it would be "fiendishly simple" and "startlingly cheap" (Levitt & Dubner 2011).

Extending Budyko's ideas, an article published in 1984 maintained that retro-fitted commercial planes could deliver the payload that would limit the need for proprietary deployment methods (Penner, Schneider, & Kennedy, 1984). Subsequently, the year 1992 saw a proposal being mooted with regard to injecting particulates, specifically ethane or propane, into the stratosphere above the poles as a means of combating ozone depletion (Cicerone, Elliott, & Turco, 1992). Irrespective of the fact all of these proposals were carried out in scientific journals, none raised the level of discourse and profile of geoengineering, and SRM in particular, as much a 1992 report by the Washington-based National Academy of Sciences, and National Academy of Engineering (hereafter referred as NAS). While devoting an entire section to geoengineering, with an emphasis on feasibility, especially for SRM, the NAS report called for a fast-track research agenda, one that could focus on "considerations of reversibility," and while providing one of the most enduring formulations of geoengineering, and SRM specifically, as a practice, the report noted: "the large-scale engineering of our environment in order to combat or counteract the effects of changes in atmospheric chemistry" (NAS, 1992, p. 433). While encompassing an array of SRM deployment methods, the NAS report (1992) did not deal with side-effects of SRM such as volcanic eruptions; and irrespective of that it continues to be regarded as pioneering work on geoengineering that speaks of the sometimes-obtuse ways with which geoengineering imageries are disseminated.

Admittedly, SRM has fallen an easy prey to criticism at the hands of skeptics and conspiracy theories, notwithstanding that has not thwarted serious research on deployment methods. Nevertheless, David Keith continues to be one of the most vocal proponents of the need for further

SRM research and in 2017, he received approval for a small-scale SRM research experiment—one of the first of its kind (Keith, Geoffrey, David St., & Kenton, 2018). The stratospheric-controlled perturbation experiment (SCoPEx) has been designed to study the effects of aerosol distribution location and size as a means of weighting the "risks and efficacy of SRM" (Dykema, Keith, Anderson, & Weisenstein, 2014, p. 17). Notwithstanding the fact that SCoPEx will deploy standard materials, such as calcium carbonate, to spawn insights on optimal aerosol deployment distribution, Keith has published research on more technologically advanced interventions, explicitly engineered nanoparticles that would vastly improve the amount of "control" over deployment and enhance the overall lifetime of the engagement—two of the main challenges of SRM (Keith, Parson, & Morgan, 2010, p. 16428). There are multiple proposals for using an array of particulates and materials, including a sunshade "near the inner Lagrange point (L1), in-line between the Earth and sun" (Angel, 2006); glass bubbles (Walter, 2011); and "land and space-based mirrors" (Weber, 2012). Such methods, as with Keith's supplication of nanotechnology, jack up see another critical issue encompassing climate engineering, especially SRM: patenting (Reynolds, Jordan, Huitema, Asselt, & Forster, 2018a, 2018b). A "land grab" on geoengineering patents is already in progress and that could restrain research, centralize deployment, and stir up possible intellectual property rights litigation (Reynolds et al., 2018a, 2018b). Remarkably, it was a concern over patenting that culminated in cancellation of a high-profile SRM experiment in the United Kingdom in 2012 (Cressey, 2012).

One of the pejorative critiques of SRM pertains to probability of wielding negative impacts on the ozone layer that staves off bulk of ultraviolet radiation from making it to the surface (HoCSTC, 2010a, 2010b). Admittedly, fast pace of ozone depletion is perturbing; nevertheless, some studies make it discernible a greater worrisome dynamic that in case a large-scale initiative were to be embarked upon, some maintain that discontinuing upkeep could be nothing short of calamitous as "there is high confidence that global surface temperatures would rise very rapidly to values consistent with the greenhouse gas forcing" (IPCC, 2013). In other words, a "start-and-stop" strategy that would generate an expiration impact, could engender, over a period spanning few decades or less, increase in surface temperatures, spurring devastating effects on multifaceted aspects of life systems (Muri et al., 2018; Trisos et al., 2018). Resultantly, SRM entails an obligation unlike anything endeavored in the history of humankind even though comparisons to nuclear-waste storage and disposal are occasionally carried out (Baum, 2014). Available evidence demonstrates that some of the major issues concerning SRM are nontechnical (Winkelmann, Levermann, Ridgwell, & Caldeira, 2015). Even the conservative estimates of the IPCC reveal that, "proposed methods will need to consider timescales extending at least up to, and likely well beyond, 2100" (Edenhofer et al., 2015, p. 4).

Irrespective of the fact that the challenges surrounding are multiple in both scale and scope, low-cost and minimal deployment constraints have given rise to concerns that unilateral deployment is a probability worth undertaking earnestly. Alluding to climatic conditions becoming frightening and urgent, Brand (2010) writes, "geoengineering schemes will suddenly jump from "plausible but dangerous" to "dangerous but we have no choice." The cost is low enough that a single nation or even a wealthy individual could set in motion a geoengineering project that would affect everyone on Earth." In an identical vein, another scholar has opined, "[...] it is a near-certainty that someone (nation or wealthy nonstate actor) will attempt to engage in geoengineering to head off utter disaster, allowing sufficient time for slower preventative solutions to take hold"

(Cascio, 2009, p. 21). Echoing similar opinion Deudney and Grove (2009) argue, "[...] actors could unilaterally select geoengineering projects for their distributional advantages, thus turning climate change into a realm of zero-sum competition." Arguing that geoengineering is not arms control, at least for the present, Parson and Keith (2013, p. 1279) have cautioned that in the eventuality of failure on the part of the states, "to build cooperation and transparency now when stakes are low, it could become as difficult and fraught as arms control, or more so, in some future of severe climate change." Elucidating the complications on a milder scale, another scholar contends that unilateral action entails the prospects of assuming the form of "directional leadership" or even "grant an actor significant leverage in international negotiations" (Rabitz, 2016, p. 106). Undoubtedly, this assortment of insights could make one surmise about the potentiality of SRM serving as a tactical and strategic maneuver to serve multiple ends, whereas some are of the view that the probable political costs would be terrible and possibly disallow implementation (Bodansky, 2011; Larson, 2016).

Irrespective of the debate over the feasibility or acceptance of unilateral SRM, concerns have been articulated over the incapability of any and all contemporary governance mechanisms and structures to deal with climate engineering, and SRM specifically (Brent, McGee, McDonald, & Rohling, 2018; HoCSTC, 2010a, 2010b; Larson, 2016). Deficiency of adequate governance; nevertheless, has not dampened the enthusiasm of the proponents of SRM from seeing the optimistic side of things. As has been opined by Schelling (1996, p. 307), "One thing that can be said for geoengineering is that it immensely reduces the complicatedness of what nations have to do internally to cope with greenhouse problems and what nations have to do internationally to cope with greenhouse problems." Maintaining a semblance of equilibrium between internal and international; local and global interests is crucial to implementing any substantive climate engineering initiative and developing potential governance structures to superintend research and potential deployment. In view of the fact that when large-scale environmental treaties like that of the Montreal Protocol, have been effectively contrived and properly regulated, governance for geoengineering would surely call for meaningful and unceasing cross-scale policy experimentation (Reynolds et al., 2018). Presently, none of the global institutions has come out in full support of climate engineering, even though the UNFCCC has called for "sinks and reservoirs of GHGs," (UNFCCC, 2015b).

Convention on biological diversity (CBD) has called upon its members or Parties to the Convention to invoke the precautionary approach and prohibit geoengineering activities at least until a number of conditions are net and this call was given as per the mandate of the tenth meeting of the Conference of the Parties to the CBD held at Nagoya, Japan in 2010, and adopted Decision X/33 that includes, in paragraph 8 (w) and (x), a section on climate-related geo-engineering and its impacts on the achievement of the objectives of the CBD. These prohibitory conditions include the following: (1) as long as there is no "science-based, global, effective, transparent control, and regulatory mechanism"; (2) in keeping with the precautionary approach and the obligations of Article 14 of the Convention; (3) until there is an adequate scientific basis to justify geoengineering; and (4) appropriate consideration of risks to the environment, biodiversity as well as social, economic, and cultural impacts (CBD, 2017; Bodle, Homan, Schiele, & Tedsen, 2012). Some commentators and scholars have termed these precautionary prohibitions as 'moratorium 'on geoengineering (ETC Group, 2010; C2G2, 2019). The 13th meeting of the Conference of Parties or COP-13 of the CBD held at Cancun (Mexico) in December 2016, reiterated its decision of the COP-10 and reaffirmed its "moratorium" on climate-related geoengineering (CBD, 2016). The issue of climate-related geoengineering came up before the United Nations Environment Assembly (UNEP) in March 2019 to consider the proposal to assess solar engineering's methods, evidence, current decision and possible future governance (Tollefson, 2019); nevertheless, in the wake of opposition from the US and Saudi Arabia, this proposal was not approved by the UNEA (Schneider, 2019). Viewed in a broad spectrum, issue of climate-related geoengineering has not been widely explored either by the IPCC or other authoritative bodies. Various reports of the IPCC that have been significantly influential in climate change policy making, have not thoroughly explored capacities and limitations of solar geoengineering and its potential in reducing impacts of climate change and managing its risks. Besides, little scholarly attention has been focused on the governance needs and potential responses that could emerge subsequent to any solar geoengineering deployment (Reynolds, 2019). While describing solar geoengineering as a complex, challenging and in many ways a novel human endeavor, Reynold (2019) emphasizes that it could "identify possible problematic outcomes and undesirable situations that could be avoided with appropriate foresight and pre-emptive action" (Reynolds, 2019).

4.5.2 Carbon dioxide reduction

CDR is the second component of geoengineering, after SRM and it is defined as the "removal and long-term sequestration of CO_2 from the atmosphere in order to reduce global warming" (NAS, 2015b, p. 33). The proposals for CDR are widely acknowledged to be an essentiality alongside intensive endeavors to minimize GHG emissions, specifically to meet the targets of Paris Agreement (Anderson 2015; Courvoisier et al., 2018; NAS, 2015b; Neslen 2015; UNFCCC, 2015b). Recent studies demonstrating the overall cost to be lower than past estimations have provided a fillip to CDR (Keith et al., 2018). CDR refers to "a cluster of technologies, practices and approaches that remove and sequester carbon dioxide from the atmosphere" (Smith & Friedmann, 2017). What differentiates CDR from other mitigation strategies is that CDR is directed at increasing the rate of negative emissions beyond carbon neutral rather than decrease net GHG emissions to zero to attain a "carbon-neutral" state. Nevertheless, in order to reach negative net carbon emissions, it is essential to accelerate rate of CDR via enhanced natural processes and development of options that capture or sequester or utilize CO_2. CDR alongside NETs, affords an assortment of methods and approaches that help remove CO_2 from the ambient air by biological, chemical or physical means and store the resulting carbon in durable reservoirs.

Options or approaches popularly used for CDR have become a defining feature of climate change mitigation scenarios that are assuming increasing emphasis in GHG emission frameworks (Clarke et al., 2014; van Vuuren et al., 2013; Fuss et al., 2014) and are seemingly consistent with the goals of Paris Agreement. In the wake of rapid pace of depletion and constriction of carbon budgets (Rogelj et al., 2016; IPCC, 2014a, 2014b), CDR options are more 'wisely used to compensate for temporary budget overshoot (Smith & Friedmann, 2017); and CDR is concerned with the management of overshoot in the eventuality of all other mitigation options being pursued. CDR options, as shown in Table 4.4, include Afforestation and reforestation; Soil carbon sequestration; biochar; Terrestrial bioenergy with carbon dioxide capture and storage (BECCS); Aquatic BECCS; Ocean fertilization; accelerated weathering (AW); and direct air capture (DAC). These options entailing technologies, practices and approaches have been managed by the people over the years

Table 4.4 Carbon dioxide removal options (Minx et al., 2018; Martin, Johnson, Stolberg, Zhang, & Young, 2017).

CDR options	Description	Readiness status
Afforestation/reforestation (AR)	Afforestation/Reforestation entails land management methodologies that comprise intentional forest management techniques for sequestering and storing CO_2 over a prolonged period.	Established
Soil carbon sequestration	Soil Carbon Sequestration consists of a land management method that is designed to enhance the amount of carbon stored in soil organic matter as well as in inorganic forms within the soil.	Demonstrated
Biochar	Biochar generates charcoal derived from biomass through a process called pyrolysis, which heats biomass to between 300°C and 800°C in a low-oxygenated environment.	Demonstrated
Terrestrial BECCS	The BECCS technique utilizes the CDR capabilities of photosynthesis through the growth of terrestrial biomass with the additional capture of CO_2 during the production of energy products.	Demonstrated
Aquatic BECCS	Aquatic BECCS absorbs CO_2 via plant growth in the ocean and then utilizes the harvested aquatic biomass to produce energy with capture and subsequent storage of CO_2.	Speculative
Ocean fertilization	Ocean fertilization decisively envisages specific nutrients into the ocean to stimulate growth in marine organisms (phytoplankton), thereby removing CO_2 from the atmosphere via photosynthesis by ocean organisms.	Speculative
Accelerated weathering	AW points to the geochemical process by which naturally occurring carbonate and silicate weathering is accelerated on land and in marine environments	Speculative
Direct air capture	DAC systems detach CO_2 directly from the atmosphere via chemical adsorption.	Speculative

Compiled from Smith, P., & Friedmann, J. (2017). Bridging the gap — Carbon dioxide removal. In UNEP (Ed.), The emissions gap report 2017 *(pp. 58–67). Nairobi: UNEP; Minx, J. C., Lamb, W. F., Callaghan, M. W., Fuss, S., Hilaire, J., Creutzig, F. . . . del Mar Zamora Dominguez, M. (2018). Negative emissions—Part 1: Research landscape and synthesis. Environment Research Letters, 13(6), 063001; Martin, D., Johnson, K., Stolberg, A., Zhang, X., & Young, C. D. (2017). Carbon dioxide removal options: A literature review identifying carbon removal potentials and costs.* University of Michigan.

carry with its wealth of knowledge and experience and they are frequently applied for the removal of carbon dioxide.

4.5.2.1 Afforestation and reforestation

Afforestation denotes planting trees on land not afforested in recent history, usually during the past 50 years or longer; while reforestation deals with the replanting of trees on more recently deforested land (Hamilton, Chokkalingam, & Bendana, 2010). Agroforestry practices comprise the incorporation of trees into agricultural systems, in conjunction with crops, livestock or both. Afforestation and reforestation along with agroforestry projects globally constitute part of various voluntary and mandatory carbon-offset trading structures (Miles & Sonwa, 2015). Potentiality of the CDR for afforestation and reforestation options is fairly significant worldwide and broad estimates show it ranging between 4 and 12 GtCO$_2$/year (Smith et al., 2016), with other recent approximations projecting even higher, at up to 28 GtCO$_2$/year (Griscom et al., 2017). The success of various ongoing projects along with the expertise garnered by forest managers worldwide over the years manifest a high level of technical readiness for afforestation and reforestation options thereby making it a viable option for option for CDR (Smith & Friedmann, 2017). Vast tracts of land would be required to attain large carbon removal rates and volumes (Kreidenweis et al., 2016), and probably large volumes of water (Trabucco, Zomer, Bossio, van Straaten, & Verchot, 2008), while the vegetation density bears a semblance of positive correlation with the intensity of precipitation sheds and wields a moderating impact on the unpredictability of water availability. Impacts of afforestation and reforestation on non-CO$_2$ GHGs surrounded by significant qualms (Benanti, Saunders, Tobin, & Osborne, 2014), and similar is the case with albedo (Zhao & Jackson, 2014), evapotranspiration, emissions of VOCs, etc.

Role of wetlands has come to be acknowledged in the removal of carbon dioxide, and wetland ecosystems provide an optimum natural mechanism for the sequestration and log-term storage of carbon dioxide (Mitsch et al., 2012). Other natural, land-based CDR solutions depend on the restoration or construction of high carbon density, anaerobic ecosystems, including "inland organic soils and wetlands on mineral soils, coastal wetlands including mangrove forests, tidal marshes and seagrass meadows, and constructed wetlands for wastewater treatment" (IPCC, 2014a, 2014b), and hereinafter, these solutions are referred as wetlands. In order to utilize potentiality of wetlands as CDR solution, it is imperative not only to preserve existing wetlands but also to ensure restoration and construction of these ecosystems along with their proper maintenance as well. According to estimates by Zedler and Suzanne (2005), up to 44%−71% of the global terrestrial biological carbon pool is stored by peatlands and coastal wetlands. Apart from the fact that the carbon stocks in peatlands and coastal wetlands are now susceptible to reversal (Parish et al., 2008), these ecosystems also possess massive carbon sequestration capacity (Page & Hooijer, 2016). Nevertheless, wetlands have received scant attention as CDR options as compared to afforestation and reforestation. Hu, Niu, Chen, Li, and Zhang (2017) estimate that by 2009 about one-third of global wetland ecosystems had been lost and suggest a number of regions where restoration work could be resumed. There are variations in the long-term sequestration rates in wetlands ranging from 0.1 to 5 tons of carbon per hectare per year, a rate that meaningfully improves when emissions avoided from restored wetlands are taken in view (Parish et al., 2008; Mitsch et al., 2012; Smith et al., 2008). Undoubtedly, making estimations about global rates and volumes of carbon sequestration is a challenging task (Adame et al., 2015); nevertheless, costs of carbon dioxide reduction costs for wetland restoration that range from US$10 to US$100/ton of carbon dioxide (Worrall et al., 2009), makes it discernible that there exist potential low-cost options for projects. Despite encumbrances that may crop up in respect of monitoring, sustaining sequestered carbon in the long term due to sink saturation, changing practices among forest managers and farmers, and creating market and policy contingencies, experience with managing forests is a testimony in support of these options.

4.5.2.2 Soil carbon sequestration

Soil carbon sequestration is a CDR option that is dependent on intentional land management targeted at enhancing the storage of carbon as soil organic matter and in labile, inorganic forms (Derek et al., 2017). Occurrence of soil carbon sequestration takes place in the wake of alteration land management practices increasing the carbon content of soil thereby ensuing in a net subtraction of carbon dioxide from the atmosphere. In view of the fact that carbon in the soil is a semblance of equilibrium of carbon inputs—absorbed from litter, residues, roots, or manure—and carbon losses, accruing largely from reparation that is increased by soil disturbance, practices that either augment inputs or decrease losses, can lead to soil carbon sequestration. Some scholars (Smith et al., 2008, 2014) have cited several land management's practices that can produce soil carbon sequestration, and some of that also entail the potential of generating soil carbon sequestration in the above-ground biomass. Significant variations characterize rates for soil carbon sequestration and these rely upon land management approaches, soil type and climate region (Smith, 2012). At the global scale, the technical probability for soil carbon sequestration is estimated at 4.8 GtCO_2eq/year (Smith, 2016). The worldwide carbon emissions mitigation probability of soil carbon sequestration is likely to vary between 1.5 and 2.6 GtCO_2eq/year, if unit costs are assumed between US$20 and US$100/ton of carbon (Smith, 2016). Nevertheless, for some systems like croplands and grazing lands, soil carbon sequestration costs can vary from minus US$45 to plus US$10/ton of carbon (Smith, 2016; Zomer et al., 2017). Estimates worked out by Smith (2016) demonstrate that subtraction of carbon dioxide via soil carbon sequestration at a rate of 2.6 GtCO_2eq/year would save US$7.7 billion, comprising US$16.9 billion of savings and US$9.2 billion of costs.

Undeniably, the ostensible energy costs for soil carbon sequestration may appear low and, in most cases, benefits could accrue to soil ecosystems and agribusinesses via soil carbon sequestration practices; nonetheless, executing these practices encompasses a significant range of potential land requirements (Amundson & Biardeau, 2018). Implementing soil carbon sequestration is faced with constraints like lack of knowledge among farmers, want of policy incentives, absence of adequate mechanism for monitoring and verification of practices, costs, and more crucially, reversibility of stored carbon (Smith & Friedmann, 2017). Well-designed pilot projects and well-concerted programs entail the potential of identifying the requisite measures that are called for in surmounting these constraints, with an emphasis on learning-by-doing and tackling key uncertainties via data acquisition and evolving practices (Smith & Friedmann, 2017).

4.5.2.3 Biochar
Biochar, a recently coined term for charcoal (Brick, 2010), is the product generated via pyrolysis of biomass into a solid, long-lasting product in the form of charcoal. This process entails heating biomass at a moderately low temperature in a low-oxygenated environment that prevents combustion. Production of biochar can be facilitated either as primary product o in the case of energy production, a secondary by-product (Shabangu, Woolf, Fisher, Angenent, & Lehmann, 2014). Owing its resistance to decomposition (Lehmann, Czimczik, Laird, & Sohi, 2015), biochar can stabilize organic matter added to soil (Weng et al., 2017). Having the ability of forming long-term carbon pools in the soil, it and afford a variety of soil fertility and soil quality cobenefits viz., improved water and nutrient retention, increased soil porosity and higher crop yields. Usage of biochar in agricultural soils has been in vogue since time immemorial because of its potential in improving water and nutrient retention (Derek et al., 2017). According to Jeffery et al. (2015), biochar provides, on a molecular level, more surface area for nutrients and beneficial microbes to stick to. Some studies have demonstrated that applying biochar to soils can decrease nitrogen leaching by 60% (Singh et al., 2010). Improvement in yields by 38%−45% by the application of biochar, as shown by some studies, can be discerned in terms of a 20% savings in fertilizer use and a 10% savings in the irrigation and seeds (Kung, McCarl, & Cao, 2013).

Biochar is prominently used these days for soil abatement, in chemical and industrial manufacturing, and, more recently, a way of appropriating carbon from the atmosphere. Undeniably, biochar can be utilized at high rates (Genesio et al., 2012), nonetheless, greater benefits of biochar can be reaped if applied in low volumes in the most responsive soils. The CDR potential via biochar is high and it has been estimated at between 1.8 and 3.3 $GtCO_2eq$/year (Woolf et al., 2010). Nevertheless, the effectiveness of biochar for CDR is uncertain. Despite the fact that biochar is an established technology, it is not yet in wider application, partly because of costs and the limited availability of infrastructure (Singh et al., 2014) because additional infrastructure in terms of pyrolysis facilities would be needed for large-scale implementation. Being a recent product, biochar phenomenon still entails a lot of scientific uncertainty regarding the consequences of its application (Laer et al., 2015) While asserting that the quantity of biomass available for biochar production is a prime factor restraining the global potential for CDR via biochar, Smith and Friedmann (2017) note that energy and water are also needed to "produce the crop feedstocks, although producing biochar can also produce power and fuels."

4.5.2.4 Bioenergy with carbon dioxide capture and storage

Bioenergy with carbon capture and storage (BECCS) is widely assumed CDR option (Creutzig, 2016). BECCS is emerging as the "best solution" to decarbonize emission-intensive industries and sectors and enable negative emissions (Consoli, 2019). The process of BECCS entails capturing CO_2, already concentrated in the biomass while the vegetation was growing, via generating an energy product. The BECCS process yields two main energy products—electricity and biofuels. BECCS is primarily based on bioenergy that refers to the energy derived from recent living biomass (in contrast to fossil) and it comprises 56 EJ (exajoules) of primary energy per year (World Energy Council, 2016), and this approximately constitutes 10% of total global primary energy use. Utility of BECCS for bioenergy comes handy because of capturing carbon from the facilities generating electricity and biofuels. Capturing of the CO_2 emissions arising from these energy products prior to the re-entry of emissions into atmosphere thereby results in negative emissions. At the cusp of 2015, there were about 15 pilot BECCS plants worldwide (Gough & Vaughan, 2015) and by the closing part of 2019, there were 18 large-scale facilities in operation, five under construction and 20 in various stages of development (Consoli, 2019). BECCS enjoys the privilege of being a major component of the integrated assessment models (IAMs) (Low & Schafer, 2020).

BECCS has been prominently highlighted in the mitigation scenarios projected by the IPCC thus far as the CDR technology (Fuss et al., 2016). Estimates made by many IAMs about the availability of sustainable bioenergy put it at 100 EJ (exajoules) per year (Creutzig et al., 2015); and fewer models accommodate estimates above 300 exajoules per year. Estimates for geological storage capacity for carbon dioxide capture are projected well above 5000 $GtCO_2$ (Ajayi, Gomes, & Bera, 2019). Nevertheless, some scholars are skeptical about the viability of the estimated capacities in all locations. (De Coninck & Benson, 2014; Global CCS Institute, 2016). Projections for the cumulative potentials of bioenergy and carbon dioxide capture and storage are estimated between 2 and 18 $GtCO_2$ per annum (Kemper, 2015; NAS 2015a). Construed in a broad perspective, attainment of this scale is bound to increase demands on land use significantly. Smith et al. (2016) opine that a level of CDR in coherence with average 2°C emissions pathways would need between 0.38 and 0.7 billion hectares of crops purpose-grown for bioenergy with carbon dioxide capture and storage. Nevertheless, assumptions under more conservative estimates could exert higher demands on land use (Monfreda, Ramankutty, & Foley, 2008). Admittedly, harnessing agricultural and forest residue as a feedback for bioenergy seldom requires competition for land use, while its removal can impact soil carbon stocks in an adverse manner (Smith et al., 2016).

Extensive use of BECCS along with the resultant potential competition for land use is a critical issue for launching large-scale BECCS deployment projects as well as policy making. Another constraint is the scant public acceptance enjoyed by carbon dioxide capture and storage and bioenergy processes (Benson et al., 2012). Besides, there prevails ambiguity with regard to whether there are massive or even any carbon reductions when accounting for displaced activities (Creutzig et al., 2015). Furthermore, absence of economic incentives and the regulatory barriers pertaining to underground storage impede large-scale implementation (De Coninck & Benson, 2014). Costly nature of many CDR approaches is another significant issue. Deployment of these approaches at a large-scale has demonstrated low level of technical readiness. The view is gaining acceptance with the scholars and policymakers alike with regard to increasing investments in developing these options that could likely yield breakthroughs in material science and manufacturing thereby entailing the potential of spurring new industries and carbon circular economy (McDonough, 2016; Center for Carbon Removal, 2017), and the example of lithium-ion batteries about three decades ago is a testimony to it (*The Economist*, 2017). According to Smith and Friedmann (2017), among the 23 countries that made commitments to undertake substantial research and development programs in the BECCS and other related areas, only the United Kingdom has made a modest investment.

4.5.2.4.1 Aquatic bioenergy with carbon dioxide capture and storage

Aquatic BECCS is said to be a hypothetical CDR option that capture CO_2 through plant growth in the ocean and then deploys the harvested aquatic biomass to produce energy with capture and subsequent storage of CO_2 (N'Yeurt, Chynoweth, Capron, Stewart, & Hasan, 2012). Irrespective of the fact that a range of aquatic species could be appropriate as a source of feedstock for Aquatic BECCS, still aquatic macroalgae has been the subject of emphasis in the bulk of literature. Aquatic macroalgae comprise wide range of kelps and seaweeds. Oceans encompassing over 70% of the earth's surface thereby, constituting a significant natural carbon sink, entail the capacity of storing roughly 5.9 $GtCO_2$/year (IPCC, 2013). CO_2 is captured from the atmosphere via diffusion and dissolution in ocean water and by means of photosynthesis by ocean organisms (Raven & Falkowski, 1999). Presently, occupying about 2% of the ocean, macroalgae have the capacity of intensifying for aquatic BECCS (N'Yeurt et al., 2012). Aquatic BECCS is considered as a speculative CDR option owing to the limited research in this regard and essential questions pertaining to its ecological impacts and costs, aquatic BECCS has come to be regarded as a speculative CDR option.

Carbon dioxide is absorbed by the aquatic BECCS process through photosynthesis, seaweed from ocean water, and is transformed into organic material that may be located in above- and below-seafloor biomass and exhales oxygen into the ocean through respiration (Chung et al., 2012). The amount of biomass growth is reliant on ambient CO_2 concentrations, nutrients, light, temperature, water motion, upwelling, salinity, and geographic location (Chynoweth, 2002). Sequester and storage CO_2 within the ocean can be facilitated by the expanding seaweed ecosystems through a process called ocean macroalgal afforestation. A variety of techniques can be deployed to harvest seaweed from the ocean floor (Roesijadi, Jones, Snowden-Swan, & Zhu, 2010), and these are exceedingly dependent on the type of macroalgae produced and the climate in which they were cultivated (Pereira, Yarish, Jørgensen, & Fath, 2008). The harvested macroalgae are carried to anaerobic digestion containers, where they are broken down to generate biogas, recyclable plant nutrients, and water (N'Yeurt et al., 2012). Biogas comprises of 60% methane and 40% CO_2 (Hughes et al., 2012). Hypothetically, methane can be used as a gas or converted to liquid fuels such as jet fuel, diesel, methanol, or gasoline. The CO_2 from the digester can be captured, compressed, and stored through various storage options (N'Yeurt et al., 2012).

Estimates about the area of seafloor usage and expansion, geographic location of macroalgal forests/ecosystems, type of macroalgae cultivated, nutrients, light, temperature, water motion, upwelling, salinity, etc., play a vital role in determining the CDR potential of aquatic BECCS (Chung et al., 2010). In other words, aquatic BECCS is confronted with some challenges and limitations as well, especially with the life cycle emissions of aquatic BECCS. The hypothetical nature of aquatic BECCS brings with it a great deal of uncertainty surrounding its overall net GHG emissions impact. Implementation of aquatic BECCS over vast ocean areas, such as the 9% of ocean surface as envisioned by N'Yeurt et al. (2012), is prone to entail an enormous shift in ocean ecosystem; and as pointed by Cotter-Cook (2016), several ecosystem shifts could take place, such as altering the structure and the function of ecosystems and associated food webs, cross-breeding among wild and introduced species, and the unplanned spread of species beyond their intended area. Besides, aquatic BECCS is limited by land availability (Alpert, Spencer, & Hidy, 1992), like most of the terrestrial CDR options.

4.5.2.5 Accelerated weathering

In common parlance, AW is construed in terms of the simulation of the damaging effects of long-term exposure of coatings, materials, and products to outdoor conditions. AW purposely increases or accelerates the geochemical weathering process wherein CO_2 gets absorbed into soil or marine sediments as carbonates or bicarbonates. Caron dioxide is taken away from the atmosphere via a reaction between CO_2, water and a mineral, generally calcium, magnesium or iron silicates, to create soluble bicarbonate ions and a solid by-product (Rau & Caldeira, 1999). Broadly, there are three approaches pertaining to CDR—aqueous carbon mineralization, enhanced carbonate and silicate weathering, and injection into basalt rock formations—and these options diverge in the range of research conducted and the level of confidence relied upon by the experts in their practical application (McLaren, 2012). In the wake of limited number of projects in operation currently, AW is reckoned with as a speculative CDR option. The aqueous mineral carbonation process is widely different from natural weathering processes, because it makes use of a reactor to enable the carbonization process with extracted minerals. Magnesium, calcium and iron silicates, often found in olivine, serpentine, wollastonite, and calcium carbonate but also industrial waste (Giannoulakis, Volkart, & Bauer, 2014; Matter et al., 2016), are the primary minerals used in this process. Aqueous carbon mineralization provides a productive use for Industrial wastes such as cement kiln dust, slag, or coal fly ash, etc., find productive use in the process of aqueous carbon mineralization and this process has elicited increasing interest because of its potential of reducing both CO_2 emissions and industrial waste (Kirchofer, Becker, Brandt, & Wilcox, 2013).

The benefits accruing from byproducts provided by AW extend beyond CDR. Most types of AW comprise carbonation of water that produces an alkaline bicarbonate solution that can be inserted into the ocean; and that bicarbonate may counteract ocean acidification (Rau & Caldeira, 1999). Utilization of the solid byproducts of aqueous carbon mineralization on land can be made as aggregate to help restore natural land contours and evade dumping more material into landfills

(Giannoulakis et al. 2014). Additionally, the usage of industrial byproducts as feedstock minerals helps in the reduction of the disposal of industrial waste (Kirchofer et al., 2013). Minerals dispersed over land with the anticipation of increased silicate weathering result in enhanced carbonate content in soil that can limit the uptake of heavy metals into plants (Wang et al., 2015). Undoubtedly, AW entails potential for CDR option; nonetheless its speculative nature coupled with the fact of wide variation in CDR estimates make AW's scale uncertain. The different magnitudes of CDR potential estimates are often attributable to the diverse AW methods, inputs, and land area used.

4.5.2.6 Ocean fertilization

Irrespective of many descriptions and statements on ocean fertilization available in a wealth of literature, there are few internationally agreed definitions of the term (CBD, 2009, p.14). Broadly, ocean fertilization is defined as "any activity undertaken by humans with the principal intention of stimulating primary productivity in the ocean, not including conventional aquaculture, or maricultural, or the creation of artificial reefs" (London Protocol, 2008). Ocean fertilization, also occasionally known as ocean nourishment, envisages insertion of nutrients into the ocean to stimulate growth of marine microscopic organisms. Insertion of a limiting nutrient purposely is prone to engender growth of phytoplankton, resulting in the sequestration and storage of additional CO_2 (Williamson et al., 2012); and iron, nitrogen, and phosphorous are regarded as limiting nutrients (Williamson et al., 2012). Undeniably, nitrogen and phosphorous are the most common limiting nutrients; nonetheless, iron is the limiting nutrient mostly used in almost one-third the earth's oceans (Wallace et al., 2010). Owing to limited demonstration projects and significant concerns pertaining to impacts on ecosystems, ocean fertilization is often construed as a speculative CDR option.

Diffusion and dissolution of carbon dioxide in ocean water and photosynthesis by ocean organisms render oceans as ideal natural sink for CO_2 (Raven & Falkowski, 1999). As per IPCC (2013), The rate of carbon sequestration of the ocean is roughly 5.9 $GtCO_2$/year. CO_2 is absorbed by phytoplankton from ocean water and stored carbon within their biomass by producing particulate organic carbon. The carbon is retained within the phytoplankton until they are taken in or remineralized by bacteria, causing the carbon to resume its original dissolved or inorganic carbon state within the ocean (Bertram, 2010). If not absorbed or remineralized, dead and living phytoplankton sink, or are forced, to deep ocean depths where they get immersed with sediments on the ocean floor, thus storing CO_2 (Denman, 2008). According to Powell (2008), only 1%–15% of the carbon gets sunken below 500 m to allow for potential burial (Powell, 2008). A limiting nutrient is added to the zones of ocean called "desolate zones," to accelerate this process; thereupon, these desolate zones are also known as high-nutrient, low-chlorophyll zones that are devoid of critical nutrient; thereafter forestalling phytoplankton or other aquatic life from growing (NASA, 2017; Williamson et al., 2012). The limiting nutrient is by and large taken to the area by an ocean vessel. Insertion of the limiting nutrient facilitates the growth of phytoplankton, enhancing primary productivity within zones that formerly lacked a high concentration of aquatic life (Bertram, 2010).

Various factors contribute to ocean fertilization and these inter alia include: the quantum of the limiting nutrient added, status of phytoplankton prefertilization, grazing pressure of zooplankton, the amount of time the atmosphere had contact with the fertilized area, stratification, mixing of water, sinking and burial rates, and light conditions (Williamson et al., 2012). A primary reason for the variation in estimates is the supposed ocean zone for ocean fertilization. In some studies, an assumed area was specified, while others exhibited estimates based on percentages of the ocean or quantities based on optimal oceanic locations. Entire ocean was assumed by Aumont and Bopp (2006) as the implementation zone for iron fertilization, while the entire North and Tropical Pacific Ocean was chosen by Jin, Gruber, Frenzel, Doney, and McWilliams (2007), Jones (2014) selected the temperate ocean, and the Southern Ocean was selected by Joos, Sarmiento, and Siegenthaler (1991), Kurz and Maier-Reimer (1993), Rickels, Rehdanz, and Oschlies (2010), and Sarmiento and Orr (1991); and Matear and Wong (1999) opted for any high-nutrient, low-chlorophyll regions. On the other hand, Peng and Broecker (1991) undertook implementation in the Antarctic (Southern) Ocean, which they supposed to be equivalent to 10% of the world's oceans, while Joos et al. (1991) presumed 16% of the world's oceans are located in the Southern Ocean. While focusing on real and unknown risks posed by ocean fertilization, like any other geoengineering method the planet; as does the inaction on climate change, focus is emphasized on pursuing research into ocean fertilization as well as also limiting channels through which to conduct beneficial research (Williamson et al., 2012; Branson, 2014).

4.5.2.7 Direct air capture

Process of direct removal of carbon dioxide from the air by its chemical reaction with other chemicals is called DAC (Jacobson, 2020). While DAC being a new concept for removal of atmospheric CO_2, it has been used air removal on the submarines and spaceships for decades (Lackner et al., 2012). An aqueous chemical sorbent or solid sorbent system constitutes primary DAC configurations. In the chemical sorbent DAC mechanism, an aqueous solution, usually comprising sodium hydroxide (NaOH), is bonded with CO_2 (Socolow et al., 2011). Ambient air blows through an absorber in this process and bonds with NaOH. The captured CO_2 solution wades through the precipitator to remove the CO_2 and regenerate the NaOH by adding calcium. The NaOH is returned to the absorber and the CO_2 is removed from the calcium by exposure to temperatures above 800°C (House et al., 2011). The CO_2 thus released is compressed and readied for storage (Socolow et al., 2011). Some studies demonstrate that solid sorbent DAC mechanisms capture more gas per unit of energy than other systems, owing to elimination of high temperatures (Lackner, 2010; Lu, Sculley, Yuan, Krishna, & Zhou, 2013). There is growing emphasis on making such DAC systems small in size, say the size of a standard shipping container, that can ensure mass production to benefit from economies of scale (Lackner, 2010).

The CDR potential of DAC considerably depends on the type of DAC mechanism deployed and temporal scope of the estimate. Estimates worked by Lackner and colleagues demonstrate that between 1 and 10 million solid sorbent DAC units could be produced per year and 1, 000 units is the conservative estimate (Lackner, 2010; Lackner et al., 2012). On the contrary, another report claims that building 80 chemical sorbent DAC systems annually would help remove 4 $GtCO_2$/year. over 100 years (Socolow et al., 2011). Keeping in view the role of the design of the DAC system contributing to the differences in the CDR potential, it has been suggested that the sequestration rate should be 1 $MtCO_2$/year. to make a significant difference, while placing emphasis on the use of an aqueous NaOH solution to capture CO_2 (Mazzotti, Baciocchi, Desmond, & Socolow, 2013; Stolaroff, Keith, & Lowry, 2008). Some studies assume air flows through contractor (Holmes, 2012; Mazzotti et al., 2013); whereas Stolaroff et al. (2008) hypothesizes a spraying mechanism to capture CO_2, an amino-modified silica adsorbent is assumed by Kulkarni and Sholl (2012) and Lackner (2010) makes use of a design that imitates a furnace filter with thin anion exchange resin.

Diverse DAC systems and system components make the range of economic estimates for DAC as the largest of any of the CDR options. Kulkarni and Sholl (2012) opine that since solid sorbent design consumes less energy owing to reduction of the energy needed for sorbet reproduction and that facilitates lower costs than other designs. Solid sorbent estimate suggested by Lackner (2010) is considerably lower than the others as it is a cost in the indefinite future. Despite the advantage, the DAC system has in terms of energy required and land use comparable to other CDR options, DAC is still regarded as a speculative CDR option because it is still in the research and development (R&D) stage with few pilot projects and its exorbitant economic costs (Derrek et al., 2017). There is an increasing consensus that DAC will not be a viable option until mid-century (Smith et al., 2016). While focusing on the implementation of DAC in the near-term, Willcox, Psarras, and Liguori (2017) suggest that it would be important to identify further opportunities for using dilute CO_2 as a feedstock whereby the energy needed for concentration of CO_2 from the air is lower.

4.6 Loss and damage associated with climate change

The notion of loss and damage associated with climate change (hereafter referred as L&D) as another option to cope with the adverse impacts of climate change has come into the reckoning and gaining traction in recent years in the discourse on Climate change. In its endeavor to combat climate change, the international community had focused on rapid reduction of GHGs through implementation of enhanced mitigation efforts from the early 1990s to the mid-2000s; nevertheless, the scientific evidence from the mid-2000s onward indicating the probability of global temperatures rising between 3°C and 4°C above the pre–industrial level within 21st century and its serious consequences for humankind and ecosystems (IPCC, 2007a, 2007b) brought shift in emphasis on adaptation along with mitigation as well (Ott, Sterk, & Watanabe, 2008). Debates over the global adaptation cost estimates (Parry et al., 2009) resulted in the establishment of the GCF at the COP-16 held at Durban in South Africa in 2011, with the avowed objective of raising a minimum of $100 billion/year by 2020 to support sustainable and climate-resilient development (IPS, 2014; GCF, 2014).

In the wake of apparent realization of the limits of adaptation to cope with the adverse impacts triggered by climate change (Barnett et al., 2015), the remnants of the adverse impacts of the vagaries of climate change that came to be known as "residual loss and damage" necessitated an extensive global understanding and agreement differentiating between adaptation and loss and damage that would culminate in recognizing that all adverse impacts of climate change could not be successfully dealt with mitigation or adaptation (Mathew, Akter, Wei-Yin, Toshio, & Maximilian, 2017). According to Perry and colleagues (2009), such remnants of the adverse effects of the climate change impacts were projected to account for two-thirds of all potential impacts across all sectors over the longer term. The acknowledgment of this realization underlined the essentiality for allocating adequate compensation and relief efforts, above and beyond the GCF, to help the victims of loss and damage in countries that were geographically and economically vulnerable to the vicissitudes of climate change (Mathew et al., 2017). Allusion to the term "loss and damage" was made at the COP-13 negotiations held in 2007 at Bali, Indonesia, wherein the Bali Action Plan (UNFCCC, 2008a) called for increased action on adaptation, including the consideration of "disaster risk reduction strategies and means to address loss and damage associated with climate change impacts in vulnerable countries" (Roberts & Huq, 2015).

The year 2008 witnessed the notion of loss and damage being acknowledged as a distinct term from adaptation in the wake of proposal mooted by the Alliance of Small Island States (AOSIS) for a Multi-Window Mechanism to address and finance the distinct concept of loss and damage due to climate change impacts (UNFCCC, 2008b). Subsequently, it was followed by the establishment of the UNFCCC Work Program on Loss and Damage in 2010 (UNFCCC, 2011a, 2011b, 2011c) and establishment of the WIM on Loss and Damage in 2013 (UNFCCC, 2013). Besides, the formation of the Loss and Damage in Vulnerable Countries Initiative took place in 2012 (ICCCAD, 2012), with the objective of understanding both the national context and the range of accessible implementation options for addressing loss and damage (Roberts & Huq, 2015). In the absence of any lifetime commitment made by the developed countries to make funds available to vulnerable countries, the loss and damage initiatives could be construed as "week attempts by the rich countries to admit liability for their contributions to climate change" (Mathew et al., 2017).

4.6.1 Defining loss and damage

Existing scientific literature offers no single widely accepted definition of the concept of L&D; therefore formulation of definition of this concept and its conceptual elaboration have continued to evolve within the UNFCCC and the academic literature with diverse groups offering assorted and "heterogeneous understanding of the terminology and concept" (Mathew et al., 2017). The concept of L&D has been defined by the UNFCCC as one of the "impacts associated with climate change in developing countries that negatively affects human and natural systems" (UNFCCC 2012). Found to be at its nascent stages, the definition delineated by the UNFCCC was construed in terms of lacking clarity. Launching of the Loss and Damage for Vulnerable Countries Initiative in 2012 is said to have helped in understanding the meaning of the concept and how it could be approached in vulnerable countries (Warner et al., 2012). Irrespective of the fact that empirical research has demonstrated that vulnerable communities around the globe are already experiencing loss and damage owing to climate change effects (Warner, van der Geest, & Kreft, 2013), and prevalence of loss and damage continues to be there; nevertheless, international agencies and forums, including that of the UNFCCC, have not come out with a widely accepted definition of loss and damage. Interestingly, it has been claimed by Bread for the World, EED-Church Development Service, and Dan Church Aid (2012:2), that "[a] precise definition of loss and damage must be quite encompassing and inclusive taking into consideration its various aspects...." Scholars are still seized of the problem in order to grasp the entire gamut of causes and impacts associated with loss and damage without having been able to reach at a satisfactory definition that can meet such standards.

Concept of L&D can be delineated on the basis of the type of impacts they trigger, such as economic and noneconomic. The UNFCCC describes economic losses as "the loss of resources, goods and services that are commonly traded in markets" (UNFCCC, 2013a). This notion of economic losses bears close proximity to the above definition of "damage" outlined by the Loss and Damage in Vulnerable Countries Initiative since economic losses are assessable and can hypothetically be restored. Conversely, noneconomic losses are not traded in markets, making them difficult to assess (UNFCCC, 2013b). Noneconomic losses comprise "losses of, inter alia, life, health, displacement and human mobility territory, cultural heritage, indigenous/local knowledge, biodiversity and ecosystem services" (BES) (UNFCCC, 2013b). Both economic and noneconomic forms of loss and damage can accrue from a single disaster induced by climate change. Differentiating between economic and noneconomic loss and damage outlines the breadth of challenges emanating from climate change and the necessity for an array of solutions including prevention (Warner et al., 2013). The manner in which academic community and climate negotiators acknowledge and deal with loss and damage is likely to impact the role of climate change adaptation and mitigation. Pinninti (2014) has opined that avoidable loss and damage allude to probable loss and damage that can be averted from ever happening through mitigation and adaptation efforts. Residual loss and damage are that which "remains once all feasible measures (especially adaptation and mitigation) have been implemented" (UNFCCC, 2012a, 2012b, p. 20). According to Pinninti (2014), residual losses and damages refer to "the portion that accrues after adjusting for the effects of [climate change adaptation] in the context of adverse impacts."

Being seized of the fact that some loss and damage associated with climate change is ineluctable and already occurring, the UNFCCC has been augmenting its efforts to address loss and damage. Apart from incorporating the suggestion of the Alliance of Small Island States (AOSIS) that promoted "an agreement on a mechanism to address the risk posed to sustainable development from the projected impacts of climate change" (UNFCCC, 2011b, p. 27), the UNFCCC, as part of the Cancun Adaptation Framework, recognized the need to strengthen international cooperation and expertise in order to understand and reduce loss and damage associated the adverse effects of climate change, including impacts related to extreme weather and slow onset events (UNFCCC, 2011a). With this development, the Loss and Damage Work Program became part of the Cancun Adaptation Framework and thereafter, the first action before the UNFCCC was to address L&D in the aftermath of Cancun summit by collecting and synthesizing desires of the Parties and select observers for the elements to be incorporated within loss and damage work program. In the wake of wide variations in opinions — with some Parties calling upon for some very specific actions while others focusing on the structure and implementation of the work program, the Small Island Developing States (SIDS), while focusing on the structure and lifespan of the work program, also drew attention to three key thematic areas: current knowledge on exposure to loss and damage, experience with various instruments to address loss and damage, and possible implementation pathways under the Convention (UNFCCC, 2011b). Amongst various conclusions reached by the UNFCCC, one of these was to implement three key thematic areas that closely resembled those suggested by the SIDS: (1) assessing the risk of loss and damage, (2) approaches to address loss and damage, and (3) the role of the Convention in enhancing the implementation of these approaches (UNFCCC, 2011c).

4.6.2 Warsaw International Mechanism for Loss and Damage

Development of the initial work program on L&D led to the adoption of specific decision at COP-18 held at Doha to create an international mechanism for loss and damage associated with climate change impacts (UNFCCC, 2013a). The decision to include an international mechanism in the Doha Decision "on loss and damage marks an important window of opportunity for the further development of such mechanisms" (Khan, Ruppel, Roschmann, & Ruppel-Schlichting, 2013, p. 846). Subsequently, the COP-19 held in Warsaw led to the official establishment of WIM for loss and damage associated with climate change impacts (hereafter referred as WIM), along with its Executive Committee (UNFCCC, 2014a). The initial two-year workplan, adopted soon after its establishment by the WIM, called upon the primary focus of the Executive Committee of the WIM until 2016, on nine action areas (UNFCCC, 2014a) that inter alia included: (1) enhancing the understanding of loss and damage impacts, (2) enhancing risk management approaches, (3) enhancing knowledge of risks of and approaches to slow onset events, (4) enhancing knowledge of non-economic losses, (5) enhancing understanding of capacity and coordination needs to address loss and damage, (6) enhancing knowledge of migration, displacement, and human mobility, (7) encourage risk management using financial instruments, (8) complement the existing work of the UNFCCC and other relevant institutions, (9) and develop a 5-year rolling workplan.

Undoubtedly, WIM has undertaken steps to address a variety of primary action areas of concern to the Parties comprehensively; nevertheless, its future actions are uncertain. There is lack of clarity in many aspects of the WIM, including its relationship with adaptation. The action area under the WIM aiming at reducing risk in order to avert loss and damage and encourage "long-term resilience of countries" (UNFCCC, 2014a), fall within the ambit of adaptation. The concept of loss and damage is distinguishable from the notion of adaptation because adaptation is "a process not an outcome" (McGray, Hammill, & Bradley, 2007, p. 7) that is aimed at preventing negative impacts from climate change, while the WIM recognizes "that loss and damage associated with the adverse effects of climate change includes, and in some cases involves more than that which can be reduced by adaptation" (UNFCCC, 2014b). The trenchant or distinct nature of the notion of loss and damage, along with the increasing urgency for taking action to deal with climate change impacts warrants exploration of the question of the status of L&D within the UNFCCC, with the probability of loss and damage being involved into another additional option in addressing climate change impacts, complementing adaptation and mitigation. As has been pointed out, "operationalizing the continuum of mitigation- adaptation and loss and damage, might require a reconsideration of the status of loss and damage" (Schafer & Kreft, 2014, p. 19). Looking upon L&D as distinct from adaptation can be instrumental in encouraging action and recognition of the impacts of climate change. Adequate and appropriate conceptualization of loss and damage associated with climate change impacts can "provide the necessary guidance for identifying the entities responsible for such change, including the private sector" (Khan et al., 2013, p. 847). In a similar vein, it has also been suggested that any approach to loss and damage, especially at global level, "must seek to increase international commitment to mitigation and adaptation, the parameters that influence the extent of residual loss and damage" (CDKN, 2012).

Nevertheless, there is lack of unanimity over letting the WIM to become an identity of its own. It is argued by Briner, Kato, Konrad, and Hood (2014) that "thorny issues" like the WIM that Parties often find cumbersome to tackle "may have been outsourced to new bodies" (Briner et al., 2014, p. 7) with the objective of permitting individual entities to narrow their resolutions. Keeping in view the controversies ranging over status of the WIM within the UNFCCC vis-à-vis adaptation, while arguing that entities within the UNFCC like the WIM are relatively at the incipient stage, Briner et al. (2014) note that they need more time to get matured before the Parties can appropriately assess their outcomes. What makes L&D as a 'thorny issue is the question of liability and compensation. Despite the fact that generation of climate change induced GHG emissions by the developed countries is an acknowledged issue within the negotiation processes of the UNFCCC; nonetheless, sharp differences "between countries and lobbying blocks" dominate the Loss and Damage negotiations, prominently with regard to issues of causality and compensation

(McNamara, 2014, p. 242). It remains somewhat a cumbersome task to establish a causal relationship between anthropogenic climate change and particular Loss and Damage. While contending that scientists can point out a relationship between gradual commencement of disasters and the frequency of extreme weather events; however, it is "almost impossible to say that an extreme event would not have happened without anthropogenic climate change" (Parker et al., 2015, p. 70).

Deliberations on liability and compensation have continued to be part of the negotiations since the inception of the UNFCCC, emphasis by the island nations and other vulnerable countries on an insurance mechanism to assist developing countries enabling them to dealing with the vagaries of climate change not only pushed the issue of liability and compensation to the forefront, albeit also proved instrumental in getting insurance and other compensations tools incorporated within action area seven of the WIM's workplan. This action area 7 emphasizes on encouraging comprehensive risk management in the diffusion of information pertaining to financial instruments and tools that address the risks of loss and damage associated with the adverse impacts of climate change to facilitate finance in loss and damage situations (UNFCCC, 2014a). Irrespective of the recognition accorded to liability and compensation under the WIM, it is yet to be seen as to far this contentious issue of loss and damage ripens into fruition, if at all. Despite the disputable claims with regard to the notion of liability, it entails the potential of limiting climate change effects, resulting in optimum outcomes. Liability can engender "an incentive to undertake cost-effective mitigation" (Doelle, 2014, p. 38). Failure on the part of countries to meet mitigation targets can spur them financing adaptation measures or repair damages, thereby generating a monetary incentive to improve mitigation efforts. This aspect has been emphasized upon by ActionAid, CARE International, Germanwatch, and WWF (2012) by emphasizing that the "precautionary principle" could be turned on by a mechanism that addresses "rehabilitation and compensation" (p. 3). Indubitably, this line of reasoning acknowledged the essentiality for each element of the UNFCCC to buttress the primary goals of the Convention; nevertheless, persuading developed nations to comply with WIM measures could seemingly prove a herculean task. While arguing that liability and compensation constitute significant elements of the loss and damage discussions, Hoffmaister, Talakai, Damptey, and Soares (2014) note that this perspective often "trivializes the complexity of the issues and inaccurately reduces the issue to one of merely determining liability and seeking compensation."

With concepts, methods and tools, along with directions for policy with regard to WIM having remained opaque and contested, the debate on the subject is concurrently broad, diffuse and somewhat confusing. Scholarly research in recent years has continued to generate actionable inputs. There is evidence on losses and damages in vulnerable countries (Warner & Van der Geest 2013); exploration and critical appraisal of definitions, logical and multiple perspectives on the narrative (Vanhala & Hestbaek 2016; Boyd, James, Jones, Young, & Otto, 2017); employable methods and models (Schinko & Mechler, 2017), swotted roles for justice and equity considerations (Huggel et al., 2016a), focused attention on noneconomic losses (Wewerinke-Singh 2018a), supportive setting out of policy and governance options (Page & Heyward, 2017; Biermann & Boas, 2017); and evaluating the role of legal responses to L&D (Wewerinke-Singh 2018b).

Executive Committee (ExCom) of the WIM since its inception has been engaged in devising work programs to inform the deliberations, and in recent years it has been subject to intense debate. While some regard it a discrete building block of negotiations under the UNFCCC alongside mitigation and adaptation, others suggest that it is said to be an integral part of the negotiations under climate change adaptation. The issue of WIM came up for discussion at the COP-24 wherein the Parties to the UNFCC were invited to integrate loss and damage in policies, plans and strategies along with the incorporation into policy options to protect displaced persons (Puig, Serdeczny, & Huq, 2018). The question of WIM governance—whether the work on loss and damage should continue to take place under the UNFCCC and Paris Agreement processes or take place solely under the Paris Agreement—was to be decided at COP-25. When the deliberations on WIM came for review at COP-25, differences between the developed and developing countries with regard to the vision of the future of the WIM could be discerned. For developed countries, minor twists to the working of the WIM's Executive Committee (ExCom) was all that was required, such as the development of a uniform technical reporting format and improving communication and outreach, while developing countries, in the wake of enhancing current and expected devastating effects, emphasized on the essentiality of considerably strengthening the ability of the WIM to facilitate work on-the-ground to address these impacts (Pierre-Nathoniel, Siegele, Roper, & Menke, 2019). Developing countries asked for: scaled-up financing from developed countries; more visible inclusion of loss and damage in the operating entities of the Financial Mechanism, enhanced capacity-building efforts; and expanded institutional arrangements under the WIM to ensure the developing countries would benefit from its work (Dagnet et al., 2019a, 2019b; Pierre-Nathoniel et al., 2019).

In view of the divided outlooks of developed and developing countries toward the future of WIM at COP-25 negotiations, reaching a consensus on the outcomes proved to be time-consuming issue. Perspective of the developing countries, presented through a coordinated G77 & China, asked for an enhanced and strengthened WIM that was able to facilitate action and support for developing countries in their efforts to implement approaches to address loss and damage associated with the adverse effects of climate change (UNFCCC, 2019a, 2019b, 2019c, 2019d). Based on the conclusions of the WIM review, the Parties to the Paris Agreement adopted the following decisions, among others, encouraging the Parties to establish loss and damage contact point and ExCom of the WIM was requested to liaise with the GCF, to clarify how developing country Parties may access funding from the GCF; and the Santiago Network was established, as part of the WIM, to "catalyze the technical assistance" required by the most vulnerable countries (UNFCCC, 2019a, 2019b, 2019c, 2019d). With regard to the operations of the WIM, the Parties to the Paris Agreement recommended that the next review of the WIM will be held in 2024 and very five years thereafter. Besides, ExCom of the WIM was requested to revise the terms of reference for and launch the expert groups on slow onset events and noneconomic losses. The ExCom of the WIM was also called upon to "establish, by the end of 2020, an expert group" that draws up a plan of action concerning finance, sources of support, collaboration, outreach and risk management (UNFCCC, 2019a, 2019b, 2019c, 2019d; Puig, Wewerinke-Singh, & Huq, 2019). Nevertheless, on the question of the governance of the WIM — whether it should operate under both the UNFCC and the Paris Agreement, as requested by developing countries, or solely under the Paris Agreement, as demanded by the US, no agreement was reached it was decided that negotiations on this issue would be resumed in 2020 (Puig et al., 2019). After about 6 years of coming into existence, the WIM has become "little more than a talk shop with minimal on-the-ground benefit to the most vulnerable" (Meva, 2019).

4.7 Ecosystems approach and climate change

Emergence of climate change as a reality that is impacting almost everything on the planet, while requiring a shift of emphasis from conventional development pathway to a sustainable development pathway, has unveiled increased scope for nature-based solutions to deal with adverse impacts of climate change. Ecosystem-based adaptation (EbA), as a subset of the nature-based solution, has come to be reckoned with as a specific type of policy mix provides impetus to sustainability transition that finally ensures BES conservation or restoration and carbon mitigation, while promoting improvements in the livelihood of the communities. EbA is increasingly gaining traction as a viable mechanism of sustainability transition, especially in the developing countries that are more eager to protect most of the planetary biodiversity and healthy ecosystems from the vagaries of climate change. Sustainability transition research is more often assisted by two approaches—research on socioecological systems (SESs) and research on socio-technical systems (STSs)—and this has spurred a great deal of scientific and public interest in large-scale societal transformation toward sustainability (Loorbach et al., 2017). Social-ecological systems are linked systems of people and nature, emphasizing that humans must be seen as a part of, not apart from, nature (Redman, Grove, & Kuby, 2004), and SES examine as to how society interacts with nature, and how such interactions can be adaptive to events like climate change and other effects (Colding & Barthel, 2019). SES research entails instances of transitions to tone down public health threat (Pant et al., 2016), to reduce susceptibility to natural disasters (Munang, Thiaw, & Alverson, 2013), contributing to sense of place (Masterson et al., 2017), resilience into energy systems (Hodbod and Adger, 2014), biocultural diversity (Mauerhofer et al., 2018), and to promoting adaptive BES conservation (Mathevet, Thompson, & Folke, 2016), and interestingly much of this focuses on the appropriateness of participatory decision-making in such systems (de Castro, Hogenboom, & Baud, 2016).

Research on STSs emphasizes on technology-focused attention on socio-technical transitions that shares some of the complex adaptive systems sensibilities of social-ecological systems research and is considered by others to provide a bridging opportunity to share lessons concerning the governance of both (Smith & Stirling, 2008). Research in STS facilitates exploration of science-society interface for transition to low-carbon systems particular sectors like agriculture, energy, industry, etc., in response to climate change and other effects (Pant, Adhikari, & Bhattarai, 2015). Sustainability transition research makes discernible many examples in many specific sectors like energy (Rogge & Reichardt, 2016), agriculture (Vogel et al., 2020), industry (Asiimwe & de Kock, 2019), transport (Geels, Kemp, Dudley, & Lyons, 2012); and these often demonstrate as to how technological innovation is a prime driver of such transitions targeting lower carbon budgets. Research on SES and STS has often been conducted independent of each other and scholarly emphasis has been on integrating both approaches in science as well as in policy making.

Dichotomy that exists between SES and STS approaches also pervades the EbA as well (Rizvi et al., 2015; GIZ, 2017). Allusions to adaptation to climate change often bring to notice images of infrastructural interventions. In this regard, Biagini, Bierbaum, and Stults (2014) have cited the example of a survey of two projects funded by the global environmental facility that found two times as much references to adaptation via physical infrastructure than to green infrastructure. Physical infrastructure is usually not cost-effective and it is adhered to with the objective of restoration or surmounting the limitations emanating from the scarcity of natural resources of one or different types (Rahman, 2014). Undeniably, EbA approaches afford cost-effective, pliable, and widely applicable adaptation alternatives (Ojea, 2015; Baig, Rizvi, Josella, & Palanca-Tan, 2016; GIZ, 2017). It seems sensible enough that if mangroves are safeguarded or restored dikes may not be essential (Martin & Watson, 2016; Fling, Elias, Faires, & Smith, 2018), and that if no deforestation occurs on hillsides (Boucher et al., 2014), there may not be any need for disputable work to be carried out (Surian, 2016). Besides, measures solely pertaining to physical infrastructure entail the potential of being confronted with obstructions in terms of long-term feasibility, efficacy or environmental soundness (Shaw, Overpeck, & Midgley, 2014; Wilbanks & Fernandez, 2014), as Lemos, Lob, Nelson, Eakin, and Bedran-Martinis (2016) cite the example of river transportation to avoid drought. Green infrastructure can be perceived and treated as a complement to development as well as man-made infrastructure (Silva & Wheeler, 2017).

Some scholarly attention has been focused on the argument that better-informed adaptive transition processes require the integration between ES and STS research (Schäpke, Omann, Wittmayer, van Steenbergen, & Mock, 2016; Loorbach, 2017). Adaptive transition is deemed as a precondition to sustainability transition in systems susceptible to climate change (EEA, 2018; Kvan & Karakiewicz, 2019). Interlinkages do exist between development and adaptation (Fankhauser, 2016) but they are not synonyms (Church & Hammill, 2019). Pursuit of sustainable development policies aim at alleviating poverty by promoting economic growth, addressing inequality through redistribution of wealth, and preventing environmental degradation via sustainable resource use and harnessing adaptation policies for addressing vulnerability and risks (Roy et al., 2018). The terminology "adaptive development" coined by Agrawal and Lemos (2015) alludes to a type of or a step in the development process that either averts or minimizes risks without casting a negative effect on human systems and natural systems, and this regard Scarano (2017) argues that the process of adaptive transition (as advocated by Pant et al., 2015) rouses adaptive development (as argued by Agrawal & Lemos, 2015) that serves as a "key step" to nurture transition from a conventional to a sustainable development paradigm (Scarano, 2017). In response to the premise that the adaptation process in question is sustainable, Brown (2011) while cautioning that not all adaptation is sustainable still coins the term "sustainable adaptation" that has seemingly found support from some other scholars as well (Eriksen et al., 2011). This notion of sustainable adaptation is termed "possibly a synonym of EbA" (Scarano, 2017). This description delineates EbA as a policy mix and as a subject in interdisciplinary science within the SES research agenda and policy arena.

Implementation of EbA policies and strategies is subject to geographical locations, whether global or local or urban or rural (Lemos et al., 2016). Similarly, scholarly literature on EbA is also somewhat divided between global and local focus, and at local level its dichotomy is in accordance with geographic setting (Scarano, 2017). A review of 132 research papers on EbA by Munroe et al. (2012) revealed that 45% were from developing countries and bulk of these studies focused on themes like urban or rural, wetlands, forests and coastal ecosystems. Another analytical review conducted by Brink et al. (2016) encompassing 110 papers in 112 cities made it discernible a strong tilt toward Europe and North America. Focus on the importance of EbA for costal vegetation, including mangroves, both in continental and small island nations has been emphasized (Martin & Watson, 2016). EbA strategies and practices entail the potential for small farmers in terms of improving the capacity of crops and livestock to uphold high crop yields and/or protect from biophysical effects of extreme weather events or increased temperatures and other vagaries of climate change (Vignola et al., 2015; Harveya et al., 2017).

Narrowing down the spatial gap between rural and urban areas, hilly mountainous regions and terrains through connectivity not only spurs sustainability at a broader spatial scale (Morton, Solecki, Dasgupta, Dodman, & Rivera-Ferre, 2014) but also envisages bridging up communication gap and enhances interface between humans and nature as well, thereby opening up Markiewicz new vistas of research to the science-policy interaction (Liu et al., 2015, 2016). Policies with regard to science range from global to local approaches. Global approaches comprise international agreements like the UNFCCC and Paris Agreement, national adaptation plans, municipal strategies for EbA, and local governance mechanism at smaller places (Scarano, 2017). Viewed in broad perspective, worldwide agreements pertaining to climate change cannot be immediately integrated into or made part of the national or subnational policies related to climate change and in the same manner local EbA strategies don't always "scale up" beyond municipal limits or a given community (Scarano, 2017). Arguing from an international perspective Ojea (2015) notes that irrespective of recent enthusiasm with EbA, there still prevail multiple constraints in mainstreaming EbA into global climate policy framework, specifically in the realms of governance, efficacy, timescale of processes, finances and scientific uncertainty. With regard to integrating EbA at national level, a review by Pramova et al. (2012) revealed that only 22% of the National Adaptation Plans of 44 developing countries included ecosystem components. Harnessing ecosystems at subnational level to help people adapt to climate change is "limited partially" by paucity of information about where the ecosystems entail the "highest potential to do so" (Scarano, 2017), and in this regard Bourne et al. (2016) suggest that that gap can be narrowed down by spatial prioritization efforts.

Interestingly, in the regions or areas where EbA policies are operational, they are characterized by variable approaches. Wamsler and Pauleit (2016), citing the examples of Sweden and Germany, note that in Swedish municipalities adaptation policies are mainstreamed via an ecosystem services perspective whereas in German municipalities the focus is on climate mitigation. Furthermore, EbA approaches to climate change are not structured or systematized or branded as such either in Germany or Sweden (Wamsler & Pauleit, 2016) or in many other places across the globe (Munroe et al., 2012). Admittedly, EbA is eliciting increased recognition and it becomes discernible from *meta*-analyses case studies the demonstrated efficacy and cost-effectiveness of EbA interventions (Reid et al., 2019); nonetheless, EbA is also concurrently confronted with recognized challenges. In the first place, EbA is faced with the identification of limits and thresholds beyond which might not hand out adaptation benefits and the extent ecosystems can provide ecosystem services under a changing climate (Roberts et al., 2012; Nalau, Becken, & Mackey, 2018). Secondly, the prevailing confusion with regard to the meaning of EbA has given rise to a range of diverse methodologies used for assessments, and the absence of coherent and relative quantitative measures of EbA success and failure makes it cumbersome to take up the case for EbA on socio-economic terms (Nalau et al., 2018). Besides, heavy dependence of EbA research on Western scientific knowledge has resulted in ignoring due focus on local and traditional knowledge with regard to coping with climate change (Doswald et al., 2014). Multisectoral nature of the EbA policy—involving both the sectors that manage ecosystems and those that benefit from ecosystem services—pose immense challenges of governance and planning (Nalau et al., 2018). Equally problematic are broad macroeconomic considerations like economic development, poverty, and access to financial resources required for implementing climate adaptation options and these together contribute to obstacles that prevent uptake of EbA at a larger scale (Nalau et al., 2018). Another contributory factor obstructing EbA implementation is differing perceptions of risks and cultural preferences (Nalau et al., 2018) and this can make prospective stakeholders entertain negative perceptions about some specific types of EbA strategies or policies (Doswald & Osti, 2011).

4.8 Conclusion

Climate change, widely recognized as a threat-multiplier and existential threat, is taking a heavy toll on humankind and ecosystems at local, national, regional and global scales and this is attested to by the plethora of reports brought out from time to time by the IPCC and outcomes of the 25 COPs held thus far under the aegis of the UNFCCC and almost identical reports made public by other international organizations dealing with issues related to climate change. Both scientific community and academic community have been engaged in in-depth exploration of the causes and outcomes of climate change impacts and concurrently devising strategies, plans and programs to tackle the adverse impacts of climate change. starting with mitigation and adaptation approaches, international community has also applied geoengineering approach comprising SRM and CDR approach and there is also growing support for loss and damage associated with climate change approach. Nevertheless, all these approaches are under implementation and have yielded some satisfactory outcomes to some extent. A combination of these approaches is deemed appropriate in coping with the adverse impacts of climate change. Admittedly, no single approach can provide a complete solution to cope with climate change impacts and each approach suffers from some type of lacuna either of governance or technology or finance or capacity building. Frankly, there is no single "silver bullet" solution to cope with the vagaries of climate change and a semblance of success can be said to be achieved by implementing an array of strategies aimed at the same target or goal.

Ironically, political response seldom matches the scientific suggestions/recommendations in terms of the priority, urgency and participation, especially at the global scale. Developed countries' reluctance to part with financial and technology corresponding to the needs of the developing countries in dealing with climate change impacts in the international negotiations have often contributed to neglect of concrete actions. Widespread awareness of the significance of climate change and its inevitable consequences under the "BAU" backdrop has increasingly led to a degree of compliance with measures aiming at reduction of GHG emissions and mitigation of climate change impacts. Strong political will is required at the global level, with developed countries agreeing to render financial and technological support to the needy developing countries to fight climate change.

References

ActionAid, CARE International, Germanwatch, & WWF. (2012). Into unknown territory: The limits to adaptation and reality of loss and damage from climate impacts.

Adame, M. F., Santini, N. S., Tovilla, C., Vázquez-Lule, A., Castro, L., & Guevara, M. (2015). Carbon stocks and soil sequestration rates of tropical riverine wetlands. *Bio-geosciences*, *12*, 3805–3818.

Adger, W. N., Huq, S., Brown, K., Conway, D., & Hulme, M. (2003). Adaptation to climate change in the developing world. *Progress in Development Studies*, *3*(3), 179–195.

Adger, W. N., Arnell, N. W., & Tompkins, E. L. (2005). Successful adaptation to climate change across scales. *Global Environmental Change*, *15*(2), 77–86.

Adger, W. N., Agrawal, S., Mirza, M., Conde, C., O'Brien, K., Pulhin, J., ... Takahashi, K. (2007). *Assessment of adaptation practices, options, constraints and capacity*. Climate change 2007: Impacts, adaptation and vulnerability. *Contribution of Working Group II to the Fourth Assessment Report of the Intergovernmental Panel on Climate Change* (pp. 717−743). Cambridge: Cambridge University Press.

Adger, W. N., Dessai, S., Goulden, M., Hulme, M., Lorenzoni, I., Nelson, D. R., ... Wreford, A. (2009a). Are there social limits to adaptation to climate change? *Climatic Change, 93*, 335−354.

Adger, W. N., Lorenzoni, I., & O'Brien, K. (Eds.), (2009b). *Adapting to climate change: Thresholds, values, governance*. Cambridge: Cambridge University Press.

Adger, W. N., Barnett, J., Brown, K., Marshall, N., & O'Brien, K. (2013). Cultural dimensions of climate change impacts and adaptation. *Nature Climate Change, 3*, 112−117.

Agrawal, A., & Lemos, M. C. (2015). Adaptive development. *Nature Climate Change, 5*, 186−187.

Ajayi, T., Gomes, J. S., & Bera, A. (2019). A review of CO_2 storage in geological formations emphasizing modeling, monitoring and capacity estimation approaches. *Petroleum Science, 16*, 1028−1063.

Aldy, J. E., & Stavins, R. N. (2010a). Introduction. In J. E. Aldy, & R. N. Stavins (Eds.), *Post-Kyoto international climate policy: Implementing architectures for agreement* (pp. 1−28). Cambridge: Cambridge University Press.

Aldy, J., & Stavins, R. (2007). *Architectures for agreement: Addressing global climate change in the post-Kyoto world*. New York: Cambridge University Press.

Alessi, M. (2015). *History of the UN climate negotiations—Part 2—From 2011 to the present. Climate Policy Info Hub*.

Alloisio, I., & Borghesi, S. (2019). Climate change mitigation. In L. F. Filho, A. M. Azul, L. Brandi, P. G. Ozuyar, & T. Wall (Eds.), *Climate action* (pp. 213−224). Champ, Switzerland: Springer.

Allwood, J. M., Bosetti, V., Dubash, N. K., Gómez-Echeverri, L., & von Stechow, C. (2014). *Glossary*. Climate change 2014: Mitigation of climate change. *Contribution of Working Group III to the Fifth Assessment Report of the Intergovernmental Panel on Climate Change*. Cambridge: Cambridge University Press.

Alpert, S. B., Spencer, D. F., & Hidy, G. (1992). Biospheric options for mitigating atmospheric carbon dioxide levels. *Energy Conversion and Management, 33*(5−8), 729−736.

Amburgey, J. W., & Thoman, D. B. (2011). Dimensionality of the New Ecological Paradigm: Issues of Factor Structure and Measurement. *Environment and Behavior, 20*(10), 1−22.

AMS (American Meteorological Society). *Policy statement on geoengineering the climate system*. (2009).

Amundson, R., & Biardeau, L. (2018). Soil carbon sequestration is an elusive climate mitigation tool. *Proceedings of the National Academy of Sciences of USA (PNAS), 115*(46), 11652−11656.

Andersen, I. (2019). Foreword. In: *Emissions gap report 2019*. Nairobi: United Nations Environment Programme.

Andreoni, J., & Samuelson, L. (2006). Building rational cooperation. *Journal of Economic Theory, 127*, 117, −154.

Angel, R. (2006). Feasibility of cooling the earth with a cloud of small spacecraft near the inner Lagrange point (L1). *Proceedings of the National Academy of Sciences, 103*(6), 17184−17189.

Arnott, J. C., Moser, S. C., & Goodrich, A. (2016). Evaluation that counts: A review of climate change adaptation indicators & metrics using lessons from effective evaluation and science-practice interaction. *Environmental Science & Policy, 66*, 383−392.

Asiimwe, M. M., & de Kock, I. H. (2019). Analysis of the extent to which industry 4.0 has been considered in sustainability or socio-technical transitions. *South African Journal of Industrial Engineering, 30*(3), 41−51.

Aumont, O., & Bopp, L. (2006). Globalizing results from ocean in situ iron fertilization studies. *Global Biogeochemical Cycles, 20*(2).

d'Auvergne, C., & Nummelin, M. (2017). Capacity- building (Article 11). In K. Daniele, C. Maria Pia, D. Meinhard, B. Jane, & H. Andrew (Eds.), *The Paris Agreement on climate change: Analysis and commentary* (pp. 277−291). Oxford: Oxford University Press.

Bach, T. (2016). Human rights in a climate changed world: The impact of COP 21, nationally determined contributions, and national courts. *Vermont Law Review*, *40*, 561−595.

Bäckstrand, K., & Lövbrand, E. (2006). Planting trees to mitigate climate change: Contested discourses of ecological modernization, green governmentality and civic environmentalism. *Global Environmental Politics*, *6*(1), 50−75.

Baig, S. P., Rizvi, A., Josella, M., & Palanca-Tan, R. (2016). *Cost and benefits of ecosystem based adaptation: The case of the Philippines*. Gland: IUCN.

Ban-Weiss, G. A., & Caldeira, K. (2010). Geoengineering as an optimization problem. *Environmental Research Letters*, *5*(3), 034009.

Barakat, S., Abeysinghe, A., Dagnet, Y., Jember, G., Jallow, B. P., More, C. H., ... Verkuijl, C. (2017). *A guide to transparency under the UNFCCC and the Paris Agreement*. London: IIED Available at:. Available from https://pubs.iied.org/pdfs/10190IIED.pdf.

Barnett, J., & O'Neill, S. (2010). *Maladaptation. Global Environmental Change*, *20*(2), 211−213.

Barnett, J. (2009). Human rights and vulnerability to climate change. In H. S. (Ed.), *Human rights and climate change* (pp. 257−271). Cambridge: Cambridge University Press.

Barrett, S. (2008). The incredible economics of geoengineering. *Environmental and Resource Economics*, *39*, 45−54.

Baum, S. D. (2014). The great downside dilemma for risky emerging technologies. *Physica Scripta*, *89*(12), 128004.

Beeler, C. (2019). Why 2020 is a key year for climate action. *Pri.org*, September 17.

Belcher, B., Suryadarma, D., & Halimanjaya, A. (2017). Evaluating policy-relevant research: Lessons from a series of theory-based outcomes assessments. *Palgrave Communications*, *3*, 17017.

Bellante, L. (2017). Building the local food movement in Chiapas, Mexico: Rationales, benefits, and limitations. *Agriculture and Human Values*, *34*(1), 119−134.

Benanti, G., Saunders, M., Tobin, B., & Osborne, B. (2014). Contrasting impacts of afforestation on nitrous oxide and methane emissions. *Agricultural and forest meteorology*, *198*, 82−93.

Benson, S. M., Bennaceur, K., Cook, P., Davison, J., de Connick, H., Farhat, K., ... Wright, I. (2012). *Carbon Capture and Storage. Global energy assessment-toward a sustainable future* (pp. 993−1068). Cambridge: Cambridge University Press.

Bertram, C. (2010). Ocean iron fertilization in the context of the Kyoto protocol and the post-Kyoto process. *Energy Policy*, *38*(2), 1130.

Bhandari, M. P. (2018). The role of international organization in addressing the climate change issues and creation of intergovernmental panel on climate change (IPCC). *Advances in Agriculture and Environmental Sciences*, *1*(1), 19−34.

Biagini, B., Bierbaum, R., & Stults, M. (2014). A typology of adaptation actions: A global look at climate adaptation actions financed through the global environment facility. *Global Environmental Change*, *25*, 97−108.

Bickerstaff, K. (2004). Risk perception research: Socio-cultural perspectives on the public experience of air pollution. *Environment International*, *30*, 827−840.

Biermann, F., & Boas, I. (2017). Towards a global governance system to protect climate migrants: Taking stock. In B. Mayer, & F. Crepeau (Eds.), *Research handbook on climate change, migration and the law* (pp. 405−419). Cheltenham: Edward Elgar Publishing.

Biesbroek, G. R., Klostermann, J. E. M., Termeer, C. J. A. M., & Kabat, P. (2013). On the nature of barriers to climate change adaptation. *Regional Environmental Change*, *13*, 1119−1129.

Bodansky, D. (2011). Governing climate engineering: Scenarios for analysis. *The Harvard Project on Climate Agreements*. Harvard Kennedy School.

Bodansky, D. (2016). The Paris climate change agreement: A new hope? *The American Journal of International Law*, *110*(2), 288−319.

Bodle, R., Homan, G., Schiele, S., & Tedsen, E. (2012). The regulatory framework for climate-related geoengineering relevant to the convention on biological diversity. In: *Part II of: Geoengineering in relation to the convention on biological diversity: Technical and regulatory matters*. In: Technical series no. 66. Montreal: Secretariat of the Convention on Biological Diversity.

Bodansky, D. (1996). May we engineer the climate? *Climatic Change*, *33*, 309−321.

BPC (Bipartisan Policy Centre Task Force on Climate Remediation Research). *Geoengineering: A national strategic plan for research on the potential effectiveness, feasibility, and consequences of climate remediation technologies*. (2011).

Bosetti, V., Carraro, C., Massetti, E., & Tavoni, M. (Eds.), (2014). *Climate change mitigation, technological innovation and adaptation: A new perspective on climate policy*. Cheltenham: Edward Elgar Publishing.

Boucher, O., Forster, P., Gruber, N., Ha-Duong, M., Lawrence, M., Lenton, T., ... Vaughan, N. (2013). Rethinking climate engineering categorisation in the context of climate change mitigation and adaptation. *WIREs Climate Change*, *5*, 23−35.

Bourne, A., Holness, S., Holden, P., Scorgie, S., Donatti, S. I., & Midgley, G. (2016). A socio-ecological approach for identifying and contextualising spatial ecosystem-based adaptation priorities at the sub-national level. *PLoS One*, *11*(5), e0155235.

Boyd, P. W., & Law, C. S. (2001). The Southern Ocean iron release experiment (SOIREE): Introduction and summary: Deep Sea research part II. *Topical Studies in Oceanography*, *48*(11−12), 2425−2438.

Boyd, W. (2010). Climate change, fragmentation, and the challenges of global environmental law: Elements of a post-Copenhagen assemblage. *University of Pennsylvania Journal of International Law*, *32*(2), 457−550.

Boyd, E., James, R. A., Jones, R. G., Young, H. R., & Otto, F. (2017). A typology of loss and damage perspectives. *Nature Climate Change*, *7*, 723−729.

Branson, M. C. (2014). Comment, a green herring: How current ocean fertilization regulation distracts from geoengineering research. *SantaClara Law Review*, *54*(1), 163−200.

Bradley, R., & Steele, K. (2015). Making climate decisions. *Philosophy Compass*, *10*(11), 799−810.

Brand, S. (2010). *Whole earth discipline: Why dense cities, nuclear power, transgenic crops, restored wildlands and geoengineering are necessary*. New York: Penguin.

Brent, K., McGee, J., McDonald, J., & Rohling, E. J. (2018). International law poses problems for negative emissions research. *Nature Climate Change*, *8*(6), 451−453.

Bread for the World, EED—Church Development Service, & DanChurchAid. *Work programme on loss and damage: Setting up an appropriate framework for identifying, prioritizing and targeting people most vulnerable towards climate change induced loss and damage*. (2012). Available at http://unfccc.int/resource/docs/2011/smsn/ngo/246.pdf.

Brick, S. (2010). *Biochar: Assessing the promise and risk to guide United States policy*. Washington, DC: National Resource Defense Council (NRDC).

Brink, E., Theodor, A. J., Ádám, D., Feller, R., Henselek, Y., Hoffmann, A., ... Wamsler, C. (2016). Cascades of green: A review of ecosystem-based adaptation in urban areas. *Global Environmental Change*, *36*, 111−123.

Briner, G., Kato, T., Konrad, S., & Hood, C. (2014). *Taking stock of the UNFCCC process and its interlinkages*. Climate Change Expert Group.

Brody, S. D., Zahran, S., Vedlitz, A., & Grover, H. (2008). Examining the relationship between physical vulnerability and public perceptions of global climate change in the United States. *Environment and Behavior*, *40*(1), 72−95.

Bronfman, N. C., Cifuentes, L. A., & Gutierrez, V. V. (2008). Participant-focused analysis: Explanatory power of the classic psychometric paradigm in risk perception. *Journal of Risk Analysis, 11*(6), 735–753.

Brousseau, E., Dedeurwaerdere, T., Jouvet, P.-A., Willinger, M., & Willinger, M. (Eds.), (2012). *Global environmental commons: Analytical and political challenges in building governance mechanisms*. Oxford: Oxford University Press.

Brown, K. (2011). Sustainable adaptation: An oxymoron? *Climate and Development, 3*, 21–31.

Bubeck, P., Botzen, W. J. W., & Aerts, J. C. J. H. (2012). A review of risk perceptions and other factors that influence flood mitigation behavior. *Risk Analysis, 32*(9), 1481–1495.

Buchholz, W., Cornes, R., & Rübbelke, D. (2012). Matching as a cure for under provision of voluntary public good supply. *Economics Letters, 117*, 727–729.

Buchner, B., Herve-Mignucci, M., Trabacchi, C., Wilkinson, J., Stadelmann, M., Boyd, R., ... Micale, V. (2013). *The global landscape of climate finance 2013. CPI Report*. Venice: Climate Policy Initiative.

Budyko, M. I. (1977a). *Climatic changes*. Washington, DC: American Geophysical Union.

Budyko, M. I. (1977b). On present-day climatic changes. *Tellus, 29*(3), 193–204.

Burke, M., Hsiang, S. M., & Miguel, E. (2015). Global non-linear effect of temperature on economic production. *Nature, 527*, 235–239.

Burton, I. (1996). The growth of adaptation capacity: Practice and policy. In J. B. Smith, N. Bhatti, G. V. Menzhulin, R. Benioff, M. Campos, B. Jallow, & ... R. K. Dixon (Eds.), *Adapting to climate change* (pp. 55–67). New York: Springer.

Burton, I., Huq, S., Lim, B., Pilifosova, O., & Schipper, E. L. (2002). From impacts assessment to adaptation priorities: The shaping of adaptation policy. *Climate Policy, 2*(2–3), 145–159.

C2G2 (Carnegie Climate Geoengineering Governance Initiative). (2019). *Geoengineering: The need for governance*. New York: C2G2.

Cameron, F. (2011). Climate change as a complex phenomenon and the problem of cultural governance. *Museum and Society, 9*(2), 84–89.

Cascio, J. *Hacking the Earth: Understanding the consequences of geoengineering*. (2009).

CBD (Convention on Biological Biodiversity). (2009). Scientific synthesis of the impacts of ocean fertilization on marine biodiversity. In: *Montreal, technical series no. 45*. Montreal: CBD Secretariat.

CBD (Convention on Biological Diversity). (2016). *Climate-related geoengineering*. UNEP/CBD/COP/13/L.4.

CBD (Convention on Biological Diversity). (2017). *COP10 Mandate*. Available at https://www.cbd.int.cop/, 27 March 2017.

CC& ES (Centre for Climate and Energy Solutions). (2013). *Outcomes of the U.N. climate change conference in Warsaw*. Available at https://www.C2es.org/documents/outcomes-of-the-u-n-climate-change-conference-in-warsaw/.

CDKN. *FEATURE: Loss and damage—From defining to understanding to action*. (2012).

Chen, A., & Murthy, V. (2019). Should we be more optimistic about fighting climate change? *Harvard Business Review Online*.

Chhetri, N., Chong, D., Conca, K., Falk, R., Gillespie, A., Gupta, A., ... Nicholson, S. (2018). Governing solar radiation management. Washington, DC: Forum for Climate Engineering Assessment, American University.

Chris, R. (2016). Systems *thinking for geoengineering policy*: How to *reduce the threat of dangerous climate change by embracing uncertainty and failure*. Routledge.

Christiansen, L., Martinez, G., & Naswa, P. (2018). *Adaptation metrics: Perspectives on measuring, aggregating and comparing adaptation results*. Copenhagen: UNEP-DTU Partnership.

Chung, I. K., Oak, J. H., Lee, J. A., Shin, J. A., Kim, J. G., & Park, K. S. (2012). Installing kelp forests/seaweed beds for mitigation and adaptation against global warming: Korean Project Overview. *ICES Journal of Marine Science: Journal du Conseil*, fss206.

Church, C., & Hammill, A. (2019). *Defining adaptation—And distinguishing it from other development investments*. Manitoba: IISD.

Chynoweth, D. P. (2002). *Review of biomethane from marine biomass*. Gainsville, FL: University of Florida.

Ciais, P., Sabine, C., Bala, G., Bopp, L., Brovkin, V., Canadell, J., ... Thornton, P. (2013). *Carbon and other biogeochemical cycles*. Climate change 2013: The physical science basis. *Contribution of Working Group I to the Fifth Assessment Report of the Intergovernmental Panel on Climate Change* (pp. 465–570). Cambridge: Cambridge University Press.

Cicerone, R. J., Elliott, S., & Turco, R. P. (1992). Global environmental engineering. *Nature, 356*, 472.

Clarke, L., Jiang, K., Akimoto, K., Babiker, M., Blanford, G., Fisher-Vanden, K., ... van Vuuren, D. P. (2014). *Assessing transformation pathways*. Climate change 2014: Mitigation of climate change. *Contribution of Working Group III to the Fifth Assessment Report of the Intergovernmental Panel on Climate Change* (pp. 413–510). Cambridge: Cambridge University Press.

Climate Action Tracker. (2011, 29 November). *Climate action tracker update: Little progress—Countries still heading for over 3°C warming*, Press Release.

Colding, J., & Barthel, S. (2019). Exploring the social-ecological systems discourse 20 years later. *Ecology and Society, 24*(1), 2.

Consoli, C. (2019). *Bioenergy and carbon capture and storage*. Melbourne: Global CCS Institute.

Creutzig, F. (2016). Economic and ecological views on climate change mitigation with bioenergy and negative emissions. *GCB Bioenergy, 8*, 4–10.

Creutzig, F., Ravindranath, N. H., Berndes, G., Bolwig, S., Bright, R., Cherubini, F., & Masera, O. (2015). Bioenergy and climate change mitigation: An assessment. *GCB Bioenergy, 7*, 5.

CRI (Climate Risk Index). (2019). *Global climate risk index 2020*. Bonn: Germanwatch e.V.

CSTCAAAS (Centre for Science Technology and Congress at the American Associate for the Advancement of Science). (2010). Congress examines geoengineering of climate system. *Issues in Science and Technology, 26*(2), 21–22.

Christensen, J., & Olhoff, A. (2019). *Lessons from a decade of emissions gap assessments*. Nairobi: United Nations Environment Programme

de Coninck, H., & Sagar, A. (2017). Technology development and transfer (Article 10). In K. Daniele, C. Maria Pia, D. Meinhard, B. Jane, & H. Andrew (Eds.), *The Paris Agreement on climate change: Analysis and commentary* (pp. 258–276). Oxford: Oxford University Press.

Craft, B., & Fisher, S. (2016). Measuring effective and adequate adaptation. *Issue Paper*. London: International Institute for Environment and Development (IIED).

Cressey, D. (2012). Cancelled project spurs debate over geoengineering patents. *Nature News, 485*(7399), 429.

Dagnet, Y., & Levin, K. (2017). Transparency (Article 13). In K. Daniele, C. Maria Pia, D. Meinhard, B. Jane, & H. Andrew (Eds.), *The Paris Agreement on climate change: Analysis and commentary* (pp. 301–318). Oxford: Oxford University Press.

Dagnet, Y., & Northrop, E. (2017). Facilitating implementation and promoting compliance (Article 15). In K. Daniele, C. Maria Pia, D. Meinhard, B. Jane, & H. Andrew (Eds.), *The Paris Agreement on climate change: Analysis and commentary* (pp. 338–351). Oxford: Oxford University Press.

Dagnet, Y., Nathan, C., Bird, N., Bouye, M., & Rocha, M. (2019a). Building capacity for the Paris Agreement's enhanced transparency framework: What can we learn from countries' experiences and UNFCCC process? *Working Paper*. Washington, DC: World Resource Institute.

Dagnet, Y., Waskow, D., Bergen, M., Levin, K., Leprince-Ringuet, N., Thwaites, J., ...Worker, J. (2019b). *COP25: What we needed, what we got, what's next*. World Resource Institute.

de Castro, F., Hogenboom, B., & Baud, M. (2016). Introduction: Environment and society in contemporary Latin America. In B. de Castro, & M. Baud Hogenboom (Eds.), *Environmental governance in Latin America* (pp. 1–25). Hampshire: Palgrave MacMillan.

De Coninck, H., & Benson, S. (2014). Carbon dioxide capture and storage: Issues and prospects. *Annual Review of Environmental Resources 2014*, *39*, 243−270.

De Groot, J. I. M., Steg, L., & Poortinga, W. (2013). Values, perceived risks and benefits, and acceptability of nuclear energy. *Risk Analysis*, *33*, 307−317.

Denton, F., Wilbanks, T. J., Abeysinghe, A. C., Burton, I., Gao, Q., Lemos, M. C., ... Warner, K. (2014). *Climate-resilient pathways: Adaptation, mitigation, and sustainable development*. Climate change2014: Impacts, adaptation, and vulnerability. Part A: Global and sectoral aspects. *Contribution of Working Group II to the Fifth Assessment Report of the Intergovernmental Panel on Climate Change* (pp. 1101−1131). Cambridge: Cambridge University Press.

Deudney, D. H., & Grove, J. (2009). *Geoengineering and world order: The emerging debate. Toronto.*

Dierwechter, Y., & Wessells, A. T. (2013). The Uneven localisation of climate action in metropolitan Seattle. *Urban Studies*, *50*(7), 1368−1385.

Doelle, M. (2014). The birth of the Warsaw loss & damage mechanism. *Carbon & Climate Law Review*, *8*(1), 35−45.

Doelle, M. (2017). Assessment of strengths and weaknesses. In K. Daniele, C. Maria Pia, D. Meinhard, B. Jane, & H. Andrew (Eds.), *The Paris Agreement on climate change: Analysis and commentary* (pp. 375−388). Oxford: Oxford University Press.

Dooling, S., & Simon, G. (2012). *Cities, nature, and development: The politics and production of urban vulnerabilities*. Vermont: Ashgate Publishing Company.

Doswald, N., Munroe, R., Roe, D., Giuliani, A., Castelli, I., Stephens, J., ... Vira, H. (2014). Effectiveness of ecosystem-based approaches for adaptation: Review of the evidence base. *Climate and Development*, *6*(2), 185−201.

Dow, K., Berkhout, F., Preston, B. L., Klein, R. J. T., Midgely, G., & Shaw, M. R. (2013a). Limits to adaptation. *Nature Climate Change*, *3*, 305−307.

Doswald, N., & Osti, M. (2011). Ecosystem-based approaches to adaptation and mitigation: Good practice examples and lessons learned in Europe. BfN, Federal Agency for Nature Conservation.

Dow, K., Berkhout, F., & Preston, B. L. (2013b). Limits to adaptation to climate change: A risk approach. Current opinion in environmental. *Sustainability*, *5*, 384−391.

Duguma, L. A., Minang, P. A., & van Noordwijk, M. (2014a). Climate change mitigation and adaptation in the land use sector: From complementarity to synergy. *Environmental Management*, *54*(3), 420−432.

Duguma, L. A., Wambugu, S. W., Minang, P. A., & van Noordwijk, M. (2014b). A systematic analysis of enabling conditions for synergy between climate change mitigation and adaptation measures in developing countries. *Environmental Science & Policy*, *42*, 138−148.

Dunlap, R. E., Liere, K. D. V., Mertig, A. G., & Jones, R. E. (2000). Measuring Endorsement of the New Ecological Paradigm: A Revised NEP Scale. *Journal of Social Issues*, *56*(3), 425−442.

Dykema, J. A., Keith, D. W., Anderson, J. G., & Weisenstein, D. (2014). Stratospheric controlled perturbation experiment: A small-scale experiment to improve understanding of the risks of solar geoengineering. *Philosophical Transactions of the Royal Society A: Mathematical, Physical and Engineering Sciences*, *372* (2031), 20140059.

Eckersley, R. (2012). Moving forward in the climate negotiations: Multilateralism or minilateralism? *Global Environmental Politics'*, *12*, 24, −42.

Edenhofer, O., Pichs-Madruga, E. R., Sokona, Y., Minx, J. C., Farahani, E., Kadner, S., Seyboth, K., et al. (Eds.), (2015). Climate change 2014: Mitigation of climate change; *Summary for Policymakers Technical Summary; Part of the Working Group III Contribution to the Fifth Assessment Report of the Intergovernmental Panel on Climate Change*. Geneva: Intergovernmental Panel on Climate Change.

EEA (European Environment Agency). (2018). *Perspectives on transition to sustainability*. Luxemburg: Publications Office of the European Union.

Eisenack, K., Moser, S. C., Hoffmann, E., Klein, R. J. T., Oberlack, C., Pechan, A., ... Termeer, C. J. A. M. (2014). *Explaining and overcoming barriers to climate change adaptation*. Nature Climate Change, 4, 867−872.

Engel, K. H., & Orbach, B. Y. (2008). Micro motives for state and local climate change initiatives. *Harvard Law and Policy Review*, 2, 119−137.

Engle, N. L. (2011). Adaptive capacity and its assessment. *Global Environmental Change*, 21(2), 647−656.

Drummond, D., Caranci, B., & Tulk, D. (2007). *Market-based solutions to protect the environment*. TD economic special report.

Erbach, G. (2015). *Negotiating a new UN climate agreement: Challenges on road to Paris*. Brussels: European Parliament.

Eriksen, S., Aldunce, P., Bahinipati, C. S., D'Almeida, Martins, R., Molefe, J. I., Nhemachena, C., ... Ulsrud, K. (2011). When not every response to climate change is a good one: Identifying principles for sustainable adaptation. *Climate and Development*, 3(1), 7−20.

Espinosa, P. (2019). *Foreword. Yearbook of global climate action*. Bonn: UN Climate Change Secretariat.

ETC Group (Action Group on Erosion, Technology and Conservation). (2010). *The geoengineering moratorium*. Montreal: ETC Group.

EPA (United States Environmental Protection Agency). *Inventory of United States greenhouse gas and sinks 1990−2015*. (2017).

ETC Group (Action Group on Erosion, Technology and Conservation). (2017). *What is wrong with solar radiation management?* ETC Group Briefing.

Falkner, R. (2015). A minilateral solution for climate change? On bargaining efficiency, club benefits and international legitimacy. Working paper of London School of Economics and Political Science.

Fankhauser, S. (2016). *Adaptation to climate change. Working paper no. 287*. London: London School of Economics.

Felgenhauer, T., & Webster, M. (2013). Multiple adaptation types with mitigation: A framework for policy analysis. *Global Environmental Change*, 23(6), 1556−1565.

Filho, W. L. (Ed.), (2015). *Climate change in the Asia-Pacific region*. Cham: Springer International Publishing.

Filho, W. L., Kalangu, B. S. J., Munishi, M. W. P., & Musiviwa (Eds.), (2017). *Climate change adaptation in Africa fostering resilience and capacity to adapt*. Cham: Springer International Publishing.

Filho, W. L., & Keenan, J. M. (Eds.), (2017). *Climate change adaptation in North America fostering resilience and the regional capacity to adapt*. Cham: Springer International Publishing.

Filho, W. L., & de Freitas, L. E. (Eds.), (2018). *Climate change adaptation in Latin America managing vulnerability, fostering resilience*. Cham: Springer International Publishing.

Filho, W. L., Trbić, G., & Filipovic, D. (Eds.), (2019). *Climate change adaptation in Eastern Europe— Managing risks and building resilience to climate change*. Cham: Springer International Publishing.

Folger, P. (2013). *Carbon capture and sequestration (CSS): A primer*. Washington, DC: Congressional Research Service.

FAO (Food and Agriculture Organization). (2015). *Disaster risk management and climate change adaptation in the CARICOM and Caribbean region: Strategy and action plan*. Rome: FAO.

FAO (Food and Agriculture Organization). (2018). Using impact evaluation to improve policy-making for climate change adaptation in the agriculture sector. *Briefing Note*. Rome: FAO.

Friedrich, J. (2017). Global Stocktake (Article 14). In K. Daniele, C. Maria Pia, D. Meinhard, B. Jane, & H. Andrew (Eds.), *The Paris Agreement on climate change: Analysis and commentary* (pp. 319−337). Oxford: Oxford University Press.

Fu, C., Zheng, Y., & Wang, W. (2014). Research perspectives on synergic relationships in addressing climate change measures. *Resources Science*, 36(7), 1535−1542.

Fuss, S., Canadell, J. G., Peters, G. P., Tavoni, M., Andrew, R. M., Ciais, P., ... Yamagata, Y. (2014). Betting on negative emissions. *Nature Climate Change, 4*(10), 850−853.

Fuss, S., Jones, C. D., Kraxner, E., Peters, G. P., Smith, P., Tavoni, M., ... Yarnagata, Y. (2016). Research priorities for negative emissions. *Environment Research Letter, 11*, 115007.

Fussel, H. M. (2007). Adaptation planning for climate change: Concepts, assessment approaches, and key lessons. *Sustainability Science, 2*, 265−275.

Fling, D., Elias, P., Faires, J., & Smith, S. (2018). *deforestation success stories: Tropical nations where forest protection and reforestation policies have worked*. Union of Concerned Scientists.

GCF (Green Climate Fund). *Background*. (2014). Available at http://www.gcfund.org/about/thefund.html.

GEF (Global Environmental Fund). (2016). *Monitoring and evaluation of climate change adaptation*. Washington, DC.: 51st GEF Council Meeting.

GEF (Global Environment Facility). (2017). *Strengthening monitoring and evaluation of climate change adaptation. GEF/STAP/LDCF.22/Inf.01*. Washington, DC: GEF.

Geels, F. W., Kemp, R., Dudley, G., & Lyons, G. (Eds.), (2012). *Automobility in transition? A socio-technical analysis of sustainable transport*. London: Routledge.

Gerber, E. R. (2013). Partisanship and local climate policy. *Cityscape (Washington, D.C.), 15*(1), 107−124.

Giannoulakis, S., Volkart, K., & Bauer, C. (2014). Life cycle and cost assessment of mineral carbonation for carbon capture and storage in European power generation. *International Journal of Greenhouse Gas Control, 21*, 140−157.

GIZ. (2015). *Impact evaluation guidebook for climate change adaptation projects*. Bonn: GIZ.

GIZ. (2017). *Valuing the benefits. Costs and impacts of ecosystem-based adaptation measures*. Bonn: GIZ.

Global CCS Institute. (2016). *The global status of CCS, 2016: Summary report*. Australia.

Goers, S., Wagner, A. F., & Wegmayr, J. (2010). New and old market-based instruments for climate change policy. *Environmental Economics and Policy. Studies, 12*(1/2), 1−30.

Goldman, M. J., Turner, M. D., & Daly, M. (2018). A critical political ecology of human dimensions of climate change: Epistemology, ontology, and ethics. *WIREs Climate Change, 9*(4), 1−15.

Goldman, S., Ungar, L., Capanna, S., & Simchak, T. (2012). *Energy efficiency: A tool for climate change adaptation*. Washington, DC: Alliance to Save Energy.

Grimmond, S. (2007). Urbanization and global environmental change: Local effects of urban warming. *Geographical Journal, 173*, 83−88.

Griscom, B. W., Adams, J., Ellis, P., Houghton, R. A., Lomax, G., Miteva, D. A., ... Fargione, J. (2017). Natural pathways to climate mitigation. *Proceedings of the National Academy of Sciences, 114*(44), 11645−11650.

Grunewald, N., & Martinez-Zarzoso, I. (2015). Did the Kyoto Protocol fail? An evaluation of the effect of the Kyoto Protocol on CO2 emissions. *Environment and Development Economics, 20*(2), 1−20.

Haasnoot, M., Brown, S., Scussolini, P., Jimenez, J. A., TVafeidis, A., & Nicholls, R. J. (2019). Generic adaptation pathways for coastal archetypes under uncertain sea-level rise. *Environmental Research Communication Letter, 1*, 071006.

Hakes, J. K., & Viscusi, W. K. (2004). Dead reckoning: Demographic determinants of the accuracy of mortality risk perceptions. *Risk Analysis, 24*(3), 651−664.

Gough, C., & Vaughan, N. E. (2015). *Synthesising existing knowledge on feasibility of BECCS. AVOID2, report WPD1a*.

Hameso, S. (2012, March 1). Development challenges in the age of climate change: The case of Sidama. In: *Paper presented at the workshop on the economy of Southern Ethiopia*. Hawassa: Ethiopian Economics Association.

Hamilton, K., Chokkalingam, U., & Bendana, M. (2010). *State of the forest carbon markets 2009: Taking root and branching out*. Washington, DC: Ecosystem Marketplace Report. Forest Trends.

Hamin, E. M., & Gurran, N. (2009). Urban form and climate change: Balancing adaptation and mitigation in the United States and Australia. *Habitat International, 33*(3), 238−245.

Harveya, C. A., Martínez-Rodrígueza, M. R., Cárdenas, J. M., Avelino, J., Rapidel, B., Vignola, R., ... Vilchez-Mendoza, S. (2017). The use of ecosystem-based adaptation practices by smallholder farmers in Central America. *Agriculture, Ecosystems and Environment, 246*, 279−290.

Hatfield, J., Takle, G., Grotjahn, R., Holden, P., Izaurralde, R. C., Mader, T., ... Liverman, D. (2014). Ch. 6: Agriculture. In J. M. Melillo, T. T. C. Richmond, & G. W. Yohe (Eds.), *Climate change impacts in the United States: The third national climate assessment* (pp. 150−174). Washington, DC: United States Global Change Research Program.

Henstra, D. (2016). The tools of climate adaptation policy: Analysing instruments and instrument selection. *Climate Policy, 16*(4), 496−521.

Head, L. (2010). Cultural ecology: Adaptation − Retrofitting a concept? *Progress in Human Geography, 34*(2), 234−242.

HoCSTC (House of Commons Science and Technology Committee). (2010a). *The regulation of geoengineering*. London: House of Commons.

Heyward, C. (2013). Situating and abandoning geoengineering: A typology of five responses to dangerous climate change. *PS: Political Science and Politics, 46*(1), 23−27.

Hisbaron, N. N. (2017). Shades of green: A comparative study of climate discourse in the Kyoto Protocol and Paris Agreement. *Unpublished MA Thesis, faculty of Humanities*. Leiden: Leiden University.

Hoffert, M., Caldeira, K., Benford, G., Criswell, D., Green, C., Herzog, H., ... Wigley, T. (2002). Advanced technology paths to global climate stability: Energy for a greenhouse planet. *Science (New York, N.Y.), 298L*, 981−987.

HoCSTC (House of Commons Science and Technology Committee). (2010b). *The regulation of geoengineering*. London: The Stationary Office.

Hoffmaister, J. P., Talakai, M., Damptey, P., & Soares, A. (2014). *Warsaw International Mechanism for loss and damage: Moving from polarizing discussions towards addressing the emerging challenges faced by developing countries*. Loss and Damage. Available at http://www.lossanddamage.net/4950.

Höhne, N., Ellermann, C., & Li, L. (2014). *Intended nationally determined contributions under the UNFCCC', discussion paper*. Ecofys. Summarized in Ogahara, J. (2014). Nationally determined contributions − Support by Germany for developing countries. In: *SB40 side event report*. Japan: Overseas Environmental Cooperation Center, (OECC).

Hood, C. (2019). *Completing the Paris 'Rulebook': Key Article 6 Issues*. Arlington, VA: Center for Climate and Energy Solutions.

House, K. Z., Baclig, A. C., Ranjan, M., van Nierop, E. A., Wilcox, J., & Herzog, H. J. (2011). Economic and energetic analysis of capturing CO2 from ambient air. *Proceedings of the National Academy of Sciences, 108*(51), 20428−20433.

Howard, A. (2017). Voluntary cooperation (Article 6). In K. Daniele, C. Maria Pia, D. Meinhard, B. Jane, & H. Andrew (Eds.), *The Paris Agreement on climate change: Analysis and commentary* (pp. 178−195). Oxford: Oxford University Press.

Houyoux, M. (2019). Analysis and use of point source emission rates from the national emission inventory. In: *Presentation at the international emissions inventory conference*, July 29−August 2, 2019.

Hu, S., Niu, Z., Chen, Y., Li, L., & Zhang, H. (2017). Global wetlands: Potential distribution, wetland loss, and status. *Science of the Total Environment, 586*, 319−327.

Huggel, C., Bresch, D., Hansen G., James R., Mechler R., Stone D., & Wallimann-Helmer I. (2016a). Attribution of irreversible loss to anthropogenic climate change. In: *EGU General assembly conference, Abstracts: 8557*.

Hughes, A. D., Black, K. D., Campbell, I., Davidson, K., Kelly, M. S., & Stanley, M. S. (2012). Does seaweed offer a solution for bioenergy with biological carbon capture and storage? *Greenhouse Gases: Science and Technology, 2*(6), 402−407.

Hurlimann, A., Barnett, J., Fincher, R., Osbaldiston, N., Mortreux, C., & Graham, S. (2014). Urban planning and sustainable adaptation to sea-level rise. *Landscape and Urban Planning, 126*, 84−93.

IAEA (International Atomic Energy Agency). (2018). *Climate change and nuclear power*. Vienna: IAEA.

ICCCAD (International Centre for Climate Change and Development). *Loss and damage in vulnerable countries initiative*. (2012).

IEA (International Energy Agency). (2019). Global energy and CO_2 status report 2018. Paris: IEA.

IMF (International Monetary Fund). (2019). *Fiscal monitor-how to mitigate climate change*. Washington, DC: International Monetary Fund.

IPBES (Intergovernmental Science-Policy Platform on Biodiversity and Ecosystem Services). (2019). *Global assessment report on Biodiversity and Ecosystem Services*. Bonn: IPBES Secretariat.

IPCC (Intergovernmental Panel on Climate Change). (2001). *Climate change 2001: Mitigation: Contribution of working group III to the third assessment report of the Intergovernmental Panel on Climate Change*. Cambridge: Cambridge University Press.

IPCC (Intergovernmental Panel on Climate Change). (2007a). *Climate change 2007: Impacts, adaptation and vulnerability: Contribution of working group II to the fourth assessment report of the Intergovernmental Panel on Climate Change*. Cambridge: Cambridge University Press.

IPCC (Intergovernmental Panel on Climate Change). (2007b). *Climate change 2007: Mitigation of climate change: Contribution of working group III to the fourth assessment report of the Intergovernmental Panel on Climate Change*. Cambridge: Cambridge University Press.

IPCC (Intergovernmental Panel on Climate Change). (2010). *Proposal for an IPCC expert meeting on geoengineering*. Submitted by the co-chairs of Working Groups I, II and III.

IPCC (Intergovernmental Panel on Climate Change). (2013). Summary for policymakers. In: *Working Group, I contribution to the IPCC fifth assessment report climate change 2013: The physical science basis*. Intergovernmental Panel on Climate Change.

IPCC (Intergovernmental Panel on Climate Change). (2014a). *Climate change 2014: Mitigation of climate change. Contribution of working group III to the fifth assessment report of the Intergovernmental Panel on Climate Change*. Cambridge: Cambridge University Press.

IPCC. (2014b). *Climate change 2014: Synthesis report. Contribution of working groups I, II and III to the fifth assessment report of the Intergovernmental Panel on Climate Change*. Cambridge: Cambridge University Press.

IPCC (Intergovernmental Panel on Climate Change). (2018). Global warming of 1.5°C. *An IPCC Special Report on the impacts of global warming of 1.5°C above pre-industrial levels and related global greenhouse gas emission pathways, in the context of strengthening the global response to the threat of climate change, sustainable development, and efforts to eradicate poverty* (pp. 445−540). Geneva: IPCC.

IPCC (Intergovernmental Panel on Climate Change). *IPCC special report on climate change and land. An IPCC special report on climate change, desertification, land degradation, sustainable land management, food security, and greenhouse gas fluxes in terrestrial ecosystems*. (2019a).

IPCC (Intergovernmental Panel on Climate Change). *IPCC special report on the ocean and cryosphere in a changing climate*. (2019b).

IPS (Institute for Policy Studies). *Green climate fund*. (2014). Available at http://climatemarkets.org/glossary/green-climate-fund.html.

Irfan, U. (2019). *The future of the Paris climate agreements is being decided this week*. vox.com.

Jacobson, M. Z. (2020). *100% Clean renewable energy and storage for everything*. Cambridge: Cambridge University Press.

Jaffe, J., & Stavins, R. (2009). Linkage of tradable permit systems in international climate policy architecture. In J. Aldy, & R. Stavins (Eds.), *Post-Kyoto international climate policy implementing architectures for agreement*. New York: Cambridge University Press.

Jain, R., & Webb, M. D. (2012). Contemporary Issues in Environmental Assessment. In R. Jain, L. Urban, H. Balbach, & D. M. Webb (Eds.), *Handbook of environmental engineering assessment* (pp. 361–448). Amsterdam: Elsevier.

Jeffery, S., Bezemer, T. M., Cornelissen, G., Kuyper, T. W., Lehmann, J., Mommer, L., & van Groenigen, J. W. (2015). The way forward in biochar research: Targeting trade-offs between the potential wins. *GCBBioenergy*, *7*(1), 1–13.

Ji, H., & Darnall, N. (2017). All are not created equal: Assessing local governments' strategic approaches towards sustainability. *Public Management Review*, *19*, 1–22.

Jin, X., Gruber, N., Frenzel, H., Doney, S. C., & McWilliams, J. C. (2007). The impact on atmospheric CO2 of iron fertilization induced changes in the ocean's biological pump. *Biogeosciences Discussions*, *4*(5), 3863–3911.

JIN (Joint International Network). (2010). Uncertainty remains after Copenhagen. *Joint Implementation Quarterly*, *15*(4), 2–3.

Jones, I. (2014). The cost of carbon management using ocean nourishment. *International Journal of Climate Change Strategies and Management*, *6*(4), 391–400.

Joos, F., Sarmiento, J. L., & Siegenthaler, U. (1991). Estimates of the effect of Southern Ocean iron fertilization on atmospheric CO2 concentrations. *Nature*, *349*, 327–329.

Jordan, J. C. (2015). Swimming alone? The role of social capital in enhancing local resilience to climate stress: A case study from Bangladesh. *Climate and Development*, *7*(2), 110–123.

Kahn, M. (2006). *Green cities: Urban growth and the environment*. Washington, DC: Brookings Institution Press.

Kahan, D. M., Peters, E., Wittlin, M., Slovic, P., Ouellette, L. L., Braman, D., & Mandel, G. (2012). The polarizing impact of science literacy and numeracy on perceived climate change risks. *Nature Climate Change*, *2*, 732–735.

Kates, R. W. (2000). Cautionary tales: Adaptation and the global poor. *Climatic Change*, *45*(1), 5–17.

Keith, D. W. (2013). *A case for climate engineering. Boston Review Books*. Cambridge, MA: The MIT Press.

Keith, D. W., Parson, E., & Morgan, M. G. (2010). Research on Global Sun Block Needed Now. *Nature*, *463*, 426, −427.

Keith, D. W., Geoffrey, H., David St., A., & Kenton, H. (2018). A process for capturing CO_2 from the atmosphere. *Joule*, *2*, 1573–1594.

Kemper, J. (2015). Biomass and carbon dioxide capture and storage: A review. *International Journal of Greenhouse Gas Control*, *40*, 401–430.

Kern, K., & Alber, G. (2008). Governing climate change in cities: Modes of urban climate governance in multilevel systems. In: *Competitive cities and climate change: OECD conference proceedings* (pp. 171–192), Milan, Italy, October 9–10, 2008. Available at http://www.oecd.org/cfe/regional-policy/50594939.pdf.

Khan, M. H. I. (2013). Legal and policy responses to loss and damage associated with climate change. In O. C. Ruppel, C. Roschmann, & K. Ruppel-Schlichting (Eds.), Climate change: International law and global governance: Volume I: Legal responses and global responsibility (1st ed., pp. 1–8). Baden-Baden: Nomos Verlagsgesellschaft mbH & Co. KG.

Kirchofer, A., Becker, A., Brandt, A., & Wilcox, J. (2013). CO2 mitigation potential of mineral carbonation with industrial alkalinity sources in the United States. *Environmental Science & Technology*, *47*(13), 7548–7554.

Klein, R. J. T., Huq, S., Denton, F., Downing, T. E., Richels, R. G., Robinson, J. B., & Toth, F. L. (2007). *Inter-relationships between adaptation and mitigation*. Climate change 2007: Impacts, adaptation and vulnerability. *Contribution of Working Group II to the Fourth Assessment Report of the Intergovernmental Panel on Climate Change* (pp. 745−777). Cambridge: Cambridge University Press.

Klein, R. J. T., Midgley, G. F., Preston, B. L., Alam, M., Berkhout, F. G. H., Dow, K., & Shaw, M. R. (2014). *Adaptation opportunities, constraints, and limits*. Climate change 2014: Impacts, adaptation and vulnerability. *Part A: Global and sectoral aspects. Contribution of Working Group II to the Fifth Assessment Report of the Intergovernmental Panel on Climate Change* (pp. 899−943). Cambridge: Cambridge University Press.

Klein, R. J. T., Adams, K. M., Dzebo, A., Davis, M., & Siebert, C. K. (2016). *Advancing climate adaptation practices and solutions: Emerging research priorities*. Stockholm: Stockholm Environment Institute (SEI).

Knopf, B. (2015). *Heaven belongs to us all − The new papal encyclical*. Energypost.eu online. Available at https://energypost.eu/heaven-belongs-us-new-papal-encyclical/.

Knuth, S., Nagle, B., Steuer, C., & Yarnal, B. (2007). Universities and climate change mitigation: Advancing grassroots climate policy in the United States. *Local. Environment, 12*(5), 485−504.

Kok, M., Lüdeke, M., Sterzel, T., Lucas, P., Walter, C., Janssen, P., & de Soya, I. (2010). *Quantitative analysis of patterns of vulnerability to global environmental change*. Den Haag/Bilthoven: Netherlands Environmental Assessment Agency (PBL).

Kolmuss, A. (2013). Doha decisions on the Kyoto surplus explained. *Carbon market watch policy brief*.

Kongsager, R., Locatelli, B., & Chazarin, F. (2016). Addressing climate change mitigation and adaptation together: A global assessment of agriculture and forestry projects. *Environmental Management, 57*, 271−282.

Krause, R. M. (2011a). An assessment of the greenhouse gas reducing activities being implemented in United States Cities. *Local Environment, 16*(2), 193−211.

Krause, R. M. (2011b). Symbolic or substantive policy? measuring the extent of local commitment to climate protection. *Environment and Planning C: Government and Policy, 29*(1), 46−62.

Krause, R. M. (2011c). Policy innovation, intergovernmental relations and the adoption of climate protection initiatives by United States Cities. *Journal of Urban Affairs, 33*(1), 45−60.

Krause, R. M. (2012). Political decision-making and the local provision of public goods: The case of municipal climate protection in the United States. *Urban Studies, 49*(11), 2399−2417.

Krause, R. M. (2013). The motivations behind municipal climate engagement: An empirical assessment of how local objectives shape the production of a public good cityscape. *Climate Change and City Hall, 15*(1), 125−141.

Kraxner, F., Nilsson, S., & Obersteiner, M. (2003). Negative emissions from BioEnergy use, carbon capture and sequestration (BECS)—The case of biomass production by sustainable forest management from seminatural temperate forests. *Biomass and Bioenergy, 24*, 285−296.

Kreidenweis, U., Florian, H., Miodrag, S., Benjamin, L. B., Elmar, K., Hermann, L.-C., & Alexander, P. (2016). Afforestation to mitigate climate change: Impacts on food prices under consideration of albedo effects. *Environment Research Letters*, 085001.

Kulkarni, A. R., & Sholl, D. S. (2012). Analysis of equilibrium-based TSA processes for direct capture of CO2 from air. *Industrial & Engineering Chemistry Research, 51*(25), 8631−8645.

Kung, C.-C., McCarl, B. A., & Cao, Z. (2013). Economics of pyrolysis-based energy production and biochar utilization: A case study in Taiwan. *Energy Policy, 60*, 317−323.

Kurz, K. D., & Maier-Reimer, E. (1993). Iron fertilization of the austral ocean—The Hamburg model assessment. *Global Biogeochemical Cycles, 7*(1), 229−244.

Kvan, T., & Karakiewicz, J. (Eds.), (2019). *Urban Galapagos transition to sustainability in complex adaptive systems*. Cham: Springer.

Lackner, K. S. (2010). Washing carbon out of the air. *Scientific American, 302*(6), 66−71.

Lackner, K. S., Brennan, S., Matter, J. M., Park, A. H. A., Wright, A., & Van Der Zwaan, B. (2012). The urgency of the development of CO2 capture from ambient air. *Proceedings of the National Academy of Sciences*, *109*(33), 13156−13162.

Laer, T. V., Smedt, D. P., Ronsse, F., Ruysschaert, G., Boeckx, P., Verstraete, W., ... Lavarsen, L. J. (2015). Legal constraints and opportunities for biochar: A case analysis of EU law. *GCB Bioenergy*, *7*, 14−24.

Lallanilla, M. (2019). *Greenhouse gasses: Causes, sources and environmental effects*. Live Science.

Landauer, M., Juhola, S., & Soderholm, M. (2015). Inter-relationships between adaptation and mitigation: A systematic literature review. *Climate Change*, *131*, 505−517.

Larson, E. J. (2016). The Red Dawn of geoengineering: First step toward an effective governance for stratospheric injections. *Duke Law & Technology Review*, *14*, 157−191.

La Viña, A. G. M., & de Leon, A. (2017). Conserving and enhancing sinks and reservoirs of greenhouse gases, including forests (Article 5). In K. Daniele, C. Maria Pia, D. Meinhard, B. Jane, & H. Andrew (Eds.), *The Paris Agreement on climate change: Analysis and commentary* (pp. 166−177). Oxford: Oxford University Press.

Lehmann, J., Czimczik, C., Laird, D., & Sohi, S. (2015). Stability of biochar in soil. In J. Lehman, & S. Joseph (Eds.), *Biochar for environmental management: Science, technology and implementation* (pp. 742−746). London: Taylor and Francis.

Lemos, M. C., Lob, Y.-J., Nelson, D. R., Eakin, H., & Bedran-Martinis, A. N. (2016). Linking development to climate adaptation: Leveraging generic and specific capacities to reduce vulnerability to drought in NE Brazil. *Global Environmental Change*, *39*, 170−179.

Levitt, S. D., & List, J. A. (2008). Science (New York, N.Y.), *319*, 909−910.

Levitt, S. D., & Dubner, S. J. (2011). *Super freakonomics global cooling, patriotic prostitutes, and why suicide bombers should buy life insurance*. New York: William Morrow.

Lin, A. C. (2013). Does geoengineering present a moral hazard? *Ecology Law Quarterly*, *40*(3), 673−712.

Lindsey, R. (2019). *Climate change: Atmospheric carbon dioxide*. Climate.gov online portal.

Lindsey, R., & Dahlman, L. A. (2020). *Climate change: Global temperature*. Climate.gov. online.

Liu, J., Hull, V., Luo, J., Yang, W., Liu, W., Viña, A., ... Zhang, H. (2015). Multiple tele-couplings and their complex interrelationships. *Ecology and Society*, *20*(3), 44.

Liu, J., Hull, V., Luo, J., Yang, W., Liu, W., Vina, A., ... Zhang, H. (2016). Human-nature interactions over distances. In J. Liu, V. Hull, W. Yang, A. Vina, X. Chen, Z. Quyang, & H. Zhang (Eds.), *Pandas and people* (pp. 218−239). Oxford: Oxford University Press.

Liverman, D. (2015). Reading climate change and climate governance as political ecologies. In T. Perreault, G. Bridge, & J. McCarthy (Eds.), *The handbook of political ecology* (pp. 303−319). Oxon.

Locatelli, B. (2011). *Synergies between adaptation and mitigation in a nutshell*. Bogor: CIFOR (Center for International Forestry Research).

Locatelli, B., Fedele, G., Fayolle, V., & Baglee, A. *Synergies between adaptation and mitigation in climate change finance*. (2015a).

Locatelli, B., Catterall, C. P., Imbach, P., Kumar, C., Lasco, R., Marín-Spiotta, E., ... Uriarte, M. (2015b). Tropical reforestation and climate change: Beyond carbon. *Restoration Ecology*. Available from https://doi.org/10.1111/rec.12209.

London Protocol. (2008). *Report of the thirtieth consultative meeting and the third meeting of contracting parties*. Available online.

Low, S., & Schafer, S. (2020). Is bio-energy carbon capture and storage (BECCS) feasible? The contested authority of integrated assessment modeling. *Energy Research and Social Science*. February, *26*, 101326.

Lu, W., Sculley, J. P., Yuan, D., Krishna, R., & Zhou, H. C. (2013). Carbon dioxide capture from air using amine-grafted porous polymer networks. *The Journal of Physical Chemistry C*, *117*(8), 4057−4061.

Luber, G., Knowlton, K., Balbus, J., Fumkin, H., Hayden, M., Hess, J., ... Ziska, L. (2014). Ch.9: Human health. In J. M. Melillo, T. (T. C.) Richmond, & G. W. Yohe (Eds.), *Climate change impacts in the United States: The third national climate assessment* (pp. 220–256). Washington, DC: United States global change research program.

Maibach, E. W., Leiserowitz, A., Roser-Renouf, C., & Mertz, C. K. (2011). Identifying like-minded audiences for global warming public engagement campaigns: An audience segmentation analysis and tool development. *PLoS One*, *6*(3), e17571.

Marchetti, C. (1977). On geoengineering and the CO_2 problem. *Climatic Change*, *1*, 59–68.

Marshall, N., & Stokes, C. J. (2014a). Identifying thresholds and barriers to adaptation through measuring climate sensitivity and capacity to change in an Australian primary industry. *Climatic Change*, *126*, 399–411.

Martin, T. G., & Watson, J. E. M. (2016). Intact ecosystems provide best defence against climate change. *Nature Climate Change*, *6*, 122–124.

Martin, D., Johnson, K., Stolberg, A., Zhang, X., & Young, C. D. (2017). *Carbon dioxide removal options: A literature review identifying carbon removal potentials and costs*. University of Michigan.

Masterson, V., Stedman, R., Enqvist, J., Tengö, M., Giusti, M., Wahl, D., & Svedin, U. (2017). The contribution of sense of place to social-ecological systems research: A review and research agenda. *Ecology and Society*, *22*(1).

Matear, R. & Wong, C. (1999). Potential to increase the oceanic CO_2 uptake by enhancing marine productivity in high nutrient low chlorophyll regions. In B. Eliasson et al. (Eds.), *Greenhouse gas control technologies*. Pergamon: Interlaken Switzerland.

Mathevet, R., Thompson, J. D., & Folke, C. (2016). Protected areas and their surrounding territory: Socioecological systems in the context of ecological solidarity. *Ecological Applications*, *26*(1), 5–16.

Mathew, L. M., & Akter, S. (2017). Loss and damage associated with climate change impacts. In C. Wei-Yin, S. Toshio, & L. Maximilian (Eds.), *Handbook of climate change mitigation and adaptation* (Second Edition, pp. 17–46). Gland: Springer.

Matter, J. M., Stute, M., Snæbjörnsdottir, S. Ó., Oelkers, E. H., Gislason, S. R., Aradottir, E. S., & Axelsson, G. (2016). Rapid carbon mineralization for permanent disposal of anthropogenic carbon dioxide emissions. *Science (New York, N.Y.)*, *352*(6291), 1312–1314.

Mauerhofer, V., Ichinose, T., Blackwell, T. B., Willig, M. R., Flint, T. G., Krause, M. S., & Penker, M. (2018). Underuse of social-ecological systems: A research agenda for addressing challenges to biocultural diversity. *Land use Policy*, *72*, 57–64.

Mayne, J. (1999). *Addressing attribution through contribution analysis: Using performance measures sensibly. Discussion paper*. Office of the Auditor General of Canada.

Mayne, J. (2008). *Contribution analysis: An approach to exploring cause and effect. Institutional learning and change brief*. Rome: Institutional Learning and Change (ILAC) Initiative.

Mazzotti, M., Baciocchi, R., Desmond, M. J., & Socolow, R. H. (2013). Direct air capture of CO_2 with chemicals: Optimization of a two-loop hydroxide carbonate system using a countercurrent air-liquid contactor. *Climatic Change*, *118*(1), 119–135.

Mbengue, M. M. (Online). Preamble. In: *Max Planck encyclopedia of public international law*. Available at http://opil.ouplaw.com/home/EPIL.

McCormick, R., & Ludwig, J. (1967). Climate modification by atmospheric aerosols. *Science (New York, N.Y.)*, *156*, 1358–1359.

McGray, H., Hammill, A., & Bradley, R. (2007). *Weathering the storm: Options for framing adaptation and development*. World Resources Institute.

McNamara, K.E. (2014). *Exploring loss and damage at the international climate change talks*, 5, 3. Springer: Heidelberg, pp. 242–246.

Mertz, O., Halsnæs, K., Olesen, J. E., & Rasmussen, K. (2009). Adaptation to climate change in developing countries. *Environmental Management, 43*, 743−752.

Meva. (2019). *How much longer can we ignore loss and damage? Climate Action Network.*

Miles, L., & Sonwa, D. J. (2015). Mitigation potential from forest-related activities and incentives for enhanced ac on in developing countries. In: *UNEP emissions gap report 2015* (pp. 44−50). Nairobi: UNEP.

Millard-Ball, A. (2012). Do city climate plans reduce emissions? *Journal of Urban Economics, 71*(3), 289−311.

Mills, G. (2007). Cities as agents of global change. *International Journal of Climatology, 27*(14), 1849−1857.

Mimura, N., Pulwarty, R. S., Duc, D. M., Elshinnawy, I., Redsteer, M. H., Huang, H. Q., . . . Kato, S. (2014). *Adaptation planning and implementation.* Climate change 2014 impacts, adaptation and vulnerability: Part A: Global and sectoral aspects. *Fifth Assessment Report of the Intergovernmental Panel on Climate Change* (pp. 869−898). Cambridge: Cambridge University Press.

Minx, J. C., Lamb, W. F., Callaghan, M. W., Fuss, S., Hilaire, J., Creutzig, F., . . . del Mar Zamora Dominguez, M. (2018). Negative emissions−Part 1: Research landscape and synthesis. *Environment Research Letters, 13*(6), 063001.

Mitsch, W. J., Bernal, B., Nahlik, A. M., Mander, U., Zhang, L., Anderson, C. J., . . . Brix, H. (2012). Wetlands, carbon, and climate change. *Landscape Ecology, 28*, 583−597.

Möhner, A. (2018). The evolution of adaptation metrics under the UNFCCC and its Paris Agreement. In L. Christiansen, G. Martinez, & P. Naswa (Eds.), *Adaptation Metrics: Perspectives on measuring, aggregating, and comparing adaptation results.* Copenhagen: UNEP DTU Partnership.

Monfreda, C., Ramankutty, N., & Foley, J. A. (2008). Farming the planet: 2. Geographic distribution of crop areas, yields, physiological types, and net primary production in the year 2000. *Global Biogeochemical Cycles, 22*(1).

Morgan, C. L. (2011). *Limits to Adaptation: A review of limitation relevant to the project 'building resilience to climate change—Coastal Southeast Asia.* Gland: IUCN.

Moosmann, L., Urrutia, C., Siemons, A., Cames, M., & Schneider, L. (2019). *International climate negotiations—Issues at stake in view of the COP 25 UN climate change conference in Madrid.* Luxembourg: Committee on the Environment, Public Health and Food Safety of the European Parliament, Policy Department for Economic, Scientific and Quality of Life Policies, European Parliament.

Morgan, J., Dagnet, Y., & Tirpak, D. (2015). *Elements and ideas for the 2015 Paris Agreement. WRI working paper.*

Morton, J. F., Solecki, W., Dasgupta, P., Dodman, D., & Rivera-Ferre, M. G. (2014). *Urban-rural interactions − Context for climate change vulnerability impacts and adaptation.* Climate change 2014: Impacts, adaptation, and vulnerability. Part A: Global and sectoral aspects. *Contribution of Working Group II to the Fifth Assessment Report of the Intergovernmental Panel on Climate Change* (pp. 153−155). Cambridge: Cambridge University Press.

Moser, S. C., & Ekstrom, J. A. (2010). A framework to diagnose barriers to climate change adaptation. *Proceedings of the National Academy of Sciences of the United States of America, 107*, 22026−22031.

Moser, S. C., & Boykoff, M. T. (Eds.), (2013). *Successful adaptation to climate change: Linking science and policy in a rapidly changing world.* New York: Routledge.

Mubaya, C. P., & Mafongoya, P. (2017). The role of institutions in managing local level climate change adaptation in semi-arid Zimbabwe. *Climate Risk Management, 16*, 93−105.

Mumpower, J. L., Liu, X., & Vedlitz, A. (2016). Predictors of the perceived risk of climate change and preferred resource levels for climate change management programs. *Journal of Risk Research, 19*(6).

Munang, R. I., Thiaw, K., & Alverson. (2013). Climate change and ecosystem-based adaptation: A new pragmatic approach to buffering climate change impacts. *Current Opinion in Environmental Sustainability, 5*(1), 67−71.

Munroe, R., Roe, D., Doswald, N., Spencer, T., Moller, I., Vira, B., ... Stephens, J. (2012). Review of the evidence base for ecosystem-based approaches for adaptation to climate change. *Environmental Evidence, 1*(2), 13.

Muri, H., Tjiputra, J., Helge Otterå, O., Adakudlu, M., Lauvset, S. K., Grini, A., ... Kristjánsson, J. E. (2018). Climate response to aerosol geoengineering: A multi-method comparison. *Journal of Climate, 31*(16), 6319−6340.

Nakićenović, N., Gilli, P. V., & Kurz, R. (1996). Regional and global exergy and energy efficiencies. *Energy., 21*(3), 223−237.

Nalau, J., Becken, S., & Mackey, B. (2018). Ecosystem-based Adaptation: A review of the constraints. *Environmental Science & Policy, 89*, 357−364.

NAS (National Academy of Sciences). (1992). *Policy implications of greenhouse warming: Mitigation, adaptation, and the science base*. Washington, DC: National Academies Press.

NAS (National Academy of Sciences). (2015a). Climate intervention: Reflecting sunlight to cool earth. *National Research Council of the NAS Committee on Geoengineering Climate*. Washington, DC: National Academies Press.

NAS (National Academy of Sciences). (2015b). Climate Intervention: Carbon dioxide removal and reliable sequestration. *National Research Council of the NAS: Committee on Geoengineering Climate*. Washington, DC: National Academies Press.

NASA (National Aeronautics and Space Administration). *Responding to climate change*. Website portal at https://climate/nasa.bov.

NASA (National Aeronautics and Space Administration). (2010). *Global warming*. NASA.

NASA. *John Martin (1935−1993)*. (2017). Available at http://earthobservatory.nasa.gov/Features/Martin/martin_4.php.

Nassiry, D., Pickard, S., & Scott, A. (2017). Implications of geoengineering for developing countries. *Working Paper 524*. London: Overseas Development Institute (ODI).

Nelson, R., Kokic, P., Crimp, S., Meinke, H., & Howden, S. M. (2010). The vulnerability of Australian rural communities to climate variability and change: Part I—Conceptualising and measuring vulnerability. *Environmental Science and Policy, 13*(1), 8−17.

Neslen, A. (2015, December 14). EU says 1.5C global warming target depends on 'negative emissions' technology. *The Guardian*.

Ngum, F., Alemagi, D., Duguma, L., Minang, P. A., Kehbila, A., & Zac, T. (2019). Synergizing climate change mitigation and adaptation in Cameroon: An overview of multi-stakeholder efforts. *International Journal of Climate Strategies and Management, 11*(1), 118−136.

NOAA (National Oceanic and Atmospheric Administration). (2020). *State of the climate: Global climate report for annual 2019*. Available from https://www.ncdc.noaa.gov/sotc/global/201913.

Noah, S. (2019). The Paris Agreement in the 2020s: Breakdown or breakup? *Ecology Law Quarterly, 46*(1).

Northrop, E., Dagnet, Y., Hohne, N., Thwaites, J., & Mogelgaard, K. (2018). Achieving the ambition of Paris: Designing the global stocktake. *Working Paper*. Washington, DC: World Resource Institute.

N'Yeurt, A. R., Chynoweth, D. P., Capron, M. E., Stewart, J. R., & Hasan, M. A. (2012). Negative carbon via ocean afforestation. *Process Safety and Environmental Protection, 90*(6), 467−474.

Oberthur, S., & Ott, H. (1999). *The Kyoto Protocol: International climate policy for the 21st century*. Berlin: Springer.

Oberthur, S., & Northrop, E. (2018). The mechanism to facilitate implementation and promote compliance with Paris Agreement: Design options. *Working Paper*. Washington, DC: World resource Institute.

OECD (Organization for Economic Cooperation and Development). (2011). OECD principles for regulatory quality and performance. In OECD (Ed.), *Regulatory policy and governance: Supporting economic growth and serving the public interest*. Paris: OECD.

OECD (Organization for Economic Cooperation and Development). (2016). *Enhancing transparency of climate finance under the Paris Agreement: Lessons from experience*. Paris: OECD.
OED (Oxford English Dictionary). (2008). *Oxford English dictionary* (Third edition). Oxford: Oxford University Press.
Ojea, E. (2015). Challenges for mainstreaming ecosystem-based adaptation into the international climate agenda. *Current Opinion in Environmental Sustainability*, *14*, 41−48.
Olson, M. (1971). *The logic of collective action*. New York: Schocken Books.
Ostrom, E. (2005). *Understanding institutional diversity*. Princeton, NJ: Princeton University Press.
Ostrom, E. (2007). Institutional rational choice: An Assessment of the institutional analysis and development framework. In P. Sabatier (Ed.), *Theories of the policy process* (Second Edition, pp. 21−64). Boulder, CO: Westview Press.
Ott, H., Sterk, W., & Watanabe, R. (2008). The Bali roadmap: New horizons for global climate policy. *Climate Policy*, *8*, 91−95.
Page, E. A., & Heyward, C. (2017). Compensating for climate change loss and damage. *Politische Studien*, *65*(2), 356−372.
Page, S. E., & Hooijer, A. (2016). In the line of fire: The peatlands of Southeast Asia. *Philosophical Transactions of Royal Society*, *B371*, 20150176.
Palutikof, J., Barnett, J., & Guitart, D. (2014a). Can we adapt to four degrees of warming? Yes, no and maybe. In P. Christoff (Ed.), *Four degrees of global warming: Australia in a hot world* (pp. 216−233). Abingdon: Routledge.
Pant, L. P., Adhikari, B., & Bhattarai, K. K. (2015). Adaptive transition for transformations to sustainability in developing countries. *Current Opinion in Environmental Sustainability*, *14*, 206−212.
Parish, F., Sirin, A., Charman, D. J., Joosten, H., Minayeva, T., Silvius, M., & Stringer, L. (2008). *Assessment on peatlands, biodiversity and climate change: Main report*. Wageningen: Global Environment Centre, Kuala Lumpur and Wetlands International.
Parker, C., Keenlyside, P., & Conway, D. (2014). *Early experiences in adaptation finance: Lessons from the four multilateral climate change adaptation funds*. Amsterdam: Climate Focus.
Parker, H. R., Cornforth, R. J., Boyd, E., James, R., Otto, F. E. L., & Allen, M. R. (2015). Implications of event attribution for loss and damage policy. *Weather.*, *70*(9), 268−273.
Parry, M., Arnell, N., Berry, P., Dodman, D., Fankhauser, S., Hope, C., ... Wheeler, T. (2009). *Assessing the costs of adaptation to climate change*. London: International Institute for Environment and Development.
Pasztor, J. (2017). The need for governance of climate geoengineering. *Ethics and International Affairs*, *31*(4), 419−430.
Parson, E. A., & Keith, D. W. (2013). End the deadlock on governance of geoengineering research. *Science (New York, N.Y.)*, *339*(6125), 1278−1279.
Peng, T., & Broecker, W. (1991). Dynamical limitations of the Antarctic iron fertilization strategy. *Nature*, *349*, 227−229.
Penner, S. S., Schneider, A. M., & Kennedy, E. M. (1984). Active measures for reducing the global climatic impacts of escalating CO_2 concentrations. *Acta Astronautica*, *11*(6), 345−348.
Pereira, R., & Yarish, C. (2008). Mass production of marine macroalgae. In S. E. Jørgensen, & B. D. Fath (Eds.), *Ecological engineering. Vol. 3 of Encyclopedia of ecology* (pp. 2236−2247). Oxford: Elsevier.
Pérez, I. S., & Kallhauge, A. C. (2017). Adaptation (Article 7). In K. Daniele, C. Maria Pia, D. Meinhard, B. Jane, & H. Andrew (Eds.), *The Paris Agreement on climate change: Analysis and commentary* (pp. 196−223). Oxford: Oxford University Press.
Pidgeon, N. (1998). Risk assessment, risk vales and the social science programme: Why we do need risk perception research. *Reliability Engineering and System Safety*, *59*, 5−15.

Pielke, R. A. (1998). Rethinking the role of adaptation in climate policy. *Global Environmental Change*, 8(2), 159–170.

Pierre-Nathoniel, D., Siegele, L., Roper, L.-A., & Menke, I. (2019). Loss and Damage at COP25 – A hard fought step in the right direction. *Climate Analytics*.

Pihl, E., Martin, M. A., Blome, T., Hebden, S., Jarzebski, M. P., Lambino, R. A., ... Sonntag, S. (2019). *10 New insights in climate science 2019*. Stockholm: Future Earth & The Earth League.

Pitt, D. R., & Randolph, J. (2009). Identifying obstacles to municipal climate protection planning. *Environment and Planning C: Government and Policy*, 27(6), 841–857.

Pitt, D. R. (2010a). The impact of internal and external characteristics on the adoption of climate mitigation policies by United States municipalities. *Environment and Planning C: Government and Policy*, 28(5), 851–871.

Pitt, D. R. (2010b). Harnessing community energy: The Keys to climate mitigation policy adoption in United States municipalities. *Local Environment.*, 15(8), 717–729.

Powell, H. (2008). Will ocean iron fertilization work? *Oceanus*, 46(1), 10, Retrieved from. Available from http://www.whoi.edu/fileserver.do?id = 30703&pt = 2&p = 35609.

Pramova, E., Locatelli, B., Djoudi, H., & Somorin, O. A. (2012a). Forests and trees for social adaptation to climate variability and change. *WIREs Climate Change*, 3(6), 581–596.

Pramova, E., Locatelli, B., Brockhaus, M., & Fohlmeister, S. (2012b). Ecosystem services in the national adaptation programmes of action. *Climate Policy*, 12(4), 393–409.

Puig, D., Serdeczny, O., & Huq, S. (2018). *Loss and damage in COP-24*. Nairobi: UNEP-DTU Available at. Available from https://unepdtu.org/wp-content/uploads/2019/02/ld-cop24-dtu-ca-icccad-final1.pdf.

Puig, D., Wewerinke-Singh, M., & Huq, S. (2019). *Loss and damage in COP-25*. Nairobi: UNEP-DTU.

Rabitz, F. (2016). Going rogue? Scenarios for unilateral geoengineering. *Futures*, 84, 98–107.

Rahman, M. (2014). *Framing ecosystem-based adaptation to climate change: Applicability in the coast of Bangladesh*. Dhaka: IUCN.

Rau, G. H., & Caldeira, K. (1999). Enhanced carbonate dissolution: A means of sequestering waste CO2 as ocean bicarbonate. *Energy Conversion and Management*, 40(17), 1803–1813.

Raven, J. A., & Falkowski, P. G. (1999). Oceanic sinks for atmospheric. *Plant, Cell & Environment*, 22(6), 741–CO755.

Raven, J., Stone, B., Mills, G., Towers, J., Katzschner, L., Leone, M., ... Hariri, M. (2018). Urban planning and design. In C. Rosenzweig, W. Solecki, P. Romero-Lankao, S. Mehrotra, S. Dhakal, & S. Ali Ibrahim (Eds.), *Climate change and cities: Second assessment report of the urban climate change research network* (pp. 139–172). New York: Cambridge University Press.

Rayner, S. (2015). To know or not to know? A note on ignorance as a rhetorical resource in geoengineering debates. In G. Matthias, & McGoey (Eds.), *Routledge international handbook of ignorance studies* (pp. 308–317). London: Routledge, Taylor & Francis Group.

Redman, C., Grove, M. J., & Kuby, L. (2004). Integrating social science into the long-term ecological research (LTER) network: Social dimensions of ecological change and ecological dimensions of social change. *Ecosystems*, 7(2), 161–171.

Reid, H., Hou, J. X., Porras, I., Hicks, C., Wicander, S., ... Roe, D. (2019). Is ecosystem-based adaptation effective? Perceptions and lessons learned from 13 project sites. *IIED Research Report*. London: IIED.

Rensburg, W. V. (2015). Climate change scepticism: A conceptual reevaluation. *Sage Open*, 5(2), 1–13.

Reynolds, J. (2018a). Governing experimental responses: Negative emissions technologies and solar climate engineering. In A. Jordan, D. Huitema, H. van Asselt, & J. Forster (Eds.), *Governing climate change: Polycentricty in action* (pp. 285–302). Cambridge: Cambridge University Press.

Reynolds, J. L., Jorge, L. C., & Sarnoff, J. D. (2018b). Intellectual property policies for solar geoengineering. *Climate Change*, 9(2), e512.

Reynolds, J. L. (2019). Solar geoengineering to reduce climate change: A review of governance proposals. *Proceedings of Royal Society, A 475*, 20190255.

Ribot, J. (2010). Vulnerability does not just fall from the sky: Toward multi-scale pro-poor climate policy. In R. Mearns, & A. Norton (Eds.), *Social dimensions of climate change: Equity and vulnerability in a warming world*. Washington, DC: The World Bank.

Rickels, W., Rehdanz, K., & Oschlies, A. (2010). Methods for greenhouse gas offset accounting: A case study of ocean iron fertilization. *Ecological Economics, 69*(12), 2495–2509.

Rippl, S. (2002). Culture theory and risk perception: A proposal for a better measurement. *Journal of Risk Research, 5*(2), 147–165.

Rizvi, A. R., Baig, S., Barrow, E., & Kumar, C. (2015a). *Synergies between climate mitigation and adaptation in forest landscape restoration*. Gland: IUCN.

Rizvi, A. R., Baig, S., & Verdone, M. (2015b). *Ecosystems based adaptation: Knowledge Gaps in making an economic case for investing in nature based solutions for climate change*. Gland: IUCN.

Roberts, D. (2019a). *The case for "conditional optimism" on climate change*. vox.com online.

Roberts, D. (2019b). *The Paris Climate Agreement is at risk of falling apart in the 2020s*. Vox online.

Roberts, E., & Huq, S. (2015). Coming full circle: The history of loss and damage under the UNFCCC. *International Journal of Global Warming, 8*(2), 141–157.

Roberts, D., Boon, R., Diederichs, N., Douwes, E., Govender, N., Mcinnes, A., ... Spires, S. (2012). Exploring ecosystem-based adaptation in Durban, South Africa: "Learning-by-doing" at the local government coal face. *Environment and Urbanization, 24*(1), 167–195.

Roesijadi, G., Jones, S. B., Snowden-Swan, L. J., & Zhu, Y. (2010). Macroalgae as a biomass feedstock: A preliminary analysis, PNNL 19944. Richland: Pacific: Northwest National Laboratory.

Rogelj, J., den Elzen, M., Höhne, N., Fransen, T., Fekete, H., Winkler, H., ... Meinshausen, M. (2016). Paris Agreement climate proposals need a boost to keep warming well below 2°C. *Nature, 534*(7609), 631–639.

Rogge, K. S., & Reichardt, K. (2016). Policy mixes for sustainability transitions: An extended concept and framework for analysis. *Research Policy, 45*, 1620–1635.

Romero Lankao, Patricia. (2007). How do local governments in Mexico City manage global warming? *Local Environment., 12*(5), 519–535.

Rosenfeld, D. (2007). New insights to cloud seeding for enhancing precipitation and for hail suppression. *Journal of Weather Modification, 39*, 61–69.

Roth, J., Echeverria, D., & Gass, P. (2019). *Current status of Article 6 of the Paris Agreement: Internationally transferred mitigation outcomes (ITMOs)*. Manitoba: International Institute for Sustainable Development.

Roy, J., Tschakert, P., Waisman, H., Abdul Halim, S., Antwi-Agyei, P., Dasgupta, P., ... Suarez Rodriguez, A. G. (2018). *Sustainable development, poverty eradication and reducing inequalities*. Global warming of 1.5°C. *An IPCC Special Report on the impacts of global warming of 1.5°C above pre-industrial levels and related global greenhouse gas emission pathways, in the context of strengthening the global response to the threat of climate change, sustainable development, and efforts to eradicate poverty* (pp. 445–540). Geneva: IPCC.

Sachs, D. J. (2015). *The age of sustainable development*. New York: Columbia University Press.

Sarewitz, D. (2011). The voice of science: Let's agree to disagree. *Nature, 478*, 7.

Sarmiento, J. L., & Orr, J. C. (1991). Three-dimensional simulations of the impact of Southern Ocean nutrient depletion on atmospheric CO2 and ocean chemistry. *Limnology and Oceanography, 36*(8), 1928–1950.

Scarano, F. R. (2017). Ecosystem-based adaptation to climate change: Concept, stability and a role for conservation science. Perspectives in ecology and conservation. *Science (New York, N.Y.), 15*(2), 65–73.

Schäpke, N., Omann, I., Wittmayer, J., van Steenbergen, F., & Mock, M. (2016). Linking transitions and sustainability: A study into social effects of transition management. In: *UFZ discussion paper, no. 11/2016*. Leipzig: Helmholtz-Zentrum fürUmweltforschung (UFZ).

Schafer, L., & Kreft, S. (2014). Loss and damage: Roadmap to relevance for the Warsaw international mechanism. Germanwatch e.V.

Scheer, D., & Renn, O. (2014). Public perception of geoengineering and its consequences for public debate. *Climatic Change*, *125*(3–4), 305–318.

Schelling, T. C. (1996). The economic diplomacy of geoengineering. *Climatic Change*, *33*(3), 303–307.

Schenk, T., Vogel, R. A. L., Maas, N., & Tavasszy, L. A. (2016). Joint fact-finding in practice: Review of a collaborative approach to climate-ready infrastructure in Rotterdam. *European. Journal of Transport and Infrastructure Research*, *16*(1), 273–293.

Schinko, T., & Mechler, R. (2017). Applying recent insights from climate risk management to operationalize the loss and damage mechanism. *Ecological Economics*, *136*, 296.

Schipper, E. L. F. (2006). Conceptual history of adaptation in the UNFCCC process. *Review of European Community and International Environmental Law*, *15*(1), 82–92.

Schipper, E. L. F., & Burton, I. (Eds.), (2008). *The Earthscan reader on adaptation to climate change*. London: Routledge.

Schneider, L. (2019). *United States and Saudi Arabia block UN efforts at climate geoengineering efforts*. DESMOG.

Schrijver, N. (2016). Managing the global commons: Common good or common sink? *Third World Quarterly*, *37*(7), 1252–1267.

Schroeder, H., & Bulkeley, H. (2009). Global cities and the governance of climate change: What is the role of law in cities? *Fordham Urban Law Journal*, *36*, 313–359.

Self, S., Zhao, J.-X., Holasek., Rick, E., Torres, R. C., & King, A. J. (1996). The atmospheric impact of the 1991 Mount Pinatubo eruption. In C. G. Newhall, & R. S. Punongbayan (Eds.), *Fire and mud: Eruptions and Lahars of Mount Pinatubo, Philippines*. Seattle, WA: Washington University Press.

Seto, K. C., Guneralp, B., & Hutyra, L. R. (2012). Global forecasts of urban expansion to 2030 and direct impacts on biodiversity and carbon pools. *Proceedings of the National Academy of Sciences*, *109*, 16083–16088.

Shabangu, S., Woolf, D., Fisher, E. M., Angenent, L. T., & Lehmann, J. (2014). Techno-economic assessment of biomass slow pyrolysis into different biochar and methanol concepts. *Fuel.*, *117*, 742–746.

Shaw, M. R., Overpeck, J. T., & Midgley, G. F. (2014). *Cross-chapter box on ecosystem-based approaches to adaptation – Emerging opportunities*. Climate change 2014: Impacts, adaptation, and vulnerability. Part A: Global and sectoral aspects. *Contribution of Working Group II to the Fifth Assessment Report of the Intergovernmental Panel on Climate Change* (pp. 35–94). Cambridge: Cambridge University Press.

Shi, L., Chu, E., Anguelovski, I., Aylett, A., Debats, J., Goh, K., ... VanDeveer, S. D. (2016). Roadmap towards justice in urban climate adaptation research. *Nature Climate Change*, *6*(2), 131–137.

Shipan, C. R., & Volden, C. (2006). Bottom-up federalism: The diffusion of antismoking policies from United States Cities to States. *American Journal of Political Science*, *50*(4), 825–843.

Siegele, L. (2017). Loss and damage (Article 8). In K. Daniele, C. Maria Pia, D. Meinhard, B. Jane, & H. Andrew (Eds.), *The Paris Agreement on Climate Change: Analysis and commentary* (pp. 224–238). Oxford: Oxford University Press.

Siegrist, M., Keller, C., & Kiers, H. A. L. (2005). A new look at the psychometric paradigm of perception of hazards. *Risk Analysis*, *25*(1), 211–222.

Silva, J. M. C., & Wheeler, E. (2017). Ecosystems as infrastructure. *Perspectives in Ecological Conservation*, *15*(1), 32–35.

Slovic, P. (1999). Trust, emotion, sex, politics, and science: Surveying the risk-assessment battlefield. *Risk Analysis*, *19*(4), 689–701.

Smit, B., Burton, I., Klein, R. J. T., & Wandel, J. (2000). An anatomy of adaptation to climate change and variability. *Climatic Change*, *45*(1), 223–251.

Smit, B., & Pilifosova, O. (2001). *Adaptation to Climate Change in the Context of Sustainable Development and Equity. Climate change 2001: Impacts, adaptation and vulnerability.* Cambridge: Cambridge University Press, Chapter 18.

Smit, B., & Wandel, J. (2006). Adaptation, adaptive capacity and vulnerability. *Global Environmental Change, 16*(3), 282−292.

Smith, A., & Stirling, A. (2008). Social-ecological resilience and socio-technical transitions: Critical issues for sustainability governance. *STEPS working paper 8.* Brighton: STEPS Centre.

Smith, J. B., & Lazo, J. K. (2001). A summary of climate change impact assessments from the United States country studies program. *Climatic Change, 50*(1−2), 1−29.

Smith, J. B., Klein, R. J. T., & Huq, S. (Eds.), (2003). *Climate change, adaptive capacity and development.* London: Imperial College Press.

Smith, P., & Friedmann, J. (2017). Bridging the gap − Carbon dioxide removal. In UNEP (Ed.), *The emissions gap report 2017* (pp. 58−67). Nairobi: UNEP.

Smith, T. F., Brooke, C., Measham, T., Preston, B., Goddard, R., Withycombe, G., Beverage, B., & Morrison, C. (2008). *Case studies of adaptive capacity: Systems approach to regional climate change adaptation strategies.* Sydney: CSIRO Climate Adaptation Flagship.

Smith, P., Gregory, P., van Vuuren, D., et al. (2010). Competition for land. *Philosophical Transactions of the Royal Society, B., 365,* 2941−2957.

Smith, P. (2012). Soils and climate change. *Current Opinion in Environmental Sustainability, 4,* 539−544.

Smith, P. (2016). Soil carbon sequestration and biochar as negative emission technologies. *Global Change Biology, 22,* 1315−1324.

Smith, P., Bustamante, M., Ahammad, H., Clark, H., Dong, H., Elsiddig, E. A., ... Tubiello, F. (2014). *Agriculture, forestry and other land use (AFOLU). Climate change 2014: Mitigation of climate change. Contribution of Working Group III to the Fifth Assessment Report of the Intergovernmental Panel on Climate Change* (pp. 811−922). Cambridge: Cambridge University Press.

Smith, P., Davis, S. J., Creutzig, F., Fuss, S., Minx, J., Gabrielle, B., & Van Vuuren, D. P. (2016). Biophysical and economic limits to negative CO2 emissions. *Nature Climate Change, 6*(1), 42−50.

Somorin, O. A., Brown, H. C. P., Visseren-Hamakers, I. J., Sonwa, D. J., Arts, B., & Nkem, J. (2011). The Congo basin forests in a changing climate: Policy discourses on adaptation and mitigation (REDD +). *Global Environmental Change, 22*(1), 288−298.

SRMGI (Solar Radiation Management Governance Initiative). (2011). *Solar radiation management: The governance of research.*

Sreeramana, A., & Shubhrajyotsna, A. (2016). Opportunities challenges for green technology in 21st century. In: *MPRA paper no. 73661.* Available at https://mpra.ub.uni-muenchen.de/73661/1/MPRA_paper_73661.pdf.

Stavins, R. N. (2011). The problem of the commons: Still unsettled after 100 Years. *American Economic Review, 101,* 81−108.

Stern, D. (2012). Modeling international trends in energy efficiency. *Energy Economics, 34*(6), 2200−2208.

Stem, T. (2018). *The Paris Agreement and its future. Paper, 5.*

Stewart, R. B. (2008). States and cities as actors in global climate regulation: Unitary vs. plural architectures. *Arizona Law Review, 50*(3), 681−707.

Stewart, R. B., Kingsbury, B., & Rudyk, B. (2009). *Climate finance: Regulatory and funding strategies for climate change and global development.* New York: New York University Press.

Stilgoe, J. (2015). *Experiment Earth: Responsible innovation in geoengineering.* London: Routledge.

Stilgoe, J. (2016). Geoengineering as collective experimentation. *Science and Engineering Ethics, 22*(3), 851−869.

Stolaroff, J. K., Keith, D. W., & Lowry, G. V. (2008). Carbon dioxide capture from atmospheric air using sodium hydroxide spray. *Environmental Science & Technology, 42*(8), 2728−2735.

Suarez, I., & Huang, J. (2015). *Addressing adaptation in a 2015 climate agreement*. Arlington, VA: The Center for Climate and Energy Solutions (C2ES), June.

Swart, R., & Raes, F. (2007). Making integration of adaptation and mitigation work: Mainstreaming into sustainable development policies? *Climate Policy*, 7(4), 288—303.

Tang, Z., Samuel, D. B., Quinn, C., Chang, L., & Wei, T. (2010). Moving from agenda to action: Evaluating local climate change action plans. *Journal of Environmental Planning and Management*, 53(1), 41—62.

Tang, Z., Wei, T., Quinn, C., & Zhao, N. (2012). Surveying local planning directors' actions for climate change. *International Journal of Climate Change Strategies and Management*, 4(1), 81—103.

The Royal Society. *Geoengineering the climate: Science, governance and uncertainty*. (2009).

Thorgeirsson, H. (2017). Objective (Article 2.1). In K. Daniele, C. Maria Pia, D. Meinhard, B. Jane, & H. Andrew (Eds.), *The Paris Agreement on Climate Change: Analysis and commentary* (pp. 131—140). Oxford: Oxford University Press.

Thornton, T. F., & Comberti, C. (2017). *Synergies and trade-offs between adaptation, mitigation and development*. Climate Change, 140(1), 5—18.

Thuy, P. T., Moeliono, M., Locatelli, B., Brockhaus, M., Di Gregorio, M., & Mardiah, S. (2014). Integration of adaptation and mitigation in climate change and forest policies in Indonesia and Vietnam. *Forests*, 5(8).

Tollefson, C., Zito, A., & Gale, F. (2012). Symposium overview: Conceptualizing new governance arrangements. *Public Administration*, 90(1).

Tollefson, J. (2019). Geoengineering debate shifts to UN environment assembly. *Nature*.

Trabucco, A., Zomer, R. J., Bossio, D. A., van Straaten, O., & Verchot, L. V. (2008). Climate change mitigation through afforestation/reforestation: A global analysis of hydrologic impacts with four case studies. *Agriculture, Ecosystems & Environment*, 126(1), 81—97.

Trisos, C. H., Amatulli, G., Gurevitch, J., Robock, A., Xia, L., & Zambri, B. (2018). Potentially dangerous consequences for biodiversity of solar geoengineering implementation and termination. *Nature Ecology & Evolution*, 2(3), 475—482.

TWN (Third World Network). (2019, 11 December). Difficult issues under Article 6 of Paris Agreement Negotiations. *Madrid News Update*.

UNDP (United Nations Development Program). (2019). *Evaluation guidelines*. New York: Independent Evaluation Office of UNDP.

UNECE (United Nations Economic Commission for Europe). (2016). The co-benefits of climate change mitigation. In: *Sustainable development brief no. 2*. Geneva: UNECE.

UNEP (United Nations Environment Program). (2013). *The emission gap report 2013: A UNEP synthesis report*. Nairobi: UNEP.

UNEP (United Nations Environment Program). (2019a). *Emissions gap report 2019*. Nairobi: UNEP.

UNFCCC (United Nations Framework Convention on Climate Change). (1992). *United Nations framework Convention on Climate Change*. FCCC/INFORMAL/84. GE.05—62220 (E) 200705. Available at https://unfccc.int/resource/docs/convkp/conveng.pdf.

UNFCCC (United Nations Framework Convention on Climate Change). (1995). *Report of the conference of the parties on its first session held in Berlin*, March 28—April 7, 1995. UNFCCC Website.

UNFCCC (United Nations Framework Convention on Climate Change). (1998). *Kyoto Protocol to the United Nations framework convention on climate change*. UNFCCC Portal Online.

UNFCCC. (2002). *UNFCCC report of the conference of the parties on its seventh session*, held at Marrakesh from October 29 to November 10, 2001. FCCC/CP/2001/13/Add.1.21.

UNFCCC (United Nations Framework Convention on Climate Change). (2005). *Decision1/CMP.1, consideration of commitments for subsequent periods for parties included in Annex I to the convention under Article 3, paragraph 9, of the Kyoto Protocol*. FCCC/KP/CMP/2005/8/Add.1.

References

UNFCCC (United Nations Framework Convention on Climate Change). (2010). *Report of the Conference of the Parties on its fifteenth session*, held in Copenhagen from December 7 to 19, 2009. FCCC/CP/2009/11/Add.1.

UNFCCC (2008a). *Report of the conference of the parties on its thirteenth session*, held in Bali from December 3 to 15, 2007. FCCC/CP/2007/6/Add.1.

UNFCCC (2008b). *UN Ad Hoc Working Group on long term cooperative action under the convention*, Fourth Session, Pozan, December 1–10, 2008.

UNFCCC (2011a). *Report of the conference of the parties on its sixteenth session*, held in Cancun from November 29 to December 10, 2010. In: *Addendum—Part Two: Action taken by the conference of the parties*. Bonn: UNFCCC.

UNFCCC (United Nations Framework Convention on Climate Change). (2011b). *Establishment of an Ad Hoc Working Group on the Durban platform for enhanced action, draft decision-/CP-17*.

UNFCCC (United Nations Framework Convention on Climate Change). (2011c). *Fact sheet: Kyoto Protocol*.

UNFCCC (United Nations Framework Convention on Climate Change). (2012a). *Report of the Conference of the Parties on its seventeenth session*, held in Durban from November 28 to December 11, 2011. UNFCCC/CP/2011/9/Add.1.

UNFCCC. *Slow onset events: Technical paper*. (2012b). Available at http://unfccc.int/resource/docs/2012/tp/07.pdf.

UNFCCC. (2013a). *Warsaw international mechanism for loss and damage associated with climate change impacts*. Draft decision -/CP.19. FCCC/CP/2013/L.15.

UNFCCC. (2013b). *Non-economic losses in the context of the work programme on loss and damage*. FCCC/TP/2013/2. Retrieved from http://unfccc.int/resource/docs/2013/tp/02.pdf.

UNFCCC (United Nations Framework Convention on Climate Change). (2013c). *Report of the Conference of the Parties serving as the meeting of the parties to the Kyoto Protocol on its eighth session*, held in Doha from November 26 to December 8, 2012. UNFCCC/KP/CMP/2012/13/Add.1.

UNFCCC. (2011a). *Report of the conference of the parties on its sixteenth session*, held in Cancun from November 29 to December 10, 2010. FCCC/CP/2010/7/Add.1.

UNFCCC. (2011b). *Views and information on elements to be included in the work programme on loss and damage*. FCCC/SBI/2011/MISC.1.

UNFCCC. (2011c). *Approaches to address loss and damage associated with climate change impacts in developing countries that are particularly vulnerable to the adverse effect of climate change to enhance adaptive capacity—Activities to be undertaken under the work programme*. FCCC/SBI/2011/L.20.

UNFCCC (United Nations Framework Convention on Climate Change). (2009b). *Fact sheet: The need for mitigation*.

UNFCCC (United Nations Framework Convention on Climate Change). (2014). *Report of the Conference of the Parties on its nineteenth session*, held in Warsaw from November 11 to 23, 2013. UNFCCC/CP/2013/10/Add.1.

UNFCCC. (2014a). *Report of the Executive Committee of the Warsaw International Mechanism for Loss and Damage associated with Climate Change Impacts*. FCCC/SB/2014/4.

UNFCCC. (2014b). *Report of the Conference of the Parties on its nineteenth session*, held in Warsaw from November 11 to 23, 2013. FCCC/CP/2013/10/Add.1. Retrieved from http://unfccc.int/resource/docs/2013/cop19/eng/10a01.pdf.

UNFCCC (United Nations Framework Convention on Climate Change). (2015a). *Report of the Conference of the Parties on its twentieth session*, held in Lima from December 1 to 14, 2014. UNFCCC/CP/2014/10/Add.1.

UNFCCC (United Nations Framework Convention on Climate Change). (2015b). *Paris Agreement*. UNFCCC.

UNFCCC (United Nations Framework Convention on Climate Change). (2016). *Climate action now, summary for policy makers 2016*. Bonn: UNFCCC.

UNFCCC (United Nations Framework Convention on Climate Change). (2016b). *Decision 1/CP.21. Adoption of the Paris Agreement. In Report of the Conference of the Parties on its twenty-first session*, held in Paris from November 30 to December 13, 2015. FCCC/CP/2015/10/Add.1.

UNFCCC (United Nations Framework Convention on Climate Change). (2017). *Technological innovation for the Paris Agreement: Implementing nationally determined contributions, national adaptation plans and mid-century strategies*. Bonn: UNFCCC Secretariat.

UNFCCC (United Nations Framework Convention on Climate Change). (2018a). *Katowice texts*.

UNFCCC (United Nations Framework Convention on Climate Change). (2018b). *Implementation of the framework for capacity building in developing countries*. FCCC/SBI/2018/5.

UNFCCC (United Nations Framework Convention on Climate Change). (2018c). *Climate aware: 2018 Talanoa dialogue*.

UNFCCC (United Nations Framework Convention on Climate Change). (n.d.). *Climate-related risks and extreme events*. UNFCCC.

UNFCCC. (n.d.). *Status of ratification of the convention*. UNFCCC.

UNFCCC. *Key points from WIM review event breakout group discussions*. (2019a).

UNFCCC. *Milestones of the WIM*. (2019b).

UNFCCC. (2019c). *Group of 77 and China submission on the review of the WIM and the report of the WIM Executive Committee*. Madrid: COP25.

UNFCCC. (2019d). *Warsaw International mechanism for loss and damage associated with climate change impacts and its 2019 review. Draft decision -/CMA.2.* FCCC/PA/CMA/2019/L.7.

van der Gaast., & Monica, A. (2015). *History of the UN climate negotiations—Part 1—From the 1980 to 2010.* Climate Policy Info Hub.

van Drooge, L., & Spaapen, J. (2017). Evaluation and monitoring of transdisciplinary collaborations. *Journal of Technology Transfer*, 1–17.

van Vuuren, D. P., Deetman, S., van Vliet, J., van den Berg, M., van Ruijven, B. J., & Koelbl, B. (2013). The role of negative CO_2 emissions for reaching 2°C—Insights from integrated assessment modelling. *Climatic Change*, *118*(1), 15–27.

Vanhala, L., & Hestbaek, C. (2016). Framing climate change loss and damage in UNFCCC negotiations. *Global Environmental Politics*, *16*, 111–129.

Vasi, I. B. (2006). Organizational environments, framing processes, and the diffusion of the program to address the global climate change among the local governments in the United States. *Sociological Forum*, *21*(3), 439–466.

Vasi, I. B. (2011). *Winds of change: The environmental movement and the global development of the wind energy industry*. New York: Oxford University Press.

Vaughan, C., & Dessai, S. (2014). Climate services for society: Origins, institutional arrangements, and design elements for an evaluation framework. *WIREs Climate Change*, *5*, 587–603.

Vaughan, N. E., & Lenton, T. M. (2011). A review of climate geoengineering proposals. *Climatic Change*, *109*(3–4), 745–790.

Verheyen, R. (2002). Adaptation to the impacts of anthropogenic climate change—The international legal framework. *Review of European Community and International Environmental Law*, *11*(2), 129–143.

Veron, J. E. N. (2011). Ocean acidification and coral reefs: An emerging big picture. *Diversity*, *3*(2), 262–274.

Vignola, R., Harvey, C. A., Bautista-Solis, P., Avelio, J., Rapidel, B., Donatti, C., & Martinez, R. (2015). Ecosystem-based adaptation for smallholder farmers: Definitions, opportunities and constraints. *Agriculture, Ecosystems and Environment*, *2011*, 125–132.

Violetti, D., & Rodriguez, H. E. (2013). Climate, positive outcomes at COP-19 at Warsaw. *Ecoscienza*, *6*, 1–3.

Virgoe, J. (2009). International governance of a possible geoengineering intervention to combat climate change. *Climatic Change, 95*(1−2), 103−119.
Vogel, C., Syndhia, M., Maria, G., Hycenth, T. N., Stefan, S., Michelle, B., & Marcos, L. (2020). Stakeholders' perceptions on sustainability transition pathways of the cocoa value chain towards improved livelihood of small-scale farming households in Cameroon. *International Journal of Agricultural Sustainability, 18*(1), 55−69.
Vogel, J., Letson, D., & Herrick, C. (2017). A framework for climate services evaluation and its application to the Caribbean agrometeorological initiative. *Climate Services, 6*, 65−76.
Voigt, C. (2019). *The 'Article 15 Committee' to facilitate implementation and promote compliance—Paris Agreement policy brief*. Norway: ECST (Roundtable on Climate Change and Sustainable Transition).
Wallace, D. W. R., Law, C. S., Boyd, P. W., Collos, Y., Croot, P., Denman, K., ... Williamson, P. (2010). *Ocean fertilization. A scientific summary for policy makers*. Paris: UNESCO.
Walsh, C. L., Dawson, R. J., Hall, J. W., Barr, S. L., Batty, M., Bristow, A. L., ... Zanni, A. M. (2011). Assessment of climate change mitigation and adaptation in cities. *Proceedings of the Institution of Civil Engineers, Urban Design and Planning, 164*(2), 75−84.
Warner, K., & Van der Geest, K. (2013). Loss and damage from climate change: Local-level evidence from nine vulnerable countries. *International Journal of Global Warming, 5*, 367.
Walter, A. G. N. (2011). Controlling the Earth's albedo using reflective hollow glass spheres. *International Journal of Global Environmental Issues, 11*(2), 91.
Wamsler, C., & Pauleit, S. (2016). Making headway in climate policy mainstreaming and ecosystem-based adaptation: Two pioneering countries, different pathways, one goal. *Climate Change, 137*(1−2), 71−87.
Wang, C., Li, W., Yang, Z., Chen, Y., Shao, W., & Ji, J. (2015). An invisible soil acidification: Critical role of soil carbonate and its impact on heavy metal bioavailability. *Scientific Reports, 5*, 12735.
Warner, K., van der Geest, K., Kreft, S., Huq, S., Harmeling, S., Koen K., & de Sherbinin, A. (2012). Evidence from the frontlines of climate change: Loss and damage to communities despite coping and adaptation. In: *Loss and damage in vulnerable countries initiative. Policy report. Report no. 9*. Bonn: United Nations University Institute for Environment and Human Security (UNU-EHS).
Warner, K., van der Geest, K., & Kreft, S. (2013). Pushed to the limit: Evidence of climate change related loss and damage when people face constraints and limits to adaptation *(No. 11)*. Bonn: United Nations University Institute of Environment and Human Security.
Waters, E., Barnett, J., & Puleston, A. (2014). Contrasting perspectives on barriers to adaptation in Australian climate change policy. *Climatic Change, 124*, 691−702.
Weber, N. M. (2012). Rainmakers, space mirrors and atmospheric vacuums: A bibliometric mapping of geoengineering research. In: *Proceedings of the 2012 iConference* (pp. 639−640). ACM Press.
Weng, Z. H., Van Zwieten, L., Singh, B. P., Tavakkoli, E., Joseph, S., Macdonald, L. M., ... Cozzolino, D. (2017). Biochar built soil carbon over a decade by stabilizing rhizo deposits. *Nature Climate Change, 7*, 371−376.
Wewerinke-Singh, M. (2018a). Climate migrants' right to enjoy their culture. In S. Behrman & A. Kent (Eds.), *Climate refugees: Beyond the legal impasse?*. Abingdon, New York: Earthscan/Routledge.
Wewerinke-Singh, M. (2018b). State responsibility for human rights violations associated with climate change. In D. Sébastien, J. Sébastien, & J. Alyssa (Eds.), *Routledge handbook of human rights and climate governance*. Abingdon: Routledge.
Whitley, S., Thwaites, J., Wright, H., & Caroline, O. (2018). *Making finance consistent with climate goals: Insights for operationalising Article 2.1c of the UNFCCC Paris Agreement*. London: Overseas Development Institute (ODI).
Wiener, J. (2009). Property and prices to protect the planet. *Duke Journal of Comparative & International Law, 19*, 515−534. Available at:. Available from http://scholarship.law.duke.edu/faculty_scholarship/2227/.

Wilbanks, T. J., & Fernandez, S. J. (2014). *Climate Change and Infrastructure, Urban Systems, and Vulnerabilities*. Washington, DC: Island Press.

Willcox, J., Psarras, P., & Liguori, S. (2017). Assessment of reasonable opportunities for direct air capture. *Environment Research Letters, 12*, 065001.

Williamson, P., Wallace, D. W., Law, C. S., Boyd, P. W., Collos, Y., Croot, P., & Vivian, C. (2012). Ocean fertilization for geoengineering: A review of effectiveness, environmental impacts and emerging governance. *Process Safety and Environmental Protection, 90*(6), 475–488.

Winkler, H. (2017). Mitigation (Article 4). In K. Daniele, C. Maria Pia, D. Meinhard, B. Jane, & H. Andrew (Eds.), *The Paris Agreement on Climate Change: Analysis and commentary* (pp. 141–165). Oxford: Oxford University Press.

Winkelmann, R., Levermann, A., Ridgwell, A., & Caldeira, K. (2015). Combustion of available fossil fuel resources sufficient to eliminate the Antarctic ice sheet. *Science Advances, 1*(8), e1500589.

Wise, R. M., Fazey, I., Stafford Smith, M., Park, S. E., Eakin, H. C., Archer Van Garderen, E. R. M., & Campbell, B. (2014). Reconceptualizing adaptation to climate change as part of pathways of change and response. *Global Environmental Change, 28*, 325–336.

World Energy Council. (2016). *World energy resources*.

Worrall, F., Evans, M. G., Bonn, A., Reed, M. S., Chapman, D., & Holden, J. (2009). Can carbon offsetting pay for upland ecological restoration? *The Science of the Total Environment, 408*(1), 26–36.

WMO (World Meteorological Organization). (2019a). *The global climate in 2015–2019*. Geneva: WMO.

WMO (World Meteorological Organization). (2019b). *Greenhouse gas concentrations in atmosphere reach yet another high*. WMO.

WMO (World Meteorological Organization). (2019c). *WMO provisional statement on the state of the global climate in 2019*. Geneva: WMO.

WMO, UNEP, IPCC, Future Earth & Future League, & GFCS. (2019). *United in science*. Available online.

Woods, N. D., & Potoski, M. (2010). Environmental federalism revisited: Second-order devolution in air quality regulation. Review of policy. *Research; Journal of Science and Its Applications, 27*(6), 721–739.

WRI (World Resource Institute). (2016). *Staying on track from Paris: Advancing the key elements of the Paris Agreement. Working paper*.

WRI (World Resource Institute). (2019). *Review of the Paris Committee on capacity building-submission by the world resource institute*. Washington, DC: WRI.

Yamin, F., Rahman, A., & Huq, S. (2005). Vulnerability, adaptation and climate disasters: A conceptual overview. *IDS Bulletin, 36*(4), 1–14.

Yang, L., Qiana, feng, Songa, D.-X., & Zhenga, K.-J. (2016). Research on urban heat-island effect. *Science Direct, 169*, 11–18.

Zahran, S., Grover, H., Brody, S. D., & Vedlitz, A. (2008a). Risk, stress, and capacity: Explaining metropolitan commitment to climate protection. *Urban Affairs Review, 43*(4), 447–474.

Zahran, S., Samuel, D. B., Vedlitz, A., Grover, H., & Miller, C. (2008b). Vulnerability and capacity: Explaining local commitment to climate-change policy. *Environment and Planning C: Government and Policy, 26*, 544–562.

Zakkour, P., Scowcroft, J., & Heidug, W. (2014). The role of UNFCCC mechanism in demonstration and deployment of CCS technologies. *Energy Procedia, 63*, 6945–6958.

Zedler, J. B., & Suzanne, K. (2005). Wetland resources: Status, trends, ecosystem services, and restorability. *Annual Review of Environmental Resources, 30*, 39–74.

Zhao, K., & Jackson, R. B. (2014). Biophysical forcings of land-use changes from potential forestry activities in North America. *Ecological Monographs, 84*(2), 329–353.

Zhao, C., Yan, Y., Chenxing, W., Mingfang, T., Gang, W., Ding, M. fang, & Yang, S. (2018). Adaptation and mitigation for combating climate change—From single to joint. *Ecosystem Health and Sustainability.*, *4*(4), 85–94.

Ziblatt, D. (2008). Why some cities provide more public goods than others: A subnational comparison of the provision of public goods in German Cities in 1912. *Studies in Comparative International Development*, *43*, 273–289.

Zihua, G., Voigt, C., & Werksman, J. (2019). Facilitating implementation and promoting compliance with the Paris Agreement Under Article 15: Conceptual challenges and pragmatic choices. *Climate Law*, *9*(1–2), 65–100.

CHAPTER 5

Toward water security

5.1 Introduction

Water, a finite source, is the most essential element for the survival of human and other biotic organisms, after air. Sustainable supply of water constitutes the bedrock of almost essential necessities of human survival such as food, energy, human health, agriculture, industry, etc., along with socioeconomic and cultural development. Owing to its pivotal role in economic development, the development and management of water resources is "chronologically locked" with the development of the "modern nation-state" (Jeffrey & Gearey, 2006). Water lies at the core of the three pillars of sustainable development — economic, social and environmental. Water resources and the essential services rendered by these resources are among the keys to attaining ambitious goals of poverty alleviation, inclusive growth, public health, food security, lives of dignity for all, along with enduring harmony with earth's basic ecosystems. Salience garnered by water issues globally in recent years seem to have proved instrumental in demonstrating increasing understanding of water's centrality as reflected in the millennium development goal (MDG) target of halving the proportion of people without sustainable access to safe drinking water between 1990 and 2010; and this period witnessed about 2.3 billion people gaining access to improved drinking-water sources like piped supplies and protected wells (WWAP, 2015). Water ranks sixth as sustainable development goal (SDG) in Agenda 2030 for SDGs.

Water has continued to be an essential and integral component of sustainable development discourse from the 1992 UN (United Nations) Conference on Environment and Development through to the 2000 MDGs to the SDGs to guide and spur the national governments and the international community to ensure sustainable and continuous supply of water. Nevertheless, this critical resource of water has seemingly come under threat from long-overlooked problems that are closely linked with its management, increased demand and climate change, the impacts of which have begun to unfold. Problems pertaining to water in terms of access to water, water stress, water scarcity, and contamination of water resources, once viewed as problems affecting only developing countries, now have come to affect the entire globe (Ray, 2010). Noting that unsustainable development pathways and governance failures have engendered substantial pressures on water resources thereby affecting quality and availability of water, Connor et al. (2015) opine that it has resulted in compromising water's ability to generate socioeconomic benefits. Besides, the planet's capacity to sustain the growing demands for freshwater is being challenged and "there can be no sustainable development unless the balance between demand and supply is restored" (Connor et al., 2015).

Problems affecting water sector are: the yawning gap between water availability and increasing demand for water, and this hiatus between supply and demand engenders the issues of water stress and scarcity and these issues are further compounded by contamination of water resources and climate change that spurs occurrence of water-induced calamities. Addressing these issues in a scientific manner entails the potential of unfolding pathways for ensuring water security.

5.2 Water availability

About 71% of Earth's surface is covered by water that is available in seas and oceans and remaining 29% is land area. These seas and oceans encompass 97.5% of water that is salty and not fit for consumption by human and other biotic organisms. Only 2.5% of the water on Earth is freshwater that can be used for human needs. Almost 70% of the world's freshwater is to be found in the cold zones of the planet in solid form, as in glaciers and perennial snow. Around 30% of freshwater is located underground whereas just under 1% is present in the form of humidity in the air and ground. Just think that only 0.3% of the planet's freshwater is easily accessible to humans, the surface water that we can see in rivers and lakes (Guppy & Anderson, 2017; UNESCO, n.d.). The total water resources in the world, estimated in the order of 43,750 km^3/year, are not evenly distributed throughout the world due to the patchwork of climates and physiographic structures; nevertheless, at the continental level, America has the largest share of the world's total freshwater resources with 45%, followed by Asia with 28%, Europe with 15.5% and Africa with 9% (FAO, 2003).

Viewed in a broad perspective, availability of surface water resources at continent level ought to remain comparatively constant as opposed to the fluctuations in demography, gross domestic product (GDP) or water demand; however, Burek et al. (2016) opine that at the subregional level, any change would be small, varying from −5% to +5%, due to climate change impacts; nonetheless, changes can be much more distinct at the country level. A large number of countries are already experiencing persistent water scarcity conditions and such countries entail the likelihood of coping with scummier surface water resources availability in the 2050s. According to Veldkamp et al. (2017), almost all countries at present that are encompassed in a belt around 10−40 degrees north, from Mexico to China and to Southern Europe are affected by water scarcity, in tandem with Australia, Western South America, and Southern Africa in the Southern Hemisphere. The period during the early-mid-2010s witnessed nearly 1.9 billion people, accounting for 27% of the then total global population, inhabiting potential strictly water-scarce regions and by 2050 this figure could increase to about 2.7−3.2 billion. Nevertheless, some 4.8−5.7 billion people, almost half of the global population, would be living in in potential water-scarce areas by 2050, if monthly variability is taken into consideration. According to Burek et al. (2016), if adaptive capacity is factored in, the possibility of 3.6−4.6 billion people going under water stress by 2050 could not be ruled out and in that eventuality bulk of such population or 91%−96% would be living in Asia, especially in Southern and eastern Asia; along with 4%−9% in Africa, mainly in the north.

Groundwater, usually contained in aquifers located in sediments and rocks, constitutes Earth's predominant reserve of freshwater, commonly with storage times spanning from decades to centuries and millennia. However, rapid pace of intensive groundwater abstraction that began from the 1950s in the wake of major advances in geological knowledge, water-well drilling, pump technology and rural electrification. Admittedly, globally groundwater withdrawals are still increasing, having reached 900 km^3/a in 2010, and provide about 36% of potable water supply, 42% of water for irrigated agriculture and 24% of direct industrial water supply. Withdrawal intensity of groundwater is said to be the highest over much of China, India, Pakistan and Iran, and parts of Bangladesh, Mexico, the United States, the EU, North Africa, and the Middle East. During the period from 2000 to 2008, estimate of rates of permanent storage depletion ranged from 100 to 145 km^3/a (Wada et al., 2016). Depletion of groundwater resource indirectly contributes to global sea-level rise by creating a transfer of water from long-term terrestrial storage to circulation in the surface hydrosphere and such an eventuality is prone to entail serious consequences for coastal areas. According to Taylor, Voss, MacDonald, Aureli, and Aggarwal (2016), this process is subject to uncertainty owing to inherent inexactitude in long-term aquifer water-balances, the average unit drainable storage of depleted aquifers and the proportion of extracted groundwater remaining in the local microclimate. As compared to current level of groundwater abstraction, a substantial surge in that abstraction amounting to 1100 km^3/year has been predicted to occur by the 2050s, comprising an increase of 39%. A comparative assessment of water withdrawal to their maximum sustainable levels enables us discern the significance of current water availability challenges. Level of global groundwater withdrawals being at about 4600 km^3/year in the 2010s were already near maximum sustainable levels (Gleick & Palaniappan, 2010; Hoekstra & Mekonnen, 2012). It has pointed in some of the past World Water Development Reports (WWDRs) that global figures often mask severe challenges at regional and global levels. According to Richey et al. (2015), almost a third of the world's biggest groundwater systems are already in distress.

Huffaker (2008) has argued that improved efficiency of irrigation water use entails the likelihood of actually contributing to an overall intensification of water depletion at basin level through increases in the total evaporation from crops and reductions in return flows. It is in this context that Ward and Pulido-Velazquez (2008) suggest that water efficiency gains in irrigation ought to be accompanied by regulatory measures on water allocations and/or irrigation areas. Attention toward limited scope of the expansion of irrigation worldwide has already been by the Comprehensive Assessment of Water Management in Agriculture (2007) albeit, with some regional exceptions, and that attention needs to shift away from surface water allocations to improving rainfed agriculture. Increasing problems of silting, available runoff, environmental concerns and restrictions have also limited the option of constructing more reservoirs. According to OECD (2016), construction of more ecosystem-friendly forms of water storage, such as natural wetlands, soil moisture and more efficient recharge of groundwater in suitable areas could prove more sustainable and cost-effective than traditional infrastructure such as dams.

5.3 Demand for water

Demand for water has constantly been increasing and global water use has recorded an increase by a factor of six over the past 100 years (Wada et al., 2016). Water has been witnessing a steady growth at a rate of about 1% per year (AQUASTAT, n.d.). Burgeoning population, economic development, rapid pace of urbanization and changing patterns of consumption along with other attendant factors have been instrumental in spurring increase in water demand over the years. The global population is expected to increase from 7.8 billion in December 2019 (Wordometere, 2019) to between 9.4 and 10.2 billion by 2050, with two-thirds of the population living in urban areas. More than half this anticipated population growth is likely to take place in Africa (+1.3 billion), with Asia (+0.75 billion) expected to be the second largest contributor to future population growth (UN-DESA, 2017). International economic growth in terms of global GDP is likely to increase by a factor of 2.5 during the period between 2017 and 2050 (OECD, n.d.), and variations in the patterns of economic growth are also anticipated during this period. This increase in population and different sectors of economy will expert increased pressures on water resources. Global demand for food and energy, both of which are water-intensive, is likely to increase nearly between 60% and 80% by 2025 (OECD, 2012). Concurrently, water cycle worldwide is getting intensified owing to global warming, with water regions primarily becoming wetter and drier regions becoming more drier (IPCC, 2014). These aspects worldwide change demonstrates the need for well-concerted planning and implementation of strategic, appropriate and sound management and countermeasures to stem the tide of depletion of water resources (Burek et al., 2016).

Global water demand that was estimated by Burek et al. (2016) around 2015 at about 4600 km^3/year is projected to increase by 20%–30% to between 5500 and 6000 km^3/year by 2050. While noting that estimations at the global scale get complicated due to availability of limited observational data and the interactions of a combination of important environmental, social, economic, and political factors, such as global climate change, population growth, land-use change, globalization and economic development, technological innovations, political stability and the extent of international cooperation, Wada et al. (2016, p. 176) point out, "Because of these interconnections, local water management has global impacts, and global developments have local impacts."

Undoubtedly, agriculture accounts for about 70% of global water withdrawals and the bulk of this water is used for irrigation; nonetheless, global estimates for annual irrigation water demand are fraught with uncertainty (WWAP, 2018). Apart from the lack of monitoring and reporting on water used for irrigation, it is also attributable to the inherently erratic nature of the practice itself. This is what makes projecting future water demand for irrigation so difficult. Making projections for future irrigation demands gets complicated in view of the given efficiency of different irrigation techniques that have a direct impact on overall water use. On the one hand, Burek et al. (2016) have projected increases in global crop irrigation water requirements for 2050 to be somewhere between 23% and 42% above the level in 2010, whereas, on the other hand, FAO (2011a) estimated a 5.5% increase in water withdrawals for irrigation from 2008 to 2050. While alluding to likely increases in irrigation water efficiency, OECD (2012) anticipated a slight decrease in water use for irrigation through the period 2000–50.

There is likelihood of food demand registering increase by 60% by 2050 (WWAP, 2018) and this increase will require more arable land and intensification of production which in turn will translate into increased water use (Leadley et al., 2014). Suggesting that these effects, including need for more land and water, can largely be avoided, FAO (2011b) points out this can be had if further intensification of agricultural production is based on ecological intensification that entails improving ecosystem services to reduce external inputs.

Global usage of water for industry currently stands at nearly 20% of the total withdrawals and this usage is dominated by energy production that accounts for 75% of the industry total and the manufacturing the remaining 25% (WWAP, 2014). Water demand for industry is projected to increase by 2050 across all the regions of the globe, with the possible exceptions of North America and Europe (Burek, et al., 2016). Africa where present industry use of water is almost negligible is projected to witness water demand for industry to increase by 800% by 2050. As per OECD (2012) estimates, global water demand for manufacturing will go up by 400% over the period 2000–50.

Global water usage for the energy sector is projected to increase by 20% of the total over the period 2010−35 (Burek et al., 2016) and by 2050 it could increase by 85% (IEA, 2012). Global water withdrawals for energy production have been projected to rise by one-fifth over the period 2010−35, whereas water consumption would increase by 85% driven by the shift toward more efficient power plants with more advanced cooling systems (that reduce water withdrawals but increase consumption) and increased production of biofuels (IEA, 2012). It has been suggested by Chaturvedi et al. (2013) that by limiting bioenergy production to nonirrigated marginal or abandoned cropland could help in lessening negative impacts on food production and prices, water use, and biodiversity.

Domestic global water use that at present roughly accounts for the remaining 10% of global water usage, is projected to witness a significant increase over the 2010−50 period in nearly all regions of the globe, with the exception of Western Europe where it remains constant. Comparatively, the African and Asian subregions are projected to register greatest increases in domestic demand to the tune of 300% increase and it could more than double in Central and South America (Burek et al., 2016). This expected growth is attributed to a projected increase in water supply services in urban settlements.

Evidently, global water demand by 2050 is prone to continue to grow significantly, with industrial and domestic demand for water will likely grow much faster than agricultural demand, although agriculture will remain the largest overall user. Rose Grant et al. (2002) had forecasted that for the "first time in world history" absolute growth in nonagricultural demand for water will exceed growth in agricultural demand, resulting in a fall in agriculture's share of total water consumption in developing countries from 86% in 1995 to 76% by 2025. According to WWAP (2018), these projections emphasize on the importance of addressing water challenges facing agriculture where agricultural demand for water, and competition for it, are both set to increase. While conceding that the demand for water by 2050 would increase dramatically albeit unequally, across all the continents, Boretti and Rosa (2019) aver those quantitative estimates are difficult to provide with accuracy as well as estimate of the WWAP (2018) "are not expected to be very accurate, and likely optimistic."

5.4 Water quality

Pollution of water resources has fast emerged as a global challenge over the years almost across the globe that is proving a potent factor in undermining economic growth as well as "the socio-environmental sustainability and health of billions of people" (FAO & IWMI, 2018). While asserting that issues like water quantity, water-use efficiency and issues pertaining to water allocation have garnered global attention, Biswas, Tortajada, and Izquierdo (2012) note that the poor management of wastewater and agricultural drainage has proved instrumental in engendering serious water quality problems across many parts of the globe, exacerbating the water crisis. Mere physical scarcity of water resources does not only contribute to water scarcity as such; nevertheless, it is also caused by the progressive decline of water quality in several basins, causing diminution in the quality of water that is safe to use (FAO & IWMI, 2018). Human settlements, industries, and agriculture are regarded as key sources of water pollution, whereas in some countries, agriculture has come to reckoned with as the biggest water polluter. Of the annual withdrawal of 3928 km^3 of freshwater, broad estimates show that only 44% is consumed, mainly through evapotranspiration by irrigated agriculture; and the remaining 56% (2212 km^3/year) is released into the environment as urban wastewater (approximately 330 km^3), industrial wastewater—including cooling water— (approximately 660 km^3) or agricultural drainage (approximately 1260 km^3) (AQUASTAT, n.d.; Mateo-Sagasta, Raschid-Sally, & Thebo, 2015).

The major regions that are susceptible to water quality threats are mostly associated with population densities and areas of economic growth (WWAP, 2018). Almost all rivers in Africa, Asia and Latin America have been subject to worsening pollution since the 1990s (UNEP, 2016a). There is likelihood of exacerbation in deterioration in water quality in ensuing decades and that could enhance threats to human health, the environment and sustainable development (Veolia/IFPRI, 2015). Bulk of all industrial and municipal wastewater, estimated nearly 80%, is released to the environment in unscientific manner and this contributes to steep decline in overall water quality entailing the potential of adversely impacting human health and ecosystems (WWAP, 2017). The most prominent water quality challenge worldwide is nutrient loading, that being region-specific, is mostly linked with pathogen loading (UNEP, 2016a). Undoubtedly, regulatory measures along with huge investments have been in place in developed countries to minimize point-source water pollution; nonetheless, challenges pertaining to water quality persist on account of under-regulated diffuse sources of pollution (WWAP, 2018). Coping with diffuse runoff of surplus nutrients from agriculture, including into groundwater, is often perceived as the widest spread water-related challenge worldwide (OECD, 2017). It was reported by almost 15% groundwater-monitoring-stations in Europe that the standard of nitrates fixed by the World Health Organization (WHO) were exceeded in drinking water (WWAP, 2018). Besides, these monitoring stations also informed that nearly 30% of rives and 40% of lakes were eutrophic or hypertrophic in 2008−11 (EC, 2013).

Water quality, apart from nutrients, is also impacted by innumerable chemicals. Increased use of chemical globally in agricultural intensification inter alia comprises 2 million tonnes per annum, with herbicides accounting for 47.5%, insecticides constituting 29.5%, fungicides comprising 17.5% and other amounting to 5.5% (De, Bose, Kumar, & Mozumdar, 2014). Attention of the international community has been drawn by a recent report of the Special Rapporteur on the right to food (UNGA, 2017) to the dire need of improving pesticide use policies. According to Sauvé and Desrosiers (2014), pollutants of emerging concerns are persistently evolving and increasing and frequently noticed at concentration higher than expected. Unbated water pollution is destined to deliver adverse impacts on human health and biodiversity (WWAP, 2017).

Water quality is impacted by the vagaries of climate change in various ways. Alterations in spatial and temporal patterns and variability of precipitation impact water flows with resultant dilution effects, while rise in temperatures spurs higher evaporation from open surfaces and soils, and decrease in water availability occurs in the wake of increased transpiration by vegetation (Hipsey & Arheimer, 2013). Future projections indicate that depletion of dissolved oxygen will be facilitated owing to higher water temperatures and it could be anticipated that higher contents of pollutants would flow into water bodies in the eventuality of extreme rain event (IPCC, 2014). Lower- and low-middle income countries are vulnerable to the anticipated occurrence of increased exposure to pollutants, mainly owing to higher population and economic growth in these countries, specifically those in Africa (UNEP, 2016a), supplemented by the lack of wastewater mechanism (WWAP, 2017). Undoubtedly, WWAP (2018) notes that given the transboundary nature of most river basins, regional cooperation will be critical to addressing projected water quality challenges; nonetheless, likelihood of pollution further reducing the amount of clean freshwater is inevitable, an aspect that has been marginally factored in WWAP 2018 (Boretti & Rosa, 2019).

5.5 Water scarcity

Notion of water scarcity, in common parlance, is often construed as the lack of access to adequate quantities of water for human and other environmental uses (White, 2012). In the wake of growing concerns about the adverse impacts of climate change and climate variability in conjunction with increasing awareness of the utility of food–energy–water nexus, there has been heightened interest about the real and perceived risk of water scarcity (Liu et al., 2017; Mekonnen & Hoekstra, 2016). The term "water scarcity" is a relative concept defined as "a gap between available freshwater supply and demand under prevailing institutional arrangements and infrastructural conditions" (FAO, 2012, p. 78). The concept of water scarcity ranges from water abundance at one extreme through several intermediate conditions, such as water stress, to an absolute lack of water at the opposite extreme (Hasan, Tarhule, Hong, & Moore, 2019). A thorough, understanding of the notion of water scarcity is significant because "it affects the views of users and policymakers on the urgency to address the water crisis, as well as their views on the most effective policies to address the water crisis" (Rijsberman, 2006). It becomes from the growing literature on the subject that scholars and policymakers have come to recognize that water scarcity is a multifaceted phenomenon that integrates aspects of the physical availability of water, including its quality, status, as well as sociocultural, economic, political, and structural dimensions (Sambu, 2011).

It is worth noting here that at its embryonic stage of evolution, the notion of water scarcity; however, was based almost entirely on the physical availability of water; and in the aftermath of the development of water stress index (WSI) in 1989 by Falkenmark and subsequent wider use of WSI as an indicator, the understanding of the concept remained no more confined merely to physical availability, rather it came to include social, economic and technological aspects as well with regard to water scarcity. The WSI is expressed by the degree water (in)sufficiency as the total renewable water resources in a country (or drainage basin) divided by the total population; and renewable water resources were denominated in flow units (where 1 flow unit = 1000 m^3) available to a country from all sources (Hasan et al., 2019). The WSI established four thresholds or indicators of renewable water resources vulnerability or stress (Falkenmark & Widstrand, 1992) (Table 5.1).

Table 5.1 Thresholds/indicator for water stress index.

Threshold/indicator (m^3 per capita)	Status
>1700	Occasional or local water stress (no stress)
1700–1000	Regular water stress (vulnerable)
1000–500	Chronic water shortage (stressed)
<500	Absolute water scarcity (scarcity)

Compiled from Falkenmark, M., & Widstrand, C. (1992). Population and water resources: A delicate balance. Population Bulletin, *47(3), 1–36.*

Table 5.2 Basic types of water scarcity and different introduced water stress indicators.

Type of water scarcity	Indicator	Reference
Physical water scarcity	Falkenmark indicator	Falkenmark (1989)
	Water resources vulnerability index	Shiklomanov (1993)
	Basic human water requirement	Gleick (1996)
	Water resources availability	Yang, Zhang, and Zehnder (2003)
	Watershed sustainability index	Chaves and Oliveira (2004)
Economical water scarcity	Physical water economic scarcity	Seckler et al. (1998)
	Green-blue water scarcity	Rockström et al. (2009a)
	Water scarcity function of water footprint	Hoekstra et al. (2011)
Social water scarcity	Social water stress index	Ohlsson (2000)
	Water use availability ratio	Alcamo and Henrichs (2002)
	Local relative water use and reuse	Vorosmarty et al. (2005)
Technological water scarcity	Water poverty index	Lawrence et. al. (2002)
	Water supply stress index	McNulty et al. (2014)

Adapted from Hasan, E., Tarhule, A., Hong, Y., & Moore, B. (2019). Assessment of physical water scarcity in Africa using GRACE and TRMM satellite data. Remote Sensing, 11(8), 904. Available at https://www.mdpi.com/2072-4292/11/8/904.

Novelty, simplicity, intuitiveness, as well as parsimonious input data requirements of the WSI earned it widespread usage and acceptance (Hasan et al., 2019). Nonetheless, the WSI has also been subject to criticism on multiple fronts (White, 2012). These criticisms spurred the development of numerous other water resources vulnerability indicators represented by four types of water scarcity—physical, economic, social, and technological (Table 5.2).

Four types of water scarcity delineated in Table 5.2 viz. physical water scarcity, economical water scarcity, social water scarcity and technological water scarcity need brief appraisals.

5.5.1 Physical water scarcity

The term "physical water scarcity" is generally used to describe the relationship between demand for water and its availability. A physical scarcity is said to take place when demand for water outstrips supply and this occurs when water resources are over-exploited (Appelgren & Klohn, 1999). If either a lack of supply or an excess of demand forms the basis of scarcity, then an increase in water availability in terms of a supply-side response or more efficient use of the available resource in terms of demand-side response, is called for to equalize the imbalance (Ohlson & Turton, 1999). Insufficient availability of water, both blue water as well as green water, to meet all demands, including the environment, constitutes the basis of physical water scarcity. Blue water alludes to liquid water in rivers, lakes, wetlands and aquifers (Rockström et al., 2009b). Kummu et al. (2016) opine that physical blue water scarcity can be essentially categorized into two aspects: water shortage (population-driven water scarcity) and water stress (demand-driven water scarcity, that is, the ratio water use to water availability). Water scarcity indicators also include economic or green water scarcity indicators (Kummu et al., 2016), where green water bears reference to rainwater ensconced in the unsaturated zone of the soil and available to plants.

According to Falkenmark et al. (2007), physical water scarcity can be further categorized into two main concepts: demand-driven scarcity (water stress) and population-driven scarcity (water shortage). Demand-driven scarcity can be appraised by exploring as to how much water is being withdrawn from rivers and aquifers, known as the use-to-availability index (Falkenmark et al., 2007). The water shortage is associated with the number of people that are to share each unit of water resources, and can be gauged by using the water crowding index, also known as the Falkenmark WSI (Falkenmark et al., 1989, 2007). It is noteworthy to mention here that there are several groupings of water demand, such as water for industrial and municipal water supply, agriculture, and environmental needs, etc. Nevertheless, assessing the role of environmental water needs in physical water scarcity has seemingly been a phenomenon of recent origin (Smakhtin et al., 2004). Indices designed to coalesce physical and social water scarcities embrace, for example, the water poverty index (Sullivan et al., 2003) and the social WSI (Ohlsson, 1998).

Several studies assessing global water scarcity from various different disciplines (Alcamo et al., 2007; Falkenmark et al., 2007; Islam et al., 2007) have dealt with as to how physical water scarcity could develop over time into the future, with a time span of a few decades ahead. The results of many of these studies have characteristically demonstrated a sharp increase in the number of people under water stress or water shortage as a consequence of increasing population and/or water use, and in some cases as a result of climatic change (Oki and Kanae, 2006).

Undoubtedly measures like building dams and abstraction of groundwater have been adhered to augment water availability in response to the prospect of water shortage; nevertheless, Kummu et al. (2010) have opined that there are already several regions wherein such measures are insufficient because "there is simply not enough water available in some regions" (Kummu et al., 2010, p. 8). The problem of water shortage was destined to be grim in the future owing to increasing population pressures (UN, 2009), higher welfare (Grubler et al., 2007), production of water-intensive biofuels (Varis, 2007a; Berndes, 2008), and climatic change (Alcamo et al., 2007). To deal with such situation, Kummu et al. (2010) suggest that there would be an increasing need "for nonstructural measures, focusing on increasing the efficiency of water use, lowering water use intensity in regions with water shortages, reforming the economic structure of countries or entire regions, and optimizing virtual water flows from regions without shortage to regions with shortage."

5.5.2 Economic water scarcity

The notion of economic water scarcity implies paucity of investment in water infrastructure or deficient human capacity to meet the demand of water in regions where people are not in a position to afford to use an adequate source of water. Factors like lack of adequate water infrastructure where the people often have to fetch water from distant places for drinking and agricultural purposes are generally contribute to economic water scarcity. Despite the stress laid on making sufficient water resources available for drinking and domestic use, nevertheless, a large portion of water made available thus gets diverted for nondrinking purposes like bathing, laundry, livestock and cleaning than for drinking and kitchen use (Madulu, 2003). Concurrently, increased emphasis on requirements of drinking water enables in tackling only a fractional part of the problem of water resources, thereby restricting the scope of solutions available (Madulu, 2003). Many regions in Africa are affected by economic water scarcity and investing in water structure in the affected regions can be helpful in poverty reduction.

Many of the developing countries relying on low-yield agriculture need investment in water retention and irrigation infrastructure to help increase food production (Duchin & López-Morales, 2012). Overcoming economic water scarcity calls for more than mere new infrastructure, it needs socioeconomic and sociopolitical sorts of interventions to help alleviate poverty and socio-inequality (Noemdoe, Jonker, & Swatuk, 2006). While referring to symptoms of economic water scarcity, Molden (2007), notes that scant infrastructure development, either small or large scale along with inadequate supply of water for agriculture and drinking; and even where infrastructure exists, the inequitable distribution of water accounts for economic water scarcity.

Undoubtedly, linkages between water scarcity and economic growth are subtle and are only gradually unfolding; nonetheless, there is sparse literature that convincingly links water scarcity to economic growth and work by Sadoff et al. (2015) is regarded seminal in this regard. While viewing water as a publicly provided input into economic production, Barbier (2004), in his study based on cross-section data for 163 countries in the 1990s, suggests that rates of freshwater utilization in the vast majority of countries didn't constrain economic growth; however, 16 countries in his sample id face conditions of extreme water scarcity. This leads to the surmise that the problem of water scarcity constraining economic growth during the 1990s was "likely quite limited" (OECD, 2016). While suggesting that the causal links between water-related investments and economic growth, evidently run in both directions, Sadoff et al. (2015, p. 45) note that water-related investments can increase economic productivity and growth, and economic growth can provide the resources to finance capital-intensive investments in water-related infrastructure. Water scarcity is a highly localized concept, subject to different interpretations by different actors; and water scarcity is dealt with by institutions and governments, it very quickly surfaces as a scarcity of social adaptive capacity (Ohlsson and Appelgren, 1998).

5.5.3 Social water scarcity

The notion of social scarcity of water alludes to a social construct of "resource management," which is ascertained by politico-economic and social power dynamics "underpinning the institutions that provide structure to social relations, security of access to bases of social power and productive wealth, and stability to the social organization of human societies" (Tapela, 2012, p. 15). Social water scarcity occurs in the wake of deficient investment, lack of skills or political will to meet growing demands for water and preventing access to the resource (WaterAid, 2013). Emphasis on the significance of a social dimension to the conceptualizations of water scarcity through developing the concept of second-order scarcity by Ohlson and Turton (1999) spurred researchers to focus on the social mechanisms that are prevalent within a given social entity that let that social entity to adapt to the circumstances forced upon it by water scarcity. Goldin (2010) feels the necessity of understanding the complexities surrounding social water scarcity as it can be a product of the interaction between resource availability, consumption patterns and the (mis)management of the resources. Emphasis by Turton (1999) on the need for discerning the social dynamics of water scarcity, along with ascertaining as to how various societies deal with this type of scarcity brings the notion of social water scarcity closely linked to water governance and management.

While referring to water scarcity is a relative concept, Appelgren and Klohn (1999) describe it partly as a social construct in that it is determined by the availability of water and consumption patterns, they further point out: "Because of the large number of factors, which influence both availability and consumption, any definition of water scarcity will vary widely from country to country and from region to region within a country...Because the concept of water scarcity is a social construct it is a matter of political and economic perception, and it may be more useful to view water scarcity as a particular mix of availability and demand" (Appelgren & Klohn, 1999, p. 362). This demonstrates that structurally induced social water scarcity pertains to the political economy and ecology of resource allocation, and the institutional frameworks and structure of water governance and management. Therefore the notion of social water scarcity alludes to a social construct of "resource management" that is ascertained by political, economic, and social power dynamics bearing out social institutions that afford structure to social relations, access to social power and social stability (Tapela, 2012; WaterAid, 2012).

For the marginalized and underprivileged communities, social water scarcity is about an inadequacy of the quality and quantity of water. Water security being a localized problem brings it in closer proximity to ensuring fair and safe access to water resources that is vital in sustaining and improving people's livelihoods (FAO, 2012). According to Marshall et al. (2009), availability of water and sanitation in townships within South Africa is not so much a problem of scarcity as it is a problem of access and control of resources. Consequently, availability of water does not merely signify its easy access to marginalized and poor communities. According to Tapela (2012), communities' cognizance of the power dynamics renders the notion of social water scarcity to be construed primarily, as the end product of dominance by overarching political, economic and social interests that determine and control the structure and nature of water resource allocation, especially in South African context. This is also obvious within the power relations between communities, municipalities and institutional actors (Rogers & Hall, 2003).

5.5.4 Technological water scarcity

In common parlance, technological water scarcity is construed as absence or lack of water-related technology to harness water for drinking and other economic productive uses. Intrinsic linkages between freshwater resources and other sectors along with ecosystems contribute in making water a complex sector (UNFCCC, 2012). Undoubtedly, deployment of technologies in water sector can be facilitated subject to fulfillment of certain requirements; nonetheless, there is no guarantee that a technology operates well in one region would also work the same way in a different country.

For instance, dam and water diversions in one location can have an impact on the water balance and microclimate in a different part of an ecosystem. In addition, with mitigation, there are significant synergies, trade-offs and cobenefits to be considered. Trade-offs are particularly significant in the water sector, where there is a conflict between the security potential of large-scale projects and the energy costs that such projects demand. Therefore climate change poses a major challenge to water managers, users and policymakers at different levels, who must examine all potential and probable impact scenarios and interrelated issues, both within and amongst regions and sectors throughout planning and implementation processes.

The use of adaptation technologies has been broadly defined as "the application of technology in order to reduce the vulnerability, or enhance the resilience, of a natural or human system to the impacts of climate change" (UNFCCC, 2005). In the water sector, site-specific solutions need to be considered within the broader context of integrated water management approaches. A lack of regard for particular contexts, alongside poor planning, as well as overemphasis on short-term outcomes, or failure to account for possible climatic consequences and adaptation limits, can result in maladaptation or "an adaptation that does not succeed in reducing vulnerability but increases it instead" (IPCC, 2001, p. 378).

Technology assumes significance in the wake of urgency for strengthening adaptive capabilities to deal with impacts of climate change on the hydrological cycle on all scales. By 2012, more than 85 countries had been subjected to conducting of Technology Needs Assessments (TNAs) for Climate Change, within the framework of the UNFCCC and its Technology Mechanism, and guided by the Technology Executive Committee (TEC) (UNFCCC, 2013). Interestingly, bulk of these countries, accounting for more than 75%, having completed their TNAs, identified the water sector as a priority sector requiring adaptation interventions. Member countries of or parties to the UNFCCC have designated water as a priority area for action, and 119 member countries of the UNFCC indicated water as a priority for adaptation action in the adaptation component of their intended nationally determined contributions (UNFCC-TEC, 2014). The role of technology in supporting adaptation to changes in water has been emphasized by the IPCC (2014). Besides, the Third Synthesis Report of the TNAs reflects the prioritization of adaptation in the water sector by 77% of Parties (UNFCCC, 2013).

An array of ways exists whereby appropriate infrastructure and technology can be harnessed for mitigating climate-related risks and vulnerabilities in water sector. UNEP-DTU, CTCN (2017) have come out with a publication that focuses on opportunities for adaptation to climate change induced hazards and building resilience by identifying water adaptation technologies that bear relevance to the water management challenges. Christiansen, Olhoff, and Trærup (2011) classify adaptation technologies into three categories: Hardware, Software and Orgware. Hardwar refers to the "hard" tachylogias such as physical infrastructure and technical equipment on the ground. Software technologies comprise approaches, processes and methodologies, including planning and decision support systems, models, knowledge transfer and building skills essential for adaptation. Orgware means organizational technologies that include: the organizational, ownership and institutional arrangements necessary for successful implementation and sustainability of adaptation solutions (Christiansen et al., 2011). Focus on both hardware and software entailing implementation of technological tools and equipment, and approaches and management strategies relevant to climate change adaptation is emphasized (UNEP-DTU, CTCN, 2017). Orgware is addressed through the dimensions of the integrated water resources management (IWRM) that are specifically significant for water technology selection an implementation.

Cognizance of the differences between these technology types, along with their synergies and complementarities is often deemed essential while considering adaptation. Hardware or hard technologies, refer to physical tools; software or soft technologies, allude to the processes, knowledge and skills needed in using the technology; and orgware or organizational technologies, refer to the ownership and institutional arrangements pertaining to a technology (Christiansen et al., 2011; UNFCCC, 2014b). In the water sector, Application of hard technologies in the water sector entails structures like ponds, wells, reservoirs and rainwater harvesting equipment, whereas "soft" technologies are those that aim at improving water-use efficiency through, for example, water recycling techniques. Institutional mechanisms, such as water-user associations and water-pricing specifications, are examples of orgware and orgware entails the potential of supporting the appropriate adoption of hardware. Undoubtedly, all types of technology are essential; nevertheless, there is a lurking concern about the application of hard technologies in isolation, their perceived impact being prioritized over soft- and org-wares (Christiansen et al., 2011; UNFCCC, 2014b). Developing countries, especially the least developed countries (LDCs), need to be encouraged and provided with adequate assistance in implementing all three technology types in a mutually supportive manner, with the avowed objective of ensuring sustainable and effective application of technologies for adaptation in the water sector to ward off water scarcity.

Technological application in LDCs is supported by various processes and institutional arrangements. Support on technology consists of TNAs that help in identifying, prioritizing and highlighting technology requirement, and technology action plans (TAPs) that are worked out on the basis of TNAs to deal with specific barriers, and identify targets, budgets and responsible stakeholders for prioritized technologies (UNFCCC, 2014a). Many countries are said to have developed TAPs especially relevant to the water sector. Cambodia is reported to have dealt with the transfer and diffusion of small dams, reservoirs, and microcatchments, along with Lebanon specifically focusing on the Water Users' Association (WUAs), and Zambia has sought to enable the implementation of boreholes and tube-wells. The Least Developed Countries Fund and the Special Climate Change Fund provide access to resources for such plans and processes. Both these funds accord priority to the water sector, allocating it 14% and 23% of their budgets, respectively (UNFCCC, 2014a). These processes and arrangements are strengthened through the work of the UNFCCC's Adaptation Committee, Technology Mechanism (including the TEC and the Climate Technology Centre and Network) and other relevant bodies under the Convention.

Availability of reliable and up-to-date data about acute and chronic water scarcity is an indispensable resource for coping with local, regional and global water scarcity, especially at a time when an estimated 4 billion are said to facing water scarcity (Mekonnen & Hoekstra, 2016) and in the wake of global population projected to reach 9 billion by 2030 (UN-DESA, 2015a,b) along with expected increase in demand for water in domestic, industrial and agriculture sectors during that period, potentially by 40% (United States-NIC, 2013). Burgeoning population in tandem with climate change driven decreases in precipitation and runoff in Africa (Gan et al., 2016) are prone to entail serious implications for future chronic and acute water scarcity. Water scarcity as a whole implies serious implications for national security (USAID, 2017a,b), food security (USAID, 2017a,b), and for attaining SDGs, through the water—energy—food nexus (McNally et al., 2019).

5.6 Water and climate change

Water and climate change are interrelated and interconnected. Water is affected by the vagaries of climate change and water also impacts climate change. Water is the primary medium through which climate change wields influence on the Earth's ecosystems and therefore people's livelihoods and well-being (UN Water, 2019). Water-related climate change impacts like droughts, floods, hurricanes and typhoons, etc., have already begun to be felt in different sectors. Rise in average temperatures and alterations in precipitation and temperature extremes are predicted to impact the availability of water resources through changes in rainfall distribution, soil moisture, glacier and ice/snow melt, and river and groundwater flows; these factors are expected to lead to further deterioration of water quality as well (UN Water, 2019). Impact of climate change on water resources becomes discernible, first and foremost, by increased variability in the water cycle that can prove instrumental in causing extreme weather events, causing diminution in the predictability of water availability, impacting water quality, along with endangering the process of sustainable development, biodiversity (Lovejoy & Hannah, 2019), access to fresh drinking water and sanitation worldwide (Climateiswater.org, n.d.).

Water plays a pivotal role in climate change mitigation, owing to the fact that bulk of efforts in reducing greenhouse gas emissions are largely dependent on safe and reliable access to water resources. Besides, emissions' reduction efforts and energy production from water contributes to assist in attaining the target of larger emission reduction. Judicious management of water resources is very vital for climate change mitigation and adaptation (Climateiswater.org, n.d.).

Direct impact of climate change is usually felt on regional hydrology, and that in turn affects global water supplies, in different ways. Rising sea level is prone to threaten coastal infrastructure and aquifers, freshwater sources fed by snowpack (Rabbani, Rahman, & Islam, 2010). Apart from impacting the melting of the glaciers and making them less reliable as water resource, rising temperatures will increase surface drying and decrease soil moisture, and shifting precipitation patterns may render arid regions vulnerable to more frequent and intense droughts while intensifying flooding in wetter regions (Kemp, Hagberg, Lammon, Mazumdar, & Bradler, 2018). Cumulative as well as individual impact of each factor is fraught with serious consequences entailing the potential of compounding the complexities generated by other. Admittedly, not all the areas will be affected equally; nonetheless, research studies demonstrate that the regions that are already the most water-insecure would be hit the hardest by climate change (WRI, 2015). Kemp and colleagues are of the view that developing effective water management policies may be a defining aspect of maintaining future political stability. Recent reports pertaining to the almost decade-long deterioration of Antarctic ice-shelves have raised the potential for sea-level rise to outdo even earlier mainstream predictions (NSIDC, 2018).

Potential of climate change impacting weather variables like precipitation and surface drying cannot be denied. Increased precipitation owing to warmer air spurs the rain to fall rather than snow, along with increased evaporation at Earth's surface. IPCC (2007) has noted that for each single degree of warming, water-holding capacity of the atmosphere increases by 7%. Occurrence of precipitation under such conditions results in heavy rains or snow fall along with entailing potential of increasing the frequency of intense floods or snow storms in areas that are already at risk; and the possibility of arid areas experiencing drought and reduced precipitation is also not ruled out. Warming temperatures adversely impact areas that are already most water-insecure, especially the Middle East and North Africa (MENA) which is already vulnerable (Arab Water Council, 2009).

5.6.1 Water and extreme weather events

Global climate change is expected to affect the frequency, intensity and duration of extreme water-related weather events like excessive rainfall, storm surges, floods, and drought are getting affected by climate change (Fussel, 2009). Extreme water-related weather events occurring in the recent past have included drought in Russia (NASA, 2010), and flooding in Sri Lanka, the Philippines, Pakistan, Australia, and Brazil (Kroeger, 2011). Quickening of the water cycle spurred by atmospheric heating is prone to make weather more extreme and variable. Altered pressure and temperature patterns, triggered by global warming, could also shift the distribution of when and where extreme water-related events generally occur (Solomon et al., 2007). While the frequency of heavy precipitation events is reported to have risen over the mid-latitude regions since 1950, even where there has been a decrease in the total reprecipitation and the area affected by drought has reported increased since the 1970s in many regions of the globe (Solomon et al., 2007). Available evidence suggests that other extreme water-related weather events such as El Niño Southern Oscillation (ENSO), hurricanes, and cyclones are becoming more frequent, intense and of greater duration (Solomon et al., 2007).

Mobilization pathogens in the environment can be triggered by excessive or heavy rainfall events resulting in the increased runoff of water from fields, transporting them into rivers, coastal waters and wells (Semenza & Menne, 2009). According to Tinker et al. (2008), such events entail the potential of enhancing raw water turbidity that has been found to be linked with gastrointestinal illness. There is also possibility of water treatment plants getting overwhelmed during periods of heavy rainfall thereby causing cross-contamination between sewage and drinking-water pipes, especially where water infrastructure is old, and sewage overflow, or bypass into local waterways (Semenza & Nichols, 2007). It is opined by Fewtrell, Kay, Watkins, and Francis (2011) that extreme precipitation events can cause increase in the risk of flooding in many areas, increasing human exposure to waterborne pathogens. According to Senhorst and Zwolsman (2005), droughts or extended dry periods sharpen the likelihood of reducing the volume of river flow and potentially enhance the concentration of effluent-derived pathogens, owing to decreased dilution by stream-receiving water.

Eruptions in the contamination of community water systems can be instrumental in causing extensive disease (Karanis, Kourenti, & Smith, 2007), specifically where the public health infrastructure is less resilient. Waterborne diseases are likely to rise with increases in extreme rainfall and deterioration in water quality following wider drought events (Pachauri & Reisinger, 2007). It is important to understand that the current effect of water-induced events on public health assumes significance in discerning future predictions, aid policy formulation, and improve adaptive capacity. It can be assumed on the basis of such events that even high-income countries are not well prepared to deal with extreme weather events (Pachauri & Reisinger, 2007). In view of the availability of limited information it is difficult to ascertain as to how different extreme water-related weather events could affect different geographical areas and pathogens.

Increased incidences of extreme weather can intensify human exposure to water contaminated by agricultural runoff, flooded water and sewage treatment systems, and standing water (habitat for toxic algal blooms as well as a breeding ground for disease vectors that increase malaria risk) while drought can negatively affect water quantity and quality (Portier et al., 2010). Drought also increases the entrainment of dust and fine particulate matter in the air caused by drought can engender an array of human health impacts, specifically for children and the elderly. According to UN-Water (2019), these impacts are felt over a range of timescales, calling for advanced planning and adaptation measures that can respond to both short-term emergencies as well as longer term stressors.

The tendencies in water availability are accompanied by anticipated changes in flood and drought risks and in such eventualities, one specific concern is that the increasing flood risk occurs in some traditionally water-scarce areas like Chile, China and India, as well as the MENA where "local coping strategies for flood events are likely to be poorly developed" (WWAP, 2018). There has been a steep rise in economic losses accruing from water-related hazards in recent years.

A recent report by the OECD notes: "The number of people at risk from floods is projected to rise from 1.2 billion today to around 1.6 billion in 2050 (nearly 20% of the world's population) and the economic value of assets at risk is expected to be around US$45 trillion by 2050, a growth of over 340% from 2010" (OECD, 2012:209). Available data demonstrates that floods have accounted for 47% of all weather-related disasters since 1995, impacting a total of 2.3 billion people. It can also be discerned from this data that incidences of floods rose to an average of 171 per year over the period 2005−14, up from an annual average of 127 in the previous decade (CRED/UNISDR, 2015). Instances of costs of flooding include 39 and 11% of GDP in the People's Democratic Republic of Korea and Yemen, respectively (CRED/UNISDR, 2015).

According to P.S. Low, the number of people affected by land degradation/desertification and drought was estimated at 1.8 billion, rendering it to be the most important category of "natural disaster" based on mortality and socioeconomic impact relative to GDP per capita (Low, 2013). The degradation of the world's land surface due to human-induced activities has already negatively affected nearly 3.2 billion people globally thereby presenting a situation that is likely to worsen further in the wake of exacerbation in the impacts of climate change (IPBES, 2018, p. 263). Land degradation is an inescapable systemic process that impacts all regions of the globe with the potential of assuming many problematic forms such as chemical contamination and pollution, salinity, soil erosion, nutrient depletion, overgrazing, deforestation and desertification (Andersson, Brogaard, & Olsson, 2011, p. 296).

Apart from being a chronic, long-term problem comparable to short-term effects of flooding, droughts are perhaps the biggest threat from climate change and fluctuations in future rainfall patterns can be anticipated to change drought occurrence and that in turn could also alter soil moisture availability for vegetation in many parts of the globe (WWAP, 2018). Chronically drought-stricken areas, also known as drylands, are home to about 3 billion people who constitute for 38% of the global population (IPCC, 2019). Moreover, more than one-fourth of humankind is confronted with a looming water crisis that is prone to further exacerbate as the effects of climate change get worsened (Sengupta & Weiyi, 2019). Ironically, a great mass of people reeling under the impact of drought and land degradation are already poor or chronically poor and quickening of these phenomena entail the likelihood of enhancing the levels of forced migration by "adding a layer of supplementary stress" to already susceptible population (Mach, 2017). Construction of additional water storage facilities by upscaling infrastructure investments can be instrumental in lessening the anticipated longer duration and severity of droughts and these measures can also yield significant trade-offs for society and environment. Increased water withdrawals to meet the growing water demand have already impacted nearly 4.2 billion people, accounting for 95% of all people affected by all disasters, causing $1.3 trillion of damage that constitutes 63% of all global disaster-related damage (UNESCAP/UNISDR, 2012).

Desertification, defined as land degradation in arid, semiarid, and dry subhumid areas, is the resultant outcome of multiple factors, including climatic variations and human activities (UNCCD, 1994). Drylands comprise arid, semiarid, and dry subhumid areas, together with hyperarid areas (UNEP, 1992). Land degradation is a "negative trend" in land condition, attributable to direct or indirect human-induced processes, including anthropogenic climate change, designated as long-term reduction or loss of at least one of the following: biological productivity, ecological integrity or value to humans (IPPC, 2019). More than 2.7 billion people are affected by desertification (IPBES, 2018, p. 36). The fast pace of land degradation is mostly the outcome of a collapsing economic system, marked by an ineffective food production mechanism, in tandem with other land degrading and polluting practices that are often harnessed to procure short-term economic gains for the benefit of a few, at the cost of multitude (UNCCD, 2017, p. 9). Available scientific literature amply demonstrates that "extreme weather and climate or slow-onset events may lead to increased displacement" (IPCC, 2019). Thus water-induced vagaries in tandem with vagaries of anthropogenic climate change result in serious complications, including desertification, land degradation, and droughts.

5.7 Securing water

Water is a finite resource and pollution/contamination of groundwater along with surface water resources exerts massive pressures on this finite resource. Moreover, fast depletion of groundwater resources without adequate renewal of underground water aquifers is contributing to water scarcity. Fast pace of melting of glaciers owing to adverse impacts of climate change that is also impacting hydrological cycle in terms of erratic rainfall, flooding, etc.; availability of water is becoming scarce to meet the growing demands of water for agriculture, industry and domestic use. This has necessitated rethinking on devising scientific pathways of securing water. The notion of water security, at the preliminary stage of its inception, was primarily used in the context of water-related issues; nevertheless, as the notion of water security started garnering salience, especially from the 1990s onwards, its usage spanned to other academic disciplines as well. The 1990s witnessed usage of the notion of water security in relation to military security, food security, and environmental security (Norman et al., 2012). From 2000s onwards, the term "water security" began to be increasingly used the wide array of scientific literature ranging from engineering and agriculture to public health and water resources (Bakker & Cook, 2011). Subsequently, the focus on water security has also begun to embrace water quality, human health, water-related hazards, sustainable development, water supply, and ecological concerns. Available literature demonstrates notion of water security encompassing a balance between the protection of resources and the enhancement of livelihoods (De Fraiture et al., 2010). According to Grey and Sadoff (2007), determinants of water security incorporate water availability for ecosystem services along with acceptable quality and quantity for human uses. Water security is presented in the literature as meeting short and long-term needs to enable access to sufficient water quality, at a fair price, for human health, safety, welfare, and productive capacity (Witter & Whiteford, 1999). In addition, water security has been linked to the political, social, and economic power of people in society (Gerlak & Mukhtarov, 2015).

5.7.1 Defining water security

Growing social and cultural significance of water and its intersection with the "already loaded term "security"" (Mason & Calow, 2012) spurred the World Economic Forum (WEF) to describe water security as an emerging "headline geopolitical issue" that may "tear into various parts of the global economic system" (WEF, 2011). The term water security was first used in political and policy circles in 2000 in the Ministerial Declaration of the 2nd World Water Forum in the Hague (Bogardi, Oswald Spring, & Brauch, 2016), and has since then gained considerable academic and political traction as is evident in the numerous publications, research, conferences and funding initiatives focusing on it (Pahl-Wostl, Gupta, & Bhaduri, 2016). From an academic perspective, natural scientists and engineers first began writing about water security in the early 1990s, with social scientists starting to follow suit a few years later (Pahl-Wostl et al., 2016).

While delineating the concept of water security Grey and Sadoff (2007, p. 547) describe it as "The availability of an acceptable quantity and quality of water for health, livelihoods, ecosystems and production, coupled with an acceptable level of water-related risks to people, environments and economies." Interpreted thus, the notion of water security implies mitigating the impacts of excess as well as scarcity; and in other words, this widely quoted definition of water security by Grey and Sadoff (2007) embraces the concept of water risk as one side of the coin—the other being availability. Professor David Grey refers to water security as "tolerable water related risk to society" (cited in Hope, 2012). Undoubtedly, this definition of water security entails the advantage of brevity; nonetheless, it is also ambiguous because it leaves the question unanswered as to what risks matter and to whom. Besides, the term "Society" that means different things to different people, "may leave room for the privileging of some interests over others" (Mason & Calow, 2012).

Admittedly, the notion of water security is not new in the water science literature; however, this term appears to have seemingly garnered greater prominence because of its linkages to other sectors that have often been reflected in a wide range of reports and conferences that have pondered over water security either in isolation or in relation to the security of other resources, especially energy and food/land (NIC, 2012; Oxford University Water Security Network, 2012). Some scholars have been of the view that there has been less problematization of water security narrative, comparable to water scarcity, irrespective of some significant interventions (Cook & Bakker, 2012) and debates about the importance of the term have had less time to evolve and polarize. Viewed in broad perspective, the definition of water security privileging availability of the resource, to some degree this underemphasizes issues of access and allocation and lines up more with the concept of physical water scarcity than with other manifestations (Mason & Calow, 2012). A broader perspective of the concept of water security was offered at the Second World Water Forum held in Hague in 2000 wherein whereby it envisaged that providing water security in the 21st century could mean ensuring that freshwater, coastal and related ecosystems are "protected and improved; that sustainable development and political stability are promoted, that every person has access to enough safe water at an affordable cost to lead a healthy and productive life and that the vulnerable are protected from the risks of water-related hazards" (WWC-WWF, 2000).

Interestingly, in the past recent years the narrative on water security has not been subject to extensive debate in the international meetings and conferences like its counterpart concept of climate security and resource security that have generally featured more prominently figured in foreign policy and defense communities' portfolios (DCDC MoD, 2010; IISS, 2011); nevertheless, there has been considerable thinking around longer-established security concepts like national and human security (Mason & Calow, 2012), and there has also been longstanding consideration of water's prospective role in conflict often in the context of water scarcity (CoFr-United States Senate, 2011). There have been some allusions in water literature of water playing potential role in localized unrest, terrorism and political oppression (Pacific Institute, 2011). The notion of water security has also been alluded to in the articulation of water's role in national and international peace and stability owing to water's strategic potential significance both as a "fugitive resource" that often transcends geographical barriers (UNDP, 2006, p. vi).

Grey and Sadoff (2007) delineate the notion of water security as the availability of an acceptable quality and quantity of water for health, livelihood, ecosystems and production coupled with an acceptable level of water-related risks to people, environments and economies. For some scholars, water security exists when humans and ecosystems are bereft of water-related harms (Horwitz & Finlayson, 2011). On the other hand, de Loë et al. (2007) and Grey & Sadoff (2007) contend that there is no unanimity on one commonly-acceptable definition of water security owing to the varies context of disciplines such as agriculture, environmental science, social sciences, and water resources, within which it is used. Appraisal of water security is also construed through the application of several approaches that are generally based on the scale at which water is viewed and the discipline under which it is assessed (Bakker & Cook, 2011). While conceding the usual emphasis on access to water to meet human needs in common approaches to water security, Longboat (2015) laments at failure of these approaches in acknowledging importance of links to aquatic ecosystems.

Viewed in a wide perspective, notion of water security serves both to revitalize old ideas and to promote new ones. Nevertheless, Lankford et al. (2013) opine that water scientists of all types are confronted with a degree of uncertainty that has come to question the very way we approach water resource futures. Some approaches to water resource futures equate this with insecurity, and see opportunities in decreasing the variability of river flows, meaning dams for agriculture or hydropower (Muller, 2012). Alternative views oppugn the paradigms of distribution of the possibly reduced or increased legitimatized "securitization" and then militarization of water resources (Cook & Bakker, 2013; Zeitoun, 2013). Suggestions by Mason (2013) for water security indicators of water availability, access, risk, ecosystem services, and institutions coincide with the considered arguments of Falkenmark (2013) that question our superficial understandings of the same for scarcity of water. Parallels can be found in Clement (2013) for his apolitical views of nature and Mirumachi (2013) for the very political and commercial nature and interests of water institutions.

Garrick and Hope (2013) mull that issues of water stress, pollution, water variability, and climate variability are best thought of and handled as risks. Besides, it has been argued by Lankford (2013) that the allocation of risk of excessive scarcity above and beyond that caused by natural distributions of rainfall and river flow can be traced to design faults in river basin architectures. On the basis of his analysis of lessons of risk management from floods in Holland, Warner and van der Geest (2013) cautions against the approach, due to its tendency to pass on residual risks to local communities. While dealing with the contested topic of corporate engagement in water security, Hepworth and Orr (2013) demonstrates that the impact of big multinationals on local and global food and water production is so great that water security practitioners cannot afford to debate that role from the sidelines. The necessity of synchronizing the traditional interests of corporations—preferential and sustained water access, permissive water quality objectives, and laissez-faire regulation—with the water security goal of improving the wider public good at all scales is also referred to in the water literature (Lankford et al., 2013). Having scrutinized literature on water law and legal frameworks, Leb and Wouters (2013) moot the suggest of developing guidelines that can serve both to appraise competing claims, and, crucially, to desecuritize water conflict issues. Water literature also focuses on approaches dealing with the potential for market-based tools to balance equity and efficiency that can work only in very well-regulated contexts (Garrick & Hope, 2013) alongside more critical persuasive views that the market should be ignored for cultural-based solutions (Boelens, 2014) and warning of the drawbacks of retaining productivity—not security, much less equity—as a guiding principle for water management.

The growing concern in academic and policy communities with regard to the state of and challenges facing the world's freshwater resources has come to be regarded as one of the main reasons for using water security terminology. This concern is seemingly in tandem with the urgent need for sustainable water and land management (Pahl-Wostl et al., 2016). More precisely, exponents use the concept to call forth water-related concerns to the level of a political priority (Fischhendler & Nathan, 2016). Moreover, use of the term water security is construed to communicate the pressing nature of water crisis at all levels, and intends to culminate in equally vital action being taken to ward off potentially violent situations, such as conflicts over scarce water resources (Funke et al., 2016), or extreme human stress such as famine or drought (Pahl-Wostl et al., 2016). Academics and scientists therefore use water security as a concept in an attempt to drive political action at various levels. Irrespective of the laudable intent of the academic community and decision-makers in using the concept of water security, little consistency becomes discernible when it comes to its different framings (Funke et al., 2019). This is shown in Table 5.3.

Cook and Bakker (2016) focus on four interrelated themes of water availability, human vulnerability to hazards, human needs and sustainability that dominate the published research on water security, in an effort to make sense of the diversity of global-level uses of the water security concept. Formulation of water security focusing on quantity and availability is often associated with water security tools and an example of this is the Maplecroft Water Index Stress (2012) as shown in Fig. 5.1, that outlines areas of water stress by calculating the ratio of domestic, industrial and agricultural water consumption against renewable water from precipitation, rivers and groundwater. Nevertheless, it also becomes discernible from the literature on water availability that a more multidimensional trend by also acknowledging the effect of nonstationarity, complexity and uncertainty in determining water security (Cook & Bakker, 2016).

Viewed in broad spectrum, framework for water-related hazards and vulnerability has encompassed a focus on an infrastructure and systems approach, such as that of the UNESCO-IHE (UN Educational, Scientific and Cultural Organization—Institute for Hydrological Education) in 2009, and this approach addresses the protection of vulnerable water systems, protection against floods and droughts, the sustainable development of water resources and safeguarding access to water functions and services (UNESCO-IHE, 2009). According to Cook and Bakker (2016), such framings have dealt with risk, vulnerability and adaptation to water security; instances of which can be discerned from studies on vulnerability and adaptation in arid climates (Few et al., 2015; Revi et al., 2015). Interestingly, the period spanning between 2011 and 2019 has witnessed the framings of water security as a human necessity to be closely associated with other kinds of securities like energy security and food security and the resultant outcome has been increasing focus on water−energy−food nexus (Brears, 2018; Smajgl & Ward, 2013; Swatuk & Cash, 2018) in supplementing the IWRM (Grigg, 2016; Rockström et al., 2014) approach along with emphasis on ecosystem adaptation approach (Bologensi, Gerlak, & Giuliani, 2018; USADI, 2017a,b).

Gender is a significant in water security. IWRM, not entirely a new concept, is regarded as a broad framework for water security and is often described as an emerging model for water resource management. Water−energy−food nexus approach along with nature-based solutions (NBSs),

Table 5.3 Timeline showing paradigm shifts in the water security concept.

Year	Paradigm shifts in the concept
1991	Savage (1991) introduces the concept of water security emphasizing on geographic concerns of water resources in arid and semiarid regions in the context of middle East.
1992	Anderson (1992) and Shuval (1992) use the concept of water security to convey geographic, political, economic and international diplomacy significance of water for and between countries (Cook & Bakker, 2012).
1995	Livingston (1995) uses notion of water security to demonstrate that market failures impact water security and linkages between water security and the design of efficient water institutions.
1997	For Allan (1997) economic systems, in general, and not hydrological or water engineering systems help attain water security in Middle Eastern economies. Amery (1997) denotes water security as a variable in Arab-Israeli conflicts.
1998	Falkenmark and Lundqvist (1998) delineate the notion of water security in terms of the management of the response and the commitment of the so-called "hydrocide" water-scarce environments.
1999	Simonovic and Fahmy (1999) adopt a modeling approach to assess water resource policies, water security in tandem with economic and social indicators. Falconer (1999) addresses the concept of water security in relation to water quality associated with human health.
2000	GWP (2000a, 2000b) attempts to define water security in terms to access to enough safe Water at affordable cost to lead a clean, healthy and productive life. Ministerial Declaration at the World Water Forum at the Hague in 2000 on the goal of providing water security in the 21st century.
2001	Falkenmark (2001) emphasizes on linking water security with environmental and food security.
2007	Grey and Sadoff (2007) provide a widely quoted definition of water security Describing it as "the availability of an acceptable quantity and quality of water for health, ecosystems and production, coupled with an acceptable level of water-related risks to the people, environments and economies."
2011	WEF (World Economic Forum) convenes a conference I 2011 at Bonn under the Title *The Water, Energy and Food Security Nexus: Solutions for a green Economy*, With the objective of presenting evidence for how a nexus approach can enhance Water, energy and food security by increasing efficiency, reducing trade-offs, building Synergies and improving governance across sectors (Holf, 2011; Kurian, 2017).
2016	UN-Water (2016) defines water security as "the capacity of a population to safeguard Sustainable access adequate quantities of acceptable quality water for sustaining Freshwater for sustaining livelihoods, human well-being, and socio-economic development, for ensuring protection against water-borne pollution and water-related disasters, and for preserving ecosystems in a climate of peace and political stability."

Adapted by the author from multiple sources referred to in the table.

specifically ecosystem-based adaptation (EbA) approach have emerged as potent tools for ensuring water security and they need brief appraisal.

5.7.2 Gender and water security

A "gender lens" provides significant insights into the differentiated risks from climate change impacts on the water cycle, due to strongly gendered norms and practices linked to water (Dickin, 2018).

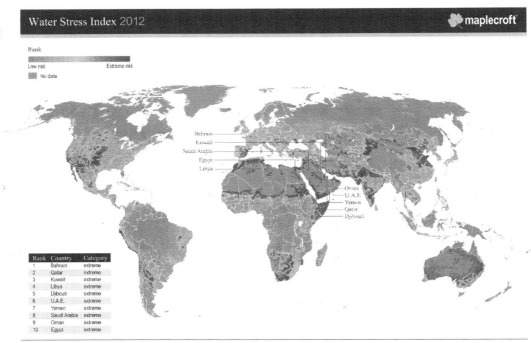

FIGURE 5.1

Water Stress Index.

From Maplecroft Water Stress Index 2012 Portal website at https://www.maplecroft.com/risk-indices/water-stress-index/.

Social protection systems along with water sector projects usually recognize gender inequalities within their respective realms and express aims relating to the empowerment of women and girls (SPIAC-B, 2019; UN Women, 2018). Undoubtedly, right to have access to "sufficient, safe, acceptable, physically accessible and affordable water" has been adopted as a basic human right by the UN General Assembly (UNGA, 2010); nevertheless, 2.1 billion people around the world still lack access to safely managed drinking water (WHO-UNICEF, 2017). Women have to share this burden of water scarcity because of their socially-assigned or family-assigned role to collect water in 8 out of 10 households who lack access to water n their premises (WHO, 2017). Consequently, women in many low- and middle-income countries have to walk for an average of six kilometers every day to fetch water (UN, 2010). According to Lenton, Wright, and Lewis (2005), in sub-Saharan Africa alone, the time cost of water collection has been estimated at 40 billion hours per year.

This responsibility of women in fetching water consumes their time and energy away from education, employment and other income-generating activities; and concurrently, it also leaves women with less time to undertake other valued activities, such as care for children or older people, and for leisure and rest (Lowe, Ludi, Sève, & Tsui, 2019). There is also likelihood of women's exposure to significant physical and psychological strain, along with risks of gender-based violence while walking to or collecting from water source locations (Graham, Hirai, & Kim, 2016). In the wake of potential for water scarcity or water insecurity set to grow worldwide as a result of the adverse impacts of ongoing climate change and burgeoning population culminating in the exacerbation of existing threats to the accessibility, availability, affordability and quality of water services; women in poor households in low- and middle-income countries are prone to bear the brunt of this growing water insecurity (Parker, Oates, Mason, & Calow, 2016). According to Lowe et al. (2019), "More arduous water collection, more limited supplies of water and the degrading quality of water for domestic and productive use all put households at risk of falling further into poverty and intensify the social and economic vulnerabilities faced by women and girls in their struggle to access this vital resource."

Thus water scarcity or insecurity is, therefore, a significant, heavily gendered, and growing driver of poverty, vulnerability and risk for women especially in many low- and middle-income countries. Accordingly, the achievement of the main objectives of social protection as well as the objective of the Agenda 2030 of "leaving no one behind" (aa), along with aims of reducing poverty and vulnerability, particularly among women and disadvantaged groups, and to enhance people's capacity to manage economic and social risks throughout their lives, remains unfulfilled. Lowe and colleagues (2019) lament at the little focus accorded within the social protection field to the importance of water security for the reduction of poverty and vulnerability, or on the role of social protection in addressing water security risks. Admittedly, both social protection experts and water experts have focused on the importance of, and potential for, addressing gender inequalities and empowering women and girls through their respective fields; nonetheless, linkages across social protection, water security, and gender equality and female empowerment domains have not been well established. Conceding that this gap is in part, the outcome of the prevalent challenge of institutional "silos," Devereux et al. (2016) argue that it also reflects the comparatively recent emergence of social protection on the development agenda worldwide and on the policy agendas of countries where water security is low. Water security and gender are inextricably linked, and the availability of clean water close to home reduces women's workloads, and the time saved in fetching water may be spent on other activities.

5.7.3 Integrated water resource management

The notion of IWRM that is often construed as an emerging model for water resource management, in reality, is not utterly new (Allan, 2003; Merrey, 2008). This notion stays on up-and-coming in the sense that it continues to develop as a management paradigm (Funke, Oelofse, Hattingh, Ashton, & Turton, 2007). IWRM is strongly associated with the recognition that water resources are complex, dynamic systems with multiactor and multiscalar interactions (Hooper, 2010; Patterson, Smith, & Bellamy, 2013; Wiek & Larson, 2012). It unveils a new normative stance for the management of water resources and pursues to redress fragmented governance regimes of the past (Metcalf, Dambacher, Rogers, & Loneragan, 2014). The concept of IWRM was established as the international guiding principle for managing increasingly scarce water and its related resources at the World Summit on Sustainable Development held in 1992 at Rio de Janeiro (UN, 1992). In 2000 Global Water Partnership (GWP) delineated IWRM as the process of promoting the coordinated development and management of water, land and related resources to optimize resultant economic and social welfare in an equitable manner without compromising the sustainability of vital ecosystems (GWP, 2000a,b). Articulation of this definition at the World Summit on Sustainable Development held at Johannesburg in 2002 provided it with legitimacy (Rahaman & Varis, 2005, p. 15). Subsequently, this definition of IWRM came to be adopted in new water policies and legislation globally (UNEP, 2012).

The explicit allocation of water to meet ecosystem needs is regarded as the major segment of environmental sustainability in water resources development. Generally, this environmental water allocation is referred to as an environmental flow. The most widely acknowledged definition of environmental flows is "the quantity, timing, and quality of water flows required to sustain freshwater and estuarine ecosystems and the human livelihoods and well-being that depend on these ecosystems" (Brisbane Declaration, 2007).

With its focus on the coordinated development and management of watershed level (GWP, 2000a,b), IWRM targets at ecological, sectoral as well as regulatory integration. Affording a congenial replacement for command-and-control approaches, IWRM envisages a shift toward participatory, bottom-up decision-making procedures that enable it to become an implementation framework in accordance with the objectives of Agenda 21 (UN, 1992) and Dublin principles (ICWE-WMO, 1992). In the wake of failed sectoral approaches, IWRM epitomizes a paradigm shift to integrated and basin-wide thinking by focusing on water bodies and their related ecosystems, water using sectors and different governance levels and taking them into consideration while making management decisions. There is need to emphasize on linking together ecological, economic and social objectives to garner sustainable outcomes. The IWRM has come to be reckoned with as one of the best practices in water resources planning to integrate water quantity and quality management for both groundwater and surface water, while incorporating a full cognizance of how the natural resources and the people of a basin are effected by multiple levels of development or by adopting new resource use policies.

Recognizing the fact that water resources are complex systems has resulted in the acknowledgment that these cannot be reduced to their component parts and managed in isolation (Rockström, et al., 2014). According to Medema et al. (2008), the characteristics of complexity and uncertainty ought to be reflected and included into the management systems of these socio-ecological systems. While summarizing this as a growing appreciation Biswas (2004:7) denotes that "water problems have become multidimensional, multisectoral, and multiregional and filled with multiinterests, multiagendas, and multicauses, and which can be resolved only through a proper multiinstitutional and multistakeholder coordination." Every aspect of the system entails a dynamic and interdependent relationship with several other elements wherein actions or processes in one part of the water system will perpetually impact actors or processes in other parts the system. This imbibes implications for governance, meaning that any effort to govern water resources essentially account for the several components that comprise the water resource system, and consequently requires that the multiple actors convoluted in the governance of each component cooperate and coordinate their efforts (Pahl-Wostl, Jeffrey, Isendahl, & Brugnach, 2011; Rijke, Farrelly, Brown, & Zevenbergen, 2013). This would imply incorporating processes and actors that are not ordinarily linked with water resource management in terms of land-use management, socioeconomic considerations and waste management. It then also necessitates that coordinated governance attempts take place at a pertinent level so as to include all of the interdependent elements of the system. Thus IWRM suggests that management endeavors function at a catchment level, or greater (Horlemann & Dombrowsky, 2011).

IWRM therefore demands for major shifts and realignments of government structures and processes (Horlemann & Dombrowsky, 2011).

While focusing on specific application of IWRM to the actors within a system Hooper et al. (2005, p. 15) observes that practically IWRM must bring together "a diverse array of people who have a "stake" in a system if it is to collaboratively manage the activities and impacts... This participatory approach produces strategies that are more coordinated, more cognizant of interconnections, and more inclusive of the diversity of goals. Furthermore, it increases support and commitment as well as the likelihood of implementation." Undoubtedly, there has been wide acceptance of IWRM by those engaged in water resource management (Horlemann & Dombrowsky, 2011); nonetheless, this approach has been subjected to criticism by some for having remained a body of ideologies and "lofty phrases" that dodge practical implementation (Medema et al., 2008; Horlemann & Dombrowsky, 2011). This approach has also been criticized on the ground that in cases where blueprints for IWRM implementation have been worked out, they are either inflexible or often ill-suited for the complex settings wherein they are acted upon (Ferguson, Brown, & Deletic, 2013). Criticism of the IWRM approach, though valid to some extent, need not entirely lessen the worth of the concept of IWRM. Even being critical of the unquestioned implementation of IWRM, Merrey (2008) acknowledges that "IWRM as a systems paradigm for understanding the problems and limitations of single-factor solutions is a critical requirement" (Merrey, 2008, p. 902). Besides, the criticism does not deny the need for some form of integration in managing water resources, rather it affirms the essentiality of translating IWRM into evocative principles that are practically implementable to improve the sustainable management of precious water resources (Patterson, et al., 2013).

IWRM needs a clear-cut assortment of management approaches for the respective water basin, ranging from participation of public and private stakeholders, the creation of adequate institutional structures to the implementation of new technologies (GWP, 2003, 2009; UNEP, 2012). The challenges of defining this "meaningful mixture" in general and especially for specific regions or river basins have been intensely described and discussed in the literature (Borchardt, Bogardi, & Ibisch, 2016; Martinez-Santos, Aldaya, & Llamas, 2014). IWRM constitutes target 6.5 of the SDG number 6 of Agenda 2030. The global average degree of implementation of IWRM in 2017/2018 was 48% corresponding to medium, low, nevertheless with great variation among countries. Undoubtedly, more than 80% of countries out of a total of 152 countries reported on IWRM implementation; nonetheless, comparisons with past surveys conducted in 2007 and 2011 reveal that while modest progress is being made, most countries will be unable to achieve indicator 6.5.1 by 2030 at current rates of implementation (UN-Water, 2018, p. 78). Calling upon governments and external support agencies to learn from experience and enhance implementation efforts to ensure accelerated progress and positive outcomes, UN-Water (2018) has emphasized on paying attention to building on IWRM monitoring and reporting to address barriers to progress on IWRM.

5.7.4 Water—energy—food security nexus

Water, energy, and food (WEF) are among the basic human needs and the growing demand for WEF in the wake of burgeoning population, rapid pace of urbanization and economic prosperity is also accelerating accordingly. By 2050, global demand for energy, water, and food is expected to rise by 61%, 55%, and 60%, respectively (WEC, 2013; WWDR, 2015). Making available the required amount of WEF in a sustainable manner at affordable prices, in the wake of growing concerns about climate change and rising inequalities, likelihood of extending human ingenuity to unprecedented scales is gaining salience. Ensuring security of water, energy and food articulates the enormity and complexity of this challenge that has emerged as a policy priority across the globe. The explicit incorporation of security concerns of WEF in the SDGs (UNGA, 2015)—SDG-6 (Water), SDG-7 (Energy), and SDG-2 (Food). This explicitly demonstrates the growing significance of WEF.

The inherent complex nature of interlinkages between energy, water, and food has come to identified with the term "nexus" and it can be denoted by water−energy−food nexus and in the process WEF security has emerged as a challenging issue to be understood and redressed. Redressal of the WEF security challenge warrants effective and efficient policies that recognize the cruciality and complexity of the nexus and are capable of reliably reconcile, articulate and resolve different nexus-related interests. Nexus delineates interconnectedness and interdependence between water, energy and food sectors that has dawn considerable attention of academia and policymakers in the past, in the present as well and will continue in the years to come also. A discernment of linkages and all the interactions within and between the nexus constituents is essential to have fully-informed analysis of the WEF nexus itself. The linkages and interactions within the WEF nexus are divided into three separate groups that have come to garner recognition much prior to water, energy and food were collectively considered as intertwined, viz., energy−water, water−food, and energy−food.

5.7.4.1 Water−energy

Water and energy are interlinked and interdependent. Gleick (1994) in his the first most comprehensive work on the water−energy nexus included both water requirement for energy production through diverse sources of energy, including renewable and nonrenewable, and energy needs for water purveying, focusing mostly on pumping, distributing, and desalinating sea water. Many scholars have assessed withdrawal and consumption of water for energy production at the extraction, processing, refining and conversion stages of energy production (Klein & Rubin, 2013; Meldrum, Nettles-Anderson, Heath, & Macknick, 2013). Focus on primary fuels has seemingly been accorded priority in the recent literature including biofuels as a potentially viable alternative energy source to minimize dependence on fossil fuel imports and meet low carbon emission targets; nevertheless, there are growing concerns as well with regard to the need of water for the large-scale implementation of bioenergy (Gerbens-Leenes, Van Lienden, Hoekstra, & Van der Meer, 2012). Keeping in view the linkages between electricity generation and water as a component of water−energy nexus, some studies have acknowledged, quantified and assessed the prevalent state of the linkage between electricity and water (Delgado, 2012; IEA, 2012). Nevertheless, Srinivasan et al. (2017) have forecast the future development of this linkage under various presumed scenarios.

Generally, requirement of water in energy generation has been assessed in the bulk of literature in terms of both withdrawal and consumption of water. Macknick and colleagues (2011) accord recognition to both water withdrawal and consumption values to be necessary indicators for water managers looking after the power plant impacts and vulnerabilities linked to water resources. However, analysis of trade-offs has been facilitated under different water availability and allocation scenarios and this has seemingly emerged as another favorite theme in water−energy discourse (Gjorgiev & Sansavini, 2018). In the wake of implied impacts of the energy-emissions on water, the association of the energy–water link with carbon emissions has assumed added importance in the scientific literature on this nexus. Harnessing of carbon-reduction technologies further exacerbate demand for water resulting in additional fuel use to compensate for energy penalties and the demands of the carbon capture system (Ramirez, Bakshi, Gibon, & Hertwich, 2013). An assessment has been made by some studies with regard to the impacts of low-carbon technologies on water consumption in power plants to meet climate objectives (Merschmann, Vasquez, Szklo, & Schaeffer, 2013), or switching to cleaner fuels (Grubert, Beach, & Webber, 2012).

Interdependence between water and energy has been examined under various scenarios with varying energy demand trends power generation fuel mixes, cooling technologies and biofuel production (IEA, 2012; Siddiqi & Anadon, 2011). Energy for water is used in municipal water pumping, treatment and distribution, wastewater treatment, recycling and, finally, disposal and energy sources analyzed include electricity (Kenway et al., 2008), energy contained in chemical production, transportation of materials, and operation (Racoviceanu et al., 2007). The water-energy nexus forms the heme of analysis by Thiede, Kurle, and Herrmann (2017) in the context of industries other than electricity. Few studies have dealt with use of energy in water sector, especially water treatment (Gleick & Cooley, 2009), technologies like reverse osmosis (McGinnis & Elimelech, 2008), and freshwater conservation like desalination (Semiat, 2008). Use of electricity in groundwater pumping and irrigation also reinforces water–energy nexus (Liang & Zhang, 2011; Nanduri & Saavedra-Antolínez, 2013).

Interlinkages between water and energy mostly deal with the use of water for energy production at different stages of the supply chain for diverse fuels and technologies, including alternative forms of energy like biofuels. Similarly, provisioning of energy for water entails energy consumption for pumping for irrigation, and different technologies and processes in the municipal water supply chain in terms of pumping, distribution, treatment, wastewater treatment, recycling, and disposal. Focus on carbon link with water–energy nexus is also receiving increasing attention.

5.7.4.2 Water–food nexus

The increasing focus on sustainable water resources for agricultural or food production demonstrates recognition of inseparable proximity between food security and water security. A broad unanimity on treating water saving in agriculture as an imperative for water security, especially in agro-based or agriculture-centric economies, gave rise to the notion of the "more crop per drop" research paradigm pioneered by Merrey (1997) that included precision irrigation technologies in tandem with water management practices like changes in agricultural practices, the introduction of less water-intensive crop varieties, increasing water-use efficiency of irrigation, etc. Introduction of the concept of water footprint by Hoekstra (2003) in relation to the management of water resources soon came to reckoned with as the cumulative virtual water content of all goods and services consumed either a single individual or by all the people in one country (Hoekstra & Hung, 2002). Linkages between national water footprints and virtual water flows with regard to international trade, especially for the crops started garnering attention of the academia as well as policymakers (Hoekstra & Chapagain, 2007). Along with the development of comparing water footprints with ecological footprints (Hoekstra, 2007), subsequent period also witnessed development of water footprint benchmark values for a massive number of crops grown worldwide (Mekonnen & Hoekstra, 2014) and this concept is now widely used at various levels.

Irrespective of plenty of knowledge available on water saving in agriculture, implementation of water-saving practices and technologies at a larger scale in food production remained minimal because of barriers like dearth of relevant knowledge and expertise, lack of inputs, inadequate fiscal resources and infrastructure along with poor policy support as being accountable for the inability to upscale such practices and technologies and make hem reach the traditional farmers (Monaghan et al., 2013). The emerging hiatus has proved instrumental in giving rise to new ideas and concepts to tide over the crisis of water scarcity for food production. There have been attempts at exploring linkages between land, food and water in order to ascertain as to how shifts in cultivated crops can prevent excessive use of water resources without adversely impacting food supply (Ren, Yang, Yang, Richards, & Zhou, 2018). In addition to the ostensible water—food linkages at the ground level, Gephart et al. (2017) report that usage of water for food production, especially freshwater for seafood production in aquaculture, keeping in view the changing dietary patters of the people, has emerged in recent literature as a concern.

Notions like trade in agricultural commodities, optimal water pricing for agriculture and changing food preferences are gaining prominence as some of the potential solutions to prevent the water crisis from impeding food security. Trade in agricultural commodities from water-rich countries to water-scarce regions is widely seen as a solution to address water scarcity. Keeping in view the direct and indirect impacts of policy measures impacting global trade, Biewald, Rolinski, Lotze-Campen, Schmitz, and Dietrich (2014) suggest adhering to measures like trade liberalization, trade barriers or agricultural subsidies. Nevertheless, cautioning against food trade as a measure to cope with water scarcity, Allan (2003) noted that in the case of developing countries, the idea of virtual water apparently seems to endanger local farming avocations in the importing regions, thereby necessitating consideration of social linkages. While optimal water pricing is suggested as a measure to encourage maximum use of water (Johansson, Tsur, Roe, Doukkali, & Dinar, 2002; Sampath, 1992). Nonetheless, it is also concurrently suggested that pricing alone is not the panacea and emphasis is laid on clearly delineated and legally enforceable water rights and responsibilities for water operators and users along with incentives for water conservation and envisaging improvement in irrigation efficiencies (Yang et al., 2003).

In the wake of globalization and increasing affluence, attention is being focused on ascertaining the impact of changing food preferences on water demand for food production and accordingly, stress is laid on water resources owing to changing food preferences (Hess, Andersson, Mena, & Williams, 2015). Other factors that have bearing on the water—food nexus are also being taken into consideration; and these, inter alia, include land use (Das, Scambos, Koenig, van den Broeke, & Lenaerts, 2015), climate (Misra, 2014), society (Rosegrant & Ringler, 1999), and the economy (Biewald et al., 2014). Improvement in supply-side efficiency to lessen food demand and resultantly water demand is emphasized by Odegard and van der Voet (2014).

5.7.4.3 Energy—food nexus

Focus on interconnectedness and interdependence between energy and food started hogging limelight in the 1970s, especially in the wake of the energy crisis that brought forth speculative perspectives about the impact of the energy crisis on food production because of increasing dependence of food supply chain on energy resources (Black et al., 2011; Pimentel et al., 1973). This conjecture got translated into reality when one of the main reasons ascribed to the augmentation in food prices worldwide from 2002 onwards was a steep hike in petroleum prices (Cassman & Liska, 2007;

Trostle, Marti, Rosen, & Wescott, 2011). While focusing on improving energy use efficiency and encouraging use of renewable energy for food production to minimize the use of fossil fuels (Goldman, 2009), emphasis was also prioritized on organic farming systems that were found to be less energy-intensive than traditional farming systems (Dalgaard, Halberg, & Porter, 2001).

A few studies have also explored different factors affecting energy productivity in agriculture over time, like price changes, size of land holdings, technological changes, resource degradation caused by farm management in the form of soil erosion, groundwater depletion, reduced genetic diversity, pest resistance, and so on. Cleveland (1995a) depicted a clear response of farmers to higher energy prices, which resulted in technical and managerial changes that finally improved energy productivity. Different types of energy like sunlight, human labor, animals, fuel and electricity are used as energy input in agriculture and this has been categorized as direct and indirect energy input (Cleveland, 1995b; Singh, Mishra, & Nahar, 2002; Ozkan, Akcaoz, & Fert, 2004). Direct energy encompasses human power, diesel, and electrical energy used in production processes, whereas indirect energy comprises pesticide and fertilizers. The excessive and unscientific use of indirect energy like fertilizers and pesticides for food production that leads to land degradation, which in turn affects crop productivity has been a matter of concern (UNCCD, 2013).

Biofuel production has come to be reckoned with as a recent emerging link between energy and food. Furtherance of energy crops, commonly called biofuels, has been high on the global agenda in recent years. Attributed to a number of reasons such as surging energy prices, increasing concerns about energy security and climate change, the eco-friendly nature of biofuels, and the income expectations of farmers and other investors (Von Braun & Pachauri, 2006), this has triggered the "food-versus-fuel" debate. Steep hike in food prices between 2005 and 2008 was attributed to biofuel production. Nevertheless, there have been divided opinions among the scholars on this issue. While Ajanovic (2011) observed no significant impact of biofuel production on feedstock prices, Rathmann, Szklo, and Schaeffer (2010) and Zhang et al. (2010) noted that production of biofuels had contributed to increasing food prices in the short term and indicated no direct long-term price relationship between fuel and agricultural commodity prices. Clearing of land for biofuel crops and the resultant loss of forests, peatlands and grasslands could exacerbate global warming and climate change (Christopher, 2008). In order to deal such a scenario, Escobar et al. (2009) suggested the establishment of international cooperation, regulations and certification mechanisms for the use of land, and the mitigation of environmental and social impacts caused by biofuel production.

Energy and food are interrelated. Energy is a valuable input in food production. Nevertheless, there are two prime concerns with regard to the impact of energy on agriculture and food. In the first instance, disproportionate use of nonrenewable fossil energy and its related effects on energy security and the environment, and secondly, replacement of food crops with energy crops — both of these entail the likelihood of resulting in hike in food prices thereby jeopardizing food security. In other words, energy security can negatively impact food security and similarly, ensuring food security can also potentially impede energy security in eventualities of energy shortages and inefficient energy use.

The key themes discernible in the literature about different interconnections of water, energy and food are summed up in Table 5.4.

Projections of future water, energy and food demands do vary, but they all agree that demand in the three sectors will significantly increase over the coming decades while the natural resources base will simultaneously be weakened through environmental degradation and climate change. All

Table 5.4 Major themes Covered in the literature on energy, water, and food interconnections.

Water, energy, and food linkages	Major themes discernible in the literature
Water—energy	• Consumption of energy generation technologies in water sector. • Consumption of water technologies by energy sector • Allocation of water for energy and other uses. • Consumption of water for biofuel production.
Water—food	• Water intensity of food crops. • Fluctuations in water footprint owing to changes in dietary-intake and patters of food trade.
Energy—food	• Energy consumption in food production and processing. • Effects of biofuels on food security.

of these threats are depicted in Fig. 5.1, which describes the relationships between water, energy, food and climate in light of global projections which indicate increasing scarcities and growing demand. This set of projections predict that by 2030, the demands for food and energy will increase by 50% and the demand for water by 30%, while we also face the challenges of adapting to and mitigating climate change (Cairns & Krzywoszynska, 2016).

The policymaking frameworks addressing natural resources management have historically been characterized by sectoral approaches and isolated policy responses, which undermine the complex relationships between sectors and resource systems. This has often resulted in segmented planning and resource stresses (Pittock, Hussey, & McGlennon, 2013). Isolated planning in the water, energy and agricultural sectors leads to unintended consequences and additional WEF resources stresses, which in turn worsens livelihoods and undermines sustainable development (Bizikova, Roy, Swanson, Venema, & McCandless, 2013). It became evident and urgent that more responsible management of WEF systems was needed to cope with the changing lifestyles and growing demand for resources and services (Liu et al., 2017).

It is noted that some literature includes the environment and/or ecosystems within the nexus (labeled as the WEFE nexus), as does the Nexus Regional Dialogues Programme. It is acknowledged that the environment and ecosystem play a fundamental role in the Nexus. In the use of the term WEF nexus in this document, a consideration of the environment and ecosystem is implicit within the consideration of each of the three sectors. Fig. 5.2 shows this integrated approach to the assessment of the WEF nexus, with ecosystems located at the center (GIZ, 2016). Furthermore, a Nexus problem is not defined as necessarily involving all three of the water, energy and food sectors; the interconnections between any two of these sectors constitute a nexus problem. Therefore better cognizance of the water—energy nexus, can be had by discerning the water—energy nexus, water—food nexus and the energy—food nexus that ultimately leads to the water—energy—food (i.e., WEF) nexus (see Fig. 5.3).

Although the systems are physically interconnected, decisions and policy planning in each sector are mostly made in isolation (Rasul, 2016; White, Jones, Maciejewski, Aggarwal, & Mascaro, 2017). Thus the nexus governance discourse postulates that to manage risks, maximize gain and optimize trade-offs in resources use, we must not only understand how these systems are physically connected but how they are institutionally linked (see, e.g., Rasul, 2016; Scott, 2017). The crucial

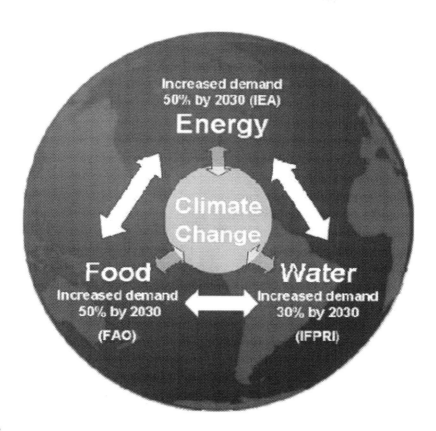

FIGURE 5.2

Water, energy, and food projections.

From Beddington, J. Food, energy, water and the climate: A perfect storm of global events?. *(2009).* <http://webarchive.nationalarchives.gov.uk/20121206120858/http://www.bis.gov.uk/assets/goscience/docs/p/perfect-storm-paper.pdf>.

role that institutions and governance processes play in enabling policy coherence and a nexus approach is underlined in several reports and frameworks.

Nexus literature acknowledges that a nexus approach requires coordination and integration across levels of government (vertical), as well as across sectors (horizontal) and emphasizes the key role of institutional relationships and effective coordination mechanisms (Scott, 2017; Weitz et al., 2017). Having "stronger institutions that are better interlinked" is identified as the key to a nexus approach (Hoff, 2011). However, considering the complexity emerging from horizontal and vertical interdependencies, the nexus researchers identify several challenges to the implementation of the nexus approach in decision-making.

The literature postulates that historically entrenched and vertically structured government departments as well as sector-based policies and regulatory mechanisms act as main barriers to the adaptation of a WEF nexus approach in decision-making (Rasul, 2016; Scott, 2017). Based on case

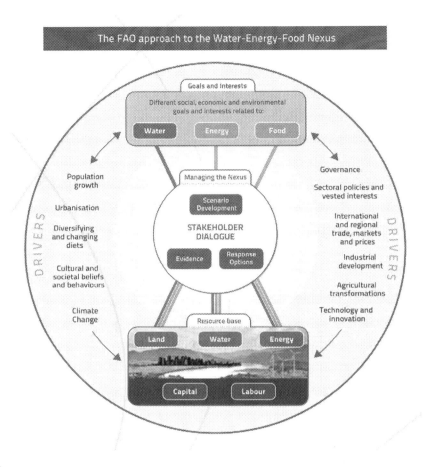

FIGURE 5.3

Interconnectedness of water—energy—food nexus.

From GIZ. Water, energy & food nexus in a nutshell. (2016). <http://www.water-energy-food.org/fileadmin/user_upload/files/2016/documents/nexus-secretariat/nexus-dialogues/Water-Energy-Food_Nexus-DialogueProgramme_Phase1_2016-18.pdf>.

studies, Scott (2017) concluded the effectiveness of the Nexus approach is determined by institutional relationships and the capacities of governing organizations to cooperate with each other.

Further key barriers to implementation highlighted in the current discourse emerge from the traditional, sector-based structures of political institutions and governance processes. The following barriers have been identified in literature:

- Lack of communication between the sectors (Weitz et al., 2017).
- Divergent sectoral institutional frameworks and interests (Weitz et al., 2017).
- Unequal distribution of power and capability between the sectors (Howarth & Monasterolo, 2016).

- A lack of willingness to cooperate and lack of trust across groups of actors belonging to different disciplines and government levels (Embid & Martin, 2017).

In 2015 the UN adopted the resolution "Transforming our world: the 2030 Agenda for Sustainable Development," which contained 17 SDGs and 169 associated targets (UN, 2015a). It should be noted that there is no specific mention of the term "Nexus" in the SDGs; however, a Nexus approach enables us to work toward simultaneously achieving water, energy and food security, thus ensuring access for basic human needs (Stephan et al., 2018). A Nexus approach allows potential trade-offs to be identified at the policy design stage, allows the identification and development of solutions that positively benefit multiple SDGs, and avoids the "silo" approach in implementation strategies (FAO, 2018).

The Paris Agreement was signed in 2015, which aims to limit the increase in global average temperatures to well below 2°C above preindustrial levels (UN, 2015b). Climate change and the WEF Nexus are inextricably linked, and this has been recognized in current political and scientific discussion. It is critical to develop effective strategies to adapt to climate change and ensure water, energy and food security for a growing global population. An overall analysis of the links between the WEF Nexus, the SDGs and the Paris Agreement recognizes the crucial role of WEF Nexus concept in achieving both the goals of the Paris Agreement and of the SDGs (FAO, 2018).

The nexus approach is increasingly used at the project level and supported by some governments, civil society, international development partners, the private sector and research (FAO, 2018). Numerous conferences and international workshops have taken place in 2018 (Nexus: Water, Food, Energy and Climate Conference at the University of North Carolina; the Reannexes Conference at Wageningen University; Water–Food Nexus High Level Panel at the World Water Forum in Brazil; various sessions at the Stockholm World Water Week; The Food–Energy–Water Nexus MiniSymposium at Monash University). Many conferences and workshops are also taking place with a focus on addressing specific nexus issues at the regional scale (Nexus Platform, n.d.).

5.7.5 Water security and nature-based solutions

NBSs are stimulated and supported by nature and their usage contributes to the improved management of water resources. An NBS can comprise conserving or rehabilitating natural ecosystems and/or the enhancement or creation of natural processes in modified or artificial ecosystems and its application can be facilitated at micro or macro scales. Acknowledgment of the role of ecosystems and the concept and application of NBS in water management has been there for many years in view of the fact that "the role of ecosystems has been entrenched in modern hydrological sciences for decades" (WWAP, 2018). According to Cohen-Shacham, Walters, Janzen, and Maginnis (2016), NBS terminology garnered salience probably around 2002. Most of the other concepts, tools, approaches or terminology already in use among different stakeholder groups are either the same or similar to or compatible to NBS. All of these approaches aim at balancing a more "technocratic, built-in-infrastructure approach" that has come to dominate water resources management, by acknowledging the potential contribution of ecosystems (WWAP, 2018).

Ecohydrology is regarded as an integrative science that deals with the interaction between hydrology and biota. It was founded based on the knowledge on the evolutionary established processes of aquatic ecosystems, and on the dual regulation between hydrology and biota. This approach aims at restoring the ecosystems' carrying capacities, so that they can become resilient on the long-term and adaptable to impacts at the entire river basin scale (Chicharo, Müller;, & Fohrer, 2015). The ecosystem approach envisages a conceptual framework for tackling ecosystem issues, adopted by the Convention on Biological Diversity (CBD, 1992) and compatible with the wise use of wetlands concept of the 1971 Ramsar Convention on Wetlands (Ramsar Convention, 2016). Ecosystem-based management and EbA or mitigation entails the conservation, sustainable management and restoration of ecosystems. Environmental flows delineate the quantities, quality and patterns of water flows needed to sustain freshwater and estuarine ecosystems and the ecosystem services provided by them. "Eco-, phyto- and bioremediation are concepts that use ecosystem restoration to reinstate a diverse system of plant communities in a particular ecosystem so that its buffering or remediation capacities are enhanced" (WWAP, 2018, p. 23). As pointed out by Cohen-Shacham et al. (2016), other concepts, tools and approaches partly related to NBS comprise ecological restoration, ecological engineering, forest landscape restoration, green or natural infrastructure, ecosystem-based disaster risk reduction and climate adaptation ecosystem services.

NBS acknowledge ecosystems as natural capital, or the stock of renewable and nonrenewable natural resources—plants, animals, air, water, soils, and minerals that coalesce to generate a flow of benefits to people (Atkinson & Pearce, 1995). The Natural Capital Protocol3 is being increasingly recognized by a wide range of stakeholders, including business, and supports the use of NBS by highlighting the flow of benefits that can be derived from using nature (NCC, 2016). Among the various approaches comprising NBS, EbA has been widely accepted as a more tenable approach to ensure water security.

5.7.5.1 Ecosystem-based adaptation and water security

EbA entails the use of biodiversity and ecosystem services to help people and communities adapt to the adverse impacts of climate change (UNEP, 2016b); and EbA approaches can assist communities adapt to water insecurity by increasing water quantity and enhance water quality among the world's 105 largest cities, one-third get their water from forested areas (USAID, 2017a,b). Healthy forests, mangroves and other natural systems wield a direct impact on water quantity by maintaining water flow, absorbing rainfall and replenishing watersheds (UNFCCC, 2011). EbA approaches that assist in the restoration and conservation of healthy ecosystems can be a cost-effective adaptation strategy to asseverate/maintain or enhance the quantity of water available for communities by recharging aquifers and improving natural water storage (Bertule et al., 2014; Talberth, Gray, Branosky, & Gartner, 2012). These approaches can also be instrumental in improving access to clean water by providing water filtration benefits alike to drainage and wastewater treatments. Forested areas entail the potential of regulating water quality and minimizing pollution by filtering runoff, preventing erosion and slowing sedimentation (Bertule et al., 2014; Talberth et al., 2012).

Runoff water from industrial areas or farmland gets filtered through wetlands and riparian buffers that also entail the potential of removing sediment before it contaminates groundwater and in an identical manner, coastal wetlands can play a crucial role in plummeting saltwater penetration in coastal aquifers (Bertule et al., 2014). EbA approaches can ameliorate the efficacy of water use and management with the help of agroforestry and conservation farming (Carabine, Venton, Tanner, & Bahadur, 2015). The low-implementation-costs of the EbA and related nature-based approaches can benefit smallholder farmers, who are extremely vulnerable to water insecurity in the face of climate change (Vignola et al., 2015). As reported by Colls, Ash, and Ninni (2009), communities in Zambia identified conservation farming as a priority-coping-strategy to assist them adapt to drought and rainfall variability. Apart from assisting in minimizing effects to water security from extreme weather events, EbA strategies can also help vulnerable communities adapt to extreme weather events. Baig, Rizvi, Pangilinan, and Palance-Tan (2015) are of the view that the conservation and restoration of mangroves, coastal marshes and coral reefs can afford protection from storm surges, floods and hurricanes/typhoons (in addition to sea-level rise) by serving as physical buffers that retain excess water, dissipate wave energy and stabilize shorelines.

Additional development benefits can accrue through EbA approaches for addressing water insecurity and that are prone to include carbon sequestration, biodiversity conservation and important goods such as wild fish, fuel and nontimber forest products. Besides, EbA can be helpful in protecting and strengthening ecosystems like coral reefs, mangrove forests and wetlands that are vital to conserving freshwater and marine biodiversity. EbA approaches that contribute to improvement of water security can also chip in benefitting water-dependent economic sectors like commercial fisheries, hydropower generation and agriculture (Munang et al., 2013; Nel et al., 2014). In comparison to conventional adaptation interventions, such as the construction of infrastructure, EbA approaches can be more cost-effective and yield positive landscape-scale impacts (Munang et al., 2013; Nel et al., 2014).

Water security can be effectively addressed by both EbA approaches and hard infrastructure; subject to the context, an adaptation strategy that includes both can be most effective (USAID, 2017a,b). Hard infrastructure options comprise construction of water treatment plants, levees, sea walls, storm water drains, pipelines, dams, canals, wells and water storage tanks. Keeping in the fact that built infrastructures also entail of potential of adversely impacting surrounding ecosystems and could lead to maladaptation in some cases (WWF & World Bank, 2013), EbA approaches are often cheaper and supply longer term results, rendering them more accessible for poorer communities; they can also yield many additional benefits for development, and there are few, if any, negative impacts from improved ecosystem management (Bertule et al., 2014; Colls et al., 2009).

Multiple benefits and cobenefits have accrued from EbA approaches as well as implementation of EbA approaches in tandem with hard infrastructure approaches. The cost of de-silting and reviving more than 6000 natural dams in the Godavari River Basin in India was estimated at $635 million versus $4 billion to construct a new dam by the researchers and the construction of the proposed dam would have displaced 250,000 people and damaged natural habitats, including 60,000 hectares of forest (Rizvi, Baig, & Verdone, 2015). In certain cases, EbA solutions also entail the potential of protecting water-related hard infrastructure. According to Rizvi et al. (2015), 16 watershed-protected-areas in Peru were estimated to provide almost $5 million in savings over a 10-year period by protecting dams and reservoirs from sedimentation; and these protected areas provided a number of other ecosystem goods and services that were valued at over $80 million annually. Harnessing EbA approaches in tandem with hard infrastructure can yield fruitful results in addressing various aspects of water security. Doswald and Estrella (2015) cite the example of a project in Bolivia where it assisted local communities in adapting to droughts by supporting the construction of hard infrastructure like water storage tanks to be complimented by EbA strategies such as the establishment of protected areas around watersheds.

Challenges confronted in enlarging the use of EbA, as pointed by USAID (2017a,b), inter alia, include: the length of time to see results for some approaches such as ecosystem restoration;

difficulty in gauging impacts and lack of institutional capacity to implement these approaches; need for better coordination among different development sectors and government ministries, availability of sufficient fiscal resources to implement approaches, and hydrological and climate information on which to base interventions. Seddon et al. (2016) view the key challenge in building the capacity of communities globally to develop robust adaptation plans that are "fine-tuned to their economic, social and environmental needs, access suitable levels of adaptation finance, and ultimately implement EbA at scale for the benefit of people and planet."

5.8 Conclusion

Water is a finite resource and it has been constantly under threat owing to multiple factors like climate change, increasing pressures on surface and groundwater resources in the wake of burgeoning population, rapid pace of urbanization and rapid profanation of economic development activities dependent on water. The problem of water insecurity is further compounded by pollution of surface water resources and exacerbated pace of depletion as well as contamination of groundwater resources. Undoubtedly, 2019 WWDR carried the theme of "Leaving no One Behind" (WWAP, 2019) and this theme is seemingly in consonance with SDG-6 of the Agenda 2030 embracing SDGs; nonetheless the progress to toward the targets of the SDG-6 had by the closing part of 2019 not been satisfactory because "Achieving target 6.1 means addressing the "unfinished business" of extending services to 844 million people who still lack even a basic water service, and progressively improving the quality of services to 2.1 billion people who lack water accessible on premises, available when needed and free from contamination" (UN-Water, 2018:11).

Water security is often regarded as a broader concept of Integrated Water Resource Management (IWRM) that is designed to seek to balance resource protection and resource use. However, even the progress in implementing the IWRM has been slow thus far (UN-Water, 2018; WWAP, 2012). IWRM is barely delineated as a perspective-way of discerning water, based on technical-scientific knowledge, while water security embraces a more inclusive and diverse way of perceiving water, with a greater emphasis on human values, ethics, and power (Gerlak & Mukhtarov, 2015). NBS have come to be reckoned with as more viable and effective pathway to address water security and among the various NBS approaches, an ecosystem-based approach (EbA) is being recognized as viable and cost-effective approach. EbA approach to water security aligns closely with the traditional Indigenous-perspective on water (Longboat, 2015). Essentiality of ensuring water security extending beyond water quality and quality to incorporate cultural and spiritual components is emphasized by Wutich et al. (2017). In sum, achieving water security is prone to variation between communities because there is no single pathway to water security.

References

Ajanovic, A. (2011). Biofuels vs food production: Does biofuels production increase food prices? *Energy*, *36*(4), 2070−2076.

Alcamo, J., & Henrichs, T. (2002). Critical regions: A model-based estimation of world water resources sensitive to global changes. *Aquatic Sciences*, *64*(4), 352−362.

References

Alcamo, J., Floke, M., & Maeker, M. (2007). Future long-term changes in global water resources driven by socio-economic and climate changes. *Hydrological Sciences Journal*, 52(2), 247–275.

Allan, T. (1997). *'Virtual water': A long-term solution for water short Middle Eastern economies*. Paper presented at the 1997 British Association Festival of Science. Available from https://www.soas.ac.uk/water/publications/papers/file38347.pdf.

Allan, T. (2003). *IWRM/IWRAM: A new sanctioned discourse? SOAS water issues study group occasional paper 50*. London: School of Oriental and African Studies, King's College.

Amery, H. A. (1997). Water security as a factor in Arab-Israeli wars and emerging peace. *Studies in Conflict and Terrorism.*, 20(1).

Anderson, E. W. (1992). The political and strategic significance of water. *Outlook on Agriculture*, 21, 247–253.

Andersson, E., Brogaard, S., & Olsson, L. (2011). The political ecology of land degradation. *Annual Review of Environment and Resources*, 36, 298–319.

Appelgren, B., & Klohn, W. (1999). Management of water scarcity: A focus on social capacities and options. *Physics and Chemistry of the Earth, Part B: Hydrology, Oceans and Atmosphere*, 24(4), 361–373.

AQUASTAT. (n.d.). *AQUASTAT website*. FAO. <https://fao.org/nr/water/aquastat/main/index.stm>.

Arab Water Council. *Vulnerability of arid and semi-arid regions to climate change: Impacts and adaptive strategies*. (2009). <http://www.preventionweb.net/files/12914_PersPap09.AridandSemiAridRegions1.pdf>.

Atkinson, G., & Pearce, D. (1995). Measuring sustainable development. In D. W. Bromley (Ed.), Handbook of environmental economics. Oxford: Wiley-Blackwell.

Baig, S., Rizvi, A., Pangilinan, M., & Palance-Tan, R. (2015). *Costs and benefits of ecosystem based adaptation: The case of the Philippines*. Switzerland: IUCN.

Bakker, K., & Cook, C. L. (2011). Water governance in Canada: Innovation and fragmentation. *International Journal of Water Resources Development*, 27(2), 275–289.

Barbier, E. B. (2004). Water and economic growth. *Economic Record*, 80(248), 1–16.

Berndes, G. (2008). Water perspectives on large-scale bioenergy. In R. S. Zalesny, S. Ronald, R. Mitchell, & J. Richards (Eds.), *Biofuels, bioenergy and bioproducts from sustainable agricultural and forest crops*. Newton Square, P.A.: US Department of Agriculture.

Bertule, M., Lloyd, G. J., Korsgaard, L., Dalton, J., Welling, R., Barchiesi, S., Smith, M., et al. (2014). *Green infrastructure guide for water management: Ecosystem-based management approaches for water-related infrastructure projects*. Nairobi: UNEP.

Biewald, A., Rolinski, S., Lotze-Campen, H., Schmitz, C., & Dietrich, J. P. (2014). Valuing the impact of trade on local blue water. *Ecological Economics*, 101, 43–53.

Biswas, A. K. (2004). Integrated water resources management: A reassessment. *Water International*, 29(2), 8.

Biswas, A. K., Tortajada, C., & Izquierdo, R. (Eds.), (2012). *Water quality management: Present situations, challenges and future perspectives*. London: Routledge.

Bizikova, L., Roy, D., Swanson, D., Venema, H., & McCandless, M. (2013). *The water–energy–food security nexus: Towards a practical planning and decision-support framework for landscape investment and risk management*. International Institute for Sustainable Development. Available online.

Black, R., Adger, W. N., Arnell, N. W., Dercon, S., Geddes, A., & Thomas, D. (2011). The effect of environmental change on human migration. *Migration and Global Environmental Change – Review of Drivers of Migration*, 21(Suppl. 1), S3–S11.

Boelens, R. (2014). Cultural politics and the Hydro-social cycle: Water, power and identity in the Andean highlands. *Geoforum.*, 57, 234–247.

Bogardi, J., Oswald Spring, U., & Brauch, H. G. (2016). Water security: Past, present and future of a controversial concept. In C. Pahl-Wostl, A. Bhaduri, & J. Gupta (Eds.), *Handbook on water security* (pp. 38–58). Cheltenham; Northampton, MA: Edward Elgar.

Bologensi, T., Gerlak, A. K., & Giuliani, G. (2018). Explaining and measuring social-ecological pathways: The case of global changes and water security. *Sustainability.*, *10*, 4378.

Borchardt, D., Bogardi, J. J., & Ibisch, R. B. (Eds.), (2016). *Integrated water resources management: Concept, research and implementation*. Switzerland: Springer Publishers.

Boretti, A., & Rosa, L. (2019). Reassessing the Projections of the World Water Development Report. *Clean Water*, *2*(15). Available from https://www.nature.com/articles/s41545-019-0039-9.pdf.

Brears, R. C. (2018). *The green economy and the water-energy-food nexus*. New York: Palgrave-Macmillan.

Brisbane Declaration (2007). The Brisbane Declaration: Environmental flows are essential for freshwater ecosystem health and human well-being. In: *10th International river symposium*, September 3–6, 2007, Brisbane. http://riverfoundation.org.au/wp-content/uploads/2017/02/THE-BRISBANE-DECLARATION.pdf.

Burek, P., Satoh, Y., Fischer, G., Kahil, M. T., Scherzer, A., Tramberend, S., ... Wiberg, D. (2016). *Water futures and solution: Fast track initiative (final report)*. Luxemburg: International Institute for Applied Systems Analysis (IIASA).

Cairns, R., & Krzywoszynska, A. (2016). Anatomy of a buzzword: The emergence of 'the water-energy-food nexus' in UK natural resource debates. *Environmental Science Policy*, *64*, 164–170.

Carabine, E., Venton, C. C., Tanner, T., & Bahadur, A. (2015). *The Contribution of Ecosystem Services to Human resilience: A rapid review*. London: Overseas Development Institute (ODI).

Cassman, K. G., & Liska, A. J. (2007). Food and fuel for all: Realistic or foolish? *Biofuels, Bioproducts and Biorefining*, *1*(1), 18–23.

CAWMA (Comprehensive Assessment of Water Management in Agriculture). (2007). *Water for food, water for life: A comprehensive assessment of water management in agriculture*. London: Earthscan/International Water Management Institute (IWMI).

CBD (Convention on Biological Diversity) (1992). Text of Convention on Biological Diversity. Available from http://www.cbd.int/convention/text/default.shtml.

Chaturvedi, V., Hejazi, M., Edmonds, J., Clarke, L., Kyle, P., Davies, E., & Wise, M. (2013). Climate mitigation policy implications for global irrigation water demand. *Mitigation and Adaptation Strategies for Global Change*, *20*(3), 1–16.

Chaves, M. M., & Oliveira, M. M. (2004). Mechanisms underlying plant resilience to water deficits: prospects for water-saving agriculture. *Journal of Experimental Botany*, *55*(407), 2365–2384.

Chicharo, L., Müller, F., & Fohrer, N. (2015). Introduction. In L. Chicharo, F. Müller, & F. Nicola (Eds.), *Ecosystem services and river basin ecohydrology* (pp. 1–6). New York: Springer.

Christiansen, L., Olhoff, A., & Trærup, S. (Eds.), (2011). *Technologies for adaptation: Perspectives and practical experiences*. Roskilde: UNEP Risø Centre.

Christopher, F. (2008). *Time to move to a second generation of biofuels*. Washington, DC: Worldwatch Institute.

Clement, F. (2013). From water productivity to water security: A paradigm shift? In B. Lankford, K. Bakker, M. Zeitoun, & D. Conway (Eds.), *Water security principles, perspectives, and practices* (pp. 148–165). New York: Routledge.

Cleveland, C. J. (1995a). Resource degradation, technical change, and the productivity of energy use in United States agriculture. *Ecological Economics*, *13*(3), 185–201.

Cleveland, C. J. (1995b). The direct and indirect use of fossil fuels and electricity in USA agriculture, 1910–1990. *Agriculture, Ecosystems & Environment*, *55*(2), 111–121.

Climateiswater.org (Portal Website). *Climate is water.* (n.d.). <http://www.climateiswater.org/>.

CoFR, United States Senate (Committee on Foreign relations, United States Senate). (2011). *Avoiding water wars: Water scarcity and Central Asia's growing importance for stability in Afghanistan and Pakistan*. Washington, DC: United States Government Printing Office.

Cohen-Shacham, E., Walters, G., Janzen, C., & Maginnis, S. (Eds.), (2016). *Nature-based solutions to address global societal challenges*. Gland: IUCN.

Colls, A., Ash, N., & Ninni, I. (2009). *Ecosystem-based adaptation: A natural response to climate change.* Gland: IUCN.

Connor, R., Miletto, M., Grego, S., Koncagul, E., Shamir, U., Gyawali, D., ... Wright, A. (2015). UN world water development report: Water for a sustainable world. Paris: UNESCO.

Cook, C., & Bakker, K. (2012). Water security: Debating an emerging paradigm. *Global Environmental Change, 22*(1), 94–102.

Cook, C., & Bakker, K. (2013). Debating the concept of water security. In B. Lankford, K. Bakker, M. Zeitoun, & D. Conway (Eds.), *Water security principles, perspectives, and practices* (2013, pp. 49–63). New York: Routledge.

Cook, C., & Bakker, K. (2016). Water security: Critical analysis of emerging trends and definitions. In C. Pahl-Wostl, A. Bhaduri, & J. Gupta (Eds.), *Handbook on water security* (pp. 19–37). Cheltenham: Edward Elgar.

CRED/UNISDR (Centre for Research on the Epidemiology of Disaster/United Nations Office for Disaster Risk Reduction). (2015). *The human costs of weather-related disasters 1995–2015.* Brussels/Geneva: CRED/UNISDR. Available online.

Dalgaard, T., Halberg, N., & Porter, J. R. (2001). A model for fossil energy use in Danish agriculture used to compare organic and conventional farming. *Agriculture, ecosystems & environment, 87*(1), 51–65.

Das, I., Scambos, T. A., Koenig, L. S., van den Broeke, M. R., & Lenaerts, J. T. M. (2015). Extreme wind-ice interaction over recovery ice stream, East Antarctica. *Geographical Letters, 42,* 8064–8070.

DCDC, MoD (Development, Concepts and Doctrine Centre, Ministry of Defence). (2010). *Global strategic trends—Out to 2040.* London: Ministry of Defence.

De, A., Bose, R., Kumar, A., & Mozumdar, S. (2014). *Targeted delivery of pesticides using biodegradable polymeric nanoparticles. Springer briefs in molecular science.* New Delhi: Springer India.

De Fraiture, C., Molden, D., & Wichelns, D. (2010). Investing in water for food, ecosystems, and livelihoods: An overview of the comprehensive assessment of water management in agriculture. *Agricultural Water Management, 97,* 495–501.

Delgado, M. (2012). Water, energy and food security in mexico city. *Resilient Cities, 2,* 105–111.

deLoë, R.; Varghese, J.; Ferreyra, C.; and Kreutzwiser, R. (2007). Water allocation and water security in Canada: Initiating a policy dialogue for 21st century. Guelph Water management Group, University of Guelph, Canada.

Devereux, S., Roelen, K., & Ulrichs, M. (2016). Where next for social protection? *IDS Bulletin, 47*(4).

Dickin, S. (2018). *Gender and water security in Burkina Faso: Lessons for adaptation.* Stockholm: SEI, SEI Policy Brief.

Doswald, N., & Estrella, M. (2015). *Promoting ecosystems for disaster risk reduction and climate change adaptation: Opportunities for integration. Discussion paper.* Nairobi: UNEP.

Duchin, F., & López-Morales, C. (2012). Do water-rich regions have a comparative advantage in food production? Improving the representation of water for agriculture in economic models. *Economic Systems Research, 24*(4), 371–389.

EC (European Commission). (2013). *Report from the Commission to the Council and the European Parliament on the implementation of Council Directive 91/676/EEC concerning the protection of waters against pollution caused by nitrates from agricultural sources based on Member State Reports for the Period 2008–2011.* Brussels: EC. Available online.

Embid, A., & Martin, L. (2017). *The nexus between water, energy and food in Latin America and the Caribbean: Planning, policy framework and the identification of priority interconnections (Original title in Spanish: El Nexo entre el agua, la energía y la alimentación en América Latina y el Caribe: Planificación, marco normativo e identificación de interconexiones prioritarias).* Bonn, Germany: GIZ. Available online.

Escobar, J. C., Lora, E., Venturini, O. J., & Ranez, E. (2009). Biofuels: Environment, technology and Food Security. *Renewable and Sustainable Energy Reviews, 13*(6-7), 1275–1287.

Falconer, I. (1999). An overview of problems caused by toxic blue-green algae (cyanobacteria) in drinking and recreational water. *Environmental Toxicology, 41*(1), 5−12.

Falkenmark, M. (1989). The massive water scarcity now threatening Africa: why isn't it being addressed? *Ambio., 18*(2), 112−118.

Falkenmark, M., Lundquist, J., & Widstrand, C. (1989). Macro-scale water scarcity requires micro-scale approaches. *Nature Resources Forum, 13*(4), 258−267.

Falkenmark, M.A.; Berntell, A.; Jägerskog, J.; Lundqvist, M.; and Tropp, H. (2007). On the verge of a new water scarcity: A call for good governance and human ingenuity. Stockholm: SIWI Policy Brief. SIWI.

Falkenmark, M. (2001). The greatest water problem: The inability to link environmental security, water security and food security. *International Journal of Water Resources Development, 17*, 539−544.

Falkenmark, M., & Lundqvist, J. (1998). Towards water security: Political determination and human adaptation crucial. *Natural Resources Forum, 22*, 37−51.

Falkenmark, M. (2013). The multiform water scarcity dimension. In B. Lankford, K. Bakker, M. Zeitoun, & D. Conway (Eds.), *Water security principles, perspectives, and practices* (2013, pp. 64−79). New York: Routledge.

Falkenmark, M., & Widstrand, C. (1992). Population and Water Resources: A Delicate Balance. *Population Bulletin, 47*(3), 1−36.

FAO & IWMI (Food and Agriculture Organization & International Water Management Institute). (2018). *More people, more food, worse water? A global review of pollution of water from agriculture.* FAO. Available from http://www.fao.org/3/ca0146en/CA0146EN.pdf.

FAO (Food and Agriculture Organization). (2003). *Review of world water resources by country.* Rome: FAO. Available from http://www.fao.org/tempref/agl/AGLW/ESPIM/CD-ROM/documents/5C_e.pdf.

FAO (Food and Agriculture Organization). (2011a). *The state of the world's land and water resources for food and agriculture: Managing systems at risk.* Rome: FAO/Earthscan.

FAO (Food and Agriculture Organization). (2011b). *Save and grow: A policy maker's guide to the sustainable intensification of the smallholder crop production.* Rome: FAO.

FAO (Food and Agriculture Organization). (2012). *The state of food and agriculture.* Rome: FAO.

FAO (Food and Agriculture Organization). (2018). Water-energy-food nexus for the review of SDG 7. *Policy Brief. #9.* Rome: FAO.

Ferguson, B., Brown, R., & Deletic, A. (2013). Diagnosing transformative change in urban water systems: Theories and frameworks. *Global Environmental Change, 23*, 264−280.

Few, R., Satyal, P., McGahey, D., Leavy, J., Budds, J., Assen, M., ... Bewket, W. (2015). Vulnerability and adaptation to climate change in semi-arid areas in East Africa. In: ASSAR working paper. ASSAR PMU, South Africa. Available online.

Fewtrell, L., Kay, D., Watkins, J., Francis, C., et al. (2011). The microbiology of urban UK flood-waters and a quantitative microbial risk assessment of flooding and gastrointestinal illness. *Journal of Flood Risk Management, 4*(2), 77−87.

Fischhendler, I., & Nathan, D. (2016). The social construction of water security discourses: Preliminary evidence and policy implications from the Middle East. In C. Pahl-Wostl, A. Bhaduri, & J. Gupta (Eds.), *Handbook on water security* (pp. 76−90). Cheltenham: Edward Elgar.

Funke, J., Prasse, C., & Ternes, T. A. (2016). Identification of transformation products of antiviral drugs formed during biological wastewater treatment and their occurrence in the urban water cycle. *Water Research, 98*, 75−83.

Funke, N., Meissner, R., Nortje, K., & Steyn, M. (Eds.), (2019). *Understanding Water Security at Local Government Level in South Africa.* London: Palgrave.

Funke, N., Oelofse, S. H. H., Hattingh, J., Ashton, P. J., & Turton, A. H. (2007). IWRM in developing countries: Lessons from the Mhlatuze Catchment in South Africa. *Physics and Chemistry of the Earth, 32*, 1237−1245.

Fussel, H. M. (2009). An updated assessment of the risks from climate change based on research published since the IPCC Fourth Assessment Report. *Climatic Change, 97*, 469–482.

Gan, T. Y., Ito, M., Hülsmann, S., Qin, X., Lu, X. X., Liong, S. Y., ... Koivusalo, H. (2016). Possible climate change/variability and human impacts, vulnerability of drought-prone regions, water resources and capacity building for Africa. *Hydrological Sciences Journal, 61*(7), 1209–1226.

Gleick, P. H. (1996). Basic water requirements for human activities: Meeting basic needs. *Water International, 21*(2), 83–92.

Garrick, D., & Hope, R. (2013). Water security risk and response: The logic and limits of economic instruments. In B. Lankford, K. Bakker, M. Zeitoun, & D. Conway (Eds.), *Water security principles, perspectives, and practices* (2013, pp. 204–219). New York: Routledge.

Gephart, J. A., Troell, M., Henriksson, P. J., Beveridge, M. C., Verdegem, M., Metian, M., ... Deutsch, L. (2017). The seafood gap in the food-water nexus literature—issues surrounding freshwater use in seafood production chains. *Advances in Water Resources, 110*, 505–514.

Gerbens-Leenes, P. W., Van Lienden, A. R., Hoekstra, A. Y., & Van der Meer, T. H. (2012). Biofuel scenarios in a water perspective: The global blue and green water footprint of road transport in 2030. *Global Environmental Change, 22*(3), 764–775.

Gerlak, A. K., & Mukhtarov, F. (2015). Ways of knowing 'water': Integrated water resources management and water security as complementary discourses. *International Environmental Agreements: Politics, Law and Economics, 15*(3), 257–272.

GIZ. *Water, energy & food nexus in a nutshell.* (2016). <http://www.water-energy-food.org/fileadmin/user_upload/files/2016/documents/nexus-secretariat/nexus-dialogues/Water-Energy-Food_Nexus-DialogueProgramme_Phase1_2016-18.pdf>.

Gjorgiev, B., & Sansavini, G. (2018). Electrical power generation under policy constrained water-energy nexus. *Applied Energy, 210*, 568–579.

Gleick, P. H., & Palaniappan, M. (2010). Peak water limits to freshwater withdrawal and use. *Proceedings of the National Academy of Sciences, 107*(25), 11155–11162.

Gleick, P. H., & Cooley, H. S. (2009). Energy implications of bottled water. *Environmental Research Letters, 4*(1), 014009.

Gleick, P. H. (1994). Water and energy. *Annual Review of Energy and the Environment, 19*, 267–299.

Goldin, J. A. (2010). Water Policy in South Africa: Trust and knowledge as obstacles to reform. *Review of Radical Economics, 42*(2), 195–212.

Goldman, M. (2009). "Water for all! The phenomenal rise of transnational knowledge and policy networks." In Kutting, G. and Lipschutz, R.D. (Eds.). *Environmental governance: Power and knowledge in a local-global world.* New York: Routledge: 145–169.

Graham, J. P., Hirai, M., & Kim, S.-S. (2016). An analysis of water collection labor among women and children in 24 sub-Saharan African countries. *PLoS One, 11*(6), e0155981.

Grey, D., & Sadoff, C. W. (2007). Sink or swim: Water security for growth and development. *Water Policy, 9* (6), 545–571.

Grübler, A., O'Neill, B., Riahi, K., Chirkov, V., Goujon, A., Kolp, P., et al. (2007). Regional, national, and spatially explicit scenarios of demographic and economic change based on SRES. *Technological. Forecasting and Social Change, 74*, 980–1029.

Grigg, N. S. (2016). *Integrated water resource management: An interdisciplinary approach.* London: Palgrave-Macmillan.

Grubert, E. A., Beach, F. C., & Webber, M. E. (2012). Can switching fuels save water? A life cycle quantification of freshwater consumption for Texas coal-and natural gas-fired electricity. *Environmental Research Letters, 7*(4), 045801.

Guppy, L., & Anderson, K. (2017). *Water crisis report*. Hamilton, Canada: United Nations University Institute for Water, Environment and Health.

GWP (Global Water Partnership). (2000b). Integrated water resources management. In: *GWP-TAC background paper #4*. http://www.gwpforum.org/gwp/library/TACNO4.PDF.

GWP (Global Water Partnership). (2000a). *Towards water security: A framework for action*. Stockholm, Sweden: GWP.

GWP (Global Water Partnership). (2003). *Toolbox integrated water resources management: Sharing knowledge for equitable, efficient and sustainable water resources management*, Stockholm.

GWP (Global Water Partnership). (2009). Lessons from integrated water resources management in practice. In: *Global Water Partnership policy brief 9*. Stockholm.

Hasan, E., Tarhule, A., Hong, Y., & Moore, B. (2019). Assessment of physical water scarcity in Africa using GRACE and TRMM satellite data. *Remote Sensing, 11*(8), 904. Available from https://www.mdpi.com/2072-4292/11/8/904.

Hepworth, N., & Orr, S. (2013). Corporate water stewardship: Exploring private sector engagement in water security. In B. Lankford, K. Bakker, M. Zeitoun, & D. Conway (Eds.), *Water security principles, perspectives, and practices* (2013, pp. 220–238). New York: Routledge.

Hess, T., Andersson, U., Mena, C., & Williams, A. (2015). The impact of healthier dietary scenarios on the global blue water scarcity footprint of food consumption in the UK. *Food Policy, 50*, 1–10.

Hipsey, M. R., & Arheimer, B. (2013). Challenges for water-quality research in the new IAHS decade on: Hydrology under societal and environmental change. In B. Arheimer, et al. (Eds.), *Understanding freshwater quality problems in a changing world* (pp. 17–29). Wallingford: International Association of Hydrological Sciences (IAHS) Press.

Hoekstra, A. Y., Chapagain, A. K., Aldaya, M. M., & Mekonnen, M. M. (2011). *The water footprint assessment manual: Setting the global standard*. London: Earthscan.

Hoff, H. (2011). Understanding the nexus. Background paper for the bonn 2011 conference: The water, energy and food security nexus. Stockholm, Sweden: Stockholm Environment Institute (SEI).

Hooper, D. U., Chapin, F. S., III, Ewel, J. J., Hector, P., Inchausti, S., Lavorel, J. H., ... Wardie, A. (2005). Effects of biodiversity on ecosystem functioning: A consensus of current knowledge. *Ecological Monographs, 75*(1), 3–35.

Hope, R. (2012). Is community water management the community's choice? Implications for water and policy development in Africa. *Water Policy, 17*(4), 664.

Horwitz, P., & Finlayson, C. (2011). Wetlands as settings for human health: Incorporating ecosystem services and health impact assessment into water resource management. *BioSciences, 61*(9), 678–688.

Hoekstra, A. Y. (Ed.), (2003). Virtual water trade: Proceedings of the international expert meeting on virtual water trade. *Value of Water Research Report. Series No. 12*. Delft, Netherlands.: UNESCO-IHE.

Hoekstra, A. Y. (2007). Human appropriation of natural capital: A comparison of ecological footprint and water footprint analysis. *Value of water research report series no. 23*. Delft, the Netherlands: UNESCO-IHE Institute for Water Education Available from. Available from https://waterfootprint.org/media/downloads/Report23-Hoekstra-2007.pdf.

Hoekstra, A. Y., & Hung, P. Q. (2002). Virtual water trade: A quantification of virtual water flows between nations in relation to international crop trade, *Value of Water Research Report Series No.11*. Delft, Netherlands: UNESCO-IHE.

Hoekstra, A. Y., & Mekonnen, M. M. (2012). The water footprint of humanity. *Proceedings of the National Academy of Sciences., 109*(9), 3232–3237.

Hoekstra, A. Y., & Chapagain, A. K. (2007). Water footprints of nations: Water use by people as a function of their consumption pattern. *Water resources management, 21*(1), 35–48.

Hooper, B. (2010). River basin organization performance indicators: Application to the Delaware River Basin commission: Supplementary file. *Water Policy, 12,* 1−24.

Horlemann, L., & Dombrowsky, I. (2011). Institutionalising IWRM in developing and transition countries: The case of Mongolia. *Environmental Earth Sciences, 65,* 1547−1559.

Howarth, C., & Monasterolo, I. (2016). Understanding barriers to decision making in the UK energy-food-water nexus: The added value of interdisciplinary approaches. *Environmental Science Policy, 61,* 53−60.

Huffaker, R. (2008). Conservation potential of agricultural water conservation subsidies. *Water Resources Research, 44*(7).

ICWE-WMO (International Conference on Water and the Environment-World Meteorological Organization). (1992). *International conference on water and the environment−development issues for the 21st century, 26−31 January 1992.* Dublin, Ireland: ICWE Secretariat, Keynote Papers.

IEA (International Energy Agency). (2012). *Water for energy: Is energy becoming a thirstier resource? Ch. 17 in World energy outlook 2012.* Paris: IEA. Available online.

IISS (International Institute for Strategic Studies). (2011). *The IISS Transatlantic dialogue on climate change and security—Report to the European Commission.* London: IISS.

IPBES (Intergovernmental Science-Policy Platform on Biodiversity and Ecosystem Services). (2018). *The IPBES assessment report on land degradation and restauration.* Bonn.

IPCC (Intergovernmental Panel on Climate Change). (2001). *Climate Change 2001: The Scientific Basis. Contribution of working group I to the third assessment report of the Intergovernmental Panel on Climate Change.* Cambridge: Cambridge University Press.

IPCC (Intergovernmental Panel on Climate Change). (2014). *Climate change 2014: Impacts, adaptation, and vulnerability. Working group II contribution to the fifth assessment report of the Intergovernmental Panel on Climate Change.* Cambridge: Cambridge University Press.

IPPC (Intergovernmental Panel on Climate Change). (2019). *Climate change and land—An IPCC special report on climate change, desertification, land degradation, sustainable land management, food security, and greenhouse gas fluxes in terrestrial ecosystems. Summary for policymakers.* Geneva: IPCC.

IPPC (Intergovernmental Panel on Climate Change). (2007). *Climate change 2007: The physical science basis. Contribution of working group I to the fourth assessment report of the Intergovernmental Panel on Climate Change.* Cambridge: Cambridge University Press.

Islam, M. S., Oki, T., Kanae, S., Hanasaki, N., Agata, Y., & Yoshimura, K. (2007). A grid-based assessment of global water scarcity including virtual water trading. *Water Resources Management, 21,* 19−33.

Johansson, R. C., Tsur, Y., Roe, T. L., Doukkali, R., & Dinar, A. (2002). Pricing irrigation water: A review of theory and practice. *Water policy, 4*(2), 173−199.

Karanis, P., Kourenti, C., & Smith, H. (2007). Waterborne transmission of protozoan parasites: A worldwide review of outbreaks and lessons learnt. *Journal of Water and Health, 5,* 1−38.

Kemp, G., Hagberg, L., Lammon, A., Mazumdar, B., & Bradler, R. (Eds.), (2018). *Water crisis security and climate change.* Washington, D.C.: Center for the National Interest.

Kenway, S. J., Priestley, A., Cook, S., Seo, S., Inman, M., Gregory, A., & Hall, M. (2008). *Energy use in the provision and consumption of urban water in Australia and New Zealand.* Sydney, Australia: Water Services Association of Australia (WSAA).

Klein, S. J., & Rubin, E. S. (2013). Life cycle assessment of greenhouse gas emissions, water and land use for concentrated solar power plants with different energy backup systems. *Energy Policy, 63,* 935−950.

Kroeger, A. (2011, 13 January). Flood recovery in Australia, Sri Lanka and the Philippines. *BBC News.* Available from https://www.bbc.com/news/world-asia-pacific-12179996.

Kummu, M., Ward, P. J., de Moel, H., & Varis, O. (2010). Is physical water · scarcity · a new phenomenon? Global assessment of water shortage · over the last two · millennia. *Environmental Research Letters, 5*(3), 034006.

Kummu, M., Guillaume, H. A., de Moel, H., Eisner, S., Florke, M., Porkka, M., ... Ward, P. J. (2016). *Scientific Reports*, 6, 38945.

Kurian, M. (2017). The water-energy-food nexus: Trade-offs, thresholds and transdisciplinary approaches to sustainable development. *Environmental Science & Policy*, 68, 97−106.

Lankford, B. (2013). Infrastructure hydro-mentalities: Water sharing, water control, and water (in)security. In B. Lankford, K. Bakker, M. Zeitoun, & D. Conway (Eds.), *Water Security Principles, Perspectives, and Practices* (pp. 254−272). New York: Routledge.

Leadley, P. W., Krug, C. B., Alkemade, R., Pereira, H. M., Sumaila, U. R., Walpole, M., ... Mumby, P.J. (2014). Progress towards the Aichi biodiversity targets: An assessment of biodiversity trends, policy scenarios and key actions. In: *CBD technical series no. 78*. Montreal, PQ: CBD (Secretariat of the Convention on Biological Diversity).

Lankford, B., Bakker, K., Zeltoun, M., & Conway, D. (Eds.), (2013). *Water security: principles, perspectives and practices*. New York: Routledge.

Lawrence, P. R., Meigh, J., & Sullivan, C. (2002). *The water poverty index: An international comparison*. Keele, Staffordshire, UK: Keele University.

Leb, C., & Wouters, P. (2013). The water · security · paradox and international law: securitization as an obstacle · to achieving water security and the role of law · in de-securitizing the world's most precious resource. In B. Lankford, K. Bakker, M. Zeitoun, & D. Conway (Eds.), *Water security: principles, perspectives and practices* (pp. 26−45). London: Earthscan.

Lenton, R., Wright, A. M., & Lewis, K. (2005). *Health, dignity, and development: What will it take? UN Millennium Project Task Force on Water and Sanitation*. New York and Stockholm: United Nations and Stockholm International Water Institute.

Liang, S., & Zhang, T. (2011). Interactions of energy technology development and new energy exploitation with water technology development in China. *Energy*, 36, 6960−6966.

Liu, J., Yang, H., Cudennec, C., Gain, A. K., Hoff, H., Lawford, R., ... Zheng, C. (2017). Challenges in operationalizing the water−energy−food nexus. *Hydrological Science Journal*, 62, 1714−1720.

Livingston, M. L. (1995). Designing water institutions: Market failures and institutional response. *Water Resources Management*, 9, 203−220.

Longboat, S. (2015). First Nations Water Security for Mother Earth. *Canadian Women Studies*, 30, 2−3.

Lovejoy, Thomas E., & Hannah, Lee (Eds.), (2019). *Climate change and biodiversity: Transforming the biosphere*. New Haven: Yale University Press.

Low, P. S. (Ed.), (2013). *Economic and social impacts of desertification, land degradation and drought*. Paris: United Nations Convention to Combat Desertification (UNCCD), White Paper I. UNCCD 2nd Scientific Conference, prepared with the contributions of an international group of scientists.

Lowe, C., Ludi, E., Sève, M. D. L., & Tsui, J. (2019). *Linking social protection and water security to empower women and girls*. London: ODI, Working paper 567.

Mach, E. *Water and migration: How far would you go for water?—A Caritas in Veritate Foundation report*. (2017). <https://environmentalmigration.iom.int/water-and-migration-how-far-would-you-go-water>.

Macknick, J., Newmark, R., Heath, G., & Hallet, K. C. (2011). Operational water consumption and withdrawal factors for electricity generation factors: A review of existing literature. *Environmental research Letters*, 7(4), 045802.

McNulty, S. G., Boggs, J. L., & Sun, G. (2014). The rise of the mediocre forest: why chronically stressed trees may better survive extreme episodic climate variability. *New Forests*, 45(3), 403−415.

Madulu, N. (2003). Linking poverty levels to water resource use and conflicts in rural Tanzania. *Physics & Chemistry of the Earth*, 28(20−27), 911.

Marshall, N. A., Marshall, P. A., Tamelander, J., Obura, D., Malleret-King, D., & Cinner, J. E. (2009). *A framework for social adaptation to climate change: Sustaining tropical coastal communities and industries*. Gland, Switzerland: IUCN.

Martinez-Santos, P., Aldaya, M. M., & Llamas, M. R. (Eds.), (2014). *Integrated water resources management in the 21st century: Revisiting the paradigm.* Leiden: CRC Press.

Mason, N. (2013). Easy as 1, 2, 3? Political and technical considerations for designing water security indicators. In B. Lankford, K. Bakker, M. Zeitoun, & D. Conway (Eds.), *Water security principles, perspectives, and practices* (2013, pp. 183−203). New York: Routledge.

Mason, N., & Calow, R. (2012). *Water security: From abstract concept to meaningful metrics.* London: ODI UK, Working paper 357.

Mateo-Sagasta, J., Raschid-Sally, L., & Thebo, A. (2015). Global wastewater and sludge production, treatment and use. In P. Drechsel, M. Qadir, & D. Wichelns (Eds.), *Wastewater − Economic asset in an urbanizing world.* The Netherlands: Springer.

McGinnis, R. L., & Elimelech, M. (2008). Global challenges in energy and water supply: The promise of engineered osmosis. *Environmental Science & Technology, 42*(23), 8625−8629.

McNally, A., Verdin, K., Harrison, L., Getirana, A., Jacob, J., Shukla, S., ... Verdin, J. P. (2019). Acute water-scarcity monitoring for Africa. *Water., 11*(10), MPDI. Available from. Available from https://www.mdpi.com/2073-4441/11/10/1968.

Medema, W., McIntosh, B. S., & Jeffrey, P. J. (2008). From premise to practice: A critical assessment of integrated water resources management and adaptive management approaches in the water sector. *Ecology and Society, 13*(2), 29.

Mekonnen, M. M., & Hoekstra, A. Y. (2014). Water footprint benchmarks for crop production: A first global assessment. *Ecological Indicators, 46,* 214−223.

Mekonnen, M. M., & Hoekstra, A. Y. (2016). Four billion people facing severe water scarcity. *Science Advances, 2*(2), e1500323.

Meldrum, J., Nettles-Anderson, S., Heath, G., & Macknick, J. (2013). Life cycle water use for electricity generation: A review and harmonization of literature estimates. *Environmental Research Letters, 8*(1), 015031.

Merrey, D. J. (1997). *Expanding the frontiers of irrigation management research: Results of research and development at the International Irrigation Management Institute, 1984 to 1995.* Colombo, Srilanka: IIMI.

Merrey, D. (2008). Is normative integrated water resource management implementable? Charting a practical course with lessons from Southern Africa. *Physics and Chemistry of the Earth, 33,* 899−905.

Merschmann, P. R. D. C., Vasquez, E., Szklo, A. S., & Schaeffer, R. (2013). Modeling water use demands for thermoelectric power plants with CCS in selected Brazilian water basins. *International Journal of Greenhouse Gas Control, 13,* 87−101.

Metcalf, S., Dambacher, J., Rogers, P., & Loneragan, N. (2014). Identifying key dynamics and ideal governance structures for successful ecological management. *Environmental Science and Policy, 37,* 34−49.

Mirumachi, N. (2013). Transboundary water security: Reviewing the importance of national regulatory and accountability capacities in international transboundary River Basins. In B. Lankford, K. Bakker, M. Zeitoun, & D. Conway (Eds.), *water security principles, perspectives, and practices* (2013, pp. 166−180). New York: Routledge.

Misra, A. K. (2014). Climate change and challenges of water and food security. *International Journal of Sustainable Built Environment, 3*(1), 153−165.

Molden, D. (2007). *Water for food, water for life: A comprehensive assessment of water management in agriculture.* London: Earthscan, and Colombo: International Water Management Institute.

Muller, M. (2012). Water management institutions for more resilient societies. *Proceedings of the ICE − Civil Engineering, 165*(6), 33−39.

Monaghan, J. M., Daccache, A., Vickers, L. H., Hess, T. M., Weatherhead, E. K., Grove, I. G., & Knox, J. W. (2013). More 'crop per drop': Constraints and opportunities for precision irrigation in European agriculture. *Journal of the Science of Food and Agriculture, 93*(5), 977−980.

Munang, R., Thiaw, I., Alverson, K., Mumba, M., Liu, J., & Rivington, M. (2013). Climate change and ecosystem-based adaptation: A new pragmatic approach to buffering climate change impacts. *Current Opinion in Environmental Sustainability.*, 5(1), 67–71.

Nanduri, V., & Saavedra-Antolínez, I. (2013). A competitive Markov decision process model for the energy–water–climate change nexus. *Applied Energy*, 111, 186–198.

NASA (National Aeronautics and Space Administration). *Severe drought in Russia.* (2010). <https://earthobservatory.nasa.gov/images/44743/severe-drought-in-southern-russia>.

NCC (Natural Capital Coalition). *Natural capital protocol.* (2016). <http://www.naturalcapitalcoalition.org/protocol>.

Nel, J. L., Le Maitre, D. C., Nel, D. C., Reyers, B., Archibald, S., Van Wilgen, B. W., Forsyth, G. G., et al. (2014). Natural hazards in a changing world: A case for ecosystem-based management. *PLoS One*, 9(5), e95942.

Nexus Platform (n.d.). The water-energy-food nexus knowledge resources platform at: <https://www.water-energy-food.org>.

NIC (National Intelligence Council, United States). (2012). *Global water security—Intelligence community assessment.* Washington, DC: Office of the Director of National Intelligence.

NIC (United States-National Intelligence Council). (2013). *Global trends 2030: Alternative worlds.* Washington, DC: Government Printing Office.

Noemdoe, S., Jonker, L., & Swatuk, L. A. (2006). Perceptions of water scarcity: The case of Genadendal and outstations. *Physics and Chemistry of the Earth*, 31(15), 771–778.

Norman, E. S., Bakker, K., & Cook, C. (2012). Introduction to the themed section: Water governance and the politics of scale. *Water Alternatives*, 5(1), 52–61.

NSIDC (National Snow and ICE Data Center). *SOTC: Ice shelves.* (2018). <https://nsidc.org/cryosphere/sotc/iceshelves.html>.

Odegard, I. Y. R., & van der Voet, E. (2014). The future of food—Scenarios and the effect on natural resource use in agriculture in 2050. *Ecological Economics*, 97, 51–59.

OECD (Organisation for Economic Co-operation and Development). (2012). *OECD environmental outlook to 2050: The consequences of inaction.* Paris: OECD Publishing.

OECD (Organisation for Economic Co-operation and Development). (2016). *Mitigating Droughts and Floods in Agriculture: Policy Lessons and Approaches.* Paris: OECD Publishing.

OECD (Organisation for Economic Co-operation and Development). (2017). *Diffuse pollution, degraded waters: Emerging policy solutions.* Paris: OECD Publishing.

OECD (Organisation for Economic Co-operation and Development). *OECD data: GDP long-term forecast (indicator).* (n.d.). <https://data.oecd.org/gdp/gdp-long-term-forecast.htm.oecd.org/gdp/gdp-long-term-forecast.htm>.

Ohlsson, L. (1998). *Water and social resource scarcity FAO Issue Paper.* Rome: Food and Agriculture Organization (FAO).

Ohlsson, L. (2000). Water conflicts and social resources scarcity. Physics and chemistry of the earth, Part B: Hydrology. *Oceans and Atmosphere*, 25(3), 213–220.

Ohlsson, L., & Appelgren, B. (1998). *Water and social resource scarcity.* Rome, Italy: FAO.

Oki, T., & Kanae, S. (2006). Global hydrological cycles and world water resources. *Science*, 313, 1068–1072.

Ohlson, L., & Turton, A. R. (1999). *The turning of screw: Social resource scarcity the bottle-neck in adaptation to water scarcity.* London: University of London, (SOAS Occasional Paper, 19).

Oxford University Water Security Network. (2012). *Water security, risk and society: International water security conference*, April 16–18, 2012, Oxford: St. Hugh's College, University of Oxford.

Ozkan, B., Akcaoz, H., & Fert, C. (2004). Energy input–output analysis in Turkish agriculture. *Renewable Energy*, 29, 2939–2951.

Pachauri, R. K., & Reisinger, A. (Eds.), (2007). *Contribution of working groups I, II and III to the fourth assessment report of the Intergovernmental Panel on Climate Change.* Geneva: IPCC. Available from. Available from https://www.ipcc.ch/site/assets/uploads/2018/02/ar4_syr_full_report.pdf.

Pacific Institute. (2011). *Pacific institute report: Setbacks and solutions of water-energy clash in U.S. intermountain west.* Oakland, California, USA: Pacific Institute.
Pahl-Wostl, C., Gupta, J., & Bhaduri, A. (2016). Water security: A popular but contested concept. In C. Pahl-Wostl, A. Bhaduri, & J. Gupta (Eds.), *Handbook on water security* (pp. 1–16). Cheltenham: Edward Elgar.
Pahl-Wostl, C., Jeffrey, P., Isendahl, N., & Brugnach, M. (2011). Maturing the new water management paradigm: Progressing from aspiration to practice. *Water Resource Management, 25*, 837–856.
Parker, H., Oates, N., Mason, N., & Calow, R. (2016). *Gender, agriculture and water insecurity.* London: Overseas Development Institute, ODI Policy Brief.
Patterson, J., Smith, C., & Bellamy, J. (2013). Understanding enabling capacities for managing the 'wicked problem' of nonpoint source water pollution in catchments: A conceptual framework. *Journal of Environmental Management, 128*, 441–452.
Pimentel, D., Hurd, L. E., Bellotti, A. C., Forster, M. J., Oka, I. N., Sholes, O. D., & Whitman, R. J. (1973). Food production and the energy crisis. *Science (New York, N.Y.), 182*(4111), 443–449.
Pittock, J., Hussey, K., & McGlennon, S. (2013). Australian Climate, Energy and Water Policies. Conflicts and synergies. *Australian Geographer, 44*(1), 3–22.
Portier, C. J., Thigpen Tart, K., Carter, S. R., Dilworth, C. H., Grambsch, A. E., Gohlke, J., ... Whung, P.-Y. (2010). *A human health perspective on climate change: A report outlining the research needs on the human health effects of climate change.* Research Triangle Park, NC: Environmental Health Perspectives/National Institute of Environmental Health Science.
Rabbani, G., Rahman, A. A., & Islam, N. (2010). Climate change and sea level rise: Issues and challenges for coastal communities in the Indian Ocean region. In D. Michel, & A. Pandya (Eds.), *Coastal zones and climate change.* Washington, DC.: The Henry L. Stimson Center.
Racoviceanu, A. I., Karney, B. W., Kennedy, C. E. J., & Colombo, A. F. (2007). Life cycle energy use and greenhouse gas emissions inventory for water treatment system. *Journal of Infrastructure Systems, 13*(4), 261–270.
Rahaman, M. M., & Varis, O. (2005). Integrated water resources management: Evolution, prospects and future challenges. *Sustainability: Science, practice and Policy, 1*, 1.
Ramirez, C. A., Bakshi, B., Gibon, T., & Hertwich, E. (2013). Assessment of low carbon energy technologies: Fossil fuels and CCS. *Energy Procedia, 37*, 2637–2644.
Ramsar Convention. (2016). *The fourth ramsar strategic plan, 2016-2024.* Glad, Switzerland: Ramsar Convention Secretariat.
Rasul, G. (2016). Managing the food, water, and energy nexus for achieving the sustainable development goals in South Asia. *Environmental Development., 18*, 14–25.
Rathmann, R., Szklo, A., & Schaeffer, R. (2010). Land use competition for production of food and liquid biofuels: An analysis of the arguments in the current debate. *Renewable Energy, 35*(1), 14–22.
Ray, B. (2010). *Water: The looming crisis in India.* New Delhi: Pentagon Press.
Ren, D., Yang, Y., Yang, Y., Richards, K., & Zhou, X. (2018). Land-water-food nexus and indications of crop adjustment for water shortage solution. *Science of The Total Environment, 626*, 11–21.
Revi, A., Bazaz, A., Krishnaswamy, J., Bendapudi, R., D'Souza, M., & Pahwa Gajjar, S. (2015). Vulnerability and adaptation to climate change in semi-arid areas in India. In: *ASSAR working paper.* ASSAR PMU, South Africa. Available from http://www.assar.uct.ac.za/sites/default/files/image_tool/images/138/South_Asia/South%20Asia%20RDS%20full%20report_update.pdf.
Richey, A. S., Thomas, B. F., Lo, M. H., Reager, J. T., Famiglietti, J. S., Voss, K., ... Rodell, M. (2015). Quantifying renewable groundwater stress with GRACE. *Water Resources Research, 51*(7), 5217–5238.
Rijke, J., Farrelly, M., Brown, R., & Zevenbergen, C. (2013). Configuring transformative governance to enhance resilient urban water systems. *Environmental Science and Policy, 25*, 62–72.

Rijsberman, F. R. (2006). Water Scarcity: fact or fiction? *Agricultural Water Management*, *80*(1-2), 5−22.

Rizvi, A. R., Baig, S., & Verdone, M. (2015). *Ecosystem based adaptation: Knowledge gaps in making an economic case for investing in nature based solutions for climate change*. Gland, Switzerland: IUCN.

Rogers, P., & Hall, A. W. (2003). *Effective water governance* (Vol. 7). Stockholm: Global Water Partnership.

Rockström, J., Falkenmark, M., Karlberg, L., Hoff, H., Rost, S., & Gerten, D. (2009a). Future water availability for food production: The potential of green water for increasing resilience to global change. *Water Resources Research*, *45*(7).

Rockström, J., Falkenmark, M., Allan, T., Folke, C., Gordon, L., Jagerskog, J., ... Varis, O. (2009b). The unfolding water drama in the Anthropocene: Towards a resilience-based perspective on water for global sustainability. *Ecohydrology*, *7*, 1249−1261.

Rosegrant, M. W., Meijer, S., & Cline, S. A. (2002). *International model for policy analysis of agricultural commodities and trade (IMPACT): Model description*. Washington, D.C.: International Food Policy Research Institute.

Rosegrant, M. W., & Ringler, C. (1999). *Impact on food security and rural development of reallocating water from agriculture*. Washington, D.C: IFPRI.

Rockström, J., Falkenmark, M., Allan, T., Folke, C., Gordon, L., Jagerskog, A., ... Varis, S. (2014). The unfolding water drama in the Anthropocene: Towards a resilience-based perspective on water for global sustainability. *Ecohydrology*, *7*(5), 1249−1261.

Sadoff, C. W., Hall, J. W., Grey, D., Aerts, J. C. J. H., Ait-Kadi, M., Brown, C., ... Wiberg, D. (2015). *Securing water, sustaining growth: Report of the GWP/OECD Task force on water security and sustainable growth*. University of Oxford. Available from https://www.water.ox.ac.uk/wp-content/uploads/2015/04/SCHOOL-OF-GEOGRAPHY-SECURING-WATER-SUSTAINING-GROWTH-DOWNLOADABLE.pdf.

Sampath, R. K. (1992). Issues in irrigation pricing in developing countries. *World Development*, *20*(7), 967−977.

Sauvé, S., & Desrosiers, M. (2014). A review of what is an emerging contaminant. *Chemistry Central Journal*, *8*(15). Available from. Available from https://bmcchem.biomedcentral.com/track/pdf/10.1186/1752-153X-8-15.

Savage, C. (1991). Middle East Waters. *Asian Affairs*, *22*, 3−10.

Scott, A. (2017). *Making governance work for water−energy−food nexus approaches. Working paper*. London: Climate and Development Knowledge Network (CDKN). Available from https://cdkn.org/wp-content/uploads/2017/06/Workingpaper_CDKN_Making-governance-work-for-water-energy-food-nexus-approaches.pdf.

Seckler, D., Barker, R., & Armsinghe, U. A. (1998). Water scarcity in the twenty-first century. *International Journal of Water Resources*, *15*(1-2), 29−42.

Seddon, N., Hou-Jones, X., Pye, T., Reid, H., Roe, D., Mountain, D., & Rizvi, A. R. R. (2016). Ecosystem-based adaptation: A win−win formula for sustainability in a warming world?. In: *Briefing, IIED-IUCN*. Available from https://pubs.iied.org/pdfs/17364IIED.pdf.

Semenza, J., & Nichols, G. (2007). Cryptosporidiosis surveillance and water-borne outbreaks in Europe. *Eurosurveillance*, *12*, E13−E14.

Semenza, J. C., & Menne, B. (2009). Climate change and infectious diseases in Europe. *Lancet Infectious Diseases*, *9*, 365−375.

Semiat, R. (2008). Energy issues in desalination processes. *Environmental Science & Technology*, *42*(22), 8193−8201.

Senhorst, H. A. J., & Zwolsman, J. J. G. (2005). Climate change and effects on water quality: A first impression. *Water Science and Technology*, *51*, 53−59.

Shiklomanov. (1993). Vital water graphics. In P. H. Gleick (Ed.), *Water in Crisis: A guide to the world's fresh water resources*. New York: Oxford University Press.

Shuval, H. I. (1992). Approaches to resolve the water conflict between Israel and her neighbors—A regional water for peace plan. *Water International, 17*, 133–143.

Siddiqi, A., & Anadon, L. D. (2011). The water-energy nexus in Middle East and North Africa. *Energy Policy, 39*(8), 4529–4540.

Simonovic, S. P., & Fahmy, H. (1999). A new modeling approach for water resources policy. *Water Resources Research, 35*(1), 295–304.

Singh, H., Mishra, D., & Nahar, N. M. (2002). Energy use pattern in production agriculture of a typical village in arid zone, India—part I. *Energy Conversion and Management, 43*(16), 2275–2286.

Smajgl, A., & Ward, J. (2013). *The water-food-energy nexus in the Mekong region: Assessing development strategies considering cross-sectoral and transboundary impacts*. New York: Springer.

Smakhtin, V., Revenga, C., & Doell, P. (2004). A pilot global assessment of environmental water requirements and scarcity. *Water International, 29*(3), 307–317.

Solomon, S., Qin, D., Manning, M., Chen, Z., Marquis, M., Averyt, K. B., & . . . Miller, H. L. (Eds.), (2007). *Climate change 2007: The physical science basis. Contribution of working group I to the fourth assessment report of the Intergovernmental Panel on Climate Change*. Cambridge: University Press, Cambridge University Press.

SPIAC-B. (Social Protection Inter-Agency Cooperation Board). (2019). *Social protection to promote gender equality and women and girls' empowerment*. Geneva: SPIAC-B (International Labour Organization).

Srinivasan, V., Sanderson, M., Garcia, M., Konar, M., Blöschl, G., & Sivapalan, M. (2017). Prediction in a socio-hydrological world. *Hydrological Sciences Journal, 62*(3), 338–345.

Stephan, R. M., Mohtar, R. H., Daher, B., Embid Irujo, A., Hillers, A., Ganter, J. C., Karlberg, L., Martin, L., Nairizi, S., Rodriguez, D. J., & Sarni, W. (2018). Water–energy–food nexus: A platform for implementing the sustainable development goals. *Water International, 43*, 472–479.

Sullivan, C., & Meigh, J. R. (2003). Considering the Water Poverty Index in the Context of Poverty Alleviation. *Water Policy, 5*(5), 513–528.

Swatuk, L. A., & Cash, C. (Eds.), (2018). *Water, energy, food and people across the global South*. New York: Palgrave-Macmillan.

Talberth, J., Gray, E., Branosky, E., & Gartner, T. (2012). *Insights from the field: Forests for water. WRI issue brief—Southern forests for the future incentives series*. Washington DC: World Resources Institute, Issue Brief 9.

Sengupta, S., & Weiyi, C. (2019). A quarter of humanity faces looming water crises. *The New York Times*. Available from http://www.nytimes.com/interactive/2019/08/06/climate/world-water-stress.htm (6 August 2019).

Tapela, B. *Social water scarcity and water use. Water research commission report*. (2012). <http://www.wrc.org.za/SiteCollectionDocuments/News%20documents/1940.dotx>.

Taylor, R., Voss, C., MacDonald, A., Aureli, A., & Aggarwal, P. (2016). *Global change and groundwater*. International Association of Hydrogeologists (IAH). Available from https://iah.org/wp-content/uploads/2016/07/IAH-Global-Change-Groundwater-14-June-2016.pdf.

Thiede, S., Kurle, D., & Herrmann, C. (2017). The water–energy nexus in manufacturing systems: Framework and systematic improvement approach. *CIRP Annals, 66*(1), 49–52.

Tinker, S. C., Moe, C. L., Klein, M., Flanders, W. D., Uber, J., Amrithrajah, A., & Singer, P. (2008). Drinking water turbidity and emergency department visits for gastrointestinal illness in Atlanta, 1993–2004. *Journal of Exposure Science and Environmental Epidemiology, 20*(1), 19–28.

Trostle, R., Marti, D., Rosen, S., & Wescott, P. (2011). *Why have food commodity prices risen again?*. ERS, USDA, Available from https://www.ers.usda.gov/webdocs/publications/40481/7392_wrs1103.pdf?v = 0.

Turton, A. (1999). *Water scarcity and social adaptive capacity: Towards an understanding of the social dynamics of water demand management in developing countries*. School of Oriental and African Studies, University of London.

UN (2009). *UN Water Annual Report.* New York: UN-Water Secretariat.
UNDESA (United Nations-Department of Economic and Social Affairs). (2017). *World Population Prospects: The 2017 Revision.* New York: UN.
UNFCCC (United Nations Framework Convention on Climate Change). (2013). *Warsaw Climate Change Conference Documents.* Bonn, Germany: UNFCCC Secretariat.
UN-Water (2016). *World water development report.* Paris: UNESCO.
UN (United Nations). (1992). *Report of the United Nations Conference on Environment and Development.* Rio de Janeiro. Available from https://www.un.org/esa/dsd/agenda21/Agenda%2021.pdf.
UN (United Nations). (2010). *The right to water fact sheet No. 35.* New York: United Nations, OHCHR, UN-Habitat, WHO.
UN Women. (2018). *Gender equality in the 2030 agenda: Gender-responsive water and sanitation systems.* New York: UN Women.
UNCCD (United Nations Convention to Combat Desertification). *United Nations convention to combat desertification in those countries experiencing serious drought and/or desertification, particularly in Africa.* (1994). <http://catalogue.unccd.int/936_UNCCD_Convention_ENG.pdf>.
UNCCD (United Nations 4Convention to Combat Desertification). (2017). *Global land outlook.* Bonn: UNCCD.
UNCCD (United Nations Convention to Combat Desertification). (2013). *A stronger UNCCD for a land-degradation neutral world, issue brief.* Bonn.
UN-DESA (United Nations Department of Economic and Social Affairs). (2015a). *World population projected to reach 9.7 billion by 2050.* New York: UN DESA.
UN-DESA (United Nations-Department of Economic and Social Affairs). (2015b). *The world's women 2015: Trends and statistics.* New York: UN-DESA. Statistics Division.
UNDP (United Nations Development Programme). (2006). *Human development report 2006, beyond scarcity: Power poverty and the global water crisis.* New York, NY: UNDP.
UNEP (United Nations Environment Program). (2012). *The UN-water status report on the application of integrated approaches to water resources management.* Nairobi: UNEP.
UNEP (United Nations Environment Programme). (2016a). *A snapshot of the world's water quality: Towards a global assessment.* Nairobi: UNEP Available from. Available from https://uneplive.unep.org/media/docs/assessments/unep_wwqa_report_web.pdf.
UNEP (United Nations Environment Programme). (1992). *World atlas of desertification.* Nairobi: UNEP.
UNEP, (United Nations Environment Programme). *Climate change adaptation: Building resilience of ecosystems for adaptation.* (2016b). <http://www.unep.org/climatechange/adaptation/what-we-do/ecosystem-based-adaptation>.
UNEP-DTU, CTCN (United Nations Environment Programme, Climate Technology Centre & Network). (2017). *Climate change adaptation technologies for water: A practitioner's guide to adaptation technologies for increased water sector resilience.* Copenhagen: CTCN.
UNESCAP/UNISDR (United Nations Economic and Social Commission for Asia and the Pacific/United Nations Office for Disaster Risk Reduction). (2012). *Reducing vulnerability and exposure to disasters. The Asia-Pacific disaster report 2012.* UNESCAP/UNISDR. http://www.unisdr.org/files/29288_apdr2012finallowres.pdf.
UNESCO (United Nations Educational, Scientific and Cultural Organization). *The properties and availability of water: A fundamental consideration for life.* (n.d.). <http://www.unesco.org/new/fileadmin/MULTIMEDIA/FIELD/Venice/pdf/special_events/bozza_scheda_DOW02_1.0.pdf>.
UNESCO-IHE (United). (2009). *Global trends in water-related disasters: An insight for policy-makers.* Paris: UNESCO. Available from http://www.ipcinfo.org/fileadmin/user_upload/drm_matrix/docs/Global%20trends%20in%20water%20related%20disasters%20~%20An%20insight%20for%20policy%20makers.pdf.

UNFCCC (United Nations Framework Convention on Climate Change). *Ecosystem-based approaches to adaptation: Compilation of information.* (2011). <http://unfccc.int/resource/docs/2011/sbsta/eng/inf08.pdf>.

UNFCCC (United Nations Framework Convention on Climate Change). (2005). Report on the seminar on the development and transfer of technologies for adaptation to climate change. In: *FCCC/SBSTA/2005/8.*

UNFCCC (United Nations Framework Convention on Climate Change). (2012). Report on the technical workshop on water and climate change impacts and adaptation strategies. In: *Subsidiary Body for Scientific and Technological Advice. UNFCCC/SBSTA/2012/4.*

UNFCCC (United Nations Framework Convention on Climate Change). (2014b). *Report of the workshop on technologies for adaptation.* Langer Eugen, Bonn, Germany, 4 March 2014. Available online.

UNFCCC (United Nations Framework Convention on Climate Change). *Presentations from the workshop on technologies for adaptation.* (2014a). <http://unfccc.int/ttclear/pages/ttclear/pages/ttclear/pages/ttclear/pages/ttclear/templates/render_cms_page?s = events_workshops_adaptationtechs>.

UNFCC-TEC (United Nations Framework Convention on Climate Change-Technology Executive Committee). (2014). *Technologies for adaptation in the water sector.* Bonn: UNFCCC-TEC, TEC Brief #5.

UNGA (United Nations General Assembly). (2010). *Resolution A/RES/64/292.* New York: UNGA.

UNGA (United Nations General Assembly). (2015). Transforming our world: The 2030 agenda for sustainable development. In: *A/RES/70/1.* Available from https://www.un.org/ga/search/view_doc.asp?symbol = A/RES/70/1&Lang = E.

UNGA (United Nations General Assembly). (2017). Report of the special rapporteur on the right to food. In: *Human Rights Council thirty-fourth session, February 27–March 24, 2017. Document A/HRC/34/48.* United Nations. https://documents-dds-ny.un.org/doc/UNDOC/GEN/G17/017/85/PDF/G1701785.pdf?OpenElement.

United Nations (UN). (2015a). *Transforming our world: The 2030 agenda for sustainable development.* Resolution adopted by the General Assembly on 25 September 2015. Available from: http://www.un.org/ga/search/view_doc.asp?symbol = A/RES/70/1&Lang = E.

United Nations (UN). (2015b). *Paris Agreement.* Available from https://unfccc.int/sites/default/files/english_-paris_agreement.pdf.

UN-Water (United Nations Water). (2019). Climate change and water. In: *UN-water policy brief.* Available from https://sustainabledevelopment.un.org/content/documents/UNWclimatechange_EN.pdf.

UN-Water (United Nations-Water). (2018). *Sustainable development goal 6: Synthesis report 2018 on water and sanitation.* New York: UN-Water.

USAID (United States Agency for International Development). (2017a). *United States government global water strategy 2017.* Washington, DC: USAID.

USAID (United States Agency for International Development). (2017b). *Ecosystem-based adaptation and water security.* Washington, DC: USAID.

Varis, O. (2007a). Water demands for bioenergy production. *International. Journal of Water Resources Development, 23,* 519–535.

Veldkamp, T. I. E., Wada, Y., Aerts, J. C. J. H., Döll, P., Gosling, S. N., Liu, J., ... Ward, P. J. (2017). Water scarcity hotspots travel downstream due to human interventions in the 20th and 21st century. *Nature Communications,* No. 15697.

Veolia/IFPRI (International Food Policy Research Institute). (2015). *The Murky future of global water quality: New global study projects rapid deterioration in water quality.* Washington, DC: IFPRI/Veolia Available from. Available from http://www.ifpri.org/publication/murky-futureglobal-water-quality-new-global-study-projects-rapid-deterioration-water.

Vignola, R., Harvey, C. A., Bautista-Solis, P., Avelino, J., Rapidel, B., Donatti, C., & Martinez, R. (2015). Ecosystem-based adaptation for smallholder farmers: Definitions, opportunities and constraints. *Agriculture, Ecosystems & Environment, 211,* 126–132.

Von Braun, J., & Pachauri, R. K. (2006). *The promises and challenges of biofuels for the poor in developing countries*. Washington DC: IFPRI.

Vorosmarty, C. J., Douglas, E. M., Gree, P. A., & Revenga, C. (2005). Geospatial indicators of emerging water stress: An application to Africa. *Ambio*, *34*(3), 230–236.

Wada, Y., Lo, M.-H., Yeh, P. J.-F., Reager, J. T., Famiglietti, J. S., Wu, R.-J., & Tseng, Y.-H. (2016). Fate of water pumped from underground and contributions to sea-level rise. *Nature Climate Change Letters*. (online), Available from. Available from https://cpb-us-w2.wpmucdn.com/blog.nus.edu.sg/dist/b/4438/files/2016/05/Wada_NCC_2016-21m7xdh.pdf.

Ward, F. A., & Pulido-Velazquez, M. (2008). Water conservation in irrigation can increase water use. *Proceedings of the National Academy of Sciences of the United States of America*, *105*(47), 18215–18220.

Warner, K., & van der Geest, K. (2013). Loss and damage · from climate · change: local-level evidence · from nine vulnerable countries. *International Journal of Global Warming*, *5*(4), 367–386.

WaterAid. (2013). *The state of the world's water 2013*. London: Water Aid. Secretariat.

WaterAid. *Water scarcity report*. (2012). <http://www.wateraid.org/uk/who-we-are/annual-reports#/annual-reports>.

WEC (World Energy Council). (2013). *World energy scenarios: Composing energy futures to 2050*. London: World Energy Council Switzerland: Project Partner Paul Scherrer Institute (PSI). Available from. Available from http://www.worldenergy.org/publications/2013/world-energy-scenarios-composing-energy-futures-to-2050.

WEF (World Economic Forum). (2011). *Water security: The water-food-energy-climate nexus, World Economic Forum water initiative*. Washington, DC: Island Press.

White, D. D., Jones, J. L., Maciejewski, R., Aggarwal, R., & Mascaro, G. (2017). Stakeholder analysis for the food-energy-water nexus in Phoenix, Arizona: Implications for nexus governance. *Sustainability*, *9*(12).

WHO (World Health Organization). (2017). *Safely managed drinking water—Thematic report on drinking water 2017*. Geneva: WHO.

WHO and UNICEF. (2017). *Progress on drinking water, sanitation and hygiene: 2017 update and SDG baselines*. Geneva: WHO and UNICEF.

Weitz, N., Strambo, C., Kemp-Benedict, C., & Nilsson, M. (2017). *Governance in the water-energy-food nexus: Gaps and future research needs*. Stockholm: SEI.

Wiek, A., & Larson, K. (2012). Water, people, and sustainability—A systems framework for analyzing and assessing water governance regimes. *Water Resource Management*, *26*, 3153–3171.

Witter, S. G., & Whiteford, S. (1999). *Water Policy: Security Issues. Vol. 11, International Review of Comparative Public Policy*. Stamford, Connecticut, USA: JAI Press.

White, C. (2012). Understanding water scarcity: Definitions and measurements. Canberra, Australia: Global Water Forum, Australian National University.

WRI (Water Resources Institute). *Ranking the world's most water-stressed countries in 2040*. (2015). <http://www.wri.org/blog/2015/08/ranking-world%E2%80%99s-most-water-stressed-countries-2040>.

Wutich, A., Budds, J., Eichellberger, L., Geere, J., Harris, L., Horney, J., ... Young, S. (2017). Advancing methods for research on household water insecurity: Studying entitlements and capabilities, socio-cultural dynamics, and political processes, institutions and governance. *Water Security*, *2*, 1–10.

WWAP (United Nations World Water Assessment Programme). (2014). *The United Nations world water development report 2014: Water and energy*. Paris: UNESCO Available from. Available from http://www.unesdoc.unesco.org/images/0022/002257/225741E.pdf.

WWAP (United Nations World Water Assessment Programme). (2017). *The United Nations world water development report 2017. Wastewater: The untapped resource*. Paris: UNESCO.

WWAP (World Water Assessment Program). (2015). *The United Nations world water development report 2015: Water for a sustainable world—Executive summary*. Paris: UNESCO. Available online.

WWAP (World Water Assessment Programme). (2018). *The United Nations world water development report 2018*. New York: UNESCO. Available online.
WWAP (World Water Assessment Programme). (2012). *UN-World water development report. Managing water under uncertainty and risk*. Paris: UNESCO.
WWAP (World Water Assessment Programme). (2019). *The United Nations world water development report 2019: Leaving no one behind*. Paris: WWAP, UNESCO.
WWC-WWF (World Water Council-World Water Forum). Ministerial declaration of the Hague on water security in the 21st century. (2000). <https://www.worldwatercouncil.org/fileadmin/world_water_council/documents/world_water_forum_2/The_Hague_Declaration.pdf>.
WWDR (2015). World Water Development Report 2015: water for a sustainable world. Paris: UNESCO.
WWF-WB (World Wildlife Fund (WWF) and The World Bank). *Operational framework for ecosystem-based adaptation: Implementing and mainstreaming ecosystem-based adaptation. Responses in the Greater Mekong sub-region*. (2013). <http://awsassets.panda.org/downloads/wwf_wb_eba_project_2014_gms_ecosystem_based_adaptation_general_framework.pdf>.
Yang, H., Zhang, X., & Zehnder, A. J. (2003). Water scarcity, pricing mechanism and institutional reform in northern China irrigated agriculture. *Agricultural Water Management*, 61(2), 143161.
Zeitoun, M. (2013). The web of sustainable water security. In B. Lankford, K. Bakker, M. Zeitoun, & D. Conway (Eds.), *Water security principles, perspectives, and practices* (pp. 11–25). New York: Routledge.
Zhang, J., Spitz, Y. H., Steele, M., Ashjian, C., Campbe, R., Berline, L., & Matrai, P. (2010). Modeling the impact of declining sea ice on the Arctic marine planktonic ecosystem. *Journal of Geographic Research*, 115, C10015.

Further reading

Ackerman, F., & Fisher, J. (2013). Is there a water–energy nexus in electricity generation? Long-term scenarios for the western United States. *Energy Policy*, 59, 235–241.
Allan, J. A. T. (2013). Food-water security: Beyond water resources and the water sector. In B. Lankford, K. Bakker, M. Zeitoun, & D. Conway (Eds.), *Water security principles, perspectives, and practices* (2013, pp. 321–335). New York: Routledge.
Carrillo, A. M. R., & Frei, C. (2009). Water: A key resource in energy production. *Energy Policy*, 37(11), 4303–4312.
Chenoweth, J., Malcolm, R., Pedley, S., & Kaime, T. (2013). Household water security and the human right to water and sanitation. In B. Lankford, K. Bakker, M. Zeitoun, & D. Conway (Eds.), *Water security principles, perspectives, and practices* (2013, pp. 307–318). New York: Routledge.
Conway, D. (2013). Water security in a changing climate. In B. Lankford, K. Bakker, M. Zeitoun, & D. Conway (Eds.), *Water security principles, perspectives, and practices* (2013, pp. 80–100). New York: Routledge.
Earle, A. (2013). The role of cities as drivers of international transboundary water management processes. In B. Lankford, K. Bakker, M. Zeitoun, & D. Conway (Eds.), *Water security principles, perspectives, and practices* (2013, pp. 101–114). New York: Routledge.
Elena, G., & Esther, V. (2010). From water to energy: The virtual water content and water footprint of biofuel consumption in Spain. *Energy Policy*, 38(3), 1345–1352.
Endo, A., & Oh, T. (Eds.), (2018). *The water-energy-food nexus human-environmental security in the Asia-Pacific ring of fire*. Singapore: Springer.
Froggatt, A. (2013). The water–energy nexus: Meeting growing demand in a resource-constrained world. In B. Lankford, K. Bakker, M. Zeitoun, & D. Conway (Eds.), *Water security principles, perspectives, and practices* (2013, pp. 115–129). New York: Routledge.

Fthenakis, V., & Kim, H. (2010). Life-cycle uses of water in United States electricity generation. *Renewable and Sustainable Energy Reviews*, *14*, 2039−2048.

Glatzel, K. (2013). *Private sector responses to water scarcity in the Hashemite Kingdom of Jordan* (Ph.D. thesis). Imperial College.

Goldin, J. A. (2004). *Prepacked trust and the water sector in Askvik, S and Bak, N. Trust in Public Institutions in South Africa*. Ashgate.

Haddadin, M. J. (2006). *Water resources in Jordan - Evolving policies for development, the environment, and conflict resolution*. Washington, DC: RFF.

Meissner, R., Funke, N., Nortje, K., & Steyn, M. (Eds.), (2019). *Understanding water security at local government level in South Africa*. Switzerland: Palgrave-Pivot.

Nexus Platform. *Nexus: The water, energy and food security resource*. (2016). <http://www.water-energy-food.org/start/>.

OECD (Organisation for Economic Co-operation and Development). (2015). *Implications of water scarcity for economic growth—Environment working paper no. 109*.

Sattler, S., Macknick, J., Yates, D., Flores-Lopez, F., Lopez, A., & Rogers, J. (2012). Linking electricity and water models to assess electricity choices at water-relevant scales. *Environmental Research Letters*, *7*(4), 045804.

Setegn, S. G. (2015). Introduction: Sustainability of integrated water resources management (IWRM). In S. G. Setegn, & M. C. Donoso (Eds.), *Sustainability of integrated water resources management water governance, climate and ecohydrology*. Switzerland: Springer.

Srinivasan, S., Kholod, N., Chaturvedi, V., Ghosh, P. P., Mathur, R., Clarke, L., ... Liu, B. (2018). Water for electricity in India: A multi-model study of future challenges and linkages to climate change mitigation. *Applied Energy*, *210*, 673−684.

Tarlock, D., & Wouters, P. (2009). Reframing the water security dialogue. *Water Law*, *20*, 5360.

UNDP (United Nations Development Programme). (2010). *Gender, climate change and community-based adaptation: A guidebook for designing and implementing gender-sensitive community-based adaptation programmes and projects*. New York: UNDP.

UNEP (United Nations Environment Programme). (2015). *Promoting ecosystems for disaster risk reduction and climate change adaptation: Opportunities for integration. Discussion paper*. Geneva: UNEP, Post-Conflict and Disaster Management Branch.

UN Water (United Nations Water). *The United Nations world water development report 2014: Water and energy*. (2014). <http://unesdoc.unesco.org/images/0022/002257/225741E.pdf> Accessed 01.08.18.

USAID (United States Agency for International Development). (2016). *United States Government global food security strategy*. Washington, DC: USAID.

Worldometers. *World population*. (n.d.). <https://www.worldometers.info/>.

Zeitoun, M., Lankford, B., Bakker, K., & Conway, D. (2013). A Battle of Ideas for Water Security. In B. Lankford, K. Bakker, M. Zeitoun, & D. Conway (Eds.), *Water security principles, perspectives, and practices*. New York: Routledge.

CHAPTER 6

Toward sustainable food security

6.1 Introduction

The global population that has reached 7.7 billion in mid-2019, is expected to reach 8.5 billion in 2030, 9.7 billion in 2050 and 10.9 billion in 2100 (UN-DESA, 2019, p. 5). In the wake of global population reaching to 9.7 billion by 2050, the question arises as to whether it would be possible to feed the growing population, especially when the people with rising incomes will be consuming more resource-intensive, animal-based foods along with simultaneous rise in greenhouse gas (GHG) emissions from agricultural production and pace of conversion of remaining forests to agricultural land. Nevertheless, a recent report by World Resource Institute (WRI) optimistically offers a five-course menu of solutions ensuring feeding every one without enhancing emissions, further deforestation or increasing poverty and this optimism is based on intensive research and modeling examining the nexus of the food system, economic development and the environment (WRI, 2018). Ranganathan, Waite, Searchinger, and Hanson (2018) have suggested that in order to feed nearly 10 billion people sustainably by 2050, there is need to close food gap, land gap and GHG mitigation gap. According to them, food gap of 56% between crop calories produced in 2010 and those required in 2050 is to be filled under "business as usual" growth. With regard to land gap, the authors suggest that a land gap of 593 Mha (million hectares) between global agricultural land area in 2010 and anticipated agricultural expansion by 2050 needs to be engulfed. They have also emphasized on narrowing down an 11-gigaton GHG mitigation gap between expected agricultural emissions in 2050 and the target level required to hold global warming level below 2°C, the essential level averting the worst climate impacts.

With global population standing at 7.7 billion people by mid-2019, more than 820 million people were still hungry in 2018 and hunger is rising in almost all subregions of Africa and, to a lesser extent, in Latin America and Western Asia. After registering declining trends in global hunger—measured by the prevalence of undernourishment (PoU)—in recent decades, this trend got reversed in 2015 and remained virtually unchanged in 2018 as well at a level slightly below 11% (FAO, 2019, p. xiv).

After decades of steady decline, the trend in world hunger—as measured by the PoU—reverted in 2015, remaining virtually unchanged in the past three years at a level slightly below 11%. Food and Agricultural Organization (FAO) has identified Prevalence of Food Insecurity, based on the Food Insecurity Experience Scale (FIES) as of two types—moderate food insecurity and severe food insecurity. Severe food insecurity is linked to the concept of hunger and the people experiencing moderate food insecurity are confronted with uncertainties about their ability of getting food and thereby being compelled to compromise quality and/or quantity of the food they consume. Taking into account all people in the world impacted by moderate levels of food insecurity in tandem with those are affected by hunger, estimates show that over two billion lack regular access to safe, nutritious and sufficient food (FAO, 2019, p. xiv).

A cogent comprehension of sustainable food security calls for brief elaboration of concepts of hunger, food security and insecurity and related notions like food sovereignty and right to food, etc.

6.2 Hunger

Humankind's fight against hunger has been a challenge throughout human history. People had been striving hard to meet the essentiality of having sufficient food necessary for their survival.

Quest for good territories endowed with fertile soils and availability of enough water resources have always constituted the basis of human settlements and development of civilizations. Undoubtedly, advent of revolutionary agricultural techniques encompassing basic agricultural implements, farming techniques, irrigation, crops' rotation, etc., have played pivotal role in ensuring a staple food supply; nonetheless, the problem of global hunger has continued to haunt humankind till date.

The notion of hunger can be construed in different ways. Hunger is a human sensation that is experienced by almost all people and according to Oxford English Dictionary it is a "feeling of discomfort or weakness caused by lack of food, coupled with the desire to eat" (Oxford English Dictionary, 2010). Even if this hunger is a common human experience, it becomes essential to differentiate between the daily feeling of hunger owing to the need to feed and hunger as a chronic problem. Construed in this sense, hunger epitomizes a problem that nowadays affects more than 820 million people in different parts of the globe (FAO, 2019). Impact of hunger is not just limited to stomach, rather it affects every part of human body, entailing the potential of causing lifelong implications, especially when it touches brain, heart, skin and bones, specifically for children under 5-years of age. The debilitating crisis engendered by hunger impacts not only people's health but also their creative faculties like educational and labor capabilities. Therefore hunger portends a threat for individuals' progress and also for the entire process of national development of a nation.

Lappé, Collins, and Rosset (1998) have opined that nevertheless chronic hunger seldom makes "the evening news" but it has emerged as a worldwide problem. As can be discerned from FAO statistics, more than 820 million people were undernourished in 2018, albeit fewer than in 1990−92, despite the population growth. Having witnessed a period of success marked by declining trends in hunger worldwide—with the total number of chronically undernourished people standing at 900 million in 2000 declining to 777 million in 2015—nowadays it on the rise again having reached up to 815 million in 2017 (FAO, 2017) and to over 820 million in 2018 (FAO, 2019). Variations worldwide are so prominent that hunger could be described as "a form of conscious racism" (Galtung, 2004). Viewed a broad spectrum, it becomes discernible that a large number of people affected by hunger worldwide lives in developing countries and are non-White. It can be ascertained from FAO statistics that 98% of the undernourished people live in developing countries, with 75% in rural areas and for 7/10 agriculture represented the main avocation and source of livelihood (The Hunger Project, 2018).

According to Windfuhr and Jonsén (2005), smallholder farmers comprising about half of the world's hungry people inhabit limited areas of land sans adequate access to productive resources. While describing "the hungry's profile," Windfuhr and Jonsén (2005) note that generally hungry people live in rural areas or marginal lands in environmentally obscure conditions; therefore are not capable of attaining self-sufficiency, 22% of the hungry are landless families, living thanks to income earned under hazardous working conditions; 8% form part of the fishing, hunting and herding communities comprising mostly women, owing to social and cultural dynamics, and so in an indirect way, are children. According to The Hunger Project (2017), there were around 520 million undernourished people in Asia and the Pacific, 243 million in sub-Saharan Africa, 41 million in Latin American and the Caribbean. As per 2014 Global Hunger Index (GHI, 2014, p. 5), levels of hunger were "extremely alarming" or "alarming" in 16 countries primarily in sub-Saharan Africa. According to 2018 Global Hunger Index (GHI, 2018), Central African Republic suffered from a level of hunger that was extremely alarming and six countries—Chad, Haiti, Madagascar, Sierra Leone, Yemen, and Zambia—suffer from levels that are alarming, while 45 countries out of the 119 countries that were ranked have serious levels of hunger.

Many factors are responsible for causing hunger. Undeniably, 2014 Global Hunger Index noted that almost 489 million of the undernourished people lived in situation of conflict and instability (GHI, 2014, p. 5); nevertheless, a 2018 UN report has listed the following factors responsible for engendering hunger: conflicts, climate change, insecurity, droughts, chronic poverty, climate disasters, weather-inducted crop production shortfalls, population displacement, escalating food prices, and the emerging use of biofuels (FSIN, 2018). During 2018, conflict and insecurity, climate shocks and economic turbulence, as the main drivers of food insecurity, continued to erode livelihoods and destroy lives. Nearly 74 million people—two-thirds of those facing acute hunger—were located in 21 countries and territories affected by conflict or insecurity. About 33 million of these people were in 10 countries in Africa; over 27 million were in seven countries and territories in West Asia/Middle East and 13 million were in three countries in South/Southeast Asia and 1.1 million in Eastern Europe (FSIN, 2019, p. 2).

Windfuhr and Jonsén (2005) have opined that some of the major factors contributing to hunger at the national level in a country, inter alia, include: marginalization, denial of access to productive resources along with inadequate land policy, budgetary allocation and unstable rural employment; whereas at international level these factors comprise: prices and dumping, and unfair competition. Nevertheless, in fact, the major cause of hunger is inaccessibility of food. Lappé et al. (1998) have identified the notion of "not-enough food belief" as the first of 12 myths of world hunger. According to Oxfam (2017), the world produces 17% more food per person today than 30 years ago. Almost similar stance is reiterated by Mercy crops (2018) when it says that there would be enough for everyone to get 2700 kcal/day if the entire world's food is evenly distributed. In the opinion of Galtung (2004), assuming an average per capita calorie requirement of 2350/day, there would be sufficient food for approximately 20% more people than the actual global population every year since the mid-1970s. Noting that there is enough food worldwide, albeit access to food is a product of political decision, Galtung (2004) further laments that such decisions are determined by political will and factors like infrastructures in terms of roads, storage facilities, etc., war and displacement, natural disasters and chronic poverty also affect access to food.

The fact of violence is attributed by Edkins (2008: 156) as contributing to denial to access to food because eruption of violence prompts the state to enforce laws of property that can in turn lead to some people's starvation. He identifies violence of three types: (1) direct violence in which the state directly intervenes to regulate production; (2) structural violence that can take the form of imposing exploitative laws of property; and (3) cultural violence that may culminate in exclusion of women or certain classes. While drawing attention to wastage of food as another factor contributing to hunger, Mercy Corps (2018) notes that up to one-third of the food produced globally is never consumed because of the wastage caused by inadequate food production mechanism, unproductive farming techniques, dearth of post-harvest storage and management facilities and enfeebled market connection, especially in developing countries.

6.2.1 Defining hunger

Hunger is defined by FAO as food deprivation or undernourishment as the consumption of fewer than about 1800 kcal a day, the minimum requirement that people need to live a healthy and productive life (FAO, 2008). In other words, it refers to the distress linked to lack of food. GHI (2014, p. 21) has identified different types of hunger—malnutrition, undernutrition, micronutrient deficiency, also called hidden hunger; undernourishment and overnutrition. Malnutrition is an abnormal physiological condition, typically owing to eating the wrong amount and/or kinds of food (FAO, 2013, p. 13). It also more broadly alludes to both overnutrition—excess intake of energy or micronutrients—undernutrition entailing deficiencies in energy, protein, and/or micronutrients. Micronutrient deficiency, also known as hidden hunger, is a form of undernutrition that takes place when intake or absorption of vitamins and minerals is too low to sustain a good health (GHI, 2014, p. 21). Undernourishment is construed as chronic calorie deficiency, with consumption of less than 1800 kcal/day—the minimum requirement most people need to lead a healthy and productive life.

Arguing that undernourishment or hidden hunger, despite being most widespread, has remained neglected or overshadowed, the 2014 GHI focused on this issue to address the challenge of hidden hunger. Mere guaranteeing food is insufficient because human body requires an adequate nutrition that includes different nutrients and according to GHI (2014, p. 2), everyone has the right to adequate food in a quantity and quality enough to satisfy one's dietary needs. Hidden hunger pertains to micronutrient deficiency or lack of minerals and vitamins. According to GHI (2014, p. 3), approximately2 billion were globally affected by hidden hunger in 2014, a number double than that of the number of people who did not have enough calories to consume. Thus hidden hunger on account of micronutrient deficiency can cause child and maternal death, physical disabilities, enfeebled immune system and diminution in creative faculties. Undoubtedly, hidden hunger is not the common hunger and it seldom generates that feeling of emptiness in the stomach; nevertheless, it is extremely dangerous because micronutrient inadequacies entail the potential of engendering, in the long run, a wide number of health problems, including impaired cognitive development, lower resistance to disease risks during childbirth for both mother and children (Biesalaski, 2013).

Prevalence of hunger, entailing a double burden of undernourishment and obesity, in many countries demonstrates as to how hunger can affect the poor in different and opposite ways. With child undernutrition recording decline continuously, the level of child obesity is witnessing increase both in low-and-middle-income countries. Malnutrition and obesity are simultaneously being experienced by many regions, such as India and Brazil. Obesity is having more weight than considered healthy. Children with obesity are at risk of developing noncommunicable diseases and psychological effects (FAO, 2019, p. 31). The global prevalence of obesity more than doubled between 1980 and 2014 (FAO, 2017, p. 19). Sedentary lifestyle along with increased consumption of processed food being part of unhealthy diets usually cause obesity and overweight. Looking at hidden hunger as distinct from other forms of hunger attests to the fact that some countries are experiencing a threefold burden of malnutrition, in the form of undernourishment, micronutrient deficiencies and obesity.

6.3 Toward food security

Food is an integral part of everyday human life. Apart from nourishing human body, food also creates livelihood for the people and substantially contributes to the global economy. However, humankind is presently faced with significant challenges of providing the people with sufficient, affordable, safe and nutritious foods in a sustainable manner. Food is evidently a basic human requirement because human body needs food energy to its vital functions going on and it on this count that food is included at the very bottom of Maslow's hierarchy of needs (Maslow, 1943). Significance of food as a foundational pillar of culture and civilizations can be denied by none (Diamond, 1997; Fraser & Rimas, 2011). Everything associated with food-related activities, ranging from farming, collection, cultivation, harvesting, preparation to consumption, etc., epitomizes a cultural act (Montanori, 2006). Most human requirement have been formulated in modern times as legitimate rights for which people can aspire and for which the entire society has an obligation to respect and provide for. With the passage of time, these rights have become legitimate and legal framework for political and social action in modern-day nation-states (Stavenhagen, 2003), and accordingly food was also accorded recognition as a human right under international law.

The right to food provides for protection of the right of all human beings to feed themselves in dignity, either by producing their food or by purchasing it, as envisaged in Article 25 of the Universal Declaration of Human Rights (United Nations, 1948) and the Article 11 of the International Covenant on Economic, Social and Cultural Rights (United Nations, 1966). The latter Covenant, a binding agreement, also specifies the fundamental right of everyone to be free from hunger and details what States individually and collectively ought to do to fulfill this obligation.

Nevertheless, buck of multidimensions of food does not stop here, because food is also a commodity, and the food industry is one of the biggest areas of economic activity globally, constituting around 10% of the global gross domestic product (GDP) (Forbes, 2007) and being one of the major and more consistent contributors to growth of all economies. The food industry, consisting of farming, food and beverage production, distribution, retail and catering, was valued between $4 and $5.7 trillion in 2008–09 (Alpen Capital, 2011, p. 5; USDA, 2009) and it was expected to increase to $7 trillion by 2014 (IMAP, 2011 citing data mining private companies such as Datamonitor and Euromonitor). The global food import bill for 2011 alone was an astonishing $1.3 trillion and international agricultural prices were likely to remain significantly higher than precrisis levels for at least the next decade (Wise & Murphy, 2012).

Nevertheless, FAO's projections for 2019 indicate that worldwide food imports will drop by 2.5% in 2019 to US$1.472 trillion (FAO, 2019a).

6.3.1 Definition of food security and its evolution

The term "food security" comprises two words, "food" and "security" that in common parlance seem to be straightforward; nevertheless, the term "food security" acquires complex connotations as it is studied and analyzed, with the constituent terms both bearing multiple values and meanings. Food is a primary physiological necessity, a foundation of culture and tradition, a tradeable commodity, and a human right (Vivero-Pol, 2013). The notion of security is also fungible, concurrently indicating the absence of real and perceived threat(s), a defensive and/or offensive stance, a present status and a future positioning-all of that is subject to gradual or abrupt change, with or without warning. Juxtaposing of these two fluid terms raises the question as to what then is meant by the resulting amalgam of "food security." Usage of these two words in tandem engenders a hybrid term that has been defined broadly enough to be focal point of multiple interpretations, mixed application and, as a result, has been the subject of decades of debate (CFS, 2012). It is perhaps better comprehended as what Bourdieu describes as pseudo-concept, wherein it entails both prescriptive and descriptive traits; by suggesting not only what to do, but also what the goal is, including both an end-state and the means to that end (Bourdieu, 2010, p. 236).

The notion of food security is concomitantly a simple definition, an analytical framework, an implementation mechanism and a goal. The concept of food security has been on the global agenda since the founding of the United Nations in the late 1940s and the interregnum period the working understanding of this concept as adjusted itself to make it explicit as to what was meant by "food security" and/or what was required to achieve it. The then American President Franklin D. Roosevelt's 1941 speech on the four freedoms-freedom from want (which included freedom from hunger), freedom of speech, freedom of worship and from fear-reiterated at the 1943 UN Conference on food and agriculture (which in turn led to the foundation of the UN FAO in 1945) is generally cited as the genesis of food security discourse (Shaw, 2007, p. 3). Elimination of hunger and ensuring food security, from the very start of discourse on food security, have often been hailed as universal values, a global concern and an urgent priority, and have, therefore, become central to the global development programs for decades now.

Incorporation of "freedom from hunger" into the conceptual substratum of the United Nations, position of food security has long been intertwined with now what has come to be called international development sector—a network of specialized institutions that looks after the global socioeconomic project, especially in the Global South. The locus assumes significance because development has been described as "a historical construct that provides a space in which poor countries are known, specified, and intervened upon" (Escobar, 1995). FAO had emerged as most prominent among the UN specialized agencies in the initial years after the founding of the UN, dealing with food-related issues and this was subsequently expanded to include many other institutions. The manner such institutional structuring looked as if food security and malnutrition was an issue exclusively pertaining to developing countries, rather than developed ones, that toward food security would suggest, as Barling, Sharpe, and Lang (2008), have pointed out, emulating the progress made in the developed countries as a means to end.

Windfuhr and Jonsén (2005, p. 31) have opined that the initial definitions of food security were developed with the specialized agencies of the UN in mind, and these definitions were subsequently expanded to incorporate national governments. Along with the evolving expansion of global governance institutions and mechanism, UN Commissions, specialized agencies of the UN, International Financial Institutions (IFIs), the NGO sector and more recently, civil society organizations have all claimed a role in food security, directly or indirectly. This perhaps gives rise to the perception that the notion of food security is apparently not sui generis, albeit appears to be as a function of the institutional context in which it exists. Acknowledgment of role of the private sector in food security governance and policy and its inclusion in the discourse on food security is a recent development.

The notion of "freedom from hunger" prior to the convening of the World Food Conference (WFC) of 1974, was prominently looked upon from a redistribution perspective entailing wider general availability of food through redistribution of existing surpluses by means of technological and trade fixes seemingly as a solution to food insecurity and hunger (Cepete, 1984; Shaw, 2007, pp. 32−36). A noticeable change in this perception starting taking place in the backdrop of the 1974 WFC, with the conference suggesting to develop both a world food production policy as well as global food security policy, operating in tandem, albeit with an emphasis on increased production in both cases (Shaw, 2007, p. 124). According to Farsund, Daugbjerg, and Langhelle (2015), it was at this juncture that the phrase "food security" itself came into global policy use. Early years of the decade of the 1980s witnessed the conceptual shift away from emphasis on production (Paulino & Mellor, 1984) in the wake of evolution of approaches to poverty and food security, influenced by Sen (1981), who had emphasized that major famines had their root cause in poor access to food, not availability or rather, that hunger was a function of access, rather than supply, what Sen termed entitlement failure (Barrett, 2002, p. 2218).

Sen (1981) tried to demonstrate that enfeebled conceptual understanding of hungry populations' inability to access food directly contributed to prolonged hunger. Describing what he had termed as "ownership bundles," Sen mooted the suggestion that ability to produce or exchange goods and services is an outcome of how individuals "own" (i.e., are able to act) within the context of land, finances, legal rights, jobs, obligations to family and other networks, etc. (Stephen Devereux quoted in Pritchett, Wollcock, & Andrews, 2010). Writing later in the closing part of the 1980s, Drèze and Sen (1989: 25) noted that there was an overcorrection with regard to the significance of food availability, and that and that entitlements (i.e., access to food) were incorrectly postulated as an "either/or" proposition vs. availability. Sen's work is often credited for paving way for acknowledgment that addressing food insecurity may not require direct agricultural production by insecure populations themselves, rather it was contingent on a much larger set of endowments (Patel, Gartaula, Johnson, & Karthikeyan, 2015).

This envisaged a shift of emphasis in the conceptual framework, suggesting that food security is a function of much more than just food, and the key lies in access, rather than availability. The observation that famines have never occurred in functioning democracies (Carolan, 2012, p. 68) not only added an additional dimension to the importance of Sen's work, albeit it also suggested that hunger was a function of political systems, and political representation was a significant factor in alleviation of hunger. In this backdrop, notion of food security got freighted with political entailments that turned it into a more subtle treatment for nondemocratic states-situating food security as a political, not simply a "technical" discourse, thereby implying that food insecurity needed political and technological solutions (Carolan, 2012, p. 68). In the wake of shift of the emphasis from availability to access, the discourse on food security got modified from production of food, to sourcing of food, along with according recognition to the entitlements of individuals and households in accessing food. In the wake of this conceptual transition in both conceptual and spatial terms, Windfuhr and Jonsén (2005, p. 29) have opined that it paved the way for consideration of food security not only at the level of the state, but at the level of the individual citizen as well. This also opened the door for important shifts in development theory, including the sustainable livelihoods approach promoted by Chambers and Conway (Erni, 2015, p. 4) and Capability theory (Nussbaum, 2011) (Table 6.1).

Table 6.1 Leading events in the evolution of food security discourse.

Period	Leading events in the global food security discourse	Key developments in conceptual approaches to food security
1970s	• World Food Summit 1974 • The term "food security" enters common usage; • Focus on food supply; • UN Plan of Action for World Food Security, 1979; • Establishment of World Food Programme in 1963; • Establishment of IFAD in 1977.	• Focus on variations in global food availability and price volatility; • "Green Revolution" continues to augment, widely introduced across Asia; • Excess global production used as food aid to secure the Cold War allegiances (Carolan, 2012, p. 70); • Increased production through introduction of technological solutions to staple food production; • Shift from global trade at nation-state level as strategy; • FAO introduces food balance sheets, 1972–74.
1980s	• Sen introduces "entitlements," access becomes key issue, shifting focus from availability; • income added to food security disclosure, FAP broadens food security definition, 1984; • major famines in Ethiopia and Sudan promote emphasis on food aid; • World Bank, poverty and hunger report, (1986).	• Analysis of food security shifts to individual, household and national level, focus pivots to access from availability; • Non-food factors (including access to land, employment or other 'other entitlement' failure) incorporated in food security analysis; • Food security definition adjusted to "access of all people at all times to enough food healthy life" (World Bank, 1986); • Geographic focus of food security on Sub-Saharan Africa and South Asia.
1990s	• 1992 International Conference on Nutrition; • GATT 1994 • World Food Summit, 1996; • Definitions of food security proliferate, estimated by Smith et al., at 200 + • UNICEF publishes "Conceptual Framework for Understanding the Causes of Malnutrition," 1990; • Global food security discourse expands to include NGOs and nonstate actors.	• WFS single definition agreed upon by 190 countries; • Updated definition of food security includes four pillars of access, availability, utilization and stability; • WFS sets target of total number of hungry people to be reduced by half by 2015; • Concerns over efficacy over food aid increase (Clay, 2002); • Nutrition security defined as separate concept; • Terms like food democracy, food citizenship & food sovereignty emerge.
2000s	• MDGs, 2000; • WFS + 5, 2001; • 2007–2009 financial, food and fuel crisis causes revival in food security as a global concern; • World Summit on Food Security, 2009; • Lancet series on maternal child health released, 2008; • Right to Food becomes voluntary; • Repositioning Nutrition as Central to Development, World Bank (2006).	• MDG Goal No. 1 and targets 1.8, 1.8A, 1.9 focus on undernutrition and malnutrition; • WFS definition adjusted to include "physical, social and economic access"; • Global trade liberalization results in food security being marketized (Zerbe, 2009 in Carolan, 2012; 71); • Focus on nutrition efforts pivots to "First 1000 days" from conception to 24 months; • Lack of progress on nutrition leads to calls for "nutritional security";

Table 6.1 Leading events in the evolution of food security discourse. *Continued*

Period	Leading events in the global food security discourse	Key developments in conceptual approaches to food security
2010s	• SUN Movement begins, 2010; • Committee for World Food Security reformulates food and nutrition security, 2012; • Sustainable Development Goals (SDGs) launch, 2015.	• Integrated Phase Classification launched, 2008; • CFS reinvigorated, increased interest in food security from global and regional groupings (G-8, ASEAN). • WFS definition of food security is adjusted to include "an environment of adequate sanitation, health services and care, allowing for a healthy active life," CFS (2012); • SDG Goal-2 includes Ending Hunger and Achieving Food Security, supported by five targets of hunger, malnutrition, small holder agriculture, sustainability and biodiversity.

Source: Adapted from Armstrong, J. W. (2018). Food Security in Lao PDR: An analysis of policy narratives in use (pp. 56–57) (Unpublished doctoral thesis). London: University of London.

With the passage of time, the definition of food security has undergone changes in consonance with the prevailing assumptions of the time. Three pre-1996 definitions of food security have been highlighted Clay (2002) as thus: "[A]availability at all times of adequate world food supplies of basic foodstuffs...to sustain a steady expansion of food consumption...and to offset fluctuations in production and prices" (Proceedings of the 1974 WFC)—"Ensuring that all people at all times have both physical and economic access to the basic food that they need" (FAO, 1983)—in the World Bank report, Poverty, and Hunger (1986), this concept of food security is elaborated in terms of: "access of all people at all times to enough food for an active, healthy life."

According to Clay (2002), the 1986 World Bank report envisages an addition of a significant temporal element, outlining chronic and transitory food insecurity, adding a timeframe to the physical elements of availability and access mentioned in the previous definitions. Equally important is the fact that appropriateness of food requirement for an active life in terms of quality and quantity was included for the first time in contrast to food intake needed for basal metabolic unction and survival only (Maxwell, 1996, p. 4). Thus food security was no longer just about staying alive, albeit was becoming about having a productive life. A shift of emphasis could be discerned in the spatial framework for food security by the mid-1990s and this shift was from global levels to household and individual level; from a food-specific focus to a more inclusive focus on livelihoods, and from objective to subjective measurement (Maxwell, 1996). Smith, Pointing, and Maxwell (1992) cataloged nearly 200 mentions about food security from both academic and gray literature that demonstrated proliferation of definitions and interpretations of the term "food security" along with denoting a massive plurality of differing definitions (Maxwell, 1996; Shaw, 2007, p. 383). It was becoming evident that there was need for clarity on the definitional aspects of food security.

6.3.1.1 Impact of the World Food Summit, 1996

The World Food Summit (WFS) held in Rome in 1996 developed a single definition for food security that entailed the consensual agreement of 190 nations. Notwithstanding subsequent attempts at envisaging adjustments, it remains the most well-known, commonly used definition, and as such, is seemingly a rational starting point from which to explore the concept. The 1996 WFS adopted Rome Declaration that defined food security: "Food security exists when all people, at all times, have physical and economic access to sufficient, safe and nutritious food to meet their dietary needs and food preferences for an active and healthy life" (FAO 1996). While envisaging a number of key aspects of food security, this definition also presents a synthesis of previous iterations on the subject. The expression "all people, at all times," denotes that it relates to entire humankind, here and now and evermore into the future. It also denotes the universality of what is suggested by the expression. The expression "physical and economic success" acknowledges that food is not merely an issue of production, albeit it also relies on the ability to procure it, needing economic and social structures like operational markets, financial income, inclusive social networks, and so on. The term "dietary needs and food preferences" denotes that there are both physiological and cultural factors to be taken into account, as well as it also suggests that despite the universal applicability of this definition, the modus operandi of attaining is contingent on social and cultural context. The final part of the definition referring to "an active and healthy life" entails that merely getting enough food to meet minimum requirements for survival is insufficient, rather it requires adequate food to sustain a full livelihood. Undoubtedly, many attempts at adjustment to this formulation have been made since 1996; nonetheless, the basic components underlying this concept still remain the foundational elements of subsequent definitions of food security.

The WFS's 1996 Rome Declaration on World Food Security envisages that food security is "a common objective-food security at the individual, household, national, regional and global levels," along with WFS Plan of Action document detailing linkages between food security and gender, food safety, access to water and education, and biodiversity protection; nevertheless, the document fails to indicate as to how its targets are to be attained. In its inability to envisage even a singular global strategy aimed at attaining food security, instead it suggests that at the national level "diverse paths to a common objective" could be applied (Sharma, 2011). Admittedly, 1996 Rome Declaration provides a normative definition for food security and a global target; nonetheless, modus operandi of attaining that target is subject to interpretation at the nation-state level. Besides, WFS's 1996 Plan of Action document was also silent on control of food production mechanisms, thereby giving rise to questions like where food should be produced, how and by whom, and who has the right to make these decisions. With the passage of time, this came to be construed as weakness, spurring the development of alternative approaches like food sovereignty and Right to Food (McKeon, 2009, p. 38).

Undeniably, adoption of the Millennium Development Goals (MDGs), agreed to by 189 countries in 2000 did not essentially representing a shift in the conceptual framework or definition of food security; nonetheless, it did provide a new momentum on food security that the WFS had not delivered. The MDGs accorded, if not to food security, at least, one very specific aspect of it—malnutrition, a top-level priority for all developing countries, establishing the first goal (MDG-1) as the eradication of poverty and hunger (Easterly, 2013, pp. 123–127; Fukuda-Parr & Orr, 2013). Initially, it seemed that the paring of poverty and hunger was in consonance with development theory and then the prevalent narratives on food security signifying as if food security was symbolic of broader poverty-related issues, and its inverse-food security as the symptom of poverty (Von Braun, 1999). Paradoxically, the MDG poverty goals were attained at the global level during the course of 2000–15, whereas the goal of malnutrition lagged far behind the target during this period. This led to the decoupling of the assumption that reduction in poverty could lead to improvement in malnutrition. Thus the assumption that poverty alleviated is equal to improved nutrition was projected to be not as closely linked as once assumed, and further research conducted in South Asia over 2000s also attested to it (Banerjee & Duflo, 2011; Deaton & Drèze, 2009).

Commitment of developing countries to the MDGs and the expectations that these countries would be assessed on their performance in the service of these goals proved instrumental in pulling food security out of the remit of the purely "technical" issue, and catapulting it to the international political fora, making it a socioeconomic benchmark against which the development efforts of a country would be assessed. Nevertheless, major global institutional stakeholders in the WFS, especially FAO, got alienated in the process of preparing MDGs and the resultant outcome was undermining of institutional commitment within FAO to the MDGs (Fukuda-Parr & Orr, 2013, p. 15, 25). While reviewing MDG-1 and the Hunger targets Fukuda-Parr and Orr opined: "... [T]he MDG target and indicators frame the problem of food insecurity as a common-sense issue of supply and production, favoring quick and measurable gains in supply and production as the key solution, and marginalizing the complex socio-economic determinants and the human development and human rights priorities of distribution, discrimination, inequitable access, and lack of voice and autonomy" (Fukuda-Parr & Orr, 2013, p. 44).

With emphasis on addressing the MDGs garnering global political momentum, attention on malnutrition was also focused. However, the corpus of knowledge around what was meant by food security continue to grow, partly as an outcome of the decoupling of poverty and hunger, and partly as reaction to the oversimplification of the MDGs. At that juncture, the need for a new definition of food security was being felt and this task was accomplished by the World Committee for Food Security in 2012.

Despite the existence of technically precise definitions of some of the terms used in the food security discourse, the usage of such terms still lacked consistency. Terms like food security and insecurity are often used interchangeably or in combination with terms like hunger, malnutrition, undernutrition and undernourishment. According to IFPRI (2011), undernutrition and malnutrition have formal definitions, but hunger does not. While eluding a single definition, hunger and its effects are socially and culturally defined or are used interchangeably with food security (Fukuda-Parr & Orr, 2013, p. 4). Each or all of these three terms—undernutrition, malnutrition, and hunger—can be construed as end states accruing from the absence of food security; nevertheless ironically, the absence of any of those three conditions does not necessarily equate to food security (Kracht, 1999).

Revitalization of the CFS in the wake of the 2007—09 food and fuel crisis (Duncan, 2014), spurred global political momentum with emphasis on revisiting the broad rubric of hunger-related policy terminology. Accordingly, the Committee for World Food Security in 2012 released a document entitled "Coming to Terms with Terminology," which had a twofold aim: to standardize terms related to food security and nutrition in common usage, and more importantly, to reevaluate the positioning of nutrition in the overall conceptual framework (CFS, 2012).

The reevaluation of nutrition signaled a tacit recognition of the shortcomings of the "four-pillars" approach to food security. The four-pillars model suggested a tantamount between one pillar and the next, such that availability was equal to access was equal to utilization. Nevertheless, improvements in food security could only be said to take place if such improvements translated into improved nutritional outcomes; that is, if fewer people were malnourished. Nonrealization of the reduction in malnutrition at the global level, despite improvements in food availability and access to food, necessitated the question that this could make sense if the problem of food insecurity was reformulated to be about more than food. Thus improvements in nutrition began to be increasingly understood not to be confined to food-related factors alone, albeit were also a function of access to sanitation, healthcare, and feeding and childcare, as envisaged in the WFS declaration. It was explicitly made clear in the CFS document that food security was not a mere function of food. The World Bank has summed it thus: "Food security, an important input for improved nutrition outcomes, is concerned with physical and economic access to food of sufficient quality and quantity in a socially and culturally acceptable manner. Nutrition security is an outcome of good health, a healthy environment, and good caring practices in addition to household-level food security" (World Bank, 2006, p. 66).

Subsequently, the CFS tried to seek rebalancing of the terminology veering around these two concepts, concluding that henceforward, the term food and nutrition security was most applicable, defined as: "Food and nutrition security exists when all people at all times have physical, social and economic access to food, which is safe and consumed in sufficient quantity and quality to meet their dietary needs and food preferences, and is supported by an environment of adequate sanitation, health services and care, allowing for a healthy and active life" (CFS, 2012). This paved way for division of policy narrative in food security with projecting nutrition as a "supra-level goal" along with food security as a subordinate strategy. Thereafter, this new definition not only moved beyond reinventing of what was proposed under food security, albeit further emphasized the significance of how food is consumed, not just accessed. In practical terms, this new definition has acquired a "reorientation" in terms of two policy sectors usually maintained as separate and distinct—agriculture as the source of food security and nutrition, previously construed as the purview of public health. On account of tying food security to nutrition, policymakers are now enjoined upon an obligation in each sector to allude to the joined-up nature of the overreaching approach.

By 2017 there were indications that even at the global level the uptake of this definition was far from complete. In 2013 Ruel described food security and nutrition security as "related but distinct," with food security "necessary but not sufficient" to ensure nutrition security. Dawe and Jaffee (2014) writing in the context of Southeast Asia presented the 1996 WFS definition as "the current definition." Such terminological gyrations reflect as to how much definition and redefinition of the term is an integral part of the narrative on food security.

6.4 Food security and climate change

Climate change wields direct impact on food systems and food security. Dealing with climate change by means of harnessing land-based technologies to procure negative emissions with the help of biomass production would progressively exert pressure on food production and food security via probable contest for land. A food system approach is helpful in discerning an explicit noesis of as to how food security, including nutrition is affected by climate change, the choices or alternatives for the food system to adapt and mitigate, synergies and trade-offs among these choices or options, along with congenial or enabling conditions for their adoption. In order to negate the adverse impact of climate change on food system and food security, including nutrition, it deems essential to appraise the role of incremental and transformational adaptation as well as exploration of the prospects of coalesce of supply-side and demand-side measures contributing to climate change mitigation.

The food system includes all the activities and actors pertaining to the production, transport, manufacturing, retailing, consumption, and waste of food, and their impacts on nutrition, health and well-being, and the environment. The existing food system, entailing production, transport, processing, packaging, storage, retail, consumption, loss and waste, is the source of food for consumption for a vast proportion of population worldwide and also supports the livelihoods of a large number of people. According to World Bank (2019), agriculture as an economic activity engenders between 1% and 60% of national GDP in many countries, with a world average of about 4% in 2017. Since the onset of the decade of the 1960s till the closing part of the year 2017, food supply per capita has recorded an increase exceeding 30% followed by increased use of nitrogen fertilizers, recording an increase of about 800%, and water resources for irrigation registering an increase of more than 100% during this period (World Bank, 2019). From the 1960s onwards, agricultural productivity has recorded a massive growth that has buttressed the development of contemporary food system that is both a big contributor to climate change and concomitantly also susceptible to the vagaries of climate change, especially in the realms of food production, transportation and marketing activities. The current emphasis on producing more food to feed the burgeoning population will require additional cropland area and this would result in generation of more GHG emissions accompanied by other environmental effects, including loss of biodiversity.

Undeniably, the activities pertaining to food system as well as actors involved in the process contribute to ensuring food security; concurrently, this process also culminates in engendering GHG emissions thereby contributing to climate change. Concomitantly, intricate interactions between climate change and food systems also contribute to food insecurity via wielding its impacts on food availability, access, utilization and stability. While adopting a food-systems approach in its Special Report on Climate Change and Land (SRCCL), released in August 2019, the Intergovernmental Panel on Climate Change (IPCC) acknowledges the interconnectedness between the demand side for and supply of food; and it emphasizes on the need for making a joint assessment with the objective of detecting the challenges of mitigation and adaptation to climate change. "Pull" from the demand side often triggers emanation of GHG emission from agriculture. Mitigation and adaptation spur envisaging modifications in the means of production, supply chain and demand practices by means of improvements in dietary options, market incentives and trade relationships with a view to develop a more durable and sound food system.

The four pillars of food security—availability, access, utilization, and stability—and their mutual interactions are vulnerable to adverse impacts of climate change (FAO, 2018a). It is interesting to ascertain as to how each of these pillars gets affected by the vagaries of climate change across the spectrum of food system activities. With some studies persistently focusing on just one pillar of food security—availability—in terms of effects on food production, more studies are concentrating on connected issues of access in terms of impacts on food prices; utilization in terms of effects on nutritional quality; and stability in terms of impacts of escalating extreme events; owing to the fact climate change wields its impact on these pillars of food security (Bailey et al., 2015). Some studies suggest that paucity of resources to invest in adaptation and diversification measures renders the low-income producers and consumers to be most affected by changing climate change (Bailey et al., 2015; UNCCD, 2017) (Table 6.2).

Table 6.2 Relationship between food security, food system and climate change.

Food security pillar	Examples of observed and projected climate change impacts	Examples of mitigation and adaptation
Availability *Production of food and its readiness for use through storage, processing distribution, sale/or purchase*	Reduced yields in crops and Livestock system Reduced yields from lack of pollinators, Pests and diseases Reduced food quality affecting availability (e.g., food spoilage and loss from mycotoxins) Disruption to food storage and transport networks from change in climate, including extremes	Development of adaptation practices Adaptation of new technologies, new and neglected varieties Enhanced by integrated practices, Better food storage Reduction of demand on by reducing Waste, modifying diets Closing of crop yield and productivity Gaps Risk management including marketing Mechanisms, financial insurance
Access *Ability to obtain food, including effects of price*	Yield reductions, changes in farmer livelihoods, limitations on ability to Purchase food Price rise and spike effects on low-income consumers, in particular women and children Due to lack of resources to purchase food Effects of increased extreme events on food supplies, disruption of agricultural trade and transportation infrastructure	Integrated agricultural practices to build resilient livelihoods Increased supply chain efficiency (e.g., reducing loss and waste) More climate-resilient food systems, shortened supply chains, dietary change, market change
Utilization *Achievement of food potential through nutrition, cooking, health*	Impacts on food safety due to increased prevalence of microorganisms and toxins Decline in nutritional quality resulting from increasing atmospheric CO_2 Increased exposure to diarrheal and other infectious diseases due to increased risk of flooding	Improved storage and cold chains Adaptive crop and livestock varieties, healthy diets, better sanitation
Stability *Continuous availability and access to food without disruption*	Greater instability of supply due to increased frequency and severity of extreme events; food price rises and spikes; instability of agricultural incomes Widespread crop failure contributing to migration and conflict	Resilience via integrated systems and Practices, diversified local agriculture, infrastructure investments, modifying markets and trade, reducing food loss and waste Crop insurance for farmers to cope with extreme events Capacity building to develop resilient systems

Table 6.2 Relationship between food security, food system and climate change. *Continued*

Food security pillar	Examples of observed and projected climate change impacts	Examples of mitigation and adaptation
Combined *Systemic impacts from interactions of all four pillars*	Increasing undernourishment as food system is impacted by climate change Increasing obesity and ill health through narrow focus on adapting limited number commodity crops Increasing environmental degradation and GHG emissions Increasing food insecurity due to competition for land and natural resources (e.g., for land-based mitigation)	Increased food system productivity and efficiency (e.g., supply-side mitigation, reducing waste, dietary change) Increased production of healthy food and reduced consumption of energy-intensive products Development of climate smart food systems by reducing GHG emissions, building resilience, adapting to climate change Governance and institutional responses (including food aid) that take into consideration gender and equity

From IPCC (Intergovernment Panel on Climate Change). (2019). IPCC Special Report on climate change, desertification, land degradation, sustainable land management, food security, and greenhouse gas fluxes in terrestrial ecosystems. Available at: https://www.ipcc.ch/site/assets/uploads/2019/11/SRCCL-Full-Report-Compiled-191128.pdf.

Attention garnered by food and its interconnectedness with environment and climate change has catapulted food in prominence since The Rio Declaration (1992) and this can be evidenced from the fact that food production was enlisted in Chapter 14 of Agenda 21 of the 1992 Rio Declaration and 2015 Paris Agreement on Climate Change (PACC) (UNFCCC, 2015) included the necessity of ensuring food security under the threat of climate change on its first page. The increasing prominence being accorded to food is also reflected in some of the recent reports of the IPCC, especially IPCC Fifth Assessment Report (IPCC, 2014a) and the Special Report on Global Warming of 1.5°C (SR15) (IPCC, 2018a).

Porter et al. (2014) in their chapter on Food Security and Food Production Systems that is included in the IPCC's Fifth Assessment Report (IPCC, 2014a), have brought to limelight new aspects by extending its emphasis beyond the impacts of climate change basically on agricultural production, inclusive of crops, livestock and agriculture, to incorporate a food-systems approach along with focusing attention on undernourished people. Prevalence of related studies had spurred the authors to principally focus on food production systems (Porter, Howden, & Smith, 2017). While emphasizing that an assortment of potential adaptation alternatives was available across all food system activities, and not just in food production, it lamented at the lack of sufficient requisite research available during that period, that is, until 2014, with regard to benefits that could probably accrue from potential innovations in food processing, packaging, transport, storage, and trade.

The chapter entitled *Agriculture, Forestry and Other Land Use (AFOLU)*, contributed by Smith et al. (2014) to the IPCC Fifth Assessment Report evaluated mitigation potential by taking into account both the supply and the demand side of land uses, by focusing on dietary changes along with food loss and waste. While focusing on crop and livestock activities within the farm precincts and land use and land-use dynamics lined to agriculture, it did not apply full food system approach to GHG emission estimates that embraces processing, transport, storage, and retail. The chapter on Rural Areas in the IPCC's Fifth Assessment Report, contributed by Dasgupta et al. (2014), showed that farm households in developing countries are susceptible to the vagaries of climate change owing to socioeconomic physiognomies and nonclimate stressors, in addition to climate risks. It was further reported that that an assortment of on-farm and off-farm climate change adaptation (CCA) measures had already been under implementation and that the local socio-cultural milieu merged as a decisive factor in the success or failure of diverse adaptation strategies for food security, especially in terms of trade, irrigation or diversification. Revi et al. (2014), while contributing his chapter on Urban Areas to the IPCC's Fifth Assessment Report, informed that food security of people residing in cities was seriously impacted by climate change via decreased supplies, including urban-produced food, and effects on infrastructure, as well as a want of access to food. Poor people inhabiting urban areas are more susceptible to swift changes of food prices due to climate change.

Some scholars have drawn attention to the fact that various climate change response options or alternatives suggested in the IPCC's Fifth Assessment Report (IPCC, 2014b, 2014c), deal with incremental adaptation or mitigation responses discretely rather than being all-encompassing of more systemic or transformational changes in numerous food systems that are extensive, in-depth and swift, necessitating social, technological, organizational and system responses (Rosenzweig & Solecki, 2018). In some instances, transformational change may call for blending of resilience and mitigation across all parts of the food system, including production, supply chains, social aspects, and dietary choices. Besides, such transformational changes in the food system need to embrace linkages to upgraded responses to land degradation, desertification, and deteriorations in quality and quantity of water resources throughout the food—energy—water nexus.

PACC, concluded in December 2015, has put forth a temperature target of limiting warming to well below 2°C and undertake efforts to limit warming to 1.5°C and concurrently Article 2 of the PACC explicitly stipulates that the agreement is within "the context of sustainable development" and actions of the countries should be "in a manner that does not threaten food production" to ensure food security (UNFCCC, 2015). Several countries party to the PACC have incorporated food systems in their mitigation and adaptation plans that are discernible from their nationally determined contributions (NDCs) for the PACC (Rosenzweig et al., 2018a). An analysis based on the appraisal of 160 NDCs submitted by the Parties to the PACC, undertaken by Richards et al. (2016) revealed that of these 103 had included agriculture mitigation; of the 113 Parties that included adaptation in their NDCs, almost 102 included agricultures among their adaptation priorities. It was further revealed that ample attention was paid to conventional agricultural practices that can be climate smart and sustainable such as crop and livestock management; however, less to the enabling services that could facilitate uptake like climate information services, insurance, credit, etc. While asserting that substantial amount of finance is required for agricultural adaptation and mitigation by least-developed countries—to the tune of $3 billion annually for adaptation and $2 billion annually for mitigation—Richards et al. (2016) lament that this could be an underestimate owing to a small sample size. Subsequently, Richards, Wollenberg, and van Vuuren (2018) reported that with regard to the mitigation side, none of the prime agricultural emitters included sector-specific contributions from the agriculture sector in their NDCs, instead most incorporated agriculture in their economy-wide targets.

A major factor with regard to the implementation of measures designed to attain the PACC goals entailing steps related to carbon dioxide removal via bioenergy. To attain the temperature target of limiting warming to well below 2°C and continuing forward steps to limit warming to 1.5°C, substantial financial investments and abrupt changes in land use will be needed to advance bioenergy with carbon capture and sequestration (BECCS), afforestation and reforestation, and biochar technologies. The global area needed for BECCS alone under the present scenarios to limit warming to 1.5°C will be in the range of 100−900 Mha, most commonly around 380−700 Mha (Rogelj et al., 2018). Therefore, attainment of the 1.5°C target is prone to culminate in major competing demands for land between climate change mitigation and food production, with cascading impacts on food security.

6.4.1 Climate change, climate drivers, and food security

In order to assess the effects of future GHG emissions, mitigation measures, and adaptation on food security, usually the common framework of the representative concentration pathways (RCPs) and the shared socioeconomic pathways is utilized (Popp et al., 2017; Riahi et al., 2017; Doelman et al., 2018). It is revealed from some recent studies that have utilized these scenario approaches that the food system externalizes costs onto human health and the environment (Springmann et al., 2018a; Swinburn et al., 2019; Willett et al., 2019), spurring need for transmuting the food system to fork out better human and sustainability results (Willett et al., 2019). Such a transmutation entails the potential of serving as a significant lever to deal with the subtle interactions between climate change and food security.

In comparison to the past research that was based on the food systems and scenario approaches along with its focus on climate change and food security that was widely followed in the IPCC's 2014 Fifth Assessment Report, the recent research trends outspread yonder production to focus on as to how climate change interacts the entire food system. New trends in recent research literature have expanded the analysis of climate change and food insecurity yonder undernutrition to encompass the overconsumption of unhealthy mass-produced food rich in sugar and fat that is risky for health in multiple but potentially dangerous ways and the role of dietary options and consumption in GHG emissions. While focusing on land-based food systems, it also somewhat deals with the role of freshwater and marine production.

Food security and in turn human health can be affected by climate change in multiple ways (Watts et al., 2018). The major mode through which climate change can affect the quantum of food, both from direct effects on yields and indirect impacts via effects of climate change on water availability and quality, pests, insects and diseases, and pollination services. Another mode is through altering patterns of carbon dioxide (CO_2) in the atmosphere in terms of affecting biomass and nutritional quality. Besides, changing climate also entails the potential of exacerbating food safety risks during transport and storage processes. Alteration in weather entails the potential of wielding direct impact on human health, especially in terms of exposure of the agricultural workforce to extreme temperatures. According to Watts et al. (2018), the people exposed to extreme temperatures becoming vulnerable to changing metabolic demands and physiological stress entail the potential for interactions with food availability because such people may need more food to cope with the situation and concurrently being impaired from producing the same food. Cumulative outcome of all these factors enhances the probability of affecting both physical health and cultural health, via altering the quantity, safety and quality of food availability for the people within their cultural ambit.

Climate drivers pertaining to temperature and precipitation along with integrated metrics that coalesce these and other variables are germane to food security and food systems. These entail the potential of impacting many aspects of the food security pillars (FAO, 2018b). Categorization of climate drivers comprises modal climate changes, seasonal changes, extreme events and atmospheric conditions. Fluctuations in climate covers or envelopes spurring shifts in cropping varieties planted exemplify modal climate change. Warming trends prolonging growing seasons represent seasonal changes. Extreme events include high temperatures impacting crucial growth periods, flooding, droughts, hurricanes, etc. Concentration of CO_2, short-lived climate pollutants (SLCPs), and dust is the stark examples of the atmospheric conditions. Fluctuating frequencies of precipitation and evaporation groundwater levels and dissolved oxygen content cumulatively are prone to affect water resources for food production (Schmidtko, Stramma, & Visbeck, 2017).

Extensive impacts, like that took place during late 2015 to early 2016 when a potent El Nino contributed to regional shifts in precipitation in the Sahel region, can be wielded by potential alterations in key modes of climate variability. Occurrence of a major drought in Ethiopia culminated in widespread crop failure necessitating food aid for more than 10 million people (United States Department of State, 2016). Climate variables such as solar radiation, wind, humidity, and (in coastal areas) salinization and storm surge in coastal areas embrace the potential of impacting agricultural production, processing and/or transport (Myers et al., 2017). Extreme climate events constitute another climate variable that bears impact on agricultural production, processing, and/or transport in terms of causing inland and coastal flooding thereby affecting the ability of the people to procure and prepare food (FAO, 2018a).

SLCPs such as ozone and black carbon have come to assume prominence owing to their significant role in impacting agricultural production via wielding direct effects on crops and indirect impacts on climate (Emberson et al., 2018). Damage is caused to plants by Ozone because it inflicts damages on cellular metabolism that impact leaf-level physiology to whole-canopy and root-system processes and feedbacks. According to Emberson et al. (2018), these effects impact leaf-level photosynthesis senescence and carbon assimilation, along with whole-canopy water and nutrient acquisition and eventually crop growth and yield. Chuwah, van Noije, van Vuuren, Stehfest, and Hazeleger (2015), having made use of atmospheric chemistry and a global integrated assessment model, have arrived at the conclusion that without a substantial reduction in air pollutant emission, high ozone concentration could result in an upsurge in crop damage of up to 20% in agricultural regions in 2050 in comparison to projections wherein alterations in ozone are not accounted for. According to Backlund, Janetos, and Schimel (2008), higher temperatures are linked to higher ozone concentrations; C3-type crops like soybeans, wheat, rice, oats, green beans, peppers, and some types of cottons, are sensitive to ozone and C4 crops are discreetly sensitive.

Methane contributes to increase in surface ozone which supplements warming-induced losses and some recent quantitative analyses now incorporate concurrently climate, long-lived pollutant like carbon dioxide (CO_2), and multiple short-lived pollutants such as methane (CH_3) and ozone (O_3) (Shindell, 2016; Shindell, Fuglestvedt, & Collins, 2017). Decrease in tropospheric ozone and black carbon entails the potential of avoiding premature deaths often caused by outdoor air pollution and it can also contribute to augmentation in annual crop yield (Shindell et al., 2012). These actions in tandem with decrease in methane can impact climate on shorter time scales in comparison to measures designed to lead to reduction in CO₂. Some scholars (Hannah et al., 2017; Shindell et al., 2017) are of the view that implementation of such actions/measures can lead to massive reduction in the risks of crossing the 2°C threshold.

6.4.2 Climate change and food availability

Availability of food is one of the pillars of food security Climate change impacts food availability through its effect on the production of food and its storage, processing, distribution, and exchange. Some studies exhibiting a firmed-up relationship between observed climate variables and crop yields, some experts believe (Innes, Tan, Van Ogtrop, & Amthor, 2015; Mavromatis, 2015), are indicative of the potential of future expected warming to wield severe impacts on crop production. A counterfactual analysis used at global scale by Iizumi and colleagues revealed that climate change between 1981 and 2010 had led to diminution in global mean yield of maize, wheat and soybeans by 4.1%, 1.8%, and 4.5% respectively in comparison to preindustrial climate even when CO₂, fertilization and agronomics are taken into account.

Drylands comprising over 40% over the earth's land area is home to over 2.5 billion people (FAO, 2011; UNEP, 2011). Nevertheless, drylands are construed as being susceptible to climate change with regard to food security, especially in developing countries; and additionally, such regions possess little capacities to deal effectively with low or decreasing crop yields (Nellemann et al., 2009; Shah, Fischer, & Van Velthuizen, 2008).

Crop yield studies with regard to India make it discernible that warming has proved instrumental in reducing wheat yields by 5.2% from 1981 to 2009, despite adaptation (Gupta, Somanathan, & Dey, 2017); that maximum daytime temperatures have registered increase along with some nighttime temperatures (Jha & Tripathi, 2017). Apparently, people living in mountainous areas of Hindu-Kush Himalayan region comprising parts of Pakistan, India, Nepal, and China, are specifically susceptible to food insecurity pertaining to climate change owing to poor infrastructure, limited access to global markets, physical isolation, low productivity, and hazard exposure to hazards, including Glacial Lake Outburst Floods (GLOFs) (Rasul et al., 2019). Ascertaining as to how climate-related changes affect food security has been facilitated through conducting surveys (Hussain, Rasul, Mahapatra, & Tuladhar, 2016) and the emerging outcomes demonstrate that the surveyed region was experiencing recurrent floods over and above protracted droughts with resultant negative impacts on agricultural yields and augmentation in food insecurity (Hussain et al., 2016).

Ascertaining as to how climate-related changes affect food security has been facilitated through conducting surveys (Hussain et al., 2016) and the emerging outcomes demonstrate that the surveyed region was experiencing recurrent floods over and above protracted droughts with resultant negative impacts on agricultural yields and augmentation in food insecurity (Hussain et al., 2016). Impacts of climate change have been examined in recent studies on a global scale with the help of applying an assortment of methodologies from a comparative perspective as well (Zhao et al., 2017a). Methodologies to examine global and local changes embrace global gridded crop model simulations (Deryng, Conway, Ramankutty, Price, & Warren, 2014), point-based crop model simulations (Asseng et al., 2015), analysis of point-based observations in the field (Zhao et al., 2016), and temperature-yield regression models (Auffhammer & Schlenker, 2014).

Outcomes emerging from the study undertaken by Zhao et al. (2017a) across diverse methods constantly demonstrated negative temperature impacts on crop yield at the global scale, commonly reinforced by analogous impacts at country and site scales. However, the approach adopted by Zhao et al. (2017a) suffers from a limitation that it is based on the supposition that yield responses to increase in temperature are linear whereas yield response differs subject to growing season temperature level. Study by Iizumi and Ramankutty (2016) has demonstrated that the projected global mean yields of maize and soybean at the end of present century would register reduction monotonically with warming, while those of rice and wheat increase with warming and level off at a warming of about 3°C (2091–2100 relative to 1850–1900). Deployment of empirical statistical models has been facilitated widely in different cropping systems, at various scales. Findings accruing from the usage of statistical models for maize and wheat tested with global climate model scenarios demonstrated that the RCP4.5 scenario brought down the size of average yield impacts, risk of major slowdowns, and exposure to critical heat extremes compared to RCP8.5 in the latter decades of the 21st century (Tebaldi & Lobell, 2018). According to Levis, Badger, Drewniak, Nevison, and Ren (2018), impacts on crops grown in the tropics are predicted to be more negative than in mid- to high-latitudes as stated in AR5 and confirmed by recent studies. Such probable negative impacts in the tropics are specifically distinct under conditions of explicit nitrogen stress (Rosenzweig et al., 2014).

An examination of biophysical impacts was undertaken by Reyer et al. (2017b) in five world regions under different warming scenarios—1°C, 1.5°C, 2°C, and 4°C warming. For the Middle East and Northern Africa region, an important correlation between crop yield decrease and temperature increase was found, irrespective of whether the effects of CO_2 fertilization or adaptation measures were taken into consideration (Waha et al., 2017). For Latin America and the Caribbean, the relationship between temperature and crop yield changes was only significant when the effect of CO_2 fertilization is considered (Reyer et al., 2017a). Some recent studies have shown that projected yield loss for West Africa is dependent on the degree of wetter or drier conditions and elevated CO_2 concentrations (Sultan & Gaetani, 2016). In a crop modeling study with RCPs 4.5 and 8.5, Faye et al. (2018b) ascertained that climate change could wield limited impacts on peanut yield in Senegal owing to the effect of elevated CO_2 concentrations.

With global warming of 1.5°C and its rise further to 2°C is projected to increase climate-related risks to food security (IPCC, 2018b). These findings of projections are based among others on studies by Rosenzweig et al. (2018a), Parkes, Defrance, Sultan, Ciais, and Wang (2018), and Faye et al. (2018a). The significance of suppositions about CO_2 fertilization was found to be important by Ren et al. (2018) and Tebaldi and Lobell (2018). Findings accruing from AgMIP-coordinated global and regional assessment results reinforce the fact that at the global scale, there are mixed results of positive and negative changes in simulated wheat and maize yields, with deteriorations in some breadbasket regions, at both 1.5°C and 2.0°C (Rosenzweig et al., 2018a). In concurrence with price changes from the global economics models, productivity diminutions in the Punjab region of Pakistan culminated in an augmentation in vulnerable households and poverty rate (Rosenzweig et al., 2018a).

6.4.2.1 Climate change and smallholder farming systems

Across the world, smallholder farmers worldwide are regarded to be unduly susceptible to climate change owing to alterations in temperature, rainfall and the frequency or intensity of extreme weather events that directly affect their crop and livestock along with their household's food security, income and well-being (Vignola et al., 2015). Peria et al. (2016) have cited the example of smallholder farmers in the Philippines, whose survival and livelihoods are principally dependent on the environment, and that they are persistently confronted with risks and bear the effects of the changing climate. Farming system approaches that have been developed by recent studies have come to regard both biophysical and economic processes affected by climate change and multiple activities. Some scholars (Rosenzweig & Hillel, 2015; Antle et al., 2015) have opined that farm households in the developing world frequently place reliance on a complex mix of crops, livestock, aquaculture, and nonagricultural activities for their livelihoods.

Smallholder farming systems have come to be acknowledged as being highly susceptible to the vagaries of climate change (Morton, 2007) owing to their dependence to a great extent on agriculture and livestock for their livelihood (Dasgupta et al., 2014). Farmers in Zimbabwe were found vulnerable owing to their marginal location, low levels of technology, and lack of other essential farming resources. Vagaries of climate change such as high frequency and severity of drought, excessive precipitation, drying up of rivers, dams and wells, and changes in timing and pattern of seasons, etc., were observed by native farmers as evidence of climate change, and indicated that protracted wet, hot, and dry weather conditions contributed to crop damage, death of livestock, soil erosion, bush fires, poor plant germination, pests, lower incomes, and corrosion of infrastructure (Mutekwa, 2009).

A survey of 600 small farmers conducted by Harvey et al. (2014b) in Madagascar demonstrated that chronic food insecurity, physical isolation and lack of access to formal safety nets increased Malagasy farmers' susceptibility to any shocks to their agricultural system, especially extreme weather events. According to a study undertaken by Paudel, Acharya, Ghimire, Dahal, and Bista (2014) in Chitwan region of Nepal, it was revealed that occurrence of extreme events and increased variability in temperature has increased the vulnerability of crops to biotic and abiotic stresses and altered the timing of agricultural operations; thereby affecting crop production.

Hailstorms, drought and dry spells were considered as climate variables of prime concern that contributed to reduction in crop yields. While evaluating vulnerability of smallholder farmers to weather extremes with regard to food security in the Peruvian Altiplan region, Sietz et al. (2012) found the pertinence of resource scarcity such as livestock, land area; diversification of activities like lack of alternative income sources, education deprivation; and income restrictions in terms of harvest failure risk; and potential role these factors play in shaping vulnerability of smallholder farmers.

Integrated methods, in tandem with the inclusion of regional economic models, take into consideration the likelihood of yield deteriorations to hike prices and thereby affecting livelihoods, up to a certain extent, in some climate change scenarios. Some scholars (Valdivia et al., 2015; Antle et al., 2015) are of the opinion that usage of regional economic models of farming systems can be facilitated to ascertain the potential for shifting to other crops and livestock, along with the role that nonfarm income can play in adaptation. Concomitantly, loss of income for smallholder farmers accruing due to climate change-related deteriorations in areas like coffee production can enhance the prospects of food insecurity for them (Hannah et al., 2017). Regional integrated assessments have been arrived at through the usage of farming system methods developed by AgMIP; for instance: sub-Saharan Africa (Kihara et al., 2015), West Africa (Adiku et al., 2015); East Africa (Rao et al., 2015), South Africa (Beletse et al., 2015), Zimbabwe (Masikati et al., 2015), South

Asia (McDermid et al., 2015), Pakistan (Ahmad et al., 2015), the Indo-Gangetic Basin (Subash et al., 2015), Tamil Nadu in India (Ponnusamy et al., 2015), and Sri Lanka (Zubair et al., 2015). These assessments make it discernible that climate change exerts pressure on smallholder farmers across sub-Saharan Africa and South Asia, with winners and losers within each area assessed. According to these assessments, temperatures are projected to increase in all locations, and diminutions in rainfall patterns are projected for the western part of West Africa and Southern Africa, whereas augmentations in rainfall are anticipated for eastern West Africa and all study regions of South Asia. In addition to these, it is also revealed from these assessments that climate change will cause yield declines in most study regions with the exception of South India and areas in central Kenya, as negative or damaging temperature effects overwhelm the positive effects of CO_2. According to Valdivia et al. (2015), these studies have used AgMIP representative agricultural pathways (RAPs) as a means to involve stakeholders in regional planning and climate resilience. Viewed from a broad perspective, RAPs are coherent with and make-up for the RCP/SSP approaches for use in agricultural model intercomparisons, improvement, and assessments of impact.

In a nutshell, it can be said that smallholder farmers are particularly susceptible to climate change because their livelihoods are often dependent principally on agriculture. Additionally, smallholder farmers frequently bear the brunt of chronic food insecurity. Smallholder farming systems are likely to be affected by vagaries of climate change in terms of exacerbated risks of pests and diseases along with extreme weather events.

6.4.3 Climate change and access to food

Access to food encompasses the ability to obtain food, including the ability to purchase food at affordable prices. Past studies have demonstrated that reduced agricultural productivity will lower down agricultural supply, spurring hike in food prices. With a view to make available an assortment of projections for prices, risk of hunger, and land-use change, Hasegawa et al. (2018) tested a combination of RCPs and SSPs with the help of a protocol-based analysis arrive at after having applied AgMIP methods. Some scholars (Valin et al., 2014; Smith et al., 2014) have opined the irrespective of varied economic models with multiple representations of the global food system along with having represented the SSPs in divergent ways such as technological change, land-use policies, sustainable diets, etc. (Stehfest et al., 2019), the combo or ensemble of participating models anticipated a 1%−29% cereal price hike in 2050 across SSPs 1, 2, and 3 owing to climate change (RCP 6.0), embracing the likelihood of impacting consumers worldwide via higher food prices; also entailing probability of varied regional effects. Nevertheless, it was also revealed that under the given present projections of demand, the median hike in cereal price was 7%.

Decline in food availability triggered by climate change is projected to spur hike in food cost and would impact consumers worldwide via higher prices and reduced purchasing power, embracing the probability of putting at risk the consumers with low-income in the wake of higher food prices (Nelson et al., 2018). Hiked food prices dampen consumer demand and that in turn not only leads to reduced energy (calories) intake worldwide (Hasegawa et al., 2018); nevertheless, also enhancing the probability of less healthy diets with lesser obtainability of key micronutrients (Nelson et al., 2018) along with the resulting surge in diet-related mortality in lower and middle-income countries (Springmann et al., 2018a). Cumulative impact of these changes would be snail-paced progress toward the obliteration of malnutrition in all its forms. The magnitude of the lower energy intake culminating in a heightened risk of hunger diverges by global economic model. Nonetheless, Hasegawa et al. (2018) have opined that all models envision a surge in the risk of hunger, with the median projection of a rise in the population at risk of inadequate energy intake by 6%, 14%, and 12% in 2050 for SSPs 1, 2, and 3, respectively, compared to a no climate change reference scenario. They have further noted that median percentage upsurge would be the equivalent of 8, 24, and 80 million, with full range 1−183 million, additional people at risk of hunger because of climate change.

Some scholars (Smith et al., 2014; Wiebe et al., 2015) are of the opinion that climate change in every likelihood can spur alterations in land use worldwide. Hasegawa and co-authors in their study have surmised that decreasing agricultural productivity largely prods the necessity for additional cropland, with 7 of 8 models projecting increasing cropland and the median increase by 2050 projected across all models of 2% compared to a no climate change reference. There is less likelihood of all regions responding to climate impacts likewise, with more indecisiveness on regional land-use change across the model ensemble than the global totals might suggest.

6.4.4 Climate change and food utilization

According to Barrett (2002), food utilization reflects concerns about whether individuals and households make appropriate use of the food to which they have access and that the nutritional value of food in term of essential micronutrient and vitamins and the ability of the body to metabolize and absorb these nutrients is an essential factor of food security. Food utilization comprises nutrient composition of food, its preparation and overall state of health. The process of food utilization is affected by food safety and quality. Food safety can be affected by climate change via altering the population dynamics of contaminating organisms that are susceptible to fluctuations in temperature and precipitation patterns, including humidity, augmentation in frequency and intensity of extreme weather events as well as changes in contaminant transport pathways. According to Tirado et al. (2010), alterations in food and farming systems, for instance, intensification to maintain supply under climate change, could also augment susceptibilities to the changing climate.

Dilapidation and putrefaction of products in storage and transport can also be impacted by fluctuating levels of humidity and temperature outside of cold chains, especially from microbial decay but also from potential alterations in the population dynamics of stored product pests such as mites, beetles, moths, etc. (Moses et al., 2015). Paterson and Lima (2010) denote that mycotoxin-producing fungi take place in explicit conditions of temperature and humidity and in that eventuality climate change is likely to impact their ambit, augmenting risks in some areas such as mid-temperate latitudes, and reducing them in others areas like that of the tropics. Battilani and colleagues (2016) have cited availability of strong evidence garnered from process-based models of particular species along with projections of future climate that demonstrate that aflatoxin contamination of maize in southern Europe will register significant increase. Besides, deoxynivalenol contamination of wheat in northwest Europe is projected to record increase by up to three times (van der Fels-Klerx et al., 2012b, 2012a). Even as the downscaled climate models make any particular inexact projection for a given geographical region (Van der Fels-Klerx et al., 2013), experimental evidence on the small-scale mulls that the combination of rising CO_2 levels, impacting physiological processes in photosynthetic organisms, and temperature changes, can be expressively greater than temperature alone (Medina et al., 2014). Risks pertaining to aflatoxins are expectedly to undergo alteration; nonetheless, detailed projections are problematic because of their dependence on local conditions (Vaughan et al., 2016).

Enteric contamination serves as the source of origin of foodborne pathogens in the terrestrial environment that can be spread by blowing wind carrying contaminated soil or flooding—the frequency of both of which are expected to rise with climate change (Hellberg & Chu, 2016). Additionally, water stored for irrigation, which may be enhanced in some regions as an adaptation strategy, can serve as a significant route for the dispersal of pathogens along with other pollutants; and contaminated water and diarrheal diseases pose serious threats to food security (Bond et al., 2018). Undeniably, there is little direct evidence in the form of modeled projections; nevertheless, the results of an assortment of reviews along with expert groups, suggest that risks from foodborne pathogens are projected to increase via multiple mechanisms (Hellberg & Chu, 2016).

The changing biology of plants altering human exposure levels can serve as a supplementary route to climate change impacts on human health. This may incorporate as to how crops sequester heavy metals under climate change (Rajkumar et al., 2013), or how they respond to changing pest pressure. According to Paterson and Lima (2011), all of these factors will engender regional differences regarding food safety impacts. This brings to surmise that food utilization through alterations in food safety and probably food access from food loss will be affected by climate change, mostly by increasing risks; however, there is little evidence to demonstrate as to how they may change for any given place.

6.4.4.1 Climate change and food quality

Food quality can be affected by climate change through two main modes. The first mode by which climate change can directly impact food quality is the direct effect of climate on plant and animal biology—by fluctuating temperatures changing he basic metabolism of plants. In the second mode, by enhancing the level of carbon dioxide climate can impact the biology via CO_2 fertilization. Direct impacts of climate change on plant and animal biology is facilitated through climate affecting an assortment of biological processes that, inter alia, include the metabolic rate in plants and ectothermic animals. Alterations in these processes can cause change in the growth rates and thereby yields. However, these alterations can also engender organisms to alter comparable investments in growth versus reproduction, thereby resulting in the alterations in assimilated nutrients. According to DaMatta et al. (2010), this could lead to reduction in protein and mineral nutrient concentration along with alteration in lipid composition. While citing the example of apples in Japan that have been subject to exposure to higher temperatures for the past over three/four decades, Sugiura and colleagues (2013) note that the resultant outcome has been, apart from early blooming, there has been change in acidity, firmness and water content, resulting in diminution in quality. Mira de Orduña (2010) has reported that warming-induced impact in other fruits like grapes, alteration in sugar composition has affected both color and aroma. According to Lara and Rostagno (2013), altering heat stress can impact yield along with meat quality.

Two processes affect the metabolism of plants and fluctuations in temperatures affect growth rates of plants along with wielding impact on yields and their nutritional quality. The first process, known as carbon dioxide fertilization or CO_2 fertilization, comprises use of CO_2 by plants in photosynthesis to form sugar; however, surging levels of CO_2, with all other things being equal, augments the process unless limited by availability of water or nitrogen. In the second process, rising level of CO_2 lets the stomata to be open for a shorter period for gas exchange, thereby plummeting water loss through transpiration. Scholarly studies on these effects comprise *meta*-analyses modeling and small-scale experiments (Franzaring et al., 2013; Mishra and Agrawal 2014; Myers et al., 2017; Ishigooka et al., 2017; Zhu et al., 2018; Loladze 2014; Yu et al., 2014).

Diminutions in protein concentration with raised carbon dioxide are linked to reduced nitrogen concentration probably triggered by nitrogen uptake not in synchronization with biomass growth, an effect that is known as "carbohydrate dilution" or "growth dilution," and by reticence of photorespiration which can make available bulk of the energy used for subsuming nitrate into proteins (Bahrami et al., 2017). Some scholars have posited other mechanisms as well (Feng et al., 2015; Bloom et al., 2014). Medek et al. (2017) inform that together, the effects on protein availability could bring as many as 150 million people into protein deficiency by 2050. Content of zinc and other nutrients in the important food crops gets lowered owing to surge in concentrations of atmospheric carbon dioxide (Myers et al., 2017).

Deficiencies of zinc and iron in human diet has emerged as a significant global public health problem and according to FAO (2013a), approximately 2 billion people suffer these deficiencies worldwide, engendering a loss of 63 million life-years annually (Myers et al., 2017). Dependence of large number of people is on C3 grain legumes as their principal dietary source of zinc and iron. A large number of diseases worldwide are caused owing to zinc deficiency, and the populations that are at highest risk of zinc deficiency receive most of their dietary zinc from crops (Myers et al., 2017). By 2050 about 138 million are projected to be placed at new risk of zinc deficiency. The people expected to be most affected by zinc deficiency reside in Africa and South Asia, with approximately 48 million living in India alone. Myers et al. (2017) have pointed that the differences between cultivars of a single crop suggest that breeding for reduced sensitivity to atmospheric CO_2 concentration could somewhat deal with these new challenges to global health. While elevated CO_2 is expected to be advantageous for crop productivity at lower temperature increases; nonetheless, it is projected to lower nutritional quality in terms of less protein, zinc, and iron.

6.4.5 Climate change and food stability

According to FAO (2006), the notion of food stability entails that to be food secure, "a population, household or individual must have access to adequate food at all times. They should not risk losing access to food as a consequence of sudden shocks (e.g., an economic or climatic crisis) or cyclical events (e.g., seasonal food insecurity). The concept of stability can therefore refer to both the availability and access dimensions of food security." Food stability pertains to people's ability to access and use food in a steady way, devoid of intervening periods of hunger. However, exacerbation in extreme events associated with climate change can hamper food stability. An analysis of PoU conducted by FAO et al. (2018) in 2017 demonstrated that the average of the PoU was 15.4% for all countries subjected to climate extremes. Simultaneously, the PoU was 20% for countries that moreover exhibited high vulnerability of agriculture production/yields to climate variability, or 22.4% for countries with high PoU vulnerability to severe drought.

Severe flooding in 2011–12 was experienced by Bangkok (Thailand) with wide-ranging disruption of the national food supply chains because they were centrally organized in the capital city (Allen et al., 2017). According to the IPCC projections, frequency, duration, and intensity of some extreme events are likely to increase in the coming decades (IPCC, 2018a, 2012). In order to test the effects of some extreme events on food security, Tigchelaar et al. (2018) used global datasets of maize production and climate variability combined with future temperature projections to quantify how yield variability will change in the world's major maize-producing and -exporting countries under 2°C and 4°C of global warming and their study demonstrated that rising instability in global grain trade and international grain prices, affecting especially the about 800 million people living in extreme poverty and who were most vulnerable to spikes in food price. Tesfaye and colleagues (2017) have projected that the range of heat-stressed areas could further expand by up to 12% in 2030 and 21% in 2050 relative to the baseline (1950–2000). In another recent study Lickley and Solomon (2018) have demonstrated that arid regions are expected to dry earlier, more critically and to a wider range than humid regions, especially the regions of sub-Saharan Africa where the population would be more susceptible to extreme weather events.

6.5 Adaptation options

Adaptation of food system to climate change, including increasing extreme events, is usually construed within a framework of autonomous, incremental, and transformational adaptation. The primary focus is accorded to regional and local considerations and adaptation options for both the supply side—comprising production, storage, transport, processing, and trade—and the demand side entailing consumption and diets, of the food system. Besides, agroecological, social, and cultural contexts are taken into consideration throughout. Some of the adverse impacts of climate change on food security can possibly be reduced or even eliminated with the help of framing effective adaptation strategies. Admittedly, persistent continuation of climate change may help adaptation to reach its limits. In the realm of food system, actions undertaken toward adaptation involve any activities initiated with a view to lessen the probability of vulnerability and augment resilience of the system to climate change. Rosenzweig and Hillel (2015) note that extended climate envelopes in some areas could alter agroecological zones, occasioning prospects for expansion toward higher latitudes and altitudes, soil and water resources.

In the wake of projections about more extreme climatic events causing more agrometeorological disasters with attendant economic and social losses, IPCC (2012) has suggested many options for adapting the food system to extreme events, with focus on measures that reduce exposure and vulnerability and enhance resilience, albeit risks cannot fully be eliminated. Harvey and colleagues (2014a) are of the view that adaptation responses to extreme events are often designed to minimize damages, tone down threats, prevent adverse impacts, or share losses, thus making the system more resilient. Given the current and anticipated climate change eventualities, entailing higher temperature, fluctuations in precipitation, flooding and extreme weather events, attainment of adaptation will necessitate both technological and nontechnological solutions. Technological solutions may involve recovering and improving orphan crops, new cultivars from breeding or biotechnology, etc., and nontechnological solutions may entail market, land management, diet change, etc. Climate determines access to food locally by its interaction with factors like market supplies over longer distances and policy drivers (Mbow et al., 2008; Howden et al., 2007), along with local agricultural productivity.

It is often acknowledged that adaptation strategies are closely associated with environmental and cultural contexts at the regional and local levels in view of the site-specific nature of climate change impacts on food system components in tandem with extensive divergence in agroecosystems types and management, and socioeconomic conditions. Development of systemic resilience facilitated through integration of climate drivers with socioeconomic drivers would decrease the impact on food security, especially in developing countries. While citing the example of Africa where improvement in food security needs evolving food systems to be more climate resilient, Mbow and colleagues (2014b) note this is to be carried out while supporting the need for augmenting the yield to feed the growing population. Godfray and Garnett (2014) are of the view that adaptation entails increasing production of food where required, modernizing demand along with waste reduction and improvement in governance.

Autonomous adaptation, also referred to as spontaneous adaptation (IPCC, 2007), in food systems is often spurred by alterations in agroecosystems, markets, or welfare changes; and it seldom comprises a conscious response to climatic stimuli. IRD (2017) has documented the instances of autonomous adaptation of rural populations in the Sahel region. According to Tripathi and Mishra (2017), while changing sowing and harvesting timing, farmers in India are also cultivating short duration varieties, intercropping, changing cropping patterns, investing in irrigation, and establishing agroforestry. While terming these measures as passive responses or autonomous adaptation, the authors further opine that the farming community does not acknowledge that these steps taken in response to perceived climatic changes.

Park et al. (2012) perceive incremental adaptation in terms of maintaining the crux and integrity of a system or process at a given scale. Incremental adaptation focuses on Improvements to existing resources and management practices are facilitated through incremental adaptation that nurses the central objective of maintaining the essence and integrity of a system or process at a given scale (IPCC, 2014a).

Transformational adaptation is perceived by Termeer, Dewulf, and Biesbroek (2017) as a process that endeavors to seek alternative livelihoods and land-use strategies required to evolve new farming systems.

Transformational adaptation modifies the vital features of a socio-ecological system either in anticipation of or in response to climate change and its impacts (IPCC, 2014a). Citing the example of limitations in incremental adaptation among smallholder rice farmers in Northwest Costa Rica, Warner and colleagues (2015) denote that the farmers shifted from cultivating rice to sugarcane production owing to diminishing market access and water scarcity. Hadarits et al. (2017) have described migration from the Oldman River Basin as a transformational adaption to climate change in the Canadian agriculture sector. The food security of farmers and consumers would be dependent on the manner of managing the transformational change in the food systems in the eventuality of high-end scenarios of climate change. Some scholars (Mockshell & Kamanda 2018; Pant et al., 2015) suggest that an integrated framework of adaptive transition, comprising management of socio-technical transitions and adaptation to socio-ecological changes, could help build transformational adaptive capacity. Necessity of overlapping phases of adaptation required to support transformational change in Africa has been suggested by Rippke et al. (2016) (Table 6.3).

Almost all pillars of food security, especially stability, are affected by risks emanating from climate change because extreme events engender strong variation to food access. Efficient management of climate-induced risks to food systems requires a wide-ranging and dynamic policy approach encompassing an assortment of drivers and scales because cooccurrence of many risks or their reinforcing each other can limit effective adaptation planning. Howden et al. (2007) are of the

Table 6.3 Synthesis of food security related adaptation options to address various climate risks.

Key climate drivers and risks	Incremental adaptation	Transformational adaptation	Enabling conditions
Extreme events and short-term climate variability Stress on water resources, drought stress, dry spells, heat extremes, flooding, shorter rainy seasons, pests	Change in variety, water management, water harvesting, supplemental irrigation during dry spells, Planting dates, pest control, feed banks, Transhumance, Other sources of revenue (e.g., charcoal, wild fruits, wood, temporary work) Soil management, composting,	Early Warning Systems Use of planning and prediction at seasonal to intra-seasonal climate risk to transition to a food safer condition. Abandonment of monoculture, diversification Crop and livestock insurance Alternate cropping, Intercropping Erosion control	Establishment of climate services Integrated water management policies, integrated land and water governance Seed banks, seed sovereignty and seed distribution policies Capacity building and extension programs
Warming trend, drying trend Reduced crop productivity due to persistent heat, long drought cycles, deforestation and land degradation with strong adverse effects on food production and nutrition quality, increased pest and disease damage	Strategies to reduce effects of recurring food challenges Sustainable intensification, agroforestry, conservation agriculture, SLM Adoption of existing drought-tolerant crop and livestock species Counter season crop production, Livestock fattening New ecosystem-based Adaptation (e.g., bee keeping, woodlots) Farmers management of natural resources Labor redistribution (e.g., mining, development projects, urban migration) Adjustments to markets and trade pathways already in place	Climate services for new agricultural programs, (e.g., sustainable irrigation districts) New technology, e.g., new farming systems, new crops and livestock breeds Switches between cropping and transhumant livelihoods, replacement of pasture or forest to irrigated/rainfed crops Shifting to small ruminants or drought resistant livestock or fish farming Food storage infrastructures, food transformation Changes in cropping area, land rehabilitation (enclosures, afforestation) perennial farming New markets and trade pathways	Climate information in local development policies. Stallholders' access to credit and production resources, National food security program based on increased productivity, diversification, transformation and trade Strengthening (budget, capacities, expertise) of local and national institutions to support agriculture and livestock breeding Devolution to local communities, women empowerment, market opportunities Incentives for establishing new markets and trade pathways

From IPCC (Intergovernmental Panel on Climate Change). (2014b). Climate change 2014: Synthesis report. Contribution of working groups I, II and III to the fifth assessment report of the Intergovernmental Panel on Climate Change. *Cambridge: Cambridge University Press; Vermeulen, S.J., Challinor, A.J., Thornton, P.K., Campbell, B.M., Eriyagama, N., Vervoort, J. ... Smith, D.R. (2013). Addressing uncertainty in adaptation planning for agriculture.* Proceedings of National Academy of Sciences U.S.A., 110, 8357–8362; Vermeulen, S.J., Dinesh, D., Howden, S.M., Cramer, C., & Thornton, P.K. (2018). Transformation in practice: A review of empirical cases of transformational adaptation in agriculture under climate change. Frontiers in Sustainable Food System, 2, 65; Burnham, M., & Ma, Z. (2016). Linking smallholder farmer climate change adaptation decisions to development. Climate and Development, 8, 289–311; Bhatta, G.D., & Aggarwal, P.K. (2015). Coping with weather adversity and adaptation to climatic variability: A cross-country study of smallholder farmers in South Asia. Climate and Development. 6 (2): 145–157.

view that facilitation of response strategies would require more than systemic reviews of risk factors ranging from the understanding by farmers of change in risk profiles to the establishment of efficient markets. Recent years have witnessed integration of CCA and disaster risk reduction (DRR) coming into the reckoning and many developing countries have embarked on CCA integrating with DRR approach. Nalau and colleagues (2016) maintain that integration of CCA with DRR is helpful in minimizing the overlap duplication of projects and programs. The Philippines is reported to have introduced legislation calling for CCA and DRR integration (Leon & Pittock, 2016).

Undoubtedly, adaptation and mitigation measures are widely regarded as measures devised to minimize the risk from climate change impacts in food systems; nevertheless, these measures can also be causes of risk themselves, especially in terms of investment risk or political risk (IPCC, 2014b). Niles et al. (2018) have expressed the opinion that adoption of agroecological practices could provide resilience for future shocks, extend farmer risk and mitigate the impact of droughts.

References

Adiku, S. G. K., MacCarthy, D. S., Hathie, I., Diancounma, M., Fredua, B. S., Amikuzono, J., ... Diarra, D. J. (2015). *Climate change impacts on West African agriculture: An integrated regional assessment (CIWARA)*. Handbook of climate change and agroecosystems: The agricultural model intercomparison and improvement project integrated crop and economic assessments, Part 2 (pp. 25–73). Singapore: World Scientific Publishing.

Ahmad, A., Ashfaq, M., Rasul, G., Wajid, S. A., Khaliq, T., Rasul, F., ... Valdivia, N. A. (2015). *Impact of climate change on the rice–wheat cropping system of Pakistan*. Handbook of climate change and agroecosystems: The agricultural model intercomparison and improvement project integrated crop and economic assessments, Part 2 (pp. 219–258). Singapore: World Scientific Publishing.

Alpen Capital. *The GCC food industry*. (2011). <http://www.alpencapital.com/downloads/GCC_Food_Industry_Report_June_2011.pdf>.

Antle, J. M., Valdvia, R. O., Boote, K., Janssen, S., Jones, J. W., Porter, C. H., ... Thorburn, P. J. (2015). *AgMIP's transdisciplinary agricultural systems approach to regional integrated assessment of climate impacts, vulnerability, and adaptation*. Handbook of climate change and agroecosystems: The agricultural model intercomparison and 18 improvement project integrated crop and economic assessments, part 1. In *ICP series on climate change impact, adaptation, and mitigation* (3, pp. 27–44). Singapore: World Scientific Publishing.

Asseng, S., Ewert, F., Martre, P., Rotter, R. P., Lobell, D. B., Cammarano, D. B., ... Zhu, Y. (2015). Rising temperatures reduce global wheat production. *Nature Climate Change*, 5, 143–147.

Auffhammer, M., & Schlenker, W. (2014). Empirical studies on agricultural impacts and adaptation. *Energy Economics.*, 46, 555–561.

Backlund, P., Janetos, A., Schimel. D. (2008). The effects of climate change on agriculture, land resources, water resources, and biodiversity in the United States. Washington, DC: United States Climate Change Science Program. <https://climatechange.lta.org/wp-content/uploads/cct/2015/03/CCSPFinalReport.pdf>.

Bailey, R., Benton, T. G., Challinor, A., Elliott, J., Gustafson, D., Hiller, B., ... Weubbles, D. J. (2015). Extreme weather and resilience of the global food system. In: *Final project report from the UK-United States Taskforce on Extreme Weather and Global Food System Resilience*. United Kingdom. <https://www.stat.berkeley.edu/~aldous/157/Papers/extreme_weather_resilience.pdf>.

Banerjee, A., & Duflo, E. (2011). *Poor economics*. New York: Public Affairs.

Barling, D., Sharpe, R., & Lang, T. (2008). *Towards a national sustainable food security policy*. London: City University.

Barrett, C. (2002). Food security and food assistance programs. In B. Gardner, & G. Rausser (Eds.), *Handbook of agricultural economics* (2, pp. 2104–2177). Elsevier BV.

Beletse, Y. G., et al. (2015). *Projected impacts of climate change scenarios on the production of maize in southern Africa: An integrated assessment case study of the Bethlehem District, Central Free State, South Africa. In* Handbook of climate change and agroecosystems: The agricultural model intercomparison and improvement project integrated crop and economic assessments. Part 2 (pp. 125–157). Singapore: World Scientific Publishing.

Biesalaski, H. K. (2013). *Hidden hunger*. Springer-Verlag.

Bourdieu, P. (2010). *Sociology is a martial art*. New York: New Press.

Von Braun, J. (1999). Food security—A conceptual basis. In U. Kracht, & M. Schultz (Eds.), *Food Security and Nutrition: The Global Challenge* (pp. 41–54). New York: St. Martin's Press.

Carolan, M. (2012). *The sociology of food*. New York: Routledge.

Cepete, M. (1984). The fight against hunger: Its history on the international agenda. *Food Policy., November*, 282–290.

CFS (Committee on World Food Security). (2012). *Coming to terms with terminology. 39th Session*. Rome: FAO.

Chuwah, C., van Noije, T., van Vuuren, D. P., Stehfest, E., & Hazeleger, W. (2015). Global impacts of surface ozone changes on crop yields and land use. *Atmospheric Environment, 106*, 11–23.

Clay, E. (2002). *Food security: Concepts and measurement*. Rome: FAO.

Dasgupta, P., Morton, J., Dodman, D., Karapinar, B., Meza, F., Rivera-Ferre, M., ... Vincent, K. (2014). *Rural areas. Climate change 2014: Impacts, adaptation, and vulnerability. Part A: Global and sectoral aspects. Contribution of working group II to the fifth assessment report of the Intergovernmental Panel on Climate Change* (pp. 613–657). Cambridge: Cambridge University Press.

Deaton, A., & Drèze, J. (2009). Food and nutrition in India: Facts and interpretations. *Economic and Political Weekly, XLIV*(7), 42–65.

Deryng, D., Conway, D., Ramankutty, N., Price, J., & Warren, R. (2014). Global crop yield response to extreme heat stress under multiple climate change futures. *Environmental. Research Letters., 9*, 34011.

Diamond, J. (1997). *Guns, germs and steel. A short history of everybody for the last 13,000 years*. New York: W. W. Norton & Company.

Drèze, J., & Sen, A. (1989). *Hunger and public action*. Oxford: Oxford University Press.

Duncan, J. (2014). *The reformed Committee on World Food Security and the global governance of food security* (Ph.D. thesis). London: City University.

Easterly, W. (2013). *The tyranny of experts*. New York: Basic Books.

Edkins, J. (2008). *Whose hunger? Concepts of famine, practices of aid*. Minneapolis, MN: University of Minnesota Press.

Emberson, L. D., et al. (2018). Ozone effects on crops and consideration in crop models. *European Journal of Agronomy, 6*, 002.

Erni, C. (Ed.), (2015). Shifting cultivation, livelihood and food security: New and old challenges for indigenous peoples in Asia. In: *Shifting cultivation, livelihood and food security* (pp. 3–41). Bangkok: FAO/IWGIA/AIPP.

Escobar, A. (1995). *Encountering development: The making and unmaking of the third world*. Princeton, NJ: Princeton University Press.

FAO. (2008). *Food security in mountains—High time for action*. Rome, Italy. Available Online.

FAO. (2011). Mountain Partnership, UNCCD, SDC, and CDE *Highlands and Drylands Mountains, a source of resilience in arid regions*. Rome: FAO Available online.

FAO. (2013). (Food and Agriculture Organization) *The state of food and agriculture*. Rome: FAO.
FAO. (2017). (Food and Agriculture Organization) *The state of food security and nutrition in the world 2017. Building resilience for peace and food security*. Rome: FAO.
FAO. (2018a). IFAD, UNICEF, WFP and WHO *The state of food security and nutrition in the world 2018: Building climate resilience for food security and nutrition*. Rome: FAO.
FAO. (2018b). (Food and Agriculture Organization) *The future of food and agriculture: Trends and challenges*. Rome: FAO.
FAO. (2019). (Food and Agriculture Organization) *The state of food security and nutrition in the world 2019: Safeguarding against economic slowdowns and downturns*. Rome: FAO.
FAO. (2019a). (Food and Agriculture Organization) *Food outlook: Biannual report on global food markets*. Rome: FAO.
Farsund, A. A., Daugbjerg, C., & Langhelle, O. (2015). Food security and trade: Reconciling discourses in the Food and Agriculture Organization and the World Trade Organization. *Food Security, 7*, 383–391.
Faye, B., et al. (2018a). Impacts of 1.5 vs 2.0 °C on cereal yields in the West African Sudan Savanna. *Environmental Research Letters, 13*, 034014.
Faye, B., Webber, H., Diop, M., Mbaye, M. L., Owusu-Sekyere, J. D., Naab, J. B., & Gaiser, T. (2018b). Potential impact of climate change on peanut yield in Senegal, West Africa. *Field Crops Research, 219*, 148–159.
Forbes. (2007, 15 November). The world's biggest industry. *Forbes*. <https://www.forbes.com/2007/11/11/growth-agriculture-business-forbeslife-food07-cx_sm_1113bigfood.html?sh=25008775373e>.
Fraser, E. D. G., & Rimas, A. (2011). *Empires of food. Feast, famine and the rise and fall of civilizations*. London: Arrow Books.
FSIN. (2018). (Food Security Information Network) *Global report on food crises 2018*. Rome: World Food Programme.
FSIN. (2019). (Food Security Information Network) *Global report on food crises 2019*. Rome: World Food Programme.
Fukuda-Parr, S., & Orr, A. (2013). *The MDG Hunger Target and the Contested Visions of Food Security*. The power of numbers: A critical review of MDG targets for human development and human rights. *Working Paper, Harvard School of Public Health & the New School*. Cambridge: Harvard.
Galtung, I. (2004). The human right to food and nutrition and the story of 840 million court cases. In: *Peace human rights, 1/2004* (pp. 111–123). <https://www.semanticscholar.org/paper/The-Human-Right-to-Food-and-Nutrition-and-the-Story-Galtung/0f59030267c4ad53f7d6c1945f9142c91c6f7101>.
GHI. (2014). (Global Hunger Index) *Global hunger index: The challenge of hidden hunger*. Bonn: International Food Policy Research Institute (IFPRI), Welt Hunger Hilfe, Concern Worldwide.
GHI. (2018). (Global Hunger Index) *Global hunger index: Forced migration and hunger*. Dublin: Welt Hunger Hilfe, Concern Worldwide.
Gupta, R., Somanathan, E., & Dey, S. (2017). Global warming and local air pollution have reduced wheat yields in India. *Climate Change, 140*, 593–604.
Hannah, L., Donatti, C. I., Harvey, C. A., Alfaro, E., Rodriguez, D. A., Bourocle, C., ... Solano, A. L. (2017). Regional modeling of climate change impacts on smallholder agriculture and ecosystems in Central America. *Climate Change, 141*, 29–45.
Harvey, C. A., Rakotobe, Z. L., Rao, N. S., Dave, R., Razafimahatratra, H., Rabarijohn, R. H., ... Mackinnon, J. L. (2014b). Extreme vulnerability of smallholder farmers to 28 agricultural risks and climate change in Madagascar. *Philosophical Transactions of the Royal Society of London. B. Biology. Sciences, 369*, 20130089.
Hussain, A., Rasul, G., Mahapatra, B., & Tuladhar, S. (2016). Household food security in the face of climate change in the Hindu-Kush Himalayan region. *Food Security, 8*, 921–937.
IFPRI. (2011). (International Food Policy Research Institute) *Global hunger index*. Washington, DC: IFPRI.
Iizumi, T., & Ramankutty, N. (2016). Changes in yield variability of major crops for 1981–2010 explained by climate change. *Environment Research Letter, 11*, 34003.

IMAP. *Food and beverage industry global report 2010. IMAP Consumers staple report.* (2011). Available online.
Innes, P. J., Tan, D. K. Y., Van Ogtrop, F., & Amthor, J. S. (2015). Effects of high-temperature episodes on wheat yields in New South Wales. *Australia. Agriculture and Forest Meteorology, 208,* 95–107.
IPCC. (2014a). (Intergovernmental Panel on Climate Change) *Climate change 2014: Mitigation of climate change: Working group III contribution to the IPCC fifth assessment report of the Intergovernmental Panel on Climate Change.* Cambridge: Cambridge University Press.
IPCC. (2014b). (Intergovernmental Panel on Climate Change) *Climate change 2014: Synthesis report. Contribution of working groups I, II and III to the fifth assessment report of the Intergovernmental Panel on Climate Change.* Cambridge: Cambridge University Press.
IPCC. (2014c). (Intergovernmental Panel on Climate Change) *Climate change 2014: Impacts, adaptation, and vulnerability: Working group II contribution to the fifth assessment report of the Intergovernmental Panel on Climate Change.* Cambridge: Cambridge University Press.
IPCC (Intergovernmental Panel on Climate Change). *Global warming of 1.5°C: An IPCC special report on the impacts of global warming of 1.5°C above pre-industrial levels and related global greenhouse gas emission pathways, in the context of strengthening the global response to the threat of climate change.* (2018a). <http://www.ipcc.ch/report/sr15/>.
Jha, B., & Tripathi, A. (2017). How susceptible is India's food basket to climate change? *Social Change, 47,* 11–27.
Kihara, J., MacCarthy, D. S., Bationo, A., Koala, S., Hickman, J., Koo, C., ... Jones, J. W. (2015). *Perspectives on climate effects on agriculture: The international efforts of AgMIP in Sub-Saharan Africa. In* Handbook of climate change and agroecosystems: The agricultural model intercomparison and improvement project integrated crop and economic assessments, part 2 (pp. 3–23). Singapore: World Scientific Publishing.
Kracht, U. (1999). Hunger, malnutrition and poverty: Trends and prospects towards the 21st century. In U. Kracht, & M. Schultz (Eds.), *Food security and nutrition: The global challenge* (pp. 55–74). New York: St. Martin's Press.
Lappé, F., Collins, J., & Rosset, P. (1998). *World hunger: 12 Myths* (Second Edition). New York: Grove Press.
Levis, S., Badger, A., Drewniak, B., Nevison, C., & Ren, X. (2018). CLM crop yields and water requirements: Avoided impacts by choosing RCP 4.5 over 8.5. *Climate Change, 146,* 501–515.
Masikati, P., Tui, H.-K., Deschemaeker, S., Crespo, K., Lennard, S., Claessens, C. J., ... Valdivia, R. O. (2015). *Crop-livestock intensification in the face of climate change: exploring opportunities to reduce risk and increase resilience in southern Africa by using an integrated multi-modeling approach.* Handbook of climate change and agroecosystems: The agricultural model intercomparison and improvement project integrated crop and economic assessments, part 2 (pp. 159–198). Singapore: World Scientific Publishing.
Maslow, A. (1943). A theory of human motivation. *Psychological Review, 50*(4), 370–396.
Mavromatis, T. (2015). Crop–climate relationships of cereals in Greece and the impacts of recent climate trends. *Theoretical and Applied Climatology, 120,* 417–432.
Maxwell, S. (1996). 'Food security: A postmodern perspective'. *Food Policy, 21*(2), 155–170.
McDermid, S. P., Dileepkumar, G., Dakshinamurthy, K. M., Nedumaran, S., Singh, P., Srinivasa, C., ... Nissanka, S. P. (2015). *Integrated assessments of the impact of climate change on agriculture: An overview of AgMIP regional research in South Asia.* Handbook of climate change and agroecosystems: The agricultural model intercomparison and improvement project integrated crop and economic assessments, part 2 (pp. 201–217). Singapore: World Scientific Publishing.
McKeon, N. (2009). *The United Nations and civil society.* London: Zed Books.
Mercy Corps. (2018). *Quick facts: What you need to know about global hunger.* Portland, OR. <https://www.mercycorps.org/blog/quick-facts-global-hunger>.
Montanori, M. (2006). *Food is culture. Arts and traditions on the table.* New York: Columbia University Press.
Morton, J. F. (2007). The impact of climate change on smallholder and subsistence agriculture. *Proceedings of the National Academy of Sciences of the United States of America, 104,* 19680–19685.

Mutekwa, V. T. (2009). Climate change impacts and adaptation in the agricultural sector: The case of smallholder farmers in Zimbabwe. *Journal of Sustainable Development in Africa, 11*(2), 237−256.

Myers, S. S., Smith, M. R., Guth, S., Golden, C. D., Vaitla, B., Mueller, N. D., ... Huybers, P. (2017). Climate change and global food systems: Potential impacts on food security and undernutrition. *Annual Review of Public Health, 38*, 259−277.

Nellemann, C., MacDevette, M., Manders, T., Eickhout, B., Svihus, B., Prins, A. G., & Kaltenborn, B. P. (2009). *The environmental food crisis: The environment's role in averting future food crises: a UNEP rapid response assessment*. UNEP/Earthprint.

Nussbaum, M. (2011). *Creating capabilities*. Cambridge: Harvard University Press.

Oxfam. (2017). *There is enough food to feed the world*. Oxfam Poster Canada, 5 September. Available online.

Oxford English Dictionary. (2010). Oxford Dictionary of English (Third Edition). Oxford, UK: Oxford Univrsity Press. Available from http://www.oed.com/.

Parkes, B., Defrance, D., Sultan, B., Ciais, P., & Wang, X. (2018). Projected changes in crop yield mean and variability over West Africa in a world 1.5 K warmer than the pre-industrial. *Earth System Dynamics, 1*, 119−134.

Patel, K., Gartaula, H., Johnson, D., & Karthikeyan, M. (2015). The interplay between household food security and well-being among small-scale farmers in the context of rapid agrarian change in India. *Agriculture and Food Security, 4*(16), 1−16.

Paudel, B., Acharya, B. S., Ghimire, R., Dahal, K. R., & Bista, P. (2014). Adapting agriculture to climate change and variability in Chitwan: Long-term trends and farmers' perceptions. *Agricultural Research, 3*(2), 165−174.

Paulino, L., & Mellor, J. (1984). The food situation in developing countries. *Food Policy, November*, 291−303.

Peria, A. S., Pulhin, J. M., Tapia, M. A., Predo, C. D., Peras, R. J. J., Evangelista, R. J. P., ... Pulhin, F. B. (2016). Knowledge, risk attitudes and perceptions on extreme weather events of smallholder farmers in Ligao City, Albay, Bicol, Philippines. *Journal of Environmental Science and Management*, ICRAF Special Issue.

Popp, A., Calvin, K., Fujimori, S., Havlik, P., Humpenoder, F., Stefest, E., ... Dietrich, J. P. (2017). Land-use futures in the shared socio-economic pathways. *Global Environmental Change, 42*, 331−345.

Porter, J. R., Xie, L., Challinor, A. J., Cochrane, K., Howden, S. M., Iqbal, M. M., ... Travasso, M. I. (2014). *Food security and food production systems. Climate change 2014 Impacts, adapt. Vulnerability. Part A Global and sectoral aspects. Contribution of working group II to fifth assessment report to the Intergovernmental Panel on Climate Change* (pp. 485−533). Cambridge: Cambridge University Press.

Porter, J. R., Howden, L. M., & Smith, P. (2017). *Considering agriculture in IPCC assessments. Nature Climate Change, 7*, 680−683.

Pritchett, L., Wollcock, M., & Andrews, M. (2010). *Capability Traps? The Mechanisms of Persistent Implementation Failure. Working Paper #234*. Washington, DC: Centre for Global Development.

Rao, K. P. C., Sridhar, G., Mulwa, R. M., Kilavi, M. N., Esilaba, A., Athanasiadis, I. N., & Valdivia, R. O. (2015). *Impacts of climate variability and change on agricultural systems in East Africa*. Handbook of climate change and agroecosystems: The agricultural model inter-comparison and improvement project integrated crop and economic assessments, part 2 (p. 75) World Scientific, −21.

Rasul, G., Saboor, A., Tiwari, P. C., Hussain, A., Ghosh, N., & Chettri, G. B. (2019). Food and nutritional security in the Hindu Kush Himalaya: Unique challenges and niche opportunities. In P. Wester, A. Mishra, A. Mukherji, & A. B. Shrestha (Eds.), *The Hindu Kush Himalaya Assessment: Mountains, climate change, sustainability and people* (pp. 301−338). Dordrecht: Springer.

Revi, A., Satterthwaite, D. E., Aragón-Durand, F., Corfee-Morlot, J., Kiunsi, R. B. R., Pelling, M., ... Solecki, W. (2014). *Urban areas. Climate change 2014: Impacts, adaptation, and vulnerability. Part A: Global and sectoral aspects. Contribution of working group II to the fifth assessment report of the Intergovernmental Panel on Climate Change* (pp. 535−612). Cambridge: Cambridge University Press.

Reyer, C. P. O., Adams, S., Albrecht, T., Baarsch, F., Boit, A., Trujillo, N. C., ... Thonike, K. (2017a). Climate change impacts in Latin America and the Caribbean and their implications for development. *Regional Environmental Change, 17*, 1601−1621.

Reyer, C. P. O., Kanta., Rigaud, K., Fernandes, E., Hare, W., Serdeczny, O., & Schellnhuber, H. J. (2017b). Turn down the heat: Regional climate change impacts on development. *Regional Environmental Change, 17*, 1563−1568.

Riahi, K., van Vuuren, D. P., Kriegler, E., Edmonds, J., O'Neill, B. C., Fujimori, S., ... Tavoni, M. (2017). The shared socioeconomic pathways and their energy, land use, and greenhouse gas emissions implications: An overview. *Global Environmental Change, 42*, 153−168.

Richards, M., Bruun, T. B., Campbell, B., Gregersen, L. E., Huyer, S., Kuntze, V., ... Vasileiou, I. (2016). How countries plan to address agricultural adaptation and mitigation. *An analysis of intended nationally determined contributions.* <https://cgspace.cgiar.org/bitstream/handle/10568/69115/CCAFS%20INDC%20info%20note-Final.pdf?sequence=3&isAllowed=y>.

Richards, M. B., Wollenberg, E., & van Vuuren, D. (2018). National contributions to climate change mitigation from agriculture: Allocating a global target. *Climate Policy, 18*, 1271−1285.

Rogelj, J., Shindell, D., Jiang, K., Fifita, S., Forster, P., Ginzburg, V., ... Vilariño, M. V. (2018). Mitigation pathways compatible with 1.5°C in the context of sustainable development. *Global warming of 1.5°C. An IPCC special report on the impacts of global warming of 1.5°C above pre-industrial levels and related global greenhouse gas emission pathways, in the context of strengthening the global response to the threat of climate change, sustainable development, and efforts to eradicate poverty.* <https://www.ipcc.ch/site/assets/uploads/sites/2/2019/02/SR15_Chapter2_Low_Res.pdf>.

Rosenzweig, C., & Coauthors. (2014). Assessing agricultural risks of climate change in the 21st century in a global gridded crop model intercomparison. *Proceedings of the National Academy of Sciences of the United States of America, 111*, 3268−3273.

Rosenzweig, C., Ruane, A. C., Antle, J., Elliott, J., Ashfaq, M., Chatta, A. A., ... Wiebe, K. (2018a). Coordinating AgMIP data and models across global and regional scales 39 for 1.5 °C and 2.0 °C assessments. *Philosophical Transactions of the Royal Society of London, 376*(2119), 20160455.

Rosenzweig, C., & Hillel, D. (2015). *Handbook of climate change and agroecosystems: The agricultural model intercomparison and improvement project (AgMIP) integrated crop and economic assessments.* London: Imperial College Press.

Rosenzweig, C., & Solecki, W. (2018). Action pathways for transforming cities. *Nature Climate Change, 8*(9), 756−759.

Schmidtko, S., Stramma, L., & Visbeck, M. (2017). Decline in global oceanic oxygen content during the past five decades. *Nature, 542*, 335−339.

Sen, A. (1981). *Poverty and famines, an essay on entitlement and deprivation.* Oxford: Clarendon Press.

Shah, M., Fischer, G., & Van Velthuizen, H. (2008). *Food security and sustainable agriculture: The challenges of climate change in Sub-Saharan Africa.* Nairobi: African Economic Research Consortium.

Sharma, R. (2011). Food sovereignty, hunger and global trade rules. African technology development. *The Journal, 8*(1/2), 10−17.

Shaw, D. J. (2007). *World food security: A history since 1945.* New York: Palgrave Macmillan.

Shindell, D., Johan, C. I., Vignati, E., van Digenen, R., Amann, M., Klimont, Z., ... Fowler, D. (2012). Simultaneously mitigating near-term climate change and improving human health and food security. *Science (New York, N.Y.), 335*, 183−189.

Shindell, D. T. (2016). Crop yield changes induced by emissions of individual climate-altering pollutants. *Earth's Future, 4*, 373−380.

Shindell, D. T., Fuglestvedt, J. S., & Collins, W. J. (2017). The social cost of methane: theory and applications. *Faraday Discussions, 200*, 429−451.

Smith, M., Pointing, J., & Maxwell, S. (1992). *Household food security, concepts and definitions: An annotated bibliography, development bibliography No. 8.* Brighton: Institute of Development Studies, University of Sussex.

Smith, P., Bustamante, Ahammad, H., Clark, H., Dong, H., Elissdig, E. A., ... Tubiello, F. (2014). Agriculture, forestry and other land use (AFOLU). *Climate change 2014: Mitigation of climate change.*

References

Contribution of working group III to the fifth assessment report of the Intergovernmental Panel on Climate Change. Cambridge: Cambridge University Press.

Springmann, M., Clark, M., Mason-D'croz, D., Wiebe, K., Leon Bodirsky, B., Lassaletta, L., ... Willett, W. (2018a). *Options for keeping the food system within environmental limits, . Nature* (562, pp. 519–525).

Stavenhagen, R. (2003). Needs, rights and social development. *Overarching Concerns Programme Paper Number 3*. Geneva: UNRISD.

Sultan, B., & Gaetani, M. (2016). Agriculture in West Africa in the twenty-first century: climate change and impacts scenarios, and potential for adaptation. *Frontiers of Plant Sciences, 7*, 1–20.

Swinburn, B. A., Kraak, V. I., Allender, S., Atkins, J. A., Baker, P. I., Bogard, J. R., ... Dietz, W. H. (2019). The global syndemic of obesity, undernutrition, and climate change: The Lancet commission report. *Lancet., 393*, 791–846.

Tebaldi, C., & Lobell, D. (2018). Estimated impacts of emission reductions on wheat and maize crops. *Climate Change, 146*, 533–545.

Termeer, C. J. A. M., Dewulf, A., & Biesbroek, G. R. (2017). Transformational change: governance interventions for climate change adaptation from a continuous change perspective. *Journal of environmental planning and management, 60*(4), 558–576.

The Hunger Project. (2017). *Know your world: Facts about hunger and poverty*. <https://thp.org/the-latest/know-your-world/>.

The Hunger Project. (2018). *2018 World hunger and poverty facts and statistics*. <https://www.worldhunger.org/world-hunger-and-poverty-facts-and-statistics/>.

The Rio Declaration. (1992). *The Rio declaration on environment and development*. <http://www.unesco.org/education/information/nfsunesco/pdf/RIO_E.PDF>.

Ranganathan, J., Waite, R., Searchinger, T., & Hanson, C. (2018). *How to sustainably feed 10 billion people by 2050, in 21 charts*. WRI. <https://www.wri.org/insights/how-sustainably-feed-10-billion-people-2050-21-charts>. Accessed 5 December 2020.

UNCCD (United Nations Convention to Combat Desertification), (2017). *Global land outlook*. Bonn. <https://www.unccd.int/sites/default/files/documents/2017-09/GLO_Full_Report_low_res.pdf>.

UN-DESA. (2019). (United Nations-Department of Economic and Social Affairs) *World population prospects 2019: Highlights*. New York: United Nations.

UNEP. (2011). (United Nations Environment Program) *Global drylands: A UN system-wide response*. Nairobi: UNEP.

UNFCCC (United Nations Framework Convention on Climate Change). *Paris Agreement*. (2015). <https://unfccc.int/sites/default/files/english_paris_agreement.pdf>.

United Nations. (1948). Universal declaration of human rights. In: *General Assembly Resolution 217 A (III). UN Doc. A/810, at 71*.

United Nations. (1966). International covenant on economic, social and cultural rights. In: *Adopted on 16 December 1966, General Assembly Resolution 2200(XXII), UN. GAOR, 21st sess., Supp. No. 16, United States Doc. A/6316 (1966), 993 UNTS 3*.

USDA (United States Department of Agriculture). (2009). *Global food markets: Global food industry structure*. United States Department of Agriculture Economic Research Service. <https://www.ers.usda.gov/topics/food-choices-health/food-consumption-demand/food-demand-analysis/>.

United States Department of State. *Briefing on announcement of new measures to address the drought in Ethiopia*. (2016). <https://2009-2017.state.gov/r/pa/prs/ps/2016/03/253954.htm>.

Vignola, R., Harvey, C. A., Bautista-Solis, P., Avelino, J., Rapidel, B., Donatti, C., & Martinez, R. (2015). Ecosystem-based adaptation for smallholder farmers: Definitions, opportunities and constraints. *Agriculture, Ecosystems and Environment, 211*, 126–132.

Vivero-Pol, J. L. (2013). *Food as a Commons: Reframing the Narrative of the Food Systems*. Available online.

Waha, K., Amann, M., Ayeb-Karlsson, S., Adams, S., Aich, V., Barrsch, S., ... Schleussner, C.-F. (2017). Climate change impacts in the Middle East and Northern Africa (MENA) region and their implications for vulnerable population groups. *Regional Environmental Change, 17,* 1623−1638.

Watts, N., Amann, M., Ayeb-Karlsson, S., Balesova, K., Bouley, T., Boykoff, M., ... Costello, A. (2018). The Lancet countdown on health and climate change: From 25 years of inaction to a global transformation for public health. *Lancet, 391,* 581−630.

Willett, W., Rockstrom, J., Loken, B., Springmann, M., lang, T., Vermuelen, S., ... Murray, C. J. R. (2019). Food in the Anthropocene: The EAT-Lancet Commission on healthy diets from sustainable food systems. *Lancet, 393,* 447−492.

Windfuhr, M., & Jonsén, J. (2005). *Food sovereignty towards democracy.* Warwickshire: ITDG Publishing.

Wise, T. A. & Murphy, S. (2012). *Resolving the food crisis: assessing global policy reforms since 2007.* Medford, MA: Global Development and Environment Institute and the Institute for Agriculture and Trade Policy. Available online.

World Bank. (1986). *Poverty and hunger: Issues and options for food security in developing countries.* Washington, DC: World Bank.

World Bank. (2006). *Repositioning nutrition as central to development.* Washington, DC: IBRD/World Bank.

World Bank. *Agriculture, forestry, and fishing, value added (% of GDP).* (2019). Available online.

WRI (World Resource Institute). (2018). *Creating a sustainable food future: A menu of solutions to feed nearly 10 billion people by 2050.*

Zhao, C., Liu, B., Piao, S., Wang, X., Lobell, D. B., uang, Y., ... Asseng, S. (2017a). Temperature increase reduces global yields of major crops in four independent estimates. *Proceedings of the National Academy of Sciences of the United States of America, 114,* 9326−9331.

Further reading

Andrews, M. S., & Clancy, K. L. (1993). The political economy of the food stamp program in the United States. In P. Pinstrup-Andersen (Ed.), *The political economy of food and nutrition polices* (pp. 61−78). Baltimore, MD: IFPRI.

Betts, R. A., et al. (2018). Changes in climate extremes, fresh water availability and vulnerability to food insecurity projected at 1.5°C and 2°C global warming with a higher-resolution global climate model. *Philosophical Transactions of Royal Society of London. A Mathematics. Engineering, Science, 376,* 20160452.

Burney, J., & Ramanathan, V. (2014). Recent climate and air pollution impacts on Indian agriculture. *Proceedings of the National Academy of Sciences of the United States of America,* 1317275111.

Cruz-Blanco, M., Santos, C., Gavilán, P., & Lorite, I. J. (2015). Uncertainty in estimating reference evapotranspiration using remotely sensed and forecasted weather data under the climatic conditions of Southern Spain. *International Journal of Climatology, 35,* 3371−3384.

FAO. *Rome declaration on World Food Security, 13 November 1996.* (1996). <http://www.fao.org/docrep/003/w3613e/w3613e00.HTM>.

Robson, D., Tiffin, R., & Wuebbles, D. J. (2015). Extreme weather and resilience of the global food system. In: *Final project report from the UK-United States Taskforce on Extreme Weather and Global Food System Resilience. United Kingdom.* <https://www.stat.berkeley.edu/~aldous/157/Papers/extreme_weather_resilience.pdf>.

Ruel, M. (2013). Food security and nutrition: Linkages and complementarities. In M. Eggersdorfer, K. Kraemer, M. Ruel, M. Van-Ameringen, H. K. Biesalski, N. Bloem, & ... V. Mannar (Eds.), *The road to good nutrition: A global perspective* (pp. 24−238). Basel: S. Karger AG.

CHAPTER 7

Sustainable smart cities

7.1 Introduction

The 21st century has witnessed cities at the center as becoming home to a growing majority of the world's population which is currently (in 2019) home to 55% of the world's population, and that figure is expected to grow to 68% by 2050, and bulk of this growth in urban population is attributed to increased migration of population from rural areas to urban areas in recent decades. Rapid pace of migration of population from rural areas to cities and towns has come to be known as the process of urbanization that can broadly be construed as an "increase in the proportion of a population living in urban areas" (OECD, 2004). Urbanization is defined as "the demographic process whereby an increasing share of the national population lives within urban settlements" (Potts, 2012) and settlements are also demarcated as urban only if most of their inhabitants obtain the majority of their livelihoods from nonfarm occupations (Arouri, Youssef, Nguyen-Viet, & Soucat, 2014). Historically, urbanization has been a prime force in human and economic development (Bairoch, 1988). Intimate boding between cities and human civilization affords an essential key to comprehend history of human civilization (Bairoch, 1988). Way for the advent of cities is said to be paved by the progress registered in the realm of agriculture during the Neolithic revolution and the subsequent period witnessed cities playing prime role as the harbinger of Industrial Revolution and that was one of the major turning points in the history of humankind (Davis, 1955). Potential of the cities in attracting specified people together to the same place at the same time helped in the emergence of cities as focal points of creativity and innovation (Grubler et al., 2013).

As a phenomenon of shifting of populations from rural to urban living patterns, the process of urbanization has occurred at different rates in different parts of the globe, taking the entire humankind toward an urbanized state. About 2% of the global population inhabited urban areas in the early 19th century (Torrey, 2004) and this was largely because cities were some of the unhealthy places to live on account of industrial pollution, overcrowding, and unhygienic conditions leading to higher rates of disease and death comparable to rural areas. Expansion of industrial base was followed by acceleration in migration from rural areas to urban areas throughout the 19th century and this consistent stream of incoming inhabitants swelled the ranks of city populations that were not only faced with the myriad fatal threats then prevalent in the urban environment but also caused far-reaching social and cultural transformations (Schwab, 2019). Undoubtedly, the formation of human settlements predates that of modern nation-state; nevertheless, it was in the wake of the aftermath of the Industrial Revolution and at the cusp of the 19th century that the urban transition actually started taking off (Farrell, 2018), and since then the world has witnessed growth in urban population from a mere 5% in 1800 to nearly 54% in the early 2010s (UN-DESA, 2014).

In 1950, nearly 30% of the global population was living in urban areas and by the year 2007 more urban areas became home to more than 50% of the world population and evolved into

"Homo Urbanus" (Crane, 2005) or city dwellers (Tibaijuka, Maseland, & Moor 2005), and in 2018 55% of the world's population was living in urban areas and by 2030, 68% of the global population is projected to be urban (UN-DESA, 2019, p. xix). According to McDonald (2008), a city of the size of Vancouver is being constructed by humankind at the rate of twice a week and by the close of the 21st century it is projected to host entire humanity (Batty, 2011).

7.2 Urbanization and sustainable development

Urbanization and the concept of sustainable development are interlinked, interconnected and interrelated. Significant changes in the patterns of production and consumption that started characterizing the global society in the aftermath of Industrial Revolution culminated in the long economic boom during the post-Second World War period of the 1950s and the 1960s that concomitantly also proved instrumental in engendering a perception that affluence and economic growth could be limitless (Du-Pisani, 2006). Irrespective of being cognizant of the looming risk of overconsumption, the neoclassical economists of that era continued to nurture the assumption that the resource crunch would be got over in the future by price adjustments and substitutable technological innovations (Du-Pisani, 2006). Subsequently, from the late 1960s and early 1970s, many scholars started realizing that the exponential industrial and commercial growth linked to growing human requirements and burgeoning population was unsustainable for the planet's limited resource base (Perez-Carmona, 2012).

Emerging patterns of consumption, manufacturing and human settlement proved to be prime factors focusing attention on the evolving sustainability crisis (Dovers, 2002). Attention to earlier scholarly works was also drawn by Du-Pisani (2006) and IISD (2012) and these early scholarly works had, apart from engendering intellectual debate, also created a growing awareness with regard to interrelatedness between these issues along with their impact within the society. Authors of these early scholarly works had tried to chronicle the evolution of intellectual debate that perhaps for the first time outlined the interrelationship between the environment, the economy and social wellbeing. Among these early scholarly works, Carson (1962) in her Silent Spring provided details as to how agricultural pesticides caused damage to both the surrounding ecology and human health. Carson's work showed how indiscriminate and unmonitored use of DDT and other pesticides damages wildlife, pollutes water sources and contaminates the food chain (Rickets, 2010), and this publication is said to be a pioneering work that led to a change in society's perception of environmental destruction. Ehrlich (1968) in his pioneering work *The Population Bomb* dealt with the impending sustainability crisis by highlighting the relationship between population, natural resource depletion and the environment. Another scholarly contribution to the impending sustainability crisis was the publication of *A Blueprint for Survival in 1972* by Goldsmith and Allen (1972). Emphasizing on the imperative for sustainable resource management, this publication called for actions to halt irreversible damage to the environment and also proved instrumental in significantly impacting attitudes toward the environment as well as modern environmentalism. Raising questions about the prevalent economic system at the outset of the 1970s and the overreliance on fossil fuel Schumacher (1973) in his Small is Beautiful, espoused for change in lifestyles as technology was seen as being unable to provide solutions to all problems.

Another development that contributed to the evolutionary discourse on sustainable development was the formation of some nongovernmental organizations (NGOs) and conclusion of a number of international agreements pertaining to different aspects of environment during the 1970s. Formation of Friends of the Earth in 1969, and Greenpeace in 1971 along with the observation of the first Earth Day in 1970 proved instrumental in bringing issues related to environment at the forefront of the public discourse (IISD International Institute for Sustainable Development, 2012). Additionally, the decade of the 1970s was also characterized with conclusion of many international agreements taking place that had direct bearing on environment-related issues and such agreements, inter alia, included: the 1971 Ramsar Convention on Wetlands of International Importance, signing in 1972 of the Convention on the Prevention of Marine Pollution by Dumping of Wastes and Other Matter and conclusion in 1973 of the Conference on International Trade in Endangered Species of Wild Animals (Quental, Lourenço, & Da Silva, 2011). Apart from these, the convening in 1972 of the United Nations Conference on the Human Environment at Stockholm came to recognized as the most significant international conference because it worked as a catalyst to generate awareness about environment and development issues confronting the planet (UN, 1972). Participants to this conference were called upon to initiate steps to prevent further ecological crises (Quental et al., 2011; UN, 1972). The outcome of the 1972 UN conference held in Stockholm came to be known as Stockholm Declaration that called for shaping our actions throughout the globe with a more prudent care for their environmental consequences. Striking a note of optimism, the Declaration stated that through fuller knowledge and wiser actions, a better future can be achieved for "ourselves and our posterity a better life in an environment more in keeping with human needs and hopes ...To defend and improve the human environment for present and future generations has become an imperative goal for mankind ..." (UN, 1972).

Another publication entitled *The Limits to Growth* by Meadows, Meadows, Randers, and Behrens III (1972) for the Club of Rome Project took into consideration linkages between economic growth and environment (Jorgensen et al., 2015). Being known as one of the most intellectually influential work of its time, this publication, apart from questioning the conventional economic growth model, also proposed ecological and economic stability for a sustainable future (Ekins, 1993; Jorgensen et al., 2015). Arguing that the limits to growth on the planet would be reached sometime within the next 100 years, if the present growth trends in global population, industrialization, pollution, food production, and resource depletion were allowed to continue unchanged (Meadows et al., 1972) warned that the resultant outcome could be a rather sudden and uncontrollable decline in both population and industrial capacity; and keeping that in view they suggested that it was possible to change "these growth trends and to establish a condition of ecological and economic stability that is sustainable far into the future" (Meadows et al., 1972, pp. 23–24). Undoubtedly, *Limits to Growth* by Meadows et al. (1972) offered an intellectually stimulating argument; nonetheless, many regarded it as both a "doomsday prediction" and a "politically unpalpable" argument (Beder, 2006).

Ward and Rene (1972) is credited to having coined the term "sustainable development" and used the same in their book *Only One Earth: The Care and Maintenance of a Small Planet*, as a compromise between the notions of development and conservation (Du-Pisani, 2006; Satterthwaite, 2006). Irrespective of the fact that the term "sustainable development" had been coined by Ward and Rene (1972) in the early 1970s; however, the term had not gained wider usage during the remaining part of the decade of the 1970s, and it was only in the aftermath of the publication of the World Conservation Strategy in 1980 that it started gaining international recognition

(Du-Pisani, 2006). This publication of World Conservation Strategy was the outcome of a collaborative effort between the International Union for the Conservation of Nature (IUCN), the United Nations Environment Programme (UNEP), and the World Wildlife Fund (WWF) under the title Living Resource Conservation for Sustainable Development (Du-Pisani, 2006; Paton, 2008). Recognizing the interrelationship between economic, social and environmental factors, this Strategy emphasized that attainment of sustainable development through the conservation of living resources identified the actions required both to improve conservation efficiency and to integrate conservation and development. It further noted: "... For development to be sustainable it must take account of social and ecological factors, as well as economic ones; of the living and nonliving resource base; and of the long-term as well as the short-term advantages and disadvantages of alternate actions ..." (IUCN, 1980, p. v-3).

In the wake of these developments, the UN established in 1983 the World Commission on Environment and Development (WCED) under the chairpersonship of Gro Harlem Brundtland (Paul, 2008). The WCED was assigned the task of finding a way to reconcile the difference between the industrial development agenda and environmental concern to resolve the clash between neo-liberalism and environmentalism (Howes, 2005). The WCED, which had also come to be known as Brundtland Commission, submitted its final reports in 1987 under the title Our Common Future, also referred as the Brundtland Report (Paul, 2008). The report comprising three main sections dealt with host of issues pertaining to development and environment. The first section deals with environmental and developmental concerns, such as poverty, inequality, pollution and the loss of biodiversity; the second part focuses on promoting sustainable development as a way to solve socioeconomic-environmental issues consequent upon economic growth (UN, 1987); and finally, the Brundtland Report proposed the formulation of an international framework to achieve sustainable development (Howes, 2005). It can thus be surmised that the normative debate on sustainability crises that had commenced in the 1960s and the 1970s had in the second half of the 1980s had seemingly been crystallized into an alternative development model known as "sustainable development." The Brundtland Commission is said to have remained the most instrumental UN project that not only institutionalized and defined the term "sustainable development" (Du-Pisani, 2006), rather the development model based on sustainable development offered by this Report has become also emerged as the alternative to the neoliberal ideology-based economic model (Lozano & Huisingh, 2011).

In the aftermath of the publication of the Brundtland Commission's Our Common Future in 1987, the concept of sustainable development not only entered into the lexicon of policy discourse, it also became a new topic for policy debate between policymakers, politicians, grassroots level activists, academics, corporate organizations, and Indigenous communities (Sneddon, Howarth, & Norgaard, 2006). The report of the Brundtland Commission advanced the argument that the issue of environmental degradation formed the core of many economic, and social issues, and that environmental crisis could influence both domestic and international peace and security (Happaerts, 2012). Thus the focus of debate whether to have growth or no growth now shifted to what type of growth was required. The Brundtland Commission defined sustainable development as: "... development that meets the needs of the present without compromising the ability of future generations to meet their own needs ..." (UN, 1987). This shift of debate from growth to sustainable development has also brought technological innovation to the forefront as a solution through which new processes and products could be adopted without damaging the environment (Beder, 2006). While emphasizing the significance of global inter- and intragenerational equity among current and future generations, the Brundtland Report also proposed that social equity, economic development and environmental protection could occur simultaneously (Du-Pisani, 2006). These notions subsequently emerged to form the three key pillars of the concept of sustainable development—the environment; the economy; and the society—and these are also known as the triple bottom line (Du-Pisani, 2006). The phrase "triple bottom line" is coined and elaborated by Elkington (1998) and it entails business prosperity, environmental quality and social justice (Sherman, 2012). In the wake of these developments, it came to realized that conventional economic development measure techniques like gross domestic product (GDP) and gross national income (GNP) did not take into consideration the environmental and human costs of development and henceforth, a new set of indicators or a benchmarking system was needed to show the full cost of development and to determine the holistic progress of society (Schlossberg & Zimmerman, 2003).

With the concept of sustainable development having started gaining salience in the aftermath of Brundtland Commission, the UN convened a number of conferences to promote the notion of sustainable development. In 1992, the UN Conference the Environment and Development, also known as the Rio Earth Summit, was held in Rio de Janeiro (UN, 1992) with the objective of outlining principles and plans to implement the concept of sustainable development (Paul, 2008). The Rio Conference proved instrumental in transforming the definition of sustainable development outlined by Brundtland Commission into a framework by drafting plans and measures to implement a sustainable development agenda, thereby making the concept germane for the common people (Conroy & Berke, 2004). The major outcomes of the conference were as follows:

1. the adoption of 27 principles on environment and development, popularly known as the Rio Declaration (UN-GA, 1992);
2. the formulation of the Agenda 21, the action plan for sustainable development;
3. the setting up of the Commission on Sustainable Development; and
4. the establishment of a funding mechanism for Agenda 21 (Boer, 1995; Paul, 2008; Stoker, 1998).

The UN member states were called upon the Rio Conference to adopt a national sustainable development strategy in order to implement he Agenda 21 action plan (Paul, 2008).

Launching of the Millennium Development Goals (MDGs) in 2000 by the United Nations was designed to aim at reducing the extent of social and environmental problems by 2015. The targeted agenda of this worldwide initiative of the MDGs, inter alia, included ambitious initiatives like poverty reduction, the spread of universal primary education, attainment of gender equality, increasing the availability of safe drinking water and ways to achieve environmental sustainability (UN, 2000). Resolving these problems was regarded to be means for minimizing the extend of various social and environmental sustainability issues (Paul, 2008; Quental et al., 2011). The World Summit on Sustainable Development held in 2002 in Johannesburg, South Africa (Paul, 2008), was greatly influenced by the MDGs initiative that was focused on dealing with the socioeconomic issues of developing countries (Drexhage & Murphy, 2010). The 2002 World Summit on Sustainable Development (UN, 2000) complimented the MDG outcomes such as attaining basic sanitation by 2015, initiatives to reduce the harmful effects of chemicals and reductions to biodiversity loss (Paul, 2008). It was observed that owing to the MDG-centric global initiatives, the global poverty level had recorded significant reduction, the number of people without clean drinking water was halved and that overall MDG targets demonstrated a number of encouraging achievements (Sachs, 2012; UN, 2015a). In the context of success of the MDG initiatives, the then UN Secretary-General noted that future progress would require an unswerving political will, and collective, long-term effort and also emphasized on the need to tackle root causes and do more to integrate the economic, social and environmental dimensions of sustainable development (UN, 2015a, p. 3).

In 2012, two decades after the 1992 Earth Summit, the UN Conference on Sustainable Development was organized at Rio de Janeiro, Brazil, on June 20–22, 2012, it also came to be known as the Rio + 20 Conference, with the avowed objective of addressing the shortcomings of the past 1992 Earth Summit (Leggett & Carter, 2012). Therefore, the Rio + 20 Conference had three major objectives: (1) to rejuvenate the global political commitment to sustainable development; (2) to assess the gaps in the previous commitment to embrace sustainable development; and (3) to address emerging challenges (Pisano Endl, & Berger 2012). By that time, the global community to a large extent had seemingly acknowledged the fact that the planet is in a new geological epoch, known as the Anthropocene, that is primarily driven by human actions (Sachs, 2012). In the background of this factor supplemented by the fact of UN recognition of the relative success of the MDG project, the Rio + 20 Conference acknowledged the essentiality of formulating more holistic goals for the future and finally culminated in the adoption by the UN of Sustainable Development Goals (SDGs) for the period from 2015, the closing phase of the MDGs, to 2030 (Osborn, Cutter, & Farooq, 2015; Sachs 2012, 2014). One of the major differences between the MDGs and the SDGs is that the MDGs were primarily focused on the developing countries whereas the SDGs are universally applicable to all member countries of the UN (Osborn et al., 2015). Launched in 2015, the 17 SDGs are based on three aspects of sustainable development—economic development; environmental stability; and social inclusion—and these encompass the most pertinent requirements of the time for both present and future generations, such as poverty, education, clean energy, sustainable consumption, and the establishment of a climate change action plan (UN, 2015b, 2015c).

Hence, since the publication of the Brundtland Commission's Our Common Future report in 1987 the concept of sustainable development had evolved, defined, structured and institutionalized into a new alternative development model. This process of evolution is shown in Table 7.1.

7.2.1 Local sustainable development

Action at multiple levels in multiple contexts is required under the sustainable development concept (Manderson, 2006). Agenda 21 adopted at the 1992 UN Conference on Environment and Development (UN, 1992) placed sustainable development within the context of different levels of action, specifically at global, national and local scales to deal with particular priorities of sustainable development. Within Chapter 28 of Agenda 21was envisaged Local Agenda 21 (LA21) that emphasized the need for a locally-relevant adaptation of Agenda 21 that takes into account of priorities of community stakeholders (Bond, Mortimer, & Cherry, 1998). Chapter 28 of Agenda 21 delineates guidelines for local governments entrusted with the task of creating LA21s that are regarded critical to attain global sustainability goals owing to the propinquity of municipalities to local stakeholders and the ability to understand and ascertain the unique context and social demands of the community (Borland et al., 2019). Local governments have come to work with an array of stakeholders, including businesses and NGOs, to frame and implement sustainable community plans, otherwise known as LA21s in response to social. Ecological, and economic challenges

Table 7.1 Process of evolution of the concept of sustainable development.

Year	Type of activity	Impact
1960s	*Key publications* Silent Spring (1962)	This publication is said to be a pioneering work that led to a change in society's perception of environmental destruction.
	The Population Bomb (1968).	This publication dealt with the impending sustainability crisis by highlighting the relationship between population, natural resource depletion and the environment.
1970s	A Blue Print for Survival (1972)	Emphasizing on the imperative for sustainable resource management, this publication called for actions to halt irreversible damage to the environment and also proved instrumental in significantly impacting attitudes towards the environment as well as modern environmentalism.
	The Limits to Growth (1972)	This publication focused on linkages between economic growth and environment.
	Only One Earth: The Care and Maintenance of a Small Plane (1972)	The term "sustainable development" was coined by the authors of this book and used for the first time.
	Small is Beautiful (1973).	This publication espoused for change in lifestyles as technology was seen as being unable to provide solutions to all problems.
1972	UN Conference on Human Environment	It was the UN's first major conference on international environmental issues, and marked a turning point in the development of international environmental politics.
1980	Publication of World Conservation Strategy by IUCN	This publication used the term sustainable development at an international level.
1987	World Commission on Environment and Development	The term sustainable development was defined by the first time at international level.
1992	UNCED, also known as Earth Summit.	Adopted Agenda 21, the action plan for sustainable development.
1994	John Elkington provided a structural definition of sustainable development	John Elkington coined the phrase "the triple bottom line" that included economic prosperity, environmental quality and social justice
2000	UN MDGs.	Addressed the socioeconomic issues of developing countries
2002	UN supported GRI	Guidelines on how to report on economic, environmental and social dimensions were released.
2012	The Rio + 20 Conference	Groundwork for SDGs that included both developing and developed counties was laid.
2015	UN adoption of SDGs	The 17 SDGs are based on three aspects of sustainable development—economic development; environmental stability; and social inclusion.

and various internationally-led sustainability programs (Clarke, 2014). According to Sachs (2012), specific priorities for sustainable development at the local level are subject to variation between and within communities all over the globe. There are various community-wide targets or goals within LA21s that comprise social, economic and environmental topics (Clarke, 2014). Local sustainable plans at the local level often comprise goals for transportation, water, waste, air quality, energy, climate change, land use, ecological diversity, food security, civic engagement, social infrastructure, housing, safety and crime, local economy, employment, poverty alleviation, noise pollution, and financial security (Clarke & MacDonald, 2012).

Delineating the definition of the term "local sustainable development," much like the concept of sustainable development, is broad and difficult to define taking into account the unique requirements of each and every community (Dale & Newman, 2006). The definition reflects the concept of sustainable development with the significant difference being the geographical scale taking place at the local level (Bridger & Lulloff, 1999). Viewed in a broad perspective, sustainable development "emphasizes an integrated response to economic, social, and environmental imperatives within a given locale, and an emphasis on intergenerational equity with regard to resource use" (Barraket, 2005, p. 77). Apart from highlighting the importance between environmental concerns, development objectives, and the social relationships (Bridger & Lulloff, 1999), the definition of the concept itself is evolving to suite the requirements of every community (Roseland, 2000), accounting for the "nested matrix of social, ecological, and economic interactions often defined by a geographical place" (Dale & Sparkes, 2011, p. 477). At the local scale, local territories allude to more than physical spaces and includes both communities and systems of relations, along with representations of managing the economy, social relations and interactions between society and the environment (ICLEI, 2012, p. 4). While acknowledging a bottom-up approach to sustainability that is dependent on local socioeconomic, environmental and economic contexts (Moallemi et al., 2019), the concept of local sustainable development also recognizes that all efforts for sustainable development need local-level action to enable progress on a large scale.

Adoption of the Agenda 2030 for Sustainable Development by the UN in September 2015 comprises 17 SDGs with 169 accompanying targets and 304 indicators (UN, 2015a, 2015b). A global framework is provided by these SDGs for collective action toward ending poverty, ensuring peace and prosperity, and protecting the planet, and all of these require transformative solutions to build the capacity and knowledge of actors for sustainable development through a revitalized global partnership for the goals (Filho et al., 2018; UN, 2015b). While addressing major systematic barriers to sustainable development, the SDGs outline goals, targets and indicators for gauging progress and enabling global actors at all levels to attain sustainable development (ICSU & ISSC, 2015). The adoption of the SDGs has paved way for a new form of governance that is dependent on cross-sector collaboration and multistakeholder initiatives to realize the SDGs (UN, 2015b) owing to the scale, scope and complexity of challenges ahead wherein no sector can manage alone (Selsky & Parker, 2005). With multistakeholder collaboration having become a key implementation mechanism for achieving the global SDGs, reporting progress on the SDGs will call for the action of governments across the globe (Willis, 2016). The role of businesses is specifically emphasized upon as relatively significant in achieving the SDGs (McGraw, Danilovich, Ma, Wilson, & Bharti Mittal, 2015), with prominent scholars like Sachs (2012) arguing that the SDGs are not achievable without them.

7.3 Sustainable urban development

The evolving and ever-changing nature of the concept of sustainable urban development (Tang & Lee, 2016), makes it sometimes to be delineated in terms of the economic sustainability of a city, that is, its potential "to reach qualitatively a new level of socio-economic, demographic and technological output which in the long run reinforces the foundations of the urban system" (Ewers & Nijkamp, 1990). This interpretation of the notion endeavors to continue economic growth and now it has come to be regarded as a relatively less strong form of sustainable development. Some have emphasized on the social sustainability by basing this concept on a wider spectrum of social principles entail futurity, equity, and participation, specifically involvement of public citizens in the land development process (Friends of the Earth, 1994). While construing this concept alongside environmental concerns, it encompasses environmental sustainability and that denotes the pursuit of urban form that synthesizes land development and preservation of nature and puts the protection of natural system into a state of vibrant equipoise (Lyle, 1994). In other words, principles aiming at attainment of urban sustainable development are usually based on environmental, economic and social considerations (Haughton & Hunter, 2003; UN-GA, 1992). Undoubtedly, the discourse on urban sustainable development has apparently witnessed added emphasis on the environment and economy; nevertheless, cities are still primarily human habitats (Tang & Lee, 2016). A potent form of sustainable development, comparable to an enfeebled form, "represents a revised form of self-reliant community development which sustains people's livelihoods using appropriate technology" (Huckle, 1996, p. 4). As cities are for people (Gehl, 2010), sustainable cities ought to be "places where people want to live and work, now and in the future. They meet the diverse needs of existing and future residents, are sensitive to their environment, and contribute to a high quality of life. They are safe and inclusive, well planned, built and run, and offer equality of opportunity and good services for all" (UK Government United Kingdom Government, 2005, p. 56). It has been discerned by McMichael (2000) that human health, wellbeing, safety, security and opportunity is likely to be influenced by the way urban settlements are planned, designed, developed and managed. Concurrently, it is also worth noting that social development and economic productivity are dependent on citizens whose mental and physical needs are fulfilled, and as such, comforts of the city inhabitants have come to play a significant role in sustainable urban development (Tang & Lee, 2016).

Apart from being a multifaceted concept that synthesizes land development and nature preservation, sustainable urban development also alludes to the capacity of nature to support its activities, the vitality of a city as a complex system, along with the quality of life of its dwellers (Tang & Lee, 2016). In other words, sustainable urban development encompasses many areas of activity like environmental protection, human development and wellbeing of the city inhabitants. Lamenting that despite all the discussions on the discourse on sustainable urban development, "no single or agreed" meaning has been produced, Tang and Lee (2016) propose to define sustainable urban development "as the capacity of any significant human settlements to maintain environmental quality and carrying capacity, to support socio-economic development and management, and to provide sufficient services and livelihoods to all current and future inhabitants." In other words, the pragmatic and full retaliation of sustainability can only occur in the overlap, or the dynamic, among the three fundamental capacities—environmental, social and economic. Various sets of frameworks, indicators, and assessment tools have been developed to facilitate a better cognizance and improved understanding of the state of or changes to, urban settlements vis-à-vis better sustainability performance (Briassoulis, 2001). A wide range of urban sustainability indicators are also in usage across different cities and regions, entailing variations in accordance with their specific requirements and goals (Brandon & Lombardi, 2005). In the wake of these developments, there have emerged various approaches in regard to the attainment of sustainable urbanization and the policies that have been harnessed in the pursuit of sustainable urbanization process have assumed an array of nomenclature like livable cities, green cities, digital cities, information & knowledge cities; resilient cities; eco-cities; low-carbon cities; and smart cities etc. (de Jong, Joss, Schraven, Zhan, & Weinjen, 2015:1). These approaches are briefly appraised below.

7.4 Livable cities

Notion of "livable cities" started gaining traction in the 1980s in the wake of city planners embarking on appraising ongoing shifts in the development patterns from the decline of urban centers to fast growing suburban areas (FHA, US Department of Transportation, 2014). Emergence of a series of reports began to question traditional growth assumptions and highlighted regions that were undertaking pioneering array of efforts to make communities more livable (Clinton-Gore Administration, 2000). As a sequel to the publication of annual surveys ranking the world's most livable cities such as the Mercer Worldwide Quality of Living Survey by Vienna-based Mercer Group and The World's Most Livable Cities by London-based Economist Intelligence Unit (EIU), the term "livable cities" has come to gain further currency. Admittedly, these surveys make use of different criteria; nonetheless, typically assess cities on factors like political stability, safety, healthcare, education, public services, transportation, housing and environmental quality (Momtaz & Elsemary, 2015). Notion of urban livability comprises an assortment of diverse issues that are underlined by a common set of guiding principles like accessibility, equity and participation that provide substance to the notion of livability. The quality of life led by people inhabiting a city is closely related to their ability to access infrastructure in the form of transportation, communication, water, and sanitation; food; clean air; affordable housing; gainful employment; and green space, including parks. According to Momtaz and Elsemary (2015), livability of a city is also determined by the access that its inhabitants have to take part in the decision-making process with regard to their requirements as well as needs. Livability can be defined as "quality of life" as experienced by the residents within a city (Timmer & Seymoar, 2005). Livability entails an urban system that enhances the physical, social and mental wellbeing and personal development of all residents of the city (Ahmed, El-Halafawy, & Ahmed Mohamed Amin, 2019).

Urban livability is about pleasant and desirable urban spaces that offer and reflect cultural and sacred enrichment (Momtaz & Elsemary, 2015), and major principles that add vitality to this theme are equity, dignity, accessibility, conviviality, participation, and empowerment (Cities PLUS, 2003). A livable city for some social groups is that where those elements have been preserved or renewed which have always been an integral part of people-friendly places. These are, as Peter Smithson once said, "relationships between streets and buildings, and buildings amongst themselves, and trees, and seasons of the year, and ornamentation, and events and other people" (Palej, 2000; Smithson, 1971). Apart from serving as a link between the past and the future, the livable city preserves the imprint of the history for the posterity (NAS National Academic Press, 2002). While regarding a livable city as a "sustainable city" that satisfies the needs of the present inhabitants without minimizing the capacity of the future generations to satisfy their needs, Momtaz and Elsemary (2015) observe that both social and physical elements must collaborate for the wellbeing and progress of the community in a livable city. While delineating a livable city as a place where common spaces constitute the centers of social life and foci of the entire community, Salzano (1997) suggests that a livable city ought to be built up or restored as a continuous network pedestrian paths and bicycle-paths do not bind together, each has its own path in all the sites of social quality and of the community life.

Various elaborations to delineate the terms "livability" and a "livable city" entail a number of dimensions, with some elaborations aligning livability with local community wellbeing; and with others trying to link livability with the physical attributes of a specific location. At the same time, it is also worth noting that livability is not just inherent in environmental characteristics; rather, it is a function of the relationship between the environment and the social life sustained by it (Hankins & Powers, 2009). This suggests, as pointed out by Wheeler (2003), that there is a social dimension to livability that reflects as to how people interact within local environments. Knox and Mayer (2009) have emphasized on subjective and relative nature of the term, with notions of what constitutes a community livable varying between groups and individuals in accordance with different and shifting perceptions, values and desires. In other words, livability means different things to different people (Balsas, 2004). And this subjective dimension may partly explain "the lack of an agreed definition of livability in the literature" (Momtaz & Elsemary, 2015). Besides, there is no agreement as to what constitutes the most appropriate index to measure livability, as no designated formula for achieving livability it depends on the context (Ahmed et al., 2019).

7.5 Eco-cities

Admittedly, a new wave of "eco-city" initiatives had started sweeping across the globe by the early 2000s (Joss, 2010); nonetheless, the idea behind term "eco-city" has a long history dating back to the 19th century. While the concept of eco-city has come to be defined in multiple ways, whereas the widely accepted elaboration of this concept was created in California that pertained to the notion of restructuring of the urban spaces in a way that could achieve a balanced development of the urban infrastructure in sync with the nature (Hu, Wadin, Lo, & Huang, 2016, p. 78). Relatively a new term in scientific literature, the term "eco-city" has been used by an array of researchers and governments to describe an assortment of ideas and concepts (Van Dijk, 2011). Other concepts like "sustainable city," "low-carbon city," "eco-community," "green city" is often used interchangeably with the term "eco-city" (Zhou & Williams, 2013). Generally, these terms denote attempts at the city level to deal with economic, social and environmental conflicts that often emanate primarily owing to the consequences of economic growth and demographic change. Eco-city planning does not constitute a singular body of thought, rather it is an assortment of various planning modes that have cropped up in response to side-effects economic development over the past many decades. Availability of surfeit of ideas and notions makes it cumbersome to work out a single precise definition of eco-city because, for example, adding a descriptor like "low carbon" as in the term "low-carbon eco-city" is prone to envisage a specific focus for Eco-city planning and performance evaluation, whereas the term "eco-city" encompasses a much wider range of environmental, social and economic issues than the reduction of CO_2 emissions (Zhou & Williams, 2013).

Undoubtedly, many eco-city notions have been available in the literature pertaining to planning theories from the past; nevertheless, these have remained only concepts until recently. As has been observed by Mark Roseland (1997), throughout the 1980s and 1990s, the term "eco-city" remained primarily a concept, "a collection of ideas and propositions about sustainable urban planning, transportation, housing, public participation and social justice, with practical examples relatively few and far between." Elements of the eco-city concept have been addressed by many different schools of thought in urban planning in the United States and the United Kingdom; nevertheless, it was found during the review of the sustainable city movement in the UK that, "notwithstanding examples of good practice, advocacy rather than research has often characterized the debate" about "sustainable city" means (Jenks & Jones, 2010). The first decade of the 21st century is said to have witnessed renewed efforts to apply sustainable city theories to projects at the neighborhood, district, and city scale (Zhou & Williams, 2013), with many of them getting started in Europe, Asia, Arica and South America; and paradoxically, relatively few of these projects taking place in the United States and the UK where bulk of the sustainable city planning theory originated (Joss, 2010).

Premalatha, Tauseef, Abbasi, and Abbasi (2013) have outlined 10 characteristics that are significant for the eco-city development. (1) Emphasis is laid on having land-use priorities such that it generates a compact, diverse, green, and safe mixed-use communities around public transport facilities; (2) eco-cities should accord transportation priorities in such a manner so as to discourage driving and stress emphasis on "access by proximity"; (3) eco-cities are called upon to restore damaged environment; (4) eco-cities should generate affordable, safe, convenient, and economically mixed housing facilities; (5) eco-cities should nurture social justice and engender improved opportunities for the underprivileged; (6) eco-cities are also called upon to support local agriculture, urban greening and community gardening; (7) eco-cities are urged to promote recycling and resource conservation while reducing pollution and hazardous waste; (8) eco-cities should support ecologically sound activities while discouraging hazardous and polluting ones; (9) eco-cities are called upon to promote simple lifestyles and not to encourage excessive consumption of material goods; and (10) eco-cities should enhance public awareness of the local environment and bioregion through educational and outreach activities (Premalatha et al., 2013, p. 661).

While lamenting that no ideal eco-city has been created yet, Register (1987, p. 3) note that certain cities around the globe entail some characteristics of an eco-city, for instance, in regard to some neighborhoods and areas but are not wholly ecological. While referring to the existence of eco healthy urban settlements of the past, especially in old European cities and towns, Register (1987, p. 3) also noted that these settlements, apart from being compact and energy-efficient, were made of local materials and were built in a manner that befitted the natural surroundings. While delineating an eco-city, Yanitsky (1982) describes the eco-city as a city that maintains a positive relationship with the environment that in turn allows for the emergence of technological and environmental aspects. The inhabitants of such cities are to provide with ample opportunities to be creative and productive, along with provisions for essential medical care to lead a healthy way of living. While espousing the cause of environmental protection in the eco-city, Yanitsky also claimed that one of the most significant characteristics of the eco-cities is their capability to be resource efficient (Su, Xu, Chen, & Yang, 2013, p. 5). Eco-cities are accountable for providing good living conditions to their inhabitants and these types of cities are prone to avail of the benefits accruing from sustainable urban planning and different management methods associated with the reduction of waste and greenhouse gas (GHG) emissions (Tsolakis & Anthopoulos, 2015, p. 1). Maintenance a semblance of equilibrium nature and economic development is critical for an eco-city along with the attainment of their sustainability and for this local governments are required to manage the local urban resources efficiently and simultaneously minimize the levels of pollution in the cities to a minimum along with emphasis on recycling to generate an eco-city environment (Amakpah, Larbi, Liu, & Zhang, 2016).

Eco-cities entail the potential of realizing long-term sustainability (Joss, 2011). Eco-cites are significant for developing countries because they can serve as catalyst for improving the standard of living of its inhabitants by providing economic security, opening up new vistas of livelihoods for its inhabitants, and elicit investments that are sine qua non for the prosperity of the city (Deb, 2010; Suzuki, Dastur, Moffatt, Yabuki, & Maruyama, 2010, p. xviii). Environmental sustainability characteristic of eco-cities makes them eligible candidate to be designated as the predecessors of the Gaia cities—Global Alliance for Incinerator Alternatives (Gaia) is an early leader in the rapidly emerging Internet of Things (IoT) marketplace offering its services for smart cities (SC4C, 2018). Eco-cities entail the potential of synthesizing various subsystems the proper management of which is significant for the creation and the efficient sustainment of an eco-city. Tsolakis and Anthopoulos (2015) have identified five management levels of an eco-city—sustainable urban growth; urban transport, GHG emissions, waste management and energy consumption—and proper management of these levels can guarantee improved living standard, employment and services to the inhabitants of eco-city (Tsolakis & Anthopoulos, 2015, pp. 2−3).

Involvement of the European Union (EU) in the development of eco-cities has envisaged a vision for the eco-city as a city that entails a sustainable community and affords accessibility for everyone, public space for daily life, green areas, bioclimate comfort, balance with nature, minimized demand for land, reduction, recycling and reuse of waste (Fook, 2014). EU's research program on developing eco-cities is characterized by short distances and the compact urban form of the eco-cities (Coplak, 2003). Other characteristics that are required in an eco-city, inter alia, include: the new balance of concentration and decentralization, network of urban quarters, power station of renewable energies, health, safety and wellbeing, sustainable lifestyle, qualified density, human scale and urbanity, strong local economy, built and managed with the inhabitants, development concentrated at suitable sites, integration into the surrounding region, minimized energy consumption, integration into global communication networks, cultural identity and social diversity (EESD, 2005, p.5).

Eco-city projects are divided by Joss (2012) into three types. While the first type alludes to new development of eco-cites, the second type represents an expansion of the cities and the third type of development of eco-cities is adopting the eco-city characteristics through retrofitting (Joss, 2012). It has also been suggested by Joss that the development of eco-cities ought to be supported by the government and implemented on a wider scale (Rapoport & Verney, 2011). Essentiality of the two concepts—Three Harmonies and Three Abilities—has been emphasized by Clarke (2014) as significant for eco-city development. The concept of Three Harmonies comprises three aspects. The first part entails people living in harmony with other people and it is social harmony; the second aspect encompasses people living in harmony with economic activities and it is called economic vibrancy; and the third aspect entails people living in harmony with the environment and it is called environmental harmony (Ghiglione & Larbi, 2015; van Dijk, 2011). In other words, Three Harmonies comprise social harmony, economic vibrancy and environmental sustainability. The concept of Three Abilities consists of three aspects—practicable, scalable and replicable. The practicable aspect pertains to the technologies that have been adopted in the eco-city and it requires that the technologies thus adopted must be affordable and commercially viable (Ghiglione & Larbi,

2015). The scalable aspect refers to the transferability to other fields of development of the models and measures that are applied to the eco-city project (Ghiglione & Larbi, 2015, p. 106). The replicable aspect epitomizes the idea that a given eco-city plan of development could be replicated in different eco-city projects within the same country as well as abroad (Clarke, 2014).

Many countries in Asia have embarked on launching development of eco-cities as an opportunity to tread on a new path of ecological development and to encourage green energy consumption in respective countries. These countries nurture the idea underlying the development of eco-cities that is relatively new and the introduction of policies and measures to implement this idea represents a complex process for the different governments (Neumann et al., 2019). Development of eco-cities in a successful manner is associated with the speed of formation and the quality of planning and concurrently, the prevalence of circular economy (CE) assumes significant role in countries where the eco-cities are being developed. Cordial cooperation between governmental and nongovernmental institutions is also a critical factor for establishing eco-cities (Hu et al., 2016, p. 77). Various challenges are to be surmounted in order to establish eco-cities; and these challenges emanating from different fields comprise issues like the lack of universal access to public services, paucity of the necessary budget and the absence of sufficient national and international governance in the urban development process (Mersal, 2017).

Attainment of sustainable urbanization is dependent on investments in green infrastructures, renewable energy and sustainable social development. Harnessing electrical power and water resources judiciously is equally significant for the development of eco-cities. Enhancing the number of parks and green fields along with renovation of the existing buildings in accordance with international green building standards are also essential in developing eco-cities. Public transport system in an eco-city is required to meet certain standards and it has to be fast, affordable and environment-friendly (ICLEI, 2020). Improved mechanism of waste and recycling system constitutes another significant factor in the development of eco-cities. Projects pertaining to the sustainable urban development also require international knowledge transfer, especially in the realms of transfer of environment-friendly technology along with adequate supply of fiscal resources. Establishment of sustainable cities requires these four conditions to be met—the provision of social development; economic development; environmental management; and urban governance. Challenges faced in the eco-city building process by developed countries differ from those of the developing countries and they are divided into social, economic and environmental sections (UN-DESA, 2013, p. 64).

Sustainable urban planning calls for the use of frameworks that comprise a number of standards and requirements imbibing various principles that are to be taken into account while creating such frameworks (World Bank, 2018a, 2018b). The use of indicators in the frameworks for sustainable cities has come to regarded as essential (Zou & Li, 2014, p. 20). The concept of sustainable planning is significant for achieving sustainable urbanization. The UN-Habitat has suggested that the provision of a clear definition of sustainable planning has become almost impossible owing to changes in the planning process that could arise overtime (Cajot, Peter, Bahu, Koch, & Marechal, 2015, p. 3368). The term "urban planning" is used loosely to allude to international interventions in the urban development process, mostly by the local government (Leyzerova, Sharovarova, & Alekhin, 2016). According, the term "planning" subsumes an assortment of mechanisms that are in fact quite distinct: regulation, collective choice, organizational design, market correction, citizen participation, and public sector action (Cajot et al., 2015, p. 3368). The eco-priority theory is designed to direct the eco-city planning. The eco-priority theory was established to guide eco-city planning when there were conflicts between different factors. The eco-priority theory advocates that eco-environmental construction and reasonable use of resources have priority among all types of socioeconomic developmental activities on the basis of a win-win situation between economic and natural processes (Xu, Lin, & Fu, 2004). While promoting the environment-friendly construction process, the eco-priority theory takes into account a number of factors. It also emphasizes on the efficient use of resources and the mutual benefits for the economy and for the natural environment. Su et al. (2013, pp. 6–7) not that the main concepts related to the eco-priority theory are: "ecological culture," "ecological efficiency," "ecological economy," "ecological behavior," "ecological allocation" and "ecological accounting."

The usage of indicators is taken into consideration while envisaging sustainability in the urban areas. The process of creating and selecting indicators is said to be confronted with challenging and problematic issues (SEP, 2018). According to Kloppand Petretta (2017, p. 92), such issues might include the following: (1) the poor availability of standardized, open and comparable data; (2) the lack of strong data collection institutions at the city scale to support monitoring for the Urban SDG; and (3) "localization"—the uptake and context specific application of the goal by diverse actors in widely different cities. The usage of indicators entails scientific, conceptual, and political functions. The scientific function deals with enhancement in the awareness of the urban areas, and the conceptual function is articulated in the public interest via and debates and discussions. Potential of the indicators in the formulation of new policies or alteration of such policies being influenced by indicators denotes their political function (Visvaldisa, Ainhoab, & Ralfsc, 2013). This is no standard or internationally acknowledged model for the indicators and this is attributed to the fact that indicators re used for different means and are often influenced by political factors. However, the urban indicators are required to be relevant. Klopp and Petretta (2017) are of the view that the essentiality for the indicators to be acceptable and practicable has been also internationally recognized (Klopp & Petretta, 2017, p. 95). Formulating indicators for the establishment of eco-cities entails the role of measuring the attainment level of the eco-projects. Differences in ecosystems and other aspects of the urban areas of the eco-projects have led to the emergence of different standards of the indicator systems. The common aspects between the eco-city frameworks comprise three categories—economy, society and nature (Su et al., 2013, p. 10). For instance, the MEP framework, developed by China's Ministry of Environmental Protection (Zhou & Williams, 2013), comprises 19 indicators that have been divided among these three categories (World Bank, 2012, pp. 37–38).

The MEP framework, especially designed for the assessment of the eco-city construction in China, is based on an index system that has been developed on a percentile scoring mechanism. The assessment aims at examining the gap that has occurred between the registered score and determined level of standards. This system is also characterized by the basic condition that suggests that the evaluation index that has been developed in regard to the quality of the eco-cities and their environmentally friendly development pattern, has to be one of the best in the region (Zhou & Williams, 2013). The structure of the indicators is also significant and is anticipated to meet the economic, social and environmental targets (Zhou & Williams, 2013, p. 11). The MEP framework, developed by Ministry of Environmental Protection (MEP) of China and named as such after it, has been in vogue since 2003 and originally it consisted of 28 indicators and subsequently after an alteration in 2007 the number of indicators was revised to 19 indicators. This alteration was facilitated to improve the framework and making it more feasible (Li & Qiu, 2015, p. 58). According to Zou and Li (2014, pp. 22−27), the 19 indicators that currently comprise the MEP framework for the assessment of eco-cities are as follows:

1. Annual net income of farmers;
2. Tertiary industry share in GDP;
3. Energy consumption per unit of GDP;
4. Water consumption per unit of industrial added value/"water efficiency of agricultural irrigation";
5. Compliance rate of enterprises should carry out cleaner production;
6. Forest coverage/percentage of the forestry and grass coverage in alpine area and grasslands;
7. Proportion of protected area in total land area;
8. Ambient air quality;
9. Water quality/coastal water quality;
10. Emissions density of key pollutants;
11. Water quality compliance rate of centralized drinking water source;
12. Centralized municipal waste water treatment/industrial water reuse rate;
13. Environmental qualities of noise;
14. Waste;
15. Urban public green area per capita;
16. Environmental protection investment shares in GDP;
17. Urbanization rate;
18. Centralized heating supply rate in heating region;
19. Public satisfaction rates on the environment.

The fulfilment of all standards and requirements in regard to the indicators of the MEP framework is essential for eco-city development in order for status of the eco-city title to be granted. No further steps are envisioned in the framework and period inspections. Once an approval is accorded to a city to be an eco-city, it will remain like that for an unlimited period of time (Li & Qiu, 2015, p. 62).

Multiple benefits accrue from the eco-city development (SEP Science for Environment Policy, 2018) and these benefits are often in the form of efficient land use, preservation and restoration, efficient transportation management, resource use efficiency, water efficiency, energy efficiency, better standards of living and profits for the private sector (Visvaldisa et al., 2013). In order to ensure preservation of the environment and the natural resources, the urban design assumes added significance for the eco-city construction. Construction of compact cities with high density population is encouraged to ensure that land is used in an efficient manner. Emphasis is also stressed on the renewal and preservation of nature and the biodiversity while embarking on the construction process of an eco-city (SEP Science for Environment Policy, 2018). Traffic jams and the working urban population traveling for long hours to reach their work place along with air pollution and high level of noise emanating from the movement of vehicles are some of the issues for the wellbeing of the urban residents. Nevertheless, such problems get diminished in a cocity where provisions for providing efficient transport system are put in place (Khanna, Fridley, & Hon, 2014; Yigitcanlar & Lee, 2014). Apart from these, several transport strategies that are beneficial for the environment are also implemented in the eco-cities, with specific emphasis on cycling, walking and an increased use of public transport (ICLEI, 2020).

Increasing use of energy and water in the urban areas leads to waste production and in the wake of expanding growth of cities, an adequate management in these areas is necessitated for the sustainable urban development. Along with specific emphasis on water use efficiency, priority needs to be accorded to adhering to measures like the reducing, recycling and reusing approaches on regular basis (Unver, Bhaduri, & Hoogeveen, 2017). The efficient use of energy is another significant factor the requires serious consideration in the eco-city construction. Energy use can be minimized by making use of efficient infrastructure facilities like streetlights. Besides, production of renewable energy and the use other environmental-friendly fuels also help in increasing energy efficiency (Mersal, 2017). Augmenting livability of the cities is the main goal of the eco-city development that becomes beneficial for the city dwellers in multiple ways. Improved living standards are helpful in spurring urban residents to perform in their workplace. The use of renewable energy and decrease in fossil-fuel-using modes of transportation are ideal eco strategies that help in preserving the environment (ICLEI, 2020). The compact urban design encompassing many green areas is creating a livable environment and entails the massive potential of minimizing air and noise pollution. Opportunities for keeping their health in good condition are afforded to the inhabitants of eco-cities by cycling, walking and enjoying the fresh air and serenity of nature in the public parks and gardens. Accrual of benefits can take place by investing in the eco-city development by construction of green buildings because by doing so a great money is expected to be saved in the near future. This estimation is based on the fact of lower operation costs involved in the green buildings (Miller, Pogue, Gough, & Davis, 2009). Apart from rendering the employees working in green building the benefit of good health enabling them perform better in their jobs, other advantages of the green building construction include reduced costs on power, and water use (Sarkar, 2016, pp. 1—2).

Viewed in a broad perspective, five key development requirements are needed in developing eco-cities and these include the following: ecological security, ecological sanitation, ecological industrial metabolism, ecological-landscape integrity, and ecological awareness (Amakpah & Liu, 2015, p. 327). Ecological security denotes the absence of threats to human life, health ease, basic rights, life-ensure sources, essential resources, social sequence; and the capability to acclimate environmental changes (Xiao, Chen, & Guo, 2002). Having been first advanced in China in the 1990s, the concept of ecological security includes natural security, economic security and social security, and they comprise a "complex false ecological secure system." (Jiang, 2011). The factor of "ecological security" takes into account the provision of: "clean air, safe and reliable water supplies, food, healthy housing and workplaces, municipal services and protection against disasters for people" (Amakpah & Liu, 2015, p. 327).

Ecological sanitation or EcoSan, is based on the concept of source control, considering human excreta as a resource instead of waste (Werner, Panesar, Rud, & Olt, 2009). The concept of source control in household entails the potential of high levels of nutrient recovery (Otterpohl, 2001). A vision of source control for household wastewater is based on the fact of different characteristics of gray, yellow and brown water. A vision of source control for household wastewater is based on the fact of very different characteristics of gray, yellow and brown water. The distinctive characteristics of the flows of household wastewater make it discernible that urine contains most of the soluble nutrients, whereas gray water, despite a very large volume compared to urine, contains only a small amount of nutrients (Gajurel & Wendland, 2007). Besides, feces that are about 10 times smaller in volume than urine, contain nutrients, high organic load and the largest part of pathogens.

Although gray water due to personal hygiene and yellow water due to contamination in sorting toilet contain pathogens, they can easily be eliminated. But feces contain as much as 100 million bacteria per gram; some of them are pathogen to human (Carr, 2001). If urine is separated and reused in agriculture, not only nutrients will be reused, but also a high level of water protection will be reached (Gajurel & Wendland, 2007). Nevertheless, the ecological sanitation advocates the existence of efficient, cost-effective eco-engineering for treating and recycling human excreta, gray water, and all wastes (Amakpah & Liu, 2015, p. 327).

Another requirement for eco-city development is "ecological-landscape integrity." Ecological integrity of a landscape is defined by Parrish, Braun, and Unnasch (2003) as the ability of an ecological system to support and maintain a community of organisms that entails species composition, diversity and functional organizational relative to those of natural habitats within a region. This framework is based on the characterization of landscape and ecological systems in terms of composition, structure and function by landscape ecologists and conservation scientists (Noss, 1990). High integrity alludes to a mechanism with natural evolutionary and ecological processes, and minimal or no influence from human activities (Parrish et al., 2003). Development of ecological indicators (EIs) is facilitated through species-specific approaches and these EIs endeavor to measure attributes of a species or community, such as population size or species diversity developing indicators of the absence of human modifications of habitat and alteration of ecological processes has come to be regarded as a complimentary and more general approach. An EI is a measurable characteristic that provides insights into the state of environment and also provides information beyond its own measurement (Noon, 2003). Indicators often serve as surrogates for properties or system responses that are too difficult or costly to measure directly (Leibowitz, Cushman, & Hyman, 1999).

Natural resource agencies and organizations are increasingly engaged in monitoring and evaluating the status and conditions of their respective lands and waters by measuring the ecological integrity of landscapes (IUCN, 2006; USDA, 2016). Many composite scoring mechanisms have been brought into use as an indicator of ecological integrity by mapping the impact of human activities on natural landscapes, including wilderness (Aplet, Thomson, & Wilbert, 2000), and the human footprint (Woolmer et al., 2008). Undoubtedly, these provide general maps of human influence and have been useful (Theobald, 2013); nevertheless, two improvements have been offered recently (Theobald, 2010, 2013). First, landscape ecologists have established that proportion of cover is a primary metric (Gardner & Urban, 2007) because no other landscape metric can be interpreted independent of it (Wickham, Riitters, Wade, & Homer, 2008), and it offers the basis for explicit interpretation required to assess landscape change (Riitters, Wickham, & Wade, 2009). Second, temporary or ad hoc scoring mechanisms like the human footprint is limited because the final score characteristically has no direct physical basis, conversion of quantitative values to ordinal categories entails the potential of violating mathematical axioms, and colinearity of individual factors results in difficulty when interpreting results (Schultz, 2001). Decision theory makes available formal methods to elicit transparent accountable indicators, such as multiple criteria analysis (Hajkowicz & Collins, 2007).

Thus the ecological-landscape integrity helps in promoting the creation of built structures, open spaces like parks and plazas, connectors such as streets and bridges, and natural features like waterways and ridgelines. Besides, ecological-landscape integrity also helps in increasing accessibility in the urban areas as far as possible along with preservation of the biodiversity. The underlying objective in achieving these twin goals in an efficient manner is to help preserve energy and natural resources along with the prevention of "automobile accidents, air pollution, hydrological deterioration, heat island effects and global warming" (Amakpah & Liu, 2015, p. 327).

Another contributing factor in the eco-city development is the ecological awareness that suggests that inhabitants of the urban areas ought to get a better perspective of as to how they can live in harmony with nature. It is imperative that the inhabitants of the urban areas get proper awareness of the urban "cultural identity and responsibility for the environment." Issues pertaining to environmental protection, with specific emphasis on the problems concerning ecological awareness of society have been receding increased focus in many countries (Piekarski, Dudziak, Stoma, Andrejko, & laska-Grzywna, 2016). Denoting a human attitude to nature that feature a great onus for its conditions, ecological awareness emerges from a thorough knowledge about nature as well as determinants to preserve it in its primary state (Kollmuss & Agyeman, 2002). Determinants of ecological awareness, inter alia, include respect for nature, strict compliance with the rules and regulations pertaining to protection of nature and counteracting environmental risks. Furthermore, ecological awareness is sometimes looked upon as the extent of urban population's interest in nature, and its protection; their opinions about environmental deterioration, and their knowledge concerning probable actions in augmenting conservation of wildlife (Malgand et al., 2014). Ecological awareness is escribed by Buger (2005) as a set of data and beliefs about nature and perception of links between the state and the character of the natural environment and the conditions and the quality of human life. It also represents attitudes of the urban residents toward the natural environment, the set of collected data and beliefs about it, and the system of values that guide human behavior (Prevot-Julliard, Clavel, Teillacdeschamps, & Julliard, 2011).

Ecological awareness has been delineated in descriptive and axiological-normative terms. The descriptive and technical approach delineates it as a combination of information and beliefs about the natural environment and the perception of links between the state and the character of the natural environment and the conditions and the quality of human life (Piekarski et al., 2016). Nevertheless, the term axiological-normative projects ecological awareness as a biocentric system, that is, understanding, experiencing, and evaluating the processes that occur in the biosphere and the entire set of interdependencies between society and nature, as well as the assessment of ecological ethics (Buger, 2005). It also reflects views and perceptions of the urban residents regarding the role of the environment in human life, anthropogenic load, the degree of overexploitation, threats and protection, including the state of knowledge about the methods and tools harnessed in managing, protecting, and shaping the environment (Prevot-Julliard et al., 2011).

Keeping in view the complex and multidimensional nature of the concept of ecological awareness, the literature on the subject differentiates its various types, sizes, shapes and components, and the basic typology permits for the identification of individual and social awareness; and as far as with regard to its dimensions, it can be delineated into real and ideal dimensions (Piekarski et al., 2016). Ecological awareness can be articulated in different forms, such as commonsense thinking, ideologies, public opinion, and/or scientific knowledge. Emotional-evaluative component of ecological awareness comprises system of values adopted by people and society; the cognitive component comprises the vision of the model of civilization and normal science; and the behavioral component embraces human needs and demands pertaining to the use of natural wealth and individual activities for environmental protection (Piekarski et al., 2016). Ecological awareness, according to Papuzinski (2006), is not only knowledge, views, and ideas about environment; rather, it is a paradigm that imbibes the values, ideas and opinions about the environment wherein an individual resides and operates the same as the all of society, Nevertheless, ecological awareness for Kalinowska (1997) is rather an ambiguous concept that alludes to different situations and mental states, and hence, various levels of this kind of consciousness may be differentiated. First, one can deal with an intuitive belief that the existence of some elements is in danger or puts the health or other requirements of an individual at risk. Second, an intuitive belief can be joined with knowledge pertaining to mechanisms of threat; and third, ecological awareness comprises an emotional reaction to the observed reality.

In the absence of a comprehensive and universally acceptable definition of the term "ecological awareness," this term has come to be described by different scientific bodies in terms of citing its individual components that are often named in diverse ways. Ecological awareness, according to Mirowski (1996), comprises ecological knowledge that is expressed via the familiarity with and understanding of means of coexistence between man and nature; and the ecological sensitivity or emotional attitude to nature; along with proecological attitude or actions undertaken to safeguard the natural environment. Environmental awareness entails broad coverage encompassing relevant aspects like geography, ecology, philosophy, envirnomics, ethics, law, political science, and more (Piekarski et al., 2016). Some scholars have opined that the prime meaning of environmental awareness includes natural concepts and values, environmental science and technology, environmental law and policy, environmental ethics, and environmental psychology (Li, Dan, & Liu, 2013). On the contrary, ecological behavior comprises actions that could be initiated by any human being and that exert a positive impact on the environment (Milfonta & Duckitt, 2010). The prime significance of ecological awareness lies in the process of molding public attitudes toward the natural environment and that makes it an important subject of research, especially in the area of the relationship between man and environment (Best, 2010; Klockner, 2013). Besides, the perception of environmental awareness by current and potential users of the environment as well as the identification and evaluation of attitudes and behaviors toward sustainable development issues and consequently the protection of the environment seem to be significant issues in this field. Thus it becomes, on the one hand, a significant factor for embarking on further analyses in this arena and on the other it could be the starting point of educational or operational activities conducted by authorities at several levels regarding the promotion of ecological behavior of both individuals and society as a whole (Piekarski et al., 2016). Ecological awareness encourages public involvement in the preservation and improvement of the urban ecosystem needs and this needs to encouraged and present in order that the eco-city project could be successfully supported (Amakpah & Liu, 2015, p. 327).

In sum, it can be said that the eco-city development projects are looked after and created in different ways and by several actors in the countries around the globe. In the eventuality of different eco-city projects being developed in one country, the possibility of their development being carried out in different ways cannot be ruled out; and in that scenario both central government and regional authorities are involved in these projects. This calls for close cooperation between the national and regional authorities along with private business and social initiatives for the eco-city development (Hu et al., 2016, p. 86). Participation of public and private actors in the governance process of eco-city development entails the structure of governance being "vertical" when it comes to connection with various types of organizations and "horizontal" when it comes to international cooperation and knowledge transfer and it is here that the governance is called upon to delineate the significance of the urban infrastructure networks and to harness the urban socio-technical systems (Bulkeley & Marvin, 2014, pp. 32−33).

7.6 Toward sustainable smart cities

Cities have emerged as the primary human living space where 55% of the world's population lived in 2018 and it is expected to increase to 68% by 2050 (UN-DESA, 2019). Cities have become places where agglomeration of economies endeavors to achieve their highest yields, generating cultural, economic and social benefits (UN-Habitat, 2001). Cities can afford multiple benefits. Cities, with human agglomeration, financial investments and resource utilization, heighten the prospects of economic development, innovation and social interaction (UN-DESA, 2013). Cities also make it possible lower unit costs so as to make available public services such as water and sanitation, healthcare, education, electricity, emergency services and public recreational areas (Satterthwaite, 2010). Nevertheless, a functioning city government is required to deliver these benefits and such a government should be able both to ensure that such accruing benefits are realized, and to adopt a sustainable framework that spurs city's growth within ecological limits (UN-DESA, 2013). However, cities along these lines are also vulnerable to challenges that endanger their endeavors to attain sustainability, for instance, via "improvement of access to, and efficiency in the use of public services, as well as reduction of their ecological footprint and financial fragility, and the building of resilience against the impact of natural hazards" (UN-DESA, 2013).

In the wake of burgeoning urban population (Rosenberger, 2018), it is becoming increasingly challenging to solve the requirements of the citizen toward improving their life quality and keep the processes in the cities effective from the operational and administrative perspectives. Given the fact of already dense population of the cities along with increasing influence of globalization, the existing problems of the cities are prone to become bigger and new ones may also emerge. Growing proximity between global trends and events and cities entail the potential of enlarging the scope of local impact affecting the economy, demographics, environment and other factors. The series of problems that emanate from the growing urbanization patterns entail the potential of reducing quality of life in urban settlement, such as inequality, pollution, ageing population, insecurity, etc. (Arouri et al., 2014). Admittedly, cities attract people looking for better social and economic opportunities; nonetheless, growth in global urbanization also brings with it increasing environmental strain (IRP, 2018). Elaborating on this aspect further, Glasmeier and Christopherson (2015) have opined that, "this will pose serious challenges for city planners, who will have to [rethink] how they provide basic city services to residents in a sustainable manner." While emphasizing that the future cities would be those that actively engage their inhabitants in the promotion of progress and improvements, rather than merely reconciling to those advancements, Smart City Expo World Congress (2016) also recognized that technology is giving rise to "a new kind of urban growth—one that is more efficient, transparent and sustainable, disrupting the form and function of our cities.". Increasing use of technology along with more frequent involvement of citizens are proving instrumental in transforming the urban landscape in novel ways and at much faster pace than ever before, thereby unfolding new opportunities for a lucrative industry segment that has come to known as smart city. There are varied estimates about the global smart city market.

According to Glasmeier and Christopherson (2015), the global smart city market would be worth US$1.565 trillion by 2020. Another assessment projects the smart city market to reach US$2.825 trillion by 2026 in comparison to US$ 345 billion in 2016 (medium.com, 2019). Another report pegs the growth of smart city market at US$3.482 trillion by 2026 (PRNewswire, 2017). In other words, smart cities are coming into the reckoning and gaining international traction.

Burgeoning global urbanization trends coupled in tandem with the quest of sustainable development have proved instrumental in the emergence of the concept of sustainable cities that has attracted the attention of many scholars, researchers and practitioners in the field as a desired goal for future urban development. The challenge has been to ensure that cities afford, for current and future generations, improved living conditions to their citizens (Al-Nasrawi, Adams, & El-Zaart, 2015). These conditions encompass four aspects—economic, technological, social, and regulatory—and Information and Communication Technology (ICT) has come to be regarded entailing high potential for solutions to many of these impediments confronting the cities while ensuring being friendly and viable to the urban environment as well (Townsend, 2013). The notion of Sustainable Smart City (SSC) has become a relatively positive connotation (Forbes, 2014), with many countries around the globe aiming at being associated with this concept as a symbol of prestige and development. Some countries claim to have developed the technical infrastructure required for an SSC and concentrated on sustainable development policies, whereas other some countries emphasize on having improvised on their e-government services implementation that is regarded to be a precondition for an SSC (Lee, Jin, & Choi, 2013). Nevertheless, these claims serve as no testimony of a SSC because there is still no assessment model to evaluate the performance of SSCs and the prevalent "models are not sensitive to the needs, resources, priorities and wider context for individual cities" (Al-Nasrawi et al., 2015).

The terms "smart city" (SC) and "sustainable smart city" have come to be used interchangeably in the literature (Ahvenniemi, Huovila, Pinto-Seppa, & Airaksinen, 2017) and in to comprehend the linkages between SC and SSC, it deems appropriate to appraise "sustainable," "smart" and "city" to have a comprehensive view of these notions.

7.6.1 Sustainable concept

Term "sustainable" is derived from the concept of sustainability that originates from the Latin word "sustinere" which means "to hold up," "to endure." Dictionary meaning of "sustainability" is "to give support" to "to keep up" (Webster's New Collegiate Dictionary, 2008, p. 1478); "Sustainable" is defined as capable of being sustained said of economic development and capable of being maintained at a set level (Chambers 21st Century Dictionary, 2007, p. 1425); and "Sustainable" is also used to referring to a way of using natural products so that no damage is caused to the environment and "Sustainability" means using the resources in a way that does not cause environmental damage (Cambridge International Dictionary of English, 2005, p. 1472). Being a normative and socially constructed notion, the term "sustainable" is used to denote development level. This means that the definition of sustainable development or sustainability cannot be, in a broader sense, based on an inductive approach, and the concept needs to be delineated on the basis of an inductive approach or top-down approach that alludes to the cognitive process of developing the definition and ought not to be "confused with the extent of participation in the process" (Höjer & Wangel, 2015). Viewed in this context, the definition of sustainable development as delineated by Brundtland Commission in its report as, "Sustainable development is development that meets the needs of the present without compromising the ability of future generations to meet their own needs" (WCED, 1987), emphasizes on two aspects—the first aspect of "needs," especially the needs of the world's poor to which overriding needs to be accorded; and second, the idea of limitations imposed by the state of technology and social organization on the ability of the environment to meet the present and future needs (Höjer & Wangel, 2015). Viewed in a wide spectrum the definition of sustainable development provided by the Brundtland Commission is seemingly global in nature and its usage at country-specific level requires an amendment in it. Nevertheless, the Swedish government has seemingly solved this dilemma by redefining a so-called "generational goal" by stating that, "the overall goal of Swedish environmental policy is to hand over to the next generation a society in which the major environmental problems in Sweden have been solved, without increasing environmental and health problems outside Sweden's borders" (Gabrys, 2014). Such an addendum can be useful not only for countries but also for their constituents such as cities and connotes the context of the city (Al-Nasrawi et al., 2015; Höjer & Wangel, 2015).

7.6.2 "Smart" concept

Usage of the term "smart" as an acronym (SMART) is credited to both Drucker (1955) and Doran (1981), nonetheless, it is difficult to ascertain whether either of these two were really the first scholars to use term SMART with reference to goals objectives in management. The acronym SMART stands for—specific, measurable, achievable, realistic, and timely—is now in common use to set goals and objectives within appraisal and performance management systems (CMI, 2011). Notion of "Specific" envisages outlining in a clear statement precisely as to what is required; notion of "Measured" entails a measure one's ability to monitor progress and to ascertain when the goal is attained; notion of "Achievable" denotes that goals can be designed to be challenging; notion of "Realistic" focuses on outcomes rather than the means of attaining them; and notion of "Timely" means time-bound strategy so as to agree the day and date by which the outcome must be achieved (CMI, 2011).

Term "smart" has come to be construed as controversial because of the controversy whether it is instrumental or a normative concept. Viewing "smart" from an instrumental perspective connotes a type of products, services and systems wherein ICT plays a pivotal role; nevertheless, there prevails disagreement over this explanation in the literature (Al-Nasrawi et al., 2015). For some scholars, the notion of "smart" represents a desired outcome rather than an instrumental concept that makes it a normative concept as is the case with sustainability (Allwinkle and Cruiskshank, 2011). Hollands (2008), later echoed by Kitchin (2014) and Allwinkle and Cruiskshank (2011), also saw smart not as instrumental but as an intended outcome that makes smart just as normative as sustainable. On the contrary, Neirotti, De Marco, Cagliano, Mangano, and Scorrano (2014, p. 25) remark on the importance of not being misled by the word smart: "the number of 'smart' initiatives launched by a municipality is not an indicator of city performance, but could instead result in an intermediate output that reflects the efforts made to improve the quality of life of the citizens."

In marketing parlance, smartness reflects a perspective of a user (Klein & Kaefer, 2008) and in the realm of urban planning, the smartness in smart growth is regarded as a normative claim and ideological dimension (Al-Nasrawi et al., 2015). The notion of smartness is increasing embraced by the governments and public agencies at all levels to differentiate their new policies, strategies and programs for targeting sustainable development, sound economic growth and better quality of their citizens (Centre on Governance, 2003). Manifestation of smartness in smart technologies needs to be focused on because technologies had permeated into commercial application of "intelligent-acting" products and services, artificial intelligence (AI), and thinking machines (Moser, 2001). Smartness in the context of technology entails the automatic computing principle, such as self-configuration, self-healing, self-protection, and self-optimization (Spangler et al., 2010). Mobile terminals, and embedded devices along with connected sensors and actuators are the sort of equipment that is often installed larger smart constructions like airports, hospitals offices of large corporations, smart homes and other smart buildings (King & Christopher, 2017). Some scholars have come to regard a smart ecosystem as a conceptual extension of smart space from the personal context to the community and large and the entire city (Jucevicius & Grumadaite, 2014).

7.6.3 "Cities" concept

The English word "city" is derived from the Latin "cīvitās," meaning a highly organized community; city-state (UN, 2014). While Mumford (1937) perceives a city as "a theater of social drama," Merriam Webster Dictionary describes it as "an inhabited place; a place larger than a village or town: a large, prominent or important center of population: a relatively permanent and highly organized center having a population with varied skills, lacking self-sufficiency in the production of food, and usually depending on manufacture and commerce to satisfy the wants of its inhabitants" (Merriam Webster Dictionary, 2004). A city is construed as an agglomeration of "population contained within the contours of contiguous territory inhabited at urban levels of residential density"—extensive level—metropolitan region: "includes both the urban agglomeration and additional surrounding areas of lower settlement density that are also under the direct influence of the city" (UN, 2014; UN-DESA, 2012). Both the notions of smart and sustainable have increasing come to be associated with cities thereby indicating the types of human structures and environments where smart solutions of problems pertaining to sustainable development can be ascertained. The concept of cities, like the notion of "smart," cannot be construed as instrumental primarily because the existence of cities is taken for granted (Al-Nasrawi et al., 2015). While taking into account an appraisal of the adequacy of cities for sustainable development, it is more significant to focus on as to what should be done to make cities more sustainable.

Owing to their potential in playing pivotal role in social and economic realms globally and wielding enormous impact on the environment, cities are reckoned with as key elements for the future (Mori & Christodoulou, 2012). Faster pace of consumption of global resources by the cities, contributing on the one hand, to their increasing economic importance; and on the hand, their poor environmental performance, has given rise to accord priority to promoting sustainability in many cases, often construed through the promotion of natural capital stocks (Al-Nasrawi, et al., 2015). Nevertheless, some recent construal of urban sustainability has increasingly emphasized on a more anthropocentric approach that focuses on the human beings as the central fact of the universe and according to that, cities should respond to human requirements through sustainable solutions for social and economic aspects (Turcu, 2013).

Combining these three different terms leads to the introduction of several definitions for the concept of cities by the international community, global agencies, governments. Academia, and private sector. Accordingly, in the available literature on smart cities' definitions the sustainability component is explicitly referred to in the definition in some cases while not in others (Al-Nasrawi et al., 2015). With a view to ensure that the sustainability component in smart cities is not left out/excluded, the International Telecommunication Union (ITU) Telecommunication Group on Smart Cities (ITU-TFG-SSC, 2014) conceptualized the new term "smart sustainable cities" that has come to be taken as "a variant of smart cities including some of the core features of eco-cities sustainable cities" (Al-Nasrawi et al., 2015).

7.6.4 Defining a smart city

Admittedly, the concept of smart city, having first emerged in the 1990s (Alawadhi et al., 2012), over the past decades has garnered international traction to help cities looking for creative solutions to their pressing urban challenges; nonetheless, there has been lack of general consensus as to what in fact a smart city means. Rather there has been a plethora of broad and vague definitions of the smart city concept, each applying a different connotation to a malleable concept. Concept of smart city has been construed by some as an alternative to convention planning modes, harnessing new technologies, especially ICT, to deal with these problems. Smart cities are often looked upon as a tool to tackle urban challenges in a progressively urbanized world (Albino, Berardi, & Dangelico, 2015; Meijer & Bolivar, 2016). Irrespective of the fact that concept of smart city has been thought of as a tool to solve the composite set of challenges confronting urban settlements; nevertheless, several scholars have pointed out that despite such numerous endeavors, there is still lack of general consensus on the definition of this concept (Albino et al., 2015; Fernandez-Anez, Fernández-Güell, & Giffinger, 2017). New approaches have been advanced suggesting incorporation of different aspects of the smart city for its contextualization and performance analysis (Lombardi, Giordano, Farouh, & Yousef, 2012). Many scholars have attempted to apply this concept of holistic smart city in their research proposals (Alawadhi et al., 2012; Fernández-Güell, Collado-Lara, Guzmán-Araña, & Fernández-Añez, 2016).

Paucity of a general consensus as to what constitutes the definition of a smart city has spurred some scholars to design conceptual and typological approaches to furnish a systematic understanding of smart city concept and policies; and in that pursuance some scholars have concentrated on the vital components of smart city, discerning the equilibrium between people, technology and institutions (Ben Letaifa, 2015) as pivotal to a city to be considered Smart. Other propositions suggesting categorization of smart city concepts and policies are seemingly mulled by various schools of thought (Kummitha & Crutzen, 2017) or a spatial approach, and moot other strategic choices sans any particular spatial allusion, such as focusing on society, innovation or business models (Angelidou, 2014). Focus on governance by scholars like Meijer and Bolivar (2016) brings them in close proximity to the ideas of Ben Letaifa (2015) and Colldahl, Frey, and Kelemen (2013). Smart city definitions have been categorized by Meijer and Bolivar (2016) in terms of technology, human resources and collaboration, embodying a fourth option that combines the three together in a holistic approach; and according to this approach, urban development should take into account the interrelationships between infrastructure, society and institutions. Interestingly, this concept of holistic smart city has found articulation in the research proposals of many scholars (Alawadhi et al., 2012; Leydesdorff & Deakin, 2010). The dearth of general consensus about definitions, according to Fernandez-Anez, Velazquez, Perez-Prada, and Monzón (2018), has resulted in the usage of the smart city conceptualization as a tool for researchers and several other stakeholders to delineate their position.

In the wake of smart city concept garnering increasing international traction (Fernandez-Anez et al., 2018; Verret & Pfeffer, 2019), the notion of smart city is seemingly more than a concept that is required to be theoretically delineated. Primarily, two key approaches may be ascertained among smart city scientists and practitioners. Viewed in a broad spectrum, the scientific literature endeavors to look beyond sector-specific approaches by suggesting a wide-ranging conceptualization of the smart city whereas, on the other hand, initiatives pertaining to smart city are worked out via sector-based approaches and projects in one or few particular areas (Fernández-Güell et al., 2016). Interestingly, execution of initiatives pertaining to smart city is still associated to these sector-specific and partial understanding, partly attributable to the limitations of governance and financing tools. Hence, there is need to engulf the gap between the theoretical extensive perspective and the sector-wide execution of the concept of smart city (Dykan, Ieromyna, Storozhylova, & Bilous, 2019). Irrespective of wide-ranging deliberations over smart city concept, little is discernible about the modus operandi of the execution of Smart Cities in effect, and this given rise to a number of penetrating questions, such as: How is the smart city concept being implemented? Are smart cities really capable of confronting current urban challenges? Is the goal of involvement of the different stakeholders of the society in implementing smart city concept being achieved? Ascertain answers to these and other related queries essentially require an understanding of different elements separately and in an extensive and interrelated manner.

Emphasis on the notion of smart city needs to be extended beyond the conceptual level and essentially to focus on how it is implemented. Generally, Smart cities are seldom built from scratch but rather via the implementation of smart city projects that ought to be tactically implanted into a smart city strategy (Angelidou, 2014). Usually, efforts are made by cities to tackle problems using a "silo" approach, that comprises municipal departments working discretely (Fernández-Güell et al., 2016; The Scottish Government, Scottish Cities Alliance, & Urban Tide, 2014). In pursuance of this sectoral approach, cities that aim at becoming smart cities often focus on augmenting specific socioeconomic aspects of daily life, such as housing, commerce, business, governance, health, education, or community (Angelidou, 2014). Follow up of this sectoral approach is mostly the outcome of the diffused disbursement of resources and financing solutions for particular types of projects. An example of this type is suggested by Eurostat (2016) wherein a project aims at optimal energy and resource efficiency, arguably by integrating technologies in the realms of transport, energy, and governance. Nevertheless, there arises a need for approach urban complexities with integrated strategies (Fernández-Güell et al., 2016), and the smart city approach is prone to progressively focus on dealing with urban problems with a holistic approach. Accordingly, smart city projects are progressively coordinated by smart city strategies in cities, as has been in the case of Vienna (City of Vienna, 2014). The significance of discerning cities as a solution for urban challenges was emphasized by the third UN-Habitat conference on Housing and Sustainable Urban Development (UN-Habitat III, 2016). Execution of smart city projects is facilitated to address multiple challenges in an increasingly urbanized world (Albino et al., 2015; Anthopoulos, Janssen, & Weerakkody, 2015). Consequently, the eventual objective of smart cities ought to be to become a solution for urban challenges. It is equally significant to confront these challenges from a multiscalar perspective, from regional to local (UN-Habitat III, 2016). Accordingly, the regional challenges framework can serve as a basis for ascertaining the challenges being confronted by a particular city. Focus of reports of international entities as well as scientific literature is centered on identifying challenges of cities or regions like the European region (European Commission, 2011; Nijkamp & Kourtit, 2013), with a view to have wide-ranging understanding of the concerned issues.

The notion of governance is progressively occupying a central place in the goal of making the smart city a comprehensive concept and has come to be regarded as one of the key elements for the successful implementation of the smart city concept (Meijer, 2016; Meijer & Bolivar, 2016). The linkages between smart governance and the essentiality for integrated approaches has also been endorsed by some scholars (Castelnovo, Misuraca, & Savoldelli, 2015). A discernment of the meaning of governance within the smart city context is considered vital for understanding the notion of smart city in its entirety (Mutiara, Yuniarti, & Pratama, 2018).

A classification of smart governance conceptualization at four levels has been proposed by Meijer and Bolivar (2016), and these four levels are— smart government, smart decision-making processes, smart administration and smart urban collaboration among various actors. This fourth level is being regarded as most transformative because involvement, collaboration and engagement of stakeholder is regarded by some scholars as a sine qua non for smart governance (Ruhlandt, 2018), and the key component of a smart city (Albino et al., 2015; Giffinger & Lü, 2015). Articulation of different visions by stakeholders of the smart city in their discourses (Fernandez, 2015; Fernandez-Anez et al., 2017) is helpful in reaching a collaborative level by incorporating different visions articulated by different stakeholders. Governance, according to Stoker (1998), is getting to implement issues confronting complexity, conflict and social change. This leads to surmise that implementing the policies is one of the indicators of good governance. Governance entails "a process of coordination of actors, social groups of institutions to achieve objectives" (Le Galès, 2004). Differentiation between the image of the smart city and its implementation has been outlined by De Santis, Allesandra, Nadia, and Anna (2014), and differences between the vision of the stakeholders in smart city development and the initiatives carried out is pointed by Alawadhi and Scholl (2013). Illation or outputs of smart governance has come to regarded as a vital element in the analysis of literature on the subject (Meijer & Bolivar, 2016; Ruhlandt, 2018), and scholars like Nam and Pardo (2011b), consider smart governance as the promotion of smart city initiatives. Smart governance can be a catalyst for the materialization of the smart city, as a consequence of the involvement of stakeholders, and as such smart city implementation should reflect the visions of different stakeholders (Mutiara et al., 2018). It can be conjectured that reducing the gap between the vision of stakeholders about smart city initiatives and the execution of the concept of smart city via projects can help in improving smart city governance strategies, as governance constitutes one of its key elements.

While fostering the smart city implementation, the EU proposed guidelines for its development vide different documents and initiatives. The view of smart city fostered by the EU evolved from a focus on ICT, energy and transportation (The European Commission, 2013) to a more extensive and integrated concept, like in the Marketplace of the European Innovation Partnership on Smart Cities and Communities (EIP-SCC), an initiative that brings together cities, industry, SMEs, banks, research and other smart city actors (European Commission, EIP-SCC European Commission, European Innovation Partnership on Smart Cities, & Communities, 2016). While emphasizing strongly on governance, policies of the EU on smart cities have focused on cementing the relationships among the stakeholders as the basis for the development and the implementation of the concept (European Commission, 2016). Establishing links between stakeholders and different groups is one of the prime roles of the Marketplace. Cities in the EU enjoy different worldwide commercial rankings on smartness in cities (IESE, 2016; Networking Society Lab, 2016). Multiple networks of cities have been set up, such as European Living Labs Network or the Marketplace itself. Involvement of stakeholders is vital for the focus of smart cities in the EU. A smart city is defined as a "multistakeholder, municipally based partnership" in a report entitled, Mapping smart cities in the EU (Manville et al., 2014). Thus EU in its concept of a smart city accords specific significance to involvement of stakeholders.

Undeniably, the smart city concept is often understood as intimately associated with the development of new technologies, albeit from a historical perspective, the scientific approach for studying cities and optimizing the management is not new (Allam & Newman, 2018). As a constantly developing notion over time, the smart city concept now, in the early 21st century, has come to be regarded, in many ways, connected with digital technology, big data, IoT and related innovative solutions that can become applicable in cities. Nevertheless, technology in itself is insufficient make a city "smart" instantly, because the notion of a smart city is a multidimensional concept entailing other elements as well (Hahnrath, 2017). Forming partnerships between various stakeholders along with application of open innovation are seemingly essential methods to help cities be more effective in development and achieve their goals (Boyes, Hallaq, Cunningham, & Watson, 2018). Remarkably, Toronto was among the top 50 smart cities globally in 2018; nonetheless, a case study there demonstrated that even through several discussions with local stakeholders, between private companies, citizens, and public officials, an agreement could not be reached, to define what "Smart" should really mean in the city context (Eden Strategy Institute & ONG&ONG Pte Ltd, 2018).

Each city is uniquely placed and requires the necessity of developing a custom strategy toward becoming a "smart" city by learning from each other and sharing best practices via pursuing platform approach and open sharing innovation practices. Apart from emphasizing on governance, special focus in many emerging strategies for smart city is centered on administrative efficiency and interoperability, improvement of services, and citizen-centric approach as added value (Pereira, Paryceka, Falcoc, & Kleinhans, 2018). The term "smart" does not necessarily refer to the size or the urban or rural geographical setting because even a community can be smart as well, especially when it has the ability to find solutions, overcome challenges and embark on innovation. Broadly speaking, transformation toward a smart city or community represents a sum total of smaller initiatives and the collaboration between private and public actors across various domains of action, rather than pursue a master plan or a single technology partnership (Gorynski, 2017). A deeper understanding of the term "smart city" can be had by looking into what a "city" means. In common parlance, a city is a relatively large and permanent human settlement (Jiang & Miao, 2014; Wirth, 1938). Cities have come to reckoned with as significant contributors to economy, education, science and other fields. As complex ecosystems that are dependent on natural systems, cities are "challenging our thinking about the development of both natural and urban environment" (Madakam, 2014).

In the literature, connotation smart city has often come to be replaced with other words like "digital," "intelligent," "future," "ubiquitous" or "virtual," and with the usage of such words the perspective of a city can undergo a change very much and while the mentioned adjectives are mostly interrelated and sometimes even used interchangeably. Nevertheless, "smart city" has seemingly come to be regarded as more user-friendly among these. While referring to interpretations that suggest that "smart" is by default also "intelligent" Albino et al. (2015) find "smart" as better for marketing purposes, as compared to other terms that are less inclusive.

Thomas Muller, the Co-Founder of bee smart city GmbH, has defined a smart city as "an ecosystem of people, processes and solutions," where "the most important driver of success is collective effort—the sum of many individual actions taken in pursuit of a shared goal" (cited in Barlow & Lévy-Bencheton, 2019). A smart city is defined by European Commission (2016) as a "place where networks and services are made more efficient with the use of digital and telecommunication technologies for the benefits of its inhabitants and business." While extending beyond technological solutions, European Commission's definition of smart city also perceives it a responsive administration that involves the citizens and addresses the needs of most social categories and groups. This bears resemblance with the ecosystem approach wherein people share their collective endeavors and the processes and solutions respond to attaining common goals (Barlow & Lévy-Bencheton, 2019). Concept of a smart city is one "where citizens, objects, utilities, etc., connect in a seamless manner using ubiquitous technologies, so as to significantly enhance the living experience in 21st century urban environments" (Northstream, 2010). While enunciating an identical definition, Harrison et al. (2010) have combined the technology and social aspects to define smart city as a city connecting the physical infrastructure, the information technology (IT) infrastructure, the social infrastructure and the business infrastructure to leverage the collective intelligence of the city. While emphasizing a strong focus on the development of innovative technological solutions, these definitions espouse for citizen-centric approach. Viewed in a broad perspective, a city is said to be smart when investments in human and social capital, traditional infrastructure and disruptive technologies foster sustainable economic growth and a high quality of life, with a judicious management of natural resources in a participatory governance mode. According to van Dijk (2011), smart solutions across all sectors of society cause emergence of smart cities and combine changing human behavior by harnessing data and innovative technology. Select Definitions of smart city are provided in Table 7.2.

In the context of a smart city, Albino et al. (2015) have pointed out two domains—the hard domain and the soft domain. In the hard domain, focus is on things like infrastructure, mobility, environment and ICT solutions, while the soft domain focuses on culture policy, education and aspects that seldom involve ICT. Since the usage of the term "smart city" across industries emphasizes focus on one of these two domains, perhaps it is one of the key factors attributable to the lack of general consensus on the definition of smart city (Albino et al., 2015). Easy accessibility of services for the needy, active participation of citizens in governance, along with a dynamic, engaged social community are some of the main advantages of a truly developed city (Deloitte, 2015; Hamilton Emily, 2016). Besides, factors like safety, mobility, economy and others that bear impact on the quality of life of the citizens also assume significance and; therefore, require sustainable solutions.

7.6.5 Dimensions of a smart city

The fact that smart cities are multidimensional has garnered ample recognition in existing research (Duan, Nasiri, & Karamizadeh, 2019; Sharma & Kumar, 2019; Ueda & Nara, 2018). Cohen and Munoz (2016) has outlined six main dimensions of fields of action of a smart city—mobility, economy, government, environment, living, and people—and strategy and actions to be adhered to by a

Table 7.2 Select definitions of smart city.

Definition	Reference
A city well performing in a forward-looking way in economy, people, governance, mobility, environment, and living, built on the smart combination of endowments and activities of self-decisive, independent and aware citizens. Smart city generally refers to the search and identification of intelligent solutions which allow modern cities to enhance the quality of the services provided to citizens.	Giffinger and Gudrun (2010).
"A city is smart when investments in human and social capital and traditional and modern communication infrastructure fuel sustainable economic growth and a high quality of life, with a wise management of natural resources, through participatory governance."	Caragliu and Del Bo (2012)
"A city connecting the physical infrastructure, the IT infrastructure, the social infrastructure and the business infrastructure to leverage the collective intelligence of the city."	Harrison et al. (2010)
Smart cities will take advantage of communications and sensor capabilities sewn into the cities' infrastructures to optimize electrical, transportation, and other logistical operations supporting daily life, thereby improving the quality of life for everyone.	Chen (2010)
"A smart city in one that is able to link physical capital with social one, and to develop better services and infrastructures. It is able to bring together technology, information and political vision into a coherent program of urban and service improvements."	Correia, Farhi, Nicolini, and Teles (2011)
"[Smart cities are about] leveraging interoperability within and across policy domains of the city. Smart city strategies require innovative ways of interacting with stakeholders, managing resources and providing services."	Nam and Pardo (2011a)
(Smart) cities as territories with high capacity for learning and innovation, which is built-in the creativity of their population, their institutions of knowledge creation, and their digital infrastructure for communication and knowledge management.	Komninos, Bratsas, Kakderi, and Tsarchopolous (2015)
"Cities are becoming smart not only in terms of the way we can automate routine functions but in ways that enable us to monitor, understand, analyze and plan the city to improve the efficiency, equity and quality of life for its citizens in real time."	Batty et al. (2012)
"Smart cities are also instruments for improving competitiveness in such a way that community and quality of life are enhanced."	Batty et al. (2012)
Smart city as a high-tech intensive and advanced city that connects people, information and city elements using new technologies in order to create a sustainable, greener city, competitive and innovative commerce, and an increased life quality	Bakici et al. (2013)
Being a smart city means using all available technology and resources in an intelligent and coordinated manner to develop urban centers that are at once integrated, habitable, and sustainable.	Barrionuevo, Berrone, and Ricart (2012)
Smart cities are the result of knowledge-intensive and creative strategies aiming at enhancing the socio-economic, ecological, logistic and competitive performance of cities. Such smart cities are based on a promising mix of human capital (e.g. skilled labor force), infrastructural capital (e.g. high-tech communication facilities), social capital (e.g. intense and open network linkages) and entrepreneurial capital (e.g. creative and risk-taking business activities).	Kourtit and Nijkamp (2012)

Table 7.2 Select definitions of smart city. *Continued*

Definition	Reference
Smart cities have high productivity as they have a relatively high share of highly educated people, knowledge-intensive jobs, output-oriented planning systems, creative activities and sustainability-oriented initiatives.	Kourtit, Nijkamp, and Arribas (2012)
Smart city [refers to] a local entity - a district, city, region or small country - which takes a holistic approach to employ[ing] information technologies with real-time analysis that encourages sustainable economic development.	IDA Singapore (2012)
A smart city is understood as a certain intellectual ability that addresses several innovative socio-technical and socio-economic aspects of growth. These aspects lead to smart city conceptions as "green" referring to urban infrastructure for environment protection and reduction of CO_2 emission, "interconnected" related to revolution of broadband economy, "intelligent" declaring the capacity to produce added value information from the processing of city's real-time data from sensors and activators, whereas the terms "innovating", "knowledge" cities interchangeably refer to the city's ability to raise innovation based on knowledgeable and creative human capital.	Zygiaris (2013)
"The idea of smart cities is rooted in the creation and connection of human capital, social capital and Information and Communication Technology (ICT) infrastructure in order to generate greater and more sustainable economic development and a better quality of life."	European Parliament (2014)
"Smart City is a city seeking to address public issues via ICT-based solutions on the basis of a multi-stakeholder, municipally based partnership."	European Parliament (2014)
Smart Cities initiatives try to improve urban performance by using data, information and information technologies (IT) to provide more efficient services to citizens, to monitor and optimize existing infrastructure, to increase collaboration among different economic actors, and to encourage innovative business models in both the private and public sectors.	Marsal-Llacuna, Colomer-Llinàs, and Meléndez-Frigola, (2015)

smart city are designed for and are about innovating the urban environment to improve these six dimensions (Pereira et al., 2018; beesmart city, 2019). These dimensions are briefly explained.

7.6.5.1 Smart mobility

In common parlance, the notion of smart mobility embraces the infrastructure and aims at providing increasingly efficient and qualitative transportation, that is, more affordable, faster and environment-friendly solutions for public and private transport, as well as adopting new forms of transportation.

In the wake of burgeoning urban population along with rapid global urbanization trends, and the reciprocity in influence of and on climate change and the environment having rendered the tasks in the realm of urban mobility increasingly demanding for traffic planners, the concept of Smart Mobility is rapidly gaining traction with scientific community worldwide (Brčić, Slavulj, Šojat, & Jurak, 2018). Growing economic prosperity of the cities and the excessive usage of private vehicles cause negative consequences for urban areas in the form of traffic congestion and delays, negative environmental impacts like noise and pollution, finally culminating in the application of negative application of the regressive investment policies in transport infrastructure sector. Confronted with the goal of establishing sustainable urban mobility, city planners throughout the globe are contemplating of smart mobility plans, and in the EU efforts in this regard have been underway for the past two decades (EU Commission, 2001, 2007, 2009, 2011). Smart mobility has come to be perceived in recent years as one of the prime options for more sustainable transport systems (Pinna, Masala, & Garau, 2017). Besides, smart mobility can be construed as "a set of coordinated actions addressed at improving the efficiency, the effectiveness and the environmental sustainability of cities" (Benevolo, Dameri, & D'Auria, 2016). Connectivity is the main aspect of smart mobility, and it along with big data, enables users to transmit all traffic information in real time while representatives of local governments of cities can concurrently carry on dynamic management (Pinna et al., 2017). In sum, urban mobility mostly pertains to traffic management in real time of passenger transport means, tracking applications and logistics, car-park management and car-sharing services, and other multiple smart mobility services (Yue, Chye, & Hoy, 2017).

Ability of the intelligent transport system (ITS) supporting urban smart mobility has been confirmed by research scientific research (Battara, Zucaro, & Tremiterra, 2017; Mangiaracina, Perego, Salvadori, & Tumino, 2017; Papa, Gargiulo, & Russo, 2017). An ITS entails the advanced mode of transportation systems that comprise many pieces of software that are helpful for safe transportation, reduce traffic congestion, decrease air pollution, enhance energy efficiency and promote the development of the associated industries (Chandra, Shivia Harun, & Reshma, 2017). According to an EU Directive (EU, 2010), ITSs "integrate telecommunications, electronics and information technologies with transport engineering to plan, design, operate, maintain and manage transport systems." ITSs encompassing modern technological and organizational transport solutions enable, among other things, traffic control, the creation of special zones of limited access and low CO_2 emissions by limiting the number of private cars in city centers (Tomaszewska & Florea, 2018). Apart from enhancing the safety of traffic passengers, ITSs also help in improving the efficacy of the transport system as well as to protect the natural environment. In other words, ITSs are advanced applications that enable various users to be better informed and make safer, more coordinated, "smarter" use of transport networks (EU, 2010).

Smart mobility entails development of logistic and transport activities, the mandatory existence of online databases and traffic optimization. While aiming at reducing negative effects of mobility, especially pollution, it also emphasizes on optimization o resource consumption (Tomaszewska & Florea, 2018). Both public and private transport systems along with heavy goods transport comprise the support system for mobility services that are pivotal for the city and its inhabitants (Czech, Biezdudnaja, Lewczuk, & Razumowskij, 2018). Management of public transportation is often designed to help the municipal authorities to make the public transportation easier to use and concurrently assist the operations' optimization with the new embedded digital characteristics like issuance of tickets, and traffic-light synchronization or traffic decongestion, efficiency enhancement, the safety and the coordination between different transport networks from the city; and the benefits accruing from, as pointed by Tomaszewska and Florea (2018), would be: real-time information about the public transportation; more efficient administration of the public transportation; possibility for citizens to access the information system online via smartphones; and even greenlighting the traffic lights on request. Major sources of air pollution comprise inefficient modes of transportation, data centers and industrial activities.

Use of fossil fuels at a large scale by transport vehicles has emerged as a major source of GHG emissions worldwide (Jacyna, Wasiak, Lewczuk, & Karon, 2017). According to International Energy Agency (IEA), the transport sector was responsible for the second-largest amount of GHG emissions in 2016, following the electricity and heat generation sector (IEA, 2017); and for the same period, 74% of transport sector emissions were attributed to road transport (Marcilio et al., 2018). Road transport will continue to grow, with a projected global increase of 60% in the total length of roads by 2050 over that in 2010 (Laurance et al., 2014). Political decisions with regard to lowering of taxes on importing second-hand cars even without taking into consideration environmental impacts contribute to augmentation in pollution levels and Tomaszewska and Florea (2018) cite the example of Romania where such measures culminated in the increase in the number of second-hand registered vehicles amounting to over 71% in 2017 comparable to the past year and led to increase in pollution level in Romania and in 2016, the European Environment Agency (EEA) identified transport as the single biggest emitter (Lewald, 2017).

For a developing country, an ITS and streamlining of traffic constitute key factors in attracting investments in businesses. Concurrently, increased traffic jams and the resultant noise and air pollution entail the potential of adversely impacting the life quality of citizens and can also cause decrease in personal and business productivity (Albalate and Fageda, 2019). There has been a corresponding mismatch between economic and population growth in urban areas on the one hand and expansion of urban areas on the other hand. The expansion of urban areas has not doubled by the development of the existing infrastructure, both of the transport and the urban utilities. Lack of road and pavement networks, unscientific urban development, inadequate laws and their poor enforcements leads to the occurrence of 1.25 million road accidents annually (WHO, 2015). It has been suggested that governments should treat road accidents as public health problem (WRI-WB, 2018). Injuries and deaths caused by road accidents wield a great economic impact. On an average, 82 developing countries pay around $220 billion per annum in the form of medical expenses and productivity losses (WRI-WB, 2018). Accordingly, there is more need for providing prevention actions by improving public transport systems, smart mobility solutions in and outside cities to minimize occurrence of accidents and road deaths affecting developing countries (Welle et al., 2018).

Facilitating implementation of smart mobility constitutes a part of an urban traffic management system connected to a public transportation management system and an information management system about urban travels, including the priority for public transportation in crossroads. The system can be connected to an air quality monitoring mechanism that permits the execution of some regulatory measures targeting pollution during traffic jams and thus contributing to smart environment (Brcic & Mlinaric 2018). Altering routes with heavy traffic can lead to decrease in pollution in inhabited areas thereby entailing the potential of enhancing the life quality of citizens and can also rehabilitate pedestrian routes in the urban transport network, making the city friendlier to its inhabitants and also contributing to smart living (Tomaszewska & Florea, 2018). It has been opined by Lewald (2017) that top priority is accorded to smarter and cleaner mobility in the future ideas pertaining to a smart city. Increasing use of public transport comparable to using personal vehicles is an effective way of reducing air pollution. While emphasizing that an intelligent public transport infrastructure must aim at reducing traffic load, Tomaszewska and Florea (2018) suggest that the people must be acquainted with civic skills as well as being familiarized with environmental changes. Citing the functional example of Shanghai Uber for bike system, van Mead (2017) suggests that use of parking charges, congestion time pricing for downtown parking areas, park-and-ride location along the metro, tram or rail line, etc., could serve as a strategic lever to control congestion along with having a positive impact on the utilization of public transport. In a similar vein, Florea & Berntzen (2017) suggest that combining the use of metro with bike-sharing from home to office and back could be the solution for replacing the travel to work with own cars, also taking into account the fact that cars remain parked 95% of the time.

Technology is harnessed by smart public transport system to provide public transport users a user-friendly experience. While making use of sensors and global positioning system (GPS) technology can help in eliciting real-time data on arrivals and departures of public transport, smart ticketing solutions may make use of smart cards or mobile phones to make ticketing efficient from the perspective of a user (Florea & Berntzen, 2017). AN intelligent traffic management in a smart city entails minimization of congestion along with making use of all available information, including elicited from citizen participation, as a stepping stone for formulating a multimodal transport system including its predictive operational control. Usage of the data collected from the public transport system can be made for the preparation of real-time situation reports and its usage can also facilitate public transport planners to adjust timetables, enhance the number of buses at peak hours on routes, create new routes and adjust fares (Tomaszewska & Florea, 2018). Mining of data from social media with regard to ascertaining citizens' perceptions pertaining to a public transport system can be undertaken. Nevertheless, the most significant challenge is as to "how can we get more people to travel by public transport, in a region where people prefer to travel by car?" (Johannessen & Berntzen, 2016). However, what is really needed is to maintain a semblance of equilibrium business goals for improving the quality of life of the people, maintenance of the city's sustainability development and the social participation and responsibility (Tomaszewska & Florea, 2018).

Keeping in view the fact that reduction in the number of cars and other personal vehicles cannot be affected immediately, the endeavors for safeguarding the environment by keeping it clean ought to be supported and encouraged by the industrial companies, especially by the automobile sector, by adhering to development of low-emission mobility strategies (Lewald, 2017). Encouraging the implementation of the polluting-emission minimization policy via the production of new vehicles and engines, including electric cars and conversion of existing cars into electric models needs to be accorded priority. Volvo and other international automobile giants are contemplating to phase out developing the new generation diesel engines and are now focusing on electric vehicles (Lowell and Huntington, 2019). Concurrently many other companies are also reportedly engaged in developing new particulate filters that are very effective in cutting emissions from existing diesel cars. Emphasis is now being focused on exploring specific IT solutions aiming at improving air quality and make mobility cleaner (beesmart.city, 2019). One of such a solution entails developing new heuristic algorithms for measuring, evaluating and minimizing CO_2, NO_x, and CO emissions in multidepot Green Vehicle Routing problems (Turkensteen, 2017) with economic and social applicability. Efforts are also underway at applying machine learning algorithms in tackling Ridesharing problems to minimize vehicle emissions, maintaining also living standards via development of infrastructure and qualitative services (Jalali, Koohi-Fayegh, El-Khatib, Hoornweg, & Li, 2017). Another IT solution being developed in this regard is development of Breath Journey software applications that coupled with GPS track with air quality for travel planning that can show the distance information, data regarding the pollution levels on the respective routes (Tomaszewska & Florea, 2018). In this manner, as pointed out by Florea and Berntzen (2017), the programming paradigm is likely to undergo change from software as a service (SaaS) to mobility as a service (MAs), making available very precise, real-time, customized, wireless mobility information services with regard to travel planning, journey time, air quality, online booking, and payment facilities.

7.6.5.2 Smart economy

Broadly, the notion of "smart economy" embraces market innovation, business climate and competitiveness to enhance the attractiveness of the smart city for start-ups, investors, and international businesses. In many cases, the notion of smart economy is often construed as innovative economy that is based in the industry equipped with state-of-the-art technology. It is largely in this context that Bakici, Almirall, and Wareham (2013) claimed that smart economy encompasses establishment of innovative clusters and mutual cooperation between enterprises, search institutions and the citizens with a view to promote, develop and implement innovation through various networks. Smart economy is described by Anttiroiko, Valkama, and Bailey (2014) as a networking economy that aims at developing new cooperation models in production, distribution and consumption. When used in the context of sustainable development, the definitions of the concept of smart economy denote resemblance to "green economy" and "green industry." Green economy is defined by UNEP as one that leads to "improved human wellbeing and social equity, while significantly reducing environmental risks and ecological scarcities" (UNEP, 2011). Green economy, while becoming a new engine of growth, also serves as a catalyst for creating decent jobs and a pivotal factor in alleviating chronic poverty. Claiming that smart economy is a green economy, Davies and Mullin (2011) aver that smart economy is instrumental in the reduction of the amount of CO_2 in industry and suggests investing in the "clean economy."

Green economy has come to represent a radical transition for more efficient, environmentally friendly and resource-saving technologies to minimize emissions and mitigate the impacts of climate change (Jänicke, 2012), and effectively deal with resource depletion and serious environmental degradation. According to Bowen and Fankhauser (2011), discourses on green economy have come to command strategic merit in reformulating a negative debate around constraints into a positive one with regard to opportunities. Having garnered traction at international, regional and national policy levels, initially as a response to 2008 financial crisis (Bina & La Camera, 2011), the concept of green economy has also come to construed as an engine for economic growth and development (Georgeson, Maslin, & Poessinouw, 2017). Regarded as an operational policy agenda designed to attain measurable progress at the environment-economy nexus (Schmalensee, 2012), green economy is also construed as a "pillar" of sustainable development implementation to facilitate the transition to low-carbon economies.

There exist several concepts of the use of "smart" environments in an assortment of subject areas such as smart transportation systems. Smart manufacturing, smart buildings, including houses, smart cities, etc., (Knyayginin et al., 2012). Smart economy is considered by many scholars specifically in the context of smart city (Galperina, 2016). Definition of smart sustainable city (SSC) outlined by ITU-FGSSC (2014) emphasizes on core themes that inter alia include (1) society, (2) economy, (3) environment, and (4) governance. Smart city concept has been hailed by Abdoullaev (2013) as the new socio-technological paradigm and advanced economic model for sustainable growth in the 21st century, that is described as the century of cities.

Emphasis on usually differentiating common characteristics of the smart economy has been focused by Bruneckiene and Sinkiene (2014, cited in Galperina, 2016) and this set of characteristics delineates its specificity. The first characteristic of innovation and knowledge economy entails emphasis on innovation execution, augmenting productivity and minimizing costs in all sectors of economy. A knowledge-based economy does not merely emphasize on new technologies or even new knowledge, rather it is the one wherein all sectors are knowledge-intensive, are responsive to new ideas and technological change, are innovative and utilize the services of highly-skilled personnel already engaged on in ongoing learning (Martinez-Fernandez, Audirac, Fol, & Cunningham-Sabot, 2012). Knowledge and skill need to be utilizable and harnessed in the production of all types of goods and services (OECD, 1999, p. 11; Smith, 1977). There two strong streams in the literature that despairingly adds to the understanding of knowledge as avital asset in innovation and growth. One stream embracing sociological perspectives focuses on the organization of knowledge as a way to optimize its efforts; and the second stream comprising organizational and innovation studies deal with the intermediaries of knowledge (Sorensen, Bloch, & Young, 2016).

Another characteristic of smart economy is learning economy that has been acknowledged as the most important process in spheres of economy. About five decades from now, an economist, Daniel Bell, had predicted the coming of a post-industrial economy herein knowledge would constitute the basis of economic power, and work and jobs would increasingly become knowledge-intensive and skill-dependent (Bell, 1973). Two decades after Bell (1973), Peter Drucker acknowledged that knowledge has become the only economic resource that matters and the world was moving into an era of "knowledge capitalism" (Drucker, 1993). Viewed in a broad spectrum, knowledge envisages a mixture of explicit and tacit elements, the latter implying that "we can know more than we can tell" (Polanyi, 1967). In accordance with knowledge-based theory of an enterprise or a firm, competitive advantage of a firm lies in in its ability to integrate and protect individual's specialized and tacit knowledge (Grant, 1996). Viewed from an economic perspective, other knowledge characteristics are also significant. For a particular firm or enterprise, knowledge can be general or specific; and knowledge can be reflected in organizational design and embedded in processes (Burton-Jones, 1999). Significantly, knowledge can only be developed, and tacit knowledge stored, in the human brain. In the realm of knowledge-intensive economy, individuals and firms ought to focus on nurturing and enhancing their "knowledge capital," and to achieve this, "they need practical techniques for identifying this knowledge assets and knowledge gaps" (Burton-Jones, 2003). Learning economy is termed as the economics of innovation and knowledge (Lundvall & OECD, 2000). Lamenting that perspective standards imposed by economics on society are "biased, incomplete, and inadequate," Lundvall (2016) cautions that the focus on rational choice, allocation of scarce resources and equilibrium only takes into account some dimensions of the modern economy, most of which are short-term and static ones. Therefore, Lundvall (2016) suggests that alternative perspectives wherein the focus is on learning as an interactive process and on processes of innovation, afford visibility and direct attention to other, at least equally significant and more dynamic, dimensions.

7.6.5.2.1 Digital economy

Digital economy is another characteristic of smart money that employs information and telecommunication technologies in a massive way in the economy, and is often described as the economy based on digital technologies. The notion of digital economy started gaining currency in the literature in the mid-1990s, and since then the definition of digital economy has over these years evolved, mirroring the swiftly changing nature of technology and its use by enterprises and consumers (Barefoot, Curtis, Jolliff, Nicholson, & Omohundro, 2018). The analyses in the early 1990s were mainly focused on the adoption of the Internet and early thinking about its impacts, and often referred to as "Internet economy" (Brynjolfsson and Kahin, 2002; Tapscott, 1996). With the expansion of the Internet use, reports from the mid-2000s onwards started reflecting increasingly on the conditions under which the "Internet economy" might emerge and expand. This phase was characterized by evolution of definition of the term "digital economy" and these definitions comprised analyses of diverse policies and technologies, on the one hand, and the growth of ICT and digitally-oriented firms and enterprises as main actors, on the other hand (OECD, 2012, 2014). In the wake of improved Internet connectivity in developing countries, in tandem with the expansion in the range of digital firms, enterprises, products and services, studies of the digital economy have embarked on embracing more extensive analyses of the situation in developing countries (UNCTAD, 2017a, 2019b; World Bank, 2016).

The discourse on digital economy in the past few years has again witnessed a shift of emphasis, with increasing focus on the way digital technologies, services, products, techniques and skill are being disseminated across economies. Usually referred to as digitalization, this process is delineated as the transition of businesses through the use of digital technologies, products and services (Brennen & Kreiss, 2014). Interestingly, digital products and services have become instrumental in facilitating change at a faster pace across a wide spectrum of sectors instead of getting being confined to those high technology sectors that been the main focal point in the past (Malecki & Moriset, 2007). Recent work, while mirroring this change, has concentrated on "digitalization" and "digital transformation"—the manner in which digital products and services are substantially disrupting traditional sectors—to explore multiple cross-sectoral digitalization trends (OECD, 2016, 2017; UNCTAD, 2017a). This has come to assume added significance for developing countries where the digital economy has started impacting the traditional sectors like agriculture, tourism and transportation. Unquestionably, significant economic changes can be spurred through the digitalization of traditional sectors rather than via the emergence of new, digitally-enabled sectors. It is significant to discern as to how investments in, and policies pertaining to, technologies or infrastructure enable or limit the emergence of digital economy so as to have cognizance of its development implications. Assessing the digital economy through the prism of certain sets of technologies is also equally significant. As emphasized by UNCTAD (2017a), the evolving economy can be linked to an increased use of advanced robotics, AI, the IoT, cloud computing, big data analytics and three-dimensional printing. Besides, interoperable systems and digital platforms are key components of the digital economy.

Digital economy is fast becoming an integral part of the entire functioning of the economic system, especially at a time when digital technologies are underpinning ever more transactions. Different technologies and economic aspects of the digital economy can be classified into three broad categories—core or foundational aspects; digital and IT; and a wider set of digitalizing sectors. Core or foundation aspects of digital economy consist of fundamental innovations, such as semiconductors and processors. It also comprises core technologies like computers and telecommunication devices. Besides, there are also enabling infrastructures that include Internet and telecoms networks. Second aspect entailing digital and IT sectors is engaged in generating key products or services that are dependent on core digital technologies, including digital platforms, mobile applications and payment services. The digital economy is impacted to a great degree by innovative services in these sectors that are contributing enormously economies, as well as enabling potential spillover effects to other sectors (UNCTAD, 2019b). The third aspect of digitalizing sectors like e-commerce make extensive use of digital products and services. Many sectors of the economy are being digitalized irrespective of the fact that change is incremental. Digitally-enabled sectors like finance, media, tourism and transportation that entail new activities or business models are undergoing transformation as a consequence of digital technologies. Growth of digitalized economy is to a large extent also depends on digitally-literate or digitally-skilled workforce, a fact less emphasized upon (UNCTAD, 2017a; UNCTAD, 2019b).

These components of digital economy are used in multiple ways as a basis or foundation for measuring the scope and impact of the digital economy (Ahmad and Ribarsky, 2018; OECD, 2014). Methodologies, at their most primary level, concentrate on measures of the core and digital sectors or appropriate proxies; prominently pertaining to the digital economy, such as digital infrastructure investments, broadband adoption; and demonstrate as to how these are associated with the growth of that economy, specifically in terms of outputs and employment in the digital and digitally-enabled sectors (OECD, 2017; UNCTAD, 2017a, 2017b). As highlighted by UNCTAD (2019b), such analyses are helpful in providing direction for policies and investments in the digital economy and also help in ascertaining potential effects on enterprises, firms, consumers and workers. It is more problematic to measure the digital economy beyond digital and digitally-enabled sectors. Viewed in a broad spectrum, effects of the digital technologies probably entail the potential of causing spillover effects and impalpable upshots, such as flexibility of an enterprise or a firm, management approaches or productivity, that are also dependent on other variables (Brynjolfsson, 1993). Attempts have been made some scholars in their studies to assess digitalization through surveys and e-commerce data, by measuring the spillover impacts from the ICT digital sectors across an economy (Barefoot et al., 2018; Knickrehm, Berthon, & Daugherty, 2016), or by examining the changing geography of global data and knowledge (Manyika et al., 2014; Ojanperä, Graham, & Zook, 2016). Nevertheless, these approaches generally are confronted with limitations owing to methodological challenges and paucity of reliable statistics (UNCTAD, 2019b).

Interestingly, suggested definitions of the digital economy are inclined to be closely associated with the components envisaged in this section of the chapter. One approach that is seemingly in consonance with some studies such as by Barefoot et al. (2018), OECD (2012a), and UNCTAD (2017a), is the definition of digital economy proposed by Bukht and Heeks (2017) that defines digital economy as "that part of economic output derived solely or primarily from digital technologies with a business model based on digital goods or services" (Bukht and Heeks, 2017, p. 17). Another approach is to consider the digital economy as comprising all the ways in which digital technologies are diffusing into the economy (Brynjolfsson & Kahin, 2002). While delineating the foundations of the digital economy in broader terms Knickrehm et al. (2016) suggest that digital economy is "the share of total economic output derived from a number of broad "digital" inputs. These digital inputs include digital skills, digital equipment (hardware, software and communications equipment) and the intermediate digital goods and services used in production. Such broad measures reflect the foundations of the digital economy" (Knickrehm et al., 2016, p. 2)

The digital economy continues to witness unprecedented growth driven by the ability to collect, use and analyze huge amounts of machine-readable information or digital data about practically everything. These digital data emanate from the digital footprints of personal, social and business activities occurring on various digital platforms. Substantial growth has been recorded by Global Internet Protocol (IP) traffic, a proxy for data flows, from about 100 GB/ day in 1992 to more than 45,000 GB/second in 2017, due to both qualitative and quantitative changes in the content of Internet traffic and by 2022, global IP traffic is projected to reach 150,700 GB/second, in the wake of more and more coming on line and by expansion of the IoT (Cisco, 2018; UNCTAD, 2019a). The IoT concerns the extension of connectivity beyond people and organizations to objects and devices, and according to one estimate, nearly 500 billion IoT devices will be deployed by 2030 (Cisco, 2016). Admittedly, the current share of digital economy is about 22.5% of the global economy; nonetheless, it still entails huge potential for future growth. Entailing the potential of creating up to $2 trillion of additional products in the global economy by 2020, the digital economy has drawn attention of the researchers toward the evolution of the global economy under the impact of digital technologies and problems and prospects for the development of global digital economy (OECD, 2017), as well as electronic fraud, digital piracy and other forms of shadow digital consumptions (Gaspareniene, Remeikiene, & Navickas, 2016; Nikonova and Dementiev, 2018; OECD, 2017). Estimates by some researchers reveal that "the digital economy is worth $11.5 trillion globally, equivalent to 15.5% of global GDP and has grown two and a half times faster than global GDP over the past 15 years" (Huawei & Oxford Economics, 2017). Noting that it is still early days in the digital era, UNCTAD (2017b) notes that there are more questions than answers about how to deal with the ensuing digital challenge. In the absence of relevant statistics and empirical evidence, along with rapid pace of technological change, decision-makers are confronted with a moving target as they try to adopt sound policies pertaining to the digital economy.

7.6.5.2.2 Competitive economy

Notion of "competitive economy," another feature of smart economy, entails the ability of an economy to compete globally by employing knowledge and innovation in the ongoing competitive battle, based on higher profits, productivity, quality, resources' cost efficiency and minimization of waste. Competitiveness is a salient feature of smart economy. The concept of competitiveness with regard to a company or an enterprise is generally accepted; nevertheless, application of this term in the context of an entire economy has been questioned, especially by Krugman (1994). The analogy between the competitiveness of companies and that of nations is contested by him as "deeply misleading" and "flatly wrong" (Krugman, 1994, p. 28). Remarkably, terms "competition" and "competitiveness" have found frequent usage both in business and public discourses pertaining to economic units, their environment and about their ability to perform in accordance with the strategic or policy goals derived from business, economic or social objectives (Listra, 2015). Despite Krugman's categorical assertion that ".... people who use the term "competitiveness" do so without a second thought" (Krugman, 1994, p. 30), the meaning of terms competition and competitiveness remains vague, and "the exact meaning depends on the problem under hand" (Listra, 2015). Even "more than two hundred years after Adam Smith we still don't know" (Boone & Mulherin 2000) answer to the question—what is competition—and Krugman (1996) claims that at least in the case of international trade of nations, the term "competitiveness" is meaningless, on the one hand, and still there is a possibility of having a precise definition of competitive markets in economics, on the other hand (Mas-Colell, Whinston, & Green 1995). Stigler (1988) defines competition as "a rivalry between individuals (or groups or nations), and it arises whenever two or more parties strive for something that all cannot obtain." A broad definition of the term "competition" is provided by

Stigler (1988), wherein defining feature of the "competition" seem to be the objectives of the competition, the goal of the analysis and dimensions of the competition (Listra, 2015).

Usage of the term "competition" is seemingly has been subject to some complications. In the first place, differing approaches adhered to by the different interested parties can give rise to the usage of the term. For example, Demsetz (1981), in his text, has analyzed economic, legal, and political dimensions of competition. Besides, frequent usage of different terminologies makes the problem more acute. Another significant complication arises from the "aims of analysis," especially when different tools are used to obtain the results even in the case of exactly the same objective of analysis (Listra, 2015); and the ideological content entails the potential of making sometimes things even more problematic (Minford, 2006) in public discourse. The existence of an array of related and partly overlapping phenomena and terms like competition, competitiveness, productivity, effectiveness, comparative advantage, and so on, can also give rise to a set of complications, a fact well illustrated by Vickers (1995). Smith (1977) describe the extent of influence cast by competition on different levels of society and elaborating it further Porter (1990, 2008) differentiates between different levels of competition as meaningful. While dealing with attempts by enterprises, firms, locations, clusters and countries to compete, Ketels (2006) also demonstrates as to how each of these emerge competitive on its own "characteristic environment" of competition. A survey of past literature on the firm's competitiveness in the international realm is provided by Buckley, Pass, and Prescott (1988); nevertheless, focus on interested approaches has been centered on by Snowdon and Stonehouse (2006) as well as Misangui, Elms, Greckhamer, and Lepine (2006).

Undoubtedly, a region is often acknowledged as a geographic entity rather than an economic unit; nonetheless, in the competition literature it still occupies a significant place. Irrespective of the fact that regional competitiveness is regarded as a contentious notion, still it has been helpful in unfolding number of phenomena in economics and good starting point in this regard has been afforded by Kitson, Martin, and Tyler (2004) and Budd and Hirmis (2004). Controversy surrounding the concept of competitiveness at the country level is attributable to Porter (2008) who explicitly spoke about as to how to improve visibly existing competitiveness; and Krugman (1994, 1996) who openly announced that concept is "full nonsense." In this regard, Listra (2015) insists that one has to keep in mind that Krugman's (1994, 1996) proof of it conveys on "assuming the objective of competition to be sustainability, not achievement." According to Listra (2015), complexity surrounding the notion of competition along with problems with the decisiveness of the focal unit of competition that is often the competitor are the major sources of the confusion with this term. The competitor can be any group of persons that can be differentiated by at least one of these criteria—the persons may be in the form of a group such as firm, via legal binding; may share common budgetary constraints; may have common preferences and/or they may entail common decision-making system and/or empowerment mechanism (Listra, 2015). The second set of decisive factors of competition emanates from the objectives the competitors set forth. The process of competition and its outcomes heavily rely on the set of objectives that may or may not be articulated clearly in the strategies of the participants. The fact that the focal unit is often market or industry is also prone to contribute to misunderstanding in this regard (Listra, 2015). In the case of economic competition, the focal unit's internal structures play role in determining a part of the outcome (Lippman & Rumelt, 1982), along with its resources that lead to the resource-based perception of the firm (Wernerfelt, 2013), and by specific form of these, capabilities and dynamic capabilities (Peteraf, Di Stefano, & Verona, 2013). According to Listra (2015), the field of competition delineates the unified framework for the analysis of competition and competitiveness in terms of six dimensions—defining criterions of focal units of competition, the objectives of competition, internal and external determinants of competitiveness, configuration of relationships influencing the competitive process, combination of static and dynamic elements in the field of competition, and the purpose of modeling. And this model enables in creating "a framework to clarify the aggregation needed when moving from the firm-level analysis to the country-level analysis" (Listra, 2015).

With growing support for national economies as entities of international competition, it is being reckoned that more effective use of production factors, and especially the nontangible assets such as innovation, cultural standards along with organizational and management skills, entail the ability of becoming basis of structural adjustment, and thus prove instrumental in changing the competitiveness of sectors of the economy (Radło, 2008, p. 75 cited in Molendowski, 2017). In the wake of this scenario, it can be assumed that countries at a similar level of development "fight for advantageous conditions for specialized workers and for the location of economic activity in innovative sectors" (Molendowski, 2017). The allegation that the notion of economic competitiveness is not applicable to large countries gets refuted in the wake of progress of globalization and the identical similarities of consumer preferences on the global scale (Mrak, 2000), because manufacturers from large countries, still supposedly placing reliance on domestic sales, are also exposed to competition in the form of innovative foreign products (Karodia, Dhiru, & David, 2014). Admittedly, economic growth in a large country doesn't directly rely on export competitiveness; nonetheless, in the era of global economy the international rivalry over competitiveness at the level of companies gets transferred to the domestic market and in that eventuality even a large industrialized economy can ill-afford to ignore competitive pressure from innovation leaders or even from more cost-effective foreign companies (Molendowski, 2017). Broadly speaking, in an open economy, the capacity of utilizing opportunities pertaining to the progress of globalization, while also confronting challenges of international competition, unfold avenues of employment in the given economy and, in effect, economic growth (Howes, 2000, p. 180). Emergence of these assumptions in the literature on competitiveness are helpful in assessing the question as to whether an appraisal of the competitiveness of national economies in an era of increasing globalization is justified; and lamentably this question still remains unanswered. Despite the "terminological chaos" in the wake of growing debate on the core phenomena of competitiveness and the attendant inconsistencies in the nomenclature, it is usually suggested that a key component of the review of the concept of international economic competitiveness is to "separate factors from results" (Molendowski, 2017).

Delineation of an economy's international competitiveness consists of ranking its competitive position and/or its competitive capacity (Molendowski, 2017). In the absence of any single, uniform approach to the interpretation of national competitiveness as a concept and to the determination of its place in other basic economic categories (Kharlamova & Vertelieva, 2013), there have emerged an array of indicators of international competitiveness of national economies and, concurrent growth of measurement methods. This alludes both to measuring the international competitive capacity of the national economy of a given country, as well as its international competitive position in a given period. It is worth mentioning that International Management Institute (IMD) publishes the results of its studies in annual reports, under the title World Competitiveness Yearbook that contains information about many dozen countries. World Bank has been publishing since 2004 its annual Doing Business reports that deal with analysis of the conditions of conducting business in the countries examined in the report. Besides, The Human Development Index (HDI) is published by the United Nations Development Programme (UNDP) annually that provides a synthetic measure of the quality of life in the countries examined in the report and this publication is often used to measure the international competitiveness of a country. Nevertheless, the Global Competitiveness Report (GCR) published by Geneva-based World Economic Forum (WEF) every year is widely acknowledged as one of the most comprehensive and often-quoted rankings of the competitiveness ranking of global economies. The ranking of each country covered in the GCR is the outcome of a yearly comparative study of the conditions of economic development of countries mentioned in the GCR. Ranking of the examined countries is facilitated on the basis of their competitiveness based on indicators formulated specifically for this purpose.

WEF's GCR for 2019 offers insights into the economic prospects of 141 countries that account for 99% of global economy. It is calculated on 103 indicators arranged in 12 "pillars": Institutions; Infrastructure; ICT adoption; Macroeconomic stability; Health; Skills; Product market; Labor market; Financial system; Market size; Business dynamism; and Innovation capability (shown in Table 7.2). Performance of a country on the overall Global Competitive Index results as well as each of its components is reported as on a as a "progress score" on a 0-to-100 scale, where 100 represents the "frontier," an ideal state where an issue ceases to be a constraint to productivity growth (WEF, 2019). These pillars are divided into four categories, namely (1) enabling environment, (2) human capital, (3) markets, and (4) innovation ecosystems, as shown in Table 7.3.

Thus smart economy, imbibing an open and transparent mechanism, apart from adding value to smart cities, also unfolds a variety of employment opportunities with labor market flexibility, diversification that promotes entrepreneurship and innovation as well as more productivity through local, regional and global linkages (Tyas et al., 2019). Embodying a high level of competitiveness globally and also at the local level, smart economy is well connected to the global economy. It facilitates an efficient and effective business environment that promotes and encourages innovation regardless of outcome; and also ensures a stable labor market with the resources and the ability to

Table 7.3 Global competitiveness index, 2019.

Category	Pillars
Enabling environment	Pillar 1: Institutions
	Pillar 2: Infrastructure
	Pillar 3: ICT adoption
	Pillar 4: Macroeconomic environment
Human capital	Pillar 5: Health
	Pillar 6: Skills
Markets	Pillar 7: Product market
	Pillar 8: Labor market
	Pillar 9: Financial System
	Pillar 10: Market size
Innovation ecosystem	Pillar 11: Business dynamism
	Pillar 12: Innovation capability

Adapted from WEF (World Economic Forum). (2019). Global competitiveness report 2019. Geneva: WEF.

adapt and make changes, if necessary. With its capabilities of guaranteeing the success and economic growth of a city, as well as the livelihoods of its citizens, smart economy has seeming emerged as an integral part of a smart city planning that entails the potential of looking ahead, innovatively and with global reach with regard to economic policy to remain attractive and competitive in the global economy (Govada, Sprujit, & Rogers, 2017; Indrawati, Ajkalhaq, & Amani, 2018).

7.6.5.3 Smart governance

Viewed in a broad framework, governance is the enabling environment that needs appropriate legal framework and efficient procedures to enable the responsiveness of the government to the needs of citizens (UN-Habitat, 2008). Governance is also construed as interaction and collaboration of diverse stakeholders in decision-making processes (Alonso & Lippez-De Castro, 2016). The fact that the notion of governance is often used to delineate the action or manner of governing a state, an organization or other groups of actors denotes that government and governance are related but different concepts (Pereira et al., 2018). Smart governance is delineated as "the capacity of employing intelligent and adaptive acts and activities of looking after and making decisions about something" (Scholl & AlAwadhi, 2016, p. 22). The concept of smart governance can also be construed as the key to smart, open and participatory government (Scholl & Scholl, 2014). In the wake of increasing role being played by these concepts in the growing discourse on smart cities, Pereira et al. (2018) opine that ICTs can be expected to play a major role in smart government as part of wider models of governance. This leads to the surmise that the adjective "smart" alludes to context- and site-embedded amalgamations of ICT, technology and innovation, as well as some type of democratic aspect (Scholl & Scholl, 2014). It has been demonstrated by some scholars that the related concepts of electronic government (e-government) and electronic governance (e-governance) also focus on the significance of ICTs in urban governance (Estevez & Janowski, 2013).

Sound management of urban problems arising from rapid pace of urbanization also requires due emphasis on establishing policies focused on wellbeing of urban citizens and this requires intelligent use of ICTs to deliver better services. Since it is one of the main objectives of smart cities to increase the quality of life in the city, it has been suggested that to manage the dynamics of smart cities, a new model of governance is required along with strong coordination by the local government to support in administering complex nature of cooperation processes with multiple stakeholders, including citizens (Testoni & Boeri, 2015). This calls for recasting the role of governments, citizens, and other social actors along with harnessing new and emergent information technologies to work out a new model of governance, including new relationship, new processes, and new government structures (Gil-Garcia, 2012). This gives rise to the question as to what should smart governance look like in the context of smart cities. While seeking answers to these and other related questions in the context of smart cities, it is found that apart from lack of clear and explicit definitions of concepts, rather some concepts overlap in many aspects, necessitating the need for more conceptual clarity and definitions that support governments and other actors in their search for smart governance and its related elements (Pereira et al., 2018).

In this context, Pereira et al. (2018) define smart governance as the ability of governments to make better decisions through the combination of ICT-based tools and collaborative governance. Construed in this sense, smart governance is the use of evidence in terms of data, people and other resources for improving decision-making and deliver results that meet the requirements and aspirations of the citizens. This is specifically significant for smart cities' initiatives that are often technology-based. "Reshaping administrative structures and processes across multiple government agencies and departments" along with "stakeholder involvement in governance" are among the major success factors in smart city initiatives (Alawadhi & Scholl, 2016, p. 2953). Envisaging improvement in the decision-making processes by creating smart governance frames for urban policies is essential to enhance the quality of public service delivery (Elisei, D'Orazio, & Prezioso, 2014). The significance of transforming the relation between government and the public along with collaborative governance together constituting key elements of smart governance leads to the concept of participatory governance that is intimately related to the new governance model, as a method, in "promoting communication, communication, interaction, collaboration, participation in decision-making and direct democracy" (Pereira et al., 2018).

Irrespective of the massive potential inherent in the smart city concept, it has seemingly been unable to reap the promising benefits accruing to the cities in the wake of associated organizational, strategic and technological challenges (Ruhlandt, 2018). Nevertheless, some scholars and practitioners have opined that bulk of the challenges for the cities to become or to be "smart" outdo the scope and capabilities of their existing organizations, institutional arrangements and governance structures (Bolívar, 2016; Caragliu & Del Bo, 2012; Gil-Garcia, Pardo, & Nam, 2015). Accordingly, increasing emphasis has been focused on governance implications of smart city investments; and this focus can be partially attributed to being a direct outcome of the "perceived failures or lack of impact resulting from smart-city investments to date" (Barns, 2018). Undoubtedly, the significance of a structured, all-embracing and practical governance has been emphasized by some scholars for the realization of smart cities (Dameri & Ricciardi, 2015); nonetheless, an ongoing discussion on as what smart city governance entails and how it is to be defined is still gathering momentum as can be discerned from Table 7.1 which shows selected definitions on smart cities. Some scholars have pointed out that lack of adequate governance mechanism is obstructing effective transformation of many cities into smart cities (Praharaj, Han, & Hawken, 2018). Cosgrave, Doddy, and Walt (2014) have tried to demonstrate as to how technology-driven developments via ICT are affecting all cities across the globe, irrespective of whether they choose to invest in or include the smart city concept into their governance agenda. Multifaceted and multilevel ecosystem of different agencies and stakeholder groups like local governments, citizens, urban planners, etc., often make city governance as immensely complex because of conflicting interests of these agencies and groups. Therefore, smart cities need an adequate governance system for connecting all forces at work, facilitating knowledge transfers, enabling decision-making in order to optimize their socioeconomic and environmental performance (Ruhlandt, 2018).

On the basis of available literature on smart cities, Tan and Taeihagh (2020) have identified eight important factors that spur the rise of smart city development, especially in developing countries. While emphasizing on the development of economic and financing capacity along with technological development, these drivers also concentrate on the strengthening of regulatory development, human capital and citizens' engagement. Emphasis is also focused on the involvement of the private sector that tends to enjoy an edge in technology and resources in envisaging a supportive ecosystem that promotes innovation as a significant driver in boosting smart city development, especially in developing countries. These drivers are summarized along with references in Table 7.4.

Table 7.4 Summary of major driving factors spurring development of smart cities.

S. no.	Driver	References
1.	Financing capacity of the government	Mishra, George, and Theorems (2019), Vadgama, Khutwad, Damle, and Patil (2015), Zhan and de Jong (2018)
2.	Building a strong regulatory environment that fosters the confidence and trust of citizens and investors	Gil-Garcia and Aldama-Nalda (2013), Chatterjee and Kar (2017), Vu and Hartley (2018), Chen, Ardila-Gomez, and Frame (2017), Joshi, Saxena, Godbole, and Shreya (2016)
3.	Technology and infrastructure readiness	Vijai and Sivakumar (2016), Wang et al. (2019), Gil-Garcia and Aldama-Nalda (2013), Elena (2015), Liu, Wang, Xie, Mol, and Chen (2012), Chen et al. (2017), Joia and Kuhl (2019), Joshi et al. (2016)
4.	Human capital	Kummitha and Crutzen (2019), Chatterjee, Kar, and Gupta (2018), Chen et al. (2017), Joia and Kuhl (2019)
5.	Stability in economic development	Lu, de Jong, and ten Heuvelhof (2018), Joshi et al. (2016)
6.	Active citizen engagement and participation	Jamal and Sen (2019), Lu et al. (2018), Viale-Pereira, Cunha, Lampoltshammer, Parycek, and Testa, 2017, Gil-Garcia and Aldama-Nalda (2013), Chatterjee, Kar, and Gupta (2017), Liu, Low, and Wang (2018), Joshi et al. (2016)
7.	Knowledge trans and participation from the private sector	Kummitha and Crutzen (2019), Lu et al. (2018), Liu et al. (2018), Chen et al. (2017)
8.	Creating a supportive ecosystem that promotes innovation and learning	Kummitha and Crutzen (2019), Peng and Bai (2018), Chen et al. (2017)

Compiled and adapted from Tan, S. Y., & Taeihagh, A. (2020). Smart city governance in developing countries: A systematic literature review. Sustainability, 12(3), 899.

7.6.5.4 Smart environment

Smart environment, generally speaking deals with improving the quality of built and natural environment, ensuring sustainability and resource management along with emphasis on reduction of waste and mitigation of emissions, etc. According to Fernandez-Anez et al. (2018), smart environment entails network and environmental monitoring, energy efficiency, urban planning and urban refurbishment, smart buildings and building renovation, resources management, environmental protection and awareness and behavioral change. Smart environment is an environment that is supported by technology that is growing at a rapid pace (Cook & Das, 2005). For Poslad (2009), smart environment is a small world where different types of smart devices are working unceasingly to make lives more comfortable. According to Rahmayanti, Oktaviani, and Syani (2018), there are three types of distinguishable smart environments: (1) virtual computing environment, (2) physical environment, and (3) human environment. With environmental problems emerging as a major problem for the cities, emphasis is stressed on conservation efforts to be carried out by providing education about the environment that ought to aim at not only to improve science but also to change positive behavior toward the environment into a habit (Bolivar, 2015). Imparting of environmental education at an early age, both through formal education and informal education, is considered as a strategic step to change the behavior and attitudes of children as young people who will replace the older generation (Chourabi et al., 2012; Van-Heur, 2010). This is facilitated to enable children better understand and care about the significance of the environment to support the survival of living things, so that the concept of smart environment in "the application of smart city can run effectively" (Rahmayanti, et al., 2018). Tackling urban climate change, waste management and CE are seeming major elements of smart environment.

7.6.5.4.1 Tackling urban climate change

Cities are being construed as ecosystems dominated by human beings and concurrently witnessing an unprecedented level of urbanization (Dirks, Gergiev, & Keeling, 2010); and accordingly increasing levels of production and the resultant increasing consumption patterns are also upsetting the environmental equilibrium causing increasing levels of pollution, increase in the levels of rates of emissions of GHGs. Thus cities are increasing becoming primarily responsible for climate change as they account for more than 70% of global CO_2 emissions (UNEP, 2018). Occupying even less than 2% of the landmass of the earth, cities consume almost three-quarters of the global natural resources (Marceau, 2008). Interestingly, UNDP (2015) identified resilient cities as one of the 17-MDGs as global cities are responsible for 60%–80% of energy consumption and 75% of carbon emissions. Thus, reduction in carbon emissions and ensuring energy-use efficiency have become integral to smart environment in future smart cities in developing countries. Unplanned or haphazard mode of rapid urbanization results in a loss of basic functionalities thereby making urban areas as nonlivable place and also create difficulty in waste management, scarcity of resources, air pollution, engendering serious concerns for human health, traffic jams and congestions, as well as yielding inadequate, degraded and ageing infrastructures (Washburn et al., 2010). Undoubtedly, urban places, apart from being significant contributors to climate change, are also severely vulnerable to the impacts of climate change (Chakravarty & Sarkar, 2018); nevertheless, concurrently they are the sources of the compelling set of opportunities toward innovation and facilitating future strategy toward climate-related actions (IPCC, 2007, 2014). It is in this context that smart environment is seen as an innovative way to operate smart cities.

The smart city approach is gaining salience as a way to solve urban problems emanating from rapid urbanization (Chakravarty & Sarkar, 2018). Nevertheless, in the literature on smart cities emphasis has been initially focused on the technological aspect of smart cities whereas the sustainability aspects have not been accorded due priority (Chakravarty & Sarkar, 2018). The utility of the application and usage of smartness in the urban context entails not only harnessing state-of-the-art ICT technologies but also embracing sustainability along with climate resilience. In this regard Nam and Pardo (2011a) have pointed out that an appraisal of various failures of urban adaptation and mitigation from the past have focused attention on the necessity for a strong information network and system as an essential intervention. Thus improved information system is integral to smart cities (Batty, Axhausen, Gianotti, & Pozdnoukhov, 2012). Reports of IPCC (2007) and UN (United Nations) (2011) have improved instrumental in broadening the concept of smart city by including the notion of sustainability and delineating its role vis-à-vis climate change mitigation. Subsequently, advanced information systems have reportedly improvised the concept by raising it to a SSC (Ahvenniemi et al., 2017; Giuseppe & Pianezzi, 2017). And this approach has come to reportedly denote the management of administrative procedures, integration of nonconventional energy in the housing, commercial and institutional sectors in the smart city and digital control.

The strategies required for the sustainability approach in creating climate-resilient cities vary from energy efficiency and technological advancement of the city's hard infrastructure, such as transport, water, waste, energy, etc., to those taking into account the soft infrastructure and people in terms of social and human capital, knowledge, inclusion, participation, social innovation and equity (Angelidou, 2014). A close scrutiny of the sustainability approach concept unfolds three broad criteria—the first criterion comprises advanced IT and system (March, 2018), the second criterion entails effectiveness of various production and consumption sectors with adequate emphasis on energy efficiency (Khansari, Mostashari, & Mansouri, 2014), and the third criterion deals with effective societal governance participation of citizens (Vanolo, 2016). These three criteria or components, also happen to be three pillars of smart city, are generally construed in a mutually exclusive manner; and in the eventuality of conceptualizing a smart city on the basis of only one component or pillar, then the task of achieving strict low-carbon targets would be rendered difficult (Deakin and Reid, 2016). In order to enable a smart city playing significant role in attaining rigorous climate change targets like 1.5°C global average temperature rise, an integration of these three pillars is necessary (Cohen & Munoz, 2016; Vollaro et al., 2014).

In the wake of multiple types of shocks and waves confronting the cities, climate change related shocks take place in tandem with other environmental, economic, and political stresses (WCRP, 2019). Accordingly, cities are required to become resilient to a broad range of shocks and stresses in order to be ready confronting climate change; and therefore, attempts designed to foster climate change ought to be incorporated into efforts to promote urban development and sustainability (Ali et al., 2019). Urban resilience to climate change often alludes to a wide range of shocks and stresses (Reed et al., 2013). Fostering urban resilience to climate change will accordingly need cities to become resistant to a wide array of overlapping and interacting shocks and stresses (Reike, Vermeulen, & Witjes, 2018). According to Turner (2010), climate resilience can be defined as "... the capacity of a system to absorb disturbance and reorganize while undergoing a change so as to still retain essentially the same function, structure and feedbacks, and therefore identity." The notion of urban climate resilience postulated in ecological theory delineates it as an adaptive system that comes back to the initial equilibrium after an external shock and the emphasis is focused on the need to bring resilience in tandem with the concept of transformation in order to embrace issues of political and socioeconomic structure and trade-offs that determine risk and vulnerability (Bahadur & Tanner, 2014).

Development of climate resilience criteria in smart cities entails the potential of equipping local governments with information to actively engage in imparting training, capacity building, and capital investment programs with enhanced efficacy (Bahadur, Tanner, & Pichon, 2016; Chakravarty & Sarkar, 2018). Principally, the local governments are called upon to identify the "hot spot" risk and vulnerability issues work out accordingly the tools and strategies for resource allocation; and thereafter, an incremental assessment of challenges could help in surmounting the obstacles (Archer, Monteith, Scott, & Gawler, 2017). These tools and strategies can be most suitable for cities that have just embarked on developing climate resilience as well as those that are already practicing well-established policies, institutional framework and requisite strategies. Available literature helps us in discerning four main principles for climate resilience (Eriksen & Lind, 2009; Roy & Pal, 2009; Roy, Cakravaty, Dasgupta, & Chakrabarty, 2018; UCS, 2016), and these include (1) recognize the context for climatic stress and vulnerability, (2) acknowledge that different stakeholders have different values and interests toward climate resiliency, (3) integrate local knowledge (traditional and cultural) with modern know-how, and (4) incorporate potential challenges into strategy development. With regard to vulnerability, there arises an essentiality of identifying the sources and components along with types of climate stress (Eakin & Wehbe, 2009; Eriksen & Lind, 2009; Leichenko, Karen, & William, 2010; Tschakert & Dietrich, 2010; Ziervogel et al., 2006). In order to identify the direct and indirect outcomes of actions/efforts, there is need to understand the fundamental social, economic, institutional and cultural conditions that contribute to a wider context for resiliency.

There is a need for being conscious of the spatial and temporal effects of such efforts. According to Chakravarty and Sarkar (2018), acknowledging the potential worth and conflicts of interest can facilitate identification of as to how prioritization of climate-resilience-related activity may be different for different groups, which may affect other groups as well as the urban economy. Strong vested interests within specific strategies entail the potential of either acting as an impediment to or strengthening the strategy. Roy et al. (2018) opine that leveraging traditional and cultural practices are least or no cost mechanisms to strategizing the resiliency. Incentive structures are said to play important role for building climate resilience (ADB, 2010), because incentives are essential to boost the investments toward climate resilience activities over existing investment portfolios rather than looking for new ways to build climate resilience. Informed strategies, legislative structure, continuous monitoring, and constant knowledge updates are often considered as key criteria for evaluating climate resilience (Schaefer, Thinh, & Greiving, 2020). Irrespective of the fact that there exist multiple different ways of measuring resilience, there are common parameters that comprise populations, neighborhoods, and systems including diversity, adaptive governance along with capacity for learning and innovation (Folke, 2016; Grieve, Kemp, Norris, & Padgett, 2017). These constitute salient characteristics of cities that are at the forefront of technological innovation and efforts to create sustainable urban infrastructure (Ernstson et al., 2010). While evaluating resilience, care needs to be taken of the fact of available evidence with regard to the broader development policies and plans comprising endeavors to promote urban resilience to climate change, including both adaptation and mitigation strategies (Bahadur & Tanner, 2014; Seto & Shepherd, 2009).

7.6.5.4.2 Urban waste management
Solid waste management (SWM) in urban areas refers to all activities and actions needed to manage waste from its inception to its final disposal; and while managing solid waste is a challenging task, it also entails the capacity to protect the environment, improve societies' quality of life and contribute to the economy as a whole. SWM varies among countries and regions, and is regarded as one of the most significant municipal services for a city to protect the environment, public health, and esthetic character (Elagroudy, Warith, Mostafa, & Zayat, 2016). In 2018, about 2.01 billion metric tons of municipal solid waste (MSW) was produced annually worldwide and the World Bank estimates that overall waste generation will increase to 3.40 billion metric tons by 2050, recording an increase by 70% (World Bank, 2018a, 2018b). The waste, generated as a result of human activities, is handled, stored, collected and disposed of, and it also entails the potential of posing risks to the environment and public health (Saxena, Srivastava, & Samaddar, 2010). Increase in quantity and complexity of generated solid waste in the cities is attributable to economic development, rapid pace of urbanization and industrialization and improved urban living standards (Gidde, Todkar, & Kokate, 2008). Sold wastes comprise domestic, commercial and industrial emanating from construction and demolition activities, along with wastes originating from agricultural, institutional and other miscellaneous activities. Sometimes domestic and commercial

wastes cannot be differentiated and are regarded together as urban wastes (Syed, 2006). MSW often combines household and commercial wastes that are generated by human activities (Rajkumar, Subramani, & Elango, 2010). The continuous indiscriminate and unscientific disposal of MSW is increasing and is linked to poverty, poor governance, urbanization, growth in population, poor standards of living and low level of environmental awareness (Rachel, Komine, Yasuhara, & Murakami, 2009), along with inadequate management of environmental knowledge (Gupta, Yadav, & Kumar, 2015). MSW comprises degradable, partially degradable and nondegradable wastes. Paper textiles, food waste, straw and yard waste fall in the category of degradable waste; whereas wood, disposable napkins and sludge comprise partially degradable waste; and nondegradable materials include leather, plastics, rubbers, glass, ash from fuel burning like coal, briquettes or woods, dust, and electronic waste (Jha, Singh, Singh, & Gupta, 2011).

Effective waste management and reduction comprise important challenges for cities in ensuring sustainable development that is compatible with environmental protection (Vilella, 2017). The crucial role cities play in waste management and reduction efforts is well acknowledged because cities are the hub of production of a wide range of waste that can be defined as "materials that are not prime products (that is, products produced for the market) for which the generator has no further use in terms of his/her own purposes of production, transformation or consumption, and of which he/she wants to dispose" (UNEP-UNITAR, 2013, p. 17). Improper and unscientific management of waste in tandem with unsustainable production and consumption patterns cause economic and social damages as "natural resources are being consumed to the point of exhaustion, generating impacts on a planetary scale, imposing huge impacts on human health and the environment and creating massive social disruption" (UNEP-UNITAR, 2013, p. 12). The UNEP has accorded significant priority to address waste management and in that direction, it has undertaken several actions to address this issue, and these actions entail partnerships, conventions, treaties, plans and provisions of technical assistance (UNDP, 2019). Keeping in view the essentiality of cities to the initiatives planned and designed to foster adequate waste management practices, local authorities and civil societies are working together in analyzing demands and address issues pertaining to waste; and accordingly, urban areas can become adequately equipped with the tools to manage waste and to promote sustainable use of resources (UNEP-UNITAR, 2013).

By ratifying the 2030 Agenda and the attendant SDGs, Member States of the United Nations renewed their commitment to the protection of the environment and fulfilment of sustainable development (UN-GA, 2015). SDG-11 envisages the goal of "making cities and human settlements inclusive, safe, resilient and sustainable," and Target 6 of SDG-11 commits the international community to reducing, by 2030 "the adverse per capita environmental impact of cities, including by paying special attention to air quality and municipal and other waste management" (UN-Habitat, 2018). Along with making waste management as a core step for the cities to reach sustainability, one of the indicators associated with target 11.6 guides urban areas to evaluate their waste management based on the proportion of urban solid waste regularly collected and with adequate final discharge out of total urban solid waste generated. However, by the closing part of 2018, 2 billion people globally did not have access to waste collection services and 3 billion people lacked access to controlled-waste disposal facilities. Amidst reports of total waste generated in the world doubling from nearly 2 billion tons in 2016 to about 4 billion tons by 2050 (UN-ECOSOC, 2019, p. 17), the objective of smart environment in the smart cities is likely to confront serious implications, if the issue of waste management is not adequately addressed.

7.6.5.5 Circular economy

Emerging smart cities in developing countries are in need of new ways of addressing the issue of waste. According to a quantitative analysis of resource requirements worldwide conducted by the Paris-based International Resource Panel (IRP) recently, it is estimated that under business-as-usual scenario material consumption by the cities worldwide will grow from 40 billion tons in 2010 to about 90 billion tons by 2050 (IRP, 2018). Huge amounts of waste are already being generated by current levels of consumption and that is exacerbating the adverse environmental impacts of increased extraction. In view of the fact that high volumes of water and energy is consumed in the upstream phases of resource extraction and manufacturing processes, it is estimated that more than two-thirds of the global energy is consumed in cities, accounting for over 70% of global CO_2 emissions (WEF, 2018). According to the World Bank (2018a, 2018b), cities are places where the highest amount of waste is generated, and cities generated 1.3 billion tons of solid waste per annum and that translated into a footprint of 1.2 kg per person per day in 2012; and by 2025 this generation of waste in the cities is expected to rise to 2.2 billion tons. Cautioning that traditional waste management and disposal practices resulting in landfill sites or pollution of the environment if not regulated properly, WEF (2018) suggests for embracing a more viable mechanism of production and consumption in the overall value chain to alleviate the burden of waste on urban areas. The CE approach that is aimed at reshaping resource use by separating growth from material extraction is seen as a viable option in managing urban waste and culminating in the creation of a more sustainable future allowing the natural environment to restore resources along with ensuring protection of environment from the adverse effects of industrial waste (WEF, 2018).

A holistic understanding of CE is provided by Kirchherr, Reike, and Hekkert (2017) when they describe it as "an economic system that is based on business models which replace the "end-of-life" concept with reducing, alternatively reusing, recycling and recovering materials in production/distribution and consumption processes, thus operating at the micro level (products, companies, consumers), meso level (eco-industrial parks) and macro level (city, region, nation and beyond), with the aim to accomplish sustainable development, which implies creating environmental quality, economic prosperity and social equity, to the benefit of current and future generations." Suggested as one of the tools in the transition to sustainability, CE is touted as a new school of thought (Murray , Murray, & Donnelly 2015, p. 377) that affords feasible and comprehensive solutions for sustainable development (Sauve, Bernard, & Bernard, 2016, p. 54). Boulding (1966), though did not use the term "circular economy," is credited with having presented an early CE theory. The concept of CE has been gaining wider traction in recent years in different fields ranging from politics to business and academia (Reike et al., 2018) in its quest of overcoming the contradictions between economic and environmental prosperity (Geissdoerfer, Savaget, Bocken, & Hultink, 2017). With the current economic model that is being construed as "linear" based on a "take-make-consume-throw away" approach of resources that has probably reached its limits, the CE constitutes an "economic system of trade and production which, at all stages of the product lifecycle, aims to increase the efficiency of resource use and reduce the impact on the environment, while developing the wellbeing of individuals" (2014 cited in Appendino, Roux, Saadé, & Peuportier, 2019). Accordingly, concept of CE is said to represent the core theme of major European plans and regulations (Petit-Boix & Leipold, 2018), such as the "Circular Economy Package" adopted in 2015 (European Commission, 2015a, 2015b, 2015c). Undoubtedly, the concept of CE has come to be studied by various disciplines ranging from economics to urban planning; nevertheless, it remains intimately linked to sustainable development issues (Kirchherr et al., 2017); and despite that no univocal and shared definition of CE has yet been developed (Appendino et al., 2019), irrespective of a wider dissemination of the concept of CE (Prieto-Sandoval, Jaca, & Ormazabal, 2018). Thus CE continues to be an evolving notion (Merli, Preziosi, & Acampora, 2018), rather ambiguous and vague (Korhonen, Nuur, Feldmann, & Birkie, 2017).

Industrial sector has been seemingly the origin basis of CE and many CE-related projects still take place in the industrial sector (Lindner, Mooij, & Rogers, 2017). Industrial eco-parks are often cited as prevalent examples of CE (Lindner et al., 2017). The Kalundborg Symbiosis industrial eco-park in Kalundborg (Denmark) comprises eight large public and private enterprises that exchange 27 separate waste streams (Kalundborg Symbiosis, 2017), envisaging links between industrial processes and matching demand with supply for resource inputs and outputs. Dell company's implementation of a closed-loop recycled plastics supply chain for its electronic products by designing for recyclability (Dell, 2017) has enabled integration of CE into consumer product design. Timberland, a shoe-making enterprise, is reported to have joined hands with a tire manufacturer to design a line of vehicle-tires that can be broken down easily and recycled into Timberland shoes (Timberland, 2018). These instances exemplify the application of CE concept at different system levels. Usage of CE at the municipal level can be seen as an occasion to pursue both economic and environmental objectives concurrently and interestingly the concept is already being integrated into municipal sustainability policies (Lindner et al., 2017). The city of Amsterdam is said to have called the integration of CE as a pillar of their sustainability policy (Circle Economy, 2016, p. 4) and the city of Vancouver has reportedly integrated CE into their Economic Commission branch (Vancouver Economic Commission, 2017).

The concept of CE has come to gain traction in the political narratives with regard to sustainability policy formulation and implementation. Many countries are said to have embraced CE as a viable economic reform model to facilitate continued economic growth (Winans, Kendall, & Deng, 2017, p. 827). Germany, Japan and China are touted as leading countries that have pioneered the concept of CE by adopting national policies in accordance with the concept. The Promotion of Effective Utilization of Resources Act 1991 was adopted by Japan and it promotes recycling waste back into the value chain (Ministry of Economy, Trade and Infrastructure, Japan, 2010, p. 10). Germany integrated CE into its national laws in 1996 in the form of a Closed Substance Cycle and Waste Management Act (German Law Archives, 2015). In 2008, China announced the enactment of the Circular Economy Promotion Law of the People's Republic of China Act, with specific emphasis on "promoting the development of the CE, improving the resource utilization efficiency, protecting and improving the environment and realizing sustainable development" (Chinalawinfo, 2017). Since the closing part of 2015, the EU has also prioritized CE approach enunciated in its document "Closing the loop—An EU action plan for the Circular Economy" that calls for a "continued, broader commitment from all levels of government, in Member States, regions and cities and all stakeholders concerned" (European Commission, 2015a, 2015b, 2015c). Other policies, like EU targets for recycling and waste management, in terms of having to reduce landfill to maximum 10% of municipal waste by 2030, are already indirectly stimulating countries and municipalities to integrate CE into their respective national policies (Lindner et al., 2017). Irrespective of traction being garnered by CE, some criticism and concerning questions have been raised scholars. Undoubtedly, CE can be seen as a "paradigm shift" from the linear economy, the complexity of this shift is criticized as being "ideal" because it requires radical transformation to the "economic order, including fundamental recasting of manufacture, retail, consumption and property rights" (Gregson, Crang, Fuller, & Holmes, 2015, p. 235). While raising the question as to how far recycling can be taken, Anderson (2007, p. 134) opines that a circular economy "cannot promote recycling in perpetuity." While conceding that the circular economy puts the environmental sustainability forward, acknowledges the need for a favorable economy context in the form of a circular model, Sauve et al. (2016, p. 55) lament that "the social objective is usually missing." Thus it seems pertinent to take these criticisms and questions into consideration while evaluating larger contribution of CE to sustainable development (Lindner et al., 2017).

7.6.5.6 Smart people

Smart people constitute another significant component of a smart city. Notion of smart people embraces mechanisms of encouraging creativity and social capital and transforming the way citizens can effectively communicate with the public and private sectors. According to Fernandez-Anez et al. (2018), smart people entail digital education, creativity, ICT-enabled working, community building and urban life management, and inclusive society. Smart people play significant role in positively determining smart city adoption (Mishra, Sen, & Kumar, 2017). Notion of smart people refers to human and social and human capital as well as citizens' participation toward city development (Anthopoulos, Janssen, & Weerakkody, 2019). For making a city to be a smart priority needs to be accorded on making its citizens to be truly innovative, inclusive and environmental conscious. Empowerment of citizens in a smart city is one of the prime factors that play a pivotal role in improving city governance (Tahir & Malek, 2016). With smart governance aiming at improving future of public services and community leadership for continuous development through innovation, smart governance initiatives have come to positively influence smart city adoption (Anthopoulos et al., 2019),

Giffinger and Gudrun (2010) acknowledge citizens as "Smart People" as a vital component of the smart city concept. It has suggested by Hollands (2008) that smart cities must start with the development of the human capital and only make use of technology when it envisages improvement in economic and political efficacy and when it enables social, cultural and urban progress. The smart people and smart communities are integral to smart cities. Smart people are citizens that are actively involved in managing, envisioning and collectively contributing to the wellbeing of their city (Walravens, 2015). Smart communities, as networks of citizens and other stakeholders linked by norms, values and goals, are in the core of smart peoples' sustained involvement (Batty et al., 2012). The involvement of citizens has been reckoned by academics and policy makers as key to succeed with smart city initiatives and urban development (Schuurman, Baccarne, De Marez, & Mechant, 2012). There are multiple ways of involving citizens—from mere informants to crucial catalysts in the innovation and coproducing process (Arnkil, Järvensivu, Koski, & Piirainen, 2010). While describing people and communities as one of the eight dimensions that constitute a smart city, Chourabi et al. (2012) also emphasize on the need to address people as part of bigger communities. Emphasis is focused by Geller (2003) on building vibrant communities and empowering citizens on a personal level. Smart people not only pursue but also generate public value themselves in these strong networks, and this engendering of value enables the creation of a social capital that is acknowledged as one of the most beneficial outputs of the bottom-up approach. (Alawadhi & Scholl, 2013; Bakici et al., 2013). In fact, social capital in terms of social knowledge

7.6 Toward sustainable smart cities

transfer, networking, trust, participation and engagement buttresses a city's: urban competitiveness (Coe, Paquet, & Roy, 2001), wealth (Caragliu & Del Bo, 2012), employment level (Shapiro, 2006), attractiveness to incoming talent (Paskaleva, 2011) and even level of happiness (Ballas, 2013).

Significance of smart people in a smart city is well illustrated from the example of London where the aim of the smart city London envisages: "To put Londoners at the core, with access to open data, leveraging London's research, technology, and creative talent, brought together through networks, to enable London to adapt and grow and City Hall to better serve Londoners needs, offering a "smarter" experience for all" (Smart London Plan, 2013). The significance of human capital for future development is emphasized therein. Authorities entrusted with the task of implementing Smart London Plan have contemplated of building smart city for smart people and by smart people, encouraging active engagement of citizens in policy development (Pozdniakova, 2018). Digital Talent Program, launched by the Mayor of London with the commencement of the Plan, incorporates various courses to tackle skills gap and encourage digital inclusion and growth in tech sectors. As per the Smart London Plan (2013), it is hoped that employment in tech sector will increase to 200,000 by 2020 (Digital Talent Program, n.d.). The work on Digital Health London, a collaborative endeavor, delivered by Med City and three Academic Health Science Networks of London, is supported by the mayor's office. The underlying idea in launching this program is to generate a platform where digital health solutions are traded and conducive conditions are created for industry (Digital Health London, n.d.). Queen Elizabeth Olympic Park is a notable example of a successful project that comprises five world class sporting venues, new low carbon dioxide emission homes, a world class cultural and education quarter along with a new media and digital hub. Besides, usage of IoT concept, weather station and solar sensors are some of other features of this park (Caprotti & Cowley, 2016; Smart London Plan, 2013).

Significance of smart people is embedded in the social sustainability segment of Stockholm Smart City Strategy that lays specific emphasis on digital inclusion, where digitalization and new technologies are deployed to engulf social divides. It also emphasizes on helping "city dwellers" to communicate, work, study experience and have an active life, based on each person's unique circumstances (SCC, 2017). While enhancing perceived safety, both in public and private spaces, emphasis is also focused on creating vibrant and safe neighborhoods. Emphasis is also focused on helping elderly to stay in home and make their life easier by using digital solutions under programs like Stockholm Digital Care, with the help of an array of digital tools (SCC, 2017). While surveys of Stockholm residents demonstrate that the people inhabiting the city are generally happy with their lives and in global terms, "Stockholm comes up high up the rankings on quality of life, innovative capacity and social trust" (CPA, 2018).

Three areas of smart people component—education, social inclusion and healthcare—are covered in the Vienna Smart City Strategy (n.d.). Focus in the education realm is on lifelong learning and enhancement in the number of young people who continue to pursue their higher levels of education. A number of events such as annual Research Festivals are being organized since 2008 by the Vienna Business Agency to engender technology-related awareness among the people, specifically youth along with the vocational orientation workshop under the title Future Jobs, etc. Attention is also focused on enhancing the role of nature and energy efficiency in the lives of the citizens starting from childhood (VBA, n.d.). Social inclusion segment embraces active inclusion of women into planning, decision-making and implementation processes, affordable and high-quality housing for all, active participation at work and corresponding fair remuneration that includes basic needs, conducive neighborly and safe living conditions for all (WSC, 2014, p. 72). Promotion of health literacy and among citizens and provisions for medical care at the highest level for all citizens form the part of healthcare segment. E-health system comprising maintenance of the electronic health record, mobile monitoring equipment, measures for data protection and data security; and mobile health systems entailing simple explanation of illness, transmitting data from sensors to doctors and booking appointments, etc. (WSC, 2014, p. 74). While citizens take active part in developing Vienna city, there are multiple ways of participating and "everyone has the possibility of voicing, discussing and implementing their own ideas and opinions regarding the city" (WSC, 2014, p. 19).

Citizens have been designated as a "Key driver for the city development" in the previous Kyiv Smart City (2017) strategy that has been revised in 2018; nevertheless, focus on citizens continues in the revised strategy that denotes "The main strategic goal of the long-term development of Kyiv is to improve the quality of life of its residents, which is determined by the economic welfare and comfort of life in the city with a rich historical tradition" (KCC, 2018, p. 16). The previous smart city strategy had set the following targets involving citizens of Kyiv (2017):

- Enhancing possibilities of citizens to manage the city and impact the decisions;
- Opening access to databases, that can be used to address the needs of citizens;
- Involving citizens, businesses, IT specialists and experts to form city development agenda using online platform and social media channels;
- Establishing direct communication between citizens and authorities through city portal and social networks;
- Development of new partnership models to realize smart city projects;
- Establishment of independent expert council to monitor project selection and concept implementation;
- Establishment of projects pull along with the platform to look for alternative financing sources;
- Establishment of clear, effective estimation system for decisions and activities of municipal authorities (including establishment of online service to conduct these estimates);
- Encouragement and development of new educational forms and qualifications to stimulate innovations and sustainable development of Kyiv;
- Support and encouragement of new innovation formats for cooperation: hackathons, innovation and educational weekends, experimental labs, etc.

Solutions involving citizens that were implemented, inter alia, included: participatory budget, e-petitions; e-portal to keep citizens up-to-date information; improvement of educational and healthcare services and implementation of "smart roads" project to track dangerous places where accidents take place (Pozdniakova, 2018). The revised strategy aims at developing the smart city standards with a view to stimulate cooperation between the city, business and community; increasing cooperation between universities, research establishments, IT industry, city authorities and public sector; creation of modern housing and utility services; security and transport; e-medicine, open government and e-democracy (KCC, 2018, p. 128).

7.6.5.7 Smart living
Smart living involves numerous characteristics that significantly enhance the quality of life of residents, such as health, culture, housing, tourism, safety, etc. Thus improving each of these features

leads to a more harmonious, satisfactory, and fulfilled life (Azkuna, 2012; Giffinger & Gudrun, 2010). According to Fernandez-Anez et al. (2018), smart living also entails tourism, culture, healthcare, security, technology accessibility and social inclusion. Viewed in broad perspective, smart living aims at improving citizens' quality of life by transforming residential areas, office, transport and energy into smart environments (Giffinger & Gudrun, 2010; Tahir & Malek, 2016). Smart living entails integrating all that contributes toward a happy and comfortable life to citizens by providing smart facilities and services enabled by the latest technology (Wolfram, 2012a, 2012b). Many of the elements of smart living have been analyzed across main parameters or dimensions of a smart city in preceding pages and here the focus is on adequate housing, social inclusion, and social protection.

7.6.5.7.1 Adequate housing

Adequate housing equipped with basic amenities is integral to smart living. Adequate housing represents a necessary articulation of a number of established human rights like the right to freedom from interference with home, privacy, and family and the right to freedom of movement (OHCHR-UN-Habitat, 2014, p. 3). Affordability is just one aspect of adequate housing and the International Federation for Housing and Planning (IFHP) defines housing as affordable if household expenditure is within 30% of household income, or 80% of the median household income, such that the household can meet other basic needs (IFHP, 2016, p. 5). While ensuring protection against forces of nature, physical safety, and ease of access for persons with disabilities, and other challenges, housing should be located away from polluted or industrial areas and should be in proximity to educational and health facilities, employment opportunities, and other social facilities. Rapid pace of urbanization that has culminated in more than half of population worldwide living in cities has also been characterized by more-than-proportionately involved rural migration to informal settlements in and around cities, known more commonly as "slums" that are densely populated urban areas characterized by poor-quality housing, a lack of adequate living space and public services, and entailing large number of informal residents with often insecure tenure (Marx, Stoker, & Suri, 2013). The UN defines a slum as a settlement where inhabitants suffer from overcrowding or "inadequate access to water, sanitation, and other basic infrastructure" (UN-Habitat, 2002, pp. 22–23) and slum dwellers are also faced with insecure tenures or poor structural quality, such as temporary or derelict housing. By 2010, at least 860 million people were living in slums and the number of slum dwellers increased by 6 million each year from 2000 to 2010 (UN-Habitat, 2012). The absolute number of people residing in slums or informal settlements grew to over 1 billion; nevertheless, the proportion of the global urban population living in slums fell from 46% to 23% between 1990 and 2016 and this progress was largely offset by internal population growth and rural—urban migration (UN-ECOSOC (United Nations-Economic and Social Council), 2019). The 2030 Agenda for Sustainable Development that includes the SDGs is seemingly one of the strongest efforts by the international community to address the spread of urban slums (UN-GA (United Nations-General Assembly), 2015, p. 14). Besides, housing and slum upgrading is also one of the seven focus areas of the UN-Habitat (2014).

Right to adequate housing has been recognized by international community as a human right and it was first codified in the Universal Declaration of Human Rights in 1948 and Article 25 of this Declaration envisages that everyone has the right to adequate housing along with other social services (UN-GA, 1948) and this Article states that everyone has the right to an adequate standard of living, including, but not limited to: decent food, clothing, housing, and other social services. Right to "an adequate standard of living... including adequate food, clothing and housing" is also incorporated in the 1966 International Covenant on Economic, Social and Cultural Rights (ICESCR), thereby placing responsibility in UN Member States to ensure the realization of this right (UN-GA, 1966). The 2011 Istanbul Program of Action for the Least Developed Countries for the Decade 2011–20 (IPoA) that was adopted at the Fourth United Nations Conference on the least developed countries (LDCs), also accorded recognition to shelters as an integral part of human and social development and a priority area for action for LDCs and the goals set by the IPoA include increasing access of slum dwellers and the rural poor to affordable housing and housing-related infrastructure through improved planning, policy, and legislation (UN (United Nations), 2011, p. 10, 27). In the wade of widespread proliferation of slums in urban areas, according priority to adequate housing as an urgent issue requiring adequate attention calls upon the UN Member States to shoulder the responsibility of ensuring that all housing needs are met with adequate structural requirements, and providing at least the minimum needs of shelter along with provisions for safe drinking water, energy, sanitation, disposal of waste, and emergency services to all inhabitants, in an inclusive and fair manner (OHCHR-UN-Habitat, 2014, pp. 6–8).

SDG-11 directly targets slums by emphasizing making cities and human settlements inclusive, safe, resilient and sustainable. There are various targets of SDG-11, like the provision of adequate, safe, affordable, and accessible housing, basic services, transport systems, and planning of sustainable urban settlements by 2030 (UN-GA (United Nations-General Assembly), 2015). Specific emphasis is focused on vulnerable segments of society such as persons with disabilities, women, children, and the elderly, particularly the development of inclusive, efficient urban policies for resilient cities. The New Urban Agenda (NUA) adopted at the Habitat III conference in October 2016 emphasizes on integrating equity and social justice into the development agenda, with the objective of eliminating urban slums, along with focusing specific emphasis on the commitment of the international community to promote age- and gender-responsive "national, subnational, and local housing policies" to support the realization of the right to adequate housing (UN-Habitat III, 2016). Nevertheless, UN report on the progress on SDGs, with specific reference to SDG-11 notes that undoubtedly substantial progress has been made in reducing the proportion of the global urban population living in slums; nonetheless, it laments that more than 1 billion people continue to live in such situations and calls for urgent action to reverse the current situation UN-ECOSOC (2019).

7.6.5.7.2 Social inclusion

Social inclusion is essential for smart living in a smart city. Social inclusion is defined by Warschauer (2003) as the extent that individuals are able to fully participate in society and control their own destinies. Even in the absence of a universally accepted definition of a smart city, it has been emphasized by some scholars that smart city ought to thrive to achieve social inclusion of all urban residents (Andrade & Doolin, 2016). Governmental emphasis on social inclusion, within political discourses, is ascribed to its economic dimension like productively being engaged in work, training among other and this at the expense of other forms of inclusion, such as cultural identity, social interaction, interpersonal networks, civic and political participation (Andrade & Doolin, 2016). Admittedly, the bulk of emphasis focused around smart city paradigm emphasizes on the use of smart technologies (Börjesson, Eriksson, & Wangel, 2015; Chourabi et al., 2012); nonetheless, a smart city should also be inclusive (Hollands, 2008). Owing to the complexity of smart urbanism, emphasis has also been focused by Luque-Ayala and Marvin (2015) on the need to seriously engage issues of social inclusion while dealing with different urban contexts. Specific priorities around smart cities accorded by European Commission aim at addressing the issue of social inclusion among other areas of low-carbon economy and climate resilience (European Commission, 2015a, 2015b, 2015c). Horizon (2020), a framework program for research and innovation, launched by European Commission at the EU level is seemingly a multifaceted program and one of the facets is to address societal challenges eliciting support for inclusive, innovative and reflective societies, a criterion for sustainable European integration (European Commission, 2016). It has also been observed that: "Inclusion is a matter not only of an adequate share in resources but equally of participation in the determination of both individual and collective life chances as well." (Stewart, 2000, p. 9).

Cities, as centers of both opportunities and concentrations of social problems, have become "attractive interiors" for migration as well as creative works, innovations and job opportunities capable of affording avenues for creating new meeting places and often acting as hubs for social networks (Abrahamsson, 2015). In the context of EU, cities are said to bring about propinquity of people, businesses and services together thereby opening doors for building a more resource efficient Europe (EEA, 2016a, 2016b). Concurrently, cities have also served as centers for severe forms of poverty, substandard housing and homeless (Abrahamsson, 2015). Rapid pace of urbanization and the resultant uneven levels of development have proved instrumental in further exacerbating these problems. According to broad estimates, nearly one-fourth of the EU population is at the risk of poverty and social exclusion and in the wake of this problems getting further accelerated as a sequel to the recent economic crisis and is undermining the overall target of Europe 2020 strategy of lifting at least 20 million people out of the risk of poverty or social exclusion by 2020 (European Commission, 2016; Eurostat, 2015). Differences in performance do exist between welfare systems of different EU countries wherein some countries have been able to reduce the risk of poverty by 60% and the least effective by less than 15% (European Commission, 2016). Poverty is often associated with other forms of inequality and exclusion within cities in areas like education, housing, health, participation and employment (Yang & Vizard, 2017), and altogether, this burden renders cities to become spaces of contestation and politicization of economic agenda that enhances marginalization, social exclusion and/or uneven development (Abrahamsson, 2015). Urban areas serve as habitat for a diverse group of people who reside side by side but with different cultures, group identity and with varied opportunities of living decent lives. Besides, there has seeming begun a process of split and widening of gap between high-income jobs that workers lack requisite qualifications and skills for and low-paid jobs on which people cannot prosper (Abrahamsson, 2015; Greene, MacDonald, & Arena, 2019). Concentration of high-skilled and ell-education workforce in specific spaces engendering differences between housing sites and often a spike in prices has been proving instrumental in creating the process of gentrification concurrently resulting in relocation of some citizens to new housing areas with lower prices and this process of gentrification is described by Wacquant (2009) as one that provides strength to segregation in the domain of societal development. Also acknowledging this fact, Abrahamsson (2015) notes that strong social tension still prevails between the affluent class of people and those who find excluded and marginalized within European cities.

Conciliation between the normative perspectives of the various actors has led to the emergence of a composite indication and according to some authors (Copeland & Daly, 2014; Frazer, Guio, Marlier, Vanhercke, & Ward, 2014), this conciliation/compromise explicitly emphasizes a glaring controversy over the social dimension of the EU: while the combination of the antipoverty and social exclusion guideline in Europe 2020 is a notable new element, it is problematic to maintain that it represents a significant step toward consolidating "Social Europe." The Commission's stance (European Commission, 2010) affirms to this in that, on the one hand it calls upon Member States to incorporate in their budgets the salient targets envisaged in EU 2020, but on the contrary, it makes evident that spending on these salient elements should occur under conditions that strictly follow the fiscal stipulations of the Stability and Growth Pact (Petmesidou, 2017). The favoring of macroeconomic stability over all other targets is evidently discernible in the Annual Growth Surveys (AGSs), especially during the first cycles of the new "Strategy" (Natali, 2014). The midterm review launched in 2015, to appraise progress with the new "Strategy," exhibited the distance from the set target and impelled stress on the antipoverty and social exclusion dimension, hitherto again in close association with macroeconomic and employment objectives. It was perhaps in this backdrop that the 2016 Annual Growth Survey (AGS) emphasized that "more effective social protection systems" were called for to face up poverty and social exclusion, "while preserving sustainable public finances and incentives to work. Any such development will have to continue to ensure that the design of in-work benefits, unemployment benefits and minimum income schemes constitutes an incentive to enter the job market. Adequate and well-designed income support, such as unemployment benefits and minimum income schemes, allow those out of work to invest in job search and training, increasing their chances to find adequate employment that matches their skills" (European Commission, 2015a, 2015b, 2015c, p. 12).

There has been an increase in income inequalities and in-work poverty in many places of the EU in recent years (Cantillon, 2011), and this indicates that a major emphasis on employment rates, as pointed out in the 2016 Annual Growth Survey (European Commission, 2015a, 2015b, 2015c), without well-concerted endeavors to improve the quality of jobs, balancing flexibility with security and encouraging equality of opportunity as well as income, seems to be ill-advised, even at times of prosperity, "let alone in a period of protracted slow recovery" (Petmesidou, 2017). Besides, academic debate in recent past years provides substantial evidence of the negative impacts of poverty and inequality on growth (Stiglitz, 2012). In the wake of controversies surrounding the social element of Europe 2020, Copeland and Daly have opined that the poverty and social exclusion as "effectively ungovernable" and their opinion is based on three grounds—(1) the "ideational coherence" characterizing the three constituent elements of the indicator, along with varying approaches to poverty and social exclusion implicit in them, (2) "insufficient political prioritization", and (3) shortfalls in the monitoring procedure. Focus on the limited role of the social dimension of the new governance structure of the EU has also been emphasized by some scholars (Urquijo, 2017). Still there is also an articulation of a more optimistic view as "to the possibility of strengthening complementarities between macroeconomic, fiscal and social coordination in the future" (Petmesidou, 2017).

7.6.5.8 Social protection

Social protection is a pivotal element of smart living systems because being an essential component of national governance it entails the potential of reducing levels of poverty and narrowing down income gaps. Social protection systems embrace three major objectives, namely (1) to guarantee access to essential goods and services for all members of a society, (2) to promote active socioeconomic security, and (3) to enhance individual and social potential for poverty reduction and societal development (Garcia & Gruat, 2003). Deemed as an investment in the social and economic development of societies and individuals, social protection not only renders help to people to deal with risks and decreases inequalities, but also enables the people to develop full potential for personal growth and meaningful contributions to their societies throughout their life (Cichon et al., 2004). Social protection refers to policies aimed at preventing and alleviating poverty, vulnerability and social exclusion throughout the lifecycle (Mathers & Slater, 2014; UN-DESA, 2018). Benefits to individuals or households are often provided by social protection systems to guarantee income security and access to health care. Measures like cash benefits, old-age pensions, in-kind transfers and disability benefits proved instrumental in mitigating the impact of the global financial crisis of 2008 among the most vulnerable, and concurrently also served as a macroeconomic stabilizer and enabled people to surmount social exclusion and poverty in both developed and developing countries (ILO, 2011). Social protection entails vast potential of stimulating demand and boost consumption, and hence contribute to economic growth. Social protection spending, during recessions, can help revive economies and stimulate employment (UN-DESA, 2018).

Concurrently, social protection systems comprise somewhat big redistributive mechanisms in developed countries, often constituting a good proportion of GDP (ILO, 2016). While emphasizing that the potential contribution to social protection to individual and societal development cannot be realized if the resources entrusted by the society to its social protection system are not with utmost care, Cichon et al. (2004) lament that too many social protection schemes—albeit well-designed—have failed because of their governance and management failed. Social protection has been catapulted to the forefront of the development agenda owing to the strong positive impacts wielded by social protection. Social protection constitutes a key component of national development strategies to alleviate poverty and inequality, promote human capital, political stability and inclusive growth (Garcia & Gruat, 2003; ILO, 2014). A global consensus on the extension of social security that has been reached among governments, employers', and workers' organizations and associations from 185 countries at all levels of development is amply reflected in the Social Protection Floors Recommendation adopted in 2012 (ILO, 2012, 2017b; Ortiz, et al., 2017). Adoption of these recommendations entails a commitment by these countries to create nationally-defined social protection floors as the prime element of their social protection mechanisms, aiming at universal protection for all. Apart from the endorsement of these recommendations by the members of the G-20, these are also integrated into the SDGs, especially in SDG-1, Target 1.3 that stipulates to "Implement nationally appropriate social protection systems and measures for all, including floors, and by 2030 achieve substantial coverage of the poor and the vulnerable Indicator" (ILO, 2017c). More than thirty developing countries are reported to have taken up the commitment and are said to have already made the vision of world with universal social protection schemes a reality; and some of them are said to have attained universality via mix of contributory social insurance and noncontributory benefits, and others have achieved universality by universal transfers, demonstrating that there are multiple ways to achieve universality (GPUSP, 2016; ILO, 2017a). Despite the fact that the need for social protection is recognized worldwide, it is lamentable that human right to social protection remains unfulfilled for a lager chunk of humanity because only 29% of the global population enjoy access to comprehensive social protection systems, whereas 71% are either covered partially or not at all (ILO, 2017a).

Elements or components of social protection instruments are often classified into three categories: (1) social assistance, (2) social insurance, and (3) labor market programs. Defined as noncontributory social protection, social assistance is often financed through taxes and targeted at low-income households and vulnerable groups (UN-DESA, 2018). Cash or in-kind social transfers like fee waivers, subsidies and child benefits, all of which are means tested, comprise the examples of social assistance. There has been rapid proliferation of cash transfers, specifically in low- and middle-income countries; and over 130 countries worldwide use direct, regular, and noncontributory cash payments as income support and poverty reduction strategies pivotal to their social protection systems (Bastagli, Hagen0Zanker, Harman, Varca, & Sturge, 2016). According to broad estimates, on an average, countries spend 1%−2% of GDP on social assistance transfers (DFID, 2011). According to Baird, Ferreira, Ozler, and Woolcock (2013), these can be unconditional or conditional on school attendance, health or job requirements. Social assistance schemes embrace nearly 31% of the world's population and have had a positive effect in reducing income inequality. Social insurance pertains to contributory programs that provide safeguard against certain life contingencies through a risk-pooling mechanism that is dependent on prior contributions (Mathers & Slater, 2014). Old-age pensions are the stark example of social insurance. According to ILO (2017a), pension schemes receive contributions from 35% of the world's labor force and make benefits available to 68% of the elderly. Unemployment benefit programs, albeit less widespread, are another example that targets the working-age population. Undoubtedly, an increasing number of countries have consolidated social protection systems to deal with development challenges, specifically under the 2030 Agenda framework that recognizes the right to social security (UN-DESA, 2018); nevertheless, only 45% of the global population is covered by at least one social protection benefit with variable coverage widely by population group; and globally, 35% of children, 22% of the unemployed and 68% of the elderly benefit (ILO, 2017a). Admittedly, there is a long way to achieve SDG 1.3; nonetheless, a number of developing countries in all regions are seemingly close to or have reached universal pension coverage as well as social protection coverage also witnessing variations across regions (OECD, 2019).

7.7 Ecosystem-based adaptation for smart cities

Burgeoning urban population worldwide along with complex patterns of economic assets, infrastructures and ensuing services renders the cities vulnerable to the vagaries of climate change (OECD, 2014; Rosenzweig, et al., 2015). Pursuit of sustainable development in urban places necessitates achieving climate adaptation and cities are reportedly undertaking increasing actions in adhering to climate adaptation (ICLEI, 2010). Congeniality of climate adaptation for urban areas is suggested by Picketts, Déry, and Curry (2013) of its being well suited to "local levels of governments, as citizens can participate in creating targeted adaptation strategies that address the important regional impacts, and these strategies will provide tangible benefits to local residents." In the similar vein, planning at municipal level is emphasized by Measham et al. (2011) as pivotal to mainstreaming adaptation actions. Ecosystem-based adaptation (EbA) entails the potential of laying a significant role in urban contexts and render assistance in dealing with spike in temperature, flash floods, and water scarcity owing to its ability to reduce soil sealing, mitigation heat-island-effect and augmenting water storage capacity in urban watersheds (Müller, Kuttler, & Barlag, 2013; USAID, 2017). Application of EbA in cities embraces approaches based on design and improvement of green and blue infrastructures like urban parks, green roofs and facades, tree-planting, rivers and ponds, etc., along with other kind of interventions that use ecosystem functions to deliver some type of adaptation to climate risks in terms of measures designed to decrease soil imperviousness (Roberts et al., 2012). Most ecosystems in the cities are "urban ecosystems" because these are ecosystems where the built infrastructure encompasses a large traction of land surface or those wherein people reside in high densities (Pickett et al., 2001). Almost all green and blue spaces in urban areas fall within the ambit of urban ecosystems and these entails relatively a low level of naturalness being profoundly managed and wholly artificial (Rocha, Zulian, Maes, & Thijssen, 2015). Green roofs are a stark example of urban ecosystems because these are exclusively managed by humans and require regular maintenance (Oberndorfer et al., 2007). Notion of EbA, with its inherent measures, is often used to denote the use of urban ecosystems to deliver services that assist in adapting to climate change (Doswald et al., 2014; Zandersen et al., 2014).

Augmentation of green infrastructure in urban areas entails the potential of decreasing the heat-island-effect and the associated health risks (Lafortezza, Davies, Sanesi, & Konijnendijk, 2013). In contrast to more stereotype or traditional infrastructure-based approaches such as levees, sea walls and irrigation systems, EbA affords the benefit of promoting "no regrets" interventions and possibly generating manifold economic, environmental and economic cobenefits that transcend the domain of climate adaptation (Jones, Hole, & Zavaleta, 2012). These cobenefits, inter alia, comprise biodiversity conservation via increased habitat conditions; mitigation through enhanced carbon sequestration, conservation of traditional knowledge, livelihood and practices of local communities; improved recreation and tourism opportunities along with enhanced food security (Demuzere et al., 2014; Munang et al., 2013c; Munang, Thiaw, Alverson, Liu, & Han, 2013b; Vignola, Locatelli, Martinez, & Imbach, 2009).

Undoubtedly, EbA approaches fall short of quantitative estimates of the adaptation potential (Jones et al., 2012); nevertheless, there is growing evidence of EbA's potential of delivering flexible, cost-effective and widely applicable options to deal with the magnitude, speed and uncertainty of climate change (Munang et al., 2013a). It is on these counts that EbA has swiftly emerged as a significant aspect of the international climate policy framework (FEBA, 2017). It is noteworthy that the European Union in its recent climate adaptation strategy (European Commission, 2013) obviously encourages the adoption of green infrastructure and ecosystem-based approaches to adaptation. In recent years, the potential role of EbA in cities has started gaining traction in the literature (Berndtsson, 2010; Bowler, Buyung-Ali, Knight, & Pullin, 2010; Müller et al., 2013), and recently, Demuzere et al. (2014) have provided an exhaustive analysis of the existing empirical evidence with regard to the contribution of green infrastructures to climate change adaptation in urban areas. However, the notion of EbA is still comparatively an emerging concept in the context of cities (Geneletti & Zardo, 2016), and there exists sparse evidence on the incorporation of EbA measures in actual urban plans and policies (Wamsler, Luederitz, & Brink, 2014). Nevertheless, some industrialized countries are reported to be addressing climate adaptation strategies and actions in their urban planning as is demonstrated by recent reviews of planning documents conducted for cities in Europe (Reckien et al., 2014), the United Kingdom (Heidrich, Dawson, Reckien, & Walsh, 2013), Australia (Baker, Peterson, Brown, & McAlpine, 2012) and North America (Zimmerman & Faris, 2011). Nonetheless, EbA finds no specific mention in these documents and papers. Multiple collections of experiences available in the gray literature either deal with urban adaptation in general, with slight allusion to EbA (EEA, 2012), or on EbA diminutive emphasis on urban areas

(Doswald et al., 2014; Naumann et al., 2011). The bulk of the case studies dealing primarily with the EbA in the literature vastly refers to natural areas, coastal zones, agriculture and forestry; nevertheless, the work of Kazmierczak and Carter (2010) provides a compilation of database of case studies dealing with EbA approaches in cities. According to Geneletti and Zardo (2016), even these case studies are not specifically related to planning and rather refer to a wide set of initiatives like incentives, schemes, physical infrastructure, delivery, guidance documents, etc., and the extent to which EbA approaches should actually be covered in planning at the urban level have remained largely undocumented.

7.8 Conclusion

The long-term projections have continued to predict that there would be further urbanization wherein the urban population that stands in 2020 at 56.2% is likely to grow to 60.4% by 2030 and almost every region is projected to become more urbanized in the ongoing decade (2020–30); nevertheless, with highly urbanized areas expected to witness a slow rate of urban growth, whereas the bulk of urban growth (about 96%) will take place in less developed areas of East Asia, South Asia and Africa with three countries—India, China and Nigeria—registering 35% of the total enhancement in urban population worldwide by 2050 (UN-Habitat, 2020). Interestingly, urban sprawl—once a phenomenon associated with the land-rich developed countries of North America and Australia—is now fast emerging as an increasingly common phenomenon happening in cities globally. The physical scale of urban regions is growing much faster comparable to their population that only results in consumption of more land for urban development but also entails serious implications for energy consumption, GHG emissions, climate change and environmental degradation. Trends emerging from a global sample of cities reveal that in 1990–2015, cities in developed countries expanded their urban land area by 1.8fold while the urban population during the same period grew by 1.twofold; therefore, implying that the increase in urban areas in relation to urban population growth increased by a ratio of 1.5 (UN-Habitat, 2020).

With one in every seven people on the Planet being a migrant, and as in 2019 the situation prevails, there are about 763 million internal migrants and 272 million are international migrants in the world and that makes migration as a driving force contributing to urbanization, whether rural-to-urban movement within countries or the clustering of international migrants in global cities. Growing levels of economic inequality and lack of affordable housing are the worrisome persistent trends confronting the burgeoning pace of urbanization. The persistent income inequalities for the past four decades have rendered about 2.9 billion people living in cities where income inequalities are pronounced, and "the nature of inequality will largely depend on what happens in cities" (UN-Habitat, 2020). Lack of affordable housing in cities worldwide has made 1.6 billion people or 20% of global population, live in inadequate housing, of which one billion people reside in urban slums and informal settlements (UN-Habitat, 2020).

Undoubtedly, SDG-11 envisages making cities and human settlements inclusive, safe, resilient and sustainable; nonetheless, the cities have experienced rising slum-dweller population, deteriorating air pollution, minimal open public spaces and limited convenient access to public transport. As per data collected in 2019 from 610 cities in 95 countries, about half of the population has convenient access to public transport. Another data collected from a sample of 911 cities from 114 countries in 2020 show that, through the 1990−2019 period, the share of urban area designated for streets and open public spaces averages only about 16% worldwide, much below the Kazimierz UN-Habitat recommendation of 30% for the same purpose. And as of March 2021, 156 countries have developed their respective national urban policies with almost half already in the implementation stage; and a further breakdown reveals that 38% of the countries are in the early stages of developing their plans, while 13% are monitoring as to how well these plans are functioning (UN-ECOSOC, 2021). This demonstrates that more needs to be done to make our cities resilient and sustainable and as such, all targets of SDG-11 are unlikely to be met by 2030. UN-Habitat's NUA to be implemented in conjunction with SDG-11 and the urban dimensions of other SDGs affords a roadmap to equitable growth and prosperity of the cities.

References

Abdoullaev, A. *Building smart cities and communities*. (2013). <http://www.eu-smartcities.eu/sites/all/files/blog/files/Building%20SMART%20CITIES%20EIP.pdf>.

Abrahamsson, H. (2015). The great transformation of our time: Towards just and socially sustainable Scandinavian cities. In E. Richard, M. Johansson, & T. Salonen (Eds.), Social transformations in Scandinavian cities: Nordic perspectives on urban marginalization and social sustainability. Lund, Sweden: Nordic Academic Press.

ADB (Asian Development BankBodyBody). (2010). *ADB climate change programs: Facilitating integrated solutions in Asia and the Pacific*. Mandaluyong, Philippines: Asian Development Bank.

Ahmad, N., & Ribarsky, J. (2018). *Towards a framework for measuring the digital economy*. Paris: OECD, OECD Working Paper.

Ahmed, N. O., El-Halafawy, A. M., & Ahmed Mohamed Amin, A. H. (2019). A critical review of urban livability. *European Journal of Sustainable Development*, 8(1), 165−182.

Ahvenniemi, H., Huovila, A., Pinto-Seppa, I., & Airaksinen, M. (2017). What are the differences between sustainable and smart cities? *Cities (London, England)*, 60, 234−245.

ADEME. (2014). *Économie Circulaire: Notions* cited in Appendino, F., Roux, C., Saadé, M., & Peuportier, B. (2019). *Circular economy in urban projects: A case studies analysis of current practices and tools*. Venice: AESOP. Available online at: https://hal.archives-ouvertes.fr/hal-02182283.

Alawadhi, S., Aldama-Nalda, A., Chourabi, H., Gil-Garcia, J. R., Leung, S., Mellouli, S., & Walker, S. (2012). Building understanding of smart city initiatives. In: *Lecture notes in computer science (including subseries lecture notes in artificial intelligence and lecture notes in bioinformatics)* (pp. 40−53). 7443 LNCS.

Alawadhi, S., & Scholl, H. J. (2016). Smart governance: A cross-case analysis of smart city initiatives. In: *2016 49th Hawaii international conference on system sciences (HICSS)* (pp. 2953−2963). IEEE Computer Society. Retrieved from https://ieeexplore.ieee.org/document/7427553.

Alawadhi, S., & Scholl, H. J. (2013). Aspirations and realizations: The smart city of Seattle. In: *2013 46th Hawaii international conference on system sciences* (pp. 1695−1703). IEEE Computer Society. Retrieved from https://www.computer.org/csdl/proceedings/2013/hicss/12OmNBzRNrw.

Albalate, D., & Fageda, X. (2019). Congestion, road safety, and the effectiveness of public policies in urban areas. *Sustainability.*, 11(5092), 1−21.

Albino, V., Berardi, U., & Dangelico, R. M. (2015). Smart cities: definitions, dimensions, performance, and initiatives. *Journal of Urban technology.*, 22(1), 3.21.

Ali, A., Razzaq, A., Mehmood, S., Zou, X., Zhang, X., Lv, Y., & Xu, J. (2019). Impact of climate change on crops adaptation and strategies to tackle its outcomes: A review. *Plants*, 8(2), 34.

Al-Nasrawi, S., Adams, C., & El-Zaart, A. (2015). A conceptual multidimensional model for assessing smart sustainable cities. *Journal of Information Systems and Technology Management*, *12*(3), 541−558.

Allam, Z., & Newman, P. (2018). Economically incentivising smart urban regeneration. Case Study of Port Louis, Mauritius. Smart. *Cities.*, *1*, 53−74.

Allwinkle, S., & Cruiskshank, P. (2011). Creating smarter cities: an overview. *Journal of Urban Technology*, *18*(2), 1−16.

Alonso, R. G., & Lippez-De Castro, S. (2016). In J. Gil-Garcia, T. Pardo, & T. Nam (Eds.), *Smarter as the new urban agenda* (Vol. 11, pp. 333−347). Cham, Switzerland: Public administration and information technology, Springer International Publishing.

Amakpah, S.-W., Larbi, M., Liu, G., & Zhang, L. (2016). Dynamics of eco-cities: A review of concepts towards operationalizing sustainable urbanization. *Journal of Environmental Accounting and Management*, *4*(1), 73−86.

Amakpah, S.-W., & Liu, G. (2015). ECO-CITIES: UE NET systems integration as new paradigm shift in sustainable energy generation and utilization. *Journal of Environmental Accounting and Management*, *3*(4), 385−394.

Andrade, A. D., & Doolin, B. (2016). Information and communication technology and the social inclusion of refugees. *MIS Quarterly*, *40*(2), 405−416.

Anderson, M. S. (2007). An introductory on the environmental economics of the circular economy. *Sustainability Science*, *2*, 133−140.

Angelidou, M. (2014). Smart city policies: A spatial approach. *Cities (London, England)*, *41*, S3−S11.

Anthopoulos, L., Janssen, M., & Weerakkody, V. (2015). Comparing smart cities with different modeling approaches. World Wide Web Conference, May 18-22, 2015, Florence, Italy. WWW 2015 Companion: 525−528.

Anthopoulos, L., Janssen, M., & Weerakkody, V. (2019). A unified smart city model (USCM) for smart city conceptualization and benchmarking. In Information Resources Management Association (IRMA) (Ed.), *Smart cities and smart spaces: Concepts, methodologies, tools, and applications* (pp. 247−264). Pennsylvania, USA: IGI Global Publishers.

Anttiroiko, A. V., Valkama, P., & Bailey, S. J. (2014). Smart cities in the new service economy: Building platforms for smart services. *AI and Society. Knowledge, Culture and Communication*, *29*(3), 323−334.

Aplet, G., Thomson, J., & Wilbert, M. (2000). Indicators of wildness: using attributes of the land to assess the context of wilderness. In S. F. McCool, D. N. Cole, W. T. Borrie, & J. O'Loughlin (Eds.), *Wilderness science in a time of change conference*. Proceedings RMRS-P-15-VOL-2 (pp. 89−98). Ogden, UT: United States Department of Agriculture, Forest Service, Rocky Mountain Research Station.

Appendino, F., Roux, C., Saadé, M., & Peuportier, B. (2019). *Circular economy in urban projects: a case studies analysis of current practices and tools*. Venice, Italy: AESOP Available at. Available from https://hal.archives-ouvertes.fr/hal-02182283/document.

Archer, D., Monteith, W., Scott, H., & Gawler, S. (2017). Developing city resilience strategies: Lessons from ICLEI-ACCCRN process. In: *Working papers series-41*. London: IIED. Retrieved from https://pubs.iied.org/pdfs/10816IIED.pdf.

Arnkil, R., Järvensivu, A., Koski, P., & Piirainen, T. (2010). *Exploring the quadruple helix*. In: Report of quadruple helix research for the CLIQ project. Tampere: Work Research Centre, University of Tampere.

Arouri, M. E. H., Youssef, A. B., Nguyen-Viet, C., & Soucat, A. *Effects of urbanization on economic growth and human capital formation in Africa. halshs-01068271*. (2014). Available online.

Azkuna, I. (2012). *Smart cities study: International study on the situation of ICT, innovation and Knowledge in cities*. Bilbao: The Committee of Digital and Knowledge-based Cities of UCLG. Available online.

Bahadur, A., Tanner, T., & Pichon, F. (2016). *Enhancing urbane climate change resilience*. Mandaluyong City, Philippines: Asian Development Bank.

Bahadur, A., & Tanner, T. (2014). Transformational resilience thinking: putting people, power and politics at the heart of urban climate resilience. *Environment and Urbanization, 26*(1), 200−214.

Baird, S., Ferreira, F. H. G., Ozler, B., & Woolcock, M. (2013). *Relative* effectiveness of conditional and unconditional cash transfers for schooling outcomes in developing countries: A systematic review. *Campbell Systematic Reviews, No.8*. Oslo: The Campbell Collaboration.

Bairoch, P. (1988). *Cities and economic development: From the dawn of history to the present*. Chicago: The University of Chicago Press.

Baker, I., Peterson, A., Brown, G., & McAlpine, C. (2012). Local government response to the impacts of climate change: an evaluation of local climate adaptation plans. *Landscape and Urban Planning, 107*(2), 127−136.

Bakici, T., Almirall, E., & Wareham, J. (2013). A smart city initiative: The case of Barcelona. *Journal of Knowledge Economy, 4*(2), 135−148.

Ballas, D. (2013). What makes a 'happy city'? *Cities (London, England), 32*, S39−S50.

Balsas, C. (2004). Measuring the livability of an urban centre: An exploratory study of key performance indicators. *Planning Practice & Research, 19*(1), 101−110.

Barefoot, K., Curtis, D., Jolliff, W., Nicholson, J. R., & Omohundro, R. (2018). *Defining and measuring the digital economy. Working paper*. Washington, DC: Bureau of Economic Analysis, United States Department of Commerce.

Barlow, M., & Le'vy-Bencheton, C. (2019). Estonia: smart nation on the Baltic Sea Available online at. Available from https://hub.beesmart.city/en/strategy/author/mike-barlow-and-cornelia-levy-bencheton.

Barns, S. (2018). Smart cities and urban data platforms: Designing interfaces for smart governance. *City, Culture and Society, 12*, 5−12.

Barraket, J. (2005). Enabling structures for coordinated action: Community organisations, social capital, and rural community sustainability. In A. Dale, & J. Onyx (Eds.), *Social capital and sustainable community development: A dynamic balance* (pp. 71−86). Vancouver, BC: University of British Columbia Press.

Barrionuevo, J. M., Berrone, P., & Ricart, J. E. (2012). Smart cities, sustainable progress. *IESE Insight, 14*, 50−57.

Bastagli, F., Hagen0Zanker, J., Harman, L., Varca, V., & Sturge, G. (2016). *Cash transfers: What does the evidence say? A rigorous review of programme impact and of the role of design and implementation features*. London: Overseas Development Institute.

Battarra, R., Zucaro, F., & Tremiterra, M. R. (2017). Smart mobility: An evaluation method to audit Italian cities. In: *5th IEEE international conference on models and technologies for intelligent transportation systems* (pp. 421−426).

Batty, M. (2011). Commentary. *Environment and Planning. A., 43*(4), 765−772.

Batty, M., Axhausen, K. W., Gianotti, F., & Pozdnoukhov, P. (2012). Smart cities of the future. *The European Physical Journal Special Topics, 214*(1), 481−518.

Beder, S. (2006). *Environmental principles and policies: An interdisciplinary approach*. Sidney: UNSW Press.

beesmart. city. *Urban mobility: Challenges and solutions in smart cities*. (2019). <https://hub.beesmart.city/solutions/en/smart-mobility/smart-mobility-challenges-and-solutions-in-smart-cities>.

Bell, D. (1973). *The coming of post-industrial society*. New York: Basic Books.

Ben Letaifa, S. (2015). How to strategize smart cities: Revealing the SMART model. *Journal of Business Research, 68*(7), 1414−1419.

Benevolo, C., Dameri, R. P., & D'Auria, B. (2016). Smart mobility in smart city. In T. Torre, A. Braccini, & R. Spinelli (Eds.), *Empowering organizations. lecture notes in information systems and organisation* (pp. 13−28). Cham, Switzerland: Springer.

Berndtsson, J. (2010). Green roof performance towards management of runoff water quantity and quality: a review. *Ecological Engineering, 36*, 351−360.

Best, H. (2010). Environmental concern and the adoption of organic agriculture. *Society and Natural Resources, 23*(5), 451−468.

Bina, O., & La Camera, F. (2011). Promise and shortcomings of a green turn in recent policy responses to the 'double crisis'. *Ecological Economics, 70*, 2308−2316.

Boer, B. (1995). Institutionalizing ecologically sustainable development: The roles of national, state, and local governments in translating grand strategy into action. *Willamette Law Review, 31*(2), 307−358.

Bolívar, M. P. R. (2016). Mapping dimensions of governance in smart cities. In: *Proceedings of the 17th international digital government research conference on digital government research* (pp. 312−324). Available online.

Bolivar, P. R. M. (Ed.), (2015). *Transforming city governments for successful smart cities*. Cham., Switzerland: Springer.

Bond, A. J., Mortimer, K. J., & Cherry, J. (1998). Policy and practice: The focus of local agenda 21 in the United Kingdom. *Journal of Environmental Planning and Management, 41*(6), 767−776.

Boone, A. L., & Mulherin, J. H. (2000). Comparing acquisitions and divestitures. *Journal of Corporate Finance, 6*(2), 117−139.

Börjesson, M. R., Eriksson, E., & Wangel, J. (2015). ICT practices in smart sustainable cities: In the intersection of technological solutions and practices of everyday life. In V. K. Johannsen, S. Jensen, V. Wohlgemuth, C. Priest, & Elina Erikson (Eds.), *Proceedings of EnviroInfo and ICT for sustainability 2015: Building the knowledge base for environmental action and sustainability* (pp. 317−324). Copenhagen: Atlantis Press.

Borland, Helen, Lindgreen, A., Maon, F., Ambrosini, V., Florencio, B. P., & Vanhamme, J. (Eds.), (2019). *Business strategies for sustainability* (1st ed.). London, UK: Routledge.

Boulding, K. E. (1966). The economics of knowledge and the knowledge of economics. *American Economic Review, 56*(1/2), 1−13.

Bowen, A., & Fankhauser, S. (2011). The green growth narrative: Paradigm shift or just spin? *Global Environmental Change, 21*, 1157−1159.

Bowler, D. E., Buyung-Ali, L., Knight, T. M., & Pullin, A. S. (2010). Urban greening to cool towns and cities: A systematic review of the empirical evidence. *Landscape and Urban Planning, 97*(3), 147−155.

Boyes, H., Hallaq, B., Cunningham, J., & Watson, T. (2018). The industrial internet of things (IIoT): an analysis framework. *Computers in Industry, 101*, 1−12.

Brandon, P. S., & Lombardi, P. (2005). *Evaluating sustainable development in the built environment*. Oxford, UK: Blackwell.

Brcic, M., & Mlinaric, D. (2018). Tracking predictive Gantt chart for proactive rescheduling in stochastic resource constrained project scheduling. *Journal of Information and Organizational Sciences, 42*(2), 179−192.

Brčić, D., Slavulj, M., Šojat, D., & Jurak, J. (2018). The role of smart mobility in smart cities. In: *5th international conference on road and rail infrastructure*, May 17−19, 2018, Zadar, Croatia. In: *CETRA papers* (pp. 1601−1606). Zagreb, Croatia: University of Zagreb.

Brennen, S., & Kreiss, D. (2014). Digitalization and digitization. *Culture Digitally, 8*. Available Online.

Briassoulis, H. (2001). Sustainable development and its indicators: Through a (planner's) glass darkly. *Journal of Environment Planning & Management, 44*, 409−427.

Bridger, J. C., & Lulloff, A. E. (1999). Toward an interactional approach to sustainable community development. *Journal of Rural Studies, 15*(4), 377−387.

Brynjolfsson, E. (1993). The productivity paradox of information technology. *Communications of the ACM, 36*(12), 66−77.

Brynjolfsson, E., & Kahin, B. (Eds.), (2002). *Understanding the digital economy*. Cambridge, MA: Massachusetts Institute of Technology.

Buckley, P., Pass, C., & Prescott, K. (1988). Measures of international competitiveness. *Journal of Marketing Management, 4*, 175−200.

Budd, L., & Hirmis, A. (2004). Conceptual framework for regional competitiveness. *Regional Studies, 38*, 1015−1028.

Buger, T. (2005). Environmental awareness of Polish community. *Instytut Gospodarki i Przestrzeni Miejskiej, Warszawa*, 300. (Original in Polish).

Bukht, R., & Heeks, R. (2017). Defining, conceptualising and measuring the digital economy. In: *GDI development informatics working papers, no. 68*. Manchester: University of Manchester. Available online.

Bulkeley, H., & Marvin, S. (2014). Urban governance and eco-cities: Dynamics, drivers and emerging lessons. In Dr. W. Hofmeister, P. Rueppel, & L. Liang Fook (Eds.), *Eco-cities: Sharing European and Asian best practices and experiences* (pp. 19−34). Singapore: KonradAdenauer Stiftung.

Burton-Jones, A. (1999). *Knowledge capitalism: Business, work and learning in the new economy*. Oxford: Oxford University Press.

Burton-Jones, A. (2003). Knowledge capitalism: The new learning economy. *Policy Futures in Education*, *1*(1), 143−159.

Cajot, S., Peter, M., Bahu, J.-M., Koch, A., & Marechal, F. (2015). Energy planning in the urban context: challenges and perspectives. *Energy Procedia*, *78*, 3366−3371.

Cambridge International Dictionary of English. (2005). Cambridge: Cambridge University Press (reprinted).

Cantillon, B. (2011). The paradox of the social investment state: employment and poverty in the Lisbon era. *Journal of European Social Policy*, *21*(5), 432−449.

Caprotti, F., & Cowley, R. (2016). *UK smart city survey*. London: King's College. Available online.

Caragliu, A., & Del Bo, C. (2012). Smartness and European urban performance: Assessing the local impacts of smart urban attributes. *Innovation: The European Journal of Social Science Research*, *25*(2), 97−113.

Carr, R. (2001). Excreta-related infections and the role of sanitation in the control of transmission. In F. Lorna, & B. Jamie (Eds.), *Water quality: Guidelines, standards and health* (pp. 89−113). London, UK: IWA Publishing.

Carson, R. (1962). *Silent spring*. New York: Houghton Mifflin.

Castelnovo, W., Misuraca, G., & Savoldelli, A. (2015). Smart cities governance: the need for a holistic approach to assess urban participatory governance. *Social Science Computer Review*, *34*(6), 724−739.

Chakravarty, D., & Sarkar, R. (2018). Conceptualizing Indian smart cities: Criteria for being climate resilient. In: *Working paper series. WPS no. 817*. Calcutta: Indian Institute of Management.

Chambers 21st Century Dictionary. (2007). Mumbai: Allied Publishers (revised ed.).

Chandra, Y. R. V. S., Shivia Harun, M., & Reshma, T. (2017). Intelligent transport system. *International Journal of Civil Engineering and Technology*, *8*(4), 2230−2237.

Chatterjee, S., Kar, A. K., & Gupta, M. P. (2017). Critical success factors to establish 5 G network in smart cities: inputs for security and privacy. *Journal of Global Information Management*, *25*(2), 15−37.

Chatterjee, S., & Kar, A. K. (2017). Effects of successful adoption of information technology enabled services in proposed smart cities of India. *Journal of Science and Technology Policy Management*, *9*, 189−209.

Chatterjee, S., Kar, A. K., & Gupta, M. P. (2018). Alignment of IT authority and citizens of proposed smart cities in India: System security and privacy perspective. *Global Journal of Flexible Systems Management*, *19*, 95−107.

Chen, Y., Ardila-Gomez, A., & Frame, G. (2017). Achieving energy savings by intelligent transportation systems investments in the context of smart cities. *Transportation Research. Part D Transport and Environment*, *54*, 381−396.

Chen, T. M. (2010). Smart grids, smart cities need better networks [Editor's Note]. *IEEE Network*, *24*(2), 2−3.

Chourabi, H., Nam, T., Walker, S., Gil-Garcia, J. R., Mellouli, S., Nahon, K., Pardo, T., & Scholl, J. R. (2012). Understanding smart cities: an integrative framework. *Proceedings of the 45th Hawaii international conference on system sciences. January 4−7, Maui, Hawaii, USA*, 2289−2297.

Chinalawinfo. *Circular economy promotion law of the People's Republic of China [Effective]*. (2017). <http://en.pkulaw.cn/display.aspx?cgid = 107971&lib = law>.

Cichon, M., Scholz, W., van de Meerendonk, A., Hagemejer, K., Bertranou, F., & Plamo, P. (2004). *Financing social protection*. Geneva: International Labour Office (ILO) and the International Social Security Association (ISSA).

Circle Economy. *Circular Amsterdam. A vision and an action agenda for the city and metropolitan area.* (2016). <https://www.amsterdam.nl/publish/pages/768044/circular-amsterdam-en-small-210316.pdf>.

Cisco. *Internet of things, at-a-glance.* (2016). <https://www.cisco.com/c/dam/en/us/products/collateral/se/internet-of-things/at-a-glance-c45-731471.pdf>.

Cisco. *Cisco visual networking index: Forecast and trends, 2017–2022. White paper.* (2018). <https://cyrek-digital.com/pl/blog/content-marketing-trendy-na-rok-2019/white-paper-c11-741490.pdf>

Cities PLUS. (2003). *A sustainable urban system: The long-term plan for greater Vancouver.* Vancouver, Canada: cities PLUS.

City of Vienna. (2014). *Smart city Wien framework strategy.* Vienna City Administration. Retrieved from https://smartcity.wien.gv.at/site/en/projekte/verkehr-stadtentwicklung/smile-die-mobilitatsplattform-der-zukunft-2/

Clarke, A. (2014). Designing social partnerships for local sustainability strategy implementation. In A. Crane, & M. M. Seitanidi (Eds.), *Social partnerships and responsible business: A research handbook* (pp. 79–102). London, UK: Routledge.

Clarke, A., & MacDonald, A. (2012). Partner engagement for community sustainability: Supporting sustainable development initiatives by reducing friction in the local economy. In: *State of knowledge report. Sustainable prosperity.* Available online.

Clinton-Gore Administration. (2000). *Building livable communities for the 21st century.* Washington, D.C.: United States Congress Library. Available online.

CMI (Chartered management Institute). (2011). *Setting smart objectives checklist 231.* Corby, England, UK: Chartered Management Institute. Available online.

Coe, A., Paquet, G., & Roy, J. (2001). E-governance and smart communities: A social learning challenge. *Social Science Computer Review, 19*(1), 80–89.

Cohen, B., & Munoz, P. (2016). Sharing cities and sustainable consumption and production: Towards an integrated framework. *Journal of cleaner production, 134,* 87–97.

Colldahl, C., Frey, S., & Kelemen, J. E. (2013). *Smart cities: Strategic sustainable development for an urban world* (p. 63). Available online.

Correia, I., Farhi, E., Nicolini, J. B., & Teles, P. (2011). Unconventional fiscal policy at the zero bound. *American Economic Review, 103*(4), 1172–1211.

Conroy, M., & Berke, P. (2004). What makes a good sustainable development plan? An analysis of factors that influence principles of sustainable development. *Environment and Planning A, 36*(8), 1381–1396.

Cook, D., & Das, S. (Eds.), (2005). *Smart environments: technology, protocols and applications.* London: Wiley.

Copeland, P., & Daly, M. (2014). Poverty and social policy in Europe 2020: Ungovernable and ungoverned. *Policy and Politics, 42*(3), 351–366.

Coplak, J. (2003). The conceptual framework of the EU Project ECOCITY. *Alfa Spectra, 10*(2), Retrieved from. Available from http://www.ecocity.szm.com/framework.html.

Cosgrave, E., Doddy, L., & Walt, N. (2014). *Delivering the smart city—Governing cities in the digital age.* London: Arup.

CPA (City Planning Association). (2018). *Stockholm city plan.* Fleminggatan, Stockholm: CPA. Retrieved from https://vaxer.stockholm/globalassets/tema/oversiktplan-ny_light/english_stockholm_city_plan.pdf

Crane, P. (2005). Nature in the metropolis. *Science (New York, N.Y.), 308*(5726), 1225, 1225.

Czech, A., Biezdudnaja, A., Lewczuk, J., & Razumowskij, W. (2018). Quantitative assessment of urban transport development—A spatial approach. *Engineering Management in Production and Services, 10*(1), 32–44.

Dale, A., & Newman, L. (2006). Sustainable community development, networks and resilience. *Environments, 34*(2), 17–27.

Dale, A., & Sparkes, J. (2011). The "agency" of sustainable community development. *Community Development Journal, 46*(4), 476–492.

Dameri, R. P., & Ricciardi, F. (2015). Smart city intellectual capital: An emerging view of territorial systems innovation management. *Journal of Intellectual Capital*, *16*(4), 860−887.

Davies, A. R., & Mullin, S. J. (2011). Greening the economy: Interrogating sustainability innovations beyond the mainstream. *Journal of Economic Geography*, *11*, 793−816.

Davis, K. (1955). The origin and growth of urbanization in the world. *American Journal of Sociology*, *60*(5), 429−437.

de Jong, M., Joss, S., Schraven, D., Zhan, C., & Weinjen, M. (2015). Sustainable-smart-resilient-low carbon, eco-knowledge cities; making sense of a multitude of concepts promoting sustainable urbanisation. *Journal of Cleaner Production*, *109*, 25−38.

Deakin, M., & Reid, A. (2016). Smart cities: under-gridding the sustainability of city-districts as energy efficient-low carbon zones. *Journal of Cleaner Production*. <https://www.napier.ac.uk/~/media/worktribe/output-679712/smart-cities-under-gridding-the-sustainability-of-city-districts-as-energy-efficient-low-1.pdf.

Deb, A. (2010). Sustainable cities in developing countries. *Building Research and Information*, *26*(1), 29−38.

Dell. (2017). *Design for environment*. Dell. Retrieved from http://www.dell.com/learn/us/en/uscorp1/corp-comm/closed-loop-recycled-content.

Deloitte. (2015). *Smart cities*. The Netherlands: Deloitte Available at. Available from https://www2.deloitte.com/content/dam/Deloitte/tr/Documents/public-sector/deloitte-nl-ps-smart-cities-report.pdf.

Demsetz, H. (1981). Economic, legal, and political dimensions of competition. In: *Discussion paper 209*. Los Angeles, CA: University of California.

Demuzere, M., Orru, K., Heidrich, O., Olazabal, E., Geneletti, D., Orru, H., ... Faehnle, M. (2014). Mitigating and adapting to climate change: Multi-functional and multi-scale assessment of green urban infrastructure. *Journal of Environmental Management*, *146*, 107−115.

De Santis, R., Allesandra, F., Nadia, M., & Anna, V. (2014). Smart city: fact and Fiction. MPRA Paper 54536, University Library of Munich, Germany. Available online at: https://mpra.ub.uni-muenchen.de/54536/1/MPRA_paper_54536.pdf.

DFID (Department for International Development). (2011). *Cash transfers evidence paper*. London: Department for International Development. Retrieved from https://webarchive.nationalarchives.gov.uk/+/http:/http://www.dfid.gov.uk/Documents/publications1/cash-transfers-evidence-paper.pdf

Digital Health. London. (n.d.). <https://digitalhealth.london/>.

Digital Talent Programme. (n.d.). <https://www.london.gov.uk/what-we-do/business-and-economy/skills-and-training/digital-talent-london>.

Dirks, S., Gergiev, C., & Keeling, M. (2010). *Smarter cities for smarter growth: how cities can optimize their systems for the talent-based economy*. NY: IBM: Global Business Services.

Doran, G. T. (1981). There's a S.M.A.R.T. way to write management's goals and objectives. *Management Review (AMA FORUM)*, *70*(11), 35−36.

Doswald, N., Munroe, R., Roe, D., Giuliani, A., Castelli, I., Stephens, J., & Reid, H. (2014). Effectiveness of ecosystem-based approaches for adaptation: Review of the evidence-base. *Climate and Development*, *6*(2), 185−201.

Dovers, S. (2002). Sustainability: Reviewing Australia's progress, 1992−2002. *International Journal of Environmental Studies*, *59*(5), 559−571.

Drexhage, J., & Murphy, D. (2010). *Sustainable development: From Brundtland to Rio 2012*. New York: United Nations & International Institute for Sustainable Development (IISD).

Drucker, P. F. (1955). *The practice of management*. Oxford, UK: Butterworth-Heinemann.

Drucker, P. F. (1993). *Post-capitalist society*. Oxford: Butterworth-Heinemann.

Duan, W., Nasiri, R., & Karamizadeh, S. (2019). Smart city concept and dimensions. *Innovative Creative Information Technology (ICIT) Journal.*, *2019*, 488−492.

Du-Pisani, J. A. (2006). Sustainable development—Historical roots of the concept. *Environmental Science*, *3*(2), 83—96.

Dykan, V., Ieromyna, M., Storozhylova, U., & Bilous, L. (2019). Implementation of smart city concept in Ukraine. *SHS Web of Conferences*, *69*, 06015, Retrieved from. Available from https://www.shs-conferences.org/articles/shsconf/pdf/2019/08/shsconf_NTI-UkrSURT2019_06015.pdf.

Eakin, H. C., & Wehbe, M. B. (2009). Linking local vulnerability to system sustainability in a resilience framework: two cases from Latin America. *Climatic Change*, *93*(3—4), 355—377.

Eden Strategy Institute & ONG&ONG Pte Ltd. (2018). Report ranks top 50 cities on leadership and governance. 6 July 2018. <https://www.edenstrategyinstitute.com/case-studies/smart-cities-world/>.

EEA (European Environment Agency). (2012). Urban adaptation to climate change in Europe challenges and opportunities for cities together with supportive national and European policies. In: *EEA technical report no. 2/2012*. European Environmental Agency.

EEA (European Environment Agency). (2016a). Urban adaptation to climate change in Europe 2016. In: *Transforming cities in a changing climate*. Retrieved from http://www.eea.europa.eu/publications/urban-adaptation-2016

EEA (European Environment Agency). *Urban environment*. (2016b). <http://www.eea.europa.eu/themes/urban>.

EESD (Energy, Environment and Sustainable Development). (2005). *Ecocity: Urban development towards appropriate structures for sustainable transport*. Oxford: EESD Available at. Available from http://www.oekostadt.at/root/img/pool/files/ecocity_project_summary.pdf.

Ehrlich, P. R. (1968). *The population bomb*. San Francisco, CA: Sierra Club Books.

Ekins, P. (1993). 'Limits to growth' and 'sustainable development': Grappling with ecological realities. *Ecological Economics*, *8*, 269—688.

Elagroudy, S., Warith., Mostafa, A., & Zayat, M. E. (2016). *Municipal solid waste management and green economy*. Berlin, Germany: Global Young Academy. Available online.

Elena, C. (2015). The making of knowledge cities in Romania. *Procedia Economics and Finance*, *32*, 534—541.

Elisei, P., D'Orazio, A., & Prezioso, M. (2014). Smart governance answers to metropolitan peripheries: Regenerating the deprived area of the Morandi Block in the Tor Sapienza Neighbourhood (Rome). In: *Proceedings REAL CORP 2014*. Retrieved from http://www.corp.at/archive/CORP2014_161.pdf.

Elkington, J. (1998). *Cannibals with forks: The triple bottom line of 21st century business*. Gabriola Island, BC: New Society Publishers.

Eriksen, S., & Lind, J. (2009). Adaptation as a political process: Adjusting to drought and conflict in Kenya's drylands. *Environmental Management*, *43*(5), 817—835.

Ernston, H., van der Leeuw, S. E., Redman, C. L., Meffert, D. J., Davis, G., Alfsen, C., & Elmqyist, T. (2010). Urban transitions: on urban resilience and human-dominated ecosystems. *Ambio*. *39* (8), 531—545, 7.

Estevez, E., & Janowski, T. (2013). Electronic governance for sustainable development—Conceptual framework and state of research. *Government Information Quarterly*, *30*(supplement 1), S94—S109.

EU (European Union). (2010). *Directive 2010/40/EU of the European Parliament and of the council on the framework for the deployment of intelligent transport systems in the field of road transport and for interfaces with other modes of transport*. Brussels: EU.

European Commission. (2010). Communication from the commission: Europe 2020. A strategy for smart, sustainable and inclusive growth. In: *COM (2010) 2020 final*. Brussels: European Commission. Retrieved from https://eur-lex.europa.eu/LexUriServ/LexUriServ.do?uri = COM:2010:2020:FIN:EN:PDF

European Commission. (2011). *Cities of tomorrow—Challenges, visions, ways forward. Information society policy link*. Brussels: Publications Office of the European Union.

European Commission. (2013). Communication from the commission to the European Parliament, the, council, the European economic and social committee and the committee of the regions. An EU strategy on adaptation to climate change (p. 216) Brussels: COM.

European Parliament (2014). Mapping Smart Cities in the EU. Available online at: https://www.europarl.europa.eu/RegData/etudes/etudes/join/2014/507480/IPOL-ITRE_ET(2014)507480_EN.pdf.

European Commission. (2015a). Annual growth survey 2016. Strengthening the recovery and fostering convergence. In: *COM (2015) 690 final*. Brussels: European Commission. Available online.

European Commission. (2015b). Closing the loop—An EU action plan for the Circular Economy. In: *COM (2015) 614 final*. Brussels. Available online.

European Commission. *Smart cities and communities. The European innovation partnership on smart cities and communities*. (2015c). <http://ec.europa.eu/eip/smartcities/index_en.htm>.

European Commission. *Employment, social affairs and inclusion*. (2016). <http://ec.europa.eu/social/main.jsp?catId = 751>.

EU Commission (European Union Commission). (2001). *White paper 2001; European transport policy for 2010: Time to decide*. Brussels: EU Commission.

EU Commission (European Union Commission). (2007). *Green Paper; Towards a new culture for urban mobility*. Brussels: EU Commission.

EU Commission (European Union Commission). (2009). *Communication from the Commission to the European Parliament. Action plan on urban mobility*. Brussels: EU Commission.

EU Commission (European Union Commission). (2011). *White Paper: Roadmap to a single European transport area—Towards a competitive and resource efficient transport system*. Brussels: EU Commission.

European Commission, EIP-SCC (European Commission, European Innovation Partnership on Smart Cities and Communities). (2016). *The marketplace of the European innovation partnership on smart cities and communities*. Brussels: European Union. Available online.

Eurostat. *People at risk of poverty or social exclusion*. (2015). <http://ec.europa.eu/eurostat/statistics-explained/index.php/People_at_risk_of_poverty_or_social_exclusion>.

Eurostat. (2016). *Urban Europe: Statistics on cities, towns and suburbs*. Luxembourg: Publications Office of the European Union. Retrieved from https://ec.europa.eu/eurostat/documents/3217494/7596823/KS-01-16-691-EN-N.pdf

Ewers, H., & Nijkamp, P. (1990). Urban sustainability. In P. Nijkamp (Ed.), *Urban sustainability* (pp. 8—10). Avebury, UK: Gower House.

Farrell, K. (2018). *Rapid urbanization an inquiry into the nature and causes of the urban transition in developing countries* (Doctoral thesis in urban and regional studies), Stockholm, Sweden: KTH Royal Institute of Technology.

FEBA (Friends of Ecosystem-based Adaptation). (2017). Making ecosystem-based adaptation effective: A framework for defining qualification criteria and quality standards *(FEBA technical paper developed for UNFCCC-SBSTA 46)*. Bonn, Germany: GIZ.

Fernandez, R. (2015). ESG integration in corporate fixed income. *Applied Corporate Finance, 27*(2), 64—72.

Fernandez-Anez, V., Fernández-Güell, J. M., & Giffinger, R. (2017). Smart city implementation and discourses: An integrated conceptual model. *The Case of Vienna Cities*, 1—12.

Fernandez-Anez, V., Velazquez, G., Perez-Prada, F., & Monzón, A. (2018). Smart city projects assessment matrix: Connecting challenges and actions in the Mediterranean region. *Journal of Urban Technology*, 1—25.

Fernández-Güell, J.-M., Collado-Lara, M., Guzmán-Araña, S., & Fernández-Añez, V. (2016). Incorporating a systemic and foresight approach into smart city initiatives: The case of Spanish cities. *Journal of Urban Technology, 23*(3), 1—25.

FHA (Federal Highway Administration, US Department of Transportation). (2014). *Highway statistics 2014*. Washington, DC: FHA. Available online.

Filho, W. L., Azeiteiro, U., Alves, F., Pace, P., Mifsud, M., Brandli, L., ... Disterheft, A. (2018). Reinvigorating the sustainable development research agenda: The role of the sustainable development goals (SDG). *International Journal of Sustainable Development and World Ecology, 25*(2), 131—142.

Florea, A., & Berntzen, L. (2017). Green IT solutions for smart city's sustainability. In: 5th *Smart cities conference slides*. Retrieved from https://administratiepublica.eu/smartcitiesconference/2017/files/SSC05-PP/DAY%202/Green%20IT%20solutions%20for%20smart%20city%20sustainability.pdf

Folke, C. (2016). Resilience. (Republished) *Ecology and Society*, *21*(4), 44.

Fook, L. L. (2014). Towards eco-cites in Europe and Asia: Sharing of best practices and experiences—An introduction. In Wilhelm Hofmeister, Patrick Rueppel, & Lye Liang Fook (Eds.), *Eco-cities: Sharing European and Asian best practices and experiences*. Singapore: Konrad Adenauer Stiftung: 7–18.

Forbes (2014). Smart Cities – A $1.5 Trillion Market Opportunity. 19 June 2014. <https://www.forbes.com/sites/sarwantsingh/2014/06/19/smart-cities-a-1-5-trillion-market-opportunity/?sh = 529e2e416053>.

Frazer, H., Guio, A.-C., Marlier, E., Vanhercke, B., & Ward, T. (2014). *Putting the fight against poverty and social exclusion at the heart of the EU agenda*. Brussels: European Social Observatory, European Social Observatory Paper Series.

Friends of the Earth. (1994). *Planning for the planet: Sustainable development policies for local and strategic plans*. London, UK: Friends of the Earth.

Gabrys, J. (2014). Programming environments—Environmentality and citizen sensing in the smart city. *Environment and Planning D. Society and Space*, *32*, 30–48.

Gajurel, D. R., & Wendland, C. (2007). *Ecological sanitation and associated hygienic risk—An overview of existing policy-making guidelines and research*. Munich, Germany: Women in Europe for a Common Future (WECF).

Galperina, H. (2016). *How to connect the other half: Evidence and policy insights from household surveys from Latin America. Global Commission on Internet Governance. Paper Series No. 34. June 2016*. London: Chatham House.

Garcia, A. B., & Gruat, J. V. (2003). *Social protection: A life cycle continuum investment for social justice, poverty reduction and development*. Geneva: ILO Available at. Available from https://www.ilo.org/public/english/protection/download/lifecycl/lifecycle.pdf.

Gardner, R. H., & Urban, D. L. (2007). Neutral models for testing landscape hypotheses. *Landscape Ecology*, *22*, 15–29.

Gaspareniene, L., Remeikiene, R., & Navickas, V. (2016). The concept of digital shadow economy: Consumers' attitude. *Procedia Economics and Finance*, *39*, 502–509.

Gehl, J. (2010). *Cities for people*. Washington, DC: Island Press.

Geissdoerfer, M., Savaget, P., Bocken, N. M. P., & Hultink, E. J. (2017). The circular economy—A new sustainability paradigm? *Journal of Cleaner Production*, *143*, 757–768.

Geller, A. L. (2003). Smart growth: A prescription for livable cities. *American Journal of Public Health*, *93*(9), 1410–1415.

Geneletti, D., & Zardo, L. (2016). Ecosystem-based adaptation in cities: An analysis of European urban climate adaptation plans. *Land Use Policy*, *50*, 38–47.

Georgeson, L., Maslin, M., & Poessinouw, M. (2017). The global green economy: A review of concepts, definitions, measurement methodologies and their interactions. *Geography and Environment*, *4*(1), e00036.

German Law Archives. *Closed Substance Cycle Waste Management Act (Kreislaufwirtschafts-und Abfallgesetz, KrW-/AbfG)—Excerpts*. (2015). <http://germanlawarchive.iuscomp.org/?p = 303>.

Ghiglione, S., & Larbi, M. (2015). Eco-cities in China: Ecological urban reality or Political nightmare? *Journal of Management and Sustainability*, *5*(1), 101–114.

Gidde, M. R., Todkar, V. V., & Kokate, K. K. (2008). Municipal solid waste management in emerging mega cities: A case study of Pune city. In: *Proceedings, GCE 2008, Indo-Italian Conference on Green and Clean Environment* (pp. 441–450). March 20–21, 2008, Pune, India.

Giffinger, R., & Gudrun, H. (2010). Smart cities ranking: An effective instrument for the positioning of the cities? *Architecture, City and Environment*, *4*(12), 7–26.

Giffinger, R., & Lü, H. (2015). *The smart city perspective: A necessary change from technical to urban innovation (First)*. Milano: Fondazione Giangiacomo Feltrinelli.

Gil-Garcia, J. R. (2012). Towards a smart state? Inter-agency collaboration, information integration, and beyond. *Information Polity, 17*(3−4), 269−280.
Gil-Garcia, J. R., & Aldama-Nalda. (2013). A smart city initiatives and the policy context: The case of the rapid business opening office in Mexico City. In: *Proceedings of the ICEGOV'13*. October 22−25, 2013, Seoul, South Korea. Retrieved from https://dl.acm.org/doi/pdf/10.1145/2591888.2591931
Gil-Garcia, J. R., Pardo, T. A., & Nam, T. (2015). What makes a city smart? Identifying core components and proposing an integrative and comprehensive conceptualization. *Information Polity, 20*(1), 61−87.
Giuseppe, G., & Pianezzi, D. (2017). Smart cities: Utopia or neoliberal ideology? *Cities (London, England), 69*, 79−85.
Glasmeier, A., & Christopherson, S. (2015). Thinking about smart cities. *Cambridge Journal of Regions, Economy and Society, 8*, 3−12.
Goldsmith, E., & Allen, R. (1972). *A blueprint for survival*. London: Penguin Books.
Gorynski, B. (2017). Smart city evolution: a review of becoming a smart city. 31 August 2017. Available online at: https://hub.beesmart.city/en/strategy/author/bart-gorynski#:~:text=Bart%20Gorynski%20is%20Managing%20Partner,simplifying%20the%20smart%20city%20business.
Govada, S. S., Sprujit, W., & Rogers, T. (2017). Smart city and framework. In V. T. M. Kumar (Ed.), *Smart economy in smart city* (pp. 187−198). Singapore: Springer.
GPUSP (Global Partnership for Universal Social Protection). (2016). Universal old-age and disability pensions in Timor-Leste, *Universal Social Protection Series*. Washington, DC: World Bank.
Grant, R. M. (1996). Prospering in dynamically competitive environments: Organizational capability as knowledge integration. *Organization Science, 7*, 375−387.
Greene, S., MacDonald, G., & Arena, O. (2019). *Technology and equity in cities-emerging challenges and opportunities*. Washington, D.C.: Urban Institute. Available online.
Gregson, N., Crang, M., Fuller, S., & Holmes, H. (2015). Interrogating the circular economy: The moral economy of resource recovery in the EU. *Economy and Society, 44*(2), 218−243.
Grieve, R., Kemp, N., Norris, K., & Padgett, C. R. (2017). Push or pull? Unpacking the social compensation hypothesis of Internet use in an educational context. *Computers and Education, 109*, 1−10.
Grubler, A., Bai, X., Buettner, T., Dhakal, S., Fisk, D. J., Ichinose, T., . . . Weisz, H. (2013). Urban energy systems. In Global Energy Assessment (Ed.), *Global energy assessment—Toward a sustainable future* (pp. 1307−1400). Cambridge, UK: Cambridge University Press.
Gupta, N., Yadav, K. K., & Kumar, V. (2015). A review on current status of municipal solid waste management in India. *Journal of Environmental Science, 37*, 206−217.
Hahnrath, J. (2017). *Foundational components of smart cities*. Berlin: IIC Smart City Forum. Available online.
Hajkowicz, S., & Collins, K. (2007). A review of multi-criteria analysis for water resource planning and management. *Water Resources Management, 21*(9), 1553−1566.
Hamilton Emily. (2016). *The benefits and risks of policymakers' use of smart city technology*.
Hankins, K. B., & Powers, E. M. (2009). The disappearance of the state from 'livable' urban spaces. *Antipode, 41*(5), 845−866.
Happaerts, S. (2012). Sustainable development and subnational governments: Going beyond symbolic politics? Environmental. *Development (Cambridge, England), 4*, 2−17.
Harrison, B., Eckman, R., Hamilton, P., Hartswick, J., Kalagnanam, J., Paraszczak., & Williams, P. (2010). Foundations for smarter cities. *IBM Journal of Research and Development, 54*(4), 1−16.
Haughton, G., & Hunter, C. (2003). *Sustainable cities*. London: Taylor and Francis.
Heidrich, O., Dawson, R. J., Reckien, D., & Walsh, C. L. (2013). Assessment of the climate preparedness of 30 urban areas in the UK. *Climate change, 120*, 771−784.
Höjer, M., & Wangel, J. (2015). Smart sustainable cities: Definition and challenges. In L. Hilty, & B. Aebischer (Eds.), *ICT innovations for sustainability* (pp. 333−349). Cham, Switzerland: Springer.
Hollands, R. (2008). Will the real smart city please stand up? *City, 12*(3), 303−320.

Horizon (2020). *Horizon 2020: Work Programme* 2018-2020. <https://ec.europa.eu/research/participants/data/ref/h2020/wp/2018-2020/main/h2020-wp1820-intro_en.pdf>.

Howes, C. (2000). U.S competitiveness and economic growth. In C. Howes, & A. Singh (Eds.), *Competitiveness matters: Industry and economic performance in the United States* (pp. 180−206). Michigan: The University of Michigan Press.

Howes, M. (2005). *Politics and the environment: Risk and the role of government and industry.* Crow's Nest, N.S.W: Allen & Unwin.

Hu, M. C., Wadin, J. L., Lo, H. C., & Huang, J. Y. (2016). Transformation toward an eco-city: Lessons from three Asian cities. *Journal of Cleaner Production, 123,* 77−87.

Huawei & Oxford Economics. *Digital spillover. Measuring the true impact of the digital economy.* (2017). <https://www.huawei.com/minisite/gci/en/digital-spillover/files/gci_digital_spillover.pdf>.

Huckle, J. (1996). Realizing sustainability in changing times. In J. Huckle, & S. Sterling (Eds.), *Education for sustainability.* London: Earthscan.

ICLEI (International Council for Local Environmental Initiatives). (2010). *Changing climate, changing communities: Guide and workbook for municipal climate adaptation.* Toronto: ICLEI.

ICLEI (International Council for Local Environmental Initiatives). (2012). Local sustainability 2012: Taking stock and moving forward. *Global Review, ICLEI Global Report.* Bonn: ICLEI.

ICLEI (International Council for Local Environmental Initiatives). (2020). *Promoting safe and sustainable cities with public transport.* Brussels, Belgium: ICLEI.

ICSU & ISSC (International Council for Science & the International Social Science Council). (2015). Review of targets for the sustainable development goals: the science perspective. <https://council.science/publications/review-of-targets-for-the-sustainable-development-goals-the-science-perspective-2015/>.

IDA Singapore. *iN2015 masterplan.* (2012). <http://www.ida.gov.sg/~/media/Files/Infocomm%20Landscape/iN2015/Reports/realisingthevisionin2015.pdf>.

IEA (International Energy Agency). *CO$_2$ emissions from fuel combustion 2017 highlights.* (2017). <https://webstore.iea.org/co2-emissions-from-fuel-combustion-highlights-2017>.

IESE (Instituto de Estudios Supriores). (2016). *IESE Cities in Motion.* Barcelona, Spain: IESE.

IFHP (International Federation of Housing and Planning). (2016). A roadmap to the past habitats and the path to habitat IV. IFHP Blog by IFHP Secretariat. <http://ifhp.org.linux4.curanetserver.dk/ifhp-blog/roadmap-past-habitats-and-path-habitat-iv>.

IISD (International Institute for Sustainable Development). (2012). *Sustainable development timeline.* Winnipeg, Canada: IISD.

ILO (International Labor Organization). (2011). *Social protection floor for a fair and inclusive globalization.* Geneva: ILO.

ILO (International Labor Organization). (2012). *The strategy of the International Labour Organization. Social security for all: Building social protection floors and comprehensive social security systems.* Geneva.

ILO (International Labor Organization). (2014). *World Social Protection Report 2014/2015, Building economic recovery, inclusive development and social justice.* Geneva: ILO.

ILO (International Labor Organization). (2016). *Delivering social protection for all.* Geneva: ILO.

ILO (International Labor Organization). *World social protection report 2017−19: Universal social protection to achieve the SDGs.* (2017a). Available online.

ILO (International Labor Organization). *Building social protection systems: International standards and human rights instruments.* (2017b). Available online.

ILO (International Labor Organization). *SDG 1.3. Social protection systems for all, including and floors, key to eradicating poverty and promoting prosperity.* (2017c). Available online.

ITU-FGSSC (International Telecommunication Union- Focus Group on Smart Cities). (2014). Smart water management in cities. Focus Group Technical Report. <https://www.itu.int/en/ITU-T/focusgroups/ssc/Pages/default.aspx>.

Indrawati, S., Ajkalhaq, N., & Amani, H. (2018). Indicators to measure smart economy: An Indonesian perspective. In: *Proceedings of the 2nd international conference on business and information management* (pp. 173−179). Barcelona, Spain, September 2018. New York: Association of Computing Machinery.

IPCC (Intergovernmental Panel on Climate Change). (2014). *Summary for policymakers*. Climate change 2014: Mitigation of climate change. *Contribution of Working Group III to the Fifth Assessment Report of the Intergovernmental Panel on Climate Change*. Cambridge, UK: Cambridge University Press.

IPCC (Intergovernmental Panel on Climate Change). (2007). *Summary for policymakers*. Climate change 2007: Mitigation. *Contribution of Working Group III to the Fourth Assessment Report of the Intergovernmental Panel on Climate Change*. Cambridge, UK: Cambridge University Press.

IRP (International Resource Panel). (2018). *The weight of cities: Resource requirements of future urbanization*. Report by the International Resource Panel. Nairobi, Kenya: United Nations Environment Programme (UNEP).

ITU-TFG-SSC (International Telecommunication Union-Task Force Group on Sustainable Smart Cities) (2014). *Technical report on smart sustainable cities: An analysis of definitions*. United Nations, International Telecommunication Union (ITU-T), Focus Group on Smart Sustainable Cities (FG-SSC). Available online.

IUCN (International Union for Conservation of Nature and Natural Resources). (1980). *World conservation strategy: Living resource conservation for sustainable development*. Gland: Switzerland.

IUCN (International Union for Conservation of Nature and Natural Resources). (2006). *Evaluating effectiveness: A framework for assessing management of protected areas* (2nd ed.). Gland, Switzerland: IUCN, Best practice protected area guidelines series No. 14.

Jacyna, M., Wasiak, M., Lewczuk, P., & Karon, G. (2017). Noise and environmental pollution from transport: Decisive problems in developing ecologically efficient transport systems. *Journal of Vibroengineering*, *19*(7), 5639−5655.

Jalali, R., Koohi-Fayegh, S., El-Khatib, K., Hoornweg, D., & Li, H. (2017). Investigating the potential of ride-sharing to reduce vehicle emissions. *Urban Planning*, *2*(2), 26−40.

Jamal, S., & Sen, A. (2019). Prospect of Faridabad as a smart city: A review. In V. R. Sharma, & Chandrakanta (Eds.), *Making cities resilient* (pp. 39−52). Cham, Switzerland: Springer International Publishing.

Jänicke, M. (2012). 'Green growth': from a growing eco-industry to economic sustainability. *Energy Policy*, *48*, 13−21.

Jenks, M., & Jones, C. (2010). Issues and concepts. In M. Jenks, & C. Jones (Eds.), *Dimensions of the sustainable city* (pp. 1−19). London: Springer.

Jha, A. K., Singh, S. K., Singh, G. P., & Gupta, P. K. (2011). Sustainable municipal solid waste management in low-income group of cities: A review. *International Society for Tropical Ecology*, *52*(1), 123−131.

Jiang, B., & Miao, Y. *The evolution of natural cities from the perspective of location-based social media*. (2014). <https://arxiv.org/ftp/arxiv/papers/1401/1401.6756.pdf>.

Jiang, X. (2011). Urban ecological security evaluation analysis based on fuzzy mathematics. *Procedia Engineering*, *15*, 4451−4455.

Johannessen, M. R., & Berntzen, L. (2016). Smart cities through implicit participation: Using gamification to generate citizen input for public transport planning. In H. J. Scholl, O. Glassey, & M. Janssen (Eds.), *Electronic government and electronic participation* (pp. 23−30). Amsterdam, The Netherlands: IOS Press.

Joia, L. A., & Kuhl, A. (2019). Smart city for development: A conceptual model for developing countries. In *Proceedings of the 15th IFIP WG 9.4 international conference on social implications of computers in developing countries ICT4D 2019* (pp. 114−203), Dar es Salaam, Tanzania, May 1−3, 2019. Available online.

Jones, H. P., Hole, D. G., & Zavaleta, E. S. (2012). Harnessing nature to help people adapt to climate change. *Nature Climate Change*, *2*(7), 504−509.

Jorgensen, S. E., Fath, B. D., Nielsen, S. N., Pulselli, F. M., Fiscus, D. A., & Bastianoni, S. (2015). *Flourishing within limits to growth: Following nature's way*. London, UK: Routledge.

Joshi, S., Saxena, S., Godbole, T., & Shreya, D. (2016). Developing smart cities: An integrated framework. *Procedia Computer Science*, *93*, 902−909.

Joss, S. (2010). Eco-cities: A global survey 2009. *WIT Transactions on Ecology and the Environment*, *129*, 239−250.

Joss, S. (2011). Eco-cities: The mainstreaming of urban sustainability—Key characteristics and driving factors. *International Journal of Sustainable Development Planning and Management*, *6*(3), 268−275.

Joss, S. (Ed.). (2012). Tomorrow's city today: Eco-City indicators, standards & frameworks. In: *Bellagio conference report*. London: University of Westminster.

Jucevicius, G., & Grumadaite, K. (2014). Smart development of innovation ecosystem. *Procedia Social and Behavioral Sciences*, *156*, 125−129.

Kalinowska (1997) cited in Piekaski, W.; Dudziak, A.; Stoma, M.; Andrejko, D.; and Dlaska-Grzywna, B. (2016). Modern assumptions and analysis of ecological awareness behaviour: an Empirical Study. *Polish Journal of Environmental Studies*, *25*(3), 1187−1195.

Kalundborg Symbiosis. *Some of the largest industrial enterprises in Denmark are partners in Kalundborg symbiosis*. (2017). <http://www.symbiosis.dk/en/partnere>.

Karodia, A. M., Dhiru, D., & David, J. E. (2014). International competitiveness, globalization and technology for developing countries. *Singaporean Journal of Business Economics and Management Studies*, *12*(9), 25−34.

Kazmierczak, A., & Carter, J. (2010). *Adaptation to climate change using green and blue infrastructure. A database of case studies* (p. 182). University of Manchester, GRaBS Project.

KCC (Kyiv City Council). (2018). Kyiv city development strategy until 2025. *Kyiv*. Ukraine: Kyiv City State Administration Available at. Available from https://dei.kyivcity.gov.ua/files/2018/1/11/Strategia.pdf.

Ketels, C. (2006). Michael Porter's competitiveness framework—Recent learnings and new research priorities. *Journal of Industry, Competitiveness and Trade*, *6*, 424−440.

Khanna, N., Fridley, D., & Hon, L. (2014). China's pilot low-carbon city initiative: A comparative assessment of national goals and local plans. *Sustainable Cities and Society*, *12*, 110−121.

Khansari, N., Mostashari, A., & Mansouri, M. (2014). Impacting sustainable behaviour and planning in smart city. *International Journal of Sustainable Land Use and Urban Planning*, *1*(2), 46−61.

Kharlamova, G., & Vertelieva, O. (2013). The international competitiveness of countries: Economic-mathematical approach. *Economics & Sociology.*, *6*(2), 39−52.

King, J., & Christopher, P. (2017). *Smart buildings: Using smart technology to save energy in existing buildings*. Washington, D.C: American Council for an Energy-Efficient Economy, Report A1701, February 2017.

Kirchherr, J., Reike, D., & Hekkert, M. (2017). Conceptualizing the circular economy: An analysis of 114 definitions. *Resources, Conservation & Recycling*, *127*, 221−232.

Kitchin, R. (2014). The real-time city? Big data and smart urbanism. *GeoJournal*, *79*(1), 1−14.

Kitson, M., Martin, R., & Tyler, P. (2004). Regional competitiveness: An elusive yet key concept? *Regional Studies*, *38*, 991−999.

Klein, C., & Kaefer, G. (2008). From smart homes to smart cities: opportunities and challenges from an industrial perspective. Next generation teletraffic and wired/wireless advanced networking . NEW 2 AN 2008. In S. Balandin, D. Moltchanov, & Y. Koucheryavy (Eds.), *Lecture Notes in Computer Science* (vol 5174). Berlin, Germany: Springer.

Klockner, C. A. (2013). A comprehensive model of the psychology of environmental behaviour — a meta-analysis. *Global Environmental Change*, *23*(5), 1028−1038.

Klopp, J. M., & Petretta, D. L. (2017). The urban sustainable development goal: Indicators, complexity and the politics of measuring cities. *Cities (London, England), 63,* 92–97.

Knickrehm, M., Berthon, B., & Daugherty, P. (2016). *Digital disruption: The growth multiplier.* Dublin: Accenture.

Knox, P. L., & Mayer, H. (2009). *Small town sustainability: Economic, social, and environmental innovation.* Basel: Birkhauser.

Knyayginin, V. N., Lipetskaya, M. S., Akim, M. E., Andreev, I. B., Andreeva, N. S., Ivanova, K. A., ... Shmeleva, S. A. (2012). *"Smart" environment, "smart" systems, "smart" production: A series of papers (green book) in the framework of the Project "Industrial and Technological Foresight of the Russian Federation"* (Vol. 4). St. Petersburg: Foundation "Center for Strategic Research" North-West.

Kollmuss, A., & Agyeman, J. (2002). Mind the gap: Why do people act environmentally and what are the barriers to pro-environmental behaviour? *Environmental Education Research, 8*(3), 239.

Komninos, N., Bratsas, C., Kakderi, C., & Tsarchopolous, P. (2015). Smart city ontologies: Improving the effectiveness of smart city applications. *Journal of Smart Cities, 1*(1), 31–46.

Korhonen, J., Nuur, C., Feldmann, A., & Birkie, S. E. (2017). Circular economy as an essentially contested concept. *Journal of Cleaner Production, 175,* 544–552.

Kourtit, K., & Nijkamp, P. (2012). Smart cities in the innovation age. *Innovation: The European Journal of Social Science Research, 25*(2), 93–95.

Kourtit, K., Nijkamp, P., & Arribas, D. (2012). Smart cities in perspective—A comparative European study by means of self-organizing maps. *Innovation: The European Journal of Social Science Research, 25*(2), 229–246.

Krugman, P. (1994). Competitiveness: A dangerous obsession. *Foreign Affairs, 73*(2), 28–44.

Krugman, P. (1996). Making sense of the competitiveness debate. *Oxford Review of the Economic Policy, 12,* 17–25.

KSC (Kyiv Smart City). *Kyiv smart city strategy.* (2017). <http://kyivsmartcity.com/Kyiv-Smart-City-Concept.pdf>.

Kummitha, R. K. R., & Crutzen, N. (2017). How do we understand smart cities? An evolutionary perspective. *Cities (London, England),* 43–52.

Kummitha, R. K. R., & Crutzen, N. (2019). Smart cities and the citizen-driven internet of things: A qualitative inquiry into an emerging smart city. *Technological Forecasting & Social Change, 140,* 44–53.

Lafortezza, R., Davies, C., Sanesi, G., & Konijnendijk, C. C. C. (2013). Green infrastructure as a tool to support spatial planning in European urban regions. *Biogeosciences and Forestry, 6,* 102–108.

Laurance, W. F., Clements, G. R., Sloan, S., O'Connell, C. S., Mueller, N. D., Goosem, M., ... Balmford, A. (2014). A global strategy for road building. *Nature, 513,* 229–232.

Le Galès, P. (2004). Governance. In L. Boussaguet, S. Jacquot, & P. Ravinet (Eds.), *Dictionnaire des politiques publiques* (pp. 242–250). Paris: Presses de Science.

Lee, S. Y., Jin, K. Y., & Choi, S. H. (2013). A study on convergence technology for building of smart city. *ICCA 2013, ASTL, 24,* 113–116.

Leggett, J. A., & Carter, N. T. (2012). *Rio + 20: The United Nations conference on sustainable development, June 2012.* Washington, DC: Congressional Research Service.

Leibowitz, S., Cushman, S., & Hyman, J. (1999). Use of scale invariance in evaluating judgment indicators. *Environmental Monitoring and Assessment, 58,* 283–303.

Leichenko, R. M., Karen, L. O. B., & William, D. S. (2010). Climate change and the global financial crisis: A case of double exposure. *Annals of the Association of American Geographers, 100*(4), 963–972.

Lewald, A. (2017). Unlocking mobility's potential. How to make mobility smarter and cleaner. *World News—Climate Change The New Economy.* Available online at https://climatechange-theneweconomy.com/tag/cop23/page/2/. Accessed 19 Novenmber 2017.

Leydesdorff, L., & Deakin, M. (2010). The Triple Helix Model and the Meta-Stabilization of Urban Technologies in Smart Cities. *A project submitted to Cornell University, New York, USA, on 17 March*, 2010. Available at: https://arxiv.org/abs/1003.3344.

Leyzerova, A., Sharovarova, E., & Alekhin, V. (2016). Sustainable strategies of urban development. *Procedia Engineering, 150*, 2055–2061.

Li, J., Dan, L. D., & Liu L. (2013). The effect of environmental awareness on consumers' green purchasing: Mediating role of green perceived value. In E. Qi, J. Shen, & R. Dou (Eds.), *The 19th international conference on industrial engineering and engineering management* (pp. 767–773). Gland, Switzerland: Springer.

Li, Y., & Qiu, L. (2015). A comparative study on the quality of China's eco-city: Suzhou vs Kitakyushu. *Habitat International, 50*, 57–64.

Lindner, P., Mooij, C., & Rogers, H. (2017). *Circular economy in cities: A strategic approach towards a sustainable society?*. Karlskrona, Sweden: Blekinge Institute of Technology.

Lippman, S., & Rumelt, R. (1982). Uncertain imitability: An analysis of interfirm differences in efficiency under competition. *The Bell Journal of Economics, 13*(2), 418–438.

Listra, E. (2015). The concept of competition and the objectives of competitors. *Procedia-Social and Behavioral Sciences, 213*, 25–30.

Liu, J., Low, S. P., & Wang, L. F. (2018). Critical success factors for eco-city development in China. *International Journal of Construction Management, 18*, 497–506.

Liu, W., Wang, C., Xie, X., Mol, A. P. J., & Chen, J. (2012). Transition to a low-carbon city: Lessons learned from Suzhou in China. *Frontiers of Environmental Science and Engineering, 6*, 373–386.

Lombardi, P., Giordano, S., Farouh, H., & Yousef, W. (2012). Modelling the smart city performance. *Innovation: The European Journal of Social Science Research, 25*(2), 137–149.

Lowell, D., & Huntington, A. (2019). *Electric vehicle market status—Manufacture commitment to future electrical mobility in the United States and worldwide*. MA, USA: MJ Bradley Associates.

Lozano, R., & Huisingh, D. (2011). Inter-linking issues and dimensions in sustainability reporting. *Journal of Cleaner Production, 19*(2), 99–107.

Lu, H., de Jong, M., & ten Heuvelhof, E. (2018). Explaining the variety in smart eco city development in China—What policy network theory can teach us about overcoming barriers in implementation? *Journal of Cleaner Production, 196*, 135–149.

Lundvall, B.-A. (2000). From the economics of knowledge to the learning economy. In OECD (Ed.), *Knowledge management in the learning economy* (pp. 1–26). Paris: OECD.

Lundvall, B.-A. (2016). *The learning economy and the economics of hope*. London: Anthem Press.

Luque-Ayala, A., & Marvin, S. (2015). Developing a critical understanding of smart urbanism? *Urban Studies, 52*(12), 2105–2116.

Lyle, J. T. (1994). *Regenerative design for sustainable development*. New York, NY, USA: John Wiley & Sons.

Madakam, S. (2014). Smart cities - six dimensions (A Scholarstical Articles Review), conference proceedings-international conference on advances in computing and information technology-ACIT 2014. Bangkok, 4–5 January 2014. Pp. 38–41.

Malecki, E. J., & Moriset, B. (2007). *The digital economy: Business organization, production processes and regional developments*. London: Routledge.

Malgand, M., Bay-Mortensen, N., Bedkowska, B., Hansen, F. N., Schow, M., Thomsen, A. A., & Hunka, A. D. (2014). Environmental awareness, the transition movement, and place: Den Selvforsynende Lands by, a Danish transition initiative. *Geoforum; Journal of Physical, Human, and Regional Geosciences, 57*, 40–47.

Manderson, A. K. (2006). A systems-based framework to examine the multi-contextual application of the sustainability concept. *Environment, Development and Sustainability, 8*(1), 85–97.

Mangiaracina, R., Perego, A., Salvadori, G., & Tumino, A. (2017). A comprehensive view of intelligent transport systems for urban smart mobility. *International Journal of Logistics Research and Applications A Leading Journal of Supply Chain Management, 20*, 39−52.

Manville, C., Cochrane, G., Cave, J., Millard, J., Pederson, K. J., Thaarup, R. K., Liebe, A., Wissner, M., Massink, R., & Kotterink, B. (2014). *Mapping smart cities in the EU*. Brussels: European Parliament.

Manyika, J., Bughin, J., Lund, S., Nottebaum, O., Poulter, D., Jauch, S., & Ramaswamy, S. (2014). Global flows in a digital age: How trade, finance, people, and data connect the world economy. Washington, DC: McKinsey Global Institute.

Marceau, J. (2008). Introduction: Innovation in the city and innovative cities. *Innovation: Management, Policy & Practice, 10*(2−3), 136−145.

March, H. (2018). The smart city and other ICT-led techno-imaginaries: Any room for dialogue with degrowth? *Journal of Cleaner Production. 197 (Part-II)*, 1694−1703.

Marcilio, G. P., Rangel, J. J., de Souza, C. L. M., Shimoda, E., da Silva, F. F., & Peixoto, T. A. (2018). Analysis of greenhouse gas emissions in the road freight transportation using simulation. *Journal of Cleaner Production, 170*, 298−309.

Marsal-Llacuna, M. L., Colomer-Llinàs, J., & Meléndez-Frigola, J. (2015). Lessons in urban monitoring taken from sustainable and livable cities to better address the smart cities initiative. *Technological Forecasting and Social Change, 90*, 611−622.

Martinez-Fernandez, C., Audirac, I., Fol, S., & Cunningham-Sabot, E. (2012). Shrinking cities: urban challenges of globalization. *International Journal of Urban and Regional research, 36*(2), 213−225.

Mathers, N., & Slater, R. (2014). *Social protection and growth: Research synthesis*. Canberra: Department of Foreign Affairs and Trade, Commonwealth of Australia.

Marx, B., Stoker, T., & Suri, T. (2013). The economics of slums in the developing countries. *The Journal of Economic Perspectives, 27*(4), 187−210.

Mas-Colell, A., Whinston, M. D., & Green, J. R. (1995). *Microeconomic theory*. New York: Oxford University Press.

McDonald, R. I. (2008). Global urbanization: Can ecologists identify a sustainable way forward? *Frontiers in the Ecology and Environment, 6*(2), 99−104.

McGraw, H., III, Danilovich, J., Ma, J., Wilson, M., & Bharti Mittal, S. (2015). Work with business to achieve sustainable development goals. *The Financial Times, 12*, 12.

McMichael, A. J. (2000). The urban environment and health in a world of increasing globalization: Issues for developing countries. *Bulletin of World Health Organization, 78*, 1117−1126.

Meadows, D. H., Meadows, D. L., Randers, J., & Behrens, W. W., III (1972). *The limits to growth*. New York: Universe Books.

Measham, T. G., Preston, B. L., Smith, T. F., Brooke, C., Gorddard, R., Withycombe, G., & Morrison, C. (2011). Adapting to climate change through local municipal planning: Barriers challenges. *Mitigation and Adaptation Strategies for Global Change, 16*, 889−909.

medium.com. *Smart cities market—global industry analysis and forecast (2017−2026)*. (2019). <https://medium.com/@surajmaximize007/smart-cities-market-global-industry-analysis-and-forecast-2017-2026-5fb303d5db85>.

Meijer, A. (2016). Smart city governance: A local emergent perspective. In J. Gil-Garcia, T. Pardo, & T. Nam (Eds.), *Smarter as the new urban agenda. Public administration and information technology* (vol 11, pp. 73−85). Cham, Switzerland: Springer.

Meijer, Albert, & Bolivar, M. P. R. (2016). Governing the smart city: a review of the literature on smart urban governance. *International Review of Administrative Sciences, 82*(2), 392−408.

Merli, R., Preziosi, M., & Acampora, A. (2018). 'How do scholars approach the circular economy? A systematic literature review'. *Journal of Cleaner Production, 178*, 703−722.

Merriam Webster Dictionary. (2004). *Dictionary, Eleventh Edition. Massachusetts, USA: Merriam-Webster Inc*. Merriam Webster.

Mersal, A. (2017). Eco-city: Challenge and opportunities in transferring a city into green city. *Procedia Environmental Sciences, 37*, 22−33.

Milfonta, T. L., & Duckitt, J. (2010). The environmental attitudes inventory: A valid and reliable measure to assess the structure of environmental attitudes. *Journal of Environmental Psychology, 30*(1), 80−94.

Miller, N. G., Pogue, P., Gough, Q. D., & Davis, S. D. (2009). Green building and productivity. *Journal of Sustainable Real Estate, 1*(1), 65−89.

Minford, P. (2006). Competitiveness in a globalised world: A comment. *Journal of International Business Studies, 37*, 176−178.

Ministry of Economy, Trade and Infrastructure, Japan. *Towards a 3R oriented, sustainable society: Legislation and trends*. (2010). Available online.

Mirowski, W. (1996). Environmental awareness and sustainability. *Issues of science studies* (p. 3) (in Polish).

Misangui, V., Elms, H., Greckhamer, T., & Lepine, J. (2006). A new perspective on a fundamental debate: A multilevel approach to industry, corporate, and business unit effects. *Strategic Management Journal, 27*, 571−590.

Mishra, A. P., Sen, A., & Kumar, A. (2017). Exploring potentials and challenges in making smart cities in India: A case study of Allahabad City, Uttar Pradesh. In P. Sharma, & S. Rajput (Eds.), *Sustainable smart cities in India* (pp. 123−142). Cham. Switzerland: Springer.

Mishra, A. K., George, H., & Theorems, M.-H. (2019). Lessons for financing smart cities in developing countries. *Environment and Urbanization Asia, 10*, 13−30.

Moallemi, E. A., Malekpour, S., Hadjikakou, M., Raven, R., Szetey, K., Moghadam, M. M., & Bryan, B. A. (2019). Local Agenda 2030 for sustainable development. *The Lancet. Planetary Health, 3*(6), e240−e241.

Molendowski, E. (2017). An internationally competitive economy: A comparison of Poland and the Visegrad Group countries in the post-accession period. *Comparative Economic Research, 20*(4), 5−21.

Momtaz, R., & Elsemary, Y. (2015). Qualitative conceptions of livability between theory and applications in Egypt. In: *International conference on IT, architecture and mechanical engineering (ICITAME'2015)*. Dubai, UAE, May 22−23, 2015. Retrieved from http://iieng.org/images/proceedings_pdf/9921E0515046.pdf

Mori, K., & Christodoulou, A. (2012). Review of sustainability indices and indicators: Towards a new city sustainability index (CSI). *Environmental Impact Assessment Review, 32*(1), 94−106.

Moser, M. A. (2001). What is smart about the smart communities' movement? *EJournal, 10/11*(1). Available at. Available from http://www.ucalgary.ca/ejournal.

Mrak, M. (2000). *Globalization: Trends, challenges and opportunities in countries in transition*. Vienna: UNIDO.

Müller, N., Kuttler, W., & Barlag, A.-B. (2013). Counteracting urban climate change: adaptation measures and their effect on thermal comfort. *Theoretical and Applied Climatology, 115*, 243−257.

Mumford, L. (1937). What is a city. In R. T. LeGates, & F. Stout (Eds.), *The city reader* (Fifth Edition, pp. 92−96). London: Routledge, in 2002 edition.

Munang, R., Thiaw, I., Alverson, K., Goumandakoye, M., Mebratu, D., & Liu, J. (2013a). Using ecosystem-based adaptation actions to tackle food insecurity. *Environmental Science Policy and Sustainable Development, 55*, 29−35.

Munang, R., Thiaw, I., Alverson, K., Liu, J., & Han, Z. (2013b). The role of ecosystem services in climate change adaptation and disaster risk reduction. *Current Opinion in Environmental Sustainability., 5*(1), 47−52.

Munang, R., Thiaw, I., Alverson, K., Mumba, M., Liu, J., & Rivington, M. (2013c). Climate change and ecosystem-based adaptation: A new pragmatic approach to buffering climate change impacts. *Current Opinion in Environmental Sustainability, 5*, 67−71.

Murray, M., Murray, L., & Donnelly, M. (2015). Systematic review protocol of interventions to improve the psychological well-being of general practitioners. *Systematic Reviews, 4*, 117.

Mutiara, D., Yuniarti, S., & Pratama, B. (2018). Smart governance for smart city. *Earth and Environmental Science, 126*, 012073.

Nam, T., & Pardo, T. A. (2011a). Conceptualizing smart city with dimensions of technology, people, and institutions. In Conference Proceedings of the 12th Annual International Conference on Digital Government Research, DGO, College Park, MD, USA. June 12−15, 2011. <https://www.researchgate.net/profile/Taewoo-Nam-3/publication/221585167_Conceptualizing_smart_city_with_dimensions_of_technology_people_and_institutions/links/0f31752f60bf009d2f000000/Conceptualizing-smart-city-with-dimensions-of-technology-people-and-institutions.pdf>.

Nam, T., & Pardo, T. A. (2011b). Smart city as urban innovation: Focusing on management, policy, and context. In Proceedings of the 5th International Conference on Theory and Practice of Electronic Governance: pp. 185−194.

NAS (National Academic Press). (2002). *Community and Quality of Life: Data Needs for Informed Decision Making*. Washington, DC: The National Academies Press.

Natali, D. (2014). Introduction: A last chance to rescue Europe? (Re-)launching social EU the only strategy against anti-euro sentiments. In D. Natali (Ed.), *Social developments in the European Union 2013*. Brussels: European Social Observatory and ETUI.

Naumann, S., Anzaldua, G., Berry, P., Burch, S., McKenna, D., Frelih-Larsen, A., ... Sanders, M. (2011). Assessment of the potential of ecosystem-based approaches to climate change adaptation and mitigation in Europe. In: *Final report to the European Commission, DG, Environment, Contract no. 70,307/2010/580412/SER/B2*. Ecologic Institute and Environmental Change Institute, Oxford University Centre for the Environment.

Neirotti, P., De Marco, A., Cagliano, A. C., Mangano, G., & Scorrano, F. (2014). Current trends in smart city initiatives−some stylised facts. *Cities (London, England), 38*, 25−36.

Networking Society Lab. (2016). *Ericsson networked society index*. Sweden: Stockholm.

Neumann, H.-M., Cai, J., Meyer, S., Jakuyte-Walangitang, D., Haindlmaier, G., Han, Y., ... Ma, E. (2019). *Bridging the planning-implementation gap in eco and smart cities. Working paper-2*. Vienna: Austrian Institute of Technology.

Nijkamp, P., & Kourtit, K. (2013). The "new urban Europe": Global challenges and local responses in the urban century. *European Planning Studies, 21*(3), 291−315.

Nikonova, Y., & Dementiev, A. (2018). Tendencies and prospects in the digital economy development in Russia. *SHS Web of Conferences, 55*, 01011.

Noon, B. (2003). Conceptual issues in monitoring ecological resources. In D. Busch, & J. Trexler (Eds.), *Monitoring ecosystems: interdisciplinary approaches for evaluating eco-regional initiatives* (pp. 27−72). Washington, DC: Island Press.

Noss, R. F. (1990). Indicators for monitoring biodiversity: A hierarchical approach. *Conservation Biology, 4*(4), 355−364.

Oberndorfer, E., Lundholm, J., Bass, B., Coffman, R. R., Doshi, H., Dunnett, N., & Rowe, B. (2007). Green roofs as urban ecosystems: ecological structures, functions, and services. *Bioscience, 57*(10), 823−833.

OECD. (2014). *Cities and climate change-policy perspectives*. Paris: OECD.

OECD (Organization for Economic Cooperation and Development). (1999). *Benchmarking Knowledge-based economies*. Paris: OECD.

OECD (Organization for Economic Cooperation and Development). (2004). *Urbanization. Glossary of statistical terms*. Paris: OECD.

OECD (Organization for Economic Cooperation and Development). (2012a). *Digital Economy, hearings*. Paris: OECD.

OECD (Organization for Economic Cooperation and Development). (2012). *Internet economy outlook 2012*. Paris: OECD Publishing.
OECD (Organization for Economic Cooperation and Development). (2016). *Ministerial declaration on the digital economy ("Cancún Declaration") from the Meeting on the digital economy: Innovation, growth and social prosperity*. Cancun, June 21–23, 2016. Available online.
OECD (Organization for Economic Cooperation and Development). (2017). *OECD digital economy outlook 2017*. Paris: OECD Publishing.
OECD (Organization for Economic Cooperation and Development). (2019). *Can social protection be an engine for inclusive growth? Development centre studies*. Paris: OECD Publishing.
OHCHR-UN Habitat. (2014). *Right to Adequate Housing. Fact Sheet No. 21*. Geneva: UN-Habitat.
Ojanperä, S., Graham, M., & Zook, M. (2016). Measuring the contours of the global knowledge economy with a digital index. In: *Paper presented at the development studies association conference*, Oxford, September 6, 2016. Available online.
Ortiz., Isabel., Durán-Valverde, F., Pal, K., Behrendt, C., & Acuña-Ulate, A. (2017). *Universal social protection floors: Costing estimates and affordability in 57 lower income countries*. Geneva: ILO.
Osborn, D., Cutter, A., & Farooq, U. (2015). *Universal sustainable development goals: Understanding the transformational challenge for developed countries (report of a study by stakeholder forum)*. London: Stakeholder Forum.
Otterpohl, R. (2001). Design of highly efficient source control sanitation and practical experiences. In P. Lens, G. Zeemann, & G. Lettinga (Eds.), *Decentralised sanitation and reuse* (pp. 164–179). London: IWA Publishers.
Palej, A. (2000). Architecture for, by and with Children: A way to teach livable city. In: *Paper presented at the international making cities livable conference*, Vienna, Austria.
Papa, R., Gargiulo, C., & Russo, L. (2017). The evolution of smart mobility strategies and behaviors to build the smart city. In: *5th IEEE international conference on models and technologies for intelligent transportation systems* (pp. 409–414).
Papuzinski, A. (2006). Ecological awareness in the light of theory and practice. *Problems of Ecology, 1*, 33. (Original in Polish).
Parrish, J. D., Braun, D. P., & Unnasch, R. S. (2003). Are we conserving what we say we are? Measuring ecological integrity within protected areas. *Bioscience, 53*(9), 851–860.
Paskaleva, K. A. (2011). The smart city: A nexus for open innovation? *Intelligent Buildings International, 3*(3), 153–171.
Paton, J. (2008). What's 'left' of sustainable development? *Journal of Australian Political Economy, 62*, 94–119.
Paul, B. D. (2008). A history of the concept of sustainable development: Literature review. *The Annals of the University of Oradea, Economic Sciences Series, 17*(2), 576–580.
Peng, Y., & Bai, X. (2018). Experimenting towards a low-carbon city: Policy evolution and nested structure of innovation. *Journal of Cleaner Production, 174*, 201–212.
Pereira, G. V., Paryceka, P., Falcoc, E., & Kleinhans, R. (2018). Smart governance in the context of smart cities: A literature review. *Information Policy, 23*, 143–162.
Perez-Carmona, A. (2012). Growth: A discussion of the margins of economic and ecological thought. In L. Meuleman (Ed.), *Trans-governance—Advancing sustainable governance* (pp. 83–161). Heidelberg: Springer.
Peteraf, M., Di Stefano, G., & Verona, G. (2013). The elephant in the room of dynamic capabilities: Bringing two diverging conversations together. *Strategic Management Journal, 34*, 1389–1410.
Petit-Boix, A., & Leipold, S. (2018). 'Circular economy in cities: Reviewing how environmental research aligns with local practices'. *Journal of Cleaner Production, 195*, 1270–1281.

Petmesidou, M. (2017). Can the European Union 2020 strategy deliver on social inclusion? Global challenges. In: *Working paper series no. 3*. Available online.

Pickett, M. L., Cadenasso, J. M., Grove, C. H., Nilon, R. V., Pouyat., Zipperer, W. C., & Costanza, R. (2001). Urban ecological systems: Linking terrestrial ecological, physical, and socioeconomic components of metropolitan areas. *Annual Review of Ecology and Systematics*, *32*(2001), 127−157.

Picketts, I. M., Déry, S. J., & Curry, J. A. (2013). Incorporating climate change adaptation into local plans. *Journal of Environmental Planning*, *57*(7), 37−41.

Piekarski, W., Dudziak, A., Stoma, M., Andrejko, D., & Ślaska-Grzywna, B. (2016). Model assumptions and analysis of ecological awareness and behaviour: An empirical study. *Polish Journal of Environmental Studies*, *25*(3), 1187−1195.

Pinna, F., Masala, F., & Garau, C. (2017). Urban policies and mobility trends in Italian Smart cities. *Sustainability.*, *9*(4), 1−21.

Pisano, U., Endl, A., & Berger, G. (2012). The Rio + 20 Conference: Objectives, processes, and outcomes. Vienna, Austria: European Sustainable Development Network (ESDN).

Polanyi, M. (1967). *The tacit dimension*. London: Routledge & Kegan Paul.

Porter, M. E. (1990). The competitive advantage of nations. *Harvard Business Review. March-April*, *1990*, 73−91.

Porter, M. E. (2008). The five competitive forces that shape strategy. *Harvard Business Review*, *86*, 79−93.

Poslad, S. (2009). *Ubiquitous computing smart devices. Smart environments and smart interaction*. London: Wiley.

Potts, D. (2012). Challenging the myths of urban dynamics in sub-Saharan Africa: The evidence from Nigeria. *World Development*, *40*(7), 1382−1393.

Pozdniakova, A. M. (2018). Smart city strategies: London-Stockholm-Vienna-Kyiv—in search of common ground and best practices. *Acta Innovations*, *27*, 31−45. Available online.

Praharaj, S., Han, J. H., & Hawken, S. (2018). Towards the right model of smart city governance in India. *International Journal of Sustainable Development and Planning*, *13*(2), 171−186.

Premalatha, M., Tauseef, S. M., Abbasi, T., & Abbasi, S. A. (2013). The promise and the performance of the world's first two zero carbon eco-cities. *Renewable and Sustainable Energy Reviews*, *25*, 660−669.

Prevot-Julliard, A. C., Clavel, J., Teillacdeschamps, P., & Julliard, R. (2011). The need for flexibility in conservation practices: Exotic species as an example. *Environmental Management*, *47*, 315−327.

Prieto-Sandoval, V., Jaca, C., & Ormazabal, M. (2018). Towards a Consensus on the circular economy. *Journal of Cleaner Production*, *179*, 605−615.

PRNewswire. (2017, January 04). *Global smart cities market to reach United States %3.584 trillion by 2026*. New York. Available online.

Quental, N., Lourenço, J. M., & Da Silva, F. N. (2011). Sustainable development policy: goals, targets and political cycles. *Sustainable Development*, *19*(1), 15−29.

Rachel, O. A., Komine, H., Yasuhara, K., & Murakami, S. *Municipal solid waste management in developed and developing countries—Japan and Nigeria as case studies*. (2009). <http://wwwgeo.civil.ibaraki.ac.jp/komine/mypapers/JGSPaper/2009/JGS2009(973)Rachel.pdf>.

Radło, M. J. (2008). Międzynarodowa konkurencyjność gospodarki. Uwagi na temat definicji, czynników i miar. In W. Bieńkowski & M. A. Weresa (Eds.), *Czynniki i miary międzynarodowej konkurencyjności gospodarek w kontekście globalizacji—wstępne wyniki badań*. Warsaw: SGH.

Rahmayanti, H., Oktaviani, V., & Syani, Y. (2018). The implementation of smart trash as smart environment concept. *E3S Web of Conferences*, *74*, 06003. Available online.

Rajkumar, N., Subramani, T., & Elango, L. (2010). Groundwater contamination due to municipal solid waste disposal—a GIS-based study in Erode city. *International Journal of Environment Science*, *1*(1), 39−55.

Rapoport, E. and Vernay A.L. (2011). Defining the eco–city: a discursive approach', Paper presented at the management and innovation for a sustainable built environment conference, International Eco–Cities Initiative, Amsterdam, The Netherlands, pp. 1–15.

Sarkar, A. N. (2016). Bench-marking, sustainability, and governance aspects of smart cities. *International Journal of Academic Library and Information Science, 4*(1), 11–31.

Reckien, D., Flacke, J., Dawson, R. J., Heidrich, O., Olazabal, M., Foley, A., & Pietrapertosa, F. (2014). Climate change response in Europe: what's the reality? analysis of adaptation and mitigation plans from 200 urban areas in 11 countries. *Climate Change, 122*(1–2), 331–340.

Reed, S. O., Friend, R., Toan, V. C., Thinphanga, P., Sutarto, R., & Singh, D. (2013). "Shared learning" for building urban climate resilience—experiences from Asian cities. *Environment and Urbanization, 25*(2), 393–402.

Register, R. (1987). *Ecocity Berkley: Building cities for a healthy future*. Berkley, CA: North Atlantic Books.

Reike, D., Vermeulen, W., & Witjes, S. (2018). The circular economy: New or refurbished as CE3.0?—Exploring controversies in the conceptualization of the circular economy through a focus on history and resource value retention options. *Resources, Conservation & Recycling, 136*, 264, 264.

Rickets, G. (2010). The roots of sustainability. *Academic Questions, 23*(1). Available online.

Riitters, K. H., Wickham, J. D., & Wade, T. G. (2009). An indicator of forest dynamics using a shifting landscape mosaic. *Ecological Indicators, 9*(1), 107–117.

Roberts, D., Boon, R., Diederichs, N., Douwes, E., Govender, N., McInnes, A., & Spires, M. (2012). Exploring ecosystem-based adaptation in Durban, South Africa learning-by-doing at the local government coal face. *Environment and Urbanization, 24*(1), 167–195.

Rocha, S. M., Zulian, G., Maes, J., & Thijssen, M. (2015). *Mapping and assessment of urban ecosystems and their services*. Brussels: European Commission. Available online.

Roseland, M. (1997). Dimensions of the eco-city. *Cities., 14*(4), 197–202.

Roseland, M. (2000). Sustainable community development: Integrating environmental, economic, and social objectives. *Progress in Planning, 54*(2), 73–132.

Rosenberger, P. (2018). *Smart cities: Individual solutions for cities*. Siemens. Retrieved from https://new.siemens.com/global/en/company/stories/mobility/hong-kong-concerted-solu-tions.html

Rosenzweig, C., Solecki, W., Romero-Lankao, P., Mehrotra, S., Dhakal, S., Bowman, T., & Ali Ibrahim, S. (2015). *ARC3.2 summary for city leaders. Urban climate change research network*. New York: Columbia University.

Roy, J., Cakravaty, D., Dasgupta, S., & Chakrabarty, D. (2018). Where is the hope? Blending modern urban lifestyle with cultural practices in India. *Current Opinion in Environmental Sustainability., 31*, 96–103.

Roy, J., & Pal, S. (2009). Lifestyles and climate change: Link awaiting activation. *Current Opinion in Environmental Sustainability., 1*(2), 192–200.

Ruhlandt, R. W. S. (2018). The governance of smart cities: A systematic literature review. *Cities (London, England), 81*, 1–23.

Sachs, J. D. (2012). From millennium development goals to sustainable development goals. *Lancet, 379*(9832), 2206–2211.

Sachs, J. D. (2014). Sustainable development goals for a new era. *Horizons. Autumn, 2014*(1), 106–119.

Salzano, E. (1997). Seven aims for the livable city. In S. H. Lennard, S. von Ungern-Sternberg, & H. L. Lennard (Eds.), *Making cities livable. International making cities livable conferences*. Newport Beach, CA: Gondolier Press.

Satterthwaite, D. (2006). *Barbara ward and the origins of sustainable development*. London, UK: International Institute for Environment and Development (IIED).

Satterthwaite, D. (2010). *Urban myths and the misuse of data that underpin them*. Helsinki: World Institute for Development Economics Research. March, UNU-WIDER Working Paper, No. 2010/28.

Sauve, S., Bernard, S., & Bernard, P. (2016). Environmental sciences, sustainable development and circular economy: Alternative concepts for trans-disciplinary research. *Environmental Development*, *17*, 48–56.

Saxena, S., Srivastava, R. K., & Samaddar, A. B. (2010). Sustainable waste management issues in India. *IUP Journal of Soil & Water Sciences*, *3*(1), 72–90.

SC4C (Smart Capacities for Cities). (2018). *GAIA catalogue of products and services*. Spain: San-Sebastian Available at. Available from http://www.gaia.es/smartcities/en/fichas/q/area = 5.

SCC (Stockholm City Council). (2017). *Strategy for Stockholm as a smart and connected city*. Stockholm. Retrieved from http://international.stockholm.se/globalassets/ovriga-bilder-och-filer/smart-city/brochure-smart-and-connected.pdf

Schaefer, M., Thinh, N. X., & Greiving, S. (2020). How can climate resilience be measured and visualized? Assessing a vague concept using GIS-based fuzzy logic. *Sustainability.*, *12*, 635.

Schlossberg, M., & Zimmerman, A. (2003). Developing statewide indices of environmental, economic, and social sustainability: A look at Oregon and the Oregon benchmarks. *Local Environment: The International Journal of Justice and Sustainability*, *8*(6), 641–660.

Schmalensee, R. (2012). From 'green growth' to sound policies: an overview. *Energy Economics*, *34*, S2–S6.

Scholl, H. J., & AlAwadhi, S. (2016). Creating smart governance: The key to radical ICT overhaul at the City of Munich. *Information Polity*, *21*(1), 21–42.

Scholl, H. J., & Scholl, M. C. (2014). Smart governance: A roadmap for research and practice. In: *iConference 2014 Proceedings*. Retrieved from https://www.ideals.illinois.edu/bitstream/handle/2142/47408/060_ready.pdf?sequence = 2

Schultz, M. T. (2001). A critique of EPA's index of watershed indicators. *Journal of Environmental Management*, *62*, 429–442.

Schumacher, E. F. (1973). *Small is beautiful*. London: Blond & Briggs.

Schuurman, D., Baccarne, B., De Marez, L., & Mechant, P. (2012). Smart ideas for smart cities: Investigating crowdsourcing for generating and selecting ideas for ICT innovation in a city context. *Journal of Theoretical and Applied Electronic Commerce Research*, *7*(3), 49–62.

Schwab, C. (2019). The transforming city in nineteenth-century literary journalism: 'Ramón de Mesonero Romanos' 'Madrid scenes' and 'Charles Dickens' 'Street sketches'. *Urban History*, *46*(2), 225–245.

Selsky, J. W., & Parker, B. (2005). Cross-sector partnerships to address social issues: Challenges to theory and practice. *Journal of Management*, *31*(6), 849–873.

SEP (Science for Environment Policy). (2018). Indicators for sustainable cities. In: *In-depth report 12*. Bristol: European Commission DG Environment by the Science Communication Unit, UWE. < https://ec.europa.eu/environment/integration/research/newsalert/pdf/indicators_for_sustainable_cities_IR12_en.pdf > .

Seto, K. C., & Shepherd, J. M. (2009). Global urban land-use trends and climate impacts. *Current Opinion in Environmental Sustainability*, *1*(1), 89–95.

Shapiro, J. M. (2006). Smart cities: Quality of life, productivity, and the growth effects of human capital. *The Review of Economics and Statistics*, *88*(2), 324–335.

Sharma, R., & Kumar, V. (2019). The multidimensional venture of developing a smart city. In: *2019 International conference on big data and computational intelligence (ICBDCI)*. Retrieved from https://ieeexplore.ieee.org/document/8686101.

Sherman, W. R. (2012). The triple bottom line: The reporting of "doing well" & "doing good". *Journal of Applied Business Research*, *28*(4), 673–681.

Smart City Expo World Congress. (2016). *Citizens for citizens report 2016*. Barcelona. Retrieved from http://media.firabcn.es/content/S078016/SCEWC_Report2016.pdf

Smart London Plan (2013). (Online). Retrieved from https://www.london.gov.uk/sites/default/files/gla_smartlondon_report_web_4.pdf

Smith, A. (1977). *An inquiry into the nature and causes of the wealth of nations*. Chicago: University of Chicago Press, [1776].
Smithson, P. (1971). Simple thoughts on repetition. *Architectural Design*, *41*, 479–481.
Sneddon, C., Howarth, R. B., & Norgaard, R. B. (2006). Sustainable development in a post-Brundtland world. *Ecological Economics*, *57*(2), 253–268.
Snowdon, B., & Stonehouse, G. (2006). Competitiveness in a globalised world: Michael porter on the microeconomic foundations of the competitiveness of nations, regions, and firms. *Journal of International Business Studies*, *37*, 163–175.
Sorensen, M. P., Bloch, C., & Young, M. (2016). Excellence in the knowledge-based economy: From scientific to research excellence. *European Journal of Higher Education*, *6*(3), 217–236.
Spangler, W. S., Kreulen, J. T., Chen, Y., Proctor, L., Alba, A., Lelescu, A., & Behal, A. (2010). A smarter process for sensing the information space. *IBM Journal of Research and Development*, *54*(4).
Stewart, A. (2000). Social inclusion: An introduction. In P. Askonas, & A. Stewart (Eds.), *Social Inclusion. Possibilities and tensions*. Great Britain: Macmillan Press Ltd.
Stigler, G. (1988). Competition. In J. Eatwell, M. Milgate, & P. Newman (Eds.), *A dictionary of economics* (pp. 531–536). London: Palgrave.
Stiglitz, J. (2012). *The price of inequality*. New York: W. W. Norton & Company.
Stoker, G. (1998). Governance as theory: Five principles. *International Social Science Journal*, *50*, 17–28.
Su, M., Xu, L., Chen, B., & Yang, Z. F. (2013). Eco-city planning theories and thoughts. In Z. F. Yang (Ed.), *Eco-cities: A planning guide* (pp. 3–14). New York: Taylor & Francis Group.
Suzuki, H., Dastur, A., Moffatt, S., Yabuki, N., & Maruyama, H. (2010). *Eco2 Cities: Ecological cities as economic cities*. Washington, DC: The World Bank.
Syed, S. (2006). Solid and liquid waste management. *Emirates Journal of Engineering Research*, *11*(2), 19–36.
Tahir, Z., & Malek, J. A. (2016). Main criteria in the development of smart cities determined using analytical method. *Planning Malaysia Journal*, *14*(5), 1–14.
Tan, S. Y., & Taeihagh, A. (2020). Smart city governance in developing countries: A systematic literature review. *Sustainability.*, *12*(3), 899.
Tang, H.-T., & Lee, Y.-M. (2016). The making of sustainable urban development: A synthesis framework. *Sustainability*, *8*, 492.
Tapscott, D. (1996). *The digital economy: Promise and peril in the age of networked intelligence*. New York: McGraw-Hill.
Testoni, C., & Boeri, A. (2015). Smart governance: Urban regeneration and integration policies in Europe. *Turin and Malmö case studies. International Journal of Scientific & Engineering Research*, *6*(3), 527–533.
The European Commission. (2013). European innovation partnership on smart cities and communities: Operational implementation plan—first public draft. *European Innovation Partnership on Smart Cities and Communities*, 111. Available online.
The Scottish Government, Scottish Cities Alliance, & Urban Tide. (2014). *Smart cities maturity model and self-assessment tool: Guidance note for completion of self-assessment tool*.
Theobald, D. M. (2010). Estimating changes in natural landscapes from 1992 to 2030 for the conterminous United States. *Landscape Ecology*, *25*(7), 999–1011.
Theobald, D. M. (2013). A general model to quantify ecological integrity for landscape assessments and United States application. *Landscape Ecology*, *28*, 1859–1874.
Tibaijuka, A. K., Maseland, J., & Moor, J. (2005). *Our urban future: making a home for Homo Urbanus*. Brown Journal of World Affairs, *11*(2), 2005, Winter/Spring.
Timberland. *International footwear catalogue*. (2018). <https://images.timberland.com/is/content/TimberlandBrand/permanent/pro-international-contact/Timberland-PRO-2018-International-Safety-Footwear-Catalog.pdf>.

Timmer, V., & Seymoar, N. (2005). The livable city. In: *The International Centre for Sustainable Cities. Vancouver working group discussion paper*. Available online.

Tomaszewska, E. J., & Florea, A. (2018). Urban smart mobility in the scientific literature — bibliometric analysis. *Engineering Management in Production and Services, 10*(2), 41−56.

Torrey, B. B. (2004). *Urbanization: An environmental force to be reckoned with*. Population Reference Bureau. Available online.

Townsend, A. (2013). *Smart cities—Big data, civic hackers and the quest for a new Utopia*. New York: Norton & Company.

Tschakert, P., & Dietrich, K. A. (2010). Anticipatory learning for climate change adaptation and resilience. *Ecology and Society, 15*(2), 11.

Tsolakis, N., & Anthopoulos, L. (2015). Eco-cities: An integrated system dynamics framework and a concise research taxonomy. *Sustainable Cities and Society, 17*, 1−14.

Turcu, C. (2013). Re-thinking sustainability indicators: Local perspectives of urban sustainability. *Journal of Environmental Planning and Management, 56*(5), 695−719.

Turkensteen, M. (2017). The accuracy of carbon emission and fuel consumption computations in green vehicle routing. *European Journal of Operational Research, 262*(2).

Turner, M. G. (2010). Disturbance and landscape dynamics in a changing world. *Ecology, 91*(10), 2833−2849.

Tyas, W. P., Nugroho, P., Sariffudin, S., Purba, N. G., Riswandha, Y., & Sitorus, G. H. (2019). Applying smart economy of smart cities in developing world: Learnt from Indonesia's home-based enterprises. *IOP Conference Series, Earth and Environmental Science, 248*, 012078.

UCS (Union of Concerned Scientists). (2016). Toward climate resilience: A framework and principles for science-based adaptation. Cambridge, MA: UCS. Available online.

Ueda, N., & Nara, F. (2018). Spatio-temporal multidimensional collective data analysis for providing comfortable living anytime and anywhere. *SIP (Industrial Technology Advances), 7*(e4), 1−17, Retrieved from. Available from https://www.cambridge.org/core/services/aop-cambridge-core/content/view/F504ECADF36841A0AE74185EBA69D198/S2048770318000045a.pdf/div-class-title-spatio-temporal-multidimensional-collective-data-analysis-for-providing-comfortable-living-anytime-and-anywhere-div.pdf.

UK Government (United Kingdom Government). (2005). *UK Government, Office of the Deputy Prime Minister. Sustainable communities: People places and prosperity*. London, UK: Office of the Deputy Prime Minister.

UN (United Nations). (1972). *Report of the United Nations Conference on the Human Environment*, Stockholm, June 5−16, 1972. A/CONF.48/14/Rev.1. Retrieved from https://www.un.org/ga/search/view_doc.asp?symbol = A/CONF.48/14/REV.1

UN (United Nations). (1987). *Report of the World Commission on Environment and Development: Our common future. A/42/427*. Retrieved from https://www.un.org/ga/search/view_doc.asp?symbol = A/42/427&Lang = E.

UN (United Nations). (1992). *United Nations conference on environment and development*, Rio de Janerio, Brazil, June 3−14, 1992. Agenda 21. Retrieved from https://sustainabledevelopment.un.org/content/documents/Agenda21.pdf.

UN (United Nations). *United Nations millennium declaration. A/RES/55/2*. (2000). <https://undocs.org/A/RES/55/2>.

UN (United Nations). (2011). *Fourth united nations conference on the least developed countries, programme of action for the least developed countries for the decade 2011−2020*. Istanbul. Retrieved from https://www.un.org/en/conf/ldc/pdf/ipoa.pdf.

UN (United Nations). (2011). *World urbanization prospects: the 2011 revision*. New York: United Nations.

UN (United Nations). (2014). Cyberschoolbus. In: *Cities of today, cities of tomorrow! "Unit 1: What is a city?*. Retrieved from http://www.un.org/cyberschoolbus/habitat/units/un01txt.a

UN (United Nations). (2015a). *The millennium development goals reports 2015*. New York: United Nations.

UN (United Nations). (2015b). *Resolution adopted by the general assembly on 25 September 2015: Transforming our world: The 2030 agenda for sustainable development*. New York: United Nations (UN).

UN (United Nations). (2015c). *Transforming our world: The 2030 Agenda for sustainable development*. New York: United Nations (UN).

UNCTAD (UN Conference on Trade and Development). (2017a). *Information economy report 2017: Digitalization, trade and development*. Geneva: UNCTAD.

UNCTAD (UN Conference on Trade and Development). (2017b). World investment report 2017: Investment and the digital economy. *UNCTAD/WIR/2017*. Geneva: UNCTAD.

UNCTAD (United Nations Conference on Trade and Development). (2019a). *Digital economy report 2019—Overview*. New York: UN Publications.

UNCTAD (United Nations Conference on Trade and Development). (2019b). *Digital economy report 2019*. New York: UN Publications.

UN-DESA (United Nations, Department of Economic and Social Affairs). (2012). *World urbanization prospects, the 2011 revision*. New York: UN.

UN-DESA (United Nations-Department of Economic and Social Affairs). (2013). *World economic and social survey 2013: Sustainable development challenges*. New York: United Nations Publications.

UN-DESA (United Nations-Department of Economic and Social Affairs). (2014). World urbanization prospects. *ST/ESA/SER.A/366*. New York: Department of Economic and Social Affairs.

UN-DESA (United Nations-Department of Social and Economic Affairs). (2018). *Promoting inclusion through social protection: Report on the world social situation 2018*. New York: UN-DESA.

UN-DESA (United Nations-Department of Economic and Social Affairs). (2019). World urbanization prospects: The 2018 revision. *ST/ESA/SER.A/420*. New York: Department of Economic and Social Affairs.

UNDP (United Nations Development Programme). (2015). *The millennium development goals report 2015*. New York: UNDP.

UN-ECOSOC (United Nations-Economic and Social Council). (2019). *Report of the Secretary General: Special edition on progress towards the sustainable development goals*. E/2019/68. Retrieved from https://undocs.org/E/2019/68

UN-ECOSOC (United Nations-Economic and Social Council). (2021). *Report of the Secretary-General: Progress towards the sustainable development goals*. E/2021/xxx.

UNEP (United Nations Environment Programme). (2011). *Towards a green economy: Pathways to sustainable development and poverty eradication. A synthesis for policy makers*. Nairobi: UNEP.

UNEP (United Nations Environment Programme). (2018). *The emissions gap report 2018*. Nairobi: UNEP.

UNEP (United Nations Environment Programme). (2019). *Global environment outlook—6. Healthy planet, healthy people*. New York: Cambridge University Press.

UNEP-UNITAR (United Nations Environment Programme-United Nations Institute for Training and Research). *Guidelines for national waste management strategies: Moving from challenges to opportunities*. (2013). Available online.

UN-GA (United Nation-General Assembly). *Universal declaration of human rights. A/RES/217 A (III)*. (1948). <https://www.un.org/en/development/desa/population/migration/generalassembly/docs/globalcompact/A_RES_217(III).pdf>.

UN-GA (United Nations-General Assembly). *International covenant on economic, social and cultural rights. A/RES/2200 (XXI)*. (1966). <https://www.un.org/en/development/desa/population/migration/generalassembly/docs/globalcompact/A_RES_2200A(XXI)_economic.pdf>.

UN-GA (United Nations-General Assembly). (1992). *Rio declaration on environment and development. A/CONF.151/26*. In: Report of the United Nations Conference on Environment and Development (Vol. I). Retrieved from https://www.un.org/en/development/desa/population/migration/generalassembly/docs/globalcompact/A_CONF.151_26_Vol.I_Declaration.pdf

UN-GA (United Nations-General Assembly). *Transforming our world—The 2030 agenda for sustainable development (A/RES/70/1)*. (2015). <https://www.un.org/en/development/desa/population/migration/generalassembly/docs/globalcompact/A_RES_70_1_E.pdf>.
UN-Habitat III. (2016). New urban agenda. In: *Conference on housing and sustainable urban development (Habitat III)*. Retrieved from http://habitat3.org/wp-content/uploads/NUA-English.pdf.
UN-Habitat (United Nations Centre for Human Settlements). (2001). *Cities in a globalizing world—Global report on human settlements*. Nairobi, Kenya: UN-Habitat.
UN-Habitat. (2002). *Enabling shelter strategies: review of experience from two decades of implementation*. Nairobi, Kenya: UN-Habitat.
UN-Habitat (United Nations Centre for Human Settlements). (2008). *State of the World's cities 2008–2009: Harmonious cities*. London: Earthscan.
UN-Habitat (2012/2013). *State of the world's cities*, (2012). Nairobi, Kenya: UN-habitat.
UN-Habitat. (2014). *State of the world's cities 2014. Re-imagining sustainable urban transitions*. Nairobi, Kenya: UN-Habitat.
UN-Habitat (United Nations Centre for Human Settlements). (2018). *SDG-11 synthesis report: Tracking progress towards inclusive, safe, resilient and sustainable cities and human settlements*. Nairobi: UN-Habitat.
UN-Habitat (United Nations Centre for Human Settlements). (2020). *World cities report 2020: The value of sustainable urbanization*. Nairobi: Kenya: UN Habitat Secretariat.
Unver, O., Bhaduri, A., & Hoogeveen, J. (2017). Water-use efficiency and productivity improvements towards a sustainable pathway for meeting future water demand. *Water Security*, *1*, 21, 27.
Urquijo, L. G. (2017). The Europeanisation of policy to address poverty under the new economic governance: The contribution of the European Semester. *Journal of Poverty and Social Justice*, *25*(1), 49–64.
USAID (United States Agency for International Development). (2017). *Ecosystem-based adaptation and water security*. Washington, D.C.: USAID.
USDA (United States Depart of Agriculture). (2016). *A citizens' guide to national forest planning*. Washington, D.C.: USDA.
Vadgama, C. V., Khutwad, A., Damle, M., & Patil, S. (2015). Smart funding options for developing smart cities: A proposal for India. *Indian Journal of Science & Technology*, *8*(34), 1–12.
van Dijk, P. (2011). Three ecological cities: examples of different approaches in Asia and Europe. In T. Wong, & B. Yuen (Eds.), *Eco-city planning: Policies, practice and design*. New York: Springer.
van Mead, N. (2017). Uber for bikes: how "dockless" cycles flooded China—and are heading overseas. *The Guardian*, March 22. Available online.
Vancouver Economic Commission. *Circular economy*. (2017). <http://www.vancouvereconomic.com/programs-initiatives/false-creek-flats/circular-economy/>.
Van-Heur, B. (2010). The built environment of higher education and research: Architecture and the expectation of innovation. *Geography Compass*, *4*(12), 1713–1724.
Vanolo, A. (2016). Is there anybody out there? The place and role of citizens in tomorrow's smart cities. *Futures*, *82*, 26–36.
VBA (Vienna Business Agency). *Helpful links and factsheets*. (n.d.). <https://viennabusinessagency.at/international/ceu-in-vienna/ceu-in-vienna/good-to-know/hilfreiche-links-und-factsheets/>.
Verret, H., & Pfeffer, K. (2019). Elaborating the urbanism in smart urbanism: Distilling relevant dimensions for a comprehensive analysis of smart city approaches. *Information, Communication & Society*, *22*(9), 1328–1342.
Viale-Pereira, G., Cunha, M. A., Lampoltshammer, T. J., Parycek, P., & Testa, M. G. (2017). Increasing collaboration and participation in smart city governance: A cross-case analysis of smart city initiatives. *Information Technology for Development.*, *23*(3), 526–553.

Vickers, J. (1995). Concepts of competition. *Oxford Economic Papers, 47*, 1−23.

Vignola, R., Locatelli, B., Martinez, C., & Imbach, P. (2009). Ecosystem-based adaptation to climate change: What role for policy-makers, society and scientists? *Mitigation and Adaptation Strategies for Global Change, 14*, 691−696.

Vijai, P., & Sivakumar, P. B. (2016). Design of IoT systems and analytics in the context of smart city initiatives in India. *Procedia Computer Science, 91*, 583−588.

Vilella, M. (2017, December 6). Zero waste cities: At the forefront of the sustainable development goals agenda. *Huffington Post*. Available online.

Visvaldisa, V., Ainhoab, G., & Ralfsc, P. (2013). Selecting indicators for sustainable development of small towns: The case of Valmiera municipality. *Procedia Computer Science, 26*, 21−32.

Vollaro, R. D. L., Evangelistia, L., Carnieloa, E., Battistaa, G., Goria, P., Guattaria., & Fanchiot, A. (2014). An integrated approach for an historical building's energy analysis in a smart cities' perspective. *Energy Procedia, 45*, 372−378.

VSC (Vienna Smart City). *Vienna smart city projects*. (n.d.). <https://smartcity.wien.gv.at/site/en/projects/>.

Vu, K., & Hartley, K. (2018). Promoting smart cities in developing countries: Policy insights from Vietnam. *Telecommunication Policy, 42*, 845−859.

Wacquant, L. (2009). *Punishing the poor: The neoliberal government of social insecurity*. London: Durham, N.C.: Duke University Press.

Walravens, N. (2015). Mobile city applications for Brussels citizens: Smart city trends, challenges and a reality check. *Telematics and Informatics, 32*(2), 282−299.

Wamsler, C., Luederitz, C., & Brink, E. (2014). Local levers for change: Mainstreaming ecosystem-based adaptation into municipal planning to foster sustainability transitions. *Global Environmental Change, 29*, 189−201.

Wang, Y., Ren, H., Dong, L., Park, H.-S., Zhang, Y., & Xu, Y. (2019). Smart solutions shape for sustainable low-carbon future: A review on smart cities and industrial parks in China. *Technological Forecasting & Social Change, 144*, 103−117.

Ward, B., & Rene, D. (1972). *Only One Earth: The Care and Maintenance of a Small Planet*. London: Penguin Books.

Warschauer, M. (2003). *Technology and Social Inclusion. Rethinking the digital divide*. Cambridge, Mass: The MIT Press.

Washburn, D., Sindhu, U., Balaouras, S., Dines, R. A., Hayes, N. M., & Nelson, L. E. (2010). *Helping CIOs understand "smart city" initiatives: Defining the smart city, its drivers, and the role of the CIO*. Cambridge, MA: Forrester Research, Inc. Available online.

WCED (World Commission on Environment and Development). (1987). Report of the World Commission on Environment and Development: Our Common Future. Available online at: https://sustainabledevelopment.un.org/content/documents/5987our-common-future.pdf.

WCRP (World Climate Research Programme). Global research and action agenda on cities and climate change science. In: *WCRP publication no. 13/2019*. Available online.

Webster's New Collegiate Dictionary. (2008). (3rd ed.). New York: Hungry Mints Inc.

WEF (World Economic Forum). (2018). *Circular economy in cities: Evolving the model for a sustainable urban future*. Geneva, Switzerland: World Economic Forum.

WEF (World Economic Forum). (2019). Global competitiveness report 2019. Geneva: WEF.

Welle, B., Sharpin, A. B., Adriazola, C., Bhatt, A., Alveano, S., Obelheiro, M., ... Bose, D. (2018). *Sustainable and safe: A vision and guidance for zero road deaths*. Washington, USA: World Resources Institute.

Werner, C., Panesar, A., Rud, S. B., & Olt, S. U. (2009). Ecological sanitation: Principles, technologies and project examples for sustainable wastewater and excreta management. *Desalination., 248*(1−3), 392−401.

Wernerfelt, B. (2013). Small forces and large firms: Foundations of the RBV. *Strategic Management Journal, 34*, 635−643.

Wheeler, S. (2003). Planning sustainable and livable cities. In R. LeGates, & F. Stout (Eds.), *The city reader*. London: Routledge.

WHO (World Health Organization). (2015). *Global status report on road safety 2015*. Geneva, Switzerland: WHO.

Wickham, J. D., Riitters, K. H., Wade, T. G., & Homer, C. (2008). Temporal change in fragmentation of continental United States forests. *Landscape Ecology, 23*, 891−898.

Willis, K. (2016). Viewpoint: International development planning and the sustainable development goals (SDGs). *International Planning Review, 38*(2), 105−111.

Winans, K., Kendall, A., & Deng, H. (2017). The history and current applications of the circular economy concept. *Renewable and Sustainable Energy Reviews, 68*, 825−833.

Wirth, L. (1938). Urbanism as a way of life. *American Journal of Sociology, 44*(1), 1−24.

Wolfram, M. (2012a). *Deconstructing smart cities: An intertextual reading of concepts and practices for integrated urban and ICT development* (pp. 171−181).

Wolfram, M. (2012b). Deconstructing smart cities: An intertextual reading of concepts and practices for integrated urban and ICT development. In: *Proceedings REAL CORP 2012 Tagungsband* (pp. 171−181), May 14−16, 2012. Retrieved from https://www.corp.at/archive/CORP2012_192.pdf

Woolmer, G., Trombulak, S. C., Ray, J. C., Doran, P. J., Anderson, M. G., Baldwin, R. F., . . . Sanderson, E. W. (2008). Rescaling the human footprint: a tool for conservation planning at an ecoregional scale. *Landscape and Urban Planning, 87*, 42−53.

World Bank. (2012). *Sustainable low-carbon city development in China*. Washington, DC: World Bank.

World Bank. (2016). *World development report 2016: Digital dividends*. Washington, DC: World Bank.

World Bank. (2018a). *What a waste 2.0 a global snapshot of solid waste management to 2050*. Washington, D.C.: World Bank.

World Bank. (2018b). *Urban sustainability framework* (1st (ed.)). Washington, DC: World Bank.

WRI-WB (World Resource Institute- World Bank). (2018). *Sustainable and Safe: A Vision and Guidance for Zero Road Accidents*. Washington, D.C.: WRI.

WSC (Wien Smart City). *Wien smart city framework strategy*. (2014). <https://smartcity.wien.gv.at/site/files/2016/12/SC_LF_Kern_ENG_2016_WEB_Einzel.pdf>.

Xiao, D. N., Chen, W. B., & Guo, F. L. (2002). On the basic concepts and contents of ecological security. *Chinese Journal of Applied Ecology, 13*, 354−358.

Xu, X., Lin, H., & Fu, Z. (2004). Probe into the method of regional ecological risk assessment—A case study of wetland in the Yellow River Delta in China. *Journal of Environmental Management, 70*, 253−262.

Yang, L., & Vizard, P. (2017). Multidimensional poverty and income inequality in EU. *CASE paper 207/LIP paper 4*. London: London School of Economics (LSE).

Yanitsky, O. (1982). Towards an eco-city: problem of integrating knowledge with practice. *International Social Science Journal, 34*(3), 469−480.

Yigitcanlar, T., & Lee, S. H. (2014). Korean ubiquitous-eco-city: a smart sustainable urban form or a branding hoax? *Technological Forecasting and Social Change, 89*, 100−114.

Yue, W. S., Chye, K. K., & Hoy, C. W. (2017). Towards smart mobility in urban spaces: Bus tracking and information application. *AIP Conference Proceedings, 1891*(1), 201−245.

Zandersen, M., Jensen, A., Termonsen, M., Buchholtz, G., Munter, B., Blemmer, M., . . . Andersen, A. H. (2014). Ecosystem based approaches to climate adaptation. Urban prospects and barriers. In: *Scientific report from DCE—Danish Centre for Environment and Energy, no. 83*. Retrieved from https://dce2.au.dk/pub/SR83.pdf

Zhan, C., & de Jong, M. (2018). Financing eco cities and low carbon cities: The case of Shenzhen international low carbon city. *Journal of Cleaner Production, 180*, 116, 125.

Zhou, N., & Williams, C. (2013). *An international review of eco-city theory, indicators, and case studies*. Berkeley, CA: Berkeley National Laboratory Available at. Available from https://china.lbl.gov/sites/all/files/lc_eco-cities.pdf.

Ziervogel, G., Nyong, A., Osman, B., Conde, C., Cortes, S., & Downing, T. (2006). Climate variability and change: Implications for household food security. In: AIACC working papers series no. 20. Washington, DC: Assessment of Impacts and Adaptations to Climate Change (AIACC).

Zimmerman, R., & Faris, C. (2011). Climate change mitigation and adaptation in North American cities. *Current Opinion in Environmental Sustainability*, *3*(3), 181–187.

Zou, X., & Li, Y. (2014). How eco are China's eco-cities? An international perspective. *IRSPSD International*, *2*(3), 18–30.

Zygiaris, S. (2013). Smart city reference model: Assisting planners to conceptualize the building of smart city innovation ecosystems. *Journal of the Knowledge Economy*, *4*(2), 217–231.

Further reading

Bairoch, P., & Goertz, G. (1986). Factors of urbanisation in the nineteenth century developed countries: A descriptive and econometric analysis. *Urban Studies*, *23*, 285–305.

Bertinelli, L., & Black, D. (2004). Urbanization and growth. *Journal of Urban Economics*, *56*, 80–96.

Bettencourt, L. M. A., Lobo, J., Helbing, D., Kuhnert, C., & West, G. B. (2007). Growth, innovation, scaling, and the pace of life in cities. *Proceedings of National Academy of Sciences*, *104*(17), 7301–7306.

Center on Governance. (2003). *Smart capital evaluation guidelines report: Performance measurement and assessment of smart capital*. Ottawa, Canada: University of Ottawa Available at. Available from http://www.christopherwilson.ca/papers/Guidelines_report_Feb2003.pdf.

Eriksen, S. H., Aldunce, P., Martins, R. D., & Sygna, L. (2011). When not every response to climate change is a good one: Identifying principles for sustainable adaptation. *Climate and Development*, *3*(1), 7–20.

Fay, M., & Opal, C. (2000). *Urbanization without growth: A not-so-uncommon phenomenon*. Washington, D.C: The World Bank, Policy Research Working Paper, No. 2412.

Fujita, M., Krugman, P., & Mori, T. (1999). On the evolution of hierarchical urban systems. *European Economic Review*, *42*(2), 209–251.

Greiving, S., Arens, S., Becker, D., Fleischhauer, M., & Hurth, F. (2017). Improving the assessment of potential and actual impacts of climate change and extreme events through a parallel modeling of climatic and societal changes at different scales. *Journal of Extreme Events*, *4*(4), 1850003.

Krugman, P. (1991). Increasing returns and economic geography. *Journal of Political Economy*, *99*, 483–499.

Kumar, A., & Kober, B. (2012). Urbanization, human capital, and cross-country productivity differences. *Economics Letters*, *117*(1), 14–17.

MacDonald, A., Clarke, A., Huang, L., & Seitanidi, M. M. (2019). Multi-stakeholder partnerships for sustainability: A resource-based view of partner implementation structure to outcomes. *Sustainability.*, *11*(3), 557.

Nuzira, F. A., & Dewancker, B. J. (2014). Understanding the role of education facilities in sustainable urban development: A case study of KSRP, Kitakyushu, Japan. *Procedia Environmental Sciences*, *20*, 632–641.

OECD (Organization for Economic Cooperation and Development). (2014). *Measuring the digital economy: A new perspective*. Paris: OECD Publishing.

UNEP (United Nations Environment Programme). (2015). *Global waste management outlook*. Nairobi: UNEP.

UN-Habitat (United Nations Centre for Human Settlements). (2011). *Cities and climate change: Global report on human settlements*. London: Earthscan.

WHO (World Health Organization). (2010). *Why urban health matters*. Paris: WHO.

CHAPTER 8

Sustaining life below water

8.1 Introduction

An ecosystem is a functional unit comprising biotic or living organisms along with their abiotic or nonliving environment, and the interactions within and between them (Allwood, Bosetti, Dubash, Gómez-Echeverri, & von Stechow, 2014). Examples of biotic or living organisms include human beings, animals, plants, trees, bacteria, fungi, protoctists, viruses, etc. Human beings constitute the part of terrestrial biotic organisms, while most of the other biotic organisms comprise part of marine ecosystem as well as terrestrial ecosystem. Nevertheless, marine biotic organisms are vitally different from the terrestrial biotic organisms in composition, characteristics and mode of living. The notion of life below water entails life in oceans, seas, rivers, estuaries, lakes, and other bodies that store water. Earth is the only planet in the entire known as universe that is endowed with liquid water on its surface and the only planet to be blessed with life that is said to have originated in the primordial ocean where it has evolved over millions of years (Convention on Biological Diversity, CBD, 2012). Comprising over 90% of all habitable space on Earth, life in ocean is found from top to bottom, from the sunny surface water of tropical seas to the cold, dark depths, thousands of meters below (Costello, Cheug, & Hauwere, 2010).

Encompassing more than 70% of the Earth's surface, oceans are home to more than 2.2. million species (Mora et al., 2013). The economic, social, and environmental significance of the oceans and seas cannot be overestimated because the market value of marine and coastal resources and industries is estimated at US$3 trillion per year, or about 5% of global GDP (UN, 2017). Primary economic services accruing from oceans comprise transportation of goods, fisheries and tourism but also potential new uses such as the generation of renewable energy and mining of materials (ICSU, 2017). Moreover, oceans are vital for human wellbeing both as a source of food and livelihood. More than three billion people rely on the seafood as the primary source of their protein, and marine fisheries provide direct or indirect employment to over 200 million people (UN (United Nations), 2017). Additionally, oceans and coastal areas are estimated to partially support a major chunk of global population (about 75% of the global population), inhabiting the wider coastal margins. Apart from hosting the largest connected ecosystem that deliver services such as climate stability, oxygen generation, nutrient cycling, and food production, oceans also play a pivotal role in climate change mitigation, absorbing about a third of carbon dioxide (CO_2) produced by anthropogenic activities (ICSU, 2017).

Admittedly, as compared to other natural resources, oceans are relatively less explored; nonetheless, oceans provide ecosystem services that are central to the very existence of both terrestrial and aquatic biospheres on planet Earth. Oceans act as a buffer to the changes in the composition of the atmosphere (Catling & Kasting, 2017), uphold the global thermal equilibrium (Wang, Han, & Sriver, 2012), control of the hydrological cycle (Gröger et al., 2007), afford a sink for all forms of wastes from the astrosphere (Goldberg, 1985; Holdgate & McIntosh, 1986; Peterson & Teal, 1986), facilitate the formation of diverse types of sedimentary and metamorphic rocks, and serve as the world's largest pool of flora and fauna (Armbrust & Palumbi, 2015). Scientific and scholarly evidence has enabled us to acknowledge the enabled value of oceans as a sink of CO_2 (Ibánhez, Araujo, & Lefèvre, 2016; Landschützer, Gruber, & Bakker, 2016; Orr & Sarmiento, 1992; Quéré et al., 2003) as well as the vital role oceans can play in global warming. Similarly, the significance of oceans has come into reckoning as never before because the value of fishery and seaweeds (Hehre & Meeuwig, 2016), salt and minerals (Loganathan, Naidu, & Vigneswaran, 2017), drinking water (Ghaffour, Missimer, & Amy, 2013), navigation (Lee & Song, 2017), petroleum resources including oils and gases under the seabed and methane hydrate in the sea floor. Nevertheless, the interactions between humans and the environment have endangered global biodiversity that is threatened by an array of pressures including overexploitation of species, habitat modification, invasive alien species and disease, pollution, and climate change (Maxwell, Fuller, Brooks, & Watson, 2016).

In marine systems, such pressures emanate from an assortment of activities like fishing, coastal development, shipping, and energy production (Halpern et al., 2015) that has left only 13% of global ocean as "wilderness" (Jones et al., 2018). Therefore, managing marine systems seemingly appears a subtle and complex attempt that entails of potential of resulting in "triple bottom line" outcomes, where conservation goals and social outcomes are optimized and overall costs are minimized (Halpern et al., 2013). Altogether, the sustainable future of humankind vastly depends on the drastic decrease of influx of pollutants into oceans and to ensure maintenance of the biodiversity therein for sustenance of ecosystem services from them (NCM, 2016). This significance is demonstrated from the declaration of sustainable development goals (SDGs) as Goal 14 clearly call to "Conserve and sustainably use the oceans, seas and marine resources for sustainable development" and outlined seven main and three associated targets in order to attain the goal (Vierros, 2017). Marine species and habitats are increasing becoming under threat from stressors—environmental and anthropogenic—with pronounced ecological consequences (Gunderson, Armstrong, & Stillman, 2016). The resultant outcomes of these stressors cause shifts in the geographical ranges (Sunday, Bates, & Dulvy, 2012), alterations in the strength and types of ecological interactions (Milazzo, Mirto, Domenici, & Gristina, 2012), and in the worst-case scenarios, collapse of population and extinction of species (McCauley et al., 2015). These stressors need to be analyzed to ascertain their causes and effects in order to formulate appropriate policies and actions that are helpful in sustaining marine life, and ecology.

8.2 Environmental stressors

The usage of term "stressor" in ecological research has often been made synonymously with other terms like "disturbance"; "pollution"; "pollutants"; or "pressures" on the supposition that the outcomes of a stressor or the equivalent descriptor imply "stress" and ought therefore to be entirely detrimental or damaging (Folt et al., 1999). Nevertheless, it is also contended that what is stressful or detrimental to one species may be nondetrimental or even advantageous to another species (Piggott, Townsend, & Matthaei, 2015). Consequently, a stressor can be construed to be "either an

8.2 Environmental stressors

abiotic or biotic factor that results from human activity and exceeds its range of normal variation and affects individual physiology, population performance, community balance or ecosystem functioning, either positively or negatively" (Townsend, Uhlmann, & Matthaei, 2008). This helps in making the differentiation between terms like "pollution" or "pollutants" that requires a substance to cause a damaging biological or ecological response (Freedman, 1995).

In common parlance, stressor can be outlined as any stimulus or factor that interferes or perturbs an organism from its natural or optimum function (Moberg & Mench, 2000; Romero, 2004). Construed in this context, climate change operates as an environmental stressor by enforcing new conditions on an organism and most prominently for aquatic organisms, ocean warming (OW) and enhanced dissolved CO_2 often prove instrumental in causing extreme changes of their natural environment (Hoegh-Guldberg & Bruno, 2010). Environmental stressors are prominently manifested through climate-related drivers that impact the ocean and often comprise so-called "slow-onset-events" such as ocean warming, acidification, deoxygenation, sea-level rise, and glacial retreat, alongside extreme weather events like marine heat waves and increased frequency of storms (Hanson, Isensee, Herr, Osborn, & Dupont, 2019). These entail the likelihood of driving biodiversity shifts in multiple ways, especially in biodiversity hotspots like coral reefs and other ecosystems that deliver significant services for coastal communities. Complex nature of interactions between organisms often renders it difficult to forecast the overall impact of climate change and ocean acidification (OA). Maintaining that different species could either struggle or thrive in response to altering ocean conditions, subject to specific tolerance limits of species or indirectly owing to changes in species interactions and habitat loss, Hanson et al. (2019) argue that currently it is not to make exact or precise predictions; nevertheless, they are certain that marine ecosystems would change and that marine biodiversity would also decrease, at least in the short term. Undoubtedly, almost all major marine ecosystems around the globe have reported footprints of climate change (Hobday & Lough, 2011; Hoegh-Guldberg & Bruno, 2010; Okey et al., 2014; Wassmann, Duarte, Agusti, & Sejr, 2011); nonetheless, neither the physical drivers of climate change nor their impacts on ocean ecosystems have demonstrated consistently or homogenously over the world oceans (Popova et al., 2016). According to Wu et al. (2012), waters of subtropical western boundary currents are getting warmer two to three times faster than the global mean for the global oceans. In the case of Arctic Ocean, polar amplification leads to faster warming than the global trend (Pithan & Mauritsen, 2014). Such amplifications, in tandem with the associated shrinking Arctic Sea ice, are prone to result in changes in the ecosystems that are usually in the opposite direction of global trends, with primary production registering an increase rather than pursuing the global trend of decline (Popova, Yool, Aksenov, Coward, & Anderson, 2014, 2016). On the other hand, upwelling zones, that are highly productive, of eastern boundary currents possess a strong sensitivity to climate change, affected by altering patterns of upsurge in events that are getting less frequent but stronger, and longer duration (Iles et al., 2012).

Hobday and Pecl (2014), on the basis of historical observation of sea surface temperature (SST), identified 24 fast-warming marine areas, or so-called hotspots, and recommended that such areas could form the basis of serving as "natural laboratories" to facilitate studying in advance the mechanistic linkages between OW and biological responses to ascertain the large-scale effects predicted for later in the 21st century. In the wake of high human dependence on marine resources in many of these hotspots, the prospects of exploring climate adaptation options cannot be easily ruled out (Miller, Ota, Sumuila, Cisneros-Montemayor, & Cheung, 2017). Alterations in ocean physical and biogeochemical parameters are expected to wield great impact on ocean ecosystems during the 21st century (Doney et al., 2012). There is also a likelihood of coastal-marine food resources undergoing change as a consequence of species-specific direct responses to drivers of climate change like supply and large-scale availability of species undergoing alteration as a response to temperature (Laffoley & Baxter, 2016); and Frusher et al. (2014) have already reported occurrence of this phenomenon in south-east Australia and another example is that of OA in the Arctic (Mathis et al., 2015). International community—from individuals, communities, industries to respective national governments—is required to focus on understanding and adapting to climate change in order to ward off such impacts to living marine resources (Frusher et al., 2014).

Nevertheless, temperature-rise is more the only climatic factor affecting ocean ecosystems because, what Gruber et al. (2019) calls "warming-up, turning sour, losing breath" has seemingly emerged as an extensively used gist of the key climatic stressors of ocean ecosystems that have come to be known as warming, acidification and deoxygenation and all these entail serious implications for marine productivity (Doney et al., 2012). It has been pointed by Doney et al. (2012) that changing ocean stratification and circulation can also yield widespread biological impacts. Alterations in climatic factors are spurred by diverse mechanisms and dissimilar aspects of global ocean dynamics and biogeochemistry (Bopp et al., 2013), and accordingly, formats or patterns of their fastest changes or hotspots do not inevitably coincide in space. Undoubtedly, warming of the ocean may not always be construed as the strongest climatic action impacting marine ecosystems (Maranon et al., 2014); nonetheless the rise of the SST entails the potential of emerging as the prime factor of climate change (Alexander et al., 2018). Usually, data pertaining to SSTs can be procured from archives of climate model simulations; however, the availability of data pertaining to temperature and other variables as a function of depth is generally restricted (Alexander et al., 2018). Almost all physiological processes in marine organisms are controlled by temperature (Deutsch, Ferrell, Seibel, Pörtner, & Huey, 2015) and as such, SSTs often constitute a significant and leading indicator and/or significant driver of fluctuations in marine ecosystem (Ottersen, Kim, Huse, Polovina, & Stenseth, 2010), including fish distribution (Pinsky, Worm, Fogarty, Sarmiento, & Levin, 2013), fish recruitment (Kristiansen, Drinkwater, Lough, & Sundby, 2011), and biodiversity (Tittensor et al., 2010). SST also greatly influences even bottom-dwelling organisms as most of them spend at least a part of their life cycle either as pelagic larvae or rely on food sources that are influenced by SST (Alexander et al., 2018). Many of the hotspots suggested by Hobday and Pecl (2014) have marine resource-dependent communities that present examples of social, economic, and ecological commonalities and contrasts that are being increasingly focused on for facilitating a large international partnership working toward minimizing coastal vulnerability (Hobday et al., 2016).

Climate change is projected to greatly impact marine ecosystems that deliver services of high social and economic value (Rogers, Sumaila, Hussain, & Baulcomb, 2014), including the primary source of protein for one in seven persons of the global population (FAO, 2012) and also help in regulating Earth's climate through the absorption and storage of atmospheric CO_2 (Kwon, Primeau, & Sarmiento, 2009). Projected impacts of climate change are likely to have far-reaching consequences for marine ecosystems impacting both their structure and functioning (Gauss et al., 2015). Four major climate drivers impacting marine ecosystem structure, functioning, and adaptive capacity have been identified by Porter et al. (2014) in IPCC's Fifth Assessment report, and these are—pH, temperature, oxygen concentration, and food availability—and all these are subject to massive perturbations in projections of future climate change scenarios (Rogers et al., 2014). Increased concentration of atmospheric CO_2 can lead to reduction in ocean pH that may result in decreased viability of calcareous organisms among other impacts (Doney et al., 2012). Ocean stratification, entailing links with warming ocean temperatures, curtails nutrient supply to photosynthetic organisms in surface water (Sigman & Hain, 2012). Warmer ocean waters also have the potential of decreasing the solubility of oxygen and exchange of subsurface waters with the atmosphere, spurring lower oceanic oxygen concentrations with the probability of negatively affecting marine organisms (Shepherd, Brewer, Oschlies, & Watson, 2017). The cumulative impact of these changes, though regionally variable, is projected to be overall global reduction in primary production that is seemingly the eventual determinant of food availability to marine ecosystems (Bopp et al., 2013). Some scholars, on the basis of observation of some cases where positive or neutral responses to potential marine stressors have been observed, have come to the surmise that uncertainty surrounding the future of the marine ecosystem is large (Boyd, Dillingham et al., 2015; Boyd, Lennartz, Glover, & Doney, 2015). According to Henson et al. (2017), "... A change in a potential stressor where the ecosystem response is unknown (and may not necessarily be negative) is termed a 'driver'."

Irrespective of the fact that there is inevitability of the projected climate change response over the coming century in these environmental drivers is great, so has been the natural variability encountered by marine organisms, thereby making it discernible that some species are endowed with the capacity to adapt or acclimate to change (Boyd et al., 2016; Doney et al., 2012; Foo & Byrne, 2016). Occurrence of natural variability takes place on timescales of a few years (international variability) to millennia (in case of glacial-inter glacial cycles), whereas climate change is an one-sided phenomenon or one-way street, and it demonstrates that associated changes in the marine environment are to some extent irreversible (Henson et al., 2017). In that event, possibility of climate change ultimately pushing marine ecosystem drivers out of the range of natural variability cannot be ruled out probably resulting in migration of species (Burrows et al., 2014), restructuring of ecological niches, envisaging novel climates (Williams et al., 2007; Williams & Jackson, 2007), and the need of for socio-economic systems to adjust to these changes so as to protect the livelihoods and human well-being (Henson et al., 2017). Besides, marine ecosystem drivers seldom vary in isolation and an array of factors can act additively or in a cooperative manner or synergistically to enhance the impact of a single driver (Frolicher, Ramseyer, Raible, Rodgers, & Dunne, 2020; Knapp et al., 2017). OA entails the potential of changing the carbon to nitrogen ratio of sinking organic material to facilitate inflow of more oxygen needed for remineralization (Riebesell et al., 2007). OA also entails the potential of acting in tandem with rising temperatures to decrease abundance of coccolithophore or calcite production (Feng et al., 2009). Concurrently, lower pH and oxygen concentration could increase temperature sensitivity in corals (Anthony, Kline, Diaz-Pulido, Dove, & Hoegh-Guldberg, 2008) and crustaceans (Findlay, Kendall, Spicer, & Widdicombe, 2010). It is opined by Henson et al. (2017) that synergistic impacts of an assortment of drivers are difficult or challenging to probe in field or lab studies owing to problems of undertaking multifactorial experiments over numerous generations. Nevertheless, Crain et al. (2008) are of the view that it can be discerned from available observational evidence that tends to demonstrate cooccurring stress can augment to or magnify the impact of a single stressor as attested to by Brennan and Collins (2015) in a microalgal species grown in 96 experiments where population growth recorded a decrease with the number of environmental drivers (Krichen, Rapaport, Floc'h, & Fouilland, 2019).

It can be discerned from some studies that manifold stressors can interact in sudden or unexpected ways, sometimes leading to a positive or neutral effect (Hale, Pigott, & Swearer, 2017; Maher, Rice, McMinds, Burkepile, & Thurber, 2019). Citing the example of warming and acidification having antagonistic impacts on sea urchin larval growth, Byrne et al. (2013) report that it results in minimal overall impact, albeit both have a negative effect on larval abnormality.

Acknowledging the lack of clarity with regard to the response of the marine ecosystem to changing drivers, Henson et al. (2017) emphasize on quantifying when, where and which groupings of stressors are likely to take place. Pace of the speed with which drivers of ecosystem stress emanate from the backdrop of natural variability assumes specific significance for ability of an organism to adapt to a changing climate (Hawkins & Sutton, 2012). The environmental niche that organisms have come to carve out roughly or unevenly resembles the surrounding conditions experienced by them (Sunday et al., 2012), and that ordains the organisms out to be resilient to the range in natural variability. Emphasizing the essentiality of organisms to adapt to at least yearly extrema in conditions during winter and summer, that is, the seasonal variability, that in almost all places of locations exceeds the international variability, Henson et al. (2017) argue that speedier the pace of exist of the system out of its natural range of variability, lesser the time the organisms will take to adapt or acclimate to the new conditions or migrate to more convenient places. Endowed with the potential of already removing about 25% of anthropogenic CO_2 emissions (Le Quéré et al., 2018) and oceans also possess the potential of removing and storing much more (Rau, McLeod, & Hoegh-Guldberg, 2012), Gattuso et al. (2018) emphasize that ocean-based solutions could significantly help in decreasing the magnitude and rate of ocean warming, OA, and lea-level rise, as well as in minimizing their effects on marine ecosystems and ecosystem services. Conceding that there might be associated risks to ocean life and people, Gattuso et al. (2018) lament that there is lack of guidance for prioritizing ocean-based interventions since there has been comparatively little research, development and deployment in this field. Significant issues in this regard pertain to determining the efficacy of a given approach in tackling changes in climate drivers and/or impacts, probable spatial and temporal scales of deployment, associated positive and negative climate, environmental, economic and societal impacts (Russell et al., 2012) and therefore the implications for ethics, equity, and governance (Williamson & Bodle, 2016). Interestingly, apart from Gattuso et al. (2018) giving a call for ocean-based solutions to tackle climate change, this idea is gaining international traction (Aldred, 2019; Hoegh-Guldberg et al., 2019) and further details in this regard follow in the concluding part of this chapter. Major environmental stressors are briefly appraised here.

8.2.1 Ocean warming

Emission of anthropogenic greenhouse gases (GHGs) has culminated in a long-term and incontrovertible warming of the planet (IPCC (Intergovernmental Panel on Climate Change), 2019a, 2019b). Bulk (more than 90%) of the excess heat is deposited within the world's oceans, where it amasses and engenders augmentation in ocean temperature (Abram et al., 2019). Since the oceans are the principal repository of the Earth's energy disequilibrium or imbalance, determining ocean heat content (OHC) is come to be acknowledged as one of the most congenial ways to quantify the rate of global warming (Cheng et al., 2018). In the wake of reports released in 2018 and 2019 (Cheng & Zhu, 2018; Cheng et al., 2019), new OHC data for the year 2019 make it discernible that the world's oceans (specifically the upper 2000 m) in 2019 were the warmest in recorded human history, especially, the ocean heat anomaly (0 − 2000 m) in 2019 was 228 Zetta Joules (ZJ, 1 ZJ = 1021 Js) above the 1981 − 2010 average and 25 ZJ above 2018 (Cheng, et al., 2020).

As the principal reservoir of heat in the climate system, the ocean has amassed more than 90% of Earth's human-induced heat since 1971, and in the fact of that enhanced anthropogenic GHGs emerging as the major driver of the observed warming (Rhein et al., 2013), this warming has been instrumental in adding over 40% of worldwide mean sea-level rise since 1993 (WCRP-GSLBG, 2018). It has also resulted in massive enhancements in the number and asperity of marine heatwaves entailing grave aftermaths for ecosystems (Frölicher, Fischer, & Gruber, 2018). OW has come to acknowledged as a key indicator of climate change (von Schuckmann et al., 2016) and as a metric for climate model evaluation (Palmer, 2017). Undeniably, replication of observed OW has occurred across many observational studies (Cheng et al., 2017; Johnson et al., 2018); nevertheless, there differentiations in the estimates, and the precise estimation of the anthropogenic alterations that spur OW is complicated by historical instrument biases (Cheng et al., 2016) as well as by scant observing system coverage, particularly prior to the launch of Argo program in 2000 (Palmer, 2017). Argo program, launched in 2000, is a global program that facilitates collection of information from inside of the ocean using a fleet of 4000 robotic instruments that drift with the ocean currents and move up and down with the surface and a mid-water-level (Roemmich et al., 2019).

Existing scientific knowledge of long-term change in OW is most robust up to the upper ocean level (0–700 m) because of the existence of the major historical measurements (Rhein et al., 2013). Since the publication of initial assessments of global ocean warming, a reliable picture of forced ocean change over longer time scales has become discernible with regard to ocean observations (Levitus et al., 2012). Many successive analyses, in the aftermath of amelioration of issues with data biases, demonstrate a clearer picture of change. Indeed, very sparse data coverage for the intermediate depths, ranging from 700 to 2000 m, is available prior to the launch of modern Argo measurements (Lyman & Johnson, 2014). Nevertheless, "pentadal" or 5-year estimates are available dating back to 1957 (Levitus et al., 2012), and these also demonstrate marked warming over the observed record, albeit at a slower pace comparable to in the upper ocean. Undeniably, all observed analyses exhibit marked and statistically robust historical warming; nonetheless, differentiations exist in their patterns and rates owing to instrumental biases, limitations of measurement coverage, and the diverse methods and processing choices deployed to reformulate global change estimates from scant observations (Boyer et al., 2016). These shortcomings or discrepancies mostly fade away in regard to the upper and intermediate ocean since the launch of the modern Argo period up to the present (Johnson et al., 2018). According to Durack et al. (2018), it is worth noting that the ice-covered polar regions and marginal seas still remain to be comprehensively sampled by Argo program, with progress going on with measurement in these regions.

Capabilities of the oceans in absorbing large volumes of energy, with comparatively low alteration in temperatures, have indisputably contributed to reducing the pace of rise in global air temperature (Dahlman & Lindsey, 2020). Nevertheless, the possibility of this absorbing effect spurring promotion of enhancement in temperature trends cannot be ruled out, irrespective of reversal in pollution trends in the future (Meehl, Washington, & Collins, 2005). Future projections with regard to OW predict an augmentation in in ocean temperatures of up to 4.3°C in some locations by the year 2100 (Mora et al., 2013). Anomalies, often construed as an established way to explain climate variability over time spans, in ocean temperature over the past many decades demonstrate variations in ocean temperature of up to 1°C; whereas, on the contrary, recent trends in OW and future projections manifest a more rapid rise in ocean temperatures comparable to fluctuations over the past 15 decades (Hegerl, Bronnimann, Shurer, & Cowan, 2018).

OW entails the potential of causing damaging effects on various marine organisms, including increased metabolism and ensuing reduced efficiency (IUCN, 2017a, 2017b) and in that context, the major issue is that organisms forced to operate at suboptimal temperatures need more energy and increased respiration to maintain homeostasis in order to fulfill their normal energy requirements and in this way it manifests as increased ingestion (Dam & Peterson, 1988); and if food is not abundantly available, suboptimal performance (Isla, Lengfellner, & Sommer, 2008). Under this scenario, there lies the probability of organisms experiencing energetic strain or migrating to new habitats. The potential for organisms to adapt to new imposed thermal conditions at the current rate of warming is unlikely as transgenerational adaptation or evolution occurs significantly slower than the expected rate of climate change (Munday, Warner, Monro, Pandolfi, & Marshall, 2013). On account of this, there is Hegel likelihood that some reduction in fitness or migration will occur, and both cases could negatively impact the organism or associated ecosystems resulting in overall strain on marine biodiversity.

The major impacts of OW occur in the form of rising sea-levels caused by both thermal expansion of the heated seawater and melting glaciers, as well as thermal stress on the oceans' ecosystems (Mimura, 2013). While rising sea-levels are a pressing issue directly affecting humans and terrestrial habitats, OW may pose a greater threat to marine ecosystems in the form of thermal stress (Vegas-Vilarrubis & Rull, 2016). Some probable outcomes of ocean heating on organisms, inter alia, include: movement of organisms toward relatively cooler poles; aberrations in the timing of key biological events spurred by temperature cues; alterations in food availability owing to inconsistent effects of warming on predators and their prey; and population declines or extinctions (Yao & Sumero, 2014). These outcomes occur because organisms have adapted and evolved to live within a limited thermal range. Existence of these ranges is attributed by Portner (2001) to optimization of organisms' physiological systems over time to operate within a specific range of temperatures. Operating outside of this range is prone to result in decreased efficiency of metabolic performance, necessitating organisms to allocate more energy toward homeostasis. This range is defined by Deutsch et al. (2015) as an upper and lower critical temperature, known as the pejus temperature, and an optimal temperature. Outside of the pejus range, aerobic performance—defined as oxygen consumption per unit respiring mass—is reduced, and subsequently worsened when moving further toward the critical range (Portner, 2001)' and outside the critical range, organisms try to temporarily change their metabolic state to endure the extreme environmental stress by procuring energy through anaerobic systems and slowing overall metabolism (Pörtner & Farrell, 2008). Apart from varying between species, thermal range also varies within species throughout different life stages making some portion of a given population more vulnerable to temperature changes (Pörtner & Farrell, 2008). A direct outcome of OW ocean is discernible in rising sea levels in the different parts of globe.

8.2.2 Global sea level rise

Numerous outcomes of global warming, especially an enhancement in the Earth's mean surface temperature and OHC (IPCC, 2013a, 2013b), melting of sea ice, loss of mass of glaciers

(Gardner et al., 2013), and ice mass loss from the Greenland and Antarctica ice sheets (Rignot, Velicogna, van den Broeke, Monaghan, & Lenaerts, 2011), are becoming discernable. On average over the past five decades, approximately 93% of heat excess amassed in the climate system due to GHG emissions has got stockpiled in the ocean, and the remaining 7% has been heating the atmosphere and continents and melting sea and land ice (von Schuckmann et al., 2016). Rise in sea levels occur owing to OW and loss of land ice mass. Lea level had remained almost constant since the end of the last deglaciation nearly 3000 years ago (Lambeck et al., 2010). Nevertheless, Global mean sea level (GMSL) started rising in the 20th century, as can be discerned from the direct observations from an in-situ tide gauges available since the mid-to-late 19th century, at a rate of 1.2—1.9 mm/yearr (Dangendorf et al., 2017); and interestingly, measurement of sea level rise is measured, since the early 1990s, by high-precision altimeter satellites and the rate has increased to $\sim$ 3 mm/year on average (Legeais et al., 2018; Nerem et al., 2018).

From 1901 to 2010, the global rise in sea level has been recorded around 200 mm, amounting to an average rate of 1.7 mm/annum (IPCC, 2013a, 2013b), and over this period the rate has recorded an increase and currently it stands at 3.2 mm/annum (Chambers et al., 2016). Currently, the thermal expansion of the world's oceans, that is, the volume of water simply owing to warming, having recoded 40% of the spike from 1993 to 2015, is seeming the largest contributor to OW (Edwards, 2017). Rest of the spike in OW is attributable to losses of land ice from glaciers (about 25%) along with 20% loss of ice sheets in Greenland and the Antarctic; along with transfer from land water sources like groundwater and snow (about 15%), with the ice sheet contributions recording an increase during this period (Edwards, 2017). Slangen et al. (2016) have estimated that nearly 70% of sea level rise from 1970 to 2005 is attributable to human activities. Shepherd et al. (2010) are of the view that owing to fact that floating ice does not add extra volume when it melts; therefore, alterations in Arctic and Antarctic Sea ice has made scant contribution to sea level rise. Undoubtedly, global sea level is expected to register to continue rising; nonetheless, projections in this regard vary widely. IPCC (Intergovernmental Panel on Climate Change) (2013a, 2013b) assessments about sea level rise during the 21st century varied from nearly 25 cm to 1 m, depending on the scenario of GHG concentrations and range of modeling uncertainties, with the largest projected contributions emanating from thermal expansion (around 30–55%), to be followed by glaciers (nearly 15–35%), deriving from up to 85% of the current volume of glaciers outside Antarctica (Church et al., 2013). Interestingly, a few individual projections are in consonance with around 1.5–2.5 m (DeConto & Pollard, 2016), primarily owing to bulk contributions from the Antarctic ice sheet; nonetheless, there is lower confidence in these estimates (Edwards, 2017).

Sea level change is driven by multiple mechanisms operating at different spatial and temporal scales (Kopp, Hay, Little, & Mitrovica, 2015). GMSL rise is mostly triggered by two factors: (1) augmented volume of seawater due to thermal expansion of the ocean as it warms and (2) increased mass of water in the ocean owing to melting ice from mountain glaciers and the Antarctic and Greenland ice sheets (Church et al., 2013). The overall amount (mass) of ocean water, and thus sea level, is also impacted to a lesser degree by alterations in global land-water storage, thereby reflecting changes in the impoundment of water in dams and reservoirs and river runoff from groundwater extraction, inland sea and wetland drainage, and global precipitation patterns, such as take place during phases of the El Niño–Southern Oscillation (ENSO) (Wada, et al., 2016). Variations in sea level and its changes globally are attributable to various reasons. First, atmosphere-ocean dynamics, spurred by ocean circulation, winds and other factors, are linked to differences in the height of the sea surface, as are the variations in density emanating from the disbursement of heat and salinity in the ocean Fasullo & Gent, 2017). Alterations in any of these factors are prone to impact sea surface height. For instance, an enfeebling of the Gulf Stream transport in the mid-to-late 2000s might have added to increased sea level rise in the ocean environment stretching to the northeaster United States coast (Ezer, 2013), a trend that would continue in the future as projected by many models (Yin & Goddard, 2013). In the opinion of Mitrovica et al. (2011), locations of land water ice melting and land water reservoir alterations covey discrete regional "static-equilibrium fingerprints" on sea level, gravitational, rotational, and crustal deformation effects. For instance, fall in sea level occurs beneath a melting ice sheet owing to the reduced gravitation of the ocean toward the ice sheet; reciprocally, it records a rise by greater that the global average far from the melting ice sheet (Sweet, Horton, Kopp, LeGrande, & Romanou, 2017). An array of factors like natural sediment compaction, compaction triggered by local extraction of groundwater and fossil fuels, and processes pertaining to plate tectonics, such as earthquakes and more gradual seismic creep entail the potential of causing vertical land movement (Wöppelmann & Marcos, 2016).

8.2.2.1 Impact of sea level rise on global areas

Coastal areas around the globe, especially the low-lying coastal areas, are vulnerable to sea level rise (Rowley, Kostelnick, Braten, Li, & Meisel, 2007). In 2010, more than 1.9 billion people lived in coastal areas, and their number is projected to reach 2.4 billion by 2050 (Kummu et al., 2016). 20 of the world's 30 megacities are located on the coasts and are expected to register increase in population faster than nonurban areas (Kummu et al., 2016). According to Grimm and Tulloch (2015), the three fastest growing coastal megacities are Lagos, Nigeria (having 4.17% population growth rate), Guangzhou, China (with 3.94% population growth rate); and Dhaka in Bangladesh (having 3.52% population growth rate). Encompassing only covering 2% of the world's land area (Bretch et al., 2013), vulnerability of low-lying coastal areas to sea level rise is likely to further accelerate in the wake of response to climate change. According to IPCC (2013a, 2013b), the average sea level rise is likely to go up to be between 28–61 cm and 52–98 cm (representative concentration pathway, RCP2.6 and RCP8.5, respectively) by the year 2100. Recent decades have witnessed sea level rise out-pacing he 20th century average, recording rates as high as 3 mm/annum since 1993 (Kopp et al., 2016), and spurred mainly by thermal expansion of ocean and loss of glaciers and ice-sheets (Church et al., 2013). The direct environmental major impacts on seal level rise are fourfold: coastal flooding; coastal erosion; exacerbated land subsidence; and saltwater intrusion (Vitousek et al., 2017). Admittedly, these impacts already exist in coastal areas; nonetheless, sea level rise further accelerates their severity. Besides, these environmental impacts also entail the potential of damaging coastal real estate, infrastructure, spurring the breakdown of coastal economies, and driving people's migration and industries from the coastal regions (Nicholls et al., 2008), including the Atlantic and Gulf shores of the coastal United States (Hauer, 2017).

The upcoming years are likely to witness population growth and increased migration from rural areas, especially low-lying coastal areas, adjacent to sea at less than 10 m height resulting in higher population density and greater pace of urbanized land, as part of the global phenomenon and that is going to be critical for developing countries of Asia and Africa (Hugo, 2011; Neumann, Ott, & Kenchington, 2017). This portends a great challenge, especially in view of the projections that seashore erosion and hinterland flooding are likely to continue to rise in coming years in various places (Barbier, 2015; Güneralp, Güneralp, & Liu, 2015; Thompson, Karunarathna, & Reeve, 2017;

Wadey, Nicholls, & Hutton, 2012). Anthropogenic activities like land use changes, hydrological or coastal infrastructures, land claim, groundwater extraction, etc., entail the potential of altering coastal response to this hazard along with impacts of global phenomena like climate change (Klein, Nicholls, & Thomalla, 2003; Sánchez-Arcilla et al., 2011; Tsu et al., 2017). Construed from this perspective, global warming could strongly alter or modify the hydrodynamic forcings that spur flooding and erosion via sea level and wave climate. In this manner, estimates by IPCC (Intergovernmental Panel on Climate Change) (2013a, 2013b) demonstrate that rise in global sea level by the year 2100 could range between 0.52 and 0.98 m with respect to the 1986–2005 period, for the highest emission RCP8.5 scenario. However, this is not a definite interval because according to Church et al. (2013), sea level rise has a 33% probability to lie outside this 5%–95% confidence interval, being the key source of uncertainty the breakdown of the Antarctica ice sheet areas. Accordingly, Jevrejeva, Moore, Grinsted, Matthews, and Spada (2014) recommend an upper limit or worst-case projection for the global sea level rise of 1.80 m by 2100, taking into account the high-end estimations of each sea level rise contributor for the RCP8.5 scenario. According to Grases, Gracia, Garcia-Leon, Lin-Ye, and Sierra (2020), this estimation would have a 5% probability of being exceeded. In contrast, 2070–99 extreme wave climate projections under the RCP8.5 scenario, demonstrate an increase of ocean wave heights focused at the tropics and at the highest latitudes of the southern oceans (Grases, et al., 2020). Furthermore, the return period of current storm events could be reduced in various coastal regions of both southern and northern hemispheres, enhancing flood and erosion risk (Wang, Feng, & Swail, 2014).

Major outcome of sea level rise is an enhancement in the exposure of coastal communities to flooding that can lead to reduction in recreational beach areas because more and more dry beach gets submerged during tidal cycles (Mcleod et al., 2010). Structures like docks and piers also entail the potential of being submerged in the event of flooding. Vulnerability of low-lying coastal areas to storm surge and flooding from torrential precipitation is always there, and in the incoming years these effects are prone to get intensified (Castrucci & Tahvildari, 2018). Massive flooding is already being caused by intense precipitation events and the quantum of rain falling in on the wettest day is projected to increase 10%–20% by 2100 (Walsh et al., 2014) and rise in sea level will result in higher water tables and decreased soil storage capacity. With water tables approaching the land surface in the wake of sea level rise, the frequency and intensity of flooding induced by precipitation is likely to increase; therefore, no delay is brooked by strong storm to spur coastal flooding (NOAA, 2012). High tides in many locations in the wake of sea level rise, land subsidence and the loss of natural barriers drive flooding. Nuisance flooding, that pertains to low levels of inundation that do not portend significant threats to public safety or cause major property damage but can disrupt day-to-day routine activities (Moftakhari, Aghakouchak, Sanders, Allaire, & Matthew, 2018), has, according to NOAA (2012), recorded a significant increase in the United States coast since the 1960s, and causes road closures and saltwater backup in storm drains and exposes infrastructure. The fact that a large number of coastal communities are already exposed to coastal flooding is attested by the fact there are 136 major port cities with more than one million population each, and 13 of these are among the top 20 most populated cities throughout the globe (Bosello & De Cian, 2014). It has been suggested by Nicholls et al. (2008) that exposure of coastal cities to flooding is projected to register an increase with burgeoning populations and the enhanced economic significance of coastal cities, specifically in developing countries. It has been also demonstrated by Hebergr (2014) that nine counties in California, in the vicinity of the Pacific Ocean, are confronted with threats owing to sea level rise and the following decades are likely to witness increase in their susceptibility.

Seal level rise also contributes to erosion of coastal and shoreline because there exists a direct relationship between sea level rise and the reduction in coastal shorelines and this fact is attested to by past studies that as seal level rises, it triggers massive erosion of coastal beaches, rocks, and low-lying areas (Zhang & Leatherman, 2011). Various studies over the past decades have related coastline retreat with seal level rise (FitzGerald, Fenster, Argow, & Buynevich, 2008), although other pertinent also been identified (Passeri et al., 2015). Entailing significant environmental effects, coastline retreat has also socio-economic implications affecting population, infrastructures and other assets (Enríquez, Marcos, Álvarez-Ellacuría, Orfila, & Gomis, 2017). The impact of sea level rise in the shoreline position is increasingly becoming a Passer matter of serious concern, especially the coastal areas that are located in densely populated regions and endowed with high urban development (Kolker, Allison, & Hameed, 2011). Many regions in the Mediterranean have economies that comprise about 14% of the total gross domestic product (GDP) of the European Union (Eurostat, 2011), and these economies are largely dependent on tourism based on beach and other seaside recreational activities. Therefore, sea level rise and its potential impacts, according to Enríquez et al. (2017), are main factors that ought to be included in coastal risk management and climate change adaptation strategies.

Land subsidence, a process that occurs when substantial amounts of groundwater are withdrawn from certain types of rocks, such as fine-grained sediments (USGS, n.d.a, n.d.b), also contributes to sea level rise with resultant consequences for coastal areas. The rock compacts owing to water that is partly responsible for holding the ground up and in the aftermath of withdrawal of water, the rock falls on itself. Land subsidence may not be noticeable too much because it can occur over large expanse of area rather than in a smaller spot like a sinkhole; and land subsidence is almost a big event that has caused immense loss worth hundreds of millions of United States dollars to the states like California, Texas and Florida of the United States over the years (USGS, n.d.a, n.d.b). As an age-old phenomenon, land subsidence is caused by shifting of Earth, in response to ice-melt, known as glacial-isostatic adjustment (GIA) and it is major driver of land subsidence, especially for much of New England, and according to Karegar, Dixon, Malservisi, Kusche, and Engelhart (2017), parts of the East Coast in the United States are witnessing as much as 1.22 mm of subsidence every year due to GIA. Another long-term process causing subsidence is compaction, which is the compression of sediments. Regarded as a natural process, compaction occurs to most sediment profiles overtime (Nawaz, Bourrie, & Trolard, 2013); and interestingly, certain parts of the United States are very prone to compaction owing to their organic-rich environments. And this a major problem for large coastal salt marshes in New Orleans and the greater Mississippi Delta in the United States, where these salt marches may undoubtedly help in defending the coast from storm surge; nonetheless, its composition leaves it vulnerable to sea level rise. According to Törnqvist et al. (2008), rates of compaction in parts of the Mississippi Delta could be as high as 10 mm/year.

Another major impact caused by sea level rise is to facilitate seawater intrusion that can be described as a mass transport of saline water into aquifer zones and estuaries where saltwater converges with freshwater and it is primarily controlled by tide and river discharge (Geyer, 1993); nevertheless, it can also be affected by wind stress (Giddings & Maccready, 2017) and vertical mixing (Prandle & Lane, 2015). Intrusion of seawater into coastal groundwater aquifers and estuaries affects not only the availability of fresh drinking water supply but can also produce estuarine circulation (Pritchard, 1956) and affect stratification (Sipson et al., 1990), thereby impacting sediment transport, producing peak estuarine turbidities (Geyer, 1993), and degrading the freshwater quality (Zhu, Gu, & Wu, 2013). There are multiple impacts of seawater intrusion in coastal aquifers (Werner et al., 2013). The progression of saltwater intrusion is sensitive to coastal topography and water management (Carrera, Hidalgo, Slooten, & Vázquez-Suñé, 2010), type of inland boundary conditions (Werner & Simmons, 2009), geological heterogeneity of coastal aquifers, initial salt distribution and relative sea surge. Undoubtedly, the multiple impacts arising out of saltwater intrusion in coastal aquifers and estuaries have been addressed in various studies; nonetheless, research combining them all is rare (Meyer, Engesgaard, & Sonnenborg, 2019), and there has been lack of multidisciplinary approaches integrating different types of observational data and numerical modeling (Werner et al., 2013). Vulnerability of coastal topography and inland boundary to saltwater intrusion has been focused on in the literature (Michael, Russoniello, & Byron, 2013). Findings by Michael et al. (2013) classified up to 70% of the global coastline as topography limited and coastal areas, where drains control the groundwater level, behave in an identical manner to topography-limited systems. Interestingly, extensive studies in recent years have been undertaken with regard to large drained coastal systems in the Dutch Polder areas that are greatly impacted by saltwater intrusions (Pauw, De Louw, & Oude Essink, 2012). Emphasis has been focused on the significance of the role of geological architecture and hydrogeological properties of coastal aquifers in exerting control on the occurrence and progression of saltwater intrusion (Carrera et al., 2010), and concurrently some modeling studies also probed this aspect by making use of synthetic geological data (Michael et al., 2013) or for particular locations (Nishikawa et al., 2009). While pointing out that the geological complexity was often simplified, particularly for regional case studies, Meyer et al. (2019) and Abarca (2006) opine that only the dominant units and structures were taken into consideration. It has been suggested that geological traits like buried valleys (Jørgensen & Sandersen, 2006; Kehew, Piotrowski, & Jørgensen, 2012; Piotrowski, 1994) can be helpful in providing preferential flow paths for saltwater (Mulligan, Evans, & Lizarralde, 2007). Some scholars (Jørgensen et al., 2012; Kaleris, Lagas, Marczinek, & Piotrowski, 2002) have been of the view that a detailed description of the coastal geology can be critical for the understanding of exchange of saltwater between sea and aquifers. Conceding that considerable research has been focused on understanding the role of geological heterogeneity on salt water intrusion by making use of two-dimensional models, Meyer et al. (2019) laments that little attention has been devoted to representing complex geological structures in the large-scale three-dimensional density-dependent flow and solute transport models and how such models impact saltwater intrusion.

8.2.3 Ocean acidification

OA denotes a suite of chemical reactions in the ocean, triggered by increasing CO_2 levels in the atmosphere (Royal Society, 2005), it is often used in the context of CO_2 emissions caused by anthropogenic activities. The ocean has assimilated approximately one-third of anthropogenic CO_2 from the atmosphere since the advent of industrial revolution (IPCC (, 2013a, 2013b; Rosane, 2019). This contributes to diminutions in pH (i.e., increase in hydrogen ion ($H+$) concentration) and carbonate ion concentration and an increase in bicarbonate ion concentration (Dickson, 2016). Absorption of CO_2 by ocean has proved instrumental in the diminution of ocean pH by nearly

0.1 unit; with pH being a measure of how acidic/basic water is and range of this measure goes from 0 to 14, with 7 being neutral and pH of less than 7 indicates acidity (USGS, n.d.a, n.d.b), over two centuries, primarily altering ocean carbonate chemistry across all ocean areas. Increase in CO_2 emissions into the atmosphere is prone to culminate in the decrease of pH and carbonate ions (Hopkins et al., 2020). Both are significant because many, if not all, marine organisms spend their energy in regulating their internal pH and carbonate ions are indispensable for shell and skeleton formation (Spalding, Finnegan, & Fischer, 2017).

Alterations in the seawater chemistry are being triggered by augmentation in atmospheric CO_2, owing to the uptake of CO_2 by the oceans (NAP, 2010). This uptake of the atmospheric CO_2 by the oceans has culminated in a shift in seawater carbonate chemistry that comprises a diminution in oceanic pH (i.e., $-\log [H+]$) and this has led to what is now known as OA; and owing to a three-fold enhancement in $H+$ (as shown in equilibrium reactions below), the pH has already decreased by 0.1 units since preindustrial times.

$$CO_2 + H_2O \leftrightarrow H_2CO_3 (\text{carbonic acid}) \tag{1}$$

$$H_2CO_3 \leftrightarrow H + + HCO_3 - (\text{bicarbonate ion}) \tag{2}$$

$$H + + HCO_3 - \leftrightarrow 2H + + CO3 2 - (\text{carbonate ion}) \tag{3}$$

Average surface ocean pH is projected to register a reduction by 0.3−0.5 units by 2100 in the wake of enhanced levels of atmospheric CO_2 (Tahil & Dy, 2015), and predictions based on the emission scenarios of the IPCC special report on emission scenarios forecast decreases in average global surface ocean pH of between 0.14 and 0.35 units over the 21st century, contributing addition to the present reduction of 0.1 units since preindustrial times (IPCC, 2014a, 2014b). Alterations in in ocean carbon chemistry are occurring at a rapid pace, in any case 100 times faster than any experienced over the past 100,000 years (CBD, 2009).

OA is described by Richard Feely as "global warming's evil twin" (Earthzine, 2015). There has been an increase in global surface temperature by nearly 0.76°C over the past 150 years, and one side impact of ocean heat absorption is increased ocean warming; accordingly, the warming corresponds to an enhancement in mean ocean temperature of 0.31°C during these 150 years (Levitus et al., 2012). Projected enhancement in global temperatures triggered by anthropogenic CO_2 emissions could probably translate into increase in OW and SSTs are predicted to register an increase by a further 3°C−5°C by the end of this century in the North Atlantic (IPCC (Intergovernmental Panel on Climate Change), 2014a, 2014b). OA and OW are worldwide phenomena that will continue to increase for many centuries (CBD, 2009; Tyrrell, 2011). Any harm or damage caused by OA and OW to the recruitment success of major organisms to an ecosystem that is prone to entail grave upshots for that ecosystem because retrieval of the main organisms is improbable except the larvae possess the capacity to manifest enough phenotypic plasticity or adaptability to the swift change in CO_2 and temperature (Eriksson, Hernroth, & Baden, 2013).

OA entails the potential of impacting the physiology, energetic apportionments for diverse metabolic processes of organisms, decrease the ability of calcifiers to engender calcium carbonate structures, and also affecting long-term fitness of some species within marine food web (Wicks & Roberts, 2012). Marine calcifying organisms seem specifically at risk in the wake of OA because additional energy would be needed to produce shells and skeletons, and in many ocean areas like polar and subpolar waters unsafe shells and skeletons face the probability of getting dissolved

(IPCC (Intergovernmental Panel on Climate Change), 2014a, 2014b). Acid—base disproportion can occur due to OA in many marine organisms like fish invertebrates and sediment fauna, nevertheless, some species can adjust or modify energetic apportionment to reimburse for enhanced energy costs of OA, albeit they may require extra food resources to accomplish this (Lloyd-Smith & Immig, 2018). Infantile or juvenile species seem more vulnerable than adults (Ishimatsu & Dissanayake, 2010), as for instance, larval oysters of the west coast of North America are already being affected with economic costs to hatcheries (Barton, Hales, Waldbusser, Langdon, & Feely, 2012) and the sensory systems and behavior of coral reef larval fish seen tender or sensitive to OA levels that will take place this century if mitigation measures are not put in place (Barth et al., 2015; Munday, Cheal, Dixson, Rummer, & Fabricius, 2014).

According to Gattuso, Mach, and Morgan (2013), there is high scientific confidence in the knowledge of the fundamental chemical processes, the cause, speed and magnitude of change and its future progression based on CO_2 emission scenarios. OA is measurable and is occurring now and at an unprecedented rate and magnitude not visible on Earth for many millennia (Benarsek et al., 2014; Ekstrom et al., 2015). OA is a simultaneous problem with a common cause to climate change. Both are spurred by augmented anthropogenic CO_2 emissions into the atmosphere at a rapid pace than natural deletion processes. The scale of future OA at the global level will tightly be contingent on the magnitude of future CO_2 emissions due to anthropogenic activities (CBD, 2014, 2010). Diminution in CO_2 emissions provides a double advantage of both decreasing OA and climate change (Pörtner et al., 2014).

There is insufficient clarity about effect of OA on marine biodiversity, food webs, biogeochemical processes, ecosystems and society as comparable to changes in ocean chemistry (CBD, 2014); however, it is probable that some will be negatively affected (IPCC, 2014a, 2014b). This portends a potential risk to human society via probable impacts on goods and services accruing from the ocean (IPCC, 2014a, 2014b). It can be discerned, for instance, from some studies that there is a decrease in growth, formation and maintenance of coral reefs with enhanced OA that will affect the goods and services accruing from them (Hilmi et al., 2015). The most noticeable of these are food and livelihoods provision from fisheries and aquaculture (Turley & Boot, 2010), storm protection from reefs and economic benefits from tourism (Turley & Boot, 2011). Estimates, as reflected in some studies, showed that an annual economic damage of OA-induced coral reef loss by 2100 were projected to be US$870 and US$528 billion, respectively for the A1 and B2 SRES Emission Scenarios, constituting a big chunk of GDP loss for the economies of many coastal regions or small islands that are dependent on the ecological goods and services of coral reefs (Hoegh-Guldberg et al., 2014). Nevertheless subsequently, Gattuso et al. (2015) showed that coral reefs are the most sensitive marine ecosystems notwithstanding the IPCC scenario (RCP2.6/RCP8.5), and currently are the most threatened ecosystems globally as well as coral reef ecosystems distinctly illustrate the close relationship between biology, ecology, human societies and global changes (Allemand & Osborn, 2019). Undoubtedly, possessing the capacity of absorbing 93% of the extra energy from the increased GHG effect, the oceans have to come a pivotal role in Earth's climate; nonetheless, the consequential warming of the ocean impacts most marine ecosystems (Pörtner et al., 2014). The amalgamation of warming and OA entail the potential of leading to the quietus of most coral-based ecosystems (Mostafal et al., 2016). Both warm water coral reefs and cold-water reefs are primary habitat builders or originators that support biodiversity as well as fisheries (Hebbeln, Portilho-Ramos, Weinberg, & Titschack, 2019), and their depletion would also represent a major loss to Earth's biological heritage (Obura, 2017). World is already experiencing the impacts of OA in different parts of the globe, especially on the west coast of North America where it is affecting valuable shellfisheries (Barton et al., 2012) and in Bay of Bengal as well (Hossain, Chowdhury, Sharifuzzaman, & Sarker, 2015), and natural wild populations of sea butterflies, also known as pteropods, in the Southern Ocean (Mannoa et al., 2017); and pteropods constitute a primary link in the food web that supports wild salmon. Apart from projections about worldwide cost of production loss with regard to mollusks exceeding over US$100 billion by 2100, models suggest that OA would often lead to reduction of fish biomass and catch and that subtle chemical or additive, incompatible and/or synergistic or interactive interactions would take place with other environmental or warming and human, pertaining to fisheries management, factors (Pörtner et al., 2014; Yu, 2019).

8.2.4 Ocean deoxygenation

Deoxygenation is "the reduction in oxygen content of the ocean due to anthropogenic effects" (Laffoley & Baxter, 2019, p. xvi). Increasing ocean deoxygenation or loss of oxygen, including expansion of oxygen minimum zones (OMZs), is a potentially significant outcome of global warming (Levin, 2018). Ocean deoxygenation is increasingly acknowledged as a key environmental threat (Breitburg et al., 2018) and a substantial part of this deoxygenation is driven by global warming (Keling et al., 2008) and that is projected to increase in ensuing decades, both in the open ocean and in the coastal waters. This trend is likely to get further exacerbated from the future enhanced anthropogenic pressures, especially from eutrophication (Fennel & Testa, 2019). Physical and biogeochemical processes administer the dynamics of oxygen in the ocean. Photosynthesis by autotrophic organisms helps ocean gin oxygen in the upper layer as well as oxygen from the atmosphere dissolving in the undersaturated waters (Gregoire, Gilbert, Oschlies, Rose, 2019). On the contrary, the ocean loses oxygen throughout the entire water column; at the surface owing to the outgassing of oxygen to the atmosphere in over saturated waters, and from the surface to depths owing to the respiration of aerobic organisms and oxidation of reduced chemical species (GOON), 2018; Gregoire et al., 2019). It has been suggested that the global ocean oxygen inventory has been disturbed since the mid-20th century (Oschlies, Brandt, Stramma, & Schmidtko, 2018). According to Gregoire et al. (2019), it can be assumed that oxygen generated by photosynthesis as a net primary production (NPP) in the upper layer of the ocean is unevenly consumed by respiration within the water column, but for a slight production of 0.002 Pmol O_2/year that resembles or corresponds to interment (Wallmann, 2000). An oxygen loss fluctuating from 0.048 Pmol O_2/year (Manning & Keeling, 2006) to 0.096 Pmol O_2/year (Schmidtko et al., 2017) ensues from an alteration of the equilibrium in the atmosphere-ocean fluctuations with a declining influx from the atmosphere to the ocean and an enhancing outflux from the ocean (Frölicher, Joos, Plattner, Steinacher, & Doney, 2009). Nevertheless, Gregoire et al. (2019) lament that the dearth of observations with adequate resolution of sampling in space and time limits the rigor of our oxygen budget for the coastal zone.

Multiple oxidation stages have facilitated evolution during the course of the Earth's history. Innumerable millennia ago, the ocean was anoxic and subsequently got slightly oxygenated in its upper layer and, ultimately adequately oxygenated in its present state (Armstrong, Frost, McCammon, Rubie, & Ballaran, 2019; Partin et al., 2013). Evidence is increasing with regard to declining oxygen in modern ocean (Breitburg, et al., 2019). Estimates put oxygen decrease nearly

1%–2% and that denotes 2.4–4.8 Pmolor 77–145 billion tons of the global oxygen inventory since the mid-20th century (Breitburg et al., 2018). This has culminated in alterations of the symmetry or equilibrium state of the ocean atmosphere attached system with the ocean freshly emerging a source of oxygen for the atmosphere. Interestingly, areas that have historically been endowed with very low oxygen concentrations are increasing numerically while new regions are now demonstrating low oxygen conditions (Gregoire, et al., 2019). Scholarly estimates show that the volume of anoxic waters have globally quadrupled since 1960 (Breitburg et al., 2018; Gregoire et al., 2019). Undeniably, there prevails uncertainty about various factors responsible for the loss of the ocean oxygen content (GOON (Global Ocean Oxygen Network), 2018); nonetheless, global warming is widely projected to contribute to ocean oxygen decline, directly because the dissolving ability of oxygen in warmer waters gets reduced, and indirectly via alterations in ocean dynamics that cause reductions in ocean ventilation (Shepherd et al., 2017). Current ocean processes are redolent of those supposed to have spurred the occurrence of ocean anoxic events (OAEs) that took place periodically during the past hundreds of millennia that culminating in causing major extinction events (Watson, 2016). Development of a full-scale OAE takes millennia of years to occur (Jenkyns, 2010); nonetheless, small oxygen inventory of the ocean that is roughly ∼0.6% of that of the atmosphere (Gregoire et al., 2019), renders it specifically sentient of trepidations of its equilibrium oxidative state.

The equilibrium between the supply and the consumption of oxygen determines the oxygen content of the ocean interior (Del Giorgio & Duarte, 2002). Bacterial respiration of labile organic matter or remineralization is the major driver of ocean consumption of oxygen (Robinson, 2019), Because of labile organic matter facilitating supply by export production (EP), an augmentation in in EP is liable to result in oxygen depletion (Danovaro, Fabiano, & Croce, 1993). On the contrary, the supply of oxygen to the ocean interior takes place via ventilation, the process whereby oxygen-rich surface waters are transported into the ocean interior through subduction (Yamamoto et al., 2015). Hence, the oxygen content of the subducted surface waters as well as their subduction rates and circulation influence the ocean interior oxygenation. Temperature and wind intensity drive the transfer of oxygen from the atmosphere to the surface of ocean, where strong winds and low temperature favor mixing and oxygen enrichment (Jaccard, Galbraith, Froelicher, & Gruber, 2014). Furthermore, the generation of dissolved oxygen by phytoplankton photosynthesis comprises another process supplying dissolved oxygen to the mixed layer. The subduction of these surface waters, ventilating the ocean interior, occurs mostly through the formation of intermediate water masses, namely sub-Antarctic Mode Water (SAMW) (Jones et al., 2016) and Antarctic Intermediate Water (AAIW) (Talley, 1996) in subpolar regions of the Southern Ocean (Sloyan & Rintoul, 2001). Of the two water masses, AAIW is volumetrically the largest and dominates the oxygen supply to the ocean interior at low latitudes (Piola & Georgi, 1982). In view of the primary role of AAIW in ventilating the ocean interior, there has been growing interest in understanding the drivers of its formation; nevertheless, large uncertainties still remain in this regard. However, it was during the 1930s that the AAIW was first described as a fresh water mass of the ocean interior (Deacon, 1933; Wust, 1935) and subsequent years have witnessed identification of multiple probable mechanisms of AAIW formation (Sloyan, Talley, Chereskin, Fine, & Holte, 2010); nonetheless, these are still under debate (Bostock, Sutton, Williams, & Opdyke, 2013; Gu et al., 2017). According to some scholars, there is one theory that the formation of AAIW is facilitated by convection during the Southern Hemisphere winter (Sloyan et al., 2010). Antarctic Surface Water

(AASW), after interacting with the atmosphere gets fresher and cooler thereafter resulting in the augmentation of the density of AASW and triggers the convection process that results in AAIW subduction. Before the subduction occurs, the northward advection at the surface is the outcome of Ekman transport, named after Swedish scientist V. Walfrid Ekman, entails ocean water at the surface moving at an angle to the wind, and the water under the surface water turning a bit more, and the water below that turning even more, and the average direction of all this turning is about a right angle from the wind direction and this average in called Ekman transport (Bravo, Ramos, Astudillo, Dewitte, & Goubanova, 2015), is usually triggered by the westerly winds (Lachkar, Orr, Dutay, & Delecluse, 2007, 2009). Besides, some modeling studies have emphasized the vital role of shallow bathymetric characteristics to explain as to why AAIW formation only takes place in a few hot spots in the Southern Ocean (Bostock et al., 2013).

Alterations in westerly wind stress are reported to be as key factor impacting AAIW ventilation (Downes, Budnick, Sarmiento, & Farneti, 2011). While, on the one hand, the severity of westerly wind controls AAIW subduction, enhanced winds lead to augmentation in subduction and increased ventilation (Downes, Bindoff, & Rintoul, 2010); whereas, on the other hand, the position of the westerlies also impacts AAIW subduction, where a southward shift in the mean position of the westerlies increases AIAW ventilation (Downes et al., 2010). Nevertheless, enhancement in temperatures and rainfall wield opposite effects to an enhancement in resilience or buoyancy that prevents AIAW ventilation (Schmidtko & Johnson, 2012). In sum, the AIAW formation is influenced by a complex combination of drivers and the resultant delivery of oxygen to the ocean interior, and the impacts of these climate change on these drivers and on the AAIW is yet to be fully ascertained (Meijers, 2014). Paleo-oceanographic studies, on the basis of reconstructing past fluctuations in oxygenation and their drivers, entail vital potential in making better predictions about future oxygen changes. No seawater from the past remains trapped in an unchanged state comparable to past atmospheric gases that can be preserved in bubbles trapped in ice sheets. Relatively, sediments constitute the best archive recording oceanic oxygenation changes (Rapp., 2013). The fluctuations of redox sensitive element concentrations in the sediments such as Uranium (U) and Rhenium (Re), have been extensively used to spot past oceanic alterations (Cole, Zhang, & Planavsky, 2017). Solubilities of U and Re in seawater diverge with the dissolved oxygen concentrations of the seawater (Morris, Bueseller, & Sims, 2011). Decline in oxygen concentration in seawater leads to reduction in U and Re from their soluble forms ($UO_2(CO_3)3)4-$, $U(OH)4$, and $ReO-4$) toward insoluble forms (UO_2 and $ReO_2.2H_2O$) (Durand et al., 2018). This results in accumulation of U and Re phases in the sediments. In other words, decrease in the concentration of dissolved oxygen in the water overlaying the sediment bed leads to increase in the concentrations of U and Re in sediments. This signal is then preserved to be used as a proxy to disclose past oceanic oxygen changes (Glud, 2008). Nevertheless, U and Re concentrations in sediments only mirror changes in oxygenation of surface waters. Another drawback that afflicts it is that the relationship between U and Re solubilities and dissolved oxygen is not linear and is proximate to a tipping point because the precipitation of the U and Re solid phases takes place only when a redox threshold is reached (Tribovillard et al., 2006). Consequently, U and Re can only be used as qualitative proxies for oxygenation change (Jorissen, Fontainer, & Thomas, 2007).

Estuary-coastal Ocean continuum, also called coastal ocean, is affected by the deoxygenation process (Rabalais, 2019). Worldwide increase of the nutrients brought by rivers from land to the ocean and resulting growth of eutrophication in multiple sites around the globe has proved instrumental in promoting algal productivity and the ensuing overconsumption of oxygen in the bottom layer that is effectually kept apart via stratification from being renewed with the oxygen in the surface waters (Gregoire, 2019). Climate warming is projected to degenerate the situation by further decreasing the ventilation of bottom waters (Hoegh-Guldberg et al., 2018). Availability of the habitat for pelagic, mesopelagic and benthic organisms is projected to be reduced in the global and coastal oceans by the deoxygenation process (Seibel & Wishner, 2019). Ocean deoxygenation casts strong effects on the rates and pathways of the split of organic matter, and the accompanying reactions envisage striking changes in the sources, sinks and cycling of an array of important elements in the environment; and furthermore this process of ocean deoxygenation affects the biologically important elements like nitrogen (N), phosphorus (P) and Iron (Fe) but also for the generation of nitrogen (C) and gases that contributes to Earth's warming by the greenhouse effect in the form of nitrous oxide (N_2O) and methane (CH_4) (Conley & Slomp, 2019). In view of the fact that mesopelagic community structure is directly dependent on the availability of oxygen for aerobic metabolism; therefore, variations in oxygen at both large and small scales are prone to influence diversity, abundance, distribution and composition of mesopelagic species, and as such there is likelihood of ocean deoxygenation leading to decrease in the minimum oxygen content in the mesopelagic zone and cause oxyclines to shift vertically (i.e., expansion of the OMZ core) in the water column (Seibel & Wishner, 2019). With climate change predicted to change oxygen concentrations throughout the open ocean, with most regions undergoing diminution owing to a slowdown in ocean ventilation and a decline in surface oxygen solubility, tinas and billfishes entail the potential of being susceptible to low ambient oxygen conditions in the wake of their high metabolic rates as well as the large differences between their resting and optimum rates, despite the fact that there are behavioral similarities among the different species (Leung, Mislan, Muhling, & Brill, 2019).

The effects of hypoxia on kelps and other macroalgae are expected to be modest in view of the fact that these are primary absorbers of CO_2 and producers of oxygen. Nevertheless, since kelps and macroalgae also respire that requires oxygen, hence hypoxia entails the potential of wielding detrimental effects on processes like NPP that supplies organic matter to support kelp food webs and ecosystems. In view of the diversity of organisms in morphology and distribution of kelps and microalgae, the effects of hypoxia are predictable to vary widely depending on the species of macroalgae and their habitat (Crowder, Ng, Frawley, Low, & Micheli, 2019). The mesopelagic communities in OMZs regions are exceptional because the fauna is renowned for their adaptations to hypoxic and suboxic environments (Parris, Ganesh, Edgcomb, DeLong, & Stewart, 2014); still, mesopelagic faunas diverge significantly to such an extent that deoxygenation and warming could lead to the augmented ascendancy of subtropical and tropical faunas most highly adapted to OMZ conditions (Koslow, 2019). Expanding OMZs are predicted to alter the structure and function of benthic communities on continental margins via modification of the taxonomic composition, body size, food-web structure, bioturbation, and carbon cycling (Levin & Gallo, 2019). In the wake of continental margins in upwelling areas being exposed to nature-induced or human-induced hypoxia over an area of 1.1. million km^2 (Levin & Sibuet, 2012), the consequential oxygen gradients deliver excellent natural laboratories for understanding adaptations, tolerances, thresholds, and ecosystem responses to ocean deoxygenation. Utilization of sublet hypoxic waters as a refuge from fish predation by zooplankton prey can be instrumental in changing pelagic predator—prey. On the other hand, evading low-oxygen bottom waters can lead to zooplankton accumulations at the interface of hypoxic waters that can be required by zooplankton predators (Roman & Pierson, 2019). Water masses with less than 2 mg/L dissolved oxygen makes mobile benthic invertebrates migrate away and the process of deoxygenation has resulted in 13-fold decrease of diversity in benthic assemblages, "abundance of benthic infauna, 25-fold; and biomass, 10-fold as dissolved oxygen approached levels of 0.05 mg/L in a seasonally severe coastal low oxygen zone" (Rabalais, 2019). Deoxygenation affects corals, seagrasses and mangroves owing to their vulnerability to hypoxia, and concurrently, they are equipped with the ability to impact oxygen concentrations in the surrounding waters, resulting in feedbacks that can affect deoxygenation rates (Altieri, Nelson, & Gedan, 2019).

Process of ocean deoxygenation affects elasmobranchs specifically that constitute part of more than 1000 species of sharks, skates and rays that essentially are water-breathers with relatively high proportion of oxygen demands because they are large-sized bodies, active predators. Behavioral responses to water hypoxic water are demonstrated by many elasmobranchs by enhanced activity associated with circumvention (Rytkonen et al., 2010). However, elasmobranchs also appear capable of enduring mild hypoxia with circulatory and/or ventilatory responses, possibly even for extended periods; nevertheless, such strategies also entail the likelihood of being not sufficient enough to withstand moderate, progressive or prolonged hypoxia or anoxia (Sims, 2019). Undoubtedly, there is paucity of detailed knowledge based on research with regard to the impacts of deoxygenation events on marine predators, especially the mammals; nonetheless, marine mammals, while not directly dependent on the oxygen in the water column, are still vulnerable to changes in distribution, availability, behavior and mortality or their gilled prey (Steingass & Naito, 2019). Prey availability is a decisive factor in marine mammal distribution and accordingly community-wide impacts on gilled species affect the behavior of marine mammals (Magera, Mills, Kaschner, Christensen, & Lotze, 2013). In the wake of increasing coastal hypoxia in critical marine mammal habitats, there are nearly 48 species of marine mammal in the hypoxia-affected regions of the Northern California Current System, Black Sea, Baltic Sea, and Gulf of Mexico (Fennel & Testa, 2019). Coastal hypoxia events can prove instrumental in envisaging alterations in the patterns of distribution, mobility, predator avoidance, and mortality of gilled animals. Intense or protracted hypoxia can lead to shifts in food webs, with the probability for affecting foraging success for marine mammals. Augmentation in global warming and the resulting diminution in availability of oxygen signal an impending pattern of deteriorating hypoxia globally (Roman, Brandt, Houde, & Pierson, 2019). These patterns entail the possibility of resulting in enhanced pressure for a variety of marine mammal species that ae even now susceptible or endangered. On the contrary, accelerated rates of predation on gilled species debilitates or spatially constricted by hypoxia could be beneficial for certain marine mammals. While noting that quantification of direct links between coastal hypoxia and marine mammals can be difficult, Steingass and Naito (2019) are sanguine about excellent potential to study species because of abundant presence of seashore marine mammals in order to begin understanding the diverse implications. Projections about anticipated effects of deoxygenation on fisheries are likely to exacerbate over the next decades, and can influence fisheries via negative effects on growth, survival and reproduction impacting biomass and movement of fish affecting their availability for harvesting (Rose et al., 2019). The intensity of the deoxygenation impacts on fisheries is expected to augment because the areas of the ocean that would demonstrate enhanced deoxygenation overlap with the coastal and oceanic regions that currently support high production of fisheries (Barange et al., 2018).

In the wake of declining content of oxygen in global ocean and its coastal waters, there has emerged one of the starkest instances of deterioration of ocean ecosystems caused by nature-induced and human-induced activities (Breitburg et al., 2018). The ocean has since the middle of the 20th century has lost an estimated billion tons of oxygen and the volume of water wherein oxygen is absolutely lacking has recorded a fourfold increase (Schmidtko et al., 2017). Bulk of this loss of oxygen is ascribable to global warming through its impacts on oxygen solubility, stratification, ocean circulation and respiration rates (IPCC (Intergovernmental Panel on Climate Change), 2019a, 2019b; Oschlies et al., 2018). Moreover, more than 500 estuaries, semi-enclosed seas and other coastal water bodies, over the same period, have reported expansion in areas of dissolved oxygen concentrations at or less than 2 mg/L or 63 μmol/L (hypoxia) as a consequence of an excessive supply of nutrients from agriculture, aquaculture, human sewage and the combustion of fossil fuels (Breitburg et al., 2018; Isensee et al., 2015). With future anthropogenic pressures projected to further accelerate these trends (Fennel & Testa, 2019), the scientific community is called upon to address these challenges in a holistic manner (Capet et al., 2020). Emphasis needs to be accorded on understanding natural variability in marine oxygenation because the observation made cover only the period spanning over the past six decades and consequently there is lack of information on longer-term variability and trends in ocean oxygenation and associated stressors or drivers (Oschlies et al., 2018). According to Breitburg et al. (2019), inaccessibility of information coupled with lack of monitoring, has probably led to an undercount of such systems, especially in developing countries. It has been predicted by numerical models that persistent warming burgeoning human populations will enhance the severity of the problem of oxygen decline in the open ocean and coastal waters (Cocco et al., 2013). Equal attention needs to be focused on understanding and predicting the response of global biogeochemical cycles to deoxygenation, with specific emphasis on as to how lower oxygen conditions impact community respiration, the nitrogen (Lam & Kuypers, 2011), and phosphorus cycles (Watson et al., 2017) across the estuarine—shelves—ocean continuum, including feed-backs on the climate system. Focus also needs to be concentrated on evaluating and mitigating the danger emanating from deoxygenation to valuable marine goods and services (Cooley, 2012) and on marine biodiversity (Vaquer-Sunyer & Duarte, 2008).

8.3 Anthropogenic stressors

Degradation of valuable marine ecosystems due to anthropogenic stressors is occurring globally at a faster pace, especially in coastal areas with high levels of human development (Barbier, et al., 2008). Rapid decline is occurring globally among corals (Pandolfi et al., 2005), seagrasses (Waycott et al., 2009), and mangroves (Valiela, Bowen, & York, 2001), specifically in part due to stressors associated with water quality, including light reduction, exposure of toxins and smothering by sediments. Anthropogenic stressors, primarily based on human activities, have come to wield impacts on the ocean that are substantial, ubiquitous (Halpern et al., 2008) and changing (Halpern et al., 2015) and the consequential combined impact of these activities relatively triggers degradation of ecosystems and even their collapse (Estes et al., 2011), and studies of individual marine ecosystems like, coral reefs, kelp forests and seagrasses, have made it discernible declining conditions worldwide owing to increasing anthropogenic stressors (Waycott et al., 2009). Multiple anthropogenic stressors like overfishing, aquaculture, pollution, climate change, coastal erosion, habitat loss, and the introduction of invasive species affect the marine environments of the entire ocean

(Jackson, 2008; Ling, Johnson, Frusher, & Ridgway, 2009). In the wake of increasing human dependence on the ocean for resources, recreation and increasing frequency of usage of the ocean as a platform for the exchange of goods in a globalized world (Steffen et al., 2011), has come to pose a number of challenges.

Anthropogenic stressors exist in the ocean in the form of pollution, such as presence of aqueous chemicals, metals, GHGs, and plastics in aquatic ecosystems and these entail the probability of further stressing organisms by meddling into their normal functioning (Bernhardt, Rosi, & Gessner, 2017). In the eventuality of the scale of stressor outdoing the ability of organism to moderately adjust to the new conditions or in case of magnitude of the stressor surpassing the ability of the organism to evolve or adapt to the new condition, the consequences of these stressors can be more pernicious (Badyaev, 2005). While reacting to these stressors, organisms embark upon a stress response that needs an enhanced energetic load, known as allostatic load, and frequently results in suboptimal metabolic performance (Kassahn, Crozier, Pörtner, & Caley, 2009). Nevertheless, some stressors entail the potential of interfering with organism in more subtle ways, take for instance the case of OA that is reported to interact with regular neurological function in some species of fish (Munday et al., 2009) probably owing to the fishes' internal pH regulation interfering with particular neurotransmitter function (Regan et al., 2016). Microplastic pollution is a stressor and its ingestion by organisms misidentified for their natural planktonic prey causes the organism to exhaust its energetic reserves (Cole, Lindeque, Fileman, Halsband, & Galloway, 2015). Usually, these stressors act to enforce some allostatic load but can often impact organisms in less obvious ways (Schulte, 2014).

Anthropogenic activities have come to be regarded as a significant set of exogenic abiotic stressors (Elliott & Whitfield, 2011) that act to change ecosystems over wide spatial and temporal scales (Arnell & Gosling, 2013). This is mainly attributable to increasing levels of CO_2 and other heat-trapping GHGs like methane (CH_4), nitrogen oxide (NO_2) and chlorofluorocarbons in the atmosphere (Karl et al., 2015). The resultant impacts of these anthropogenic activities are discernible in the forms of global warming, eutrophication, augmented stratification, alterations in the ocean chemistry, fluctuations in the patterns of precipitations, circulation and freshwater inputs, and cumulative outcome of these impacts is projected to be significantly important in the future (Rajaratnam, Romano, Tsiang, & Diffenbaugh, 2015). Apart from CO_2 related effects as portent of threats to marine systems, supplementary endogenic regional pressures (Elliott & Whitfield, 2011) like increasing pollution, destruction of exogenic organic habitat, overfishing, invasive species, etc., also threaten marine systems, acting concurrently in tandem with human-induced climate change (Bulling et al., 2010). Hence, none of these stressors act in isolation but rather have composite impacts (Boyd, Dillingham et al., 2015; Boyd, Lennartz et al., 2015). Instead of a single factor of stressor yielding serious consequences for an organism or community of species, the additive synergistic or combined action of various factor stressors has the likelihood of causing more harm as compared to simply the totality of the individual threats (Darling, McClanahan, & Côté, 2013). One type of stress like rising mean sea temperatures may make marine and freshwater organisms less able to deal with the physiological demands of adapting (Yao & Somero, 2014) to enhancing frequencies of freshwater perturbation that could occur with changing precipitation patterns and their resilience could be further enfeebled by exposure to industrial contaminants. Given the fact that anthropogenic and natural stressors seldom deliver unvarying or consistent effects at various levels of biological organization (Wu et al., 2014), coupled with the nonlinear response of both individual and entire ecosystems to gradients of stressor intensity (Smith, Knapp, & Collins, 2009), it is difficult to predict the ultimate impact to be wielded by a single stressor or a series of stressors.

Stressors entail the potential of having noninteractive or composite impacts, subject to the context within which these stressors interact (Halpern et al., 2009). Additive model has often been used with a view to ascertain the combined ecological impacts of anthropogenic activities are undertaken (Piggott et al., 2015). Stressor impacts within the framework of additive model can be classified into three categories—either as additive or equal to the rich individual effects; or synergistic or greater than the sum of their individual effects; or antagonistic or less than the sum of their individual effects (Guisan, Edwards, & Hastie, 2002). There still prevails a sort of mixed consensus, irrespective of substantial current interest evinced in mechanistic underpinnings of these different kinds of effects, on which effects, according to Darling and Côté (2008), are most prevalent in "real" world marine system. In the opinion of Van den Brink et al. (2016), this is mostly attributable to twin factors of difficulty in establishing causality and the dearth of factorial studies that have been carried out in this regard at the community and ecosystem level. Signal or evidence of nonadditive effects, synergistic, or antagonisitic, is common in nature (Villar-Argaiz, medina-Sanchez, Bidanda, & Carrillo, 2018), specifically in coastal marine systems like estuaries that are persistently exposed to an array of natural and anthropogenic stressors (Ellis et al., 2015; Hewitt, Ellis, & Thrush, 2015). Some studies, in wake of high probability of composite effects in these systems, have focused on the measures being devised by scientists and policymakers who are progressively looking to execute or implement strategies that take into account the net effects of manifold human stressors (Brooks & Crowe, 2019), specifically those that engender unpredictable nonadditive effects or "ecological surprises" defined by Lindenmayer, Likens, Krebs, and Hobbs (2010) as unexpected findings about the natural environment.

8.3.1 Marine microplastic pollution

Marine plastic pollution, one of the major anthropogenic stressors, is currently assuming serious proportions with an estimated 6.4 million to 12 million metric tons (MMT) of plastics being dumped in the oceans annually (IMO, 2018; Routley, 2018). Plastic used for packaging is specifically constitutes bulk part of this marine litter owing to its sheer abundance and difficulty in recycling it (Geyer, Jambeck, & Law, 2017). Of all the marine litter that is dumped annually in the ocean worldwide, approximately 70%—80% of it is nondegradable plastics (Auta, Emenike, & Fauziah, 2017; Plastics Europe, 2016). The low rate of recycling is termed as a market failure by Newman et al. (2015) because the price of plastic products seldom reflects the true cost of the disposal. The society at large is made to bear the brunt of the cost of recycling and disposal of plastic products rather than by the producer or consumer (Newman et al., 2015). According to *GRID-Arendal*, known as GRIDA, a nonprofit environmental communications center based in Norway, this drawback "in our system allows for the production and consumption of large amounts of plastic at a very low 'symbolic' price. Waste management is done "out of sight" from the consumer, hindering awareness of the actual cost of a product throughout its life" (GRIDA, 2018a, para. 1). Single-use plastics comprising plastic shopping bags, plastic straws, plastic wrapping plastic bottles, and plastic cutlery is used only for minutes (or even seconds) before being thrown away (Routley, 2018), comparable to many other applications of plastics, is specifically problematic in terms of recycling and disposable. Out of all the plastics used by people worldwide per year, almost half of it is used once only or is a single-use plastic (Routley, 2018). According to Routley (2018), an estimated 500 billion plastic shopping bags are consumed per annum globally and that equals around one million plastic bags being used every minute, or 150 bags annually per capita worldwide, with each of these plastic bags having an average lifespan of just 15 minutes. It has been opined by Geyer et al. (2017) that the massive scale at which single-use plastics is in daily usage and the worldwide turnover of the plastics enhances the probability of some of it mismanaged substantially in comparison to other types of plastic applications with longer timespan, such as car parts, electronic parts and furniture.

International community has been making endeavors to address the issue of marine plastic pollution, albeit gradually until around 1994. Nevertheless, with the London Convention on Dumping in 1972 commenced the first big international effort to address the issue of plastics in ocean (IMO, 1990). The subsequent 20–25 years of international endeavors veered largely around targeting dumping of waste at the sea rather than emphasizing on land-based sources of marine plastic pollution. Up until the year 1994, no international endeavor to single out plastic as the key cause of concern of marine plastic debris. The focus of the international community underwent a change in the aftermath of the Third International Conference on Marine Debris in 1994 (Faris, 1994), and the establishment of Global Program of Action for the Protection of the Marine Environment from Land-based Activities (GPA) in 1995 (UNEP, 1995). More and more international efforts acknowledged around year 2000 that it was important to focus on both sea- and land-based sources of plastic in oceans. With the convening of the Fifth International Conference of Marine Debris in 2011, international efforts have gathered momentum and focus has now turned over to plastics and microplastics from land-based sources. The Fifth International Conference on Marine Debris held in Honolulu, Hawaii, USA, adopted the Honolulu Strategy that envisaged one of the first comprehensive outlines on as to how to tackle marine plastics pollution (Chen, 2015). Goal A of the Honolulu strategy aims at reducing the amount and impact of land-based sources of marine debris introduced into the sea the, through seven specific strategies (Chen, 2015; Honolulu Strategy, 2011). With the adoption of the SDGs by the United Nations in 2015 (UN, 2015) and inclusion of SDG-14 in the SDGs for the protection and sustainability of the ocean and marine wildlife, a strong signal has been sounded with regard to increased attention dedicated to plastics pollution in the 21st century. SDG 14 reads: "Conserve and sustainably use the oceans, seas and marine resources for sustainable development," and the first chapter of SDG 14 tackles plastics pollution via marine debris (UN, 2015, p. 23). SDG 14.1 reads: "By 2025, prevent and significantly reduce marine pollution of all kinds, in particular from land-based activities, including marine debris and nutrient pollution" (UN, 2015, p. 23).

The fact that only a few countries are responsible for the bulk of the plastics in oceans has seemingly emerged as the main driver for international community in providing impetus to their endeavors in tackling the issue. Pointing out that marine plastics pollution has many of the same characteristics as other global issues, Borrelle et al. (2017) notes that as with many other contaminants like GHGs and ozone-depleting substances, "plastics is not constrained by national boundaries, because it migrates via water and air currents and settles in benthic sediments. More than 50% of the ocean's area sits beyond national jurisdiction, including the infamous "garbage patches" in oceanic gyres where plastic accumulates" (Borrelle et al., 2017, pp. 1–2). Admittedly, many countries have already introduced domestic policies and legislation targeting the issue of combating marine plastics pollution from land-based sources (Borrelle et al., 2017; European Commission, 2008); nonetheless, it is essential that the biggest sources of marine plastics pollution—many of which are in East Asia, also act and envisage domestic and regional policies to deal with the issue of the release of plastics into the oceans from their respective territories (Walnum, 2018). Many European countries that have already policies with regard to combating plastics pollution are now emphasizing upon other countries, especially in East Asia, to legislate similar targeted policies to deal with marine plastics pollution (Borrelle et al., 2017), and interestingly, many of these policies veer around diminution in consumption in single-use plastics, and mitigation of leakages via recycling and waste management policies (European Commission, 2008; UNEP, 2017). Recently for over past couple of years, there has seemingly emerged a semblance of consensus in the international community on the necessity of combating marine plastics pollution at the source on land. The UNEP in February 2017 seemingly "declared war on plastics pollution" by launching a massive campaign for "targeting sources of marine litter, micro-plastics in cosmetics and excessive waste of single-use plastics by 2022" (Corben, 2017, p. para. 10).

The UN Ocean Conference held in July 2017 called upon all member countries to reduce consumption of plastics and four countries—China, Thailand, Indonesia, and the Philippines—committed to reduce plastics consumption (Corben, 2017). This was followed up in December 2017, where member countries of the UNEP reiterated their commitment to the combat against marine plastics pollution by signing a resolution on marine debris and microplastics the Third United Nations Environment Assembly (UNEA) (UNEP, 2017). This shows that the international community has been steadfast in its commitment to get a binding and specific international agreement on marine plastics pollution in recent years.

Research findings of the late 1960s with regard to monitoring and assessment of the volume of plastics released into the marine environment underwent a quick change when the subsequent research started to demonstrate and acknowledge that marine plastic pollution delivered negative consequences for marine animals and ecosystems other than just annoyance (Barboza & Gimenez, 2015; Ryan, 2015). Earlier research on marine plastics pollution was limited in scope because of its perception of the issue as one that yields negative consequences for the marine environment thereby failing to grasp the bigger issue of ingestion, enlargement, microplastics, and toxins moving up the ladder of the food chain and by merely focusing on local and regional impacts thereby failing to identify the issue in a global context (Barboza & Gimenez, 2015; Ryan, 2015). Irrespective of the fact that the first report of the effect of plastic on marine animals came in the 1960s (GRIDA, 2018b; Ryan, 2016); nevertheless, the subtleties and complexities surrounding issue did not get full focus until the late 1980s (Gall & Thompson, 2015; Laist, 1987), and summing up the emergence of the issue and its potential consequences, Laist (1987) wrote: "The deceptively simple nature of the threat, the perceived abundance of marine life, and the size of the oceans have, until recently, caused resource managers to overlook or dismiss the proliferation of potentially harmful plastics debris as being insignificant. However, developing information suggests that the mechanical effects of these materials affect many marine species in many ocean areas, and that these effects justify recognition of persistent plastic debris as a major form of ocean pollution" (Laist, 1987, p. 319). However, the full gamut of direct and indirect negative consequences of the issue of marine plastic pollution is yet to be fully explored (Prata et al., 2019). Despite varied estimates, between 400 and 600 marine species are reported to have been shown to have either ingested or gotten entwined in plastics since the commencement of research starting monitoring the impacts plastics have on the marine environment (Green et al., 2015). Currently, it is widely recognized that marine plastics pollution entails serious consequences specifically in the domains of marine wildlife, human health, and economic implications (Avio, Gorbi, & Regoli, 2017; Ryan, 2016). And these require brief appraisal.

Marine wildlife, comprising fish, marine birds, marine mammals and turtles, is the prime victim of marine plastics pollution (Visbeck et al., 2014). As early as in 1987, Laist (1987) had taken note of emerging consequences emanating from plastics on marine animals and arguing in this regard, he stated: "The accumulating debris poses increasingly significant threats to marine mammals, seabirds, turtles, fish, and crustaceans. The threats are straightforward and primarily mechanical. Individual animals may become entangled in loops or openings of floating or submerged debris or they may ingest plastic materials. Animals that become entangled may drown, have their ability to catch food or avoid predators impaired, or incur wounds from abrasive or cutting action of attached debris. Ingested plastics may block digestive tracts, damage stomach linings, or lessen feeding drives" (Laist, 1987, p. 319). In other words, individual animals either ingest plastics mistaking it as food or get entangled in plastic debris. Big quantities of plastics, both microplastic and microplastic, is being discovered in the stomachs of dead sea birds, fish and marine mammals (Laist, 1997; Ryan, 2016). Decomposition of plastic does not occur naturally within the lifespan of these animals, and any plastics ingested would stay in animal's stomach until the animal dies (Laist, 1997). Serious consequences for animal's health in the form of false satiation of stomach is full, pathological stress, blocked enzyme production, reduced growth rate and reproductive complications are borne out by ingested plastics (Auta et al., 2017). It has also been demonstrated by research studies that plastics could block parts or even the entire stomach of the animal thereby hindering the ability of the animal to ingest food (Auta et al., 2017; Gall & Thompson, 2015; Ryan, 2016). Plastic pollution affects marine animals differently subject to their size, diet and in area of ocean they inhabit (Routley, 2018). According to Routley (2018), an estimated 66% of global fish stock has ingested some degree of plastics, and plastics ingested by fish can cause intestinal injury and death (Auta et al., 2017). Besides, plastics ingested by fish also entails the potential of transferring toxins and microplastics up the food chain to bigger fish, marine mammals, and even humans (Auta et al., 2017).

Among the marine wildlife animals, turtles have been reported both to ingest and to get entangled in plastics (Gall & Thompson, 2015; Laist, 1997). Some recent studies have demonstrated that out of the seven species of sea turtles worldwide, plastic ingestion has been reported in all seven of the species of sea turtles and plastic entanglement in six species (Wilcox, Puckridge, Schuyler, Townsend, & Hardesty, 2018). Plastic bags that look almost similar to jellyfish in water, a favorite food for turtles, have specifically been dominant, and along with it other plastic debris like Styrofoam, plastic ropes and plastic pellets have also been discovered inside the stomach of dead sea turtles (Duncan et al., 2019; Gall & Thompson, 2015). Sea turtles ingesting plastic are prone to have "blockage in the gut, ulceration, internal perforation and death; even if their organs remain intact, turtles may suffer from false sensations of satiation and slow or halt reproduction" (CBD, n.d., para. 1). Seabirds have also fallen an easy prey to plastic ingestion and plastic debris. Seabirds primarily rely on the coastline and upper layer of the ocean to pick up food and their vulnerability of exposure to the ingestion of marine plastics debris has been reported (Kühn et al., 2015). Reports regarding plastic ingestion by seabirds were one of the first signal of the negative impact of marine plastic pollution on marine wildlife and the scientists were already alarmed at the outset of the 1960s (Ryan, 2015). Kuhn et al. (2015) reported that out of 406 species of birds monitored, entanglement was found in 103 species (25%), and ingestion in 164 species (40%). According to Wilcox, Van Sebille, and Hardesty (2015), plastic pollution also entails the likelihood of starvation of seabirds to death owing to reduction of the storage volume of the stomach along with affecting their reproductive abilities (CBD, n.d.; Kuhn et al., 2015). Big marine mammals like seals and whales have been found to be falling a prey to both ingestion and getting themselves entangled in marine plastic debris (Auta et al., 2017). While reporting about tests conducted over dead marine mammals, some scholars have stated that out of the 115 known species of marine mammals in the world, 30 species accounting for 45% of the total had ingested plastics and 52 species or 30% had got themselves entangled in (Routley, 2018). Ingestion of plastics by marine mammals can enhance toxicological stress and clog up the stomach ultimately leading to starvation for all species of marine mammals (Auta et al., 2017). Furthermore, microplastics have also been reported to having clogged up the breathing filter of whales, decreasing the mammal's ability to filter food or krill from the water (Besseling et al., 2015). Bigger mammals on top of the marine food chain are prone to have contamination levels up to 3 billion times higher than their surrounding environment owing to Bestselling the multiplication effect up the food chain (WWF, n.d.)

In recent years, attention has been focused on the impacts of pieces of microplastics reaching human through consumption of seafood (Galloway, 2015; Ryan, 2015), and findings of this research suggest that toxics released from plastics once inside the human body could change hormones, reduce fertility, decrease cognitive abilities and enhance the risk of cancer for humans (Naturvernforbundet, 2018). Humans being at the top of the food chain consume fish and mussels thereby enlarging the scope of receiving the transfer of toxics and microplastic particles (Galloway, 2015). Specifically concerned about the multiplication effect of toxins released from microplastics and delivered through the food chain, researchers and scientific organizations engaged in monitoring the impact of marine plastic pollution nursing apprehensions that humans could end up with a large quantity of microplastics in their bodies via seafood consumption (Galloway, 2015). Large mammals and humans are vulnerable to have contamination levels of toxics and chemicals released by microplastics many times more comparable to other marine smaller animals (WWF, n.d; Galloway, 2015). Apart from the argument by Peixoto et al. (2019) that even common commercial table salt can be a transferring agent of microplastics to humans, some researchers have gone one step ahead by arguing that effects that nanoplastics or particles smaller than 100 nm in diameter, can impact human health in future (Lambert & Wagner, 2016). Undeniably, economic costs of marine plastics pollution is not so huge as comparable to its impacts on marine wildlife or human health (Newman, et al., 2015); nonetheless, according to broad estimates, the economic cost inflicted by marine plastics pollution on society varies between $8 billion and $13 billion per annum (GRIDA, 2018b; UNEP, 2014), and bulk of this cost is incurred from the necessity of cleaning up and removal of marine plastic pollution to ensure successful functioning of coastline business operations (GRIDA, 2018b).

With increasing recognition that marine plastics pollution is a serious threat to marine environment and human health (Law, 2017; Ryan, 2015), researchers and scholars have come out with an array of suggestions to combat the issue both at source and through removal plastics that is already present in the ocean. While Loehr et al. (2017) emphasizes on the need for global action on the lines suggested in SDG-14, Thevenon and De Sousa (2017) argue that employment of economic tools is key to deal with the issue. Emphasis is also focused on specifically targeting single-use or any plastic products with a short useable lifespan (Steensgard et al., 2017; Xanthos & Walker, 2017). These suggestions are often directed toward policymakers and other decision-makers by adhering to measures and tools to limit plastics reaching the ocean (Xanthos & Walker, 2017). Other tangible suggestions emphasize on providing economic incentives (Oosterhuis et al., 2014), taxation (Wagner et al., 2017), monitoring (Thevenon & De Sousa, 2017), public education (Le Guern, 2018). Emphasis has also been stressed on Extended Producer Responsibility (Perella, 2017), and waste management strategies (Newman et al., 2015). While espousing for a holistic approach wherein marine plastics pollution is construed as a consequence of human mismanagement of land and sea resources, (Haward, 2018; Hu, 2012; Thevenon & De Sousa, 2017; Vince & Hardesty, 2017), some scholars propose the creation of circular plastic economy to tackle the current emerging issue of plastics in ocean (Eriksen, Thiel, Prindiville, & Kiessling, 2018). There have been suggestions with regard to the creation of an overreaching national strategy that addresses marine plastic pollution to mitigate against plastics reaching the ocean (IUCN, 2017a, 2017b; HELCOM. (Baltic Marine Environment Protection Commission), 2015); however, in disapproval of this suggestion (Routley, 2018) argues that most of the world's countries have ocean coastlines, and with so many jurisdictions and varying degrees of environmental scrutiny, truly curbing the flow of plastic isn't realistic in the near term" (Routley, 2018: para. 6). Recent years have been characterized by growing interest from the public, media, policymakers, and international organizations on the topic on combating marine plastic pollution and most prominent has been the establishment of the Joint Group of Experts on the Marine Aspects of Marine Environment Protection (GESAMP), comprising experts nominated by nine United Nations associated agencies (IMO, FAO, UNESCO-IOC, UNIDO, WMO, IAEA, UN, UNEP, UNDP), and it publishes research on effective tools to combat marine plastics pollution (GESAMP, 2015). GESAMP has a pool of 500 scientists from 50 countries, who contribute with science-driven publications, workshops and reports on marine plastics pollution (GESAMP, 2018). The UNEP has also developed a toolkit for policymakers that lists and explains policy measures to combat certain types of marine plastics pollution, with particular focus on single-use plastics (UNEP, 2017). The Honolulu Strategy further outlines government measures that can be used in the fight against marine plastics pollution (Honolulu Strategy, 2011). Amidst these salutary developments, one can hope that concrete solutions in combating marine pollution will be found soon.

8.3.2 Oil-spills pollution

Oil spill is an anthropogenic stressor that has been persistently threatening the marine environment despite the adequate technical developments in the safety of extraction and transport of crude oil and gas. These oil spills engender grievous and decade long destruction for marine and coastal ecosystems and the organisms that sustain them (Joye, 2015). Recent decades have witnessed both small and large incidents of oil spills worldwide. In the 1990s, there were 358 incidents of oil spills and in the 2000s, there were 181 incidents of oil spill, and the ensuing decade witnessed 62 spills of 7 tons and over resulting in the loss of 164, 000 tons of oil and 91% of this oil was split in just 10 incidents and during 2019, one large spill exceeding 700 tons and two medium spills ranging between 7 and 700 tons were recorded (ITOPF, 2020). According to Fingas (2013, p. 225), washing ballast tanks are accountable for 36,000 metric tons or 11.2 million gallons of oil entering the oceans worldwide every year due to human-induced activities and nontank vessels. Release of oil

8.3 Anthropogenic stressors

into the ocean via natural seeps takes place at a low pace to which deep sea organisms are adapted to in comparison to the sudden faster pace of release of bulk amount of oil that happens during an oil spill or extraction incident (Blackburn, Mazzacano, Fallon, & Black, 2014: 152). While noting that upgraded technological standards for oil production have minimized large spills ranging from 7 to 700 tons and larger than 700 tons of oil during the decade of 2000s, Fingas (2013, p. 225) laments at the fact of neglect of taking note of smaller spills representing less than 7 tons of oil and accounting for about 80% of all recorded spill being let to go unnoticed and unreported. Exposure of marine organisms to the accumulation of dissolved and dispersed oil is at peak during preliminary stages of the spill (Gros et al., 2014; NRC (National Research Council), 2005; Reddy et al., 2013), and accordingly, foremost toxicity mechanisms smaller spills arise from instant exposure that vanishes from sight during the first few days without the oil been the oil been essentially degraded from the sea surface (Short et al., 2007). In view of mounting global demand for oil and gas and depleting on-land oil resources, focus has now shifted to increasing marine oil exploration. While describing that world's oil and gas resources are lying beneath the oceans and advances in drilling, exploration and production technologies allowing production in water more than 10,000 feet deep and more than 100 miles offshore, Allison and Mandler (2018) note that spills are rarebit damage sensitive ocean and coastal environments. Some scholars have noted oil toxicity at its peak during the preliminary stages of a spill (Gros et al., 2014); therefore, even smaller spills entail the probability of delivering protracted effects and posing a risk to ecosystem health and biota (Yuewen & Adzigb, 2018). It has been reported in the recent past that prime shipping routes on the North Sea have been recurrently polluted with oil, either from inadvertent spills or may be from natural oil spills (Gros et al., 2014).

Crude oil consists of an amalgamation of hydrocarbons and 10% of molecules with heteroatoms such as Sulfur, Oxygen and Nitrogen (Tissot & Welte, 1984, p. 699; Liu & Kujawinski, 2015). The amalgamation of hydrocarbon is at variation from trivial uneven and explosive monocyclic hydrocarbons to big nonvolatile polycyclic aromatic hydrocarbons (PAHs) (Fingas, 2013). Final impact and state of hydrocarbons in the marine environment is regulated by their solubility, molecular weight and availability (NRC (National Research Council), 2005). Toxicity in crude oil is primarily from aromatic hydrocarbons, specifically PAHs (Neff, 2002). Irrespective of their explosive nature, monocyclic aromatic hydrocarbons are less tenacious and therefore, are not accumulated in water, sediments and tissues of marine organism (NRC, 2003). It becomes discernible from some recent observations that toxicity is attributable not only to the prevalence of PAHs but even the presence of small oil droplets also wields severe impacts (Gonzalez-Doncel, Gonzalez, Fernandez-Torija, Navas, & Tarazona, 2008). This fact was established when Atlantic haddock (*Melanogrammus aeglefinus*) demonstrated uttermost sensitivity to dispersed oil that was supposed and associated to direct interaction with oil droplets (Sørhus et al., 2017; Sørensen et al., 2017). Photooxidation and microbial degradation have currently come to reckoned as important process for breaking down or eliminating oil from the marine environment (Das & Chandran, 2011; Rathi & Yadav, 2019; Wolfe et al., 1996). Concurrently, some studies have also demonstrated the risks for aquatic organisms emanating from the by-products from these biodegradation processes (Bellas et al., 2013).

Composition of marine ecosystems embraces an array of animals of different types varying from microorganism, invertebrates like fish birds, mammals, and turtles to invertebrates such as copepods, mollusks, crustaceans, and echinoderms (Yuewen & Adzigb, 2018). These organisms fall an easy prey to variable degree of effect in the event of an oil-spill incident. The toxicological effects of oil, like increased mortality or as sublethal injury, weakened feeding and reproduction and avoiding predators, on fish communities have been focused in the literature (Tarnecki & Patterson, 2015), along with oil-spill effects on estuarine communities, mammals, birds and turtles (Haney, Geiger, & Short, 2014; USF&WS (United States Fish & Wildlife Service), 2011). deep-water corals (White et al., 2012), plankton (Almeda, Baca, Hyatt, & Buskey, 2014), foraminifera (Schwing et al., 2015), and microbial communities (Bñlum et al., 2012).

Oil spills can be instrumental in envisaging changes in fish metabolism (Cohen, Nugegoda, & Gagnon, 2001), and affecting fishes in multiple ways; including increased mortality (Fodrie et al., 2014), kill or cause sublethal damage to fish eggs and larvae in the form of morphological deformities, decreased feeding and growth rates, enhance vulnerability to predators and starvation (Sorhus et al., 2016), habitat degradation, loss of hatching ability of eggs, maculating of gill structures, diminished reproduction, growth, development, feeding, respiration (Blackburn et al., 2014). Endowed with the capability of smelling oil, fish can escape it (Davis, Moffat, & Shepherd, 2002) and concurrently fish are able to avoid drifting oil better in the open sea than in shallow shoreline water where oil frequently moves toward them and the accumulation of oil in the water becomes higher. Early development stages of fish are regarded as the most vulnerable to the effects of oil (Short, 2003) Fish spawning and nursery grounds entail a high potential of becoming exposed to oil effects in coastal waters (Lecklin, Ryömä, & Kuikka, 2011). Exposure of fish eggs and larvae to oil spills can cause mortality and or serious developmental disorders as malformation in internal orders diminution in reproductive capability, and curving of the notochord (Sorhus et al., 2016). There is also likelihood of sublethal effects like edemas, changes in progeny, neuronal cell death, failed inflation of the swim bladder, and anemia being caused by exposure to embryonic PAHs (Incardona et al., 2009). Undoubtedly, fish stock is said to be at risk and vulnerable to large oil spills (Fodrie et al., 2014); nonetheless, this vulnerability is mostly experienced at the early stages, that is, egg and larvae (Bellas et al., 2013) owing to their under-developed membrane and body structure, and detoxification structure (Langangen et al., 2017). According to Sørhus et al. (2016), these effects were owing to the effect crude oil had on the genes, in terms of change in expression, that control the ion, amount of water and purpose or morphogenesis of specific tissues and organs. The effects are pertaining to the changes in composition, structure and life history of marine fish eggs and larvae. The normal development and role of the heart can be interrupted by the PAH that can further trigger pace and contractility flaws at the scale of the emerging heart and failure of circulation (Sørhus et al., 2017). It was revealed by Barron et al. (2005) that when juvenile pink salmon (Oncorhynchus gorbuscha) were exposed to crude oil, they displayed astonished reaction, slow and low movement, loss of stability and balance, melanosis and inconsistent swimming. Nevertheless, these responses were not intensified by phototoxicity owing to its high skin pigmentation (Incardona et al., 2014). It has been opined by Rahikainen et al. (2017) that since the oil affects the early development of fish, adverse effects on population dynamics might be detectable only after several years.

Seabirds are prone to be adversely affected when exposed to oil spills. High vulnerability of seabirds to oil spills is because they spend most time on the shoreline for forage for food where they get easily exposed to drifting oil (Henkel, Sigel, & Taylor, 2012). Among the detrimental effects of exposure of seabirds to oil are impact on their health and behavior and in an event of consumption of oil by seabirds can be harmful to their lungs, liver and kidney. One of the common impacts of oil is the ensnaring of their feathers that causes change in feather microstructure

(O'Hara & Marandin, 2010), leading to loss of floating and flying ability owing to compressed plumage thereby allowing water to contact skin (Leighton, 1993) causing anemia, pneumonia, intestinal irritation, kidney damage, altered blood chemistry, decreased growth, impaired osmoregulation, decreased reproduction capability and viability of eggs, abnormal parenting behavior, and decreased growth of the offspring (Albers, 2003), and hypothermia can eventually lead to death, specifically during cold weather (Gaston, 2004). Additional adverse effects arise, for example, from the toxicity of PAHs: birds get oil into their system while preening, drinking oiled sea water, consuming contaminated food or inhaling toxic vapors (Briggs, Yoshida, & Gershwin, 1996). Some toxic effects may not be evident immediately or may not cause death. These sublethal effects have an impact on birds' physiology and behavior, and can have population-level impacts through diminished health and reproductive fitness (Finch, Wooten, & Smith, 2011; Haney et al., 2014).

Among the marine mammals, dolphins, sea turtles and whales are primarily known to breathe at sea surface (Helm, Costa, O'Shea, Wells, & Williams, 2015; Rosenberger, Adrianne, MacDuffee, Rosenberger, & Ross, 2017) and as such are exposed to oil spills causing severe damage to their eyes and adenoidal tissue damage, low immunity, lung and adrenal diseases (IPIECA-IOGP, 2015; ITOPF, 2011). In an eventuality of an oil spill, sea otters are prone to have their fur soiled and that obstructs insulation and water repellence and in the process of cleanup likelihood of sea otters consuming oil is increased and that can cause tissue damages (Helm et al., 2015; IPIECA-IOGP, 2015). Undoubtedly, little research is available in the realm of the effects of oil spills on mammals; nonetheless, various responses of marine biota to oil spills have been documented following the Deepwater horizon oil spill (NOAA, 2010; Frasier, Solson-Berga, Stokes, & Hilderbrand, 2020). It is revealed that ingestion of oil by marine mammals causes respiratory irritation, inflammation, emphysema/pneumonia, gastrointestinal inflammation, ulcers, bleeding, diarrhea, and may cause damage to organs (Frasier et al., 2020; Helm et al., 2015). It was assumed by Carmichael, Graham, Aven, Worthy, and Howden (2012) that stress from bacterial attacks and decreased diet resources on account of oil spill were instrumental in causing death of bottlenose dolphins. Study of live bottlenose dolphins exposed to oil revealed lung and adrenal diseases along with other poor health disorders (Helm et al., 2015). Some studies confirm the susceptibility of the dolphins to bacterial infections and other type of health hazards that can probably lead to death (Helm et al., 2015). Findings by NAP (2010) reveal sea turtles experiencing esophageal papillae in their throats from oil ingestion and owing to the ability of sea turtles inhaling for extended times, ingested oil could be highly absorbed into their bodies. Direct linkages between sea turtle mortality and oil pollution because of oil spills are not established by some studies (Drabeck, Chatfield, & Richards-Zawacki, 2014; Hart, Lamont, Sartain, & Fujisaki, 2014), still NAP (2010) insists that survival of sea turtles intensely smeared in oil is improbable without medical attention.

Adverse impacts of oil spills on marine organisms are now well recognized. Every situation involving oil-spill accident is unique and is subject to the specific conditions and circumstances in that area, and on the characteristics of the spill (NRC, 2003). The recovery of the affected habitats and species following an oil spill is subject to a large extent on the type of ecosystem, the susceptibility of the species and the climate of the region where oil spill takes place. Usually, the recovery is said to move at a faster pace in warmer climate and on rocky shores in comparison to cold climates and marshes (UNEP, n.d.). Studies suggest that the oil spills that occurred in the Arab Gulf Region in 1991 were rather rapidly accommodated by 1992 and it is believed that it could have been the outcome of warm waters of the Gulf (GESAMP, 2001 p. 23). Undoubtedly, may studies have dealt with short-term effects of oil marine spills; nevertheless, emphasizing on the need for studying long-term effects of marine oil spills, Yuewen and Adzigb (2018) suggest the need for "research data on focusing more on the level of impacts, where long term monitoring will be done to grasp the full knowledge behind the oil spill impacts on marine organisms, the impacts of cleanup exercise, dispersants and other response methods to oil spill."

8.3.3 Overfishing

With its ability to represent more than 15% of the blue economy sector (Link & Watson, 2019) along with exports being valued at about $164 billion (FAO, 2020, p. 8), and providing about 59.51 million jobs globally (FAO, 2020, p. 7), fisheries have become a significant part of the global economy. Apart from trade and livelihoods, fish provide constant source of protein for a part of population worldwide and in 2017, fish consumption comprised 17% of the global population's intake of animal proteins, and 7% of all proteins consumed, and globally, fish provided more than 3.3 billion people with 20% of their average per capita intake of animal proteins, reaching 50% or more in developing and small-island developing states (SIDS). (FAO (Food & Agriculture Organization), 2020: 5). In 2018, global fish production was estimated to have reached nearly 179 million tons, with a total first sale value worth $401 billion, of which 82 million tons worth $250 billion came from aquaculture production. Bulk potion of the global fish production is used for human consumption, equivalent to than an average estimated supply of 20.5 kg per capita and remaining part of the fish production, about 22 million tons, are destined for nonfood purposes like production of fishmeal and fish oil (FAO, 2020, p. 5).

The high seas embracing 47% of the Earth's surface or 64% of the world's oceans (Costello & Chaudhary, 2017), encompass areas beyond national jurisdiction (ABNJ), as defined by the United Nations Convention on the Law of the Seas (UNCLOS) as incorporating the water columns beyond the exclusive economic zones (EEZs) of coastal countries and the "area" that is that is the seabed beyond the limits of the continental shelf (Biodiversity, n.d.; Kimball, 2005). Apart from providing multiple biodiversity benefits to society, the high seas also assume immense significance owing to the commercial potential of marine genetic resources (MGR) and mineral deposits along with fisheries (Vierros, Suttle, Harden-Davies, & Burton, 2016). In the wake of depleting fish resources in the coastal areas, the focus on the high seas fisheries is gathering momentum that is further spurred by improvement in navigation technology and fishing gear in recent decades. High seas capture fisheries—comprising mostly pelagic and deep-sea fish—that stood at 450,000 tons in 1950 increased to almost 5.2 million tons by 1989 (Dunn et al., 2018), and it recorded an average production of 4.3 million tons annually between 2009 and 2014, or to some extent over 4% of total annual marine catch (Schiller, Bailey, Jacquet, & Sala, 2018). China, Taiwan, and Japan rank at the top among the largest fishing countries in terms of revenue, recording about 45% of the $7.6 billion in high seas catch-landed value in 2014 (Sala et al., 2018). Undoubtedly, a couple of years back some studies demonstrated that fish capture from high seas played a little role in global food security (Schiller et al., 2018; Teh et al., 2016); nevertheless, a recent study (Bell et al., 2019), while establishing a link between human health and high seas fisheries suggests that in some countries, especially in Pacific Island nations, food commodities such as canned tuna from fish captured in EEZs and on the high seas are assuming increasing significance in narrowing down food nutrition gaps.

While 60% of fish population is fully fished, nearly 30%−35% of fish populations are fished unsustainably (Link & Watson, 2019). The consequences of unsustainable fisheries extend beyond simple status of fish populations and economic viability of fisheries to global food security; cultural survival, and even national security (Blanchard et al., 2017; Jennings et al., 2017). Unsustainable fishing not only impacts fish populations, fleet and fishery systems but also culminates in much wider implications for marine ecosystems (Link, 2018). Link (2018) has emphasized on the essentiality of adopting a broader, more systematic mechanism of detecting and delineating overfishing in order to concurrently addressing many of related challenges, before it consecutively impacts fish population after fish population, fishery after fishery, and marine ecosystem functioning. Even in the high seas, unsustainable fishing is likely to continue to remain one of the major threats to biodiversity. While bottom trawling specifically has proved instrumental in bringing in declines in species diversity in ABNJ (Norse et al., 2012; Wright et al., 2015), particularly around seamounts and deep-coral ecosystems as these are consisted of enormously slow-growing and long-lived organisms and this renders them specifically susceptible to fishing impacts (Clark et al., 2015). Pelagic fisheries on the high seas are also reported to have considerably led to the overexploitation of a number of targeted highly migratory stock such as tuna and billfishes (White & Costello, 2014). Declining trends are reported in a number of bycatch species like marine turtles, seabirds, sharks, and marine mammals (Oliver, Braccini, Newman, & Harvey, 2015). Contrary economic incentives like fishing subsidies in tandem with weak regulatory control and these spur depletion of fisheries stock at a faster pace and if this scenario is allowed to continue unabated there is likelihood of further aggravation of existing challenges (Norse et al., 2012).

Overfishing entails taking more fish than the annual yield that a fish stock can engender, and it renders the fish stock and marine ecosystem more vulnerable to change even without a stressor such as climate change (Sumaila & Tai, 2020). Overfishing has come to be widely acknowledged as a direct pressure and major risk to marine environments and ocean health, considerably plummeting fish biomass in the ocean (Halpern et al., 2015). Human-induced anthropogenic activities have exerted plentiful and far-reaching impacts on the global ocean (Halpern et al., 2015), and overfishing, another anthropogenic stressor, has already had long-lasting effects on marine ecosystems and still remains as one of the greatest threats to marine health (Gattuso et al., 2018; Halpern et al., 2015). Marine ecosystem is often affected by overfishing (Coll, Libralato, Tudela, Palomera, & Pranovi, 2008), and is also perceived as a catalyst for spurring ecosystem regime shifts (Daskalov et al., 2007). Overfishing wields negative effects on many indicators of ocean health, including biodiversity, food security, and coastal livelihoods along with economies (Halpern et al., 2012). Overfishing can directly be instrumental in affecting diminution in fish biomass, impacting biodiversity and the sustainability of fisheries, as well as accelerating the effects of destructive fishing gear on marine ecosystems (Sumaila & Tai, 2020). Unsustainable overfishing operations that often illegal, unreported or unregulated and often carried with destructive fishing gear by means of bottom trawls and dynamite, are bound to negatively impact benthic substate (Bailey and Sumaila, 2015).

Broad estimates based on recent reports with regard to state of fisheries in European waters make it discernible that between 40% and 70% of fish stocks are presently at an unsustainable level—either due to overfishing or at their lower biomass limits (Froese et al., 2018; STECF (Scientific, Technical, & Economic Committee for Fisheries), 2019). According to estimates over 90% of fish stocks get overexploited in the Mediterranean Sea (Colloca, Scarcella, & Libralato,

2017), and high levels of overfishing along with continuing reductions in catch being reported from the Black Sea (Tsikliras, Dinouli, Tsiros, & Tsalkou, 2015). On the contrary, Norwegian Sea and Barents Sea in Norther European side have fared better on overfishing score and this attributed to historically sound management mechanism for fisheries and the bulk of stocks in northern European waters with a biomass that can generate maximum sustainable yield (MSY) and not presently subject to overfishing (Froese et al., 2018). With a view to manage European fishing fleets and for conserving fish stocks, European Union framed a set of rules, known as the Common Fisheries Policy (CFP) and first introduced in the 1970s and after going through successive updates it went into effect on January 1, 2014 (EU, n.d.). At that stage, all EU member states committed to ending overfishing by 2015 oy by 2020 at the latest, to restore and protect all EU stocks in a sustainable way. However, during the second week of January 2020, it was reported that EU failed to end overfishing by 2020 and interesting in a tweet on August 30, 2019, the EU proposed to continue overfishing past the deadline of January 1, 2020 (Nicolas, 2020).

Apart from impacting marine ecosystems, overfishing also causes damage to marine food chain. Flows of biomass from low to high trophic levels of a food web and their changes over time comprises a major aspect of ecosystem functioning, incorporating the impacts of natural and human perturbations (Gascuel & Pauly, 2009). Flow of a biomass depends on species' characteristics like trophic levels, production or consumption rates and their interactions as well as traits of the physical environment such as temperatures (Odum, 1969). Ecosystem stressors, especially fishing, can spur natural choice in exploited fish populations (Maureaud et al., 2017), as the constant targeting of fish above a certain size has been shown to opt for individuals with lower asymptotic size and earlier suppuration (Olsen et al., 2004). According to Perry et al. (2010), such selection events influence the structure and diversity of communities in the long term, and thus biomass flows in the food web. Pauly, Christensen, Dalsgaard, Froese, and Torres (1998) have informed that movement in bulk of large individuals from higher trophic levels and high value fish out of the marine ecosystem from the highest trophic levels and most valuable species at the time they are fishing culminates in serial depletion and fishing down marine food webs. The resultant outcome is the enfeeblement of fishing stocks and their exposure to vulnerability to all types of stressors inclusive of climate change (Sumaila & Tai, 2020). Natural and human elements of ocean health get affected as a consequence of climate-related impacts on marine ecosystems, entailing the probability of envisaging alterations in species' distribution and abundance that can lead to increase in local invasions and extinctions, reallocating marine biodiversity and its composition (Pecl et al., 2017; Sunday et al., 2017). Afterward, it will impact goods and services of marine ecosystem, including food security, and benefits accruing to dependent coastal communities (Sumaila et al., 2019). Moreover, the enhanced variance in environmental change will also augment the variability—and reduce the predictability and reliability—of goods and services to human society (IPCC, 2014a, 2014b). Thus, the cumulative impact of overfishing and climate change is baneful for fish stock and marine ecosystems (Sumaila & Tai, 2020). Therefore, putting an end to overfishing is seen as a way out to ensure more effective conservation and sustainable use of marine fish and ecosystems.

Overfishing as an indirect stressor causes habitat degradation, especially when destructive fishing gear is used and some studies show that overfishing has already caused habitat loss (Halpern et al., 2015). Planetary functions like carbon storage, coastal protection/erosion necessitate the urgency for maintaining the integrity of ocean biodiversity. Envisaging improvements in the conditions of marine habitats, especially corals, seamounts, mangroves, seagrass, can be beneficial to other components of the ecosystem including fish stocks and enhance resilience to other pressures, specifically climate change (Gaines et al., 2018). At a time when pressures and stressors are negatively impacting fish stocks and marine ecosystem health, resilience is seen as capable of warding off or counteracting these negative effects (Halpern et al., 2012). Integrated combo of marine diversity and habitat can be instrumental in chipping in biomass to fisheries and nonfood products such as ornamental fish used on aquarium (Livengood & Chapman, 2007), and these are the portents of implications for global food security and livelihoods and economies of the coastal communities. While pointing out implications for marine life from habitat loss and impact of overfishing on other aspects of ocean health like coastal protection and carbon storage, Sumaila and Tai (2020) emphasize on the need for eliminating overfishing to reduce habitat degradation that could enhance the health of marine ecosystems and the fish stocks they sustain.

8.3.4 Greenhouse gases

The climate system is a dynamic network of physical, chemical and energy processes evolving over timescales spanning over days to millennia, and entailing interactions between the hydrosphere, lithosphere, biosphere, atmosphere, and cryosphere (Rose, 2015). These processes embrace, but are not limited to: precipitation, pressure (wind), temperature, photosynthesis, weathering, and movement of energy or heat [radiative forcing (RF)] (Kärnbränslehantering, 2005). Vicissitudes in the system are impelled by both natural and human perturbations to the planetary energy budget (Myhre et al., 2013). Energy to the system is forced externally by solar output, orbital variation, and volcanic activity (Myhre et al., 2013). Theoretically, radiative energy is supplied daily by solar output that is either reflected back to space or absorbed by water, land surfaces, and the atmosphere (Rose, 2015). At night land surfaces reemit radiative energy which either escapes to space or is trapped by GHGs in the atmosphere that are themselves radiative forcers on the climate system (Rose, 2015). This radiative energy in the climate system is otherwise called heat and though its only source is solar output, radiative energy flows between and is stored in the internal climate system components (Trenberth & Stepaniak, 2004). Conceiving the planetary land surface as a uniform body, constant upsurge in radiative energy translates into temperature increase at the surface by a simple relationship whereby temperature change is proportional to RF and climate sensitivity (Baggenstos et al., 2019). Due to the significant difference in heat capacity between the land and the ocean, the ocean can assimilate far greater quantities of radiative energy than the land (Dahlman & Lindsey, 2020). *Per se*, the ocean both does not reemit radiative energy to the same degree as land surfaces, and it is the driver of the thermal inertia of the climate system (Resplandy et al., 2019). This means that with any change to the radiative balance of the system, there is an insulated effect on the temperature of the ocean and as a result on global temperatures (Zickfeld & Herrington, 2015). Whereas the ocean can absorb energy from the atmosphere, the capacity to do so decreases as temperate increases (Ehlert & Zickfeld, 2017). Such changes to the radiative balance in the prevalent climate system can be evaluated by alterations in the internal cycles of the climate system. Internal changes to vegetation, land surface, sea and land ice, ocean, and atmospheric processes all influence the system. Some studies have demonstrated that the most dominant forcers on internal components are the movement of energy between the atmosphere and ocean along with associated feedbacks from the other components (Kren, Pilewskie, & Coddington, 2017) and the carbon cycle (Ehlert & Zickfeld, 2017).

The carbon cycle, affected by temperature fluctuations, as well as its own internal and external mechanisms, is the movement of carbon—organic, inorganic, dissolved inorganic, and CO_2—between the constituents of the climate system (Isson et al., 2020). For the reason that CO_2 is a GHG, alterations in the atmospheric concentration of CO_2 lead to changes in the radiative balance of the climate system and thus result into changes in surface-air temperature (SAT) and sea-surface temperature (SST) (Hartmann et al., 2013). Natural procedures of plant and animal respiration, and tectonic outgassing cause augmentations in the atmospheric concentration of CO_2 (Heinze et al., 2015), while photosynthesis, ocean dissolution, and silicate weathering bring about diminutions in atmospheric CO_2 (Ciais et al., 2013), and connected increases in the land's biosphere and soil and ocean carbon pools. These sinks function on yearly, centennial, and multiple-millennia timescales (Willeit, Ganopolski, Dalmonech, Foley, & Fuelner, 2014) and the steadiness of natural input and output can be stirred up by emitting CO_2 into the atmosphere and enhancing SAT and SST (Ciais et al., 2013). In the opinion of Ciais et al. (2013), human output of CO_2 beyond the effectiveness of the short-term sinks induces a portion the emissions to collect in the atmosphere. Climate modeling research has made it discernible that increasing CO_2 emissions and SAT change are roughly linearly related (Matthews, Zickfeld, Knutti, & Allen, 2018; Tokarska, Gillett, Weaver, Arora, & Eby, 2016). The TCRE, also known as the transient climate response to cumulative CO_2 emission, occurs because of amalgamation of the nonlinear processes of CO_2 RF and exclusion from the atmosphere by sinks (MacDougall & Friedlingstein, 2015). An increase in cumulative emissions renders enhancing SAT to reduce the effectiveness biosphere sinks as well as also renders increasing SST to decrease the efficiency of ocean carbon dissolution, culminating in a larger airborne fraction of emissions (MacDougall & Knutti, 2016). Nevertheless, this larger fraction reveals a lesser per-unit RF owing to impregnation of absorption bands (MacDougall & Knutti, 2016). Hence, each unit of CO_2 emissions spurs almost the same climate response (Leduc, Matthews, & Elia, 2015).

Absorption of heat by the ocean does not stop with CO_2 emissions: instead, the ocean continues to absorb heat from the atmosphere for many centuries following ZEC (Katz, 2015), though at a reduced rate over time (Frölicher, Winton, & Sarmiento, 2014; Zickfeld, Solomon, & Gilford, 2017). Since rise in thermal sea-level occurs owing to thermal expansion of the ocean that is caused by increases in ocean heat (Church et al., 2013), sea-level rise will also continue for many centuries after CO_2 emissions end (Ehlert & Zickfeld, 2017; Zickfeld et al., 2017). The level of ocean temperature change, at the time of ending CO_2 emissions, is decided by the level of atmospheric temperature change that is linked to the cumulative CO_2 emissions (Matthew, Solomon, & Pierrehumbert, 2012). Ocean temperatures change, because of its being associated with atmospheric temperature change, and with its thermal sea-level rise, are dependent on the collective CO_2 emissions before zero emissions, and can be increased by additional warming from non-CO_2 forcers prior to emissions stop (Mimura, 2013). As compared to the tropics, poles are reported to be influenced greatly by the higher levels of global warming (Partanen, Keller, Korhinen, & Matthews, 2016; Harvey, 2020), meaning under high warming scenarios the ocean should have a reduced potential to uptake carbon and possibly heat (Hoegh-Guldberg et al., 2018). This impact was examined by Frölicher and Paynter (2015) as well as by Ehlert and Zickfeld (2017) by making use of an idealized or flawless scenario where CO_2 concentration is registering an increase by 1% per year until 2°C temperature change was reached or until concentration was doubled, respectively. In 2°C analysis by Frölicher and Paynter (2015), the rate of ocean heat absorption was shown to decrease faster than the decline in heat uptake from land and ocean CO_2 uptake, culminating in additional warming after emissions were zeroed. In this case, the time until atmosphere-ocean thermal equilibrium, that is, the point at which temperatures stabilized, was observed at roughly 1000 years after zeroed emissions (Frölicher & Paynter, 2015). Over that time period, the warming commitment from past emissions was reported at 0.5°C, leading to peak warming of approximately 2.5°C above pre-industrial temperature (Frölicher & Paynter, 2015). However, findings by Ehlert and Zickfeld (2017) showed that at higher amounts of cumulative emissions, the ocean uptake of CO_2 appears to be insufficient to counteract the continued warming from thermal inertia, resulting in a positive zero-emissions commitment.

Admittedly, many factors are attributable to global climate change (IPCC, 2013a, 2013b, pp. 8−9); nonetheless, the preponderance of scientific evidence upholds that the prime cause is anthropogenically − induced GHG emissions (UCS, 2018). The IPCC Third Assessment Report stated that "most of the observed warming over the last 50 years is likely to have been due to the increase in GHG concentrations" (IPCC, 2001, p. 51). This assessment was reinforced by the IPCC Fourth Assessment Report, stating that "most of the observed increase in global average temperatures since the mid-20th century is very likely due to the observed increase in anthropogenic GHG concentrations" (ICCT, 2007, p. 39). The Working Group I report of the IPCC Fifth Assessment Report asserted that "it is extremely likely that human influence has been the dominant cause of the observed warming since the mid-20th century" (IPCC, 2001, p. 12). The scientific evidence provided by these IPCC reports has been strengthened and GHGs are regarded as the largest contribution to this phenomenon (IPCC, 2013a, 2013b). In the period 1970−2004, global GHG emissions from human activities increased by 70%, and CO_2 increased by 80%. It represented 77% of total GHG emissions in 2004 indicating that it is the most important anthropogenic GHG (ICCT, 2007, p. 36). However, CH_4 and N_2O accounted for only 14.3% and 7.9% of total GHG emissions, respectively, in 2004. To the extent climate change is concerned, oceans have been treated "both as victims of the problem and as part of the solution" (Freestone, 2009, p. 383). On the positive side, the oceans can transfer the heat between surface waters of the ocean and the lower atmosphere so as to adjust the global climate and weather; they also serve as a vital sink for absorbing anthropogenic GHG emissions (Currie & Wowk, 2009, p. 388). On the negative side, scientific data from all continents and most oceans had laid bare the fact that the climate changes resulting from GHG emissions have wreaked havoc on marine systems, leading to global marine-species redistribution and marine biodiversity reduction in sensitive regions, OA, and other future risks (IPCC, 2014a, 2014b: 16−17; Hoegh-Guldberg et al., 2018). The impacts from GHG emissions have provided further reasons to combat climate change (IPCC, 2018). However, the transboundary nature of GHG emissions renders its regulation a challenging task. Given the transboundary nature of GHG emissions, the regulation of this GHG issue needs to be conducted globally. Nevertheless, differing interests from various countries have made it a regulatory challenge to reach a consensus in relation to the reduction of GHG emissions (Simlinger & Mayer, 2019).

GHG emissions may originate from various sources, including land-based sources and marine shipping sources, such as automobile exhaust and shipping discharges. As a gaseous constituent in the atmosphere, GHG emissions often travel with the wind from the territory of or in other places under the jurisdiction or control of one country, to another place under the jurisdiction or control of another country or a place beyond the limits of national jurisdiction (IPCC, 2015). Hence, GHG emissions are often transboundary in nature (IPCC, 2014a, 2014b). The impacts of GHG emissions on the marine environment in a transboundary context include but are not limited to the following four aspects (IPCC, 2014a, 2014b). First, GHG emissions may gradually lead to the rise of ocean temperature so as to alter the dynamics of the marine environment. As a result of excessive emissions from GHGs, atmospheric CO_2 concentrations have increased from 280 to 380 ppm since the beginning of the industrial revolution, which is estimated to have led to a 0.74°C + 0.18°C global temperature rise during the past 100 years (Meinshausen et al., 2017). It is "virtually certain" that the upper ocean (0−700 m depths) warmed from 1971 to 2010, and "likely" that the ocean warmed from depths between 700 and 2000 m over the period 1957−2009 (Johnson et al., 2020). In the wake of prevalence of such circumstances, species distribution, polar systems, and global and regional weather patterns may be changed. (Currie & Wowk, 2009, p. 389). Some of the carbon stored in the form of methane hydrates from the seabed may ultimately be released once the deep ocean warms (Woody, 2019) and the dynamics, structure and biodiversity of marine ecosystems is also likely to shift (Nicholls & Cazenave, 2010, p. 1517).

Second, GHG emissions also entail the potential of causing sea-level rise and produce adverse impacts. Observations indicate that sea levels have risen by an average of 1.7 + 0.3 mm/year since 1950, and that this rate increased to 3.3 + 0.4 mm/year from 1993 to 2009, suggesting that the seal level rise is not only happening but that it is also accelerating (Nicholls & Cazenave, 2010, p. 1517). The 2007 IPCC Fourth Assessment Report concluded that under global warming conditions caused by excessive GHG emissions and ignoring the contribution from melting sea ice, global sea level will rise by 18 −59 cm in this century and reach 59 cm by 2099 (ICCT, 2007, p. 45). Even though, the IPCC's predictions on sea level rise are regarded as "remarkably conservative" and "wildly optimistic," largely owing to the IPCC's methodology of not taking into account the potential melting land ice (Schofield, 2007, p. 406). The 2013 IPCC Fifth Assessment Report strengthened this trend and asserted that more than 95% of the ocean area will experience sea level rise by the end of the 21st century (IPCC, 2013a, 2013b). It was further confirmed that global sea-level rise rates are "very likely" accelerating (IPCC, 2013a, 2013b). In 2019, global mean sea level was 87.61 mm centimeters above the 1993 average—the highest annual average in the satellite record (1993−present). From 2018 to 2019, global sea level rose 6.1 mm (Lindsey, 2020). Owing to the sea level rise, certain coastal hazards like flooding of coastal land, storm surges, erosion, destruction of infrastructure, settlements, and facilities are prone to occur. Apart from ensuring the safety of coastal residents requiring them to move to safer places as and when such occurrences happen, so as to avoid larger losses and seek more secure shelter (Boehm, 2006), and some low-lying coastal States, for instance the Maldives and Tuvalu, are even facing the risk of disappearance (Hinkel et al., 2019). In 2009, the Carteret Islanders of Papua New Guinea became the world's first entire community to be displaced by climate change (Connell, 2016). Vanuatu communities have also been displaced from the Torres Islands as a result of sea-level rise (Hinkel et al., 2019; IPCC (Intergovernmental Panel on Climate Change), 2014a, 2014b). Moreover, sea level rise may impact maritime jurisdictional claims by coastal States due to the alterations in their baselines (Hinkel et al., 2019; Sefrioui, 2017).

Third, GHG emissions are prone to play important role in causing OA and affect marine ecosystems negatively (Allsopp, Page, Johnston, & Santillo, 2009, p. 158). Since the ocean's role as a carbon sink, mounting CO_2 levels have led to heightened concentration of CO_2 into the surface water of the ocean and as the CO_2 dissolves into the seawater and acts as a weak acid, the carbonates in the ocean are reduced (Gattusso et al., 2015; NOAA, 2020). This chemical process is known as

OA. Acidification of the deep ocean could take place as one of the potential side effects of a process known as ocean fertilization (Williamson & Bodle, 2016), an activity designed to mitigate the deleterious effects of excess GHG emissions (Warner, 2009, p. 427). Up to now, the oceans have absorbed one-third of all anthropogenically sourced CO_2 (Warner, 2009, p. 426; Gruber et al., 2019). The mean surface ocean pH has dropped 0.1 units from 8.2 to 8.1 since 1750, and this number is anticipated to drop an extra 0.3—0.4 units by the end of the century (Doney et al., 2012). Measured on a logarithmic scale, pH is a dimensionless measure of the acidity of water or any solution. Given that pure water's pH is 7, acid solutions' pH is smaller than 7 while basic solutions' pH is larger than 7 (ICCT, 2007, p. 85).

Another active contributor of GHG emissions to marine environment is shipping and ships are instrumental in emitting about 3% of global CO_2 and GHG emissions (CO_2-eq), emitting roughly 1 billion tons of CO_2 and GHGs annually, on average from 2007 to 2012 (Smith et al., 2015). In the wake of phenomenal increase in global marine trade, ship emissions are likely to register further increase in both absolute terms and in shipping's share global CO_2 and GHG emissions. Estimates by Smith et al. (2015) project that ship CO_2 emissions will record increase 50%—250% from 2012 to 2050, and according to CE Delft (2017), emissions will register an increase 20%—120% over the same period for global temperature rise scenarios less than 2°C. In any case, shipping emissions are expected to increase in all situations. Undeniably, some sectors may adhere to measures designed to reduce GHG emissions; nonetheless, shipping will be responsible for an increasingly large share of global climate pollution in the wake of growing international marine trade and if no remedial action to reduce shipping emissions is taken, the shipping sector worldwide could be accountable 17% of global CO_2 emissions in 2050 (Cames, Graichen, Siemons, & Cook, 2015). Similar indications have been articulated by International Maritime Organization (IMO) in its fourth GHG Report released in early August 2020 that states that the GHG emissions—including CO_2, CH_4, and N_2O, expressed in CO_{2e}—of total shipping (international, domestic, and fishing) have increased from 977 million tons in 2012 to 1076 million tons in 2018 (9.6% increase). Noting that the share of shipping emissions in global anthropogenic GHG emissions has increased from 2.76% in 2012 to 2.89% in 2018, the IMO report GHG emissions are projected to grow from about 90% of 200t emissions in 2018 to 90%—130% of 2008 emissions by 2050, and these projected enhancements in GHG emissions are in a stark contrast to the IMO GHG reduction targets that set a goal to reduce the total annual GHG emissions from international shipping by at least 50% by 2050 compared to 2008, in terms of a climate change strategy adopted by the IMO in 2018 (DieselNet, 2020).

Buhaug et al. (2009) have classified GHG emissions emanating from ships into four categories—emissions from exhaust gases, emissions from refrigeration, cargo emissions, and other emissions. Emissions of exhaust gases are the main emissions coming from ships and these emanate from sources like main engines, auxiliary engines, boilers, and incinerators. Emissions from refrigeration are generally emitted through two main channels—leaks during operations and during maintenance of refrigeration and air-conditioning equipment or during the dismantling process (Buhaug et al., 2009). Cargo emissions include various emissions and leakages, in particular leaks of refrigerant from refrigerated containers and trucks, and volatile compounds emissions [CH_4 and nonmethane volatile organic compounds (NMVOCs)] from liquid cargoes. Other emissions comprise emissions from testing and maintenance of firefighting equipment, and other equipment (Buhaug et al., 2009).

GHG emissions emanating from international shipping wield adverse impacts on the environment, involving atmospheric composition, human health and climate (IMO, 2013), and are by their nature transboundary and chronically accumulative (Winnes, Styhre, & Fridell, 2015). First, most ships' emissions, excluding those in ports and in the vicinity of coastlines, are emitted in or transported to the marine boundary layer (MBL), also known as marine atmospheric boundary layer (MABL), is where the ocean and atmosphere exchange large amounts of heat, moisture, and momentum, primarily via turbulent transport (Winnes et al., 2015). It is that part of the atmosphere that has direct contact and, hence, is directly influenced by the ocean (Rahim, Islam, & Kuruppu, 2016; Semedo, Saetra, Rutggerson, Kahma, & Pettersson, 2009, p. 2256), and where they affect atmospheric composition negatively (Eyring et al., 2010, pp. 4744−4745; ICCT, 2007; OECD, 2008). Usually, a ship emits locally at relatively high concentrations (Eyring et al., 2010, p. 4744). Therefore, once GHG emissions from shipping are injected into the atmosphere, they mix with the ambient air and become diluted (Eyring et al., 2010, p. 4744, ICCT, 2007; OECD, 2008). In the interim, these emissions are chemically transformed prior to new secondary species, such as ozone, are produced, and some of them are subsequently removed from the atmosphere by wet and dry deposition (Eyring et al., 2010, pp. 4744−4745; ICCT, 2007; OECD, 2008). As a result, the atmosphere's composition will be affected (Rahim et al., 2016; Winnes et al., 2015). Second, GHG emissions from international shipping may also indirectly influence human health by means of the formation of ground-level ozone and particulate matter (Eyring et al., 2010, pp. 4752−4754; ICCT, 2007; OECD, 2008). Approximately 70% of the emissions from oceangoing shipping take place within 400 km of the coastline along the main seaborne trade routes (Corbett, Fischbeck, & Pandis, 1999, p. 3465; Leaper, 2019; Winnes et al., 2015). It can thus be inferred that most of the emissions from international shipping also occur within this distance from coastlines. If this is the case, these emissions may be transported hundreds of kilometers inland, bringing about air quality problems and impacting on human health (Leaper, 2019; OECD, 2008; Winnes et al., 2015). Lastly, ship emissions have an effect on climate—changing clouds and RF (Capaldo, Corbett, Kasibhatla, & Fischbeck, 1999; Fuglestvedt, et al., 2009). As a residual product remaining at the end of the crude oil refining chain, heavy fuel oil (HFO) has been widely used by various ships due to its competitive price (Santos, Loh, Bannwart, & Trevisan, 2014) However, the by-products of combustion from HFO, including CO_2, black carbon (BC), nitrogen oxides (NO_x), Sulfur dioxide (SO_2) and carbon monoxide (CO), produce significant impacts on climate by means of various physical and chemical interactions (Perera, 2018).

International shipping is the prime source of GHG emissions that pollutes the marine environment and release of these emissions into the atmosphere can potentially be dangerous to human health, marine ecosystems and organisms. IMO is currently reported to be engaged in two-pronged strategy, adopted in April 2018, toward addressing GHG emissions from international shipping through regulatory work, supported by capacity-building initiatives. In the first place, IMO has adopted regulations to deal with air pollutants from ships and has accordingly adopted mandatory energy-efficiency measures to reduce emissions of GHGs from international shipping, under Annex VI of IMO's pollution prevention treaty, known as MARPOL (IMO, 2018). The second step taken by IMO as part of its two-pronged strategy is that it is engaged in capacity-building projects worldwide to support the implementation of those regulations and boost innovation and technology transfer (IMO, 2018). Undoubtedly, regulating GHG emissions from international shipping is a priority area that requires diligence and sincere commitments of all stakeholders; nonetheless, it cannot be achieved overnight, albeit with the cooperation of all concerned IMO may be able to bring down the levels of GHG emissions soon to ensure healthy marine environment.

8.3.5 Land-based sources of marine pollution

Land-based sources, also called marine debris, constitute anthropogenic stressor that greatly influences marine ecosystems. Land-based marine litter is a growing problem that enforces progressively grave threat to the environment. Marine litter comprises any insistent, manufactured or processed solid material cast-off, disposed of or abandoned in the marine and coastal environment. Composition of marine litter includes items that have been made or used by people and intentionally rejected or inadvertently lost into the sea and on beaches, including such materials transported into the marine environment from land by rivers, wind or draining or sewage systems. (Galgani et al., 2010; UNEP, 2009). Albeit, all litter originates from land, the sources of marine litter are generally classified sea- and land-based (ARCADIS (Consultants in Infrastructure Water Environment Buildings), 2012; UNEP, 2009) depending on the type of activities through which litter enters into water. It is estimated that up to 80% of the marine litter is from land-based sources (Eunomia, 2016). Direct outfalls of land-based marine pollution (LBMP) usually comprise all inputs that are released directly into the marine environment via human activities in coastal settlements, nearby cities, and native industries and the consequential discharge and disposal of urban and industrial waste, litter, untreated sewage, and coastal agricultural installations like fish farms (Tibbetts, 2015). Inputs by rivers to LBMP include all inputs that are collected in an entire river catchment area and finally culminating in being discharged into the sea through the river. A river catchment area generally encompasses a rather substantial area, both inland and on the coast and that means that human inputs can enter the ocean from far within the land via being washed off by rainfall into a river or stream (Clark, 2001) and examples of such inputs include organic waste like sewage, pesticides and fertilizers from agriculture, as well as petroleum, plastics, oil-spills, and litter from municipal sewage washed off from roads and landfills (Jambeck et al., 2015).

Land-based sources of marine pollution comprise as much as 80% of the man-made inputs into the oceans and these constitute as one of the biggest grave threats to the marine environment (Tanaka, 2017). Marine litter—a type of land-based marine pollutant, has proven to be specifically endemic in the marine environment (Tanaka, 2017; Wang, Zhang, & Mu, 2017). Marine litter is a composite or collective term delineating all human solid wastes that have found its way into the marine environment (Tanaka, 2017). Marine litter is either deliberately released into the marine environment through dumping garbage into the sea from a ship or released into the sea inadvertently through being washed off the land into rivers and water ways, and thereby being transported into the sea (Bergmann, Gutow, & Klages, 2015; Wang et al., 2017). Interestingly, about 80% of marine litter comes from land-based sources via River Inputs and Direct Outfalls, and only 20% emanates from sea-based sources like shipping or offshore inputs (Haward, 2018). What makes marine litter more problematic than liquid pollutants like chemicals, oil or fertilizers reaching the marine environment is the fact that marine litter in majority of cases does not get diluted in the ocean. Instead, marine litter either floats on top of the surface of the water, or sinks to the bottom (Wang et al., 2017). Effective combat on marine litter is confronted with an uphill task in the wake of somewhat vast expanse of the catchment areas of rivers and streams as well as densely populated some coastal economies (Auta et al., 2017; Routley, 2018). As a way out of this predicament is to emphasize on the essentiality of addressing the issue via combating the most pressing issue first, as suggested by many studies, the most pressing issue of marine pollution at present is plastics that has already been dealt with in this chapter (Auta et al., 2017; Xanthos & Walker, 2017).

Pollutants from human activities on land identified as having the potential to affect marine ecosystems and organisms include: wastewater, agricultural runoff, nutrients, turbidity, sedimentation, heavy metals, salinity, pharmaceuticals and personal care products, hydrocarbons and other organics, biocides such as pesticides, herbicides and fungicides, and pathogens (Caraco, M. Novotney, & Gadbois, 2009). According to estimates of Food and Agriculture Organization (FAO), annual global freshwater withdrawals stand at 3928 km^3 (WWDR (World Water Development Report), 2017), and more than half of it being released into the environment as wastewater in the form of municipal and industrial effluent and agriculture drainage, and less than half is being mainly consumed by agriculture via evaporation in irrigated cropland (Krushelnytska, 2018). Bulk of the wastewater, nearly over 80%, released into the environment is not scientifically treated and interestingly, developed countries treat 70% of wastewater generated by them whereas this percentage ration declines to 38% in upper middle-income countries and to 28% in lower income countries (Sato, Qadir, Yamamoto, Endo, & Zahoor, 2013). Hazards to environment and health arising from the release of untreated wastewater into the marine environment depend on the type of pollutants. Excessive nutrients like phosphorous and nitrogen, including ammonia, in wastewater lead to eutrophication, or overfertilization of receiving waters—freshwater or marine—that can be toxic to aquatic organisms, boost excessive algae blooms, cause diminution in available oxygen, harm spawning grounds, alter habitat, and lead to a decline in certain species (Freeman et al., 2020; Krushelnytska, 2018).

Agricultural activities in the form of runoff water have come to be regarded as a primary contributor to an augmentation in pollutant delivery to marine ecosystems (NOAA, n.d.). Industrialized agriculture is one of the largest sources of nitrogen (N) and phosphorus (P) pollution in the form of animal manure, inefficient nutrient application, bad irrigation practices, and soil erosion (Krushelnytska, 2018; Lindwall, 2019). The use of N-based fertilizers is projected to double or even triple within the next 50 years (Beman, Arrigo, & Matson, 2005). Nonetheless agricultural runoff is highest among developed countries, marine nitrogen pollution is ever more widespread because of agriculture intensification worldwide. The level of land-based dissolved inorganic nitrogen (DIN) export from watersheds to large marine ecosystems (LMEs) varies globally across a large range of magnitudes (Lee, Rosalynn, Seitzinger, & Mayorga, 2015). Fertilizer was the major source of DIN to LMEs in most of Europe and Asia, while manure was the primary source in most of Central and South America. The smallest loads are exported to many polar and Australian LMEs, while the largest loads are exported to northern tropical and subtropical LMEs. The LMEs receiving the largest loads of land-based DIN are the North Brazil Shelf, Bay of Bengal, Guinea Current, South China Sea, East China Sea, and Gulf of Mexico LME (Krushelnytska, 2018; Lee et al., 2015).

Nutrients are conceived of as a pollutant when rarified concentrations wield a negative impact on marine and estuarine biological systems (Chislock, Doster, Zitomer, & Wilson, 2013). Augmentations in dissolved inorganic nutrients have been shown to contributing to reduction in coral calcification rates, diminution in fertilization success and probably engender the growth of macroalgae that vie for space with corals (Malone & Newton, 2020). The profusion of clionid sponges, a contender and predator for corals enhances the rarified nutrient concentration (Chislock et al., 2013). It has been reported by Ward-Paige, Risk, and Sherwood (2005) that deteriorations in coral cover have been linked to enhanced density of clionid sponges along coral reef track in the Florida Keys. Turbidity, being linked to physical forces acting on the seabed as well as terrestrial runoff, fluctuates naturally in time and space (Shi & Wang, 2010). Turbidity mostly consists of dissolved inorganic matter including light-absorbing colored organic matter and sediment load with clung or adhered organic components like bacteria (Bulmer, Townsend, Drylie, & Lohrer, 2018). Augmentation in turbidity tends to decrease water clarity that subsequently leads to reduced light availability to seagrass (SAFMC, 2009; SFWMD, 2009b) and coral zooxanthellae photosynthesis (SAFMC, 2009). Turbidity entails the potential of envisaging decrease in the recruitment, survival and settlement of coral larvae (Bulmer et al., 2018).

Other major land-based pollutants affecting marine ecosystems and organisms are salinity, sedimentation and heavy metals. Alterations in salinity are often referred to as being instrumental in spurring lethal and sublethal effects on corals (Gegner & Voolstra, 2019), oysters (Olive, Chang, Inn, Hwai, & Yasin, 2019; PBC, 2008; SFWMD, 2012) and seagrass (Lirman & Cropper, 2003; Oscar, Barak, & Winters, 2018; SFWMD, 2009a, 2012) in general and southeast Florida in particular. Salinity variations have also been reported as being held responsible for the loss of oyster and seagrass habitats in the SEFCRI region including St. Lucie River (SFWMD, 2009a, 2012), Lake Worth Lagoon (PBC, 2008; SFWMD, 2009b) and Biscayne Bay watersheds (Caccia & Boyer, 2005). Worrisome concerns are being expressed over the damaging role salinity changes play in the survival of already stressed coral organisms or other benthic fauna prevalent in marginal environment (Chui & Ang, 2017; Camp et al., 2018). The survival of benthic organisms including oysters, seagrass and corals in marginal habitats is seemingly a fertile realm of research necessary for their perseverance in such environments (Ridge, Rodriguez, & Fodrie, 2017). Sedimentation entails the potential of engendering mortality of filter feeding animals such as corals (Sarkis, 2017), sponges (SAFMC, 2009) and oysters (Housego & Rosman, 2016) by constraining feeding (Rogers et al., 2014), and by physically restricting or burial of sessile organisms (SAFMC, 2009). Sedimentation can reduce growth rates of oysters (Housego & Rosman, 2016) and corals (Sarkis, 2017) through shading or abrasion or scuff. Sublethal effects of sedimentation on corals comprise enhanced respiration and mucus production, as well as diminutions photosynthesis, reproduction, and larval survival rates (Sarkis, 2017). Sedimentation can also harmfully impact seagrass and other soft bottom estuarine habitats by direct burial and engender turbidity that reduces light penetration and decreases photosynthetic production in coastal waters (Donatelli, Fagherazzi, & Leonardi, 2018; SAFMC. (South Atlantic Fisheries Management Council), 2009). Heavy metals are reckoned with casting lethal and sublethal impacts on marine fauna like corals (Joseph et al., 2011), mollusks (Spicer & Weber, 2004; Wang & Guangyuan, 2017), and crustaceans (Barbieri, Passos, & Garcia, 2005; Mazzei et al., 2014; Spicer & Weber, 2004). Rarified metal accumulations have demonstrated to cast a negative impact on coral fecundity, reproduction and recruitment (Joseph et al., 2011). Copper, lead, zinc, cadmium, and nickel harmfully affected fertilization success on Scleractinia coral, with copper being the most toxic (Reichelt-Brushett & Harrison, 2005). Heavy metals entail the likelihood of turning toxic if the bioavailable metal concentrations overdo limen values, which vary for different organisms (SAFMC, 2009; Tchounwou, Yedjou, Patlolla, & Sutton, 2012).

Currently ongoing and emerging policy measures veering around regulating and managing for collective impacts to the oceans envisage a dire need to discern and understand as to how and how fast, cumulative effects are changing. Halpern et al. (2019) have opined that enlargement of existing uses of the ocean and emerging new ones — including offshore energy, ocean farming, and ocean mining—necessitates an understanding of what else is affecting those locations, how those new uses will contribute to existing impacts, and censoriously whether the collective impact of these ocean uses is changing, and how rapidly. Interestingly, both the European Union's Marine Strategy Framework Directive (MSFD) (EC, 2008) and the United Nation's SDG-14 focus on evaluating and bringing down collective pressures to the oceans, and the upcoming renegotiation of the Aichi Biodiversity Targets in 2020 (CBD, 2010) will benefit from a deeper understanding of the pace and pattern of change in cumulative impacts. Furthermore, the accelerating rate of creating marine protected areas (MPAs) to meet CBD targets of 10% of the ocean within protected area by 2020 (CBD (Convention on Biological Diversity), 2010), and the push worldwide to produce very large MPAs (Toonen et al., 2013), could likewise profit from "detailed maps of where and how fast cumulative impacts are changing, as this information is critical to siting and managing effective protected areas" (Halpern et al., 2019).

8.4 Ecosystem-based management for oceans

EBM is based or predicated on making use of the natural ecosystem boundaries as a framework instead of being confined by political or administrative boundaries (AORA, 2018). In the context of oceans, EBM focuses on the maintenance or augmentation of ecological structure and function, and the benefits that accrue to the society from healthy oceans (Link & Browman, 2017). EBM inevitably entails a degree of coordination across countries that share ocean ecosystems, and among national agencies and departments that shoulder responsibilities pertaining to ocean health and utilization of marine resource (Rudd et al., 2018). Adequate conceptualization and efficient implementation of EBM entails the potential of facilitating systematic, holistic perspectives on ocean management. Per se, EBM has managed to garner significant national and international interest and attention among governments and agencies entrusted with the onus for the sustainable utilization and management of ocean and coastal resources (AORA, 2017; European Parliament, 2018). The notion of EBM had started garnering increasing interest from the policymakers, academia and those dealing with ocean affairs during the first decade of the 21st century and this interest was manifested in the publication of a book (McLeod & Leslie, 2009a, 2009b), a number of articles in scientific journals (Tallis et al., 2010), issuance of a consensus statement (United States-S&PE, 2005), a web page (EBMTN, 2018), and an interim framework for coastal and marine spatial planning in the United States (IOPTF, 2009), epitomizing "the maturation of the movement to spread the word of EBM (Hoaglaned, 2010). Issuance of scientific consensus statement on EBM in 2005 described EBM as an integrated place-based management approach that focuses on the maintenance of Hoagland the integrity or augmenting the resilience of an entire ecosystem, including its structure, functioning, processes and dynamics (McLeod & Leslie, 2009a, 2009b, p. 4).

EBM explicitly takes note of both intrasystem and intersystem linkages, focuses on collective impacts, and is adaptive (Hoaglaned, 2010). While intra-system linkages would consist of biological relationships like predator prey, competition, commensalism, among many others, as well as dependencies on environmental conditions; whereas intersystem linkages would embrace connections among distinct ecosystems in the air, in the water, and on the land (Hoaglaned, 2010). According to web-based EBMTN (Ecosystem-Based Management Tools Network) (2018), multiple stakeholders are engaged by EBM in a "collaborative process to define problems and find solutions." Guerry (2005) is in favor of comparing EBM with the failed management approaches that focus on single species or single sectors. The governance of large-scale environments like coastal ocean have often been subject to criticism as being too centralized as well as also too fragmented (IOPTF, 2009, pp. 1–2). Pikitch, Santora, Babcock, and Bakun (2004) citing arguments of some commentators note that single-species fishery management "myopic alignments" of special interests with captured agencies, also called "single sector," have been the "norm" and resultantly policies and politics have yielded ineffective and unfair results. EBM now has come to be acknowledged now as a science-based solution to the governance problems of large-scale environments or ecosystems (Wasson et al., 2015).

In contrast to ecosystem-based science that engenders knowledge of ecosystem structures, functions and processes, EBM is seemingly construed as an integrated management approach entailing human activities to maintain health, productivity and resilience of ecosystem to generate ecosystem services (Foley et al., 2010). In planning, marine ecosystem-based science provides a discernment of the marine ecosystems and helps in the identification of the scale of activities that are or may be initiating impacts to these ecosystems (Halpern, 2009). From a management perspective, EBM needs a process for identifying the range or scale of environmental protection and conservation measures that are required to minimize these impacts (Cormier et al., 2017). From a pragmatic point of view, marine activities and the pressures they engender are managed or regulated by an array of standards, codes of practice, and guidelines that are implemented through their specific environmental protection regulatory frameworks (Boyes & Elliott, 2014). Indeed, this approach has been the hallmark of fisheries management and pollution legislation for decades (Bell, McGillivray, Pedersen, Lees, & Stokes, 2013; Garcia, Rice, & Charles, 2014). Viewed in a broad perspective, bulk of these frameworks are worked out to manage localized impacts within the footprint of individual marine activities (Cormier, Elliott, & Kannen, 2018; Cormier, Stelzenmüller et al., 2018; Lodge & Verlaan, 2018). Fragmentary or piecemeal management approaches by species, by activity or by concern are prone to be subject of criticism in EBM (McLeod, Lubchenco, Palumbi, & Rosenberg, 2005). Even though management strategies that emphasize on ecosystem structures, functions and processes are preferred as a more holistic approach; nonetheless, such conservation strategies are not essentially effective at minimizing all probable pressures and still need efficacious environmental protection strategies that are designed in accordance with the pressures engendered by particular activities (Cormier et al., 2017).

In the wake of enactment of legislation based on ecosystem management such as the Canadian Oceans Act or the European Wild Bird and Habitat directives (Canada, 2019), management attention shifted from conservation strategies to minimizing impacts to ecosystem structures, functions and processes via MPAs and environmental quality guidelines (De Santo & Jones, 2007). Nevertheless, incorporation of integrated coastal and ocean management into the Canadian Ocean Act and specifically, the European MSFD envisaged the essentiality of managing and reducing the accumulated pressures produced by human activities in addition to conservation strategies (Piet et al., 2019). According to some scholars (Cormier & Lonsdale, 2019; Cormier, Elliott et al., 2018; Cormier, Stelzenmüller et al., 2018), this unwittingly envisages some confusion with regard to the roles of environmental protection and conservation strategies. Individual activities and their resultant pressures that can wield influence on any local habitats and species are regulated by environmental protection strategies, whereas conservation strategies prefer spatial prohibitions of marine activities to attain wider ecosystem objectives (Cormier, Elliott, & Rice, 2019). Assuming that the anticipated outcomes of these two strategies are not identical, it is obvious that those entrusted the task of or involved in marine planning processes are called upon to ascertain complex and subtle differences between sector-based environmental protection regulatory frameworks and marine conservation regulatory frameworks. Besides, there is also a need for discernment of the efficacy of these regulatory frameworks as being appropriate for the purpose of attaining broader ecosystem and conservation objectives. Provided that many environmental protection regulatory frameworks are used independently to manage marine activities (Boyes, Elliott, Murillas-Maza, Papadopoulou, & Uyarra, 2016; Cavallo, Elliott, Quintino, & Touza, 2018), integration has amply to do with arraying the outcomes of these frameworks with the ecosystem and conservation resultants laid down by conservation strategies to deal with the shortfalls or gaps of current integrated management practices (van Hoof, 2015; van Leeuwen, van Hoof, & van Tatenhove, 2012). Provided that spatial conservation strategies alone like MPAs are unable to deal with pressures outside a MPA (Mach et al., 2017), improvements would be required to the efficacy of current environmental protection regulatory frameworks (Cormier et al., 2017, 2019).

The European MSFD (European Commission, 2008) seeks to apply an ecosystem-based approach to the management of human activities with a view to ensure that their collective pressures are kept within levels compatible with the achievement of good environmental status (Cormier, Elliott et al., 2018; Cormier, Stelzenmüller et al., 2018). In quest of the management logic of the controls envisaged in the program of measures of the MSFD, environmental protection regulatory frameworks for particular controls would have to execute input controls and spatial and temporal disbursement or distribution controls to meet the anticipated results of the output controls to uphold and maintain and attain good environmental status (Cormier, Elliott et al., 2018; Cormier, Stelzenmüller et al., 2018). In case efficacy of these regulatory frameworks are unable to meet the remaining or residual pressure allowed by the output control, it would imply then that mitigation and remediation tools would be required. However, this would also connote that mitigation and remediation or redressal tools would be required if the efficacy of these regulatory frameworks are unable to cope with the residual pressure allowed by the output control. This in way lessens the challenges faced to ascertain the good environmental status as well as to fix the limina for the output controls (Berg, Fürhaupter, Teixeira, Uusitalo, & Zampoukas, 2015). It does, however, emphasize the need to develop tools to analyze the environmental protection regulatory frameworks as well as the science to carry out efficiency studies of the standards, codes of practice and guidelines used in such frameworks (Cormier, Elliott et al., 2018; Cormier, Stelzenmüller et al., 2018). The operational execution of the ecosystem-based approach to management is not only about integration of policies, it is also about integration of the environmental protection and conservation regulatory frameworks (Ansong, Gissi, & Calado, 2017).

Increasing use of ocean resources by humans (Gelcich, 2014; Martin, Maris, & Sinberloff, 2016) have necessitated the urgency of addressing competing uses as well as has also impelled multiple international endeavors to contemplate options for managing numerous ecosystem services. The concept of EBM has come to be acknowledged as foremost among the options for managing multiple ecosystem services (Leslie, Rosenberg, & Eagle, 2008). While integrating ecological, social and governance objectives the CBD describes EBM as: "a strategy for the integrated management of land, water and living resources that promotes conservation and sustainable use in an equitable way" (CBD, 2011). As there is no single universally acceptable and applicable definition for EBM (O'Hagan, 2020) for different reasons, most definitions emphasize that EBM is an integrated approach that takes into consideration links among living and nonliving resources, involving the management of species, other natural commodities/services, and humans as components of the ecosystem (Steenberg, Duinker, & Nitoslawski, 2019). Irrespective of variance the terminology and particular emphasis used, such as EBM, ecosystem approach to fisheries, ecosystem management, ecosystem-based ocean planning, etc., the overall distinction of EBM approaches is that they embrace the interactions among ecosystem components, humans and the cumulative impacts of numerous activities, encouraging conservation and sustainable use of resources (Arkema, Abramson, & Dewsbury, 2006). During the past decade, environmental policy discussions around the world have increasingly encouraged an ecosystem approach to managing the oceans (Arriagada, 2018). EBM is currently dominating policy debates, global organizations such as the United Nations and FAO have come out with a series of recommendations that has made EBM globally recognized as the best practice for ocean governance (Gelcich et al., 2009; UNEP, 2016). The 12 key guiding principles developed by the CBD (2007) for EBM implementation have come to be regarded as vital to define different elements that are required to be incorporated in an EBM approach (Clarke & Jupiter, 2010; Long et al., 2015). Concurrently, in the wake of the EBM having become part of international conventions, it has been able to garner support from scientific consensus statements (Sardà, O'Higgins, Cormier, & Tintore, 2014) and is becoming objective of many national level policies (Leslie, Sievanen, Crawford, & Gruby, 2015). Furthermore, EBM guiding principles are part of the majority of the 20 Aichi biodiversity targets subscribed by the CBD conference of the parties (Leadley et al., 2014). In fact, target 6 explicitly states that "by 2020 all fish and invertebrate stocks and aquatic plants are managed and harvested sustainably, legally and applying ecosystem-based approaches, so that overfishing is avoided" (CBD, 2011). Thus, EBM is internationally recognized as a best practice for ocean governance, acknowledged by multiple stakeholders, and to which 198 countries have formally committed (Rudd et al., 2018; UNEP, 2016).

8.4.1 Ecosystem-based approach versus ecosystem-based management

The ecosystem-based approach (EBA)/ EBM is a multidimensional concept that started finding mention in environmental literature in the first half of the 20th century (Grumbine, 1994, p. 28), though its underlying philosophies have been practiced for centuries in many cultures (Long et al., 2015). An early iteration of the approach in international law is found in the 1980 Convention on the Conservation of Antarctic Marine Living Resources (CAMLR, 2019), which requires conservation or harvesting activities to be carried out with regard to the maintenance of the ecological relationship between harvested and other species as well as to the marine ecosystem as a whole (Langlet & Rayfuse, 2018b, p. 2). The terms "EBA" and EBM' are being used as interchangeable terms (EC, 2018a, 2018b; Liebernecht, 2020). In the 1990s, when environmental conservation literature progressively developed holistic principles and approaches for managing humans as part of ecosystems, the concept of ecosystem approach also garnered ample traction (Waylen, et al., 2014). In the wake of the parties to the CBD agreeing in 1995 that ecosystem approach "should be the primary framework of action to be taken under the Convention," this approach garnered recognition as a policy concept. In 2000, the parties to the CBD at COP-5 adopted vide Decision V/6, a definition according to which the ecosystem approach "is a strategy for the integrated management of land, water and living resources that promotes conservation and sustainable use in an equitable way." It further "requires adaptive management to deal with the complex and dynamic nature of ecosystems and the absence of complete knowledge or understanding of their functioning"

(CBD (, 2000). This definition was followed up with the elaboration and subsequent endorsement by the CBD parties of the 12 so-called "Malawi Principles for the Ecosystem Approach" (CBD, 2000). The ecosystem approach was endorsed by the parties to the Helsinki Convention through a joint statement with OSPAR in 2003 (OSPAR, 2003). The statement defines the ecosystem approach as "the comprehensive integrated management of human activities based on the best available scientific knowledge about the ecosystem and its dynamics, in order to identify and take action on influences which are critical to the health of marine ecosystems, thereby achieving sustainable use of ecosystem goods and services and maintenance of ecosystem integrity" (OSPAR (Commission for the Protection of the Marine Environment of the North-East Atlantic), 2003, p. 16).

Apart from politically endorsed statements and definitions, there is ample scientific literature that deals with as to how the approach should be defined and what it means in practice to apply an ecosystem approach in different kinds of natural resource management (Long et al., 2015; Tallis et al., 2010). The problem gets with further complicated in the wake of the fact there are a number of concepts with identical or overlapping meaning, such as "EBM," "ecosystem approach to management," and "ecosystem-based marine management." These all come with different definitions in different contexts. Nevertheless, Arkema et al. (2006: 528) have opined that an analysis of definitions of "ecosystem management" (EM), "EBM" and "ecosystem-based fisheries management" (EBFM) provided in the scientific literature found no statistically relevant differences in how they related to notions such as the inclusion of humans in ecosystems, complexity, ecosystem goods and services, or the precautionary approach. None of these terms thus seem to have a distinct or established content that distinguishes it from the concept of "ecosystem approach." This is substantiated by a review of references to the ecosystem approach between 1957 and 2012 by Waylen et al. (2014, p. 1215) that identified three primary uses of the term, one being "as an alternative to ecosystem management or EBM"; and the other primary uses were "in reference to an integrated and equitable approach to resource management as per the CBD; and as a term signifying a focus on understanding and valuing ecosystem services" (Waylen et al., 2014, p. 1215).

Murawski (2007, p. 682) in his study has inferred that the many definitions of concepts like ecosystem approach to management and EBM "invariably share a number of common characteristics involving broadening stakeholder involvement, evaluation of multiple simultaneous drivers or 'pressures' on ecosystems, and specifying that EAM/EBM is geographically based vs. being primarily species or single-issue driven." He further goes on to say that an ecosystem approach to management has some defining characteristics, in comparison to more narrow management approaches, namely that it is: '(1) geographically specified, (2) adaptive in its development over time as new information becomes available or as circumstances change, (3) takes into account ecosystem knowledge and uncertainties, (4) recognizes that multiple simultaneous factors may influence the outcomes of management (particularly those external to the ecosystem), and (5) strives to balance diverse societal objectives that result from resource decision making and allocation. Additionally, because of its complexity and emphasis on stakeholder involvement, the process of implementing EAM needs to be (6) incremental and (7) collaborative' (Murawski, 2007, p. 682). The dearth of an explicitly clear definition of EBM cherishing overall backing in terms of its usage may be the reason as to why many scholars defer the issue of definition and concentrate on the main challenges, and steps that EBM needs to address (WWF, 2017, p. 15). Leaving the issue of definitions aside, it is the practical application aspect of ecosystem approach that poses a real challenge for the managers and they have to cope with it as well (Österblom et al., 2010, p. 1290). It has been opined that in practical terms, the differing contexts wherein an ecosystem approach is needed or executed, renders a universal definition of tangible management measures that would comprise the operationalization of the ecosystem approach impracticable and perhaps obstructive (Langlet & Westholm, 2019). Nevertheless, it has been suggested that general guidelines as envisaged in the Malawi Principles for the ecosystem approach (CBD (Convention on Biological Diversity), 2000), can be utilizable by making available an overreaching frame of reference that can be augmented by pragmatic experience of the problems confronted and the lessons imbibed in the course of the execution of the approach in particular contexts and situations (Langlet & Rayfuse, 2018a, p. 447).

Viewed in a broader context, the notion of ecosystem approach is bereft of sustentative guidance that permits maintain equilibrium between conservation and sustainable utilization of natural resources (Douvere, 2010, p. 765) and it entails concepts and principles that too wide-ranging and composite (Ansong et al., 2017, p. 66). Undeniably, ambiguous and imprecise nature of notions like "ecosystem health" and "ecosystem integrity" that are associated with the ecosystem approach have been subjected to criticism for their lack of substance; nonetheless, it is this ambiguity and imprecision that has been acknowledged as a significant factor responsible for the wide-ranging acceptance of the ecosystem approach (Engler, 2015, p. 295). It has been opined by Langlet and Westholm (2019) that it is this relative vagueness that makes it essential to identify and manage potential value conflicts "as well as inconsistent perceptions and expectations in the practical application" of the ecosystem approach. While commenting on the proximity between EBA and EBM, Langlet and Witholm (2019) say: "In fact, many definitions of the ecosystem approach contain provisions indicating that management is to be "within the limits of" or respecting the functions of the ecosystem, thus signaling that ecosystems impose restrictions on the volume and nature of human activities affecting them, although the detailed nature of those limits may be contested."

With all its potential worth and utility, nevertheless, there has been little substantive proof of the efficacy of EBM (Tallis et al., 2010). Besides, apart from scientific complexities, the requirement of setting up of novel institutions for governance present significant challenges (Leslie et al., 2008). Being skeptical about EBM, Berghofer et al. (2008) opine that underneath EBM's characteristically scientific façade, there may be "poor justification" for some of its primary general principles. For some scholars, there are an array of concerns that the pace of adopting and implementing EBM is inadequate, irrespective of its promise (Leslie & McLeod, 2007; Link & Browman, 2017)' especially in the wake of exacerbating pace of environmental change (Barange et al., 2018), technological advance (Sutherland et al., 2017), and the emerging international significance of the "blue growth" agenda (Visbeck et al., 2014). There prevails currently a common perception that within and across jurisdictions there is uneven or excessive legislative complications (Boyes & Elliott, 2014; Boyes et al., 2016), comparatively a low level of policy cohesiveness or configuration pertaining to ocean and coastal EBM, and that more arrayed legislation is required to expedite EBM adoption and help maintain healthy oceans (Marshak et al., 2017). Research has also focused on enfeebled implementation of EBM (Link & Browman, 2014, 2017; Marshak et al., 2017), until present times, there has been little research focus on the legal and policy mandates required to support efficacious EBM (Boyes et al., 2016), and how those might impact EBM implementation.

References

Abarca, E. (2006). *Seawater intrusion in complex geological environments*. Technical University of Catalonia, UPC PhD Thesis 154. Available from http://www.tdx.cat/TDX-0222107−105833.

Abram, N., Gattuso, J.P., Prakash, A., Cheng, L., Chidichimo, M.J., Enomoto, H., ... & von Schuckman, K. (2019). Framing and context of the report. In *IPCC special report on the ocean and cryosphere in a changing climate*. Available from https://www.ipcc.ch/sites/assets/upload/sites/3/2019/11/05_SROCC_SPM_FINAL.pdf.

Albers, P. (2003). Petroleum and individual polycyclic aromatic hydrocarbons. In D. J. Hoffman, B. A. Rattner, & G. A. Burton (Eds.), *Handbook of ecotoxicology* (pp. 341−371). Boca Raton FL: CRC Press.

Aldred, J. (2019). *Oceans hold solutions to help tackle climate change*. September 23. Available online from https://chinadialogueocean.net/10234-solutions-to-climate-change-ocean/.

Alexander, M. A., Scott, J. D., Friedland, K. D., Mills, K. E., Nye, J. A., Pershing, A. J., & Thomas, A. C. (2018). Projected sea surface temperatures over the 21st century: Changes in the mean, variability and extremes for large marine ecosystem regions of Northern Oceans. *Elementa Science of the Anthropocene*, 6(1), 9.

Allemand, D., & Osborn, D. (2019). Ocean acidification impacts on coral reefs: From sciences to solutions. *Regional Studies in Marine Science*, 28, 100558.

Allison, E., & Mandler, B. (2018). *Off-shore oil and gas: Tachlogical and environmental challenges in increasing deep water*. Petroleum and the Environment. Part13/24. Available at: https://www.americangeosciences.org/sites/default/files/AGI_PetroleumEnvironment_web.pdf.

Allsopp, M., Page, R., Johnston, P., & Santillo, D. (Eds.), (2009). *State of the world's oceans*. Dordrecht: Springer.

Allwood, J. M., Bosetti, V., Dubash, N. K., Gómez-Echeverri, L., & von Stechow, C. (2014). *Glossary. Climate change 2014: Mitigation of climate change. Contribution of working group III to the fifth assessment report of the intergovernmental panel on climate change* (pp. 1251−1274). Cambridge, UK.: Cambridge University Press.

Almeda, R., Baca, S., Hyatt, C., & Buskey, E. J. (2014). Ingestion and sublethal effects of physically and chemically dispersed crude oil on marine planktonic copepods. *Ecotoxicology (London, England)*, 23, 988−1003.

Altieri, A. H., Nelson, H. R., & Gedan, K. B. (2019). The significance of ocean deoxygenation for tropical ecosystems − corals, seagrasses and mangroves. In D. Laffoley, & J. M. Baxter (Eds.), *Ocean deoxygenation: Everyone's problem—Causes, impacts, consequences and solutions* (3, pp. 401−429). Gland, Switzerland: IUCN.

Ansong, J., Gissi, E., & Calado, H. (2017). An approach to ecosystem-based management in maritime spatial planning process. *Ocean & Coastal Management*, 141, 65−81.

Anthony, K. R. N., Kline, D. I., Diaz-Pulido, G., Dove, S., & Hoegh-Guldberg, O. (2008). Ocean acidification causes bleaching and productivity loss in coral reef builders. *Proceedings of National Academy of Sciences of the United States of America*, 105, 17442−17446.

AORA (Atlantic Ocean Research Alliance). (2017). *Working group on the ecosystem approach to ocean health and stressors*. Reykjavik, Iceland: AORA Office.

AORA (Atlantic Ocean Research Alliance). (2018). *Working group on the ecosystem approach to ocean health and stressors*. Mandates for Ecosystem-based Ocean Governance across Canada, the EU, and the United States. London, UK: AORA. Available online from: https://www.un-ilibrary.org/content/journals/15643913/54/2/41.

ARCADIS (Consultants in Infrastructure Water Environment Buildings). (2012). *Case study on the plastic cycle and its loopholes in the four European regional seas areas*. European Commission DG Environment Framework contract. ENV.D.2./ETU/2011/0041. Brussels: ARCADIS.

Arkema, K. K., Abramson, S. C., & Dewsbury, B. L. (2006). Marine ecosystem-based management: From characterization to implementation. *Frontiers in Ecology and the Environment*, 4(10), 525–532.

Armbrust, E., & Palumbi, S. (2015). Uncovering hidden worlds of ocean biodiversity. *Science (New York, N.Y.)*, 348(6267), 865–867.

Armstrong, K., Frost, D. J., McCammon, C. A., Rubie, D. C., & Ballaran, T. B. (2019). Deep magma ocean formation set the oxidation state of Earth's mantle. *Science (New York, N.Y.)*, 365(6456), 903–906.

Arnell, N. W., & Gosling, S. N. (2013). The impacts of climate change on river flow regimes at the global scale. *Journal of Hydrology*, 486, 351–364.

Arriagada, R. A. (2018). Assessing the implementation of marine ecosystem-based management into national policies: Insights from agenda setting and policy responses. *Parine Policy*, 92, 1–17.

Auta, H., Emenike, C., & Fauziah, S. (2017). Distribution and importance of microplastics in the marine environment: A review of the sources, fate, effects, and potential solutions. *Environment International*, 102, 165–176.

Avio, C. G., Gorbi, S., & Regoli, F. (2017). Plastics and microplastics in the oceans: From emerging pollutants to emerged threat. *Marine Environmental Research*, 128, 2–11.

Badyaev, A. V. (2005). Stress-induced variation in evolution: from behavioural plasticity to genetic assimilation. *The Royal Society Proceedings B, Biological Science*, 272(1566), 877–886.

Baggenstos, D., Haberli, M., Schmitt, J., Shackleton, S. A., Birner, B., Severinghaus, J. P., ... Fischer, H. (2019). Earth's radiative imbalance from the Last Glacial Maximum to the present. *Proceedings of National Academy of Sciences of the United States of America*, 116(30), 14881–14886.

Barange, M., Bahri, T., Beveridge, M. C. M., Cochrane, K. L., Funge-Smith, S., & Poulain, F. (Eds.), (2018). *Impacts of climate change on fisheries and aquaculture: Synthesis of current knowledge, adaptation and mitigation options*. Rome: FAO, Fisheries and Aquaculture Technical Paper No. 627.

Barbier, E. B. (2015). Climate change impacts on rural poverty in low-elevation coastal zones. *Estuarine Coastal and Shelf Science*, 165, a1–a13.

Barbier, E. B., Koch, E. W., Silliman, B. R., Hacker, S. D., Wolanski, E., Primavera, J., ... Reed, D. J. (2008). Coastal ecosystem-based management with nonlinear ecological functions and values. *Science (New York, N.Y.)*, 319, 321–323.

Barbieri, E., Passos, E. A., & Garcia, C. A. B. (2005). Use of metabolism to evaluate the sublethal toxicity of mercury on Farfantepeneus brasiliensis larvae. *Journal of Shellfish Research*, 24(4), 1229–1233.

Barboza, L. G. A., & Gimenez, B. C. G. (2015). Microplastics in the marine environment: current trends and future perspectives. *Marine Pollution Bulletin*, 97(1–2), 5–12.

Barth, P., Berenshtein, I., Bnson, M., Roux, N., Banaigs, B., & Lecchini, D. (2015). From the ocean to a reef habitat: How do the larvae of coral reef fishes find their way home? A state of art on the latest advances. *Vie et Milieu*, 65(2), 91–100.

Barton, A., Hales, B., Waldbusser, G. G., Langdon, C., & Feely, R. A. (2012). The Pacific oyster, Crassostrea gigas, shows negative correlation to naturally elevated carbon dioxide levels: Implications for near-term ocean acidification effects. *Limnology and Oceanography*, 57(3), 698–710.

Bell, J. D., Sharp, M. K., Havice, E., Batty, M., Charlton, K. E., Russell, J., ... Gillett, R. (2019). Realizing the food security benefits of canned fish for Pacific Island countries. *Marine Policy*, 100, 183–191.

Bell, S., McGillivray, D., Pedersen, O. W., Lees, E., & Stokes, E. (2013). *Environmental law* (Ninth. (ed.)). Oxford, UK: Oxford University Press.

Bellas, J., Saco-Alvarez, L., Nieto, O., Bayona, J. M., Albaiges, J., & Beiras, R. (2013). Evaluation of artificially-weathered standard fuel oil toxicity by marine invertebrate embryogenesis bioassays. *Chemosphere*, 90, 1103–1108.

Beman, M. J., Arrigo, K. R., & Matson, P. A. (2005). Agricultural runoff fuels large phytoplankton blooms in vulnerable areas of the ocean. *Nature*, 434(7030), 211–214.

Benarsek, N., Feely, R. A., Reum, J. C. P., Peterson, B., Menkel, J., Alin, S. R., & Hales, B. (2014). Limacina helicina shell dissolution as an indicator of declining habitat suitability owing to ocean acidification in the California current ecosystem. *Proceedings of the Royal Society, Series B, 281*, 20140123.

Berg, T., Fürhaupter, K., Teixeira, H., Uusitalo, L., & Zampoukas, N. (2015). The marine strategy framework directive and the ecosystem-based approach—Pitfalls and solutions. *Marine Pollution Bulletin, 96*, 18–28.

Bergmann, M., Gutow, L., & Klages, M. (2015). *Marine anthropogenetic litter.* Bremerhaven, Germany: Springer International.

Bernhardt, E. S., Rosi, E. J., & Gessner, M. O. (2017). Synthetic chemicals as agents of global change. *Frontiers in Ecology and Environment, 15*(2), 84–90.

Besseling, E., Foekema, E. M., Van Franeker, J. A., Leopold, M. F., Kühn, S., Rebolledo, E. B., ... Koelmans, A. A. (2015). Microplastic in a macro filter feeder: humpback whale Megaptera novaeangliae. *Marine Pollution Bulletin, 95*(1), 248–252.

Biodiversity, A.-Z. (n.d.). *Areas beyond national jurisdiction (ABNJ).* Available online at: https://www.biodiversitya-z.org/content/areas-beyond-national-jurisdiction-abnj.

Blackburn, M., Mazzacano, C. A. S., Fallon, C., & Black, S. H. (2014). *Oil in our oceans: A review of the impacts of oil spills on marine invertebrates.* Portland, USA: The Xerces Society for Invertebrate Conservation.

Blanchard, J. L., Watson, R. A., Fulton, E. A., Cottrell, R. S., Nash, K. S., Bryndum-Bunchholz, A., ... Jennings, S. (2017). Linked sustainability challenges and trade-offs among fisheries, aquaculture and agriculture. *Nature Ecology & Evolution, 1*, 1240–1249.

Bñlum, J., Borglin, S., Chakraborty, R., Fortney, J. L., Lamendella, R., Mason, O. U., ... Jansson, J. K. (2012). Deep-sea bacteria enriched by oil and dispersant from the deepwater horizon spill. *Environmental Microbiology, 14*, 2405–2416.

Boehm, P. (2006). *Global warming: Devastation of an atoll. The independent.* 30 August 2006. Available online from: https://www.independent.co.uk/climate-change/news/global-warning-devastation-of-an-atoll-413922.html.

Borrelle, S. B., Rochman, C. M., Liboiron, M., Bond, A. L., Lusher, A., Bradshaw, H., & Provencher, J. F. (2017). Opinion: Why we need an international agreement on marine plastic pollution. *Proceedings of the National Academy of Sciences, USA, 114*(38), 9994–9997.

Bosello, F., & De Cian, E. (2014). Climate change, sea level rise, and coastal disasters. A review of modeling practices. *Energy Economics., 46*, 593–605.

Bostock, H. C., Sutton, P. J., Williams, M. J., & Opdyke, B. N. (2013). Reviewing the circulation and mixing of Antarctic intermediate water in the South Pacific using evidence from geochemical tracers and Argo float trajectories. *Deep Sea Research Part I: Oceanographic Research Papers, 73*, 84–98.

Boyd, P. W., Cornwall, C. E., Davison, A., Doney, S. C., Fourquez, M., Hurd, C. L., ... McMinn, A. (2016). Biological responses to environmental heterogeneity under future ocean conditions. *Global Change Biology, 22*, 2633–2650.

Boyd, P. W., Dillingham, P. W., McGraw, C. M., Armstrong, E. A., Feng, Y. Y., Hurd, C. L., ... Nunn, B. L. (2015). Physiological responses of a Southern Ocean diatom to complex future ocean conditions. *Nature Climate Change, 6*, 207–213.

Boyd, P. W., Lennartz, S. T., Glover, D. M., & Doney, S. C. (2015). Biological ramifications of climate-change mediated oceanic multi-stressors. *Nature Climate Change, 5*, 71–79.

Boyer, T., Domingues, C. M., Good, S. A., Johnson, G. C., Lyman, J. M., Ishii, M., ... Bindoff, N. L. (2016). Sensitivity of global upper-ocean heat content estimates to mapping methods, XBT bias corrections, and baseline climatologies. *Journal of Climate, 29*(13), 4817–4842.

Boyes, S. J., & Elliott, M. (2014). Marine legislation—The ultimate "horrendogram": International law, European directives & national implementation. *Marine Pollution Bulletin, 86*, 39–47.

Boyes, S. J., Elliott, M., Murillas-Maza, A., Papadopoulou, N., & Uyarra, M. C. (2016). Is existing legislation fit-for-purpose to achieve good environmental status in European seas? *Marine Pollution Bulletin, 111*, 18−32.

Bravo, L., Ramos, M., Astudillo, O., Dewitte, B., & Goubanova, K. (2015). Seasonal variability of the Ekman transport and pumping in the upwelling system off central-northern Chile (∼ 30o S) based on a high-resolution atmospheric regional model (WRF). *Ocean Science Discussions, 12*, 3003−3041.

Breitburg, D., Conley, D. J., Isensee, K., Levin, L. A., Limburg, K. E., & Williamson, P. (2019). What can we do? Adaptation and solutions to declining ocean oxygen. In D. Laffoley, & J. M. Baxter (Eds.), *Ocean deoxygenation: Everyone's problem—Causes, impacts, consequences and solutions* (pp. 545−562). Gland, Switzerland: IUCN.

Breitburg, D., Levin, L. A., Oschlies, A., Grégoire, M., Chavez, F. P., Conley, D. J., ... Zhang, J. (2018). Declining oxygen in the global ocean and coastal waters. *Science (New York, N.Y.), 359*, eaam7240.

Brennan, G., & Collins, S. (2015). Growth responses of a green alga to multiple environmental drivers. *Nature Climate Change, 5*, 892−897.

Briggs, K. T., Yoshida, S. H., & Gershwin, M. E. (1996). The influence of petrochemicals and stress on the immune system of seabirds. *Regulatory Toxicology and Pharmacology, 23*, 145−155.

Brooks, P. R., & Crowe, T. P. (2019). Combined Effects of multiple stressors: New insights into the influence of timing and sequence. *Frontiers in Ecological Evolution., 2019*, 000387.

Bulling, M. T., Hicks, N., Murray, L., Paterson, D. M., Raffaelli, D., White, P. C., & Solan, M. (2010). Marine biodiversity−ecosystem functions under uncertain environmental futures. *Philosophical Transactions of the Royal Society of London B: Biological Sciences, 365*(1549), 2107−2116.

Bulmer, R. H., Townsend, M., Drylie, T., & Lohrer, A. M. (2018). Elevated turbidity and the nutrient removal capacity of seagrass. *Frontiers in Marine Science, 5*, 462.

Burrows, M. T., Schoeman, D. S., Richardson, A. J., Molinos, J. G., Hoffman, A., Buckley, L. B., ... Poloczanska, E. S. (2014). Geographical limits to species-range shifts are suggested by climate velocity. *Nature, 507*, 492−495.

Byrne, M., Foo, S., Soars, N. A., Wolfe, K. D. I., Nguyen, H. D., Hardy, N., & Dworjanyn, S. A. (2013). Ocean warming will mitigate the effects of acidification on calcifying sea urchin larvae (Heliocidaris tuberculata) from the Australian global warming hot spot. *Journal of Experimental Marine Biology and Ecology, 448*, 250−257.

Caccia, V. G., & Boyer, J. N. (2005). Spatial patterning of water quality in Biscayne Bay, Florida as a function of land use and water management. *Marine Pollution Bulletin, 50*, 1416−1429.

Cames, M., Graichen, J., Siemons, A., & Cook, V. (2015). *Emission reduction targets for international aviation and shipping*. European Parliament. Available from http://www.europarl.europa.eu/RegData/etudes/STUD/2015/569964/IPOL_STU(2015)569964_EN.pdf.

CAMLR (Convention of Antarctic Marine Living Resources). (2019). *Conference on the conservation of Antarctic marine living resources*, Canberra, Australia, 7−20 May 1980. CCAMLR website online, 23 October 2019.

Camp, E. F., Schoepf, V., Mumby, P. J., Hardtke, L. A., Rodolfo-Metalpa, R., Smith, D. J., & Suggett, D. J. (2018). The future of coral reefs subject to rapid climate change: Lessons from natural extreme environments. *Frontiers in Marine Science, 5*, 4.

Canada. (2019). Oceans Act. S.C., 1996, c.31. Available from https://laws-lois.justice.gc.ca/PDF/O-2.4.pdf.

Capaldo, K., Corbett, J. J., Kasibhatla, P., & Fischbeck, P. (1999). Effects of ship emissions on sulphur cycling and radiative climate forcing over the ocean. *Nature, 400*(6746), 743−746.

Capet, A., Cook, P., Garcia-Robledo, E., Hoogakker, B., Paulmier, A., Rabouille, C., & Vaquer-Sunyer, R. (2020). Editorial: Facing marine deoxygenation. *Frontiers in Marine Science, 7*, 46.

Caraco, D., M. Novotney, N., & Gadbois, T. C. (2009). *LBSP Project 21—Review of programs in Southeast Florida that generate pollution and provide recommendations to reduce landbased pollution impacting the Southeast Florida coral reefs*. Ellicott City, MD: Center for Watershed Protection.

Carmichael, R. H., Graham, W. M., Aven, A., Worthy, G., & Howden, S. (2012). Were multiple stressors a 'perfect storm' for Northern Gulf of Mexico bottlenose dolphins (*Tursiops truncatus*) in 2011? *PLoS One, 7*, e41155.

Carrera, J., Hidalgo, J. J., Slooten, L. J., & Vázquez-Suñé, E. (2010). Computational and conceptual issues in the calibration of seawater intrusion models. *Hydrogeology Journal, 18*, 131−145.

Castrucci, L., & Tahvildari, N. (2018). Modeling the impacts of sea level rise on storm surge inundation in flood-prone urban areas of Hampton Roads, Virginia. *Marine Technology Society Journal, 52*(2), 92−105.

Catling, D. C., & Kasting, J. F. (2017). Formation of Earth's atmosphere and oceans. In D. C. Catling, & J. F. Kasting (Eds.), *Atmospheric evolution on inhabited and lifeless worlds* (pp. 171−197). Cambridge, UK: Cambridge University Press.

Cavallo, M., Elliott, M., Quintino, V., & Touza, J. (2018). Can national management measures achieve good status across international boundaries?—A case study of the Bay of Biscay and Iberian coast sub-region. *Ocean & Coastal Management, 160*, 93−102.

CBD (Center for Biological Diversity). (n.d.). *Ocean plastics pollution—A global tragedy for our oceans and sea life*. Available online at: https://www.biologicaldiversity.org/campaigns/ocean_plastics/.

CBD (Convention on Biological Diversity). (2000). *Malawi principles for the ecosystem approach*. UNEP/CBD/ COP/5/23, Decision V/6 Ecosystem Approach (2000).

CBD (Convention on Biological Diversity). (2007). *Ecosystem approach principles*. CBD website online, 2 July 2007. Available at: https://www.cbd.int/ecosystem/principles.shtml.

CBD (Convention on Biological Diversity). (2009). *Scientific synthesis of the impacts of ocean acidification on marine biodiversity*. Montreal, Canada: Secretariat of the Convention on Biological Diversity, Technical Series No. 46.

CBD (Convention on Biological Diversity). (2010). *Strategic plan for biodiversity 2011−2020 and the Aichi targets*. Montreal: CBD Available at. Available from https://www.cbd.int/undb/media/factsheets/undb-factsheet-sp-en.pdf.

CBD (Convention on Biological Diversity). (2011). *Ecosystem approach*. CBD Website Online. Available from http://www.cbd.int/ecosystem/.

CBD (Convention on Biological Diversity). (2012). *Marine biodiversity—One ocean, many worlds of life*. Montreal: CBD Secretariat.

Chambers, D. P., Cazenave, A., Champollion, N., Dieng, H., Llovel, W., Forsberg, R., ... Wada, Y. (2016). *Evaluation of the global mean sea level budget between 1993 and 2014, . Surveys in geophysics*. (38, pp. 395−428). Switzerland: Springer. (1): 1−19.

Cheng, L., Abraham, J., Goni, G., Boyer, T., Wijffels, S., Cowley, R., ... Zhu, J. (2016). XBT science: Assessment of instrumental biases and errors. *Bulletin of the American Meteorological Society, 97*(6), 924−933.

Cheng, L., Abraham, J., Zhu, J., Trenberth, K. E., Boyer, T., Locarnini, R., ... Mann, M. E. (2020). Record-setting ocean warmth continued in 2019. *Advances in Atmospheric Sciences, 37*(2), 137−142.

Cheng, L., Trenberth, K. E., Fasullo, J., Boyer, T., Abraham, J., & Zhu, J. (2017). Improved estimates of ocean heat content from 1960 to 2015. *Science Advances, 3*(3), e1601545.

Cheng, L. J., Trenberth, K. E., Fasullo, J., Abraham, J., Boyer, T. P., Von Schuckmann, K., & Zhu, J. (2018). Taking the pulse of the planet. *EOS (Rome, Italy), 99*, 14−16.

Cheng, L. J., & Zhu, J. (2018). 2017 was the warmest year on record for the global ocean. *Advances in the Atmospheric Sciences, 35*(3), 261−263.

Cheng, L. J., Zhu, J., Abraham, J. P., Trenberth, K. E., Fasullo, J. T., Zhang, B., ... Song, X. (2019). 2018 continues record global ocean warming. *Advances in the Atmospheric Sciences, 36*, 249–252.

Chislock, M. F., Doster, E., Zitomer, R. A., & Wilson, A. E. (2013). Eutrophication: Causes, consequences, and controls in aquatic ecosystems. *Nature Education Knowledge, 4*(4), 10.

Chui, A. P. Y., & Ang, P., Jr. (2017). High tolerance to temperature and salinity change should enable Scleractinia coral Platygyra acuta from marginal environments to persist under future climate change. *PLoS One, 12*(6), e0179423.

Church, J. A., Clark, P. U., Cazenave, A., Gregory, J. M., Jevrejeva, S., Levermann, A., ... Unnikrishnan, A. S. (2013). *Sea Level Change. Climate change 2013: The physical science basis. Contribution of working group I to the fifth assessment report of the intergovernmental panel on climate change* (pp. 1137–1216). Cambridge, UK: Cambridge University Press.

Ciais, P., Sabine, C., Bale, G., Bopp, L., Brovkin, V., Canadell, J., ... Thornton, P. (2013). *Carbon and other biogeochemical cycles. Climate change 2013: The physical science basis. Contribution to working group I to the fifth assessment report of the international panel on climate change* (pp. 465–570). Cambridge, UK: Cambridge University Press.

Clark, M. R., Althaus, F., Schlacher, T. A., Williams, A., Bowden, D. A., & Rowden, A. A. (2015). The impacts of deep-sea fisheries on benthic communities: A review. *ICES Journal of Marine Science, 73*(suppl_1), i51–i69.

Clarke, P., & Jupiter, S. (2010). *Principles and practice of ecosystem-based management: A Guide for conservation practitioners in the tropical Western Pacific*. Suva, Fiji: Wildlife Conservation Society. Available at: https://www.reefresilience.org/pdf/EBMguide_2010.pdf.

Cocco, V., Joos, F., Steinacker, M., Frölicher, T., Bopp, L., Dunne, J., ... Tjiputra, J. (2013). Oxygen and indicators of stress for marine life in multi-model global warming projections. *Biogeosciences., 10*, 1849–1868.

Cohen, A., Nugegoda, D., & Gagnon, M. M. (2001). Metabolic responses of fish following exposure to two different oil spill remediation techniques. *Ecotoxicology and Environmental Safety, 48*, 306–310.

Cole, D. B., Zhang, S., & Planavsky, N. J. (2017). A new estimate of detrital redox-sensitive metal concentrations and variability in fluxes to marine sediments. *Geochimica et Cosmochimica Acta, 2015*, 337–353.

Cole, M., Lindeque, P., Fileman, E., Halsband, C., & Galloway, T. S. (2015). The impact of polystyrene microplastics on feeding, function and fecundity in the marine copepod *Calanus helgolandicus*. *Environmental Science and Technology, 49*(2), 1130–1137.

Coll, M., Libralato, S., Tudela, S., Palomera, I., & Pranovi, F. (2008). Ecosystem overfishing in the ocean. *PLoS One, 3*(12), e3881.

Colloca, F., Scarcella, G., & Libralato, S. (2017). Recent trends and impacts of fisheries exploitation on mediterranean stocks and ecosystems. *Frontiers in Marine Science, 4*, 244.

Conley, D. J., & Slomp, C. P. (2019). Ocean deoxygenation impacts on microbial processes, biogeochemistry and feedbacks. In D. Laffoley, & J. M. Baxter (Eds.), *Ocean deoxygenation: Everyone's problem—Causes, impacts, consequences and solutions* (pp. 249–262). Gland, Switzerland: IUCN.

Connell, J. (2016). Last days in the Carteret Islands? Climate change, livelihoods and migration on coral atolls. *Asia Pacific Viewpoint, 57*(1), 3–15.

Cooley, S. R. (2012). How human communities could "feel" changing ocean biogeochemistry. *Current Opinions in Environmental Sustainability., 4*, 258–263.

Corben, R. (2017, June 22). *Asia's booming plastics industry prompts ocean pollution fears*. Available online at: https://www.voanews.com/a/asia-plastics-industry/391158.

Corbett, J. J., Fischbeck, P. S., & Pandis, S. N. (1999). Global Nitrogen and sulfur inventories for oceangoing ships. *Journal of Geophysical Research, 104*(3), 3457–3470.

Cormier, R., Elliott, M., & Kannen, A. (2018). *IEC/ISO 31010 Bow-tie analysis of marine legislation: A case study of the marine strategy framework directive.* ICES Cooperative Research Report. No. 342. Available online from: https://www.ices.dk/sites/pub/Publication%20Reports/Cooperative%20Research%20Report%20(CRR)/CRR342.pdf.

Cormier, R., Elliott, M., & Rice, J. (2019). Putting on a bow-tie to sort out who does what and why in the complex arena of marine policy and management. *Science of the Total Environment, 648*, 293–305.

Cormier, R., Kelble, C. R., Anderson, M. R., Allen, J. I., Grehan, A., & Gregersen, Ó. (2017). Moving from ecosystem-based policy objectives to operational implementation of ecosystem-based management measures. *ICES Journal of Marine Science, 74*, 406–413.

Cormier, R., & Lonsdale, J. (2019). Risk assessment for deep sea mining: An overview of risk. *Marine. Policy., 2*, 56.

Cormier, R., Stelzenmüller, V., Creed, I. F., Igras, J., Rambo, H., Callies, U., & Johnson, L. B. (2018). The science-policy interface of risk-based freshwater and marine management systems: From concepts to practical tools. *Journal of Environmental Management, 226*, 340–346.

Costello, J. M., Cheug, A., & Hauwere, A. D. (2010). Surface area and the seabed area, volume, depth, slope, and topographic variation for the world's seas, oceans, and countries. *Environmental Science and Technology, 44*(23), 8821–8828.

Costello, M. J., & Chaudhary, C. (2017). Marine biodiversity, biogeography, deep-sea gradients, and conservation. *Current Biology, 27*, R511–R527.

Crowder, L. B., Ng, C. A., Frawley, T. H., Low, N. H. N., & Micheli, F. (2019). The significance of ocean deoxygenation for kelp and other macroalgae. In D. Laffoley, & J. M. Baxter (Eds.), *Ocean deoxygenation: Everyone's problem—Causes, impacts, consequences and solutions* (pp. 309–322). Gland, Switzerland: IUCN.

Dahlman, L.A., & Lindsey, R. (2020). *Climate change: Ocean heat content.* 13 February. Available online from https://www.climate.gov/news-features/understanding-climate/climate-change-ocean-heat-content.

Dam, H. G., & Peterson, W. T. (1988). The effect of temperature on the gut clearance rate constant of planktonic copepods. *Journal of Experimental Marine Biology and Ecology, 123*(1), 1–14.

Dangendorf, S., Marcos, M., Wöppelmann, G., Conrad, C.P., Frederikse, T., & Riccardo Riva, R. (2017). Reassessment of 20th century global mean sea level rise. *Proceedings of National Academy of Sciences, USA*. 114: 5946–5951.

Danovaro, R., Fabiano, N., & Croce, N. D. (1993). Labile organic matter and microbial biomasses in deep-sea sediments (Eastern Mediterranean Sea). *Deep Sea research Part-1: Oceanographic Research, Papers. 40*(5), 953–965.

Darling, E. S., & Côté, I. M. (2008). Quantifying the evidence for ecological synergies. *Ecology Letters, 11*(12), 1278–1286.

Darling, E. S., McClanahan, T. R., & Côté, I. M. (2013). Life histories predict coral community disassembly under multiple stressors. *Global Change Biology, 19*(6), 1930–1940.

Das, N., & Chandran, P. (2011). Microbial degradation of petroleum hydrocarbon contaminants: An overview. *Biotechnology Research International, 2011*, 941810.

Davis, H. K., Moffat, C. F., & Shepherd, N. J. (2002). Experimental tinting of marine fish by three chemically dispersed petroleum products, with comparisons to the baer oil spill. *Spill Science & Technology Bulletin, 7*, 257–278.

De Santo, E. M., & Jones, P. J. S. (2007). Offshore marine conservation policies in the North East Atlantic: Emerging tensions and opportunities. *Marine Policy, 31*, 336–347.

Deacon, M. (1933). *A general account of the hydrology of the South Atlantic Ocean.* Cambridge, UK: Cambridge University Press.

DeConto, R. M., & Pollard, D. (2016). Contribution of Antarctica to past and future sea-level rise. *Nature, 531*(7596), 591–597.

Del Giorgio, P. A., & Duarte, C. M. (2002). Respiration in the open. *Nature, 420*, 379–384.

Delft, C. E. (2017). Update of maritime greenhouse gas emission projections. *Prepared for BIMCO. Submitted as MEPC 71/INF*, 34.

Deutsch, C., Ferrell, A., Seibel, B., Pörtner, H. O., & Huey, R. B. (2015). Climate change tightens a metabolic constraint on marine habitats. *Science (New York, N.Y.), 348*, 1132–1135.

Dickson, A. J. (2016). Introduction to CO_2 chemistry in sea water. IAEA. Available online.

DieselNet (2020). IMO releases final report of the Fourth GHG Study. Online, 4 August 2020. Available at: https://dieselnet.com/news/2020/08imo.php.

Donatelli, C., Fagherazzi, S., & Leonardi, N. (2018). Seagrass impact on sediment exchange between tidal flats and salt marsh, and the sediment budget of shallow bays. *Geographical Research Letters, 45*(10), 4933–4943.

Doney, S. C., Ruckelshaus, M., Duffy, J. E., Barry, J. P., Chan, F., Chad, A. E., ... Talley, L. D. (2012). Climate change impacts on marine ecosystems. *Annual Review of Marine Science, 4*, 11–37.

Douvere, F. (2010). The importance of marine spatial planning in advancing ecosystem-based sea use management. *Marine Policy, 32*, 762–771.

Downes, S. M., Bindoff, N. L., & Rintoul, S. R. (2010). Changes in the subduction of Southern Ocean water masses at the end of the twenty-first century in eight IPCC models. *Journal of Climate, 23*, 6526–6541.

Downes, S. M., Budnick, A. S., Sarmiento, J. L., & Farneti, R. (2011). Impacts of wind stress on the Antarctic Circumpolar Current fronts and associated subduction. *Geophysical Research Letters, 38*(11), GL047668.

Drabeck, D. H., Chatfield, M. W. H., & Richards-Zawacki, C. L. (2014). The status of Louisiana's diamond-back terrapin (Malaclemys terrapin) populations in the wake of the Deepwater horizon oil spill: insights from population genetic and contaminant analyses. *Journal of Herpetology, 48*, 125–136.

Duncan, E. M., Broderick, A. C., Fuller, W. J., Galloway, T. S., Godfrey, M. H., Hamann, M., ... Santillo, D. (2019). Microplastic ingestion ubiquitous in marine turtles. *Global Change Biology, 25*(2), 744–752.

Dunn, D. C., Jablonicky, C., Crespo, G. O., McCauley, D. J., Kroodsma, D. A., Boerder, K., ... Halpin, P. N. (2018). Empowering high seas governance with satellite vessel tracking data. *Fish and Fisheries, 19*(4), 729–739.

Durack, P. J., Glecker, P. J., Purkey, S. G., Johnson, G. C., Lyman., John, M., & Boyer, T. P. (2018). Ocean warming: From the surface to deep in observations and models. *Oceanography, 31*(2), 41–51.

Durand, A., Chase, Z., Noble, T. L., Bostock, H., Jaccard, S. L., Townsend, A. T., ... Jacobsen, G. (2018). Reduced oxygenation at intermediate depths of the southwest Pacific during the last glacial maximum. *Earth and Planetary Sciences, 491*, 48–57.

Earthzine (2015). *Ocean acidification, global warming's 'evil twin'*. May 29. Available online at: https://earthzine.org/ocean-acidification-global-warmings-evil-twin/.

EBMTN (Ecosystem-Based Management Tools Network). (2018). *About ecosystem-based management*. Available online, 12 November 2018. Available from https://toolkit.climate.gov/tool/ecosystem-based-management-ebm-tools-network.

EC (European Commission) (2008). *Directive 2008/56/EC of the European Parliament and of the Council*. 17 June 2008. Available from https://eur-lex.europa.eu/legal-content/EN/TXT/PDF/?uri = CELEX: 32008L0056&from = EN.

EC (European Commission). (2018a). *A European strategy for plastics in a circular economy*. Brussels: EU Secretariat. Available from https://ec.europa.eu/environment/circular-economy/pdf/plastics-strategy-brochure.pdf.

EC (European Commission). (2018b). *Policy brief: Implementing the ecosystem-based approach in maritime spatial planning*. Brussels: European Commission Directorate.

Edwards, T. (2017). *Future of the sea: Current and future impacts of sea level rise on the UK*. London: Government Office of Science.

Ehlert, D., & Zickfeld, K. (2017). What determines the warming commitment after cessation of CO_2 emissions? *Environmental Research Letters*, *12*(1), 015002.

Ekstrom, J. A., Suatoni, L., Cooley, S. R., Pendleton, L. H., Waldbusser, G. G., Cinner, J. E., ... Gledhill, D. (2015). Vulnerability and adaptation of United States shellfisheries to ocean acidification. *Nature Climate Change*, *5*, 207–214.

Elliott, M., & Whitfield, A. K. (2011). Challenging paradigms in estuarine ecology and management. *Estuarine. Coastal and Shelf Science*, *94*(4), 306–314.

Ellis, J. I., Hewitt, J. E., Clark, D., Taiapa, C., Patterson, M., Sinner, J., ... Thrush, S. F. (2015). Assessing ecological community health in coastal estuarine systems impacted by multiple stressors. *Journal of Experimental Marine Biology and Ecology*, *473*, 176–187.

Engler, C. (2015). Beyond rhetoric: navigating the conceptual tangle towards effective implementation of the ecosystem approach to oceans management. *Environmental Reviews*, *23*(3), 288–320.

Enríquez, A. R., Marcos, M., Álvarez-Ellacuría, A., Orfila, A., & Gomis, D. (2017). Changes in beach shoreline due to sea level rise and waves under climate change scenarios: Application to the Balearic Islands (western Mediterranean). *Natural Hazards and Earth System Sciences*, *17*, 1075–1089.

Eriksen, M., Thiel, M., Prindiville, M., & Kiessling, T. (2018). Microplastic: What are the solutions. In M. Wagner, & S. Lambert (Eds.), *Freshwater microplastics* (pp. 273–298). Cham, Switzerland: Springer.

Eriksson, S. P., Hernroth, B., & Baden, S. P. (2013). Stress biology and immunology in nephrops norvegicus. In L. J. Magnus, & J. P. Mark (Eds.), *Advances in marine biology* (pp. 149–200). Cambridge, Massachusetts: Academic Press.

Estes, J. A., Terborgh, J., Brashares, J. S., Power, M. E., Berger, J., Bond, W. J., ... Wardle, D. A. (2011). Trophic downgrading of planet earth. *Science (New York, N.Y.)*, *333*, 301–306.

EU (European Union) (n.d.). *The Common Fisheries Policy: origins and development*. Available online from: https://www.europarl.europa.eu/factsheets/en/sheet/114/the-common-fisheries-policy-origins-and-development#:~:text=A%20common%20fisheries%20policy%20(CFP,and%20stable%20jobs%20for%20fishers.

Eunomia. (2016). *Plastics in the marine environment*. London: Eunomia Research & Consulting Ltd Available at. Available from https://www.eunomia.co.uk/reports-tools/plastics-in-the-marine-environment/.

European Commission. (2008). Directive 2008/56/EC of the European Parliament and of the Council establishing a framework for community action in the field of marine environmental policy (Marine Strategy Framework Directive). *Official Journal of the European Union*, *L164*, 19–40.

European Parliament. (2018). *International Ocean Governance: An Agenda for the Future of Our Oceans in the Context of the 2030 Sustainable Development Goals*. Brussels: European Parliament.

Eurostat (2011). *The Mediterranean and Black Sea basins*. Available from http://ec.europa.eu/eurostat/en/web/products-statistics-in-focus/-/KS-SF-11–014.

Ezer, T. (2013). Sea level rise, spatially uneven and temporally unsteady: Why the United States East Coast, the global tide gauge record, and the global altimeter data show different trends. *Geophysical Research Letters*, *40*, 5439–5444.

FAO (Food and Agricultural Organisation). (2012). *The state of world fisheries and aquaculture*. Rome: Fisheries and Aquaculture Department of FAO.

FAO (Food and Agriculture Organization). (2020). *The state of world fisheries and aquaculture 2019, sustainability in action*. Rome: FAO.

Faris, J. (1994). *Seas of debris: A summary of the third international marine debris*. Alaska, USA: Alaska Fisheries Science Center.

Fasullo, J. T., & Gent, P. R. (2017). On the relationship between regional ocean heat content and sea surface height. *Journal of Climate*, *30*(22), 9295–9311.

Feng, Y. Y., Hare, C. E., Leblanc, K., Rose, J. M., Zhang, Y., DiTullio, G. R., ... Hutchins, D. A. (2009). Effects of increased pCO_2 and temperature on the North Atlantic spring bloom. I. The phytoplankton community and biogeochemical response. *Marine Ecology Progress Series*, *388*, 13–25.

Fennel, K., & Testa, J. M. (2019). Biogeochemical controls on coastal hypoxia. *Annual Review of Marine Science*, *11*, 105−130.

Finch, B. E., Wooten, K. J., & Smith, P. N. (2011). Embryotoxicity of weathered crude oil from the Gulf of Mexico in mallard ducks (*Anas platyrhynchos*). *Environmental Toxicology and Chemistry*, *30*, 1885−1891.

Findlay, H. S., Kendall, M. A., Spicer, J. I., & Widdicombe, S. (2010). Post-larval development of two intertidal barnacles at elevated CO_2 and temperature. *Marine Biology*, *157*, 725−735.

Fingas, M. F. (2013). *The basics of oil spill cleanup* (3rd Edn.). Boca Raton: CRC Press.

FitzGerald, D. M., Fenster, M. S., Argow, B. A., & Buynevich, I. V. (2008). Coastal impacts due to sea-level rise. *Annual Review of Earth & Planetary Sciences*, *36*, 601−647.

Foley, M. M., Halpern, B. S., Micheli, F., Armsby, M. H., Caldwell, M. R., Crain, C. M., ... Steneck, R. S. (2010). Guiding ecological principles for marine spatial planning, . *Marine Policy* (34, pp. 955−966).

Foo, S. A., & Byrne, M. (2016). Acclimatization and adaptive capacity of marine species in a changing ocean. *Advances in Marine Biology*, *74*, 69−116.

Frasier, K. E., Solson-Berga, A., Stokes, L., & Hilderbrand, J. A. (2020). Impacts of the deep-water horizon oil spills on marine mammals and sea turtles. In S. Murawski, C. Ainsworth, S. Gilbert, D. Hollander, C. Paris, M. Schlüter, & D. Wetzel (Eds.), *Deep oil spills: Facts, fate and effects* (pp. 431−462). Cham, Switzerland: Springer.

Freedman, B. (1995). *Environmental ecology: the ecological effects of pollution, disturbance, and other stresses*. Cambridge, MA: Academic Press.

Freeman, S., Booth, A. M., Sabbah, I., Tiller, R., Dierking, J., Klun, K., ... Angel, D. L. (2020). Between source and sea: The role of wastewater treatment in reducing marine microplastics. *Journal of Environment Management*, *266*, 110642.

Freestone, D. (2009). Climate change and the oceans. *Carbon & Climate Law Review*, *3*(4), 383.

Froese, R., Winker, H., Coro, G., Demirel, N., Tsikliras, C., Dimarchopoulou, D., ... Matz-Luck, N. (2018). Status and rebuilding of European fisheries. *Marine Policy*, *93*, 159−170.

Frölicher, T. L., Fischer, E. M., & Gruber, N. (2018). Marine heatwaves under global warming. *Nature*, *560* (7718), 350−364.

Frölicher, T. L., Joos, F., Plattner, G.-K., Steinacher, M., & Doney, S. C. (2009). Natural variability and anthropogenic trends in oceanic oxygen in a coupled carbon cycle−climate model ensemble. *Global Biogeochemical Cycles*, *23*, GB1003.

Frölicher, T. L., & Paynter, D. J. (2015). Extending the relationship between global warming and cumulative carbon emissions to multi-millennial timescales. *Environmental Research Letters*, *10*(7), 075002.

Frolicher, T. L., Ramseyer, L., Raible, C. C., Rodgers, K. B., & Dunne, J. (2020). Potential predictability of marine ecosystem drivers. *Biogeosciences.*, *17*, 2061−2083.

Frölicher, T. L., Winton, M., & Sarmiento, J. L. (2014). Continued global warming after CO_2 emissions stoppage. *Nature Climate Change*, *4*(1), 40−44.

Frusher, S. D., Hobday, A. J., Jennings, S. M., Creighton, C., D'Silva, D., Haward, M., ... van Putten, E. I. (2014). The short history of research in amarine climate change hotspot: From anecdote to adaptation in south-east Australia. *Reviews in Fish Biology and Fisheries*, *24*, 593−611.

Fuglestvedt, J., Berntsen, T., Eyring, V., Isaksen, I., Lee, D. S., & Sausen, R. (2009). Shipping emissions: From cooling to warming of climate—and reducing impacts on health. *Environmental Science & Technology*, *43*(24), 9057−9062.

Galgani, F., Fleet, D., van Franeker, J., Katsanevakis, S., Maes, T., Mouat, J., ...& Janssen, C. (2010). *Marine strategy framework directive task team 10 report marine litter*. Luxemburg: JRC (EC Joint Research Centre) Scientific and Technical Reports. Available from https://edepot.wur.nl/174619.

Gall, S. C., & Thompson, R. C. (2015). The impact of debris on marine life. *Marine Pollution Bulletin*, *92*(1−2), 170−179.

Galloway, T. S. (2015). Micro-and nano-plastics and human health. In M. Bergmann, L. Gustow, & M. Krages (Eds.), *Marine anthropogenic litter* (pp. 343–366). Cham. Switzerland: Springer.

Garcia, S. M., Rice, J., & Charles, A. T. (2014). *Governance of marine fisheries and biodiversity conservation.* Chichester, UK: John Wiley & Sons, Ltd.

Gardner, A. S., Moholdt, G., Cogley, J. G., Wouters, B., Arendt, A. A., Wahr, J., ... Paul, F. (2013). A reconciled estimate of glacier contributions to sea level rise: 2003 to 2009. *Science (New York, N.Y.), 340*, 852–857.

Gascuel, D., & Pauly, D. (2009). EcoTroph: Modelling marine ecosystem functioning and impact of fishing. *Ecological Modeling, 220*(21), 2885–2898.

Gaston, A. (2004). *Seabirds—A natural history.* London: Black Publishers Ltd.

Gattuso, J.-P., Mach, K. J., & Morgan, G. (2013). Ocean acidification and its impacts: an expert survey. *Climatic Change, 117*(4), 725–738.

Gattuso, J.-P., Magnan, A., Billé, R., Cheung, W. W. L., Howes, E. L., Joos, F., ... Turley, C. (2015). Contrasting futures for ocean and society from different anthropogenic CO_2 emissions scenarios. *Science (New York, N.Y.), 349*(6243), aac4722.

Gattuso, J. P., Magnan, A. K., Bopp, L., Cheumg, W. W. L., Duarte, C. M., Hinkel, J., ... Rau, G. H. (2018). Ocean solutions to address climate change and its effects on marine ecosystems. *Frontiers in Marine Science, 5*, 337.

Gegner, H., & Voolstra, C. (2019). A salty coral secret: How high salinity helps corals to be stronger. *Frontiers for Young Minds, 7*, 38.

Gelcich, S. (2014). Towards polycentric governance of small-scale fisheries: insights from the new 'Management Plans' policy in Chile. *Aquatic Conservation, 24*, 575–581.

Gelcich, S., Defeo, O., Iribarne, O., Del Carpio, G., Dubois, R., Horta, S., ... Castilla, J. C. (2009). Ecosystem based management in the Southern Cone of South America: stakeholder perceptions and lessons for implementation. *Marine Policy, 33*(5), 801–806.

GESAMP (Joint Group of Experts on the Marine Aspects of Marine Environment Protection). (2015). *Sources, fate and effects of microplastics in the marine environment: a global assessment.* London: IMO Available at. Available from https://ec.europa.eu/environment/marine/good-environmental-status/descriptor-10/pdf/GESAMP_microplastics%20full%20study.pdf.

GESAMP (Joint Group of Experts on the Marine Aspects of Marine Environment Protection). (2018). *Science for a sustainable ocean.* Available online from http://www.gesamp.org.

GESAMP (Joint Group of Experts on the Scientific Aspects of Marine Environmental Protection). (2001). *Protecting the ocean from land-based activities.* London: IMO, Report 71.

Geyer, R., Jambeck, J. R., & Law, K. L. (2017). Production, use, and fate of all plastics ever made. *Science Advances, 3*(7), e1700782.

Geyer, W. R. (1993). The importance of suppression of turbulence by stratification on the estuarine turbidity maximum. *Estuaries and Coasts, 16*, 113–125.

Ghaffour, N., Missimer, T. M., & Amy, G. L. (2013). Technical review and evaluation of the economics of water desalination: Current and future challenges for better water supply sustainability. *Desalination., 309*, 197–207.

Giddings, S. N., & Maccready, P. (2017). Reverse estuarine circulation due to local and remote wind forcing, enhanced by the presence of a long-coast estuaries. *Journal of Geophysical Research: Oceans, 122*, 10184–10205.

Goldberg, E. D. (1985). The oceans as waste space. *Ocean Yearbook. (Online), 5*, 150–161.

Gonzalez-Doncel, M., Gonzalez, L., Fernandez-Torija, C., Navas, J. M., & Tarazona, J. V. (2008). Toxic effects of an oil spill on fish early life stages may not be exclusively associated to PAHs: Studies with Prestige oil and medaka (*Oryzias latipes*). *Aquatic Toxicology, 87*, 280–288.

GOON (Global Ocean Oxygen Network). (2018). *The ocean is losing its breath: Declining oxygen in the world's ocean and coastal waters*. Paris: IOC-UNESCO, IOC Technical Series, No. 137.

Grases, A., Gracia, V., Garcia-Leon, M., Lin-Ye, J., & Sierra, J. P. (2020). Coastal flooding and erosion under a changing climate: Implications at a low-lying coast (Ebro Delta). *Water., 12*, 346.

Gregoire, M., Gilbert, D., Oschlies, A., & Rose, K. (2019). What is deoxygenation? In D. Laffoley, & J. M. Baxter (Eds.), *Ocean deoxygenation: Everyone's problem - Causes, impacts, consequences and solutions* (pp. 1−21). Gland, Switzerland: IUCN. (2019).

GRIDA. (2018a). *How plastic moves from the economy to the environment*. Available online from http://www.grida.no/resources/6908.

GRIDA. (2018b). *Marine plastics global policy timeline*. Available online from http://www.grida.no/resources/6912.

Grimm, M., & Tulloch, J. (Eds.), (2015). *The megacity state: The world's biggest cities shaping our future*. Munich: Allianz SE. Available from https://www.allianz.com/content/dam/onemarketing/azcom/Allianz_com/migration/media/press/document/Allianz_Risk_Pulse_Megacities_20151130-EN.pdf.

Gröger, M., Maier-Reimer, E., Mikolajewicz, U., Schurgers, G., Vizcaíno, M., & Winguth, A. (2007). Changes in the hydrological cycle, ocean circulation, and carbon/nutrient cycling during the last interglacial and glacial transition: Changes in the hydrological cycle. *Paleoceanography, 22*, PA 420.

Gros, J., Nabi, D., Würz, B., Wick, L. Y., Brussaard, C. P., Huisman, J., ... Arey, J. S. (2014). First day of an oil spill on the open sea: Early mass transfers of hydrocarbons to air and water. *Environmental Science and Technology, 48*, 9400−9411.

Gruber, N., Clement, D., Carter, B. R., Feely, R. A., va Heyven, S., Hoppema, M., ... Wanninkhof, R. (2019). The oceanic sink for anthropogenic CO_2 from 1994 to 2007. *Science (New York, N.Y.), 363*(6432), 1193−1199.

Gu, S., Liu, Z., Zhabg, I., Remper, J., Joos, F., & Oppo, D. W. (2017). Coherent response of antarctic intermediate water and atlantic meridional overturning circulation during the last deglaciation: reconciling contrasting neodymium isotope reconstructions from the Tropical Atlantic. *Paleoceanography and Paleoclimatology, 32*(10), 1036−1053.

Guerry, A. D. (2005). Icarus and daedalus: Conceptual and tactical lessons for marine ecosystem based management. *Frontiers in Ecology and the Environment, 3*(4), 202−211.

Guisan, A., Edwards, T. C., Jr., & Hastie, T. (2002). Generalized linear and generalized additive models in studies of species distributions: Setting the scene. *Ecological Modeling, 157*, 89−100.

Gunderson, A. R., Armstrong, E. J., & Stillman, J. H. (2016). Multiple stressors in a changing world: The need for an improved perspective on physiological responses to the dynamic marine environment. *Annual Review of Marine Science, 8*, 357−378.

Güneralp, B., Güneralp, I., & Liu, Y. (2015). Changing global patterns of urban exposure to flood and drought hazards. *Global Environmental Change, 31*, 217−225.

Hale, R., Pigott, J. J., & Swearer, S. E. (2017). Describing and understanding behavioral responses to multiple stressors and multiple stimuli. *Ecology and Evolution, 7*(1), 38−47.

Halpern, B. S. (2009). A global map of human impact on marine ecosystems. *Science (New York, N.Y.), 319*, 948−953.

Halpern, B. S., Frazier, M., Afflerbach, J., Lowndes, J. S., Michell, F., O'hara, C., ... Selkoe, K. A. (2019). Recent pace of change in human impact on the world's ocean. *Scientific Reports, 9*, 11609.

Halpern, B. S., Frazier, M., Potapenko, J., Casey, K. S., Koenig, K., Longo, C., ... Walbridge, S. (2015). Spatial and temporal changes in cumulative human impacts on the world's ocean. *Nature Communications, 6*, 7615.

Halpern, B. S., Kappel, C. V., Selkoe, K. A., Micheli, F., Ebert, C. M., Kontgis, C., ... Teck, S. J. (2009). Mapping cumulative human impacts to California current marine ecosystems. *Conservation Letters, 2*(3), 138−148.

Halpern, B. S., Klein, C. J., Brown, C. J., Beger, M., Grantham, H. S., Mangubhai, S., ... Possingham, H. P. (2013). Achieving the triple bottom line in the face of inherent trade-offs among social equity, economic return, and conservation. *Proceedings of National Academy of Science*, *110*, 6229–6234.

Halpern, B. S., Longo, C., Hardy, D., McLeod, K. L., Samhouri, J. F., Katona, S. K., ... Zeller, D. (2012). An index to assess the health and benefits of the global ocean. *Nature*, *488*, 615–620.

Haney, J. C., Geiger, H. J., & Short, J. W. (2014). Bird mortality from the deepwater horizon oil spill. I. Exposure probability in the offshore Gulf of Mexico. *Marine Ecology Progress Series*, *513*, 225–237.

Hanson, L., Isensee, K., Herr, D., Osborn, D., & Dupont, S. (2019). Impacts of Climate Change and Ocean Acidification on Marine Biodiversity. In CBD (Ed.), *Background briefs for 2020 ocean pathways week*. Montreal, Canada: CBD Secretariat.

Hart, K. M., Lamont, M. M., Sartain, A. R., & Fujisaki, I. (2014). Migration, foraging, and residency patterns for northern Gulf loggerheads: Implications of local threats and international movements. *PLoS One*, *9*, e103453.

Hartmann, D. L., Klein Tank, A. M. G., Rusticucci, M., Alexander, L. V., Brönnimann, S., Charabi, Y., ... Zhai, P. M. (2013). *Observations: Atmosphere and surface. Climate change 2013: The physical science basis. Contribution of working group I to the fifth assessment report of the intergovernmental panel on climate change* (pp. 159–254). Cambridge, UK: Cambridge University Press.

Harvey, C. (2020). *Why is the South Pole warming so Quickly? It's complicated*. Scientific American online from: https://www.scientificamerican.com/article/why-is-the-south-pole-warming-so-quickly-its-complicated/#:~:text=When%20the%20winds%20are%20stronger,over%20the%20last%20few%20decades.

Hauer, M. E. (2017). Migration induced by sea-level rise could reshape the United States population landscape. *Nature Climate Change*, *7*, 321.

Haward, M. (2018). Plastic pollution of the world's seas and oceans as a contemporary challenge in ocean governance. *Nature Communications*, *9*(1), 667.

Hawkins, E., & Sutton, R. (2012). Time of emergence of climate signals. *Geophysical Research Letters*, *39*(1), L01702.

Hebbeln, D., Portilho-Ramos, R. D. C., Weinberg, C. C., & Titschack, J. (2019). the fate of cold-water corals in a changing world: A geological perspective. *Frontiers in Marine Science*, *2019*, 00119.

Hegerl, G. C., Bronnimann, S., Shurer, A., & Cowan, T. (2018). The early 20th century Warming: Anomalies, causes and consequences. *Wires Climate Change*, *9*, e552.

Hehre, E. J., & Meeuwig, J. J. (2016). A global analysis of the relationship between farmed seaweed production and herbivorous fish catch. *PLoS One*, *11*, e0148250.

Heinze, C., Meyer, S., Goris, N., Anderson, L., Steinfeldt, R., Chang, N., ... Bakker, D. C. E. (2015). The ocean carbon sink—Impacts, vulnerabilities and challenges. *Earth System Dynamics*, *6*, 327–358.

HELCOM (Baltic Marine Environment Protection Commission). (2015). *Regional action plan for Marine Litter in the Baltic Sea*. Available from http://www.helcom.fi/Lists/Publications/Regional%20Action%20Plan%20for%20Marine%20Litter.pdf.

Helm, R. C., Costa, D. P., O'Shea, T. J., Wells, R. S., & Williams, T. L. (2015). Overview of effects of oil spills on marine mammals. In M. F. Fingas (Ed.), *Handbook of oil spill science and technology* (pp. 455–475). London: John Wiley.

Henkel, J. R., Sigel, B. J., & Taylor, C. M. (2012). Large-scale impacts of the deepwater horizon oil spill: Can local disturbance affect distant ecosystems through migratory shorebirds? *Bioscience*, *62*, 676–685.

Henson, S. A., Beaulieu, C., Ilyina, T., John, J. G., Long, M., Seferian, R., ... Sarmiento, J. L. (2017). Rapid emergence of climate change in environmental drivers of marine ecosystems. *Nature Communications*, *8*, 1462.

Hewitt, J. E., Ellis, J. I., & Thrush, S. F. (2015). Multiple stressors, nonlinear effects and the implications of climate change impacts on marine coastal ecosystems. *Global Change Biology*, *22*(8), 2665–2675.

Hilmi, N., Allemand, D., Kavanagh, C., Laffoley, D., Metian, M., Osborn, D., & Reynaud, S. (Eds.), (2015). *Bridging the gap between ocean acidification impacts and economic valuation: Regional impacts of ocean acidification on fisheries and aquaculture*. Gland, Switzerland: IUCN.

Hinkel, J., van de Wal, R., Magnan, A.K., Abd-Algawad, A., Cai, R., Cifuentes-Jara, M., ... & Sebesvari, Z. (2019). Sea level rise and implications for low lying islands, coasts and communities. In *IPCC special report on the ocean and cryosphere in a changing climate*. Available from https://report.ipcc.ch/srocc/pdf/SROCC_FinalDraft_Chapter4.pdf.

Hoaglaned, P. (2010). A (social) scientific look at ecosystem-based management. *Ocean & Coastal Law Journal*, *15*(1), 167–175. Available at. Available from http://digitalcommons.mainelaw.maine.edu/oclj/vol15/iss1/22.

Hobday, A. J., Cochrane, K., Downey-Breedt, N., Howard, J., Canela, S. J., Byfield, V., ... Putten, I. V. (2016). Planning adaptation to cli-mate change in fast-warming marine regions with seafood-dependent coastal com-munities. *Reviews in Fish Biology and Fisheries*, *26*, 249–264.

Hobday, A. J., & Lough, J. M. (2011). Projected climate change in Australian marine and freshwater environments. *Marine and Freshwater Research*, *62*, 1000–1014.

Hobday, A. J., & Pecl, G. T. (2014). Identification of global marine hotspots: Sentinels for change and vanguards for adaptation action. *Reviews in Fish Biology and Fisheries*, *24*, 415–425.

Hoegh-Guldberg, O., Cai, R., Brewer, P. G., Fabry, V. J., Hilmi, K., Jung, S., ... Sundby, S. (2014). The ocean. *In* Climate change 2014: Impacts, adaptation, and vulnerability. Contribution of working group II to the fifth assessment report of the intergovernmental panel on climate change (pp. 1655–1731). Cambridge, UK: Cambridge University Press.

Hoegh-Guldberg, O., Caldeira, K., Chopin, T., Gaines, S., Haugan, P., Hemer, M., ... Tyedmers, P. (2019). *The ocean as a solution to climate change: Five opportunities for action*. Washington, DC: World Resources Institute.

Hoegh-Guldberg, O., Jacob, D.; Taylor, M.; Bindi, M.; Brown, S.; Camilloni, I., ... & Zhou, G. (2018). Impacts of 1.5ºC Global Warming on Natural and Human Systems. In *Global Warming of 1.5°C. An IPCC Special Report on the impacts of global warming of 1.5°C above pre-industrial levels and related global greenhouse gas emission pathways, in the context of strengthening the global response to the threat of climate change, sustainable development, and efforts to eradicate poverty*. Switzerland: IPCC: 175–311. Available from https://www.ipcc.ch/site/assets/uploads/sites/2/2019/06/SR15_Chapter3_Low_Res.pdf.

Holdgate, M. W., & McIntosh, P. T. (1986). The oceans as a waste disposal option—Management, decision making and policy. In G. Kullenberg (Ed.), *The role of the oceans as a waste disposal option* (pp. 1–18). Netherlands: Springer.

Honolulu Strategy (2011). *The Honolulu strategy: A global framework for prevention and management of marine debris*. UNEP and NOAA. Available online.

Hopkins, F. E., Suntharalingam, P., Gehlen, M., Andrews, O., Archer, S. D., Bopp, L., ... Williamson, P. (2020). The impacts of ocean acidification on marine trace gases and the implications for atmospheric chemistry and climate. *Proceedings of Royal Society. A*, *476*, 20190769.

Hossain, M. S., Chowdhury, S. R., Sharifuzzaman, S. M., & Sarker, S. (2015). *Vulnerability of the Bay of Bengal to ocean acidification*. Dhaka, Bangladesh: IUCN.

Housego, R. M., & Rosman, J. H. (2016). A Model for understanding the effects of sediment dynamics on oyster reef development. *Estuaries and Coasts*, *39*, 495–509.

Hu, N. T. A. (2012). Integrated oceans policymaking: An ongoing process or a forgotten concept? *Coastal Management*, *40*(2), 107–118.

Hugo, G. (2011). Future demographic change and its interactions with migration and climate change. *Global Environmental Change.*, *21*(Suppl. 1), s21–s33.

Ibánhez, J. S. P., Araujo, M., & Lefèvre, N. (2016). The overlooked tropical oceanic CO_2 sink. *Geophysical Research Letters*, *43*(8), 3804–3812.

ICCT (International Council on Clean Transportation). (2007). *Air pollution and greenhouse gas emissions from ocean-going ships*. Washington, D.C.: ICCT.

ICSU (International Council for Science). (2017). *A guide to SDG interactions: From Science to Implementation*. Paris: ICSU.

IMO (International Maytime Organization). (1990). *London dumping convention: The first decade and beyond*. London, UK: IMO Secretariat.

IMO (International Maritime Organization). (2013). *Report of the expert workshop on the update of GHG emissions estimate for international shipping (update-EW)*. Note by the Secretariat, MEPC 65th Session, Agenda Item 5, IMO Doc MEPC 65/5/2 (4 March 2013).

IMO (International Maritime Organization). (2018). *Adoption of the initial IMO strategy on reduction of the emissions from ships and existing IMO activity related to reducing GHHG emission in the shipping sector*. Note by the IMO to the UNFCCC Talanoa Dialogue. Available from https://unfccc.int/sites/default/files/resource/250_IMO%20submission_Talanoa%20Dialogue_April%202018.pdf.

Incardona, J., Carls, M., Day, H., Sloan, C., Bolton, J., Colloer, T., & Scholz, N. (2009). Cardiac arrhytmia is the primary response of embryonic Pacific herring (*Clupea pallasi*) exposed to crude oil during weathering. *Environmental Science & Technology, 43*, 201−207.

Incardona, J. P., Gardner, L. D., Linbo, T. L., Brown, T. L., Esbaugh, A. J., mager, E. M., ... Scholz, N. L. (2014). Deepwater horizon crude oil impacts the developing hearts of large predatory pelagic fish. *Proceedings of National Academy of Science, USA, 111*, E1510−E1518.

IOPTF (Interagency Ocean Policy Task Force). (2009). *Interim framework for effective coastal and marine spatial planning*. Available from http://www.whitehouse.gov/sites/default/files/microsites/091209-Interim-CMSP-Framework-Task-Force.pdf.

IPCC (Intergovernmental Panel on Climate Change. (2001). *Climate Change 2001: Synthesis report. A contribution of working groups I, II, and III to the third assessment report of the intergovernmental panel on climate change*. Cambridge, UK: Cambridge University Press.

IPCC (Intergovernmental Panel on Climate Change). (2013a). Climate change 2013: The physical science basis, contribution of working group I to the *Fifth Assessment Report of the Intergovernmental Panel on Climate Change*. Cambridge, U.K.: Cambridge University Press.

IPCC (Intergovernmental Panel on Climate Change). (2013b). *Summary for policymakers. Climate change 2013: The physical science basis. Contribution of working group I to the fifth assessment report of the intergovernmental panel on climate change* (pp. 3−29). Cambridge, UK: Cambridge University Press.

IPCC (Intergovernmental Panel on Climate Change). (2014a). *Climate change 2014: Impacts, adaptation, and vulnerability. Part A: Global and sectoral aspects. Contribution of working group II to the fifth assessment report of the intergovernmental panel on climate change*. Cambridge, UK: Cambridge University Press.

IPCC (Intergovernmental Panel on Climate Change). (2014b). *Climate change 2014: Synthesis report. Contribution of working groups I, II and III to the fifth assessment report of the intergovernmental panel on climate change*. Geneva, Switzerland: IPCC Available at. Available from https://www.ipcc.ch/site/assets/uploads/2018/05/SYR_AR5_FINAL_full_wcover.pdf.

IPCC (Intergovernmental Panel on Climate Change) (2015). *Summary for policy makers, technical summary, part of the working group III Contribution to the fifth assessment report of the intergovernmental panel on climate change*. Available from https://www.ipcc.ch/site/assets/uploads/2018/03/WGIIIAR5_SPM_TS_Volume-3.pdf.

IPCC (Intergovernmental Panel on Climate Change). (2018). Global Warming of 1.5°C. In *IPCC Special Report on the impacts of global warming of 1.5°C above pre-industrial levels and related global greenhouse gas emission pathways, in the context of strengthening the global response to the threat of climate change, sustainable development, and efforts to eradicate poverty*. Available from https://www.ipcc.ch/site/assets/uploads/sites/2/2019/06/SR15_Full_Report_High_Res.pdf.

IPCC (Intergovernmental Panel on Climate Change). (2019a). *IPCC special report on the ocean and cryosphere in a changing climate*. Available from https://www.ipcc.ch/sites/assets/upload/sites/3/2019/11/05_SROCC_SPM_FINAL.pdf.

IPCC (Intergovernmental Panel on Climate Change). (2019b). Summary for policymakers. In *IPCC special report on the ocean and cryosphere in a changing climate*. Available from https://www.ipcc.ch/sites/assets/upload/sites/3/2019/11/05_SROCC_SPM_FINAL.pdf.

IPIECA-IOGP (International Petroleum Industry Environmental Conservation Association-International Association of Oil and Gas Producers). (2015) *Impacts of oil spills on marine ecology*. Good Practice Guide Series, Oil Spill Response Joint Industry Project (OSR-JIP), IOGP Report Number: 525.

Isensee, K., Levin, L., Breitburg, D., Gregoire, M., Garçon, V., & Valdés, L. (2015). *The ocean is losing its breath. Ocean and climate, scientific notes*. 20−28. Available online at: https://ocean-climate.org/wp-content/uploads/2020/02/4.-The-Ocean-is-losing-its-breath-Scientific-notes-2016.pdf.

Ishimatsu, A., & Dissanayake, A. (2010). Life threatened in acidic coastal waters. In A. Ishimatsu, & H. J. Lie (Eds.), *Coastal environmental and ecosystem issues of the east China Sea* (pp. 283−303). Nagasaki, Japan: Terrapub & Nagasaki University.

Isla, J. A., Lengfellner, K., & Sommer, U. (2008). Physiological response of the copepod Pseudocalanus sp. in the Baltic Sea at different thermal scenarios. *Global Change Biology*, *14*(4), 895−906.

Isson, T. T., Planavsky, N. I., Coogan, L. A., Stewart, R. M., Ague, I. I., Bolton, E. W., ... Kump, L. R. (2020). Evolution of the global carbon cycle and climate regulation on Earth. *Global Biogeochemical Cycles*, *34*(2), e2018GB006061.

ITOPF (International Tanker Owners Pollution Federation). (2011) *Effects of oil pollution on the marine environment*. Technical Information Paper (TIP) No. 13, International Tanker Owners Pollution Federation.

ITOPF (International Tanker Owners Pollution Federation). (2020). *Oil tanker spill statistics 2019*. london: itopf, available online.

IUCN (International Union for Conservation of Nature). (2017a). *National marine plastic litter policies in EU Member States: An overview*. Brussels, Belgium: IUCN.

IUCN (International Union for Conservation of Nature). (2017b). *Ocean warming: Issue brief*. Cham, Switzerland: IUCN Secretariat.

Jaccard, S. L., Galbraith, E. D., Froelicher, T. L., & Gruber, N. (2014). Ocean (de)oxygenation across the last deglaciation, insights for the future. *Oceanography*, *27*, 26−35.

Jackson, J. B. (2008). Ecological extinction and evolution in the brave new ocean. *Proceedings of National Academy of Science, USA*, *105*(Suppl 1), 11458−11465.

Jenkyns, H. C. (2010). Geochemistry of oceanic anoxic events. *Geochemistry, Geophysics, Geosystems*, *11*(3), 002788.

Jennings, S., Stentiford, G. D., Leocadio, A. M., Jeffery, K. R., Metcalfe, J. D., katisiadki, I., ... Verner-Jeffreys, D. W. (2017). Aquatic food security: Insights into challenges and solutions from an analysis of interactions between fisheries, aquaculture, food safety, human health, fish and human welfare, economy and environment. *Fish Fish*, *17*, 893−938.

Jevrejeva, S., Moore, J. C., Grinsted, A., Matthews, A. P., & Spada, G. (2014). Trends and Acceleration in Global and Regional Sea Levels since 1807. *Global and Planetary Change*, *113*, 11−22.

Johnson, G. C., Lyman, J. M., Boyer, T., Cheng, L., Domingues, C. M., Gilson, J., ... Wijffels, S. E. (2018). Global oceans: Ocean heat content. State of the climate in 2017. *Bulletin of the American Meteorological Society*, *99*(8), S72−S77.

Johnson, G. C., Lyman, L. M., Boyer, T., Cheng, L., Domingues, C. M., Gilson, J., ... Wijffels, S. E. (2020). Ocean heat content. In Jessica Blunden, & Derek S. Arndt (Eds.), *State of the climate in 2019: Global oceans* (101, pp. S140−S144). Bulletin of American Meteorology Society. (8).

Jones, D. C., Meilers, A. J. S., Shuckburg, E., Sallee, J.-B., Haynes, P., McAufield, E. K., & Mazloff, M. R. (2016). How does Subantarctic M ode Water ventilate the Southern Hemisphere subtropics? *JGR Oceans*, *121*(9), 6558−6582.

Jones, K. R., Klein, C. J., Halpern, B. S., Venter, O., Grantham, H., Kuempel, C. D., ... Watson, J. E. M. (2018). The location and protection status of earth's diminishing marine wilderness. *Current Biology, 28* (15), 2506−2512, e3.

Jørgensen, F., & Sandersen, P. B. E. (2006). Buried and open tunnel valleys in Denmark—Erosion beneath multiple ice sheets. *Quaternary Science Reviews, 25,* 1339−1363.

Jørgensen, F., Scheer, W., Thomsen, S., Sonnenborg, T. O., Hinsby, K., Wiederhold, H., ... Auken, E. (2012). Transboundary geophysical mapping of geological elements and salinity distribution critical for the assessment of future sea water intrusion in response to sea level rise. *Hydrology and Earth System Sciences, 16,* 1845−1862.

Jorissen, F. J., Fontainer, C., & Thomas, E. (2007). Paleoceanographical proxies based on deep-sea benthic foraminiferal assemblage characteristics. In C. Hillaire-Marcel, & A. D. Vernal (Eds.), *Proxies in late Cenozoic Paleoceanography* (Vol.1, pp. 263−325). Amsterdam, Netherlands: Elsevier.

Joseph, B., Raj, J. S., Edwin, T. B., Sankarganesh, P., Jeevitha, M. V., Ajisha, S. U., & Rajan, S. (2011). Toxic effect of heavy metals on aquatic environment. *International Journal of Biological and Chemical Sciences, 4*(4).

Joye, S. B. (2015). Deepwater horizon, 5 years on. *Science (New York, N.Y.), 349,* 592−0593.

Kaleris, V., Lagas, G., Marczinek, S., & Piotrowski, J. A. (2002). Modelling submarine groundwater discharge: An example from the western Baltic Sea. *Journal of Hydrology, 265,* 76−99.

Karegar, M. A., Dixon, T. H., Malservisi, R., Kusche, J., & Engelhart, S. E. (2017). Nuisance flooding and relative sea-level rise: The importance of present-day land motion. *Scientific Reports, 7*(1), 1−9.

Karl, T. R., Arguez, A., Huang, B., Lawrimore, J. H., McMahon, J. R., Menne, M. J., ... Zhang, H. M. (2015). Possible artifacts of data biases in the recent global surface warming hiatus. *Science (New York, N.Y.), 348*(6242), 1469−1472.

Kärnbränslehantering, A. B. S. (2005). *Climate and climate-related issues for the safety assessment SR-Can.* Stockholm: SNFWM Available at. Available from https://www.osti.gov/etdeweb/servlets/purl/20843902.

Kassahn, K. S., Crozier, R. H., Pörtner, H. O., & Caley, M. J. (2009). Animal performance and stress: Responses and tolerance limits at different levels of biological organization. *Biological Reviews, 84*(2), 277−292.

Katz, C. (2015). *How long can oceans continue to absorb Earth's excess heat?* Yale Environment 360 online. Available from https://e360.yale.edu/features/how_long_can_oceans_continue_to_absorb_earths_excess_heat.

Kehew, A. E., Piotrowski, J. A., & Jørgensen, F. (2012). Tunnel valleys: Concepts and controversies—A review. *Earth−Science Reviews, 113,* 33−58.

Kimball, L. A. (2005). *The international legal regime of the high seas and the seabed beyond the limits of national jurisdiction and options for cooperation for the establishment of marine protected areas (MPAs) in marine areas beyond the limits of national jurisdiction.* Montreal: CBD Secretariat.

Klein, R. J. T., Nicholls, R. J., & Thomalla, F. (2003). Resilience to natural hazards: How useful is this concept? *Environmental Hazards, 5,* 35−45.

Knapp, S., Schweiger, A., Kalberg, A., Asmus, H., Asmus, R., Brey, T., ... Krause, G. (2017). Do drivers of biodiversity change differ in importance across marine and terrestrial systems—Or is it just different research communities' perspectives? *Science of the Total Environment, 574,* 191−203.

Kolker, A. S., Allison, M. A., & Hameed, S. (2011). An evaluation of subsidence rates and sea-level variability in the northern Gulf of Mexico. *Geophysical Research Letters, 38*(21), 1−6.

Kopp, R. E., Hay, C. C., Little, C. M., & Mitrovica, J. X. (2015). Geographic variability of sea-level change. *Current Climate Change Reports, 1,* 192−204.

Kopp, R. E., Kemp, A. C., Bittermann, K., Horton, B. P., Donnelly, J. P., Gehrels, W. R., ... Rahmstorf, S. (2016). Temperature-driven global sea-level variability in the common era. *Proceedings of National Academy of Sciences, USA., 113*(11), E1434−E1444.

Koslow, A. J. (2019). The significance of ocean deoxygenation for continental margin mesopelagic communities. In D. Laffoley, & J. M. Baxter (Eds.), *Ocean deoxygenation: Everyone's problem - Causes, impacts, consequences and solutions* (pp. 323–339). Gland, Switzerland: IUCN.

Kren, A. C., Pilewskie, P., & Coddington, O. (2017). Where does Earth's atmosphere get its energy? *Journal of Space Weather and Space Climate, 7*, A10.

Krichen, E., Rapaport, A., Floc'h, E. L., & Fouilland, E. (2019). Demonstration of facilitation between microalgae to face environmental stress. *Scientific Reports, 9*, 16076.

Kristiansen, T., Drinkwater, K. F., Lough, R. G., & Sundby, S. (2011). Recruitment variability in North Atlantic cod and match-mismatch dynamics. *PLoS One, 6*(3), e17456.

Krushelnytska, O. (2018). *Solving Marine Pollution: Successful models to reduce wastewater, agricultural runoff and marine litter.* Washington DC: World Bank Group.

Kummu, M., De Moel, H., Salvucci, G., Viviroli, D., Ward, P. J., & Varis, O. (2016). Over the hills and further away from coast: Global geospatial patterns of human and environment over the 20th–21st centuries. *Environmental Research Letters, 11*(3), 034010.

Kwon, E. Y., Primeau, F., & Sarmiento, J. L. (2009). The impact of remineralization depth on the air-sea carbon balance. *Nature Geoscience, 2*, 630–635.

Lachkar, Z., Orr, J. C., Dutay, J. C., & Delecluse, P. (2007). Effects of mesoscale eddies on global ocean distributions of CFC-11, CO_2, and Delta C-14. *Ocean Science, 3*, 461–482.

Lachkar, Z., Orr, J. C., Dutay, J.-C., & Delecluse, P. (2009). On the role of mesoscale eddies in the ventilation of Antarctic intermediate water. *Deep Sea Research Part I: Oceanographic Research, 909–925*, Papers. 56.

Laffoley, D., & Baxter, J. M. (Eds.), (2016). *Explaining ocean warming: Causes, scale, effects and consequences.* Gland, Switzerland: IUCN.

Laffoley, D., & Baxter, J. M. (Eds.), (2019). *Ocean deoxygenation: Everyone's problem—Causes, impacts, consequences and solutions.* Gland, Switzerland: IUCN.

Laist, D. W. (1987). Overview of the biological effects of lost and discarded plastic debris in the marine environment. *Marine Pollution Bulletin, 18*(6), 319–326.

Laist, D. W. (1997). Impacts of marine debris: entanglement of marine life in marine debris including a comprehensive list of species with entanglement and ingestion records. In J. M. Coe, & D. B. Roger (Eds.), *Marine Debris* (pp. 99–139). New York, NY: Springer.

Lam, P., & Kuypers, M. M. M. (2011). Microbial nitrogen cycling processes in oxygen minimum zones. *Annual Review of Marine Science, 3*, 317–345.

Lambeck, K., Woodroff, C. D., Antonioli, F., Anzidei, M., Gehrels, W. R., Laborel, J., & Wright, A. J. (2010). Paleoenvironmental records, geophysical modelling and reconstruction of sea level trends and variability on centennial and longer time scales. In J. A. Church, P. L. Woodworth, T. Aarup, & W. S. Wilson (Eds.), *Understanding sea level rise and variability* (pp. 61–121). Hoboken, N.J.: Blackwell.

Lambert, S., & Wagner, M. (2016). Characterization of nanoplastics during the degradation of polystyrene. *Chemosphere, 145*, 265–268.

Landschützer, P., Gruber, N., & Bakker, D. C. E. (2016). Decadal variations and trends of the global ocean carbon sink: decadal air-sea CO_2 flux variability. *Global biogeochemical cycles, 30*(10), 1396–1417.

Langangen, Ø., Olsen, E., Stige, L. C., Ohlberger, J., Yaragina, N. A., Vikeba, F. B., ... Hijermann, N. O. (2017). The effects of oil spills on marine fish: Implications of spatial variation in natural mortality. *Marine Pollution Bulletin, 119*, 102–109.

Langlet, D., & Rayfuse, R. (2018a). Challenges in implementing the ecosystem approach: Lessons learned. In D. Langlet, & R. Rayfuse (Eds.), *The ecosystem approach in ocean planning and governance* (pp. 445–461). Gland, Switzerland: Brill Publishers.

Langlet, D., & Rayfuse, R. (2018b). The ecosystem approach in ocean planning and governance: An introduction. In D. Langlet, & R. Rayfuse (Eds.), *The ecosystem approach in ocean planning and governance* (pp. 1−14). Leiden, Netherlands: Brill Publishers, Brill 2018b.

Langlet, D. & Westholm, A. (2019). *Synthesis report on the ecosystem approach to maritime spatial planning.* Available from http://www.panbalticscope.eu/wp-content/uploads/2019/12/PBS-Synthesis-Report.pdf.

Law, K. L. (2017). Plastics in the marine environment. *Annual Review of Marine Science*, 9, 205−229.

Le Guern, C. (2018). *When the mermaids cry: The great plastic tide.* Available on line at: http://plastic-pollution.org/.

Le Quéré, C., Andrew, R. M., Friedlingstein, P., Sitch, S., Pongratz, J., Manning, A. C., ... Zhu, D. (2018). Global carbon budget 2017. *Earth System Science Data*, 10, 405−448.

Leadley, P. W., Krug, C. B., Alkemade, R., Pereira, H. M., Sumaila, U. R., Walpole, M., ... Mumby, P. J. (2014). *Progress towards the Aichi biodiversity targets: An assessment of biodiversity trends, policy scenarios and key actions.* Montreal, Canada: Secretariat of the Convention on Biological Diversity.

Leaper, R. (2019). The role of slower vessel speeds in reducing greenhouse gas emissions. *Underwater Noise and Collision Risk to Whales. Frontiers in Marine Science*, 6, 505.

Lecklin, T., Ryömä, R., & Kuikka, S. (2011). A Bayesian network for analyzing biological acute and long-term impacts of oil spill in the Gulf of Finland. *Marine Pollution Bulletin*, 62, 2822−2835.

Leduc, M., Matthews, H. D., & Elia, R. D. (2015). Quantifying the limits of a linear temperature response to cumulative CO_2 emissions. *Journal of Climate*, 28(24), 9955−9968.

Lee, C. Y., & Song, D. P. (2017). Ocean container transport in global supply chains: Overview and research opportunities. *Transportation Research Part B Methodological*, 95, 442−474.

Lee, Y., Rosalynn., Seitzinger, S., & Mayorga, E. (2015). Land-based nutrient loading to LMEs: A global watershed perspective on magnitudes and sources. *Environmental Development.*, 17(Suppl-1), 220−229.

Legeais, J.-F., Ablain, M., Zawadzki, L., Zuo, H., Johannessen, J. A., Scharffenberg, M. G., ... Benveniste, J. (2018). An improved and homogeneous altimeter sea level record from the ESA climate change initiative. *Earth System Science Data.*, 10, 281−301.

Leighton, F. A. (1993). The toxicity of petroleum oils to birds. *Environmental Review*, 1, 92−103.

Leslie, H. M., & McLeod, K. L. (2007). Confronting the challenges of implementing marine ecosystem-based management. 5. *Frontiers in Ecology and the Environment*, 5(10), 540−548.

Leslie, H. M., Rosenberg, A. A., & Eagle, J. (2008). Is a new mandate needed for marine ecosystem-based management? *Frontiers in Ecology and the Environment*, 6(1), 43−48.

Leslie, H. M., Sievanen, L., Crawford, T. G., & Gruby, R. (2015). Learning from ecosystem-based management in practice. *Coastal Management*, 43(5), 471−495.

Leung, S., Mislan, K. A. S., Muhling, B., & Brill, R. (2019). The significance of ocean deoxygenation for open ocean tunas and billfishes. In D. Laffoley, & J. M. Baxter (Eds.), *Ocean deoxygenation: Everyone's problem - Causes, impacts, consequences and solutions* (pp. 277−308). Gland, Switzerland: IUCN.

Levin, L. A. (2018). Manifestation, drivers, and emergence of open ocean deoxygenation. *Annual Review of Marine Sciences*, 10, 229−260.

Levin, L. A., & Gallo, N. D. (2019). The significance of ocean deoxygenation for continental margin benthic and demersal biota. In D. Laffoley, & J. M. Baxter (Eds.), *Ocean deoxygenation: Everyone's problem - Causes, impacts, consequences and solutions* (pp. 341−361). Gland, Switzerland: IUCN.

Levin, L. A., & Sibuet, M. (2012). Understanding continental margin biodiversity: A new imperative. *Annual Review of Marine Science*, 4, 79−112.

Levitus, S., Antonov, J. I., Boyer, T. P., Baranova, O. K., Garcia, H. E., Locarnini, R. A., ... Zweng, M. M. (2012). World ocean heat content and thermosteric sea level change (0−2000 m), 1955−2010. *Geophysical Research Letters*, 39, L10603.

Lindenmayer, D. B., Likens, G. E., Krebs, C. J., & Hobbs, R. J. (2010). Improved probability of detection of ecological "surprises.". *Proceedings of the National Academy of Sciences, USA., 107*(51), 21957−21962.

Lindsey, R. (2020). *Climate change: Global sea level.* Climate.gov website, 14 August 2020. Available from https://www.climate.gov/news-features/understanding-climate/climate-change-global-sea-level.

Lindwall, C. (2019). *Industrial agricultural pollution.* Available online from https://www.nrdc.org/stories/industrial-agricultural-pollution-101.

Ling, S. D., Johnson, C. R., Frusher, S. D., & Ridgway, K. R. (2009). Overfishing reduces resilience of kelp beds to climate-driven catastrophic phase shift. *Proceedings of National Academy of Science. USA., 106* (52), 22341−22345.

Link, J. S. (2018). System-level optimal yield: increased value; less risk; improved stability, and better fisheries. *Canadian Journal of Fisheries and Aquatic Sciences. Sci, 75*, 1−16.

Link, J. S., & Browman, H. I. (2014). Integrating what? Levels of marine ecosystem-based assessment and management. *ICES Journal of Marine Sciences, 71*, 1170−1173.

Link, J. S., & Browman, H. I. (2017). Operationalizing and implementing ecosystem-based management. *ICES Journal of Marine Science, 74*, 379−381.

Link, J. S., & Watson, R. R. (2019). Global ecosystem overfishing: Clear delineation within real limits to production. *Science Advances., 5*(6), eaav0474.

Lirman, D., & Cropper, W. P. (2003). The influence of salinity on seagrass growth, survivorship, and distribution within Biscayne Bay, Florida: Field, experimental, and modeling studies. *Estuaries, 26*, 131−141.

Liu, Y., & Kujawinski, E. B. (2015). Chemical composition and potential environmental impacts of water-soluble polar crude oil components inferred from ESI FT-ICR MS. *PLoS One, 10*(9), e0136376.

Livengood, E., & Chapman, F. (2007). The ornamental fish trade: An introduction with perspectives for responsible aquarium fish ownership. *University of Florida IFAS Extension EDIS, 16*, 1−8. Available at. Available from https://journals.flvc.org/edis/article/view/116719.

Lloyd-Smith, M. & Immig, J. (2018). *Ocean pollutants guide: Toxic threats to human health and marine life.* Available from https://ipen.org/sites/default/files/documents/ipen-ocean-pollutants-v2_1-en-web.pdf.

Lodge, M. W., & Verlaan, P. A. (2018). Deep-sea mining: international regulatory challenges and responses. *Elements, 14*, 331−336.

Loganathan, P., Naidu, G., & Vigneswaran, S. (2017). Mining valuable minerals from seawater: a critical review. *Environmental Science, Water Research and Technology, 3*(1), 37−53.

MacDougall, A. H., & Friedlingstein, P. (2015). The origin and limits of the near proportionality between climate warming and cumulative CO2 emissions. *Journal of Climate, 28*(10), 4217−4230.

MacDougall, A. H., & Knutti, R. (2016). Enhancement of non-CO_2 radiative forcing via intensified carbon cycle feedbacks. *Geophysical Research Letters, 43*(11), 5833−5840.

Mach, M. E., Wedding, L. M., Reiter, S. M., Micheli, F., Fujita, R. M., & Martone, R. G. (2017). Assessment and management of cumulative impacts in California's network of marine protected areas. *Ocean & Coastal Management, 137*, 1−11.

Magera, A. M., Mills, F. J. E., Kaschner, K., Christensen, L. B., & Lotze, H. K. (2013). Recovery Trends in Marine Mammal Populations. *PLoS One, 8*(10), e77908.

Maher, R. L., Rice, M. M., McMinds, R., Burkepile, D. E., & Thurber, R. B. (2019). Multiple stressors interact primarily through antagonism to drive changes in the coral microbiome. *Scientific Report, 9*, 6834.

Malone, T. C., & Newton, A. (2020). The globalization of cultural eutrophication in the coastal ocean: Causes and consequences. *Frontiers in Marine Science, 7*, 670.

Manning, A. C., & Keeling, R. F. (2006). Global oceanic and land biotic carbon sinks from the Scripps atmospheric oxygen flask sampling network. *Tellus, 58B*, 95−116.

Mannoa, C., Bednarsekb, N., Tarlinga, G. A., Pecka, V. L., Comeauc, S., Adhikarid, D., ... Ziveri, P. (2017). Shelled pteropods in peril: Assessing vulnerability in a high CO2 Ocean. *Earth Science Reviews*, *169*, 132−145.

Maranon, E., Cermeno, P., Huete-Ortega, M., Lopez-Sandoval, D. C., Mouri∼no-Carballido, B., & Rodrıguez-Ramos, T. (2014). Resource supply overrides temperature as a control-ling factor of marine phytoplankton growth. *PLoS One*, *9*, e99312.

Marshak, A. R., Link, J. S., Shuford, R., Monaco, M. E., Johannesen, E., & Bianchi, G. (2017). International perceptions of an integrated, multi-sectoral, ecosystem approach to management. *ICES Journal of Marine Sciences*, *74*, 414−420.

Martin, J.-L., Maris, V., & Sinberloff, D. S. (2016). The need to respect nature and its limits challenges society and conservation science. *Proceedings of the National Academy of Sciences, USA.*, *113*(22), 6105−6112.

Mathis, J. T., Cooley, S. R., Lucey, N., Colt, S., Ekstrom, J., Hurst, T., ... Feely, R. A. (2015). Ocean acidification risk assessment for Alaska's fishery sector. *Progress in Oceanography*, *136*, 71−91.

Matthew, H. D., Solomon, S., & Pierrehumbert, R. (2012). Cumulative carbon as a policy framework for achieving climate stabilization. *Philosophical Transactions of the Royal Society A.*, *370*, 4365−4379.

Matthews, H. D., Zickfeld, K., Knutti, R., & Allen, M. R. (2018). Focus on cumulative emissions, global carbon budgets and the implications for climate mitigation targets. *Environmental Research Letters*, *13*(1), 010201.

Maureaud, A., Gascuel, D., Colléter, M., Palomares, M. L. D., Du Pontavice, H., Pauly, D., & Cheung, W. W. L. (2017). Global change in the trophic functioning of marine food webs. *PLoS One*, *12*(8), e 0182826.

Maxwell, S. L., Fuller, R. A., Brooks, T. M., & Watson, J. E. M. (2016). Biodiversity: The ravages of guns, nets and bulldozers. *Nature*, *536*, 143−145.

Mazzei, V., Longo, G., Brundo, M. V., Sinatra, F., Copat, C., Conti, G. O., & Ferrante, M. (2014). Bioaccumulation of cadmium and lead and its effects on hepatopancreas morphology in three terrestrial isopod crustacean species. *Ecotoxicology and Environmental Safety*, *110*, 269−279.

McCauley, D. J., Pinsky, M. L., Palumbi, S. R., Estes, J. A., Joyce, F. H., & Warner, R. R. (2015). Marine defaunation: Animal loss in the global ocean. *Science (New York, N.Y.)*, *347*(6219), 1255641.

Mcleod, E., Hinkel, J., Vafeidis, A. T., Nicholls, R. J., Harvey, N., & Salm, R. (2010). Sea-level rise vulnerability in the countries of the coral triangle. *Sustainability Science.*, *5*(2), 207−222.

McLeod, K. L., & Leslie, H. M. (2009a). Why ecosystem-based management? *Proceedings of the National Academy of Sciences, USA.*, *115*(14), 3−12.

McLeod, K. L., & Leslie, H. M. (Eds.), (2009b). *Ecosystem-based management of the oceans*. Washington, D. C.: Island Press.

McLeod, K.L., Lubchenco, J., Palumbi, S., & Rosenberg, A.A. (2005). *Scientific consensus statement on marine ecosystem-based management*. United States Commission on Ocean Policy. Washington, D.C.

Meehl, G. A., Washington, W. M., & Collins, W. D. (2005). How much more global warming and sea level how much more global warming and sea level rise? *Science (New York, N.Y.)*, *307*, 1769−1773.

Meijers, A. (2014). The Southern Ocean in the coupled model intercomparison project phase 5. *Philosophical Transactions of The Royal Society: A mathemetiacal Physical and Engineering Sciences*, *372*(2019), 0296.

Meinshausen, M., Vogell, E., Nauels, A., Lorbacher, K., Meinshausen, N., Etheridge, D. M., ... Weiss, R. (2017). Historical greenhouse gas concentrations for climate modeling (CMIP6). *Geoscientific Model Development.*, *10*, 2057−2116.

Meyer, R., Engesgaard, P., & Sonnenborg, T. O. (2019). Origin and dynamics of saltwater intrusion in a regional aquifer: Combining 3D saltwater modeling with geophysical and geochemical data. *Water Resources Research*, *55*(3), 1792−1813.

Michael, H. A., Russoniello, C. J., & Byron, L. A. (2013). Global assessment of vulnerability to sea-level rise in topography-limited and recharge-limited coastal groundwater systems. *Water Resources Research, 49*, 2228−2240.

Milazzo, M., Mirto, S., Domenici, P., & Gristina, M. (2012). Climate change exacerbates interspecific interactions in sympatric coastal fishes. *Journal of Animal Ecology, 82*, 468−477.

Miller, D. D., Ota, Y., Sumuila, R. U., Cisneros-Montemayor, A. M., & Cheung, W. W. L. (2017). Adaptation strategies to climate change in marine systems. *Global Change Biology, 24*(1), e1−e14.

Mimura, N. (2013). Sea-level rise caused by climate change and its implications for society. *Proceedings of the Japan Academy. Series B Physical and Biological Sciences, 89*(7), 281−301.

Mitrovica, J. X., Gomez, N., Morrow, E., Hay, C., Latychev, K., & Tamisiea, M. E. (2011). On the robustness of predictions of sea level fingerprints. *Geophysical Journal International, 187*, 729−742.

Moberg, G. P., & Mench, J. A. (2000). *The biology of animal stress: Basic principles and implications for animal welfare.* Wallingford: CABI.

Moftakhari, H. R., Aghakouchak, A., Sanders, B. F., Allaire, M., & Matthew, R. A. (2018). What is nuisance flooding? Defining and monitoring an emerging challenge. *Water Resources Research, 54*(7), 4218−4227.

Mora, C., Wei, C. L., Rollo, A., Amaro, T., Baco, A. R., Billett, D., ... Yasuhara, M. (2013). Biotic and Human Vulnerability to Projected Changes in Ocean Biogeochemistry over the 21st Century. *PLoS Biology, 11*(10).

Morris, S., Bueseller, K., & Sims, K. W. W. (2011). Re-evaluating the 238 U-salinity relationship in seawater: Implications for the 238 U− 234 Th disequilibrium method. *Marine Chemistry, 127*(1−4), 31−39.

Mostafal, K. M. G., Liu, C.-Q., Zhai, W. D., Minella, M., Vione, D., Gao, K., ... Sakugawa, H. (2016). Reviews and syntheses: Ocean acidification and its potential impacts on marine ecosystems. *Biogeosciences., 13*, 1767−1786.

Mulligan, A. E., Evans, R. L., & Lizarralde, D. (2007). The role of paleochannels in groundwater/seawater exchange. *Journal of Hydrology, 335*, 313−329.

Munday, P. L., Cheal, A. J., Dixson, D. L., Rummer, J. L., & Fabricius, K. E. (2014). Behavioural impairment in reef fishes caused by ocean acidification at CO2 seeps. *Nature Climate Change, 4*, 487−493.

Munday, P. L., Dixson, D. L., Donelson, J. M., Jones, G. P., Pratchett, M. S., Devitsina, G. V., & Døving, K. B. (2009). Ocean acidification impairs olfactory discrimination and homing ability of a marine fish. *Proceedings of National Academy of Sciences, 106*(6), 1848−1852.

Munday, P. L., Warner, R. R., Monro, K., Pandolfi, J. M., & Marshall, D. J. (2013). Predicting evolutionary responses to climate change in the sea. *Ecology Letters, 16*(12), 1488−1500.

Murawski, S. A. (2007). Ten myths concerning ecosystem approaches to marine resource management. *Marine Policy, 31*, 681−690.

NAP (National Academic Press). (2010). *Ocean acidification: A national strategy to meet the challenges of a changing ocean.* Washington, D.C: NAP.

Nawaz, M. F., Bourrie, G., & Trolard, F. (2013). Soil compaction impact and modelling. A review. *Agronomy for Sustainable Development, 33*, 291−309.

NCM (Nordic Council of Ministers). (2016). *Classification of marine ecosystem services. Marine ecosystem services: Marine ecosystem services in Nordic marine waters and the Baltic Sea—Possibilities for valuation* (pp. 23−38). Copenhagen: Nordic Council of Ministers.

Nerem, R. S., Beckley, B. D., Fasullo, J., Hamlington, B. D., Masters, D., & Mitchum, G. T. (2018). Climate change driven accelerated sea level rise detected in the Altimeter Era. *Proceedings of National Academy of Sciences, USA, 115*, 2022−2025.

Neumann, B., Ott, K., & Kenchington, R. (2017). Strong sustainability in coastal areas: a conceptual interpretation of SDG 14. *Sustainability Science., 12*, 1019−1035.

Nicholls, R.J., Hanson, S., Herweijer, C., Patmore, N., Hallegatte, S., Corfee-Morlot, J., ... & Muir-Wood, R. (2008). *Ranking port cities with high exposure and vulnerability to climate extremes*. Paris: OECD Environment Working Papers.

Nicolas, E.S. (2020). *EU failed to end overfishing by 2020: Lost opportunity?* Available online at: https://euobserver.com/environment/147099.

Nishikawa, T., Siade, A. J., Reichard, E. G., Ponti, D. J., Canales, A. G., & Johnson, T. A. (2009). Stratigraphic controls on seawater intrusion and implications for groundwater management in Dominguez Gap area of Los Angeles, California, USA. *Hydrogeology Journal*, *17*, 1699−1725.

NOAA (National Oceanic and Atmospheric Administration). (n.d.). *What is the biggest source of pollution in the Ocean?* Available online from https://oceanservice.noaa.gov/facts/pollution.html.

NOAA (National Oceanic and Atmospheric Administration). (2010). *Oil and sea turtles: Biology, planning, and response*. Washington, D.C.: NOAA Office of Response and Restoration.

NOAA (National Oceanic and Atmospheric Administration). (2012). Service assessment hurricane/post-tropical cyclone sandy. Washington, DC.: NOAA.

NOAA (National Oceanic and Atmospheric Administration). (2020). *Ocean acidification*. NOAA Website, April 2020. Available online.

Norse, E. A., Brooke, S., Cheung, W. W., Clark, M. R., Ekeland, I., Froese, R., ... Watson, R. (2012). Sustainability of deep-sea fisheries. *Marine Policy*, *36*(2), 307−320.

NRC (National Research Council). (2003). *Oil in the sea III: Inputs, fates, and effects*. The National Academies Press.

NRC (National Research Council). (2005). *Oil spill dispersants: Efficacy and effects*. Washington, DC: The National Academies Press.

Obura, D. (2017). Refilling the coral reef glass. *Science (New York, N.Y.)*, *357*(6357), 1215.

Odum, E. P. (1969). The strategy of ecosystem development. *Science (New York, N.Y.)*, *164*(3877), 262−270.

Olive, G., Chang, J. L., Inn, L. V., Hwai, A. T. S., & Yasin, Z. (2019). The effects of salinity on the filtration rates of juvenile tropical oyster *Crassostrea iredalei*. *Tropical Life Sciences Research.*, *27*(Suppl-1), 45−51.

Oliver, S., Braccini, M., Newman, S. J., & Harvey, E. S. (2015). Global patterns in the bycatch of sharks and rays. *Marine Policy*, *54*, 86−97.

Olsen, E. M., Heino, M., Lilly, G. R., Morgan, M. J., Brattey, J., Ernande, B., & Dieckmann, D. (2004). Maturation trends indicative of rapid evolution preceded the collapse of northern cod. *Nature*, *428*(6986), 932−935.

Oscar, M. A., Barak, S., & Winters, G. (2018). The tropical invasive seagrass, halophila stipulacea, has a superior ability to tolerate dynamic changes in salinity levels compared to its freshwater relative, *Vallisneria americana*. *Frontiers in. Plant Sciences*, *9*, 950.

Oschlies, A., Brandt, P., Stramma, L., & Schmidtko, S. (2018). Drivers and mechanisms of ocean deoxygenation. *Nature Geoscience*, *11*, 467−473.

OSPAR (Commission for the Protection of the Marine Environment of the North-East Atlantic). (2003). Statement on the ecosystem approach to the management of human activities. In *Annual report 2002−2003*. Volume-2. London: OSPAR Commission: 16−22. Available from https://www.ospar.org/documents?v = 6955.

Österblom, H., Gardmark, A., Bergstrom, L., Muller-karulis, B., Folke, C., Lindegren, M., ... Mollmann, C. (2010). Making the ecosystem approach operational—Can regime shifts in ecological- and governance systems facilitate the transition?'. *Marine Policy*, *34*(6), 1290−1299.

Ottersen, G., Kim, S., Huse, G., Polovina, J. J., & Stenseth, N. C. (2010). Major pathways by which climate may force marine fish populations. *Journal of Marine Systems*, *79*(3−4), 343−360.

Palmer, M. D. (2017). Reconciling estimates of ocean heating and Earth's radiation budget. *Current Climate Change Reports*, *3*(1), 78−86.

Pandolfi, J., Jackson, J. B. C., Baron, N., Bradbury, R. H., Guzman, H. M., Hughes, T. P., ... Sala, E. (2005). Are United States coral reefs on the slippery slope to slime? *Science (New York, N.Y.)*, *307*, 1725−1726.

Parris, D. J., Ganesh, S., Edgcomb, V. P., DeLong, E. F., & Stewart, F. J. (2014). Microbial eukaryote diversity in the marine oxygen minimum zone off northern Chile. *Frontiers in Microbiology*, *5*, 543.

Partanen, A.-I., Keller, D. P., Korhinen, H., & Matthews, H. D. (2016). Impacts of sea spray geoengineering on ocean biogeochemistry. *Geophysical Research Letters*, *43*(14), 4300−4308.

Partin, C. A., Bekker, A., Planavsky, N. J., Scott, C. T., Gill, B. C., Li, C., ... Lyons, T. W. (2013). Large-scale fluctuations in Precambrian atmospheric and oceanic oxygen levels from the record of U in shales. *Earth and Planetary Science Letters*, 284−293, #69−370.

Passeri, D. L., Hagen, S. C., Medeiros, S. C., Bilskie, M. V., Alizad, K., & Wang, D. (2015). The dynamic effects of sea level rise on low-gradient coastal landscapes: A review. *Earth's Future*, *3*, 159−181.

Pauly, D., Christensen, V., Dalsgaard, J., Froese, R., & Torres, F. (1998). Fishing down marine food webs. *Science (New York, N.Y.)*, *279*(5352), 860−863.

Pauw, P., De Louw, P. G. B., & Oude Essink, G. H. P. (2012). Groundwater salinisation in the Wadden Sea area of the Netherlands: Quantifying the effects of climate change, sea-level rise and anthropogenic interferences. *Netherlands Journal of Geosciences-Geologie En Mijnbouw*, *91*, 373−383.

PBC (Palm beach County). (2008). *Lake worth lagoon management plan*. West Palm Beach, FL: Palm Beach County Department of Environmental Resources Management.

Pecl, G. T., Araújo, M. B., Bell, J. D., Blanchard, J., Bonbrake, T. C., Chen, I.-C., ... Williams, S. E. (2017). Biodiversity redistribution under climate change: Impacts on ecosystems and human well-being. *Science (New York, N.Y.)*, *355*, 6332.

Peixoto, D., Pinheiro, C., Amorim, J., Oliva-Teles, L., Guilhermino, L., & Vieira, M. N. (2019). Microplastic pollution in commercial salt for human consumption: A review. *Estuarine, Coastal and Shelf Science*, *219*, 161−168.

Perella, M. (2017). *Extended producer responsibility: The answer to cutting waste in the UK?* Available online.

Perera, F. (2018). Pollution from fossil-fuel combustion is the leading environmental threat to global pediatric health and equity: Solutions exist. *International Journal of Environmental Research and Public Health*, *15*(1), 16.

Perry, R. I., Cury, P., Brander, K., Jennings, S., Möllmann, C., & Planque, B. (2010). Sensitivity of marine systems to climate and fishing: Concepts, issues and management responses. *Journal of Marine Systems*, *79*(3−4), 427−435.

Peterson, S. & Teal, J. (1986). *Scientific basis for the role of the oceans as a waste disposal option*. Marine Policy.10.

Piet, G. J., Culhane, F., Jongbloed, R., Robinson, L. A., Rumes, B., & Tamis, J. (2019). An integrated risk-based assessment of the North Sea to guide ecosystem-based management. *Science of the Total Environment*, *654*, 694−704.

Piggott, J. J., Townsend, C. R., & Matthaei, C. D. (2015). Reconceptualizing synergism and antagonism among multiple stressors. *Ecology and Evolution*, *5*(7), 1538−1547.

Pikitch, E. K., Santora, C., Babcock, E. A., & Bakun, A. (2004). Ecosystem-based fishery management. *Science (New York, N.Y.)*, *305*, 345−346.

Pinsky, M. L., Worm, B., Fogarty, M. J., Sarmiento, J. L., & Levin, S. A. (2013). Marine taxa track local climate velocities. *Science (New York, N.Y.)*, *341*(6151), 1239−1242, PMID: 24031017.

Piotrowski, J. A. (1994). Tunnel-valley formation in northwest Germany—Geology, mechanisms of formation and subglacial bed conditions for the Bornhöved tunnel valley. *Sedimentary Geology*, *89*, 107−141.

Pithan, F., & Mauritsen, T. (2014). Arctic amplification dominated by temperature feed-backs in contemporary climate models. *Nature Geoscience*, *7*, 181−184.

Plastics Europe. (2016). *World plastics production 1950–2015*. Available online.
Popova, E., Yool, A., Byfield, V., Cochrane, K., Coward, A. C., Salim, S., ... Roberts, M. J. (2016). From global to regional and back again: common climate stressors of marine ecosystems relevant for adaptation across five ocean warming hotspots. *Global Change Biology, 22*, 2018–2053.
Popova, E. E., Yool, A., Aksenov, Y., Coward, A. C., & Anderson, T. R. (2014). Regional variability of acidification in the Arctic: a sea of contrasts. *Biogeosciences., 11*, 293–308.
Porter, J. R., Xie, L., Challinor, A. J., Cochrane, K., Howden, M., Iqbal, M. M., ... Travasso, M. I. (2014). Food security and food production systems. *In* Climate change 2014: Impacts, adaptation and vulnerability. Working group II Contribution to the IPCC 5th assessment report (pp. 411–484). Cambridge, UK: Cambridge University Press.
Portner, H. O. (2001). Climate change and temperature-dependent biogeography: Oxygen limitation of thermal tolerance in animals. *The Science of Nature, 88*(4), 137–146.
Pörtner, H. O., & Farrell, A. P. F. (2008). Physiology and climate change. *Science (New York, N.Y.), 322*(5902), 690–692.
Pörtner, H.-O., Karl, D., Boyd, P. W., Cheung, W., Lluch-Cota, S. E., Nojiri, Y., ... Zavialov, P. (2014). *Ocean systems. Climate change 2014: Impacts, adaptation, and vulnerability. Contribution of working group II to the fifth assessment report of the intergovernmental panel on climate change* (pp. 411–484). Cambridge, UK: Cambridge University Press.
Prandle, D., & Lane, A. (2015). Sensitivity of estuaries to sea level rise: Vulnerability indices. *Estuarine Coastal and Shelf Science, 160*, 60–68.
Prata, J. C., Silva, A. L. P., da Costa, J. P., Mouneyrac, C., Walker, T. R., Duarte, A. C., & Rocha-Santos, T. (2019). Solutions and integrated strategies for the control and mitigation of plastic and microplastic pollution. *International Journal of Environment Research and Public Health, 16*(13), 2411.
Quéré, C. L., Aumont, O., Bopp, L., Bousquet, P., Ciais, P., Francey, R., ... Rayner, P. J. (2003). Two decades of ocean CO_2 sink and variability. *Tellus B: Chemical and Physical Meteorology, 55*(2), 649–655.
Rabalais, N. N. (2019). The significance of ocean deoxygenation for estuarine and coastal benthos. In D. Laffoley, & J. M. Baxter (Eds.), *Ocean deoxygenation: Everyone's problem - Causes, impacts, consequences and solutions*. Gland, Switzerland: IUCN: 379–399.
Rahikainen, M., Hoviniemi, K.-M., Mäntyniemi, S., Vanhatalo, J., Helle, I., Lehtiniemi, M., ... Kuikka, S. (2017). Impacts of eutrophication and oil spills on the Gulf of Finland herring stock. *Canadian Journal of Fisheries and Aquatic Sciences, 74*(8), 1218–1232.
Rahim, M. M., Islam, M. T., & Kuruppu, S. (2016). Regulating global shipping corporations' accountability for reducing greenhouse gas emissions in the seas. *Marine Policy, 69*, 159–170.
Rajaratnam, B., Romano, J., Tsiang, M., & Diffenbaugh, N. S. (2015). Debunking the climate hiatus. *Climatic Change, 133*(2), 129–140.
Rapp, D. (2013). Ocean sediment data. In D. Rapp (Ed.), *Ice ages and interglacials* (pp. 171–189). Berlin, Germany: Springer.
Rathi, V., & Yadav, V. (2019). Oil degradation taking microbial help and bioremediation: A review. *Journal of Biomediation and Biodegradation, 10*(2). Available online.
Rau, G. H., McLeod, E. L., & Hoegh-Guldberg, O. (2012). The need for new ocean conservation strategies in a high-carbon dioxide world. *Nature Climate Change, 2*, 720–724.
Regan, M. D., Turko, A. J., Heras, J., Andersen, M. K., Lefevre, S., Wang, T., ... Nilsson, G. E. (2016). Ambient CO_2, fish behaviour and altered GABAergic neurotransmission: Exploring the mechanism of CO_2-altered behaviour by taking a hypercapnia dweller down to low CO_2 levels. *Journal of Experimental Biology, 219*(1), 109–118.
Reichelt - Brushett, A. J., & Harrison, P. (2005). The effect of selected trace metals on the fertilization success of several Scleractinia coral species. *Coral Reefs, 24*, 524–534.

Resplandy, L., Keeling, R. F., Eddebbar, Y., Brooks, M., Wang, R., Bopp, L., ... Oschlies, A. (2019). Quantification of ocean heat uptake from changes in atmospheric O_2 and CO_2 composition. *Scientific Reports, 9*, 20244.

Ridge, J. T., Rodriguez, A. B., & Fodrie, F. J. (2017). Evidence of exceptional oyster-reef resilience to fluctuations in sea level. *Ecology and Evolution, 7*(23), 10409−10420.

Riebesell, U., Shultz, K., Bellerby, R., Botros, M., Fritsche, P., Meyerhofer, M., ... Zoliner, E. (2007). Enhanced biological carbon consumption in a high CO_2 ocean. *Nature, 450*, 545−548.

Rignot, E. J., Velicogna, I., van den Broeke, M. R., Monaghan, A. J., & Lenaerts, J. T. M. (2011). Acceleration of the contribution of the Greenland and Antarctic ice sheets to sea level rise. *Geophysical Research Letters, 38*, L05503.

Robinson, C. (2019). Microbial respiration, the engine of ocean deoxygenation. *Marine Biogeochemistry., 2018*, 00533.

Roemmich, D., Alford, M. H., Clauster, H., Johnson, K., King, B., Moum, J., ... Yasuda, I. (2019). On the future of argo: A global, full-depth, multi-disciplinary array. *Frontiers in Marine Science, 6*, 439.

Rogers, A. D., Sumaila, U. R., Hussain, S. S., & Baulcomb, C. (2014). *The high seas and us: understanding the value of high-seas ecosystems*. Oxford, UK: Global Ocean Commission.

Roman, M. R., Brandt, S. B., Houde, E. D., & Pierson, J. J. (2019). Interactive effects of hypoxia and temperature on coastal pelagic zooplankton and fish. *Frontiers in Marine Science, 6*, 139.

Roman, M. R., & Pierson, J. J. (2019). The significance of ocean deoxygenation for estuarine and coastal plankton. In D. Laffoley, & J. M. Baxter (Eds.), *Ocean deoxygenation: Everyone's problem - Causes, impacts, consequences and solutions*. Gland, Switzerland: IUCN: 363−378.

Rosane, O. (2019). *Oceans absorb almost 1 / 3 of global CO_2 emissions, but at what cost?* World Economic Forum. 19 Mach. Available online from: https://www.weforum.org/agenda/2019/03/oceans-absorb-co2-challenges-emerge/.

Rose, B.E.J. (2015). *Components of climate system*. Available online from http://www.atmos.albany.edu/facstaff/brose/classes/ATM623_Spring2015/Notes/Lectures/Lecture04%20−%20Climate%20system%20components.html.

Rose, K. A., Gutiérrez, D., Breitburg, D., Conley, D., Craig, J. K., Froehlich6, H. E., ... Prema, D. (2019). Impacts of ocean deoxygenation on fisheries. In D. Laffoley, & J. M. Baxter (Eds.), *Ocean deoxygenation: Everyone's problem—Causes, impacts, consequences and solutions*. Gland, Switzerland: IUCN: 519−544.

Rosenberger, J., Adrianne, L., MacDuffee, M., Rosenberger, A. G. J., & Ross, P. S. (2017). Oil spills and marine mammals in British Columbia, Canada: Development and application of a risk-based conceptual framework. *Archives of Environmental Contamination and Toxicology, 73*, 131.

Routley, N. (2018, April 25). This shocking chart shows the true impact of plastic on our planet. Available online from: https://www.weforum.org/agenda/2018/04/visualizing-the-prolific-plastic-problem-in-our-oceans.

Rowley, R. J., Kostelnick, J. C., Braten, D., Li, X., & Meisel, J. (2007). Risk of rising sea level to population and land area. *Eos, Transactions, American Geophysical Union, 88*, 105−107.

Royal Society. (2005). *Ocean acidification due to increasing atmospheric carbon dioxide*. Policy document 12/05. Available from http://eprints.uni-kiel.de/7878/1/965_Raven_2005_OceanAcidificationDueTo Increasing_Monogr_pubid13120.pdf.

Rudd, M. A., Dickey-Collas, M., Ferretti, J., Johannesen, E., Macdonald, N. M., McLaughlin, R., ... Link, J. S. (2018). Ocean ecosystem-based management mandates and implementation in the North Atlantic. *Frontiers in Marine Science, 5*, 485.

Russell, L. M., Rasch, P. J., Mace, G. M., Jackson, R. B., Shepherd, J., Liss, P., ... Morgan, M. G. (2012). Ecosystem impacts of geoengineering: A review for developing a science plan. *Ambio, 41*(4), 350−369.

Rytkonen, K. T., Renshaw, G. M. C., Ashton, K. J., Williams-Pritchard, G., Leder, E. H., & Nikinmaa, M. (2010). Elasmobranch qPCR reference genes: a case study of hypoxia preconditioned epaulette sharks. *BMC Molecular Biology, 11*, 27.

SAFMC (South Atlantic Fisheries Management Council) (2009). *Fishery ecosystem plan of the south Atlantic region*. Available online from: https://safmc.net/download/SAFMC_HabitatPolicy_ClimateVariabilityFisheries_Final_Dec2016.pdf.

Sala, E., Mayorga, J., Costello, C., Kroodsma, D., Palomares, M. L. D., Pauly, D., ... Zeller, D. (2018). The economics of fishing the high seas. *Science Advances*, *4*(6), eaat2504.

Sánchez-Arcilla, A., Mösso, C., Sierra, J. P., Mestres, M., Harzallah, A., Senouci, M., & El Rahey, M. (2011). Climate drivers of potential hazards in Mediterranean coasts. *Regional Environmental Change*, *11*(8), 617–636.

Santos, R. G., Loh, W., Bannwart, A. C., & Trevisan, O. V. (2014). An overview of heavy oil properties and its recovery and transportation methods. *Brazilian Journal of Chemical Engineering*, *31*(3), 571–590.

Sardà, R., O'Higgins, T., R. Cormier, A. D., & Tintore, J. (2014). A proposed ecosystem-based management system for marine waters: Linking the theory of environmental policy to the practice of environmental management. *Ecology and Society*, *19*(4), 51.

Sarkis, S. (2017). *Threatened Corals see effect of sedimentation stress in early life stages.* NOAA. Available online.

Sato, T., Qadir, M., Yamamoto, S., Endo, T., & Zahoor, A. (2013). Global, regional, and country level need for data on wastewater generation, treatment, and use. *Agricultural Water Management*, *130*(2013), 1–13. Available online from: https://www.sciencedirect.com/science/article/abs/pii/S0378377413002163.

Schiller, L., Bailey, M., Jacquet, J., & Sala, E. (2018). High seas fisheries play a negligible role in addressing global food security. *Science Advances*, *4*(8), eaat8351.

Schmidtko, S., & Johnson, G. C. (2012). Multidecadal warming and shoaling of antarctic intermediate water. *Journal of Climate*, *25*, 207–221.

Schulte, P. M. (2014). What is environmental stress? Insights from fish living in a variable environment. *Journal of Experimental Biology*, *217*, 23–34.

Schwing, P. T., Romero, I. C., Brooks, G. R., Hastings, D. W., Larson, R. A., & Hollander, D. J. (2015). A decline in benthic foraminifera following the deep water horizon event in the northeastern Gulf of Mexico. *PLoS One*, *10*, e0128505.

Sefrioui, S. (2017). Adapting to sea level rise: A law of the sea perspective. In G. Andreone (Ed.), *The future of the law of the sea* (pp. 3–22). Cham. Switzerland: Springer.

Seibel, B. A., & Wishner, K. A. (2019). The significance of ocean deoxygenation for mesopelagic communities. In D. Laffoley, & J. M. Baxter (Eds.), *Ocean deoxygenation: Everyone's problem—Causes, impacts, consequences and solutions*. Gland, Switzerland: IUCN: 265–276.

Semedo, A., Saetra, O., Rutggerson, A., Kahma, K. K., & Pettersson, H. (2009). Wave-induced wind in the marine boundary layer. *Journal of the Atmospheric Sciences*, *66*(8), 2256–2271.

SFWMD (South Florida Water Management District. (2009a). *St. Lucie river watershed protection plan.* West Palm Beach, FL.: South Florida Water Management District.

SFWMD (South Florida Water Management District). (2009b). *Lake worth lagoon watershed and stormwater loading analysis.* West Palm Beach, FL.: South Florida Water Management District.

SFWMD (South Florida Water Management District). (2012). *St. Lucie river watershed protection plan update.* West Palm Beach, FL: South Florida Water Management District, Appendix 10–1, South Florida Environmental Report 2012.

Shepherd, A., Wingham, D., Wallis, D., Giles, K., Lexon, S., & Sundal, A. V. (2010). Recent loss of floating ice and the consequent sea level contribution. *Geophysical Research Letters*, *37*, 13.

Shepherd, J. G., Brewer, P. G., Oschlies, A., & Watson, A. J. (2017). Ocean ventilation and deoxygenation in a warming world: Introduction and overview. *Philosophical Transactions A Mathematics, Physics, Engineering Science*, *375*(2102), 20170240.

Shi, W., & Wang, M. (2010). Characterization of global ocean turbidity from moderate resolution imaging spectroradiometer ocean color observations. *Journal of Geophysical Research, Oceans*, *115*, C11.

Short, J. (2003). Long-term effects of crude oil on developing fish: Lessons from the Exxon Valdez oil spill. *Energy Sources*, *25*, 509–517.

Short, J. W., Irvine, G. V., Mann, D. H., Maselko, J. M., Pella, J. J., Lindeberg, M. R., ... Rice, S. D. (2007). Slightly weathered Exxon Valdez oil persists in Gulf of Alaska beach sediments after 16 years. *Environment Science & Technology, 41*, 1245–1250.

Sigman, D. M., & Hain, M. P. (2012). The biological productivity of the ocean. *Nature Education Knowledge, 3*(10), 21.

Simlinger, F., & Mayer, B. (2019). Legal responses to climate change induced loss and damage. In R. Mechler, L. Bouwer, T. Schinko, S. Surminski, & J. Linnerooth-Bayer (Eds.), *Loss and damage from climate change. Climate risk management, policy and governance* (pp. 179–203). Cham, Switzerland: Springer.

Sims, D. W. (2019). The significance of ocean deoxygenation for Elasmobranchs. In D. Laffoley, & J. M. Baxter (Eds.), *Ocean deoxygenation: Everyone's problem—Causes, impacts, consequences and solutions* (pp. 431–441). Gland, Switzerland: IUCN.

Slangen, A. B. A., Church, J. A., Agosta, C., Fettweis, X., Marzeion, B., & Kristin, R. (2016). Anthropogenic forcing dominates global mean sea-level rise since 1970. *Nature Climate Change, 6*(7), 701–705.

Sloyan, B. M., Talley, L. D., Chereskin, T. K., Fine, R., & Holte, J. (2010). Antarctic intermediate water and subantarctic mode water formation in the Southeast Pacific: The role of turbulent mixing. *Journal of Physical Oceanography, 40*, 1558–1574.

Smith, M. D., Knapp, A. K., & Collins, S. L. (2009). A framework for assessing ecosystem dynamics in response to chronic resource alterations induced by global change. *Ecology, 90*(12), 3279–3289.

Smith, T.W.P., Jalkanen, J.P., Anderson, B.A., Corbett, J.J., Faber, J., Hanayama, S., & Pandey, A. (2015). *Third IMO greenhouse gas study 2014*. Available online from: https://greenvoyage2050.imo.org/wp-content/uploads/2021/01/third-imo-ghg-study-2014-executive-summary-and-final-report.pdf.

Sørensen, L., Sørhus, E., Nordtug, T., Incardona, J. P., Linbo, T. L., Giovanetti, L., ... Meir, S. (2017). Oil droplet fouling and differential toxicokinetics of polycyclic aromatic hydrocarbons in embryos of Atlantic haddock and Cod. *PLoS One, 12*, e0180048.

Sørhus, E., Incardona, J. P., Furmanek, T., Goetz, G. W., Scholz, N. L., Meir, S., ... Jentoft, S. (2017). Novel adverse outcome pathways revealed by chemical genetics in a developing marine fish. *eLife., 6*, e20707.

Sørhus, E., Incardona, J. P., Karlsen, Ø., Linbo, T., Sørensen, L., Nordtug, T., ... Meir, S. (2016). Crude oil exposures reveal roles for intracellular calcium cycling in Haddock craniofacial and cardiac development. *Scientific Reports, 6*, 31058.

Spalding, C., Finnegan, S., & Fischer, W. W. (2017). Energetic costs of calcification under ocean acidification. *Global Biogeochemical Cycles, 31*(5), 866–877.

Spicer, J. I., & Weber, R. E. (2004). Respiratory impairment in crustaceans and molluscs due to exposure to heavy metals. *Comparative Biochemistry and Physiology Part C: Comparative Pharmacology., 100*(3), 339–342.

STECF (Scientific, Technical and Economic Committee for Fisheries). (2019). *Monitoring the performance of the common fisheries policy (STECF-Adhoc-19-01)*. Luxembourg: Publications Office of the European Union.

Steenberg, S., Duinker, P., & Nitoslawski, S. (2019). Ecosystem-based management revisited: Updating the concepts for urban forests. *Landscape and Urban Planning, 186*, 24–35.

Steingass, S., & Naito, Y. (2019). The significance of ocean deoxygenation for ocean megafauna. In D. Laffoley, & J. M. Baxter (Eds.), *Ocean deoxygenation: Everyone's problem - Causes, impacts, consequences and solutions*. Gland, Switzerland: IUCN: 449–459.

Sumaila, U. R., & Tai, T. C. (2020). End overfishing and increase the resilience of the ocean to climate change. *Frontiers in Marine Science, 7*, 523.

Sumaila, U. R., Tai, T. C., Lam, V. W. Y., Cheung, W. W. L., Bailey, M., Cisneros-Montemayor, A. M., ... Gulati, S. S. (2019). Benefits of the Paris Agreement to ocean life, economies, and people. *Science Advances, 5*, eaau3855.

Sunday, J. M., Bates, A. E., & Dulvy, N. K. (2012). Thermal tolerance and the global redistribution of animals. *Nature Climate Change*, *2*, 686−690.

Sunday, J. M., Fabricius, K. E., Kroeker, K. J., Anderson, K. M., Brown, N. E., Barry, J. P., ... Hartley, C. D. G. (2017). Ocean acidification can mediate biodiversity shifts by changing biogenic habitat. *Nature Climate Change*, *7*, 81−85.

Sutherland, W. J., Barnard, P., Broad, S., Clout, M., Connor, B., & Côté, I. M. (2017). A 2017 horizon scan of emerging issues for global conservation and biological diversity. *Trends in Ecology and Evolution*, *32*, 31−40.

Sweet, W. V., Horton, R., Kopp, R. E., LeGrande, A. N., & Romanou, A. (2017). *Sea level rise,* . In Climate science special report: fourth national climate assessment (Volume I, pp. 333−363). Washington, D.C.: United States Global Change Research Program.

Tahil, A. S., & Dy, D. T. (2015). Effects of reduced pH on the growth and survival of post-larvae of the donkey's ear abalone, *Haliotis asinina* (L.). *Aquaculture International*, *23*, 141−153.

Talley, L. D. (1996). Antarctic intermediate water in the South Atlantic. In W. G. Berger, W. H. Sielder, & D. J. Webb (Eds.), *The South Atlantic: Present and past circulation* (pp. 219−238). Berlin: Springer.

Tallis, H., Levin, P. S., Ruckelshaus, M., Lester, S. E., McLeod, K. L., Fluharty, D. L., & Halpern, B. S. (2010). The many faces of ecosystem-based management: Making the process work today in real places. *Marine Policy*, *34*, 340−348.

Tanaka, Y. (2017). Land-based marine pollution. In A. Nollkaemper, & I. Plakokefalos (Eds.), *International Law of the Sea* (pp. 294−315). Cambridge: Cambridge University Press.

Tarnecki, J. H., & Patterson, W. F. (2015). Changes in red snapper diet and trophic ecology following the deepwater horizon oil spill. *Marine and Coastal Fisheries*, *7*, 135−147.

Tchounwou, P. B., Yedjou, C. G., Patlolla, A. K., & Sutton, D. J. (2012). Heavy metal toxicity and the environment. In A. Luch (Ed.), Molecular, clinical and environmental toxicology. *Experientia Supplementum* (vol 101, pp. 133−164). Basel: Springer.

Teh, L. S., Lam, V. W., Cheung, W. W., Miller, D., Teh, L. C., & Sumaila, U. R. (2016). Impact of high seas closure on food security in low-income fish dependent countries. *PLoS One*, *11*(12), e0162985.

Thevenon, F., & De Sousa, J. M. (2017). Tackling marine plastic pollution: Monitoring, policies, and sustainable development solutions. In P. Nunes, L. E. Svensson;, & A. Markandya (Eds.), Handbook on the economics and management of sustainable oceans (pp. 353−380). Northampton, MA: Edward Elgar Publishing.

Thompson, D. A., Karunarathna, H., & Reeve, D. E. (2017). Modelling extreme wave overtopping at Aberystwyth promenade. *Water.*, *9*, 663.

Tissot, B. P., & Welte, D. H. (1984). *Petroleum formation and occurrence*. Verlag: Springer.

Tittensor, D. P., Mora, C., Jetz, W., Lotze, H. K., Ricard, D., Berghe, E. V., & Worm, B. (2010). Global patterns and predictors of marine biodiversity across taxa. *Nature*, *466*(7310), 1098−1101.

Tokarska, K. B., Gillett, N. P., Weaver, A. J., Arora, V. K., & Eby, M. (2016). The climate response to five trillion tonnes of carbon. *Nature Climate Change*, *6*(9), 851−855.

Toonen, R. J., Wilhelm, T. A., Maxwell, S. M., Wagner, D., Bowen, B. W., Sheppard, C. R. C., ... Friedlander, A. M. (2013). One size does not fit all: The emerging frontier in large-scale marine conservation. *Marine Pollution Bulletin*, *77*, 7−10.

Törnqvist, T. E., Wallace, D. J., Storms, J. E. A., Wallinga, J., Van Dam, R. L., Blaauw, M., ... Snijders, E. M. A. (2008). Mississippi Delta subsidence primarily caused by compaction of Holocene strata. *Nature Geoscience*, *1*(3), 173−176.

Townsend, C. R., Uhlmann, S. S., & Matthaei, C. D. (2008). Individual and combined responses of stream ecosystems to multiple stressors. *Journal of Applied Ecology*, *45*(6), 1810−1819.

Trenberth, K. E., & Stepaniak, D. P. (2004). The flow of energy through the earth's climate system. *Quarterly Journal of the Royal Meteorological Society*, *130*, 2677−2701.

Tsikliras, A. C., Dinouli, A., Tsiros, V. Z., & Tsalkou, E. (2015). The Mediterranean and Black Sea fisheries at risk from overexploitation. *PLoS One, 10*, 1−19.

Turley, C., & Boot, K. (2010). *Environmental consequence of ocean acidification: a threat to food security*. Nairobi: UNEP Emerging Issues Bulletin.

Turley, C., & Boot, K. (2011). The ocean acidification challenges facing science and society. In J.-P. Gattuso, & L. Hanson (Eds.), *Ocean acidification* (pp. 249−271). Oxford: Oxford University Press.

Tyrrell, T. (2011). Anthropogenic modification of the oceans. *Philosophical Transactions of the Royal Society, 369*, 887−908.

UCS (Union of Concerned Scientists) (2018). *Scientists agree: Global Warming is happening and humans are the primary cause*. UCS Website, 4 January 2018. Available online from https://www.ucsusa.org/resources/global-warming-happening-and-humans-are-primary-cause.

UN (United Nations). (2015). *Transforming our world: The 2030 agenda for sustainable development*. A/RES/70/1. 21 October 2015. Available at: https://www.un.org/ga/search/view_doc.asp?symbol = A/RES/70/1&Lang = E.

UN (United Nations). (2017). *Oceans: Facts and figures*. Available Online at: http://www.un.org/sustainable-development/oceans/.

UNEP. (2009). *Marine litter: A global challenge*. Nairobi: Nairobi: UNEP Available at. Available from http://www.unep.org/pdf/unep_marine_litter-a_global_challenge.pdf.

UNEP (United Nations Environment Programme). (n.d.). *Global Marine Oil Pollution Information Gateway*. Available online from http://oils.gpa.unep.org/facts/sources.htm.

UNEP (United Nations Environment Programme). (1995). *Global programme of action for the protection of marine environment from land-based activities*. UNEP (OCA)/LBA/IG.2//7, 5 December 1995. Nairobi: UNEP.

UNEP (United Nations Environment Programme). (2016). *Regional oceans governance making regional seas programmes, regional fishery bodies and large marine ecosystem mechanisms work better together*. Nairobi: UNEP.

UNEP (United Nations Environment Programme). (2017). *Draft resolution on marine litter and microplastics (UNEP/EA.3/L.20. Resolution)*. Nairobi: UNEP.

UNEP (United Nations Environment Programme). (United Nations Environment Programme). (2014). *Plastic waste causes financial damage of US $13 billion to marine ecosystems each year as concern grows over microplastics*. Available online.

USF&WS (United States Fish and Wildlife Service) (2011). *Deepwater horizon response consolidated fish and wildlife collection report*. Available from http://www.fws.gov/home/dhoilspill/pdfs/Consolidated%20Wildlife%20Table%2001252011.pdf.

USGS (Unites States geological Survey) (n.d.a). *Land subsidence*. Available from https://www.usgs.gov/special-topic/water-science-school/science/land-subsidence?qt-science_center_objects = 0#qt-science_center_objects.

USGS (United States Geological Survey). (n.d.b) *pH and water*. Available from https://www.usgs.gov/special-topic/water-science-school/science/ph-and-water?qt-science_center_objects = 0#qt-science_center_objects.

Valiela, I., Bowen, J. L., & York, J. K. (2001). Mangrove forests: one of the world's threatened major tropical environments at least 35% of the area of mangrove forests has been lost in the past two decades, losses that exceed those for tropical rain forests and coral reefs, two other well-known threatened environments. *Bioscience, 51*, 807−815.

Van den Brink, P. J., Choung, C. B., Landis, W., Mayer-Pinto, M., Pettygrove, V., Scanes, P., ... Stauber, J. (2016). New approaches to the ecological risk assessment of multiple stressors. *Marine and Freshwater Research, 67*(4), 429−439.

van Hoof, L. (2015). Fisheries management, the ecosystem approach, regionalisation and the elephants in the room. *Marine Policy, 60*, 20−26.

van Leeuwen, J., van Hoof, L., & van Tatenhove, J. P. M. (2012). Institutional ambiguity in implementing the European Union marine strategy framework directive. *Marine Policy, 36*, 636−643.

Vegas-Vilarrubis, T., & Rull, T. (2016). Undervalued impacts of sea-level rise: Vanishing Deltas. *Frontiers in Ecology and Evolution*, *4*, 77.

Vierros, M. (2017). Global marine governance and oceans management for achievement of SDG 14. UN Chronicle. *LIV* (1−2). Available Online from: https://www.un-ilibrary.org/content/journals/15643913/54/2/41.

Vierros, M., Suttle, C. A., Harden-Davies, H., & Burton, G. (2016). Who owns the ocean? Policy issues surrounding marine genetic resources. *Limnology and Oceanography Bulletin*, *25*, 29−35.

Villar-Argaiz, M., medina-Sanchez, J. M., Bidanda, B. A., & Carrillo, P. (2018). Predominant non-additive effects of multiple stressors on autotroph C: N:P ratios propagate in freshwater and marine food webs. *Frontiers in Microbiology*, *9*, 69.

Vince, J., & Hardesty, B. D. (2017). Plastic pollution challenges in marine and coastal environments: from local to global governance. *Restoration Ecology*, *25*(1), 123−128.

Visbeck, M., Kronfeld-Goharani, U., Neumann, B., Rickels, W., Schmidt, J., & van Doorn, E. (2014). Securing blue wealth: The need for a special sustainable development goal for the ocean and coasts. *Marine Policy*, *48*, 184−191.

Vitousek, S., Barnard, P. L., Fletcher, C. H., Frazer, N., Erikson, L., & Storlazzi, C. D. (2017). Doubling of coastal flooding frequency within decades due to sea-level rise. *Scientific Reports*, *7*(1), 1399.

Wada, Y., Lo, M.-H., Yeh, P. J. F., Reager, J. T., Famiglietti, J. S., Wu, R.-J., & Tseng, Y.-H. (2016). Fate of water pumped from underground and contributions to sea-level rise. *Nature Climate Change*, *6*, 777−780.

Wadey, M. P., Nicholls, R. J., & Hutton, C. (2012). Coastal flooding in the Solent: An integrated analysis of defenses and inundation. *Water.*, *4*, 430−459.

Wallmann, K. (2000). Phosphorus imbalance in the global ocean? *Global Biogeochemical Cycles*, *24*, GB4030.

Walnum, N. (2018, May 18). *Marianergropa er verdens dypeste havområde. Her ble det funnet en plastpose i 1998*. In Norwegian Language. (*Record Diver found plastic bag at the deepest point of the sea*). Available from https://www.dagbladet.no/nyheter/marianergropa-er-verdens-dypeste-havomrade-her-ble-det-funnet-en-plastpose-i-1998/69800975.

Walsh, J. D., Wuebbles, K., Hayhoe, J., Kossin, K., Kunkel, G., Stephens, P., . . . Willis, J. (2014). Our changing climate. *In* Climate change impacts in the United States: The third national climate assessment (pp. 19−67). Washington, DC: Global Change Research Program.

Wang, J., Zhang, W., & Mu, J. (2017). *Marine debris (micro-and macro-) monitoring and research in China*. Available from https://webgate.ec.europa.eu/maritimeforum/system/files/EU-.

Wang, J. W., Han, W., & Sriver, R. L. (2012). Impact of tropical cyclones on the ocean heat budget in the Bay of Bengal during 1999: 1. Model configuration and evaluation: Tropical cyclones and ocean heat budget, 1. *Journal of Geophysical Research Oceans*, *117*, C1.

Wang, W.-X., & Guangyuan, L. (2017). Heavy metals in bivalve mollusks. In a Cartus, & D. Schrenk (Eds.), *Chemical contaminants and residues in food*. New Delhi: Woodhead Publishers: 553−594.

Wang, X. L., Feng, Y., & Swail, V. R. (2014). Changes in global ocean wave heights as projected using multi-model CMIP5 simulations. *Geophysical Research Letters*, *41*, 1026−1034.

Ward-Paige, C. A., Risk, M., & Sherwood, O. (2005). Clionid sponge surveys on the Florida Reef Tract suggest land-based nutrient inputs. *Marine Pollution Bulletin*, *51*, 570−579.

Warner, R. (2009). Marine snow storms: Assessing the environmental risks of ocean fertilization. *Carbon & Climate Law Review*, *3*(4), 426−436.

Wassmann, P., Duarte, C. M., Agusti, S., & Sejr, M. K. (2011). Footprints of climate change in the Arctic marine ecosystem. *Global Change Biology*, *17*, 1235−1249.

Wasson, K., Suarez, B., Akhavan, A., McCarthy, E., Kenneth, J. K., Johnson, S., . . . Feliz, D. (2015). Lessons learned from an ecosystem-based management approach to restoration of a California estuary. *Marine Policy*, *58*, 60−70.

Waycott, M., Durate, C. M., Carruthers, T. J. B., Orth, R. J., Dennison, W. C., Olyarnik, S., ... Williams, S. L. (2009). Accelerating loss of seagrasses across the globe threatens coastal ecosystems. *Proceedings of National Academy of Science, USA, 106*, 12377–12381.

Waylen, K. A., Hastings, E. J., Banks, E. A., Holstead, K. L., Irvine, R. J., & Blackstock, K. L. (2014). The need to disentangle key concepts from ecosystem-approach jargon. *Conservation Biology, 28*(5), 1215–1224.

WCRP-GSLBG (World Climate Research Programme-Global Sea Level Budget Group). (2018). Global sea-level budget 1993–present. *Earth System Science Data, 10*(1), 1551–1590.

Werner, A. D., Bakker, M., Post, V. E. A., Vandenbohede, A., Lu, C., Ataie-Ashtiani, B., ... Barry, D. A. (2013). Seawater intrusion processes, investigation and management: Recent advances and future challenges. *Advances in Water Resources, 51*, 3–26.

White, C., & Costello, C. (2014). Close the high seas to fishing? *PLoS Biology, 12*(3), e1001826.

White, H. K., Hsing, P. Y., Cho, W., Shank, T. M., Cordes, E. E., Quattrini, A. M., ... Fisher, C. R. (2012). Impact of the deepwater horizon oil spill on a deep-water coral community in the Gulf of Mexico. *Proceedings of National Academy of Science, USA., 109*, 20303–20308.

Wilcox, C., Puckridge, M., Schuyler, Q. A., Townsend, K., & Hardesty, B. D. (2018). A quantitative analysis linking sea turtle mortality and plastic debris ingestion. *Scientific Reports, 8*, 12536.

Wilcox, C., Van Sebille, E., & Hardesty, B. D. (2015). Threat of plastic pollution to seabirds is global, pervasive, and increasing. *Proceedings of the National Academy of Sciences, USA., 112*(38), 11899–11904.

Willeit, M., Ganopolski, A., Dalmonech, D., Foley, A. M., & Fuelner, G. (2014). Time-scale and state dependence of the carbon-cycle feedback to climate. *Climate Dynamics, 42*, 1699–1713.

Williams, J. W., & Jackson, S. T. (2007). Novel climates, no-analog communities, and ecological surprises. *Frontiers in Ecology and the Environment, 5*(9), 475–482.

Williamson, P. & Bodle, R. (2016). *Update on climate geoengineering in relation to the convention on biological diversity: Potential impacts and regulatory framework*. CBD Technical Series No. 84. Montreal: Secretariat of the CBD.

Winnes, H., Styhre, L., & Fridell, E. (2015). Reducing GHG emissions from ships in port areas. *Research in Transportation Business & Management, 17*, 73–82.

Wolfe, D. A., Krahn, M. M., Casillas, E., Sol, S., Thomas, T. A., Lunz, J., & Scott, K. J. (1996). Toxicity of intertidal and subtidal sediments contaminated by the Exxon Valdez oil spill. In S. D. Rice, R. B. Spies, D. A. Wolfe, & B. A. Wright (Eds.), *Proceedings of the Exxon Valdez oil spill symposium* (pp. 121–139). Bethesda: American Fisheries Society.

Woody, T. (2019). *Huge amounts of greenhouse gases lurk in the oceans, and could make warming far worse*. nationalgeographic. com, 17 December 2019. Available online.

Wöppelmann, G., & Marcos, M. (2016). Vertical land motion as a key to understanding sea level change and variability. *Reviews of Geophysics, 54*, 64–92.

Wu, L., Cai, W., Zhang, L., Nakamura, H., Timmermann, A., Joyce, T., ... Giese, B. (2012). Enhanced warming over the global subtropical western boundary currents. *Nature Climate Change, 2*, 161–166.

Wu, P. P.-Y., Mengersen, K., McMahon, K., Kendrick, G. A., Chartrand, K., York, P. H., ... Caley, J. (2014). Timing anthropogenic stressors to mitigate their impact on marine ecosystem resilience. *Nature Communications., 8*, 1263.

Wust, G. (1935). The stratosphere of the Atlantic Ocean, The German Meteor expedition 1925–1927. Meteor. 6: 109–288. Also available. In W. J. Emery (Ed.), *The stratosphere of the Atlantic Ocean, scientific results of the German Atlantic expedition of the research vessel meteor, 1925–1927*. New Delhi: Amerind Publishing Co.

WWDR (World Water Development Report). (2017). *World water development report 2017: Wastewater: the untapped resource*. Paris: UNESCO.

WWF (World Wide Fund). (2017). *Delivering ecosystem-based marine spatial planning in practice—Assessing the integration of an ecosystem-based approach into UK and Ireland Marine Spatial Plans*. Gland, Switzerland: WWF.

Xanthos, D., & Walker, T. R. (2017). International policies to reduce plastic marine pollution from single-use plastics (plastic bags and microbeads): A review. *Marine Pollution Bulletin, 118*, 17–26.

Yamamoto, A., Abe-Ouchi, A., Shigemitsu, A., Oka, A., Takahashi, K., Ohgaito, R., & Yamanaka, Y. (2015). Global deep ocean oxygenation by enhanced ventilation in the Southern Ocean under long-term global warming. *Global Biogeochemical Cycles, 29*(10), 1801–1815.

Yin, J., & Goddard, P. B. (2013). Oceanic control of sea level rise patterns along the East Coast of the United States. *Geophysical Research Letters, 40*, 5514–5520.

Yu, L. (2019). Assessing the economic impacts of ocean acidification on Asia's Mollusk mariculture. *International Journal of Maritime Affairs and Fisheries, 17*(2), 17–30.

Yuewen, D., & Adzigb, L. (2018). Assessing the impact of oil spills on marine organisms. *Journal of Oceanography and Marine Research, 6*(1), 1000179.

Zhang, K., & Leatherman, S. (2011). Barrier island population along the United States Atlantic and Gulf Coasts. *Journal of Coastal Research, 27*(2), 356–363.

Zhu, J. R., Gu, Y. L., & Wu, H. (2013). Determination of the period not suitable for taking domestic water supply to the Qingcaosha Reservoir near Changjiang River Estuary. *oceanologia et limnologia sinica, 44*, 1138–1145.

Zickfeld, K., & Herrington, T. (2015). The time lag between a carbon dioxide emission and maximum warming increases with the size of the emission. *Environmental Research Letters, 10*(3), 031001.

Zickfeld, K., Solomon, S., & Gilford, D. M. (2017). Centuries of thermal sea-level rise due to anthropogenic emissions of short-lived greenhouse gases. *Proceedings of the National Academy of Sciences, USA, 114*(4), 657–662.

Further reading

Allen, M., Babiker, M., Chen, Y., de Coninck, H., Connors, S., van Diemen, R., ... & Zickfeld, K. (2018). IPCC, 2018: Summary for Policymakers. In: Global warming of 1.5°C. *In IPCC Special Report on the impacts of global warming of 1.5°C above pre-industrial levels and related global greenhouse gas emission pathways, in the context of strengthening the global response to the threat of climate change, sustainable development, and efforts to eradicate poverty*. Available from https://www.ipcc.ch/site/assets/uploads/sites/2/2019/05/SR15_SPM_version_report_LR.pdf.

Balseiro, A., Espí, A., Márquez, I., Pérez, V., Ferreras, M. C., Marin, J. G. E., & Prieto, J. M. (2005). Pathological features in marine birds affected by the Prestige's oil spill in the north of Spain. *Journal of Wildlife Diseases, 41*, 371–378.

Bergmann, M., Gutow, L., & Klages, M. (2016). *Marine Anthropogenic Litter*. Bremerhaven, Germany: Springer International.

Bouin, M.-N., Simeoni, P., Crawford, W. C., Calmant, S., Bore, J.-M., Kanas, T., & Pelletier, B. (2011). Comparing the role of absolute sea-level rise and vertical tectonic motions in coastal flooding, Torres Islands (Vanuatu). *Proceedings of the Academy of Sciences, USA, 108*(32), 13019–13022.

Bowerman, N. H. A., Frame, D. J., Huntingford, C., Lowe, J. A., Smith, S. M., & Allen, M. R. (2013). The role of short-lived climate pollutants in meeting temperature goals. *Nature Climate Change, 3*(12), 1021–1024.

Boyes, S. J., & Elliott, M. (2016). Brexit: The marine governance horrendogram just got more horrendous!. *Marine Pollution Bulletin, 111*, 41–44.

Boyes, S. J., Warren, L., & Elliott, M. (2003). *Regulatory responsibilities & enforcement mechanisms relevant to marine nature conservation in the United Kingdom*. Institute of Estuarine and Coastal Studies, Hull, UK: University of Hull.

Brander, L. M., Rehdanz, K., Tol, R. J., & Van Beukering, P. J. H. (2012). the economic impact of ocean acidification on coral reefs. *Climate Change Economics, 3*(1), 1250002.

Bryant, R. L. (1992). Political ecology: an emerging research agenda in third-world studies. *Political Geography, 11*, 12–36.

Chiswell, S. M., & Sutton, P. J. H. (1998). A deep eddy in the Antarctic intermediate water north of the Chatham Rise. *Journal of Physical Oceanography, 28*, 535–540.

Collins, W. J., Fry, M. M., Yu, H., Fuglestvedt, J. S., Shindell, D. T., & West, J. J. (2013). Global and regional temperature-change potentials for near-term climate forcers. *Atmospheric Chemistry and Physics, 13*, 2471–2485.

Collins, W. J., Webber, C. P., Cox, P. M., Huntingford, C., Lowe, J., Sitch, S., ... Powell, T. (2018). *Environmental Research Letters, 13*(3), 054003.

Cormier, R. (2019a). Ecosystem approach for management of deep-sea mining activities. In R. Sharma (Ed.), *Environmental issues of deep-sea mining: Impacts, consequences and policy perspectives* (pp. 381–402). Cham, Switzerland: Springer International Publishing.

Cormier, R. (2019b). *Waste assessment guidance: Bow-tie analysis and cumulative effects assessment*. London: Scientific Group of the London Protocol, IMO.

Cormier, R., Savoie, F., Godin, C., & Robichaud, G. (2016). Bowtie analysis of avoidance and mitigation measures within the legislative and policy context of the fisheries protection program. *Canada Manuscript Report on Fisheries and Aquatic Sciences*, 3093. Available from http://oaresource.library.carleton.ca/wcl/2016/20160527/Fs97-4-3093-eng.pdf.

DFO (Fisheries and Oceans Canada). (2009). *The role of the Canadian government in the oceans sector*. Ottawa: Fisheries and Oceans Canada Available at. Available from https://waves-vagues.dfo-mpo.gc.ca/Library/337909.pdf.

Domingues, C. M., Church, J. A., White, N. J., Gleckler, P. J., Wijffels, S. E., Barker, P. M., & Dunn, J. R. (2008). Improved estimates of upper-ocean warming and multi-decadal sea-level rise. *Nature, 453*(7198), 1090–1093.

Estes, J. A., Doak, D. F., Springer, A. M., & Williams, T. M. (2013). Causes and consequences of marine mammal population declines in southwest Alaska: a food-web perspective. *Philosophical Transactions of Royal Society London B. Biological Sciences, 364*(1524), 1647–1658.

Fock, H. O. (2011). Natura 2000 and the European common fisheries policy. *Marine Policy, 35*, 181–188.

Frame, D. J., Macey, A. H., & Allen, M. R. (2014). Cumulative emissions and climate policy. *Nature Geoscience, 7*(10), 692–693.

Galgani, F., Hanke, G., & Maes, T. (2015). Global distribution, composition and abundance of marine litter. In M. Bergmann, L. Gutow, & M. Klages (Eds.), *Marine anthropogenic litter* (pp. 29–56). Cham, Switzerland: Springer.

Garcia, S. M., Zerbi, A., Aliaume, C., Do Chi, T., & Lasserre, G. (2003). The ecosystem approach to fisheries. Issues, terminology, principles, institutional foundations, implementation and outlook. Rome: FAO, FAO Fisheries Technical Paper. No. 443.

Gelich, S., Reyes-Mendy, F., Arriagada, R., & Castillo, B. (2018). Assessing the implementation of marine ecosystem-based management into national policies: Insights from agenda setting and policy responses. *Marine Policy, 92*, 40–47.

Giambelluca, T. W., Leon, L. R., Hawkins, E., & Trauernicht, C. (2017). Global risk of deadly heat. *Nature Climate Change, 7*, 501–506.

Gobler, C. J., & Baumann, H. (2016). Hypoxia and acidification in ocean ecosystems: coupled dynamics and effects on marine life. *The Royal Society Biological Letters*, *12*(5), 20150976.

Heberger, M. (2012). *The impacts of sea level rise on the San Francisco Bay*. Sacramento, CA: California Energy Commission.

Horta e Costa, B., Claudet, J., Franco, G., Erzini, K., Caro, A., & Gonçalves, E. J. (2016). A regulation-based classification system for marine protected areas (MPAs). *Marine Policy.*, *72*, 192−198.

Hsu, T.-W., Shih, D.-S., Li, C.-Y., Lan, Y.-J., & Lin, Y.-C. (2017). A study on coastal flooding and risk assessment under climate change in the mid-western coast of Taiwan. *Water.*, *9*, 390.

Incardona, J., Collier, T., & Scholz, N. (2003). Defects in cardiac function precede morphological abnormalities in fish embryos exposed to polycyclic aromatic hydrocarbons. *Toxicology and Applied Pharmacology*, *196*, 191−205.

IPCC (Intergovernmental Panel on Climate Change). (2007). *Climate change 2007: Synthesis report, contribution of working groups I, II and III to the fourth assessment report of the intergovernmental panel on climate change*. Cambridge, UK: Cambridge University Press.

IUCN (International Union for Conservation of Nature). (2018). *Marine plastics, issues brief*. Gland, Switzerland: IUCN.

Keeling, R. F., & Garcia, H. E. (2003). The change in oceanic O_2 inventory associated with recent global warming. *Proceedings of the National Academy of Sciences, USA*, *99*(12), 7848−7853.

Kelaher, B. P., Page, A., Dasey, M., Maguire, D., Read, A., Jordan, A., & Coleman, M. A. (2015). Strengthened enforcement enhances marine sanctuary performance. *Global Ecology and Conservation*, *3*, 503−510.

Kittinger, N., Koehn, J. Z., Cornu, L. E., Ban, N. C., Gopnik, M., Armsby, M., ... Crowder, L. B. (2014). A practical approach for putting people in ecosystem-based ocean planning. *Frontiers in Ecology and the Environment*, *12*(8), 448−456.

Les, A. C., Gouhier, T. C., Menge, B. A., Stewart, J. S., Haupt, A. J., & Lynch, M. C. (2012). Climate-driven trends and ecological implications of event-scale upwelling in the California Current System. *Global Change Biology*, *18*, 783−796.

Levy, H., II (1971). Normal atmosphere: Large radical and formaldehyde concentrations predicted. *Science, New Series*, *173*(3992), 141−143.

Lieberknecht, L.M. (2020) *Ecosystem-based integrated ocean management: A framework for sustainable ocean economy development*. A report for WWF-Norway by GRID-Arendal.

Matthews, H. D., Landry, J.-S., Partanen, A.-I., Allen, M., Eby, M., Forster, P. M., ... Zickfeld, K. (2017). Estimating carbon budgets for ambitious climate targets. *Current Climate Change Reports*, *3*, 69−77.

Millar, R., Allen, M., Rogelj, J., & Friedlingstein, P. (2016). The cumulative carbon budget and its implications. *Oxford Review of Economic Policy*, *32*(2), 323−342.

Millar, R. J., Fuglestvedt, J. S., Friedlingstein, P., Rogelj, J., Grubb, M. J., Matthews, H. D., ... Allen, A. R. (2017). Emission budgets and pathways consistent with limiting warming to 1.5 degrees C. *Nature Geoscience*, *10*(10), 741.

Oude Essink, G. H. P., Van Baaren, E. S., & De Louw, P. G. B. (2010). Effects of climate change on coastal groundwater systems: A modeling study in the Netherlands. *Water Resources Research*, *46*, W00F04.

Palmer, M. D., Roberts, C. D., Balmaseda, M., Chang, Y.-S., Chepurin, G., Ferry, N., ... Xue, Y. (2017). Ocean heat content variability and change in an ensemble of ocean re-analyses. *Climate Dynamics.*, *49*(3), 909−930.

Piet, G., Delacamara, G., Gomez, C.M., Lago, I.M., Rouillard, J., Martin, R., & van Duinen, R. (2017). *Making ecosystem-based management operational*. Deliverable 8.1, European Union's Horizon 2020 Framework Programme for Research and Innovation grant agreement No. 642317. Available from https://aquacross.eu/sites/default/files/D8.1_Making%20ecosystem-based%20management%20operational_v2_13062018.pdf.

Ricketts, P. J., & Hildebrand, L. (2011). Coastal and ocean management in Canada: Progress or paralysis? *Coastal Management, 39*, 4−19.

Ries, J. (2013). Biodiversity and ecosystems: Acid ocean cover up. *Nature Climate Change, 1*, 294−295.

Rogelj, J., McCollum, D. L., O'Neill, B. C., & Riahi, K. (2013). 2020 emissions levels required to limit warming to below 2 degrees C. *Nature Climate Change, 3*(4), 405−412.

Seinfeld, J. H., & Pandis, S. N. (2016). *Atmospheric chemistry and physics: From air pollution to climate change*. New York: John Wiley & Sons.

Simmons, C. T., Fenstemaker, T. R., & Sharp, J. M. (2001). Variable-density groundwater flow and solute transport in heterogeneous porous media: Approaches, resolutions and future challenges. *Journal of Contaminant Hydrology, 52*, 245−275.

Smith, S. M., Lowe, J. A., Bowerman, N. H. A., Gohar, L. K., Huntingford, C., & Allen, M. R. (2012). Equivalence of greenhouse-gas emissions for peak temperature limits. *Nature Climate Change, 2*(7), 535−538.

UNEP (United Nations Environment Programme). (2005). *Marine Litter—An analytical overview*. Nairobi: UNEP.

UNFCCC (United Nations Framework Convention on Climate Change). (2015). *Adoption of the Paris Agreement. United Nations Off*. Geneva, Switzerland: United Nations Office Available from. Available from https://unfccc.int/sites/default/files/english_paris_agreement.pdf.

Venn-Watson, S., Colegrove, K. M., Litz, J., Kinsel, M., Terio, K., Saliki, J., ... Rowles, S. (2015). Adrenal gland and lung lesions in Gulf of Mexico common bottlenose dolphins (*Tursiops truncatus*) found dead following the deepwater horizon oil spill. *PLoS One, 10*, e0126538.

Whitney, C. K., Gardner, J., Ban, N. C., Vis, C., Quon, S., & Dionne, S. (2016). Imprecise and weakly assessed: Evaluating voluntary measures for management of marine protected areas. *Marine. Policy., 69*, 92−101.

Zickfeld, K., Eby, M., Matthews, H. D., & Weaver, A. J. (2009). Setting cumulative emissions targets to reduce the risk of dangerous climate change. *Proceedings of the National Academy of Sciences, USA, 106*(38), 16129−16134.

CHAPTER 9

Preserving life on Earth

9.1 Introduction

The term "Earth system" alludes to Earth's interacting physical, chemical, and biological processes and this Earth system comprises the land, oceans, atmosphere and poles. It also incorporates the planet's natural cycles—the carbon, water, nitrogen, phosphorus, sulfur and other cycles—and deep Earth processes (Steffen et al., 2004). Biotic life too is an integral part of the Earth system and this biotic life affects the carbon, nitrogen, water, oxygen, and many other cycles and processes. With human society becoming an integral part of this Earth system, along with social and economic systems of the human society getting embedded within the Earth system, in many instances, the human systems have now emerged as the major drivers of change in the Earth system (Steffen et al., 2004). The Earth system has been defined as "the suite of interacting physical, chemical and biological global-scale cycles and energy fluxes that provide the life-support system for life at the surface of the Planet" (Steffen, Crutzen, & McNeill, 2007, p. 615). Humans have been relying on that life-support system, and over the past many millennia, the Earth system had been relatively stable until the advent of the industrial revolution, and specifically since the mid-20th century burgeoning population growth and industrialization (Steffen et al., 2018), human activities have proved instrumental in vastly changing most of the processes and subsystems in the Earth system at an accelerated pace that, if allowed to continue uninterruptedly, would perhaps destabilize the resilience of major components of Earth system operationalization (IPCC, 2018; Steffen et al., 2018). Earth is presently undergoing unparalleled rapid pace of change on account of anthropogenic climate change and unsustainable use of natural resources (Corlett, 2015). Increased warming in tandem with alterations in the frequency and duration of drought (Dai, 2013) and large storms (Dettinger, Udall, & Georgakakos, 2015) have seemingly impacted all levels of ecological organization from individual behavior (Root et al., 2003) to ecosystem process (Grimm et al., 2013; Walther et al., 2002). This can be prominently discerned in terrestrial landscapes where changing temperatures are projected to alter habitat ranges for mobile organisms, while some species may enlarge their habitats, others may pop off or fade away owing to shrinking habitats (Bentz et al., 2010). Nonmobile species, like those observed in coral reefs, are incapable of avoiding human pressures and resultantly their abundance is reduced worldwide in the wake of enhanced water temperatures and acidity in oceans (Hoegh-Guldberg et al., 2007). Vagaries of climate change and exploitation and unsustainable use of natural resources, such as forests, freshwater ecosystems, genetic resources, wildlife, land use, etc., have endangered the survival of human life and other biotic organisms and species on land and this warrants an urgency to understand as to how these perturbations will cause multivariate responses across physical, thermal, chemical, and biological gradients at innumerable scales.

9.2 Freshwater ecosystems

Freshwater ecosystems, a subset of Earth's aquatic ecosystems, embrace lakes and ponds, rivers, streams, springs, bogs, and wetlands. Freshwater ecosystems are indispensable to environmental health, economic wealth and human well-being (Grill et al., 2019). The integrity of freshwater ecosystems is of prime significance, as the freshwater supply ranks in the top 10 (listed as fourth) of the most pressing concerns for human survival (WEF, 2019). Humans make use of freshwater ecosystems for drinking and irrigation purposes, waste disposal, transportation, energy production, fisheries, and for harvesting plants and minerals (Strayer & Dudgeon, 2010). All these services provided by freshwater ecosystems have high economic value. Encompassing nearly 1% of Earth's surface, lakes and rivers host 9.5% of all animal species (He, Burgess, Gao, & Li, 2019a; He et al., 2019b) turning freshwater ecosystems areas of high biodiversity. All the same, this biodiversity is confronted with plentiful threats responsible for large declines in abundance and spatial distribution of many species. According to Freyhof and Brooks (2011), 44% of freshwater mussels, 37% of fish, 23% of amphibians, 19% of reptiles, 15% of mammals and dragonflies, 13% of birds, 9% of butterflies, and 7% of aquatic plants are threatened in Europe. As per 2018 Living Planet Index of the World-Wide Fund for Nature (WWF), number of vertebrates in freshwater that account for about one-third of all vertebrates, declined in their abundance and spatial distribution by 83% compared to 1970 data (WWF, 2018). This specifically presents an alarming position if compared to the declines reported for marine and terrestrial ecosystems with values around 36% and 38%, respectively (He et al., 2019a,b; Reid et al., 2019). Numerous freshwater species are reported to have suffered distinct range contractions, like the European sturgeon (*Acipenser sturio*) that was uprooted from all major European rivers, and is now confined to the Garrone River in France (Limam, 2020). Other species such as the baiji (*Lipotes vexillifer*) and the Chinese paddlefish (*Psephurus gladius*) are considered extinct as a consequence of human activities and intrinsic factors like long life cycle and lower fecundity of their biology (He et al., 2019a,b). Apart from human activities, many freshwater species are limited to small isolated areas and this situation is responsible for high rates of autochthony and increased risk of extinction (Abell et al., 2008). Undoubtedly, given the remarkable biodiversity, the vulnerability and the level of endemism present in freshwater ecosystems these areas should, in theory, be considered a conservation priority (Flitcroft, Cooperman, Harrison, Juffe-Bignoli, & Boon, 2019); nevertheless regrettably, that has not been the case since, for instance, conservation literature tends to be biased toward terrestrial systems with 81% of research articles focused on these ecosystems (Di Marco et al., 2017).

Species are registering decline at an alarming rate under the prevalent biodiversity crisis (IPBES, 2019) principally owing to human activities and global consumption of natural resources (WWF, 2018). This has proved instrumental in causing chemical, physical and biological alterations in all ecosystems, being freshwater ecosystems no exception (Carpenter, Stanley, & Vander Zanden, 2011). Priority is accorded to human well-being when it comes to decision-making and policy, resulting in loss of biodiversity that supports some of these same priorities (Darwall et al., 2018). Overexploitation, climate change, introduction of nonnative species, pollution, changes in water flow and destruction and fragmentation of habitats pose the biggest threats to freshwater ecosystems (Grzybowski & Glińska-Lewczuk, 2019). Overfishing or overexploitation is mostly affecting fishes, amphibians and reptiles (Dudgeon, 2019), but even macroinvertebrates like some freshwater bivalves can be highly explored (Lopes-Lima et al., 2017; Zieritz et al., 2018). The conjunction between the increasing water temperature and the higher global atmospheric temperature contributes to the depletion of dissolved oxygen in water, and also affects the physiology, phenology and distribution of many species (Carpenter et al., 2011) and can even cause direct mortality (Nogueira, Lopes-Lima, Varandas, Teixeira, & Sousa, 2019).

Freshwater ecosystems are also greatly affected by the introduction of invasive species (Strayer & Dudgeon, 2010), impacting ecosystems structure and functioning as well as their biotic and abiotic interactions (Sousa et al., 2019). Undoubtedly, some species are introduced accidentally by humans; nonetheless, many are deliberately introduced by humans to serve their needs, such as aquaculture, sport fishing, and pest control (Sousa, Novais, Costa, & Strayer, 2014). Alterations in the land use of riparian areas and watersheds, mostly in the wake of agricultural expansion, changes the input of water, sediments, nutrients and other chemicals resulting in eutrophication and pollution that impacts natural communities, modifies biogeochemical cycles and disturbs the habitat for many species (Strayer & Dudgeon, 2010). Uneven availability of freshwater in time and space spurs the urgency for controlling water reserves mostly for agriculture, industry and domestic use through the construction of dams and levees, thereby erecting physical obstacles and that in turn alter the connectivity of rivers, capture about 25% of global sediment load (Tickner et al., 2020; Vörösmarty & Sahagian, 2000), can vastly change the natural river flow, harm fish and other species migration and aggravate the occurrence of infectious diseases caused by standing-water related parasites (Carpenter et al., 2011). There are close to 2.8 million dams worldwide that are fragmenting rivers and altering their courses and this number is prone to increase in the near future especially in areas such as Amazon, Mekong, Congo, and the Balkans basins (Grill et al., 2019). All these threats entail the potential of operating synergistically, as flow modification can be aggravated by climate change in terms of more frequent floods and droughts (Vörösmarty, Green, Salisbury, & Lammers, 2010), or invasive species that are more likely to thrive in human modified environments (Gaertner et al., 2017). Apart from these persistent threats, Reid et al. (2019) describe new emerging threats that have been identified affecting freshwater biodiversity and are related to e-commerce, algal blooms, engineered nanomaterials, microplastic pollution, urban light and noise, increase salinization, and decrease of calcium.

Response of freshwater ecosystems to anthropogenic influence is of significant concern because freshwater ecosystems provide valuable ecosystem services (Matthews, 2016). Freshwater ecosystems provide water for drinking and irrigation purposes, support up to 12% of the world's known species, making freshwater ecosystems a global biodiversity hotspot (Mittermeier, Farrell, Harrison, Upgren, & Brooks, 2010) and a scarce resource representing only about 0.01% of all water on Earth (Shiklomanov, 1993), with increasing anthropogenic demands (Vörösmarty et al., 2000). The use of freshwater ecosystems as potent instrument of change over long-term and interannual basis has become widely acknowledged because of their integration of multiple ecological processes across landscapes (Martin-Ortega, Ferrier, Gordon, & Khan, 2015). Long term decreases to the duration of global biodiversity ice cover on lakes and rivers (Benson et al., 2012), has been shown to change nutrient levels and biological activity (Preston et al., 2016). Drought and dam construction in rivers have seemingly regulated flow regimes and caused reductions to biodiversity (Poff & Zimmerman, 2010). Future conservation endeavors of freshwater in a time of global change call for a thorough understanding of the biological changes triggered by climate and water use drivers (Malhi et al., 2020).

Freshwater ecosystems are seemingly under threat from an array of stressors that either individually or in combination (Reid et al., 2019), entail the potential of putting pressures on their biological communities (Ormerod, Dobson, Hildrew, & Townsend, 2010). Freshwater ecosystems, including climate change, are impacted by an assortment of human-induced environmental changes (Benateau, Gaudard, Stamm, & Altermatt, 2019; Vörösmarty et al., 2000), urbanization (McDonald et al., 2011), agricultural intensification (van Soesbergen, Sassen, Kinsey, & Hill, 2019; Stoate et al., 2009) and water abstraction (Jiménez Cisneros et al., 2014; Bunn & Arthington, 2002). The biological consequences/implications of these changes are variable, ranging from the formation of new or novel ecosystems and species interactions (Hobbs et al., 2006), through to the complete breakdown or collapse of biological communities and diminutions in the provision of ecosystem services (Dobson et al., 2006). Undoubtedly, the nature and sternness of impacts mirror the frequency and intensity of anthropogenic activities (Chester & Robson, 2013); nonetheless, abiotic and biotic factors intercede the extent to which anthropogenic disturbances induce significant ecosystem-specific responses that are general or specific to any given environment (Nõges et al., 2016). Water quality is a significant factor in the conservation of freshwater biodiversity (Martinuzzi et al., 2014), arising from pollution by both organic and inorganic chemicals (Schwarzenbach et al., 2006, 2010). Presently, over 7700 chemicals are produced and used in significant quantities worldwide (EPA, 2018), with many of these synthetic chemicals entering and subsequently threatening freshwater systems (Bernhardt, Rosi, & Gessner, 2017). Nevertheless, the dispersal and quantity of Rosa chemicals in the environment is both spatially and temporally variable (Malaj et al., 2014). The combination of chemical pollutants in any given freshwater ecosystem is an outcome of contemporary releases of current-use and emerging chemicals like personal-care products, pharmaceuticals, systemic pesticides as well as the recirculation of persistent chemicals like brominated flame retardants, organochlorine pesticides, etc. (Kortenkamp, 2007). Undeniably, improvements in wastewater treatment has proved instrumental in reducing the levels of gross pollution across freshwaters and led to widespread recovery of biological communities (Vaughan & Ormerod, 2012); nevertheless, mixtures of xenobiotic chemicals remain pervasive and continue to generate significant potential for ecological effects across large regions of the globe (Bernhardt et al., 2017).

9.2.1 Climate change and freshwater ecosystems

Climate change has spurred significant changes in several climate trends (IPCC, 2007), at a remarkable speed (Woodward et al., 2010). Hence, climate change has been projected as the greatest emerging threat to global biodiversity and to the functioning of ecosystems (Sintayehu, 2018). Freshwater ecosystems afford various features that render them specifically vulnerable to climate change; for instance, freshwater ecosystems are relatively isolated and physically fragmented within large terrestrial landscapes (Woodward, Perkins, & Brown, 2010). Therefore, freshwater species cannot easily dissolve or diffuse while the environment is changing (Woodward, et al., 2010). Furthermore, freshwater ecosystems are confronted with innumerable anthropogenic pressures while they are heavily explored for goods and services (Carpenter et al., 2011), and irrespective of the fact that a small percentage of the Earth surface is covered by freshwater ecosystems, they have been estimated to support almost 6% of all described species (Collen et al., 2014). Accordingly, Woodward et al. (2010) have claimed that "freshwater biodiversity is disproportionately at risk on a global scale." Freshwater systems located in higher latitudes and altitudes have already recorded impacts of climate change because these regions have been experiencing higher rates of warming (IPCC, 2007; Woodward et al., 2010). These specific systems have been perceived as "sentinel systems" because they can provide early warming evidence on the impacts of climate change in freshwater systems. The impacts of climate change have already been recorded in freshwater systems standing in higher latitudes and altitudes, since these regions have been experiencing higher rates of warming (Layer, Hildrew, Monteith, & Woodward, 2010). Indeed, many Arctic lakes and ponds have been registering regime shifts in their biological communities (Smol et al., 2005), and that has been linked to long-term warming. Freshwater ecosystems discern the impact of warmer temperatures that decrease the duration of the ice season, because the warmer temperatures affect the melting of glaciers, permafrost and other ice sheets (Koç et al., 2009). Furthermore, a warmer climate also triggers changes in evaporation and precipitation ratios (Carpenter et al., 2011). All these factors make up significant causes of limnological changes, provided they aggravate transformations in the seasonality and magnitude of hydrologic flows (Carpenter et al., 2011). Salinity is another climate change-related driver of freshwater transformation (Mengel et al., 2016). Sea level rise contributes to the rise in salinity in freshwater ecosystems and this salinity entails severe impacts on freshwater biodiversity, as it affects the survival and reproduction of freshwater species (Mimura, 2013).

Global environmental change has seemingly put freshwater ecosystems particularly at risk, causing greater declines in biodiversity than either their terrestrial or marine counterparts (WWF, 2018). This risk is attributable to a combination of three factors. In the first place, freshwaters are disproportionately exposed to multiple, severe environmental and anthropogenic stressors, resulting from the implicit dependence of human populations on freshwater resources (Strayer & Dudgeon, 2010). Secondly, freshwater ecosystems are hotspots of biological diversity—reflecting large species richness per unit area or water volume (Strayer & Dudgeon, 2010). Thirdly, freshwater organisms appear specifically sensitive to environmental change due to the exceptional environmental conditions present in these systems, such as the restricted oxygen availability in the subaqueous environment (Forster, Hirst, & Atkinson, 2012). With increases predicted in the anthropogenic pressures placed on freshwater resources (Bunn & Arthington, 2002), the biodiversity and functioning of freshwater systems is at considerable risk (Davis et al., 2015). In other words, climate change is transforming the composition, biodiversity and functioning of freshwater ecosystems (Cañedo-Argüelles et al., 2013). An in-depth discernable understanding on the capacity of freshwater organisms to deal with environmental changes is vital to better figure out the resilience of freshwater ecosystems under related environmental pressure and hence to a conscientious development of effective protection/management actions toward the conservation of these valuable resources (Bastawrous & Hennig, 2012; Woodward et al., 2010).

9.3 Conserving forests

Encompassing 31% of the global land area, forests provide habitats for 80% of amphibian species, 75% of bird species and 68% of mammal species, and nearly 60% of all vascular plants are found in tropical forests (FAO, 2020a,b,c). Apart from providing more than 86 million green jobs the millions of people around the globe who count on them for food security and their livelihood, over 90% of those living in extreme poverty are dependent on forests for wild food, firewood or part of their livelihood (FAO, 2020a,b,c). For long, humans have relied on forest ecosystems and the supply of ecosystem services along with their associated benefits. Forest ecosystems deliver critical and diverse services and values to human society such as food, fiber, timber, flood protection, clean water, and climate regulation (Sands, Norton, & Weston, 2013), and these benefits are widely referred to as ecosystem services (Birch et al., 2010). Supply of these services is critical for human survival, health and well-being (MEA, 2005) as well as economic prosperity (TEEB, 2010). Besides, forest ecosystem services are also critical to the survival of the bulk of world's poor people (FAO, 2020a,b,c). Apart from being a primary habitat, for a wide range of species, forests also support biodiversity maintenance and conservation along with sequestration and storage of carbon from the atmosphere, contributing to regulation of the global carbon cycle and climate change mitigation (Pugh et al., 2019; Jenkins & Schaap, 2018). Healthy forest ecosystems facilitate soil conservation and help in stabilizing stream flows and water runoff thereby preventing land degradation and desertification and decreasing the risks of natural disasters like droughts, floods and landslides (Jenkins & Schaap, 2018). Apart from these, forests also serve as sites of esthetic, recreational and spiritual value in many cultural and societal contexts (Cooper, Brady, Steen, & Bryce, 2016). The increasing demands for these ecosystem services in the wake of burgeoning population and the resultant increased demand for agricultural lands for ensuring food security (Tallis, Kareiva, Marvier, & Chang, 2008), along with rapid pace of industrialization and urbanization have exacerbated pressures on forests culminating in conversion of forests to agriculture and land uses pertaining to urban population growth (Ahrends et al., 2010). Forest fragmentation and diminutions in habitat patch sizes have resulted in "deleterious" edge effects (Laurance et al., 2007) and also contributing to decreasing plant and animal population sizes (Laurance, Laurance, & Hilbert, 2008) that in turn entail the likelihood of lowering population viability and genetic variation (Aguilar, Quesada, Ashworth, Herrerias-Diego, & Lobo, 2008).

The adverse impacts of forest fragmentation and isolation are projected to be further aggravated by other anthropogenic threats like forest fires (Laurance et al., 2011), especially in the context of global climate change (Aragao & Shimabukuro, 2010). Apart from forest loss and fragmentation, other factors like enigmatic or cryptic deforestation (Puryavaud, Davidar, & Laurance, 2010), in terms of the selective logging and internal degradation of forests, have changed forest structure and plant communities, thereby crippling biodiversity, regeneration capacity and vivacity and vitality of forests (Aerts et al., 2011). The concurrent decrease in both forest quantity and quality is projected to trigger gigantic extinction of multiple species inhabiting forest habitats (Wright & Muller-Landau, 2006). Loss of forest biodiversity entails the potential of seriously impairing the functioning of forest ecosystems in terms of the activities, processes or properties of forests, such as decomposition of organic matter, soil nutrient cycling and water retention, and accordingly the capability of forest to provide the ecosystem services (Duffy, 2009). In other words, loss of forests entails both local and global consequences because human and natural ecosystems rely on stable global carbon and hydrologic cycles and the ability of the forests to mitigate impacts of the vagaries of climate change (Seymour & Busch, 2016; Watson et al., 2018). In order to sustain healthy forest ecosystems, it is seemingly essential to address the problems of deforestation, droughts, desertification, forest fires, and fragmentation of forests among others.

9.3.1 Deforestation

In common parlance, notion of "deforestation" is construed in terms of the conversion of forested areas to nonforest land use such as arable land, urban use, logged area or wasteland. According to FAO, "deforestation is the conversion of forest to another land use or the long-term reduction of tree canopy cover below the 10% threshold" (FAO, 2007). There are both proximate/direct causes and underlying/indirect causes of deforestation (MEA, 2005). Proximate/direct causes comprise human activities or immediate actions that directly affect forest cover and loss of carbon, such as agricultural expansion—both commercial and subsistence—infrastructure expansion, and wood extraction; while on the one hand, underlying/indirect causes are complicated interactions of primary socioeconomic, political, cultural, and technological processes that are often far away from their area of impact (Kissinger, Herold, & De Sy, 2012), and these causes either operate at local level or wield an indirect impact from the national or global level, and these are related to markets, commodity prices, population growth, domestic markets, national policies governance and local circumstances, etc. (Obersteiner et al., 2009). Increasing demand for agricultural and forest products is getting exacerbated by global market pressures, dietary preferences and loss and waste along agricultural value chains and that in turn, drive deforestation and forest degradation (IPCC, 2019). The essentiality of making food available to global population is generally regarded as the prime cause of loss of forests and resultant loss of biodiversity. Combination of population pressure and poverty in Africa has emerged as the main threat to forest conservation (FAO, 2020a,b,c), propelling poor farmers to convert forests to cropland (Lung & Schaab, 2010) and to procure wood-fuel in an unsustainable manner. In other regions, changes in consumption patterns of affluent segments of population are said to be driving deforestation. Besides, an array of political and socioeconomic forces interacting at the global to local levels is also contributing to deforestation (Carr, Suter, & Barbier, 2005). An analysis conducted by Hosonuma et al. (2012) with regard to national data of 46 tropical and subtropical countries comprising about 78% forest area in those climatic regions demonstrated that large-scale commercial agriculture is the most common driver of deforestation, accounting for about 40% of it. This analysis also made it discernible that the deforestation drivers differed prominently between regions and even within countries.

Global population that currently in 2020 stands at 7.8 billion is projected to reach 9.9 billion by 2050 thereby registering an increase of more than 25% from the current level (WPDS, 2020). A burgeoning population entails a growing demand for agricultural and forest products (Rademaekers, Eichler, Berg, Obersteiner, & Havlik, 2010). Given that the rural birth rates are stabilized, increased per capita consumption of food, especially meat consumption, along with rapid pace of urbanization and industrialization as an ongoing process and more rapidly in the near future reinforces the trend of deforestation being spurred by global commodity markets more and less by local population (DeFries, Rudel, Uriarte, & Hansen, 2010). In 2009, FAO (2009) had predicted a 70% increase in the demand for food by 2050, with a required increase of 49% in the quantity of cereals produced and an increase of 85% in the volume of meat to be produced, and almost nearly all that additional food was expected to be consumed by developing countries as per increases in population and living standard (Foresight, 2011). Admittedly, growing emphasis on increasing agricultural yield has been in vogue for some decades; nonetheless, the likelihood of further intensification of agricultural land leading to more deforestation in some circumstances cannot be ruled out (Boucher et al., 2011). With eucalyptus pulp and logs dominating the market a decade back, FAO (2010a,b) pointed toward indications that production was moving away from primary forests to plantations. Concurrently, during the same period, it was argued that timber production had not been a significant driver of deforestation in Africa (Fisher, 2010), yet it could be growing in importance (Laporte et al., 2007). Illegal logging has been adversely impacted in the wake of import controls in the United States and European Union (EU). A number of countries are taking part in the EU Voluntary Partnership Agreements under the European Forest Law Enforcement, Governance and Trade (FLEGT) Action Plan enacted in 2003 (EU, n.d.a). trends emerging from an in-depth study of this Action Plan by Lawson and MacFaul (2010) demonstrate that the action taken by concerned governments, civil society and the private sector in the previous decade with regard to illegal logging and the related had been extensive and commanded a considerable impact. Still, there is need to exercise more vigilance to discourage illegal logging.

Increasing emphasis on renewable energy resources has propelled many countries to focus on adoption of policies and support measures for alternative energy resources, including the biofuel consumption, and accordingly EU has adopted Renewable Energy Directive (RED) (EU, n.d.b) and United States has come out with Renewable Fuel Standard (DOE, n.d.). Other increased energy consuming countries like Brazil, China and India have also fixed their respective targets. According to OECD/FAO (2011), it is estimated that by 2020, 21% of the increase in global coarse grains production above current levels, 29% of the global vegetable oil production's increase, and 68% of global sugar cane production's increase is likely to go to biofuels. And such a scenario surely entails the probability of expansion of land to be used for biofuel production thereby exerting increased pressure on forests in the near future (Lapola et al., 2010). During the past two decades, it has been reported that there have been important regional shifts in the patterns of fuel wood use worldwide, with wood for energy having recorded significant augmentation in the case of Africa and Latin America along with recording a decline in Asia by nearly half (FAO, n.d.), and it also demonstrates a pattern of enhanced development and the availability of alternative sources to household in these countries (Klenk, Mabee, Gong, & Bull, 2012). According to Hofstad, Kohlin, and Namaalwa (2009), fuel wood remains a major source of domestic energy for some time and domestic fuel wood use is expected to remain comparatively stable for some years in the near future, and this means that fuel wood will continue to exert pressure on forests thereby contributing to deforestation to some extent.

Mining also contributes to deforestation both within and beyond lease boundaries. Within leased areas, forests are kept ready for mineral extraction, processing and development of infrastructure (Alvarez-Berríos & Aide, 2015). Nevertheless, in off-lease areas effects of mining are hypothetically more wide ranging and their pathways more composite (Ferreira et al., 2014; Duran, Rauch, & Gaston, 2013). Deforestation entails the probability of expanding to more distant areas, say more than 10 km, yonder lease boundaries (Duran et al., 2013), owing to the cumulative effects of land-use displacement (Scheuler et al., 2011), urban expansion (Mwitwa, German, Muimba-Kankolongo, & Puntodewo, 2012), growth in commodity supply chains (Sonter, Barrett, Moran, & Soares-Filho, 2015), and complications arising out of mine-waste discharge (Edwards et al., 2014) and spills (Edwards & Laurance, 2015). Minor sector has come to play a significant role in global economy not only for the minerals and metals it produces but in the development of future technology (Maddox, Howard, Knox, & Jenner, 2019), and in the wake of rise in global demand broad estimates show that at least 50% of mineral and metal requirements will have to be met by mining for foreseeable future (Nassar, 2018). Accounting for about a quarter of global GDP and indirectly contributing up to 15% of employment, minor sector has come to play a prominent role in the economies of more than 80 countries, especially those in the lower- to middle-income group (ICMM, 2016). Impacts of mining can be discerned at and near mine sites in terms of land clearance, displacement of people and the production of big volumes of waste (eLAW, 2010). To generate 9 million tons of refined metals that are produced, the waste material engendered from the mining process alone is comparable to approximately 9 tons per year for every person on Earth (Franks, 2015). Nevertheless, indirect effects of mining can be even more permeating, taking place far away from the mining site, including the impacts of associated infrastructure and the induction of people that are often linked with large-scale mining projects (Sonter et al., 2017). As demands for mineral and metals grow worldwide, there will be more exploitation of forests for mineral abstraction and this is bound to further lead to deforestation (Rademaekers et al., 2010).

9.3.2 Impact of droughts

A vital role is played by drought events in influencing tropical forests these droughts cast massive effects on tree mortality and growth (Bonal, Burban, Stahl, Wagner, & Hérault, 2016). Mortality of tropical trees has been largely observed in natural droughts that are frequently associated with multiyear climate cycles explained by El Niño-Southern Oscillation (ENSO) events (Slik, 2004). Nevertheless, the ENSO drought event that occurred in 2015–16 has come to reckoned as the strongest since 1979 and is ascribed to sea surface warming in the Central Equatorial Pacific (Jiménez-Muñoz et al., 2016), and is an event that entails the potential of increasing tree mortality in coming years. Some drought events are linked to an increase in sea surface temperatures (SSTs) in the tropical North Atlantic Ocean and are often considered to be 1-in-100-year events (Malhi et al., 2008). Two major warming Atlantic SST-linked were identified in 2005 and 2010 vastly impacting the Amazon basin (Marengo, Tomasella, Alves, Soares, & Rodriguez, 2011), resulting in peaks of fire activity, dried lakes, the low levels in many rivers in southwestern Amazon. These droughts proved instrumental in contributing to augmentation in tree mortality in lowlands forests with major alterations in forest composition and structure (Esquivel-Muelbert et al., 2018). High tree mortality is reported to be one of the short-term responses of forests to drought, with evidence of lagged mortality in one or more years altering forests from carbon sinks into sources in some tropical regions (Corlett, 2016). Large trees, with stems larger than 40 cm in diameter, were strongly affected by the droughts (Phillips et al., 2010). Hydraulic failure and carbon starvation have been proposed as mechanisms inducing mortality in large vs. smaller trees, with hydraulic failure being the most supported hypothesis (Rowland et al., 2015). There is increasing interest in understanding of responses of forest ecosystem in lowland Amazonian systems (Gatti et al., 2014), with advances in linking drought effects to functional characteristics (Kraft, Metz, Condit, & Chave, 2010). In view of the fact that predicting tree mortality in response to drought is difficult, it has been suggested that wood density—a significant functional characteristic in plant life history—can be used as a proxy to predict mortality risk in tropical lowland forest (Kraft et al., 2010).

There is seemingly increasing interest of the scientific community with regard to responses and reactions of forests to drought across diverse biomes to ascertain which growth features or functional characteristics better delineate different species-specific responses to such climate extremes (Gazol, Camarero, Anderegg, & Vicente-Serrano, 2017a). As annual growth rings of trees provide short- (say about years) and long-term (say about decades) information with regard to how trees respond to drought (Gazol, Sangüesa-Barreda, Granda, & Camarero, 2017c). Drought effects on forests can last for years because dry spells decrease forest productivity and tree growth and also often leads to the enfeeblement of trees by deteriorating their vigor (Peltier, Fell, & Ogle, 2016). It can be discerned from some recent studies that forest resilience to drought—the capacity to resist and recover after a drought—is primarily dependent on drought intensity; nonetheless, such resilience can also be relying on climate types (Zhang, Shao, Jia, & Wei, 2017), tree species or specific functional traits (Greenwood et al., 2017). Moreover, variations in the strategies of tree species and individuals can occur while dealing with drought (Gazol et al., 2017a,b), leading to modification in forest growth responses to drought that can be reflected in tree ring width variability. Owing to their close linkages to wide-ranging climatic conditions, droughts do not always share the same climate conditions, and, thereby, they can affect forest growth differently across different climates and biomes (Anderegg, Anderegg, & Berry, 2013).

A vital role is played by drought events in influencing tropical forests these droughts cast massive effects on tree mortality and growth (Bonal et al., 2016). Mortality of tropical trees has been largely observed in natural droughts that are frequently associated with multiyear climate cycles explained by El Niño-Southern Oscillation (ENSO) events (Slik, 2004). Nevertheless, the ENSO drought event that occurred in 2015—16 has come to reckoned as the strongest since 1979 and is ascribed to sea surface warming in the Central Equatorial Pacific (Jiménez-Muñoz et al., 2016), and is an event that entails the potential of increasing tree mortality in coming years. Some drought events are linked to an increase in sear surface temperatures (SSTs) in the tropical North Atlantic Ocean and are often considered to be 1-in-100-year events (Marengo, Nobre, Tomasella, Cardoso, & Oyama, 2008; Marengo et al. 2011). Two major warming Atlantic SST-linked were identified in 2005 and 2010 vastly impacting the Amazon basin (Marengo et al., 2011), resulting in peaks of fire activity, dried lakes, the low levels in many rivers in southwestern Amazon. These droughts proved instrumental in contributing to augmentation in tree mortality in lowlands forests with major alterations in forest composition and structure (Esquivel-Muelbert et al., 2018). High tree mortality is reported to be one of the short-term responses of forests to drought, with evidence of lagged mortality in one or more years altering forests from caron sinks into sources in some tropical regions (Corlett, 2016). Large trees, with stems larger than 40 cm in diameter, were strongly affected by the droughts (Phillips et al., 2010). Hydraulic failure and carbon starvation have been proposed as mechanisms inducing mortality in large versus smaller trees, with hydraulic failure being the most supported hypothesis (Rowland et al., 2015). There is an increasing interest in understanding of responses of forest ecosystem in lowland Amazonian systems (Gatti et al., 2014), with advances in linking drought effects to functional characteristics (Kraft et al., 2010). In view of the fact that predicting tree mortality in response to drought is difficult, it has been suggested that wood density—a significant functional characteristic in plant life history—can be used as a proxy to predict mortality risk in tropical lowland forest (Kraft et al., 2010).

There is seemingly increasing interest of the scientific community with regard to responses and reactions of forests to drought across diverse biomes to ascertain which growth features or functional characteristics better delineate different species-specific responses to such climate extremes (Gazol et al., 2017a). As annual growth rings of trees provide short- (say about years) and long-term (say about decades) information with regard to how trees respond to drought (Gazol et al., 2017c). Drought effects on forests can last for years because dry spells decrease forest productivity and tree growth and also often leads to the enfeeblement of trees by deteriorating their vigor

(Peltier et al., 2016). It can be discerned from some recent studies that forest resilience to drought—the capacity to resist and recover after a drought—is primarily dependent on drought intensity; nonetheless, such resilience can also be relying on climate types (Zhang et al., 2017), tree species or specific functional traits (Greenwood et al., 2017). Moreover, variations in the strategies of tree species and individuals can occur while dealing with drought (Gazol et al., 2017a,b), leading to modification in forest growth responses to drought that can be reflected in tree ring width variability. Owing to their close linkages to wide-ranging climatic conditions, droughts do not always share the same climate conditions, and, thereby, they can affect forest growth differently across different climates and biomes (Anderegg et al., 2013, 2015, 2016; Phillips et al., 2009, 2010; Williamson et al., 2000; Condit et al., 2004; Marengo et al., 2008, 2011; Lewis, Brando, Phillips, van der Heijden, & Nepstad, 2011; Choat et al., 2012; Chao et al., 2008; Vicente-Serrano et al., 2013; Camarero, Gazol, Sangüesa-Barreda, Oliva, & Vicente-Serrano, 2015; Gazol & Camarero, 2016; Lewis et al., 2011; Camarero et al., 2015; Stanturf, Palik, & Dumroese, 2014a; Chazdon, 2008; Ahrends et al., 2010; Lamb, Stanturf, & Madsen, 2012; Stanturf, Palik, Williams, Dumroese, & Madsen, 2014b; Asner et al., 2006, 2008; Murdiyarso, Skutsch, Guariguata, & Kanninen, 2008; Hosonuma et al., 2012; Lund, 2009; Putz & Nasi, 2009; Putz & Redford, 2010).

9.3.3 Forest degradation

Forest degradation has come to be acknowledged as an issue of widespread global concern. An array of human activities that are driven by a. assortment of macroeconomic, demographic technological, institutional and political factors often contribute to forest degradation (Bustamante et al., 2018). the process of forest degradation can occur over a long period and could only be discerned slowly (Sasaki & Putz, 2009), denoting that degradation of forest is facilitated over time (Thompson, Kamenik, & Schmidt, 2005). This process of degradation, in bulk of the cases, entails a decrease in biomass and alterations to the structure and species composition or biodiversity of the forest, along with in its natural regeneration. These alterations in the biotic constituents of the system can also trigger changes in soil and water, and in the interactions between these constituents, eventually impacting forest functioning and lessening the provision of ecosystem goods and services (Modica, Merlino, Solano, & Mercurio, 2015). Accordingly, forest degradation denotes the situation entailing long-term and grievous environmental alterations, and does not encompass short-term alterations or variations like those linked to forest management for silvicultural targets (FAO, 2011). Unsustainable manipulation or extraction in terms of disproportionate harvesting of forest products, overgrazing, wildfires, and the proliferation of invasive species or pests are the major causes that lead to degradation of forests (Bustamante et al., 2016). Trends emerging from an analysis conducted by Hosonuma et al. (2012) in respect of the direct drivers of degradation in 46 developing countries reveal that the major causes are harvesting for timber (52%), harvesting for firewood and charcoal (31%), uncontrolled forest fires (9%), and grazing (7%). Taking into account this multifariousness of these drivers, the process of degradation can be expeditious, or may happen more gradually, as in the case of selective harvesting of the best trees in a forest and that may be carried out in few or through various occasions (Vásquez-Grandón, Donoso, & Gerding, 2018). Broad estimates demonstrate that a large number of the global forests have been subjected to degradation. It is in this backdrop that forest degradation has come to reckoned with as a grave environmental, social and economic problem (FAO, 2011; Vásquez-Grandón, et al., 2018). The impacts of forest degradation are damaging both to the ecosystems as well as to the society because they entail the likelihood of adversely affecting millions of people that are dependent—wholly or partially—on the goods and services that are produced by forests on a local, regional or global scale (FAO, 2011; Vásquez-Grandón, et al., 2018).

In the wake of multiple adverse impacts of forest degradation, the notions of "degraded forest" and "forest degradation" have come to be delineated from diversified perspectives, in line with the interests, targets and objectives of the multifariousness of programs, international conventions and global policies that deal with biodiversity, climate change and forest management (FAO, 2011; Vásquez-Grandón, et al., 2018). Presently, there are more than 50 definitions of the concepts of "degraded forest" and "forest degradation" that are in vogue (Vásquez-Grandón, et al., 2018), with significance varying from soil degradation (Hudson & Alcantara-Ayala, 2006) to, recently, loss of carbon stock and mitigation of climate change (Morales-Barquero et al., 2014). While glancing through the proceedings of the international negotiations and deliberations on issues pertaining to forest degradation, one can infer that the avowed objective is to a consensus on these definitions (Chazdon et al., 2016; Vásquez-Grandón, et al., 2018). Nevertheless, the generic definition of forest degradation delineated by FAO in 2002 defined the term as: "reduction in the capacity of a forest to provide goods and services" (FAO, 2002) is still widely used globally. In this regard, Vásquez-Grandón et al. (2018) have opined that this definition of forest degradation as envisaged by FAO (2002) is used to compare and monitor statistics across different countries but excludes aspects such as a reference state or different degrees of degradations. However, it has been pointed that mere interpretation of degradation as a loss of traits or functions, although intuitive, is insufficient to decide the scale of degradation of a forest (Ghazoul, Burivalova, Garcia-Ulloa, & King, 2015). Dealing with difficulties in arriving at a clear and practical definition of a degraded forest, Vásquez-Grandón et al. (2018) attribute this in part to the innate level of "uncertainty surrounding the degradation process" and further add that it also emanates from the "high degree of variability in the capacity of forests to recover to their original undisturbed state (i.e., resilience)." It has been argued by Sasaki and Putz (2009) that in the wake of urgency for arriving at agreements on dealing with the challenges presented by climate change, it is warranted that the definition of forest degradation ought to incorporate the entire gamut of biophysical and social conditions under which forests thrive and the multiple ways in which forests can degrade and this endeavor needs to be followed by easily monitorable parameters (Thompson et al., 2005). Admittedly, reversal of the process of forest degradation is a worldwide challenge; nevertheless, it requires local solutions and answers and this endeavor needs to be followed up by easily monitorable parameters (Thompson et al., 2005).

Along with deforestation, forest degradation has major consequences for human societies and biodiversity, and significantly contributes to greenhouse gas (GHG) emissions (CBD, 2002; Mery et al., 2010). Deforestation is an obvious ecosystem change but forest degradation is more difficult to discern and quantify (Sasaki & Putz, 2009). Blay (2012: 267) differentiates between deforestation, which is "the removal of forest cover and conversion to other land uses such as agriculture," and degradation, which is defined as "a permanent decline in the productive capacity of the land." Degradation not only implies disappearing vegetation but also the deterioration of biophysical conditions of a site that may be barriers to the forests' self-recovery (Lemenih & Bongers, 2010). In its publication Good Practice Guidelines (IPCC, 2006), the IPCC does not identify forest degradation by name by the guidelines it "can be estimated as the effect on emissions and removals of human interventions on land continuing to be used as forests" (GFOI, 2013, p. 19), and further that "[forest] degradation is interpreted here as the processes leading to long-term loss of carbon without land-use change, otherwise there would be deforestation" (GFOI, 2013, p. 61). Natural disturbance is also said to be a contributor to forest degradation. Natural disturbance, in common parlance, is caused by natural, but not regularly recurring activity, and includes fire, windfall and waterlogging. Forest degradation and natural disturbance are defined based on whether the cause is natural or anthropogenic, but that distinction is not always clear. For instance, fire is a natural process but rarely occurs naturally in the Amazon. Instead, fires are normally started for vegetation management and can spread into surrounding forests (Cochrane & Schulze, 1999; Nelson, 1994). Logging of the forest promotes high-risk fire conditions or tree mortality due to windthrow in exposed canopies (Brando et al., 2014). Consequently, the difference between degradation and natural disturbance is often unclear and the causes overlapping.

Undeniably, many studies have focused on the process of forest degradation in the context of Brazilian Amazon (Tyukavina et al., 2017), nonetheless, the large inconsistencies in the estimates about area of degradation reflect differences in estimation procedures, definitions of land cover and change, and input data, and they also owing to well-documented difficulties in monitoring (Mitchell, Rosenqvist, & Mora, 2017). With regard to the total area impacted by natural disturbance in the Amazon being uncertain; however, subregional estimates suggest that impacts of degradation and natural disturbance on an average of 14%–123% of the area of deforestation and as much as 500% of the area due to fires in severe drought years have been impacted (Aragão et al., 2018; Tyukavina et al., 2017). Carbon emissions from forest degradation and natural disturbance have even been shown to Aragon be larger than those from deforestation, even if they are not incorporated in large-area climate models or national GHG inventories (Pearson et al., 2017). It is essential to minimize these uncertainties for according priority to mitigation measures and for understanding the impact of forest disturbance on biodiversity and on climate change. trends emerging from research findings of Baccini et al. (2017) have proved partly triggering debate over the role of tropical forest disturbance in the carbon cycle because the finding by Baccini et al. (2017) determined that disturbance in tropical forests result in a net source of carbon emissions into the atmosphere. These results differ from previous studies that identified tropical forests to be a net carbon sink and Baccini et al. (2017) point out that a significant difference in their study as compared to previous studies has been the inclusion of degradation and natural disturbance that accounted for 69%of the total emissions. Undeniably, Hansen, Potapov, and Tyuavina (2019) have described about limitations in this approach; nevertheless, it is worth noting that the estimated emissions from degradation and natural disturbance were significantly greater than previously reported (Pearson et al., 2017) and in the light of this it can be surmised that the total emissions from degradation and natural disturbance remain subject to debate and; therefore, there prevails uncertainty with regard to the role of forests in the global carbon cycle, especially in the Amazon region.

9.3.4 Wild and forest fires

Forest fires, also called wildfires, are a global phenomenon that entail serious consequences for the environment, population and property and increased frequency and severity of wildfires in recent years has been recognized to be associated with climate change (Williams et al., 2019). Several studies have identified an increase in the frequency and severity of wildfires in the boreal region

(Veraverbeke et al., 2017; Hanes et al., 2019). Forest fires are a significant issue impacting a variety of environment, ecosystem and climate functions and structures. They are one of the most pertinent factors affecting vegetation succession, vegetation composition and carbon budgets globally. The high volume of smoke emissions emanating from forest fires are of special significance at global scale because this proves instrumental in altering atmospheric composition, deteriorates air quality and contributes to global warming and climate change (Fuchs, Stein, Strunz, Strobl, & Frey, 2015). Forests fires emit mainly Carbon Dioxide that comprises around a 90%, and is the main GHG (Carrielo & Anderson, 2007). Africa alone accounts for 40% of global annual carbon emissions, with the highest incidence of vegetation fires in the world (Tsela et al., 2014). The atmospheric chemistry change is a highly complex factor to include in the emission models. Estimations are primarily based on the amount of biomass consumed, derived from burned area mapping, pre-fire biomass information and knowledge of the degree of fires combustion completeness (Bastarrika, Chuvieco, & Martín, 2011). There are many socioeconomic implications arising out of uncontrolled fires. These implications are specifically acute in developed regions such as Australia, Greece, Portugal, Russia, or the State of California in the United States as a result of the large numbers of severe fires that have taken place in recent years due to the growing urbanization of forested areas (Bastarrika et al., 2011). The increasing incidents of wildfire also entail the potential of significantly impacting freshwater resources, as alterations from wildfire to water catchments can temporarily reduce evapotranspiration (Bond-Lamberty, Peckham, Gower, & Ewers, 2009), increase snow accumulation (Pomeroy, Granger, Pietroniro, Toth, & Hedstrom, 1999), and cause snowmelt to occur earlier in the spring (Gleason, McConnell, Arienzo, Chellman, & Calvin, 2019). These collective effects can increase runoff, erosion, organic material input, and trace metal levels in water bodies.

Existing in the confluence of temporal and spatial scales, fire events are experienced as a local phenomenon, and yet are recognized as a global-scale environmental process with importance and impacts at the Earth system level (Rabin, Malyshev, Magi, Shevliakova, & Pacala, 2017). In terms of hazard management and action planning, fires are studied on individual basis as individual fires that take place over days to weeks; nonetheless, fire danger and change in regimes are studied on a seasonal basis, with implications for the atmosphere and biosphere on timescales of hundreds to millions of years (Koele et al., 2017; Goulart et al., 2017). Essentially, the incidence of fire depends on the availability of fuel, oxygen and ignition; however, the nature of the fire regime including frequency, seasonality, size, intensity, and ecosystem effects, is dependent on a number of processes that, inter alia, include fuel connectivity, fuel type, resistance of fuel to fire, density and dryness, conditions such as topography, temperature, moisture, wind, anthropogenic factors of ignition, management and suppression, and natural factors such as lightning ignition. Significant impact, both on a global scale and Koala a local level, is wielded by forest fires, at global-scale forest fire affects global vegetation composition and dynamics through mortality, regrowth, and species diversity (Barlow & Peres, 2008), atmospheric chemistry through emissions of gases, including CO_2, carbon monoxide, sulfur dioxide, and nitrous oxide, the carbocycle via the release of carbon from vegetation, the hydrological cycle through decrease in evapotranspiration and release of particulate matters as well as changing the surface albedo from burnt area, black carbon, and altering vegetation cover (Hantson et al., 2016). Forest fire is claimed to be the single most important disturbance of vegetation worldwide (Hantson et al., 2016). Annual burnt area globally reaches nearly 350 million hectares or 3.4 million km^2 per year, and ensuing CO_2 emissions in the past have been as much as 50% of fossil fuel emissions (Jolly et al., 2015a,b; Bowman et al., 2009). For instance, in 1997, carbon emissions from forest fires in Indonesia alone was 0.81 and 3 GtC (van der Werf et al., 2004; GFED data), compared to 6.55 GtC of total global fossil fuel emissions that year (Le, Nkonya, & Mirzabaev, 2016). In 2016, total fossil fuel emissions had risen to approximately 10.0 GtC/year (Global Carbon Project, 2016) and forest fire emissions were slightly lower, being 1.869 GtC in 2016 and projected to be at 1.822 GtC in 2017 (GFED, 2018), accounting for nearly 20% of global emissions. There are also social and economic dimension of forest fires. In 2017, forest fires in California, United States, were estimated to have affected over 10,878,000 people (USCB, 2017), killed 43 people (Cal Fire), and estimates of economic cost for this fire event alone were estimated to be $71.1 billion and $347.8 billion (Thomas, Butry, Gilbert, Webb, & Fund, 2017). Apart from impacting air quality, and destruction of human lives and infrastructure, forest fire is also said to be a vital process for successful ecosystem functioning in terms of maintaining savanna ecosystems (Andela et al., 2017), vegetation succession in terms of seed release and resprouting in lodgepole pine Eucalyptus, Banksia; post-fire flowering in Australian grass tree and small-scale farming [waste disposal, land clearance, disease and pest control (Brandt, 1966)], increasing soil fertility (Santín & Doerr, 2016), soil carbon (Koele et al., 2017), and encouraging new growth for grazing animals (EMP, 2019).

9.3.5 Forest fragmentation

Forest fragmentation is described by Snyder (2014) as the breaking of large, adjoining, forested areas into smaller pieces of forest; and characteristically these pieces are separated by roads, agriculture, utility corridors, subdivisions, or other human development. Forest fragmentation has been defined by Thomson (2003, cited in Rahman, Jashimuddin, Kamrul, & Kumar, 2016a) as the conversion of large areas of "contiguous native" forest to other kinds of vegetation and/or land use leaving residual tracts or patches of forest that are varied in size and isolation. Habitat fragmentation is one of the foremost causes of biodiversity erosion, especially in tropical forests (Tabarelli et al., 1999). Trends emerging from the study undertaken by Gibson et al. (2013), within 25–26 years of fragmentation half of the native species are said to have vanished. Between the span of 2000 and 2010, there were approximately 13 million hectares of forest lost globally each year either via destruction by natural causes or owing to the fact that land was converted to other land uses (Rahman et al., 2016a). Over 60% of economies in the Asia region between 2000 and 2011 was either extended or retained the area meant for agriculture and five of the 10 economies with the maximum rate of deforestation in 2012 were located in Southeast Asia (Rahman et al., 2016a; Estoque et al., 2019). In the South Asia region, the rate of deforestation was 0.18% (ADB, 2010). High population density and small land area persistently wields pressure on forests for meeting the basic needs, such as food, shelter, fuel and even livelihood of the native populace (Rahman, Islam, & Islam, 2016b). With the rising demand for increased food production worldwide, agriculture is in for rapid expansion (Tscharntke et al., 2012), specifically in tropical forests (Green, Cornell, Scharlemann, & Balmford, 2005), and this fast becoming rapid expansion of agriculture is one of the prime activities diminution and fragmentation of natural forests globally (Moran & Catterall, 2014), and encompasses nearly 40% of the soil throughout the globe (Foley, 2005). Accordingly, agricultural landscapes have come to be construed as domains of immense significance for the analysis and development of strategies that assimilate biodiversity and ecosystem services along with food production (Balmford et al., 2018). Tropical regions, being endowed with world's highest biodiversity, also bear the brunt for increasing agricultural land (FAO, 2016b,c). This necessitates the urgency for developing production models that not only accord priority to increased production efficacy but also the externalities insinuated on the biodiversity and ecosystem services of a landscape (Balmford et al., 2018; Fahrig et al., 2015).

Edge effect that is produced by erecting borders between forest and agricultural ecosystems, is one of the major consequences of fragmentation (Laurance et al., 2011) the erection of this hasty border constitutes the prime cause of changes in the microclimate, the soil fertility and the prevalence of chemical substances from the surroundings in the borders and inside or interior of the fragmentation (Murcia, 1995). This wields short-term and long-term impacts on species composition (Andrade et al., 2015), frequently supporting the spread of generalist and pioneer species (Ribeiro, Nunes-Freitas, Fidalgo, & Uzêda, 2019), and hardly exotic species (Lambert, Dudley, & Robbins, 2014). Edge effects are usually more severe in fragments of smaller area and with higher edge ration (Ribeiro et al., 2019). Nevertheless, the severity and types of edge impacts are directly linked to traits and usage of intensity in the environment (Chabrerie, Jamoneau, Gallet-Moron, & Decocq, 2013). Fertilization and pH corrections that necessitate a frequent use of chemical inputs are the two major practices that are adhered to for intensification of agriculture (Tscharntke et al., 2012), and some studies make it discernible that augmentation of these practices in farm fields triggers alterations in soil chemical traits on nearby forest fragments through contamination by fertilizers (Ribeiro et al., 2019). This contamination takes place through fertilized soil particles in the crops being carried out by wind to the adjoining fragments. Such alterations in fertility levels of soils in forest fragments entail the likelihood of significantly affecting the floristic composition in forest fragments owing to the wide impact of soil chemical traits on vegetation composition and spatial distribution (Bentos, Nascimento, Vizcarra, & Williamson, 2017). Growth of trees in forest environments are being limited by some nutrients (Alvarez-Clare & Mack, 2015), and the persistent augmentation in fertility levels in forest fragment soils could lead to changes in soil chemical relationships (Alvarez-Clare & Mack, 2015), and may also cause species losses owing to competition and mortality (Sardans & Penuelas, 2012). It also becomes discernible from some studies that some species that are pioneers or innovators demonstrate more efficacy in making use of nutrient-surpluses, with an augmentation in growth rates (Alvarez-Clare & Mack, 2015). According to Ribeiro et al. (2019), alterations in soil fertility conditions in tandem with other effects linked with forest fragmentation may promote production of pioneer species to the disadvantage of late "successional" species.

Of the total global forest cover, about 30% of that has been cleared, with another 20% having been degraded in recent years and bulk of the rest has been fragmented, leaving only about 15% intact (Rahman et al., 2016a). Having fallen prey to fragmentation, these fragmented forests have been divided into small habitats (Numata & Cochrane, 2012). According to Haddad et al. (2015), approximately 20% of the world's remaining forest is within 100 m of an edge in intimate contiguity agricultural, urban or other altered environments where effects on forest ecosystems are most serious, and more than 70% of the global forests are within 1 km of a forest edge. Microclimatic differences that trigger buffer effect or edge effect are prone to lead to alteration in species composition, change in species richness and also entailing the potential of pushing species with small population to extinction and parasitic disturbance (Zhu, Xu, Wang, & Li, 2004). Some studies have also discussed about forest fragmentation spurring abiotic alteration that impacts habitats of both floral and faunal community (Laurance et al., 2011). Likelihood of extinction cascades taking place cannot be ruled out in landscapes with low vegetation cover, low landscape connectivity, degraded native vegetation and intensive land use in altered areas, specifically if keystone species or entire operational groups of species are lost (Fischer & Lindenmayer, 2007). Human-induced land use change entails the potential of trigger massive-scale destruction and fragmentation of native vegetation (Bennett & Saunders, 2010). This surmise is also reinforced by Abdullah and Nakagoshi (2007) when they argue that human-induced land-use alteration is a significant decisive factor leading to forest fragmentation. Nevertheless, there are other number of activities or events like road construction, logging, conversion to agriculture or wildfires that contribute to forest fragmentation, and finally, cause of fragmentation is either anthropogenic or natural in origin (Wade, Riitters, Wickham, & Jones, 2003).

Forests play indispensable role as sanctuaries of biodiversity, multiple species and when as providers of ecosystem services. In the wake of growing demands of burgeoning human population, mere conservation of remaining intact forests along with their biodiversity, functions and ecosystem services is seemingly an improbable and insufficient solution. In order to meet the growing demands or ecosystem services made available by forests, especially the multiple provisioning services on which the people are heavily dependent for livelihoods and products like timber, medicines, thatch, fiber and meat, a massive program of forest restoration seems to the only solution that will be efficacious in the long run (Holl & Aide, 2011). Planting short-rotation single- or multiple-species plantations on degraded soils, restoring plantings in secondary forests or assisted-regeneration in selective logged forests are some of measures that can help in forest restoration (Lamb, 2010). These and other related measures designed to restore forests embrace management interventions that aim at restoring ecosystems that have been subjected to degradation, damage or destruction on account of anthropogenic activities (Birch et al., 2010). Therefore, ecological restoration is a significant approach that entails the potential of enhancing levels of biodiversity in human-altered ecosystems (Brudvig, 2011) and may also in mitigating the impact of climate change (Harris, 2009). Recent years have witnessed focus of forestry shifting from sustained yield management to sustainable forest management (SFM) (Hahn & Knoke, 2010). SFM is seemingly emerged as the prevailing forest management paradigm worldwide (MacDicken et al., 2015), and it emphasizes on balancing economic, ecological and socio-cultural functions of forest. There is no globally acceptable definition of SFM. Nevertheless, on the European level, SFM is defined as "The stewardship and use of forests lands in a way, and at a rate, that maintains their biodiversity, productivity, regeneration capacity, vitality and their potential to fulfill, now and in the future, relevant ecological, economic and social functions at local, national and global levels, and that does not cause damage to other ecosystems" (MCPFE, 1993). Inclusion of forests as SDG-15 in United Nation's Sustainability Development Goals of the 2030 Agenda for Sustainable Development that aims to "protect, restore and promote sustainable use of terrestrial ecosystems, sustainably manage forests, combat desertification, halt and reverse land degradation and halt biodiversity loss" (UN, 2017), has proved instrumental in bringing on the global level the significance of forests and their sustainable management.

9.4 Sustainable land use

Human activities have been instrumental in altering the land surface since prehistoric times, and these activities have been in the form of converting natural vegetation to crops, pastures and savanna through the use of fires and deforestation, using wood harvesting for fuel and construction of infrastructure that has been on the rise in the wake of burgeoning population levels (Williams, 2006 in Wilkenskjeld, Kloster, Pongratz, & Reick, 2014). Land-use activities have affected over 50% of the land surface over the last three centuries, 25% of global forest area has been lost, and agriculture now accounts for around 30% of the land surface (Turner et al., 1995), and that continues to grow at a rate of 7 million hectares per year (). These land-use transitions in the form of cutting of forests and planting of crops, etc., have not only envisaged important primary changes to ecosystems, but also have left behind secondary land in various stages of regrowth (Hurtt et al., 2011). Vegetation impacts the surface fluxes of radiation, heat and moisture, where conversion to crops or pastureland can reduce the aerodynamic roughness and alter evaporation patterns, soil moisture and latent heat (Betts, 2005). Land-use and land-cover change (LUCC) entails an alteration in the use and management of land by humans and a corresponding change in terrestrial surface and subsurface (biota, soil, topography, surface and groundwater, and human structures characteristics) of the Earth (Robinson, Brown, French, & Reed, 2013). Most land-use activities essential for human development involve the appropriation of natural resources and ecosystem services that negatively impact the Earth's ecosystem both locally and globally (Foley, 2005). Global rise in cropland and fertilizer use from 1960s to 1990s proved instrumental in causing environmental damages, including water quality degradation, arable land loss, and soil erosion (Foley, 2005). Broad estimate show that land-use activities associated with agriculture and timber harvesting resulted in a net loss of about 7–11 million km^2 of forest over the previous 300 years (Goldewijk & Ramankutty, 2004), and the resultant outcome has been that nearly 40% of the global terrestrial surface is occupied by croplands and pastures (Foley, 2005).

Apart from affecting the Earth's ecosystems, LUCC also significantly impacts global carbon cycle. Agriculture, forestry, and other land-use (AFOLU) sector accounted for about 25% of anthropogenic GHG emissions on an average annually between 1990 and 2011 (Tubiello et al., 2014) and agricultural practices contributed the most to the global anthropogenic non-CO$_2$ GHGs, the annual value of which accounted for about 10%–12% of global anthropogenic emissions (Smith et al., 2014). In the wake of adverse impacts wielded by LUCC on the Earth's ecosystem and carbon cycle, there is growing demand for devising well-concerted plans for preparing land management strategies that require a trade-off among factors changing land management; for instance, maintaining an equilibrium in economic development among various sectors, social benefits of diverse groups of people, and environment effects on different regions (Houghton, 2010). Multicriteria decision analysis (MCDA) is touted as an effective instrument that can assimilate multiple variables and importance of these variables into a decision process, by offering a structured framework to support decision-making problems through exploring objectives and concerns of multiple stakeholders (Velasquez & Hester, 2013). MCDA's capabilities of dealing with quantitative and qualitative data, multiple stakeholders' conflicts, and sensitivity and robustness of various choices make it a widely used application for decision analysis, particularly in Climate change analysis and ecosystem management (Giove, Brancia, Satterstrom, & Linkov, 2009). Apart from being both a cause and consequence of climate change (Settele et al., 2014), LUCC also directly affects the land surface. Regional and global climate is Settee also affected by land-use changes when the latter changes GHG concentrations in the atmosphere, the surface-energy budget, surface and cloud albedo, wind profiles and altering the natural carbon cycle via removal of carbon from natural vegetation and enhancing carbon emissions (Pielke et al., 2002) or vice versa through reforestation.

Having been instrumental in changing other basic cycles as well, including the nitrogen and water cycle, alterations in land use can prove critically significant factors in the climate system thereby entailing the potential in enhancing or dampening the global thermodynamics of climate change.

Past land-cover alterations have helped in the augmentation of in atmospheric CO_2 content and thereby to global warming, with high probability of biogeochemical impact. The mean biogeochemical warming has been premediated from observation-based estimates ($+0.25°C \pm 0.10°C$) (Li et al., 2017; Avitabile et al., 2016), or assessed from peppy worldwide vegetation models ($+0.24°C \pm 0.12°C$) (Peng et al., 2017; Arneth et al., 2017; Pugh et al., 2015; Hansis, Davis, & Pongratz, 2015) and global climate models ($+0.20°C \pm 0.05°C$) (Simmons & Matthews, 2016). The extent of these simulated biogeochemical impacts could; however, be taken lightly as they are seemingly no answerable for an array of processes like land management, nitrogen/phosphorus cycles, alterations in the emission of CH_4, N_2O and non-GHG emissions from land (Arneth et al., 2017; Pongratz et al., 2018). While accounting for those compounds, Mahowald, Ward, Doney, Hess, and Randerson (2017) and Ward, Mahowald, and Kloster (2014) in their respective studies have discerned a global net positive radiative forcing in response to past anthropogenic land-cover alterations, denoting a net surface warming. Nevertheless, in the first instance, the assessed biophysical radiative forcing in those studies only sheds light on alterations in albedo and not on alterations in "turbulent fluxes." In the second place, the cumulative assessments also rely on other various key modeling assessments like climate sensitivity, CO_2 fertilization triggered by land-use emissions, probable synergistic impacts, and cogency of radiative forcing concept for land forcing. Interestingly, most of the other studies based on different modeling do not shed light on the evolution of vegetation in areas that are not managed. Nonetheless, it has been reported by studies based on observation and numerical data about greening of the land in boreal regions followed on from both extended growing season and poleward migration of tree lines (Lloyd, Rupp, Fastie, & Starfield, 2003). This greening event augments global warming through a diminution of surface albedo (Forzieri, Alkama, Miralles, & Cescatti, 2017).

9.4.1 Land degradation

Land is a significant resource permitting the production of food, the conservation of biodiversity, enabling the natural management of water systems and serving as a carbon entrepot (FAO, 1999; EP, 2008). M misappropriation of land resources have led to emergence of land degradation and desertification and both have critical environmental problems facing the world today (UNCCD, 2017). Land degradation wields negative impacts on agricultural productivity and ecological function and finally affecting human sustenance and quality of life (Masoudi & Amiri, 2015). On account of environmental factors on various scales of time and space, about 25% of the global biomass has been subjected to degradation (Quyet, 2014), trying to make out land degradation requires a multiscale approach (Masoudi & Amiri, 2015), and this approach assumes significance with regard to land management goals and few studies are said to have examined land degradation with multiscale approach (Masoudi & Jokar, 2017). Accruing from an array of parameters, including climate change and human-induced activities in arid, semiarid and dry subhumid regions (UNEP, 1992), land degradation continues to be a global problem of immense magnitude (UNCED, 1992; UNEP, 2007; UNCCD, 2017). The concept of land degradation has been defined in the literature in multiple different ways with divergent emphases on biodiversity, ecosystem functions and ecosystem services (Montanarella, Scholes, & Brainich, 2018). Nevertheless, Special Report on Climate Change and Land (SRCCL) released by the IPCC in 2019 has defined land degradation as a "negative trend in land condition, caused by direct or indirect human-induced processes, including anthropogenic climate change, expressed as long-term reduction or loss of at least one of the following: biological productivity, ecological integrity or value to humans" (IPCC, 2019). SRCCL's definition is primarily derived from UNCCD's definition of land degradation elaborated in 1994. Article 1 of the UNCCD says: "land degradation in arid, semiarid and dry subhumid areas resulting from various factors, including climatic variations and human activities. Land degradation in arid, semiarid and subhumid areas is reduction or loss of the biological or economic productivity and integrity of rainfed cropland, irrigated cropland, or range, pasture, forest, and woodlands resulting from land uses or productivity and from a process or combination of processes, including processes arising out from human activities and habitation patterns, such as (1) soil erosion caused by wind and/or water; (2) deterioration of the physical, chemical, biological, or economic properties of soil; and (3) long-term loss of natural vegetation" (UNCCD, 1994, Article 1). The SRCCL has complemented the more detailed UNCCD definition of land degradation by expanding its scope to all regions, and not just limiting it to drylands, with a view to provide an operational definition that stresses on the relationship between land degradation and climate (IPCC, 2019). By focusing on the three aspects—biological productivity, ecological integrity and value to humans—the SRCCL definition is coherent with the notion of land degradation neutrality (LDN) that aims and maintaining or enhancing the land-based natural capital, and the ecosystem services accruing from it (Cowie et al., 2018).

Many factors contribute to land degradation and these, inter alia, include the following: soil erosion, soil contamination, soil salinization, decline in soil organic matter, soil sealing, landslides, soil compaction, and loss of soil biodiversity (EP, 2007). Soil erosion is the fading away of the land surface by physical forces like rainfall, flowing water, wind, ice, change in temperature, gravity or other natural or anthropogenic factors that facilitate abrasion, detachment and removal of soil or geological material from one point of Earth's surface to be deposited at other point. Entailing many processes, the overall effect of land erosion is of particles being carried away from one location to be deposited elsewhere. Soil erosion, though taking place naturally, is often aggravated by anthropogenic activities (Adornado, Yoshida, & Apolinares, 2009). Wind, rainfall and associated runoff processes, susceptibility of soil to erosion and traits of land cover and management affect soil erosion (Panagos et al., 2015). Possible negative impacts of soil erosion like nutrient loss, river and reservoir siltation, deterioration water quality and decreases in soil productivity make it critical to have understanding and mitigating erosion and associated degradation (Bagherzadeh, 2014). Soil contamination, another contributor to land degradation, is caused by the prevalence of xenobiotic chemicals or other change in the nature of soil environment. Caused by industrial activity, agricultural chemicals or inappropriate disposal of waste, soil contamination can be local or diffuse (Van-Camp et al., 2004). Diffuse contamination, comprising heavy metals, acidification, nutrient surplus, etc., is usually caused by contaminants transported over far-off places, and as such soil contamination portends a worrisome threat to agricultural productivity, food safety and human health, apart from leading to land degradation (FAO, 2018). Soil salinization also contributes to land degradation. It is the process wherein water-soluble salts like chloride, sulfate, sodium, magnesium, carbonates and bicarbonates get accumulated in the soil to the extent that soil fertility gets severely restricted (Shrivastava & Kumar, 2015). There are primary and secondary processes of salinization.

Primary process entails accumulation of salts via natural processes like physical or chemical weathering and transport from saline geological deposits or groundwater; and secondary salinization occurs due to human interventions such as use of salt-rich irrigation water or other unscientific irrigation practices and/or inadequate drainage conditions (Kibblewhite et al., 2008). Soil organic matter embraces all living soil organisms together with the remains of dead organisms in their various degrees of decomposition and it constitutes an important ingredient of soil owing to its influence on soil structure and stability, water retention, cation exchange capacity, soil ecology and biodiversity and also as a source of plant nutrients (EU, 2009). Crop residue, animal and green manures, compost and other organic materials are the sources of organic matter. The organic carbon content of a soil consists of heterogeneous mixtures of both simple and complex substances containing carbon. A decline in organic matter results in a decrease in fertility and loss of structure, reduction in water infiltration capacity, leading to enhanced runoff and erosion, and loss of carbon content limiting the soil's ability to provide nutrients for sustainable plant production, and all these together exacerbate overall land degradation (Van-Camp et al., 2004).

Soil sealing also leads to land degradation. Soil sealing comprises covering of the soil surface with impervious matter as a consequence of Kibble white urban development and infrastructure construction and it is also used to describe an alteration in the nature of soil resulting in impermeability, primarily attributable to compaction by agricultural machinery (Kibblewhite et al., 2008). This brings soil sealing and land consumption intimately proximate to each other, especially when natural, seminatural and cultivated land gets covered by built surfaces and structures thereby leading to degradation in soil functions or causing their loss (EP, 2008). Soil degradation is also attributable to landslides. A landslide is described by Geological Survey of the United States as the movement of a mass of rock, debris, artificial fill or Earth down a slope, under the force of gravity (USGS, 2004). Soil functioning is threatened by landslides in two ways: In the first instance, via the removal of soil from its in situ position; and in the second instance the deposition of colluvium on in situ down slope from the area where the soil mass "failed" (EP, 2009; Gan & Wang, 2018). Soil compaction also contributes to soil degradation. The multiple forces of soil compression by agricultural machinery can cause soil particles to soil surface compacted closer together into a smaller volume and this compression of particles together leads to reduction in pore space—the space between particles—thereby narrowing down the space available in the soil for water and air (AARD, 2010). Soil compaction leads to reduction in soil biological activity and soil productivity for agriculture and forest cropping, culminating in decreased water infiltration capacity and increased erosion risk (EP, 2009). The diminution in pore volume that results from compaction is primarily owing to a decrease in micropores that makes available connectivity for water and gas via the soil profile (Kibblewhite et al., 2008).

There exists very intimate relationship between soil and biodiversity because soils have come to be acknowledged as significant reservoir of biodiversity worldwide and being home to nearly one-fourth of all soil surface living organisms on the planet (Decaëns, Jiménez, Gioia, Measey, & Lavelle, 2006). Soil affords multiple functions and services supporting life on the planet in terms of engendering biomass, biogeochemical cycling, relating water movement, climate and pollution (Adhikari & Hartemink, 2016; Chen et al., 2020). Nevertheless, the affordability of soils in delivering these services vastly relies on their biodiversity (Bardgett & Van Der Putten, 2014), and this dependence is up to such an extent that soil biodiversity has come to acknowledged as the basis of soil security (McBratney, Field, & Koch, 2014). The soil biota is not only endowed with its own remarkable capacity to stand up against events that engender disturbance or alteration but also possesses certain capacity to recover from these perturbations (Tibbett, Fraser, & Duddigan, 2020). Arguing that the capacity to recover from change is thought to be a primary trait of biodiversity, Tibbett et al. (2020) have opined that soils with higher biodiversity are considered to possess an innate resistance and resilience to change. Soils endowed with lower resistance to a disturbance and diminution in capacity to recover are prone to suffer loss of biodiversity (Downing, van Nes, Mooij, & Scheffer, 2012). Generally, fall in biodiversity is thought to be the diminution of forms of life inhabiting soils, both in terms of quantity and variety (Jones, Hiederer, Rusco, & Montanarella, 2005). Several soil threats have been identified by Gardi, Jeffery, and Saltelli (2013) that negatively impact soil biodiversity, including land-use change, intensive exploitation by humans and diminution in soil organic matter. Owing to close proximity between soil and biodiversity, there also exists multifaceted relationship between soil erosion and biodiversity, and erosion is now widely acknowledged as one of the main threats to biodiversity. Given the fact that approximately 2.8 tons of soil are lost per hectare annually worldwide (ESDAC, 2018), soil loss necessarily affects the organisms inhabiting this ecosystem. Concomitantly, the wide range of species living in soil affects its aggregate stability and water infiltration via their movement and feeding activities, and this brings to the surmise that soil biodiversity surely plays a key role in soil loss processes as well (Orgiazzi & Panagos, 2018).

9.4.2 Desertification

Desertification is increasing becoming a worrisome issue globally. Desertification is caused by both natural and anthropogenic processes and has had multiple definitions over time (Capozzi, 2018). Ding et al. (1998) described desertification as a reduction in the productivity of biological life over time. The United Nations Convention to Combat Desertification (UNCCD, 1999) identified desertification as land degradation in arid, semiarid and dry, humid areas caused by various factors, including climatic changes and human activities. Findings by Veron et al. (2018) demonstrate that two main definitions coexist within recent literature: "desertification as land degradation in arid, semiarid and subhumid areas resulting from various factors including climatic variations and human activities," and "desertification as a persistent reduction in the capacity of ecosystems to supply services... over extended periods of time." Pourghasemi et al. (2019) define desertification simply as a land degradation process due to climate change and unsustainable land management practices. In China, desertification has been defined as the process whereby extreme economic activities of humans disrupt the ecological balance of arid and semiarid lands and result in the formation of blown sand and sand dunes (Duan et al., 2019). Although the formation of dunes is not a necessary condition for desertification worldwide, the concept of ecological balance is essential to the definition. Desertification is defined as land degradation in arid, semiarid and subhumid areas, accruing from multiple factors, including variations in climate and human activities (UNCCD, 1994). As a negative trend in land condition, land degradation is triggered by direct or indirect human-induced processes, including anthropogenic climate change, articulated as long-term diminution or loss of at least one of the following: biological productivity, ecological integrity or value to human (Evans et al., 2019). Arid, semiarid, and dry and subhumid areas in tandem with hyperarid areas, comprise drylands (UNEP, 1992), and house about three billion people (van der Esch et al., 2017).

The distinction between desertification and land degradation is not process-based, rather it is geographic. Land degradation can take place anywhere across the globe while when it takes place in drylands, it is regarded as desertification; nonetheless, desertification is not confined to irretrievable forms of land degradation, nor can be compared to expansion of desert, rather it epitomizes all forms and levels of land degradation taking place in drylands (Evans et al., 2019). The geographic categorization of drylands is usually facilitated on the basis of the aridity index (AI) that delineates the ratio of average annual precipitation to the potential evapotranspiration. Currently, drylands encompass approximately 46% of the global land areas (Koutroulis, 2019; Prăvălie, 2016). Interestingly, hyperarid areas, where the AI is below 0.05, are often counted as drylands but are included in the definition of desertification (UNCCD, 1994). Admittedly, deserts are precious ecosystems (UNEP, 2006) and they are geographically located in drylands and susceptible to climate change; nonetheless, they are not thought to be vulnerable to desertification. Aridity is a long-term climatic lineament marked by low average precipitation or water availability (Türkeş, 1999). Aridity is distinguishable from drought which is a temporary climatic event (Maliva & Missimer, 2012); furthermore, droughts are not confined merely to drylands and they can happen both in drylands and humid areas (Wilhite, Sivakumar, & Pulwarty, 2014). Assuming that future climate is marked by significant augmentation in CO_2, the utility of currently applicable AI thresholds gets confined under climate change. Usage of other variables like precipitation, soil moisture and primary productivity to detect drylands, instead of the AI, entails no clear signal that the extent of drylands will alter overall under climate change (Lemordant, Gentine, Swann, Cook, & Scheff, 2018). Therefore, there is likelihood that some dryland areas will expand, while others will shrink.

A major portion of dryland areas, nearly 70%, is situated in Africa and Asia, and the largest land use/cover, in terms of drylands, if deserts are not included, are grasslands followed by forests and croplands, and "other lands" category comprises bare soil, ice, rock, and other land areas that are not included within other five categories (). Grasslands are defined by FAO (2016) as permanent pastures and meadows persistently used for more than 5 years. Seasonal migratory grazing in drylands very often brings in nonpermanent pasture systems, as a result of which some of the areas under "other land" denomination is also used as nonpermanent pastures (Fetzel et al., 2017). Drylands are home to about 3 billion people, approximately accounting for 38% of the total global population (Koutroulis, 2019; van der Esch et al., 2017), with the highest number of those people inhabiting the drylands of South Asia, followed by sub-Saharan Africa and Latin America (van der Esch et al., 2017). In 2007, Reynolds et al., 2007 indicated that about 250 million people were directly affected by desertification and some recent estimates demonstrated that about 400 million people lived in those dryland regions in 2015, experiencing tremendous losses in biomass productivity between the 1980s and the 2000s (Le et al., 2016). Projections indicate that the population in drylands is likely to rise about twice as fast as nondrylands, reaching 4 billion people by 2050 (van der Esch et al., 2017). Bulk of the population, say 90%, in drylands live in developing countries (UN-EMG, 2011) and people inhabiting drylands are greatly susceptible to desertification and climate change (Huang et al., 2017; Lawrence, Maxwell, Rew, Ellis, & Bekkerman, 2018) because of their predominant dependence on agriculture as their livelihood and agricultural sector is most vulnerable to climate change (Rosenzweig et al., 2014). Climate change is anticipated to have considerable impacts on all types of agricultural livelihood systems in drylands (CGIAR-RPDS, 2014). Assessing areas at risk of desertification and identifying appropriate mitigation measures are of primary importance (Ladisa, Todorovic, & Liuzzi, 2012). The causes of desertification include deforestation, overgrazing, unsustainable agricultural practices, and poorly managed irrigation. Further factors such as population growth, economic instability, and national policies may augment the potential for these unsustainable land management practices to emerge, and consequently, both natural and human-caused climate change can aggravate desertification (Wiesman, 2009). Desertification is one of the fundamental problems around the world that is progressively rising.

It has been opined by many scholars that desertification is caused primarily by human-induced activities (Jabbar & Zhou, 2013; Boudjemline & Semar, 2018). Human-induced factors are prime cause of aggravating land degradation, including population growth, unsustainable economic development, reduced environmental and ecological awareness, overconversion and improper conversion of pastures for overcultivation and sowing, improper management of water resources, overgrazing, firewood, uncontrolled collection of herbal medicines, deforestation and displacement cultivation, terrace clipping on slopes and mountainous areas, salinization/alkalization, oil exploration and mineral mining. Take the instance of Libya, where the growing population pressure on natural resources, especially on water, soil and vegetation, and excessive use of natural resources by entities or individuals have been instrumental in contributing to desertification (Heshmati & Squires, 2013). Desertification occurs because arid land ecosystems are extremely susceptible to overexploitation and inappropriate land-use practices (Maestre, Reynolds, Huber-Sannwald, Herrick, & Stanford, 2006). Desertification can trigger a general climate tendency toward more drought, or initiate an alteration in the local climate (Wiesman, 2009). Besides, it also becomes discernible from the trends emerging from the research carried out on global climate that annual rainfall decreases in the Mediterranean, average temperatures increase, and an increase in intra-year variability and in the same season (Boudjemline & Semar, 2018). As per climate forecasts, in terms of RCP 4.5 (RCP 8.5), the rate of drought is projected to increase by 11% before 2100 and that means that 50% of the total land surface will be covered by dry areas (Lu et al., 2018). Climate change along with irregular rainfall continues to enhance the vulnerability of arid areas in East Africa. Nevertheless, combating drought is considered to have essential role in reducing climate change and that are significant points of biodiversity and support various livelihoods (Winowiecki, Vågen, Kinnaird, & O'Brien, 2018). As a consequence of climate change impacting desertification, vegetation conditions in arid, semiarid and dry humid environments have undergone drastic alterations (Huang et al., 2017).

Loss of fertile farmland due to desertification is a growing serious concern worldwide. With around one-third of world's total surface area being at the risk of desertification that entails the likelihood of affecting directly the livelihoods of about one billion people at risk. Desertification is already rendering a sizeable portion of fertile land useless for planting each year. Careful selection of crops and sustainable management of water resources have become indispensable in resisting the severe conditions of droughts (Wiesman, 2009). Population growth and the resultant increasing food requirements of increasing population have proved instrumental in exerting more pressure on the land and spread of agricultural production into marginal drought areas, including traditional pastures (Winowiecki, et al., 2018). The resultant effect is discernible in the decrease in productivity of forests, natural pastures and agricultural lands. Alterations in physical and chemical properties and their resultant impact on soil productivity entails the likelihood of leading diminution in arable land and the net outcome would be sharp decline in agricultural productivity and more humans to feed (Heshmati & Squires, 2013). Ecological functioning of a region is impacted by unsustainable alterations in land use, probably triggering diminution in both productivity and biodiversity (Jabbar & Zhou, 2013). In the Egyptian context Ismael (2015) points out in that region, agriculture is influenced by desertification, soil depletion, climate change and water scarcity. Desertification has come to be acknowledged as a severe threat to agriculture, forests and other vegetation. Jiang et al. (2019) point out there are some significant alterations in agriculture and animal husbandry that entail the potential of causing desertification to move in different directions like: diminution in irrigation and expansion of irrigated agriculture resulting in an augmentation in in the volume of irrigation water, giving up of vast plantation areas, breakdown of animal husbandry and exit from far-off pastures. Forests play pivotal role in safeguarding agricultural crops by protecting them from wind, sunlight and evapotranspiration, and in this regard Vorovencii (2015) has reported that surface areas and new ones are presently reforested and ecological reconstruction projects are being implemented in order to minimize the risk of desertification and to avert the adverse consequences of desertification that, inter alia, include: loss of agricultural areas, reduction of forest biomass, pastures and fertile plains, loss of surface water and groundwater in forests, salinization of land, and alterations in the water quality (Pashaei, Rashki, & Sepehr, 2017). According to Pasternak and Schlissel (2012), Mediterranean rainforest system is said to be probably the oldest and most sustainable rainforest system in the world, and it is primarily based on fruit trees, specifically olive trees. There are said to be some tress and shrub species that are resistant to arid and semiarid conditions and probably entail the potential of stabilizing nitrogen, and such trees and shrub species can necessarily contribute to organic matter content and soil fertility, an in this context Caliskan and Boydak (2017) cite the example of Russian olive as a nitrogen-fixing species that can make significant contribution to nitrogen-pool.

9.5 Conserving mountain ecosystems

Mountains encompass 24% of the Earth's land surface (UNEP-WCMC, 2002). Rugged and topsy-turvy terrain, a low-temperature climate regime, steep slopes along with institutional remoteness are some of the common traits characterizing high mountain regions, and these features are usually associated with physical and social-ecological processes that, while not distinct to mountain regions, symbolize many aspects of these regions. Location of mountains at higher elevation relative to the neighboring landscape makes them often reflect cryosphere components, such as glaciers, snow cover, and permafrost, with a tremendous impact on nearby lowland areas even located even far away from the mountains (Huggel, Carey, Clague, & Kääb, 2015). This enables the mountain cryosphere to play a major role in large parts of the globe. It was reported at the outset of 2000s that mountain regions are home to 12% of the global human population (Hillstrom & Hillstrom, 2003). Another 12% of the population is inhabiting the areas in the immediate proximity (Meybeck, Green, & Vörösmarty, 2001). As per the gridded population data, approximately 671 million people or 10% of the global population lived in high mountain regions in 2010 (Jones & O'Neill, 2016) and a distance of less than 100 km from glaciers and permafrost located in mountain regions and this population is projected to grow to 736–844 million across the shared socioeconomic pathways by 2050 (Gao, 2019). Nevertheless, many people living outside of mountain regions and not enumerated in these numbers are also prone to be affected by the mountain cryosphere. Almost all of the world's major rivers originate in mountains and more than half the world's mountain regions play pivotal role in supplying water to downstream regions (Viviroli, Dürr, Messerli, Meybeck, & Weingartner, 2007).

Mountains deliver multiple ecosystem services across the spectrum and also serve as repositories of biological and cultural diversity. The vital ecosystem services supplied by mountains, also entailing tangible economic value, include: water, power, tourism, minerals, medicinal herbs and plants, and fibers (Macchi, 2010). Besides these, mountains also influence the climate change of their surrounding areas and serve as carbon sinks. Nevertheless, fragility of mountain ecosystems renders them vulnerable to both natural and anthropogenic drivers of change that range from volcanic and seismic events and flooding to global climate change and loss of vegetation and soils on account of inappropriate agricultural and forestry practices and extractive industries (Herrera, n.d.). Enhanced warming projected at high elevations puts mountain ecosystems at increased risk of falling a prey to the adverse impacts of the vagaries of climate change (Pepin et al., 2015). Enhancing air temperatures (IPCC, 2014) and drought frequency (Mann & Gleick, 2015) along with higher magnitude, more common and warmer extreme precipitation events (Dettinger et al., 2015) have resulted from climate change and will continue in Western United States mountain watersheds. Snow droughts (Harpold, Dettinger, & Rajagopal, 2017) and earlier snowmelt-caused runoff is projected to increase in mountain ecosystems (Sadro, Sickman, Melack, & Skeen, 2018; Harpold & Brooks, 2018). Mountain lakes are extremely susceptive of climate change (Thompson et al., 2005) and coalesce change from atmospheric, terrestrial and aquatic environments serving as indicators of climate change (Adrian et al., 2009). There prevails seemingly an agreement to some extent among many scholars that aquatic systems in mountains are not well understood owing to their general inaccessibility (Hock et al., 2019); nevertheless, in view of the fact that higher elevations are projected to have increased warming (Pepin et al., 2015), mountain lakes are extremely sensitive to change (Thompson et al., 2005). Currently, the available literature shows that our understanding of climate and water use change on freshwaters is focused on lowland systems, while mountain ecosystems entail the likelihood of acting as the best assessors or investigators of environmental change (Williamson et al., 2000).

Mountains signify distinct areas to explore climate change and evaluate climate-related impacts. One factor that makes mountains unique is that, as the climate quickly with altitude changes over comparatively short horizontal distances, so do vegetation and hydrology (Whiteman, 2000). High biodiversity is exhibited by mountains owing to the complex topography in alpine environments (Winkler et al., 2016). Climate projections predict that global warming will not be consistent or uniform but will vary amply between different regions; specifically, climate change will be greater over land and at high altitudes and elevations (Gobiet et al., 2014). It has been explicitly highlighted by IPCC (2014) about the high sensitivity of mountain regions to climate change. Apart from increasingly threatening mountain ecosystems, climate change is also causing biodiversity loss, degradation of habitats, deterioration of freshwater quality and landscape modifications (Woodward et al., 2010) and all these portend a grave threat to the ecological integrity of terrestrial and freshwater ecosystems and the services rendered by them (Huss et al., 2017). All glaciers in the mountain areas of the World except those in Antarctica, Greenland, the Canadian and Russian Arctic, and Svalbard, comprise nearly 170,000 glaciers encompassing an area of 250,000 km^2 (RGI Consortium, 2017) with a total ice volume of 87 ± 15 mm sea level equivalent (Farinotti et al., 2019). These glaciers serve as source of freshwater not only the mountain population but also for the downstream areas as well. The runoff per unit area generated in mountains is on average nearly twofold as high as in lowlands (Viviroli et al., 2011) making mountains a vital source of freshwater in sustaining ecosystems and supporting livelihoods in and far-off areas beyond the mountain ranges. Availability of snow, glaciers and permafrost on the mountains usually exercise a powerful rheostat on the amount, timing and biogeochemical qualities of runoff. Freshwater availability is affected by the changes to the mountain cryosphere owing to climate change and it entails direct consequences for human population and ecosystems. An augmentation in average winter runoff over the past decades has been reported in Western Canada (Moyer, Moore, & Koppes, 2016), in the European Alps (Bard et al., 2015) and Norway (Fleming & Dahlke, 2014), owing to increased precipitation falling as rain under hot conditions. Runoff in summer conditions has been reported as getting reduced in basins, for instance in Western Canada (Brahney, Weber, Foord, Janmaat, & Curtis, 2017) and the European Alps (Bocchiola, 2014), but getting enhanced in High Mountain Asia (Reggiani & Rientjes, 2015; Engelhardt et al., 2017). Projections point out toward a continuous enhancement in winter runoff in many snow and/or glacier-fed rivers over the 21st century (Hock et al., 2019).

Climate change is already exercising increasing influence in mountain ecosystems. There can be variations in the response of mountain ecosystems keeping in consonance with the rate of climate change, the ecological domain and the biogeographical region (Müller, Baessler, Schubert, & Klotz, 2010). About 20% of the native flora of the European continent is contained by high mountains of Europe (Väre et al., 2003) and serve as hub of plant diversity, accommodating highly specialized vascular plants (Barthlott, Mutke, Rafiqpoor, Kier, & Kreft, 2005) and many native species (Stanisci, Carranza, Pelino, & Chiarucci, 2011). Climate change is thought to be one of the major threats to plant diversity above the tree-line. Model projections based on the basis of usage of climate change scenarios have predicted a dramatic diminution of appropriate habitats for high-elevation herbaceous plants (Engler et al., 2011) irrespective of the fact that thermal microhabitat assortments afford alpine species both shelter habitats and serve as thresholds as atmospheric temperatures increase (Scherrer, Körner, 2011). The gradual growth rates of long-lived alpine plants may result in a tardy reduction in the ranges of species, creating an "extinction debt" (Dullinger et al., 2012). Mountain forests are specifically susceptible to climate change owing to their long rotation cycles that may impede their adaptation capacity (Lindner et al., 2010). Nevertheless, benefits can also accrue to forests from global climate change, as enhancing concentrations of CO_2 and nitrogen deposition should accelerate photosynthesis rates and forest growth (Sperry et al., 2019). Assessments of the impacts of extensive climate-induced ecological changes occurring in high mountain ecosystems in the recent past decades were examined in an orderly manner only in earlier IPCC assessments (Fischlin et al., 2007) but not in the fifth Assessment Report (AR5) of the IPCC (Settele et al., 2014).

Two of the most palpable alterations comprise range shifts of plants and animals in Central Europe and the Himalayas but also for other mountain regions (You et al., 2018; He et al., 2019), and enhancements in species richness on mountain summits (Fell et al., 2017; Steinbauer et al., 2018) of which some have augmented in recent decades (Steinbauer et al., 2018); nonetheless, decelerating over the past 10 years in Austria (Lamprecht, Semenchuk, Steinbauer, Winkler, & Pauli, 2018). Undoubtedly, many alterations in freshwater ecosystems have been directly ascribed to changes in the cryosphere (Milner et al., 2017); nevertheless, extricating the direct impact of atmospheric warming from the effect of concurrent cryospheric change and independent biotic processes has been often challenging for terrestrial ecosystems (Frei et al., 2018; Lamprecht et al.,

2018). Alterations in climate in high mountains exert further stress on biota that are already affected by land use and its resultant change, direct exploitation and pollutants (Díaz et al., 2019; Wester et al., 2019). There is a need for species to alter their behaviors, including seasonal aspects, and distributional ranges to follow or track appropriate climate conditions (Settele et al., 2014). Climate change scenarios transcending mean global warming of 1.5°C comparable to preindustrial levels have been assessed to result in casting major effects on species richness, community structure and ecosystem functioning in high mountain regions (Hoegh-Guldberg et al., 2018). The extent and separation of mountain habitats (Cotto et al., 2017) that may at variance in accordance with the topography of mountain ridges (Graae et al., 2018), impacts disapprovingly the survival of species as they migrate across mountain ranges, enhancing at large the risks for many species from climate change (Dobrowski & Parks, 2016).

High mountains are specifically vulnerable to hazards arising from snow, ice and permafrost because these elements exercise key controls of mountain slope stability (Haeberli & Whiteman, 2015). Recent decades have witnessed increasing frequency of rocks detaching and falling from steep slopes within zones of degrading permafrost, especially in North America, New Zealand and Europe (Coe, Bessette-Kirton, & Geertsema, 2017). It emerges from the future projections that there would be an overall reduction in snow depth and snow cover duration at lower elevation; nevertheless, the anticipation of occasional large snow precipitation events taking place is predicted to remain possible throughout most of the 21st century (Hock et al., 2019). It has been estimated by Castebrunet, Eckert, Giraud, Durand, and Morin (2014) that there would be an overall 20% and 30% reduction in natural avalanche activity in the French Alps for the mid and end of the 21st century, respectively, under A1B scenario, relative to the reference period 1960–90; and interestingly, similar conclusions have been reached by Katsuyama, Inatsu, Nakamura, and Matoba (2017) for Northern Japan and by Lazar and Williams (2008) with regard to North America. The hazards occurring in high mountains also adversely affect lives and livelihoods of the people living in high mountains along with impacting those living in neighboring areas as well. Floods occurring due to glacier lake outburst over the past two centuries are said to have caused at least 400 deaths in Europe, 5745 deaths in South America and 6300 deaths in Asia (Carrivick & Tweed, 2016); nevertheless, these number are thought to be vastly skewed by individual large events taking place in Yungay, Peru (Carey, 2005) and Kedarnath, India (Allen, Rastner, Arora, Huggel, & Stoffel, 2016). Countrywide economic impacts accruing from glacier floods have been very grave in Nepal and Bhutan (Carrivick & Tweed, 2016). Besides, the dislocation of transportation corridors can affect trading of goods and services (Khanal, Hu, & Mool, 2015), and economic losses incurred on account of glacier floods can wield significant long-lasting impacts (IHCAP, 2017).

Climate-related alterations in high mountains has seemingly emerged as a new and developing area of research, with specific gaps in terms of methodically assessing their cost-benefits and long-term efficaciousness as "fit-for-purpose" solutions in the context of mountains. Undeniably, adaptation measures have reportedly been suggested in many studies (Adler, Huggel, Orlove, & Nolin, 2019); nonetheless, "improved inter-compatibility of successful adaptation cases, including the transferability of evidence for how adaptation can address both climate change and sustainable development goals (SDGs) in different mountain regions" (Hock et al., 2019), hold key to support a sound basis for future projections of adaptation to changes in the high mountains (Adler et al., 2019; McDowell, Huggel, Frey, Cramer, & Ricciardi, 2019).

9.6 Protecting biodiversity

Biodiversity is fundamental to human life on Earth; nonetheless, concurrently, there is unequivocal increasing evidence that biodiversity is being destroyed by human-induced activities at an unprecedented rate. The report of the Intergovernmental Science Policy Platform on Biodiversity and Ecosystem Services (IPBES) released in May 2019 has warned that around one million animal and plant species are threatened with extinction, many within decades, more than ever before in human history (Diaz et al., 2019). It is further reported that there has been a fall of at least 20%, mostly since 1900, in the average richness of native species in most major land-based habitats along with threat of extinction to more than 40% of amphibian species, almost 33% of reef-forming corals and more than a third of all marine animals (Diaz et al., 2019). Unsustainable pursuit of human development not only portends a threat to biological systems but also causes disturbances and decline in ecosystem services. Loss of biodiversity means losing species of animals, plants and bacteria and all other living organisms, including specific genes and physical environments (Adebayo, 2019). Describing biodiversity loss as "burning the library of life," Carrington (2018) alludes to losing evolutionary experienced garnered and developed over millions of years. Undeniably, 168 countries had signed the UN's Convention on Biological Diversity (CBD) in 1993 (CBD, 2020); nevertheless, current extinction rate is, standing between 100 and 1000 times, greater relative to the background rates and definition of "extinction" (Lovejoy, 2017). The increasing extinction rate has seemingly led researchers to believe that world has entered a "sixth mass extinction" phase (Baronsky et al., 2011; Saltre & Bradshaw, 2019). A mass extinction is characterized by losing over 75% of the estimated number of species living at a specific time (Baronsky et al., 2011). The increasing extinction rate is not only rendering the world poorer each day by loss of species but extinction rates are projected to increase in the future (Diaz et al., 2019).

The CBD defines biodiversity as: "The variability among living organisms from all sources including, inter alia, terrestrial, marine and other aquatic ecosystems and the ecological complexes of which they are part; this includes diversity within species, between species and of ecosystems" (UN, 1992, p. 3) The Earth is home to an estimated 10 million species of living organisms such as plants, animals and microbes (Diaz et al., 2019) and the degree of variability among them and their habitats is described by the term biodiversity. A sound, healthy and productive natural environment with a high level of biodiversity is a vital element for the functioning of ecosystems, and sequentially, for human survival (Pimentel et al., 1997). Together with the irreversible nature of the loss of biodiversity (Dirzo & Raven, 2003; Roe, Seddon, & Elliott, 2019), a general and well-established scientific consensus is that it is also accelerating due to human involvement (Díaz et al., 2019; MA, 2005). Ecosystems and species become threatened when land and water resources are exploited in the name of development. Simultaneously, humans are strongly reliant on ecosystem services provided by nature in ways not yet fully comprehended (Lubchenco, 1998; Nicholson et al., 2009). Winter, Pflugmacher, Berger, and Finkbeiner (2018) have opined that evaluating as to how to qualitatively and quantitatively measure the anthropogenic impact on the loss of biodiversity has become increasingly important. The 2018 Living Planet Report has reported a decline of 60% in wildlife population between 1970 and 2014, and this alarming rate of decline was estimated by Pimm, Russell, Gittleman, and Brooks (1995) to be in the range of 100–1000 times greater than the standard background rate of extinction prior to major human impact on the environment;

nevertheless, subsequent research has even suggested a higher estimate of 10,000 times (De Vos, Joppa, Gittleman, Stephens, & Pimm, 2015). It is in the wake of these alarming estimates of rapid loss of biodiversity that has led some researches to refer to the "Sixth mass Extinction" (Baronsky, et al., 2011).

Biodiversity plays a vital role in the functioning of ecosystems that heavily rely on the functional characteristics, variety and abundance of the existing organisms (Hooper et al., 2005). Loss of biodiversity entails the potential of not only to vastly change an ecosystem's properties but also sequentially to directly affect how it can deliver goods and services to humankind via ecosystem services (Díaz, Fargione, Chapin, & Tilman, 2006). When loss of biodiversity takes place in an ecosystem initially, its impact on the ecosystem services is relatively miniscule; nonetheless, the magnitude of that impact is expected to speed up with continuous losses of biodiversity and may compete with some of the major global change stressors like global warming and ozone depletion (Cardinale et al., 2012). In 2005, Millennium Ecosystem Assessment defined "ecosystem service" as "the benefits that people obtain from ecosystems" with instances being food, nutrient cycling, water purification, etc., and concurrently it also estimated that "approximately 60% of the ecosystem services...are being degraded or used unsustainably..." (MEA, 2005). Since 2017, a new and identical concept, "Nature's Contribution to People (NCP)" has been presented by the IPBES (Diaz et al., 2019) that endeavors to incorporate more insights and knowledge from the humanities and other scientific disciplines that ecosystem services thus far excluded. Concurrently, arguments are advanced by some experts that new NCP approach calls for more validation as well as operational guidance from the further research (Ellis et al., 2019; Kadykalo et al., 2019).

Loss of biodiversity is caused by both direct and indirect drivers. Direct drivers include: habitat change, climate change, invasive alien species (IAS), overexploitation of species and pollution. Indirect drivers, inter alia, are demographic, economic, socio-political, cultural and religious, and scientific and technological (MEA, 2005; Burkmar & Bell, 2015). Undeniably, there have always been alterations in biodiversity and ecosystem services owing to natural causes; nevertheless, the anthropogenic indirect drivers have come to rule the roost in causing biodiversity loss. Bulk of the onus for biodiversity loss is to be shared by human-induced activities such as hunting and habitat loss through deforestation, agricultural expansion and intensification, industrialization and urbanization (Cebaloss, Ehrlich, & Dirzo, 2017), and that together claimed a 30%−50% encroachment on natural ecosystems at the end of the 20th century (Vitousek, Mooney, Lubchenco, & Melillo, 1997). Human-induced anthropogenic activities entail the potential of simultaneously impacting the biodiversity, productivity and stability of Earth's ecosystems; however, there is no consensus on the causal relationship linking these variables. An analysis of data from 12 multiyear experiments by Hautier et al. (2015), with these experiments manipulating significant anthropogenic drivers, including plant diversity, nitrogen, carbon dioxide, fire, herbivory and water, demonstrates that each driver affects ecosystem productivity. Nevertheless, the stability of ecosystem productivity, as is revealed from the analysis, gets altered only by those drivers that change biodiversity, with a given diminution in plant species numbers resulting in a quantitatively identical reduction in ecosystem stability irrespective of which driver triggered the biodiversity loss. This analysis also suggests that alterations in biodiversity brought in by diverse environmental changes could be a decisive factor "determining how global environmental changes affect ecosystem stability" (Hautier et al., 2015). Now it is being acknowledged that relative research endeavors on different drivers are not well-aligned with their assessed impact, and an array of driver interactions are hardly taken into

9.6 Protecting biodiversity

account and that brings to surmise that research on drivers of biodiversity loss needs "urgent realignment to match predicted severity and inform policy goals" (Major et al., 2018).

Land use has come to be acknowledged as the prime cause of biodiversity loss (Souza, Teixeira, & Ostrmann, 2015; García-Vega & Newbold, 2020). Land is an indispensable and precious natural resource for the survival of human beings and multitude of biotic organisms, and also for the smooth functioning of terrestrial ecosystems (FAO, 2020a,b,c). Some of the most significant land use (LU) practices for humans include shelter, food production, freshwater balance and climate regulation (Foley, 2005). One-and-a-half-decade ago Millennium Ecosystem Assessment (MEA, 2005) had come to the conclusion that LU is one of the prime drivers of biodiversity loss and this conclusion has been reinforced by recent research (Diaz et al., 2019; Maier, Lindner, & Francisco, 2019). It has been argued that land-use changes have had the greatest relative negative impact on terrestrial ecosystems in the past half century (Diaz et al., 2019). Focus on this issue has been emphasized in one of the SDGs, vide SDG-15 dealing with "Life on Land" (Sachs et al., 2020; UN, 2015a, 2015b). Nevertheless, unprecedented decline of biodiversity at such a rapid pace is prone to whittle away the progress of many global societal and environmental goals. The trajectories currently in vogue make it discernible that most of the SDGs, including SDG-15, will not be achieved by their deadline of 2030 (Diaz et al., 2019; Sachs et al., 2020). In the process of using land, a major part of the terrestrial surface has been transformed by humans, primarily through conversion of natural land to agricultural land (Olofsson & Hickler, 2008). Currently 40% of the global land surface is occupied by croplands and pastures (Zabel, Putzenlechner, & Mauser, 2014), with total global areas having increased by 110% and 59%, respectively, from 1850 to 2015 (Houghton & Nassikas, 2017). The emissions from agricultural production, together with lost carbon storage at various LU activities, accounted for 24% of the total anthropogenic GHG emissions in 2010 (IPCC, 2019). The total global area of cropland has more than doubled during the period from 1850 to 2015, and the area of forest land shrank by 17% (Houghton & Nassikas, 2017). It was estimated by Searchinger et al. (2014) that the demand for food is expected to have a 50% increase by 2050. With human activities playing a part in nearly 40% of global terrestrial net primary productivity (Vitousek, Ehrlich, Ehrlich, & Matson, 1986) and major demographic stresses like population growth, urbanization, climate change, etc., the pressure on land and ecosystems is destined to intensify as demand soars (Searchinger et al., 2014).

The largest portion of total carbon emissions is shared by deforestation that is primarily caused by land-use alteration, accounting for 77% since 1850, and 85% between 2006 and 2015 (Houghton & Nassikas, 2017); leading drivers include swidden agriculture practices, frontier settlements and logging operations (Lambin et al., 2001). Forest land plays a pivotal role in climate mitigation owing of its ability to sequester carbon (Houghton & Nassikas, 2017), and as such, the destruction of native forests is a key driver for the loss of biodiversity globally (Panfil & Harvey, 2016). On an average, there is loss of approximately 0.6% of global forest cover every year (FAO, 2015). Nevertheless, there has been diminution in the rate of deforestation globally since its peak in the 1990s, after several hundreds of years of steady climbs (Houghton & Nassikas, 2017). The same downward trend has been found for swidden areas worldwide (van Vliet et al., 2012). Due to this unprecedented rate of deforestation, the looming mass extinction of species has been discussed widely in many case studies. While there is a dearth of empirical *meta*-analysis for the normalized global state of biodiversity loss, many seem to agree that up to half of all species are under threat (Diaz et al., 2019).

Alterations in land use may apparently seem to have only local effects, but they have global impacts as well, and as such, require both global and local policy responses (Hertel, West, & Villoria, 2019; Brown et al., 2014). There has emerged institutional framework in the aftermath of 1992 Earth Summit to address in the form of United Nations Framework Convention on Climate Change (UNFCCC) and the CBD under the aegis of the UN to address climate-related issues and ensuring conservation of biodiversity. The main long-term objective of UNFCCC is to "...stabilize GHG concentration in the atmosphere at a level that would prevent dangerous anthropogenic interferences with the climate system" (UNFCCC, 1992, p. 4). The CBD's three main objectives are: "the conservation of biodiversity, sustainable use of biodiversity, and the fair and equitable sharing of the benefits arising from the use of genetic resources" (UN, 1992, p. 3). Under the aegis of the CBD in 2010, 20 Aichi Biodiversity targets were agreed upon to be achieved by 2020 (Smith, 2015; CBD, 2010). Establishment of Kyoto Protocol in 1998, while emphasizing on envisaging concrete binding commitment targets toward emission limitations and reductions during certain commitment periods (UNFCCC, 1998), also required transparent and verifiable annual reporting of GHG emissions and removal of its sinks to the UNFCCC secretariat, including those accruing from, among others Land Use, Land Use Change, and Forestry (LULUCF) activities (EU, 2014; UNFCCC, 2013). LULUCF is one of the GHG inventory sectors that addresses emissions and removal of GHG, and its main categories include land-use changes between forest land, cropland, grassland, wetland, settlement, harvested wood products and others, with impact categories for carbon dioxide (CO_2), methane (CH_4), nitrous oxide (N_2O), nitrogen oxides (NOx), carbon monoxide (CO) and nonmethane volatile organic compounds (NMVOC) (Clilverd et al., 2019). Paris Agreement on Climate Change concluded in 2015 (UN, 2015a,b) requires all signatory parties to the agreement to hold the increase of global average temperature well below 2°C above preindustrial level, and pursue the effort to limit it to 1.5°C. Another significant milestone in terms of multilateral climate negotiations (Rajamani, 2016), Paris Agreement calls upon all signatory parties to communicate and maintain their domestic mitigation and adaptation ambitions in terms of nationally determined contributions (NDCs) to be submitted every five years from 2015 onwards, with each successive one representing a higher level of ambition than the previous.

The 17 SDGs presented as part of the UN 2030 Agenda for Sustainable Development (Rosa, 2017), and unanimously adopted in 2015 by all UN member states (UN-DPI, 2015), address the universal, global challenges currently confronting humankind. There has been an increasing amount of literature in recent years, linking climate change mitigation endeavors to some societal objectives, especially those among SDGs (Campagnolo & Davide, 2019; Iyer et al., 2018), and this finding is seemingly consistent with the findings of Northrop et al. (2016), who found climate actions all across all the NDCs aligned with more than 90% of SDG targets. The close alignment between NDCs and SDGs affords an opportunity for both to be implemented together, in a synergic and mutually supportive manner (Iyer et al., 2018). The key SDGs pertaining to LULUCF that affect biodiversity and ecosystem services comprise SDG 12 Responsible Production and Consumption and SDG-15 Life on Land. Despite slowing deforestation and more resources in conservation, the 2020 targets for SDG-15 are unlikely to be met (UN, 2019). Many proposed indicators—in terms of a measurement of value that can represent the state of something—for biodiversity assessment exist, although a general consensus for the most appropriate one is lacking (Petchey, O'Gorman, & Flynn, 2009). In the wake of global nature of biodiversity loss that is sought to be addressed, Ahmed, van Bodegom, and Tukker (2019) have opined that this would be one of the most

9.6 Protecting biodiversity

"irreplaceable and essential factors to incorporate quantitatively." The Nagoya Protocol on Access to genetic resources (CBD, 2011) aims at creating a legally binding framework that promotes transparency for CBD's third objective that emphasizes on sharing all benefits from the utilization of genetic resources equally and fairly.

The relationship between mitigation measures and adaptation measures is a significant factor in addressing climate and biodiversity issues. Mitigation tackles the problem of how to lower the physical emission of GHG, whereas adaptation addresses as to how people and ecosystems anticipate alterations in order to lessen potential damage (Vermeulen et al., 2013). Admittedly, both approaches aim at the same goal of reducing impacts of climate change; nonetheless, they differ in their methods and ought to be analyzed together, albeit many of climate change projects focus on one or the other approach (Kongsager, 2018). Irrespective of the fact that Kongsager (2018) has argued that all mitigation projects entail the potential of contributing to adaptation in some form; however, the most sustainable action accrues from a combination of both mitigation and adaptation (Duguma, Minang, & van Noordwijk, 2014). Nonetheless, adherence to drastic mitigation measures to bring climate change process to a standstill entails the risk of rendering the people in the poorest parts of the globe to suffer (Gordijn & ten Have, 2012). Besides, putting of the development potential of those regions that have made least contribution to climate change is prone to give rise to many questions about justice, compromise and responsibility (Gordijn & ten Have, 2012). IPCC's Special Report on Global warming of 1.5°C has reported with high confidence that the Earth will rise 1.5°C in mean temperature in the middle of this century if it continues to increase at its current rate (IPCC, 2018). Even if all emissions are brough to a naught at this juncture, the impact of climate change will still continue for centuries as a consequence of the emissions that already exist in the atmosphere (Hoegh-Guldberg et al., 2018). Moreover, many of its effects will be irreversible, such as the loss of ecosystems and biological species (IPCC, 2019). In view of continued inevitability of climate change, it becomes important to focus on the adaptation approach in the policy measures of accommodating the changing environment. The annual reporting mechanism of GHG inventories mandated by Kyoto Protocol is regarded as a tool of mitigation since it aims at documenting and tracking the progress of climate change goals. Similar mechanism seems important to track the status of biodiversity and ecosystem services that are irreplaceable and essential for human survival (Vermeulen et al., 2013). Apply a similar reporting mechanism for biodiversity and ecosystem services as it exists for the GHG can be helpful in enabling early detection of alterations in biodiversity and ecosystem services, and therefore, related aspects of the adaptation process could be furthered.

Recent years have witnessed increasing emphasis on biodiversity offsetting (BO) schemes, as an attempt to counter the decline of ecosystem, as a vital factor in mitigating negative impacts (Rainey et al., 2015; OECD, 2016a,b). BO is seen as a novel policy instrument for financing biodiversity conservation by aiming at compensating for the biodiversity losses that take place in one place from economic activities, by requiring the developers involved in the exploitation of natural resources, to finance the costs of environmental protection or restoration activities somewhere else (Koh, Hahn, & Boostra, 2019). Biodiversity offsets aim at being an incentive by companies to work with ecological conservation by implementing a system of internalizing external environmental costs, often called "polluter pays principle" (Koh, Hahn, & Ituarte-Lima, 2017a,b). According to Bennett, Gallant, and Ten Kate (2017), policies for offsetting biodiversity losses are already underway in at least 33 countries around the globe, collectively restoring and protecting 8.3 million ha of land. This extensive use of BO has reportedly been spurred by three major drivers. First factor or driver is the adoption of legislation and policies promoting compensation by national governments, the European Commission (2011) and the CBD (2008). Secondly, global financial institutions require biodiversity offsets to be considered as a condition of sanctioning funds for the projects (World Bank, 2017); and finally, the voluntary commitments from firms preemptively overseeing business risks (BBOP, 2012). It has been argued that with the help of BO, both the biodiversity conservation and human development could be achieved simultaneously (Koh et al., 2017a, 2017b). Nevertheless, ascribing a monetary value to things such as biodiversity and ecosystem services has been subjected to criticism as being the expansion via commodification (Hahn et al., 2015) of the market economy into the realm of nature. It is also argued that economic valuation methods and market principles are not ideologically neutral, rather they embrace an individualistic, competitive logic that incentivizes self-regarding behavior (Koh et al., 2019). Besides, Sullivan and Hannis (2015) opine that the market logic also leads alteration in human relationships with nonhuman nature. In the backdrop this criticism, Koh et al. (2019) suggest that governments thinking of adopting BO policies based on the recommendations of the CBD (CBD, 2016) can; therefore, "design biodiversity offset policies with a high or low market involvement to match their political-economic culture" (Koh et al., 2019).

9.7 Genetic resources

The Nagoya Protocol applies to the genetic resources of all organisms, excluding humans, within all the geographical areas of the contract parties. The Nagoya Protocol does not determine the genetic resources specifically, but in accordance with the CBD, genetic material and genetic resources are defined as follows: "Genetic material" refers to any material of plant, animal, microbial or other origin containing functional units of heredity. "Genetic resources" refer to genetic material of actual or potential value. Decree number 511/2014 of the European Parliament and the Council (EU, 2014) uses these same definitions.

Life on Earth is buoyed by communities "of plants, animals, and microorganisms interacting with each other within ecosystems, and with the physical environment". (Wilson, Carpenter, & Wilson MA, 1999). Scientists and researchers have evinced interest alike in plants, animals and microorganisms because they contain scientifically demonstrable value and that also explains as to why they have been able to attract the attention of business and industry in developed and developing countries (Wilson et al., 1999). A "gene" is described as being a very small or tiny portion or piece of matter located in in cells and that is found in chromosomes is made of DNA, an acronym of "deoxyribonucleic acid" that is the chemical at the center of the cells of living things, that controls the structure and purpose of each cell and carries the genetic information during reproduction (CALD, 2008, p. 412). Those DNA contain information for constructing proteins received by each animal or plant from its parents, and manage its physical development, behavior and so on (Glawzeski, 2000, pp. 300–301; Comfort, 2012). "Genetic material" means "any material of plant, animal, microbial or other origin containing functional unit of heredity" (CBD, 1992, p. 3). "Genetic resources" is a term identified by the CBD meaning "all genetic material of actual or potential value" (CBD, 1992, p. 3). Primarily, this term encompasses all living organisms—plants, animals and microbes—that carry genetic material potentially useful to humans, and genetic resources can be had from the wild, domesticated or cultivated sources; sources that are taken from "natural environments (in situ) or human-made collections (ex situ), for example, botanical gardens, gene-banks, seed banks and microbial cultural collections" (EU, n.d.c).

9.7.1 Types of genetic resources

Genetic resources identified for food, agriculture and forestry include both wild species and domesticated forms that reflect the major areas of use, such as crop production, animal husbandry, forestry, fisheries and microorganisms, and they grouped as: Plant Genetic Resources; Animal genetic Resources; Forest genetic Resources; and Aquatic and Marine Genetic Resources (EU, n.d.d; FAO, 2019a,b).

9.7.1.1 Plant genetic resources

Plant genetic resources are the most valuable and indispensable basic raw materials to meet the current and future needs of crop production programs. There are about 382,000 species of vascular plants worldwide (RBG Kew, 2017), out of which a little over 6000 have been cultivated for food (IPK, 2017). Of these, lesser than 200 species, as of 2014, had significant production levels worldwide, with only nine (sugar cane, maize, rice, wheat, potatoes, soybeans, oil-palm fruit, sugar beet and cassava) accounting for over 66% of all crop production by weight (FAO, 2017a,b). Given that an indicator entailing extensive application for monitoring within-species diversity is yet to be developed, the genetic diversity within crop species can be wide ranging and thus making precise extent of quantification of such diversity problematic. Genetic erosion and genetic susceptibility at within-species level are significant worrisome dimensions of concern. Genetic erosion within species has been defined as "the loss of individual genes and the loss of particular combinations of genes (i.e., of gene complexes) such as those manifested in locally adapted landraces" (FAO, 1997). The term "genetic erosion" is occasionally used in a narrow sense, alluding to the loss of genes or alleles, and sometimes in a broader sense, indicating the loss of varieties. Genetic vulnerability has been defined as "the condition that results when a widely planted crop is uniformly susceptible to a pest, pathogen or environmental hazard as a result of its genetic constitution, thereby creating a potential for widespread crop losses" (FAO, 1997). It can be discerned from the evidence available in the country reports, prepared for the Second Report on the State of the World's Plant Genetic Resources for the Food and Agriculture (Second SoW-PGRFA) (FAO, 2010a,b) that altogether the diversity prevalent in farmers' fields has been reduced and that potential threats to diversity are getting increased, with situation vastly varying depending on the country, location, type of production system, etc. There is substantial concurrence that, overall, the change over from traditional production systems applying farmers' varieties/landraces to "modern" production systems subject to dependence on officially-released varieties are contributing to genetic erosion (CGRFA, 2019). The situation is seemingly complex in view of the fact that most of farmers' varieties/landraces are reported to have either disappeared or have become rarer (Casanas, Simo, Casalas, & Prohens, 2017). The possibility of Many farmers who make use of planting modern varieties also continuing to maintain traditional varieties cannot be ruled out (FAO, 2019a,b).

Some research studies reflecting trends in genetic diversity within released verities also make it discernible the existence of a complex situation with some reporting no diminution, or even increases in diversity overtime (Bhandari, Bhanu, Srivastava, Singh, & Shreya, 2017). Undeniably, applying newly-adopted varieties could add genetic diversity to an agricultural system; nonetheless, the possibility of their completely substituting the original ones in some cases cannot be ruled out. While this scenario not only makes assessment of the balance of diversity difficult, it also renders it difficult to make definitive statements about trends in genetic susceptibility, as can be evidenced from the fact that more than half the country reports prepared for the Second SoW-PGRFA "indicate the presence of significant genetic vulnerability" (FAO, 2019a,b). While some regions have reported decrease in the diversity of crop wild relatives, in places where the climate conditions are changing but species migration is prevented by ecogeographical barriers, the crops and their species especially appear to be under threat (Aguirre-Gutierrez, van Treuren, Hoekstra, & van Hintum, 2017).

9.7.1.2 Animal genetic resources

Animal genetic resources are defined as genetic diversity in domesticated animal species entailing "economic or other socio-cultural values and found among species, among animal breeds within the species and in cryo-conserved material (embryos and semen). Genetic diversity refers to differences in allele frequencies and allele combinations among breeds of farm animal species and the spectrum of genetic variation within the breeds" (Kantanen et al., 2015). There is relatively small number of animal species domesticated for use in food and agriculture. The Global Databank for Animal Genetic Resources, hosted by FAO, records data on 38 species (CGRFA, 2019). An assessment of the status and trends of animal genetic resources for food and agriculture worldwide are made largely on the basis of summary statistics on breed risk status that entail the proportions of the world's breeds that are categorized as being at risk, not at risk, extinct or of unknown risk status according to the classification system used by FAO (FAO, 2020a,b,c). FAO has been publishing global data of this kind since 1993 and the most recent being Status and trends of animal genetic resources—2019 (CGRFA, 2019). This approach entails some limitations. In the first instance, this approach treats all breeds equally irrespective of their significance to the overall diversity of the species or importance in terms of other possible conservation criteria. Secondly, it facilitates registration of changes only when breeds move from one risk-status category to another. And thirdly, lack of regular updating of data on the size and structure of breed populations poses a foremost pragmatic restriction in the monitoring of risk status in many countries, specifically in the developing regions of the world (FAO, 2020a,b,c).

FAO is proposed "custodian" UN agency for 21 SDG indicators across SDGs 2, 5, 6, 12, 14, and 15. Accordingly, FAO is also "custodian" of indicator 2.5.2. that deals with "proportion of local breeds classified as being at risk, not at risk or at unknown level of risk of extinction" (FAO, 2017a,b, p. 23). As per data monitored by FAO, as of March 2018, of the total 8, 803 recorded breeds, 7745 were classed as local breeds, that is, reported present only in one country. A total of 594 local breeds were reported as extinct; and among extant local breeds, 26% were categorized as being at risk, 7% as not at risk and the remaining 67% as being of unknown risk status (CGRFA, 2019). It can be discerned from a comparison of data from 2006 to 2014 that there has been a slight reduction, say from 29% to 26%, in the proportion of local breeds categorized as being at risk of extinction (CGRFA, 2019). Nevertheless, FAO also suggests that the apparent trend is required to be interpreted with caution in view of the limitations in the state of reporting (FAO, 2020a,b,c). As per the Red List of Threatened Species prepared by the International Union of Conservation of Nature (IUCN), among the extant species considered as having been the wild ancestors of major livestock species, the most gravely at risk are the African wild ass (*Equus africanus*) and the wild Bactrian camel (*Camelus ferus*), both of these are categorized as Critically Endangered (IUCN, 2012). The wild water buffalo (*Bubalus arnee*) and the banteng (*Bos javanicus*) are classified as Endangered; the Indian bison (*Bos gaurus*), wild yak (Bos mutus), mouflon (*Ovis orientalis*), wild goat (*Capra aegagrus*) and swan goose (*Anser cygnoides*) are categorized as Vulnerable, and the European rabbit (*Oryctolagus cuniculus*) is classified as Near Threatened (IUCN, 2012). On the whole, it is seemingly apparent that a higher proportion of livestock wild relative species are threatened with extinction in comparison to mammalian and bird species in general. According to McGowan (2010), as of 2010, 25% of species in order Galliformes (chicken relatives), 83% of species in tribe Bovini (cattle relatives), 44% of species in subfamily Caprinae (sheep and goat relatives) and 50% of species in family of Suidae (pig relative) were classified as threatened.

9.7.1.3 Forest genetic resources

The notion of "forest genetic resources" refers to the "heritable materials maintained within and among tree and other woody plant species that are of actual or potential economic, environmental, scientific or societal value" (FAO, 2014). Admittedly, uncertainty pervades with regard to the total number of extant trees; nonetheless, it is estimated to be around 60,000 (Beech, Rivers, Oldfield, & Smith, 2017). Approximately 8000 species of trees, scrubs, palms and bamboo are listed by the Commission for Genetic Resources for Food and Agriculture (CGRFA) for its Report of the World's Forest Genetic Resources (SoW-FGR) (FAO, 2014) as based on the country reports submitted for SoW-FGR, and of this total number, nearly 2400 are actively managed for the products and/or services they deliver. More than 700 species are currently incorporated in the tree-breeding programs worldwide. The monitoring of the status and trends of forest genetic resources is facilitated at ecosystem, species and intraspecific levels. Nonetheless, these monitoring efforts are faced with many methodological and other constraints. While many countries face difficulties in assessing their primary forest area, factors like forest degradation, forest restoration and species composition are also difficult to monitor precisely (FAO, 2020a,b,c). Newton et al. (2015) have reported about launching of an initiative in 2015 known as Global Tree Assessment (GTA), an initiative launched by Botanic Gardens Conservation International (BGCI) and the IUCN Species Survival Commission Global Tree Specialist Group (IUCN-SSCGTSG) and it aims at providing conservation assessments of trees species across the globe by 2020 (BGCI, n.d.).

Factors such as conversion of forests to agriculture, unsustainable harvesting of trees for wood and nonwood products grazing and browsing, forest fires, and invasive species along with climate change have emerged as portents threatening and eroding forest genetic resources (FAO, 2014). Vast areas of land once covered by forests have been converted to other land uses in many parts of the globe. Despite the significant reduction in the rate of annual loss of forests over recent decades, the global forest area still continues to shrink. Undeniably, in the wake of absence of any reliable global monitoring system in place for monitoring intraspecific diversity in tree species, the FAO's report on State of World Forest Genetic Resources (SoW-FGR) makes available an overview of knowledge in this regard wherein schemes for genetic monitoring have been envisaged at global (Namkoong et al., 2002) and regional levels (Aravanopoulos et al., 2015); nevertheless, these schemes have not been put into practice and only a handful of countries have tested such schemes practically (Konnert, Maurer, Degan, & Kätzel, 2011). In view of the loss of intraspecific diversity in economically significant tree species having been a matter of key concern in forest management for decades, forest management practices can wield generic impacts on tree population and that situation calls necessity of making an assessment on a case-by-case basis (FAO, 2020a,b,c). Wickneswari et al. (2014) have opined that the range of the impact is dependent on the management system and the stand structure, as well as on the demography, biological traits and ecology of the species. For example, silvicultural interventions, such as the thinning of stands in temperate forests generally have limited generic consequences (Lefevre, 2004), and numerous silvicultural systems maintain generic diversity in tree population rather well (Geburek & Müller, 2005). Nevertheless, any alteration in evolutionary processes within tree populations as a sequel to or on account of forest management practices is prone to cast a more profound impact on the generic diversity of ensuing generations of trees (Lefevre et al., 2014).

9.7.1.4 Aquatic and marine genetic resources

Aquatic species live in freshwater and marine species live in saltwater. Thus, life in freshwater and saltwater survives in different environments. The fish are unable to survive in each other's environment because of the variance in physiology. Aquatic species, especially fish, survive in freshwater and live in streams, rivers, lakes, wetlands, etc., that entail salinity of less than 0.05%. Contingent upon the species, fish can survive in temperatures varying from 5 and 24 degrees Celsius, and they can also adapt to an alteration in habitat, such as the rise and fall in water levels, temperature and oxygenation levels (Wood, 2017). On the other hand, marine life alludes to fish living in seas and oceans, also known as saltwater fish, because they can only survive in waters entailing high levels of salinity, and coral reef and seagrass bed constitute habitats for marine species (Wood, 2017). Habitat of the fish is determined by their physiology. The way their organs operate exhibits that freshwater fish retain more salt within their bodies than in the water they live in and that is why they can survive in water containing less than 0.05% salinity, whereas saltwater fish lose water to their surroundings via the process of osmosis because they need to drink plenty of water to maintain a sound and healthy body and this what makes them survive in seas and oceans (Wood, 2017). Aquatic resources denote lakes, rivers, streams, springs, seeps, reservoirs, ponds, riparian areas and wetlands; and the fauna that reside within them, and aquatic resources comprise permanent, seasonal, flowing, standing, natural or man-made water bodies (Law Insider, n.d.). Aquatic genetic resources (AqGRs) for food and agriculture include DNA, genes, chromosomes, tissues, gametes, embryos and other early life history stages, individuals, strains, stocks and communities of organisms of actual or potential value for food and agriculture (FAO, 2019a,b: xxix). Even though, the term "marine genetic resources" (MGRs) has never been formally described (Vierros, Suttle, Harden-Davies, & Burton, 2016), it suggests a subset of "genetic resources," that have been defined under the CBD as "genetic material of actual or potential value" (CBD, 1992:3).

9.7.1.4.1 Aquatic genetic resources

Among the AqGRs worldwide, there are more than 31,000 species of finfish, 52,000 species of aquatic molluscs, 64,000 species of aquatic crustaceans and 14,000 species of aquatic plants (Leveque et al., 2008; WoRMS, 2018). Global capture fisheries harvested over 1800 species, including finfish, crustaceans, molluscs, echinoderms, coelenterates and aquatic plants in 2016 (FAO, 2020a,b,c). Among these enumerable global capture fisheries, were also included numerous genetically distinct stocks and phenotypes. Data with regard to fisheries and aquaculture available with FAO is based on the reports submitted by its member countries and it constitutes a valuable source of information on multiple aspects of aquatic species. Nevertheless, information is not often submitted at the species level that makes specifically problematic in regard to enumeration of inland fisheries, where more than half of overall production is not "designated" by species (Bartley et al., 2016). As per available data with regard to aquaculture, it can be discerned that more species are being farmed than ever before, particularly as more marine fish are being farmed currently in captivity (). The information based on country reports prepared for the State of the World's Aquatic Genetic Resources (SoW-AqGR) reveals about the farming of 694 species and other taxonomic groups (CGRFA, 2019). Data available with FAO, as of 2016, shows that approximately 598 species were used in aquaculture, and included 369 finfish spenfish (inclusive of hybrids); 104 mollusc species; 64 crustacean species; 7 amphibian and reptile species (excluding alligators, caimans or crocodiles); 9 other aquatic invertebrate species; and 40 species of aquatic algae (FAO, 2018a). Substantial variations across the world are discernible at the level of monitoring of species and populations harvested in marine and inland fisheries as well as those raised in aquaculture. As compared to monitoring of diversity in terrestrial livestock and crop sectors, the monitoring of diversity at intraspecies level in respect of aquatic sector still remains underdeveloped (FAO, 2020a,b,c).

9.7.1.4.2 Marine genetic resources

Seas and oceans contain a rich diversity of biological molecules within million species of plants, animals and bacteria (Bollmann et al., 2010, p. 114). Greiber (2011, p. 1) notes that acquiring genetic material from marine organisms represents a major source of diversity and novelty, and the exploitation of these genetic resources has received considerable attention in recent decades. In the wake of massive expansions in technological capabilities and the developments of advanced methods, scientists and bioprospecting companies are now appropriately equipped to explore a greater part of the marine biodiversity (OECD, 2016a,b). Admittedly, abundant diversity available in seas and oceans affords ample opportunities within scientific research and development that probably entails the potential of extending far beyond current available knowledge; nonetheless, owing to the difficulties, such as access, technology and work hours needed to develop novel products from marine organisms, this huge marine genetic diversity has until rather recently remained nearly unexploited (Broggiato, Arnaud-Haond, Chiarolla, & Greiber, 2014, p. 177). Increase in knowledge and augmentation in capacity for collecting and identifying biomolecules via intricate screening processes having been advanced to such a degree in recent decades that the potential for exploitation and exploration of marine organisms is now considered to be feasible. Apart from the prospects of the potential of garnering massive monetary gains a new drug represents, it is also an essentiality in order to confront the future challenges associated with the drug resistance of pathogenic bacteria, viruses, parasites and fungi (McIntosh, 2018). MGRs have also demonstrated their vital significance into fields of basic research, such as taxonomy, and barcoding to explore, identify and determine marine species into the taxonomic hierarchy (Fedder, 2013, pp. 15—16). Undeniably, findings from the utilization of MGRs is prone to yield many benefits for a range of sectors within scientific research; nevertheless, there are not many instances of a straight development path to be carved out of the sampling and collecting of marine organisms that leads all the way toward a commercial product derived from MGRs, despite often including various attempts

(Mostarda et al., 2016). Furthermore, two major benefits accruing from the research of MGRs, inter alia, include: expansion of basic scientific understanding of marine biology and biochemistry; and the economic incomes and success by way of marketing commercial products derived from the collection and sampling of genetic material (Kate & Laird, 2000). Besides the increasing emphasis in international law and science on integrated ecosystem-based management, marine biodiversity and advanced ecosystem processes, further scientific research on MGRs constitute a vital component of the advancement of knowledge that is essential to understand the complexities in the global oceans (Broggiato et al., 2018, p. 11).

An assessment of the state of the global marine fisheries is facilitated by FAO via the analysis of over 400 stocks of fish. Categorization of species targeted by marine fisheries is in accordance with whether these are overfished or optimally fished in a sustainable manner or underfished. As of 2015, 33.1% of fish stocks were estimated to be overfished, 59.9% to be optimally fished in a sustainable manner and the remaining 7% to be underfished (FAO, 2018a). The share of fish stocks within optimally fished in a sustainable manner or underfished decreased from 90% in 1974 to 66.9% in 2015 (FAO, 2018a). An equivalent analysis for inland fisheries is not made available by FAO. Monitoring of the inland capture fishery resources is rendered difficult due to various factors, including the dispersed character of the sector, involvement of multitude of people in the sector, the seasonal and subsistence nature of many small-scale inland fisheries, local consumption of the bulk of the catch along with informal trading locally, along with the fact that local populations are prone to be impacted by activities other than fishing, including stocking from aquaculture and diversion of water for other uses like agriculture and hydropower development (FAO, 2012). In the absence of any dedicated program of FAO to address the state of inland fisheries, Committee on Fisheries under the aegis of FAO had recommended for "the development of an effective methodology to monitor and assess the status of inland fisheries, to underpin their value, to give them appropriate recognition and to support their management ... [and] requested that FAO develop this assessment methodology, including broader ecosystem considerations that impact inland fisheries" (FAO, 2016a). FAO is reported to have developed a partnership on "Hidden Harvest-2" initiative to improve the information baseline on small-scale baseline and the outcome of this program "will also provide important baseline information and deliver methodologies for further work in the context of road map leading up to the International Year of Artisanal Fisheries and Aquaculture in 2022" (Committee on Fisheries, 2018). Many of the conservation and trade organizations are reportedly assessing the status of most of the marine and aquatic species. Over 1, 300 marine species including, plants, fish, molluscs, crustaceans and other invertebrates, along with 5200 wetland species have been classified by the IUCN Red List in Endangered, Threatened or Vulnerable categories (IUCN, 2019). Amongst the freshwater fish species, excluding broader range of biodiversity, out of 5785 species about which assessment for The IUCN Red List by the close of 2011 was made, 60 species were considered extinct, 8 Extinct, 8 Extinct in the Wild and 1679, about 29.3%, as threatened (Carrizo, Smith, & Darwall, 2013). Assuming that the 1062 species categorized as data deficient are threatened at the same scale for which data is available, thereafter, the proportion of threatened species would account for 36.1% of the total (Carrizo et al., 2013). Nevertheless, not all the fish species that have been assessed by IUCN are used for food and agriculture. CITES (The Convention on International Trade in Endangered Species of Wild Fauna and Flora) maintains data on status of species that are internationally traded, and many species used in food and agriculture, such as tunas, sturgeons, and sharks, are listed in CITES Appendices (CITES, 2019).

9.7.1.5 Access and benefits sharing

The notion of access and benefit sharing (ABS) of genetic resources is seemingly designed to enable fair disbursement of benefits between the users like universities and biotech companies, and the providers, mostly the biodiversity rich countries, so as to facilitate both open new avenues for innovation and generate incentives for biodiversity conservation (CBD, 2010a; Sirakaya, 2019). Facilitation of accessibility to genetic resources is indispensable for research pertaining to conservation of plant genetic resources along with research and development to agricultural products and evolved crops that can be able to adapt to new weather conditions as a consequence of climate change (CGRFA, 2015). Accordingly, enablement of access to genetic resources in general along with benefit sharing accruing from that accessibility is a prime element for sustainable development thereby ensuring secured research, over and above environmental sustainability and uninterrupted resource supply (CBD, 2011). ABS is rapidly gaining international traction as a system under public international law

Bulk of the crop plants were developed over millennia in various regions of the world and a major chunk of those regions falls in the territories that are now known as developing countries. Farmers and breeders in these countries have traditionally relied on the notion of "open access" to genetic resources from these products (Ranganathan, 2007). In fact, genetic resources among the world's most important resources that used for significant multiple purposes, such as in the life-support systems, ecosystem services and cultural objects as well as production inputs and goods (Ranganathan, 2007). Genetic resources entail monetary value in terms of food, medicines, chemicals, fibers, structural material, fuel, and other purposes (FAO, 1997; Guruswamy, 2017, pp. 127–129). Apart from continuing to be the key component for the improvement of agricultural crops, genetic resources are reliable source for the supply of traditional largely plant-based treatment for primary healthcare and for other products like pharmaceutical crops protection products and perfume for the consumption of the bulk of population (CBD, 2010b). The immense significance of genetic resources and their multiple uses gave rise to issues pertaining to controlling, utilization and safeguarding genetic resources. Undeniably, the management and protection of genetic resources have often constituted a significant challenge confronting that are poor in genetic resources as well as those countries that are rich in genetic resources (Ogwu, Osawaru, & Abana, 2014). Construed in a historical perspective, the "open access" approach to these genetic resources incurred as a result not only their depletion but also the impoverishment of the countries that delivered biological resources, and added to that, the environmental degradation that contributed to extinction of species (Oli, 2009).

Genetic resources had already by the early 1990s come to be regarded as "green gold" that entailed high monetary value due to their genetic potential (Ishaq Khan, 2007: 94). Admittedly, almost all the developed countries are technologically well-equipped; nonetheless, most of the genetic resources are commercialized outside the countries where they were originally accessed (Ishaq Khan, 2007, p. 96). Developed countries possess technology but fall short of genetic resources while developing countries abound with genetic resources and lack appropriate technology (Rubenstein, Heisey, Shoemaker, Sullivan, & Frisvold, 2005). At the outset of the 1990s, it was hoped that if a share of benefits arising out of their utilization went back to "gene" rich countries, this could generate incentives for conservation and minimize destructive activities that were of least economic benefit (Richerzhagen, 2010, pp. 205–208). Endowed with abundant genetic resources, the countries of the South even though accepted the significance of looking after biodiversity, concomitantly, they regarded outside endeavors to impose a limit on internal development as an "intrusion on sovereignty," especially in the wake of their already being confronted with conditions of real poverty, the call for delimitation of vast tracts of land for conservation or protection of biological resources or for so-called "sustainable development" suddenly appeared for these countries to be something being "externally imposed" to obstruct their onward march on the path of economic development (Richerzhagen, 2010, pp. 135−136). Consequently, the countries of the South, possessing abundant genetic resources and lacking in technology, raised the bidding by letting the "gene poor" and "technology rich" North countries realize that the South would embrace sustainable development only if "...(t)he North would assume the costs" (Richerzhagen, 2010, p. 136). It was realized that issues pertaining to genetic resources could be a matter of difficulty between the interests of developed and developing countries, and there was a need to find an appropriate equilibrium between the interests of developed and developing countries (Pena-Neira, 2009). The management of genetic resources being an issue of global concern because it constitutes a significant issue confronting both use and provider countries, called for a global solution that was acceptable to all stakeholders.

Developments in the realms of science and technology, over time, facilitating the utilization of natural resources in ways other than the known, proved instrumental in spurring the need to modernize the existing laws and engendered new expectations in natural-resources-rich countries with regard to benefits they could garner from the new utilization ways of their biodiversity. With the coming into being of the CBD under the aegis of the UN in 1992 along with incorporation of certain provisions in respect of genetic resources, an obligation was also established to share in a fair and equitable manner the benefits derived from the use of genetic resources (CBD, 1992, Article 15.1: 9). The mention of the genetic resources in the CBD treaty text as the object of access, and the results of research and developments as benefits to distribution (CBD, 1992, Article 15.7: 10), envisaged a distinction respecting the obligations derived from the utilization of biodiversity. It is noteworthy that even prior to the establishment of the CBD, the FAO had established rules with regard to the use of plant-related genetic resources in 1983 entitled International Undertaking on Plant Genetic Resources (IUPGR) (FAO, 1983). Vide Article 1 of the IUPGR, "plant genetic resources are a heritage of mankind and consequently should be available without restriction" (FAO: 1983, p. 1). This statement reflected the way property rights over genetic resources were traditionally perceived; nevertheless, over time, this view underwent a change and in 1991, the recognition of sovereign rights over plant genetic resources were incorporated in the IUPGR (FAO, 1997, p. 3).

Admittedly, the CBD in its 1992 treaty text included these changes; nonetheless, the CBD could be seen as the first international instrument destined to acknowledge sovereign rights of countries over all genetic resources within their respective frontiers. Nevertheless, some scholars, such as Mgboji (2003), have opined that it would be erroneous to argue that CBD and FAO created a new regime of state sovereignty over biodiversity because they simply reaffirmed an inherent preexisting right of state jurisdiction over plant life forms. Therefore, it can be surmised that the CBD did not create sovereignty rights over nature, rather extended those existent rights to the use of Goji "new" elements of biodiversity, in the form of genetic resources (GRs) and traditional knowledge (TK). One of the CBD's objective is the achievement of ABS (CBD, 1992, Article 1: 3), has also been discussed in other international fora as well. Accordingly, there are currently three different ABS systems. In the first place, there is FAO system as envisaged in the International Treaty on

Plant Genetic Resources for Food and Agriculture (ITPGRFA) (FAO, 2009), and it is designed to rule on the use of plant genetic resources for food and agriculture. Secondly, there is World Health Organization's (WHO) system that is devised in the Pandemic Influenza Preparedness (PIP) Framework (WHO, 2011) that was created to regulate the use of genetic resources for the development of vaccines in a pandemic situation. And finally, there is the CBD system that is developed in the Nagoya Protocol (Nagoya Protocol, 2011), and it aims to govern the utilization of genetic resources for all purposes different from those regulated by the FAO and the WHO. These three systems entail a set of rules that embrace the international ABS system that were created to complement each other and be mutually supportive.

The underlying rationale for ABS is that access to genetic resources is subject to the prior informed consent (PIC) of the Party making available the resources and, where made available, shall be upon mutually agreed terms (MATs) (Nagoya Protocol, 2011, p. 6; Grieber et al., 2012). Each country, on the grounds of the exercise of sovereign rights through its respective national laws, can decide about the conditions under which ABS will operate in the concerned country's territory (CBD, 1992, pp. 9–10). This stipulates that compliance with ABS obligations ought to be achieved by each country at a national level. This aspect, known as the bilateral national approach to compliance with ABS obligations, has been acknowledged as one of the impediments for the realization of benefit sharing. The subject of Genetic resources is not the only subject-matter of ABS. The knowledge of indigenous local communities (ILCs) pertaining to the use and management of biodiversity, also known as traditional knowledge (TK), and when associated with genetic resources, also falls within the purview of the ABS (CBD, 1992, p. 6). Issues in respect of TK are clarified vide Article 3 of the NP (Nagoya Protocol, 2011, p. 5). Significantly, an obligation is envisaged in the NP for the countries to take into account customary law in the framing of national law and policy (Nagoya Protocol, 2011, p. 9). Undeniably, this could be construed as a step forward in the protection of traditional knowledge and the rights of the indigenous local communities; nonetheless, some authors, such as Tobin (2013) seem less sure when it is pointed out that as the negotiations for the agreement on one or more instruments for the protection of TK advanced within the WIPO Intergovernmental Committee on Intellectual Property, Genetic Resources, Traditional Knowledge and Folklore, allusions to customary law started disappearing from the draft instrument. Deliberations on the scope of the ABS not only entail GR and TK; rather, issues such as byproducts, derivatives and information are also under debate.

Advances in science and technology have been accompanied by economic expectation, especially with regard to new forms of utilization of GR and TK. It is in this backdrop, as opined by Oberthur et al. (2011), that such advances also envisaged an economic incentive to introduce patent protection for biodiversity-based inventions, as a result, the ABS debate was placed into World Intellectual Property Organization (WIPO) forum where the relationship between the CBD and the Agreement on Trade-Related Aspects of Intellectual Property Rights (TRIPS) has been discussed for many years. In order to have a better understanding of this debate, it seems significant to emphasize that, irrespective of the recognition of the right to benefit from the use of GR and TK, neither the CBD nor the NP establish or recognize states' property rights over their GR or of ILCs on their TK, nor the TRIPS (Phillips, 2016). According to Tyedt and Schei (2014), acknowledgment of intellectual property rights (IPRs) over TK in global forums like TRIPS and the WIPO have often met with failure. In the opinion of Oguamanam (2008), this could have been triggered by the lack of political will on the part of industrialized countries who are the direct beneficiaries of nonstop appropriation of local knowledge, the informal nonscientific way TK is generated, the asymmetrical relationship of power between industrialized countries and ILCs, and because IPRs are construed as a capitalist instrument that is unsuitable for societies that operate in a largely communal model outside or on the fringes of the contested paradigm of the market economy framework. This explanation helps to elucidate why, in the wake of lack of recognition of IPRs over GR and TK, and the not so appropriate integration of the ABS rules with the IP system (Oberthur et al., 2011), there have been some claims where IPRs over inventions using GR and/or TK could be attained reportedly without complying with ABS obligations, specifically with regard to benefit sharing. This is also seen as another impediment in the realization of ABS. Irrespective of the ostensible simplicity of ABS in terms of mutual exchange of resources between two countries, many scholars are in agreement that this system does not work properly (Vogel, Muller, & Angerer, 2015; Dutfield, McManis, & Ong, 2018).

This clarification helps to explain why, given the lack of recognition of IPRs over GR and TK, and the poor integration of the ABS rules with the IP system, there have been some claims where IPRs over inventions using GR and/or TK could be obtained allegedly without complying with ABS obligations (PIC, MAT, and benefit sharing). As could be predicted, this is perceived as another obstacle to ABS realization. Despite the apparent simplicity of ABS (mutual exchange of resources between two parties), it seems that the majority of authors agree that this system does not properly work. ABS is delineated as a complex matter by Oguamanam and Jain (2017) because it contains controversial aspects such as mandatory obligations under the patent system to disclose the origin of the resources and proof that prior consent was obtained. Undeniably, the goals of ABS have managed to garner broad support in the global diplomacy space, as has the "innovative if unproven approach for creating incentives through ABS for the biodiversity conservation" (Laird et al., 2020); however, the real devil lies in the details. In the recent past it had become evident that commercial demand for genetic resources was not enough to incentivize biodiversity conservation. Even the transactions that have been materialized under the aegis of the CBD have yet to yield desired benefits for conservation (Laird & Kate, 2002; Wynberg & Laird, 2009; Prathapan, Pethivagoda, Bawa, Raven, & Rajan, 2018). Inability of the CBD in commencing work in earnest on digital sequence information (DSI) is lamented at by Laird and Wynberg (2018). Relative to DSI, the ABS mechanism was designed to physical samples, and the term "DSI itself remains a negotiated placeholder term, the planning and scope of which remain in dispute" (Laird et al., 2020). Keeping in view the increasing international traction being garnered by the ABS mechanism, Laird and colleagues (2020) emphasize on the urgent need for a global institutional and conceptual framework for ethical research and commercialization, and the environmental and social implications of scientific advances, and they also call for "a new approach for ethically sharing the benefits of science and technology" (Laird et al., 2020).

9.8 Conservation of wildlife

Undeniably, in common parlance, the term "wildlife" often refers to native or wild animals (Fryxell, Sinclair, & Caughley, 2014, p. 1); nonetheless, wildlife has been defined as "a collective term for wild animals and plants living in their natural habitat, not under human control" (Rees, 2013, p. 1208). In this backdrop, then the question arises as to how wildlife is related to "conservation." In this regard, Hambler and Canney (2013, p. 2) has endeavored to define conservation as "the protection of wildlife from irreversible harm." The notion of wildlife conservation has evolved over many centuries in multiple forms across the globe (Hambler & Canney, 2013, p. 2). Advocacy for the preservation of "wildlands and forests" throughout the English-speaking world was supported by John Evelyn in his influential book entitled "Sylva" published in 1664 (Cited in McKusick, 2013, p. 110). And subsequently, the then President of the United States Theodore Roosevelt popularized the term "conservation" in the United States as "the wise use of the Earth and its resources" (Hambler & Canney, 2013, p. 2). In the wake of increasing understanding of the natural world, there arose environmental movements in Europe, including Australia, in the nineteenth century in the conservation and public health campaigns and organizations (Rootes, 2008). The notion that biodiversity augmented ecological stability reportedly spurred a substantial body of scientific research that consequently galvanized the impetus for wildlife conservation (Mainka & Trivedi, 2002; Mikkelson, 2009, p. 1; CBD, 2014). Change in land use, especially for agriculture and urban development is regarded as the prime cause of biological degradation, depletion of wildlife ecosystems along with loss of wildlife species (Johnson, 2006; McShane et al., 2011; Hambler & Canney, 2013). Australia is reported to have experienced implications of the effects of change in land use in terms of more mammal extinction than any other continent since European colonization (Dickman, Pimm, Cardillo, Macdonald, & Service, 2007). Apart from making available clean air and water, wildlife ecosystems also provide cultural and "esthetic" fulfillment, a supply of renewable resources and economic benefits via ecotourism opportunities (Hambler & Canney, 2013, p. 11). Accordingly, destruction of wildlife ecosystems is prone to threaten human ability to sustain life (Lindsay et al., 2008).

9.8.1 Threats to wildlife species

Wildlife species and ecosystems are under constant threat from multiple factors like human–wildlife conflict, habitat loss, illegal trafficking in wildlife, invasive species, climate change, and pollution (Prem, 2020; DES-Australia, 2018). A brief understanding of these factors is helpful in appraising the magnitude of the problems confronting wildlife.

9.8.1.1 Human–wildlife conflict

In the wake of ruthless competition for natural resources coupled with shrinking and fragmenting habitats, the question of coexistence between humans and potentially dangerous predators has been instrumental in bringing the challenge of coexistence with wildlife into limelight (Chapron et al., 2014). Undeniably, conservationists are said to be equipped with toolkit for the pragmatic mitigation of conservation conflicts; nonetheless, those toolkits are seemingly inadequate to deal with basal cultural and social dimensions (Macdonald, Loveridge, Rabinowitz, Macdonald, & Loveridge, 2010). This could be attributable to the fact that the major emphasis of the bulk of existing research on human–predator interactions being focused on conflictual relations and particularly on human–wildlife conflict, and besides that, much of the conservation research has been "driven by natural scientists concentration on the biology and behavior of predators and prey and the impacts of predators on prey" (Pooley et al., 2017). It has been pointed by Madden and McQuinn (2015) that in recent times, social science methods have been commandeered to improve the human cost–benefit ratio of cohabiting with such animals. Even within the ambit of research on interactions between humans and predators, people and predators are prone to be studied distinctly and with different ontologies, epistemologies, and methodologies (Ghosal & Kjosavik, 2015).

Research focusing on the social dimensions of conservation points toward quantitative social science, bringing out ideas from social psychology and economics, designed to exploring and changing beliefs and attitudes impacting unwarranted behavior, usually to protect wildlife rather than humans (Hayman, Harvey, Mazzotti, Israel, & Woodward, 2014). Studies focusing on the roles of culture and value in human−wildlife coexistence empirically are occasional and the humanities are almost completely nonexistent in the field (Mekonen, 2020). It is almost an established fact that while ostensibly conservation conflicts entail adverse human−wildlife relations, construed in a broader context they often ruminate adverse human−human relations, wherein the views of conversationists come into conflict with those of others with seemingly discordant goals, and in both cases, one party is understood to proclaim its interests at the expense of another's (Draheim, Madden, McCarthy, & Parsons, 2015; Redpath, Bhatia, & Young, 2015a; Redpath, Gutierrez, Wood, & Young, 2015b). It is in this backdrop that it has been emphasized for the adoption of a broader interdisciplinary approach to human−wildlife conflicts (Agelici, 2016); nevertheless, notions about the utility of such a broader approach are still mostly conceptual or retrospectively applied (Pooley et al., 2017). Recent years have been characterized by endeavors at undertaking individual studies that explore human−predator interactions in vernal, interdisciplinary and more integrated ways; nevertheless, these kinds of studies are widely scattered across different disciplines (Ghosal & Kjosavik, 2015).

Human−wildlife conflict is generally construed as conflict that takes place between people and wildlife (Woodroffe, Thirgood, & Rabinowitz, 2005); actions by humans or wildlife that wield a negative effect on the other (Conover, 2002); threats posed by wildlife to human life, economic security, or recreation (Treves & Karanth, 2003); or the perception that wildlife poses a threat to human safety, health, food, and property (Peterson, Birckhead, Leong, Peterson, & Peterson, 2011). The term wildlife is broadly described as nondomesticated plants and animals (Reidinger & Miller, 2013), even though domesticated and feral animals are sometimes included in the human−wildlife conflict literature. Wildlife damage-control or management is outlined as the science and art of lessening the negative consequences of wildlife while maintaining or enhancing their positive aspects (Reidinger & Miller, 2013), and is usually construed as synonymous with human−wildlife conflict mitigation (Reidinger & Miller, 2013). Various scholars mention that the notion of human−wildlife conflict is convoluted by inherent tensions from human−human conflicts over conservation and use of resources (Redpath et al., 2015a,b). Another impediment is that human interactions with wildlife are generally constructed negatively even if significant positive benefits—recreational, educational, psychological, and ecosystem services are accruing or available (Soulsbury & White, 2015). Consequently, a growing convergence surrounds the phrase human−wildlife conflict and coexistence to imply the acknowledgment of both problems and solutions (Dickman et al., 2007), even though some authors raise the question as to whether coexistence is more exactly cooccurrence (Harihar et al., 2013).

Significant repercussions accrue from human−wildlife conflict for human health, security and safety, and welfare as well as for biodiversity and ecosystem health. Impacts of human−wildlife conflict on humans can be both direct or indirect. Human injury or death can occur when animals bite, claw, gore, or else launch direct attack on people; during the occurrence of collisions between animals and automobiles, trains, buses, planes, boats and ships, and other vehicles, and from the transmission of a zoonotic disease or parasite (Conover, 2002). Direct material and economic loss and damage to crops, livestock, game species, and property can accrue as a consequence of human–wildlife conflict (Loveridge, Wang, Frank, & Seidensticker, 2010). Indirect impacts of conflict that are usually more difficult to gauge, inter alia, comprise opportunity costs to farmers, and rangers entrusted with the task of guarding crops or livestock, reduced psychological well-being, disruptions of livelihoods along with insecurity of food (Hoare, 2012; Barua, Baghwat, & Sushrut, 2013). It is suggested by Soulsbury and White (2015) that there are variations on a continuum in human–wildlife interactions from positive to negative, in intensity from little to severe, and in frequency from rare to common. Undeniably, attacks by giant predators like tigers, lions and sharks on humans are relatively not so common in present times; nonetheless, such attacks, as and when occur, can be lethal and trigger strong public resentment (Woodroffe et al., 2005). On the other hand, conflict between people and common garden pests or birds like geese could be more common but provoke less concern (Nyhus, 2016). Entailing the probability of highly variable conflict frequency within and amongst the geographic regions, some households or farms within reach of a local community may happen to suffer slight damage whereas neighborhoods could witness additional killing events or sufferings, especially in the eventuality of a predator killing more animals in a single attack (Dickman et al., 2007), or more properties could be saved from damage as compared to others. Extinction constitutes the most extreme biological impact. While informing that multiple terrestrial and marine vertebrate species have become extinct in recorded history, Dirzo et al. (2014) note that populations of many remaining species declined in abundance. According to Ripple et al. (2014), the diminution of large, predatory animals has brought forth cascading ecological repercussions for other species and ecosystem services, and many of these declines are linked to conflict with humans (Nyhus, 2016).

Approaches designed to tackle negative impacts of human–predator conflicts, inter alia, include lethal control or translocations of presumed problem animals, facility of information about predator behavior and technical fixes for preventing damage (McManus, Dickman, Gaynor, Smuts, & Macdonald, 2015), and working out financial mechanism to counterbalance impacts (Dickman et al., 2007). Endeavors at discerning the behavior of harmful predators determine the most efficacious mechanisms for minimizing attacks, and these include safeguarding livestock or making available safe water-collection points to avoid crocodile attacks, and sensitizing local communities about adhering to methods to lessen their susceptibility, etc., have had success to some extent (Marker & Boast, 2015). Concomitantly, such approaches have also met with failure in certain places for an array of reasons, including inability to elicit participation of local people, high opportunity costs of effectual methods of safeguarding livestock, and aversion to apparent breaches of freedom of behavior (Barua et al., 2013), or as a consequence of epistemological incongruities over as to what spurs predator attacks (Pooley, 2016). Actual damage is generally minimal in cases where negative attitudes to predators are articulated as objections to the damage caused by them. In certain cases, it is the apprehension of an attack by an animal to be presumed as targeting humans, especially in the case of sharks, and helps in promoting such attitudes (Neff, 2012). In other cases, it can be people's dislike or scunner of a risk they presume is thrust upon them by an outside authority, for instance, conservationist authorities reintroducing predators to a specific area (Dickman & Hazzah, 2016). It is suggested by Kansky and Knight (2014) that impalpable costs, such as psychological costs of danger, are the most significant variables delineating attitudes to carnivores—importantly more so than palpable costs, in terms of direct monetary losses. Cavalcanti, Marchini, Zimmermann, Gese, and Macdonald (2010) have pointed that maltreatment of jaguars (known as *Panthera onca*) in Brazil is scarcely related to the monetary effects of livestock depredation than conservationists anticipated and bore more relevance to the cultural and social perceptions of probable threat and the entertainment and status associated with jaguar hunting.

It can be discerned from the suite of surmises, analyses and advice on impact diminution seldom essentially help in the resolution of conservation conflicts. According to Pooley et al. (2017), and underresearched subject is why scientifically sound mitigation measures are so frequently ignored or discontinued. Suggestions based on the evaluative research on conservation-conflict mitigation indicate that the peripheral or superficial impacts of predation generally hide diversity of rudimentary issues pertaining to different epistemologies, historical contexts, and ascertain differences that are "beyond the competencies" of natural scientists to tackle (Madden & McQuinn, 2015; Dickman & Hazzah, 2016). It is in this backdrop that Pooley et al. (2017) have drawn attention to the existence of important gaps and shortfalls in the current understanding of and approaches to "mitigating the more intractable of these human–predator conflict scenarios" (Pooley et al., 2017).

9.8.1.2 Habitat loss

The notion of habitat alludes to an area that is endowed with resources and conditions present to generate occupancy by a given organism (Hall, Krausman, & Morrison, 1997), and these resources and conditions embrace food, water and any specific factors required by a species for survival and reproductive activities (Leopold, 1933; NRC, 1995). Splitting of a large expanse of habitat into a number of smaller patches of smaller total area results in habitat fragmentation, and these smaller patches are separated from each other by a matrix of habitats unlike the original habitat (Wilcove, McLellan, & Dobson, 1986). Habitat fragmentation is explained by alterations in habitat configuration and can be unattached from or over and above the effects of habitat loss—a diminution in habitat abundance (Fahrig, 2003). Human contribution to habitat loss and fragmentation takes place in an array of ways. Intensity of land use has been on the increase in the wake of burgeoning human population worldwide and the resultant increase in demand for food production (Donald & Evans, 2006). Alteration of the landscape matrix by humans either for economic productivity or urban development has been facilitated in ways and manners that affect the spatial density, diversity, and quality of wildlife habitat (Radeloff et al., 2005). In the eventuality of habitat patches becoming smaller and more isolated, the resultant impact is discernible in terms of the ability of wildlife being affected to disperse—a factor that further gets complicated by erection of physical barriers like fencing and roadways that dissect the landscape (White, Michalak, & Lerner, 2007). In order to deal with such situations, it devolves on the land managers to have cognizance of the knowledge of the habitat requirements for a specific species or group of species of interest so that informed decisions are made as to how landscape changes would influence ecological communities and then adhere to undertaking adequate steps to ensure the management of those habitats in a way that delivers benefits to wildlife (Krausman, 1999).

9.8.1.3 Wildlife trafficking

Wildlife trade that has come to be regarded as phenomenon that has emerged in the aftermath of the Second World War (Mackenzie, 1988), and it is emerging as a major cause for the global decline of wild fauna and flora. Global trade in wildlife and their products, estimated at approximately $12 billion annually at the cusp of the 21st century (Menon & Kumar, 1998; Oldfield, 2002), of which ≥30% was illegal (Oldfield, 2002), is thought to be second only to narcotics and illegal arms trade (Reeve, 2002). Consumer demands for wildlife have recorded increase and diversified from the closing part of the 1980s. Fascinatingly, wild animals and their products have been witnessing phenomenal hike in price tags since the 1990s owing to their augmented sparseness and other factors spurring their demands (Hillstrom & Hillstrom, 2003). It can be discerned from some studies in the recent past (Hillstrom & Hillstrom, 2003; Thapar, 2003) that there has been augmentation of demands for imported wildlife in Europe, North America, China, South Korea, and Japan. In recent times, trafficking in wildlife has become one of the most pressing threats confronting many endangered and threatened species. Growing demand for consumption, status and traditional medicine has fueled the price tags of illicit wildlife trade that is estimated between $7 and $23 billion per year making wildlife trafficking one of the world's most lucrative businesses (GEF, 2020). While referring to complicated illegal trafficking structures that facilitate the killing, transport and sale of prized species like pangolins, rhino and elephants, Rosen (2020) notes that some resourceful and wealthy traffickers use their connections to facilitate the poaching and illicit trade in wildlife. According to recent report, there could anywhere be from five to more than 15 people involved across the trafficking chain (UNODC, 2020, pp. 68–69). There were approximately 180,000 seizures from 149 countries representing almost 6000 species between 1999 and 2018 (UNODC, 2020, p. 32). According to Rosen (2020), no country can be said to be immune from this crime that impacts biodiversity, human health, national security and socioeconomic development.

9.8.1.4 Wildlife, climate Change, and invasive alien species

There are multiple ways that climate change is impacting wildlife (Hance, 2017). Augmentations in temperature and alterations in precipitation can wield direct impact on species depending on their physiology and tolerance of environmental changes (Somero, 2010). Besides, climate change can also alter a species' food supply or its reproductive timing, indirectly impacting its fitness (Milligan, Holt, & Lloyd, 2009). A comprehension of these interactions is a significant move in devising management strategies to enable species survive the changing climate. Impact of climate change is almost severe on amphibians, birds, mammals and reptiles. The adverse impacts of climate change on amphibians have been focused on by many researchers and offered differing results suggesting variations in risks among taxa (Laurance, 2008; Blaustein et al., 2010). Rare species or species that possess limited dispersal are prone to have limited movements, and may not be in apposition to change their distribution to put up changes in the locations of appropriate habitats, in comparison to species with continental distributions may possess inherent resiliency to a broad assortment of conditions and possess better survival capacity as a whole. Some scholars suggest manipulation of hydroperiod or moisture regimes at sites as a prominent tool to mitigate the effects of climate change on any wildlife group, and implementation of this tool is facilitated by an array of methods, including: irrigation, site excavation, vegetation management, riparian buffer creation down wood recruitment and litter supplementation (Shoo et al., 2011).

Bird populations, that seemingly deliver significant ecosystem services with repercussions for human health and well-being, including pest control, sanitation, seed dispersal and pollination (Sekercioglu, Daily, & Ehrlich, 2004) are faced with conservation challenges worldwide, and about 12% of these species a facing a high-risk of extinction in the near future (Vie, Hilton-Taylor, & Stuart, 2009). Strategies designed to mitigate the adverse impacts of climate change on bird populations embrace maintenance of the resilience of the habitats of the birds by alleviating compound stressors that are prone to interact with climate change and amplify its impact (King & Finch,

2013). The impacts of climate change on mammals can be discerned sometimes directly via the study of their biology and physiology. For instance, once it becomes evident that wolverines need snow to dens (Copeland et al., 2010), and persistent spring snow defines the southern extent of their range (Copeland et al., 2010), the probable effects of reduced snowpack are straightforward (McKelvey et al., 2011). Uncertainties pervade in predicting mammalian responses to changing climate and these uncertainties don't preclude active management to conserve mammals, they merely change the nature of that management (McKelvey, Perry, & Mills, 2013).

Native wildlife is vastly threatened from IAS, whether introduced accidently or intentionally. A rapid rate of growth and reproduction is often recorded by alien invasive species and they are prone to spread across ecosystems in an aggressive manner (Phillips et al., 2010). Invasive species are one of the leading threats to native wildlife, putting 42% of threatened or endangered species at risk (Team Wanderdust, 2019). The damage caused by invasive species to the cane toad in Australia is one of the glaring examples of the adverse impacts of IAS on wildlife (Jolly et al., 2015a,b). Invasive plants grab rapidly land from native plant species eventually resulting in the reduction of the number of native animals in the area. Admittedly, most studies have only looked at the effects of one invasive plant on one animal species and little is known about the consequences of invasive plants (Fletcher et al., 2019).

9.9 Preventing invasive aliens species on land and water

Many scholars and researches have come to acknowledge in the recent past decades that IAS are a grave source of threat to the local biodiversity, ecosystem services, environmental quality (Jones & McDermott, 2018; Bartz & Kowarik, 2019) and human health (Stone, Witt, Walsh, Foster, & Murphy, 2018; Jones, 2019). Even the IPBES (2019) has projected that 20% of the Earth's surface, embracing also the global biodiversity hotspots are at risk owing to IAS. In comparison to low-income countries, high-income countries have registered 30 times higher number of invasions by IAS (Seebens et al., 2018). As compared to Asia-Pacific/African regions, IAS hotspots are relatively in high-income countries of European Union (EU), Australasia and North America (IPBES, 2019). This trend can be ascribed to burgeoning trade and transport activities in high-income countries (Rai & Singh, 2020). In the wake of advent of fresh and novel ecosystems, the likelihood of potential threats to the environment and human health cannot be ruled out (Rai & Singh, 2020). Progress in the biomedical sector, especially catering to protect human health, is getting hampered in the wake of recent alterations in the global environment attributable to land use, biotic invasions of flora and fauna induced by climate change (Ebi, Frumkin, & Hess, 2017). Apart from IAS, native biodiversity is also threatened by other global threats, especially the climate change. While alluding to ongoing debate among invasion ecologists as to whether invasive alien plant species are the first/second-most grave threat to biodiversity, Rai and Singh (2020) report that only 27.3% are in favor of this surmise or they should be further ranked below (Young & Larson, 2011). Notwithstanding the fact that these rankings with regard to biodiversity threats/extinctions could be region-specific (Rai & Singh, 2020), most of the invasion ecologists share the common view that this worldwide problem of IAS is getting exacerbated by the anthropogenic perturbations (Young & Larson, 2011). Furthermore, IAS has already been declared as major driver of biodiversity loss (IPBES, 2019).

IAS are recording invasions at an unprecedented rate in recent times, and this increase in invasions is to such an extent that their potential to disrupt ecosystem interactions, and the magnitude of those consequences, qualifies IAS as a major driver of global environmental change (IUCN, 2017). Resilience of ecosystems in most of the geographical realms is challenged by IAS. Evidence on effects of IAS has been quantified on species, community dynamics and overall structure and function of ecosystems (Hulme et al., 2013). The search for determining as to which of the habitats are most susceptible to plant invasions has been accorded priority since the launching of invasion biology as a focused arena of science (Pyšek & Chytrý, 2014). Available scientific literature attests to the presence of plethora of literature on patterns of alien plants for diverse habitat types and embrace an array of partial scales from particular landscape types (Moravcová et al., 2015), broad regional scales (Giorgi et al., 2016) and at continental scales (Hejda et al., 2009). Irrespective of the fact that there exist differences across scales, habitats and regions with regard to the subtle and complex interactions between environmental factors and their impacts on species (González-Moreno, Diez, Ibáñez, Font, & Vilà, 2014), these studies assume significant role in developing generalizations on drivers of geographic patterns o alien species (Pyšek & Chytrý, 2014). Notwithstanding the geographical differences, climate is often a significant driver of invasive species abundant at the continental scale (Richardson & Pysek, 2012).

Composition and structure of riparian zone species is an illustration of the outcomes of complex landscape-scale interactions, as they are an assortment of plant communities constituted in response to changing environmental, hydrological and habitat characteristics of the river (Hood & Naiman, 2000). Given that the frequency and intensity of inundation and vulnerability of riparian plants to spoilage by flooding entails a dominant effect on the dispersal of riparian vegetation (Hupp & Bornette, 2003), the interactivity between landform traits and flow regulate the disturbance regime, water and moisture availability (Parsons, McLoughlin, Kotschy, Rogers, & Rountree, 2005). It has suggested that in order to have an adequate understanding of the dynamics of riparian invasions and their resultant impacts, it is essential to ascertain the factors that affect their distribution (Garófano-Gómez et al., 2011). Extreme climate events like massive floods, droughts or heat waves can facilitate invasion further because they can disrupt ecological and evolutionary processes (Diez et al., 2012). Invasions by extreme climatic events are facilitated by two pathways—abrupt mortality of residence species or decreasing capacity to access resources—and that eventually leads to diminution in invasion resistance (Havel, Kovalenko, Thomaz, Amalfitano, & Kats, 2015). While alluding to the good progress recorded in the realm of bio-geomorphic impacts of invasive species, Fei et al. (2014) lament at the scant attention focused on the distribution patterns of alien plants in relation to riparian landforms.

Being construed as one of the main drivers of biodiversity loss and environmental degradation, IAS represent a huge threat to ecosystems as well as to human health and livelihoods (Rendekova, Micieta, Hrabovsky, Eliasova, & Miskovic, 2019). Owing to their damage potential, IAS are regulated in a wide array of legal instruments in a categorical manner. A three-step hierarchical mechanism of measures comprising their prevention, eradication and control is generally applied by countries to minimize the spread and impact of IAS (CBD, 1992, p. 6; EC, 2013). Nevertheless, this rigid treatment of IAS in the present regulation entails multiple considerations to be taken into account, especially because its applicability and topicality would be subjected to question in the wake of ongoing process of climate change that constitutes one of the biggest ecological challenging impacts of the century (IPCC, 2014); however, climate change is exacerbating these movements

(Teillard et al., 2016). As one of the greatest drivers of biodiversity loss, climate change is already forcing perceptible alterations in ecosystem balances that has to adapt to new circumstances via shifting habitats, distributions and life cycles (Hoegh-Guldberg et al., 2018), and consequently, species entail the likelihood of resorting to a series of survival adaptation measures varying from alterations in their biology to the requirement of moving to new ecosystems so as to escape extinction (Council of Europe, 2008; Baumsteiger & Movie, 2017).

Movements of species of animals and plants have been facilitated by humans around the globe for many centuries and this has been done both not only deliberately for agriculture, horticulture and the pet trade but also inadvertently via stowaways and contaminants (Hulme et al., 2008; Faulkner, Robertson, Rouget, & Wilson, 2015). These species being relocated in new environments by humans are termed as "alien" species in that ecosystem (Blackburn et al., 2011). A small proportion of those species that attain the ability to survive in new or novel environments are seemingly capable of surmounting the impediments to their reproduction via getting appropriate mates and thereafter becoming "naturalized," and spread to distant places from their original point of introduction, thereby retaining the prospects of becoming "invasive" (Blackburn et al., 2011). Most of the invasive species possess negative impacts in the environments of their initial introduction and such negative impacts, inter alia, include triggering the extinction of other species (Clavero & García-Berthou, 2005) and affecting the ecosystem services like causing diminution in the supply of water (Le Maitre, van Wilgen, Chapman, & McKelly, 1996). Pimentel, Lach, Zuniga, and Morrison (2000) maintain that the costs of coping with or remediating these impacts can be very high. Horticulture has come to be reckoned with as a prime global pathway for the introduction of plant species (Foxcroft, Parsons, McLoughlin, & Richardson, 2008; Foxcroft, Richardson, & Wilson, 2008; Niemiera, Holle, & Inderjit, 2009), and it is possibly instrumental in fostering strong positive correlation between human population density and alien plant species abundance (Aronson et al., 2014a,b). The horticultural trade, by way of business transaction mechanism, is able to surmount the first impediment in the introduction-naturalization-invasion continuum, as plants are sourced and carried away to distant places across the globe, where in those new sites these alien plant species are generally taken care of and their survival encouraged by those who are looking after their floral investments. Gardeners characteristically change or control edaphic factors like soil fertility, acidity and moisture content to suit the species or suite of species they are attempting to cultivate (Soares et al., 2015). Some of the features that apparently aim for a tempting horticultural specimen are identical to those that are worrisome for invasion biologists, such as vegetative reproduction, drought tolerance, ability to re-sprout and prolific flowering and/or fruiting (Marco, Lavergne, Dutoit, & Valerie, 2010).

The significant role of the horticultural trade in boosting preference for alien plant species across the globe over indigenous plants within gardens has been focused on in many scholarly studies (Ööpik, Bunce, & Tischler, 2013; Cubino, Subiros, & Lozano, 2015). Propagule insistence also gets enhanced because some species are recurrently introduced in abundance via this pathway over the period (Zenni, 2014). This introduction trend is expectedly aggravated in the wake of increasing trend toward online trading that makes many more species probably available to potential buyers (Lenda et al., 2014). Irrespective of the magnitude of introduction of alien plants into urban areas via the horticultural trade, many prospective invasive species have not yet spread far and wide, thereby leaving behind high level of "invasive debt" (Rouget et al., 2016); in the eventuality of no introduction of new species, those that are already planted entail the potential of turning into invasive in the future (Cubino et al., 2015). Plants introduced for horticulture are primarily located in urban areas first but they sometimes manage to stretch out from urban areas to nearby natural and seminatural environments owing to their being highly disturbed, affording opportunities for recruitment (Alston & Richardson, 2006)—the so-called "weed-shaped hole" (Buckley, Bolker, & Rees, 2007). Likewise, movements of humans within and out of urban areas can contribute to the distribution of propagules to the adjacent natural spaces, specifically seeds that can be transported to even distant places through vehicles (von der Lippe et al., 2013), and this demonstrates that urban spaces can serve as pivotal points for launching of invasions into neighboring areas (von der Lippe & Kowarik, 2008; Marco et al., 2010). Cities are collectivities of urbanization that is characterized by dense human population and many cities are endowed with ports of trade and likely entry as well as gardens, thereby possessing better prospects of affording much greater alien plant species abundance relative to rural towns and are often also the first place in country to which a plant is introduced (Dodd, McCarthy, Ainsworth, & Burgman, 2016; Padayachee et al., 2017). In the wake of burgeoning expansion of urban space across all regions of the globe, there is rapid growth of urban population as well; with about 55% of global population living in urban areas in 2016 (UN-DESA, 2016), and this number is going to reach 68% by 2050 (UN-DESA, 2018), and this increasing trend in urbanization entails the probability of concomitant increase in the risks of introduction of invasive species.

9.9.1 Defining invasive alien species

There is seemingly a lack of a legally binding definition of IAS and most of the instruments regulating IAS also do not provide any such definition as well. While the CBD treaty text vide Article 8 (h) refers to "prevent the introduction of, control or eradicate those alien species which threaten ecosystems, habitats or species" (CBD, 1992, p. 6); nevertheless, an entry in the CBD website dated 20 March 2009 alludes to IAS as "plants, animals, pathogens and other organisms that are nonnative to an ecosystem and which may cause economic or environmental harm or adversely affect human health. In particular, they impact adversely upon biodiversity, including decline or elimination of native species—through competition, predation or transmission of pathogens—and the disruption of local ecosystems and ecosystem functions" (CBD, 2009). A more detailed definition of IAS is given by IUCN by denoting that the "animals, plants or other organisms introduced by man out of their natural range of distribution, where they become established and disperse, generating a negative impact on the local ecosystem and species" (CMS, 2014, p. 7).

A series of common traits that delineate IAS can be discerned from the definitions of IAS by CBD and IUCN. In the first place, IAS are always outside their natural range of distribution; and secondly, IAS have to be introduced by a human in the new ecosystem, and this introduction can be deliberate or not (UNCLOS, 1982, Article 196: 101; CBD, 2002). Nevertheless, the words envisaged in the CBD delineation of IAS, "whose introduction and/or spread outside their natural past or present distribution" indicate that spread outside natural distribution is referred to as an alternative or a substitute, "and/or," to human introduction, entails the likelihood of arrival of species to a new ecosystem even without human endeavors. There is lack of a consensual agreement on as to what extent the human factor is essential so as to designate a species an IAS, specifically in reference to anthropogenic environmental change, that both impels the movement of species and makes it easier for species to settle in the new ecosystems (Warren, 2007). Lastly,

IAS have to engender an adverse impact on the new ecosystem and its native species or, at any rate, pose a threat, inter alia, to its biological diversity human livelihoods, economic activities and health (IUCN, 2017). In the absence of a universally acceptable definition of IAS, the one made available by CBD is in frequent usage. Nevertheless, the impacts of IAS find detailed mention in the Convention of Migratory Species (CMS, 2014), that makes use of the definition of IAS delineated by IUCN along with the Aichi Biodiversity Target 9 that defines IAS in identical terms as the one used in CBD, as "those alien species which threaten ecosystems, habitats or species" (CBD, 2012). Interestingly, neither the Law of the Seas treaty (UNCLOS, 1982) nor the Convention on International Trade in Endangered Species of Wild Fauna and Flora (CITES, 1973) provide any definition of IAS, despite usage of the term. However, the international Convention for the Control and Management of Ships' Ballast Water and Sediments (BWM Convention, 2004), vide its Article 1 (8) endeavors to provide a definition of IAS; nonetheless, the definition of IAS made available by BWM Convention differs from the IAS definition delineated by CBD and IUCN because of the particular incorporation of the term "pathogens" in it, and it even goes to the extent of applying a different terminology for mentioning about IAS, referring to them as "harmful aquatic organisms and pathogens" (CWM Convention, 2004, p. 3), whereas the IUCN alludes to "other organisms."

A distinction between alien species (AS) and IAS is facilitated by the IUCN, and it defines alien species as: "Alien species (nonnative, nonindigenous, foreign, exotic) means a species, subspecies, or lower tax on occurring outside of its natural range (past or present) and dispersal potential (i.e., outside the range it occupies naturally or could not occupy without direct or indirect introduction or care by humans) and includes any part, gametes or propagule of such species that might survive and subsequently reproduce" (Pagad, Genovesi, Carnevali, Scaleral, & Clout, 2015). This distinction assumes significance because of the least probability of an alien species causing any adverse impact to the new ecosystem and as such prospects of its being treated as invasive are very dim. Besides, a nonnative species entailing least probability of surviving in the new ecosystem may not be affording any chance of affecting the ecosystem in any way and as such may not be accounted in the category of invasive species. However, it can also be inferred that an alien species could become invasive empirically or a posteriori in the eventuality of a change in the conditions of an ecosystem or it becomes enfeebled or disrupted on account of an array of factors, such as climate change (Early et al., 2016). In the eventuality of a species becoming invasive to an ecosystem then that species entails the remote possibility of being able to settle and become native unless a certain amount of time has lapsed in between and even thereafter its status of naturalization still would be at stake (Carthey & Banks, 2012), assuming the massive adverse impact that is caused by it along with its long-term consequences for an ecosystem.

9.9.2 Invasive alien species and climate change

Climate change wields immense impact on IAS, especially from the climate-induced extreme events like hurricanes, floods and droughts that entail the capability of transporting IAS to new destinations thereby decreasing the resistance of habitats to invasions (IUCN, 2017). The species possessing high resistance capability and expansion rate along with good adaptability to new environments, entail better prospects of becoming invasive. Introduction of IAS in new ecosystem and its interaction with it is prone to disrupt ecosystem equilibrium and predator−prey relationships, thereby emerging as one of the prime causes of biodiversity loss (Raatz, van Velzen, & Gaedke, 2019), along with causing harm to marine industries, human health, and other ecosystem services and dependent native communities. Island territories in the wake of isolation of their ecosystems are particularly vulnerable to IAS (IUCN, 2018). IAS entail the potential of transforming the ecosystems and their resultant consequences, including the risk of biotic homogenization (Olden & Rooney, 2006). As a sequel to global warming that is the direct consequences of anthropogenic activities and the acute effects of global warming on the oceans that serve as major CO_2 accumulators (DeVries et al., 2019), a sort of direct disruption of marine ecosystem equilibrium is taking place, along with habitat degradation and biodiversity loss; and as such climate change is said to have added a new dimension to marine IAS.

It is to be noted that climate change is capable of triggering a wide range of consequences in ocean conditions and each of them spurs a different ecological response. The major ones to be taken into account with regard to IAS are five (Capdevila-Arguelles & Zilletti, 2008). In the first place, increase in the ocean water temperature is an outcome of the enhanced air temperature and GHGs concentrations along with other factors like ocean ice melting (IUCN, 2017a). In the second place, rising sea levels cause a series of alterations in ocean currents and ocean circulation. Thirdly, a diminution in the ocean salinity is driven by an increase of storm frequency and altered rainfall amounts (NASA, 2020). In the fourth place is the acidification of the water and other chemical changes like an augmentation of the presence of CO_2 and a decrease of the water pH. And, there are altered patterns of primary production (Knapp, Carroll, Fahey, & Monson, 2014; Christopher et al., 2006). Nevertheless, the long-term consequences of these alterations are prone to be subjected to variations from species to species, from ecosystem to ecosystem, and they are often unpredictable (Minteer & Collins, 2012). Furthermore, the collective action of all these elements in a specific marine area is yet unclear (Harvey, Gwynn-Jones, & Moore, 2013). Nonetheless, there is growing scientific evidence of climate change affecting and modifying that the impact IAS wield on a given ecosystem, usually causing a discrete increase in the chances of success (Capdevila-Arguelles & Zilletti, 2008).

Ample scientific and scholarly attention has been focused on the probable responses of alien species to the impacts of climate change in the ocean (Ballard et al., 2018). Alien species that get used to warmer ecosystems entail the potential of becoming more abundant as well as being able to enlarge their ranges to presently nonwarmer latitudes that could become available to them in the future as a result of the rise in temperatures (Malhi et al., 2020). Arrival of these new species to higher latitudes entail the likelihood of exerting pressure over the native species thereby contributing to latter's extinction or forcing them to seek out shelter in even higher latitudes. Furthermore, increase in sea temperature can be instrumental in causing stress on the species, resulting in massive death toll of species that can result in empty spaces in an ecosystem that can possibly be occupied by alien species, a scenario that leads to a situation that, inter alia, gives rise to the question of positive effects of alien species (Gallardo et al., 2019). Lastly, as a sequel to the effects of climate change in the water, native species perforce move to the north wherein they become alien to the new ecosystem and potentially invaders as well (Stayer & Dudgeon, 2010).

9.9.3 Regulatory framework for invasive alien species

Regulatory framework for dealing with IAS, inter alia, includes legal instruments, such as UNCLOS, the CBD, BWM Convention, the CMS agreement, CITES along with soft law like Aichi

Biodiversity Target number 939, etc. These are briefly analyzed to assess the extent of regulatory mechanism in place to prevent IAS. UNCLOS, CBD, or CMS go further and put in place mechanisms of eradication, control and containment of IAS to implement if prevention fails. Section 9.1 of Part XII of UNCLOS regulates the introduction of IAS in an ecosystem and it envisages general provision for the protection and preservation of the marine environment (UNCLOS, 1982, p. 98). Nevertheless, the only article of UNCLOS that deals with IAS particularly is Article 196 and it is said to be the first provision of an international legal instrument to address IAS (UNCLOS, 1982, p. 99). No definition of IAS or alien species is provided by UNCLOS. Furthermore, the manner in which the UNCLOS regulates IAS, vide two provisions in Article 196 that encompass two different matters, these may seem little confusing and require interpretation (Proelss, Maggio, Blitza, & Daum, 2017, p. 1319). IAS are regulated by the CBD vide Article 8 (h) very briefly and no definition of IAS is provided by it (CBD, 1992, p. 6). This provision enjoins upon the States to "prevent the introduction of, control or eradicate those alien species which threaten ecosystems, habitats or species" (CBD, 1992, p. 6). From the wording of Article 8 (h) it appears that for regulating a nonnative species, it is not sufficient to be alien but it needs to threaten the ecosystem, habitat or species of that new ecosystem (Rai & Singh, 2020). Furthermore, the wording of Article 8 (h) also seems to suggest that the potential for causing harm is sufficient for alien species to be regarded as IAS under CBD, thereby precipitating the prevention, control or eradication requirements of Article 8 (h), and concomitantly, the context of this requirement is Article 8 that encompasses the in situ conservation of the species that means the conservation of the ecosystems and the maintenance, or recuperation, of the species that are dependent on it in their natural surroundings (Pallewatta, Reaser, & Gutierrez, 2003), as per Article 2 (CBD, 1992, p. 3). This demonstrates that IAS is contemplated by the CBD as a direct obstacle in the fulfillment of that objective of in situ conservation (CBD, 2002).

Among the 20 targets envisaged in the Aichi Biodiversity Targets, formulated in the context of the Strategic Plan 2011−20 and set on the context of five strategic goals to halt the loss of biodiversity by 2050, the Aichi Biodiversity Target 9 deal with IAS as follow: "By 2020, IAS and pathways are identified and prioritized, priority species are controlled or eradicated and measures are in place to manage pathways to prevent their introduction and establishment" (CBD, 2010). This target bears close proximity to the SDG-15.8 of SDGs that stipulates: "By 2020, introduce measures to prevent the introduction and significantly reduce the impact of IAS on land and water ecosystems and control or eradicate the priority species" (UN, 2015a,b). Nevertheless, the CBD in the extended technical rationale of the Aichi Biodiversity Target 9 provides a definition of Pollenate IAS and a detailed list of the harmful impacts that can have in an ecosystem as follow: "Invasive alien species are those alien species which threaten ecosystems, habitats or species (Article 8 (h)). They are a major threat to biodiversity and ecosystem services, as identified by most Parties in their fourth national reports. They often have a particularly detrimental effect in island ecosystems. In some ecosystems, such as many island ecosystems, invasive alien species are the leading cause of biodiversity loss. In addition, invasive alien species can pose a threat to food security, human health and economic development. Increasing trade and travel means the threat is likely to increase unless additional action is taken" (CBD, 2012).

An endeavor to provide answer to one of the most common ways of unintentional introduction of IAS on an ecosystem is made through Ballast Water Management Convention (BWMC), adopted under the aegis of the International Meteorological Organization (IMO). Article 1 (2) of the BWMC describes ballast water as the use of water to maintain the ship trim, floatability, draught and stability when is empty of its cargo (BWMC, 2004, p. 2). In view of the fact that ships take water from the sea in which they are located when they leave their cargo and the water already taken is often released in entirely a different place with completely different ecosystems, species, etc., and in this eventuality, introductions of alien species are common if there is no preventive mechanism in place. The BWM Convention adopted under the aegis of IMO in 2004 came into force in 2017. Some amendments to the BWM Convention adopted in April 2018 were entered into force in October 2019. These amendments formalize an implementation schedule to ensure ships manage their ballast water to meet a specified standard, known as D-2 Standard, which specifies that ships can only discharge ballast water that meets the following criteria: less than 10 viable organisms per cubic meter which are greater than or equal to 50 μm in minimum dimension; less than 10 viable organisms per milliliter which are between 10 and 50 μm in minimum dimension; less than 1 colony-forming unit (cfu) per 100 mm of Toxicogenic *Vibrio cholerae*; less than 250 cfu per 100 mm of *Escherichia coli*; and less than 100 cfu per 100 mm of Intestinal Enterococci (IMO, 2019).

Because ships take water from the sea in which they are located when they leave their cargo and release it in a completely different place, with completely different ecosystems, species, etc., alien species introductions are common if no mechanism of prevention is in place. The report that resulted from the sixth meeting of the Conference of the Parties to the CBD encourages the IMO, in Part III (7), "to complete the preparation of an international instrument to address the environmental damage caused by the introduction of harmful aquatic organisms in ballast water and to develop as a matter of urgency, mechanisms to minimize hullfouling as an invasion pathway." In its Preamble, the BWM Convention highlights its intention of developing Article 196(1) UNCLOS with the objective of addressing one of the main pathways of introduction of IAS. The BWM Convention defines IAS in Article 1(8) using the term "harmful aquatic organisms and pathogens" in the following terms aquatic organisms or pathogens which, if introduced into the sea including estuaries, or into freshwater courses, may create hazards to the environment, human health, property or resources, impair biological diversity or interfere with other legitimate uses of such areas. Again, the main characteristic for invasiveness is the potential hazard to the environment and the consequences the alien species introduced through ballast water may cause. The BWM Convention requires states to implement a plan on ballast water management to minimize and eliminate the transfer of harmful aquatic organisms and pathogens, as well as the cooperation among them to cooperate to agree on and continuously develop international standards. The D-2 Standard is designed to ensure that viable organisms are not released into new sea areas, and make mandatory the Code for Approval for Ballast Water Management Systems that sets out how ballast water management systems used to attain the D-2 standard have to be assessed and approved. It further envisages that compliance with the D-2 standard would be phased-in overtime for individual ships, up to September 8, 2024, and overtime, more and more will be compliant with the D-2 standard (IMO, 2019). It is hoped that these amendments with be helpful in halting IAS in marine ecosystems.

The conservation and sustainable use of migratory animals and their habitats is envisaged in the Convention on the CMS (1979), and it also makes available a classification of their status in Appendix 1 that provides a list of the endangered migratory species along with a list of species that have an unfavorable conservation status vide Appendix II. The CMS vide Article III (4)(c) stipulates the obligation for the Parties to the Convention "to the extent feasible and appropriate, to prevent, reduce or control factors that are endangering or are likely to further endanger the species, including strictly controlling the introduction of, or controlling or eliminating, already introduced exotic species" (CMS, 1979). Furthermore, the Parties to the Convention are also required to cooperate in the prevention, early detection and rapid response against IAS, and calls for collaboration among governments, economic sectors and nongovernmental and international organizations to prevent its international movement (CMS, 1979).

In pursuance of Articles IV and V of the CMS (CMS, 1979), emphasis is focused on cooperating in reaching a favorable state of conservation of Albatrosses and Petrels and this resulted in concluding an Agreement on Conservation of Albatrosses and Petrels in 2001 (ACAP, 2001). Parties to this agreement are enjoined upon by the obligation to "Eliminate or control nonnative species detrimental to albatrosses and petrels" vide Article III (1) (b) (ACAP, 2001). This provision pursues the same argument as the one emphasized in the CMS in the sense that it alludes to "nonnative" species and not to their invasiveness, as well as having a trigger for the obligation of eliminating or controlling such alien species that wield the detrimental effect on albatrosses and petrels in particular. Operation of this Agreement is facilitated through an Action Plan for the attainment and maintenance of a favorable conservation status of the albatrosses and petrels, in accordance with the obligations envisaged in Article VI, that includes conservation of the habitats, research and collation of information, among others (Cooper et al., 2006). In order to encourage the adoption of best-practice mitigation measures to reduce seabird mortality in specifically longline fisheries in international waters outside national jurisdiction, the ACAP has been working with tuna Regional Fishery Management Organizations (tREMOs), the Commission for the Conservation of Antarctic Marine Living Resources (CCAMLR) and other relevant fisheries management organizations (ACAP, 2015). In 2019, Advisory Committee of ACAP announced that a conservation crisis continues to be faced by its 31 listed species, with thousands of albatrosses, petrels and shearwaters dying annually as a consequence of fisheries operations, and with a view to increase awareness with regard to this crisis, ACAP has inaugurated a World Albatross Day in 2020, to be observed annually on June 19, the date on which ACAP agreement was opened for signature (ACAP, 2020).

A regime of export and import permits, established under the Convention on International Trade in Endangered Species of Wild Fauna and Flora (CITES), granted by a national Scientific and Management Authority designated in accordance with Article IX of CITES, of species threatened with extinction or that may become threatened of a regulation on their export and import were no to exist (CITES, 1973, p. 7). Appendix I, II & III contain a list of species, in accordance with the level of danger of the species as a result of trade. Trade in species assumes significance in the context of IAS, particularly because the movement of living species and the introduction of alien species in a new ecosystem, either intentionally or by accident, as could be the case of plant or animal introduced without permits or that escapes, can be a probable entry path of IAS (STDF, 2013). An assortment of recommendations has been issued by the Conference of the Parties of CITES with regard to this, calling upon the parties to consider the risks and species problems that invasive species can trigger when creating national legislation pertaining to trade of living animals or plants. Considering the regulations of the receptor country when exporting potential invasive species to it (CITES, 2007). The Parties are also called upon to coordinate CITES and CBD to further augment the cooperation among parties on the issue of introduction of potential invasive species (CBD, 2012a). Despite its limitations, CITES has come to be reckoned as the most powerful international convention on biodiversity conservation. The possibility of monitoring trade in endangered species and consequential imposition of automatic sanctions on the noncomplying parties envisages a powerful tool to prevent or at least mitigate illegal trade in endangered species (Lavorgna, Rutherford, Vaglica, Smith, & Sajeva, 2018). Morgera, Tsioumani, and Matthias (2014) have opined that by permitting the CITES secretariat to evaluate as to how Parties' legislations permit the implementation of the Convention and thereafter recommending the suspension of trade in CITES species for those not complying with requisite legislation, is prone to further augment the strength of CITES.

Interestingly, scattered nature of legal instruments dealing with IAS require these different instruments to be taken into account in order to address IAS, and that seems to be a difficult task (CBD, 2001; Pallewatta et al., 2003). Nevertheless, similarities can be drawn among various legal instruments dealing with IAS, specifically in the type of measures that States shall implement to safeguard the ecosystems from their harmful impacts. There is a dire need of generating a hierarchical system of measures against IAS comprising prevention, early detection, eradication, containment and control and this can be facilitated by considering all available legal instruments dealing with IAS in a wider global perspective (Early et al., 2016).

References

AARD (Alberta Agriculture and Rural Development). (2010). Agricultural soil compaction: Causes and management. *Agdex*, 510−511. Available from https://www1.agric.gov.ab.ca/$department/deptdocs.nsf/all/agdex13331/$file/510-1.pdf?OpenElement.

Abdullah, S. A., & Nakagoshi, N. (2007). Forest fragmentation and its correlation to human land use change in the state of Selangor, Peninsular Malaysia. *Forest Ecology and Management*, *241*, 39−48.

Abell, R., Thieme, M. L., Revenga, C., Bryer, M., Kottelat, M., Bogutskaya, N., ... Petry, P. (2008). Freshwater ecoregions of the world: A new map of biogeographic units for freshwater biodiversity conservation. *Bioscience*, *58*, 403−414.

ACAP (Agreement on Conservation of Albatrosses and Petrels). (2001). *Text of the agreement on the conservation of albatrosses and petrels*, concluded at Cape Town, South Africa, January 29−February 2, 2001. Available from https://www.gc.noaa.gov/documents/acap_annexes.pdf.

ACAP (Agreement on Conservation of Albatrosses and Petrels). (2015). *Agreement on the conservation of albatrosses and petrels: Achievements in the first ten years. 2004−2014*. Hobart: ACAP Secretariat. Available online.

ACAP (Agreement on Conservation of Albatrosses and Petrels). (2020). *About ACAP*. ACAP. Available online. Updated on 17.09.20.

ADB (Asian Development Bank) (2010). Focused action. Priorities for addressing climate change in Asia and the Pacific. Manila, Philippine: ADB Secretariat.

Adebayo, O. (2019). Loss of biodiversity: The burgeoning threat to human health. *Annals of Ibadan Postgraduate Medicine*, *17*(1), 1−3. Available online.

Adhikari, K., & Hartemink, A. E. (2016). Linking soils to ecosystem services—A global review. *Geoderma*, *262*, 101−111.

Adler, C., Huggel, C., Orlove, B., & Nolin, A. (2019). Climate change in the mountain cryosphere: Impacts and responses. *Regional Environmental Change*, *19*(5), 1225−1228.

Adornado, H. A., Yoshida, M., & Apolinares, H. (2009). Erosion vulnerability assessment in REINA, Quezon Province, Philippines with raster-based tool built within GIS environment. *Journal of Agriculture Research*, *18*, 24−31.

Adrian, A., O'Reilly, C. M., Zagaresse, H., Baines, S. B., Hessen, H. O., Keller, W., ... Winder, M. (2009). Lakes as sentinels of climate change. *Limnology and Oceanography, 54*(6), 2283−2297.

Aerts, R., Hundera, K., Berecha, G., Gijbels, P., Baeten, M., Van Mechelen, M., ... Honnay, O. (2011). Semiforest coffee cultivation and the conservation of Ethiopian Afromontane rainforest fragments. *Forest Ecology Management, 261*, 1034−1041.

Agelici, F. M. (Ed.), (2016). *Problematic wildlife: A cross-disciplinary approach*. London: Springer.

Aguilar, R., Quesada, M., Ashworth, L., Herrerias-Diego, Y., & Lobo, J. (2008). Genetic consequences of habitat fragmentation in plant populations: Susceptible signals in plant traits and methodological approaches. *Molecular Ecology, 17*(24), 5177−5188.

Aguirre-Gutierrez, J., van Treuren, R., Hoekstra, R., & van Hintum, T. J. L. (2017). Crop wild relatives range shifts and conservation in Europe under climate change. *Diversity and Distribution, 23*(7), 739−750.

Ahmed, D. A., van Bodegom, P. M., & Tukker, A. (2019). Evaluation and selection of functional diversity metrics with recommendations for their use in life cycle assessments. *The International Journal of Life Cycle Assessment, 24*, 485−500.

Ahrends, A., Burgess, N. D., Milledge, S. A. H., Bulling, M. T., Fisher, B., Smart, J. C. R., ... Lewis, S. L. (2010). Predictable waves of sequential forest degradation and biodiversity loss spreading from an African city. *Proceedings of the National Academy of Sciences of the United States of America, 107*(33), 14556−14561.

Allen, S. K., Rastner, I. P., Arora, M., Huggel, C., & Stoffel, M. (2016). Lake outburst and debris flow disaster at Kedarnath, June 2013: Hydrometeorological triggering and topographic predisposition. *Landslides, 13*(6), 1479−1491.

Alston, K. P., & Richardson, D. M. (2006). The roles of habitat features, disturbance, and distance from putative source populations in structuring alien plant invasions at the urban/wildland interface on the Cape Peninsula, South Africa. *Biological Conservation, 132*, 183−198.

Alvarez-Berríos, N. L., & Aide, T. M. (2015). Global demand for gold is another threat for tropical forests. *Environmental Research Letters, 10*, 014006.

Alvarez-Clare, S., & Mack, M. C. (2015). Do foliar, litter, and root nitrogen and phosphorus concentrations reflect nutrient limitation in a lowland tropical wet forest? *PLoS One, 10*(4), 1−16.

Andela, N., Morton, D., Giglio, L., Chen, Y., van der Werf, G., Kasibhatla, P., ... Randerson, J. (2017). A human-driven decline in global burned area. *Science (New York, N.Y.), 356*, 1356−1361.

Anderegg, L. D. L., Anderegg, W. R. L., & Berry, J. A. (2013). Not all droughts are created equal: Translating meteorological drought into woody plant mortality. *Tree Physiology, 33*, 672−683.

Anderegg, W. R. L., Schwalm, C., Biondi, F., Camarero, J. J., Koch, G., Litvak, M., ... Pacala, S. (2015). Pervasive drought legacies in forest ecosystems and their implications for carbon cycle models. *Science (New York, N.Y.), 349*, 528−532.

Anderegg, W. R. L., Klein, T., Bartlett, M., Sack, L., Pellegrini, A. F., Choat, B., & Jansen, S. (2016). Meta-analysis reveals that hydraulic traits explain cross-species patterns of drought-induced tree mortality across the globe. *Proceedings of the National Academy of Sciences of the United States of America, 113*, 5024−5029.

Andrade, E. R., Jardim, J. G., Santos, B. A., Melo, F. P. L., Talora, D. C., Faria, D., & Cazetta, E. (2015). Effects of habitat loss on taxonomic and phylogenetic diversity of understory Rubiaceae in Atlantic forest landscapes. *Forest Ecology and Management, 349*, 73−84.

Aragao, L., & Shimabukuro, Y. E. (2010). The incidence of fire in Amazonian forests with implications for REDD. *Science, 328*(5983), 1275−1278.

Aragão, L. E. O. C., Anderson, L. O., Fonseca, M. G., Rosan, T. M., Vedovato, L. B., Wagner, F. H., ... Saatchi, S. (2018). 21st Century drought-related fires counteract the decline of Amazon deforestation carbon emissions. *Nature Communications., 9*, 1−12.

References

Aravanopoulos, F. A., Tollefsrud, M. M., Graudal, L., Koskela, J., Kätzel, R., Soto, A., ... Bozzano, N. (2015). *Development of genetic monitoring methods for genetic conservation units of forest trees in Europe. European Forest Genetic Resources Programme (EUFORGEN)*. Rome: Biodiversity International. Available online.

Arneth, A., Sitch, S., Pongratz, J., Stocker, B. D., Ciais, P., Poulter, B., ... Zaehie, S. (2017). Historical carbon dioxide emissions caused by land-use changes are possibly larger than assumed. *Nature Geoscience, 10*, 79−84.

Aronson, M. F., La Sorte, F. A., Nilon, C. H., Katti, M., Goddard, M. A., Lepczyk, C. A., ... Winter, M. (2014a). A global analysis of the impacts of urbanization on bird and plant diversity reveals key anthropogenic drivers. *Proceedings of the Royal Society B Biological Sciences, 281*, 20133330.

Aronson, M. F. J., Handel, S. N., La Puma, I. P., & Clemants, S. E. (2014b). Urbanization promotes nonnative woody species and diverse plant assemblages in the New York metropolitan region. *Urban Ecosystems, 18*, 31−45.

Asner, G. P., Broadbent, E. N., Oliveira, P. J. C., Keller, M., Knapp, D. E., & Silva, J. N. M. (2006). Condition and fate of logged forests in the Brazilian Amazon. *Proceedings of the National Academy of Sciences of the United States of America, 103*, 12947−12950.

Asner, G. P., Hughes, R. F., Vitousek, P. M., Knapp, D. E., Kennedy-Bowdoin, T., Boardman, J., ... Green, R. O. (2008). Invasive plants transform the three-dimensional structure of rain forests. *Proceedings of the National Academy of Sciences of the United States of America, 105*, 4519−4523.

Avitabile, V., Herold, M., Heuvelink, A. G. P., Jong, D. B., & Laurin, V. G. (2016). An integrated pan-tropical biomass map using multiple reference datasets. *Global Change Biology, 22*, 1406−1420.

Baccini, A., Walker, W., Carvalho, L., Farina, M., Sulla-Menashe, D., & Houghton, R. A. (2017). Tropical forests are a net carbon source based on aboveground measurements of gain and loss. *Science (New York, N.Y.), 358*, 230−234.

Bagherzadeh, A. (2014). Estimation of soil losses by USLE model using GIS at Mashhad plain, Northeast of Iran. *Arabian Journal of Geosciences, 7*, 211−220.

Ballard, S., Porro, J., & Tromsdroff. (2018). *The roadmap to a low-carbon urban water utility: An International Guide to the WaCClim approach*. London: IWA Publishing Alliance House.

Balmford, A., Amano, T., Bartlett, H., Chadwick, D., Collins, A., Edwards, D., ... Eisner, R. (2018). The environmental costs and benefits of high-yield farming. *Nature Sustainability, 1*(9), 477−485.

Bard, A., Bernard, B., Lang, M., Giuntoli, L., Korck, J., Koboltschntg, G., ... Volken, D. (2015). Trends in the hydrologic regime of Alpine rivers. *Journal of Hydrology, 529*(3), 1823−1837.

Bardgett, R. D., & Van Der Putten, W. H. (2014). Belowground biodiversity and ecosystem functioning. *Nature, 515*, 505−511.

Barlow, J., & Peres, C. A. (2008). Fire-mediated dieback and compositional cascade in an Amazonian forest. *Philosophical Transactions of the Royal Society of London B: Biological Sciences, 363*(1498), 1787−1794.

Baronsky, A. D., Matzke, N., Tomiya, S., Wogan, G. O. U., Swartz, V., Quental, T. B., ... Ferrer, E. A. (2011). Has the Earth's sixth extinction already arrived? *Nature, 471*(7336), 51−57.

Barthlott, W., Mutke, J., Rafiqpoor, M. D., Kier, G., & Kreft, H. (2005). Global centers of vascular plant diversity. *Nova Acta Leopold NF, 342*, 61−83.

Bartley, D. M., Leonard, N. J., Youn, S., Taylor, W. W., Baigún, C., Barlow, C., ... Valbo-Jorgensen, J. (2016). Moving towards effective governance of fisheries and freshwater resources. In W. W. Taylor, D. M. Bartley, C. I. Goddard, N. L. Leonard, & R. L. Welcomme (Eds.), *Freshwater, fish and the future* (pp. 251−280). Rome: FAO. Available from http://www.fao.org/3/a-i5711e.pdf.

Bartz, R., & Kowarik, I. (2019). Assessing the environmental impacts of invasive alien plants: A review of assessment approaches. *NeoBiota., 43*, 69−99.

Barua, M., Baghwat, S. A., & Sushrut, A. (2013). The hidden dimensions of human–wildlife conflict: Health impacts, opportunity and transaction costs. *Biological Conservation*, *157*, 309–316.

Bastarrika, A., Chuvieco, E., & Martín, M. P. (2011). Mapping burned areas from Landsat TM/ETM + data with a two-phase algorithm: Balancing omission and commission errors. *Remote Sensing of Environment*, *115*(4), 1003–1012. Available from http://www.sciencedirect.com/science/article/pii/S0034425710003433.

Bastawrous, A., & Hennig, B. D. (2012). The global inverse care law: a distorted map of blindness. *British Journal of Ophthalmology*, *96*(10), 1357–1358.

Baumsteiger, J., & Movie, P. B. (2017). Assessing extinction. *Bioscience*, *67*(4), 357–366.

BBOP (Business and Biodiversity Offsets Program). (2012). *Standard on Biodiversity Offsets*. Washington, DC: BBOP. Available online.

Beech, E., Rivers, M., Oldfield, S., & Smith, P. P. (2017). Global tree search: The first complete global database of tree species and country distributions. *Journal of Sustainable Forestry*, *36*(5), 454–489.

Benson, B. J., Magnuson, J. J., Jensen, O. P., Card, V. M., Hodgkins, G., Korhonen, J., . . . Granin, N. G. (2012). Extreme events, trends, and variability in Northern Hemisphere lake-ice phenology (1855–2005). *Climatic Change*, *112*, 299–323.

Benateau, S., Gaudard, A., Stamm, C., & Altermatt, F. (2019). *Climate change and freshwater ecosystems: Impacts on water quality and ecological status. Hydro-CH2018 Project*. Bern: Federal Office for the Environment (FOEN).

Bennett, A. F., & Saunders, D. A. (2010). Habitat fragmentation and landscape change. *Conservation Biology for All*, *93*, 1544–1550.

Bennett, G., Gallant, M., & Ten Kate, K. (2017). State of biodiversity mitigation 2017 markets and compensation for global infrastructure development. *Forest Trends*. Available from https://www.forest-trends.org/wp-content/uploads/2018/01/doc_5707.pdf.

Bentos, T. V., Nascimento, H. E. M., Vizcarra, M. A., & Williamson, G. B. (2017). Forest ecology and management effects of light-gaps and topography on Amazon secondary forest: Changes in species richness and community composition. *Forest Ecology & Management*, *396*, 124–131.

Bentz, B. J., Régnière, J., Fettig, C. J., Hansen, E. M., Hayes, J. L., Hicke, J. A., . . . Seybold, S. J. (2010). Climate change and bark beetles of the Western United States and Canada: Direct and indirect effects. *Bioscience*, *60*, 602–613.

Bernhardt, E. S., Rosi, E. J., & Gessner, M. O. (2017). Synthetic chemicals as agents of global change. *Frontiers in Ecology and the Environment*, *15*, 84–90.

BGCI (Botanical Gardens Conservation International). *Global tree assessment*. (n.d.). <https://www.bgci.org/our-work/projects-and-case-studies/global-tree-assessment/>.

Bhandari, H. R., Bhanu, A. N., Srivastava, K., Singh, A. N., & Shreya, A. H. (2017). Assessment of genetic diversity in crop plants—An overview. *Advances in Plants & Agriculture Research*, *7*(3), 279–286.

Birch, J. C., Newton, A. C., Aquino, C. A., Cantarello, E., Echeverria, C., Kitzberger, T., . . . Garavito, N. T. (2010). Cost-effectiveness of dryland forest restoration evaluated by spatial analysis of ecosystem services. *Proceedings of the National Academy of Sciences of the United States of America*, *107*(50), 21925–21930.

Betts, R. (2005). Integrated approaches to climate-crop modelling: Needs and challenges. *Philosophical Transactions of the Royal Society B-Biological Sciences*, *360*, 2049–2065.

Blackburn, T. M., Pyšek, P., Bacher, S., Carlton, J. T., Duncan, R. P., Jarosik, V., . . . Richardson, D. M. (2011). A proposed unified framework for biological invasions. *Trends in Ecology and Evolution*, *26*, 333–339.

Blaustein, A. R., Walls, S. C., Bancroft, B. A., Lawler, J. J., Searle, C. L., & Gervasi, S. S. (2010). Direct and indirect effects of climate change on amphibian populations. *Diversity*, *2*, 281–313.

Blay, D. (2012). Restoration of deforested and degraded areas in Africa. In J. A. Stanturf, P. Madsen, & D. Lamb (Eds.), *A goal-oriented approach to forest landscape restoration* (pp. 267–319). New York: Springer.

Bocchiola, D. (2014). Long term (1921–2011) hydrological regime of Alpine catchments in Northern Italy. *Advances in Water Resources, 70*, 51–64.

Bollmann, M., Bosch, T., Colijn, F., Ebinghaus, R., Froese, R., Gussow, K., … Weinberger, F. (2010). *World ocean review: Living with the oceans*. Hamburg: Maribus GmbH.

Bonal, D., Burban, B., Stahl, C., Wagner, F., & Hérault, B. (2016). The response of tropical rainforests to drought—Lessons from recent research and future prospects. *Annals of Forest Science, 73*, 27–44.

Bond-Lamberty, B., Peckham, S. D., Gower, S. T., & Ewers, B. E. (2009). Effects of fire on regional evapotranspiration in the central Canadian boreal forest. *Global Change Biology, 15*(5), 1242–1254.

Boudjemline, F., & Semar, A. (2018). Assessment and mapping of desertification sensitivity with MEDALUS model and GIS—Case study: Basin of Hodna, Algeria. *Journal of Water and Land Development, 36*(1), 17–26.

Bowman, D. M. J. S., Balch, J. K., Artaxo, P., Bond, W. J., Carlson, J. M., Cochrane, M. A., … Harrison, S. P. (2009). Fire in the Earth system. *Science, 324*, 481–484.

Brando, P. M., Balch, J. K., Nepstad, D. C., Morton, D. C., Putz, F. E., Coe, M. T., … Soares-Filho, B. S. (2014). Abrupt increases in Amazonian tree mortality due to drought-fire interactions. *Proceedings of the National Academy of Sciences of the United States of America, 111*, 6347–6352.

Brahney, J., Weber, F., Foord, V., Janmaat, J., & Curtis, P. J. (2017). Evidence for a climate-driven hydrologic regime shift in the Canadian Columbia Basin. *Canadian Water Resources Journal, 42*(2), 179–192.

Broggiato, A., Arnaud-Haond, S., Chiarolla, C., & Greiber, T. (2014). Fair and equitable sharing of benefits from the utilization of marine genetic resources in areas beyond national jurisdiction: Bridging the gaps between science and policy. *Marine Policy, 49*, 176–185.

Broggiato, A., Vanagt, T., Lallier, L., Jaspars, M., Burton, G., & Muyldermans, D. (2018). Mare geneticum: Balancing governance of marine genetic resources in international waters. *The International Journal of Marine and Coastal Law, 33*(1), 3–33.

Boucher, D., Elias, P., Lininger, K., May-Tobin, C., Roquemore, S., & Saxon, E. (2011). *The root of the problem: What's driving tropical deforestation today?* Cambridge, MA: Union of Concerned Scientists.

Brandt, C. (1966). Agricultural burning. *Journal of the Air Pollution Control Association, 16*(2), 85–86.

Brown, D. G., Polsky, C., Bolstad, P., Brody, S. D., Hulse, D., Kroh, R., … Thomson, A. (2014). Land use and land cover change. In J. M. Melillo, T. C. Richmond, & G. W. Yohe (Eds.), *Climate change impacts in the United States: The third national climate assessment* (pp. 318–332). Washington, DC: United States Global Change Research Program.

Brudvig, L. A. (2011). The restoration of biodiversity: Where has research been and where does it need to go? *American Journal of Botany, 98*(3), 549–558.

Buckley, Y. M., Bolker, B. M., & Rees, M. (2007). Disturbance, invasion and re-invasion: Managing the weed-shaped hole in disturbed ecosystems. *Ecological Letters, 10*, 809–817.

Bunn, S. E., & Arthington, A. H. (2002). Basic principles and ecological consequences of altered flow regimes for aquatic biodiversity. *Environmental Management, 30*, 492–507.

Burkmar, R., & Bell, C. (2015). *Drivers of biodiversity loss*. Shrewsbury: Field Studies Council.

Bustamante, M. M., Roitman, I., Aide, T. M., Alencar, A., Anderson, L. O., Aragão, L., … Vieira, I. C. G. (2016). Toward an integrated monitoring framework to assess the effects of tropical forest degradation and recovery on carbon stocks and biodiversity. *Global Change Biology, 22*, 92–109.

Bustamante, M., Helmer, E. H., Schill, S., Belnap, J., Brown, L. K., Brugnoli, E., … Thompson, L. (2018). Direct and indirect drivers of change in biodiversity and nature's contributions to people. *The IPBES regional assessment report on biodiversity and ecosystem services for the Americas* (pp. 295–435). Bonn:

Secretariat of the Intergovernmental Science-Policy Platform on Biodiversity and Ecosystem Services (IPBES).

BWM Convention (Ballast Water Management Convention) (2004). Text of the International Convention for the Control and Management of Ships' Ballast Water and Sediments. BMW/CONF/36. 16 February 2004. London: IMO.

BWMC (Ballast Water Management Convention). *Text of the ballast water management convention adopted at the international convention for the control and management of ships' ballast water and sediments. BWM/CONF/36.* (2004). <http://library.arcticportal.org/1913/1/International%20Convention%20for%20the%20Control%20and%20Management%20of%20Ships%27%20Ballast%20Water%20and%20Sediments.pdf> Accessed 16.02.04.

CALD (Cambridge Advanced Learners' Dictionary). (2008). *Cambridge advanced learners' dictionary* (Third Edition). Cambridge: Cambridge University Press.

Caliskan, S., & Boydak, M. (2017). Afforestation of arid and semiarid ecosystems in Turkey. *Turkish Journal of Agriculture and Forestry, 41*(5), 317–330.

Camarero, J. J., Gazol, A., Sangüesa-Barreda, G., Oliva, J., & Vicente-Serrano, S. M. (2015). To die or not to die: Early-warning signals of dieback in response to a severe drought. *Journal of Ecology, 103*, 44–57.

Campagnolo, L., & Davide, M. (2019). Can the Paris deal boost SDGs achievement? An assessment of climate mitigation co-benefits or side-effects on poverty and inequality. *World Development, 122*, 96–109.

Cañedo-Argüelles, M., Kefford, B. J., Piscart, C., Prat, N., Schäfer, R. B., & Schulz, C. J. (2013). Salinisation of rivers: An urgent ecological issue. *Environmental Pollution, 173*, 157–167.

Capdevila-Arguelles, L., & Zilletti, B. (2008). A perspective on climate change and invasive alien species. In: *A report prepared for the 2nd meeting of the group of experts on biodiversity and climate change.* Convention on the Conservation of European Wildlife and Natural Habitats. T-PVS/Inf. 5 rev. Available from https://rm.coe.int/168074629c.

Capozzi, R. (2018). The best leaders get the culture right: Proven levers to change a Corporate Culture. Investments and Wealth Monitor. May-June 2018: 56–63. Greenwood Village, CO, USA. Available at: https://investmentsandwealth.org/getattachment/1525cb6e-97a2-4f5c-a6e3-96df551e740d/IWM18MayJun-ProvenLeversChangeCorpCulture.pdf.

Cardinale, B. J., Duffy, E., Gonzalez, A., Hooper, D. U., Perrings, C., Venail, P., ... Naeem, S. (2012). Biodiversity loss and its impact on humanity. *Nature, 486*(7401), 59–67.

Carey, M. (2005). Living and dying with glaciers: People's historical vulnerability to avalanches and outburst floods in Peru. *Global & Planetary Change, 47*(2–4), 122–134.

Carpenter, S. R., Stanley, E. H., & Vander Zanden, M. J. (2011). State of the World's Freshwater Ecosystems: Physical, chemical, and biological changes. *Annual Review of Environment and Resources, 36*(1), 75–99.

Carr, D. L., Suter, L., & Barbier, A. (2005). Population dynamics and tropical deforestation: State of the debate and conceptual challenges. *Population and Environment, 27*(1), 89–113.

Carrington, D. (2018, 12 March). What is biodiversity and why does it matter to us?. *The Guardian.* Available from https://www.theguardian.com/news/2018/mar/12/what-is-biodiversity-and-why-does-it-matter-to-us.

Carrizo, F., Smith, K. G., & Darwall, W. R. T. (2013). Progress towards a global assessment of the status of freshwater fishes (Pisces) for the IUCN Red list: Application to conservation programmes in zoos and aquariums. *International Zoo Yearbook, 47*(1), 46–64.

Carthey, A. J. R., & Banks, P. B. (2012). When does an alien become a native species? A vulnerable native mammal recognizes and responds to its long-term alien predator. *PLoS One, 7*(2), e31804.

Carrivick, J. L., & Tweed, F. S. (2016). A global assessment of the societal impacts of glacier outburst floods. *Global & Planetary Change, 144*, 1–16.

Carrielo, F., & Anderson, L. O. (2007). Multitemporal analysis of the spectral response of scars of burned areas using Landsat/ETM sensor. In: *Geoscience and remote sensing symposium, 2007. IGARSS 2007. IEEE international*, Barcelona, Spain, July 23−28, 2007. Available from https://ieeexplore.ieee.org/document/4423681.

Casanas, F., Simo, J., Casalas, J., & Prohens, J. (2017). Toward an evolved concept of landrace. *Frontiers in Plant Science*, *8*, 145.

Castebrunet, H., Eckert, N., Giraud, G., Durand, Y., & Morin, S. (2014). Projected changes of snow conditions and avalanche activity in a warming climate: The French Alps over the 2020−2050 and 2070−2100 periods. *The Cryosphere*, *8*(5), 1673−1697.

Cavalcanti, S. M. C., Marchini, S., Zimmermann, A., Gese, E. M., & Macdonald, D. W. (2010). Jaguars, livestock, and people in Brazil: Realities and perceptions behind the conflict. In D. W. Macdonald, & J. A. Loveridge (Eds.), *The Biology and Conservation of Wild Felids* (pp. 383−402). Oxford: Oxford University Press.

CBD (Convention on Biological Diversity). (1992). Text of the convention on biological convention. New York: United Nations. Available from https://www.cbd.int/doc/legal/cbd-en.pdf.

CBD (Convention on Biological Diversity). (2001). Review of the efficiency and efficacy of existing legal instruments applicable to invasive alien species. In: *CBD technical series no. 2*. Montreal, Canada: CBD Secretariat. Available from https://www.cbd.int/doc/publications/cbd-ts-02.pdf.

CBD (Convention on Biological Diversity). *Guiding principles for the implementation of Article 8 (h). UNEP/CBD/COP/6/20*. (2002). <https://www.cbd.int/doc/decisions/cop-06-dec-23-en.pdf>.

CBD (Convention on Biological Diversity). *COP 9 decision IX/11: Review of implementation of Articles 20 and 21*. (2008). <https://www.cbd.int/decisions/cop/9/11>.

CBD (Convention on Biological Diversity). (2009). *What are invasive alien species?*. CBD. Available from https://www.cbd.int/idb/2009/about/what/#:~:text=Invasive%20alien%20species%20are%20plants,or%20adversely%20affect%20human%20health.

CBD (Convention on Biological Diversity). *Strategic plan for biodiversity 2011−2020 and the Aichi targets*. (2010). <https://www.cbd.int/doc/strategic-plan/2011-2020/Aichi-Targets-EN.pdf>.

CBD (Convention on Biological Diversity). (2010a). *Introduction to access and benefit sharing*. Montreal: Secretariat of CBD. Available from https://www.cbd.int/abs/infokit/all-files-en.pdf.

CBD (Convention on Biological Diversity). (2010b). *Uses of genetic resources*. Montreal: CBD. Available from https://www.cbd.int/abs/infokit/factsheet-uses-en.pdf.

CBD (Convention on Biological Diversity). (2011). *Factsheet—Access and benefit sharing*. Montreal: Secretariat of CBD. Available from https://www.cbd.int/abs/infokit/revised/web/factsheet-abs-en.pdf.

CBD (Convention on Biological Diversity). (2012). *Aichi target 9: Technical rationale extended (provided in document COP/10/INF/12/Rev. 1)*. CBD. Available from https://www.cbd.int/sp/targets/rationale/target-9/.

CBD (Convention on Biological Biodiversity). (2012a). *Partners-CITES*. CBD. Available from https://www.cbd.int/invasive/collaboration.shtml?org=cites.

CBD (Convention on Biological Diversity). (2014). *Global biodiversity outlook-4*. Montreal: CBD Secretariat. Available from https://www.cbd.int/gbo/gbo4/publication/gbo4-en-hr.pdf.

CBD (Convention on Biological Diversity). *Biodiversity offsets: A user guide*. (2016). <https://www.cbd.int/financial/doc/wb-offsetguide2016.pdf>.

CBD (Convention on Biological Diversity) (2020). *History of the convention*. CBD. Available from https://www.cbd.int/history/. Updated 25.09.20.

Cebaloss, G., Ehrlich, P. R., & Dirzo, R. (2017). Biological annihilation via the ongoing sixth mass extinction signaled by vertebrate population losses and declines. *Proceedings of the National Academy of Sciences of the United States of America*, *14*, E6089−E6096.

CGRFA (Commission on Genetic Resources for Food and Agriculture). (2019). *Draft voluntary guidelines for the conservation and sustainable use of farmers' varieties/landraces. CGRFA-15/19/9/9.2/Inf.1*. Rome: FAO. Available from http://www.fao.org/3/ca5601en/ca5601en.pdf.

CGRFA (Commission on Genetic Resources for Food and Agriculture). (2015). *Voluntary guidelines to support the integration of genetic diversity into national climate change adaptation planning*. Rome: FAO. Available from http://www.fao.org/3/a-i4940e.pdf.

Chabrerie, O., Jamoneau, A., Gallet-Moron, E., & Decocq, G. (2013). Maturation of forest edges is constrained by neighbouring agricultural land management. *Journal of Vegetation Science*, 24(1), 58–69.

Chao, K. J., Phillips, O. L., Gloor, E., Monteagudo, A., Torres-Lezama, A., & Martinez, R. V. (2008). Growth and wood density predict tree mortality in Amazon forests. *Journal of Ecology*, 96, 281–292.

Chapron, G., Kaczensky, P., Linnell, J. D. C., von Arx, M., Huber, D., Andren, H., ... Boitani, L. (2014). Recovery of large carnivores in Europe's modern human-dominated landscapes. *Science (New York, N.Y.)*, 346, 1517–1519.

CGIAR-RPDS (Consultative Group on International Agriculture Research). (2014). *Annual report 2014. Pathways to lasting impact for rural dryland 38 communities in the developing world*. Amman, Jordan. Available from http://drylandsystems.cgiar.org/sites/default/files/DS_annual_report_2014.pdf.

Chazdon, R. L. (2008). Beyond deforestation: Restoring forests and ecosystem services on degraded lands. *Science (New York, N.Y.)*, 320, 1458–1460.

Chazdon, R. L., Brancalion, P. H. S., Laestadius, L., Bennett-Curry, A., Buckingham, K., Kumar, C., ... Wilson, S. J. (2016). When is a forest a forest? Forest concepts and definitions in the era of forest and landscape restoration. *Ambio*, 45, 538–550.

Chen, X. D., Dunfield, K. E., Fraser, T. D., Wakelin, S. A., Richardson, A. E., & Condron, L. M. (2020). Soil biodiversity and biogeochemical function in managed ecosystems. *Soil Research*, 58, 1–20.

Chester, E. T., & Robson, B. J. (2013). Anthropogenic refuges for freshwater biodiversity: Their ecological characteristics and management. *Biological Conservation*, 166, 64–75.

Choat, B., Jansen, S., Brodribb, T. J., Cochard, H., Delzon, S., Bhaskar, R., ... Zanne, A. E. (2012). Global convergence in the vulnerability of forests to drought. *Nature*, 491, 752–755.

Christopher, M., Peck, H., & Towill, D. (2006). A taxonomy for selecting global supply chain strategies. *The International Journal of Logistics Management*, 17(2), 277–287.

CITES (Convention on International Trade in Endangered Species of Wild Flora and Fauna). *Text of the convention on international trade in endangered species of wild flora and fauna*. (1973). <https://www.cites.org/sites/default/files/eng/disc/CITES-Convention-EN.pdf>.

CITES (Convention on International Trade in Endangered Species of Wild Flora and Fauna). (2007). *Fourteenth meeting of the conference of the parties to CITES*, The Hague, Netherlands, June 3–15, 2007. Available from https://cites.org/eng/cop/14/inf/index.php.

CITES (Convention on International Trade in Endangered Species of Wild Fauna and Flora). *Appendices I, II, & III*. (2019). <https://cites.org/sites/default/files/eng/app/2019/E-Appendices-2019-11-26.pdf>.

Clavero, M., & García-Berthou, E. (2005). Invasive species are a leading cause of animal extinctions. *Trends in Ecology and Evolution*, 20, 110.

Clilverd, H., Buys, G., Thomson, A., Malcolm, H., Henshall, P., & Matthews, R. (2019). *Mapping carbon emissions & removals for the land use, land use change & forestry sector*. London: Department of Business, Energy & Industrial Strategy.

CMS (Convention of Migratory Species). (1979). *Convention on the conservation of migratory species of wild animals*. Bonn. Available from https://www.cms.int/sites/default/files/instrument/CMS-text.en_.PDF.

CMS (Convention on Migratory Species). (2014). Glossary: Invasive alien species-IUCN. In: *Report of the review of the impact of invasive alien species protected under the convention on migratory species (CMS)*.

UNEP/CMS/COP11/Inf.32. Available from https://www.cms.int/sites/default/files/document/COP11_Inf_32_Report_Review_of_Impact_of_IAS_Eonly.pdf.

Cochrane, M. A., & Schulze, M. D. (1999). Fire as a recurrent event in tropical forests of the eastern Amazon: Effects on forest structure, biomass, and species composition. *Biotropica*, *31*, 2−16.

Coe, J. A., Bessette-Kirton, E. K., & Geertsema, M. (2017). Increasing rock-avalanche size and mobility in Glacier Bay National Park and Preserve, Alaska detected from 1984 to 2016 Landsat imagery. *Landslides*, *15*(3), 393−407.

Collen, B., Whitton, F., Dyer, E. E., Baillie, J. E. M., Cumberlidge, N., Darwall, W. R. T., . . . Bohn, M. (2014). Global patterns of freshwater species diversity, threat and endemism. *Global Ecology and Biogeography*, *23*(1), 40−51.

Comfort, N. (2012). *The science of human perfection: How genes became the heart of American medicine*. New Haven, CT: Yale University Press.

Condit, R., Aguilar, S., Hernandez, A., Perez, R., Lao, S., Angehr, G., . . . Foster, R. B. (2004). Tropical forest dynamics across a rainfall gradient and the impact of an El Nino dry season. *Journal of Tropical Ecology*, *20*, 51−72.

Cooper, J., Baker, G. B., Double, M. C., Gales, R., Papworth, W., Tasker, M. L., & Waugh, S. M. (2006). The agreement on the conservation of Albatrosses and Petrels: Rationale, history, progress and the way forward. *Marine Ornithology*, *34*, 1−5.

Cooper, N., Brady, E., Steen, H., & Bryce, R. (2016). Aesthetic and spiritual values of ecosystems: Recognizing the ontological and axiological plurality of cultural ecosystem 'services'. *Ecosystem Services*, *21*(Part-B), 2018−2229.

Committee on Fisheries. (2018). *Progress in improving the information baseline and assessment of the contribution of small-scale fisheries in marine and inland waters*, thirty-third session, Rome, July 9−12, 2018. COFI/2018/Inf.18. Rome: FAO. Available from http://www.fao.org/3/MW788EN/mw788en.pdf.

Conover, M. R. (2002). *Resolving human-wildlife conflicts: The science of wildlife damage management*. Boca Raton, FL: CRC Press.

Copeland, J. P., McKelvey, K. S., Aubry, K. B., Landa, A., Persson, J., Inman, R. M., . . . May, R. (2010). The bioclimatic envelope of the wolverine: DO climatic constraints limit their geographic distribution? *Canadian Journal of Zoology*, *88*, 233−246.

Corlett, R. T. (2015). The Anthropocene concept in ecology and conservation. *Trends in Ecology & Evolution*, *30*, 36−41.

Corlett, R. T. (2016). The impacts of droughts in tropical forests. *Trends in Plant Science*, *21*(7), 584−593.

Cotto, O., Wessely, J., Georges, D., Klonner, G., Schmid, M., Dullinger, S., . . . Guillaume, F. (2017). A dynamic eco-evolutionary model predicts slow response of alpine plants to climate warming. *Nature Communications*, *8*, 15399.

Council of Europe. (2008). A perspective on climate change and invasive alien species. In: *Second meeting of the group of experts on biodiversity and climate change*. Convention on the Conservation of European Wildlife and Natural Habitats. T-PVS/Inf 5 rev. Available from https://rm.coe.int/168074629c.

Cowie, A. L., Orr, B. J., Sanchez, V. M. C., Chasek, P., Crossman, N. D., Erlewein, A., . . . Welton, S. (2018). Land in balance: The scientific conceptual framework for Land Degradation Neutrality. *Environmental Science & Policy*, *79*, 25−35.

CWM Convention (International Convention for the Control and Management of Ships' Ballast Water and Sediments). *Adoption of the final act on international convention for the control and management of ships' ballast water and sediments. BWM/CONF/36*. (2004). http://library.arcticportal.org/1913/1/International%20Convention%20for%20the%20Control%20and%20Management%20of%20Ships%27%20Ballast%20Water%20and%20Sediments.pdf.

Cubino, J. P., Subiros, J. V., & Lozano, C. B. (2015). Propagule pressure from invasive plant species in gardens in low-density suburban areas of the Costa Brava (Spain). *Urban Forestry & Urban Greening, 14*, 941−951.

Dai, A. (2013). Increasing drought under global warming in observations and models. *Nature Climate Change, 3*(1), 52−58.

Darwall, W., Bremerich, V., De Wever, A., Dell, A. I., Freyhof, J., Gessner, M. O., ... Weyl, O. (2018). The alliance for freshwater life: A global call to unite efforts for freshwater biodiversity science and conservation. *Aquatic Conservation: Marine and Freshwater Ecosystems, 28*, 1015−1022.

Davis, J., O'Grady, A. P., Dale, A., Arthington, A. H., Gell, P. A., Driver, P. D., ... Page, T. J. (2015). When trends intersect: The challenge of protecting freshwater ecosystems under multiple land use and hydrological intensification scenarios. *Science of the Total Environment, 534*, 65−78.

Decaëns, T., Jiménez, J. J., Gioia, C., Measey, G. J., & Lavelle, P. (2006). The values of soil animals for conservation biology. *European Journal of Soil Biology, 42*, S23−S38.

DeFries, R. S., Rudel, T., Uriarte, M., & Hansen, M. (2010). Deforestation driven by urban population growth and agricultural trade in the twenty-first century. *Nature Geoscience, 3*(3), 178−181.

DES-Australia (Department of Environment & Science, Australia). (2018). *Threats to wildlife.* <https://environment.des.qld.gov.au/wildlife/animals/threats> Updated on 23.03.18.

Dettinger, M., Udall, B., & Georgakakos, A. (2015). Western water and climate change. *Ecological Applications, 25*, 2069−2093.

De Vos, J. M., Joppa, L. N., Gittleman, J. L., Stephens, P. R., & Pimm, S. L. (2015). Estimating the normal background rate of species extinction: Background rate of extinction. *Conservation Biology, 29*, 452−462.

DeVries, T., Le Quere, C., Andrews, O., Berthet, S., Hauck, J., Ilyina, T., ... Seferian, R. (2019). Decadal trends in the ocean carbon sink. *Proceedings of the National Academy of Sciences of the United States of America, 116*(24), 11646−11651.

Di Marco, M., Chapman, S., Althor, G., Kearney, S., Besancon, C., Butt, N., ... Watson, J. E. (2017). Changing trends and persisting biases in three decades of conservation science. *Global Ecology and Conservation, 10*, 32−42.

Díaz, S., Fargione, J., Chapin, F. S., & Tilman, D. (2006). Biodiversity Loss Threatens Human Well-Being. *PLoS Biology, 4*, e277.

Díaz, S., Settele, J., Brondizio, E.S., Ngo, H. T., Gueze, M., Agard, J., ... Zayas, N. (2019). *Summary for policymakers of the global assessment report on biodiversity and ecosystem services of the Intergovernmental Science-Policy Platform on Biodiversity and Ecosystem Services (advance unedited version).* Available from https://ipbes.net/sites/default/files/2020-02/ipbes_global_assessment_report_summary_for_policymakers_en.pdf.

Diaz, S., Settle, J., Brondizio, E., Ngo, H. T., Arneth, A., Agard, J., ... Zayas, C. A. (2019). Pervasive human-driven decline of life on Earth points to the need for transformative change. *Science, 366*(6471), eaax3100.

Dickman, C., Pimm, S., & Cardillo, M. (2007). The pathology of biodiversity loss: The practice of conservation. In D. Macdonald, & K. Service (Eds.), *Key topics in conservation biology* (pp. 1−16). Blackwell Publishing Ltd.

Dickman, A. J., & Hazzah, L. (2016). Money, myths and man-eaters: Complexities of human-wildlife conflict. In F. M. Agelici (Ed.), *Problematic wildlife* (pp. 339−356). London: Springer.

Diez, J. M., D'Antonio, C. M., Dukes, J. S., Grosholz, E. D., Olden, J. D., Sorte, C. J. B., ... Miller, L. P. (2012). Will extreme climatic events facilitate biological invasions? *Frontiers in Ecology and the Environment, 10*(5), 249−257.

Ding, M., Zhang, M., Wong, J. L., Rogers, N. E., Ignarro, L. J., & Voskuhl, R. R. (1998). Cutting edge: antisense knockdown of inducible nitric oxide synthase inhibits induction of experimental autoimmune encephalomyelitis in SJL/J mice. *Journal of Immunology, 160*(6), 2560−2564.

Dirzo, R., & Raven, P. H. (2003). Global state of biodiversity and loss. *Annual Review of Environment Resources*, *28*, 137−167.

Dirzo, R., Young, H. S., Galetti, M., Ceballos, G., Isaac, N. J. B., & Collen, B. (2014). Defaunation in the Anthropocene. *Science (New York, N.Y.)*, *345*, 401−406.

Dobson, A., Lodge, D., Alder, J., Cumming, G. S., Keymer, J., McGlade, J., ... Xenopoulos, M. A. (2006). Habitat loss, trophic collapse, and the decline of ecosystem services. *Ecology*, *87*, 1915−1924.

Dodd, A. J., McCarthy, M. A., Ainsworth, N., & Burgman, M. A. (2016). Identifying hotspots of alien plant naturalization in Australia: Approaches and predictions. *Biological Invasions*, *18*, 631−645.

DOE (Department of Energy, United States). *Renewable fuel standard*. (n.d.). <https://afdc.energy.gov/laws/RFS.html>.

Dobrowski, S. Z., & Parks, S. A. (2016). Climate change velocity underestimates climate change exposure in mountainous regions. *Nature Communications*, *7*, 1−8.

Donald, P. F., & Evans, A. D. (2006). Habitat connectivity and matrix restoration: The wider implications of agri-environment schemes. *Journal of Applied Ecology*, *43*, 209−218.

Downing, A. S., van Nes, E. H., Mooij, W. M., & Scheffer, M. (2012). The resilience and resistance of an ecosystem to a collapse of diversity. *PLoS One*, *7*, 1−7.

Draheim, M. M., Madden, F., McCarthy, J.-B., & Parsons, E. C. M. (Eds.), (2015). *Human-wildlife conflict: Complexity in the marine environment*. Oxford: Oxford University Press.

Duan, Y., Edwars, J. S., & Dwiedi, Y. K. (2019). Artificial intelligence for decion-making in the era of big data—Evolution, challenges and research agenda. *International Journal of Information Management*, *48*, 63−71.

Dudgeon, D. (2019). Multiple threats imperil freshwater biodiversity in the Anthropocene. *Current Biology*, PR960−PR967.

Duffy, J. E. (2009). Why biodiversity is important to the functioning of real-world ecosystems. *Frontiers in Ecology and the Environment*, *7*(8), 437−444.

Duguma, L. A., Minang, P. A., & van Noordwijk, M. (2014). Climate change mitigation and adaptation in the land use sector: From complementarity to synergy. *Environmental Management*, *54*(3), 420−432.

Dullinger, S., Gattringer, A., Thuiller, W., Moser, D., Zimmermann, N. E., Guisan, A., ... Hulber, K. (2012). Extinction debt of high-mountain plants under twenty-first-century climate change. *Nature Climate Change*, *2*, 619−622.

Duran, A. P., Rauch, J., & Gaston, K. J. (2013). Global spatial coincidence between protected areas and metal mining activities. *Biological Conservation*, *160*(272−278), 88−405.

Dutfield, G. (2018). If we have never been modern, they have never been traditional: 'traditional knowledge', biodiversity, and the flawed ABS paradigm. In C. R. McManis, & B. Ong (Eds.), *Routledge handbook of biodiversity and the law* (pp. 276−290). New York: Routledge.

Early, R., Bradley, B. A., Dukes, J. S., Lawler, J. J., Olden, J. D., Blumenthal, D. M., ... Tatem, A. J. (2016). Global threats from invasive alien species in the twenty-first century and national response capacities. *Nature Communications*, *7*, 12485.

Ebi, K. L., Frumkin, H., & Hess, J. (2017). Protecting and promoting population health in the context of climate and other global environmental changes. *Anthropocene*, *19*, 1−12.

EC (European Commission). *Our life insurance, our natural capital: An EU biodiversity strategy to 2020*. (2011). <https://www.eea.europa.eu/policy-documents/our-life-insurance-our-natural>.

Edwards, D. P., Sloan, S., Weng, L., Dirks, P., Sayer, J., & Laurance, W. F. (2014). Mining and the African environment. *Conservation Letters*, *7*, 302−311.

Edwards, D. P., & Laurance, W. F. (2015). Preventing tropical mining disasters. *Science (New York, N.Y.)*, *350*, 1482.

eLAW. (2010). *Guidebook for evaluating mining project EIAs*. Eugene, OR: Environmental Law Alliance Worldwide. Available from https://www.elaw.org/files/mining-eia-guidebook/Full-Guidebook.pdf.

Ellis, R., Skehen, P., Li, S., Shintani, N., & Lambert, C. (2019). *Task-based language teaching: Theory and practice*. Cambridge, UK: Cambridge University Press.

Engler, R., Randin, C. F., Thuiller, W., Dullinger, S., Zimmermann, N. E., Araújo, M. B., ... Guisan, A. (2011). 21st century climate change threatens mountain flora unequally across Europe. *Global Change Biology, 17*, 2330–2341.

Engelhardt, M., Leclercq, P., Eidhammer, T., Kumar, P., Landgren, O., & Rasmussen, R. (2017). Melt water runoff in a changing climate (1951–2099) at Chhota Shigri Glacier, Western Himalaya, Northern India. *Annals of Glaciology, 58*(75), 47–58.

ESDAC (European Soil Data Centre). *Soil biodiversity and soil erosion*. (2018). <https://esdac.jrc.ec.europa.eu/themes/soil-biodiversity-and-soil-erosion>.

Estoque, R. C., Ooba, M., Avitabile, V., Hijoka, Y., DasGupta, R., Togawa, T., & Murayam, Y. (2019). The future of Southeast Asia's forests. *Nature Communications., 10*, 1829.

EC (European Commission). *Invasive alien species—Framework the identification of invasive alien species of EU concern. ENV.B.2/ETU/2013/0026*. (2013). <https://ec.europa.eu/environment/nature/invasivealien/docs/Final%20report_12092014.pdf>.

EP (European Parliament). (2008). *Land degradation and desertification*. Brussels: European Parliament, Policy Department. Available from https://www.europarl.europa.eu/RegData/etudes/etudes/join/2009/416203/IPOL-ENVI_ET(2009)416203_EN.pdf.

EP (European Parliament). (2007). Text of European Parliament resolution on climate change. 14 February 2007. Available at: https://www.europarl.europa.eu/doceo/document/TA-6-2007-0038_EN.html?redirect.

EP (European Parliament). (2009). Text adopted: Combating climate change. 11 March 2009, Strasburg. Available at: https://www.europarl.europa.eu/doceo/document/TA-6-2009-0121_EN.html.

EPA (Environmental Protection Agency, USA). (2018). *Toxic Substances Control Act (TSCA) chemical substances inventory*. Washington, DC.

EU (European Union). *What is the EU FLEGT action plan?*. (n.d.a). <http://www.euflegt.efi.int/flegt-action-plan>.

EU (European Union). *Renewable energy directive*. (n.d.b). <https://ec.europa.eu/energy/topics/renewable-energy/renewable-energy-directive/overview_en>.

EU (European Union). *Glossary of key terms in the context of "access & benefit-sharing"*. (n.d.c.). <https://ec.europa.eu/environment/nature/biodiversity/international/abs/pdf/Glossary%20for%20Europa.pdf>.

EU (European Union). *Genetic resources*. (n.d.d). <https://biodiversity.europa.eu/topics/genetic-resources#:~:text=Animals%2C%20plants%2C%20micro%2Dorganisms,grouped%20under%20the%20concept%20Agrobiodiversity>.

EU (European Union). *Organic matter decline. Fact sheet no. 3*. (2009). <https://esdac.jrc.ec.europa.eu/projects/SOCO/FactSheets/ENFactSheet-03.pdf>.

EU (European Union). *Commission Implementing Regulation (EU) No 749/2014 of 30 June 2014 on structure, format, submission processes and review of information reported by Member States pursuant to Regulation (EU) No 525/2013 of the European Parliament and of the Council*. (2014). <https://eur-lex.europa.eu/legalcontent/EN/TXT/?uri=uriserv:OJ.L_.2014.203.01.0023.01.ENG>.

Esquivel-Muelbert, A., Baker, T. R., Dexter, K. G., Lewis, S. L., Brienen, R. J. W., Feldpausch, T. R., ... Phillips, O. L. (2018). Compositional response of Amazon forests to climate change. *Global Change Biology, 25*(1), 39–56.

Evans, J., Garcia-Oliva, F., Hussein, I. A. G., Iqbal, M. M., Kimutai, J., Knowles, T., ... Weltz, M. (2019). Desertification. In: *Climate change and land: An IPCC special report on climate change, desertification, land degradation, sustainable land management, food security, and greenhouse gas fluxes in terrestrial ecosystems*. Available from https://www.ipcc.ch/site/assets/uploads/2019/08/2d.-Chapter-3_FINAL.pdf.

FAO (Food and Agriculture Organization). *Wood energy consumption patterns*. (n.d.). <http://www.fao.org/3/w7744e/w7744e08.htm>.

FAO (Food and Agriculture Organization). *International undertaking on plant genetic resources. Resolution 8/83 (1983)*. (1983). <http://www.fao.org/wiews-archive/docs/Resolution_8_83.pdf>.

FAO (Food and Agriculture Organization). (1997). *The state of the world's plant genetic resources for food and agriculture*. Rome: FAO. Available from http://www.fao.org/3/a-w7324e.pdf.

FAO (Food and Agriculture Organization). (1999). *Prevention of land degradation, enhance of carbon sequestration, and conservation of biodiversity through land use change and sustainable land management with a focus on Latin America and the Caribbean*. Rome: FAO. Available from http://www.fao.org/3/a-bc909e.pdf.

FAO (Food and Agriculture Organization). (2002). Proceedings: Second expert meeting on harmonizing forest-related definitions for use by various stakeholders. Rome: FAO.

FAO (Food and Agriculture). (2007). *The state of the world's animal genetic resources for food and agriculture*. Rome: FAO. Available from http://www.fao.org/3/a1250e/a1250e.pdf.

FAO (Food and Agriculture Organization). *International treaty on plant genetic resources for food and agriculture (ITPGRFA)*. (2009). <http://www.fao.org/3/a-i0510e.pdf>.

FAO (Food and Agriculture Organization). (2010a). What wood-fuels can do to mitigate climate change. *Forestry paper 162*. Rome: FAO.

FAO (Food and Agriculture Organization). (2010b). *The second report on the state of the world's plant genetic resources for food and agriculture*. Rome: FAO. Available from http://www.fao.org/3/i1500e/i1500e.pdf.

FAO (Food and Agriculture Organization). (2011). *Assessing forest degradation: Towards the development of globally applicable guidelines*. Rome: FAO.

FAO. (Food and Agriculture Organization). (2012). *The state of world fisheries and aquaculture*. Rome: FAO. Available from http://www.fao.org/3/a-i2727e.pdf.

FAO (Food and Agriculture Organization). (2014). *The state of the world's forest genetic resources*. Rome: FAO. Available from http://www.fao.org/3/a-i3825e.pdf.

FAO (Food and Agriculture Organization). (2015). Global forest resources assessment 2015: How are the world's forests changing? (Second Edition). Rome: FAO. Available from http://www.fao.org/3/a-i4793e.pdf.

FAO (Food and Agriculture Organization). (2016a). *Report of the 32nd session of the committee on fisheries (Rome, 11–15 July 2016)*, Fortieth session, Rome, July 3–8, 2017. Committee on Fisheries. C 2017/23. Available from http://www.fao.org/3/a-mr484e.pdf.

FAO (Food and Agricultural Organization). (2016b). *State of the world's forests 2016. Forests and agriculture: Land-use challenges and opportunities*. Rome: FAO. Available from http://www.fao.org/3/a-i5588e.pdf.

FAO (Food and Agriculture Organization). (2016c). Global forest resources assessment 2015. How are the world's forests changing? (Second edition). Rome: FAO. Available from http://www.fao.org/3/ai4793e.pdf.

FAO (Food and Agriculture Organization). (2017). *FAO and the SDGs indicators: Measuring up to the agenda 2030 for sustainable development*. Rome: FAO. Available from http://www.fao.org/3/a-i6919e.pdf.

FAO (Food and Agriculture Organization). (2018). *Soil pollution: A hidden reality*. Rome: FAO. Available from http://www.fao.org/3/I9183EN/i9183en.pdf.

FAO (Food and Agriculture Organization). (2018a). *The state of world fisheries and aquaculture 2018—Meeting the sustainable development goals*. Rome: FAO. Available from http://www.fao.org/3/i9540en/I9540EN.pdf.

FAO (Food and Agriculture Organization). (2019a). *The state of the world's aquatic genetic resources for food and agriculture*. Rome: FAO. Available from http://www.fao.org/3/ca5256en/CA5256EN.pdf.

FAO (Food and Agriculture Organization). (2019b). *The state of the world's biodiversity for food and agriculture*. Rome: FAO. Available from http://www.fao.org/3/CA3129EN/ca3129en.pdf.

FAO (Food and Agriculture Organization). *Risk status of animal genetic resources*. (2020a). <http://www.fao.org/dad-is/risk-status-of-animal-genetic-resources/en/> Updated 21.09.20.
FAO (Food and Agriculture Organization). *The future of our land: Facing the challenge*. (2020b). <http://www.fao.org/3/a-x3810e.pdf>.
FAO (Food and Agriculture Organization). (2020c). *The state of the world's forests 2020. Forests, biodiversity and people*. Rome: FAO.
Fahrig, L. (2003). Effects of habitat fragmentation on biodiversity. *Annual Review of Ecology, Evolution, and Systematics*, *34*, 487–515.
Fahrig, L., Girard, J., Duro, D., Pasher, J., Smith, A., Javorek, S., ... Tischendorf, L. (2015). Farmlands with smaller crop fields have higher within- field biodiversity. *Agriculture, Ecosystems and Environment*, *200*, 219–234.
Farinotti, D., Round, V., Huss, M., Compagno, L., & Zekollari, H. (2019). Large hydropower and water-storage potential in future glacier-free basins. *Nature*, *575*, 341–344.
Faulkner, K. T., Robertson, M. P., Rouget, M., & Wilson, J. R. U. (2015). Understanding and managing the introduction pathways of alien taxa: South Africa as a case study. *Biological Invasions*, *18*, 73–87.
Fedder, B. (2013). *Marine genetic resources, access and benefit sharing—Legal and biological perspectives*. London: Routledge.
Fei, S., Philips, J., & Shouse, M. (2014). Biogeomorphic impacts of invasive species. *Annual Review of Ecology, Evolution and Systematics*, *45*, 69–87.
Fell, J., Grodzicki, M., Martin, R., & O'Brien, E. (2017). A role for systemic asset management companies in solving Europe's non-performing loan problems. *European Economy*, *1*, 71–85.
Ferreira, J., Aragao, L. E. O. C., Barlow, J., Barreto, P., Berenguer, E., Bustamante, M., ... Zuanon, J. (2014). Brazil's environmental leadership at risk. *Science (New York, N.Y.)*, *346*, 706–707.
Fetzel, T., Havlik, P., Herroro, M., & Erb, K.-H. (2017). Seasonality constraints to livestock grazing intensity. *Global Change Biology*, *23*(4), 1636–1647.
Fisher, B. (2010). African exception to drivers of deforestation. *Nature Geoscience*, *3*, 375–376.
Fischer, J., & Lindenmayer, D. B. (2007). Landscape modification and habitat fragmentation: A synthesis. *Global Ecology and Biogeography*, *16*, 265–280.
Fischlin, A., Midgley, G. F., Price, J., Leemans, R., Gopal, B., Turley, C., ... Velichko, A. (2007). *Ecosystems, their properties, goods and services. Climate change 2007: Impacts, adaptation and vulnerability. Contribution of working group II to the fourth assessment report of the intergovernmental panel of climate change* (pp. 211–272). Cambridge: Cambridge University Press.
Fleming, S. W., & Dahlke, H. E. (2014). Modulation of linear and nonlinear hydroclimatic dynamics by mountain glaciers in Canada and Norway: Results from information-theoretic polynomial selection. *Canadian Water Resources Journal*, *39*(3), 324–341.
Fletcher, R. A., Brooks, R. K., Lakoba, V. T., Sharma, G., Heminger, A. R., Dickinson, C. C., & Barney, J. N. (2019). Invasive plants negatively impact native, but not exotic, animals. *Global Change Biology*, *25*(11), 3694–3705.
Flitcroft, R., Cooperman, M. S., Harrison, I. J., Juffe-Bignoli, D., & Boon, P. J. (2019). Theory and practice to conserve freshwater biodiversity in the Anthropocene. *Aquatic Conservation*, *29*(7), Special Issue.
Foley, J. A. (2005). Global consequences of land use. *Science (New York, N.Y.)*, *309*(5734), 570–574.
Foresight. (2011). *The future of food and farming (2011) final project report*. London: The Government Office for Science. Available from https://assets.publishing.service.gov.uk/government/uploads/system/uploads/attachment_data/file/288329/11-546-future-of-food-and-farming-report.pdf.
Forster, J., Hirst, A. G., & Atkinson, D. (2012). Warming-induced reductions in body size are greater in aquatic than terrestrial species. *Proceedings of the National Academy of Sciences of the United States of America*, *109*, 19310–19314.

Forzieri, G., Alkama, R., Miralles, D. G., & Cescatti, A. (2017). Satellites reveal contrasting responses of 32 regional climate to the widespread greening of Earth. *Science (New York, N.Y.), 356*(6343), 1180–1184.

Foxcroft, L. C., Parsons, M., McLoughlin, C. A., & Richardson, D. M. (2008). Patterns of alien plant distribution in a river landscape following an extreme flood. *South African Journal of Botany, 74*, 463–475.

Foxcroft, L. C., Richardson, D. M., & Wilson, J. R. U. (2008). Ornamental plants as invasive aliens: Problems and solutions in Kruger National Park, South Africa. *Environmental Management, 41*, 32–51.

Franks, D. M. (2015). *Environment. Mountain movers: Mining, sustainability and the agents of change*. London: Routledge.

Frei, E. R., Bianchi, E., Bernareggi, G., Bebi, P., Dawes, M. A., Brown, C. D., ... Rixen, C. (2018). Biotic and abiotic drivers of tree seedling recruitment across an alpine treeline ecotone. *Scientific Reports, 8*(1), 10894.

Freyhof, J., & Brooks, E. (2011). *European red list of freshwater fishes*. Luxembourg: Publications Office of the European Union. Available online.

Fryxell, J., Sinclair, A., & Caughley, G. (2014). *Wildlife ecology, conservation, and management* (3rd ed.). West Sussex: Wiley-Blackwell.

Fuchs E., Stein E., Strunz G., Strobl C., & Frey C. (2015). Fire monitoring—The use of medium resolution satellites (AVHRR, MODIS, TET) for long time series processing and the implementation in user driven applications and services. In: *The International Archives of the Photogrammetry, Remote Sensing and Spatial Information Sciences. Volume XL-7/W3, 2015. 36th International symposium on remote sensing of environment*, Berlin, Germany, May 11–15, 2015. Available from https://elib.dlr.de/95769/1/Final-Paper-isprsarchives-XL-7-W3-797-2015.pdf.

Gaertner, M., Wilson, J. R., Cadotte, M. W., MacIvor, J. S., Zenni, R. D., & Richardson, D. M. (2017). Non-native species in urban environments: Patterns, processes, impacts and challenges. *Biological Invasions, 19*, 3461–3469.

Gallardo, B., Bacher, S., Bradley, B., Comin, F. A., Gallien, L., Jeschke, J. M., ... Vila, M. (2019). InvasiBES: Understanding and managing the impacts of invasive alien species on biodiversity and ecosystem services. *NeoBiota., 50*, 109–122.

Gan, F., & Wang, T. (2018). Water and soil loss from landslide deposits as a function of gravel content in the Wenchuan earthquake area, China, revealed by artificial rainfall simulations. *PLoS One, 13*(5), e0196657.

Gao, J. (2019). *Global population projection grids based on shared socioeconomic pathways (SSPs), downscaled 1-km grids, 2010–2100*. Palisades, NY. Available from https://catalog.data.gov/dataset/global-population-projection-grids-based-on-shared-socioeconomic-pathways-ssps-downsc-2010.

Garófano-Gómez, V., Martínez-Capel, F., Peredo-Parada, M., Marín, E. J. O., Mas, R. M., Costa, R. M. S., & Pinar-Arenas, J. L. (2011). Assessing hydro-morphological and floristic patterns along a regulated Mediterranean river: The Serpis River (Spain). *Limnetica, 30*, 307–328.

Gatti, L. V., Gloor, M., Miller, J. B., Doughty, C. E., Malhi, Y., Domingues, L. G., ... Lloyd, J. (2014). Drought sensitivity of Amazonian carbon balance revealed by atmospheric measurements. *Nature, 506*, 76–80.

García-Vega, D., & Newbold, T. (2020). Assessing the effects of land use on biodiversity in the world's drylands and Mediterranean environments. *Biodiversity & Conservation, 29*, 393–408.

Gardi, C., Jeffery, S., & Saltelli, A. (2013). An estimate of potential threats levels to soil biodiversity in EU. *Global Change Biology, 19*, 1538–1548.

Gazol, A., & Camarero, J. J. (2016). Functional diversity enhances silver fir growth resilience to an extreme drought. *Journal of Ecology, 104*, 1063–1075.

Gazol, A., Camarero, J. J., Anderegg, W. R. L., & Vicente-Serrano, S. M. (2017a). Impacts of droughts on the growth resilience of Northern Hemisphere forests. *Global Ecology and Biogeography, 26*, 166–176.

Gazol, A., Ribas, M., Gutiérrez, E., & Camarero, J. J. (2017b). Aleppo pine forests from across Spain show drought-induced growth decline and partial recovery. *Agriculture and Forest Meteorology, 232*, 186–194.

Gazol, A., Sangüesa-Barreda, G., Granda, E., & Camarero, J. J. (2017c). Tracking the impact of drought on functionally different woody plants in a Mediterranean scrubland ecosystem. *Plant Ecology*, *218*, 1009−1020.

Geburek, T., & Müller, F. (2005). How can silvicultural management contribute to genetic conservation? In T. Geburek, & J. Turok (Eds.), *Conservation and management of forest genetic resources in Europe* (pp. 651−669). Zvolen: Arbora Publishers.

GEF (Global Environment Facility). *Illegal wildlife trade*. (2020). <http://www.thegef.org/topics/illegal-wildlife-trade>.

Ghazoul, J., Burivalova, Z., Garcia-Ulloa, J., & King, L. A. (2015). Conceptualizing forest degradation. *Trends in Ecology & Evolution*, *30*(10), 622−632.

Ghosal, S., & Kjosavik, D. J. (2015). Living with leopards: Negotiating morality and modernity in Western India. *Society & Natural Resources*, *28*, 1092−1107.

Gibson, R., Tanner, C., & Wagner, A. F. (2013). Preferences for truthfulness: Heterogeneity among and within individuals. *American Economic Review*, *103*(1), 532−548.

Giorgi, F., & Gutowski, W. J. (2016). Coordinated experiments for projections of regional climate change. *Current Climate Change Reports*, *2*, 202−210.

Giove, S., Brancia, A., Satterstrom, F. K., & Linkov, I. (2009). Decision support systems and environment: Role of MCDA. In A. Marcomini, G. W. Suter, & A. Critto (Eds.), *Decision support systems for risk-based management of contaminated sites* (pp. 53−73). New York: Springer Science Business Media.

Glawzeski, J. (2000). *Environmental law in South Africa*. Oxford: Butterworth Publishers.

Global Carbon Project. *Global Carbon Project historical emissions*. (2016). <http://www.globalcarbonproject.org/carbonbudget/17/data.htm>.

GFED (Global Fire Emissions Databases). *Global fire emissions databases, version4 (GFEDv4)*. (2018). <https://daac.ornl.gov/VEGETATION/guides/fire_emissions_v4.html> Revised 24.02.18.

GFOI (Global Forest Observation Initiative). (2013). *Integrating remote-sensing and ground-based observations for estimation of emissions and removals of greenhouse gases in forests* (p. 226). Methods and Guidance from the Global Forest: Pub: Group on Earth Observations Initiative 2. Available from https://redd.unfccc.int/uploads/2_77_redd_20140218_mgd_report_gfoi.pdf.

Gleason, K. E., McConnell, J. R., Arienzo, M. M., Chellman, N., & Calvin, W. M. (2019). Fourfold increase in solar forcing on snow in western United States burned forests since 1999. *Nature Communications*, *10*(1), 1−8.

Goldewijk, K. K., & Ramankutty, N. (2004). Land cover change over the last three centuries due to human activities: Assessing the differences between two new global data sets. *GeoJournal*, *61*, 335−344.

González-Moreno, P., Diez, J. M., Ibáñez, I., Font, X., & Vilà, M. (2014). Plant invasions are context-dependent: Multiscale effects of climate, human activity and habitat. *Diversity and Distributions*, *20*, 720−731.

Gordijn, B., & ten Have, H. (2012). Ethics of mitigation, adaptation and geoengineering. *Medicine, Health Care, and Philosophy*, *15*, 1−2.

Goulart, A. C., Macario, K. D., Scheel-Ybert, R., Alves, E. Q., Bachelet, C., Pereira, B. B., ... Feldpausch, T. R. (2017). Charcoal chronology of the Amazon forest: A record of biodiversity preserved by ancient fires. *Quaternary Geochronology*, *41*, 180−186.

Graae, B. J., Vandvik, V., Armbruster, W. J., Eiselhardt, W. L., Svenning, J.-C., Hylander, K., ... Lenoir, J. (2018). Stay or go−how topographic complexity influences alpine plant population and community responses to climate change. *Perspectives in Plant Ecology Evolution.*, *30*, 41−50.

Green, R. E., Cornell, S. J., Scharlemann, J. P. W., & Balmford, A. (2005). Farming and the fate of wild nature. *Science (New York, N.Y.)*, *307*, 550−555.

Greenwood, S., Ruiz-Benito, P., Martínez-Vilalta, J., Lloret, F., Kitzberger, T., Allen, C. D., ... Jump, A. S. (2017). Tree mortality across biomes is promoted by drought intensity, lower wood density and higher specific leaf area. *Ecology Letters*, *20*, 539−553.

Greiber, T. (2011). *Access and benefit sharing in relation to marine genetic resources from areas beyond national jurisdiction—A possible way forward*. Bonn: Federal Agency for Nature Conservation & IUCN.

Grieber, T., Moreno, S. P., Ahren, M., Carrasco, J. N., Kamau, E. C., Medaglia, J. C., ... Welch, F. C. (2012). *An explanatory guide to the Nagoya Protocol on access and benefit-sharing*. Gland: IUCN.

Grill, G., Lehner, B., Thieme, M., Geenen, B., Tickner, D., Antonelli, F., ... Zarfl, C. (2019). Mapping the world's free-flowing rivers. *Nature, 569*, 215—221.

Grimm, N. B., Chapin, F. S., Bierwagen, B., Gonzalez, P., Groffman, P. M., Luo, Y., ... Williamson, C. E. (2013). The impacts of climate change on ecosystem structure and function. *Frontiers in Ecology and the Environment, 11*, 474—482.

Grzybowski, M., & Glińska-Lewczuk, K. (2019). Principal threats to the conservation of freshwater habitats in the continental biogeographical region of Central Europe. *Biodiversity Conservation, 28*, 4065—4097.

Guruswamy, L. D. (2017). *International environmental law in a nutshell*. St. Paul, MN: West Academic Publishing.

Haddad, N. M., Brudvig, L. A., Clobert, J., Davies, K. F., Gonzalez, A., Holt, R. D., ... Townshend, J. R. (2015). Habitat fragmentation and its lasting impact on Earth's ecosystems. *Science Advances, 1*(2), e1500052.

Haeberli, W., & Whiteman, C. (2015). *Snow and ice-related hazards. Risks, and Disasters*. Amsterdam: Elsevier.

Hanes, C. C., Wang, X., Jain, P., Parisien, M. A., Little, J. M., & Flannigan, M. D. (2019). Fire-regime changes in Canada over the last half century. *Canadian Journal of Forest Research, 49*(3), 256—269.

Hahn, W. A., & Knoke, T. (2010). Sustainable development and sustainable forestry: Analogies, differences, and the role of flexibility. *European Journal of Forest Research, 129*, 787—801.

Hahn, T., McDermott, C., Ituarte-Lima, C., Schultz, M., Green, T., & Tuvendal, M. (2015). Purposes and degrees of commodification: Economic instruments for biodiversity and ecosystem services need not rely on markets or monetary valuation. *Ecosystem Services, 16*, 74—82.

Hall, L. S., Krausman, P. R., & Morrison, M. L. (1997). The habitat concept and a plea for standard terminology. *Wildlife Society Bulletin, 25*, 173—182.

Hambler, C., & Canney, S. M. (2013). *Conservation*. Cambridge: Cambridge University Press.

Hance, J. (2017, 5 April). Climate change impacting 'most' species on Earth, even down to their genomes. *The Guardian*. Available from https://www.theguardian.com/environment/radical-conservation/2017/apr/05/climate-change-life-wildlife-animals-biodiversity-ecosystems-genetics.

Hansen, M. C., Potapov, P., & Tyuavina, A. (2019). Comment on "Tropical forests are a net carbon source based on aboveground measurements of gain and loss". *Science (New York, N.Y.), 363*(6423), eaar3629.

Hansis, E., Davis, S. J., & Pongratz, J. (2015). Relevance of methodological choices for accounting of land 9 use change carbon fluxes. *Global Biogeochemical Cycles, 29*, 1230—1246.

Hantson, S., Arneth, A., Harrison, S., Kelley, D., Prentice, I., Rabin, S., ... Yue, C. (2016). The status and challenge of global fire modelling. *Biogeosciences., 13*, 3359—3375.

Harihar, A., Chanchani, P., Sharma, R. K., Vattakaven, J., Gubbi, S., Pandav, B., & Noon, B. (2013). Conflating "co-occurrence" with "coexistence". *Proceedings of the National Academy of Sciences of the United States of America, 110*, E109.

Harpold, A. A., Dettinger, M. D., & Rajagopal, S. (2017). Defining snow drought and why it matters. *EOS—Earth and Space Science News, 98*, EO068775.

Harpold, A. A., & Brooks, P. D. (2018). Humidity determines snowpack ablation under a warming climate. *Proceedings of the National Academy of Sciences of the United States of America, 115*, 1215, LP-1220.

Harris, J. A. (2009). Soil microbial communities and restoration ecology: Facilitators or followers? *Science (New York, N.Y.), 325*(5940), 573—574.

Harvey, B. P., Gwynn-Jones, D., & Moore, P. J. (2013). Meta-analysis reveals complex marine biological responses to the interactive effects of ocean acidification and warming. *Ecology and Evolution, 3*(4), 1016—1030.

Hautier, Y., Tilman, D., Isbell, F., Seabloom, E. W., Borer, T., & Reich, P. B. (2015). Anthropogenic environmental changes affect ecosystem stability via biodiversity. *Science (New York, N.Y.)*, *348*(6232), 336−339.

Havel, J. E., Kovalenko, K. E., Thomaz, S. M., Amalfitano, S., & Kats, L. B. (2015). Aquatic invasive species: Challenges for the future. *Hydrobiologia*, *750*(1), 147−170.

Hayman, R. B., Harvey, R. G., Mazzotti, F. J., Israel, G. D., & Woodward, A. R. (2014). Who complains about alligators? Cognitive and situational factors influence behavior toward wildlife. *Human Dimensions of Wildlife*, *19*, 481−497.

He, X., Burgess, K. S., Gao, L. M., & Li, D. Z. (2019a). Distributional responses to climate change for alpine species of Cyananthus and Primula endemic to the Himalaya-Hengduan Mountains. *Plant Diversity*, *41*, 26−32.

He, F., Zarfl, C., Bremerich, V., David, J. N. W., Hogan, Z., Kalinkat, G., . . . Jähnig, S. C. (2019b). The global decline of freshwater megafauna. *Global Change Biology*, *25*(11), 3883−3892.

Hejda, M., Pysek, P., & Jarosik, V. (2009). Impact of invasive plants on the species richness, diversity and composition of invaded communities. *Journal of Ecology*, *97*(3), 393−403.

Herrera, B. (n.d.). *Mountain ecosystems*. IUCN. Available from https://www.iucn.org/commissions/commission-ecosystem-management/our-work/cems-specialist-groups/mountain-ecosystems.

Hertel, T. W., West, T. A. P., & Villoria, N. B. (2019). A review of global-local-global linkages in economic land-use/cover change models. *Environmental Research Letters*, *14*(5), 053003.

Heshmati, G. A., & Squires, V. R. (2013). *Combating desertification in Asia*. Africa and the Middle East. New York: Springer.

Hillstrom, K., & Hillstrom, L. C. (2003). *Asia: A continental overview of environmental issues*. CA: ABC-CLIO.

Hoare, R. (2012). Lessons from 15 years of human elephant conflict mitigation: Management considerations involving biological, physical and governance issues in Africa. *Pachyderm*, *51*, 60−74.

Hobbs, R. J., Arico, S., Aronson, J., Baron, J. S., Bridgewater, P., Cramer, V. A., . . . Zobel, M. (2006). Novel ecosystems: Theoretical and management aspects of the new ecological world order. *Global Ecology and Biogeography*, *15*, 1−7.

Hock, R. G., Rasul, C., Adler, B. C., Gruber, S., Hirabayashi, Y., Jackson, M., . . . Steltzer, H. (2019). High mountain areas. In: *IPCC special report on the ocean and cryosphere in a changing climate*. Available from https://www.ipcc.ch/site/assets/uploads/sites/3/2019/11/06_SROCC_Ch02_FINAL.pdf.

Hoegh-Guldberg, O., Mumby, P. J., Hooten, A. J., Steneck, R. S., Greenfield, P., Gomez, E., . . . Hatziolos, M. E. (2007). Coral reefs under rapid climate change and ocean acidification. *Science (New York, N.Y.)*, *318*, 1737−1742.

Hoegh-Guldberg, O., Jacob, D., Taylor, M., Bindi, M., Brown, S., Camilloni, I., . . . Zhou, G. (2018). Impacts of 1.5°C global warming on natural and human systems. In: *Global warming of 1.5°C. An IPCC special report on the impacts of global warming of 1.5°C above pre-industrial levels and related global greenhouse gas emission pathways, in the context of strengthening the global response to the threat of climate change, sustainable development, and efforts to eradicate poverty*. Available from https://www.ipcc.ch/site/assets/uploads/sites/2/2019/02/SR15_Chapter3_Low_Res.pdf.

Hofstad, O., Kohlin, G., & Namaalwa, J. (2009). How can emissions from wood-fuel be reduced? In A. Angelsen, M. Brockhaus, M. Kanninen, E. Sills, W. D. Sundelin, & S. Wertz-Kanounnikoff (Eds.), *Realising REDD + : National strategy and policy options* (pp. 237−248). Bogor: Center for International Forestry Research. Available from https://www.cifor.org/publications/pdf_files/Books/BAngelsen090219.pdf.

Holl, K. D., & Aide, T. M. (2011). When and where to actively restore ecosystems? *For Ecology and Management*, *261*(10), 1558−1563.

References

Hood, G. W., & Naiman, R. J. (2000). Vulnerability of riparian zones to invasion by exotic vascular plants. *Plant Ecology, 148*, 105−114.

Hooper, D. U., Chapin, F. S., Ewel, J. J., Hector, A., Inchausti, P., Lavorel, S., ... Wardle, D. A. (2005). Effects of biodiversity on ecosystem functioning: A consensus of current knowledge. *Ecological Monographs, 75*, 3−35.

Hosonuma, N., Herold, M., De Sy, V., De Fries, R. S., Brockhaus, M., Verchot, L., ... Romijn, E. (2012). An assessment of deforestation and forest degradation drivers in developing countries. *Environmental Research Letters, 7*(4), 044009.

Houghton, R. A. (2010). How well do we know the flux of CO_2 from land-use change? *Tellus, B 62*, 337−351.

Houghton, R. A., & Nassikas, A. A. (2017). Negative emissions from stopping deforestation and forest degradation globally. *Global Change Biology, 24*(1), 350−359.

Huang, J., Li, Y., Fu, C., Chen, F., Fu, Q., Dai, A., ... Wang, G. (2017). Dryland climate change: Recent progress and challenges. *Reviews of Geophysics, 55*(3), 719−778.

Hudson, P. F., & Alcantara-Ayala, I. (2006). Ancient and modern perspectives on land degradation. *Catena, 65*, 102−106.

Huggel, C., Carey, M., Clague, J. J., & Kääb, A. (Eds.), (2015). *The high-mountain cryosphere: Environmental changes and human risks*. Cambridge: Cambridge University Press.

Hulme, P. E., Bacher, S., Kenis, M., Klotz, S., Kühn, I., Minchin, D., ... Vilà, M. (2008). Grasping at the routes of biological invasions: A framework for integrating pathways into policy. *Journal of Applied Ecology, 45*, 403−414.

Hulme, P. E., Pyšek, P., Jarošík, V., Pergl, J., Schaffner, U., & Vilà, M. (2013). Bias and error in understanding plant invasion impacts. *Trends in Ecology and Evolution, 28*, 212−218.

Hupp, C. R., & Bornette, G. (2003). Vegetation as a tool in the interpretation of fluvial geomorphic processes and landforms in humid temperate areas. In G. M. Kondolf, & H. Piégay (Eds.), *Tools in fluvial geomorphology*. Chichester: John Wiley and Sons, Ltd,: 269−288.

Hurtt, G., Chini, L., Frolking, S., Betts, R., Feddema, J., Fischer, G., ... Wang, Y. (2011). Harmonization of land-use scenarios for the period 1500−2100: 600 years of global gridded annual land-use transitions, wood harvest, and resulting secondary lands. *Climatic Change, 109*, 117−161.

Huss, M., Bookhagen, B., Huggel, C., Jacobsen, D., Bradley, R. S., Clague, J. J., ... Winder, M. (2017). Toward mountains without permanent snow and ice. *Earth's Future, 5*(5), 418−435.

ICMM (International Council on Mining and Metals). (2016). *Role of mining in national economies: Mining contribution index* (3rd ed., supplement). London: ICMM. Available online.

IMO (International Meteorological Organization). (2019). *Addressing invasive species in ships' ballast water treaty enter into force*. IMO. Available from https://www.imo.org/en/MediaCentre/PressBriefings/Pages/21-BWM-Amendments-EIF-.aspx.

IPBES (Intergovernmental Science-Policy Platform on Biodiversity and Ecosystem Services). (2019). *Summary for policymakers of the global assessment report on biodiversity and ecosystem services of the intergovernmental science-policy platform on biodiversity and ecosystem services*. Bonn: IPBES secretariat.

IHCAP (Himalayas. Indian Himalayas Climate Adaptation Programme). (2017). *Mountain and lowland linkages: A climate change perspective in the Himalayas*. Indian Himalayas Climate Adaptation Programme (IHCAP). Available from http://ihcap.in/?media_dl = 872.

IPCC (Intergovernmental Panel on Climate Change). *Guidelines for national greenhouse gas inventories, volume-4: Forest land*. (2006). <https://www.ipcc-nggip.iges.or.jp/public/2006gl/pdf/4_Volume4/V4_04_Ch4_Forest_Land.pdf>.

IPCC (Intergovernmental Panel on Climate Change). (2007). *Climate change 2007: The physical science basis. Contribution of working group I to the fourth assessment report of the intergovernmental panel on climate change*. Cambridge: Cambridge University Press.

IPCC (Intergovernmental Panel on Climate Change). (2014). *Climate change 2014: Synthesis report. Contribution of working groups I, II and III to the fifth assessment report of the international panel on climate change*. Cambridge: Cambridge University Press.

IPCC (Intergovernmental Panel on Climate Change). *Global warming of 1.5°C. An IPCC Special Report on the impacts of global warming of 1.5°C above pre-industrial levels and related global greenhouse gas emission pathways, in the context of strengthening the global response to the threat of climate change, sustainable development, and efforts to eradicate poverty*. (2018). <https://www.ipcc.ch/site/assets/uploads/sites/2/2019/06/SR15_Full_Report_High_Res.pdf>.

IPCC (Intergovernmental Panel on Climate Change). *Climate change and land: An IPCC special report on climate change, desertification, land degradation, sustainable land management, food security, and greenhouse gas fluxes in terrestrial ecosystems*. (2019). <https://www.ipcc.ch/site/assets/uploads/2019/11/SRCCL-Full-Report-Compiled-191128.pdf>.

IPK (Leibniz Institute of Plant Genetics and Crop Plant Research). *Mansfeld's world database of agriculture and horticultural crops*. (2017). <https://mansfeld.ipk-gatersleben.de/apex/f?p = 185:3>.

Ishaq Khan, Y. (2007). Traditional knowledge, genetic resources and developing countries in Asia: The concerns, (2007–2008) at 94. *Wake Forest Intellectual Property Law Journal*, 8(1), 82–127. Available from http://ipjournal.law.wfu.edu/files/2009/09/article.8.81.pdf.

Ismael, H. (2015). Evaluation of present-day climate-induced desertification in El-Dakhla Oasis, Western Desert of Egypt, based on integration of MEDALUS method, GIS and RS techniques. *Present Environment and Sustainable Development*, 9(2), 47–72.

IUCN (International Union of Conservation of Nature). (2012). *IUCN red list categories and criteria: Version 3.1* (2nd ed.). Cambridge. Available from https://www.iucn.org/km/content/iucn-red-list-categories-and-criteriaversion-31-second-edition.

IUCN (International Union of Conservation of Nature). (2017). Invasive alien species and climate change. *IUCN Issue Brief*. Gland: IUCN. Available from https://www.iucn.org/sites/dev/files/ias_and_climate_change_issues_brief_final.pdf.

IUCN (International Union of Conservation of Nature). (2017a). *Issues brief: Ocean warming*. Gland: IUCN. Available from https://www.iucn.org/sites/dev/files/ocean_warming_issues_brief_final.pdf.

IUCN (International Union of Conservation of Nature). (2018). *Guidelines for invasive species planning and management on islands*. Gland: IUCN.

IUCN (International Union for Conservation of Nature). (2019). *The IUCN red list of threatened species*. Gland: IUCN Secretariat. Available from https://www.iucn.org/sites/dev/files/content/documents/iucn_brochure_english_screen.pdf.

Iyer, G., Calvin, K., Clarke, L., Edmonds, J., Hultman, N., Hartin, C., ... Pizer, W. (2018). Implications of sustainable development considerations for comparability across nationally determined contributions. *Nature Climate Change*, 8, 124–129.

Jabbar, M. T., & Zhou, J. X. (2013). Environmental degradation assessment in arid areas: A case study from Basra Province, southern Iraq. *Environmental Earth Sciences*, 70(5), 2203–2214.

Jenkins, M., & Schaap, B. (2018). *Forest ecosystem services, background analytical study*. New York: UN Forum on Forests.

Jiang, L., Bao, A., Jiapaer, G., Guo, H., Zheng, G., Gafforov, K., ... De Maeyer, P. (2019). Monitoring land sensitivity to desertification in Central Asia: Convergence or divergence? *Science of The Total Environment*, 658, 669–683.

Jiménez Cisneros, B. E., Oki, T., Arnell, N. W., Benito, G., Cogley, J. G., Döll, P., ... Mwakalila, S. S. (2014). *Freshwater resources. Climate change 2014: Impacts, adaptation, and vulnerability. Part A: Global and sectoral aspects. Contribution of working group II to the fifth assessment report of the intergovernmental panel on climate change* (pp. 229–269). Cambridge: Cambridge University Press.

Jiménez-Muñoz, J. C., Mattar, C., Barichivich, J., Santamaría-Artigas, A., Takahashi, K., Malhi, Y., ... van der Schrier, G. (2016). Record-breaking warming and extreme drought in the Amazon rainforest during the course of El Niño 2015–2016. *Scientific Reports*, *6*, 33130.

Jolly, M., Cochrane, M., Freeborn, P., Holden, Z., Brown, T., Williamson, G., & Bowman, D. (2015a). Climate-induced variations in global wildfire danger from 1979 to 2013. *Nature Communications*, *6*, 7537.

Jolly, C. J., Shine, R., & Greenlees, M. J. (2015b). The impact of invasive cane toads on native wildlife in southern Australia. *Ecology and Evolution*, *5*(18), 3879–3894.

Jones, R. J. A., Hiederer, R., Rusco, E., & Montanarella, L. (2005). Estimating organic carbon in the soils of Europe for policy support. *European Journal of Soil Science*, *56*, 655–671.

Jones, B., & O'Neill, B. C. (2016). Spatially explicit global population scenarios consistent with the shared socioeconomic pathways. *Environmental Research Letters*, *11*(2016), 084003.

Jones, B. A. (2019). Tree shade, temperature, and human health: Evidence from invasive species-induced deforestation. *Ecological Economics*, *156*, 12–23.

Jones, B. A., & McDermott, S. M. (2018). Health impacts of invasive species through an altered natural environment: Assessing air pollution sinks as a causal pathway. *Environmental and Resource Economics*, *71*(1), 23–43.

Kadykalo, A. N., López-Rodriguez, M. D., Ainscough, J., Droste, N., Ryu, H., Ávila-Flores, G., ... Harmáčková, Z. V. (2019). Disentangling 'ecosystem services' and 'nature's contributions to people'. *Ecosystems and People*, *15*, 269–287.

Kansky, R., & Knight, A. T. (2014). Key factors driving attitudes towards large mammals in conflict with humans. *Biological Conservation*, *179*, 93–105.

Kantanen, J., Løvendahl, P., Strandberg, E., Eythorsdottir, E., Li, M.-H., Kettunen-Præbel, A., ... Meuwissen, T. (2015). Utilization of farm animal genetic resources in a changing agro-ecological environment in the Nordic countries. *Frontiers in Genetics*, *6*, 52.

Kate, K. T., & Laird, S. A. (2000). Biodiversity and business: Coming to terms with the 'grand bargain'. *International Affairs*, *76*, 241–264.

Katsuyama, Y., Inatsu, M., Nakamura, K., & Matoba, S. (2017). Global warming response of snowpack at mountain range in northern Japan estimated using multiple dynamically downscaled data. *Cold Regions Science & Technology*, *136*, 62–71.

Kibblewhite, M. G., Jones, R. J. A., Baritz, R., Huber, S., Arrouays, D., Micheli, E., & Stephens, M. (2008). *ENVASSO final report part I: Scientific and technical activities*. ENVASSO Project (Contract 022713) coordinated by Cranfield University, UK, for Scientific Support to Policy, European Commission 6th Framework Research Program. Available from https://www.dafne.at/prod/dafne_plus_common/attachment_download/cadb6ff31fc3ff4741304a4f7a378fe9/ENV_D11_FinalRep_prt2bk.pdf.

Khanal, N. R., Hu, J.-M., & Mool, P. (2015). Glacial lake outburst flood risk in the Poiqu/Bhote Koshi/Sun Koshi river basin in the Central Himalayas. *Mountain Research and Development*, *35*(4), 351–364.

King, D., & Finch, D. M. (2013). *The effects of climate change on terrestrial birds of North America*. Washington, DC: United States Department of Agriculture, Forest Service, Climate Change Resource Center. Available from http://www.fs.usda.gov/ccrc/topics/wildlife/birds.

Kissinger, G., Herold, M., & De Sy, V. (2012). *Drivers of deforestation and forest degradation: A synthesis report for REDD+ policymakers*. Vancouver: Lexeme Consulting.

Klenk, L. N., Mabee, W., Gong, Y., & Bull, G. (2012). *Deforestation, forest management and governance. Encyclopedia of life sciences*. Hoboken: John Wiley & Sons, Online edition.

Knapp, A. K., Carroll, C. J. W., & Fahey, T. J. (2014). Patterns and controls of terrestrial primary production in a changing world. In R. Monson (Ed.), *Ecology and the environment* (pp. 205–246). New York: Springer.

Koç, N., Njåstad, B., Armstrong, R., Corell, R. W., Jensen, D. D., Leslie, K. R., & ... Winther, J.-G. (Eds.), (2009). *Melting snow and ice: A call for action*. Tromsa: Centre for Ice, Climate and Ecosystems.

Koele, N., Bird, M., Haig, J., Marimon-Junior, B. H., Marimon, B. S., Phillips, O. L., ... Feldpausch, T. R. (2017). Amazon Basin forest pyrogenic carbon stocks: First estimate of deep storage. *Geoderma., 306*, 237–243.

Koh, N. S., Hahn, T., & Ituarte-Lima, C. (2017a). Safeguards for enhancing ecological compensation in Sweden. *Land Use Policy, 64*, 186–199.

Koh, N. H., Hahn, T., & Boostra, W. J. (2019). How much of a market is involved in a biodiversity offset? A typology of biodiversity offset policies. *Journal of Environmental Management, 232*, 679–691.

Koh, N. S., Han, T., & Ituarte-Lima, C. (2017b). Safeguards for enhancing ecological compensation in Sweden, Stockholm universitet, Naturvetenskapliga fakulteten & Stockholm Resilience Center *Land Use Policy, 64*, 186.

Kongsager, R. (2018). Linking climate change adaptation and mitigation: A review with evidence from the land-use sectors. *Land, 7*, 158.

Konnert, M., Maurer, W., Degan, B., & Kätzel, R. (2011). Genetic monitoring in forests—Early warning and controlling system for ecosystemic changes. *iForest, 4*, 77–81.

Kortenkamp, A. (2007). Ten years of mixing cocktails: A review of combination effects of endocrine-disrupting chemicals. *Environmental Health Perspectives, 115*(Suppl), 98–105.

Koutroulis, A. G. (2019). Dryland changes under different levels of global warming. *Science of The Total Environment, 655*, 482–511.

Kraft, N. J. B., Metz, M. R., Condit, R. S., & Chave, J. (2010). The relationship between wood density and mortality in a global tropical forest data set. *The New Phytologist, 188*, 1124–1136.

Krausman, P. R. (1999). Grazing behavior of livestock and wildlife. In: *Idaho Forest, Wildlife & Range Experiment Station. Bulletin. #70* (pp. 85–90). Moscow, ID: University of Idaho. Available from https://digital.lib.uidaho.edu/digital/collection/fwres/id/42/.

Ladisa, G., Todorovic, M., & Liuzzi, G. T. (2012). A GIS-based approach for desertification risk in Apulia region, SE Italy. *Physics and Chemistry of the Earth, Parts A/B/C, 49*, 103–113.

Laird, S. A., & Kate, K. T. (2002). In S. Pagiola, J. Bishop, & N. Landell-Mills (Eds.), *Selling forest services: Market-based mechanisms for conservation and development* (pp. 151–172). Earthscan.

Laird, S. A., & Wynberg, R. (2018). *Fact finding and scoping study on digital sequence information on genetic resources in the context of the convention of biological diversity and the Nagoya Protocol*. Montreal: CBD. Available from https://www.cbd.int/doc/c/b39f/4faf/7668900e8539215e7c7710fe/dsi-ahteg-2018-01-03-en.pdf.

Laird, S., Wynberg, R., Rourke, M., Humphries, F., Muller, M. R., & Lawson, C. (2020). Rethink the expansion of access and benefit sharing. *Science (New York, N.Y.), 367*(6483), 1200–1202.

Lamb, D. (2010). *Regreening the bare hills: Tropical forest restoration in the Asia-Pacific region*. Verlag: Springer.

Lamb, D., Stanturf, J., & Madsen, P. (2012). What is forest landscape restoration? In J. Stanturf, D. Lamb, & P. Madsen (Eds.), *Forest landscape restoration: Integrating natural and social sciences* (pp. 3–23). Dordrecht: Springer.

Lambert, A. M., Dudley, T. L., & Robbins, J. (2014). Nutrient enrichment and soil conditions drive productivity in the large-statured invasive grass Arundo donax. *Aquatic Botany, 112*, 16–22.

Lambin, R. F., Turner, B. L., Giest, H. J., Agbola, S. B., Angelsen, A., Bruce, J. W., ... Xu, J. (2001). The causes of land-use and land cover change: moving beyond the myths. *Global Environmental Change, 11*, 261–269.

Lamprecht, A., Semenchuk, P. R., Steinbauer, K., Winkler, M., & Pauli, H. (2018). Climate change leads to accelerated transformation of high-elevation vegetation in the central Alps. *New Phytologist*, *220*(2), 447–459.

Lapola, D. M., Schaldach, R., Alcamo, J., Bondeau, A., Koch, J., Koelking, C., & Priess, J. A. (2010). Indirect land-use changes can overcome carbon savings from biofuels in Brazil. *Proceedings of the National Academy of Sciences of the United States of America*, *107*, 3388–3393.

Laporte, N., Stabach, J., Grosch, R., & Lin, T. S. (2007). Expansion of industrial logging in Central Africa. *Science*, *316*(5830), 1451.

Laurance, W. F., Nascimento, H. E. M., Laurance, S. G., Andrade, A., Ewers, R. M., Harms, K. E., ... Ribeiro, J. E. (2007). Habitat fragmentation, variable edge effects, and the landscape-divergence hypothesis. *PLoS One*, *2*(10), e1017.

Laurance, W. F., Laurance, S. G., & Hilbert, D. W. (2008). Long-term dynamics of a fragmented rainforest mammal assemblage. *Conservation Biology*, *22*(5), 1154–1164.

Laurance, W. F. (2008). Global warming and amphibian extinctions in Eastern Australia. *Australia Ecology*, *33*, 1–9.

Laurance, W. F., Camargo, J. L. C., Luizao, R. C. C., Laurance, S. G., Pimm, S. L., Bruna, E. M., ... Lovejoy, T. E. (2011). The fate of Amazonian forest fragments: A 32-year investigation. *Biological Conservation*, *144*(1), 56–67.

Lavorgna, A., Rutherford, C., Vaglica, V., Smith, M. J., & Sajeva, M. (2018). CITES, wild plants, and opportunities for crime. *European Journal on Criminal Policy and Research*, *24*, 269–288.

Law Insider. *Definition of aquatic resources*. (n.d.). <https://www.lawinsider.com/dictionary/aquatic-resources#:~:text=Aquatic%20resources%20means%20wetlands%2C%20streams,or%20man%2Dmade%20water%20bodies>.

Lawrence, P. G., Maxwell, B. D., Rew, L. J., Ellis, C., & Bekkerman, A. (2018). Vulnerability of dryland agricultural regimes to economic and climatic change. *Ecology and Society*, *23*(1), art34.

Lawson, S., & MacFaul, L. (2010). *Illegal logging and related trade: Indicators of the global response*. Chatham House: The Royal Institute of International Affairs.

Layer, K., Hildrew, A., Monteith, D., & Woodward, G. (2010). Long-term variation in the littoral food web of an acidified mountain lake. *Global Change Biology*, *16*(11), 3133–3143.

Lazar, B., & Williams, M. (2008). Climate change in western ski areas: Potential changes in the timing of wet avalanches and snow quality for the Aspen ski area in the years 2030 and 2100. *Cold Regions Science & Technology*, *51*(2–3), 219–228.

Le, Q. B., Nkonya, E., & Mirzabaev, A. (2016). Biomass productivity-based mapping of global land degradation hotspots. In E. Nikoya, A. Mirzabaev, & J. von Braun (Eds.), *Economics of land degradation and improvement—A global assessment for sustainable development* (pp. 55–84). Cham: Springer International Publishing.

Lefevre, F. (2004). Human impacts on forest genetic resources in the temperate zone: An updated review. *Forest Ecology and Management*, *197*(1–3), 257–271.

Lefevre, F., Boivin, T., Bontemps, A., Courbet, F., Davi, H., Durand-Gillmann, M., ... Pichot, C. (2014). Considering evolutionary processes in adaptive forestry. *Annals of Forest Science*, *71*(7), 723–739.

Le Maitre, D. C., van Wilgen, B. W., Chapman, R. A., & McKelly, D. H. (1996). Invasive plants and water resources in the Western Cape Province, South Africa: Modelling consequences of a lack of management. *Journal of Applied Ecology* (33, pp. 161–172).

Lemenih, M., & Bongers, F. (2010). The role of plantation forests in fostering ecological restoration: Experiences from East Africa. In F. Bongers, & T. Tennigkeit (Eds.), *Degraded forests in Eastern Africa* (pp. 171–219). London: Earthscan.

Lemordant, L., Gentine, P., Swann, A. S., Cook, B. I., & Scheff, J. (2018). Critical impact of vegetation physiology on the continental hydrologic cycle in response to increasing CO_2. *Proceedings of the National Academy of Sciences of the United States of America*, *115*(16), 4093–4098.

Lenda, M., Skórka, P., Knops, J. M. H., Moron, D., Sutherland, W. J., Kuszewska, K., & Woyciechowski, M. (2014). Effect of the internet commerce on dispersal modes of invasive alien species. *PLoS One*, *9*, e99786.

Leopold, A. (1933). *Game management*. New York: Charles Scribner's Sons.

Lewis, S. L., Brando, P. M., Phillips, O. L., van der Heijden, G. M. F., & Nepstad, D. (2011). The 2010 Amazon drought. *Science (New York, N.Y.)*, *331*(6017), 554.

Li, W., Ciais, P., Peng, S., Yue, C., Wang, Y., Thurner, M., ... Zeachle, S. (2017). Land-use and land-cover change carbon emissions between 1901 and 2012 constrained by biomass observations. *Biogeosciences.*, *14*, 5053−5067.

Limam, A. *World's largest freshwater fish being hunted to extinction in Europe*. (2020). Available online.

Lindner, M., Maroschek, M., Netherer, S., Kremer, A., Barbati, A., Garcia-Gonzalo, J., ... Marchetti, M. (2010). Climate change impacts, adaptive capacity, and vulnerability of European forest ecosystems. *Forest Ecology and Management*, *259*, 698−709.

Lindsay, D., Barr, K., Lance, R., Tweddale, S., Hayden, T., & Leberg, P. (2008). Habitat fragmentation and genetic diversity of an endangered, migratory songbird, the golden-cheeked warbler (*Dendroica chrysoparia*). *Molecular Ecology*, *17*(9), 2122−2133.

Lloyd, A. H., Rupp, T. S., Fastie, C. L., & Starfield, A. M. (2003). Patterns and dynamics of tree-line advance on the Seward Peninsula, Alaska. *Journal of Geophysical Research Atmospheres*, *107*(D2), ALT 2−1-ALT 2−15.

Lopes-Lima, M., Sousa, R., Geist, J., Aldridge, D. C., Araujo, R., Bergengren, J., ... Zogaris, S. (2017). Conservation status of freshwater mussels in Europe: State of the art and future challenges. *Biological Reviews*, *92*, 572−607.

Lovejoy, T. (2017). Amazon approaches deforestation tipping point, Blog Post in Brazil Institute, Wilson Center. Available at: https://www.wilsoncenter.org/blog-post/amazon-approaches-deforestation-tipping-point.

Loveridge, A. J., Wang, S. W., Frank, L. G., & Seidensticker, J. (2010). People and wild felids: Conservation of cats and management of conflicts. In D. W. Madonald, & A. J. Loveridge (Eds.), *Biology and conservation of wild fields* (pp. 161−195). New York: Oxford University Press.

Lu, W., Ridgwell, A., Thomas, E., Hardistry, D. S., Luo, G., Algeo, T. J., ... Lu, Z. (2018). Late inception of a resiliently oxygenated upper ocean. *Science*, *361*(6398), 174−177.

Lubchenco, J. (1998). Entering the century of environment: A new social contract for science. *Science (New York, N.Y.)*, *279*(5350), 491.

Lund, H. G. (2009). What is a degraded forest? Gainesville, VA: Forest Information Services. Available from http://home.comcast.net/∼gyde/2009forestdegrade.doc.

Lung, T., & Schaab, G. (2010). A comparative assessment of land cover dynamics of three protected forest areas in tropical eastern Africa. *Environmental monitoring and assessment*, *161*(1), 531−548.

MA (Millennium Ecosystem Assessment). (2005). *Ecosystems and human well-being: Synthesis*. Washington, DC: Island Pres.

Macchi, M. (2010). *Mountains of the world−ecosystem services in a time of global and climate change*. Kathmandu: ICIMOD.

Macdonald, D. W., Loveridge, A. J., & Rabinowitz, A. (2010). Felid futures: Crossing disciplines, borders, and generations. In D. W. Macdonald, & A. J. Loveridge (Eds.), *The biology and conservation of wild felids* (pp. 600−649). Oxford: Oxford University Press.

Mackenzie, J. M. (1988). *The empire of nature: Hunting, conservation and British imperialism*. Manchester: Manchester University Press.

Madden, F., & McQuinn, B. (2015). Understanding social conflict and complexity in marine conservation. In M. M. Draheim, F. Madden, J.-B. McCarthy, & C. Parsons (Eds.), *Human-wildlife conflict* (pp. 3−16). Oxford: Oxford University Press.

Maddox, T., Howard, P., Knox, J., & Jenner, N. (2019). *Forest-smart mining, identifying factors associated with the impacts of large-scale mining (LSM) on forests*. Washington, DC: World Bank.

Maestre, F. T., Reynolds, J. F., Huber-Sannwald, E., Herrick, J., & Stanford, S. M. (2006). Understanding global desertification: Biophysical and socioeconomic dimensions of hydrology. In P. D'Odorico, & A. Porporato (Eds.), Dryland ecohydrology. (pp. 315–332). Dordrecht: Springer.

Mahowald, N. M., Ward, D. S., Doney, S. C., Hess, P. G., & Randerson, J. T. (2017). Are the impacts of land 22 use on warming underestimated in climate policy? *Environmental Research Letters*, *12*, 094016.

Maier, S., Lindner, J., & Francisco, J. (2019). Conceptual framework for biodiversity assessments in global value chains. *Sustainability.*, *11*, 1841.

Mainka, S. A., & Trivedi, M. (Eds.), (2002). *Links between biodiversity conservation, livelihoods and food security: The sustainable use of wild species for meat*. Gland: IUCN. Available from https://portals.iucn.org/library/sites/library/files/documents/SSC-OP-024.pdf.

Major, T., Doropoulos, C., Schwarzmueller, F., Gladish, D. W., Kumaran, N., Merkel, K., . . . Gagi, V. (2018). Global mismatch of policy and research on drivers of biodiversity loss. *Nature Ecology & Evolution*, *2*, 1071–1074.

Malaj, E., Peter, C., von der, Ohe., Grote, M., Kunhe, R., Mondy, C. P., . . . Schafer, R. B. (2014). Organic chemicals jeopardize the health of freshwater ecosystems on the continental scale. *Proceedings of the National Academy of Sciences of the United States of America*, *111*(26), 9549–9554.

Malhi, Y., Roberts, J. T., Betts, R. A., Killeen, T. J., Li, W., & Nobre, C. A. (2008). Climate change, deforestation, and the fate of the Amazon. *Science (New York, N.Y.)*, *319*, 169–172.

Malhi, Y., Franklin, J., Seddon, N., Solan, M., Turner, M. G., Field, C. B., & Knowlton, N. (2020). Climate change and ecosystems: Threats, opportunities and solutions. *Philosophical Transactions of the Royal Society of London. B Biological Sciences*, *375*(1794), 0104.

Maliva, R., & Missimer, T. (2012). Aridity and drought. In R. Maliva, R. G. Missimer, & M. Thomas (Eds.), *Arid land water evaluation and management* (pp. 21–39). Cham: Springer.

Mann, M. E., & Gleick, P. H. (2015). Climate change and California drought in the 21st century. *Proceedings of the National Academy of Sciences of the United States of America*, *112*(13), 3858–3859.

Marco, A., Lavergne, S., Dutoit, T., & Valerie, B.-M. (2010). From the backyard to the backcountry: How ecological and biological traits explain the escape of garden plants into Mediterranean old fields. *Biological Invasions*, *12*, 761–779.

Marengo, J. A., Nobre, C. A., Tomasella, J., Cardoso, M. F., & Oyama, M. D. (2008). Hydro-climatic and ecological behaviour of the drought of Amazonia in 2005. *Philosophical Transactions of the Royal Society of London. B-Biological Sciences*, *363*, 1773–1778.

Marengo, J. A., Tomasella, J., Alves, L. M., Soares, W. R., & Rodriguez, D. A. (2011). The drought of 2010 in the context of historical droughts in the Amazon region. *Geophysical Research Letters*, *38*(12), L12703.

Marker, L. L., & Boast, L. K. (2015). Human–wildlife conflict 10 years later: Lessons learned and their application to cheetah conservation. *Human Dimensions of Wildlife*, *20*, 302–309.

Martin-Ortega, J., Ferrier, R. C., Gordon, I. J., & Khan, S. (2015). *Freshwater ecosystem services: A global perspective*. Cambridge: Cambridge University Press.

Martinuzzi, S., Januchowski-Hartley, S. R., Pracheil, B. M., McIntyre, P. B., Plantinga, A. J., Lewis, D. J., & Radeloff, V. C. (2014). Threats and opportunities for freshwater conservation under future land use change scenarios in the United States. *Global Change Biology*, *20*, 113–124.

Masoudi, M., & Amiri, E. (2015). A new model for hazard evaluation of vegetation degradation using DPSIR framework, a case study: Sadra region, Iran. *Polish Journal of Ecology*, *63*(1), 1–9.

Masoudi, M., & Jokar, P. (2017). A new model for desertification assessment using geographic information system (GIS)—A case study, Runiz Basin, Iran. *Polish Journal of Ecology*, *65*, 236–246.

Matthews, N. (2016). People and freshwater ecosystems: Pressures, responses and resilience. *Aquatic Procedia*, *6*, 99–105.

McDowell, G., Huggel, C., Frey, H., Cramer, H., & Ricciardi, V. (2019). Adaptation action and research in glaciated mountain systems: Are they enough to meet the challenge of climate change? *Global Environmental Change, 54*, 19−30.

McIntosh, J. (2018). Antibiotic resistance: What you need to know. *Medical News Today*. Available from https://www.medicalnewstoday.com/articles/283963.php.

McKelvey, K. S., Copeland, J. P., Schwartz, M. K., Littell, J. S., Aubry, K. B., Squires, J. R., ... Mauger, G. S. (2011). Climate change predicted to shift wolverine distributions, connectivity, and dispersal corridors. *Ecological Applications, 21*, 2882−2897.

McKelvey, K. S., Perry, R. W., & Mills, L. S. (2013). *The effects of climate change on mammals*. Washington, DC: United States Department of Agriculture, Forest Service, Climate Change Resource Center. Available from http://www.fs.usda.gov/ccrc/topics/wildlife/mammals.

McManus, J. S., Dickman, A. J., Gaynor, D., Smuts, B. H., & Macdonald, D. W. (2015). Dead or alive? Comparing costs and benefits of lethal and non-lethal human-wildlife conflict mitigation on livestock farms. *Oryx, 49*, 687−695.

Menon, V., & Kumar, A. (1998). Wildlife Crime. New Delhi: Natraj Publishers.

Meybeck, M., Green, P., & Vörösmarty, C. (2001). A new typology for mountains and other relief classes: An application to global continental water resources and population distribution. *Mountain Research and Development, 21*, 34−45.

Mittermeier, R. A., Farrell, T. A., Harrison, I. J., Upgren, A. L., & Brooks, T. M. (2010). *Fresh water: The essence of life*. Arlington, VA: CEMEX & ILCP.

MacDicken, K. G., Sola, P., Hall, J. E., Sabogal, C., Tadoum, M., & de Wasseige, C. (2015). Global progress toward sustainable forest management. *Forest Ecology and Management, 352*, 47−56.

McBratney, A., Field, D. J., & Koch, A. (2014). The dimensions of soil security. *Geoderma., 213*, 203−213.

McDonald, R., Green, P., Balk, D., Fekete, B. M., Reyenga, C., Todd, M., & Montgomery, M. F. (2011). Urban growth, climate change, and freshwater availability. *Proceedings of the National Academy of Sciences of the United States of America, 108*(15), 6312−6317.

McGowan, P. J. K. (2010). Conservation status of wild relatives of animals used for food. *Animal Genetic Resources, 47*, 115−118, Full text is available from. Available from http://www.fao.org/3/i1823t/i1823t12.pdf.

McKusick, J. (2013). John Evelyn: The forestry of imagination. *Wordsworth Circle, 44*(2), 110−114.

McShane, T., Hirsch, P., Trung, T., Songorwa, A., Kinzig, A., Monteferri, B., & Pulgar-Vidal, M. (2011). Hard choices: Making trade-offs between biodiversity conservation and human well-being. *Biological Conservation, 144*(3), 966−972.

MCPFE (Ministerial Conference on the Protection of Forests in Europe). (1993). Resolution H1: General guidelines for the sustainable management of forest in Europe. In: *Second Ministerial Conference on the Protection of Forests in Europe*, Helsinki, Finland, June 16−17, 1993.

Mekonen, S. (2020). Coexistence between human and wildlife: The nature, causes and mitigations of human wildlife conflict around Bale Mountains National Park, Southeast Ethiopia. *BMC Ecology, 20*, 51.

Mengel, M., Levermann, A., Frieler, K., Robinson, A., Marzeion, B., & Winkelmann, R. (2016). Future sea level rise constrained by observations and long-term commitment. *Proceedings of the National Academy of Sciences of the United States of America, 113*(10), 2597−2602.

Mery, G., Katila, P., Galloway, G., Alfaro, R.I., Kanninen, M., Lobovikov, M., & Varjo, J. (Eds.). (2010). Forests and society: Responding to global drivers of change. In: *IUFRO world series*, vol. 25. Vienna: Austria International Union of Forest Research Organizations. Available from http://www.iufro.org/science/special/wfse/forests-society-global-drivers/.

Mgboji, I. (2003). Beyond rhetoric: State sovereignty, common concern, and the inapplicability of the common heritage concept to plant genetic resources. *Leiden Journal of International Law, 16*(4), 821−837.

Mikkelson, G. (2009). Diversity-stability hypothesis. *Encyclopedia of environmental ethics and philosophy, 1*, 255−256.

Milligan, S. R., Holt, W. V., & Lloyd, R. (2009). Impacts of climate change and environmental factors on reproduction and development in wildlife. *Philosophical Transactions of the Royal Society of London. B Biological Science, 364*(1534), 3313−3319.

Milner, A. M., Khamis, K., Battin, T. J., Brittain, J. E., Barrand, N. E., Fureder, L., . . . Brown, L. E. (2017). Glacier shrinkage driving global changes in downstream systems. *Proceedings of the National Academy of Sciences of the United States of America, 114*(37), 9770−9778.

Minteer, B. A., & Collins, J. P. (2012). Species conservation, rapid environmental change, and ecological ethics. *Nature Education Knowledge, 3*(10), 14.

MEA (Millennium Ecosystem Assessment). (2005). *Ecosystems and human well-being: Synthesis*. Washington, DC: Island Press.

Mimura, N. (2013). Sea-level rise caused by climate change and its implications for society. *Proceedings of the Japan Academy, Series B, 89*(7), 281−301.

Mitchell, A. L., Rosenqvist, A., & Mora, B. (2017). Current remote sensing approaches to monitoring forest degradation in support of countries measurement, reporting and verification (MRV) systems for REDD + . *Carbon Balance and Management, 12*, 9.

Modica, G., Merlino, A., Solano, F., & Mercurio, R. (2015). An index for the assessment of degraded Mediterranean forest ecosystems. *Forest Systems, 24*, e037.

Montanarella, L., Scholes, R., & Brainich, A. (2018). *The IPBES assessment report on land degradation and restoration*. Bonn. Available from https://ipbes.net/sites/default/files/2018_ldr_full_report_book_v4_pages.pdf.

Moracova, L., Pysek, P., Krinke, L., Mullerova, J., Perglova, I., & Pergi, J. (2015). Long-term survival in soil of seed of the invasive herbacaes plant: Heracleum mantegazziannum. *Presila, 90*(3), 225−234.

Morales-Barquero, L., Skutsch, M., Jardel-Peláez, J. E., Ghilardi, A., Kleinn, C., & Healey, R. J. (2014). Operationalizing the definition of forest degradation for REDD + , with application to Mexico. *Forests, 5*, 1653−1681.

Moran, C., & Catterall, C. P. (2014). Responses of seed-dispersing birds to amount of rainforest in the landscape around fragments. *Conservation Biology, 28*(2), 551−560.

Morgera, E., Tsioumani, E., & Matthias, B. (2014). *Unravelling the Nagoya protocol. A commentary on the Nagoya Protocol on access and benefit-sharing to the convention on biological diversity*. Leiden: Koninklijke Brill nv.

Mostarda, E., Japp, D., Heinecken, C., Stiassny, C. P., Ebert, B., Xavier, J., & Tracy, D. (2016). *Marine species biological data collection manual*. Rome: FAO. Available from http://www.fao.org/3/a-i6353e.pdf.

Moyer, A. N., Moore, R. D., & Koppes, M. N. (2016). Streamflow response to the rapid retreat of a lake-calving glacier. *Hydrological Processes, 30*(20), 3650−3665.

Müller, F., Baessler, C., Schubert, H., & Klotz, S. (Eds.), (2010). *Long-term ecological research : Between theory and application*. Cham: Springer.

Murdiyarso, D., Skutsch, M., Guariguata, M., & Kanninen, M. (2008). Measuring and monitoring forest degradation for REDD: Implications of country circumstances. In: *Cifor Infobriefs. Number. 16*. Available from https://ris.utwente.nl/ws/files/119576843/Murdiyarso2008measuring.pdf.

Murcia, C. (1995). Edge effects in fragmented forests: Implications for conservation. *Trends in Ecology and Evolution, 10*, 58−62.

Mwitwa, J., German, L., Muimba-Kankolongo, A., & Puntodewo, A. (2012). Governance and sustainability challenges in landscapes shaped by mining: Mining-forestry linkages and impacts in the Copper Belt of Zambia and the DR Congo. *Forest Policy Economics, 25*, 19−30.

Nagoya Protocol. (2011). *Nagoya protocol on access to genetic resources and the fair and equitable sharing of benefits arising from their utilization to the convention on biological diversity*. Montreal:

Secretariat of the CBD. Text and Annex. Available from https://www.cbd.int/abs/doc/protocol/nagoya-protocol-en.pdf.

Namkoong, G., Boyle, T., El-Kassaby, Y. A., PalmbergLerche, C., Eriksson, G., Gregorius, H.-R., ... Prabhu, R. (2002). Criteria and indicators for sustainable forest management: Assessment and monitoring of genetic variation. In: *Forest Genetic Resources working papers FGR/37E*. Rome: FAO. Available from http://www.fao.org/3/a-y4341e.pdf.

NASA (National Aeronautics and Space Administration). (2020). *NASA science-salinity*. NASA. Available from <https://science.nasa.gov/earth-science/oceanography/physical-ocean/salinity> Updated 29.10.20.

Nassar, N. T. (2018). Quantifying and characterizing in-use stocks of non-fuel mineral commodities. In: *Symposium on the availability of raw materials from secondary sources: A key aspect of circular economy*. Geneva, Palais des Nations: USGS.

Neff, C. (2012). Australian beach safety and the politics of shark attacks. *Coastal Management*, 40, 88−106.

Nelson, B. W. (1994). Natural forest disturbance and change in the Brazilian Amazon. *Remote Sensing Reviews*, 10, 105−125.

Newton, A., Oldfield, S., Rivers, M., Mark, J., Schatz, G., Tejedor Garavito, N., ... Miles, L. (2015). Towards a global tree assessment. *Oryx*, 49(3), 410−415.

Nicholson, E., Mace, G. M., Armsworth, P. R., Atkinson, G., Buckle, S., Clements, T., ... Milner-Gulland, E. J. (2009). Priority research areas for ecosystem services in a changing world. *Journal of Applied Ecology*, 46(6), 1139−1144.

Niemiera, A. X., & Holle, B. V. (2009). Invasive plant species and the ornamental horticulture industry. In Inderjit (Ed.), *Management of invasive weeds* (pp. 167−187). Dordrecht: Springer.

Nõges, P., Argillier, C., Borja, Á., Garmendia, J. M., Hanganu, J., Kodeš, V., ... Birk, S. (2016). Quantified biotic and abiotic responses to multiple stress in freshwater, marine and ground waters. *Science of the Total Environment*, 540, 43−52.

Nogueira, J. G., Lopes-Lima, M., Varandas, S., Teixeira, A., & Sousa, R. (2019). Effects of an extreme drought on the endangered pearl mussel Margaritifera margaritifera: A before/after assessment. *Hydrobiologia*, 1−11.

Northrop, E., Biru, H., Lima, S., Buoye, M., & Song, R. (2016). Examining the alignment between the intended nationally determined contributions and sustainable development goals. Working paper. Washington, DC: World Resources Institute.

NRC (National Research Council). (1995). *Science and the endangered species act*. Washington, DC: National Academic Press.

Numata, I., & Cochrane, M. A. (2012). Forest Fragmentation and Its Potential Implications in the Brazilian Amazon between 2001 and 2010. *Open Journal of Forestry*, 2, 265−271.

Nyhus, P. J. (2016). Human-wildlife conflict and coexistence. *The Annual Review of Environment and Resources*, 41, 143−171.

Obersteiner, M., Huettner, M. M., Kraxner, F., McCallum, I., Aoki, K., Bottcher, H., ... Reyers, B. (2009). On fair, effective and efficient REDD mechanism design. *Carbon Balance and Management*, 4, 11.

Oberthur, S., Pozarowska, J., Rabitz, F., Gerstetter, C., Lucha, C., McGlade, K., & Tedsen, E. (2011). *Intellectual property rights on genetic resources and the fight against poverty. Study for the European Parliament*. Brussels: European Parliament.

OECD/FAO (Organization for Economic Cooperation & Development—Food and Agriculture Organization). (2011). *OECD/FAO agricultural outlook 2011−2020*. Paris: OECD Publishing and FAO.

OECD (Organization for Economic Cooperation and Development). (2016a). *Biodiversity offsets, effective design and implementation*. Paris: OECD.

OECD (Organization for Economic Cooperation and Development). (2016b). *The Ocean Economy in 2030*. Paris: OECD.

Oguamanam, C. (2008). Local knowledge as trapped knowledge. *The Journal of World Intellectual Property*, *11*(1), 29−41.

Oguamanam, C., & Jain, V. (2017). Access and benefit sharing, Canadian and aboriginal research ethics policy after the Nagoya Protocol: Digital DNA and transformations in biotechnology. *Journal of Environmental Law and Practice*, *3*(1), 79−89.

Ogwu, M. C., Osawaru, M., & Abana, C. M. (2014). Challenges in conserving and utilizing plant genetic resources (PGR). *International Journal of Genetic and Molecular Biology*, *6*(2), 16−23.

Olden, J. D., & Rooney, T. P. (2006). On defining and quantifying biotic homogenization. *Global Biology and Diversity*, *15*, 113−120.

Oldfield, S. (Ed.), (2002). *The trade in wildlife: Regulation for conservation*. Cambridge: Fauna and Flora International, Resource Africa and TRAFFIC International.

Oli, K. P. (2009). Access and benefit sharing from biological resources and associated traditional knowledge in the HKH region—Protecting community interests. *International Journal of Biodiversity and Conservation*, *1*(5), 105−118.

Olofsson, J., & Hickler, T. (2008). Effects of human land-use on the global carbon cycle during the last 6,000 years. *Vegetation History & Archaeobotany*, *17*, 605−615.

Orgiazzi, A., & Panagos, P. (2018). Soil biodiversity and soil erosion: It is time to get married: Adding an earthworm factor to soil erosion modelling. *Global Ecology and Biogeography*, *27*(10), 1155−1167.

Ööpik, M., Bunce, R. G. H. B., & Tischler, M. (2013). Horticultural markets promote alien species invasions: An Estonian case study of herbaceous perennials. *NeoBiota.*, *17*, 19−37.

Ormerod, S. J., Dobson, M., Hildrew, A. G., & Townsend, C. R. (2010). Multiple stressors in freshwater ecosystems. *Freshwater Biology*, *55*, 1−4.

Pagad, S., Genovesi, P., Carnevali, L., Scaleral, R., & Clout, M. (2015). IUCN SSC Invasive Species Specialist Group: Invasive alien species information management supporting practitioners, policy makers and decision takers. *Management of Biological Invasions*, *6*(2), 127−135.

Padayachee, A. L., Irlich, U. M., Faulkner, K. T., Gaertner, M., Procheş, Ş., Rouget, M., & Wilson, J. R. U. (2017). How do invasive species travel to and through urban environments? *Biological Invasions*, *19*, 3557−3570.

Pallewatta, N., Reaser, J. K., & Gutierrez, A. (Eds.). (2003). Prevention and management of invasive alien species. In: *Proceedings of a workshop on forging cooperation throughout south and southeast Asia*. Cape Town: Global Invasive Species Programme. Available from http://issg.org/pdf/publications/GISP/Resources/SEAsia-1.pdf.

Panagos, P., Borrelli, P., Poesen, J., Ballabio, C., Lugato, E., Meusburger, K., & Alewell, C. (2015). The new assessment of soil loss by water erosion in Europe. *Environmental Science and Policy*, *54*, 438−447.

Panfil, S. N., & Harvey, C. A. (2016). REDD + and biodiversity conservation: A review of the biodiversity goals, monitoring methods, and impacts of 80 REDD + projects: Biodiversity conservation in REDD + projects. *Conservation Letters*, *9*, 143−150.

Parsons, M., McLoughlin, C. A., Kotschy, K. A., Rogers, K. H., & Rountree, M. W. (2005). The effects of extreme floods on the biophysical heterogeneity of river landscapes. *Frontiers in Ecology and the Environment*, *3*, 487.

Pashaei, M., Rashki, A., & Sepehr, A. (2017). An integrated desertification vulnerability index for Khorasan-Razavi, Iran. *Natural Resources and Conservation*, *5*, 44−55.

Pasternak, D., & Schlissel, A. (Eds.), (2012). *Combating desertification with plants*. Cham: Springer.

Pearson, N., Griffiths, P., Biddle, S. J. H., Johnston, J. P., & Haycraft, E. (2017). Individual, behavioral and home environmental factors associated with eating behaviors in young adolescents. *Appetite*, *112*, 35−43.

Peltier, D. M. P., Fell, M., & Ogle, K. (2016). Legacy effects of drought in the southwestern United States: A multi-species synthesis. *Ecological Monographs*, *86*, 312−326.

Pena-Neira, S. (2009). *Balancing rights and obligations in sharing benefits from natural genetic resources: Problems, discussions and possible solutions*. Anuario mexicano de derecho internacional. Available from http://www.scielo.org.mx/scielo.php?script = sci_arttext&pid = S1870-46542009000100005.

Peng, S., Ciais, P., Maignan, F., Li, W., Chang, J., Wang, T., & Yue, C. (2017). Sensitivity of land use change 22 emission estimates to historical land use and land cover mapping. *Global Biogeochemcal Cycles, 31*(23), 626–643.

Pepin, N., Bradley, R. S., Diaz, H. F., Baraer, M., Caceres, E., Forsythe, N., ... Yang, D. Q. (2015). Elevation-dependent warming in mountain regions of the world. *Nature Climate Change, 5*, 424–430.

Petchey, O. L., O'Gorman, E. J., & Flynn, D. F. (2009). A functional guide to functional diversity measures. In S. Naeem, D. E. Bunker, A. Hector, M. Loreau, & C. Perrings (Eds.), *Biodiversity, ecosystem functioning, & human wellbeing* (pp. 49–59). Oxford: Oxford University Press.

Peterson, M. N., Birckhead, J. L., Leong, K., Peterson, M. J., & Peterson, T. R. (2011). Rearticulating the myth. of human–wildlife conflict. *Conservation Letters, 3*, 74–82.

Phillips, F.-K. (2016). Intellectual property rights in traditional knowledge: Enabler of sustainable development. *Utrecht Journal of International and European Law, 32*(83), 1–18.

Phillips, O. L., Aragão, L. E. O. C., Lewis, S. L., Fisher, J. B., Lloyd, J., López González, G., ... Torres-Lezama, A. (2009). Drought sensitivity of the Amazon rainforest. *Science.* (323, pp. 1344–1347).

Phillips, O. L., van der Heijden, G., Lewis, S. L., Lopez-Gonzalez, G., Aragao, L., Lloyd, J., ... Vilanova, E. (2010). Drought-mortality relationships for tropical forests. *New Phytologist, 187*, 631–646.

Pielke, R. A., Marland, G., Betts, R. A., Chase, T. N., Eastman, J. L., Niles, J. O., ... Running, S. W. (2002). The influence of land-use change and landscape dynamics on the climate system: Relevance to climate-change policy beyond the radiative effect of greenhouse gases. *Philosophical Transactions of the Royal Society of London Series A-Mathematical Physical and Engineering Sciences, 360*(1797), 1705–1719.

Pimentel, D., Wilson, C., McCullum, C., Huang, R., Dwen, P., Flack, J., ... Cliff, B. (1997). Economic and environmental benefits of biodiversity. *Bioscience, 47*, 747–757.

Pimentel, D., Lach, L., Zuniga, R., & Morrison, D. (2000). Environmental and economic costs of nonindigenous species in the United States. *Bioscience, 50*, 53.

Pimm, S. L., Russell, G. J., Gittleman, J. L., & Brooks, T. M. (1995). The future of biodiversity. *Science (New York, N.Y.), 269*, 347–350.

Poff, N. L., & Zimmerman, J. K. H. (2010). Ecological responses to altered flow regimes: A literature review to inform the science and management of environmental flows. *Freshwater Biology, 55*, 194–205.

Pomeroy, J. W., Granger, R., Pietroniro, A., Toth, B., & Hedstrom, N. (1999). Classification of the boreal forest for hydrological processes. *Ninth international Boreal forest research association conference*, 49–59. Available from https://citeseerx.ist.psu.edu/viewdoc/download?doi = 10.1.1.712.7324&rep = rep1&type = pdf.

Pongratz, J., Dolman, H., Don, A., Erb, K.-H., Fuchs, R., Herold, M., ... Naudts, K. (2018). Models meet data: Challenges and opportunities in implementing land management in Earth system models. *Global Change Biology, 24*(4), 1480–1487.

Pooley, S. (2016). A cultural herpetology of Nile crocodiles in Africa. *Conservation and Society, 14*, 391–405.

Pooley, S., Barua, M., Beinart, W., Dickman, A., Holmes, G., Lorimer, J., ... Milner-Gulland, E. J. (2017). An interdisciplinary review of current and future approaches to improving human–predator relations. *Conservation Biology, 31*(3), 513–523.

Pourghasemi, S. R., Gayen, A., Panahi, M., Rezaie, F., & Blasche, K. (2019). Multi-hazard probability assessment and mapping in Iran. *Science of the Total Environment, 692*, 556–571.

Prathapan, K. D., Pethiyagoda, R., Bawa, K. S., Raven, K. H., & Rajan, P. D. (2018). 172 Signatories from 35 countries on 'when the cure kills—CBD limits biodiversity research. *Science (New York, N.Y.), 360*(6396), 1405–1406.

Prăvălie, R. (2016). Drylands extent and environmental issues. A global approach. *Earth-Science Reviews, 161*, 259–278.

Prem, H. T. (2020). What are the biggest threats to wildlife and why? *Posted in the Animals in Disasters blog*. Available online.

Preston, D. L., Caine, N., McKnight, D. M., Williams, M. W., Hell, K., Miller, M. P., ... Johnson, P. T. J. (2016). Climate regulates alpine lake ice cover phenology and aquatic ecosystem structure. *Geophysical Research Letters, 43*, 5353−5360.

Proelss, A., Maggio, A. R., Blitza, E., & Daum, O. (Eds.), (2017). *United Nations convention on the law of the sea: A commentary*. Verlag: C.H. Beck.

Pugh, T. A. M., Arnetyh, A., Olin, S., Ahlstrom, A., Bayer, A. D., Goldewijk, K. K., ... Schurgers, G. (2015). Simulated carbon emissions from land-use change are substantially 43 enhanced by accounting for agricultural management. *Environmental Research Letters, 10*, 124008.

Pugh, T. A. M., Lindeskog, M., Smith, B., Poulter, B., Arneth, A., Haverd, V., & Calle, L. (2019). Role of forest regrowth in global carbon sink dynamics. *Proceedings of the National Academy of Sciences of the United States of America, 116*(10), 4382−4387.

Puryavaud, J-P., Davidar, P., & Laurance, W. F. (2010). Cryptic destruction of India's native forests. *Conservation Letters*. 3, 390−394.

Putz, F. E., & Nasi, R. (2009). Carbon benefits from avoiding and repairing forest degradation. In A. Angelsen (Ed.), *Realising REDD + : National strategy and policy options* (pp. 249−362). Bogor: CIFOR.

Putz, F. E., & Redford, K. H. (2010). The importance of defining 'forest': Tropical forest degradation, deforestation, long-term phase shifts, and further transitions. *Biotropica., 42*, 10−20.

Pyšek, P., & Chytrý, M. (2014). Habitat invasion research: Where vegetation science and invasion ecology meet. *Journal of Vegetation Science, 25*, 1181−1187.

Quyet, M. V. (2014). *Multi-level assessment of land degradation. The case of Vietnam* (Ph.D. Thesis). ETH Zurich Research Collection. Available from https://www.research-collection.ethz.ch/bitstream/handle/20.500.11850/92195/eth-47053-02.pdf.

Raatz, M., van Velzen, E., & Gaedke, U. (2019). Co-adaptation impacts the robustness of predator−prey dynamics against perturbations. *Ecology and Evolution, 9*(7), 3823−3836.

Rabin, S. S., Malyshev, S. L., Magi, B. I., Shevliakova, E., & Pacala, S. W. (2017). A fire model with distinct crop, pasture, and non-agricultural burning: Use of new data and a model-fitting algorithm for FINALv1. *Geoscience Model Development (Cambridge, England), 11*, 815−842. Available from https://pdfs.semanticscholar.org/d2a7/a5d9188a2df2e67b7b170795f35a5eecfa32.pdf.

Radeloff, V. C., Hammer, R. B., Stewart, S. I., Fried, J. S., Holocomb, S. S., & McKeefry, J. F. (2005). The wildland-urban interface in the United States. *Ecological Applications, 15*, 799−805.

Rademaekers, K., Eichler, L., Berg, J., Obersteiner, M., & Havlik, P. (2010). *Study on the evolution of some deforestation drivers and their potential impacts on the costs of an avoiding deforestation scheme*. Rotterdam: Prepared for the European Commission by ECORYS and IIASA. Available from https://ec.europa.eu/environment/enveco/biodiversity/pdf/deforestation_drivers_annexes.pdf.

Rahman, M. F., Jashimuddin, M., Kamrul, I., & Kumar, N. T. (2016a). Land use change and forest fragmentation analysis: A geoinformatics approach on Chunati Wildlife Sanctuary, Bangladesh. *Journal of Civil Engineering and Environmental Sciences, 2*(1), 20−29.

Rahman, M. F., Islam, K., & Islam, K. N. (2016b). Industrial symbiosis: A review on uncovering approaches, opportunities, barriers and policies. *Journal of Civil Engineering and Environmental Sciences, 2*, 11−19.

Rai, P. K., & Singh, S. K. (2020). Invasive alien plant species: Their impact on environment, ecosystem services and human health. *Ecological Indicators, 111*, 106020.

Rainey, H. J., Pollard, E. H. B., Duston, G., Ekstrom, J. M. M., Livingstone, S. R., Temple, H. J., & Pilgrim, J. D. (2015). A review of corporate goals of no net loss and net positive impact on biodiversity. *Oryx, 49*(2), 232−238.

Rajamani, L. (2016). Ambition and differentiation in the 2015 Paris Agreement: Interpretative possibilities and underlying politics. *International & Comparative Law Quarterly, 65*(2), 493−514.

Ranganathan, R. (2007). Plant genetic resources for food and agriculture: A common heritage of mankind. *Panel discussion on making intellectual property work for development.* Available from http://www.iccwbo.org/uploadedFiles/ICC/policy/intellectual_property/pages/R_Ranganathan26April07.pdf.

RBG Kew (Royal Botanic Gardens Kew). (2017). *The state of the world's plants 2017.* Kew. Available from https://stateoftheworldsplants.org/2017/report/SOTWP_2017.pdf.

Redpath, S. M., Bhatia, S., & Young, J. C. (2015a). Tilting at wildlife: Reconsidering human−wildlife conflict. *Oryx, 49*(2), 222−225.

Redpath, S. M., Gutierrez, R. J., Wood, K. A., & Young, J. C. (Eds.), (2015b). *Conflicts in conservation: Navigating toward solutions.* Cambridge: Cambridge University Press.

Rees, P. (2013). *Dictionary of zoo biology and animal management.* West Sussex: John Wiley & Sons.

Reeve, R. (2002). *Policing international trade in endangered species: The CITES treaty and compliance.* London: The Royal Institute of International Affairs.

Reggiani, P., & Rientjes, T. H. M. (2015). A reflection on the long-term water balance of the Upper Indus Basin. *Hydrology Research, 46*, 446−462.

Reid, A. J., Carlson, A. K., Creed, I. F., Eliason, E. J., Gell, P. A., Johnson, P. T., ... Cooke, S. J. (2019). Emerging threats and persistent conservation challenges for freshwater biodiversity. *Biological Reviews, 94*, 849−873.

Reidinger, R. F., Jr., & Miller, J. E. (2013). *Wildlife damage management: Prevention, problem solving, and conflict resolution.* Baltimore, MD: Johns Hopkins University Press.

Rendekova, A., Micieta, K., Hrabovsky, M., Eliasova, M., & Miskovic, J. (2019). Effects of invasive plant species on species diversity: Implications on ruderal vegetation in Bratislava City, Slovakia, Central Europe. *Acta Societatis Botanicorum Poloniae, 88*(2), 3621.

Reynolds, J. F., Smith, D. M. S., Lambin, E. F., Turner, B. L., Mortimore, M., Batterbury, S. P. J., ... Walker, B. (2007). Global desertification: Building a science for dryland development. *Science (New York, N.Y.), 316*(5826), 847−851.

Ribeiro, J. C. T., Nunes-Freitas, A. F., Fidalgo, E. C. C., & Uzêda, M. C. (2019). Forest fragmentation and impacts of intensive agriculture: Responses from different tree functional groups. *PLoS One, 14*(8), e0212725.

Richardson, D. M., & Pysek, P. (2012). Naturalization of introduced plants: Ecological drivers of biogeographical patterns. *New Phytologist, 196*(2), 383−396.

Richerzhagen, C. C. (2010). *Protecting biological diversity: The effectiveness of access and benefit-sharing regimes.* New York: Routledge.

Ripple, W. J., Estes, J. A., Beschta, R. L., Wilmers, C. C., Ritchie, E. G., Hebblewhite, M., ... Wirsing, A. J. (2014). Status and ecological effects of the world's largest carnivores. *Science (New York, N.Y.), 343*, 6167.

Robinson, D. T., Brown, D. G., French, N. H. F., & Reed, B. C. (2013). Linking land use and carbon cycle. In D. G. Brown, D. T. Robinson, N. H. F. French, & B. C. Reed (Eds.), *Land use and the carbon cycle* (pp. 3−23). New York: Cambridge University Press.

Roe, D., Seddon, N., & Elliott, J. (2019). Biodiversity loss is a development issue: A rapid review of evidence. *IIED Issue Paper.* London: IIED.

Root, T. L., Price, J. T., Hall, K. R., Schneider, S. H., Rosenzweig, C., & Pounds, J. A. (2003). Fingerprints of global warming on wild animals and plants. *Nature, 421*, 57−60.

Rootes, C. (2008). The environmental movement. In M. Klimke, & J. Scharloth (Eds.), *1968 in Europe* (pp. 295−305). New York: Palgrave Macmillan.

Rosa, W. (2017). Transforming our world: The 2030 agenda for sustainable development. In W. Rosa (Ed.), *A new era in global health.* New York: Springer Publishing Company.

Rosen, T. (2020). The evolving war on illegal wildlife trade. *Brief # 3.* Canada: IISD. Available from https://www.iisd.org/system/files/2020-10/still-one-earth-wildlife-trade.pdf.

Rosenzweig, C., Elliott, J., Deryng, D., Ruane, A. C., Müller, C., Arneth, A., ... Jones, J. W. (2014). Assessing agricultural risks of climate change in the 21st century in a global gridded crop model intercomparison. *Proceedings of the National Academy of Sciences of the United States of America, 111*(9), 3268−3273.

Rouget, M., Robertson, M. P., Wilson, J. R. U., Hui, C., Essl, F., Renteria, J. L., & Richardson, D. M. (2016). Invasion debt—Quantifying future biological invasions. *Diversity and Distribution, 22,* 445−456.

Rowland, L., da Costa, A. C. L., Galbraith, D. R., Oliveira, R. S., Binks, O. J., Oliveira, A. A. R., ... Meir, P. (2015). Death from drought in tropical forests is triggered by hydraulics not carbon starvation. *Nature, 528,* 119−122.

Rubenstein, K. D., Heisey, P., Shoemaker, R., Sullivan, J., & Frisvold, G. (2005). Crop genetic resources: An economic appraisal. *Economic Information Bulletin, 2*(2005). Available from https://www.ers.usda.gov/webdocs/publications/44121/17452_eib2_1_.pdf?v = 3443.3.

Sachs, J., Schmidt-Traub, G., Kroll, C., Lafortune, G., Fuller, G., & Woelm, F. (2020). *The sustainable development goals and COVID-19. Sustainable development report 2020.* Cambridge: Cambridge University Press.

Sadro, S., Sickman, J. O., Melack, J. M., & Skeen, K. (2018). Effects of climate variability on snowmelt and implications for organic matter in a high-elevation lake. *Water Resources Research, 54,* 4563−4578.

Saltre, F., & Bradshaw, J. A. (2019, 13 November). Are we in the sixth mass extinction on Earth? The signs are all there. *Theconversation.com.* Available from https://theconversation.com/what-is-a-mass-extinction-and-are-we-in-one-now-122535.

Sands, R., Norton, D. A., & Weston, C. J. (2013). The environmental value of forests. In S. Roger (Ed.), *Forestry in a global context* (2nd Edition, pp. 55−76). Wallingford: CABI International.

Santín, C., & Doerr, S. H. (2016). Fire effects on soils: The human dimension. *Philosophical Transactions of the Royal Society of London B: Biological Sciences, 371*(1696), 20150171.

Sardans, J., & Penuelas, J. (2012). The role of plants in the effects of global change on nutrient availability and stoichiometry in the plant-soil system. *Plant Physiology, 160*(4), 1741−1761.

Sasaki, N., & Putz, F. E. (2009). Critical need for new definitions of "forest" and "forest degradation." In global climate change agreements. *Conservation Letters, 2,* 226−232.

Scherrer, D., & Körner, C. (2011). Topographically controlled thermal-habitat differentiation buffers alpine plant diversity against climate warming. *Journal of Biogeography, 38,* 406−416.

Scheuler, R. S., Jackson, S. E., & Tarique, I. (2011). Framework for global talent management: HR actions for dealing with global talent challenges. In H. Scullion, & D. Collings (Eds.), *Global talent management* (pp. 17−37). New York: Routledge.

Schwarzenbach, R. P., Egli, T., Hofstetter, T. B., von Gunten, U., & Wehrli, B. (2010). Global water pollution and human health. *Annual Review of Environment and Resources, 35,* 109−136.

Schwarzenbach, R. P., Escher, B. I., Fenner, K., Hofstetter, T. B., Johnson, C. A., von Gunten, U., & Wehrli, B. (2006). The challenge of micropollutants in aquatic systems. *Science (New York, N.Y.), 313,* 1072−1077.

Shiklomanov, I. (1993). World fresh water resources. In P. H. Gleick (Ed.), *Water in crisis: A guide to the world's fresh water resources* (pp. 13−24). New York: Oxford University Press.

Searchinger, T., Hanson, C., Ranganathan, J., Lipinski, B., Waite, R., Winterbottom, R., ... Chemineau, P. (2014). *Creating a sustainable food future. A menu of solutions to sustainably feed more than 9 billion people by 2050. World resources report 2013−14: Interim findings.* Washington, DC: World Resources Institute (WRI).

Seebens, H., Blackburn, T. M., Dyer, E. E., Genovesi, P., Hulme, P. E., Jeschke, J. M., ... Essl, F. (2018). Global rise in emerging alien species results from increased accessibility of new source pools. *Proceedings of National Academy of Sciences, U.S.A., 115*(10), E2264−E2273.

Sekercioglu, C. H., Daily, G. C., & Ehrlich, P. R. (2004). Ecosystem consequences of bird declines. *Proceedings of the National Academy of Sciences of the United States of America, 101,* 18042−18047.

Settele, J., Scholes, R. J., Bunn, S. E., Betts, R. A., Bunn, S. E., & Winter, M. (2014). *Terrestrial and inland water systems. In IPCC's climate change 2014 impacts, adaptation and vulnerability, part A: Global and sectoral aspects. Fifth assessment report of the working groups of the Intergovernmental Panel on Climate Change* (pp. 271–360). Cambridge, UK: Cambridge University Press.

Seymour, F., & Busch, J. (2016). *Why forests? Why now? The science. Economics, and politics of tropical forests and climate change.* Washington, DC: Center for Global Development; Brookings Institutional Press.

Shoo, L. P., Olson, D. H., McMenamin, S. K., Murray, K. A., Sluys, M. V., Donnelly, M. A., ... Hero, J.-M. (2011). Engineering—A future for amphibians under climate change. *Journal of Applied Ecology, 48*, 487–492.

Shrivastava, P., & Kumar, R. (2015). Soil salinity: A serious environmental issue and plant growth promoting bacteria as one of the tools for its alleviation. *Saudi Journal of Biological Sciences, 22*(2), 123–131.

Simmons, C. T., & Matthews, H. D. (2016). Assessing the implications of human land-use change for the transient climate response to cumulative carbon emissions. *Environment Research Letters, 11*(3), 025001.

Sintayehu, D. W. (2018). Impact of climate change on biodiversity and associated key ecosystem services in Africa: A systematic review. *Ecosystem Health and Sustainability, 4*(9), 225–239.

Sirakaya, A. (2019). Balanced Options for Access and Benefit-Sharing: Stakeholder Insights on Provider Country Legislation. *Frontiers in Plant Science, 10*, 1175.

Slik, J. W. F. (2004). El Niño droughts and their effects on tree species composition and diversity in tropical rain forests. *Oecologia, 141*, 114–120.

Smith, P., Bustamante, M., Ahammad, H., Clark, H., Dong, H., Elsiddig, E. A., ... Tubiello, F. (2014). *Agriculture, forestry and other land use (AFOLU). Climate change 2014: Mitigation of climate change. Contribution of working group III to the fifth assessment report of the intergovernmental panel on climate change* (pp. 811–922). Cambridge: Cambridge University Press.

Smith, C. (2015, 29 January). More needs to be done to halt global biodiversity loss and meet Aichi targets. *Weekly News Alert Online.* Available from https://ec.europa.eu/environment/integration/research/newsalert/pdf/Aichi_biodiversity_targets_unlikely_to_be_met_by_2020_401na5_en.pdf.

Smol, J. P., Wolfe, A. P., Birks, H. J. B., Douglas, M. S. V., Jones, V. J., Korhola, A., ... Weckstrom, J. (2005). Climate-driven regime shifts in the biological communities of arctic lakes. *Proceedings of the National Academy of Sciences of the United States of America, 102*(12), 4397–4402.

Snyder, M. (2014). What is forest fragmentation and why is it a problem? *Northern Woodlands.* Available from https://northernwoodlands.org/articles/article/forest-fragmentation#:~:text=Forest%20fragmentation%20is%20the%20breaking,subdivisions%2C%20or%20other%20human%20development.

Soares, M. P., Reys, P., Pifano, D. S., de Sa, J. L., da Silva, P. O., Santos, P. M., & Silva, S. G. (2015). Relationship between edaphic factors and vegetation in savannas in the Brazilian Midwest. *Journal of Science of Solo, 39*(3). Available from https://www.scielo.br/scielo.php?pid=S0100-06832015000300821&script=sci_arttext.

Somero, G. (2010). The physiology of climate change: How potentials for acclimatization and genetic adaptation will determine 'winners' and 'losers'. *Journal of Experimental Biology, 213*(6), 912–920.

Sonter, L. J., Barrett, D. J., Moran, C. J., & Soares-Filho, B. S. (2015). Carbon emissions due to deforestation for the production of charcoal used in Brazil's steel industry. *Nature and Climate Change, 5*, 359–363.

Sonter, L. J., Herrera, D., Barrett, D. J., Galford, G. L., Moran, C. J., & Soares-Filho, B. S. (2017). Mining drives extensive deforestation in the Brazilian Amazon. *Nature Communications., 8*(1), 1013.

Soulsbury, C. D., & White, P. C. L. (2015). Human-wildlife interactions in urban areas: A review of conflicts, benefits and opportunities. In A. Taylor, & P. White (Eds.), *Wildlife research: Interactions between humans and wildlife in urban areas* (pp. 541–553). Australia: CSIRO.

Sousa, R., Novais, A., Costa, R., & Strayer, D. L. (2014). Invasive bivalves in fresh waters: Impacts from individuals to ecosystems and possible control strategies. *Hydrobiologia*, *735*, 233−251.

Sousa, R., Nogueira, J. G., Ferreira, A., Carvalho, F., Lopes-Lima, M., Varandas, S., & Teixeira, A. (2019). A tale of shells and claws: The signal crayfish as a threat to the pearl mussel *Margaritifera margaritifera* in Europe. *Science of the Total Environment*, *665*, 329−337.

Souza, D. M., Teixeira, R. F. M., & Ostrmann, O. P. (2015). Assessing biodiversity loss due to land use with life cycle assessment: Are we there yet? *Global Change Biology*, *21*(1), 32−47.

Sperry, J. S., Venturas, M. D., Todd, H. N., Trugman, A. T., Anderegg, W. R. L., Wang, Y., & Tai, X. (2019). The impact of rising CO_2 and acclimation on the response of United States forests to global warming. *Proceedings of the National Academy of Sciences of the United States of America*, *116*(51), 25734−25744.

Stanisci, A., Carranza, M. L., Pelino, G., & Chiarucci, A. (2011). Assessing the diversity pattern of cryophilous plant species in high elevation habitats. *Plant Ecology*, *212*, 595−600.

Stanturf, J. A., Palik, B. J., & Dumroese, R. K. (2014a). Contemporary forest restoration: A review emphasizing function. *Forest Ecology and Management*, *331*, 292−323.

Stanturf, J. A., Palik, B. J., Williams, M. I., Dumroese, R. K., & Madsen, P. (2014b). Forest restoration paradigms. *Journal of Sustainable Forestry*, *33*, S161−S194.

Stayer, D. L., & Dudgeon, D. (2010). Freshwater biodiversity conservation: recent progress and future challenges. *Journal of the North American Benthological Society*, *29*(1), 344−350.

STDF (Standards and Trade Development Facility). (2013). *International trade and invasive alien species*. Geneva: WTO Secretariat. Available from https://www.standardsfacility.org/sites/default/files/STDF_IAS_EN_0.pdf.

Strayer, D. L., & Dudgeon, D. (2010). Freshwater biodiversity conservation: Recent progress and future challenges. *Journal of the North American Benthological Society*, *29*, 344−358.

Steffen, W., Sanderson, A., Tyson, P. D., Jäger, J., Matson, P. A., Moore, B., III, ... Wasson, R. J. (2004). *Global change and the earth system: A planet under pressure*. Heidelberg: Springer.

Steffen, W., Crutzen, P. J., & McNeill, J. R. (2007). The Anthropocene: Are humans now overwhelming the great forces of nature. *Ambio*, *36*(8), 614−622.

Steffen, W., Rockström, J., Richardson, K., Lenton, T. M., Folke, C., Liverman, D., & Donges, J. F. (2018). Trajectories of the earth system in the Anthropocene. *Proceedings of the National Academy of Sciences of the United States of America*, *115*(33), 8252−8259.

Steinbauer, M. J., Grytnes, J.-A., Jurasisnki, G., Kulonen, A., Lenoir, J., Pauli, H., ... Wipf, S. (2018). Accelerated increase in plant species richness on mountain summits is linked to warming. *Nature*, *556*(7700), 231−234.

Stoate, C., Báldi, A., Beja, P., Boatman, N. D., Herzon, I., van Doorn, A., ... Ramwell, C. (2009). Ecological impacts of early 21st century agricultural change in Europe—A review. *Journal of Environmental Management*, *91*, 22−46.

Stone, C. M., Witt, A. B. R., Walsh, G. C., Foster, W. A., & Murphy, S. T. (2018). Would the control of invasive alien plants reduce malaria transmission? A review. *Parasites Vectors*, *11*, 76.

Sullivan, S., & Hannis, M. (2015). Nets and frames, losses and gains: Value struggles in engagements with biodiversity offsetting policy in England. *Ecosystem Services*, *15*, 162−173.

Tallis, H., Kareiva, P., Marvier, M., & Chang, A. (2008). An ecosystem services framework to support both practical conservation and economic development. *Proceeding of the National Academy of Sciences, USA*, *105*, 9457−9464.

Team Wanderdust. *These are the 5 biggest threats to the world's wildlife*. (2019). <https://www.wanderlust.co.uk/content/biggest-threats-to-wildlife/> Updated 11.10.19 October 2019.

TEEB (The Economics of Ecosystems and Biodiversity). (2010). *The economics of ecosystems and biodiversity: Ecological and economic foundations*. London; Washington, DC: Earthscan.

Teillard, F., Anton, A., Dumont, B., Finn, J. A., Henry, B., Souza, D. M., ... White, S. (2016). A review of indicators and methods to assess biodiversity—Application to livestock production at global scale. *Livestock Environmental Assessment and Performance (LEAP) Partnership*. Rome: FAO. Available from http://www.fao.org/3/a-av151e.pdf.

Thapar, V. (2003). *Battling for survival: India's wilderness over two centuries*. New Delhi: Oxford University Press.

Thomas, D., Butry, D., Gilbert, S., Webb, D., & Fund, J. (2017). *The costs and losses of wildfires; A literature survey*. United States Department of Commerce, National Institute of Standards and Technology. Available from https://nvlpubs.nist.gov/nistpubs/SpecialPublications/NIST.SP.1215.pdf.

Thomson, J. J. (2003). Causation: Omissions. Philosophy and Phenomenological. *Research*, 66(1), 81–103.

Thompson, R., Kamenik, C., & Schmidt, R. (2005). Ultra-sensitive alpine lakes and climate change. *Journal of Limnology*, 64, 139–152.

Tibbett, M., Fraser, T. D., & Duddigan, S. (2020). Identifying potential threats to soil biodiversity. *PeerJ.*, 8, e9271.

Tobin, B. (2013). Bridging the Nagoya compliance gap: The fundamental role of customary law in protecting of indigenous peoples' resource and knowledge rights. *Law Environment and Development Journal*, 9(2), 142–152.

Treves, A., & Karanth, K. U. (2003). Human-carnivore conflict and perspectives on carnivore management worldwide. *Conservation Biology*, 17, 1491–1499.

Tscharntke, T., Tyliankis, J. M., Rand, T. A., Didham, R., & Westphal, C. (2012). Landscape moderation of biodiversity patterns and processes: Eight hypotheses. *Biological Reviews*, 87, 661–685.

Tsela, P., Wessels, K., Botai, J., Archibald, S., Swanepoel, D., Steenkamp, K., & Frost, P. (2014). Validation of the two standard MODIS satellite burned-area products and an empirically-derived merged products in South Africa. *Remote Sensing*, 6(2), 1275–1293. Available from https://www.mdpi.com/2072-4292/6/2/1275/htm.

Tubiello, S.F.N., Savatore, M., Gole, R.D.C., Ferrara, A., Rossi, S., Biancalani, R., ... Flammini, A. (2014). Agriculture, forestry and other land use emissions by sources and removals: 1990–2011 analysis. Working Paper Series, FAO Statistical Division. Rome: FAO.

Türkeş, M. (1999). Vulnerability of Turkey to desertification with respect to precipitation and aridity conditions. *Turkish Journal of Engineering and Environmental Sciences*, 23(5), 363–380. Available from http://journals.tubitak.gov.tr/engineering/abstract.htm?id = 3523.

Turner, B. L., Skole, D. L., Sanderson, S., Fischer, G., Fresco, L. O., & Leemans, R. (1995). Land-use and land-cover change. Science/research plan. In: *IGBP report no. 35 and HDP report no. 7*. Stockholm and Geneva: IGBP.

Tyukavina, A., Hansen, M. C., Potapov, P. V., Stehman, S. V., SmithRodriguez, K., Okpa, C., & Aguilar, R. (2017). Types and rates of forest disturbance in Brazilian Legal Amazon, 2000–2013. *Science Advances*, 3, e1601047.

UN (United Nations). *Text of convention on biological diversity*. (1992). <https://www.cbd.int/doc/legal/cbd-en.pdf>.

UN (United Nations). *Paris Agreement*. (2015a). <https://sustainabledevelopment.un.org/content/documents/17853paris_agreement.pdf>.

UN (United Nations). *Transforming our world: The 2030 agenda for sustainable development. A/RES/70/1.* (2015b). <https://www.un.org/ga/search/view_doc.asp?symbol = A/RES/70/1&Lang = E>.

UN (United Nations). (2017). *Forests, desertification and biodiversity*. United Nations Sustainable Development. Available from http://www.un.org/sustainabledevelopment/biodiversity/.

UN (United Nations). *Sustainable development goals report 2019*. (2019). <https://unstats.un.org/sdgs/report/2019/The-Sustainable-Development-Goals-Report-2019.pdf>.

UN-DPI (United Nations-Department of Public Information). (2015). *Historic new sustainable development agenda unanimously adopted by 193 UN members*. United Nations Sustainable Development. Available

References

from https://sustainabledevelopment.un.org/content/documents/8371Sustainable%20Development%20Summit_final.pdf.

UNCCD (United Nations Convention to Combat Desertification). *United Nations convention to combat desertification in those countries experiencing serious drought and/or desertification, particularly in Africa.* (1994). <https://www.unccd.int/sites/default/files/relevant-links/2017-01/UNCCD_Convention_ENG_0.pdf>.

UNCCD (United Nations Convention to Combat Desertification). (2017). *Global land outlook.* Bonn, Germany: UNCCD Secretariat. Available from https://www.unccd.int/sites/default/files/documents/2017-09/GLO_Full_Report_low_res.pdf.

UNCED (United Nations Conference on Environment and Development). (1992). *Managing fragile ecosystems: Combating desertification and drought.* United Nations Conference on Environment and Development. Agenda 21, Chapter 12.

UNCLOS (United Nations Convention on the Law of the Sea). *Text of the UN convention on the law of the sea.* (1982). <https://www.un.org/depts/los/convention_agreements/texts/unclos/unclos_e.pdf>.

UN-DESA (United Nations-Department of Economic and Social Affairs). (2018). *World cities in 2018-data booklet.* New York: UN-DESA. Available from https://www.un.org/en/events/citiesday/assets/pdf/the_worlds_cities_in_2018_data_booklet.pdf.

UN-DESA (United Nations, Department of Economic and Social Affairs). *The world's cities in 2016−data booklet.* ST/ESA/SER.A/392. (2016). Available from http://www.un.org/en/development/desa/population/publications/pdf/urbanizati on/the_worlds_cities_in_2016_data_booklet.pdf.

UN-EMG (United Nations-Environment Management Group). (2011). *Global drylands: A UN system-wide response.* New York: United Nations. Available from https://www.zaragoza.es/contenidos/medioambiente/onu//issue07/1107-eng.pdf.

UNEP (United Nations Environment Programme). (1992). *World atlas of desertification.* Nairobi: UNEP.

UNEP (United Nations Environment Programme). (2006). *Global deserts outlook.* Nairobi: UNEP. Available from http://wedocs.unep.org/handle/20.500.11822/9581.

UNEP (United Nations Environment Programme). (2007). *Global environmental outlook GEO-4.* Nairobi: UNEP.

UNEP-WCMC (United nations Environment Programme-World Conservation Monitoring Centre). (2002). *Mountain watch: Environmental change and sustainable development in mountains.* Nairobi: UNEP. Available from http://www.unep-wcmc.org/mountains/mountainwatchreport/.

FAO (Food and Agriculture organization) (2017). Food and Agriculture data for 2017. FAOSTAT eb Portal. Available at: http://fao.org/faostat/en/#home.

UNFCCC (United Nations Framework Convention on Climate Change). *Text of the United Nations framework convention on climate change.* (1992). <https://unfccc.int/resource/docs/convkp/conveng.pdf>.

UNFCCC (United Nations Framework Convention on Climate Change). *Kyoto Protocol to the United Nations framework convention on climate change.* (1998). <https://unfccc.int/resource/docs/convkp/kpeng.pdf>.

UNFCCC (United Nations Framework Convention on Climate Change). *Revision of the UNFCCC reporting guidelines on annual inventories for parties included in Annex I to the Convention.* (2013). <https://unfccc.int/resource/docs/2013/cop19/eng/10a03.pdf#page = 2>.

UNODC (United Nations Office on Drugs and Crime). (2020). *World wildlife crime report 2020.* Vienna: United Nations Office on Drugs and Crime. Available online.

USCB (United States Census Bureau). *Impacts of 2017 Californian fires.* (2017). <https://www.census.gov/topics/preparedness/events/wildfires/2017-cawildfires.html>.

USGS (United States Geological Survey). (2004). Landslide types and processes. In: *United States Geological Survey fact sheet 2004−3072, Version 1.0.* Available from http://pubs.usgs.gov/fs/2004/3072/pdf/fs2004-3072.pdf.

Van-Camp, L., Bujarrabal, B., Gentile, A.-R., Jones, R. J. A., Montanarella, L., Olazabal, C., & Selvaradjou, S.-K. (2004). Reports of the technical working groups established under the thematic strategy for soil protection. *EUR 21319 EN/4.* Luxembourg: Office for Official Publications of the European Communities.

van der Esch S., ten Brink, B., Stehfest, E., Bakkenes, M., Sewell, A., Bouwman, A., ... Mantel, S. (2017). Exploring future changes in land use and land condition and the impacts on food, water, climate change and biodiversity. In: *Scenarios for the UNCCD Global Land Outlook*. The Hague. Available from https://www.pbl.nl/sites/default/files/downloads/pbl-2017-exploring-future-changes-in-land-use-and-land-condition-2076b_1.pdf.

van der Werf, G. R., Randerson, J. T., Collatz, G. J., Giglio, L., Kasibhatla, P. S., Arellano, A. F., Jr., ... Kasischke, E. S. (2004). Continental-scale partitioning of fire emissions during the 1997 to 2001 El Niño/La Niña period. *Science (New York, N.Y.)*, *303*, 73−76.

van Soesbergen, A., Sassen, M., Kinsey, S., & Hill, S. (2019). Potential impacts of agricultural development on freshwater biodiversity in the Lake Victoria basin. *Aquatic Conservation*, *29*(7), 1052−1062.

van Vliet, N., Mertz, O., Heinimann, A., Langanke, T., Pascual, U., Schmook, B., ... Ziegler, A. D. (2012). Trends, drivers and impacts of changes in swidden cultivation in tropical forest-agriculture frontiers: A global assessment. *Global Environmental Change*, *22*, 418−429.

Vare, H., Lampinen, H., Humpheries, C. J., & Williams, H. (2003). Taxonomic diversity of vascular plants in the European Alpine areas. In L. Nagy, G. Grabberr, C. Korner, & D. B. A. Thompson (Eds.), *Alpine biodiversity in Europe. Berlin, Germany* (pp. 133−148). Springer.

Vásquez-Grandón, A., Donoso, P. J., & Gerding, V. (2018). Forest degradation: When is a forest degraded? *Forests*, *9*, 726.

Vaughan, I. P., & Ormerod, S. J. (2012). Large-scale, long-term trends in British river macroinvertebrates. *Global Change Biology*, *18*, 2184−2194.

Velasquez, M., & Hester, P. T. (2013). An analysis of multi-criteria decision-making methods mark. *International Journal of Operations Research*, *10*(2), 56−66.

Veraverbeke, S., Rogers, B. M., Goulden, M. L., Jandt, R. R., Miller, C. E., Wiggins, E. B., & Randerson, J. T. (2017). Lightning as a major driver of recent large fire years in North American boreal forests. *Nature Climate Change*, *7*(7), 529−534.

Vermeulen, S. J., Challinor, A. J., Thorton, P. K., Campbell, B. M., Eriyagama, N., Vervoort, J. M., ... Smith, D. R. (2013). Addressing uncertainty in adaptation planning for agriculture. *Proceedings of the National Academy of Sciences of the United States of America*, *110*(21), 8357−8362.

Veron, J., Stafford-Smith, M., DeVantier, L., & Turak, E. (2015). Overview of distribution patterns of zooxanthellate Scleractinia. *Frontiers of Marine Science*, *1*, 81.

Vicente-Serrano, S. M., Gouveia, C., Camarero, J. J., Beguería, S., Trigo, R., López-Morenoa, J. I., ... Sanchez-Lorenzo, A. (2013). Response of vegetation to drought time-scales across global land biomes. *Proceedings of the National Academy of Sciences of the United States of America*, *110*, 52−57.

Vie, J.-C., Hilton-Taylor, C., & Stuart, S. N. (Eds.), (2009). *Wildlife in a changing world—An analysis of the 2008 IUCN red list of threatened species*. Gland: IUCN.

Vierros, M., Suttle, C. A., Harden-Davies, H., & Burton, G. (2016). Who owns the ocean? Policy issues surrounding marine genetic resources. *Limnology & Oceanography Bulletin*, *25*, 29−35.

Vitousek, P. M., Ehrlich, P. R., Ehrlich, A. H., & Matson, P. A. (1986). Human appropriation of the products of photosynthesis. *Bioscience*, *36*, 368−373.

Vitousek, P. M., Mooney, H. A., Lubchenco, J., & Melillo, J. M. (1997). Human domination of Earth's ecosystems. *Science (New York, N.Y.)*, *277*, 494−499.

Viviroli, D., Dürr, H. H., Messerli, B., Meybeck, M., & Weingartner, R. (2007). Mountains of the world, water towers for humanity: Typology, mapping, and global significance. *Water Resource Research: A Journal of Science and its Applications*, *43*, W07447.

Viviroli, D., Archer, D. R., Buytaert, W., Fowler, H. J., Greenwood, G., Hamlet, A. F., ... Woods, R. A. (2011). Climate change and mountain water resources: Overview and recommendations for research, management and policy. *Hydrology & Earth System Sciences*, *15*(2), 471−504.

Vogel, J. H., Muller, M. R., & Angerer, K. (2015). *Submission of views in preparation for the Expert Meeting on the need for and modalities of a global multilateral benefit-sharing mechanism and the first meeting of the Compliance Committee of the Nagoya Protocol: Annex to collective submission of the IUCN Joint SSC-WCEL Global Specialist Group on ABS, Genetic Resources and Related Issues (ABSSG): Annex.* Available from https://www.cbd.int/abs/submissions.shtml.

von der Lippe, M., Bullock, J. M., Kowarik, I., Knopp, T., & Wichmann, M. (2013). Human-mediated dispersal of seeds by the airflow of vehicles. *Plos One, 8*(8), 10.

von der Lippe, M., & Kowarik, I. (2008). Do cities export biodiversity? Traffic as dispersal vector across urban–rural gradients. *Diversity and Distribution, 14*, 18–25.

Vörösmarty, C. J., & Sahagian, D. (2000). Anthropogenic disturbance of the terrestrial water cycle. *Bioscience, 50*, 753–765.

Vörösmarty, C. J., McIntyre, P. B., Gessner, M. O., Dudgeon, D., Prusevich, A., Green, P., ... Davies, P. M. (2000). Global threats to human water security and river biodiversity. *Nature, 467*, 555–561.

Vörösmarty, C. J., Green, P., Salisbury, J., & Lammers, R. B. (2010). Global water resources: Vulnerability from climate change and population growth. *Science (New York, N.Y.), 289*, 284–288.

Vorovencii, I. (2015). Assessing and monitoring the risk of desertification in Dobrogea, Romania, using Landsat data and decision tree classifier. *Environmental Monitoring and Assessment, 187*(4), 204.

Wade, T. G., Riitters, K. H., Wickham, J. D., & Jones, K. B. (2003). Distribution and causes of global forest fragmentation. *Conservation Ecology, 7*, 7.

Walther, G.-R., Post, E., Convey, P., Menzel, A., Parmesan, C., Beebee, T. J. C., ... Bairlein, F. (2002). Ecological responses to recent climate change. *Nature, 416*, 389–395.

Ward, D. S., Mahowald, N. M., & Kloster, S. (2014). Potential climate forcing of land use and land cover change. *Atmospheric Chemistry & Physics, 14*(23), 12701–12724.

Warren, C. R. (2007). Perspectives on the 'alien' vs 'native' species debate: A critique of concepts, language and practice. *Progress in Human Geography, 31*(4), 427–446.

Watson, J. E., Evans, T., Venter, O., Williams, B., Tulloch, A., Stewart, C., ... Lindenmayer, D. (2018). The exceptional value of intact forest ecosystems. *Nature, Ecology and Evolution, 2*, 599–610.

Wester, P., Mishra, A., Mukherji, A., & Shrestha, A. B. (Eds.), (2019). *The Hindu Kush Himalaya assessment: Mountains, climate change, sustainability and people.* Cham, Switzerland: Springer.

White, P., Michalak, J., & Lerner, J. (2007). Linking conservation and transportation: Using the state wildlife action plans to protect wildlife from road impacts. *Defenders of Wildlife.* Available from https://defenders.org/sites/default/files/publications/linking_conservation_and_transportation.pdf.

Whiteman, C. D. (2000). *Mountain meteorology: Fundamentals and applications.* New York: Oxford University Press.

Wiesman, Z. *Desert olive oil cultivation: Advanced biotechnologies.* (2009). <https://ebookcentral.proquest.com>.

Wickneswari, R., Rajora, O. P., Finkeldey, R., Aravanopoulos, F., Bouvet, J. M., Vaillancourt, R. E., ... Vinson, C. (2014). Genetic effects of forest management practices: Global synthesis and perspectives. *Forest Ecology and Management, 333*, 52–65.

Wilcove, D. S., McLellan, C. H., & Dobson, A. P. (1986). Habitat fragmentation in the temperate zone. In M. E. Soule (Ed.), *Conservation biology: The science of scarcity and diversity* (pp. 237–256). Sunderland, MA: Sinauer Associates.

Winkler, M., Lamprecht, A., Steinbauer, K., Hülber, K., Theurillat, J.-P., Breiner, F., ... Pauli, H. (2016). The rich sides of mountain summits—A pan-European view on aspect preferences of alpine plants. *Journal of Biogeography, 43*, 2261–2273.

Winowiecki, L. A., Vågen, T. G., Kinnaird, M. F., & O'Brien, T. G. (2018). Application of systematic monitoring and mapping techniques: Assessing land restoration potential in semi-arid lands of Kenya. *Geoderma., 327*, 107–118.

WEF (World Economic Forum). (2019). *The global risk report.* Geneva: World Economic Forum.

Winter, L., Pflugmacher, S., Berger, M., & Finkbeiner, M. (2018). Biodiversity impact assessment (BIA +) - methodological framework for screening biodiversity. *Integrated Environmental Assessment & Management, 14*, 282–297.

WPDS (World Population Data Sheet). *2020 World population data sheet shows older populations growing, total fertility rates declining*. (2020). <https://www.prb.org/2020-world-population-data-sheet/#: ~ : text = The%202020%20Data%20Sheet%20identifies,2020%20population%20of%207.8%20billion>.

Wilhite, D. A., Sivakumar, M. V. K., & Pulwarty, R. (2014). Managing drought risk in a changing climate: The role of national drought policy. *Weather and Climate Extremes, 3*, 4−13.

Wilkenskjeld, S., Kloster, J., Pongratz, T. R., & Reick, C. H. (2014). Comparing the influence of net and gross anthropogenic land-use and land-cover changes on the carbon cycle in the MPI-ESM. *Biogeosciences, 11*, 4817−4828.

Williams, M. (2006). *Deforesting the Earth from pre-history to global crisis*. Chicago: University of Chicago Press.

Williamson, G. B., Laurance, W. F., Oliveira, A. A., Delamonica, P., Gascon, C., Lovejoy, T. E., & Pohl, L. (2000). Amazonian tree mortality during the 1997 El Nino drought. *Conservation Biology, 14*, 1538−1542.

Wilson, M., Carpenter, S., & Wilson MA, C. S. R. (1999). Economic valuation of freshwater ecosystem services in the United States: 1971−1997. *Ecological Applications, 9*, 772−783.

Wood, L. (2017). *What is the difference between Freshwater and Marine life?*. animals.mom.com. Available from https://animals.mom.com/what-is-the-difference-between-freshwater-marine-life-6980623.html.

Woodroffe, R., Thirgood, S., & Rabinowitz, A. (Eds.), (2005). *People and wildlife: Conflict or coexistence?* Cambridge: Cambridge University Press.

Woodward, G., Perkins, D. M., & Brown, L. E. (2010). Climate change and freshwater ecosystems: Impacts across multiple levels of organization. *Philosophical Transactions of the Royal Society of London. B Biological Sciences, 365*(1549), 2093−2106.

Woodward, G., Dybkjær, J. B., Ólafsson, J. S., Gíslason, G. M., Hannesdóttir, E. R., & Friberg, N. (2010). Sentinel systems on the razor's edge: Effects of warming on Arctic geothermal stream ecosystems. *Global Change Biology, 16*(7), 1979−1991.

World Bank. *The world bank environmental and social framework*. (2017). <http://documents1.worldbank.org/curated/en/383011492423734099/pdf/The-World-Bank-Environmental-and-Social-Framework.pdf>.

WHO (World Health Organization). (2011). *pandemic influenza preparedness framework (PIPF) for the sharing of influenza viruses and access to vaccine and other benefits*. Paris: WHO.

WoRMS (World Register of Marine Species). *World register of marine species*. (2018). <http://www.marine-species.org/> Updated April 2018.

Wright, S. J., & Muller-Landau, H. C. (2006). The future of tropical forest species. *Biotropica., 38*, 287−301.

WWF (World Wide Fund for Nature). (2018). *Living planet report, 2018: Aiming higher*. Gland: WWF.

Wynberg, R., & Laird, S. A. (2009). Indigenous peoples. In R. Wynberg, D. Schroeder, & R. Chennells (Eds.), Consent and benefit sharing: Lessons fron the San-Hoodia case (pp. 69−86). Cham: Springer.

You, B., & Sun, Y. (2018). Innovative strategies for electrocatalytic water splitting. *Accounts of Chemical Research, 51*, 1571−1580.

Young, A. M., & Larson, B. M. H. (2011). Clarifying debates in invasion biology: A survey of invasion biologists. *Environmental Research, 111*, 893−898.

Zabel, F., Putzenlechner, B., & Mauser, W. (2014). Global agricultural land resources—A high resolution suitability evaluation and its perspectives until 2100 under climate change conditions. *PLoS One, 9*(9), e107522.

Zhang, Q., Shao, M., Jia, X., & Wei, X. (2017). Relationship of climatic and forest factors to drought- and heat-induced tree mortality. *PLoS One, 12*, e0169770.

Zenni, R. D. (2014). Analysis of introduction history of invasive plants in Brazil reveals patterns of association between biogeographical origin and reason for introduction. *Austral Ecology, 39*, 401−407.

Zhu, H., Xu, Z. F., Wang, H., & Li, B. G. (2004). Tropical rain forest fragmentation and its ecological and species diversity changes in southern Yunnan. *Biodiversity & Conservation, 13*, 1355−1372.

Zieritz, A., Bogan, A. E., Froufe, E., Klishko, O., Kondo, T., Kovitvadhi, U., ... Zanatta, D. T. (2018). Diversity, biogeography and conservation of freshwater mussels (Bivalvia: Unionida) in East and Southeast Asia. *Hydrobiologia, 810*, 29−44.

Further reading

Asner, G. P., Kellner, J. R., Kennedy-Bowdoin, T., Knapp, D. E., Anderson, C., & Martin, R. E. (2013). Forest canopy gap distributions in the Southern Peruvian Amazon. *Public Library of Science One*, 8, 1−10.

Bellard, C., Jeschke, J. M., Leroy, B., & Mace, G. M. (2018). Insights from modeling studies on how climate change affects invasive alien species geography. *Ecology and Evolution*, 8(11), 5688−5700.

Beniston, M., & Fox, D. G. (1996). Impacts of climate change on mountain regions. *Climate change 1995— Impacts, adaptations and mitigation of climate change: Scientific-technical analysis. Contribution of working group II to the second assessment report of the intergovernmental panel on climate change* (pp. 191−213). Cambridge: Cambridge University Press.

Corcoran, E. C., Nellemann, E., Baker, R., Bos, D., & Osborn, H. S. (2010). *Sick water? The central role of wastewater management in sustainable development*. Norway: UNEP-UN Habitat.

Drechsel, P., Qadir, M., & Wichelns, D. (2015). *Wastewater: Economic asset in an urbanizing world*. Cham: Springer.

ENP (Exmoor National Park, USA). *Swaling on Exmoor, 2018−2019 annual report*. (2019). <https://www.exmoor-nationalpark.gov.uk/_data/assets/pdf_file/0022/260158/Swaling-Report-2019.pdf>.

Evangelista, A., Frate, L., Carranza, M. L., Attorre, F., Pelino, G., & Stanisci, A. (2016). Changes in composition, ecology and structure of high-mountain vegetation: A re-visitation study over 42 years. *AoB Plants*, 8, 1−11.

Foley, J. A., Ramankutty, N., Brauman, K. A., Cassidy, E. S., Gerber, J. S., Johnston, M., ... Zaks, D. P. M. (2011). Solutions for a cultivated planet. *Nature*, 478(7369), 337−342.

Giam, X. (2017). Global diversity loss from tropical deforestation. *Proceedings of the National Academy of Sciences of the United States of America*, 114(23), 5775−5777.

Halewood, M. (Ed.), (2016). *Farmers' crop varieties and farmers' rights*. London: Earthscan.

Houghton, R. A. (2002). Magnitude, distribution and causes of terrestrial carbon sinks and some implications for policy. *Climate Policy*, 2, 71−88.

Jackson, R., Carpenter, S. R., Dahm, C. N., McKnight, D. M., Naiman, R. J., Postel, S. L., & Running, S. W. (2001). Water in a changing world. *Ecological Applications*, 11(4), 1027−2045.

Le Quéré, C., Andrew, R. M., Friedlingstein, P., Sitch, S., Pongratz, J., Manning, A. C., ... Zhu, D. (2018). Global carbon budget 2017. *Earth System Science Data*, 10, 405−448. Available from https://research-information.bris.ac.uk/ws/portalfiles/portal/149328928/essd_10_405_2018.pdf.

Lévêque, C., Oberdorff, T., Paugy, D., Stiassny, M. L. J., & Tedesco, P. A. (2008). Global diversity of fish (Pisces) in freshwater. *Hydrobiologia*, 595(1), 545−567.

Mekonnen, M. M., & Hoekstra, A. Y. (2016). Four billion people facing severe water scarcity. *Science Advances*, 2(2), e1500323.

Milla, S., Depiereux, S., & Kestemont, P. (2011). The effects of estrogenic and androgenic endocrine disruptors on the immune system of fish: A review. *Ecotoxicology (London, England)*, 20, 305−319.

Olson, D. M., Dinerstein, E., Wikramanayake, E. D., Burgess, N. D., Powell, G. V., Underwood, E. C., ... Kassem, K. R. (2001). Terrestrial ecoregions of the world: A new map of life on Earth: A new global map of terrestrial ecoregions provides an innovative tool for conserving biodiversity. *Bioscience*, 51, 933−938.

Good practice guidance for land use, land-use change and forestry. In J. Penman, M. Gytarsky, T. Hiraishi, T. Krug, D. Kruger, R. Pipatti, & ... F. Wagner (Eds.), *Intergovernmental panel on climate change, national greenhouse gas inventories program*. Available from https://www.ipcc-nggip.iges.or.jp/public/gpglulucf/gpglulucf_files/GPG_LULUCF_FULL.pdf.

Phalan, B., Onial, M., & Balmford, A. (2011). Green RE. Reconciling food production and biodiveristy. *Science (New York, N.Y.)*, 333, 1289−1291.

Puyravaud, J. P., Davidar, P., & Laurance, W. F. (2010). Cryptic loss of India's native forests. *Science (New York, N.Y.)*, 329(5987), 32.

Rockström, J., Gordon, L., Folke, C., Falkenmark, M., & Engwall, M. (1999). Linkages among water vapour flows, food production and terrestrial ecosystem services. *Conservation Ecology*, *3*(2), 5.

Schewe, J., Heinke, J., Gerten, D., Haddeland, I., Arnell, N. W., Clark, D. B., ... Kabat, P. (2014). Multimodel assessment of water scarcity under climate change. *Proceedings of the National Academy of Sciences of the United States of America*, *111*(9), 3245–3250.

Schlenker, W., & Lobell, D. B. (2010). Robust negative impacts of climate change on African agriculture. *Environmental Research Letters*, *5*(1), 014010.

Schueler, V., Kuemmerle, T., & Schroder, H. (2011). Impacts of surface gold mining on land use systems in Western Ghana. *Ambio*, *40*, 528–539.

Tvedt, M. W., & Schei, P. J. (2014). The term 'genetic resources' flexible and dynamic while providing legal certainty?. In S. Oberthr, & G. K. Rosendal (Eds.), *Global governance of genetic resources: Access and benefit sharing after the Nagoya Protocol* (pp. 19–32). New York: Routledge.

CHAPTER 10

Mainstreaming ecosystem-based adaptation

10.1 Introduction

Unprecedented global challenges are currently confronting humankind that warrant a primal metamorphosis of society in order to deal with degradation of functions that back up life and ensure human development (Rockstrom et al., 2009). Accumulation of sustainability challenges like climate change or loss of biodiversity at a global level are often delineated by numerous scales and facets and their causes and effects pertain to regional and local dynamics (Jerneck et al., 2011). For that reason, the korero on climate change has come to emphasize, over and above to the mitigation of greenhouse gas (GHG) emissions, on adaptation to adverse impacts so as to expansively address the worldwide challenge and facilitate a transition toward sustainability (Crane & Landis, 2010; IPCC, 2014a). Dearth of passable responses to climate change at national and global levels has enabled the regional and local setting to be progressively regarded as an efficacious swivel to accost the underlying processes of this sustainable challenge (McCormick, Anderberg, Coenen, & Neij, 2013; Zborel et al., 2012). Pivotal role of local governments in supervising wide-ranging responses to climate change is acknowledged by Rauken, Mydske, and Winsvold (2014), and "acting...to incorporate climate change adaptation (CCA) into their development plans and polices and infrastructure investments" (IPCC, 2014b, p. 6), and in this regard, spatial planning is a primary pathway for adaptation (McDonald, 2011), and also focuses on respective governance arrangements (Agrawala & Van Aalst, 2008).

Ecosystem-based adaptation (EbA) has garnered international traction as a wide-ranging approach capable of reducing the adverse impacts of climate change along with delivering numerous benefits while tailored to place-based features (Chong, 2014). Benefits accruing from EbA inter alia include: GHG emissions, improvement in and protection of livelihood, generation and conservation of recreation of areas, support for biodiversity, improvement of human well-being along with the potential of being more cost-effective than optional adaptation approaches (Doswald et al., 2014; IPCC, 2012). EbA has come to be regarded as an efficacious way of reassessing the existing paradigms of combating risk and natural disasters that have for decades been commanded by hard or technical solutions and gray infrastructures (Jones, Hole, & Zavaleta, 2012). Undeniably, the concept of EbA is still at evolving stage (Doswald et al., 2014); nonetheless, systematic incorporation of ecosystem services (ESSs) into municipal planning deal with the inherent linkages between nature and human well-being and, eventually, entails the potential of reconciling human-environment systems and engendering sustainability transitions (Chong, 2014; IPCC, 2014a; Wu, 2014).

Given that modus operandi for orderly mainstreaming and institutionalization of EbA entails a little theory (Andrade et al., 2011) renders it somewhat blurred as to how local authorities can facilitate apt integration of this approach into their respective development plans and policies (IPCC,

2014a). Besides, insufficient evidence available in the literature with regard to the degree to which EbA is already applied in urban planning practice (Turnpenny, Russel, & Jordan, 2014) is further compounded by the fact of its integration into prevalent structures and processes coupled with the fact of the driving forces or obstacles to further integration. In this backdrop, it becomes imperative to ascertain potential pathways to sustainably mainstream EbA into urban governance and planning by taking cues from in-depth studies undertaken on municipalities in different countries, with specific focus on modus operandi of integration of EbA into municipal practice, ascertainment of key characteristics of current mainstreaming strategies and assessing their ability to promote sustainability transition and transformative adaptation.

10.2 Concept of mainstreaming

The concept of mainstreaming, as a term has been increasingly in usage since the 1990s by social scientists and policymakers to focus on critical cross-cutting but neglected issues to bring them centerstage, aims at bringing to the forefront a significant issue that is not the concern or business of a specific sector but is the common concern of all sectors or what is often referred to as "everybody's business" (UN-ESCAP, 2017). The fundamental philosophy underlying the notion of mainstreaming is that by bringing the critical issues to the forefront, it becomes the essential principle of governance that pervades all sectors and all levels, even going beyond the public sector to embrace the private sector, the corporate world, academia, media, civil society and communities. In the absence of a universally accepted definition of the concept of mainstreaming, the term "mainstreaming" has been conceptualized and operationalized differently by different scholars. In the context of climate change, mainstreaming denotes "incorporation of the challenges posed by climate change into the work of city authorities/municipalities by formulating effective responses to it, which—to become sustainable—then need to be anchored in existing planning processes and frameworks, and policy across all sectors and levels" (Wamsler & Ing, 2007, p. 4). According to Runhaar et al. (2016), the mainstreaming approach aims "to integrate climate adaptation as an objective in existing policy domains. This means that synergies between existing policy objectives and climate adaptation are established and that existing resources are used to address climate adaptation. As opposed to the dedicated approach, mainstreaming focuses on performance- based decision-making—that is, actors focus on to what extent climate adaptation is required and feasible within the given context."

Since the notion of nature-based solutions for CCA not only renders support but also builds on the concept of ESSs and can be used in an identical manner with EbA (Naumann, Kaphengst, McFarland, & Stadler, 2014); hence, both the concepts are entrenched in theory and practice of CCA and ESSs (Chong, 2014). In the wake of increasing necessity being felt for mainstreaming EbA into municipal planning (Ojea, 2015), the concept of climate adaptation mainstreaming is also garnering similar emphasis. Nevertheless, the notion of climate adaptation mainstreaming has evolved from risk reduction mainstreaming that has been vociferously promoted since the World Conference on Disaster Risk Reduction (DRR) in 2005 and builds on expertise garnered from mainstreaming other cross-cutting issues (Wamsler & Pauleit, 2016). Besides, notion of climate adaptation mainstreaming also emanates from environmental policy integration for sustainable development (Van Asselt, Rayner, & Persson, 2015), and more particularly climate policy integration or mainstreaming that has been gaining traction since 1997 (Collier, 1997). Construed in a broad framework, climate policy integration/mainstreaming alludes to the incorporation of climate considerations in sector policy and practice (Berkhout et al., 2015). Mainstreaming, often framed in technical language, is driven by the need to alter the predominant paradigm at numerous levels of governance (Wamsler, 2014). While altering the rules of the game, mainstreaming also calls into question ideas, attitudes or activities that are considered mainstream or normal (Picciotto, 2002). Sequentially, it gets related to the concepts of sustainable transformation and transformative adaptation (IPCC, 2012; Pelling, O'Brien, & Matyas, 2014), where the latter is acknowledged for its potential to deal with underlying causes of risk and failed approaches to sustainable development (Revi et al., 2014).

The burgeoning themes from the literature on mainstreaming CCA and other cross-cutting issues, such as environmental policy, risk reduction, HIV/AIDS, gender, etc., can be categorized into certain strategic activities (Nunan, Campbell, & Foster, 2012) and be designated as a mainstreaming framework that incorporates and consolidates these strategies, including normative, operational, organizational and procedural factors at different policymaking stages (Persson, 2004). These different mainstreaming strategies are exhaustive and complementary because they entail both the kind and scale of integration, linking direct and indirect approaches to adaptation (Uittenbroek, Janssen-Jansen, & Runhaar, 2013) and are all needed for attaining sustainable transformation (Wamsler, 2014). An efficacious mainstreaming of CCA is a potent tool to achieve urban resilience and this attested to by the fact that the bulk of vulnerable populations rely on climate-sensitive livelihood sources such as urban agriculture, live-in disaster-prone areas like informal settlements in flood plains and unsustainable landscapes and possess limited adaptive capacities. Furthermore, in spatial or land-use planning, mainstreaming of climate adaptation can be facilitated into other sectors like infrastructure that is climate proof or green infrastructure, into agriculture with the help of use of drought resistant crops and efficient technologies, into educational systems by reviewing curriculum to enhance risk perception and adaptation response, and into water management in the context of drought hazards as well as excessive precipitation (Wamsler, 2014). There also exists a nexus between spatial planning and other concerns of natural resources and ecology, infrastructure and economy that are all susceptible or exposed to climate hazards and disasters. Nevertheless, Araos et al. (2016) have opined that spatial planning can as well play a crucial role in promoting CCA.

10.2.1 Categories of mainstreaming climate change adaptation

Categories of mainstreaming CCA, inter alia, include: Programmatic mainstreaming; managerial mainstreaming; organizational mainstreaming; regulatory mainstreaming; and directed mainstreaming (Runhaar et al., 2016; Runhaar, Wilk, Persson, Uittenbroek, & Wamsler, 2018; Wamsler, Luederitz, & Brink, 2014). Nevertheless, this categorization of mainstreaming includes these main categories of mainstreaming but is not limited to them.

10.2.1.1 Programmatic mainstreaming

Programmatic mainstreaming embraces the modification of the implementing body's sector work by integrating aspects related to adaptation into on-the-ground operations, projects or programs. Global Environment Facility (GEF) has embarked upon a programmatic approach and it is known as GEF programmatic approach that can be defined as "a long-term and strategic arrangement of individual yet interlinked project aimed at achieving large-scale impacts on the global environment" (Global Environment Facility, n.d., p. 7). The GEF programmatic approach is ordained to observe the following principles: (1) Be country-owned and build on national priorities designed to support sustainable development, as identified in the context of national and regional planning framework; (2) Emphasize the GEF's catalytic role and leverage additional financing from other sources [e.g., donors, private sector, nongovernmental organizations (NGOs)]; (3) Be based on an open and transparent process of multistakeholder representation—from dialoge to implementation—in accordance with the GEF's policy on public involvement; (4) Be cost-effective and seek to maximize global environmental benefits (Global Environment Facility, n.d., p. 7).

10.2.1.2 Managerial mainstreaming

Managerial mainstreaming entails the modification of managerial and working structures, including internal formal and informal norms and job descriptions, the configuration of sections or departments, as well as personnel and financial assets, so that aspects pertaining to adaptation are adequately addressed and institutionalized in a better way.

10.2.1.3 Organizational mainstreaming

Organizational mainstreaming comprises both inter- and intraorganizational mainstreaming. While laying emphasis on promoting collaboration and networking of different departments, individual sections, and or different stakeholders like other governmental and nongovernmental bodies, educational and research organizations, public and private sector, etc., it aims at creating or generating a shared understanding and knowledge between the departments, thereby, helping in enhancing positive coherence, developing competence and steering collective issues of adaptation.

10.2.1.4 Regulatory mainstreaming

Regulatory mainstreaming refers to the modification of formal and informal planning procedures, including planning strategies and frameworks, regulations, policies and legislation and related instruments that lead to the integration of adaptation. This enables in the inking of the procedures to policy to facilitate the generation of adaptation approaches that confide within the laws, plans, and regulations. Besides, regulatory mainstreaming of CCA entails a reference to the explicit measures to make climate change one of the foci of any urban plan and planning framework (Wamsler & Brink, 2014).

10.2.1.5 Directed mainstreaming

Directed mainstreaming alludes to top-down support to mainstreaming adaptation at the local planning levels. Higher level support is extended to redirect the emphasis on aspects pertaining to mainstreaming adaptation, such as providing topic-specific funding, promotion of new projects, supporting staff education or directing responsibilities. Directed mainstreaming can be seen as a vital component of all of the other categories or strategies as it is concerned with as to how related changes occur or come about, both top-down versus bottom-down (Wamsler, 2014).

Application of mainstreaming framework can be facilitated to overall adaptation or particular aspects of it, like ecosystem-based (EB) approaches, as well as other cross-cutting subjects like climate change mitigation. Nevertheless, application of mainstreaming framework to adaptation enjoins upon that all strategies need an exhaustive approach for climate risk reduction, including measures of hazard reduction and avoidance, vulnerability reduction, preparedness for response and preparedness for recovery (Wamsler & Brink, 2014). It is equally noteworthy that application of mainstreaming framework can be facilitated to single department or other implementing bodies at all levels; nonetheless, their collaboration and networking with other stakeholders is critical since governance and problem scales are usually mismatched. Mainstreaming framework aims at addressing the implementing body's own challenges like risks to premises and assets, with a view to ensure that adaptation work can continue even during adverse events or circumstances. This mainstreaming framework envisages local pathways to sustainably mainstream EbA into municipal planning and in the process analyze their linkages or discrepancies with local processes for climate policy integration/mainstreaming.

10.3 Mainstreaming ecosystem-based adaptation into policy process

EbA is deeply ingrained in the theory and practice of ESSs and CCA planning (Chong, 2014), and as such, ESSs are "the conditions and processes through which natural ecosystems, and the species that make them up, sustain and fulfil human life" (Daily, 1997, p. 41). Besides, ESSs include, but not limited to, natural processes that facilitate regulation of local climate erosion, soil retention, water infiltration, and natural hazards (Larondelle, Haase, & Kabisch, 2014). The ESSs concept, which is often developed to integrate ecological principles into economic planning or considerations and local decision-making (TEEB, 2010), is regarded to be an efficacious way of advancing sustainable urban planning (Ahern, Cilliers, & Niemelä, 2014). Keeping this in view, ESS planning is a place-specific or place-based approach that emphasizes on the creation, restoration, and conservation of ecological structures to provide society with precise or specific services from nature (Staes, Vrebos, & Meire, 2010). On the contrary, focus of the CCA is centered on the modification of human-environment characteristics to tone down the adverse impacts of climate extremes and variability (Wamsler, Brink, & Rivera, 2013). Accordingly, CCAS planning facilitates assessment and modification of activities, policies and built environment in consistent with the current and projected impacts of climate change and related societal vulnerabilities (Dannevig, Rauken, & Hovelsrud, 2012).

Emphasis on the need for mainstreaming ESSs and CCA—the two conceptual constituents of EbA—is stressed in the scientific literature (Clar, Prutsch, & Steurer, 2013). Integrating or mainstreaming framework is closely akin to the concepts of sustainability transitions (Forrest & Wiek, 2014), sustainable transformation (McCormick et al., 2013), and transformative adaptation, and the latter is acknowledged for its potential to deal with root causes of risk and failures in sustainable development approaches (Revi et al., 2014). A systematic exploration of the potential ways whereby EbA can be integrated into urban planning requires applying a multiple mainstreaming framework. Classification of mainstreaming approaches is dependent upon the fact as to whether these approaches are based on horizontal or vertical integration, and that manifests the quality of governance relations between actors (Rauken et al., 2014). Horizontal integration can be construed in terms of processes that are implemented by less powerful bodies like departments, and particularly, conditions are specified by a single actor who inspires or coordinates mainstreaming but who does not possess enough authority to exercise top-down control (Nunan et al., 2012). Nevertheless, vertical integration or mainstreaming can be delineated as one that is implemented by powerful governmental entities like city councils, and firm guidance from central or core legislative powers or actors during the implementation process (Jacob & Volkery, 2004).

Moreover, emerging mainstreaming themes from the available literature can be categorized under six strategic activities (Wamsler, 2015) as follows: (1) The commencement of new on-the-ground activities that directly concentrate on the topic under consideration (Wamsler, 2014), (2) aligning on-the-ground activities of various departments to integrate the topic under consideration (Wamsler, 2014), (3) strategic cooperation between relevant internal and external stakeholders (Wamsler, 2014), (4) the alteration of organizational working structures (Roberts & O'Donoghue, 2013), (5) the emendation and formulation of policies, regulations and instruments (Wamsler, 2014), and (6) directed instructions to support the integration of the topic under consideration (Wamsler, 2014).

EbA entails the potential of delivering manifold benefits to society and ecosystems in terms of reduction in disaster and climate risks through natural infrastructure, providing livelihood capitals, maintaining ecosystem's equilibrium, etc., and occurrence of this potential can take lace simultaneously in a cost-effective manner without any major trade-offs (Colls, Ash, & Ikkala, 2009). Besides, EbA also provides multiple nonmarketable social, environmental and economic cobenefits that comprise prime forces of local development via strengthening local adaptive capacities (Huq, 2016). There are numerous intervention areas where EbA can be instrumental in promoting adaptation in areas such as agriculture and food security, forestry, biodiversity, DRR, urban planning, coastal and water resource management (Travers, Elrick, Kay, & Vestergaard, 2012). In other words, EbA offers an array of flexible, convenient and cost-effective adaptation intervention to ward-off undefined capital investment along with helping in promoting soft adaptation methods (Jones et al., 2012). Inclusion of EbA in development and climate policies is prone to spur promotion of sustainable adaptation and development for communities, decision-makers and ecosystems as well as paving an "in-built" pathway for the attainment of sustainable development goals (SDGs) (Szabo et al., 2015). It has been reported that bulk of global supplies of ESSs, about 60%, are either being already degraded or subject of unsustainable management (Munang, Thiaw, Alverson, Liu, & Han, 2013). It is in this backdrop and in pursuance of an integrated adaptation and development approach that EbA entails the potential of playing a significant role for realizing the SDGs embracing ecosystem management, CCA and DRR in comparison to the single disciplinary approach of adaptation, development and ecosystem management (Doswald et al., 2014).

In many instances, specifically like that of coastal fishing communities in small island states, EbA represents the only viable adaptation options to be implemented (Jones et al., 2012). Thus Huq, Bruns, and Huq (2017) have emphasized that EbA deservedly needs better attention at political agenda setting level, and it is required to be mainstreamed at the climate and development policy and practice levels. Better prospects of systematic integration of adaptation into ecosystem management and development activities can be augmented by mainstreaming EbA at policy level and concurrently, it also helps in ensuring essential financial, human, technological and knowledge supports (Carabine, Venton, Tanner, & Bahadur, 2015). Undeniably, EbA approaches and methods are still in their embryonic stages (Pramova, Locatelli, Brockhaus, & Fohlmeister, 2012); nevertheless, national governments, international organizations and NGOs have embarked upon initiating the integration of EbA into policy and management albeit, it has not yet emerged as a standard practice (Guerry et al., 2015). From a theoretical perspective, EbA can be "embedded into national, regional and local policy and practice by adopting an integrated, participatory and ecosystem-based approach to territorial planning" (Wertz-Kanounnikoff, Locatelli, Wunder, & Brockhaus, 2011) for integrated development and adaptation, recent years have witnessed a steady increase in the number of pilot EbA projects as well as scientific studies; nevertheless, that cannot be said to the only criterion or a guarantee of an automatic inclusion of EbA into the policy and practice (Huq & Hugé, 2012). Making interventions at the policymaking and planning areas across scales is said to be one of the key cornerstones of unlocking the potentials of EbA (IUCN (International Union for Conservation of Nature), 2008). In this regard, Huq et al. (2017) have suggested that a required assessment of the current extent of EbA inclusion in the development and adaptation process is direly needed because once the extent is identified it would be easy to underline and address potential scopes, entry points, strengths and weaknesses along with challenges.

In the wake of availability of limited studies in the literature portraying EbA internalization in multilevel policy contexts (Pramova et al., 2012; Wamsler et al., 2014), it is difficult to assess as to what extent EbA is being integrated and mainstreamed into climate and development policies. Assessments thus far available make it discernible that EbA is still a peripheral adaptation constituent (Pramova et al., 2012), construed as an environmental subissue (Wamsler et al., 2014), and predominant in developed countries (Ruckelshaus et al., 2013). It has been pointed out that as a sequel to nonverification of findings in a national policy context owing to the lack or absence of country-level studies pertaining to EbA implementation, the resultant outcome is that potential gaps, challenges and entry points have remained unexplored (Huq et al., 2017). In an identical manner, little is known with regard to which extent EbA approaches are already integrated/streamlined into policies and practices and how best it could be integrated (Wamsler, 2015). Pivotal role is assigned to local government entities like municipalities in facilitating implementation of EbA through extensive community participation leading to sustainable adaptation (Wamsler et al., 2014). Nevertheless, role of national policies and the policymaking process assumes prime significance because of being a provider of overreaching guidance to their local development adaptation practices (Huq et al., 2017).

10.4 Entry points for mainstreaming ecosystem-based adaptation

Apart from promoting integrated approach to climate change, mainstreaming EbA into development policy and climate change policy also helps in vigorously addressing both the climate extreme events and issues pertaining to sustainability (Ojea, 2015). With regard to EbA in the context of climate change, the avowed objective of "mainstreaming" seemingly is to internalize the goals of DRR, ensuring healthy and productive ecosystems along with engendering "no-regret" practices as part of overall development interventions (Vignola, McDaniels, & Scholz, 2013). Operationalization of mainstreaming strategies and norms on a policy cycle comprises three stages: (1) agenda setting, (2) policy formulation, and (3) implementation and evaluation (Benson, Gain, & Rouillard, 2015). Admittedly, there is lack of normative guidance that can help separate governance agencies and entities in accordance with policy cycle stages. Nevertheless, the agenda setting stage entails problems, conflicts, emerging global and national issues and political will (Gain, Giupponi,

& Benson, 2015). Developing legislative, regulatory and programmatic directions and policies designed to address the problems fall within the ambit of policy formulation stage. The policy implementation stage involves formulation of I implementation guidelines along with allocation of essential resources required for implementing the policies. Hence, policy mainstreaming characteristically denotes that EbA considerations are efficaciously required to occur at all three stages horizontally and/or vertically to either coalesce current policies into a single policy or include the necessary concerns into existing ones (Gain et al., 2015). Concurrently, the examination of policy cycle also takes into account as to how diverse mainstreaming norms and practices can effectually adjust with different levels to open up new entry points for further levels of mainstreaming (Wamsler & Pauleit, 2016).

CCA, inclusive of the ecosystems and the services delivered by these ecosystems, is as much an institutional, political and economic issue as it is a technical issue. A major aspect of mainstreaming is to ascertain suitable entry points for integrating EbA into solid as well as complex policy and planning frameworks and decision-making processes. Entry points afford windows of opportunity in terms of situations or processes that help garner the interest of policymakers, stakeholders or the public at large for integrating EbA into ongoing national and subnational processes and reining in synergies with other approaches (GIZ, 2018e). They entail the potential of taking place at all levels. Entry points may pertain to problem awareness, political will, and policy solutions designed to overcome the problems. Problem awareness refers to examples of crisis over food and water, climate change impacts disasters, etc. Political will includes government actors or civil society/interest groups/voters that are driving forces to address the problems. In an identical manner, policy solutions that are designed to surmount the problems, inter alia, include: strategies and concrete actions at multiple levels by several actors. In other words, manifold things have to come together to facilitate successful mainstreaming of EbA.

Altogether, potential entry points are often governance processes that may comprise the development, revision and/or strengthening of policy instruments, including planning instruments, command and control instruments, economic and fiscal instruments, educational and awareness-raising measures, voluntary measures and institutions. Development or policy instruments include: development plans, sector plans, nationally determined contributions (NDCs), national adaptation plan (NAP) processes, National Biodiversity Strategies Action Plan (NBSAP), watershed plans, strategic environmental assessments, and land-use plans. Command and control instruments comprise climate change and environmental laws, standards, and environmental impact assessments. Investment programs, funds, taxes and fees, etc., fall within the purview of economic and fiscal instruments. Educational and awareness-raising measures include environmental education, extension programs. Besides, voluntary measures comprise voluntary environmental agreements and standards. Moreover, institutional instruments include climate change task forces, watershed committees, and land-use associations (GIZ, 2017a).

10.4.1 Mainstreaming ecosystem-based adaptation in national adaptation plan

The process of NAP was established under the 2010 Cancun Adaptation framework under the aegis of United Nations Framework Convention on Climate Change (UNFCCC) and it is designed to assist countries reduce vulnerability, build adaptive capacity and mainstream adaptation in development planning. Being a continuous, progressive and iterative process, NAP follows a country-specific, gender-sensitive, participatory and fully transparent approach. Owing to its increasing significance in the ongoing international climate debate, the NAP process has come to occupy a pivotal role in planning and implementing EbA measures, particularly in countries where ecosystems are instrumental in playing a role in facilitating risk reduction for the communities and the general public (GIZ, 2017a).

The UNFCCC has envisaged technical guidelines for the NAP process, especially for the least developed countries that are more vulnerable to climate change impacts, and these guidelines help the NAP process as entry point for EbA integration (UNFCCC, 2012). As per these guidelines, the NAP lays down strategic framework that aims at reduction of vulnerabilities of the communities and people and helps in the maintenance of their livelihoods in the context of climate change. Keeping in view the fact that the most vulnerable communities and the people are mostly dependent on ecosystems across sectors and scales, the EbA is required to be a primary element in the NAP process. Since the NAP facilitates identification of major medium-term and long-term goals under the NDCs; hence it cannot be regarded as a mere "wish list" rather it needs to be construed as a prioritized suite of strategies funded by national and international sources (UNFCCC, 2012). Countries vulnerable to climate change are specifically called upon to prioritize their role in the NAP process owing to the manifold social, environmental and economic benefits accruing from ecosystems and the services these ecosystems deliver that often go beyond adaptation (GIZ, 2017a). Close linkages between the Nap and the adaptation goals under the NDCs make it imperative that the NAP process ought to serve as a concrete process for attaining NDC adaptation goals that embrace the use of ecosystems and their services in bulk of all NDCs.

As a process for adaptation that is very country-specific and dependent on governance and policy structures, the NAP can vary from top-down to bottom-up approaches and as such it is called upon to focus on traditional and indigenous knowledge on ecosystem management in the context of adaptation wherever feasible o viable. Keeping in view the need to build NAP upon existing and ongoing initiatives and policy frameworks, EbA can be instrumental in building bridges between NAPs and existing policy processes that support conservation and the sustainable management of ecosystems and their services. While emphasizing close links between NAP and subnational policy and management framework, decisions pertaining to ecosystems and their management, including their role in the context of adaptation, GIZ (2017a) also lays stress on integrating EbA in the local development and budgeting frameworks to ensure sustainability. Effective implementation of NAP requires involvement of an array of actors such as government, civil society, private sector; sectors like agriculture, marine resources, water and sanitation, etc., and scales from local to landscape or even biomes. Being a multisector and landscape approach, EbA entails the potential of supporting complex policy-settings like the NAP process that calls for a significant time investment to enhance ownership and acceptance in society. An appropriate mechanism of monitoring and reporting to convey impacts is another imperative of NAP because setting up of monitoring and evaluation (M&E) systems is helpful in monitoring EbA impacts and facilitates their communication as essential part of the NAP; concurrently, it is equally important to keep in mind challenges to the M&E systems that inter alia include lack of data and lack of coordination across sectors and scales, and accordingly, it also calls for applying realistic time horizons for showing adaptation benefits so that the NAP is adequately equipped with adequate capacity development, specifically at local level to bridge information gaps (GIZ, 2017a).

10.4.2 Mainstreaming ecosystem-based adaptation into sectoral adaptation plan

Given that certain sectors like agriculture, forestry, fisheries, water resource management, human health and protected areas are specifically sensitive to climate variability; hence, each of these sectors is required to take into consideration adaptation measures in their policy planning processes. Often, emphasis on policies at the sector level is focused on to ensure that CCA priorities determined at the national level in the NAP get operationalized (OECD, 2009). Given that adaptation solutions are highly sector-specific, each sector is called upon to work out a sectoral plan with the objective of delineating climate adaptation measures and this affords an opportunity for integrating EbA in the policy framing process for the requisite processes. Sectors like agriculture, water resources, protected areas and DRR are particularly vulnerable to climate change owing to their dependence on ecosystems and the services they deliver, and as such these sectors can be crucial in building resilience for both ecosystems and communities. Apart from being a priority sector, water resource management is also of multisectoral significance because of its close linkages with agriculture, food security and health, etc. Accordingly, policy development on CCA and DRR for various sectors affords a potential entry point for integrating EbA via water resource management actions (GIZ, 2019).

With a view to ensure security of water resources, South Africa has in KwaZulu-Natal launched "uMngeni Ecological Infrastructure" project that aims at integrating watershed planning and management. As a collaboration of public–private partnership, this project proceeds with the principles of EbA through harnessing the ability of ecosystems to deliver services to downstream communities in a resilient manner. Another project "Building resilience in the Greater uMngeni Catchment," led by the district municipality, is designed to enhance the resilience of vulnerable communities via interventions like early warning systems, climate-smart agriculture and climate proofing settlements (GIZ, 2018b).

Protected area management has come to play a significant role in building resilience to climate change for both ecosystems and communities, minimizing climate risks and ensuring the provision of the provision of ESS both within these areas and further afield. Dudley et al. (2010) have opined that protected areas ought to be a significant part in the efforts for scaling up EbA and mainstreaming in NDCs and NAPs. There are 68 Nature Protected Areas in Peru that are extremely vulnerable to climate impacts and the government of Peru has adopted the master Plan for Protected Areas that serves as a guide for the planning and management of all protected areas in that country. Under this Master Plan, EbA is going to be the part of this Master Plan so that EbA is mainstreamed into the planning and updating processes of each protected area ensuring alignment of EbA objectives with conservation objectives and an action plan will be prepared by each protected area with identified EbA measures that in turn will be incorporated in the Peru's NAP (GIZ, 2018a).

Fruitful results have come out from case studies, especially from Peru (GIZ, 2018a) and South Africa (GIZ, 2018b) with regard to adoption of ecosystem approaches in addressing DRR. The adoption of ecosystem-based (EB) approaches to address DRR is known as ecosystem-based disaster risk reduction (Eco-DRR). Measures like reforestation of riverbanks and riverine flood control via natural infrastructure, such as flood bypass, riparian buffers, wetland construction, etc., constitute integral part of the Eco-DRR. These measures are required to be part of disaster risk management planning as effective prevention and postdisaster reconstruction (known as build back better)

10.4 Entry points for mainstreaming ecosystem-based adaptation   **613**

solutions (GIZ, 2019). It can be discerned from the case studies, especially from Peru (GIZ, 2018a) and South Africa (GIZ, 2018b) that mainstreaming Eco-DRR is often launched and initiated at the local and regional level and is a highly participative and collaborative process. In 2004, the city of Durban in South Africa is said to have embarked on a Municipal Climate Protection Program, representing one of the earliest municipal scale programs worldwide to include Eco-DRR approaches (GIZ, 2018b). Similarly, in Peru, the Regional Government of Lima and the Council of Water Resources of the Chancay-Huaral Basin encouraged restoration of urban riparian areas to lessen the impact of extreme rainfall and flood events; and it is further revealed that the enabling conditions are prevalent to mainstream EbA in the Regional Climate Change Strategy of Lima and the Regional Development Plan, considering the national budget allocation for nature infrastructure projects (GIZ, 2018a).

10.4.3 Mainstreaming ecosystem-based adaptation into local and community planning processes

The key emphasis in local planning processes is focused on distribution and management of natural resources, and as such, the avowed objective is to safeguard and strengthen rural livelihoods contribution to poverty alleviation and thereby promoting economic development at all scales. National and sectoral plans are often confronted with some sort of constraints to effectively respond to local needs with regard to specific aspects of climate adaptation; therefore nature-based solutions need to be included into locally-driven rural planning processes to ensure that local communities get access to the multiple benefits accruing from ESSs (GIZ, 2019). Involvement of multiple stakeholders like local and indigenous populace, civil society and local activists, etc., in the local policy planning process becomes imperative for making planning at local level to be more effective and result-oriented. Village action plans along with their microprojects, apart from being key planning tools for local communities aiming at promoting the sustainable management of a territory and natural resources available therein, also help in addressing a particular problem that takes place at local level, such as, food security, water quality or availability. As such, local planning process is crucial to integrate EbA as potential solution for tackling these challenges in the wake of climate change. Introduction of EbA measures can be facilitated either as stand-alone microprojects or in tandem with other measures as part of the village action plans; and these action plans generate ownership of processes that in turn can guarantee sustainability of projects and initiatives in the long run. The example of Peru demonstrates as to how integrating EbA as part of comanagement agreements with local population helped in creating ownership of processes and enabled the replication of those processes in other protected areas as well (GIZ, 2018a). The example from Vietnam demonstrates as to how local communities affected by rising sea levels and storms successfully launched mangrove restoration measures (GIZ, 2017b).

In Peru, it has been reported that the comanagement agreements, also often described as "life plans," have helped in promoting a vision of integral development to guide the land-use planning in a territory owned by the community. Peru government's project EbA Amazonia entailed the objective of reducing the vulnerability to climate change of indigenous communities that are dependent on fragile ecosystems as the Amarakaeri Communal Reserve, Madre de Dios and Amazonas. The initiative proved instrumental in promoting EbA measures to ensure food security and minimize deforestation. The activities launched under this initiative were developed in close collaboration with the indigenous populace and integrated into microprojects. It has been observed that, "The form integral part of the comanagement agreement with the Communal Reserve. Replication of this success story is essential for the effective mainstreaming of EbA across all Communal Reserves and to become part of the Master Plan for Nature protected Areas in Peru," it is also hoped that apart from the local benefits, "this will contribute to the Regional Climate Change Strategy of Madre de Dios and the national agenda for climate change" (GIZ, 2018a).

In the case of Vietnam, where over 700 ha of mangroves and tidal flats in Phu Long Commune, Cat Ba Island, are the main source of subsistence and income generation for local people, this area also serves as the buffer zone for Cat Ba National Park (NP) where mangrove degradation was a big problem, particularly in the context of increasingly intense storms and rising sea levels. Smallholder farmers in this region were vulnerable not only to these extreme events, but also to saltwater intrusion and the subsequent salinization of the soils. Furthermore, water pollution and aquatic diseases seriously damaged shrimp farms (GIZ, 2017b). In the beginning, the local people sought to restore the mangrove belt by themselves to reduce the impact of storms and improve the aquatic resources; nevertheless, subsequently, they received the technical support of Cat Ba NP as well as policy support from the local authorities in terms of low-interest loans for aquaculture. The resultant impact of these measures is discernible from the fact that mangrove rehabilitation and a change of aquaculture toward more sustainable management—from intensive aquaculture to semi-intensive and extensive farming—has taken place. (GIZ, 2017b).

10.5 Mainstreaming ecosystem-based adaptation in G-20 countries

The "Group of Twenty" or "G-20" comprises 19 countries and the Europe Union, and these 19 countries include Argentina, Australia, Brazil, Canada, China, France, India, Indonesia, Italy, Japan, Mexico, Russia, Saudi Arabia, South Africa, South Korea, Turkey, the United Kingdom, United States, and the Germany. Collectively, the G-20 economies constitute around 90% of the gross world product, 80% of global trade, two-thirds of the world population and nearly half of the world land area (Schmarek & Harmeling, 2018). Established in 1999 with the objective of promoting international financial stability, G-20 member countries have increasingly come to realize that climate change is a significant underlying factor impacting their stability and development; and accordingly, these countries are called upon to play significant role in addressing the vagaries of climate change to ensure stability and uninterrupted pace of development. G-20 countries are characterized by variability in their vulnerability to climate change as well as variability in their preparedness in addressing climate change effects. According to the Global Adaptation Index, pioneered by the United States-based Notre Dame university and known as Notre Dame Global Adaptation Initiative (ND-GAIN) that summarizes a country's vulnerability to climate change and other global challenges in combination with its readiness to address those challenges, India, South Africa, Brazil, Mexico and Saudi Arabia are less prepared (with a ND-GAIN preparedness value of around 0.4), whereas the United States, Australia, Canada, Japan, and South Korea are more prepared for climate change (with a ND-GAIN preparedness value of around 0.7) (ND-GAIN, 2019). This amply demonstrates significant potential of G-20 countries to exchange experiences and expertise to enhance the preparedness of countries that lag in preparedness to address climate change impacts. In 2018, the G-20 countries formed the Climate Sustainability Working Group (CSWG) for promoting EB approaches in G-20 countries and a couple of meetings of the CSWG held in 2018 have emphasized on the significance of promoting EB approaches, including for adaptation and resilient infrastructure thereby affording an opportunity for the member countries to discuss the prevalent state of the EB approaches and comprehend their importance (G-20, 2018).

In order to identify the extent to which EB approaches were integrated into the NAPs and strategies of G-20 countries, Prabhakar, Scheyvens, and Takahashi (2019) reviewed these NAPs to assess whether the plans and strategies acknowledged EB approaches and whether they identified specific approaches, priority sectors, mainstreaming strategies, timeframes for implementation and funding. Findings demonstrated that most G-20 countries recognized the significance of EB approaches in their respective plans and strategies and EB approaches constituted a guiding principle of adaptation for some countries like Brazil, Italy, Mexico, and United States. Apart from demonstrating that almost all G-20 countries promote EB approaches linked with ecosystem protection and conservation as well as most countries identified urban areas, coastal areas, mountainous areas, and marine areas as priority areas for EB approaches, it is also revealed from the findings that while China and Japan emphasized on the importance of EB approaches for resilient infrastructure, Japan, Indonesia, Germany, and South Africa stressed on the identification of land-use planning and other related spatial approaches as necessary entry points for the integration of EB approaches into natural resource management (Prabhakar et al., 2019). Almost all G-20 countries have gone beyond according mere recognition to EB approaches in their respective NAPs and strategies by delineating mechanisms for integrating EB approaches into their developmental sectors, and in this regard, it is noteworthy that Brazil and South Africa in their respective NAPS specified actions to integrate the notion of EB approaches into sectoral adaptation strategies by evolving suitable methodologies or integration, by appropriating a working group that channelizes the strategy, and by arraying a timeframe and precise indicators for gauging progress.

Emphasis on the necessity for collaboration among different agencies to promote EB approaches is revealed in the adaptation strategy of South Africa where emphasis is also stressed on the significance of EB approaches in other national strategies, as has been the case with Brazil and Japan. Nevertheless, South Africa has developed the Strategic Framework and Overreaching Implementation Plan for Ecosystem-Based Adaptation, 2016–21 which entails four areas of work to promote EbA. These four areas are as follows: (1) effective coordination, learning and communication to mobilize capacity and resources for EB; (2) research, M&E to provide evidence for the contribution of EB approaches to a climate-resilient economy and society; (3) integration of EB approaches into policies, plans, and decision-making to support an overall CCA strategy; and (4) implementation of projects to demonstrate the ability of EB approaches to deliver a wide range of cobenefits (DEA-SA, 2017). With a view to implement this 2016–21 EbA strategy, South Africa, apart from developing guidelines, also established a coordination mechanism along with implementing a pilot project.

While portraying an optimistic picture, based on the review of NAPs and related strategies being adhered to by G-20 countries, of the expectation of these countries of generating a wide range of benefits by having recognized EB approaches, along with the fact that some G-20 countries have also identified indicators and targets and developed guidelines for mainstreaming as well as allocating modern funding for implementation of EB approaches, Prabhakar et al. (2019) lament that these countries are still confronted with important challenges, specifically in promoting EB approaches colligated with mainstreaming, financing, M&E, and institutional and policy bottlenecks. While referring to these bottlenecks as related to capacity and attitudinal factors and insufficient research and evidence on the efficacy of EB approaches, Prabhakar et al. (2019) suggest that in order to augment mainstreaming and understand costs and benefits in the short to long term, it is significant that research and M&E efforts gather tangible evidence in the efficacy of EB approaches.

10.6 Mainstreaming ecosystem-based adaptation in Sweden

A study, based on the investigation of four coastal municipalities in southern Sweden—Malmö, Helsingborg, Lomma, and Kristianstad—is undertaken by Wamsler and Brink (2014) to examine the major traits of existing mainstreaming strategies, with a view to enhance knowledge with regard to the potential ways to mainstream EbA into municipal planning. The study is based on both vertical and horizontal integration processes and results emerging from this study demonstrate that, even though ESS planning and climate change planning together lay down the conceptual foundation for EbA, related activities are usually implemented distinctly and are seldom exhaustive. The study endeavors to show as to how combined mainstreaming strategies can strengthen and complement each other and how effective leadership in the integration processes entails the ability to pay-off for a lack of guidance or supporting legislation from higher decision-making authorities. Trends emerging from the study make it discernible that undoubtedly, all of the investigated mainstreaming strategies have been applied in practice; nonetheless, there are variations in the importance accorded to particular strategies and specific activities.

While providing insights into the practices adhered to by the civil servants in promoting ecosystem-based planning and making use of green and blue infrastructure for climate adaptation, the analysis arrived at by Wamsler et al. (2014) also focuses on a gap between the rapidly emerging concept of EbA and its practical implementation; synergies between strategies that can help to overcome barriers to mainstreaming; ambiguities with regard to mainstreaming concept and potential drawback; and mainstreaming strategies as potential lever for change in sustainability transitions. While emphasizing that the mainstreaming needs strategies that focus on context-specific characteristics and harness network governance, it is also argued that systematic application of mainstreaming strategies entails the potential of moving sustainability into the core of municipal decision-making; thereby, enhancing the probability of leading to the combination of governance dimension, involving a diversity of actors to engender knowledge along with encouraging goal-oriented and learning-by-doing approaches, while keeping in view the local context (Wamsler et al., 2014). Concurrently, Wamsler and Brink (2014) conclude that systemic mainstreaming of sustainability issues is a promising avenue for launching and promoting sustainability transitions and also entails the potential of addressing criticism heaped on other mainstreaming topics or approaches.

Another study by Wamsler and Pauleit (2016) provides a comparative analysis of local pathways to sustainably mainstream EbA applied in 12 municipalities in Germany and Sweden, and the emerging results demonstrate that progress is apparently similar in the two countries, there are perceptible differences with regard to the local triggers, the significance accorded to specific mainstreaming strategies and their link to climate policy integration, While noting that EB approaches to CCA are not labeled or systematized in either country, the study ascertains that the mainstreaming of EbA is enabled through municipalities' level of experience in mainstreaming other issues. It also becomes evident from this study by Wamsler and Pauleit (2016) that while in Sweden, adaptation mainstreaming is driven by the ESS concept that is often disconnected from the integration of climate integration policy, whereas in Germany, structures and planning processes set up for climate change mitigation are steering the path. While emphasizing on the need for more explicit consideration, both in research and practice, with regard to systematic adaptation and its potential linkages and disconnects with climate mitigation policy integration, Wamsler and Pauleit (2016) call for more research on the role of inclusive, not fragmented, climate policy integration to guarantee sustainable planning and transformation. "Sustainable change will, however, remain elusive as long as our understanding of mainstreaming remains naïve, it is organizations themselves that need to change, rather than only 'mainstreaming' change in selected on-the-ground measures" (Wamsler & Pauleit, 2016).

10.7 Mainstreaming ecosystem-based adaptation in Bangladesh

Huq et al. (2017) provide an analysis of the extent to which mainstreaming/integration of EbA to climate change in the policymaking process has taken place in Bangladesh. The study focuses on three-stage hybrid policymaking cycle—agenda setting, policy formulation, and policy implementation—wherein the contribution of EbA can be horizontally or vertically mainstreamed and mainstreamed. After examining a total of nine national and the sectoral development and climate change policies along with 329 CCA projects belonging to different policymaking stages, Huq et al. (2017) surmise that the role of EbA is marginally taken into account as an adaptation component in most of the reviewed policies in the case of Bangladesh. It is further revealed that at the horizontal mainstreaming stage of policy formulation and implementation, EbA is largely ignored and priority is accorded to structural adaptation policies and projects such as, large-scale concrete dams, and embankments. The resultant outcome is tangible from the fact that EbA's role to adapt sectors like urban planning, biodiversity management and DRR is either ignored or left unchecked, and the implementation stage gets overwhelming priorities and investments to adhere to hard adaptation measures such that only 38 projects were related to EbA.

It also becomes discernible from analysis by Huq et al. (2017) that CCA in Bangladesh is predominantly construed as a structural and technical approach at the expense of ignoring its sociocultural attachments with humans and nature and this has left EbA approach as an underappreciated measure throughout the decision-making stages. Besides, there also exist huge differences in adaptation priorities and understanding at national and sectoral levels in Bangladesh. An examination of sectoral policies and projects reveals that EbA is certainly not a top priority in sectoral adaptation activities. According to Huq et al. (2017), it looked as if policymakers of Bangladesh, both at national and sectoral levels, remain uninformed and unaware with regard to the potential of the EbA approach, as a result of which potential scope of EbA in DRR, urban planning and water management is either ignored or undervalued. Arguing that the EbA mainstreaming in Bangladesh is considerably offset by dominant structural adaptation ideologies, the expert and bureaucracy dependent policymaking process, and the lack of adaptive integration capacities at institutional level, Huq et al. (2017) emphasize that this needs to be adequately addressed for CCA.

10.8 Mainstreaming ecosystem-based adaptation in India

Mainstreaming of EbA to climate change, in the Indian context in particular and in the context of other countries in general, is closely linked to the gravity of adverse impacts wielded by the vagaries of climate change and the policy measures devised in response to deal with those challenges in the form of national CCA plan and its percolation to the provincial and local levels. India, being a populous and tropical developing country, is confronted with challenges of immense magnitude arising out of the vagaries of climate change in terms of extreme weather events, flashfloods, torrential rains, droughts, heat waves, wildfires, landslides, soil erosion, air pollution and water pollution, depletion of groundwater resources, pollution of surface water resources, retreating of Himalayan glaciers, etc. Between 2014 and 2018, India has been the fifth most rains, droughts vulnerable of 181 countries to the effects of climate change, according to Global Climate Risk Index (GCRI) 2020 released in early December 2019. Nevertheless, in 2019, India was the seventh most vulnerable of the 180 countries to the impacts of climate change (GCRI (Global Climate Risk Index), 2021). India's overall ranking on the GCRI between 2014 and 2019 is shown in Table 10.1.

According to GCRI (2020), India suffered 2081 deaths in 2018 owing to extreme weather events caused by climate change in terms of cyclones, heavy-rainfall, floods and landslides. Overall economic losses accruing to India on account of climate change in 2018 were to the tune of nearly $37 billion, the second highest in the world, and that amounted to losing about 0.36 per unit of GDP. It is noticeable from Table 10.1 that between 2014 and2018, India's overall ranking on the GCRI had remained fluctuating over the period of five years but it had always been the country with one of the five highest economic losses due to climate change and it also had the deaths due to extreme weather events in 4 of 5 years.

According to GCRI (2021), between 2000 and 2019, over 475,000 people lost their lives worldwide on account of more than 11,000 extreme weather events globally and incurred losses

Table 10.1 India's ranking on the global climate risk index, 2014–19.

Assessment year	Total countries assessed	Index (rank)	Fatalities (rank)	Fatalities (per 100,000 inhabitants) (rank)	Economic loses (rank)
2014	138	10	1	36	2
2015	135	4	1	24	1
2016	182	6	1	30	3
2017	124	14	2	29	4
2018	181	5	1	34	1
2019	180	7	NA	NA	NA

N.B. NA: Not available.
From Global Climate Risk Index reports, 2015 to 2021, Bonn: Germanwatch.

amounting to $2.56 trillion in purchasing power parities. In the case of India, flooding caused by heavy rains was responsible for 1800 deaths across 14 states and led to the displacement of 1.8 million people, and in overall, 11.8 million people were affected by the intense monsoon and economic losses were estimated to the tune of nearly $10 billion. During 2019, India was visited by eight tropical cyclones and six of these eight cyclones got intensified to become "very severe"; and cyclone Fani impacted almost 28 million people, killing about 90 people in India and Bangladesh and causing economic losses to the tune of $8.1 billion (GCRI, 2021) (Fig. 10.1).

Recent years have witnessed some research studies, a couple of reports by the Intergovernmental Panel on Climate Change (IPCC) and even the Government of India's report in 2020 emphasizing on the urgency for making strenuous efforts to reach the targets the Paris Agreement on Climate Change (referred here as Paris Agreement) that binds every signatory nation to reach an agreed level for cutting the greenhouse emissions and by doing so India along with other countries will be placed in a better position to effectively deal with adverse impacts of climate change. A recent report by Government of India warns that India is likely to see an estimated rise of temperature of about 4.4°C by the end of the 21st century. Noting that India has witnessed an increase in the frequency of droughts and floods in the recent past few decades, the report while warning that the frequency of warm days and nights is projected to rise 55% and 70%, respectively, also points out that while humid regions of central India have become drought-prone zones, flood risk has also risen in the east coast (MoES-GoI, 2020). While alluding to the data of past three decades (1986–2015) during which the temperatures of the warmest day and the coldest night have recorded a rise by about 0.63°C and 0.4°C, respectively, the report warns that by the closing part of the 21st century; these temperatures are projected to rise by approximately 4.7°C and 5.5°C, respectively, relative to the corresponding temperatures in the recent past (1976–2005). Asserting that human-induced climate change have influenced vagaries of climate change in the past, the report warns that this phenomenon is expected to continue apace during the remaining eight decades of

FIGURE 10.1

An aerial view of destruction caused by cyclone Fani in Odisha, India.

Courtesy Wikimedia Commons.

the 21st century and in this regard, the report suggests that in order to improve the accuracy of future climate projections, particularly that of regional forecasts, it is essential to develop strategic approaches for improving the knowledge of earth system processes, and to contribute enhancing observation systems and climate models (MoES-GoI, 2020).

Warning signals about India facing devastating climate change effects, including killer heat waves and severe flooding in the remaining eight decades of the 21st century have been sounded by another research study by Almazroui, Saeed, Saeed, Islam, and Ismail (2020). Noting that India is the most densely populated region in the world, with relatively high sensitivity and low resilience to changes in its climate, all of which makes India very exposed and vulnerable to any changes that could occur in the remaining period of the 21st century, Almazroui et al. (2020) also forecast killer heat waves over the plains along with severe flooding likely to occur with annual rainfall over India projected to increase under all emission scenarios in the remaining decades of this century. Basing their analysis on global climate models, the researchers have observed a potentially large increase of more than 6°C under a high emission scenario over northwestern India, comprising the complex Karakorum and Himalayan mountain ranges; and the resultant increased warming entails the potential of further accelerating the snow and glaciers melt over the Indian Himalayan region (IHR) and that could have serious implications for crops, ecosystems and populations living downstream. Accordingly, the study calls for urgent steps to reduce GHG emissions to avert associated risks to India's population, their livelihoods, ecosystems and nation's economy.

10.8.1 India's move toward climate change policy

India moved toward framing a national climate policy in the wake of the 2007 Fourth Assessment Report of the IPCC that warned of a "dangerous" increase in the frequency and intensity of extreme weather events, especially in tropical and subtropical countries and called upon countries to step up action on climate change (IPCC, 2007). Accordingly, the Government of India established the Prime Minister's Council on Climate Change (PMCCC) in 2007 itself. In 2008, the PMCCC formulated the National Action Plan on Climate Change (NAPCC) that proposed constituting eight missions to deal with CCA and mitigation. These eight missions (as shown in Fig. 10.1) are as follows: (1) The National Solar Mission, (2) the National Mission for Enhanced Energy Efficiency, (3) the National Mission for Sustainable Habitat, (4) the National Water Mission, (5) the National Mission for Sustaining the Himalayan Ecosystem, (6) the National Mission for a Green India, (7) the National Mission for Sustainable Agriculture, (8) and the National Mission for Strategic Knowledge on Climate Change (PMCCC-GoI, 2008). While incorporating India's vision of ecologically sustainable development and steps to be adhered to in implementing it, the NAPCC is based on the awareness that climate change action ought to proceed concurrently on several intimately interrelated domains like energy, industry, agriculture, water, forests, urban spaces, and the fragile mountain environment (Fig. 10.2).

Undoubtedly, the NAPCC comprises India's response to climate change based on its own resources; nonetheless concurrently, it also recognizes that it is closely linked to the parallel multilateral endeavors, based on the Go principles and provisions of the UNFCCC, emphasizing on the establishment of a global climate change regime. Nevertheless, India's expectations that the ongoing multilateral negotiations under the aegis of the UNFCCC would yield an agreed outcome, based on the principle of Common but Differentiated Responsibility and Respective Capabilities (CBDR),

10.8 Mainstreaming ecosystem-based adaptation in India   621

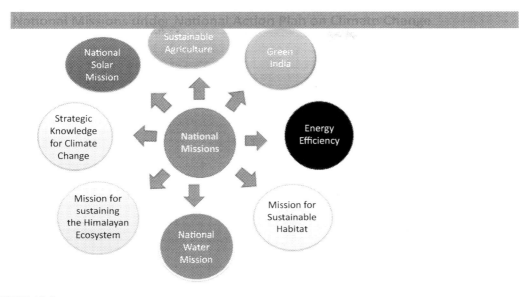

FIGURE 10.2

National missions under national action plan on climate change.

that would enable developing countries like India, via international financial support and technology transfer, to accelerate its shift toward a future of renewable and clean energy, have not been fully met with. An evaluation of Go the NAPCC along with an assessment of the implementation of eight missions established under the aegis of the NAPCC was taken by Delhi-based Center for Science and Environment (CSE) and its report published in 2018 informs that the NAPCC's approach lacks specificity and that most missions (some of which were set up after a gap of some years) have recorded "uncertain" progress along with being confronted with problems like financial constraints, lack of interministerial coordination, paucity of technical expertise and project clearance delays and these have been instrumental in impeding the efficient implementation of these missions (CSE, 2018).

10.8.2 From water–energy–food nexus approach to ecosystem-based adaptation

Emphasis on water–energy–food nexus (w-e-f nexus) approach to deal with water and climate change-related problems started gaining traction from 2008 onwards in India, especially in the aftermath of the publication of the NAPCC by the Government of India and follow-up of the recommendations envisaged in the NAPCC by different states in India. Interestingly, India Water Foundation (IWF), a registered nonprofit civil society, was also founded in 2008 with its headquarters at New Delhi under the leadership of Dr. Arvind Kumar, with the initial objective of engendering heightened awareness amongst the people about water and environment-related issues in the Asia-Pacific region regarding the vital role water and environment play in human lives, their impact on health, economic growth, livelihoods of the people and calamities that wreak havoc due to nonjudicious harnessing of these natural resources. As water is an essential component of power generation and food production, therefore, IWF is also engaged in ensuring environmental security, water security, energy security and food security which are essential for sustainable development (IWF, 2013a).

Soon after its establishment, IWF had started taking part in water-related activities in Meghalaya, especially in the realms of capacity building and generating awareness amongst the people of Meghalaya about water conservation and keeping water resources free from pollution. Nevertheless, in the aftermath of signing of the Memorandum of Understanding (MoU) in August 2012 between IWF and the Meghalaya Basin Development Authority (MBDA), Government of Meghalaya, IWF became a knowledge partner in managing water resources in Meghalaya in 2012, and some inputs provided by Dr. Arvind Kumar, president of IWF, that proved helpful in the finalization of Integrated Basin Development and Livelihoods Promotion Program (IBDLP)—a flagship program of Meghalaya government to improve the quality of life of the people of the state—IWF was called upon to play increased role in facilitating management of water resources in the state.

IBDLP is based on four pillars—Knowledge Management, Natural Resources Management, Entrepreneurship Development, and Good Governance—that aim at ensuring all-round development of Meghalaya and improvement of the lives of the people of Meghalaya. The IBDLP is implemented in a mission-mode (Govt. of Meghalaya, 2012). At the initial stage of implementation, there were nine missions during 2013–14, which were subsequently increased keeping in view of the requirements of the pace of realizing SDGs. Integrated Water Resources Management (IWRM) approach was emphasized under the Meghalaya Water Mission, one of the components of the IBDLP, to manage water-related problems in the state; nevertheless, Dr. Kumar soon realized the limitations of IWRM approach and stressed on the need to take environment-related issues into consideration while addressing water-related issues. Noting that water is the primary medium through which climate change influences Earth's ecosystems and thereafter the livelihoods of the people and well-being of societies, and that variations in climate change affect availability and distribution of rainfall, snowmelt, river flows, and groundwater, and that inadequate management of water resources could jeopardize progress on poverty reduction and sustainable development in all socioeconomic and environmental dimensions in general and in Meghalaya in particular, it was emphasized that adaptation to climate change is closely linked to water and its role in sustainable development; therefore essential adaptation measures dealing with climate variability and built upon existing land and water management practices entailed the potential of creating resilience to climate change, enhancing water security and contributing to sustainable development (Kumar, 2013a).

Thereafter, emphasis on environment-plus approach along with IWRM was emphasized by IWF in Meghalaya that entailed a holistic approach focusing on capacity building of sector and actor through sensitization, incentivization, and galvanization; establishment of a nodal agency as a hub for knowledge sharing and networking in water and environment sectors, assimilation and dissemination of water and environment-related knowledge, intersectoral approach, equal emphasis on soft approach along with hard approach; and from sectoral to collective approach in water and environment sectors (IWF, 2013b). This proved to be precursor to the subsequent emphasis on w-e-f nexus approach that was emphasized by IWF in Meghalaya. Realizing the limitations of IWRM approach that addresses all sectors of development from a water management perspective while leaving other related sectors like food and energy untouched, that the process of IWRM being translated into action remains chiefly driven by the "water/environment sector," with other sectors consulted but not substantially involved; and also conceding that the water demands and water quality requirements of all sectors may get addressed in the bulk IWRM planning; nevertheless, the policy and strategy developments within sectors seldom adequately consider the vital cross-cutting role of water; therefore the emphasis is on the water, energy and food security nexus approach aims to figure out the true involvement of ownership of the food and energy sectors, an aspect that is often falling back in IWRM approach. There exists no antagonism between the Nexus approach and IWRM principles, rather the nexus approach lays emphasis on exploring the opportunities of coordinated and integrated actions between water, energy and food and aims to do away with the "silo" thinking between the three development areas.

Interestingly, the ground work for w-e-f nexus approach had been laid down by the convening of the Climate Summit for a Living Himalayas in 2011 at Bhutan that brought together government of Bangladesh, Bhutan, India and Nepal. The Framework of Cooperation adopted at this summit aimed at building regional resilience to the negative impacts of climate change in the Himalayas with the themes of ensuring food security and securing livelihoods; securing the natural freshwater systems in the region, ensuring sustainable use of biodiversity and ensuring energy security (WWF, 2011). With along with IWRM approach, w-e-f nexus approach was also implemented in Meghalaya to deal with water-related issues as well as coping with issues pertaining to climate change. In the meanwhile, EbA approach to climate change had started gaining international traction by 2014 and emphasis on integrating EbA into national and provincial and local levels in India also started gathering momentum and accordingly IWF focused its attention on integrating EbA into the water end environment-related policies in Meghalaya.

10.8.3 Mainstreaming/integrating ecosystem-based adaptation into Meghalaya

The state of Meghalaya is situated in the northeast region of India and it extends for about 300-km in length and about 100-km in width. Having an area of 22,429 km^2. And lying in the vicinity of Bangladesh, Nepal, and Bhutan, Meghalaya has a landscape that is mostly rolling plateau with south-facing slopes being extremely steep, with the hill rising to 2000 m. It is bounded on the north and east by the state of Assam and on the south and west by Bangladesh (see map below). Being a compact and bounded state, Meghalaya is cool despite its proximity to tropics. Undoubtedly, it abounds with lakes and waterfalls; nonetheless, the state of Meghalaya lies in in a severe earthquake belt and has faced some of them in the past centuries. It is also vulnerable to water-induced and environment-induced disasters because of its fragile geo-environmental setting and economic underdevelopment, thereby posing potential threat to neighboring states as well as neighborhood countries as well (Kumar, 2021) (Fig. 10.3).

In September 2014 Government of Meghalaya adopted the State Action Plan on Climate Change (SAPCC) for Meghalaya and the prime objective of the SAPCC was to strategize adaptation and mitigation initiatives toward emission stabilization, develop resilience of ecosystems, climate proofing of the livelihood sector and diversification of the economy by reducing the dependency on natural resources. With the SAPCC serving as a guiding document to take the climate agenda forward and infuse it into the development planning of the state, Government of Meghalaya has, in line with the NAPCC, has prioritized the sectors of agriculture, forestry, water resources, sustainable energy, habitat, mining and health, for adaptation planning in respect of

624 Chapter 10 Mainstreaming ecosystem-based adaptation

FIGURE 10.3

Map showing location of Meghalaya (Kumar, 2018).

From Kumar, A. (2018). Learning from the implementation of EbA interventions in the restoration and conservations of ecosystems: Lessons learned from EbA implementation in Meghalaya, India. In: A PPT presentation made at the sixth Asia-pacific climate change forum, on October 17–19, 2018 at Asian Development Bank headquarters, Manila, Philippines.

climate change (Govt. of Meghalaya, 2014). Meghalaya's SAPCC noted that climate change-related knowledge was "very limited" at the state level and thereafter emphasized on the need for the localized indicators, investments in generating local dataset for modeling, forecasting and tracking. Emphasizing on the need for identifying the drivers and indicators of climate change, the SAPCC also noted that owing to uniqueness in geographic location of the state, the profile of either drivers or indicators of climate change could be different from elsewhere and the report called upon the state machinery to take equal interest in assessing these drivers and indicators as "they are the nucleus of all studies related to climate change" (Govt. of Meghalaya, 2014, pp. 102–103).

Undoubtedly, communities worldwide have been adapting to climate variability for centuries; nonetheless, their coping mechanisms are being outpaced by the fast-changing climate. In response to global climate change impacts, most countries have focused on "hard" or "gray" infrastructure options such as embankments for flood control or new reservoirs to cope with water shortages. These options can be costly to build and maintain, and generally do not take the benefits accruing from ecosystems and the services delivered by them. In view of the limitations articulated in Meghalaya's SAPCC and the urgency of making the state capable of coping with the vagaries of climate change while pursuing its path of sustainable development uninterruptedly, IWF under the leadership of Dr. Arvind Kumar espoused the case of integrating EbA to climate change in state's policies. EbA entails conservation, sustainable management and restoration of ecosystems as cost-effective solutions that can easily help people adapt to the impacts of climate change and the stark examples of such nature-based solutions to climate change, inter alia, include: sustainable agriculture, integrated water resource management and sustainable forest management, etc. Harnessing the power of nature can bring in benefits to human communities and natural systems. Though primarily an adaptation approach, EbA entails the potential of contributing to contributing to climate change mitigation by reducing GHG emissions that transpire from habitat loss and ecosystem degradation.

Meghalaya is rich in important minerals like coal and limestone, and has a large forest cover, good soil, is biodiversity-rich and possesses abundant water bodies (Figs. 10.3 and 10.4); however, piecemeal approaches to water, energy, or biodiversity in the past resulted sectoral solution and lacked cross-cutting outcomes with other interconnected sectors. While laying emphasis on mainstreaming EbA in the policymaking process of different sectors at provincial and local levels, IWF also stressed on capacity building of all stakeholders in order to equip them with sufficient knowledge and capacities to harness EbA to cope with climate change impacts.

Priority was accorded to water because water is a socioeconomic asset to tackle the vagaries of climate change and socioeconomic development indicators that facilitate interconnectedness with other sectors such as agriculture, energy, environment, etc.; connected policy areas, economic sectors, and societies, tools for cooperation and building trust along with improving livelihoods of the native people. While engaging the stakeholders in building their capacities in harnessing EbA, they were first acquainted with the ESSs and subsequently the services delivered by them with the help of Fig. 10.5.

Acquaintance of the stakeholders—policymakers, those entrusted with the task of implementing the policies and programs and those for whom policies are meant to be beneficial or the expected beneficiaries—with the ecosystems and the services delivered by them facilitates the task of

FIGURE 10.4

Mawphlang sacred forest in eastern Meghalaya.

Courtesy Wikimedia Commons.

626 Chapter 10 Mainstreaming ecosystem-based adaptation

FIGURE 10.5

Nohkalikai waterfalls in Meghalaya.

Courtesy Wikimedia Commons.

adapting to EbA somewhat easy. Thereafter, IWF has been emphasizing on familiarizing the stakeholders with the principles of EbA on the lines as shown in Figs. 10.6 and 10.7.

While according priority to implementing EbA approach that seeks to promote societal resilience via ecosystem management and conservation, and recognizes the centrality of ecosystems in the adaptation process, IWF has embarked on multiple competence and capacity-building programs of all stakeholders in Meghalaya from time to time to enable them to harness EbA in managing water and other natural resources in the state in a sustainable manner. These training and capacity-building programs have been carried out as part of the Meghalaya Government's flagship program of IBDLP. Through coordination with state government agencies like MBDA and international agencies like JICA, GIZ, IFAD World Bank, etc., as development partners, IWF has been able to facilitate convergence so that cross-cutting partnerships are promoted alongside community-led initiatives to elicit essential fiscal help and technical expertise to fulfill the objectives envisaged in the IBDLP and other subsequent development plans to ensure uninterrupted pace of sustainable development of Meghalaya.

Multiple benefits have started accruing in different sectors in Meghalaya in the aftermath of integrating EbA in water and other related sectors. Owing to the fact that EbA addresses crucial links between climate changes, biodiversity, ESSs and sustainable resource management, the adaptation of EbA has demonstrated clear linkages between water and human development and their cumulative impact on people's lives and livelihoods. Integration of EbA in water sector in Meghalaya has resulted in creating alternative water-related livelihoods for the native communities, and that has been successful in addressing the problem of unemployment, especially in rural areas, creation of new jobs, ensuring sustainable mining sans polluting water resources, alleviation of poverty, enhancing gender mainstreaming along with resulting in enhanced socioeconomic-environmental indicators of the state of Meghalaya. Emphasis on according equal priority to IWRM, w-e-f nexus approach along with EbA approach by IWF for managing natural resources in

10.8 Mainstreaming ecosystem-based adaptation in India

FIGURE 10.6

Ecosystem services.

the state in a holistic state has helped in creating a "Brand Meghalaya." EbA has enabled the local communities to garner multiple benefits accruing from the restoration of man-made wetlands, forest conservation and sustainable forest management that have been helpful in carbon-sequestration, improved water quality, reduction in risks from natural hazards, biodiversity conservation, improvement in alternative livelihoods and poverty reduction (Fig. 10.8).

Moreover, harnessing of EbA along with IWRN and w-e-f approaches has brought Meghalaya in the vicinity of realizing specific SDGs, specifically of improved livelihoods in terms of SDG-1, increased water, energy and food security in terms of SDG-2, SDG-6, and SDG-7; enhanced climate change resilience in terms of SDG-13, and augmented biodiversity conservation and ESs in terms of SDG-13 and SDG-15. Besides, sacred groves in Meghalaya are proving to be a shining example of traditional forest management. These traditional cultural practices serve as water catchments, and water conservation is perhaps the well-documented ecological service provided by these sacred groves. Tribes in Meghalaya used bamboo-based drip-irrigation to bring water to seasonal crops and successfully prevented water wastage and increased crop-yield with less water (Kumar, 2021). It is worth mentioning that appropriate sanitation strategies have made Mawlynnong in Meghalaya the "cleanest village in Asia." EbA has met with success in Meghalaya and neighboring states like Assam, Sikkim, and other underdeveloped aspirational states, are propelled to adopt EbA approach in their development policies (Fig. 10.9).

Nevertheless, mainstreaming EbA in Meghalaya, like anywhere else, is also confronted with some stumbling blocks. Our experience in Meghalaya has demonstrated that restoration efforts are likely to fail if the sources of degradation persist; therefore, in that case, it is essential to identify and eliminate or remediate ongoing stresses wherever possible. There has also been a limitation in terms of environmental financing, institutional convergence, lack of integration of horizontal and vertical development sectors. Undoubtedly, Meghalaya has prospered over the years by registering double-digit growth in its economy; nonetheless, it has taken almost a decade to bring visible impact at the grassroots level due to lack of effective awareness and regular capacity-building

628   Chapter 10 Mainstreaming ecosystem-based adaptation

FIGURE 10.7

Principles of ecosystem-based adaptation.

FIGURE 10.8

Integrated land and water resources modeling at Umiam Lake in Meghalaya connecting water resources-related ESS with climate change.

endeavors. Since EbA addresses the crucial links between climate change, biodiversity and sustainable resources' management, it is imperative that effective mitigation and adaptation strategies are required to focus on ecosystem restoration as part of an explicit response to climate change.

Success stories from Meghalaya in the wake of mainstreaming of EbA leading to providing alternative livelihoods have been acknowledged by United Nations Social Development Network (USDN) and published online at UNSDN portal. The first story deals with East Jaintia Hills District of Meghalaya, an area where people are mostly engaged in mining activities such as lime stone and coal mining. Coal mining has been the main source of livelihood in the entire District and also a part of the state. It is an area where financial and livelihood activities revolve around extraction of nonrenewable resources or fossils fuels which contribute to high return on investment. Undoubtedly, the people have been making more investments in coal mining businesses rather than any other activities, thus driving out any other activities due to the scale of returns on investment; nevertheless, this mining business has created several problems and negative impacts on the environment. For example, soil degradation due to mining has become a serious problem in East Jaintia Hills District. Therefore, in view of these adverse impacts on the environment, IWF, in cooperation with MBDA authorities emphasized on the need to introduce schemes for the reclamation of land and to provide sustainable livelihoods to the poor and achieve rural development. Capacity-building programs, professional training and technical/financial support by various institutions under the aegis of the IBDLP have helped many people from the district to improve their livelihoods and therefore an augmentation in their incomes.

The first story revolves round Bijoy Wann, a resident of Moopyniein, Wapungskur Village of East Jaintia Hills District, was first drawn toward a career in mining, but in the aftermath of undergoing capacity-building programs, he has not only changed his mind but also has shown keen interest in skill upgrading, training and bank linkages on poultry/piggery farming. He owns a big plot of land which is feasible for poultry farming (as shown in Figs. 10.9 and 10.10).

FIGURE 10.9

Indigenous rice seed exchange enhance awareness of local varieties protecting them, utilizes and enhances local farmers' knowledge and technical expertise along with promoting biodiversity protection.

Author at project site somewhere in Meghalaya.

FIGURE 10.10

Bijoy Wann with his site for Poultry hut.

Photo made available by India Water Foundation.

He has started constructing the poultry hut. Bijoy has been facilitated by the local branch of Meghalaya Rural Bank (MRB), where he decided to take a general credit card loan. With this loan he bought materials (like cements, blocks, etc.) and has started working. He has an ambition of earning his livelihood from livestock activities only. During the past 2 years, his earnings have tripled to about $1,500 per year.

The second success story veers round Donald Shylla, a resident of Sonapyrdi village, East Jaintia Hills District, who has emerged as one of the progressive farmers of the village. He wants to become an entrepreneur, and set an example for others. Capacity-building programs and technical/financial assistance under the IBDLP has enabled him to enhance his earnings through a rubber plantation. He has two plots of land; one in Wahdiekyiad near LumTongseng village and the other at Wahlakhar in Sonapyrdi. In Wahdiekyiad he has planted 600–700 rubber trees since 2005. Through rubber extraction he has earned about $3000 per year. At his other plot at Wahlakhar, which is close to Lukha River in Sonapyrdi village, he has started conducting multiple activities such as building fish ponds, cultivating vegetables, fruits, etc., in order to generate some income from that land (Figs. 10.10 and 10.11).

He has three fish ponds, which produce up to 400–500 kg of fish very year. He generates an additional income of $3500 per annum through fish and around $1000 per year from the sale of vegetables like tomatoes, potatoes, pumpkins, beans, cucumbers, cabbages, and capsicums. He acknowledges that capacity building and programs under IBDLP have helped him augment his income (UNSDN, 2016).

IWF has projected Meghalaya Model, comprising IWRM, w-e-f and EbA, in its presentations both at national and international fora. A presentation on "Climate Change and its Impacts on Water Sector in India: A Case Study of Meghalaya" was made by IWF at the Third Asia-Pacific

10.8 Mainstreaming ecosystem-based adaptation in India

FIGURE 10.11

Donald Shylla (sitting) near his fish pond.

Photo provided by India Water Foundation.

Adaptation Forum 2013 held on March 18–20, 2013 at Incheon in Republic of South Korea. Besides, focus on Green Mission of Meghalaya and the role of EbA vis-à-vis climate change in urban planning was stressed by IWF in its presentation on "New Trends and Innovations and Assigning New Responsibilities to the Collaborating Partners in Urban Planning Toward Resilient Cities" made at the Regional Policy Dialogue on Sustainable Urbanization in South Asia organized at New Delhi, India on December 17–18, 2014, under the aegis of the UN-ESCAP. Meghalaya was also the focal point of a presentation made by IWF on "Harnessing the Water–Energy–Food Nexus for Trans-boundary Basin Management Cooperation: A Case Study of Meghalaya's CCA" at the Fifth Workshop on Adaptation to Climate Change in Trans-boundary Basins, convened on October 14–15, 2014 at Palais des nations, Geneva, Switzerland.

Meghalaya was also in limelight in the presentation made by IWF at SIWI International Water Week at the latter's session on Freshwater Ecosystems and Human Development held at Stockholm, Sweden, on August 23–28, 2015; wherein IWF made presentation on "Ecosystem-based Adaptation—Sustainable Water Use in Urban Areas: A case Study of Meghalaya's Green Mission in Urban/Peri-Urban Areas." Focus on Meghalaya with regard to integrating EbA and lessons learned therefrom was again centered in IWF presentation entitled "Learning from EbA Interventions in the Restoration and Conservation of Ecosystems: Lessons Learnt from EbA in Meghalaya, India," made at the Sixth Asia-Pacific Climate Change Forum, held on October 17–19, 2018 at Asian Development Headquarters in Manila, Philippines. Besides, some takeaway inputs, partly based on the successful implementation of EbA in Meghalaya have been included in

IWF's submission to the UNFCCC recently. In this way, IWF has tried to catapult Brand Meghalaya as a model for successful implementation and mainstreaming EbA at a global scale.

10.8.4 Integrating ecosystem-based adaptation into Arunachal Pradesh

A study by Saikia et al. (2020) focuses on climate change impacts and EbA in Eastern Himalayan forests of Arunachal Pradesh, a state located in north-east region of India. Noting that high variations in altitudinal gradient precipitation and land surface temperature (LST) have resulted in high vegetation diversity and density, the researchers highlight that the varying relief zones implies all major types of climatic zones in the region. The subtropical zone (800–1800 m) occupies major geographical area comprising about 28.8%, followed by alpine (>2800 m; constituting about 27.3%); temperate (1800–2800 m, comprising about 23%), and tropical zone (<800 m, accounting for 20.9%). Significant variations in the precipitation pattern and reduction in cumulative precipitation with varied intensity during 1998–2013 period can be discerned from this study. It is further revealed that the Tropical Rainfall Measuring Mission (TRMM)-based standard anomaly of precipitation with reference to mean precipitation exhibits higher negative anomaly in recent years in comparison to past years, whereas CRUTS 4.00-based standard anomaly of long-term observations of precipitation patterns during the 1901–2015 period exhibits episodic variability with increasing unevenness in recent years. The spatiotemporal annual LST patterns for the period 2001–16 indicates that northern periphery in latitudinal direction have low (5°C–10°C) to very low (<5°C) LST, whereas central · to south parts have moderate (15°C–20°C) to very high (>25°C) LST (Saikia et al., 2020). While dealing with climate change impacts on the Himalayan forests biomes that are being further accelerated by various anthropogenic disturbances, including illegal timber felling, forest fire, jhum cultivation and hydroelectric projects, Saikia et al. (2020) have endeavored to explore the prospects of harnessing EbA and Eco-DRR approaches to determine significant ecosystems services, reduce community vulnerability, and enhance resilience of people as well as the environment to climate change.

10.8.5 Integrating ecosystem-based adaptation into the Indian Himalayan Region

Degradation of the mountain ecosystems in the IHR triggered by climate change and other human activities is increasingly impacting provisioning of multiple and essential ESSs, and given the fact of direct dependence of human well-being on these ESSs, it becomes imperative to adopt reliable and suitable adaptation measures to enhance resilience of communities to climate change-related effects. It is in this backdrop that the Indian Himalayan Climate Adaptation Program (IHCAP) has undertaken a project that, while emphasizing on the need for adopting an EbA approach to augment the resilience of the communities dependent on natural resources toward climate change, also draws on the lessons from earlier studies undertaken by IHCAP to assess vulnerability of biodiversity and ecosystems in Kullu district of Himachal Pradesh in India (IHCAP, 2016). The Great Himalayan National Park (GHNP), along with the Sainj Wildlife Sanctuary, Tirthan Wildlife Sanctuary and an eco-zone, encompass a total of 1171 km^2 of protected area in Kullu. The biodiversity-rich GHNP comprises 832 floral and 386 faunal species. Nevertheless, 34 of the 47 medicinal plants, categorized as threatened plants in Himachal Pradesh, are found in the GHNP. Apart from being a rich repository of biodiversity, the GHNP provides a wide variety of ecosystem goods and services to the region. Taking into account the vulnerabilities confronting the GHNP and the high dependency of local communities on medicinal plants, the IHCAP (2017) suggests an example of EbA response in terms of propagating medicinal plants in the eco-zone of the GHNP that may be helpful in enhancing the resilience of the ecosystem in GHNP by reducing the dependence of communities on forest resources. The adaptation response can entail promotion of medicinal plants in privately-owned agricultural land or wasteland of the eco-zone of the GHNP. Resultantly, in the wake of communities having better access to medicinal plants from the agricultural or wasteland, these interventions may be helpful in minimizing dependence of the local communities on forest resources in GHNP, and accordingly owing to reduced pressure, these interventions are prone to lead to augmented quality of forest resources and biodiversity thereby "making the forest more resilient to the changing climate" (IHCAP, 2017).

Noting that the enhanced resilience of forests entails the potential of assisting in increased sustainability of ESSs provided by the GHNP, IHCAP (2017) suggests that promotion of medicinal plants on barren and agricultural land in the eco-zone of GHNP may help in reducing the water and soil runoff thereby minimizing the exposure to natural hazards such as landslides and flash floods. Emphasizing on the essentiality of implementing EbA to restore degraded ecosystems and assuring the future provisioning of ESSs in the wake of high vulnerability of ecosystems and communities in IHR, IHCAP (2017) makes specific recommendations in the context of EbA on following counts. In the first instance, integrated approach is recommended because EbA allows an integrated approach entailing an array of ecosystem management, restoration, and conservation activities for achieving the objectives of adaptation to climate change. The other recommendation veers round flexibility because one of the greatest advantages of implementing EbA approach is its flexibility in implementation because this approach permits accommodating adjustments and incremental implementation depending on the level and degree of climate change across the lifespan of the project. Besides, relatively low costs involved in the implementation of EbA interventions widen the scope of adjustments in plans in order to address the uncertainties pertaining to climate change. Another recommendation relates to iterative approach because from an implementation perspective, EbA permits an iterative process with regular monitoring to address the uncertainties involved in climate change.

10.9 Emerging lessons of mainstreaming ecosystem-based adaptation

As a deliberate process, occurrence of EbA mainstreaming is required to take place at various levels of governance, including civil society and private sector. Entailing the ability of targeting manifold outcomes like policies, plans and legislation, EbA can follow different pathways to attain these outcomes. Various case studies of EbA mainstreaming analyzed by GIZ (2019) and IIED (2019) reflect on some emerging lessons pertaining to the policy processes, institutional arrangements, stakeholder engagement and communication as steppingstones for spurring mainstreaming of EbA. In the first place, it is essential to have an understanding of policy processes and institutional arrangements at all levels to be able to explore potential windows of opportunity for mainstreaming EbA. Availability of information on climate variability and change is essential to take adaptive decisions. Making available adequate information with the necessary level of detail in a user-friendly manner is warranted because in view of the uncertainty of information pertaining to climate change, especially its impacts on the functioning of the ecosystems and the services delivered by them, is prone to create a problematic situation. While arguing that scenario-thinking to address uncertainty as well as co-production of information and knowledge management furnishes a good basis for informed and transparent decision-making, GIZ (2019) further suggests that coconstruction and of an information and knowledge base from indigenous, local and scientific sources can be helpful in envisaging vigorous and locally appropriate solutions to build the resilience of natural and societal systems.

It is equally important to lay stress on integrating EbA criteria and norms into the national budget allocation process because by doing so will help in enabling mainstreaming outcomes at various levels. The example of Peru makes it discernible as to how public financing for EbA can be earmarked via national budgets across sectors and multiple scales, ranging from local to regional and national levels (GIZ, 2018a). Aside from ecosystem-related sectors like water, agriculture and forestry, Peru has reportedly made the case for the potential integrating biodiversity and ecosystems conservation in other sectors, such as infrastructure, thus moving from gray to green or hybrid adaptation measures. Equal emphasis needs to be focused on ensuring a semblance of equilibrium between top-down and bottom-up approaches for mainstreaming EbA. From a pragmatic perspective, integrating climate change into national and sectoral policies and strategies ought to place reliance on a mix of top-down and bottom-up approaches, because this has been attested to by experiences garnered from ongoing projects, especially in South Africa, that opting for a synergetic affords opportunities and multiple benefits. While launching EbA initiatives at the local level, joint analysis of feedback and lessons learned can be integrated into policy processes and that can enable replication and scaling-up of the approach. For example, most of the ongoing EbA programs in South Africa, within both the rural and urban areas, have often been developed in a bottom-up, gradual manner, leveraging funding and the capacity opportunities as and when they emerge, and afforded opportunity to imbibe lessons learned to support scaling up (GIZ, 2018b).

Another emerging lesson from the ongoing and past EbA mainstreaming projects warrants that EbA needs to be made an integral part of land-use policy. There arises a dire need for defining and embedding EbA in the land-use policy planning process, especially under the prevailing circumstances worldwide, and the need to have a clear, effective and efficient law or order that aims at mainstreaming EbA in the development of cities and communities, especially in the wake of rapid pace of urbanization globally. The case of EbA mainstreaming in the Philippines amply demonstrates as to how mainstreaming has grown from land-use planning in a "ridge-to-reef" approach, to incorporate legal aspects, the definition of responsibilities and training on all levels (GIZ, 2018c). Besides, interagency collaboration across governance scales is also called for in delivering essential policy change for EbA. The example of Mexico shows that the role of institutions and policy processes at different governance scales is necessary for achieving mainstreaming and this exemplified from the interagency collaboration, especially in the case of Inter-Ministerial Committee on Climate Change in Mexico (CICC$) that has recommended to make EbA a cross-cutting concept (GIZ, 2018d).

Another emerging lesson emphasizes on the critical role of collaboration and institutional leadership as an enabling factor for EbA mainstreaming in policymaking processes Undoubtedly, a typical arrangement for facilitating coordination of adaptation strategies is often assigned to the Ministry of Environment for assuming overall responsibility for climate change; nevertheless,

10.9 Emerging lessons of mainstreaming ecosystem-based adaptation

experience suggests that such an arrangement leads to enfeebled ministerial coordination, whereas a powerful central agency or ministry like the Ministry of Finance can facilitate better coordination in implementation by sectoral ministries along with reviewing legislation and holding concerned implementation agencies accountable for their actions and results. The example of Peru amply illustrates that the established collaboration between the Ministry of Economy and Finance and the Ministry of Environment is better positioned to efficaciously mainstream EbA (GIZ, 2018a).

Generation of knowledge capital needed for effective EbA requires new institutional partnerships and resources, and keeping in view the transdisciplinary character of EbA, it is essential for institutions to work outside their traditional "comfort zones." For example, South Africa shows as to how a research partnership fostered between eThekwini Municipality and the University of KwaZulu-Natal has focused on advancing knowledge on biodiversity conservation and management within the context of global environmental change and includes an internship program designed to build human capital for the municipality in these areas (GIZ, 2018b). In the case of the Philippines, creation of the "People's Survival Fund" as an annual fund for local government units and community organizations to implement CCA projects is expected to better equip vulnerable communities to deal with the adverse impacts of climate change (GIZ, 2018d).

For the effective EbA mainstreaming, increased collaborative governance and planning is necessary. Being inherently an interdisciplinary and cross-sectoral approach, EbA requires a multistakeholder process wherein EbA mainstreaming can be spurred and supported at different governance levels and by different stakeholders and in that sense, different sets of actors can hold "key roles" in EbA governance (GIZ, 2019). Collaborative governance has come to play a crucial role to help achieving longstanding change, both at the policy level and on the ground, and concurrently, it has proven to be extremely beneficial to identify and work with an agency or body that has the mandate to bring all actors together. For example, South Africa illustrates where substantial advances in EbA planning have been recorded in a number of catchments based on a governance model between state and private institutions, and these catchments usually represent a critical level of governance and can afford a good opportunity to create new or support existing entities to lead participatory planning and EbA development (GIZ, 2018b).

Another emerging lesson to be learned from the past and ongoing EbA mainstreaming projects is that participation of private sector is essential in making EbA mainstreaming effective and successful. Cost-effectiveness, direct and short-term benefits of EbA mainstreaming are seemingly sufficient factors to attract private sector. Aside from the allocation of public funding for EbA projects, some countries may require the mobilization of new and additional financial resources to bridge the financial gap and at that juncture, involvement of private sector in the context of innovative direct financing models for EbA can be helpful in engulfing the fiscal gap. In this regard, GIZ (2019) cautions that since businesses usually privilege decisions that are efficient and cost-effective, it is especially important to provide concrete evidence that EbA measures actually yield the results that they are designed to generate.

Role of effective communication is another lesson that emerges from the past and ongoing EbA mainstreaming projects. Communicating in simple messages, apart from being a potent tool to heighten awareness and interest, can also be helpful in eliciting more resources. In recent years, as GIZ (2019) informs, the conservation community worldwide has increased its communication and awareness-raising endeavors by harnessing the concept of "natural solutions" in an attempt to buy-in from other sectors for the conservation of natural resources for multiple purposes. In an identical manner, successful implementation of EbA measures ordains that practitioners and policymakers communicate effectively with an assorted range of stakeholders and this process involves communicating complex concepts and notions like uncertainty and probabilistic information, ESSs and their values, and often very technical scientific data in a simple style and language that can be understood by everyone without any ambiguity. In the Philippines, the "National Integrated Climate Change Database Information and Exchange System" coalesces data from public and private sources, and that data are important for coherent policymaking and investment decision-making (GIZ, 2018c).

10.10 Barriers to mainstreaming ecosystem-based adaptation

Effective EbA mainstreaming in development planning calls for an enabling environment at all levels of EbA implementation; however, while implementing EbA is usually well understood at the national level and at the levels of field implementation; nevertheless, barriers do crop up at the time of rendering policies into plans, programs, and budgets. Accordingly, it becomes important to have brief overview of such barriers so that adequate measures are devised to overcome these barriers by clearly articulating the major objectives for EbA to be embedded in national planning processes in order to generate an enabling framework for local-level implementation and to seek increased access to financial resources.

Absence of a common language and methods for EbA serves as a major barrier owing to the fact that different agencies engaged in prioritizing and promoting ecosystems management and adaptation are used to adhered to making usage of distinct terminology and diverse methodologies, which often entail the potential of hindering coordination. Besides, overlapping institutional mandates also serve as a barrier because while at the national level mandates on CCA and ecosystem management are often clear; nonetheless, at the regional and local levels responsibilities of institutions often overlap. In bottom-up cases entailing potential to mainstream EbA there is no clear indication as to who heads the articulation of initiatives from local to regional level and subsequently to national level to inform policy processes. Ignoring the fact that the effective EbA mainstreaming is a resource-intensive and long-term process is bound to give rise to problems, especially when budget timelines and budgets overlook this factor. Oftentimes project leaders—either institutional or community stakeholders are left with no option but to embark on the adaptation planning in a "vacuum." As it has been noted, "The planning and implementation process in many cases is a multiyear participatory initiative, which needs to secure trust and establish long-term relationships among a range of actors. This can be a demotivating factor for initiating such processes" (IIED, 2019).

Limited horizontal coherence of policies at different levels also serves as a hindrance, especially in the wake of prevalence of limited alignment between policies and sectoral action plans in respect of adaptation measures as well as DRR. Lack of articulation with policies across governance levels is another obstacle because a great difficulty is prone to arise in the absence of clear delineation policies and plans at local, provincial and national level. Integration of EbA into the NAP is required to be further articulated and aligned with provincial and local planning process in order to be implemented in an effective manner, and it often proves to be a challenging task (GIZ, 2019).

10.10 Barriers to mainstreaming ecosystem-based adaptation 637

M&E of EbA measures being a challenging task can also become another hindrance in hassle-free implementation of EbA because it is difficult to find indicators for M&E of EbA measures, since processes rely on time frames, actions and planning across sectors and regional or administrative entities, with complex interactions and interdependencies. Furthermore, the situation gets more complicated in view of the very limited availability of information with regards to baselines or construction and monitoring of indicators. Moreover, another difficulty arises in the wake of limited capacity of national-level institutions to support the validation and implementation of EbA measures because these national entities are often unable to adequately respond to the need for technical assistance such as revision and approval of EbA project proposals pursuing public investment. This is amply demonstrated from the examples of Peru (GIZ, 2018a; Reid, Podvin, & Segura, 2018) and South Africa (GIZ, 2018b) where EbA was expected to be supported by an array of policies and plans, there; however, the human resources needed for implementation were often insufficient and there also existed a specific gap in skill relating to EbA M&E.

Both weak governance mechanism and weak institutional framework are an anathema to EbA mainstreaming because these are prone to lead to limited creation and enforcement of relevant legislation and management regimes at the local level. Lack of capacity is also reported, especially at higher government levels, specifically in terms of incorporating EbA into national adaptation policy and planning processes. Corruption is said to have contributed to government support for mining in Peru (Reid, Savadogo, & Somda, 2018), poor enforcement of grazing regulations in Kenya (Reid & Orindi, 2018) and illegal natural resource extraction in Bangladesh (Reid & Ali, 2018), all of which proved instrumental in undermining EbA implementation. Weak community organizations and weak traditional leadership can cause lack of the technical skills needed to implement EbA, and perhaps led stakeholders in some cases to think that this had made implementation challenging in certain cases. It was reported with regard to the project site in Burkina Faso that there were no strong local organizations (Reid, Scorgie, Muller, & Bourne, 2018).

Insufficient collaboration between and among the concerned institutions and sectors can also serve as a handicap in EbA implementation, especially in view of the given fact that EbA is characteristically a multisectoral endeavor that requires collaboration across a range of government levels. Nonetheless, when governments tend to be structured according to sector and political rivalry or instability, hindrance in collaboration is bound to arise. Given the fact that local government departments or technical services often work independently from each other and under that scenario, stakeholders often see participation of local people or communities as insufficient in the spaces where decisions are taken. Under such circumstances, the possibility of governance at provincial or regional government levels is often prone to get fragmented and siloed. Another outcome of such a situation is that collaboration at the national level between agencies responsible for climate change, disaster prevention and relief are often inadequate and some such sites identified by IIED (2019) need transboundary collaboration, and the example of Bangladesh is cited where hilsa fish travel through river systems and ocean waters under the jurisdictions of Indian and Myanmar, respectively (Reid & Ali, 2018).

Knowledge gaps and inadequate knowledge sharing with regard to mainstreaming EbA entails the potential of impeding implementation of EbA at different levels. It has been reported by IIED (2019) that many stakeholders considered that government was required to improve its understanding of EbA and it was noted by that community understanding of the benefit accruing from environmental protection and EbA was limited, while others emphasized on the need for a stronger scientific-evidence-base on EbA, especially in terms of Orinda quantitative socioeconomic assessments and economic cost-benefit analyses. The process of M&E—and securing robust evidence of impacts—can be rendered difficult in view of knowledge gaps relating to mainstreaming EbA. While contending that a comprehensive evaluation of the full gamut of socio-economic benefits accruing from EbA is a challenging task and benefits emerging from EbA are often undervalued, it is suggested that this lacuna can be addressed by interpreting the science behind EbA and making it more accessible, especially for the policymakers, and focusing more emphasis on inclusion of EbA in national curriculum and higher education (IIED, 2019).

Enfeebled policy framework and weak legal support are also barriers to the mainstreaming of EbA. Admittedly, many countries have policies and strategies dealing with climate change and DRR; nevertheless, EbA is usually poorly integrated into these policy frameworks and as such policy support for climate change and EbA gets rendered insufficient. It has been reported that Bangladesh has no policy or strategy for addressing climate change impacts in the fisheries sector as well as no national-level policy or strategy recognizing and facilitating EbA (Reid & Ali, 2018). Prevalence of weak policy framework and lack of legal support in other areas like water extraction and use and payments for ESSs schemes, can also undermine the EbA mainstreaming. While noting that stakeholders at various sites witnessed government policies as undermining local agency by failing to support user rights on communal land or limiting the devolution and decentralized governance, IIED International Institute for Environment and Development (2019) suggests that policies can also be top-down and ill-suited to local conditions. Prevalence of high-levels of poverty and poor infrastructure can also be instrumental in obstructing the mainstreaming of EbA. Nursing this notion, many stakeholders at several sites considered that these obstacles limited the scope of potential benefits of EbA. Examples are cited about poor transport networks that limited access to market; prevalence of high levels of unemployment and illiteracy, limited mobile phone coverage along with poor water supplies, etc., that altogether reportedly joined hands in facilitating reduced adaptive capacity. At the same time, it is also argued that limited funding pouring in for EbA at local, regional and national levels often constrains EbA implementation even when plans and policies according priority to EbA are well in place; concurrently, it is also lamented that with some donors supporting EbA and that support may not be coming via government agencies or through government channels, can also undermine nationally-determined adaptation priorities (IIED, 2019).

10.11 Conclusion

Increasing emphasis on integrating/mainstreaming EbA into climate change adaption and development policies and strategies at national, regional and local levels is gaining traction globally (GIZ, 2019; IIED, 2019; Terton & Greenwalt, 2021). The UNFCCC (2017) has argued that EbA has exhibited potential to enhance social and ecological resilience to climate change and adaptive capacity in the long-term, and while strongly supporting UNFCCC's this view, IIED (2019) has also tried to demonstrate that EbA can provide an array of strong, long-lasting, and wide-reaching adaptation-related benefits, social eco-benefits and ecosystem-related benefits. The UNFCCC's suggestion that "countries should consider EbA in their approach to adaptation, including in national adaptation plans" (UNFCCC, 2017) also finds support from the IIED (2019). Keeping in view uncertainties linked with climate change and the potential impacts it may wield on ecosystems and societies, can in now be a cause for any inaction on tackling the menace of climate change. Development in the technology and the experiences, expertise, and knowledge amassed in recent decades has seemingly enabled humankind to put in place adequate strategies to deal with the vagaries of climate change. And these strategies, apart from being inclusive and adaptive, also entail the potential of being learned rapidly and improved upon as well.

There is no gainsaying in the fact that by sharing success stories, challenges and failures with regard to CCA entail brighter chances of vastly improving prospects of developing good, efficacious strategies for mainstreaming of EbA. Improvement in the ecosystem resilience by reducing the impacts of nonclimatic stressors can be facilitated through CCA and risk reduction via nature-based solutions and in this regard, mainstreaming EbA objectives into national strategies that regulate and coordinate the activities of the different sectors has emerged as a critical step. In the eventuality of a particular sector coming under pressure in tandem with occurrence of a felt-need for change, the possibility of good moments for mainstreaming taking place gets enhanced. Maintenance of ecosystem values and services for human development into an uncertain future warrants immediate and well-concerted action irrespective of the choice of type of entry points. A specific focus needs to be emphasized on appropriate management of the "green capital" keeping in view the vitality of resilient ecosystems for EbA. Restoration of deteriorated ecosystems will increasingly play a role in the context of EbA projects. With regard to monitoring of progress and outcomes of adaptation actions for the ecosystems and their functions in the context of EbA, GIZ (2019) suggests ensuring following essentials: ensuring effective use of resources for management and restoration of ecosystems; ensuring that desired results are being achieved; ensuring that actions are not having maladaptive outcomes either for communities or for biodiversity; and that knowledge is developed for future planning. Paice and Chambers (2016) have emphasized that implementation and monitoring should incorporate of the process outcomes and a formal research component to ensure rigor and confidence in the research because such a measure can entail the potential of contributing to bridge existing knowledge gaps and thereby help in providing essential information for other management entities in development and implementation of apposite strategies. In view of the exacerbated pace of the vagaries of climate change, there is dire need of integrating/mainstreaming EbA into CCA plans, policies, and strategies at national, regional, and local levels along with ensuring regular reflection on EbA projects along with their biodiversity pillars as well as also ensuring frequent and regular exchange of fresh knowledge and expertise garnered through ongoing EbA mainstreaming projects in order to facilitate overcoming of existing barriers.

References

Agrawala, S., & Van Aalst, M. (2008). Adapting development cooperation to adapt to climate change. *Climate Policy*, 8(2), 183–193.

Ahern, J., Cilliers, S., & Niemelä, J. (2014). The concept of ecosystem services in adaptive urban planning and design: A framework for supporting innovation. *Landscape and Urban Planning*, 125, 254–259.

Almazroui, M., Saeed, S., Saeed, F., Islam, N. M., & Ismail, M. (2020). Projections of precipitation and temperature over the South Asian countries in CMIP6. *Earth Systems and Environment*, 4, 297–320.

Andrade, A., Córdoba, R., Dave, R., Girot, P., Herrera-F, B., Munroe, R., ... Vergara, W. (2011). *Draft principles and guidelines for integrating ecosystem-based approaches to adaptation in project and policy design: A discussion document*. Gland: IUCN.

Araos, M., Berrang-Ford, L., Ford, J. D., Austin, S. E., Biesbroek, R., & Lesnikowski, A. (2016). Climate change adaptation planning in large cities: A systematic global assessment. *Environmental Science and Policy*, 66, 375–382.

Benson, D., Gain, A. K., & Rouillard, J. (2015). Water governance in a comparative perspective: From IWRM to a "Nexus". Approach? *Water Alternatives*, 8, 756–773.

Berkhout, F., Bouwer, L. M., Bayer, J., Bouzid, M., Cabeza, M., Hanger, S., ... van Teeffelen, A. (2015). European policy responses to climate change: Progress on mainstreaming emissions reduction and adaptation. *Regional Environmental Change*, 15(6), 949–959.

Carabine, E., Venton, C. C., Tanner, T., & Bahadur, A. (2015). *The contribution of ecosystem services to human resilience a rapid review*. London: Overseas Development Institute.

Chong, J. (2014). Ecosystem-based approaches to climate change adaptation: Progress and challenges. *International Environmental Agreements: Politics, Law & Economics*, 14, 391–405.

Clar, C., Prutsch, A., & Steurer, R. (2013). Barriers and guidelines for public policies on climate change adaptation: A missed opportunity of scientific knowledge-brokerage. *Natural Resources Forum*, 37, 1–18.

Collier, U. (1997). Sustainability, subsidiarity and deregulation: New directions in EU environmental policy. *Environmental Politics*, 6(2), 1–23.

Colls, A., Ash, N., & Ikkala, N. (2009). *Ecosystem-based adaptation: A natural response to climate change*. Gland: IUCN.

Crane, R., & Landis, J. (2010). Planning for climate change: Assessing progress and challenges. *Journal of American Planning Association*, 76(4), 389–401.

CSE (Centre for Science and Environment). (2018). *Coping with climate change: An analysis of India's national action plan on climate change*. New Delhi: CSE.

Daily, G. C. (1997). *Nature's services. Societal dependence on natural ecosystems*. Washington, DC: Island Press.

Dannevig, H., Rauken, T., & Hovelsrud, G. (2012). Implementing adaptation to climate change at the local level. *Local Environment*, 17, 597–611.

DEA-SA (Department of Environmental Affairs-Government of South Africa). (2017). *National climate change adaptation strategy, 2016–2021*. Pretoria, South Africa.

Doswald, N., Munroe, R., Roe, D., Giuliani, A., Castelli, I., Stephens, J., ... Reid, H. (2014). Effectiveness of ecosystem-based approaches for adaptation: review of the evidence-base. *Climate & Development*, 6(2), 185–201.

Dudley, N., Stolton, S., Belokurov, A., Krueger, L., Lopoukhine, N., MacKinnon, K., & ... Sekhran, N. (Eds.), (2010). *Natural solutions: Protected areas helping people cope with climate change*. Gland: IUCN.

Forrest, N., & Wiek, A. (2014). Learning from success toward evidence-informed sustainability transitions in communities. *Environmental Innovation and Societal Transitions*, 12, 66–88.

G-20 (Group of Twenty Web Portal). (2018). *Chair's summary of the first and second meeting of the G-20 climate sustainability working group*. Buenos Aires.

Gain, A. K., Giupponi, C., & Benson, D. (2015). The water-energy-food (WEF) security nexus: The policy perspective of Bangladesh. *Water International*, 40, 895–910.

GCRI (Global Climate Risk Index). (2020). *Global climate risk index 2020*. Bonn: Germanwatch.

GCRI (Global Climate Risk Index). (2021). *Global climate risk index 2021*. Bonn: Germanwatch.

GIZ (Deutsche Gesellschaft für Internationale Zusammenarbeit). (2017a). *Learning brief: Entry points for mainstreaming ecosystem-based adaptation*. Bonn: GIZ.

GIZ (Deutsche Gesellschaft für Internationale Zusammenarbeit). (2017b). *Factsheet: Strategic mainstreaming ecosystem-based adaptation in Vietnam*. Bonn: GIZ.

GIZ (Deutsche Gesellschaft für Internationale Zusammenarbeit). (2018a). *Entry points for mainstreaming ecosystem-based adaptation. The case of Peru*. Bonn: GIZ.

GIZ (Deutsche Gesellschaft für Internationale Zusammenarbeit). (2018b). *Entry points for mainstreaming ecosystem-based adaptation. The case of South Africa.* Bonn: GIZ.

GIZ (Deutsche Gesellschaft für Internationale Zusammenarbeit). (2018c). *Entry points for mainstreaming ecosystem-based adaptation. The case of Philippines.* Bonn: GIZ.

GIZ (Deutsche Gesellschaft für Internationale Zusammenarbeit). (2018d). *Entry points for mainstreaming ecosystem-based adaptation. The case of Mexico.* Bonn: GIZ.

GIZ (Deutsche Gesellschaft für Internationale Zusammenarbeit) (2018e). Mainsteaming the ecosystem-based adaptation approach into policy planning. Bonn: GIZ.

GIZ (Deutsche Gesellschaft für Internationale Zusammenarbeit). (2019). *Emerging lessons for mainstreaming ecosystem-based adaptation: Strategic entry points and processes.* Bonn: GIZ.

Global Environment Facility (n.d.). *Adding value and promoting higher impact through the GEF's programmatic approach.* Washington, DC.

Govt. of Meghalaya (Government of Meghalaya, India). (2012). *Integrated basin development and livelihoods promotion program: A report to citizens.* Shillong: Govt of Meghalaya.

Govt. of Meghalaya (Government of Meghalaya, India). (2014). *State action plan on climate change.* Shillong: Govt. of Meghalaya.

Guerry, A. D., Polasky, S., Lubchenco, J., Chaplin-Kramer, R., Daily, G. C., Griffin, R., ... Vira, B. (2015). Natural capital and ecosystem services informing decisions: From promise to practice. *Proceedings of National Academy of Sciences, United States of America, 112*, 7348–7355.

Huq, N. (2016). Small scale fresh water ponds in rural Bangladesh: Navigating roles and services. *International Journal of Water, 10*, 73–85.

Huq, N., Bruns, A., & Huq, S. (2017). Mainstreaming ecosystem services based climate change adaptation (EbA) in Bangladesh: Status, challenges and opportunities. *Sustainability, 9*, 926.

Huq, N., & Hugé, J. (2012). "Greening" integrated water resources management policies for tackling climate change impacts: A call for sustainable development. In H. Leal Filho (Ed.), *Climate change and the sustainable use of water resources* (pp. 173–183). Berlin/Heidelberg: Springer.

IHCAP (Indian Himalayas Climate Adaptation Program). (2016). *Climate vulnerability, hazards and risk: An integrated pilot study in Kullu District, Himachal Pradesh' synthesis report.* New Delhi: IHCAP.

IHCAP (Indian Himalayas Climate Adaptation Program). (2017). *Ecosystem-based adaptation: An integrated response to climate change in the Indian Himalayan Region.* New Delhi: IHCAP.

IIED (International Institute for Environment and Development). (2019). Is ecosystem-based adaptation effective? Perceptions and lessons learned from 13 project sites. In: *IIED research report.* London: IIED.

IPCC (Intergovernmental Panel on Climate Change). (2007). *Climate Change 2007: Synthesis report. Contribution of working groups I, II and III to the fourth assessment report of the Intergovernmental Panel on Climate Change.* Geneva: IPCC.

IPCC (Intergovernmental Panel on Climate Change). (2012). *Managing the risks of extreme events and disasters to advance climate change adaptation. A special report of working groups I and II of the Intergovernmental Panel on Climate Change.* Cambridge: Cambridge University Press.

IPCC (Intergovernmental Panel on Climate Change). (2014a). *Climate Change 2014: Impacts, adaptation, and vulnerability. Summary for policymakers.* Cambridge: Cambridge University Press.

IPCC (Intergovernmental Panel on Climate Change). (2014b). Chapter 8: Urban areas. Working group II contribution to the fifth assessment report of the Intergovernmental Panel on Climate Change. In: *Climate change 2014: Impacts, adaptation, and vulnerability.* Cambridge: Cambridge University Press.

IUCN (International Union for Conservation of Nature). (2008). *The ecosystem approach: Learning from experience.* Gland: IUCN.

IWF (India Water Foundation). (2013a). *Annual report.* New Delhi: IWF.

IWF (India Water Foundation). (2013b). *India water foundation brochure.* New Delhi: IWF.

Jacob, K., & Volkery, A. (2004). Institutions and instruments for government self-regulation: Environmental policy integration in a cross-country perspective. *Journal of Comparative Policy Analysis: Research and Practice, 6*, 291–309.

Jerneck, A., Olsson, L., Ness, B., Anderberg, S., Baier, M., Clark, E., ... Persson, J. (2011). Structuring sustainability science. *Sustainability Science, 6*, 69–82.

Jones, H. P., Hole, D. G., & Zavaleta, E. S. (2012). Harnessing nature to help people adapt to climate change. *Nature Climate Change, 2*(7), 504–509.

Kumar, A. (2013). Climate change and its impact on water sector in India: A case study of Meghalaya. In: *A PPT presentation at the third Asia-Pacific adaptation forum*, Incheon, Republic of Korea, March 18–20, 2013.

Kumar, A. (2018). Learning from the implementation of EbA interventions in the restoration and conservations of ecosystems: Lessons learned from EbA implementation in Meghalaya, India. In: *A PPT presentation made at the sixth Asia-pacific climate change forum*, on October 17–19, 2018 at Asian Development Bank headquarters, Manila, Philippines.

Kumar, A. (2021). *Ecosystem-based adaptation approach: The case of Meghalaya*. India Water Portal. https://www.indiawaterportal.org/article/ecosystem-based-approach-case-meghalaya. Accessed 18.01.21.

Larondelle, N., Haase, D., & Kabisch, N. (2014). Mapping the diversity of regulating ecosystem services in European cities. *Global Environmental Change, 26*, 119–129.

McCormick, K., Anderberg, S., Coenen, L., & Neij, L. (2013). Advancing sustainable urban transformation. *Journal of Clean Production, 50*, 1–11.

McDonald, J. (2011). The role of law in adapting to climate change. *Wiley Interdisciplinary Review of Climate Change, 2*(2), 283–295.

MoES-GoI (Ministry of Earth Sciences-Government of India). (2020). *Assessment of climate change over the Indian region—A report*. Singapore: Springer.

Munang, R., Thiaw, I., Alverson, K., Liu, J., & Han, Z. (2013). The role of ecosystem services in climate change adaptation and disaster risk reduction. *Current Opinion in Environmental Sustainability, 5*, 47–52.

Naumann, S., Kaphengst, T., McFarland, K., & Stadler, J. (2014). *Nature-based approaches for climate change mitigation and adaptation. The challenges of climate change—partnering with nature*. Bonn: German Federal Agency for Nature Conservation (BfN).

ND-GAIN (Notre Dame-Global Adaptation Initiative). *ND-GAIN country index*. (2019). <https://gain.nd.edu/our-work/country-index/>.

Nunan, F., Campbell, A., & Foster, E. (2012). Environmental mainstreaming: The organizational challenges of policy integration. *Public Administration & Development, 32*(3), 262–277.

OECD (Organization for Economic Cooperation and Development). (2009). *Policy guidance on integrating climate change adaptation into development cooperation*. Paris: OECD Secretariat.

Ojea, E. (2015). Challenges for mainstreaming ecosystem-based adaptation into the international climate agenda. *Current Opinion in Environmental Sustainability, 14*, 41–48.

Paice, R., & Chambers, J. (2016). *Climate change adaptation planning for protection of coastal ecosystems*. Queensland: National Climate Change Adaptation Research Facility, CoastAdapt Information Manual 10. Gold Coast.

Pelling, M., O'Brien, K., & Matyas, D. (2014). Adaptation and transformation. *Climate Change, 133*, 113–127.

Persson, A. (2004). *Environmental policy integration: an introduction. PINTS—Policy integration for sustainability, background paper*. Stockholm: Stockholm Environment Institute (SEI).

Picciotto, R. (2002). The logic of mainstreaming: A development evaluation perspective. *Evaluation, 8*, 322–339.

PMCCC-GoI (Prime Minister's Council on Climate Change-Government of India). (2008). *National action plan on climate change*. New Delhi: PMCCC.

Prabhakar, S. V. R. K., Scheyvens, H., & Takahashi, Y. (2019). *Ecosystem-based approaches in G20 countries: Current status and priority actions for scaling up.* Hayama: Institute for Global Environmental Strategies (IGES), IGES Discussion Paper.

Pramova, E., Locatelli, B., Brockhaus, M., & Fohlmeister, S. (2012). Ecosystem services in the national adaptation programmes of action. *Climate Policy, 12,* 393−409.

Rauken, T., Mydske, P. K., & Winsvold, M. (2014). Mainstreaming climate change adaptation at the local level. *Local Environment, 20*(4), 408−423.

Reid, H., & Ali, L. (2018). Ecosystem-based approaches to adaptation: Strengthening the evidence and informing policy. In: *Research results from the Incentive-based Hilsa Conservation Programme, Bangladesh. Project report.* London: IIED.

Reid, H., Podvin, K., & Segura, E. (2018). Ecosystem-based approaches to adaptation: Strengthening the evidence and informing policy. In: *Research results from the Mountain EbA project, Peru. Project report.* London: IIED.

Reid, H. & Orindi, V. (2018). Ecosystem-based approaches to adaptation: Strengthening the evidence and informing policy. In: *Research results from the Supporting Counties in Kenya to Mainstream Climate Change in Development and Access Climate Finance project, Kenya. Project report.* London: IIED.

Reid, H., Savadogo, M., & Somda, J. (2018). Ecosystem-based approaches to adaptation: strengthening the evidence and informing policy. In: *Research results from the Ecosystems Protecting Infrastructure and Communities project, Burkina Faso. Project report.* London: IIED.

Reid, H., Scorgie, S., Muller, H., & Bourne, A. (2018). Ecosystem-based approaches to adaptation: strengthening the evidence and informing policy. In: *Research results from the Climate Resilient Livestock Production on Communal Lands project, South Africa. Project report.* London: IIED.

Revi, A., Satterthwaite, D., Aragón-Durand, F., Corfee-Morlot, J., Kiunsi, R., Pelling, M., ... Sverdlik, A. (2014). Toward transformative adaptation in cities: The IPCC's fifth assessment. *Environment & Urbanization, 26,* 11−21.

Roberts, D., & O'Donoghue, S. (2013). Urban environmental challenges and climate change action in Durban, South Africa. *Environment and Urbanization, 25,* 299−319.

Rockstrom, J., Steffen, W., Noone, K., Persson, A., Chapin, F. S., Lambin, E. F., ... Foley, J. A. (2009). A safe operating space for humanity. *Nature, 461,* 472−475.

Ruckelshaus, M., McKenzie, E., Tallis, H., Guerry, A., Daily, G., Kareiva, P., ... Bernhardt, J. (2013). Notes from the field: Lessons learned from using ecosystem service approaches to inform real-world decisions. *Ecological Economics, 115,* 11−21.

Runhaar, H., Wilk, B., Persson, Å., Uittenbroek, C., & Wamsler, C. (2018). Mainstreaming climate adaptation: Taking stock about "what works" from empirical research worldwide. *Regional Environmental Change, 18* (4), 1201−1210.

Runhaar, H. A. C., Uittenbroek, C. J., van Rijswick, H. F. M. W., Mees, H. L. P., Driessen, P. P. J., & Gilissen, H. K. (2016). Prepared for climate change? A method for the ex-ante assessment of formal responsibilities for climate adaptation in specific sectors. *Regional Environmental Change, 16*(5), 1389−1400.

Saikia, P., Kumar, A., Diksha., Lal, P., Nikita., & Khan, M. L. (2020). Ecosystem-based adaptation to climate change and disaster risk reduction in Eastern Himalayan Forests of Arunachal Pradesh, Northeast India. In S. Dhyani, A. Gupta, & M. Karki (Eds.), *Nature-based solutions for resilient ecosystems and societies. Disaster resilience and green growth* (pp. 391−408). Singapore: Springer.

Schmarek, C., & Harmeling, S. (2018). *G20 and climate change: time to lead for a safer future.* Geneva: Care International.

Staes, J., Vrebos, D., & Meire, P. (2010). A framework for ecosystem services planning. In P. H. Liotta, W. G. Kepner, J. M. Lancaster, & D. A. Mouat (Eds.), *Achieving environmental security: ecosystem services and human welfare* (pp. 53−72). Amsterdam: IOS.

Szabo, S., Renaud, F. G., Hossain, M. S., Sebesvári, Z., Matthews, Z., Foufoula-Georgiou, E., & Nicholls, R. J. (2015). Sustainable development goals offer new opportunities for tropical delta regions. *Environment: Science and Policy for Sustainable Development, 57*(4), 16–23.

TEEB (The Economics of Ecosystems and Biodiversity). (2010). *Mainstreaming the economics of nature: a synthesis of the approach, conclusions and recommendations of TEEB*. Geneva: TEEB.

Terton, A., & Greenwalt, J. (2021). *Building resilience with nature: Maximizing ecosystem-based adaptation through national adaptation plan processes*. Canada: IISD.

Travers, A., Elrick, C., Kay, R., & Vestergaard, O. (2012). *Ecosystem-based adaptation guidance: moving from principles to practice*. Nairobi: UNEP.

Turnpenny, J., Russel, D., & Jordan, A. (2014). The challenge of embedding an ecosystem services approach: Patterns of knowledge utilisation in public policy appraisal. *Environment and Planning C: Government and Policy, 32*, 247–262.

Uittenbroek, C. J., Janssen-Jansen, L. B., & Runhaar, H. (2013). Mainstreaming climate adaptation into urban planning: Overcoming barriers, seizing opportunities and evaluating the results in two Dutch case studies. *Regional Environmental Change, 13*(2), 399–411.

UN-ESCAP (United Nations-Economic and Social Commission for Asia and the Pacific). (2017). *Mainstreaming disaster risk reduction for sustainable development: A guidebook for the Asia-Pacific*. Bangkok: UNESCAP.

UNFCCC (United Nations Framework Convention on Climate Change). (2012). *National Adaption Plans: Technical guidelines for national adaptation plan process*. Bonn: United Nations Climate Change Secretariat.

UNFCCC (United Nations Framework Convention on Climate Change). (2017). Adaptation planning, implementation and evaluation addressing ecosystems and areas such as water resources. In: *Synthesis report by the secretariat for the subsidiary body for scientific and technological advice*, 46th session, May 8–18, 2017. Bonn: UNFCCC Secretariat.

UNSDN (United Nations Social Development Network). (2016). *Economic inclusion and environmental sustainability: Case study of Meghalaya*. UNSDN Portal. Available online.

Van Asselt, H., Rayner, T., & Persson, Å. (2015). Climate policy integration. In K. Bäckstrand, & E. Lövbrand (Eds.), *Research handbook on climate governance* (pp. 388–399). Cheltenham: Edward Elgar.

Vignola, R., McDaniels, T. L., & Scholz, R. W. (2013). Governance structures for ecosystem-based adaptation: Using policy-network analysis to identify key organizations for bridging information across scales and policy areas. *Environmental Science and Policy, 31*, 71–84.

Wamsler, C. (2014). *Cities, disaster risk and adaptation*. London: Routledge.

Wamsler, C. (2015). Mainstreaming ecosystem-based adaptation: Transformation toward sustainability in urban governance and planning. *Ecology and. Society, 20*(2), 30.

Wamsler, C., & Brink, E. (2014). Adaptive capacity: From coping to sustainable transformation. In S. Eriksen, T. Inderberg, K. O'Brien, & L. Sygna (Eds.), *Climate change adaptation and development: Transforming paradigms and practices* (pp. 54–82). London: Routledge.

Wamsler, C., Brink, E., & Rivera, C. (2013). Planning for climate change in urban areas: From theory to practice. *Journal of Cleaner Production, 50*, 68–81.

Wamsler, C., & Ing, D. (2007). *Operational framework for integrating risk reduction for organisations working in settlement development planning*. Lund: Lund University.

Wamsler, C., Luederitz, C., & Brink, E. (2014). Local levers for change: Mainstreaming ecosystem-based adaptation into municipal planning to foster sustainability transitions. *Global Environmental Change, 29*, 189–201.

Wamsler, C., & Pauleit, S. (2016). Making headway in climate policy mainstreaming and ecosystem-based adaptation: Two pioneering countries, different pathways, one goal. *Climate Change, 137*, 71–87.

Wertz-Kanounnikoff, S., Locatelli, B., Wunder, S., & Brockhaus, M. (2011). Ecosystem-based adaptation to climate change: What scope for payments for environmental services? *Climate and Development*, *3*(2), 143–158.

Wu, J. (2014). Urban ecology and sustainability: The state-of-the-science and future directions. *Landscape and urban planning*, *125*, 209–221.

WWF (World Wide Fund for Nature). (2011). *WWF's living Himalayas initiative factsheet*. Gland: WWF. Available online.

Zborel, T., Holland, B., Thomas, G., Baker, L., Calhoun, K., & Ramaswami, A. (2012). Translating research to policy for sustainable cities. *Journal of Industrial Ecology*, *16*, 786–788.

Further reading

Kumar, A. (2013b). Transboundary basin management cooperation in Himalayan Region of South Asia: Challenges and opportunities. In: *Paper presented at from Rio + 20 to real results: Strengthening of regional cooperation in North and Central Asia in order to improve the efficiency of the water resources*, Conducted under the Auspices of UN-ESCAP in Almaty on November 18–20, 2013.

UNISDR (United Nations International Strategy for Disaster Reduction). (2005). Hyogo framework for action 2005–2015: Building resilience of nations and communities to disasters. In: *World conference on disaster reduction*, Kobe, Hyogo, Japan.

CHAPTER 11

Conclusion

The contemporary world stands at a "zeitenwende"—a German word meaning a turning point of the age—where humankind may find itself on the verge of prime discontinuities in environmental and global affairs that may entail the potential of making the recent past more of an unpredictable guide to developments that humankind is expected to witness in the near future. Humankind along with both biotic and abiotic organisms of the planet Earth have been visited, especially from 2020 onward, by unprecedented global pandemic, now known as COVID-19, and extreme weather and climate disruptions in the forms of superstorms, cyclones, flooding, heatwaves, droughts, and wildfires; fueled by anthropogenic climate change, the cumulative impact of these developments has been instrumental in affecting the lives of innumerable people, destroying their livelihoods and forcing many millions from their homes. The havoc wreaked by the ongoing COVID-19 pandemic since 2020 along with natural disasters and extreme weather events as well as climate disruptions, fueled by anthropogenic climate change, have together reminded the world of its fragility and brought forth the inherent risks of high levels of interdependence in ensuing years and decades, wherein the world is projected to confront more intense and cascading global challenges ranging from disease to climate change to the disruptions from new technologies and fiscal crises, and cumulative impact of these challenges "will repeatedly test the resilience and adaptability of communities, states, and the international system, often exceeding the capacity of existing systems and models. This looming disequilibrium between existing and future challenges and the ability of institutions and systems to respond is likely to grow and produce greater contestation at every level" (NIC, 2021).

11.1 Natural disasters versus human-induced disasters

With the world at the cusp of the inauguration of the third decade into the 21st century, new shapes and sizes are being taken on by disaster risk with every passing year. These disasters, never waiting their turn, entail increased interconnected risks and embrace risk drivers and consequences that are multiplying and cascading, colliding in unexpected ways, in the absence of a commensurate systemic response with national and local strategies for disaster risk reduction appropriate for the purpose. This is attested to by the increasing number of disasters taking place worldwide along with claiming enhanced number of human lives, affecting a greater number of people and causing increased economic losses globally in 2000–19 as compared to 1980–99, as shown in Table 11.1.

It can be discerned from Table 11.1 that the period of two decades between 1980 and 1999 witnessed occurrence of 4212 disasters, claiming 1.19 million human lives and affecting about 3.25 billion people worldwide and the economic losses incurred on account of these disasters amounted

Table 11.1 Disaster impacts: 1980–99 versus 2000–19.

Period	Reported disasters	Total deaths (in millions)	Affected people (in billions)	Economic losses (in US$ trillions)
1980–99	4212	1.19	3.25	1.63
2000–19	7348	1.23	4.03	2.97

From UNDRR (United Nations Office for Disaster Risk Reduction). (2020). Human cost of disasters: An overview of last twenty years, 2000–2019. Geneva, Switzerland: UNDRR.

Table 11.2 Total disaster events by type: 1980–99 versus 2000–19.

Period	Drought	Earthquakes	Extreme Temperatures	Flood	Landslide	Mass movement (dry)	Storm	Volcanic activity	Wildfire
1980–99	263	445	130	1389	254	27	1457	84	163
2000–19	338	552	432	3254	376	13	2043	102	238

From UNDRR (United Nations Office for Disaster Risk Reduction). (2020). Human cost of disasters: An overview of last twenty years, 2000–2019. Geneva, Switzerland: UNDRR.

to US$1.63 trillion. Nevertheless, the number of disasters occurring in the period 2000–19 recorded a phenomenal increase by standing at 7348 along with increase in number of total deaths amounting to 1.23 million as well as an increase in the number of affected people standing at 4.03 billion. Quantum of economic losses incurred during 2000–19 on account of disasters also witnessed an increase by standing at US$2.97 trillion as compared to 1980–99.

A comparative analysis of disaster-related events by type occurring during 1980–99 and 2000–19, as shown in Table 11.2, demonstrates that the number of droughts that stood at 263 in 1980–99 had increased to 338 in 2000–19 and almost similar pattern of increase was witnessed in the occurrence of earthquakes during the period under review.

Nevertheless, phenomenal increase in events pertaining to extreme temperatures was reported in 2000–19 when the number of such occurrences stood at 432 as compared to occurrence of 130 such events in 1980–99. An identical pattern of phenomenal increase was reported in occurrence of floods and storms during the period under review. While an increase in volcanic activity and wildfires was also reported, a diminution was reported in events related to mass movement in 2000–19 as compared to 1980–99. The resultant impacts of the global temperature in 2019 standing at 1.1°C above the preindustrial level are being felt in the enhanced frequency of extreme weather events including heatwaves, droughts, flooding, winter storms, hurricanes and wildfires, despite the improvements made in terms of early warnings, disaster preparedness and response.

Bulk of the disasters, accounting for 79% of the total disasters, occurring worldwide between 1970 and 2019, entailed weather, water, and climate related hazards, and these disasters accounted for 56% of deaths and 75% of overall economic losses accruing from disasters. There has been continuous increase in the percentage of disasters associated with weather, climate, and water related events in recent decades, with the period between 2010 and 2019 recording an increase by 9% in such events compared to the previous decade and by almost 14% with respect to 1991–2000 (WMO, 2020). Small Island Developing States (SIDS) and Least Developed Countries (LDC) have been badly affected due to weather, climate, and water related hazards; and between 1970 and 2019, SIDS have incurred a loss of US$153 billion owing to these events and in the meanwhile 1.4 million people in LDC have lost their lives, thereby accounting for 70% of the total deaths, on account of these events (WMO, 2020).

As available evidence demonstrates that bulk of the disasters are caused by weather, climate, and water related events, water has come to be recognized as the prime factor amply impacting both weather and climate related events. Climate change affects, and in turn also affected by, water resources worldwide. While reducing the predictability of availability of water and affecting water quality, climate change also exacerbates the occurrence of extreme weather events, threatening globally sustainable socioeconomic development and biodiversity. This, in turn, entails serious implications for water resources, and, as such, climate change accelerates the pace of ever-growing challenges linked to the sustainable management of water. On the contrary, the way water is managed impacts the drivers of climate change. Nevertheless, water is the primary medium through which climate change affects ecosystems, human societies, and economies in multiple ways. In certain cases, these impacts are clearly visible, such as through the increasing frequency and intensity of storms, floods, and droughts. Water-related impacts of climate change also embrace negative impacts on food security, human health, energy production, livelihoods, and biodiversity (WWDR, 2020).

The term "water resources" and "aquatic resources" are sometimes used interchangeably; however, the term "water resources" is a genus and the term "aquatic resources" is its species. Aquatic resources comprise those areas where the presence and movement of water is a prime process impacting their development, structure, and functioning; and aquatic resources may include, but are not limited to, vegetated and nonvegetated wetlands or aquatic sites like mudflats, deep-water habitats, lakes, and streams (Law Insider, n.d.). Encompassing nearly 70% of the Earth's surface, aquatic ecosystems support a vast array of ecosystem services. Irrespective of their significance, aquatic ecosystems are becoming increasingly vulnerable to anthropogenic stressors like contaminants and climate change impacts. Increasing interactions between climate change and human activity, such as nutrient enrichment, microplastic and organic pollution extraction, salinization, and catchment modifications, have proved instrumental in rendering aquatic ecosystems increasing vulnerable in recent decades. These numerous stressors and their interactions can sometimes spur dramatic ecological changes across all trophic levels, and even where "critical transition" takes place relatively briskly, they are often the result of multiple, smaller changes over much longer timescales (Roberts, Bishop, & Adams, 2020). Aquatic ecosystems worldwide are increasingly becoming one of the most threatened ecosystems due to overexploitation, water pollution, flow modifications, habitat destruction or degradation, and invasion of exotic species, and as a result of which, declines in aquatic biodiversity are occurring at a faster scale than those in the most affected terrestrial ecosystems. As biodiversity is linked to ecosystem services and the resilience of ecosystem to environmental change, fluctuations or alterations in biodiversity entail the potential of having significant consequences for the services that are derived by humans from ecosystems (Forio & Goethals, 2020).

This brings us to have a brief overview of the current status of and future outlook for global warming, biodiversity, food security, water security, sustainable cities, sustaining oceans, global

650 Chapter 11 Conclusion

climate change governance, and impact of COVID-19 and finally assess the future prospects of ecosystem-based adaptation (EbA).

11.2 Tackling global warming

IPCC (2013) had predicted that global mean temperatures will continue to rise over the 21st century if greenhouse gas (GHG) emissions continue unabated, and the global mean temperature for 2020 was $1.2°C \pm 0.1°C$ above the 1850–1900 baseline (shown in Fig. 11.1), and that places 2020 as one of the three warmest years on record worldwide. This assessment of World Meteorological Organization is based on five global temperature datasets—HardCRUT analysis, NOAA Global Temp, GISTEMP, ERA5, and JRA-55—as indicated in Fig. 11.1. All five of these datasets have placed 2020 as one of the three warmest years on record, and the expanse of the five estimates of the annual mean varies between $1.15°C$ and $1.28°C$ above preindustrial levels.

Nevertheless, Paris Agreement on Climate Change emphasizes on holding the global average temperature to well below $2°C$ above preindustrial levels and to embark on endeavors to limit the temperature increase to $1.5°C$ above preindustrial levels (UNFCCC, 2015). The year 2016 is regarded as the warmest year on record to date, which began with an extraordinarily strong El Nino, a phenomenon contributing to elevated global temperatures, and irrespective of neutral or

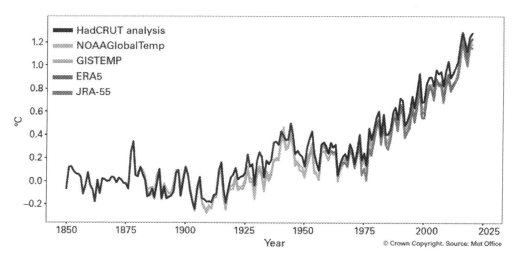

FIGURE 11.1

Global annual mean temperature difference from preindustrial conditions (1850–1900) for five global temperature datasets.

Source: Adapted from WMO (World Meteorological Organization) (2021). State of the global climate 2020: Unpacking the indicators. Geneva, Switzerland: WMO. Available online from https://public.wmo.int/en/our-mandate/climate/wmo-statement-state-of-global-climate.

comparatively weak El Nino conditions early in 2020 and La Nina conditions developing by late September 2020, the warmth of 2020 was comparable to that of 2016. Undoubtedly, the overall warmth of 2020 is evident; nonetheless, variations in temperature anomalies across the globe were also visible, wherein most land areas were reported warmer than the long-term average (1981–2010), and one such area in northern Eurasia recorded temperatures of more than 5°C above average. Other significant areas, inter alia, included few areas of south-western United States, the northern and western parts of South America, parts of Central America, and wider areas of Eurasia, inclusive of some parts of China. And for Europe, 2020 has been the warmest year on record. As far as areas recording below average temperatures in 2020 are concerned, these included western Canada, limited areas of Brazil, northern India, and south-eastern Australia (WMO, 2021).

As far as warmth over the ocean is concerned, unusual warmth became discernible during 2020 in parts of the tropical Atlantic and Indian Oceans. The pattern of sea-surface temperature anomalies in the Pacific—being characteristic of La Nina, entailing cooler-than-average surface waters in the eastern equatorial Pacific surrounded by a horseshoe-shaped band of warmer-than-average waters, was observed in 2020 most notably in the North-East Pacific from Japan to Papua New Guinea (WMO, 2021).

11.2.1 Greenhouse gases

Global warming, both on land and over the oceans, is primarily triggered by emissions of GHGs. Atmospheric concentrations of GHG mirror an equilibrium between emission from anthropogenic activities and natural sources, and sinks in the biosphere and ocean. Augmented levels of concentration of GHGs in the atmosphere primarily on account of human activities have been the prime driver of climate change since the mid-20th century. GHG concentration in 2019 attained new highs, as shown in Fig. 11.2, with worldwide average mole fractions of carbon dioxide (CO_2) at 410.5 ± 0.2 parts per million (ppm), methane (CH_4) at 1877 ± 2 parts per billion (ppb) and nitrous oxide (N_2O) at 332.0 ± 0.1 ppb, respectively, 148%, 260%, and 123% of preindustrial levels. The enhancement in CO_2 from 2018 to 2019 (2.6 ppm) was higher that the increase from 2017 to 2018, being at 2.3 ppm, and the average annual increase over the past decade, being 2.37 per year. For methane (CH_4), the increase from 2018 to 2019 was mildly lower than the increase from 2017 to 2018, yet still higher than the average annual increase over the past decade (WMO, 2021).

Top row: Globally averaged mole fraction (measure of concentration), from 1984 to 2019, of CO_2 in ppm (left), CH_4 in ppb (center) and N_2O in ppb (right). The dark black line is the monthly mean mole fraction with the seasonal variations removed; the blue dotted line shows the monthly averages. Bottom row: The growth rates representing increases in successive annual means of mole fractions are shown as gray columns for CO_2 in ppm per year (left), CH_4 in ppb per year (center), and N_2O in ppb per year (right). In the case of nitrous oxide (N_2O), the increase from 2018 to 2019 was also lower than that discerned from 2017 to 2018 and proximate to the average growth rate over the previous decade.

The unabated pace of rise in temperature in excess of 3°C during the 21st century—an increase far beyond the Paris Agreement goals of limiting warming to well below 2°C and pursuing 1.5°C was being construed in 2019 as heading for a catastrophe. However, in the wake of outbreak of COVID-19 global pandemic in early 2020, adoption of green recovery measures, to some extent, had given the hope that the world was nearing close to pathway to 2°C and growing commitments

652 Chapter 11 Conclusion

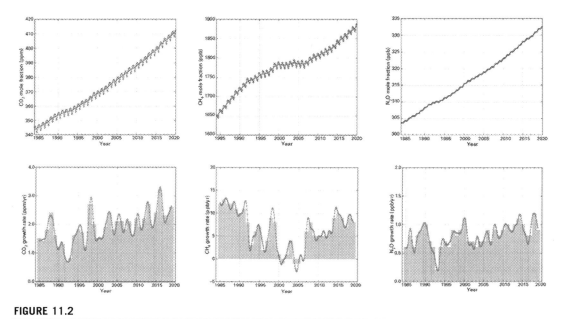

FIGURE 11.2

Greenhouse gases in the atmosphere.

Source: *From WMO (World Meteorological Organization) (2021).* State of the global climate 2020: Unpacking the indicators. *Geneva, Switzerland: WMO. Available online from https://public.wmo.int/en/our-mandate/climate/wmo-statement-state-of-global-climate.*

to net-zero emissions by 2050 (UNEP, 2020a). Nevertheless, "temporary" reduction in emissions in 2020 is likely to lead to only a slight decrease in the yearly growth rate of CO_2 concentration in the atmosphere and that will be practically indistinguishable from the natural interannual variability spurred primarily by terrestrial biosphere and real-time data from specific locations indicate that levels of CO_2, CH_4, and N_2O continued to increase · in 2020 (WMO, 2021).

It has been explicitly made clear by the IPCC Special Report on Global Warming of 1.5°C that limiting warming to 1.5°C above preindustrial levels implies reaching net-zero CO_2 emissions globally by around 2050, with concurrent deep · reductions in emissions of non-CO_2 forcers (IPCC, 2018). This requires strong political will, especially on the part of major global emitters like the United States, China, India, Russia, and other countries to render cooperation and adhere to the commitments envisaged in the Paris Agreement in letter and spirit. Undoubtedly, return of the United States under Biden Administration to the Paris Agreement and Biden Administration's announcement of reducing GHG emission at least by 50% by 2030 can be seen as positive developments; nevertheless, the growing geopolitical and geo-economic rivalry between the United States and China on the one hand, and between the United States and Russia on the other hand, outbreak of the second wave of COVID-19 from March 2021 onward, with India and Brazil being worst-affected countries in particular along with many other countries worldwide, realization of the goals of the Paris Agreement along with sustainable development goals (SDGs) by 2030 seems somewhat an improbable proposition.

11.3 Global outlook for biodiversity

With around one million plant and animal species being threatened with extinction, many within the decades, more than ever before in human history (IPBES, 2019), and none of the Aichi Biodiversity Targets achieved (CBD, 2020a), future outlook in coming decades seems very bleak, unless concerted efforts are not adhered to urgently at every level from local to global to conserve, restore, and use nature sustainably. Substantial decline and change occurring at unprecedented scales in the richness, abundance, composition, and distribution of the Earth's biological biodiversity has triggered a sort of "Biodiversity Crisis" of immense magnitude (Fajardo et al., 2021) (Fig. 11.3).

Over the years, the Earth has seemingly continued its trajectory toward declining biodiversity primarily due to global failures to reach the Aichi Biodiversity Targets on overall lack of investments, resources, knowledge, and accountability toward biodiversity conservation. The national goals adopted in each participating country, member of the CBD, didn't always align with the Aichi targets, and the resultant sum of national successes has proved insufficient to reach the overall Aichi targets by the stipulated period of 2020 (Xu et al., 2021). Having realized in 2020 that most of the Aichi targets of the Strategic Plan for Biodiversity 2011–20 were not on track to be achieved by 2020 and the fact that this failure could jeopardize achievement of the SDGs and ultimately the planet's life-support system, the global community under the auspices of the CBD started focusing on the successes and failures in the context of the implementation of the Strategic Plan and negotiating global biodiversity framework for the post-2020 era.

Accordingly, it was decided to convene the 2020 UN Biodiversity Conference to be held in 2020 in Kunming, China. However, this plan has got disrupted in the wake of outbreak of

FIGURE 11.3

Rain forests. Rainforest in Borneo (Indonesia).

Source: Wikimedia Commons.

COVID-19 global pandemic in 2020 and surge in its second wave from March 2021 onward, while online consultations are continuing. The fifteenth meeting of the COnference of Parties of the CBD has been postponed at least till the end of 2021 and it was perhaps expected to adopt a framework and set the level of ambition for negotiating the same globally for the next decade; nevertheless, one may see the proposed meeting being at crossroads, and failure to adopt a new framework—and implement it—may lead to irreversible loss in the near future (Tsioumani & Dsioumanis, 2020).

Subsequently, in early January 2020, the CBD Secretariat released the zero draft of the post-2020 global biodiversity framework that builds on the Strategic Plan for Biodiversity 2011–20 and aims at bringing about a "transformation" in society's relationship with biodiversity and to ensure that, by 2050, the shared vision of living in harmony with nature is fulfilled. In August 2020 the CBD published an updated zero draft of the post-2020 global biodiversity framework that contains reworded targets and updated language. The draft framework presents five long-term goals for 2050 related to the CBD's 2050 Vision for Biodiversity, and each of these goals entails an associated outcome for 2030. The five goals address net loss and ecosystem resilience, reductions in the percentage of species threatened with extinction, the maintenance and enhancement of genetic diversity, the benefits of nature to people, and the fair and equitable sharing of the benefits from the use of genetic resources and associated genetic knowledge (CBD, 2020b). A total of 20 action-oriented targets are incorporated in the draft framework and these are designed to contribute to achieving those five goals, and these 20 targets are presented under the categories of reducing threats to biodiversity, meeting people's needs through sustainable use and benefit-sharing, and tools and solutions for implementation and mainstreaming. A preliminary list of indicators in the proposed draft framework may be used to assess progress toward the goals and targets. It considers elements of guidance on goals, SMART targets, indicators, baselines, and monitoring frameworks, relating to the drivers of biodiversity loss and for achieving transformational change.

As per proposed draft, the new framework will be based on "a theory of change" with elements of resource mobilization, mainstreaming, digital sequence information, sustainable use capacity-building, national planning, reporting processes, issues associated with responsibility and transparency, and indicators. The framework's theory of change assumes that transformative actions are taken to put in place tools and solutions for implementation and mainstreaming, reducing the threats to biodiversity, and ensuring that biodiversity is used sustainably in order to meet people's needs. And these are actions are supported by enabling conditions and adequate means of implementation, including financial resources, capacity, and technology. Fig. 11.4 shows this theory of change.

The theory of change is complimentary to and supportive of the 2030 Agenda for Sustainable Development. It also takes into account the long-term strategies and targets of other agreements, including the biodiversity-related and Rio Conventions, to ensure synergistic delivery of benefits from all the agreements for the planet and people (CBD, 2020b).

The new framework is expected to be implemented using a rights-based approach and recognizing the principle of intergenerational equity. The theory of change for the framework acknowledges the need for appropriate recognition of gender equality, women's empowerment, youth, gender-responsive approaches, and the full and effective participation of indigenous peoples and local communities in the implementation of this framework. It is to be seen whether the draft framework will be adopted and approved in its existing shape or with some modifications by the 15th meeting of the Conference of Parties to the CBD as and when it is held in the near future.

11.3 Global outlook for biodiversity

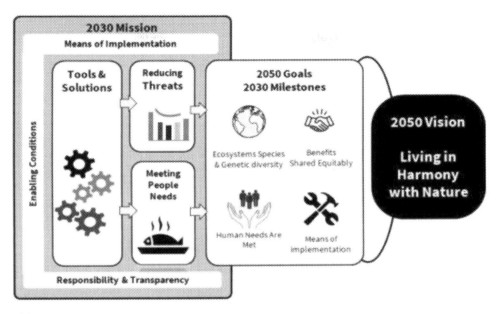

FIGURE 11.4

Theory of change as contemplated in post-2020 global biodiversity framework.

Source: *CBD (Convention on Biological Diversity). (2020b).* Update of the zero draft of the post-2020 global biodiversity framework. CBD/POST2020/PREP/2/1. Montreal, Canada: CBD Secretariat.

International Union for Conservation of Nature (IUCN), while stating its position on the post-2020 Zero Draft Biodiversity Framework has suggested that the new draft, while avoiding duplication, should enhance complementarity with existing frameworks, in particular the 2030 Agenda for Sustainable Development. While emphasizing that new framework ought to be structured to reflect the pathway from where we are now to the changes, we'd like to see in 2050, IUCN wants the new framework to have focused, concrete, and measurable targets (IUCN, 2021). While expecting the new framework to be a truly global framework, IUCN notes that the new framework should reflect the CBD objectives as well as three components of biodiversity—species, ecosystems, and genes—in coherent specific outcome goals. While emphasizing that the new framework should integrate nature-based solutions to safeguard and maintain ecosystems, IUCN posits that all voices need to embraced in the new framework so that indigenous people and local communities, regional and city governments, the private sector, NGOs, women, youth, and society at large must not only be invited to debate but the framework should also incentivize their explicit contributions toward the global goals (IUCN, 2021).

While identifying an array of existing and proposed post-2020 biodiversity targets that risk being severely compromised due to climate change, even if other barriers to their achievement were removed, Arneth et al. (2020) suggest that the next set of biodiversity targets should explicitly address climate change-related risks since many aspirational goals will not be feasible under even lower-end projections of future warming; and adopting more flexible and dynamic approaches to conservation, rather than static goals, would allow us to respond flexibly to changes in habitats, genetic resources, species composition, and ecosystem functioning and leverage biodiversity's capacity to contribute to climate change mitigation and adaptation.

11.4 Global outlook for life on Earth

Global outlook for life on the Earth at current juncture seems bleak, especially at a time when the world is reeling from the deepest global disruption and health crisis of a lifetime in the wake of available unequivocal and alarming evidence that nature is unraveling and that planet earth is flashing red warning signs of vital natural systemic failure due to fast declining pace of diversity owing to humanity's increasing dstruction of nature, and all these cumulatively are having catastrophic impacts on almost all aspects of human lives (WWF, 2020). Life on the Earth is abundantly represented by freshwater ecosystems, forests, mountain ecosystems, land use, genetic resources, and wildlife.

Freshwater ecosystems extend direct support to terrestrial ecosystems, like mountain ecosystems and forests, large marine ecosystems, and coastal zones; and as such freshwater ecosystems are indispensable for sustainable development and human well-being. Besides, freshwater ecosystems play a pivotal role for the health of economies and societies globally. Concurrently, freshwater ecosystems are also confronted with threats such as enhanced levels of pollution, rapid pace of urbanization and industrialization, rising food and energy production, water-related disasters, and displacement of people (UNEP, 2017) (Fig. 11.5).

Freshwater habitats, including rivers, streams, lakes, and other wetlands are among the most threatened ecosystems on the Earth, with their biodiversity declining far faster than that of oceans

FIGURE 11.5

Aquatic ecosystem.

Source: *Image courtesy: Wikimedia Commons.*

or forests. With almost 90% of global wetlands have been lost, and of the 3741 monitored populations, representing 944 species of mammals, birds, amphibians, reptiles, and fishes, have registered decline by an average of 84% and most these declines are seen in freshwater amphibians, reptiles, and fishes; and they are recorded across all regions, especially Latin America and Caribbean (WWF, 2020). Freshwater ecosystems are significant components of the global carbon cycle, such that under the influence of global change factors organic carbon burial decreased and mineralization rate increased. Climate change poses a grave threat to different freshwater ecosystems by raising water temperatures, changing precipitation patterns and enhancing water flow conditions, enhancing species invasion, and increasing extreme events (Peng, Biao, Yi-Xuan, Qiu-Jin, & Hong-Xiu, 2020).

In order to deal with the threats to freshwater ecosystems, United Nations Environment Program (UNEP) in 2017 came out with Freshwater Strategy (2017–21) providing actionable guidance to support countries' implementation of sustainable freshwater management practices globally. Built on several concept and pillars, this strategy, while emphasizing on the essentiality of freshwater to the environment as a whole, underpins adherence to effective ecosystem-based management and EbA. The Freshwater Strategy (2017–21) was expected to provide global leadership, contributing to the topics of immediate and pressing concerns and actively following other closely related processes. To be operationalized through a combination of ongoing and new initiatives in support of UNEP member countries, the strategy also defined an array of key areas with example activities including direct provision of expertise, development and dissemination of tools and technologies, and a range of awareness and knowledge sharing efforts (UNEP, 2017). The duration of this strategy is to expire by the end of 2021 and almost nothing is known about the successes or failures of this strategy and it is only after the UNEP reviews the position and comes out with some fresh strategy, In the meanwhile, freshwater ecosystems are likely to continue to face the problems.

Encompassing 31% of the global land area that is not equally distributed around the globe, forests harbor bulk of the Earth's terrestrial biodiversity implying that global biodiversity is extremely dependent on the way human interact with and use the world's forests. Apart from providing habitats for 80% of amphibian species, 75% of bird species, and 68% of mammal species, forests are also home to about 60% of vascular plants, especially in tropical forests. Besides, mangroves and nurseries for numerous species of fish and shellfish (FAO, 2020a, 2020b). Deforestation, degradation of forest land, droughts, and forest fires have continued to significantly contribute to loss of forests, and the resultant loss of biodiversity. Approximately, some 420 million hectares of forest have been lost since 1990 through conversion of two other land uses; nevertheless, the rate of deforestation has decreased between 2015 and 2020, which was estimated at 10 million hectares per annum, down from 16 million hectares per year in the 1990s. There has been increase in the number of wild and forest fire incidents over the years. During 2000–19 there occurred 238 incidents of wild and forest fires worldwide as compared to 163 such incidents occurring during 1980–99 (UNDRR, 2020) (Fig. 11.6).

In April 2017 UN Strategic Plan for Forests, 2017–30 was adopted and it featured a set of six goals and 26 associated targets to be achieved by 2030. Building on the vision of Agenda 2030 and recognizing that real change requires decisive, collective action, within and beyond the UN System, it also included a target to increase forests area by 3% by 2030, signifying an increase of 120 million hectares. In April 2021 the first evaluation of two where the world stands in implementing the UN Strategic Plan for Forests 2030 was presented in the Global Forest Goals Report 2021

FIGURE 11.6

Forests.

Source: *Wikimedia Commons.*

(UN-DESA, 2021). While providing a snapshot of actions being taken for forests, this report emphasizes that it is essential to meet the deadline 2030. While noting that the world had been making progress in key areas such as increasing global forest area via afforestation and restoration, these advances are also under threat from the deteriorating state of natural environment. Taking note of the impacts of COVID-19 pandemic that has proved instrumental in aggravating the challenges facing the countries in managing forests, the report estimates that world gross product fell by an estimated 4.3% in 2020, the sharpest contraction of global output since the Great Depress of the 1930s, the report informs more than just a health crisis, COVID-19 is driving losses of lives and livelihoods, extreme poverty, inequality, and food insecurity. Pointing out that forests and forest-dependent people are both a casualty and an important part of solution for the resilient recovery from the COVID-19 recovery along with a response to the climate and biodiversity losses that rooted in the forests, the report suggests that safeguarding and restoring forests are among the environmental actions that can reduce the risks of future zoonotic disease outbreaks (UN-DESA, 2021).

Global land use plays a significant role in impacting life on the Earth. Global land resources are currently under greater pressures than at any other time in human history, and the burgeoning population, coupled with growing levels of consumption, is placing ever-larger demands on these finite land resources, and the resultant impact is accelerated competition among land uses and its provisioning of goods and services. The rapid pace of land degradation, droughts, and desertification are huge pressures on land. There has been increase in occurrence of droughts in recent years, and during 2000–19 the number of droughts taking place stood at 338 as compared to 268 drought that took place during 1980–99 (UNDRR, 2020) (Fig. 11.7).

According to IPCC (2019a), risks from desertification are projected to get augmented due to climate change and the number of dry land population vulnerable to multiple impacts related to water,

FIGURE 11.7

Drought dry riverbed in California.

Source: *Wikipedia Commons.*

energy, and land sectors like water stress, drought intensity, and habitat degradation is also projected to increase. About half of the vulnerable population resides in South Asia, followed by Central Asia, West Africa, and East Asia. Conceding that land is finite in quantity, the first edition of Global Land Outlook (UNCCD, 2017) suggests that with changes in consumer and corporate behavior and the adoption of more efficient planning and sustainable practices, there will be sufficient land available in the long-term to meet both the demand for essentials and the need for a wide range of goods and services. The Global Land Outlook has highlighted six response pathways—multifunctional landscape approach, resilience building, farming for multiple benefits, managing the rural−urban interface, no net loss, and creating an enabling environment—and these pathways can be followed by producers and consumers, governments, and corporations to stabilize and reduce pressures on land resources. The second edition of Global Land Outlook is likely to be brought out by the UNCCD sometime in 2021.

Mountains comprise another important component of life on the Earth. Encompassing about a quarter of the Earth's land surface, mountains constitute a significant source of freshwater, centers of biological and cultural diversity along with traditional and indigenous knowledge. Being home to a quarter to global population, high mountains consist of all mountain regions where glaciers, snow, or permafrost are important characteristics of the landscape, and by supplying freshwater to half of humankind, mountains are often referred to as water towers of the world. According to IPCC (2019b), current trends in cryosphere-related changes in high mountain ecosystems are expected to continue and their impacts to intensify. Decline in snow cover, glaciers, and permafrost is projected to continue in almost all regions throughout the 21st century, thereby emphasizing the urgency for addressing the hydro-climatic changes in high mountains, their impacts and their downstream effects. Lamenting that the available scientific evidence is very heterogeneous across and within mountain regions and where new observations, learning, thinking, and experiences are acquired by international research initiatives with limited or no engagements with local scientific or operational communities, Adler, Pomeroy, and Nitu (2020) suggest the need for substantially strengthening the scientific understanding of social-ecological systems in high mountains. They also emphasize the need for amplified knowledge of the ecosystem services and goods provided by the cryosphere and other critical systems in mountain regions and their human uses.

Wildlife, comprising both plants and animals, forms another important constituent of life on the Earth. As part of the world's ecosystems, wildlife provides balance and stability to nature's processes. Burgeoning human population and the resultant increasing levels of consumption are exerting additional pressures on wildlife and animal resources and some of these being pushed on the brink of extinction. The introduction of alien species, climate change, pollution, and poaching are other threats to wildlife. While there exist appropriate institutional mechanism at international level to deal with wildlife-related issues to ensure conservation and safety of wildlife; nevertheless, trafficking and unsustainable trade in wildlife commodities have continued unabated thereby causing unprecedented declines in some of the world's rare and lesser-known species, and value of illegal wildlife trade is estimated between US$7 and US$23 billion annually, making wildlife illegal trade as the fourth largest illegal trade (Rosen, 2020).

Illegal trade in wildlife entails the potential of leading to the spread of zoonoses, such as SARS-CoV-2 that caused the COVID-19 pandemic. In the wake of spread of COVID-19 in 2020, calls to shut down wildlife markets led to further proliferation of illegal wildlife trade. New challenges emerged for The Convention on International Trade in Endangered Species of Wild fauna and Flora (CITES) that is the only international convention addressing legal and illegal trade in wild species of wildlife flora and fauna. In the wake of spread of COVID-19 pandemic, the response of CITES has been that matters pertaining to zoonotic diseases are outside the mandate of CITES and as such it has no competence "to make comments regarding the recent news on the possible link between human consumption of wild animals and COVID-19" (CITES, 2020). While focusing on a new treaty to curb illegal wildlife trade, it is concurrently emphasized that "new treaty" should address "One Health," the human health, veterinary, and wildlife issues and the relationship with wildlife trade. According to Rosen (2020), the first step would be to consider if CITES can further adapt to better curb wildlife trade and address the challenges brought by COVID-19, and prior to a new treaty is considered, "countries are more likely to explore what CITES can offer in the meantime" (Rosen, 2020).

Overall outlook for life on the Earth does not seem so good in coming years, especially in the wake of disruptions caused by the global outbreak of COVID-19 pandemic in 2020 and its continuing surge in some countries in 2021, and these disruptions in socioeconomic, political, health, and economic sectors have not only diverted the focus from addressing climate-related issues, albeit resources meant for addressing climate-related issues have to diverted to deal with the pandemic.

11.5 Global outlook for life below water

Life below water is mainly concerned with existing conditions of the oceans and seas wherein the marine biodiversity, marine ecology, and the marine organisms are impacted by ocean warming,

11.5 Global outlook for life below water

sea-level rise, acidification, deoxygenation, and marine pollution from plastics and land-based sources. Oceans are known sources of sinks for carbon dioxide, and as bulk of the excess of energy that gets accumulated in the Earth system owing to the increasing concentration of GHGs is taken up by the ocean and this addition energy warms the ocean and the resultant thermal expansion of the water thereafter leading to sea-level rise. Concentration of CO_2 in the oceans leads to impacting ocean chemistry, lowering the average pH of the water and this process is known as ocean acidification, and the cumulative impact of these changes wields a broad range of effects in the open ocean and coastal areas. Ocean heat content (OHC) is a measure of heat accumulation in the Earth system as about 90% of it is stored in the ocean. A positive Earth Energy Imbalance—that is driving global warming through an accumulation of energy in the form of heat in the Earth system—indicates that the Earth's climate system is still responding to the current forcing and that there is likelihood of occurrence of more warming even if the forcing does not record any further increase (WMO, 2021).

It is now possible to measure OHC changes to a depth of 2000 m with the deployment of Argo network of the profiling floats. In 2019 the 0−2000−m depth layer of the global ocean continued to warm, attaining a new record high and it is expected that it will continue to warm in the future. According to WMO (2021), a preliminary analysis based on three datasets suggests that 2020 exceeded that record, with heat storage at intermediate depth (700−2000 m) increased at a comparable rate to the rate of heat storage. All datasets relied upon by WMO (2021) agree that ocean warming rates demonstrate a particular strong increase over the past two decades.

Rise in global sea level is attributable mostly to a combination of meltwater from glaciers and ice sheets and thermal expansion of seawater as it warms. Global mean sea level has risen about 21−24 cm since 1880, with about a third of that coming in the last two and a half decades. Since early 1993 the altimetry-based global mean rate of sea-level rise on average has accounted for 3.3 ± 0.3 mm/year. In 2019 global mean level was 86.7 mm above the 1993 average—the highest average in the satellite record (Lindsey, 2021). Global mean sea level continued to rise in 2020. The sea level continued to rise in 2020 nonuniformly at the regional scale. The strongest regional trends over the period—from January 1993 to June 2020—were visible in the southern hemisphere: east of Madagascar in the Indian Ocean, east of New Zealand in the Pacific Ocean, and east of Rio de la Plata/South America in the South Atlantic Ocean. Regional seal-level tends are dominated by variations in OHC (WMO, 2021).

The near-surface layer of the oceans gets affected by extreme heat and in common parlance, this situation is called a marine heatwave (MHW). MHWs are extreme climatic events that entail the potential of having devastating impacts on ecosystems, causing abrupt ecological changes and socioeconomic consequences. Satellite retrievals of sea-surface temperature can be used to monitor MHWs, and an MHW is categorized as moderate, strong, severe, or extreme. According to WMO (2021), much of the ocean experienced at least one "strong" MHW at some point in 2020; with MHWs absent in the Atlantic Ocean south of Greenland and in the eastern equatorial Pacific Ocean; the Laptev Sea experiencing a particularly intense MHW from June to December 2020, with sea-ice extent being usually low in that region, and adjacent land areas experiencing heatwaves during the summer. Nearly, one-fifth of the global ocean was experiencing an MHW on any given day in 2020.

Approximately 23% of the annual emissions of anthropogenic carbon dioxide into the atmosphere is absorbed by the ocean and it helps in minimizing the impacts of climate change.

Nevertheless, absorption of CO_2 by the ocean facilitates interaction with seawater leading to lowering its pH and this process, known as ocean acidification, impacts many marine organisms and ecosystem services, threatening food security by endangering fisheries and aquaculture. Apart from being a specific problem in the polar oceans, it also affects coastal protection by enfeebling coral reefs that shield coastlines. A decline in the pH of the ocean affects the capacity of the ocean to absorb CO_2 from the atmosphere, culminating in diminishing the capacity of the ocean to moderate climate change. Efforts at the global level have been underway to collect and compare ocean acidification observation data and these show an increase of variability with minimum and maximum pH values are highlighted, and a decline in average pH at the available observing sites between 2015 and 2019. The increase in the amount of available data highlights the variability and the trend in ocean acidification, as well as the need for sustained long-term observations to better characterize the natural variability in ocean carbonate chemistry (WMO, 2021). According to European Environment Agency, the average surface open ocean pH is projected to decline further in the range of 0.04–0.29 pH units by 2081–2100 relative to 2006–2015, depending on future CO_2 emissions (EEA, 2021).

Deoxygenation means loss of oxygen, and since 1950 the open ocean oxygen content has registered decrease by 0.5%–3%, and oxygen minimum zones that are permanent features of the open ocean are expanding; nevertheless, the trend of deoxygenation in the global coastal ocean is still uncertain (WMO, 2021). Recent decades have witnessed increase in the number of hypoxic sites in the global coastal ocean in response to worldwide eutrophication. However, IUCN (2019) projections predict that there will be distinct regional differences in the intensity of oxygen loss as well as variations in ecological and biogeochemical impacts. While reporting consensus across models that oxygen loss at mid and high latitudes will be strong and driven by both reduced solubility and increased respiration effects, projections are more ambiguous in the tropics where models suggest that there will be compensation between oxygen decline owing to reduced solubility and oxygen increase caused by reductions in cumulative respiration (IUCN, 2019). Suggesting that an exhaustive assessment of deoxygenation in the open and coastal ocean would benefit from building a consistent, quality-controlled, open-access global ocean oxygen dataset and atlas complying with FAIR data principles, WMO (2021) notes that an effort in this regard has been initiated by the Global Ocean Oxygen Network (GO$_2$NE), the International Ocean Carbon Coordination Project (IOCCP), the National Oceanic and Atmospheric Administration (ONAA) and the German Collaborative Research Center 754 (SFB, 754).

Marine pollution has emerged as a grave threat to the oceans. Bulk of the marine pollution, around 80%, originates from land-based pollution and that, inter alia, includes untreated sewage, agricultural runoff, oils and heavy metals from industry, and sediment washed in from earthworks and logging. Coastal areas are specifically vulnerable to marine pollution. Most of the coastal areas worldwide abound with coral reefs, mangroves, seagrass, meadows, salt marshes, and tidal flats and all of these are complex ecosystems that support food production and livelihoods. Oceans and coasts provide significant ecosystem services, such as regulating climate, maintaining the food web, and enabling marine transportation. The value of maritime ecosystem services is estimated about US$2.5 trillion per annum, thus making it the world's seventh largest economy (UNEP-GEF, 2018). And degradation of coastal and ocean environment is estimates somewhat to cost US$250940 billion yearly in the decline of fisheries, storm damage in coastal areas, loss of revenue earnings from tourism, and other impacts (UNEP-GEF, 2018). Land-based sources of marine pollution are a huge and complex suite of problems that no overreaching regulatory or programmatic approach can properly address and global responses have become more specific and specialized overtime, focusing on specific pollutants, like the 2001 Stockholm Convention on POPs and the 2013 Minamata Convention on Mercury (Paul, 2021).

In 1995 Global Program of Action for the Protection of the Marine Environment from Land-based Activities (GPA) was launched as a unique intergovernmental mechanism to deal with land-based polluting activities by the member countries party to GPA. In February 2017 UNEP launched the Clean Sea campaign that has succeeded in garnering support and commitments from 60 countries for actions like establishing national recycling facilities and facilitating the use of single-use plastics. In January 2019 GPA in collaboration with UNEP launched a five-year project to coordinate action around wastewater pollution, nutrient management, and marine litter. This project identified nine source categories of marine and coastal pollution: sewage, persistent organic pollutants (POPs), radioactive substances, heavy metals, oils (hydrocarbons), nutrients, sediment mobilization, litter, and physical alteration and destruction of habitat. In order to address these pollution sources, it proposed a logical sequence of problem assessment, priority setting, management strategies, evaluation, and financing. Establishment of a clearing house mechanism was proposed to improve access to information and expertise on each of the source categories. In the wake of global responses having become more specialized over the years, focusing on specific pollutants, there have been changes in the policy architecture that have proved instrumental in decreasing the relevance of the GPA and "have likely reduced the effectiveness of the GPA as a global program, as energy and donor interest flows to these new instruments" (Paul, 2021).

Marine litter has emerged as one of the gravest threats to marine environment and ecology. There are many types of marine litter, including manufactured materials such as plastics, paper, and used wood that end up in the marine environment. Apart from being a major source of marine pollution, marine litter is also threatening the livelihoods of over 3 billion people worldwide who are dependent on the marine environment, the shipping sector, telecommunication undersea cables, rescue missions, tourism, marine wildlife, and food chains. In terms of environment damage, the cost of marine litter is estimated to be nearly US$13 billion per annum (Abalansa, Mahrad, Vondolia, Icely, & Newton, 2020). Among the many different types of marine litter, plastic has come to be regarded as the most harmful to the marine environment and marine wildlife. Plastics comprise macro, micro, mega, meso, and nano-plastics. Availability of plastics in abundance as a persistent type of litter, accounting for 60%−80% of the litter in the marine environment is attributable to multiple factors like its multipurpose usage and resultant high consumption rate, its physical features like its durability and corrosion resistance, and low recycling rates and poor waste management practices on the part of many countries. The most worrisome aspect of plastic is that it persists in the environment for hundreds of years and breaks down into the microplastics that find their way into the food chain (Dauvergne, 2018). Thus marine plastic litter has become a global issue.

With more than 8 billion tons of plastic in circulation worldwide and more plastic produced each day, a dire need for a multipronged approach is felt to address this mounting crisis of marine plastic litter. The 1989 Basel Convention on the Control of Transboundary Movements of hazardous Wastes and Their Disposal took the first major legally binding action to control plastic waste. In 2019 annexes to the 1989 Basel Convention were amended to control the transboundary movement of plastic waste, classifying certain plastic as hazardous and subjecting their movement to the prior informed consent procedure and effective from January 2021, any state exporting certain types of plastic waste will require the importing state's prior informed consent and countries will have right to refuse the importation of any such waste outright (Kantai, 2020). The Basel Convention also established a plastic waste partnership to minimize the generation of Kentia plastic waste along with promoting its environmentally sound management. These steps initiated under the Basel Convention are expected to encourage recycling of plastic waste, with the implicit hope that more recycled plastic in the supply chain will discourage fresh plastic production.

UNEP has also been seized of the issue of plastic pollution and in 1995, UNEP established the Global Program of Action on the Protection of the Marine Environment from Land-based Activities (GPA) and over the years was the launch of the Global Partnership on Marine Litter in 2012; nonetheless, in the wake of mounting concerns worldwide about the increasing quantum of plastic litter in environment, UNEP in 2017 established an expert group on marine litter and microplastics and the expert group took into consideration a variety of options to control marine litter, including negotiating a new treaty on plastic pollution or ramping up existing voluntary measures like G-20's Osaka Blue Ocean Vision and G-7's Ocean Plastic Charter. Nonetheless, the limitation of these existing measures is that these comprise only a small group of countries belonging to G-20 group or G-7 group, thereby leaving behind a vast majority of countries. On the other hand, a new treaty would be characterized by its inclusivity owing its global nature wherein every country in the world would have an option to join global action against plastic pollution. The UNEP expert group is expected to present its report soon and many hope that this will initiate an intergovernmental negotiating process toward a universal and inclusive treaty on plastic pollution (Kantai, 2020). In the wake of the outbreak of COVID-19 global pandemic, plastic pollution has increased worldwide due to the increased demand for personal protective equipment (PPE) like masks and gloves. Many countries have recoded an increase in plastic packaging for food as food providers distribute food to allay fears around food safety as it relates to the pandemic. Undoubtedly, the pandemic has slowed down the gains registered over the past few years to manage plastic pollution. Since the second wave of the COVID-19 has surged in many countries in the first half of 2021, the problem of plastic pollution is likely to receive scant attention for the time being. In the meanwhile, progress toward negotiating a global treaty on plastic pollution can be continued.

Overfishing is another problematic issue that is worrisome for the sustainable management of marine environment and ecology because overfishing is one of the most significant drivers of declines in ocean wildlife populations. In common parlance, overfishing takes place when fish is caught faster than stock can replenish. Overfishing is closely tied to bycatch, the capture of unwanted sea life, while fishing for a different species, and this a portent marine threat that engenders the unnecessary loss of billions of fish along with innumerable sea turtles and cetaceans. By 2018 global fish production was estimated to have reached around 179 million tons, with a total fish sale value estimated at US$ 401 billion, and of which 82 million tons came from aquaculture production (FAO, 2020a, 2020b). Undoubtedly, fish production is projected to increase to 204 million tons in 2030 as compared to 2018 level, thereby recording an increase of 15%%; nevertheless, concurrently there has been no slowing down of global appetite for fish and fish products, with global consumption per capita has in 2018 having reached 20.5 kg., is forecast to rise by one kilo per person by 2030, and a warning signal is sounded by FAO that failure to apply effective fisheries management measures is likely to threaten both food security and livelihoods (FAO, 2020a, 2020b).

Over the years, a continuous declining trend is discernible in the state of marine fishery resources, as per FAO's long-term monitoring of assessed marine fish stock, the quantity of fish stocks that are within biologically sustainable levels got reduced from 90% in 1974 to 65.8% in 2017, recording a 1.1% decrease since 2015, with 59.6% classified as being maximally sustainably fished stocks and 6.2% underfished stocks. The percentage of stocks fished at biologic unsustainable levels got augmented from 10% in 1974 to 34.2% in 2017, and these fish stocks of the world's marine fisheries were classified as overfished, and worried over these increasing trends in overfishing, FAO favors further effort and solid actions to combat overfishing. Overfishing, delineated as stock abundance fished to below the level that can produce maximum sustainable yield, not only engenders negative impacts on biodiversity and ecosystem functioning but also reduces fish production that entails the potential of subsequently leading to negative social and economic consequences (FAO, 2020a, 2020b). Under the prevailing circumstances amidst the outbreak of COVID-19 in 2020 and by the middle of 2021 with second wave of COVID-19 having adversely impacts many countries worldwide, achieving SDG target 14.4 of ending overfishing of marine fisheries by 2020 has become a remote possibility for the time being. While suggesting that the United Nations Fish Stocks Agreement, in force since 2001, should be used as the legal basis for management measures for the high sea fisheries (FAO, 2020a, 2020b).

Undoubtedly, COVID-19 pandemic has led to environmental recovery in some ecosystems from a global "anthropause"; nonetheless, such evidence for natural resources with extraction or production value, especially fisheries, is limited (Stokes, Lynch, Valbo-Jorgesen, & Smidt, 2020). There has been indirect impact on fisheries sector of measures undertaken to contain pandemic in terms of national level lockdowns and cross-country travel, transportations and supply chains. According to UN News (2020), more than 90% of small-scale fishers in parts of the Mediterranean and the Black Sea, had been forced to stop because they were unable to sell their catch. Aquaculture production for export had been hit by disruptions in international transport and distribution channels to tourism and restaurant industry had shrunk. Other factors affecting fisheries in the wake of COVID-19 included migrant laborer issues and risks linked to crowded fresh markets (UN News, 2020). Nevertheless, as of June 2021, with some of the countries in Europe and the United States having inoculated their respective population with anti-COVID-19 vaccines along with declining trends in these countries of pandemic infections and the resultant opening of economy, the fisheries sector is likely to gather momentum.

11.6 COVID-19 pandemic and climate change

No individual in the present globalized and interconnected world is immune from socioeconomic and health-related impacts of the ongoing COVID-19, going on since its outbreak in early January 2020, and the vagaries of climate change that are with us for past many decades. When the COVID-19 outbreak was declared a global pandemic on March 12, 2020 by WHO, the climate change was already globally at the forefront of political deliberations and agendas with the thinking that it was a critical juncture to embark upon a decisive action to protect the future of the planet Earth. Nevertheless, in the wake of accelerating impact of the global pandemic assuming new proportions, the world's spotlight started moving away from climate change, and the governments and the scientific community, apart from providing healthcare facilities to combat the pandemic, also focused on finding an effective vaccine to contain the pandemic. In the meanwhile, the scientific community has also focused on the similarities and differences between COVID-19 and climate change that require brief description.

Apparently, nothing similar becomes discernible immediately between the COVID-19 pandemic and climate change; nonetheless, a closer inspection reveals a number of shred factors. In the first instance, both crises are ascribed to massive needless loss of life. The COVID-19 pandemic is acknowledged to affect people, especially the elderly and those with underlying health conditions, causing serious respiratory disease, and climate change affects air quality, drinking water, food supply, and shelter, all factors that are closely aligned to health. The COVID-19 pandemic, since its outbreak in January 2020 till mid-May 2021, has already claimed about 3.4 million human lives (WHO, 2021) and is likely to continue to do so, while climate change is expected to claim approximately 250,000 deaths annually between 2030 and 2050 (WHO, 2018). Second, both crises share the similarity of impacting certain demographic groups more so than others. The vulnerable and disadvantaged segments of the society pay a heavier price in both crises, with people reeling under appalling poverty suffering from the vagaries of climate change and the pandemic more so than the rich, and regrettably the existing disparities between the rich and the poor in terms of healthcare and exposure to factors that adversely affect health—and these disparities are amply highlighted in the pandemic and climate change crises (Moore, 2021). Another similarity between these two crises is that both these crises have proved instrumental in pushing regional healthcare systems worldwide to the limit by pointing chinks their health armory, especially in the wake of both climate change and COVID-19 pandemic having resulted in a substantial number of people being hospitalized, forcing countries to reassess as to how they manage their respective healthcare systems.

In terms of differences between COVID-19 pandemic and climate change, pandemic outbreak occurs within a short duration and affects humankind globally, while climate change is a localized event, albeit with global repercussions. COVID-19 pandemic unfolded within months to affect humanity worldwide, whereas climate change emerges relatively gradually, and its impact has often been contested. Differences also exist in terms of solutions to these crises. COVID-19 can be overcome within a short span, whereas solutions to climate change remain complex, long-term, and heavily contested; still addressing climate change calls for a "multigenerational commitment to population-wide lifestyle changes" (Trembath & Wang, 2020).

Scientific research has enabled in uncovering some linkages between climate change and outbreak of the COVID-19 pandemic. An evidence reveals that climate change may have staged a causal role in the emergence of the virus responsible for the COVID-19 pandemic, severe acute respiratory syndrome coronavirus 2 or SARS-Cov-2 (Beyer, Manica, & Mora, 2021). Another research has been to link the climatic changes that take place as a result of climate change directly to COVID-19 with the emphasis that the number of bat species present is linked with the number of coronaviruses in a specific environment (Watts et al., 2021). Climate change facilitates rise and fall in temperature, increase or decrease in atmospheric carbon dioxide, and expansion or contraction of cloud cover, and these processes in turn directly affects the growth of plants and trees and this enables climate change in affecting natural habitats and ecosystems through altering environmental factors. Even slight adjustments also entail the potential of wielding a great impact on the species living within an ecosystem. Nevertheless, COVID-19 is not the only infectious disease linked to climate change. With scientific community having been focused on highlighting the link between environmental conditions and epidemic diseases for many years (Parham et al., 2015), now at this juncture it can be hoped that this link between climate change and pandemic contagious diseases will be prioritized, forcing "policy-makers to consider the wide reaching impact of climate change and make calculated strategies to prevent further environmental damage and revers, where possible, the damage that has already occurred" (Moore, 2021).

Aligning government responses to the COVID-19 pandemic and climate change entails the potential of allowing for the overall improvement of public health, as well as fostering a sustainable economic future for regions globally, and aligning responses also affords a chance to protect the planet Earth's biodiversity and further damage to diverse ecosystems (Moore, 2021). Emphasizing on the need for aligning responses to both COVID-19 pandemic and climate change in order to address both these crises optimally, a recent report published in The Lancet notes that because of the common factors of the pandemic and climate change, converging responses is rational because both are linked to human activity and both contribute to the degradation of the environment (Watts et al., 2021). While human-induced climate change spurs outbreak of diseases, the illegal wildlife trade is also a significant factor contributing in the spread of zoonotic diseases such as SARS-Cov-2, especially because trafficking in wildlife trade cause humans to mix with different species in places where they not expected to fueling zoonosis (Moore, 2021). In the wake of havoc wreaked by COVID-19 pandemic and climate change in terms of loss of human lives and resultant socioeconomic and health-related consequences, one can hope in the ensuing years more well-concerted strategies would be implemented in rectifying human behavior so that it wields lesser and lesser impact on the environment and accordingly on the infectious diseases.

The COVID-19 pandemic epitomizes what has been called a super-wicked problem (Roberts, 2020) because such problems are characterized by the emergent need of finding a solution; where the solution often lies with those who are causing it; where the apex authority to tackle the problem is weak; and where rectifying the actions taken at that juncture "can store up problems for future generation" (Cole & Dodds, 2021). Pandemics share these similarities with climate change, and both these crises are complex and urgent, and if left undetected or unchecked, it entails the potential of continuing to exert huge burdens on the future health and well-being of humankind and the biosphere. Given the fact that neither of the two crises respects geographical frontiers, the effect of the pandemics and climate change within countries is vastly dependent on the intersection of laws, policies, and social factors (Cole & Bickersteth, 2018). There is a dire of collective human action to resolve these "super-wicked" problems. Experience of recent years in problem solving at the global level, as has been the case with majority of countries struggling to meet the targets of the UN's Agenda 2030 and the goals of the UNFCCC, is sufficient to serve as a reminder that some new approaches are called for. People cannot be expected to socially distance from the havoc wreaked by the vagaries of climate change in this Anthropocene era and that has been implicated in the emergence and spread of COVID-19 (Frontera, Martin, Vlachos, & Sgubin, 2020). While calling for more openness and sharing about threats emerging within countries' own borders, Cole and Dodds (2021) emphasize on the need for providing early warning to the world, with more collaboration and resource sharing to tackle it; and ensuring more protection to the natural environment "on whose health our own health so strongly depends".

11.7 Outlook for ecosystem-based adaptation

Increasing traction is being garnered by climate change and biodiversity loss as major risks affecting ecosystems' ability to deliver the services that support economies and human well-being. While biodiversity buttresses ecosystem processes and functions that deliver other critical ecosystem services (CBD, 2019), climate change directly impacts biodiversity, accordingly altering the composition and function of ecosystems at an extraordinary speed and, therefore threatening the services delivered by ecosystems, specifically those that regulate climate and disease control (Dasgupta, 2020). It is significant to note that ecosystems and biodiversity play a crucial role in facilitating supporting efforts to minimize the negative effects of climate change. Apart from facilitating storage and removal of carbon from the atmosphere, ecosystems and biodiversity also provide important natural buffers to hazardous events, and thus the efficacy of most ecosystem-based mitigation and adaptation actions places vital reliance on the functional provision of ecosystem services (Kapos, Wicander, Salvaterra, Dawkins, & Hicks, 2019), as they are themselves climate-sensitive and ought to remain within safe biophysical limits (Seddon et al., 2020).

It is in this backdrop that nature-based solutions (NBS) are mulled as "the best way to achieve human well-being, tackle climate change and protect our living planet" (UNEP, 2020b). The growing political recognition of the climate change and biodiversity crises and the emphasis on the need for protecting and enhancing nature's multiple benefits had never seemed more crucial, with 2020 being widely referred to as the "Super year for Nature" (UNEP, 2020c). The COVID-19 pandemic that has been with us since early 2020 has amply demonstrated as to how human impacts on natural habitats, biodiversity loss, and ecosystems degradation are making "virus spillover" events much more likely (Johnson et al., 2020) and ordains us to reconsider human relationship with nature. Widely acknowledged as an umbrella term, NBS encompass multiple ecosystem-based approaches, like EbA, ecosystem-based disaster risk reduction (EcoDRR), and ecosystem-based mitigation; however, EbA has garnered wider recognition as a means to protect, restore, and enhance ecosystem services to reduce climate change risks and improve resilience of people (Terton & Greewalt, 2021). Inherent strength of EbA is its potential of engendering economic returns and delivering multiple benefits, such as improved health, biodiversity protection food security, and alternative livelihood opportunity, all of which can help build resilience to climate change.

With over a decade of operationalization of EbA in global, regional, and local contexts, it has seemingly emerged as an essential approach to adaptation that is efficacious in generating ecological, social, and economic resilience. There is a dire need of putting EbA at the heart of national development and climate strategies of each country in order to fully maximize and provide EbA at the scale and pace required, and the National Adaptation Plan (NAP) provides an appropriate opening to do just this. By enabling countries to strategically integrate adaptation into their respective decision-making, planning, and budgeting, the NAP process endeavors to make adaptation part of standard development practice. According to UNFCCC (2019), about 120 countries have initiated and/or are undertaking activities pertaining to the NAP process, affording an important opportunity to scale up EbA, and it demonstrates a strong commitment to nature and to address the biodiversity loss (Terton & Greewalt, 2021).

During the period of its being in operation at global, regional, and local scales spanning over a decade, implementation of EbA has been confronted with some impediments that have inhibited its successful implementation. While some of the constraints have been similar to the broader reporting on adaptation constraints (Leal Filho & Nalau, 2018), there has been an array of other EbA-specific impediments that have come to notice. In terms of governance constraints, EbA is constrained by lack of people's participation and codevelopment of projects (Wamsler & Paulet, 2016), and lack of attention to differing governance arrangements, such as traditional versus government; and governance arrangements are reflections of power relationships and influence greatly as to what priorities are put forward and how particular policy issues are dealt with (Djoudi, Brockhaus, & Locatelli, 2013). Institutional fragmentation, organizational silos, and problems across vertical and horizontal governance (Wamsler & Paulet, 2016) often impede implementation of EbA, specifically where multiple jurisdictions cover same geographical area such as watershed. Multilevel governance creates an increasingly complex array of stakeholders, priorities, and values that ought to be accommodated within the decision-making processes on adaptation (Juhola, 2016). With regard to EbA, there also exist social and cultural constraints, as has been noted by Lukasiewicz, Pittock, and Finlayson (2016) that changing land use of an area as an EbA strategy entails the potential of changing the landscape and invoke resistance among stakeholders due to culturally preferred landscapes. While reviewing the constraints impeding implementation of EbA, Nalau, Becken, and Mackey (2018) suggest that the time is ripe for further theoretical development of this concept that could include further reflexive examination of how robust the evolving "heuristics" are in practice that are now used to define EbA as a research issue and policy problem because such reflections could offer useful advice for policy, research, and practice for the communities enabling them to take climate adaptation forward that delivers multiple ecological and social benefits.

In the wake of the COVID-19 pandemic that has been with us since early 2021 and its second wave along with various variants has afflicted many countries till the closing of May 2021 and is likely to continue to do so in near future as well, the resultant national lockdowns have led to severe economic and social shocks in many countries. As with climate change, early warming signals clearly indicate that the socioeconomic and health impacts of the COVID-19 have disproportionately affected the most vulnerable countries and population groups (Kebede, Stave, & Kattaa, 2020). Initially, there indications that the pandemic and the ensuing stringent measures implemented to manage it would have significant implications for adaptation processes and would perhaps continue to do so long after the pandemic had passed; nevertheless, as the situation prevails in the middle of 2021, portraying a comprehensive and robust picture of global adaptation processes will be impacted by the pandemic is neither feasible nor possible because the available evidence is fragmented and anecdotal with robust data and analyses usually lacking (UNEP, 2021). Nonetheless, available evidence, though paltry, demonstrates as to how the COVID-19 pandemic is affecting the diverse dimensions of the global adaptation processes in the short term as well as how it is likely to alter the outlook for these processes in the longer term. In the short term, the dire need of managing the direct public health impacts of the virus and its subsequent economic consequences has seen adaptation and related topics, such as climate mitigation and environmental sustainability, "fall down" the political agenda at all levels of governance (Hammill, 2020). At the same time, ongoing and scheduled adaptation planning and implementation processes at the global, national, and local levels have witnessed ample proportions of the human and financial resources, including bilateral and multilateral support, previously kept apart for them being reallocated toward endeavors to manage the impacts of the pandemic virus (Johnson, Vera, & Zühr, 2020). Concurrently, the

COVID-19 pandemic restrictions and reallocation of resources have reportedly impeded important adaptation planning meetings and stakeholder consultations (Hammill, 2020) as well as requiring implementer and funders to adopt new modes of operating to deal with rapidly changing priorities and operational realities (Adaptation Fund, 2020).

Long-term socioeconomic implications of the COVID-19 pandemic can be projected to entail lasting implications for worldwide adaptation processes even after the pandemic will be over. Most probably, the stringent negative effects of the pandemic on the global economy entail the likelihood of reducing the availability of adaptation finance (Quevedo, Peters, & Cao, 2020). In the wake of the pandemic, reduced pressure on public finance is projected to be disproportionately felt in developing and LDC, where governments are likely to face being concurrently hit by reductions in domestic tax revenues and external finance (OECD, 2020). In the wake of the prevailing situation in mid-2021, the clouds of uncertainty hovering the global economic outlook are prone to cast doubt on the feasibility/viability of many countries' long-term plans, as many of the assumptions upon which these plans and strategies are built upon, for example, the availability of domestic budget resources, borrowing headroom, access to international climate finance and economic growth, etc., may seem no longer sound. Concurrently, alterations in national and donor priorities as an outcome of the probable side-effects of a global recession, for instance, widespread business failure, and high rate of unemployment, could see budgets allocated for implementing climate actions and plans coming under threat or redirected toward adaptation actions that are regarded as more likely to achieve outcomes linked to stimulating economic growth (UNEP, 2021). Thus total outlook for resources, biodiversity loss, life on planet Earth, and life below water seems bleak in the immediate near future and whatever progress had been made in recent years in these realms has seemingly been washed away by the disruptions caused by the outbreak of the COVID-19 pandemic since early 2020 and its second wave continue to surge in 2021, thereby making the goals and targets of Agenda 2030 as well as Paris Agreement on Climate Change that has become operative in 2021, as unattainable within the stipulated period. Even after overcoming the havoc wreaked by the pandemic, it will take some years for the global community to recover from the losses in the health and socioeconomic realms, and to resume the unfinished agenda on climate change.

References

Abalansa, S., Mahrad, B. E. L., Vondolia, G. K., Icely, J., & Newton, A. (2020). The marine plastic litter issue: A social-economic analysis. *Sustainability*, *12*, 8677.

Adaptation Fund (2020). *Report on the Adaptation Fund's response to the Covid-19 pandemic and adaptive measures to mitigate its impact on the fund's portfolio*. AFB/EFC.26.b/4. Bonn, Germany: Adaptation Fund Board.

Adler, C., Pomeroy, J., & Nitu, R. (2020). High mountain summit: Outcomes and outlook. *WMO Bulletin*, *69*(1). Online at https://public.wmo.int/en/resources/bulletin/high-mountain-summit-outcomes-and-outlook.

Arneth, A., Shin, Y.-J., Leadley, P., Rondnini, C., Bukvareva, E., Kolb, M., ... Saito, O. (2020). Post-2020 biodiversity targets need to embrace climate change. *Proceedings of the National Academy of Sciences, United States of America*, *117*(49), 32882–32891.

Beyer, R., Manica, A., & Mora, C. (2021). Shifts in global bat diversity suggest a possible role of climate change in the emergence of SARS-CoV-1 and SARS-CoV-2. *Science of the Total Environment*, *767*, 145413.

CBD (Convention on Biological Diversity). (2019). *Voluntary guidelines for the design and effective implementation of ecosystem-based approaches to climate change adaptation and disaster risk reduction and supplementary information*. CBD Technical Series No. 93. Montreal, Canada: CBD Secretariat.

CBD (Convention on Biological Diversity) (2020a). *Global biodiversity outlook 5*. Montreal, Canada: CBD Secretariat.

CBD (Convention on Biological Diversity). (2020b). *Update of the zero draft of the post-2020 global biodiversity framework*. CBD/POST2020/PREP/2/1. Montreal, Canada: CBD Secretariat.

CITES (Convention on International Trade in Endangered Species of Wild Fauna and Flora). (2020). *CITES Secretariat's statement in relation to COVID-19*. Press release. Available online from https://cites.org/eng/CITES_Secretariat_statement_in_relation_to_COVID19.

Cole, J., & Bickersteth, S. (2018). What's planetary about health? An analysis of topics covered in The Lancet Planetary Health's first year. *Lancet Planet Health*, 2(7), e283−e284.

Cole, J., & Dodds, K. (2021). Unhealthy geopolitics: Can the response to COVID-19 reform climate change policy? *Bulletin of World health Organization*, 99, 148−154.

Dasgupta, P. (2020). *The Dasgupta review: Independent review on the economics of biodiversity (interim report)*. London: Government of UK.

Dauvergne, P. (2018). Why is the global governance of plastic failing the oceans? *Global Environmental Change*, 51, 22−31.

Djoudi, H., Brockhaus, M., & Locatelli, B. (2013). Once there was a lake: Vulnerability to environmental changes in northern Mali. *Regional Environmental Change*, 13(3), 493−508.

EEA (European Environment Agency). (2021). *Ocean acidification*. Available online from https://www.eea.europa.eu/ims/ocean-acidification.

Fajardo, P., Beauchesne, D., Carbajal-López, A., Daigle, R. M., Fierro-Arcos, L. D., Goldsmit, J., ... Christofoletti, R. A. (2021). Aichi target 18 beyond 2020: Mainstreaming traditional biodiversity knowledge in the conservation and sustainable use of marine and coastal ecosystems. *PeerJ*, 9, e9616.

FAO (Food and Agriculture Organization). (2020a). *The state of the world's forests 2020: Forests, biodiversity and people*. Rome: FAO.

FAO (Food and Agriculture Organization). (2020b). *The state of world fisheries and aquaculture 2020: Sustainability in action*. Rome, Italy: FAO

Forio, M. A. E., & Goethals, P. L. M. (2020). An integrated approach of multi-community monitoring and assessment of aquatic ecosystems to support sustainable development. *Sustainability*, 12, 5603.

Frontera, A., Martin, C., Vlachos, K., & Sgubin, G. (2020). Regional air pollution persistence links to COVID-19 infection zoning. *Journal of Infection*, 81(2), 318−356.

Hammill, A. (2020). *How COVID-19 is reinforcing the need for climate adaptation in vulnerable countries*. NDC partnership portal. Available online from https://ndcpartnership.org/news/how-covid-19-reinforcing-need-climate-adaptation-vulnerable-countries.

IPBES (Intergovernmental Science-Policy Platform on Biodiversity and Ecosystem Services). (2019). *Summary for policymakers of the global assessment report on biodiversity and ecosystem services of the Intergovernmental Science-Policy Platform on Biodiversity and Ecosystem Services*. Bonn, Germany: IPBES secretariat.

IPCC (Intergovernmental Panel on Climate Change) (2013). *Climate change 2013: The physical science basis*. Contribution of working group I to the fifth assessment report of the Intergovernmental Panel on Climate Change. Cambridge: Cambridge University Press.

IPCC (Intergovernmental Panel on Climate Change). (2018). *Global warming of 1.5°C*. An IPCC Special Report on the impacts of global warming of 1.5°C above pre-industrial levels and related global greenhouse gas emission pathways, in the context of strengthening the global response to the threat of climate change, sustainable development, and efforts to eradicate poverty. Available online from https://www.ipcc.ch/site/assets/uploads/sites/2/2019/06/SR15_Full_Report_High_Res.pdf.

IPCC (Intergovernmental Panel on Climate Change) (2019a). *Climate change and land: An IPCC special report on climate change, desertification, land degradation, sustainable land management, food security, and greenhouse gas fluxes in terrestrial ecosystems.* Available online from https://www.ipcc.ch/site/assets/uploads/2019/11/SRCCL-Full-Report-Compiled-191128.pdf.

IPCC (Intergovernmental Panel on Climate Change). (2019b). *IPCC special report on the ocean and cryosphere in a changing climate.* Available online from https://www.ipcc.ch/site/assets/uploads/sites/3/2019/12/SROCC_FullReport_FINAL.pdf.

IUCN (International Union for Conservation of Nature). (2019). *Ocean deoxygenation: Everyone's problem—Causes, impacts, consequences and solutions.* Gland, Switzerland: IUCN.

IUCN (International Union for Conservation of Nature). (2021). IUCN position on updated zero draft post-2020 global biodiversity framework. Gland: IUCN Office. Available online at: <https://www.iucn.org/sites/dev/files/iucn_position_on_the_updated_zero_draft_of_the_post-2020_global_biodiversity_framework_-_april_2021.pdf>.

Johnson, C. K., Hitchens, P. L., Pandit, P. S., Rushmore, J., Evans, T. S., Young, C. C. W., & Doyle, M. M. (2020). Global shifts in mammalian population trends reveal key predictors of virus spillover risk. *Proceedings of the Royal Society B: Biological Sciences, 287*(1924), 20192736.

Johnson, Z., Vera, I., & Zühr, R. (2020). *How are donor countries responding to COVID-19? Early analyses and trends to watch.* Donor Tracker Webinar. Available online from https://donortracker.org/insights/how-are-donor-countries-responding-covid-19-early-analyses-and-trends-watch.

Juhola, S. (2016). Barriers to the implementation of climate change adaptation in land use planning: a multi-level governance problem? *International Journal of Climate Change Strategies and Management, 8*(3), 338–355.

Kantai, T. (2020). *Confronting the plastic pollution pandemic. IISD Brief # 8.* Winnipeg, Canada: IISD.

Kapos, V., Wicander, S., Salvaterra, T., Dawkins, K., & Hicks, C. (2019). *The role of the natural environment in adaptation: Background paper for the Global Commission on Adaptation.* Washington, D.C.: Global Commission on Adaptation.

Kebede, T. A., Stave, S. E., & Kattaa, M. (2020). *Rapid assessment of the impacts of COVID-19 on vulnerable populations and small-scale enterprises in Iraq.* Geneva, Switzerland: International Labor Organization.

Law Insider (n.d.). *Aquatic resources definition.* Available online at: <https://www.lawindier.com/dictionary/aquatic-resources>.

Leal Filho, F. W., & Nalau, J. (2018). *Limits to climate change adaptation.* Berlin, Germany: Springer International.

Lindsey, R. (2021). *Climate change: Global sea level.* Climate.gov. Available online from https://www.climate.gov/news-features/understanding-climate/climate-change-global-sea-level.

Lukasiewicz, A., Pittock, J., & Finlayson, M. (2016). Institutional challenges of adopting ecosystem-based adaptation to climate change. *Regional Environmental Change, 16*(2), 487–499.

Moore, S. (2021). *Climate Change and COVID-19.* News-Medical. Available online from https://www.news-medical.net/health/Climate-Change-and-COVID-19.aspx.

Nalau, J., Becken, S., & Mackey, B. (2018). Ecosystem-based adaptation: A review of the constraints. *Environmental Science and Policy, 89*, 357–364.

NIC (National Intelligence Council). (2021). *Global trends 2040: A more contested world.* Washington, D.C.: NIC.

OECD (Organization for Economic Co-operation and Development). (2020). *Impact of the coronavirus (COVID-19) crisis on development finance.* Paris: OECD.

Parham, P. E., Waldock, J., Christophides, G. K., Hemming, D., Agusto, F., Evans, K. J., ... Michael, E. (2015). Climate, environmental and socio-economic change: Weighing up the balance in vector-borne disease transmission. *Philosophical Transactions of the Royal Society of London. Series B: Biological Sciences, 370*(1665), 20139551.

References

Paul, D. (2021). Protecting the marine environment from land-based activities. *IISD Brief # 9*. Winnipeg, Canada: IISD.

Peng, X., Biao, L., Yi-Xuan, H., Qiu-Jin, G., & Hong-Xiu, W. (2020). Responses of freshwater ecosystems to global change: Research progress and outlook. *Chinese Journal of Plant Ecology, 44*(5), 565−574.

Quevedo, A., Peters, K., & Cao, Y. (2020). *The impact of COVID-19 on climate change and disaster resilience funding: Trends and signals*. London and Zurich: ODI and Zurich Flood Resilience Alliance.

Roberts, A. (2020). Pandemics and politics. *Survival, 62*(5), 7−40.

Roberts, L. R., Bishop, I. J., & Adams, J. K. (2020). Anthropogenically forced change in aquatic ecosystems: Reflections on the use of monitoring, archival and palaeolimnological data to inform conservation. *Geo: Geography and Environment, 2*, 89.

Rosen, T. (2020). *The evolving war on illegal wildlife trade. Brief# 3*. Winnipeg, Canada: IISD.

Seddon, N., Chausson, A., Berry, P., Girardin, C., Smith, A., & Turner, B. (2020). Understanding the value and limits of nature-based solutions to climate change and other global challenges. *Philosophical Transactions of the Royal Society B, 375*(1794).

Stokes, G. L., Lynch, A. J., Valbo-Jorgesen, J., & Smidt, S. J. (2020). COVID-19 pandemic impacts on global inland fisheries. *Proceedings of the National Academy of Sciences, USA, 117*(47), 29419−29421.

Terton, A., & Greewalt, J. (2021). *Building resilience with nature: Maximizing ecosystem-based adaptation through national adaptation plan processes*. Winnipeg, Canada: IISD.

Trembath, A. & Wang, S. (2020). *Why the COVID-19 response is no model for climate action*. The Breakthrough Institute. Available online from https://thebreakthrough.org/issues/energy/covid-19-climate.

Tsioumani, A., & Dsioumanis, E. (2020). *Biological diversity: Protecting the variety of life on Earth*. Winnipeg, Canada: IISD.

UN News (2020). *As consumption rises, here's why sustainable fisheries management matters*. UN News online from https://news.un.org/en/story/2020/06/1065842.

UNCCD (United Nations Convention to Combat Desertification). (2017). *Global Land Outlook*. Bonn, Germany: UNCCD Secretariat

UN-DESA (United Nations, Department of Economic and Social Affairs). (2021). *The Global Forest Goals Report 2021*. Washington, D.C: UN-DESA Secretariat.

UNDRR (United Nations Office for Disaster Risk Reduction). (2020). *Human cost of disasters: An overview of last twenty years, 2000−2019*. Geneva, Switzerland: UNDRR.

UNEP (United nations Environment Programme). (2017). *Freshwater strategy*, 2017−2021. Nairobi: UNEP.

UNEP (United Nations Environment Programme). (2020a). *Emissions gap report*. Nairobi: UNEP.

UNEP (United Nations Environment Program). (2020b). *Spotlight on nature and biodiversity*. UNEP Portal. Available Online from https://www.unep.org/news-and-stories/news/spotlight-nature-and-biodiversity.

UNEP (United Nations Environment Program). (2020c). *Agenda Item 2: Synergies between the 2020 "super year for nature" and UNEA-5 (UNEP/EA.5/BUR.1/3)*. Meeting of the Bureau of the Environment Assembly.

UNEP (United Nations Environment Program). (2021). *Adaptation Gap Report 2020*. Nairobi: UNEP.

UNEP-GEF (United Nations Environment Programme-Global Environment Facility). (2018). *From source to sea: Protecting our oceans through partnership and investments*. UNEP/GPA/IGR.4/INF/5.

UNFCCC (United Nations Framework Convention on Climate Change). (2015). *Paris Agreement*. Bonn, Germany: UNFCCC Secretariat.

UNFCCC (United Nations Framework Convention on Climate Change). (2019). *Progress in the process to formulate and implement national adaptation plans (FCCC/SBI/2019/INF.15)*. Subsidiary Body for Implementation. Item 12 of the Provisional Agenda: National Adaptation Plans. Bonn, Germany: UNFCCC Secretariat.

Wamsler, C., & Paulet, S. (2016). Making headway in climate policy mainstreaming and ecosystem-based adaptation: Two pioneering countries, different pathways, one goal. *Climate Change, 137*(1), 71−87.

Watts, N., Amann, M., Arnell, N., Ayeb-karlsson, S., Beagley, J., belesova, K., ... Costello, A. (2021). The 2020 report of The Lancet Countdown on health and climate change: Responding to converging crises. *The Lancet, 397*(10269), 129−170.

WHO (World Health Organization). (2018). *Climate change and health-factsheet.* WHO portal. Available online from https://www.who.int/health-topics/climate-change/4#tab=tab_1.

WHO (World Health Organization). (2021). *WHO Coronavirus (COVID-19) dashboard.* WHO portal online at https://covid19.who.int/.

WMO (World Meteorological Organization). (2020). *2020 State of climate services.* Geneva, Switzerland: WMO.

WMO (World Meteorological Organization). (2021). *State of the global climate 2020: Unpacking the indicators.* Geneva, Switzerland: WMO. Available online from https://public.wmo.int/en/our-mandate/climate/wmo-statement-state-of-global-climate.

WWDR (World Water Development Report). (2020). *Water and climate change.* Paris: UNESCO.

WWF (World Wide Fund for Nature). (2020). *Living planet report 2020—Bending the curve of biodiversity loss.* Gland, Switzerland: WWF.

Xu, H., Cao, Y., Yu, D., Cao, M., He, Y., Gil, M., & Pereira, H. M. (2021). Ensuring effective implementation of the post-2020 global biodiversity targets. *Nature Ecology & Evolution, 5,* 411−418.

Index

Note: Page numbers followed by "*f*" and "*t*" refer to figures and tables, respectively.

A

Abrupt climate change, 55−57
Accelerated weathering, 194−195
Access and benefit sharing (ABS), 543−546
Access to food, climate change and, 310−311
Ad hoc Working Group on Long-term Cooperative Action (AWG LCA), 156
Adaptation
   climate change, 169−178
      delineating, 172−173
      evaluating, 176−178
      food system, 314−317
      limits and barriers to, 175−176
      under Paris Agreement, 173−175
   in EbA approach, 110−117
      adaptive capacity, 114−116
      costs and benefits of, 125−126
      limits to, 116−117
      need for, 113−114
   vs. mitigation, 178−182, 179*t*
Adaptive capacity, 114−116
Aerosol, 45−46
Afforestation, 189−190
Agricultural drought, 73
Agriculture, forestry and other land-use (AFOLU), 165, 303−304, 520−521
Aichi Biodiversity Targets, 558, 653
Alien species (AS), 556
Amazonian systems, 511, 515
Animal genetic resources, 538−539
Antarctic Intermediate Water (AAIW), 433−434
Antarctic Surface Water (AASW), 433−434
Anthropocene, and global warming, 59−60
Anthropocene Working Group (AWG), 60
Anthropogenic drivers, of climate change, 60−68
Anthropogenic stressors, 437−460
Aquatic bioenergy with carbon dioxide capture and storage, 193−194
Aquatic genetic resources (AqGRs), 540−542
Aquatic resources, 649
Arctic Oscillation (AO) index, 79
Areas beyond national jurisdiction (ABNJ), 448
Asia-Pacific region, 621−622
Association of Small Island States (AOSIS), 199
Atlantic and Indian Oceans, 651
Atlantic Meridional Overturning Circulation (AMOC), 46−47
Atmosphere, 44−46
Atmospheric drought, 73
Autonomous adaptation, 315

B

Bali Plan of Action, 156
Ballast Water Management Convention (BWMC), 558−559
Basel Convention, 663−664
Biochar, 191−192
Biodiversity
   ecosystem services and, 123−124
   global outlook for, 653−656, 655*f*, 656*f*
   protecting, 531−536
Biodiversity Crisis, 653
Biodiversity offsetting (BO), 535−536
Bioenergy with carbon capture and storage (BECCS), 192−193
Bioenergy with carbon dioxide capture and storage, 192−194
Biofuel, 263
Biosphere, 49−51
Biotic life, 503
Blueprint for Survival in 1972, 326
Brand Meghalaya, 626−627
Brundtland Commission, 328−329
Brundtland Report, 328
Business-as-usual (BAU), 158

C

Carbon dioxide removal (CDR), 183−184, 188−197, 189*t*
   accelerated weathering, 194−195
   afforestation, 189−190
   biochar, 191−192
   bioenergy with carbon dioxide capture and storage, 192−194
   direct air capture, 196−197
   ocean fertilization, 195−196
   reforestation, 189−190
   soil carbon sequestration, 190−191
Carbon geoengineering, 183−184
Carbon intensity of human well-being (CIWB), 62
Center for Science and Environment (CSE), 620−621
Circular economy, 374−376
Cities concept, in SSC, 348−349
Cities for Climate Protection (CCP), 147
Cleanest village in Asia, 627
Climate
   definition of, 42−44
   and weather, 43−44
Climate change, 39, 143, 419−421, 506−507, 529−530, 556−557, 665−668

Climate change (*Continued*)
  abrupt climate change, 55–57
  and access to food, 310–311
  adaptation, 169–178
    delineating, 172–173
    evaluating, 176–178
    of food system, 314–317
    limits and barriers to, 175–176
    under Paris Agreement, 173–175
  adverse impacts of, 105–106
  anthropogenic drivers of, 60–68
    consumption, 67–68
    economic growth, 61–64
    urbanization, 64–67
  climate drivers, 305–307
  definition of, 51–55
  ecosystems approach and, 203–206
  extreme weather events, 68–81
    droughts, 72–74
    extreme cold events, 69–70
    extreme heat events, 70–72
    extreme snow/ice storms, 74–76
    tropical cyclones, 76–80
    wildfires, 81
  and food availability, 307–310
  and food quality, 312–313
  and food security, 300–314, 302t
  and food stability, 313–314
  and food utilization, 311–313
  geoengineering and, 182–197
    carbon dioxide reduction, 188–197
    solar radiation management, 185–188
  vs. global warming, 58–59
  impacts of, 41, 82–83
  linkages between mitigation and adaptation, 178–182, 179t
  loss and damage associated with, 197–202
    defining, 198–199
    Warsaw International Mechanism for, 199–202
  mitigation, 144–169
    climate risk, 145–147
    excludable benefits, 149
    harnessing, 150–153
    internal determinants, 147–149
    international framework for, 153–169
  and smallholder farming systems, 308–310
  threat multiplier effect, 39–40
  water and, 247–251
Climate change adaptation (CCA), 16, 603–608, 610, 612, 616–618, 636
  and DRR, 19–21
Climate Summit for a Living Himalayas in 2011, 623
Climate Sustainability Working Group (CSWG), 614–615
Climate system, 44–51
  atmosphere, 44–46
  biosphere, 49–51
  cryosphere, 47–48
  hydrosphere, 46–47
  lithosphere, 48–49
Climate Technology Center Network (CTCN), 166
Climatological drought, 73
Clouds, 45–46
CO_2 fertilization, 50
Commission for Genetic Resources for Food and Agriculture (CGRFA), 539
Common but Differentiated Responsibility and Respective Capabilities (CBDR), 620–621
Common Fisheries Policy (CFP), 449–450
Community-Based Disaster Risk Reduction (CBDRR), 13–14
Competitive economy, 363–367
Conference of the Parties (COP), 52–53
Conference on International Trade in Endangered Species of Wild Animals, 327
Conserving forests, 507–519
Conserving mountain ecosystems, 527–530
Convention of Migratory Species (CMS), 555–558, 560
Convention on Biodiversity (CBD), 107
Convention on Biological Diversity (CBD), 267–268
Convention on International Trade in Endangered Species of Wild Fauna and Flora (CITES), 560–561, 660
Convention on the Prevention of Marine Pollution by Dumping of Wastes and Other Matter, 327
COP-1, 20, 155–156, 161t
COP-3, 161t
COP-5, 463–464
COP-7, 161t
COP-10, 187–188
COP-12, 161t
COP-13, 166, 170–171, 187–188, 197–198
CoP-15, 40, 113, 143, 156
COP-16, 156, 161t, 197
COP-17, 52–53, 156–157
COP-18, 157, 161t
COP-19, 158, 161t
COP-20, 158–160, 161t
COP-21, 158–160, 161t
COP-22, 161t
COP-23, 161t
COP-24, 161t
COP-25, 161t, 201–202
Cost—benefit analysis (CBA), 129–130
  usage example, 130–131
Cost-effective analysis, 129
Coupled Model Inter-comparison Project Phase 5 (CMIP5), 70
COVID-19 pandemic, 649–654, 657–658, 660, 664–670
Critical transition, 649
Cryosphere, 47–48
Cultural Theory, 146

Cultural violence, 291–292

D

Dansgaard—Oescher (D—O) events, 55–56
Deforestation, 509–511
Deleterious edge effects, 507–508
Desertification, 250–251, 524–527
Digital economy, 360–363
Digital sequence information (DSI), 546
Direct air capture, 196–197
Direct violence, 291–292
Directed mainstreaming, 606–607
Disaster, 10–12
    definition of, 2–9
        toward consensual, 5–7
    hazard-disaster linkages, 7–10
    risk reduction. *See* Disaster risk reduction (DRR)
    and social systems, 9–10
Disaster management, defined, 12–13
Disaster Research Center (DRC), 2
    and climate change adaptation, 19–21
Disaster risk reduction (DRR), 11–14, 604–605, 608–610, 617–618, 636–638
    deployment of EbA for, 24–25
    ecosystem-based adaptation for, 19–25
    Hyogo Framework, 14–15
    Sendai Framework, 15–19
Dissolved inorganic nitrogen (DIN), 458
Droughts, 72–74
    impact of, 511–513

E

Earth Day, 327
Earth Energy Imbalance, 660–661
Earth, global outlook for life on, 656–660, 658f
Earth, preserving life on
    conserving forests, 507–519
        deforestation, 509–511
        forest degradation, 513–515
        forest fragmentation, 517–519
        impact of droughts, 511–513
        wild and forest fires, 515–517
    conserving mountain ecosystems, 527–530
    freshwater ecosystems, 504–507
        climate change and freshwater ecosystems, 506–507
    genetic resources, 536–546
        genetic resources, types of, 537–546
    land and water, preventing invasive aliens species on, 552–561
        defining invasive alien species, 555–556
        invasive alien species and climate change, 556–557
        regulatory framework for invasive alien species, 557–561
    protecting biodiversity, 531–536
    sustainable land use, 519–527
        desertification, 524–527
        land degradation, 521–524
    wildlife, conservation of, 546–552
        wildlife species, threats to, 547–552
Earth's biological heritage, 431–432
Earth System Models (ESMs), 50–51
East Jaintia Hills District, 629–630
Eco-city projects, 335–345
    types, 337–338
Ecohydrology, 267–268
Ecological awareness, 343–344
Ecological-landscape integrity, 342–343
Ecological sanitation (EcoSan), 341–342
Economic growth, 61–64
Economic water scarcity, 243–244
Ecosystem services (ESSs), 117–124, 603–605, 607–608, 616–617, 626–627, 632–633, 635–636
    and biodiversity, 123–124
    classification of, 119–123
        provisioning services, 120–122, 120t
        regulating services, 121t, 122–123
        sociocultural services, 122t, 123
Ecosystem-based adaptation (EbA)
    concept of, 107–110
    costs and benefits, 125–132, 131t
    defining, 107–110
    deployment for DRR, 24–25
    for DRR, 19–25
    vs. ecosystem-based DRR, 23
    ecosystem services, 117–124
        classification of, 119–123
        provisioning services, 120–122, 120t
        regulating services, 121t, 122–123
        sociocultural services, 122t, 123
    ingredients, 110–117
    from water—energy—food nexus approach to, 621–623
    and water security, 268–270
Ecosystem-based DRR (Eco-DRR), 22–25
    mechanism of, 22
    vs. ecosystem-based approach, 23
Ecosystem-based management (EBM), 460–465
    ecosystem-based approach *vs.*, 463–465
Edge effects, 518
El Niño—Southern Oscillation (ENSO), 425–426, 511
Energy efficiency, 153
Energy—food nexus, 262–267, 264t
Environmental degradation, 22
Environmental stressors, 418–437
European Forest Law Enforcement, 509–510
European Innovation Partnership on Smart Cities and Communities (EIP-SCC), 351–352
European Union (EU), in eco-city development, 337

European Union's Marine Strategy Framework Directive (MSFD), 459−462
"Everybody's business", 604
Exclusive economic zones (EEZs), 448
Expert Team on Climate Change Detection and Indices (ETCCDI), 70
Export production (EP), 433−434
Extreme cold events, 69−70
Extreme heat events, 70−72
Extreme snow/ice storms, 74−76
Extreme weather events, 68−81
    droughts, 72−74
    extreme cold events, 69−70
    extreme heat events, 70−72
    extreme snow/ice storms, 74−76
    tropical cyclones, 76−80
    wildfires, 81

F

Feely, Richard, 430
Food and Agricultural Organization (FAO), 289, 458, 462−463, 509−510, 514, 538−542, 664−665
Food availability, climate change and, 307−310
Food Insecurity Experience Scale (FIES), 289
Food quality, climate change and, 312−313
Food security, 293−300
    and climate change, 300−314, 302*t*
    climate drivers, 305−307
    definition of, 293−300
    evolution of, 296*t*
    World Food Summit (1996), 297−300
Food stability, climate change and, 313−314
Food utilization, climate change and, 311−313
Forest degradation, 513−515
Forest fragmentation, 517−519
Forest genetic resources, 539−540
Framework's theory, 654
Freshwater ecosystems, 504−507
Friends of the Earth, 327

G

G-7's Ocean Plastic Charter, 664
G-20's Osaka Blue Ocean Vision, 664
Garbage patches, 440−441
Gender, and water security, 254−257
Genetic erosion, 537
Genetic resources, 536−546
Geoengineering, 182−197
    carbon dioxide reduction, 188−197
    definitions of, 183*t*
    solar radiation management, 185−188
GHGs emissions, 151−152, 151*f*
Glacial Lake Outburst Floods (GLOFs), 307
Glacial-isostatic adjustment (GIA), 428

Global Alliance for Incinerator Alternatives (Gaia), 337
Global Atmospheric Research Program (GARP), 52
Global biodiversity, 657
Global climate change negotiations, timeline of, 161*t*
Global Climate Risk Index (GCRI) 2020, 618
Global Competitive Index (2019), 366*t*
Global Competitiveness Report (GCR), 365−366
Global Environment Facility (GEF), 605−606
Global Forest Goals Report 2021, 657−658
Global Hunger Index, 290−291
Global Land Outlook, 658−659
Global mean sea level (GMSL), 424−426
Global Network of Civil Society Organizations for Disaster Reduction (GNDR), 18−19
Global population, 289
Global Risk Report (2006), 39
Global sea level rise, 424−429
Global Tree Assessment (GTA), 539
Global warming (GW), 57−60
    and anthropocene, 59−60
    vs. climate change, 58−59
Global warming's evil twin, 430
Great Himalayan National Park (GHNP), 632−633
Green economy, 359
Greenhouse gas (GHG), 44−45, 113, 422−425, 431−432, 438−441, 451−456, 514−516, 520−521, 533−535, 557, 603, 620, 650−652, 660−661
    and climate change, 105
    ozone, 44−45
Greenhouse Gas Removal (GGR), 183−184
Greenpeace, 327
Gross domestic product (GDP), 61−62, 417, 428, 431−432, 510−511
Groundwater, 236−237

H

Habitat loss, 550
Hazard
    definition of, 1, 7
    natural, discernment of, 8
Hazardous air pollutants (HAP) emissions, 149
Hidden Harvest-2, 542
Hidden hunger. *See* Micronutrient deficiency
Homo economicus model, 145−146
Human Development Index (HDI), 365−366
Human—wildlife conflict, 547−550
Hunger, 290−293
    definition of, 290, 292−293
    factors responsible for, 291
    types of, 292
Hydrological drought, 73
Hydrosphere, 46−47
Hyogo Framework for Action (HFA), 14−15

Index

I

India Water Foundation (IWF), 621–623, 625–627, 630–632
India's move toward climate change policy, 620–621, 621f
Indian Himalayan Climate Adaptation Program (IHCAP), 632–633
Indian Himalayan region (IHR), 620, 632–633
Information and Communication Technology (ICT), 346
Integrated Basin Development and Livelihoods Promotion Program (IBDLP), 622, 626, 629–630
Integrated Water Resources Management (IWRM), 246, 257–259, 622–623, 626–627, 630–631
Intellectual property rights (IPRs), 545–546
Intelligent transport system (ITS), 356
Intergovernmental Panel on Climate Change (IPCC), 40, 53, 156–157, 301, 420–421, 430–432, 453–454, 514–515, 521–522, 529, 535, 619–620
   Fifth Assessment Report (2014), 106, 169–170, 303–304
   Fourth Assessment Report (2007), 106, 170–171
   Third Assessment Report of, 44
Intergovernmental Science Policy Platform on Biodiversity and Ecosystem Services (IPBES), 531
Inter-Ministerial Committee on Climate Change in Mexico (CICC$), 634
International Council of Scientific Unions (ICSU), 52
International Covenant on Economic, Social and Cultural Rights (ICESCR), 293, 379–380
International Decade for Natural Disaster Reduction, 13
International Energy Agency (IEA), 356–357
International Federation for Housing and Planning (IFHP), 379
International framework, for climate change mitigation, 153–169
   Kyoto Protocol, 155–160
   Paris Agreement on Climate Change, 160–169
International Management Institute (IMD), 365–366
International Maritime Organization (IMO), 455
International Meteorological Organization (IMO), 558–559
International Strategy for Disaster Reduction (ISDR), 13
International Treaty on Plant Genetic Resources for Food and Agriculture (ITPGRFA), 544–545
International Undertaking on Plant Genetic Resources (IUPGR), 544
International Union of Conservation of Nature (IUCN), 327–328, 538–539, 555–556, 655
Internationally Transferred Mitigation Outcomes (ITMOs), 165
Internet economy, 360–361
Intrusion on sovereignty, 543–544
Invasive alien species (IAS), 532–533, 552–554
   and climate change, 556–557
   defining, 555–556
   regulatory framework for, 557–561

IUCN Species Survival Commission Global Tree Specialist Group (IUCN-SSCGTSG), 539

K

Kyoto Protocol, 155–160, 535

L

Land and water, preventing invasive aliens species on, 552–561
Land degradation neutrality (LDN), 521–524
Land surface temperature (LST), 632
Land use (LU), 533
Land Use, Land Use Change, and Forestry (LULUCF), 152–153, 534–535
Land-based marine pollution (LBMP), 457
Land-use and land-cover change (LUCC), 519–521
Large marine ecosystems (LMEs), 458
Least Developed Countries (LDCs), 105–106, 648–649
Life below water, global outlook for, 660–665
Life on Earth, 536–537
Lithosphere, 48–49
Livable cities, 334–335
Local sustainable development, 330–332

M

Mainstreaming, concept of, 604–607
Mainstreaming ecosystem-based adaptation
   Bangladesh, mainstreaming ecosystem-based adaptation in, 617–618
   G-20 countries, mainstreaming ecosystem-based adaptation in, 614–616
   India, mainstreaming ecosystem-based adaptation in, 618–633, 618t, 619f
      Arunachal Pradesh, integrating ecosystem-based adaptation into, 632
      ecosystem-based adaptation, from water—energy—food nexus approach to, 621–623
      Indian Himalayan region, integrating ecosystem-based adaptation into, 632–633
      India's move toward climate change policy, 620–621
      Meghalaya, mainstreaming/integrating ecosystem-based adaptation into, 623–632, 624f, 625f, 626f, 627f, 628f, 629f, 630f
   mainstreaming, concept of, 604–607
      mainstreaming climate change adaptation, categories of, 605–607
   mainstreaming ecosystem-based adaptation
      barriers to, 636–638
      emerging lessons of, 633–636
   mainstreaming ecosystem-based adaptation, entry points for, 609–614
      local and community planning processes, mainstreaming ecosystem-based adaptation into, 613–614

Mainstreaming ecosystem-based adaptation (*Continued*)
    national adaptation plan, mainstreaming ecosystem-based adaptation in, 610–611
    sectoral adaptation plan, mainstreaming ecosystem-based adaptation into, 612–613
    mainstreaming ecosystem-based adaptation into policy process, 607–609
    Sweden, mainstreaming ecosystem-based adaptation in, 616–617
Malnutrition, 292–293
Managerial mainstreaming, 606
Marine atmospheric boundary layer (MABL), 455–456
Marine boundary layer (MBL), 455–456
Marine genetic resources (MGRs), 448, 540–542
Marine heatwave (MHW), 661
Marine microplastic pollution, 439–444
Marine pollution, land-based sources of, 457–460
Marine protected areas (MPAs), 459–462
Maximum sustainable yield (MSY), 449–450
Meghalaya Basin Development Authority (MBDA), 622, 626, 629
Meghalaya Rural Bank (MRB), 630
Meghalaya Water Mission, 622
Memorandum of Understanding (MoU), 622
Meteorological drought, 73
Micronutrient deficiency, 292
Millennium Development Goals (MDGs), 235, 298–299, 329–330
Million metric tons (MMT), 439–440
Ministry of Environmental Protection (MEP) framework, 339–340
Mitigation, climate change, 144–169
    vs. adaptation, 178–182, 179*t*
    climate risk, 145–147
    excludable benefits, 149
    harnessing, 150–153
    internal determinants, 147–149
    international framework for, 153–169
Monitoring and evaluation (M&E) systems, 611, 615–616, 636–638
Multicriteria analysis, 129
Multicriteria decision analysis (MCDA), 520–521
Municipal solid waste (MSW), 372–373

N

The Nagoya Protocol, 536
Nairobi Work Program (NWP), 127
National Action Plan on Climate Change (NAPCC), 620–622
National adaptation plan (NAP), 610–612, 615–616, 636–637, 668
National Adaptation Programs of Action (NAPA), 105–106
National Research Council (NRC), 56–57
Nationally determined contributions (NDCs), 534, 611
Natural disasters *vs.* human-induced disasters, 647–650, 648*t*, 650*f*

Natural solutions, 635–636
Nature-based solutions (NBS), 668
Nature's Contribution to People (NCP), 532
Negative Emissions Technologies (NETs), 183–184
Net primary production (NPP), 46, 432, 435–436
1995 Global Program of Action, 663
Nongovernmental organizations (NGOs), 51
North Atlantic Oscillation (NAO), 79
Notre Dame Global Adaptation Initiative (NDGAIN), 614–615

O

Obesity, 292–293
Occurrence of ocean anoxic events (OAEs), 432–433
Ocean acidification (OA), 419–422, 429–432, 438, 454–455
Ocean deoxygenation, 432–437
Ocean fertilization, 195–196
Ocean heat content (OHC), 422–425, 660–661
Ocean warming (OW), 419, 422–425, 430
Oil-spills pollution, 444–448
Organizational mainstreaming, 606
Overfishing, 448–451
Overnutrition, 292
Oxygen minimum zones (OMZs), 432, 435–436
Ozone (O_3), 44–45

P

Pacific-North American (PNA) pattern, 79
Paris Agreement on Climate Change (PACC), 60, 113, 144, 160–169, 303–304, 619–620, 650–651, 670
    adaptation under, 173–175
People's Survival Fund, 635
Personal protective equipment (PPE), 664
Peru, 613–614
Physical water scarcity, 242–243
Plant genetic resources, 537–538
"Polluter pays principle", 535–536
The Population Bomb, 326
Pollution of water resources, 239–240
Pressure valve effect, 149–150
Prime Minister's Council on Climate Change (PMCCC) in 2007, 620
Programmatic mainstreaming, 605–606
Provisioning services, in ecosystem services, 120–122, 120*t*
Psychometric paradigm, 146

R

Ramsar Convention on Wetlands, 267–268, 327
Rationalizing Biodiversity Conservation in Dynamic Ecosystems project (RUBICODE) project, 118
Reforestation, 189–190
Regulating services, in ecosystem services, 121*t*, 122–123
Regulatory mainstreaming, 606
Renewable Energy Directive (RED), 510
Resilience, 10–12

Index

Ridge-to-reef approach, 634
Risk, definition of, 2
Roosevelt, Franklin D., 294

S

SARSCoV-2, 660
Sea surface temperatures (SSTs), 419–420, 430, 452, 511
Second Report on the State of the World's Plant Genetic Resources for the Food and Agriculture (Second SoW-PGRFA), 537
Second World War, 550–551
Sendai Framework for Disaster Risk Reduction (SFDRR), 15–19
Severe acute respiratory syndrome coronavirus 2/SARS-Cov-2, 666–667
Severe convective storms (SCSs), 75–76
Short-lived climate pollutants (SLCPs), 306
Sixth Asia-Pacific Climate Change Forum, 631–632
Sixth mass Extinction, 531–532
Small Island Developing States (SIDS), 648–649
Smallholder farming systems, and climate change, 308–310
Smart city
    definition of, 349–353, 354t
    dimensions of, 353–384
        circular economy, 374–376
        smart economy, 359–367
        smart environment, 369–373
        smart governance, 367–368
        smart living, 378–382
        smart mobility, 355–359
        smart people, 376–378
        social protection, 382–384
    ecosystem-based adaptation for, 384–386
    factors spurring development of, 369t
Smart economy, 359–367
    competitive economy, 363–367
    digital economy, 360–363
Smart environment, 369–373
    tackling urban climate change, 370–372
    urban waste management, 372–373
Smart governance, 367–368
Smart living, 378–382
    adequate housing, 379–380
    social inclusion, 380–382
Smart mobility, 355–359
Smart people, 376–378
SMART (Specific, Measurable, Achievable, Realistic, and Timely) acronym, 347–348
Snowball effect, 149–150
Social inclusion, 380–382
Social protection systems, 382–384
Social water scarcity, 244–245
Sociocultural services, in ecosystem services, 122t, 123
Socioeconomic, and environmental drought, 73
Soil carbon sequestration, 190–191
Soil sealing, 523
Solar geoengineering, 183–184
Solar radiation management (SRM), 183–188
Solid waste management (SWM), 372–373
Special Report on Climate Change and Land (SRCCL), 301, 521–522
Spontaneous adaptation. *See* Autonomous adaptation
State Action Plan on Climate Change (SAPCC), 623–624
State of the World's Aquatic Genetic Resources (SoWR-AGRFA), 540–541
Structural violence, 291–292
Sub-Antarctic Mode Water (SAMW), 433–434
Sunlight Reflection Methods (SRM), 183–184
Super year for Nature, 668
Surface-air temperature (SAT), 452
Sustainable development goals (SDGs), 235, 267, 418, 440, 533–535, 538–539, 558, 608, 627, 652
Sustainable development, defined, 328–329
Sustainable forest management (SFM), 519
Sustainable land use, 519–527
Sustainable lifestyle, 68
Sustainable smart city (SSC), 345–384. *See also* Smart city
    cities concept, 348–349
    smart concept, 347–348
    sustainable concept, 346–347
Sustainable urban development, 332–334
Sustaining life below water
    anthropogenic stressors, 437–460
        greenhouse gases, 451–456
        marine microplastic pollution, 439–444
        marine pollution, land-based sources of, 457–460
        oil-spills pollution, 444–448
        overfishing, 448–451
    ecosystem-based approach *vs.* ecosystem-based management, 463–465
    environmental stressors, 418–437
        global sea level rise, 424–429
        ocean acidification, 429–432
        ocean deoxygenation, 432–437
        ocean warming, 422–424
    oceans, ecosystem-based management for, 460–465

T

Tackling global warming, 652f, 653f
Tackling urban climate change, 370–372
Technological water scarcity, 245–247
Technology action plans (TAPs), 247
Technology Executive Committee (TEC), 166, 246
Technology Needs Assessments (TNAs), 246
Theory of change, 654
Third United Nations Environment Assembly (UNEA), 441
Threat multiplier effect, 39–40
Three Abilities, 337–338
Three Harmonies, 337–338
Transformational adaptation, 315

682 Index

Tropical cyclone, 76–80
  extratropical cyclones, 78–80
Tropical Rainfall Measuring Mission (TRMM), 632
2020 UN Biodiversity Conference, 653–654
2030 Agenda for Sustainable Development, 654–655

U

UN Strategic Plan for Forests, 657–658
UN's Convention on Biological Diversity (CBD), 531–532, 534–536, 540, 544–546, 555–559
Undernourishment, 292
Undernutrition, 292
United Nations Conference on the Human Environment, 327
United Nations Convention on the Law of the Seas (UNCLOS), 448
United Nations Convention to Combat Desertification (UNCCD), 107, 521–522, 524, 534
United Nations Development Programme (UNDP), 365–366
United Nations Environment Program (UNEP), 107, 327–328, 440–441, 443–444, 657, 663–664
  Emissions Gap Report (2013), 113–114
  Emissions Gap Report (2017), 114
United Nations Framework Convention on Climate Change (UNFCCC), 19, 40, 51–52, 105–106, 155, 610–611, 620–621, 631–632
  Cancun Adaptation Framework, 22
  First Conference of the Parties (COP-1), 155–156
United Nations International Children Emergency Fund (UNICEF), 62–63
United Nations Office for Disaster Risk Reduction (UNDRR), 19
United Nations Social Development Network (USDN), 629
United Nations vide Resolution 44/236, 13
United States Mayors Climate Protection Agreement (USMCPA), 150
United States Presidential/Congressional Commission on Risk Assessment and Risk Management, 2
Universal Declaration of Human Rights, Article 25, 293
Unprecedented global challenges, 603
Urban livability, 334–335
Urban planning, 338–339
Urban waste management, 372–373
Urbanization, 64–67, 325
  and sustainable development, 326–332
    evolution of, 331*t*
    local sustainable development, 330–332
US Climate Change Science Program (USCCSP), 57

V

Violence, 291–292
Vulnerability, 10–12
  reduction of, 21

W

Wapungskur Village of East Jaintia Hills District, 629
Warsaw International Mechanism (WIM) for loss and damage, 199–202
  Executive Committee, 201–202
Water, 235
  availability, 236–237
  and climate change, 247–251
    extreme weather events, 248–251
  demand for, 237–239
  quality, 239–241
  scarcity, 241–247, 242*t*
    economic, 243–244
    physical, 242–243
    social, 244–245
    technological, 245–247
  security, 251–270
    defining, 251–255, 255*t*
    ecosystem-based adaptation and, 268–270
    and gender, 255–257
    integrated water resource management, 257–259
    and nature-based solutions, 267–270
    water—energy—food security nexus, 259–267, 264*t*, 266*f*
Water—energy nexus, 260–261, 264*t*
Water/environment sector, 622–623
Water—food nexus, 261–262, 264*t*
Water management drought, 73
Water resources, 649
Water stress index (WSI), 241, 241*t*
Water vapor, in atmosphere, 45–46
Weather
  climate and, 43–44
  definition of, 43
Wild and forest fires, 515–517
Wildfires, 81
Wildlife
  climate change/invasive alien species, 551–552
  conservation of, 546–552
Wildlife species, 547–552
Wildlife trafficking, 550–551
World Climate Conference (WCC), 52
World Commission on Environment and Development (WCED), 328
World Conference on Disaster Risk Reduction (WCDRR), 14–16
World Economic Forum (WEF), 39, 365–366
World Food Conference (WFC), 295
World Food Summit (WFS) (1996), 297–300
World Health Organization (WHO), 665–666
World Intellectual Property Organization (WIPO), 545–546
World Meteorological Organization (WMO), 42–43
World Resource Institute (WRI), 289
World Wildlife Fund (WWF), 327–328
World's most livable cities, 334
World-Wide Fund for Nature (WWF), 504

Printed in the United States
by Baker & Taylor Publisher Services